# A
# BIBLIOGRAPHY
# OF
# MULTIVARIATE
# STATISTICAL
# ANALYSIS

# A
# BIBLIOGRAPHY
## OF
# MULTIVARIATE
# STATISTICAL
# ANALYSIS

**T. W. Anderson**
*Stanford University*

**Somesh Das Gupta**
*University of Minnesota*

**George P. H. Styan**
*McGill University*

A Halsted Press Book

## JOHN WILEY & SONS

NEW YORK · TORONTO

OLIVER & BOYD
Tweeddale Court
14 High Street
Edinburgh EH1 1YL
A Division of Longman Group Limited

Published in the U.S.A. and
Canada by Halsted Press
A Division of John Wiley & Sons, Inc.
New York

Anderson, Theodore Wilbur.

  A bibliography of multivariate statistical analysis.

  " A Halsted Press book."
  1. Multivariate analysis—Bibliography.
I. Das Gupta, Somesh, joint author. II. Styan, George
Peter Hansbenno, joint author. III. Title.
Z6654.M8A53     016.5195'3     72-5950
ISBN 0-470-02650-2

First published 1972

© 1972 T. W. Anderson, S. Das Gupta & G. P. H. Styan

All rights reserved.
No part of this publication may be reproduced,
stored in a retrieval system, or transmitted
in any form or by any means—electronic,
mechanical, photocopying, recording or otherwise—
without the prior permission of the Copyright Owners
and the Publishers except for any purpose of the
United States Government. Any request should be
addressed in the first instance to the Publishers.

Printed in Great Britain by
T. & A. Constable Ltd., Hopetoun Street, Edinburgh

MATH.-SCI.

ZQA
278
.A53

1327446

# Contents

TO
ALL WHO MADE
*ABOMSA*
POSSIBLE

# Preface

The preparation of this bibliography began in 1963 at Columbia University when the three co-authors were studying, researching and writing multivariate statistical analysis. The bibliography contained in the book on multivariate analysis by the senior author formed a modest beginning. Originally our interest was in the mathematical theory of statistical inference concerning the multivariate normal distribution; the scope of the bibliography has extended, however, to include inference and model building based on other parametric distributions as well as a large literature of nonparametric methods.

At the outset we read papers to increase our own knowledge of the field. In organizing this knowledge we found it helpful to classify the papers according to their contents. This led eventually to our subject-matter codes for the papers as described in Chapter V. As we proceeded it became clear that we should search the literature systematically for papers in the field. Our expectation is that this bibliography will serve students and researchers to find papers in any aspect of multivariate statistical analysis. A subject-matter code corresponding to a topic of interest may be found in the index at the end of Chapter V. Chapter IV lists the authors who have written under each code and the associated number of entries in Chapter III, which is the core of the bibliography: 6093 research papers, reprints, translations and amendments arranged alphabetically by author. Chapter II identifies the journals and collections in which the research papers were published. Selected books in the general area of multivariate statistical analysis and those concentrating on related special topics are given separately in Chapter I.

Research papers (and their reprints and translations) are included if published in 1966 or earlier, and books, if published in 1970 or earlier. Amendments are given whenever known; they include corrections, acknowledgements of priority and updates.

The ingredients of multivariate statistical analysis treated in this bibliography are identified in Chapter V. For the most part we have tried to be comprehensive in scope, but in some areas where it was necessary to be selective we have confined our attentions to multivariate analysis dealing specifically with statistical inference or important developments in the mathematical theory.

The listing of papers is meant to include all research papers on statistical methods based on the multivariate normal distribution. Furthermore, it purports to include all papers on other multivariate distributions of continuous variables. Discrete multivariate distributions are also covered but not completely. We have included those papers which we considered to be primarily in a multivariate setting. The multinomial distribution, for example, has only selective coverage. Measures of association include the product-moment correlation coefficient among others. For multiple regression and time series analysis we have concentrated on models with more than one response or dependent variable. Among the nonparametric methods are included those based on nonparametric correlation coefficients, but the literature is so large that we had to be selective. It was difficult to draw the line in the direction of model building and statistical analysis in various substantive fields, particularly psychology and economics.

Additions and corrections to Chapters II and III are included in the Addenda at the end of this book.

The listing of books is intended to include all books which deal comprehensively with multivariate statistical analysis or which concentrate on a related special topic. We have been selective, however, with books in psychometrics and econometrics, as well as textbooks which only deal with multivariate analysis in part.

Bibliographies which we searched are listed in Chapter I under Section 7. Particularly helpful was the three-volume set by Kendall & Doig (1962, 1965, 1968). In addition, many entries were found by searching:

*Jahrbuch über die Fortschritte der Mathematik.* [Vol. 1-60, 1868-1934. Pub. Walter de Gruyter & Co., Berlin & Leipzig]

*Mathematical Reviews.* [Vol. 1-22, 1940-1961. Vol. 23, 1962- two volumes per year. Author

Index Vol. 1-20 and Vol. 21-28. Pub. American Mathematical Society, Providence, R. I.]

*Statistical Theory and Method Abstracts.* [Vol. 1, 1959-1960. Vol. 2, 1961- one volume per year. Vol. 1-4 entitled *International Journal of Abstracts. Statistical Theory and Method.* Pub. Longman Group, Journals Division, Edinburgh, for the International Statistical Institute]

*Zentralblatt für Mathematik und ihre Grenzgebiete.* [Vol. 1, 1931-. Pub. Springer-Verlag, Berlin]

We also searched many bibliographies published as articles in journals or collections; these are listed in Chapter III under subject-matter code *1.2.*

All journals and collections which had at least ten entries when the above searching was completed were searched. Authors who then had at least ten entries were contacted or bibliographies of their writings searched (cf. §III.6). We also searched 'Specific non-normal multivariate distributions', a 15-page manuscript by Donald B. Owen, 'Selected references on clustering and pattern recognition', a 6-page manuscript by Herbert Solomon, and 'Permuted Index of Source Titles: 1 & 2', some 8 side-inches of computer printout by John W. Tukey. We are grateful to these authors for making these unpublished materials available to us.

The listing of papers was generated by the CDC 6600 computer at the University of Minnesota, using a computer program which interpreted the entries from coded input and sorted them into author order. For input, each author and the abbreviation of each journal or collection were coded; this ensured that each author and each journal or collection would be cited consistently. This computerization was most helpful in our work, especially in the searching stages, since it enabled us to obtain listings of the current file in any one of a number of different orders. Several programs were developed to aid in updating the files as corrections and additions were found.

Our bibliographic procedures and style are explained in the introductions to each of Chapters I, II and III. We were guided by:

*Bibliographic Procedures & Style; a Manual for Bibliographers in the Library of Congress.* By Blanche Prichard McCrum & Helen Dudenbostel Jones. Pub. The Library of Congress, Washington, D.C., 1954. 133 pp. [Reprinted 1966 with list of abbreviations]

*Guide to Reference Books.* Eighth Edition. By Constance M. Winchell. Pub. American Library Association, Chicago, 1967. 741 pp.

*Manual of Foreign Languages for the Use of Librarians, Bibliographers, Research Workers, Editors, Translators, and Printers.* Fourth Edition, revised and enlarged. By Georg F. von Ostermann. Pub. Central Book Company, Inc., New York.

We owe a tremendous debt of gratitude to David G. Doren for the computer programming which led to two-thirds of this book. The layout and arrangement of the entries in Chapter III is almost entirely due to him. Thanks are also due to the University of Minnesota Computer Center for very generous support of computer time on its CDC 6600, and especially to James K. Foster.

Our sincere appreciation goes to editorial assistants Judy Hebert, Joan Widerkehr, Cheryl Schiffman, and Leslie Ann Hathaway who each devoted about two years to *Abomsa*; in 1964 Judy created this acronym, a term of endearment for all who have participated in this project. We are also grateful to Peter Trenholme and Evelyn Matheson Styan. Peter's computer programs for the IBM 7094 at Columbia University during 1964-1967 formed the basis for the programs which eventually created Chapter III. Evelyn married into the project in 1967 and has since spent many days searching for references in libraries across North America and improving *Abomsa* in other ways.

We are particularly grateful to Heather Benson, as well as Pauline Mullins and Mary Ann Nikoriak for cheerfully and patiently preparing the difficult and tedious typescript. Special thanks are also due to Henry Koro for his expertise in photographing the entire manuscript.

There are numerous other colleagues to thank for assistance of various kinds. They include Arlene Grossman, Charles Johansson, Kathleen Dooley Kjellström, Irena Murray, Lolly Schiffman, Honora Shaughnessy, Jim Swearingen, and Peter Wachter as well as Gertrude Battell, Judi Campbell, Katherine Cane, Charles Cotterill, Warren Dent, Mr. Dhar, Colleen Doren, Gerry DuChaine, Pamela Oline Gerstman, N. C. Giri, Jean-Guy Hébert, Abdul Jhavary, Peter Keating, Lillian Kit, Monique Andrée Legault, Dennis Lienke, Pi-erh Lin, Jacqueline Marin, William Reed, Carol Hallett Robbins, Harold Sackrowitz, William Samborsky, Issie Scarowsky, Richard Stanley, Takakazu Sugiyama, Denys Voaden, Laurel Ward, and Lynn Zmistowski. We also wish to thank Alison G. Doig, Maurice G. Kendall, Ingram Olkin and John W. Tukey for guidance in the early stages of this bibliography.

An important factor in preparing this bibliography was the generous assistance of many of the authors whose papers we cite, as well as the help of librarians at Columbia University, The Library of Congress, Linda Hall Library, McGill University, The New York Public Library, Stanford University, and University of Minnesota. In addition, we are grateful to Bernard Lindgren and Edward Rosenthall for their generous provision of facilities at the University of Minnesota and McGill University. We also greatly appreciate the assistance and encouragement of our colleagues at Columbia, McGill, Minnesota, and Stanford. To those others who have helped us and who are not mentioned we extend special thanks.

The beginning of the preparation of this bibliography at Columbia University and its continuation at Stanford University was made possible by financial support from the Probability and Statistics Program of the Office of Naval Research under Contracts Nonr-266(33), NR 042-034; Nonr-4195(00), NR 042-233; Nonr-4259(08), NR 042-034; Nonr-225(53), NR 042-002; N 00014-67-A-0112-0030, NR 042-034. The bibliography was completed with additional financial support from the National Science Foundation and the Graduate School at the University of Minnesota, as well as from the National Research Council, Ottawa and the Social Sciences Research Fund at McGill University.

*Stanford, California*
*Minneapolis, Minnesota*
*Montreal, Quebec*

*February*, 1972

T. W. ANDERSON
SOMESH DAS GUPTA
GEORGE P. H. STYAN

# Cyrillic Alphabet

Transliteration as followed by *Mathematical Reviews*, Volume 30 (1965), p. 1207, with alternative forms in lower case.

## Cyrillic—Latin

| | | | | |
|---|---|---|---|---|
| А, а | A | Р, р | R |
| Б, б | B | С, с | S |
| В, в | V | Т, т | T |
| Г, г | G | У, у | U |
| Д, д | D | Ѳ, Ф, ф | F |
| Ђ, Е, е | E (ie͡) | Х, х | H (kh) |
| Ё, ё | E | Ц, ц | C (ts, t͡s) |
| Ж, ж | Ž(zh) | Ч, ч | Č (ch) |
| З, з | Z | Ш, ш | Š (sh) |
| И, и | I | Щ, щ | ŠČ (shch) |
| І, Й, й | Ĭ(i, j) | Ъ, ъ | ″ |
| К, к | K | Ы, ы | Y |
| Ѧ, Л, л | L | Ь, ь | ′ |
| М, м | M | Э, э | È (ė, e′) |
| Н, н | N | Ю, ю | JU (yu, i͡u) |
| О, о | O | Я, я | JA (ya, i͡a) |
| П, п | P | Ѵ, ѵ | Ẏ |

## Latin—Cyrillic

| | | | | |
|---|---|---|---|---|
| A | А | N | Н |
| B | Б | O | О |
| C | Ц | P | П |
| Č, ch | Ч | R | Р |
| D | Д | S | С |
| E | Е, Ё | Š, sh | Ш |
| È | Э | ŠČ, shch | Щ |
| F | Ф | T | Т |
| G | Г | ts, t͡s | Ц |
| H | Х | U | У |
| I | И | V | В |
| Ĭ | Й | Y | Ы |
| i͡a | Я | Ẏ | Ѵ |
| i͡e | Е | ya | Я |
| i͡u | Ю | yu | Ю |
| JA | Я | Z | З |
| JU | Ю | Ž, zh | Ж |
| K | Ќ | ′ | Ь |
| kh | Х | ″ | Ъ |
| L | Л | | |
| M | М | | |

CHAPTER I  BOOKS

# I.1   INTRODUCTION

We have found only eleven books which deal comprehensively with
multivariate statistical analysis (Section *1* of §I.2).  Several other
books, however, concentrate on special topics such as correlation
(Section *2A*), classification, cluster analysis and selection *(2B)*, factor
analysis *(2C)*, and multiple time series *(2F)*.  There are many books in
psychometrics and econometrics which deal with multivariate analysis to
some extent.  Almost all statistics texts have some material on the
product-moment correlation coefficient and the multivariate normal distribution.

In §I.2 we have listed 213 books classified according to the following scheme:

> *1.* Multivariate Statistical Analysis
> *2.* Special Topics in Multivariate Analysis
> > *2A.* Correlation *(4.1-4.9)*
> > *2B.* Classification, Cluster Analysis and Selection
> > *(6.1-6.7)*
> > *2C.* Factor Analysis *(15.1-15.3)*
> > *2D.* Psychometrics other than Factor Analysis
> > *(15.4-15.5)*
> > *2E.* Econometrics *(16.2)*
> > *2F.* Multiple Time Series *(16.1-16.5)*
> > *2G.* Miscellaneous
> *3.* Partial Contents: Multivariate Analysis
> *4.* Related to Multivariate Analysis
> *5.* Tables
> *6.* Computer Programs
> *7.* Bibliography

The codes which follow the headings above of Sections *2A* to *2F* are
subject-matter codes used for the annotation of papers in Chapter III;
these codes are identified in Chapter V.

Sections *1*, *2A*, *2B*, and *2F* are intended to be exhaustive.  In
Sections *2C*, *2D*, *2E*, and *2G* we have been selective, including only those
books which have substantial material on the theory of multivariate
analysis.  Section *3* includes statistics books which have some chapters
on multivariate analysis; the relevant chapter headings are cited.  Other
books which we consider useful for multivariate analysis are listed in
Section *4*; our selection here is highly subjective.  In Sections *5* and *6*
are listed tables and computer programs dealing primarily with multivariate
analysis; also cited are some books which concentrate on univariate statistics but which include multivariate criteria as well.  Section *7* contains
the statistical bibliographies which we have consulted for the preparation
of this *Bibliography*; included for completeness are some unpublished works
as well as some bibliographies which are only peripheral to multivariate
analysis.

    Books are listed if published in 1970 or earlier; amendments, how-
ever, are listed even if published after 1970.  The authors and titles
are transcribed from the title pages whenever possible.  Characters in
the Cyrillic alphabet are transliterated according to the scheme in *Mathematical Reviews*, Volume 30 (1965), p.1207 (and in the Preface of this
*Bibliography*).  The year in which the book was first published is given
in parentheses.  When there are various editions, we list the latest one
and give details of previous editions in square brackets.  Translations
are listed as known.  A Dewey or Library of Congress (LC) call number is
sometimes given.  References to abstracts or reviews often appear at the
end of the entry in parentheses as follows:

     JFM  *Jahrbuch über die Fortschritte der Mathematik*
      MR  *Mathematical Reviews*
     ZMG  *Zentralblatt für Mathematik und ihre Grenzgebiete*.

The Preface contains further information on these journals.  References
are by page number, except for *Mathematical Reviews* from Volume 20 onwards, which are by serial number.

# I.2 AN ANNOTATED LIST OF THE BOOKS

## 1. MULTIVARIATE STATISTICAL ANALYSIS

ANDERSON, T. W. (1958) *An Introduction to Multivariate Statistical Analysis*. John Wiley & Sons, Inc., New York. xii + 374 pp. [Russian translation (1963) *Vvedenie v Mnogomernyĭ Statističeskiĭ Analiz*. Gosudarstvennoe Izdatel'stvo Fiziko-matematičeskoĭ Literatury, Moskva. 500 pp.] (MR19, p.992)

COOLEY, William W., & LOHNES, Paul R. (1962) *Multivariate Procedures for the Behavioral Sciences*. John Wiley & Sons, Inc., New York. x + 211 pp.

DEMPSTER, A. P. (1969) *Elements of Continuous Multivariate Analysis*. Addison-Wesley Publishing Company, Reading, Mass. xii + 388 pp.

DuBOIS, Philip H. (1957) *Multivariate Correlational Analysis*. Harper & Brothers, Publishers, New York. xv + 202 pp. (MR19, p.990)

HOPE, Keith (1968) *Methods of Multivariate Analysis, with Handbook of Multivariate Methods Programmed in Atlas Autocode*. University of London Press, Ltd., London. 288 pp. [Paperback edition *Methods of Multivariate Analysis*. Unibooks, London. 165 pp.]

KENDALL, M. G. (1957) *A Course in Multivariate Analysis*. Charles Griffin & Company, Ltd., London. 185 pp. [Number 2 of Griffin's Statistical Monographs & Courses. Copyright 1968] (MR19, p.1093)

MILLER, Kenneth S. (1964) *Multidimensional Gaussian Distributions*. John Wiley & Sons, Inc., New York. viii + 129 pp. (MR30-1564)

MORRISON, Donald F. (1967) *Multivariate Statistical Methods*. McGraw-Hill Book Company, New York. xiii + 338 pp. (MR35-3811)

ROY, S. N. (1957) *Some Aspects of Multivariate Analysis*. John Wiley & Sons, Inc., New York and Indian Statistical Institute, Calcutta. viii + 214 pp. [Indian Statistical Series No. 1] (MR19, p.1093)

SEAL, Hilary L. (1964) *Multivariate Statistical Analysis for Biologists*. Methuen & Co., Ltd., London and John Wiley & Sons, Inc., New York. xi + 207 pp.

SIOTANI [= SHIOYA], Minoru, & ASANO, Choichiro (1966) *Tahenryo Kaiseki Ron*. Kyoritsu Shuppan Co., Ltd., 4-6-19, Kohinata, Bunkyo-ku, Tokyo, Japan. [11] + 236 pp. [= *Statistical Multivariate Analysis*. In Japanese. Joho Kagaku Koza (= Lectures on Information Science) A·5·3, edited by Toshio Kitagawa]

## 2. SPECIAL TOPICS IN MULTIVARIATE ANALYSIS

### 2A. Correlation.

BAGGALEY, Andrew R. (1964) *Intermediate Correlational Methods*. John Wiley & Sons, Inc., New York. [3] + 211 pp. (MR30-2656)

BONFERRONI, C. E., & BRAMBILLA, F. (1942) *Studi sulla Correlazione e sulla Connessione*. Università Commerciale "Luigi Bocconi", Milano. 368 pp. [Istituto di Statistica, Serie A, Vol. I. LC:QA276.M47. Contents:- C.E. Bonferroni: I. Di una estensione del coefficiente di correlazione; II. Di un coefficiente di correlazione simultanea; III. Nuovi indici di connessione fra variabili statistiche. F. Brambilla: Teoria statistica della correlazione e della connessione]

de MONTESSUS de BALLORE, Vicomte Robert (1932) *La Méthode de Corrélation suivie de la Table des Carrés des Nombres Entiers de 1 à 1000*. Gauthier-Villars & Cie., Paris. 77 pp. (ZMG5, p.405)

EZEKIEL, Mordecai (1941) *Methods of Correlation Analysis*. Second Edition. John Wiley & Sons, Inc., New York. xix + 531 pp. [First published 1930, xiv + 427 pp.] (JFM56, p.1092)

GINI, Corrado (1955) *Variabilità e Concentrazione*. Seconda Edizione aggiornata a cura di E. Pizzetti e T. Salvemini. Libreria Eredi Virgilio Veschi, Roma. xxviii + 738 + [3] pp. [Memorie di Metodologia Statistica Volume Primo. Università degli Studi di Roma, Facoltà di Scienze Statistiche, Demografiche ed Attuariali]

HAGGARD, Ernest A. (1958) *Intraclass Correlation and the Analysis of Variance*. Dryden Press, Inc., New York. xvii + 171 pp.

KENDALL, Maurice G. (1970) *Rank Correlation Methods*. Fourth Edition. Charles Griffin & Company, Ltd., London. viii + 202 pp. [First published 1948. Second Edition 1955; Third Edition 1962, vii + 199 pp.]

QUENOUILLE, M. H. (1952) *Associated Measurements*. Butterworths Scientific Publications, London. x + 242 pp. (MR14, p.568)

RISSER, R., & TRAYNARD, C.-E. (1958) *Les Principes de la Statistique Mathématique. Livre II: Corrélation. Séries Chronologiques*. Deuxième Édition, revue et augmentée. Gauthier-Villars & Cie., Paris. xi + 418 pp. [Traité du Calcul des Probabilités et de ses Applications. Tome I, Les Principes de la Théorie des Probabilités: fasc. IV] (MR19, p.894)

*2A.* Correlation. (cont.)

ROMANOVSKIĬ, Vsevolod Ivanovič (1927) *Èlementy Teorii Korreljacii.*
Izdan'e 2, Pererabotannoe i Dopolnennoe. Opytno-issledovatel'skiĭ,
Institut Vodnogo Knozjaistva, Tashkent. [= *Elements of the Theory
of Correlation.* Second Edition, revised and extended. In Russian.
LC:QA273.R6]

SALVEMINI, Tommaso (1959) *Regressione e Correlazione.* Edizioni Scienti-
fiche "Einaudi", Torino. vii + 112 pp.

SLUCKIĬ, E. E. (1912) *Teorija Korreljacii i Èlementy Učenija o Krivyh
Raspredelenija.* Isvestija Kievskogo Kommerčeskogo Instituta Kn. XVI,
Kiev. iv + 208 pp. [= *Correlation Theory and Elements of Teaching
on Curved Distributions.* In Russian]

TRELOAR, Alan E. (1942) *Correlation Analysis.* Burgess Publishing Co.,
Minneapolis, Minn. 64 pp. (MR4, p.220)

TSCHUPROW, A. A. [ČUPROV, Aleksandr Aleksandrovič] (1939) *Principles of
the Mathematical Theory of Correlation.* William Hodge & Company,
Ltd., London and Nordeman Publishing Co., Inc., New York. x + 194 pp.
[Translation of the German edition (1925) *Grundbegriffe und Grund-
probleme der Korrelationstheorie.* B. G. Teubner, Leipzig. v + [1] +
153 pp. Russian edition *Èlementy Matematičeskoĭ Teorii Korreljacii*]
(MR1, p.151)


*See also* Section *3* ANDERSON (1935), ELDERTON (1948), FISHER (1970);
Section *5* DAVID (1938), MINER (1922).

## 2B. Classification, Cluster Analysis and Selection.

FU, K. S. (1968) *Sequential Methods in Pattern Recognition and Machine Learning.* Academic Press, New York. xi + 227 pp. [Volume 52 in Mathematics in Science and Engineering; a Series of Monographs and Textbooks]

ROZEBOOM, William W. (1966) *Foundations of the Theory of Prediction.* Dorsey Press, Homewood, Ill. viii + 628 pp.

RULON, Phillip J., TIEDEMAN, David V., TATSUOKA, Maurice M., & LANGMUIR, Charles R. (1967) *Multivariate Statistics for Personnel Classification.* John Wiley & Sons, Inc., New York. xi + 406 pp.

SEBESTYEN, George S. (1962) *Decision-making Processes in Pattern Recognition.* The Macmillan Co., New York. viii + 162 pp. (MR27-6620)

SOKAL, Robert R., & SNEATH, Peter H. A. (1963) *Principles of Numerical Taxonomy.* W. H. Freeman and Company, San Francisco. xvi + [1] + 359 pp.

TATSUOKA, Maurice M. (1970) *Discriminant Analysis; the Study of Group Differences.* Institute for Personality and Ability Testing, 1602-04 Coronado Drive, Champaign, Ill. iv + 57 pp. [Number 6 of Selected Topics in Advanced Statistics; an Elementary Approach]

TRYON, Robert C., & BAILEY, Daniel E. (1970) *Cluster Analysis.* McGraw-Hill Book Company, New York. xviii + 347 pp.


*See also* Section *2D* HORST (1968), MAGNUSSON (1967), THORNDIKE (1949); Section *3* BECHHOFER, KIEFER & SOBEL (1968), RAIFFA & SCHLAIFER (1961); Section *5* BOSE & CHAUDHURI (1966); Section *7* HODGES (1950), POSTEN (1962).

## 2C. Factor Analysis.

ADCOCK, C. J. (1954) *Factorial Analysis for Non-Mathematicians.* Melbourne University Press, Melbourne. 88 pp.

BURT, Cyril (1940) *The Factors of the Mind; an Introduction to Factor Analysis in Psychology.* University of London Press, Ltd., London. xiv + 509 pp. [Macmillan Co., New York, 1941]

CATTELL, Raymond B. (1952) *Factor Analysis; an Introduction and Manual for the Psychologist and Social Scientist.* Harper & Brothers, Publishers, New York. xiii + 462 pp.

FRUCHTER, Benjamin (1954) *Introduction to Factor Analysis.* D. Van Nostrand Company, Inc., Princeton, N. J. xii + 280 pp.

HARMAN, Harry H. (1967) *Modern Factor Analysis.* Second Edition, revised. The University of Chicago Press, Chicago. xx + 474 pp. [First published 1960, xvi + 471 pp.] (MR28-2610)

HENRYSSON, Sten (1957) *Applicability of Factor Analysis in the Behavorial Sciences; a Methodological Study.* Almqvist & Wiksell, Stockholm. 156 pp. [Stockholm Studies in Educational Psychology No. 1]

HOLZINGER, Karl J. (1930) *Statistical Résumé of the Spearman Two-Factor Theory.* The University of Chicago Press, Chicago. iv + 44 pp.

HOLZINGER, Karl J., & HARMAN, Harry H. (1941) *Factor Analysis; a Synthesis of Factorial Methods.* The University of Chicago Press, Chicago. xii + 417 pp. (MR4, p.18)

HORST, Paul (1965) *Factor Analysis of Data Matrices.* Holt, Rinehart & Winston, Inc., New York. xix + 730 pp. (MR33-8062)

JÖRESKOG, K. G. (1963) *Statistical Estimation in Factor Analysis; a New Technique and its Foundation.* Almqvist & Wiksell, Stockholm. 145 pp.

KELLEY, Truman Lee (1928) *Crossroads in the Mind of Man; a Study of Differentiable Mental Abilities.* Stanford University Press, Stanford, Calif. vii + 238 pp.

LAWLEY, D. N., & MAXWELL, A. E. (1963) *Factor Analysis as a Statistical Method.* Butterworths, London. viii + 117 pp. [Butterworths Mathematical Texts No. 2]

SPEARMAN, Charles (1927) *The Abilities of Man; their Nature and Measurement.* Macmillan and Co., Limited, London. viii + 415 + xxxiii pp. [Macmillan Co., New York, vi + 416 + xxxiv pp., 1927. Reprinted 1932. French translation (1936) *Les Aptitudes de l'Homme; leur Nature et leur Mesure*] (JFM53, p.521)

*2C*. Factor Analysis.  (cont.)

STEPHENSON, William (1953)  *The Study of Behavior; Q-technique and its
    Methodology*.  The University of Chicago Press, Chicago.  ix + 376 pp.

THOMSON, Godfrey H. (1951)  *The Factorial Analysis of Human Ability*.
    Fifth Edition.  University of London Press, Ltd., London.  xv + 383 pp.
    [First published 1939.  Second edition 1946, Third 1948, Fourth 1950.
    Fifth Edition reprinted 1956]

THURSTONE, L. L. (1933)  *The Theory of Multiple Factors*.  Edwards
    Brothers, Inc., Ann Arbor, Mich.  vii + 65 pp.  (JFM59, p.1207)

THURSTONE, L. L. (1933)  *A Simplified Multiple Factor Method and an
    Outline of the Computations*.  The University of Chicago Press,
    Chicago.

THURSTONE, L. L. (1935)  *The Vectors of Mind; Multiple-Factor Analysis
    for the Isolation of Primary Traits*.  The University of Chicago
    Press, Chicago.  xv + 266 pp.  (JFM62, p.1380)

THURSTONE, L. L. (1947)  *Multiple-Factor Analysis; a Development and
    Expansion of "The Vectors of Mind"*.  The University of Chicago Press,
    Chicago.  xix + 535 pp.  (MR9, p.47)

TRYON, Robert Choate (1939)  *Cluster Analysis; Correlation Profile and
    Orthometric (Factor) Analysis for the Isolation of Unities in Mind
    and Personality*.  Edwards Brothers, Inc., Ann Arbor, Mich.  viii +
    122 pp.

ÜBERLA, K. (1968)  *Faktorenanalyse; eine systematische Einführung für
    Psychologen, Mediziner, Wirtschafts- und Sozialwissenschaftler*.
    Springer-Verlag, Berlin.  xii + 399 pp.


*See also*  Section *2D* MAGNUSSON (1967); *2G* HOTELLING (1933).

## 2D. Psychometrics other than Factor Analysis.

COOMBS, C. H. (1964) *A Theory of Data.* John Wiley & Sons, Inc., New York. xviii + 585 pp.

GUILFORD, Joy Paul (1954) *Psychometric Methods.* Second Edition. McGraw-Hill Book Company, New York. ix + 597 pp.

GULLIKSEN, Harold (1950) *Theory of Mental Tests.* John Wiley & Sons, Inc., New York. xix + 486 pp.

HORST, Paul (1968) *Psychological Measurement and Prediction.* Brooks/Cole Publishing Company, Belmont, Calif. xii + 455 pp.

LAZARSFELD, Paul F., & HENRY, Neil W. (1968) *Latent Structure Analysis.* Houghton Mifflin Company, Boston. ix + 294 pp.

LONG, Max F. (1959) *Psychometric Analysis.* Huna Research Publications, Vista, Calif. 118 pp.

LORD, Frederic M., & NOVICK, Melvin R. [With contributions by Allan Birnbaum] (1968) *Statistical Theories of Mental Test Scores.* Addison-Wesley Publishing Company, Reading, Mass. xvii + 568 pp.

MAGNUSSON, David (1967) *Test Theory.* Addison-Wesley Publishing Company, Reading, Mass. x + 270 pp. [English translation of the second Swedish edition of *Testteori,* Almqvist & Wiksell/Gebers Förlag AB, Stockholm, 1966; first published 1961]

NUNNALLY, Jim (1967) *Psychometric Theory.* McGraw-Hill Book Company, New York. xii + 640 pp.

PFANZAGL, J. (1968) *Theory of Measurement.* Physica-Verlag, Würzburg. 235 pp.

THORNDIKE, Robert L. (1949) *Personnel Selection.* John Wiley & Sons, Inc., New York. viii + 358 pp.

TORGERSON, Warren S. (1958) *Theory and Methods of Scaling.* John Wiley & Sons, Inc., New York. xiii + 460 pp.

*See also* Section *2B* ROZEBOOM (1966).

## 2E. Econometrics.

CHRIST, Carl F. (1966) *Econometric Models and Methods*. John Wiley & Sons, Inc., New York. xxiii + 705 pp.

DAVIS, Harold Thayer (1941) *The Theory of Econometrics*. The Principia Press, Inc., Bloomington, Ill. 480 pp. (MR3, p.11)

DHRYMES, Phoebus J. (1970) *Econometrics: Statistical Foundations and Applications*. Harper & Row, Publishers, Inc., New York. xii + 592 pp.

FÉRON, Robert (1968) *Modèles Statiques de l'Économétrie*. Editions Eyrolles, Paris. 142 pp.

FISHER, Franklin M. (1966) *The Identification Problem in Econometrics*. McGraw-Hill Book Company, New York. xi + 203 pp.

FISHMAN, George S. (1968) *Spectral Methods in Econometrics*. Harvard University Press, Cambridge, Mass. xi + 212 pp.

FISK, P. R. (1967) *Stochastically Dependent Equations; an Introductory Text for Econometricians*. Charles Griffin & Company, Ltd., London. viii + 181 pp. [Number 21 of Griffin's Statistical Monographs & Courses]

GOLDBERGER, Arthur S. (1964) *Econometric Theory*. John Wiley & Sons, Inc., New York. xi + 399 pp. [Spanish translation (1969) *Teoria de la Econometria*. Limusa-Wiley, Mexico]

GUITTON, H. (1959) *Statistique et Économétrie*. Dalloz, Paris. 544 pp. [Spanish translation *Estadística y Econometria*]

HAAVELMO, Trygve (1944) *The Probability Approach to Econometrics*. The University of Chicago Press, Chicago. viii + 118 pp. [Separate imprint of entry no. 2271 in Chapter III] (MR6, p.93)

HUANG, David S. (1970) *Regression and Econometric Methods*. John Wiley & Sons, Inc., New York. xiii + 274 pp.

JOHNSTON, J. (1963) *Econometric Methods*. McGraw-Hill Book Company, New York. xiii + 300 pp. [Second Edition 1972, x + 437 pp.]

KLEIN, Lawrence R. (1953) *A Textbook of Econometrics*. Row, Peterson and Co., Evanston, Ill. ix + 355 pp. [Spanish translation *Manuel de Econometria*. Aguilar, Madrid] (MR18, p.265)

2E. Econometrics. (cont.)

KOOPMANS, Tjalling (1937) *Linear Regression Analysis of Economic Time Series*. De Erven F. Bohn, n. v., Haarlem. xi + [1] + 150 pp. [Number 20 of Publicatië, Nederlandsch Ekonomisch Instituut]

LANGE, Oskar (1962) *Introduction to Econometrics*. Second Edition, revised and enlarged. Pergamon Press Ltd., Oxford and Państwowe Wydawnictwo Naukowe, Warszawa. 433 pp. [Translation of the Polish second edition (1961) *Wstęp do Ekonometrii*, Warszawa. First published 1958 (Warszawa); 1959 (London), 384 pp. Serbo-Croat translation 1960, Sarajevo. Italian translation 1962. Spanish translation (1962) *Introducción a la Econometria*. Fondo de Cultura Economica, Mexico. 356 pp.] (MR25-1948)

LESER, C. E. V. (1966) *Econometric Techniques and Problems*. Charles Griffin & Company, Ltd., London. viii + 119 pp. [Number 20 of Griffin's Statistical Monographs & Courses]

MALINVAUD, Edmond (1970) *Statistical Methods of Econometrics*. Second Edition. Rand McNally & Company, Chicago and North-Holland Publishing Company, Amsterdam. xvi + 744 pp. [Studies in Mathematical and Managerial Economics, Vol. 6. First published 1966, xiv + 631 pp. Translation of *Méthodes Statistiques de l'Econométrie*. Deuxième Édition. Dunod, Paris, 1969. xiv + 782 pp. Finance et Économie Appliqué, Vol. 16. First published 1964, xiv + 634 pp. Spanish translation (1967) *Metodos Estadísticos de la Econometría*, Ariel, Barcelona]

MENGES, Günter (1961) *Ökonometrie*. Betriebswirtschaftlicher Verlag Dr. Th. Gabler, Wiesbaden. 251 pp. [Verlags-Nr. 607. Die Wirtschaftswissenschaften 34, Lfg. Reihe B: Volkswirtschaftslehre, Beitrag Nr. 20]

THEIL, H. [Assisted by J. S. Cramer, H. Moerman, & A. Russchen] (1961) *Economic Forecasts and Policy*. Second Revised Edition. North-Holland Publishing Company, Amsterdam. xxxii + 567 pp. [Contributions to Economic Analysis No. XV. First published 1958, xxxi + 562 pp.]

TINBERGEN, Jan (1951) *Econometrics*. The Blakiston Co., Philadelphia. 258 pp.

TINBERGEN, Jan (1968) *Statistical Testing of Business-Cycle Theories. I. A Method and its Application to Investment Activity. II. Business Cycles in the United States of America 1919-1932*. Agathon Press, Inc., New York. 244 + [3] pp. [First published League of Nations, Geneva, 1939]

*2E.* Econometrics.  (cont.)

TINTNER, Gerhard (1952)  *Econometrics.*  John Wiley & Sons, Inc., New
    York.  xiii + 370 pp.  [Reprinted by Science Editions, 1965.  Japan-
    ese translation 1961, Bun Ga Do Publishers, Tokyo]

VALAVANIS, Stefan (1959)  *Econometrics; an Introduction to Maximum Like-
    lihood Methods.*  Edited, from manuscript, by Alfred H. Conrad.
    McGraw-Hill Book Company, New York.  xvii + 223 pp.  [Author also
    known as Valavanis-Vail]

WALTERS, A. A. (1968)  *An Introduction to Econometrics.*  Macmillan Co.,
    New York.  377 pp.

WOLD, Herman O. A.  [In association with Lars Juréen] (1953)  *Demand
    Analysis; a Study in Econometrics.*  John Wiley & Sons, Inc., New York.
    xvi + 358 pp.  [First published 1952, Gebens, Stockholm.  Spanish
    translation (1956) *Análisis de la Demanda; un Estudio de Econometría.*
    Selecciones Gráficas, Madrid.  427 pp.]  (MR16, p.274)

WONNACOTT, Ronald J., & WONNACOTT, Thomas H. (1970)  *Econometrics.*
    John Wiley & Sons, Inc., New York.  ix + 445 + [8] pp.


*See also*  Section *2F* GRANGER (1964);  Section *2G* FRISCH (1934), REIERSØL
    (1945);  Section *6* KLEIN, EVANS & HARTLEY (1969).

## 2F. Multiple Time Series.

GRANGER, C. W. J.   [In association with M. Hatanaka] (1964)   *Spectral Analysis of Economic Time Series*.   Princeton University Press, Princeton, N. J.   xviii + 299 pp.   (MR30-4368)

HANNAN, E. J. (1970)   *Multiple Time Series*.   John Wiley & Sons, Inc., New York.   xi + 536 pp.

QUENOUILLE, M. H. (1957)   *The Analysis of Multiple Time-Series*.   Charles Griffin & Company, Ltd., London.   105 pp.   [Number 1 of Griffin's Statistical Monographs & Courses.   Copyright and Second impression (with minor corrections), 1968]   (MR19, p.1205; MR39-3669)

ROBINSON, Enders A. (1967)   *Multichannel Time Series Analysis with Digital Computer Programs*.   Holden-Day, Inc., San Francisco.   viii + 298 pp.

*See also*   Section *2E* KOOPMANS (1937); Section *3* BARTLETT (1966), JENKINS & WATTS (1968), WHITTLE (1963), WIENER (1949); Section *4* DOOB (1953); Section *7* WOLD (1965).

## 2G. Miscellaneous.

BHARUCHA-REID, A. T. (undated)  *A Survey on the Theory of Random Equations.*
The Institute of Mathematical Sciences, Madras, India.  72 pp.
[Matscience Report 31, mimeographed.  Cover = *Lectures on Theory of*
*Random Equations.*  Dewey:  519.B469s]

DELTHEIL, R. (1926)  *Probabilités Géométriques.*  Gauthier-Villars & Cie.,
Paris.  123 pp.  [Traité du Calcul des Probabilités et de ses Applications.  Tome II, Les Applications de la Théorie des Probabilités
aux Sciences Mathématiques et aux Sciences Physiques:  fasc. II]
(JFM52, p.526)

FRISCH, Ragnar (1934)  *Statistical Confluence Analysis by Means of*
*Complete Regression Systems.*  Universitetets Økonomiske Institutt,
Oslo.  192 pp.  [Publikasjon Nr. 5.  Separate impression from *Nordic*
*Statistical Journal,* Vol. 5, 1933]  (ZMG11, p.219)

HOTELLING, Harold (1933)  *Analysis of a Complex of Statistical Variables*
*into Principal Components.*  Warwick and York, Baltimore.  48 pp.
[Separate imprint of entry no. 2587 in Chapter III.  Dewey:  519.4.H79]

KAMALOV, M. K. (1958)  *Raspredelenie Kvadratičnyh Form v Vyborkah iz*
*Normal'noĭ Sovokupnosti.*  Akademija Nauk Uzbekskoi SSR, Institut
Matematiki i Mehaniki im. V. I. Romanovskogo, Tashkent.  289 pp.
[= *Distribution of Quadratic Forms in Samples of a Normal Population.*
In Russian]  (MR22-10038)

KENDALL, M. G., & MORAN, P. A. P. (1963)  *Geometrical Probability.*
Charles Griffin & Company, Ltd., London.  125 pp.  [Number 10 of
Griffin's Statistical Monographs & Courses]  (MR30-4275)

McQUIE, Robert, CASSADAY, George, CHAPMAN, Robert, & MONTWEILER, William
(1968)  *Multivariate Analysis of Combat.*  U. S. Department of
Commerce, Washington, D. C.  109 pp.  [Report AD673-294 available
from Clearinghouse for Federal Scientific and Technical Information,
Springfield, Va.  Abstract in *U. S. Government Research and Develop-*
*ment Reports,* October 25, 1968 (#20, p.111, Section 15G:  Operations,
Strategies and Tactics).  " ... the methods combine mean values,
data plots and correlation coefficients with factor analysis and
discriminant functions ... applied to data from ... 12 German tank
battles in World War 2."]

MARDIA, K. V. (1970)  *Families of Bivariate Distributions.*  Charles
Griffin & Company, Ltd., London.  ix + 109 pp.  [Number 27 of
Griffin's Statistical Monographs & Courses]

MEHTA, M. L. (1967)  *Random Matrices and the Statistical Theory of Energy*
*Levels.*  Academic Press, New York.  x + 259 pp.

REIERSØL, Olav (1945)  *Confluence Analysis by means of Instrumental Sets*
*of Variables.*  Almqvist & Wiksell, Stockholm.  119 pp.  [Separate
imprint of entry no. 4501 in Chapter III]  (MR7, p.317)

### 3. PARTIAL CONTENTS: MULTIVARIATE ANALYSIS

ACTON, Forman S. (1966) *Analysis of Straight-Line Data*. Dover Publications, Inc., New York. xiii + 267 pp. [First published 1959, John Wiley & Sons, Inc., New York] (MR21-4512)

  *Chapter 4* Samples from bivariate normal populations. *5* Regression with both $x$ and $y$ in error.

ANDERSON, Oskar N. (1935) *Einführung in die Mathematische Statistik*. Julius Springer, Wien. v + 314 pp. (ZMG12, p.111)

  *Kapitel IV* Korrelationstheorie und verwandte Forschungsgebiete.

BARTLETT, M. S. (1966) *An Introduction to Stochastic Processes with Special Reference to Methods and Applications*. Second Edition. Cambridge University Press, Cambridge. xvi + 362 pp. [First published 1955] (MR16, p.939)

  *Section 6.5* Multivariate and multidimensional stationary processes. *Chapter 7* Prediction and communication theory. *Section 9.3* Multivariate autoregressive series. *Section 9.4* Multidimensional series.

BECHHOFER, Robert E., KIEFER, Jack, & SOBEL, Milton (1968) *Sequential Identification and Ranking Procedures with Special Reference to Koopman-Darmois Populations*. The University of Chicago Press, Chicago. xvii + 420 pp. [Statistical Research Monographs Volume III]

  *Chapter 5* Some specific ranking problems and associated procedures. *6* Some special analytical results for the basic ranking and identification procedures applied to Koopman-Darmois populations for general k. *11* Koopman-Darmois populations. *14* Some analytical results concerning the PCS-function, ASN-function, and RE-function of the basic sequential ranking procedure for Goal I.

CHAKRAVARTI, I. M., LAHA, R. G., & ROY, J. (1967) *Handbook of Methods of Applied Statistics. Volume I, Techniques of Computation, Descriptive Methods, and Statistical Inference*. John Wiley & Sons, Inc., New York. xiv + 460 pp. (MR37-998a)

  *Part I* Techniques of computation: *§3* Numerical solution of equations, *§5* Matrix calculations. *Part II* Descriptive methods: *§5* Bivariate data: regression and correlation. *§6* Multivariate data: multiple regression and correlation. *Part III* Statistical inference: *§6* Tests of parameters of normal populations, *§8* Tests of models, *§9* Multivariate analysis.

3. Partial Contents: Multivariate Analysis.  (cont.)

CRAMÉR, Harald (1946)  *Mathematical Methods of Statistics*.  Princeton
University Press, Princeton, N. J.  xvi + 575 pp.  [First published
1945, Almqvist & Wiksell, Uppsala.  Princeton Mathematical Series No.9]
(MR8, p.39)

   *Chapter 3* Point sets in $n$ dimensions.  *8-9* Theory of measure and
integration in $R_n$.  *11* Matrices, determinants and quadratic forms.
*21-24* Variables and distributions in $R_n$.  *29* Exact sampling distributions.  *31* Tests of significance for parameters.


CRAMÉR, Harald (1970)  *Random Variables and Probability Distributions*.
Third Edition.  Cambridge University Press, Cambridge.  ix + 118 pp.
[First published 1937.  Second Edition 1962.  Cambridge Tracts in
Mathematics and Mathematical Physics No. 36]   (MR29-2879)

   *Third Part* Distributions in $R_k$:   §*IX* General properties, characteristic functions:  §*X* The normal distribution and the central
limit theorem.


ELDERTON, William Palin (1953)  *Frequency Curves and Correlation*.   Fourth
Edition.  Cambridge University Press, Cambridge.  xi + 272 pp.
[First published 1906, Second Edition 1927, Charles and Edwin Layton,
London.  Third Edition 1938.  Fourth Edition reprinted 1954, Dover
Publications, Inc., New York]

   *Chapter VII* Correlation.  *VIII* Theoretical distributions.
Spurious correlation.  *IX* Correlation of characters not quantitatively
measurable.  *XIII* Partial correlation.


ELDERTON, William Palin, & JOHNSON, Norman Lloyd (1969)  *Systems of
Frequency Curves*.  Cambridge University Press, Cambridge.  vii +
216 pp.  [Revision of ELDERTON (1953) with "practically all the
material on correlation removed"]

   *Chapter 8* Frequency surfaces.


EZEKIEL, Mordecai, & FOX, Karl A. (1959)  *Methods of Correlation and
Regression Analysis:  Linear and Curvilinear*.  Third Edition.  John
Wiley & Sons, Inc., New York.  xv + 548 pp.  [Revision of EZEKIEL
(1941) in Section *2A*]  (MR24-A3762)

   *Chapter 21* Measuring the relation between one variable and two
or more others operating jointly.  *24* Fitting systems of two or more
simultaneous equations.

3. Partial Contents: Multivariate Analysis.  (cont.)

FISHER, Sir Ronald A. (1970)  *Statistical Methods for Research Workers.*
   Fourteenth Edition, revised and enlarged by E. A. Cornish.  Oliver
   and Boyd, Edinburgh.  xiii + 362 pp.  [First published 1925.  Second-
   Thirteenth Editions, 1928-1958.  French translation (1947) *Les
   Méthodes Statistiques adaptées à la Recherche Scientifique.*  Presses
   Universitaires de France, Paris.  ix + 324 pp.  Japanese translation
   (1952) *Kenkyû-sya no tame no Tôkeiteki-hôhô.*  Sôbun-sya, Tokyo.
   Spanish translation (1949) *Métodos Estadísticos para Investigadores.*
   Aguilar, Madrid.  322 pp.  Also German, Italian and Russian trans-
   lations]  (JFM51, p.414)

   *Chapter IV* Tests of goodness of fit, independence and homogeneity;
with table of $\chi^2$.  *VI* The correlation coefficient.  *VII* Intraclass
correlations and the analysis of variance.


FRASER, D. A. S. (1968)  *The Structure of Inference.*  John Wiley & Sons,
   Inc., New York.  x + 344 pp.  (MR38-3946)

   *Chapter 3* Linear models.  *4* Conditional analysis.  *5* Marginal
analysis.


GRAYBILL, Franklin A. (1961)  *An Introduction to Linear Statistical
   Models. Volume I.*  McGraw-Hill Book Company, New York.  xiii + 463 pp
   (MR23-A3612)

   *Chapter 1* Mathematical concepts.  *3* The multivariate normal distribution.  *4* Distribution of quadratic forms.  *7* Computing techniques.


HANNAN, E. J. (1965)  *Group Representations and Applied Probability.*
   Methuen & Co., Ltd., London.  iii + 71 pp.  [Methuen's Review Series
   in Applied Probability Vol.3.  Separate imprint of entry no. 2330 in
   Chapter III]  (MR35-2307b)

   *Section 8.2* Multivariate analysis.  *10.2* The distribution of the
eigenvalues of the covariance matrix.


JENKINS, Gwilym M., & WATTS, Donald G. (1968)  *Spectral Analysis and its
   Applications.*  Holden-Day, Inc., San Francisco.  xviii + 525 pp.
   (MR37-6000)

   *Chapter 5* Introduction to time series analysis.  *6* The spectrum.
*8* The cross correlation function and cross spectrum.  *9* Estimation
of cross spectra.  *10* Estimation of frequency response functions.
*11* Multivariate spectral analysis.

3. Partial Contents: Multivariate Analysis.  (cont.)

JOHNSON, Norman L., & KOTZ, Samuel (1969)  *Distributions in Statistics:*
*Discrete Distributions.*  Houghton Mifflin Company, Boston.  xvi +
328 pp.

    *Chapter 11* Multivariate discrete distributions.


JOHNSON, Norman L., & LEONE, Fred C. (1964)  *Statistics and Experimental*
*Design in Engineering and the Physical Sciences.  Volume II.*  John
Wiley & Sons, Inc., New York.  ix + 399 pp.  (MR30-2581)

    *Chapter 17* Multivariate observations.


KENDALL, Maurice G. (1952)  *The Advanced Theory of Statistics.  Volume I.*
Fifth Edition.  Charles Griffin & Company, Ltd., London.  xi + 457 pp.
[First published 1943.  Second Edition 1945, Third 1947, Fourth 1948]
(MR6, p.89)

    *Chapter 11* Approximations to sampling distributions.  *13* Associa-
tion and contingency.  *15* Partial and multiple correlation.  *18* Rank
correlation.


KENDALL, Maurice G. (1951)  *The Advanced Theory of Statistics.  Volume II.*
Third Edition.  Charles Griffin & Company, Ltd., London.  vii + 521
pp.  [First published 1946.  Second Edition 1947]  (MR8, p.473)

    *Chapter 28* Multivariate analysis.


KENDALL, Maurice G., & STUART, Alan (1969)  *The Advanced Theory of Statis-*
*tics.  Volume 1, Distribution Theory.*  Third Edition.  Charles Griffin
& Company, Ltd., London.  xii + 439 pp.  [First published 1958.
Second Edition 1963.  Russian translation (1966) *Teorija Raspredelenii.*
Izdat. "Nauka", Moskva.  587 pp.]  (MR23-A2247, 35-3767)

    *Chapter 13* Approximations to sampling distributions -(2).  *15*
The multivariate normal distribution and quadratic forms.  *16* Distributions associated with the normal.


KENDALL, Maurice G., & STUART, Alan (1967)  *The Advanced Theory of Stat-*
*istics.  Volume 2, Inference and Relationship.*  Second Edition.
Charles Griffin & Company, Ltd., London.  ix + 690 pp.  [First published 1961]

    *Chapter 26* Statistical relationship:  linear regression and
correlation.  *27* Partial and multiple correlation.  *29* Functional
and structural relationship.

3. Partial Contents: Multivariate Analysis.  (cont.)

KENDALL, Maurice G., & STUART, Alan (1968)  *The Advanced Theory of Statistics. Volume 3, Design and Analysis, and Time-Series.*  Second Edition.  Charles Griffin & Company, Ltd., London.  x + 557 pp.  [First published 1966]

    *Chapter 41* Multivariate distribution theory.  *42* Tests of hypotheses in multivariate analysis.  *43* Canonical variables.  *44* Discrimination and classification.  *48* The sampling theory of serial correlation.

KOERTS, J., & ABRAHAMSE, A. P. J. (1969)  *On the Theory and Application of the General Linear Model.*  Rotterdam University Press, Rotterdam.  ix + 185 pp.

    *Chapter 3* Estimators of disturbances.  *5* The distribution of the test statistic.  *8* The correlation coefficient.  *9* A computer program for calculating the distribution of quadratic forms in normal variables and for calculating BLUS residuals.

KULLBACK, Solomon (1968)  *Information Theory and Statistics.*  Dover Publications, Inc., New York.  xv + 399 pp.  [First published 1959, John Wiley & Sons, Inc., New York.  xvii + 395 pp.  The Dover edition is an unabridged replication of the Wiley edition with a new preface and numerous corrections and additions]  (MR21-2325)

    *Chapter 3* Inequalities of information theory.  *6* Multinomial populations.  *8* Contingency tables.  *9* Multivariate normal populations.  *11* Multivariate analysis; the multivariate linear hypothesis.  *12* Multivariate analysis; other hypotheses.  *13* Linear discriminant functions.

MILLER, Rupert G., Jr. (1966)  *Simultaneous Statistical Inference.*  McGraw-Hill Book Company, New York.  xv + 272 pp.  (MR35-6282)

    *Chapter 5* Multivariate techniques.  *Section 6.2* Multinomial populations.

PLACKETT, R. L. (1960)  *Principles of Regression Analysis.*  Oxford University Press, London.  ix + 173 pp.  (MR28-2604)

    *Chapter 1* Linear equations.  *2* Quadratic forms in normal variables.

*3.* Partial Contents: Multivariate Analysis.  (cont.)

RAIFFA, Howard, & SCHLAIFER, Robert (1961)  *Applied Statistical Decision Theory*.  Division of Research, Graduate School of Business Administration, Harvard University, Boston.  xxviii + 356 pp.  (MR22-8618)

   *Chapter 5B* Selection of the best of several processes.  *8* Multivariate normalized density functions.  *12* Independent multinormal process.  *13* Normal regression process.

RAO, C. Radhakrishna (1952)  *Advanced Statistical Methods in Biometric Research*.  John Wiley & Sons, Inc., New York.  xvii + 390 pp.  (MR14, p.388)

   *Chapter 1* Algebra of vectors and matrices.  *2* Theory of distributions.  *4* The general theory of estimation and the method of maximum likelihood.  *6* Tests of homogeneity of variances and correlations.  *7* The tests of significance in multivariate analysis.  *8* Statistical inference applied to classificatory problems.  *9* The concept of distance and the problem of group constellations.

RAO, C. Radhakrishna (1965)  *Linear Statistical Inference and its Applications*.  John Wiley & Sons, Inc., New York.  xviii + 522 pp.  [Second corrected printing 1968]  (MR36-4668)

   *Chapter 1* Algebra of vectors and matrices.  *3* Continuous probability models.  *6* Large sample theory and methods.  *8* Multivariate analysis.

SEBER, G. A. F. (1966)  *The Linear Hypothesis: a General Theory*.  Charles Griffin & Company, Ltd., London.  viii + 115 pp.  [Number 19 of Griffin's Statistical Monographs & Courses]

   *Chapter 9* Multivariate linear hypotheses.

SPRENT, Peter (1969)  *Models in Regression and Related Topics*.  Methuen & Co., Ltd., London.  x + 173 pp.

   *Chapter 2* Bivariate normal and least squares regression.  *3* Lawlike relationships in the presence of random variation.  *4* Regression and functional relationship with heterogeneous and correlated departures.  *6* Multidimensional functional relationships and canonical analysis.

WHITTLE, P. (1963)  *Prediction and Regulation by Linear Least-Square Methods*.  The English Universities Press, Ltd., London and D. Van Nostrand Company, Inc., Princeton, N. J.  x + 147 pp.  (MR28-650)

   *Chapter 9* Multivariate processes.

*3.* Partial Contents: Multivariate Analysis.  (cont.)

WIENER, Norbert (1949)  *Extrapolation, Interpolation and Smoothing of
    Stationary Time Series, with Engineering Applications.*  The M. I. T.
    Press, Cambridge, Mass.  ix + 163 pp.  (MR11, p.118)

 *Chapter IV* The linear predictor and filter for multiple time
series.


WILKS, Samuel S. (1943)  *Mathematical Statistics.*  Princeton University
    Press, Princeton, N. J.  xi + 284 pp.  [Errata 1950, 2 pp. Japanese
    translation (1951) *Sûritôkei-gaku.*  Kasuga-syuppan-sya, Tokyo]
    (MR5, p.41)

 *Chapter II* Distribution functions.  *III* Some special distributions.
*V* Sampling from a normal population.  *VI* On the theory of statistical
estimation.  *X* On combinatorial statistical theory.  *XI* An introduction
to multivariate statistical analysis.


WILKS, Samuel S. (1962)  *Mathematical Statistics.*  John Wiley & Sons, Inc.,
    New York.  xvi + 644 pp.  [Second printing with corrections 1963]
    (MR26-1949)

 *Chapter 2* Distribution functions.  *3* Mean values and moments of
random variables.  *5* Characteristic functions and generating functions.
*7* Some special continuous distributions.  *12* Parametric statistical
estimation.  *18* Multivariate statistical theory.


WILLIAMS, E. J. (1959)  *Regression Analysis.*  John Wiley & Sons, Inc.,
    New York.  ix + 214 pp.  (MR22-3066)

 *Chapter 9* Simultaneous regression equations.  *10* Discriminant
functions.  *11* Functional relations.


YULE, G. Udny, & KENDALL, M. G. (1950)  *An Introduction to the Theory of
    Statistics.*  Fourteenth Edition, revised and enlarged.  Charles
    Griffin & Company, Ltd., London.  xxiv + 701 pp.  [First published
    1911.  Second-Tenth Editions, 1912-1932.  With M. G. Kendall:  Elev-
    enth-Thirteenth Editions, 1937-1944.  Portuguese translation (1948)
    *Introdução à Teoria da Estatística.*  Instituto Brasileiro de Geo-
    grafia e Estatística, Rio de Janeiro.  681 pp.  Spanish translation
    (1947) *Introducción a la Estadística Matemática.*  Aguilar, Madrid.
    Also Czech translation 1926 and Polish, 1921]  (MR12, p.35)

 *Chapter 9* Correlation and regression.  *10* Normal correlation.
*11* Further theory of correlation.  *12* Partial correlation.  *13* Correlation and regression:  some practical problems.  *14* Miscellaneous
theorems involving the correlation coefficient.

## 4. RELATED TO MULTIVARIATE ANALYSIS

BECKENBACH, Edwin F., & BELLMAN, Richard (1965) *Inequalities*. Second
    Revised Printing. Springer-Verlag, Berlin. xi + 198 pp. [First
    published 1961. Ergebnisse der Mathematik und ihrer Grenzgebiete.
    Neue Folge. Band 30] (MR28-1266)

DOOB, J. L. (1953) *Stochastic Processes*. John Wiley & Sons, Inc., New
    York. vii + 654 pp. (MR15, p.445)

DWYER, Paul S. (1951) *Linear Computations*. John Wiley & Sons, Inc., New
    York. xi + 344 pp. (MR13, p.283)

GALTON, Francis (1889) *Natural Inheritance*. Macmillan & Co., Limited,
    London. ix + 259 pp.

GAUSS, Karl Friedrich [= Carolo Friderico] (1828) *Theorematis Fundamen-
    talis in Doctrina de Residuis Quadraticis. Demonstrationes et Ampli-
    ficationes Novae*. Typis Dieterichianis, Gottingae.
    [In Latin. See also entry nos. 1865-1870 in Chapter III. English
    translation (1957) *Gauss's Work (1803-1826) on the Theory of Least
    Squares*. Hale F. Trotter, OOR Project No. 264, Contract No.
    DA36-034-ORD2001. 183 pp. Mimeographed. (Dewey: 519.8.G274g)
    French translation (1855) *Méthode des Moindres Carrés*. Mallet-
    Bachelier, Paris. 167 pp. German translation (1877) *Abhandlung
    zur Methode der kleinsten Quadrate*. Druck & Verlag von P. Stankiewicz'
    Buchdruckerei, Berlin. 208 pp.]

GAUSS, Karl Friedrich (1901) *Sechs Beweise des Fundamentaltheorems
    über quadratische Reste*. Verlag von Wilhelm Engelmann, Leipzig.
    111 pp. [Ostwald's Klassiker der Exakten Wissenschaften Nr. 122.
    German translation of six *Commentationes*]

GRAYBILL, Franklin A. (1969) *Introduction to Matrices with Applications
    in Statistics*. Wadsworth Publishing Company, Inc., Belmont, Calif.
    [v] + 372 pp.

GUTTMAN, Irwin (1970) *Statistical Tolerance Regions: Classical and
    Bayesian*. Charles Griffin & Company, Ltd., London. x + 150 pp.
    [Number 26 of Griffin's Statistical Monographs & Courses]

KARLIN, Samuel, & STUDDEN, William J. (1966) *Tchebycheff Systems:
    with Applications in Analysis and Statistics*. Interscience Pub-
    lishers, New York. xviii + 586 pp. [Pure and Applied Mathematics:
    A Series of Texts and Monographs, Volume XV] (MR34-4757)

KENDALL, M. G. (1961) *A Course in the Geometry of n Dimensions*. Charles
    Griffin & Company, Ltd., London. viii + 63 pp. [Number 8 of
    Griffin's Statistical Monographs & Courses] (MR25-476)

*4.* Related to Multivariate Analysis.  (cont.)

KENDALL, Maurice G., & BUCKLAND, William R. (1960)  *A Dictionary of
    Statistical Terms*.  Second Edition.  Oliver and Boyd, Edinburgh.
    xi + 575 pp.  [Published for the International Statistical Insti-
    tute.  Third Edition, revised and enlarged, 1971. 166 pp.  Spanish
    translation (1959) *Diccionario de Terminos Estadisticos*.  Talleres
    Graficos Fenner, S. R. L., Rosario, Argentina. 482 pp.]  (MR25-5560)

KNUTH, Donald E. (1969)  *The Art of Computer Programming.  Volume 1,
    Fundamental Algorithms*.  Addison-Wesley Publishing Company, Reading,
    Mass.  xxi + 634 pp.  [*Volume 2, Seminumerical Algorithms*.  xi +
    624 pp.]

LANCASTER, H. O. (1969)  *The Chi-squared Distribution*.  John Wiley &
    Sons, Inc., New York.  xiv + 356 pp.

LEHMANN, E. L. (1959)  *Testing Statistical Hypotheses*.  John Wiley &
    Sons, Inc., New York.  xiii + 369 pp.  (MR21-6654)

LIGHTHILL, M. J. (1958)  *Introduction to Fourier Analysis and Genera-
    lised Functions*.  Cambridge University Press, Cambridge.  viii +
    79 pp.  [Reprinted 1959, 1960]  (MR22-5888)

LUKACS, Eugene (1970)  *Characteristic Functions*.  Second Edition, revised
    and enlarged.  Charles Griffin & Company, Ltd., London.  x + 350 pp.
    [First published 1960, Number 5 of Griffin's Statistical Monographs
    & Courses. 216 pp.]  (MR23-A1392)

RAHMAN, N. A. (1967)  *Exercises in Probability and Statistics, for Math-
    ematics Undergraduates*.  Charles Griffin & Company, Ltd., London.
    xi + 307 pp.

ROBINSON, Enders A. (1959)  *An Introduction to Infinitely Many Variates*.
    Charles Griffin & Company, Ltd., London. 132 pp.  [Number 6 of
    Griffin's Statistical Monographs & Courses]  (MR22-5060)

SOMMERVILLE, D. M. Y. (1958)  *An Introduction to the Geometry of n
    Dimensions*.  Dover Publications, Inc., New York.  xviii + 196 pp.
    [First published 1929, Methuen & Co., Ltd., London.  The Dover
    edition is an unabridged and unaltered republication of the Methuen
    edition]  (MR20-6672)

WALKER, Helen M. (1929)  *Studies in the History of Statistical Method,
    with Special Reference to Certain Educational Problems*.  The Williams
    & Wilkins Company, Baltimore.  viii + 229 + [3] pp.  (JFM55, p.622)

## 5. TABLES

BARK, L. S., BOL'ŠEV, L. N., KUZNECOV, P. I., & ČERENKOV, A. P. (1964) *Tablicy Raspredelenija Releja-Raĭsa.* Vyčislitelnyi Tsentr. Matematičeskii Institut V. A. Steklova. Akademija Nauk SSSR. Moskva. xxviii + 246 pp. [= *Tables of the Rayleigh-Rice Distribution.* In Russian] (MR31-4142)

BOSE, P. K., & CHAUDHURI, S. B. (1966) *On some Problems associated with $D^2$-statistics and p-statistics.* Asia Publishing House, London. xi + 58 pp.

DAVID, F. N. (1938) *Tables of the Ordinates and Probability Integral of the Distribution of the Correlation Coefficient in Small Samples.* Cambridge University Press, Cambridge. xxxviii + 55 pp. [Issued by the *Biometrika* Office, London. Reprinted 1954] (ZMG19, p.74)

DAVID, F. N., KENDALL, M. G., & BARTON, D. E. (1966) *Symmetric Function and Allied Tables.* Cambridge University Press, Cambridge. x + 278 pp. (MR34-2099)

DUNNETT, Charles W. (1958) *Tables of the Bivariate Normal Distribution with Correlation $1/\sqrt{2}$.* Lederle Laboratories, Pearl River, N. Y. 16 pp. [Mimeographed. On microfilm with *Mathematics of Computation* file on Unpublished Mathematical Tables. Abstract: Volume 14, p.79, 1960]

GROENEWOUD, C., HOAGLIN, D. C., & VITALIS, J. A. (1967) *Bivariate Normal Offset Circle Probability Tables with Offset Ellipse Transformations.* Cornell Aeronautical Laboratory, Inc., Buffalo, New York. xxii + 1320 pp.

KRISHNA IYER, P. V., & SIMHA, P. S. (undated) *Tables of Bivariate Random Normal Deviates.* Defence Science Laboratory, Delhi, India. iii + 120 pp. (MR34-8527)

MIJARES, Tito A. (1964) *Percentage Points of the Sum $V_1^{(s)}$ of s roots (s = 1 - 50). A Unified Table of Tests of Significance in Various Univariate and Multivariate Hypotheses.* The Statistical Center, University of the Philippines, Manila. vii + 241 pp.

MINER, John Rice (1922) *Tables of $\sqrt{1-r^2}$ and $1-r^2$ for Use in Partial Correlation and in Trigonometry.* The Johns Hopkins Press, Baltimore. 49 pp.

NATIONAL BUREAU OF STANDARDS, United States (1959) *Tables of the Bivariate Normal Distribution Function and Related Functions.* U. S. Government Printing Office, Washington, D. C. xiv + 258 pp. [Applied Mathematics Series No. 50. LC: QA47.U56. no:50]

5. Tables.  (cont.)

OWEN, D. B. (1957)  *The Bivariate Normal Probability Distribution*.  U. S.
     Department of Commerce, Washington, D. C.  133 pp.  [Accession No.
     SC 3831(TR).  Available from Clearinghouse for Federal Scientific
     and Technical Information, Springfield, Va.]

OWEN, D. B. (1962)  *Handbook of Statistical Tables*.  Addison-Wesley Pub-
     lishing Company, Reading, Mass.  xii + 580 pp.  (MR28-4608)

     *Section 8* Multivariate normal and $t$-distributions.  *13* Rank corre-
     lation.  *19* Product moment correlation coefficient.

PEARSON, E. S., & HARTLEY, H. O. (1966)  *Biometrika Tables for Statisti-
     cians, Volume I*.  Third Edition.  Cambridge University Press, Cam-
     bridge.  xvi + 264 pp.  [First published 1954.  Second edition 1958]
     (MR34-8535)

     *Table 13* Percentage points for the distribution of the correlation
     coefficient, $r$, when $\rho=0$.  *14* The $z$-transformation of the correlation
     coefficient, $z = \tanh^{-1}r$.  *15* Charts giving confidence limits for the
     population correlation coefficient, $\rho$, given the sample coefficient,
     $r$.  Confidence coefficients, 0·95 and 0·99.  *44* Distribution of
     Spearman's rank correlation coefficient, $r_s$, in random rankings.  *45*
     Distribution of Kendall's rank correlation coefficient, $t_k$, in random
     rankings.  *46* Distribution of the concordance coefficient, $W$, in
     random rankings.

PILLAI, K. C. Sreedharan (1957)  *Concise Tables for Statisticians*.  The
     Statistical Center, University of the Philippines, Manila.  viii + 50 pp.

     *Table 14* Values of the correlation coefficient for different
     levels of significance.  *15* Correlation coefficient $r$, for values of
     $z$.  *16* Percentage points of the largest characteristic root, $\theta_s$, of a
     matrix in multivariate analysis.  *17* Percentage points of the sum of
     the roots, $V^{(s)}$, of a matrix in multivariate analysis.

PILLAI, K. C. Sreedharan (1960)  *Statistical Tables for Tests of
     Multivariate Hypotheses*.  The Statistical Center, University of the
     Philippines, Manila.  viii + 46 pp.

SMIRNOV, N. V., & BOL'ŠEV, L. N. (1962)  *Tablicy dlja Vyčislenija Funkcii
     Dvumernogo Normal'nogo Raspredelenija*.  Izdat. Nauk SSSR, Moskva.
     204 pp.  [= *Table for Evaluating a Function of a Two-dimensional
     Normal Distribution*.  In Russian]  (MR27-911)

*See also*  Section 7 GREENWOOD & HARTLEY (1962).

## 6. COMPUTER PROGRAMS

CLYDE, Dean J. (1969) *Multivariate Analysis of Variance on Large Com-*
*puters*. Clyde Computing Service, Box 166, Coconut Grove, Miami,
Florida. 43 pp.

CLYDE, Dean J., CRAMER, Elliot M., & SHERIN, Richard J. (1966)
*Multivariate Statistical Programs*. Biometric Laboratory, University
of Miami, Coral Gables, Florida. 61 pp.

DIXON, W. J. *editor* (1967) *BMD: Biomedical Computer Programs*. Second
Edition. University of California Press, Berkeley & Los Angeles.
x + 600 pp. [University of California Publications in Automatic
Computation No. 2. First published 1965. Third Printing (of the
Second Edition, revised, 1970]

    *Class M* Multivariate analysis: *BMD01M* Principal component analy-
sis. *02M* Regression on principal components. *03M* Factor analysis.
*04M* Discriminant analysis for two groups. *05M* Discriminant analysis
for several groups. *06M* Canonical analysis. *07M* Stepwise discriminant analysis.

I. B. M. (1968) *System/360 Scientific Subroutine Package (360A-CM-03X)*
*Version III. Programmer's Manual*. Fourth Edition. International
Business Machines Corporation, White Plains, New York. 454 pp.
First published 1966. [Manual No. H20-0205-3]

    Correlation and regression, *pp.32-48*. Discriminant analysis,
*pp.52-54 & 425-429*. Factor analysis, *pp.55-59 & 429-433*. Matrices,
*pp.98-170*. Canonical correlation, *pp.418-421*

JONES, Kenneth J. (1964) *A Multivariate Statistical Analyzer. (A System*
*of Fortran II programs to be run under the 7090-4 Fortran Monitor*
*System)* Harvard Cooperative Society, 1400 Massachusetts Ave.,
Cambridge, Mass. iii + 180 pp.

KLEIN, Lawrence R., EVANS, Michael K., & HARTLEY, M. (1969) *Econometric*
*Gaming: A Kit for Computer Analysis of Macroeconomic Models*. Macmillan Co., New York. 44 pp.

SONQUIST, John A., & MORGAN, James N. (1964) *The Detection of Inter-*
*action Effects. A Report on a Computer Program for the Selection of*
*Optimal Combinations of Explanatory Variables*. Survey Research Center,
Institute for Social Research, The University of Michigan, Ann Arbor,
Mich. xi + 292 pp. [Monograph No. 35]

*See also* Section *4* KNUTH (1969); Section *7* YOUDEN (1965, 1968).

## 7. BIBLIOGRAPHY

AMERICAN MATHEMATICAL SOCIETY (1966) *Index to Translations selected by the American Mathematical Society.* Providence, R. I. iii + 90 pp.

ARNIM, Max (1944-1952) *Internationale Personalbibliographie, 1800-1943.* 2 Bände. Hiersemann, Leipzig. [Band III, *1944-1959 und Nachträge*, von Gerhard Bock & Franz Hodes, 1961-1963. Hiersemann, Stuttgart. 659 pp.]

BOWLEY, Arthur Lyon (1928) *F. Y. Edgeworth's Contributions to Mathematical Statistics.* Royal Statistical Society, London. vii + 139 pp. [Updated by H. G. Johnson: "F. Y. Edgeworth: A Bibliography", i + 34 pp., mimeographed. cf. paper by M. G. Kendall in *Biometrika*, Vol. 55 (1968), p.275] (JFM54, p.573)

BUCKLAND, William R., & FOX, Ronald A. (1963) *Bibliography of Basic Texts and Monographs on Statistical Methods, 1945-1960.* Second Edition. Oliver and Boyd, Edinburgh. vii + 297 pp. [Published for the International Statistical Institute. First published 1951, The Hague] (MR31-4098)

CAPPELLI EDITORE (1959) *Bibliografie con Brevi Cenni Biografici.* Bologna. 550 pp. [Biblioteca di "Statistica" -2. Available from Casa Editrice Licinio Cappelli, 40124 Bologna, via Farini 6]

d'OTTONE R., Horacio *compiler* (1960) *Bibliography of Statistical Textbooks and other Teaching Material.* Second Edition. Pan American Union, Washington, D. C. xii + 120 pp. [= *Bibliografia de Tratados y Demas Material de Enseñanza de Estadística.* In English and Spanish. Published for the Inter American Statistical Institute, Document no. 4074ab-5/2/60-3000]

GREENWOOD, J. Arthur, & HARTLEY, H. O. (1962) *Guide to Tables in Mathematical Statistics.* Princeton University Press, Princeton, N. J. lxii + 1014 pp. (MR27-4299)

HODGES, Joseph L., Jr. (1950) *Discriminatory Analysis. I. Survey of Discriminatory Analysis.* USAF School of Aviation Medicine, Randolph Field, Texas. 115 pp. [Report Number 1, Project Number 21-49-004 under Contract No. AF41(128)-8. Distribution No. 1514. Mimeographed]

HOUSEHOLDER, A. S. (1969) *KWIC Index for Matrices in Numerical Analysis. Volume I. Primary Authors A-J.* Oak Ridge National Laboratory, Oak Ridge, Tennessee. vii + 125 pp. [Report No. ORNL-4418 Volume I. Contract No. W-7405-eng-26, Mathematics Division. Available from Clearinghouse for Federal Scientific and Technical Information, Springfield, Va.]

7. Bibliography.  (cont.)

HOUSEHOLDER, A. S. (1970)  *KWIC Index for Matrices in Numerical Analysis.
Volume II. Primary Authors K-Z*.  Oak Ridge National Laboratory, Oak
Ridge Tennessee.  vii + 151 pp.  [Report No. ORNL-4418 Volume II.
Contract No. W-7405-eng-26, Mathematics Division.  Available from
Clearinghouse for Federal Scientific and Technical Information,
Springfield, Va.]

INSTITUT INTERNATIONAL de STATISTIQUE [= INTERNATIONAL STATISTICAL
INSTITUTE] (1952)  *Bibliographie sur la Méthode Statistique et ses
Applications*.  Paris.  49 pp.

JOINER, Brian L., LAUBSCHER, N. F., BROWN, ELEANOR S., & LEVY, BERT (1970)
*An Author and Permuted Title Index to Selected Statistical Journals*.
U. S. Government Printing Office, Washington, D. C.  iv + 506 pp.
[National Bureau of Standards Special Publication 321]

KENDALL, Maurice G., & DOIG, Alison G. (1962)  *Bibliography of Statistical
Literature 1950-1958*.  Oliver and Boyd, Edinburgh.  xii + 297 pp.
(MR26-3134)

KENDALL, Maurice G., & DOIG, Alison G. (1965)  *Bibliography of Statistical
Literature 1940-1949*.  Oliver and Boyd, Edinburgh.  190 pp.

KENDALL, Maurice G., & DOIG, Alison G. (1968)  *Bibliography of Statistical
Literature Pre-1940; with Supplements to the Volumes for 1940-49 and
1950-58*.  Oliver and Boyd, Edinburgh.  356 pp.

LANCASTER, H. O. (1968)  *Bibliography of Statistical Bibliographies*.
Oliver and Boyd, Edinburgh.  ix + 103 pp.  [Published for the Inter-
national Statistical Institute.  Updated in *Revue de l'Institut
International de Statistique*:  "A second list", Vol. 37 (1969), pp.57-
67.  "A third list", Vol. 38 (1970), pp.258-267.  "A fourth list",
Vol. 39 (1971), pp.64-73]

MATUSITA, K., & HUDIMOTO, H. (1958)  *Statistical Bibliography in Japan*.
The Institute of Statistical Mathematics, Tokyo.  58 pp.  (MR24-A3018)

MAUNDER, W. F. *editor* (1970)  *Bibliography of Index Numbers.  An International Team Project*.  The Athlone Press, University of London.
xxvii + 215 pp.  [Published for The International Statistical Insti-
tute]

MORANT, G. M. *editor* [Assisted by B. L. Welch] (1939)  *A Bibliography of
the Statistical and Other Writings of Karl Pearson*.  Biometrika Office,
University College London.  viii + 119 pp.  [Printed at the University
Press, Cambridge]  (JFM65-1092)

7. Bibliography.  (cont.)

OPERATIONS RESEARCH GROUP, Engineering Administration Department, Case
    Institute of Technology (1963)  *A Comprehensive Bibliography on
    Operations Research, 1957-1958.*  John Wiley & Sons, Inc., New York.
    xiii + 403 pp.  [Publications in Operations Research No. 8]

PATIL, Ganapati P., & JOSHI, Sharadchandra W. (1968)  *A Dictionary and
    Bibliography of Discrete·Distributions.*  Oliver and Boyd, Edinburgh.
    xii + 268 pp.  [Published for the International Statistical Insti-
    tute]

POSTEN, Harry O. (1962)  *Bibliography on Classification, Discrimination,
    Generalized Distance and Related Topics.*  International Business
    Machines Corporation, Thomas J. Watson Research Center, Yorktown
    Heights, New York.  20 pp.  [Research Report RC-743.  Updated by:
    "Addenda", 5 pp. "Second Addenda", 5 pp.  Both undated.  All mimeo-
    graphed]

SAVAGE, I. Richard (1962)  *Bibliography of Nonparametric Statistics.*
    Harvard University Press, Cambridge, Mass.  [2] + 284 pp.
    (MR24-A2490)

SOCIETÀ ITALIANA di STATISTICA (1957)  *Bibliografia sui Metodi Statis-
    tici e le loro Applicazioni; elenco dei Principali Scritti di
    Autori Italiani.*  Roma.  56 pp.

TSAO, Chia Kuei *compiler* (1961)  *Bibliography of Mathematics published
    in Communist China during the Period 1949-1960.*  American Mathe-
    matical Society, Providence, R. I.  83 pp.  [Contemporary Chinese
    Research Mathematics, Volume 1]

WOLD, Herman O. A. (1965)  *Bibliography on Time Series and Stochastic
    Processes.*  Oliver and Boyd, Edinburgh.  xv + 516 pp.  [Published for
    the International Statistical Institute]  (MR34-8467)

YOUDEN, W. W. (1965)  *Computer Literature Bibliography 1946 to 1963.*
    U. S. Government Printing Office, Washington, D. C.  iv + 463 pp.
    [National Bureau of Standards Miscellaneous Publication 266]
    (MR31-4208)

YOUDEN, W. W. (1968)  *Computer Literature Bibliography; Volume 2 1964-
    1967.*  U. S. Government Printing Office, Washington, D. C.  iv +
    381 pp.  [National Bureau of Standards Miscellaneous Publication 309]

## I.3 AUTHOR INDEX

SHERIN, Richard J.
    *see 6*, CLYDE, Dean J.
SHIOYA, Minoru
    *see 1*, SIOTANI, Minoru
SIMHA, P. S.
    *see 5*, KRISHNA IYER, P. V.
SIOTANI, Minoru, *1*
SLUCKIĬ, E. E., *2A*
SMIRNOV, N. V., *5*
SNEATH, Peter H. A.
    *see 2B*, SOKAL, Robert R.
SOBEL, Milton
    *see 3*, BECHHOFER, Robert E.
SOCIETÀ ITALIANA di STATISTICA, *7*
SOKAL, Robert R., *2B*
SOMMERVILLE, D. M. Y., *4*
SONQUIST, John A., *6*
SPEARMAN, Charles, *2C*
SPRENT, Peter, *3*
STEPHENSON, William, *2C*
STUART, Alan
    *see 3*, KENDALL, Maurice G.
STUDDEN, William J.
    *see 4*, KARLIN, Samuel

TATSUOKA, Maurice M., *2B*
    *see also 2B*, RULON, Philip J.
THEIL, H., *2E*
THOMSON, Godfrey H., *2C*
THORNDIKE, Robert L., *2D*
THURSTONE, L. L., *2C*
TIEDEMAN, David V.
    *see 2B*, RULON, Philip J.
TINBERGEN, Jan, *2E*
TINTNER, Gerhard, *2E*

TORGERSON, Warren S., *2D*
TRAYNARD, C.-E.
    *see 2A*, RISSER, R.
TRELOAR, Alan E., *2A*
Trotter, Hale F.
    *see 4*, GAUSS, Karl Friedrich
TRYON, Robert Choate, *2B*, *2C*
TSAO, Chia Kuei, *7*
TSCHUPROW, A. A., *2A*

ÜBERLA, K., *2C*
U. S. NATIONAL BUREAU OF STANDARDS
    *see 5*, NATIONAL BUREAU OF
        STANDARDS, United States

VALAVANIS, Stefan, *2E*
VITALIS, J. A.
    *see 5*, GROENEWOUD, C.

WALKER, Helen M., *4*
WALTERS, A. A., *2E*
WATTS, Donald G.
    *see 3*, JENKINS, Gwilym M.
Welch, B. L.
    *see 7*, MORANT, G. M.
WHITTLE, P., *3*
WIENER, Norbert, *3*
WILKS, Samuel S., *3*
WILLIAMS, E. J., *3*
WOLD, Herman O. A., *2E*, *7*
WONNACOTT, Ronald J., *2E*
WONNACOTT, Thomas H., *2E*

YOUDEN, W. W., *7*
YULE, G. Udny, *3*

# CHAPTER II   JOURNALS AND COLLECTIONS OF PAPERS

II.1     Introduction.

II.2     An Annotated List of the Journals and
         Collections.

II.3     Journals and Collections arranged by
         Number of Papers.

II.4     Duplicated Publications.

II.5     Reference Sources.

# II.1 INTRODUCTION

Chapter III contains a listing of papers published in 819 journals and collections; the abbreviations used are identified in §II.2. Collections of papers identify books in which each chapter is a research paper with separate authorship. These include Festschriften, Readings, Collected Works, and Conference and Symposium Proceedings. The number at the end of each entry in §II.2 indicates the number of papers (including reprints, translations and amendments) in Chapter III; these frequencies are arranged numerically in §II.3.

For each journal the full title is followed by the place of publication in parentheses; information on volume numbers, years, and previous, subsequent or alternate names is given in square brackets for many journals. For each collection the full title is followed by the editorship in full whenever known; the publisher, and place and year of publication (as well as details of series, if any) appear in square brackets for most collections. For conferences and symposia, the locations where and the dates when they were held are given whenever known; the ordinal number of the conference or symposium is cited in the entry in Chapter III when no separate volume number has been allocated. The place of publication is given in its original language if this employs the Latin alphabet, and usually in English, otherwise.

The abbreviations follow closely the style used by *Mathematical Reviews* (cf. Volume 9 (1948), pp. 723-733 and Volume 22 (1961), pp. 2527-2543); some abbreviations have been expanded to achieve greater clarity. For most collections an indication of the editorship (or person honoured by a Festschrift) is given in parentheses at the end of the abbreviation. We have tried to abbreviate a particular word the same way in all cases; no two publications, however, are given the same abbreviation and no two abbreviations identify the same publication. Punctuation within an abbreviation is kept to a minimum with periods included only to indicate abbreviations of words. Accents and diacritical marks are not used in the abbreviations (which are all in capital letters) but are always given elsewhere when relevant; transliteration of Cyrillic characters follows the scheme used by *Mathematical Reviews* (cf. Volume 30 (1965), p. 1207 and the Preface of this *Bibliography*). When the title of a publication varies due to language or similar reason (but not due to series), then we try to use the same abbreviation each time (e.g., BULL. INST. INTERNAT. STATIST. -- in some years *Bulletin de l'Institut International de Statistique* and in others, *Bulletin of the International Statistical Institute*). When the title of a publication changes due to series, a new abbreviation is adopted; most indications of series are given within the abbreviation.

Reprints of papers are listed as separate entries in Chapter III except for those publications which have been duplicated in bulk; versions in the first-named publications in §II.4 are not cited. Translations are also listed separately in Chapter III; for DOKL. AKAD. NAUK SSSR NOV. SER. (Vol. 1-56) and *Doklady Akademii Nauk SSSR. Serija A*, however, we cite only the versions in C.R. ACAD. SCI. URSS NOUV. SER. and C.R. ACAD. SCI. URSS SER. A.

Whenever possible we have personally inspected the title page of the publications; for those which proved hard to find the Library of Congress call number is quoted when known with the prefix LC. In some cases, however, we have been guided by listings of journals in reference books (see §II.5).

ABHANDL. DEUTSCHEN AKAD. WISS. BERLIN KL. MATH. PHYS. TECH.
*Abhandlungen der Deutschen Akademie der Wissenschaften zu Berlin.
Klasse für Mathematik, Physik und Technik.* (Berlin) ... 2

ACAD. REPUB. POP. ROMINE BUL. STI. MAT. FIZ. *Academia Republicii Populare Romîne. Buletinul Ştiinţelor Matematică şi Fizica.* (Cluj) ... 2

ACAD. ROY. BELG. BULL. CL. SCI. SER. 5 *Académie Royale de Belgique: Bulletins de la Classe des Sciences, 5ème Série.* (Bruxelles) ... 11

ACAD. ROY. BELG. CL. SCI. MEM. COLLECT. IN-8. *Académie Royale de Belgique. Classe des Sciences. Mémoires. Collection in-8°.* (Bruxelles) ... 1

ACAD. SERBE SCI. PUBL. INST. MATH. *Académie Serbe des Sciences. Publications de l'Institut Mathématique.* (Beograd) ... 1

ACAD. SINICA SCI. REC. *Academia Sinica. Science Record.* (Peking) [Vol. 1-4, 1957-1960. Then *Scientia Sinica*] ... 1

ACTA COMMENT. SER. A *Acta et Commentationes Universitatis Tartuensis (Dorpatensis). A. Mathematica, Physica, Medica.* (Tartu, Estonia, USSR) [= *Eesti Vabariigi Tartu Ülikooli Toimetused*] ... 1

ACTA CRYST. *Acta Crystallographica.* (København) [Vol. 1-23, 1948-1967; Vol. 24, 1968-, *Section A, B*] ... 4

ACTA GENET. STATIST. MED. *Acta Genetica et Statistica Medica.* (Basel) ... 1

ACTA LITT. SCI. REG. UNIV. HUNGAR. SECT. SCI. MATH. *Acta Litterarum ac Scientiarum Regiae Universitatis Hungaricae Francisco-Iosephinae. Sectio Scientiarum Mathematicarum.* (Szeged) [Vol. 1-9, 1922-1941. Then ACTA UNIV. SZEGED. ACTA SCI. MATH.] ... 1

ACTA MATH. *Acta Mathematica.* (Stockholm) ... 8

ACTA MATH. ACAD. SCI. HUNGAR. *Acta Mathematica Academiae Scientiarum Hungaricae.* (Budapest) [Magyar Tudományos Akadémia. Was *Hungarica Acta Mathematica*] ... 4

ACTA MATH. SINICA *Acta Mathematica Sinica.* (Peking) [Translated as CHINESE MATH. ACTA] ... 5

ACTA SOC. SCI. FENN. NOVA SER. A *Acta Societatis Scientiarum Fennicae. Nova Series A.* (Helsinki) ... 1

ACTA UNIV. SZEGED. ACTA SCI. MATH. *Acta Universitatis Szegediensis.*
*Acta Scientiarum Mathematicarum.* (Szeged) [Vol. 10, 1941-. Was
ACTA LITT. SCI. REG. UNIV. HUNGAR. SECT. SCI. MATH.] ... 1

ACTUAL. SCI. INDUST. *Actualités Scientifiques et Industrielles.*
(Paris) [Conférences Internationales de Sciences. Mathématiques.
LC:Q111.A3] ... 2

AGRIC. EXPER. STATION AUBURN UNIV. BULL. *Agricultural Experiment*
*Station, Auburn University, Auburn, Alabama. Bulletin.* (Auburn)
... 1

AKAD. NAUK SSSR SIBIRSK. OTDEL. INST. MAT. VYC. SISTEMY SB. TRUDOV
*Akademija Nauk SSSR. Sibirskoe Otdelenie. Institut Matematiki.*
*Vyčislitel'nye Sistemy. Sbornik Trudov.* (Novosibirsk) ... 1

AKAD. NAUK UZBEK. SSR TRUDY INST. MAT. MEH. *Akademija Nauk Uzbekskoi*
*SSR. Trudy Instituta Matematiki i Mehaniki im.V. I. Romanovskogo.*
(Tashkent) ... 3

AKTUAR. VEDY *Aktuárské Vědy: Pojistná Matematika; Matematická*
*Statistika.* (Praha) [Jednota Českých Matematiků a Fyziků v Praze]
... 6

AKUST. ZUR. *Akustičeskii Žurnal. Akademija Nauk SSSR.* (Moscow)
[Translated as SOVIET PHYS. ACOUST.] ... 3

ALLGEMEIN. STATIST. ARCH. *Allgemeines statistisches Archiv. Organ der*
*Deutschen Statistischen Gesellschaft.* (München) ... 4

AMER. ANTHROP. *American Anthropologist.* (Washington, D.C.) [Vol.
1-11, 1888-1898; *New Series* Vol. 1, 1899-.] ... 1

AMER. ECON. REV. *American Economic Review.* (Ithaca, N.Y.) ... 2

AMER. J. EPIDEM. *American Journal of Epidemiology.* (Baltimore)
[Vol. 1-80, 1921-1964 as *American Journal of Hygiene*] ... 2

AMER. J. MATH. *American Journal of Mathematics.* (Baltimore) ... 11

AMER. J. PHYS. ANTHROP. *American Journal of Physical Anthropology.*
(Washington, D.C.) [Vol. 1-29, 1918-1942; *New Series* Vol. 1,
1943-.] ... 2

AMER. J. PSYCHOL. *The American Journal of Psychology.* (Ithaca, N.Y.)
... 14

AMER. J. SCI. *The American Journal of Science.* (New Haven, Conn.)
[Vol. 1, 1818-. Vol. 101-150 (1871-1895) also numbered Vol. 1-50
*Third Series, The American Journal of Science and Arts*] ... 2

AMER. J. SOCIOL. *American Journal of Sociology*. (Chicago) ... 3

AMER. MATH. MONTHLY *The American Mathematical Monthly*. (Lancaster,Pa.) [Vol. 72(2),1965 is paginated separately as *Part II, Computers and Computing, Number 10 of the Herbert Ellsworth Slaught Memorial Papers*] ... 42

AMER. NATUR. *American Naturalist*. (Boston) ... 6

AMER. SOCIOL. REV. *American Sociological Review*. (Menasha,Wis.) ... 10

AMER. STATIST. *The American Statistician*. (Washington,D.C.) [Was *American Statistical Association. Bulletin*] ... 4

AN. INST. SUPER. CI. ECON. FINAN. *Anais do Instituto Superior de Ciências Económicas e Financeiras. Universidade Técnica*. (Lisboa) ... 1

AN. REAL SOC. ESPAN. FIS. QUIM. SER. A *Anales de la Real Sociedad Española de Física y Química. Serie A - Física*. (Madrid) ... 1

AN. SOC. CI. ARGENT. *Anales de la Sociedad Científica Argentina*. (Buenos Aires) [Vol. 130, 1965-, in two series] ... 6

AN. UNIV. BUCURESTI SER. STI. NATUR. *Analele Universității "C.I.Parhon" Bucureşti. Seria Ştiinţelor Naturii*. (Bucureşti) [Became AN. UNIV. BUCURESTI SER. STI. NATUR. MAT. FIZ.] ... 1

AN. UNIV. BUCURESTI SER. STI. NATUR. MAT. FIZ. *Analele Universității Bucureşti. Seria Ştiinţele Naturii. Matematică-Fizică*. (Bucureşti) [Was AN. UNIV. BUCURESTI SER. STI. NATUR.] ... 1

ANALYST *The Analyst*. (Des Moines,Iowa) [Vol. 1-10, 1874-1883. Then *Annals of Mathematics*] ... 3

ANN. ACAD. SCI. FENN. SER. B *Annales Academiae Scientiarum Fennicae. Series B*. (Helsinki) ... 1

ANN. ECOLE POLYTECH. DELFT *Annales de l'École Polytechnique de Delft*. (Leiden) ... 2

ANN. EUGENICS *Annals of Eugenics*. (London) [Vol. 1-18, 1925-1954. Then ANN. HUMAN GENET.] ... 33

ANN. FAC. ECON. COMM. PALERMO *Annali della Facoltà di Economia e Commercio*. (Palermo) [Università di Palermo] ... 2

ANN. HUMAN GENET. *Annals of Human Genetics.* (London) [Vol. 19, 1954-. Was ANN. EUGENICS] ... 2

ANN. HYDROGRAPHIE *Annalen für Hydrographie.* (Berlin) ... 1

ANN. INST. H. POINCARE *Annales de l'Institut Henri Poincaré.* (Paris) [Université, Faculté des Sciences. Vol. 1-18, 1930-1963. Then *Section A* and SECT. B] ... 2

ANN. INST. H. POINCARE SECT. B *Annales de l'Institut Henri Poincaré. Section B. Calcul des Probabilités et Statistique.* (Paris) [Université, Faculté des Sciences. Vol. 1, 1964-.] ... 2

ANN. INST. STATIST. MATH. TOKYO *Annals of the Institute of Statistical Mathematics.* (Tokyo) ... 103

ANN. MATH. PURES APPL. *Annales de Mathématiques Pures et Appliquées.* (Paris) [Vol. 1-22, 1810-1831. Then J. MATH. PURES APPL.] ... 1

ANN. MATH. SER. 2 *Annals of Mathematics. Second Series.* (Princeton, N.J.) [Vol. 1, 1900-. Was "First Series", 1884-1899 and ANALYST, 1874-1883] ... 18

ANN. MATH. STATIST. *The Annals of Mathematical Statistics.* (Baltimore) ... 636

ANN. NEW YORK ACAD. SCI. *Annals of the New York Academy of Sciences.* (New York) ... 4

ANN. SCI. ECOLE NORM.SUPER. *Annales Scientifiques de l'École Normale Supérieure.* (Paris) [*Serie 3*, Vol. 1, 1884-.] ... 1

ANN. SOC. POLON. MATH. *Annales de la Société Polonaise de Mathématique.* (Kraków) [=*Rocznik Polskiego Towarzystwa Matematycznego.* Vol. 1-25, 1922-1952] ... 1

ANN. SOC. SCI. BRUXELLES SECT. A *Annales de la Société Scientifique de Bruxelles. Section A. Sciences Mathématiques.* (Bruxelles) [Vol. 47-56, 1927-1936] ... 1

ANN. STATIST. SER. 2A *Annali di Statistica, Serie IIa.* (Roma) [Vol. 1-3, 1878; 4-5, 1879; 6, 1881; 7-10, 1879; 11-17, 1880; 18-25, 1881. LC: HA1360.A2; Italy. Istituto Centrale di Statistica. Ministero di Agricoltura, Industria e Commercio. Direzione Generale della Statistica. Pub. Regia Tipografia D. Ripamonti; Tipografia dei fratelli Bencini] ... 3

ANN. STATIST. SER. 3A  *Annali di Statistica, Serie 3a.*  (Roma) [Some years no volume numbers; vol. 4 = 1883.  LC: Z7554.I8B6.  Publisher as SER. 2A]  ... 3

ANN. UNIV. LYON NOUV. SER. I SCI. MED.  *Annales de l'Université de Lyon. Nouvelle Série. I. Sciences, Médecine.*  (Lyon) [No.1-50, 1899-1934. Then ANN. UNIV. LYON TROISIEME SER. SCI. SECT. A]  ... 1

ANN. UNIV. LYON TROISIEME SER. SCI. SECT. A  *Annales de l'Université de Lyon. Troisième Série. Sciences. Section A: Sciences Mathématiques et Astronomie.*  (Lyon) [Vol. 1, 1936-.  Was ANN. UNIV. LYON NOUV. SER. I SCI. MED.]  ... 6

ANN. VERSICHERUNGSW.  *Annalen des gesamten Versicherungswesens.*  (Leipzig)  ... 1

ANNEE PSYCHOL.  *Année Psychologique.*  (Paris)  ... 2

ANWEND. MATRIZ. WIRTSCH. STATIST. PROBLEME (ADAM ET AL.)  *Anwendungen der Matrizenrechnung auf wirtschaftliche und statistische Probleme.* Edited by Ing. Dr. A. Adam, Dr. F. Ferschl, Dr. A. Klamecker, Dr. A. Klingst, Dr. Ing. O. Pichler, Dr. J. Roppert, Dr. H. Scholz, Dipl.-Math. K. Wenke & Priv.-Doz. Dr. W. Wetzel. [Einzelschriften der Deutschen Statistischen Gesellschaft Nr. 9.  Physica-Verlag, Würzburg, 1959]  ... 2

APL. MAT.  *Aplikace Matematiky.*  (Praha) [Československá Akademie Věd]  ... 3

APPL. STATIST.  *Applied Statistics.*  (London) [Vol. 1-13, 1952-1964. Then also *Journal of the Royal Statistical Society. Series C*]  ... 17

APPL. STATIST. METEOROL.  *Applied Statistics in Meteorology.*  (Tokyo) [= *Kisho to Tokei, Oyo Suikeigaku Zasshi*]  ... 2

ARCH. ANTHROP.  *Archiv für Anthropologie.*  (Braunschweig) [1936-, as *Archiv für Anthropologie und Völkerforschung*]  ... 1

ARCH. ANTROP. ETNOLOGIA  *Archivio per l'Antropologia e la Etnologia.* (Firenze)  ... 1

ARCH. DE PSYCHOL. GENEVA  *Archives de Psychologie.*  (Genève)  ... 2

ARCH. ELEK. UBERTR.  *Archiv der elektrischen Übertragung.*  (Wiesbaden)  ... 1

ARCH. GES. PSYCHOL. *Archiv für die gesamte Psychologie.* (Leipzig)
... 1

ARCH. MATH. *Archiv der Mathematik.* (Basel) ... 3

ARCH. MATH. PHYS. *Archiv der Mathematik und Physik.* (Greifswald) ... 1

ARCH. MATH. WIRTSCH. *Archiv für Mathematische Wirtschafts- und Sozialforschung.* (Leipzig) ... 1

ARCH. OF PSYCHOL. (COLUMBIA) *Archives of Psychology.* (New York) [Vol. 1-41, no. 300, 1908-1945. Monograph Series edited by R.S. Woodworth, Columbia University. Mainly doctoral dissertations] ... 2

ARCH. OIKON. KOIN. EPIST. *Archion Oikonomikōn Kai Koinōnikōn Epistemōn.* (Athens) ... 1

ARCH. PAD. *Archiv für Pädagogik.* (Leipzig) ... 1

ARCH. RASS. GES. BIOL. *Archiv für Rassen- und Gesellschafts-biologie, einschliesslich Rassen- und Gesellschafts-hygiene.* (Berlin) ... 1

ARCH. VERZEKERINGS-WETENSCH. AANVERW. VAKKEN *Archief voor de Verzekerings-Wetenschap en aanverwante vakken.* ('s-Gravenhage) [Vol. 1-17, 1895-1918] ... 2

ARCHIMEDE *Archimede; rivista per gli insegnanti e i cultori di matematiche pure e applicate.* (Firenze) [Vol. 1, 1949 = Vol. 48, 1949 of *Bollettino di Matematica*] ... 1

ARK. ASTRONOM. *Arkiv för Astronomi.* (Stockholm) [Vol. 1, 1950-. Was part of ARK. MAT. ASTRONOM. FYS.] ... 3

ARK. MAT. *Arkiv för Matematik.* (Stockholm) [Vol. 1, 1949-. Was part of ARK. MAT. ASTRONOM. FYS.] ... 6

ARK. MAT. ASTRONOM. FYS. *Arkiv för Matematik, Astronomi och Fysik.* (Stockholm) [Vol. 1-36, 1903-1949. Then ARK. MAT., ARK. ASTRONOM., and *Arkiv för Fysik*] ... 7

ASSOC. ACTUAIR. BELG. BULL. *Association Royale des Actuaires Belges. Bulletin.* (Bruxelles) ... 1

ASTRONOM. NACHR. *Astronomische Nachrichten.* (Kiel) ... 2

ASTRONOM. ZUR. *Astronomičeskiĭ Žurnal. Akademija Nauk SSSR.*
(Moscow) [Vol. 1-4, 1924-1927 as *Russkiĭ Astronomičeskiĭ Žurnal.*
Translated as *Soviet Astronomy*] ... 2

ASTROPHYS. J. *The Astrophysical Journal.* (Chicago) ... 4

ATTI CONG. UN. MAT. ITAL. *Atti del Congresso dell'Unione Matematica
Italiana.* [1st Congress, 1937. 2nd, 1940 (Bologna), Pub. 1942.
3rd, 1948 (Pisa), Pub. 1951. 4th, 1951 (Taormina), Pub. 1953. 5th,
unknown. 6th, 1959 (Napoli). Edizioni Cremonense Casa Editrice
Perrella, Roma] ... 4

ATTI REALE ACCAD. NAZ. LINCEI SER. 6 *Atti della Reale Accademia
Nazionale dei Lincei. Serie Sesta. Rendiconti. Classe di Scienze
Fisiche, Matematiche e Naturali.* (Roma) [Vol. 1-29, 1925-1939.
Then SER. 7] ... 3

ATTI REALE ACCAD. NAZ. LINCEI SER. 7 *Atti della Reale Accademia
d'Italia [Nazionale dei Lincei]. Serie Settima. Rendiconti della
Classe di Scienze Fisiche, Matematiche e Naturali.* (Roma) [Vol.
1-5, 1939-1943. Was SER. 6] ... 1

ATTI REALE ACCAD. SCI. TORINO PUBBL. ACCAD. CL. SCI. FIS. MAT. NATUR.
*Atti della Reale Accademia delle Scienze di Torino. Pubblicati dagli
Accademici Segretari delle due Classi. Classe di Scienze Fisiche,
Matematiche e Naturali.* (Torino) ... 3

ATTI REALE IST. VENETO SCI. LETT. ARTI *Atti del Reale Istituto Veneto
di Scienze, Lettere ed Arti. Classe di Scienze Matematiche e
Naturali.* (Venezia) ... 7

ATTI RIUN. SCI. SOC. ITAL. STATIST. *Atti della Riunione Scientifica
della Società Italiana di Statistica.* (Roma) ... 15

ATTI SOC. ITAL. PROG. SCI. *Atti della Società Italiana per il Progresso
delle Scienze.* (Roma) ... 2

AUSGEWAHLTE SCHR. (ANDERSON REPRINTS) *Oskar Anderson: Ausgewählte
Schriften.* Edited by Hans Kellerer, Werner Mahr, Gerda Schneider,
& Heinrich Strecker. [J.C.B. Mohr/Paul Siebeck, Tübingen, 1963.
Two volumes] ... 7

AUSTRAL. COUNC. SCI. INDUST. RES. J. *Australian Council of Science and
Industry. Research Journal.* (Melbourne) [= *Commonwealth
Scientific and Industrial Research Organization. Journal*] ... 1

AUSTRAL. J. APPL. SCI. *Australian Journal of Applied Science.* (Melbourne) ... 2

AUSTRAL. J. BOT. *Australian Journal of Botany.* (Melbourne) ... 2

AUSTRAL. J. PHYS. *Australian Journal of Physics.* (East Melbourne) ... 5

AUSTRAL. J. PSYCHOL. *Australian Journal of Psychology.* (Victoria) ... 1

AUSTRAL. J. SCI. *The Australian Journal of Science.* (Sydney) ... 1

AUSTRAL. J. STATIST. *The Australian Journal of Statistics.* (Sydney) ... 20

AUTOMAT. REMOTE CONTROL *Automation and Remote Control.* (New York) [Vol. 17, 1956-. Translation of AVTOMAT. I TELEMEH.] ... 1

AVH. NORSKE VIDENS. AKAD. OSLO *Avhandlinger Utgitt av det Norske Videnskaps-Akademi i Oslo.* (Oslo) ... 1

AVTOMAT. I TELEMEH. *Avtomatika i Telemehanika. Akademija Nauk SSSR.* (Moscow) [Vol. 1, 1934-. Translated as AUTOMAT. REMOTE CONTROL] ... 1

BAROMETRO ECON. ITAL. *Barometro Economico Italiano.* (Roma) [1929-1933 as *Barometro Economico*] ... 2

BEHAV. SCI. *Behavioral Science.* (Ann Arbor, Mich.) ... 1

BELL SYSTEM TECH. J. *The Bell System Technical Journal.* (New York) ... 6

BER. MATH. TAGUNG TUBINGEN *Bericht über die Mathematiker-Tagung Tübingen von 23. bis 27. September 1946.* [Mathematischen Institut der Universität Tübingen] ... 1

BERNOULLI (1713), BAYES (1763), LAPLACE (1813), ANNIV. VOLUME *Bernoulli (1713), Bayes (1763), Laplace (1813) Anniversary Volume. Proceedings of an International Research Seminar, Statistical Laboratory, University of California, Berkeley, California, 1963.* Edited by Jerzy Neyman & Lucien M. LeCam. [Springer-Verlag, New York, 1965] ... 1

BIOGRAPH. MEM. FELLOWS ROY. SOC. *Biographical Memoirs of the Fellows of the Royal Society.* (London) [Was OBIT. NOTICES FELLOWS ROY. SOC.] ... 1

BIOGRAPH. MEM. NAT. ACAD. SCI. U.S.A.  *Biographical Memoirs of the National Academy of Sciences, U.S.A.*  (Washington, D.C.) ... 1

BIOM. BULL. (NEURO-ENDOCRINE RES.)  *Biometric Bulletin.*  (Worcester, Mass.)  [No. 1-4, 1936-1938. Memorial Foundation for Neuro-endocrine Research] ... 1

BIOM. PRAX.  *Biometrie-Praximetrie.*  (Bruxelles) ... 2

BIOM. Z.  *Biometrische Zeitschrift.*  (Berlin) ... 8

BIOMETRICS  *Biometrics.*  (Blacksburg, Va.) ... 63

BIOMETRIKA  *Biometrika.*  (London) ... 468

BIOSCIENCE  *Bioscience.*  (Washington, D.C.) ... 1

BIOTYPOLOGIE  *Biotypologie.*  (Paris) ... 7

BJULL. SREDNE-ASIAT. GOS. UNIV.  *Bjulleten'. Srednyǐ-Aziatskiǐ Gosudarstvennyǐ Universitet.*  (Tashkent)  [LC:Q60.T33] ... 3

BOL. FAC. INGEN. AGRIMENS. MONTEVIDEO  *Boletín de la Facultad de Ingeniería y Agrimensura de Montevideo.*  (Montevideo) ... 1

BOL. SOC. MAT. SAO PAULO  *Boletim de Sociedade de Matemática de São Paulo.*  (São Paulo) ... 1

BOLL. UN. MAT. ITAL.  *Bollettino dell'Unione Matematica Italiana.*  (Bologna) ... 1

BRITISH J. EDUC. PSYCHOL.  *The British Journal of Educational Psychology.*  (Birmingham) ... 8

BRITISH J. MATH. STATIST. PSYCHOL.  *British Journal of Mathematical and Statistical Psychology.*  (London)  [Vol. 18, 1965-. Was BRITISH J. STATIST. PSYCHOL.] ... 7

BRITISH J. PSYCHOL.  *The British Journal of Psychology.*  (London) ... 72

BRITISH J. PSYCHOL. STATIST. SECT.  *The British Journal of Psychology, Statistical Section.*  (London)  [Vol. 1, pp.1-72, 1947; pp.73-222, 1948; Vol. 2-5, 1949-1952. Then BRITISH J. STATIST. PSYCHOL.] ... 33

BRITISH J. SOC. MED.  *British Journal of Social Medicine.*  (London) ... 1

BRITISH J. STATIST. PSYCHOL. *The British Journal of Statistical Psychology*. (London) [Vol. 6-17, 1953-1964. Was BRITISH J. PSYCHOL. STATIST. SECT. Became BRITISH J. MATH. STATIST. PSYCHOL.] ... 35

BULL. ACAD. POLON. SCI. CL. 3 *Bulletin de l'Académie Polonaise des Sciences. Classe Troisième*. (Warszawa) [Vol. 1-5, 1953-1957] ... 1

BULL. AMER. MATH. SOC. *Bulletin of the American Mathematical Society*. (Providence, R.I.) ... 25

BULL. AMER. METEOROL. SOC. *The Bulletin of the American Meteorological Society*. (Easton, Pa.) ... 3

BULL. AMER. SOC. TEST. MATER. *Bulletin of the American Society for Testing Materials*. (Philadelphia) ... 1

BULL. CALCUTTA MATH. SOC. *Bulletin of the Calcutta Mathematical Society*. (Calcutta) ... 10

BULL. CENTRE ETUDES RECH. PSYCHOTECH. *Bulletin du Centre d'Études et Recherches Psychotechniques*. (Paris) ... 2

BULL. DEPT. EDUC. RES. UNIV. TORONTO *Bulletin of the Department of Educational Research of the University of Toronto*. (Toronto) ... 1

BULL. FUKUOKA GAKUGEI UNIV. SECT. III *Bulletin of Fukuoka Gakugei University. Section III. Natural Sciences*. (Fukuoka) ... 1

BULL. GEODESIQUE *Bulletin Géodésique*. (Paris) [*Nouvelle Série* No. 1, 1946-. Organe de l'Association Internationale de Géodésie] ... 1

BULL. INST. INTERNAT. STATIST. *Bulletin de l'Institut International de Statistique*. ('s Gravenhage) [= *Bulletin of the International Statistical Institute*; Vol. 31 (5) duplicated as *Econometrica*, Vol. 17 (Supplement)] ... 69

BULL. INTERNAT. POLSKA AKAD. UMIEJET. (KRAKOW) SER. A *Bulletin International. Polska Akademia Umiejętności. Wydział Matematyczno-przyrodniczy*. (Kraków) [= *Bulletin International de l'Académie Polonaise des Sciences et des Lettres. Classe des Sciences Mathématiques et Naturelles. Série A: Sciences Mathématiques*] ... 2

BULL. JAPAN STATIST. SOC. *Bulletin of the Japan Statistical Society*. (Tokyo) [= *Nippon Tôkei Gakukai Kaiko*] ... 1

BULL. MATH. BIOPHYS. *The Bulletin of Mathematical Biophysics*. (Colorado Springs, Colo.) [Supplement to PSYCHOMETRIKA] ... 4

BULL. MATH. STATIST. *Bulletin of Mathematical Statistics.* (Tokyo)
[= *Tôkei Sûri Kenkyû*] ... 12

BULL. PHILOS. SOC. WASHINGTON *Bulletin of the Philosophical Society of
Washington.* (Washington, D.C.) ... 1

BULL. RES. COUNC. ISRAEL *Bulletin of the Research Council of Israel.*
(Jerusalem) ... 1

BULL. RES. INST. UNIV. TRAVANCORE SER. B *Bulletin of the Research
Institute. University of Travancore, Trivandrum. Series B,
Statistics.* (Trivandrum, Kerala, India) ... 1

BULL. SCI. ECOLE POLYTECH. TIMISOARA *Bulletin Scientifique de l'École
Polytechnique de Timişoara.* (Timişoara) ... 1

BULL. SCI. MATH. SER. 2 *Bulletin des Sciences Mathématiques. Série 2.*
(Paris) ... 6

BULL. SOC. BELG. STATIST. *Bulletin de la Société Belge de Statistique.*
(Bruxelles) [44 rue de Louvain. No volume numbers] ... 1

BULL. SOC. MATH. FRANCE *Bulletin de la Société Mathématique de France.*
(Paris) ... 3

BULL. SOC. MATH. GRECE *Bulletin de la Société Mathématique de Grèce.*
(Athens) [= *Deltion tes Hellenikes Mathematikes Hetereias*] ... 1

BULL. SOC. ROY. SCI. LIEGE *Bulletin de la Société Royale des Sciences
de Liège.* (Liège) ... 1

BULL. STATIST. INST. NAT. STATIST. BRUXELLES *Bulletin de Statistique
publié par l'Institut National de Statistique.* (Bruxelles) ... 1

BULL. STATIST. SOC. N.S.W. *Bulletin of the Statistical Society of
New South Wales.* (Sydney) [Became AUSTRAL. J. STATIST.] ... 1

BULL. TRIMEST. INST. ACTUAIR. FRANC. *Bulletin Trimestriel de l'Institut
des Actuaires Français.* (Paris) ... 2

BULL. UNIV. TEXAS *Bulletin of the University of Texas.* (Austin,
Texas) ... 1

BULL. ZOOL. SOC. SAN DIEGO *Bulletins of the Zoological Society of
San Diego.* (San Diego, Calif.) ... 1

C.R. ACAD. BULGARE SCI. *Comptes Rendus de l'Académie Bulgare des
Sciences.* (Sofia) [Doklady Bolgarskoĭ Akademii Nauk] ... 1

C.R. ACAD. SCI. PARIS   *Comptes-Rendus Hebdomadaires des Séances de l'Académie des Sciences.* (Paris) [Vol. 262, 1966-, as *Série A, B, C, D*] ... 91

C.R. ACAD. SCI. PARIS SER. A   *Comptes-Rendus Hebdomadaires des Séances de l'Académie des Sciences. Série A. Sciences Mathématiques.* (Paris) [Vol. 262, 1966-.] ... 8

C.R. ACAD. SCI. URSS NOUV. SER.   *Comptes Rendus de l'Académie des Sciences de l'Union des Républiques Soviétiques Socialistes. Nouvelle Série.* (Moscow) [1933-1947. Articles mainly in French, German or English. Translation of DOKL. AKAD. NAUK SSSR NOV. SER. Vol. 1-56 (not cited). 55 vols. published in all, called Vol. 1-49, 51-56. Vol. 50 never published: last part of Vol. 49 dated December 30, 1945, followed by first part of Vol. 51, dated January 10, 1946] ... 6

C.R. ACAD. SCI. URSS SER. A   *Comptes Rendus de l'Académie des Sciences de l'Union des Républiques Soviétiques Socialistes. Série A.* (Leningrad) [1922-1933. Articles mainly in French, German or English. Either translation of or identical to *Doklady Akademii Nauk SSSR. Serija A.* 158 (or 160) unnumbered issues; 25 in 1928] ... 1

C.R. ASSOC. FRANC. AVANCE. SCI. CONF.   *Association Française pour l'Avancement des Sciences. Conférences. Comptes Rendus.* [49ème Session, 1925 (Grenoble), Pub. 1926; 52ème, 1928 (La Rochelle); 55ème, 1931 (Nancy)] ... 3

C.R. CONF. INTERNAT. PSYCHOTECH.   *Comptes Rendus. Conférence Internationale de Psychotechnique.* [VIIIème Conférence 11-15, Sept. 1934 (Praha), Pub. 1935] ... 1

C.R. SEANCES SOC. PHYS. HIST. NATUR. GENEVE   *Compte Rendu des Séances de la Société de Physique et d'Histoire Naturelle de Genève. Supplément aux Archives des Sciences Physiques et Naturelles.* (Genève) ... 1

CAHIERS CENTRE ETUDES RECH. OPERAT.   *Cahiers du Centre d'Études de Recherche Opérationelle.* (Bruxelles) ... 3

CAHIERS INST. SCI. ECON. APPL. SER. E   *Cahiers de l'Institut de Science Économique Appliquée. Série E.* (Paris) [No. 2 = Supplément No. 138] ... 4

CAHIERS SEM. ECONOMET.   *Cahiers du Séminaire d'Économétrie.* (Paris) ... 4

CALCUTTA MATH. SOC. GOLDEN JUBILEE COMMEM. VOLUME *Calcutta Mathematical Society Golden Jubilee Commemoration Volume (1958-1959). (Jubilee Year - 1958)* [Part I = pp. 1-248, Part II = pp. 249-497. Pub. Golden Jubilee Celebration Committee, Calcutta Mathematical Society, Calcutta, undated: Part I, circa 1959 (LC received 1961), Part II, circa 1961 (LC received 1963). LC: QA3.C54] ... 6

CALCUTTA STATIST. ASSOC. BULL. *Calcutta Statistical Association Bulletin.* (Calcutta) ... 38

CANAD. J. MATH. *Canadian Journal of Mathematics.* (Toronto) [= *Journal Canadien de Mathématiques*] ... 3

CANAD. J. PHYS. *Canadian Journal of Physics.* (Ottawa) [Vol. 29, 1951-. Was CANAD. J. RES. SECT. A] ... 2

CANAD. J. PSYCHOL. *Canadian Journal of Psychology.* (Toronto) [= Revue Canadienne de Psychologie] ... 1

CANAD. J. RES. *Canadian Journal of Research.* (Ottawa) [Vol. 1-12, 1929-1935. Then SECT. A and *Section B, C, D*] ... 3

CANAD. J. RES. SECT. A *Canadian Journal of Research. Section A. Physical Sciences.* (Ottawa) [Vol. 13-28, 1935-1950. Was CANAD. J. RES. Became CANAD. J. PHYS.] ... 2

CANAD. MATH. BULL. *Canadian Mathematical Bulletin.* (Montreal) [= *Bulletin Canadien de Mathématiques*] ... 3

CASOPIS PEST. MAT. *Časopis pro Pěstování Matematiky. Československá Akademie Věd.* (Praha) [Vol. 76, 1951-. Was part of CASOPIS PEST. MAT. FYS.; Translated as CZECH. MATH. J.] ... 2

CASOPIS PEST. MAT. FYS. *Časopis pro Pěstování Matematiky a Fysiky. Část Vědecká.* (Praha) [Vol. 1-75, 1872-1950. Then CASOPIS PEST. MAT. and *Československý Časopis pro Fysiku*] ... 2

CHARACTER AND PERSONALITY *Character and Personality; an international quarterly of psychodiagnostics and allied studies.* (Durham, N.C.) [Became *Journal of Personality*] ... 2

CHEM. AND ENGRG. NEWS *Chemical and Engineering News.* (Washington, D.C.) ... 1

CHINESE MATH. ACTA *Chinese Mathematics.* (Providence, R.I.) [Vol. 1, 1962-. Translation of ACTA MATH. SINICA, Vol. 10-.] ... 4

COLLECTED PAPERS 70-TH ANNIV. NIHON UNIV. NATUR. SCI. *The Collected Papers for 70th Anniversary of Nihon University.* (Tokyo) [Vol. 3 *Natural Science*, 1960] ... 1

COLLOQ. ANAL. STATIST. BRUXELLES *Colloque sur l'Analyse Statistique.* [15-17 Déc. 1954 (Bruxelles). Pub. Centre Belge de Recherches Mathématiques. Georges Thone, Liège; Masson & Cie., Paris, 1955] ... 1

COLLOQ. INTERNAT. CENTRE NAT. RECH. SCI. *Colloques Internationaux du Centre National de la Recherche Scientifique.* [Vol. 13, "Le Calcul des Probabilités et ses Applications", 1949. Vol. 65, "L'Analyse Factorielle et ses Applications", 1955 (Vol. LVIII on flyleaf is a misprint!). Vol. 87, "Le Calcul des Probabilités et ses Applications", 1958 (Pub. 1959). Pub. C.N.R.S., 13 quai Anatole France, Paris 7ème] ... 15

COMMENT. MATH. HELVETIA *Commentarii Mathematici Helvetici.* (Zürich) ... 2

COMMENT. SOC. REG. SCI. GOTTING. REC. CL. MATH. *Commentationes Societatis Regiae Scientiarum Gottingensis. Recentiores. Classis Mathematicae.* (Göttingen) ... 4

COMMUN. ASSOC. COMP. MACH. *Communications of the Association for Computing Machinery.* (Baltimore) ... 2

COMPLEX SYSTEMS CONTROL (NAUKOVA DUMKA, KIEV) *Complex Systems of Control.* [Academy of Sciences of the Ukrainian SSR, Institute of Cybernetics. Naukova Dumka, Kiev, 1965] ... 1

COMPUT. INFORMATION SCI. (TOU + WILCOX) *Computer and Information Sciences.* Edited by Julius T. Tou & Richard H. Wilcox. [Spartan Books, Washington, D.C., 1964. LC:Q320.C6] ... 2

COMPUT. J. *The Computer Journal.* (London) ... 9

COMUN. ACAD. REPUB. POP. ROMINE *Comunicările Academiei Republicii Populare Romîne.* (Bucureşti) ... 2

CONG. INTERNAT. PHILOS. SCI. *Congrès International de Philosophie des Sciences.* [1949 (Paris). Actual. Sci. Indust. No. 1146. Pub. Herman & Cie., Paris, 1951. Vol. IV = *Calcul des Probabilités*] ... 1

CONTRIB. BOYCE THOMPSON INST. *Contributions. Boyce Thompson Institute for Plant Research, Inc.* (Yonkers, N.Y.) ... 1

CONTRIB. MATH. PSYCHOL. (FREDERIKSEN + GULLIKSEN) *Contributions to Mathematical Psychology.*  Edited by Norman Frederiksen & Harold Gulliksen.  [Holt, Rinehart & Winston, Inc., New York, 1964]  ... 3

CONTRIB. MATH. STATIST. (FISHER REPRINTS) *Contributions to Mathematical Statistics.*  [R.A. Fisher Reprints.  John Wiley & Sons, Inc., New York, 1950.  Reprints of 43 selected papers, with notes by the author]  ... 3

CONTRIB. ORDER STATIST. (SARHAN + GREENBERG) *Contributions to Order Statistics.*  Edited by Ahmed E. Sarhan & Bernard G. Greenberg.  [John Wiley & Sons, Inc., New York, 1962]  ... 3

CONTRIB. PROB. STATIST. (HOTELLING VOLUME) *Contributions to Probability and Statistics.  Essays in Honor of Harold Hotelling.*  Edited by Ingram Olkin, Sudhish G. Ghurye, Wassily Hoeffding, William G. Madow, & Henry B. Mann.  [Stanford University Press, Stanford, Calif., 1960]  ... 14

CONTRIB. STATIST. (MAHALANOBIS VOLUME) *Contributions to Statistics, presented to Professor P.C. Mahalanobis on the occasion of his 70th birthday.*  Edited by C.R. Rao.  [One of two volumes (cf.  ESSAYS ECONOMET. PLANNING).  Undated.  Statistical Publishing Society, Calcutta, circa 1964.  Pergamon Press, Oxford, circa 1965]  ... 10

COPEIA  *Copeia; published to advance the science of cold-blooded vertebrates.*  (New York)  ... 1

COWLES COMMISS. MONOG.  *Cowles Commission for Research in Economics.  Monograph.*  [No. 10, "Statistical Inference in Dynamic Economic Models", edited by Tjalling C. Koopmans, 1950.  No. 13, "Activity Analysis of Production and Allocation", edited by Tjalling C. Koopmans, 1951.  No. 14, "Studies in Econometric Methods", edited by William C. Hood & Tjalling C. Koopmans, 1953. Pub. John Wiley & Sons, Inc., New York]  ... 17

CSIRO DIV. MATH. STATIST. TECH. PAPER  *Commonwealth Scientific and Industrial Research Organization, Division of Mathematical Statistics, Technical Paper.*  (Melbourne)  ... 4

CURRENT SCI.  *Current Science.*  (Bangalore)  ... 3

CYBERNETICA  *Cybernetica.*  (Namur)  ... 1

CZECH. MATH. J.  *Czechoslovak Mathematical Journal.*  (Praha)  [= *Čehoslovackiĭ Matematičeskiĭ Žurnal.*  Vol. 1, 1951-, corresponds to Vol. 76, 1951-, of CASOPIS. PEST. MAT. but has different content]  ... 4

DE ECONOMIST  *De Economist.*  (Haarlem) [Nederlandsch Economisch
    Institut, Pub. De Erven F. Bohn N.V.]  ... 1

DEFENCE SCI. J.  *Defence Science Journal.*  (Delhi) ... 2

DEUTSCHE MATH.  *Deutsche Mathematik.*  (Leipzig) ... 1

DEUTSCHE STATIST. ZENTRALBLATT  *Deutsches statistiches Zentralblatt.*
    (Berlin) ... 1

DEUX CONG. MATH. HONGR.  *Deuxième Congrès Mathématique Hongrois.*
    [Second Hungarian Mathematical Congress, Zweiter Ungarischer
    Mathematischer Kongress. 24-31 August 1960 (Budapest).  Pub.
    Akadémiai Kiadó, Budapest, 1961.  Two vols. Vol. 2 Part IV:  Calcul
    des Probabilités et Statistiques Mathématiques] ... 4

DOKL. AKAD. NAUK ARMJAN. SSR  *Doklady.  Akademija Nauk Armjanskoĭ SSR.*
    (Yerevan, Armenia, USSR) ... 2

DOKL. AKAD. NAUK SSSR  *Doklady Akademii Nauk SSSR.*  (Moscow) [Vol.
    100, 1955-.  Was DOKL. AKAD. NAUK SSSR NOV. SER.  Vol. 130, 1960-
    translated as SOVIET MATH. DOKLADY] ... 19

DOKL. AKAD. NAUK SSSR NOV. SER.  *Doklady Akademii Nauk SSSR. Novaja
    Serija.*  (Moscow) [Vol. 1-99, 1935-1954.  Vol. 1-49, 51-56, not cited,
    translated as C.R. ACAD. SCI. URSS NOUV. SER.  Vol. 50 not published.
    Vol. 100, 1955- as DOKL. AKAD. NAUK SSSR] ... 15

DOKL. AKAD. NAUK UZBEK. SSR  *Doklady Akademii Nauk Uzbekskoĭ SSR.*
    (Tashkent) ... 3

DOPOVIDI AKAD. NAUK UKRAIN. RSR  *Dopovidi Akademiĭ Nauk Ukraïns'koĭ
    Radjans'koĭ Socialističnoĭ Respubliki.*  (Kiev) ... 3

DRAPERS CO. RES. MEM. BIOM. SER.  *Drapers' Company Research Memoirs.
    Biometric Series.*  (London) ... 3

DUKE MATH. J.  *Duke Mathematical Journal.*  (Durham, N.C.) ... 18

EARLY STATIST. PAPERS (PEARSON REPRINTS)  *Karl Pearson's Early
    Statistical Papers.*  [Cambridge University Press, Cambridge, 1956.
    Eleven papers reprinted for the Biometrika Trustees and first issued
    in a collected edition 1948] ... 5

ECOLOGY  *Ecology.*  (Brooklyn, N.Y.) ... 1

ECON. ESSAYS (CASSEL VOLUME)  *Economic Essays in Honour of Gustav
    Cassel.*  [George Allen & Unwin, Ltd., London, 1933] ... 1

ECONOMET. MODEL BUILDING (WOLD) *Econometric Model Building. Essays on the Causal Chain Approach.* Edited by Herman O.A. Wold. [Contributions to Economic Analysis, Vol. 36, North-Holland Publishing Company, Amsterdam, 1964] ... 6

ECONOMETRICA *Econometrica.* (Chicago) ... 88

EDUC. PSYCHOL. MEAS. *Educational and Psychological Measurement.* (Chicago) ... 51

EGYPTIAN STATIST. J. *Egyptian Statistical Journal.* (Cairo) [= *Al-Majallah Al-isha'hsa'iya Al-misriyah.* Text chiefly in Arabic] ... 2

EINZELSCHR. DEUTSCHEN STATIST. GES. *Einzelschriften der Deutschen Statistischen Gesellschaft.* (München) ... 1

EKON. TIDSKR. *Ekonomisk Tidskrifts.* (Uppsala) ... 1

ELECTRONICS *Electronics.* (New York) ... 1

ENGRG. CYBERNETICS *Engineering Cybernetics.* (New York) [Translation of of selected articles from IZV. AKAD. NAUK SSR TEH. KIBERNET.] ... 1

ENSEIGNEMENT MATH. *L'Enseignement Mathématique.* (Genève) [Vol. 1-40, 1899-1954] ... 2

ERFAHRUNGSBER. DEUTSCHEN FLUGWETTERD. *Erfahrungsberichte des Deutschen Flugwetterdienstes.* (Berlin) ... 1

ERGEB. MATH. KOLLOQ. *Ergebnisse eines mathematischen Kolloquiums.* (Leipzig) ... 7

ERKENNTNIS *Erkenntnis.* (Leipzig) [= *Journal of Unified Science*] ... 1

ESSAYS ECON. ECONOMET. (HOTELLING VOLUME) *Essays in Economics and Econometrics: A Volume in Honor of Harold Hotelling.* Edited by Ralph W. Pfouts. [Studies in Economics and Business Administration. The University of North Carolina Press, Chapel Hill, N.C., 1960] ... 1

ESSAYS ECONOMET. PLANNING (MAHALANOBIS VOLUME) *Essays on Econometrics and Planning, presented to Professor P.C. Mahalanobis on the occasion of his 70th birthday.* Edited by C.R. Rao. [One of two volumes (cf. CONTRIB. STATIST.). Undated. Statistical Publishing Society, Calcutta, circa 1964. Pergamon Press, Oxford, circa 1965] ... 2

ESTADIST. CHILENA SANTIAGO *Estadística Chilena.* (Santiago) ... 1

ESTADIST. ESPAN. *Estadística Española.* (Madrid) ... 5

ESTADISTICA *Estadística.* (Washington, D.C.) ... 1

ESTUDOS MAT. ESTATIST. ECONOMET. *Estudos de Matemática, Estatística e Econometria.* (Lisboa) [Vol. VII, 1962-1963. Curso de Matemáticas Superiores, Prof. Mira Fernandes. Instituto Superior de Ciências Económicas e Financeiras, Lisboa] ... 1

EUGENICS LAB. MEM. *Eugenics Laboratory Memoirs.* (London) [Francis Galton Laboratory for National Eugenics] ... 3

EVOLUTION *Evolution; international journal of organic evolution.* (Lancaster, Pa.) ... 1

EXPERIENTIA *Experientia.* (Basel) ... 2

FAC. SCI. STATIST. DEMOG. ATTUAR. IST. STATIST. CALC. PROB. *Facoltà di Scienze Statistiche, Demografiche e Attuariali. Istituto di Statistica e Istituto di Calcolo delle Probabilità.* (Roma) ... 3

FALL JOINT COMP. CONF. PROC. *Fall Joint Computer Conference, Proceedings.* [Spartan Books, Washington, D.C., 1962-. Was *Eastern Joint Computer Conference, Proceedings*] ... 3

FESTSK. STEFFENSEN *Festskrift til Professor, Dr. Phil J.F. Steffensen fra Kolleger og Elever paa hans 70 Aars Fødselsdag 28. Februar 1943.* [Den Danske Aktuarforening, København, 1943] ... 1

FIRST REPUBL. MATH. CONF. OF YOUNG RES. (AKAD. NAUK UKRAIN. SSR, KIEV) *First Republican Mathematical Conference of Young Researchers. Part II.* [April 2-4, 1964. Pub. Akademija Nauk Ukrainskoĭ SSR, Institut Matematiki, Kiev, 1965] ... 1

FOOD TECH. *Food Technology.* (Chicago) ... 2

FORSAKRINGSMAT. STUD. TILL. F. LUNDBERG *Försäkringsmatematiska Studier Tillägnade Filip Lundberg.* [Pub. Stockholm, 1946] ... 1

GACETA MAT. *Gaceta Matemática.* (Madrid) ... 1

GANITA *Ganita.* (Lucknow) ... 5

GAZETA MAT. SER. A *Gazeta Matematică. Series A.* (București) [Vol. 69, 1964-.] ... 1

GENETICS *Genetics; a periodical record of investigations bearing on heredity and variation.* (Princeton, N.J.) ... 9

GENUS *Genus.* *Comitato italiano per lo studio dei problemi della popolazione. Società Italiana di genetica ed eugenica.* (Roma) ... 1

GEOL. FOREN. I STOCKHOLM FORHANDL. *Geologiska Föreningens i Stockholm. Förhandlingar.* (Stockholm) ... 2

GIORN. ECON. ANN. ECONOMIA *Giornale degli Economisti e Annali di Economia.* (Bologna) [*Serie Nuova*, Vol. 1-26, 1939-1967] ... 7

GIORN. ECON. ANN. ECONOMIA SER. 2 *Giornale degli Economisti e Annali di Economia. Serie Segunda.* (Bologna) ... 1

GIORN. ECON. RIV. STATIST. SER. 3 *Giornale degli Economisti e Rivista di Statistica. Serie Terza.* (Bologna) ... 1

GIORN. IST. ITAL. ATTUARI *Giornale dell'Istituto Italiano degli Attuari.* (Roma) ... 14

GLASNIK MAT. FIZ. ASTRONOM. SER. 2 *Glasnik Matematičko-Fizički i Astronomski. Serija II. Periodicum Mathematico-Physicum et Astronomicum.* (Zagreb) ... 1

GROWTH *Growth; a journal for studies of development and increase.* (Menasha, Wis.) ... 7

HANDB. MATH. PSYCHOL. (LUCE, BUSH, + GALANTER) *Handbook of Mathematical Psychology.* Edited by R.D. Luce, R.R. Bush & E. Galanter. [John Wiley & Sons, Inc., New York, 1963. Vol. 1 = Chap. 1-8; Vol. 2 = Chap. 9-14; Vol. 3 = Chap. 15-21] ... 1

HANDB. MULTIVARIATE EXPER. PSYCHOL. (CATTELL) *Handbook of Multivariate Experimental Psychology.* Edited by Raymond B. Cattell. [Rand McNally & Co., Chicago, 1966] ... 13

HANDB. SOC. PSYCHOL. (LINDZEY) *Handbook of Social Psychology.* Edited by Gardner Lindzey. [Addison-Wesley Pub. Co., Inc., Cambridge, Mass., 1954] ... 1

HANDB. VERERBUNGSWISS. *Handbuch der Vererbungswissenschaft.* (Berlin) ... 1

HANDEL. NEDERL. NATUUR- EN GENEESK. CONG. *Handelingen van het Nederlandsch Natuur- en Geneeskundig Congres.* (Haarlem) ... 1

HARVARD EDUC. REV. *Harvard Educational Review.* (Cambridge, Mass.) ... 7

HEREDITAS GENET. ARK. LUND  *Hereditas; genetiskt arkiv.*  (Lund)  ... 1

HERMATHENA  *Hermathena;  a series of papers on Literature, Science, and Philosophy.*  (Dublin)  ... 1

HUMAN BIOL.  *Human Biology;  a record of research.*  (Baltimore)  ... 3

HUMAN RELATIONS  *Human Relations; studies towards the integration of the social sciences.*  (London)  ... 1

IBM J. RES. DEVELOP.  *IBM Journal of Research and Development.*  (New York)  ... 1

IBM SYSTEMS J.  *IBM Systems Journal.*  (New York)  ... 4

ICC BULL.  *ICC Bulletin, International Computation Centre.*  (Roma)  ... 1

IDOJARAS  *Az Időjárás.  A Magyar Meteorológiai Társaság Folyóirata.*  (Budapest)  ... 2

IEEE INTERNAT. CONV. REC.  *IEEE International Convention Record.*  (New York)  [Vol. 11, 1963-. Was *IRE International Convention Record*]  ... 1

IEEE TRANS. CIRCUIT THEORY  *IEEE Transactions on Circuit Theory.*  (New York)  [Was *IRE Transactions on Circuit Theory*]  ... 2

IEEE TRANS. ELECTRONIC COMPUT.  *IEEE Transactions on Electronic Computers.*  (New York)  [Vol. EC-12, 1963-.  Was IRE TRANS. ELECTRONIC COMPUT.]  ... 4

IEEE TRANS. INFORMATION THEORY  *IEEE Transactions on Information Theory.*  (New York)  [Vol. IT-9, 1963-.  Was IRE TRANS. INFORMATION THEORY]  ... 2

IFO-STUDIEN  *IFO-Stüdien.*  (Berlin)  [Institut für Wirtschaftsforschung]  ... 1

ILLINOIS J. MATH.  *Illinois Journal of Mathematics.*  (Urbana, Ill.)  ... 2

INC. STATIST.  *The Incorporated Statistician.*  (London) [Vol. 1-11, 1950-1961.  Then STATISTICIAN]  ... 4

INDIAN ANTHROP. ESSAYS MEM. MAJUMDAR  *Indian Anthropology: Essays in Memory of D.N. Majumdar.*  Edited by T.N. Madan & Gopāla Sarana. [Asia Publishing House, Bombay, 1962]  ... 1

INDIAN J. AGRIC. SCI.  *Indian Journal of Agricultural Science.*
   (Calcutta)  ... 1

INDIAN J. ENT.  *The Indian Journal of Entomology.*  (New Delhi)  ... 1

INDUST. MATH.  *The Journal of the Industrial Mathematics Society.*
   (Detroit)  [= *Industrial Mathematics*]  ... 1

INDUST. MILANO  *L'Industria: rivista di economia politica.*  (Milano)
   [No volume numbers: pagination consecutive per year]  ... 3

INDUST. PSYCHOTECH.  *Industrielle Psychotechnik; Angewandte Psychologie in
   Industrie - Handel - Verkehr - Verwaltung.*  (Berlin)  ... 2

INDUST. QUAL. CONTROL  *Industrial Quality Control.*  (Buffalo, N.Y.)
   ... 11

INFORMATION AND CONTROL  *Information and Control.*  (New York)  ... 6

INGEN. SKR.  *Ingeniørvidenskabelige Skrifter.*  (København)  ... 1

IN MEM. N.I. LOBATSCHEVSKII  *In Memoriam N.I. Lobatschevskii.*  [2 Volumes.
   Title *sic*.  Collection des mémoires présentés à la Société Physico-
   Mathématique de Kasan... (Vol. 1)... pour la fête de l'inauguration
   du monument de Lobatchefsky (1/13 Septembre 1896), par Mm. Hermite,
   Halsted, Girardville, Laisant, Lemoine, Neuberg, Ocagne.  Extrait
   du *Bulletin de la Société Physico-Mathématique de Kasan, 2 Série,*
   Vol. VI, 1897.  (Vol. 2)... à l'occasion de la célébration du
   centenaire de la découverte de la Géométrie Non-Euclidienne par N.I.
   Lobatcheffsky *(sic)*, par les savants de divers pays.  Pub. Glavnauka,
   1927.  LC: QA3.I6]  ... 2

INTERNAT. ECON. REV.  *International Economic Review.*  (Osaka)  ... 2

INTERNAT. J. ENGRG. SCI.  *International Journal of Engineering Science.*
   (New York)  ... 1

INTERNAT. J. OPINION ATTITUDE RES.  *International Journal of Opinion and
   Attitude Research.*  (Mexico)  ... 1

INZEN. SB. AKAD. NAUK SSSR  *Inženernyĭ Sbornik. Akademija Nauk SSSR.*
   (Moscow)  [Otdelenie Tehničeskih Nauk, Institut Mehaniki.  English
   title is *Engineering Review*]  ... 1

IOWA STATE COLL. J. SCI.  *Iowa State College Journal of Science.*
   (Ames, Iowa)  [Vol. 1-33, 1926-1959; Vol. 34, 1959-, as *Iowa State
   Journal of Science*]  ... 2

IOWA STATE COLL. OFF. PUBL.  *Iowa State College of Agriculture and Mechanic Arts. Official Publication.*  (Ames, Iowa) ... 1

IRE TRANS. AUTOMAT. CONTROL  *IRE Transactions on Automatic Control.* (New York) [Became *IEEE Transactions on Automatic Control*] ... 1

IRE TRANS. ELECTRONIC COMP.  *IRE Transactions on Electronic Computers.* (New York) [Vol. EC-2 - EC-11, 1953-1962.  Vol. PGEC-1 = *Transactions of the IRE Professional Group on Electronic Computers.*  Became IEEE TRANS. ELECTRONIC COMPUT.] ... 1

IRE TRANS. INFORMATION THEORY  *IRE Transactions on Information Theory.* (New York) [Vol. IT-1 - IT-8, 1955-1962.  Was TRANS. IRE PROF. GROUP INFORMATION THEORY.  Became IEEE TRANS. INFORMATION THEORY] ... 19

IRE TRANS. MED. ELECTRONICS  *IRE Transactions on Medical Electronics.* (New York) [Became *IEEE Transactions on Bio-medical Engineering*] ... 1

ISIS  *Isis; an international review devoted to the history of science and civilization.*  (Cambridge, Mass.) ... 1

ISRAEL J. MATH.  *Israel Journal of Mathematics.*  (Jerusalem) ... 1

IZV. AKAD. NAUK AZERBAIDZAN. SSR SER. FIZ.-TEH. MAT. NAUK  *Izvestija Akademii Nauk Azerbaĭdžanskoĭ SSR.  Serija Fiziko-Matematičeskih i Techničeskih Nauk.*  (Baku, Azerbaĭdžan, USSR) ... 1

IZV. AKAD. NAUK  SER. 6  *Izvestija Akademii Nauk. VI Serija.*  (Moscow) [Semi-monthly (except July and August) in 21 vols. 1907-1927. Title varies.  Vol. 1-10: *Izvestija Imperatorskoĭ Akademii Nauk. VI Serija.*  Vol. 11-18 (No. 9/11): *Izvestija Rossiĭskoĭ Akademii Nauk. VI Serija.*  Vol. 18 (No. 12/15) -19 (No. 16/17): *Izvestija Akademii Nauk Sojuza Sovetskih Socialističeskih Respublik (SSSR). VI Serija.*  Vol. 19 (No. 18) -21: *Izvestija Akademii Nauk SSSR. VI Serija*] ... 3

IZV. AKAD. NAUK SSSR SER. MAT.  *Izvestija Akademii Nauk SSSR.  Serija Matematičeskaja.*  (Moscow) [Vol. 1, 1937-.  Bimonthly, except 4 issues in 1937.  Title varies.  1937-1938: *Izvestija Otdelenija Matematičeskih i Estestvennyh Nauk.  Serija Matematičeskaja*] ... 11

IZV. AKAD. NAUK SSSR TEH. KIBERNET.  *Izvestija Akademii Nauk SSSR. Tehničeskaja Kibernetika.*  (Moscow) [Translated as ENGRG. CYBERNETICS] ... 1

IZV. AKAD. NAUK UZSSR. SER. FIZ. MAT. NAUK  *Izvestija Akademii Nauk UzSSR. Serija Fiziko-Matematičeskih Nauk.* (Tashkent) [Vol. 1, 1957-.  UzSSR Fanlar Akademijasining Ahboroti]  ... 5

IZV. ASSOTS. NAUC. INST. UNIV.  *Izvestija Assotsiatsija Naučno-Issledovatel'skih Institutov. Universitet.* (Moscow) [Vol. 1-2, 1928-1929]  ... 1

IZV. VYSS. UCEBN. ZAVED. MAT.  *Izvestija Vysših Učebnyh Zavedeniĭ Matematika. Ministerstvo Vysšego Obrazovanija SSSR.* (Kazan)  ... 2

J. ABNORM. SOC. PSYCHOL.  *Journal of Abnormal and Social Psychology.* (Boston) [Vol. 16-69, 1921-1964.  Then *Journal of Abnormal Psychology*]  ... 2

J. ACOUST. SOC. AMER.  *The Journal of the Acoustical Society of America.* (New York)  ... 5

J. ADV. RES.  *Journal of Advertising Research.* (New York)  ... 3

J. AGRIC. RES.  *Journal of Agricultural Research.* (Washington, D.C.)  ... 4

J. AGRIC. SCI.  *Journal of Agricultural Science.* (Cambridge)  ... 2

J. ALLERGY  *Journal of Allergy.* (St. Louis, Mo.)  ... 1

J. AMER. MED. ASSOC.  *The Journal of the American Medical Association.* (Chicago)  ... 2

J. AMER. SOC. AGRON.  *Journal of the American Society of Agronomy.* (Washington, D.C.) [1913-1948.  Then *Agronomy Journal*]  ... 2

J. AMER. STATIST. ASSOC.  *Journal of the American Statistical Association.* (Washington, D.C.) [Vol. 18, 1922-.  Was QUART. PUBL. AMER. STATIST. ASSOC.  Vol. 23-29, 1928-1934 included a supplement *Proceedings* cited as part (SP) in Chapter III.  Vol. 18 covers 1922-1923.  Thereafter one volume per year]  ... 251

J. AND PROC. ASIAT. SOC. BENGAL  *Journal and Proceedings of the Asiatic Society of Bengal.* (Calcutta) [*New Series* Vol. 1-30, 1905-1934]  ... 2

J. ANIMAL ECOL.  *The Journal of Animal Ecology.* (London)  ... 1

J. ANIMAL SCI.  *Journal of Animal Science.* (Madison, Wis.) [Vol. 1, 1942-.  Was PROC. AMER. SOC. ANIMAL PROD.]  ... 1

J. ANNAMALAI UNIV.  *Journal of the Annamalai University.* (Chidambaram, Madras, India)  ... 2

J. APPL. PHYS. *Journal of Applied Physics.* (Lancaster, Pa.) ... 8

J. APPL. PROB. *Journal of Applied Probability.* (London) ... 11

J. APPL. PSYCHOL. *Journal of Applied Psychology.* (Worcester, Mass.) ... 15

J. ASSOC. COMPUT. MACH. *Journal of the Association for Computing Machinery.* (Baltimore) ... 11

J. ASTRONAUTICAL SCI. *Journal of the Astronautical Sciences.* (New York) ... 1

J. AUSTRAL. MATH. SOC. *The Journal of the Australian Mathematical Society.* (Brisbane) [Vol. 1, p.1-240 is 1959; p.240- is 1960] ... 8

J. BASIC ENGRG. *Journal of Basic Engineering.* (New York) ... 2

J. BRITISH INST. RADIO ENGRS. *Journal of the British Institution of Radio Engineers.* (London) ... 1

J. CHEM. PHYS. *The Journal of Chemical Physics.* (Lancaster, Pa.) ... 2

J. CLIN. PSYCHOL. *Journal of Clinical Psychology.* (Brandon, Vt.) [Separates published as *Journal of Clinical Psychology. Monograph Supplement*] ... 5

J. COLL. ARTS SCI. CHIBA UNIV. *Journal of the College of Arts and Sciences, Chiba University.* (Tokyo) [= *Chiba Daigaku Bunrigakubu Kiyō*] ... 1

J. ECOL. *Journal of Ecology.* (London) ... 1

J. ECOLE POLYTECH. *Journal de l'École Polytechnique.* (Paris) ... 1

J. EDUC. PSYCHOL. *Journal of Educational Psychology.* (Washington, D.C.) ... 108

J. EDUC. RES. *Journal of Educational Research.* (Madison, Wis.) ... 12

J. EXPER. EDUC. *Journal of Experimental Education.* (Madison, Wis.) ... 27

J. EXPER. PSYCHOL. *Journal of Experimental Psychology.* (Washington, D.C.) ... 11

J. FARM ECON. *Journal of Farm Economics.* (Lancaster, Pa.) [Vol. 1-49, 1919-1967. Then *American Journal of Agricultural Economics*] ... 3

J. FORESTRY *Journal of Forestry.* (Washington, D.C.) ... 1

J. FRANKLIN INST. *Journal of the Franklin Institute.* (Philadelphia) ... 3

J. GENERAL PSYCHOL. *The Journal of General Psychology; experimental, theoretical, clinical, and historical psychology.* (Worcester, Mass.) ... 4

J. GENET. *Journal of Genetics.* (Cambridge) ... 1

J. GENET. PSYCHOL. *Journal of Genetic Psychology.* (Provincetown, Mass.) ... 1

J. GEOL. *Journal of Geology.* (Chicago) ... 1

J. GEOPHYS. RES. *Journal of Geophysical Research.* (Washington, D.C.) [Was TERREST. MAGNET. ATMOSPH. ELEC.] ... 2

J. HEREDITY *Journal of Heredity.* (Washington, D.C.) ... 1

J. HOKKAIDO GAKUGEI UNIV. SECT. B *Journal of the Hokkaido Gakugei University: Section B.* (Sapporo) [= *Hokkaidō Gakugei Daigaku Kiyō dai-2-bu*] ... 2

J. INDIAN MATH. SOC. *The Journal of the Indian Mathematical Society.* (Madras) [Vol. 1-20, 1909-1933. Then J. INDIAN MATH. SOC. NEW SER.] ... 2

J. INDIAN MATH. SOC. NEW SER. *The Journal of the Indian Mathematical Society. New Series.* (Madras) [Vol. 1, 1934-. Was J. INDIAN MATH. SOC.] ... 2

J. INDIAN SOC. AGRIC. STATIST. *Journal of the Indian Society of Agricultural Statistics.* (New Delhi) ... 26

J. INDIAN STATIST. ASSOC. *Journal of the Indian Statistical Association.* (Bombay) ... 30

J. INDUST. ENGRG. CHEM. *The Journal of Industrial and Engineering Chemistry.* (Washington, D.C.) ... 1

J. INST. ACTUAR. *Journal of the Institute of Actuaries.* (Cambridge) ... 3

J. INST. ELEC. COMMUN. ENGRS. JAPAN *Journal. Institute of Electrical and Communication Engineers of Japan.* (Tokyo) [= *Denki Tsūshin Gakkai Zasshi*] ... 1

J. KARNATAK UNIV. *Journal of the Karnatak University.* (Dharwar) ... 3

J. LONDON MATH. SOC. *The Journal of the London Mathematical Society.* (London) ... 13

J. MADRAS UNIV. SECT. B *Journal of the Madras University. Section B. Contributions in Mathematics, Physical and Biological Sciences.* (Madras) [Vol. 15, 1943-.] ... 1

J. MAHARAJA SAYAJIRAO UNIV. BARODA *Journal of the Maharaja Sayajirao University of Baroda.* (Baroda) ... 2

J. MARKET. RES. *Journal of Marketing Research.* (Chicago) ... 3

J. MATH. ANAL. APPL. *Journal of Mathematical Analysis and Applications.* (New York) ... 4

J. MATH. AND MECH. *Journal of Mathematics and Mechanics.* (Bloomington, Ind.) ... 2

J. MATH. AND PHYS. *Journal of Mathematics and Physics.* (Cambridge, Mass.) ... 1

J. MATH. PHYS. *Journal of Mathematical Physics.* (New York) ... 2

J. MATH. PSYCHOL. *Journal of Mathematical Psychology.* (New York) ... 8

J. MATH. PURES APPL. SER. 1 *Journal de Mathématiques Pures et Appliquées. Série 1, Recueil Mensuel de Mémoires sur les Diverses Parties des Mathématiques.* (Paris) ... 1

J. MATH. PURES APPL. SER. 9 *Journal de Mathématiques Pures et Appliquées. Neuvième Série.* (Paris) ... 3

J. MATH. SOC. JAPAN *Journal of the Mathematical Society of Japan.* (Tokyo) [Vol. 1, 1948-. Was PROC. PHYS. MATH. SOC. JAPAN] ... 1

J. METALS *Journal of Metals.* (New York) ... 1

J. METEOROL. SOC. JAPAN *Journal of the Meteorological Society of Japan.* (Tokyo) [= *Kishō Shūshi. Second Series* Vol. 1, 1923-.] ... 1

J. NARA GAKUGEI UNIV. NATUR. SCI. *The Journal of Nara Gakugei University. Natural Science.* (Nara) [= *Nara Gakugei Daigaku Kiyō*] ... 1

J. OPERATIONS RES. SOC. JAPAN *Journal of the Operations Research Society of Japan.* (Tokyo) [= *Keiei Kaguku Kyokai*] ... 1

J. OPT. SOC. AMER. *Journal of the Optical Society of America.* (Lancaster, Pa.) ... 4

J. OSAKA INST. SCI. TECH. PART I *Journal of the Osaka Institute of Science and Technology. Part I. Mathematics and Physics.* (Osaka) ... 1

J. PHILOS. *Journal of Philosophy.* (New York) ... 1

J. PSYCHOL. *Journal of Psychology; the general field of psychology.* (Worcester, Mass.) ... 4

J. REINE ANGEW. MATH. *Journal für die reine und angewandte Mathematik.* (Berlin) ... 5

J. RES. NAT. BUR. STANDARDS *Journal of Research of the National Bureau of Standards.* (Washington, D.C.) [Vol. 1-62, 1928-1959. Then *Section A, B, C, D*] ... 3

J. RES. NAT. BUR. STANDARDS SECT. B *Journal of Research of the National Bureau of Standards. Section B. Mathematics and Mathematical Physics.* (Washington, D.C.) [Vol. 63, 1959-.] ... 13

J. RES. NAT. BUR. STANDARDS SECT. D *Journal of Research of the National Bureau of Standards. Section D. Radio Propagation.* (Washington, D.C.) [Vol. 63-69, 1959-1965] ... 1

J. ROY. ANTHROP. INST. *Journal of the Royal Anthropological Institute of Great Britain and Ireland.* (London) [Vol. 1-95, 1871-1965. Then *Man*] ... 1

J. ROY. STATIST. SOC. *Journal of the Royal Statistical Society.* (London) [Vol. 50-110, 1887-1947. Was J. STATIST. SOC. LONDON. Became J. ROY. STATIST. SOC. SER. A] ... 45

J. ROY. STATIST. SOC. SER. A *Journal of the Royal Statistical Society. Series A. General* (London) [Vol. 111, 1948-. Was J. ROY. STATIST. SOC.] ... 17

J. ROY. STATIST. SOC. SER. B *Journal of the Royal Statistical Society. Series B. Methodological.* (London) [Vol. 10, 1948-. Was J. ROY. STATIST. SOC. SUPP.] ... 113

J. ROY. STATIST. SOC. SUPP. *Supplement to the Journal of the Royal Statistical Society.* (London) [Vol. 1-9, 1934-1947. Then J. ROY. STATIST. SOC. SER. B] ... 10

J. SCI. HIROSHIMA UNIV. SER. A *Journal of Science of the Hiroshima University. Series A. Mathematics, Physics, Chemistry.* (Hiroshima) ... 3

J. SOC. HONGR. STATIST. *Journal de la Société Hongroise de Statistique.* (Budapest) [=*Magyar Statisztikai Társaság*] ... 2

J. SOC. INDUST. APPL. MATH. *Journal of the Society for Industrial and Applied Mathematics.* (Philadelphia) [Vol. 1-13, 1953-1965. Then SIAM J. APPL. MATH.] ... 17

J. SOC. INDUST. APPL. MATH. SER. B NUMER. ANAL. *Journal of the Society for Industrial and Applied Mathematics. Series B. Numerical Analysis.* (Philadelphia) [Vol. 1-2, 1964-1965. Then SIAM J. NUMER. ANAL.] ... 2

J. SOC. PSYCHOL. *Journal of Social Psychology; political, racial and differential psychology.* (Worcester, Mass.) ... 2

J. SOC. STATIST. PARIS *Journal de la Société de Statistique de Paris.* (Paris) ... 9

J. STATIST. SOC. LONDON *Journal of the Statistical Society.* (London) [Vol. 1-49, 1838-1886. Then J. ROY. STATIST. SOC.] ... 1

J. TEXTILE INST. *The Journal of the Textile Institute.* (Manchester) ... 1

J. WASHINGTON ACAD. SCI. *Journal of the Washington Academy of Sciences.* (Washington, D.C.) ... 1

JAHRESBER. DEUTSCHEN MATH. VEREIN. *Jahresbericht der Deutschen Mathematiker Vereinigung.* (Stuttgart) ... 1

JAPAN. J. GENET. *The Japanese Journal of Genetics.* (Tokyo) [= *Idengaku Zasshi*] ... 1

JAPAN. J. MATH. *Japanese Journal of Mathematics.* (Tokyo) ... 1

JAPAN. J. PSYCHOL. *Japanese Journal of Psychology.* (Tokyo) ... 1

KAZAN. GOS. UNIV. UCEN. ZAP. *Kazanskiǐ Ordena Trudovogo Krasnogo Znameni Gosudarstvennyǐ Universitet im. V.I. Ul'janova-Lenina. Učenye Zapiski.* (Kazan) ... 1

KIEV. GOS. UNIV. MAT. SB. *Kievskiǐ Gosudarstvennyǐ Universitet. Matematičeskiǐ Sbornik.* (Kiev) [Russian version of KIIV. DERZ. UNIV. NAUK. ZAP.] ... 1

KIIV. DERZ. UNIV. NAUK. ZAP. *Kiïvs'kiǐ Deržavniǐ Universitet im. T.G. Ševčenka. Naukovi Zapiski.* (Kiev) [Ukrainian version of KIEV. GOS. UNIV. MAT. SB.] ... 1

KODAI MATH. SEM. REP. *Kodai Mathematical Seminar Reports. Bulletin of the Tokyo Institute of Technology. Series C.* (Tokyo) ... 1

KON. AKAD. WETENSCH. AMSTERDAM PROC. SECT. SCI. *Koninklijke Akademie van Wetenschappen te Amsterdam. Proceedings of the Section of Sciences.* (Amsterdam) [Vol. 1-37, 1898-1934. Then KON. NEDERL. AKAD. WETENSCH. PROC. SECT. SCI. English translation from VERSLAG AFD. NATUURK. KON. AKAD. WETENSCH.] ... 15

KON. NEDERL. AKAD. WETENSCH. PROC. SECT. SCI. *Koninklijke Nederlandse Akademie van Wetenschappen. Proceedings of the Section of Sciences.* (Amsterdam) [Vol. 38-53, 1935-1950. Was KON. AKAD. WETENSCH. AMSTERDAM PROC. SECT. SCI. Became KON. NEDERL. AKAD. WETENSCH. SER. A. Duplicated as (1939-1950) *Indagationes Mathematicae ex actis quibus titulus*] ... 9

KON. NEDERL. WETENSCH. PROC. SER. A *Koninklijke Nederlandse Akademie van Wetenschappen. Proceedings. Series A. Mathematical Sciences.* (Amsterdam) [Vol. 54, 1951-. Title varies: 1951-1954 *Proceedings of the Section of Sciences. Series A. Mathematical Sciences.* Was KON. NEDERL. AKAD. WETENSCH. PROC. SECT. SCI. Duplicated as *Indagationes Mathematicae ex actis quibus titulus*] ... 12

KONG. EXPER. PSYCHOL. BER. *Kongress für experimentale Psychologie. Bericht.* (Leipzig) ... 1

KOZGAZDASAGI SZEMLE *Közgazdasági Szemle.* (Budapest) ... 1

KUMAMOTO J. SCI. SER. A *Kumamoto Journal of Science. Series A. Mathematics, Physics and Chemistry.* (Kumamoto) ... 1

KUNGL. SVENSKA VETENSKAPSAKAD. HANDL. *Kungl. Svenska Vetenskapsakademiens Handlingar.* (Stockholm) [*Ny Följd* Vol. 1-63, 1855-1923] ... 1

KUNGL. TEK. HOGSK. HANDL. STOCKHOLM *Kungl. Tekniska Högskolans Handlingar.* (Stockholm) ... 2

KWART. STATYST. GLOWNY URZAD STATYST. *Kwartalnik Statystyczny.*
(Warszawa) [= *Revue Trimestrielle de Statistique.* Główny Urząd
Statystyczny Rzeczypospolitej Polskiej (Office Central de Statis-
tique de la République Polonaise). LC: HA1451.A245] ... 2

LANDBOUWK. TIJDSCHR. WAGENINGEN *Landbouwkundig Tijdschrift te
Wageningen.* (Groningen) ... 2

LAND ECON. *Land Economics.* (Madison, Wis.) ... 1

LANGUAGE *Language.* (Austin, Texas) ... 1

LAS CIENCIAS (MADRID) *Las Ciencias.* (Madrid) ... 1

LA SCUOLA IN AZIONE *La Scuola in Azione.* (Milano) [Ente Nazionale
Idrocarburi – E.N.I. Scuola Enrico Mattei Di Studi Superiori Sugli
Idrocarburi] ... 2

LATVIJAS PSR ZINATNU AKAD. VESTIS FIZ. TEH. ZINATNU SER. 6 *Latvijas
PSR Zinātņu Akadēmijas Vēstis. Fizikas un Tehnisko Zinātņu.
Sērija 6.* (Riga, Latvia, USSR) [Izvestija Akademii Nauk SSSR
Latviĭskoĭ SSR] ... 1

LIMIT THEOR. STATIST. INFER. (IZDAT FAN, TASHKENT) *Limit Theorems and
Statistical Inference.* [Academy of Sciences of the Uzbek SSR.
V.I. Romanovskiĭ Institute of Mathematics, Izdat. "Fan" Uzbek. SSR,
Tashkent, 1966] ... 1

LITOVSK. MAT. SB. *Litovskiĭ Matematičeskiĭ Sbornik. Vysšie Učebnye
Zavedenija Litovskoĭ SSR.* (Vilnius, Lithuania, USSR) ... 8

LUNDS UNIV. ARSSKR. N. F. AVD. 2 *Lunds Universitets Årsskrift. Ny
Fö+ Andra Avdelningen 2. Medecin samt Matematiska och
Naturvetenskapliga Ämnen.* (Lund) [*Acta Universitatis Lundensis.
Nova Series* Vol. 1–59, 1905–1963. Duplicated as (1931–1963)
*Kungl. Fysiografiska Sällskapets i Lund Förhandlingar*] ... 6

L'VOV GOS. UNIV. UCEN. ZAP. SER. MEH. MAT. *L'vovskiĭ Gosudarstvennyĭ
Universitet imeni Ivana Franko. Učenye Zapiski. Serija Mehaniko-
Matematičeskaja.* (L'vov) ... 1

MAGYAR TUD. AKAD. MAT. KUTATO. INT. KOZL. *A Magyar Tudományos Akadémia
Matematikai Kutató Intézetének Közleményei.* (Budapest) [Vol.
1–9, 1956–1964. Then STUDIA SCI. MATH. HUNGAR.] ... 10

MAGYAR TUD. AKAD. MAT. OSZT. KOZL. *A Magyar Tudományos Akadémia Matematikai és Fizikai Todományok Osztályának Közleményei.* (Budapest)
... 2

MANAG. SCI. SER. A SCI. *Management Science. Series A, Sciences.*
(Baltimore) [= *Journal of the Institute of Management Sciences*]
... 1

MAN IN INDIA *Man in India; a quarterly record of anthropological
science with special reference to India.* (Ranchi, Bihar, India)
... 1

MAT. ELEM. SER. 4 *Matematica Elemental.* $4^a$ *Serie.* (Madrid) [Revista
publicada por el instituto "Jorge Juan" de Matematicas y la real
sociedad matematica española. Consejo Superior de Investigaciones
cientificas patronato "Alfonso el Sabio"] ... 1

MAT.-FYZ. CASOPIS SLOVEN. AKAD. VIED *Matematicko-Fyzikálny Časopis.
Slovenská Akadémia Vied.* (Bratislava) ... 1

MAT. SB. *Matematičeskiǐ Sbornik.* (Moscow) [Vol. 1-42, 1866-1935.
Then MAT. SB. NOV. SER.] ... 3

MAT. SB. NOV. SER. *Matematičeskiǐ Sbornik. Nova Serija.* (Moscow)
[Vol. 43, 1936-. Was MAT. SB.; Translated as *Mathematics of the
USSR - Sbornik.* Vol. 1, 1967 = Vol. 72 of Russian] ... 4

MAT. TIDSSKR. A *Matematisk Tidsskrift. A. Elementary.* (København)
[1919-1952. Then NORDISK MAT. TIDSKR.] ... 1

MAT. TIDSSKR. B *Matematisk Tidsskrift. B. Advanced.* (København)
[1919-1952. Then *Mathematica Scandinavica*] ... 1

MATH. ANN. *Mathematische Annalen.* (Leipzig, Berlin) ... 7

MATH. COMPUT. *Mathematics of Computation.* (Washington, D.C.) [Vol. 14,
1960-. Was MATH. TABLES AID COMPUT.] ... 7

MATH. COMPUT. SCI. BIOL. MED. *Mathematics and Computer Science in
Biology and Medicine.* (London) [Proceedings of Conference held
by Medical Research Council in association with the Health Depart-
ments. July 1964 (Oxford), Pub. 1965] ... 2

MATH. GAZETTE *The Mathematical Gazette.* (London) ... 1

MATH. JAPON. *Mathematica Japonicae.* (Osaka) ... 2

MATH. MAG. *Mathematics Magazine.* (Buffalo) ... 5

MATH. METH. DIGIT. COMPUT. (RALSTON + WILF) *Mathematical Methods for
Digital Computers.* Edited by Anthony Ralston & Herbert S. Wilf,
[John Wiley & Sons, Inc., New York, 1960] ... 2

MATH. METH. SOC. SCI. (ARROW, KARLIN, + SUPPES) *Mathematical Methods in the Social Sciences, Proceedings of the First Stanford Symposium.* Edited by Kenneth J. Arrow, Samuel Karlin, & Patrick Suppes. [Stanford University Press, Stanford, Calif., 1960] ... 1

MATH. NACHR. *Mathematische Nachrichten.* (Berlin) ... 4

MATH. NOTAE *Mathematicae Notae, Boletín del Instituto de Matemática.* (Rosario) ... 1

MATH. NOTES EDINBURGH *Mathematical Notes.* (Edinburgh) [Vol. 1-30, 1907-1937. Then *The Edinburgh Mathematical Notes,* Vol. 31-41, 1939-1957. Merged into PROC. EDINBURGH MATH. SOC. SER. 2] ... 3

MATH. PSYCHOL. (MILLER) *Mathematics and Psychology.* Edited by George A. Miller. [John Wiley & Sons, Inc., New York, 1964] ... 1

MATH. STUDENT *The Mathematics Student.* (Madras) ... 4

MATH. TABLES AIDS COMPUT. *Mathematical Tables and Other Aids to Computation.* (Washington, D.C.) [Vol. 1-13, 1944-1959. Then MATH. COMPUT.] ... 4

MATH. THINK.MEAS. BEHAV. (SOLOMON) *Mathematical Thinking in the Measurement of Behavior.* Edited by Herbert Solomon. [The Free Press, Glencoe, Ill., 1960] ... 1

MATH. THINK. SOC. SCI. (LAZARSFELD) *Mathematical Thinking in the Social Sciences.* Edited by Paul F. Lazarsfeld. [The Free Press, Glencoe, Ill., 1954] ... 3

MATH. Z. *Mathematische Zeitschrift.* (Berlin) ... 3

MATHEMATICA, CLUJ *Mathematica.* (Cluj) [Vol. 1-16, 1929-1940. Seminarului de Matematici al Universităţii Cluj. Became MATHEMATICA, TIMIŞOARA] ... 3

MATHEMATICA, TIMISOARA *Mathematica.* (Timişoara) [Vol. 17-23, 1941-1948. Seminarul de Analiză Matematică Facultatea de Ştiinţe, Timişoara. Was MATHEMATICA, CLUJ. Became *Mathematica;* Societatea de Ştiinţe Matematice şi Fizice din R.P.R. Filiala Cluj. Vol. 1 (24), 1959-.] ... 1

MEAS. ECON. (GRUNFELD VOLUME) *Measurement in Economics. Studies in Mathematical Economics and Econometrics in Memory of Yehuda Grunfeld.* Edited by Carl F. Christ, Milton Friedman, Leo A. Goodman, Zvi Griliches, Arnold C. Harberger, Nissan Liviatan, Jacob Mincer, Yair Mundlak, Marc Nerlove, Don Patinkin, Lester G. Telser & Henri Theil. [Stanford University Press, Stanford, Calif., 1963] ... 1

MEAS. PREDICT. (STOUFFER ET AL.) *Measurement and Prediction.* Studies in Social Psychology in World War II, Volume IV. Edited by Samuel A. Stouffer, Louis Guttman, Edward A. Suchman, Paul F. Lazarsfeld, Shirley A. Star & John A. Clausen. [Princeton University Press, Princeton, N.J. & Oxford University Press, London, 1950. Paperback edition: Science Editions, John Wiley & Sons, Inc., New York, 1966] ... 3

MED. AND BIOL. (IGAKU TO SEIBUTSUGAKU) *Medicine and Biology.* (Tokyo) [= *Igaku to Seibutsugaku*] ... 1

MEDD. LUNDS ASTRONOM. OBS. SER. 1 *Meddelande från Lunds Astronomiska Observatorium. Series I.* (Lund) ... 1

MEDD. LUNDS ASTRONOM. OBS. SER. 2 *Meddelande från Lunds Astronomiska Observatorium. Series II.* (Lund) ... 3

MEDED. LANDBOUWHOGESCHOOL WAGENINGEN *Mededelingen van de Landbouwhoge-school te Wageningen.* (Wageningen) ... 2

MEDIZIN. DOKUMENT. *Medizinische Dokumentation.* (Bielefeld) [= *Medical Documentation.* Became *Methods of Information in Medicine*] ... 1

MEM. AMER. MATH. SOC. *Memoirs of the American Mathematical Society.* (Lancaster, Pa.) ... 2

MEM. AMER. PHILOS. SOC. *Memoirs of the American Philosophical Society.* (Philadelphia) ... 1

MEM. COLL. SCI. UNIV. KYOTO SER. A *Memoirs of the College of Science. University of Kyoto. Series A. Mathematics.* (Kyoto) ... 1

MEM. FAC. ENGRG. KOBE UNIV. *Memoirs of the Faculty of Engineering. Kobe University.* (Kobe) [= *Kōbe Daigaku Kōgakubu Kenkyū Hōkoku*] ... 1

MEM. FAC. SCI. KYUSHU UNIV. SER. A *Memoirs of the Faculty of Science, Kyūshū University. Series A. Mathematics.* (Fukuoka) ... 6

MEM. INDIAN METEOROL. DEPT. *Memoirs of the Indian Meteorological Department.* (Calcutta, Simla) [Occasional Discussions and Compilations of Meteorological Data relating to India and the Neighbouring Countries. LC: QC990.I35] ... 7

MEM. INST. IMP. FRANCE (ANNEE 1810) *Mémoires de la Classe des Sciences Mathématiques et Physiques de l'Institut Impérial de France, Année 1810.* (Paris) ... 1

MEM. PRES. ACAD. ROY. SCI. INST. FRANCE  *Mémoires Présentés par Divers Savants à l'Académie Royale des Sciences de l'Institut de France.* (Paris) ... 1

MEM. PROC. MANCHESTER LIT. PHILOS. SOC.  *Memoirs and Proceedings of the Manchester Literary and Philosophical Society.* (Manchester) [Vol. 41, 1897-.] ... 1

MEM. REALE ACCAD. SCI. TORINO  *Memorie della Reale Accademia delle Scienze di Torino.* (Torino) [= *Mémoires de l'Académie Impériale des Sciences, Littérature et Beaux-Arts de Turin pour les Années 1811-1812*] ... 1

MEM. ROY. ASTRONOM. SOC.  *Memoirs of the Royal Astronomical Society.* (London) ... 1

MEM. ROY. METEOROL. SOC.  *Memoirs of the Royal Meteorological Society.* (London) [Vol. 1-4 (no. 1-40), 1926-1939. Then merged into QUART. J. ROY. METEOROL. SOC.] ... 1

MEM. SCI. MATH.  *Mémorial des Sciences Mathématiques.* (Paris) ... 1

MESSENGER MATH.  *The Messenger of Mathematics.* (Cambridge) [*New Series* Vol. 1-58, 1871-1929. Then merged with QUART. J. PURE APPL. MATH. to form QUART. J. MATH. OXFORD SER.] ... 1

MET. VYC. (LENINGRAD)  *Metody Vyčislenii.* (Leningrad) ... 1

METEOROL. MAG. LONDON  *Meteorological Magazine.* (London) ... 1

METEOROL. MONOG.  *Meteorological Monographs.* (Boston) ... 1

METEOROL. Z.  *Meteorologische Zeitschrift.* (Berlin) ... 7

METEOROS.  *Meteoros. Revista de meteorologia y geofisica del Servicio Meteorologico Nacional.* (Buenos Aires) ... 2

METRIKA  *Metrika; Zeitschrift für theoretische und angewandte Statistik.* (Würzburg) [Vol. 1, 1958-. Was MITT. MATH. STATIST. and STATIST. VIERTELJAHRESSCHR.] ... 20

METROECONOMICA  *Metroeconomica. Rivista Internazionale di Economica.* (Bologna) ... 1

METRON  *Metron. Rivista Internazionale di Statistica.* (Roma) [Vol. 1, 5-14, 16-19 paginate afresh within part number(s)] ... 48

MICHIGAN MATH. J.  *The Michigan Mathematical Journal.* (Ann Arbor, Mich.) ... 2

MITT. MATH. STATIST. *Mitteilungsblatt für Mathematische Statistik.* (München) [Vol. 1-9, 1949-1957. Then merged with STATIST. VIERTELJAHRESSCHR. to form METRIKA] ... 18

MONATSHEFTE MATH. *Monatshefte für Mathematik.* (Wien) [Vol. 52, 1948-. Was MONATSHEFTE MATH. PHYS.] ... 2

MONATSHEFTE MATH. PHYS. *Monatshefte für Mathematik und Physik.* (Wien) [Vol. 1-51, 1890-1944. Then MONATSHEFTE MATH.] ... 3

MONOG. GOTTING. APUD HEN. DIET. *Monograph. Gottingae. Apud Henricum Dietrich.* [Typis Dieterichianis, Göttingen. Reprints (1823, 1828) from COMMENT. SOC. REG. SCI. GOTTING. REC. CL. MATH.] ... 2

MONTHLY NOTICES ROY. ASTRONOM. SOC. *Monthly Notices of the Royal Astronomical Society.* (London) ... 6

MONTHLY NOTICES ROY. ASTRONOM. SOC. GEOPHYS. SUPP. *Monthly Notices of the Royal Astronomical Society. Geophysical Supplement.* (London) ... 1

MONTHLY WEATHER REV. *Monthly Weather Review.* (Washington, D.C.) ... 1

MULTIVARIATE ANAL. PROC. INTERNAT. SYMP. DAYTON (KRISHNAIAH) *Multivariate Analysis: Proceedings of an International Symposium held in Dayton, Ohio, June 14-19, 1965.* Edited by Paruchuri R. Krishnaiah. [Academic Press Inc., New York, 1966] ... 30

MULTIVARIATE BEHAV. RES. *Multivariate Behavioral Research.* (Fort Worth, Texas) ... 3

NACHR. GES. WISS. GOTTING. MATH. PHYS. KL. *Nachrichten von der Gesellschaft der Wissenschaften zu Göttingen. Mathematisch-Physikalische Klasse.* (Göttingen) [Akademie der Wissenschaften. Göttingen] ... 1

NAGOYA MATH. J. *Nagoya Mathematical Journal.* (Nagoya) ... 1

NAT. CONV. TRANS. AMER. SOC. QUAL. CONTROL *National Convention Transactions. American Society for Quality Control.* (New York) ... 1

NATURE *Nature.* (London) ... 11

NATURWISS. *Die Naturwissenschaften.* (Berlin) ... 1

NATURWISS. WOCHENSCHR. JENA *Naturwissenschaftliche Wochenschrift.* (Jena) ... 1

NATUURWETENSCH. TIJDSCHR. (ANTWERP) *Natuurwetenschappelijk Tijdschrift.* (Antwerpen) ... 1

NAUCN. DOKL. VYSS. SKOLY *Naučnye Doklady Vyssheĭ Školy. Serija Radio-Elektronika.* (Moscow) ... 1

NAVAL RES. LOGISTICS QUART. *Naval Research Logistics Quarterly.* (Washington, D.C.) ... 3

NEDERL. TIJDSCHR. PSYCHOL. GRENSGEB. *Nederlands Tijdschrift voor de Psychologie en haar Grensgebieden.* (Amsterdam) ... 1

NEW STATIST. TABLES *New Statistical Tables.* (London) [Issued by the Biometrika Office, University College, London, and printed by the University Press, Cambridge. Reprints from BIOMETRIKA] ... 2

NEW YORK STATIST. *The New York Statistician.* (New York) ... 1

NEW ZEALAND J. SCI. *New Zealand Journal of Science.* (Wellington) [Vol. 1, 1958-. Was part of NEW ZEALAND J. SCI. TECH.] ... 1

NEW ZEALAND J. SCI. TECH. SECT. A *New Zealand Journal of Science and Technology. Section A. Agricultural Section.* (Wellington) [Vol. 20-38, 1918-1937] ... 3

NORDIC STATIST. J. *Nordic Statistical Journal.* (Stockholm) [Vol. 1-4, 1929-1932. Also issued as Vol. 8-11 of NORDISK STATIST. TIDSKR.] ... 4

NORDISK MAT. TIDSKR. *Nordisk Matematisk Tidskrift.* (Oslo) [Vol. 1-11, 1953-1963. Was MAT. TIDSSKR. A and *Norsk Matematisk Tidsskrift*] ... 4

NORDISK PSYKOL. MONOG. NR. 3 (UPPSALA SYMP. PSYCHOL. FACT. ANAL.) *Nordisk Psykologi's Monografiserie Nr. 3.* Uppsala Symposium on Psychological Factor Analysis, 17-19 March 1953. [Almqvist & Wiksell, 1953] ... 5

NORDISK STATIST. TIDSKR. *Nordisk Statistisk Tidskrift.* (Stockholm) [Vol. 1-11, 1922-1932. Vol. 8-11 also issued as Vol. 1-4 of NORDIC STATIST. J.] ... 2

NORDISK TIDSSKR. INFORMATIONS-BEHANDLING *Nordisk Tidsskrift for Informations-Behandling.* (København) ... 2

NORDISK TIDSSKR. TEK. OKON. *Nordisk Tidsskrift for Teknisk Økonomi.* (København) ... 1

NUCLEAR PHYS. *Nuclear Physics*. (Manchester) ... 3

NUMER. MATH. *Numerische Mathematik*. (Berlin) ... 8

NUOVO CIMENTO *Il Nuovo Cimento*. (Bologna) ... 2

OBIT. NOTICES FELLOWS ROY. SOC. *Obituary Notices of the Fellows of the Royal Society*. (London) [Vol. 1-9, 1932-1954. Then BIOGRAPH. MEM. FELLOWS ROY. SOC.] ... 2

OCCUP. PSYCHOL. LONDON *Occupational Psychology*. (London) ... 2

OFF. REP. 1940 MEET. AMER. EDUC. RES. ASSOC. *Official Report of 1940 Meeting, American Educational Research Association*. [St. Louis, Mo., Feb. 24-27, 1940] ... 1

OPERATIONS RES. *Operations Research*. (Baltimore) ... 6

OSAKA J. MATH. *Osaka Journal of Mathematics*. (Osaka) [Vol. 1, 1964-. Was OSAKA MATH. J.] ... 1

OSAKA MATH. J. *Osaka Mathematical Journal*. (Osaka) [Vol. 1-16, 1949-1963. Then OSAKA J. MATH.] ... 5

OSIRIS *Osiris; studies on the history and philosophy of science, and on the history of learning and culture*. (Bruges) ... 1

PACIFIC J. MATH. *Pacific Journal of Mathematics*. (Berkeley, Calif.) ... 7

PACIFIC SCI. *Pacific Science; a quarterly devoted to the biological and physical sciences of the Pacific region*. (Honolulu) ... 1

PAKISTAN J. SCI. *Pakistan Journal of Science*. (Lahore) ... 1

PAM. ALEKS. ALEKS. ANDRONOVA *Pamjati Aleksandra Aleksandroviča Andronova*. (Moscow) [Izdat. Akad. Nauk SSSR. Mem. vol. A.A. Andronov, 1955] ... 1

PAPERS MICHIGAN ACAD. SCI. ARTS LETT. *Papers of the Michigan Academy of Science, Arts, and Letters*. (Ann Arbor, Mich.) ... 1

PAPERS PRES. COLLOQ. STATIST. DESIGN LAB. EXPER. *Papers presented at the Colloquium in Statistical Design of Laboratory Experiments*. [Vol. 4028 qv = NAVORD Report 4028. U.S. Naval Ordnance Laboratory, White Oak, Md.] ... 1

PEKING UNIV. J. *The Peking University Journal. Natural Science Reports.* (Peking) [= *Acta Scientiarum Naturalium Universitatis Pekinensis.* = *Pei-Ching ta Hsueh Hsueh Pao. Tzu Jan k'o Hsueh.* Vol. 1, 1955-.] ... 2

PHILOS. MAG. SER. 4 *The London, Edinburgh, and Dublin Philosophical Magazine and Journal of Science, Fourth Series.* (London) [Vol. 1-50, 1851-1875. Then SER. 5] ... 2

PHILOS. MAG. SER. 5 *The London, Edinburgh, and Dublin Philosophical Magazine and Journal of Science, Fifth Series.* (London) [Vol. 1-50, 1876-1900. Then SER. 6] ... 13

PHILOS. MAG. SER. 6 *The London, Edinburgh, and Dublin Philosophical Magazine and Journal of Science, Sixth Series.* (London) [Vol. 1-50, 1901-1925. Then SER. 7] ... 10

PHILOS. MAG. SER. 7 *The London, Edinburgh, and Dublin Philosophical Magazine and Journal of Science, Seventh Series.* (London) [Vol. 1-46, 1926-1955. Then *Eighth Series*] ... 15

PHILOS. TRANS. ROY. SOC. LONDON SER. A *Philosophical Transactions of the Royal Society of London. Series A. Mathematical and Physical Sciences.* (London) [Vol. 178, 1887-.] ... 7

PHILOS. TRANS. ROY. SOC. LONDON SER. B *Philosophical Transactions of the Royal Society of London. Series B. Biological Sciences.* (London) [Vol. 178, 1887-.] ... 1

PHYS. REV. *The Physical Review; a journal of experimental and theoretical physics.* (New York) [*Series 2, Vol. 1, 1913-.*] ... 3

PHYSICA *Physica.* (Amsterdam) ... 1

PLANT PHYSIOL. *Plant Physiology.* (Lancaster, Pa.) ... 1

POLSKA AKAD. NAUK STUD. MATH. *Polska Akademia Nauk Studia Mathematica.* (L'vov) ... 3

POLSKA AKAD. NAUK ZASTOS. MAT. *Polska Akademia Nauk. Instytut Matematyczny Zastosowania Matematyki.* (Warszawa) ... 8

PONT. ACAD. SCI. ACTA *Pontificia Academia Scientiarum. Acta.* (Roma) ... 1

PORTUGAL. MATH. *Portugaliae Mathematica.* (Lisboa) ... 1

PROB. STATIST. (CRAMER VOLUME) *Probability and Statistics. The Harald Cramér Volume.* Edited by Ulf Grenander. [John Wiley & Sons, Inc., New York; Almqvist & Wiksell, Stockholm, 1959] ... 9

PROBLEMS MEAS. CHANGE (HARRIS) *Problems in Measuring Change.* Edited by Chester W. Harris. [University of Wisconsin Press, Madison, Wis., 1963] ... 6

PROBLEMS OPERATIONS RES. (MECNIEREBA) *Problems in Operations Research.* [Academy of Sciences of the Georgian SSR. Institute of Cybernetics. Izdat. "Mecniereba", Tbilisi, 1966] ... 1

PROBLEMY KYBERNETIKY (PRAGUE) *Problémy Kybernetiky.* (Praha) [Nakladatelství Československé Akademie Věd] ... 1

PROC. AMER. ACAD. ARTS SCI. *Proceedings of the American Academy of Arts and Sciences.* (Boston) ... 4

PROC. AMER. MATH. SOC. *Proceedings of the American Mathematical Society.* (Lancaster, Pa.) ... 24

PROC. AMER. SOC. ANIMAL PROD. *Proceedings of the American Society of Animal Production.* (Madison, Wis.) [Vol. 1-33, 1908-1940. Then J. ANIMAL SCI.] ... 1

PROC. AMER. SOC. HORT. SCI. *Proceedings. American Society for Horticultural Science.* (Geneva, N.Y.) ... 3

PROC. BERKELEY SYMP. MATH. STATIST. PROB. *Proceedings of the Berkeley Symposium on Mathematical Statistics and Probability, August 13-18, 1945; January 27-29, 1946.* Edited by Jerzy Neyman. [University of California Press, Berkeley & Los Angeles, 1949] ... 7

PROC. CAMBRIDGE PHILOS. SOC. *Proceedings of the Cambridge Philosophical Society.* (Cambridge) ... 59

PROC. CANAD. MATH. CONG. *Proceedings of the Canadian Mathematical Congress.* (Toronto) ... 1

PROC. EDINBURGH MATH. SOC. SER. 2 *Proceedings of the Edinburgh Mathematical Society. Second Series.* (Edinburgh) ... 12

PROC. EDUC. RES. FORUM *Proceedings. Educational Research Forum, Endicott, New York, August 26-31, 1940.* [IBM Corp., Endicott, N.Y., 1941] ... 3

PROC. FED. AMER. SOC. EXPER. BIOL. *Proceedings of the Federation of American Societies for Experimental Biology.* (Baltimore) ... 1

PROC. FIFTH BERKELEY SYMP. MATH. STATIST. PROB. *Proceedings of the Fifth Berkeley Symposium on Mathematical Statistics and Probability, June 21–July 18, 1965; December 27, 1965 – January 7, 1966.* Edited by Lucien M. LeCam & Jerzy Neyman. [University of California Press, Berkeley & Los Angeles, 1967] ... 1

PROC. FIRST IBM CONF. STATIST. *IBM – Proceedings of the First IBM Conference on Statistics.* Edited by Carl F. Kossack. [IBM Dept. Education, Poughkeepsie, N.Y.] ... 1

PROC. FOURTH BERKELEY SYMP. MATH. STATIST. PROB. *Proceedings of the Fourth Berkeley Symposium on Mathematical Statistics and Probability, June 20–July 30, 1960.* Edited by Jerzy Neyman. [University of California Press, Berkeley & Los Angeles, 1961] ... 9

PROC. GLASGOW MATH. ASSOC. *Proceedings of the Glasgow Mathematical Association.* (Glasgow) [Vol. 1–7, 1952–1965. Then *Glasgow Mathematical Journal*] ... 1

PROC. IBM SCI. COMPUT. SYMP. STATIST. *Proceedings of the IBM Scientific Computing Symposium, October 21–23, 1963.* [Held at Yorktown Heights, N.Y. Pub. IBM Data Processing Division, White Plains, N.Y., 1965] ... 5

PROC. IMP. ACAD. TOKYO *Proceedings of the Imperial Academy.* (Tokyo) [Vol. 1, 1912–. After Feb. 1945, PROC. JAPAN ACAD.] ... 2

PROC. INDIAN ACAD. SCI. SECT. A *Proceedings of the Indian Academy of Sciences. Section A.* (Bangalore) ... 5

PROC. INDUST. COMPUT. SEM. IBM NEW YORK *Proceedings, Industrial Computation Seminar, September 1950.* [International Business Machines Corp., New York, N.Y., 1951] ... 1

PROC. INST. ELEC. ENGRS. LONDON *The Proceedings of the Institution of Electrical Engineers. Part C. Institution Monographs.* (London) ... 1

PROC. INST. RADIO ENGRS. *Proceedings of the Institute of Radio Engineers.* (New York) [Vol. 1–50, 1913–1962. Then *Proceedings of IEEE*] ... 4

PROC. INST. STATIST. MATH. TOKYO *Proceedings of the Institute of Statistical Mathematics.* (Tokyo) [Vol. 1, 1953–. Was RES. MEM. INST. STATIST. MATH. TOKYO. All articles in Japanese; some with English summaries] ... 30

PROC. INTERNAT. CONG. LOGIC METH. PHILOS. SCI. *Logic, Methodology and Philosophy of Science. Proceedings of the 1960 International Congress.* Edited by Ernest Nagel, Patrick Suppes & Alfred Tarski. [Stanford University Press, Stanford, Calif., 1962] ... 1

PROC. INTERNAT. CONG. MATH. *Proceedings of the International Congress of Mathematicians.* [8th Congress, 1928 (Bologna), Pub. 1932. 9th, 1932 (Zürich). 11th, 1950 (Cambridge, Mass.), Pub. 1952. Also called *Atti* ... and *Verhandlungen* ...] ... 7

PROC. INTERNAT. CONG. PSYCHOL. *Proceedings. International Congress of Psychology.* [6th Congress, 1909 (Genève)] ... 1

PROC. INTERNAT. SYMP. CLASS. CONTAG. DISCRETE DISTNS. (PATIL) *Classical and Contagious Discrete Distributions, Proceedings of the International Symposium, McGill University, Montreal, 1963.* Edited by Ganapati P. Patil. [Statistical Publishing Society, Calcutta, 1965] ... 1

PROC. INTERNAT. WOOL TEXTILE ORG. *International Wool Textile Organization. Technical Committee Proceedings.* (London) ... 1

PROC. INVIT. CONF. TEST. PROBLEMS *Proceedings of the Invitational Conference on Testing Problems.* [Educational Testing Service, Princeton, N.J.] ... 1

PROC. JAPAN ACAD. *Proceedings of the Japan Academy.* (Tokyo) [Vol. 1, 1912-. Until Feb. 1945, PROC. IMP. ACAD. TOKYO] ... 3

PROC. LONDON MATH. SOC. *Proceedings of the London Mathematical Society.* (London) [Vol. 1-35, 1865-1903. Then SER. 2] ... 1

PROC. LONDON MATH. SOC. SER. 2 *Proceedings of the London Mathematical Society. Second Series.* (London) [Vol. 1-54, 1903-1951. Then SER. 3] ... 5

PROC. LONDON MATH. SOC. SER. 3 *Proceedings of the London Mathematical Society. Third Series.* (London) ... 1

PROC. MATH. PHYS. SOC. EGYPT *Proceedings of the Mathematical and Physical Society of Egypt.* (Cairo) ... 3

PROC. NAT. ACAD. SCI. INDIA SECT. A *Proceedings of the National Academy of Sciences, India. Section A. Physical Sciences.* (Allahabad) ... 3

PROC. NAT. ACAD. SCI. U.S.A. *Proceedings of the National Academy of Sciences of the United States of America.* (Washington, D.C.) ... 19

PROC. NAT. INST. SCI. INDIA *Proceedings of the National Institute of Sciences of India.* (New Delhi) ... 1

PROC. PAKISTAN STATIST. CONF. *Proceedings of the Pakistan Statistical Conference.* [1st Conference, 1950 (University Panjab, Lahore). Pub. Panjab University Press, Lahore, 1951] ... 2

PROC. PHYS. MATH. SOC. JAPAN SER. 3 *Proceedings of the Physico-Mathematical Society of Japan. Series 3.* (Tokyo) [= *Nippon Sugakubuturigakkwai.* Vol. 1-26, 1919-1944. Then J. MATH. SOC. JAPAN and *Journal of the Physical Society of Japan*] ... 14

PROC. PHYS. SOC. *Proceedings of the Physical Society of London.* (London) ... 2

PROC. RES. FORUM *Proceedings of the Research Forum, Endicott, New York, August 26-30, 1946.* [IBM Corp., Endicott, N.Y., 1947] ... 2

PROC. ROY. IRISH ACAD. A *Proceedings of the Royal Irish Academy. Section A.* (Dublin) ... 1

PROC. ROY. SOC. EDINBURGH *Proceedings of the Royal Society of Edinburgh.* (Edinburgh) [Vol. 1-60, 1832-1940. Then SECT. A and *Section B*] ... 18

PROC. ROY. SOC. EDINBURGH SECT. A *Proceedings of the Royal Society of Edinburgh. Section A. Mathematical and Physical Sciences.* (Edinburgh) [Vol. 61, 1942-.] ... 16

PROC. ROY. SOC. LONDON *Proceedings of the Royal Society of London.* (London) [Vol. 1-75, 1800-1905. Then SER. A and SER. B] ... 15

PROC. ROY. SOC. LONDON SER. A *Proceedings of the Royal Society of London. Series A. Mathematical and Physical Sciences.* (London) [Vol. 76, 1906-.] ... 15

PROC. ROY. SOC. LONDON SER. B *Proceedings of the Royal Society of London. Series B. Biological Sciences.* (London) [Vol. 76, 1906-.] ... 4

PROC. ROY. SOC. N. S. W. *Proceedings of the Royal Society of New South Wales.* (Sydney) ... 1

PROC. SECOND BERKELEY SYMP. MATH. STATIST. PROB. *Proceedings of the Second Berkeley Symposium on Mathematical Statistics and Probability, July 31-August 12, 1950.* Edited by Jerzy Neyman. [University of California Press, Berkeley & Los Angeles, 1951] ... 6

PROC. SIXTH ALL-UNION CONF. THEORY PROB. MATH. STATIST. (VILNIUS) *Proceedings of the Sixth All-Union Conference on the Theory of Probability and Mathematical Statistics, Vilnius, 5-10 Sept. 1960; Palanga, 12-14 Sept. 1960.* [Gosudarstv. Izdat. Političesk. i Naučn. Lit. Litovosk. SSR, Vilnius, 1962] ... 2

PROC. SOCIAL STATIST. SECT. AMER. STATIST. ASSOC. *Proceedings of the Social Statistics Section. American Statistical Association.* (Washington, D.C.) ... 1

PROC. STATIST. TECH. MISSILE EVAL. SYMP. *Proceedings of the Statistical Techniques in Missile Evaluation Symposium.* Edited by Boyd Harshbarger. [Held at Dept. of Statistics, V.P.I., Blacksburg, Va.,1958] ... 1

PROC. STEKLOV INST. MATH. *Proceedings of the Steklov Institute of Mathematics.* (Providence, R.I.) [Translation of TRUDY MAT. INST. STEKLOVA] ... 2

PROC. SYMP. SPECTRAL THEORY DIFFER. PROBLEMS *Proceedings of the Symposium on Spectral Theory and Differential Problems.* [Mathematics Department, Oklahoma Agricultural and Mechanical College, Stillwater, Oklahoma, 1951] ... 1

PROC. SYMP. TIME SERIES ANAL. BROWN UNIV. (ROSENBLATT) *Proceedings of the Symposium on Time Series Analysis held at Brown University, June 11-14, 1962.* Edited by Murray Rosenblatt. [SIAM Series in Applied Mathematics, John Wiley & Sons,Inc., New York, 1963] ... 9

PROC. THIRD BERKELEY SYMP. MATH. STATIST. PROB. *Proceedings of the Third Berkeley Symposium on Mathematical Statistics and Probability, December 1954; July & August 1955.* Edited by Jerzy Neyman. [University of California Press, Berkeley & Los Angeles, 1956] ... 9

PRZEGLAD GEOGRAFICZNY *Przegląd Geograficzny.* (Warszawa) [= *Revue Polonaise de Géographie.* Articles in Polish and French] ... 1

PSYCHOL. BULL. *Psychological Bulletin.* (Washington, D.C.) ... 24

PSYCHOL. MONOG. *The Psychological Monographs.* (Princeton, N.J.) [Vol. 1-80, 1895-1966. Vol. 1-11 called *The Psychological Review. Monograph Supplements.* All carry a sequence number which is not cited] ... 1

PSYCHOL. PRINCIPLES SYSTEM DEVELOP. *Psychological Principles in System Development.* [Holt, Rinehart, & Winston, Inc., New York, 1962] ... 1

PSYCHOL. REC.  *The Psychological Record.*  (Bloomington,Ind.)  ... 2

PSYCHOL. REP.  *Psychological Reports.* (Louisville, Ky.) [Separates published as *Psychological Reports. Monograph Supplement*] ... 8

PSYCHOL. REV.  *Psychological Review.*  (Princeton, N.J.)  ... 23

PSYCHOL. SCALING THEORY APPL. (GULLISKEN + MESSICK)  *Psychological Scaling: Theory and Applications.*  Edited by H.Gulliksen & S.Messick. [John Wiley & Sons, Inc., New York, 1960]  ... 2

PSYCHOL. STUDY SCI. (KOCH)  *Psychology; A Study of a Science.*  Study I. Conceptual and Systematic.  Edited by Sigmund Koch.  [Volume 3: Formulations of the Person and the Social Context.  McGraw-Hill Book Co., Inc., New York, 1959]  ... 1

PSYCHOMET. MONOG.  *Psychometric Monographs.*  (Chicago) [No. 1-15, 1938-1967.  No.16, 1969- as *Psychometrika. Monograph Supplement.*  Published for the Psychometric Society by the University of Chicago Press]  ... 6

PSYCHOMETRIKA  *Psychometrika; a journal devoted to the development of psychology as a quantitative rational science.*  (Chapel Hill, N.C.) ... 387

PUBL. FAC. SCI. UNIV. CHARLES  *Publications de la Faculté des Sciences de l'Université Charles.*  (Praha) [= *Spisy Vydávané Přírodovědeckou Fakultou Karlovy University*] ... 1

PUBL. INST. STATIST. UNIV. PARIS  *Publications de l'Institut de Statistique de l'Université de Paris.*  (Paris) ... 20

PUBL. RES. INST. MATH. SCI. KYOTO UNIV. SER. A  *Publications of the Research Institute for Mathematical Sciences. Series A.*  (Kyoto) ... 1

PUBL. SCI. TECH. MINIST. AIR  *Publications Scientifiques et Techniques du Ministère de l'Air.*  (Paris) ... 1

PUBL. UN. MAT. ARGENT.  *Publicación de la Unión Matemática Argentina.* (Buenos Aires) [Vol. 1-7, 1936-1941. Then REV. UN. MAT. ARGENT.] ... 1

PUBLIC OPINION QUART.  *The Public Opinion Quarterly.*  (Princeton, N.J.) ... 2

QUART. APPL. MATH.  *Quarterly of Applied Mathematics.*  (Providence, R.I.) ... 12

QUART. J. ECON.  *Quarterly Journal of Economics.*  (Cambridge, Mass.) ... 5

QUART. J. MATH. OXFORD SECOND SER.  *The Quarterly Journal of Mathematics. Oxford Second Series.*  (Oxford) [Vol. 1, 1950-.  Was QUART. J. MATH. OXFORD SER.] ... 8

QUART. J. MATH. OXFORD SER.  *The Quarterly Journal of Mathematics. Oxford Series.*  (Oxford) [Vol. 1-20, 1930-1949.  Was MESSENGER MATH. and QUART. J. PURE APPL. MATH.  Became QUART. J. MATH. OXFORD SECOND SER.] ... 2

QUART. J. MECH. APPL. MATH.  *The Quarterly Journal of Mechanics and Applied Mathematics.*  (London) ... 2

QUART. J. PURE APPL. MATH.  *The Quarterly Journal of Pure and Applied Mathematics.*  (London) [Vol. 1-50, 1857-1927.  Then merged with MESSENGER  MATH. to form QUART. J. MATH. OXFORD. SER.] ... 3

QUART. J. ROY. METEOROL. SOC.  *Quarterly Journal of the Royal Meteorological Society.*  (London) ... 4

QUART. PUBL. AMER. STATIST. ASSOC.  *Quarterly Publications of the American Statistical Association.*  (Boston) [Vol. 1-16, 1888-1919. One volume covers two years; Vol. 17, 1920-1921 titled *Quarterly Publication.*  Then J. AMER. STATIST. ASSOC.] ... 12

QUESTOES MET. INST. ANTROP. COIMBRA  *Questões de Método.  Contribuições Para o Estudo da Antropologia Portugesa.  Universidade de Coimbra.*  (Coimbra) ... 2

RADIOTEHNIKA  *Radiotehnika.*  (Moscow) [Vol. 1, 1946-. Vol. 12-16, 1957-1961, translated as *Radio Engineering;* 1962-, translated as *Telecommunications and Radio Engineering*] ... 2

RAD JUGOSLAV. AKAD. ZNAN. UMJET. ODJEL MAT. FIZ. TEH. NAUKE  *Rad Jugoslavenske Akademije Znanosti i Umjetnosti. Odjel za Matematičke, Fizičke i Tehničke Nauke.*  (Zagreb) ... 1

READINGS MATH. PSYCHOL. (LUCE, BUSH, + GALANTER)  *Readings in Mathematical Psychology.*  Edited by R.D. Luce, R.R. Bush,& E. Galanter.  [John Wiley & Sons, Inc., N.Y., 1963.  Two volumes] ... 4

READINGS MATH. SOCIAL SCI. (LAZARSFELD + HENRY)  *Readings in Mathematical Social Science.*  Edited by Paul F. Lazarsfeld & Neil W. Henry. [Science Research Associates, Inc., Chicago, 1966] ... 7

REC. INDIAN MUSEUM  *Records of the Indian Museum.  A Journal of Indian Zoology.*  (Calcutta) ... 3

RECENT ADVANC. MATRIX THEORY (SCHNEIDER) *Recent Advances in Matrix Theory Proceedings of an Advanced Seminar Conducted by the Mathematics Research Center, U.S.Army, at the University of Wisconsin, Madison, October 14-16, 1963.* Edited by Hans Schneider. [Publication No. 12 of the Mathematics Research Center, U.S.Army, University of Wisconsin, The University of Wisconsin Press, Madison & Milwaukee, 1964] ... 3

REND. ACCAD. SCI. FIS. MAT. NAPOLI SER. 4 *Rendiconto dell'Accademia delle Scienze Fisiche e Matematiche. Società Reale di Napoli. Serie 4.* (Napoli) ... 1

REND. CIRCOLO MAT. PALERMO *Rendiconti del Circolo Matematico di Palermo.* (Palermo) ... 1

REND. IST. LOMBARDO SCI. LETT. CL. SCI. MAT. NATUR. *Rendiconti. Istituto Lombardo di Scienze e Lettere. Classe di Scienze Matematiche e Naturali.* (Milano) ... 1

REND. MAT. E APPL. REG. UNIV. ROMA REALE IST. NAZ. ALTA MAT. *Rendiconti di Matematica e delle sue Applicazioni. Regia Università di Roma e Reale Istituto Nazionale di Alta Matematica.* (Roma) [*Serie 5,* Vol. 1, 1940-.] ... 2

REND. SEM. FAC. SCI. UNIV. CAGLIARI *Rendiconti del Seminario della Facoltà di Scienze della Università di Cagliari.* (Cagliari) ... 1

REND. SEM. MAT. FIS. MILANO *Rendiconti del Seminario Matematico e Fisico di Milano.* (Milano) ... 1

REP. BRITISH ASSOC. ADVANC. SCI. *Reports of the British Association for the Advancement of Science.* (London) ... 2

REP. BRITISH ASSOC. DUNDEE *Reports of the British Association, Dundee.* (Dundee) ... 1

REP. STATIST. APPL. RES. UN. JAPAN. SCI. ENGRS. *Reports of Statistical Application Research, Union of Japanese Scientists and Engineers.* (Tokyo) ... 12

REP. THIRD ANNUAL CONF. ECON. STATIST. *Report of the Third Annual Research Conference on Economics and Statistics, Colorado Springs.* [Cowles Commission for Research in Economics] ... 1

REP. UNIV. ELECTRO-COMMUN. PHYS. SCI. ENGRG. *Reports of the University of Electro-Communications, Physical Sciences and Engineering.* (Tokyo) ... 1

RES. BULL. EDUC. TEST. SERVICE *Research Bulletin, Educational Testing Service.* (Princeton, N.J.) [Multilith reports] ... 2

RES. MEM. INST. STATIST. MATH. TOKYO  *Research Memoirs of the Institute of Statistical Mathematics.*  (Tokyo) [= *Kokyūroku.* Vol. 1-8, 1944-1953.  Early volumes handwritten!  Became PROC. INST. STATIST. MATH. TOKYO] ... 25

RES. PAPERS STATIST. FESTSCHR. NEYMAN (DAVID)  *Research Papers in Statistics.  Festschrift for J. Neyman.*  Edited by F.N. David. [John Wiley & Sons, Ltd., London, 1966] ... 8

RES. QUART. AMER. ASSOC. HEALTH PHYS. EDUC. REC.  *Research Quarterly of the American Association for Health, Physical Education, and Recreation.*  (Ann Arbor, Mich.) ... 1

REV. BRASIL ESTATIST.  *Revista Brasileira de Estatística.*  (Rio de Janeiro) ... 2

REV. CANAD. BIOL.  *Revue Canadienne de Biologie.*  (Montréal) ... 1

REV. ECON. STATIST.  *Review of Economic Statistics.*  (Cambridge, Mass.) [1919-1947; 1948-, as *Review of Economics and Statistics*] ... 1

REV. EDUC. RES.  *Review of Educational Research.*  (Washington, D.C.) ... 7

REV. FAC. CI. UNIV. LISBOA SER. 2A CI. MAT.  *Revista da Faculdade de Ciências.  Universidade de Lisboa.  $2^a$ Série A, Ciências Matemáticas.* (Lisboa) ... 2

REV. GENERALE SCI. PURES APPL.  *Revue Générale des Sciences Pures et Appliquées et Bulletin de l'Association Française pour l'Avancement des Sciences.*  (Paris) ... 1

REV. INST. INTERNAT. STATIST.  *Revue de l'Institut International de Statistique.*  ('s-Gravenhage) [= *Review of the International Statistical Institute*] ... 27

REV. MAT. HISP. AMER. SER. 2  *Revista Matemática Hispano-Americana. Segunda Serie.*  (Madrid) [Vol. 1-13, 1926-1938.  Later SER. 4] ... 6

REV. MAT. HISP. AMER. SER. 4  *Revista Matemática Hispano-Americana. Cuarta Serie.*  (Madrid) [Vol. 1, 1941-.  Was SER. 2] ... 3

REV. MATH. PURES APPL.  *Revue de Mathématiques Pures et Appliquées.* (Bucureşti) [Vol. 1-8, 1956-1963.  Then *Revue Roumaine de Mathématiques Pures et Appliquées*] ... 1

REV. MATH. UN. INTERBALK.  *Revue Mathématique de l'Union Interbalkanique.* (Athens) ... 1

REV. MODERN PHYS. *Reviews of Modern Physics.* (New York) ... 3

REV. SCI. INSTRUMENTS *The Review of Scientific Instruments.* (Rochester, N.Y.) ... 1

REV. SCI. PARIS *Revue Scientifique.* (Paris) ... 1

REV. STATIST. APPL. *Revue de Statistique Appliquée.* (Paris) [Institut de Statistique. Université de Paris] ... 7

REV. UN. MAT. ARGENT. *Revista de la Unión Matemática Argentina.* (Buenos Aires) [Was PUBL. UN. MAT. ARGENT.] ... 1

REV. UNIV. NAC. LITORAL FAC. CI. ECON. COM. POLIT. (ROSARIO) SER. 3 *Revista. Universidad Nacional de Litoral. Facultad de Ciencias Económicas, Comerciales y Políticas. Serie 3.* (Rosario) ... 1

RICERCA SCI. *La Ricerca Scientifica. Consiglio Nazionale delle Ricerche.* (Roma) ... 1

RIV. ITAL. ECON. DEMOG. STATIST. *Rivista Italiana di Economia, Demografia e Statistica.* (Roma) [Vol. 1, 1947-. Vol. 1-3, 1947-1950 as *Rivista Italiana di Demografia e Statistica*] ... 7

RIVEON LEMATEMATIKA *Riveon Lematematika.* (Jerusalem) ... 1

ROCZNIK TOWARZ. NAUK. WARSZAWSK. *Rocznik Towarzystwa Naukowego Warszawskiego.* (Warszawa) [= *Annuaire de la Société des Sciences et des Lettres de Varsovie.* (Année XXIV = 1931, Pub. 1932)] ... 1

RUSS. ASTRONOM. ZUR. *Russkiĭ Astronomičeskiĭ Žurnal.* (Moscow) [Vol. 1-4, 1924-1927. Then ASTRONOM. ZUR.] ... 3

S. AFRICAN J. AGRIC. SCI. *South African Journal of Agricultural Science.* (Pretoria) [= *Suid-Afrikaanse Tydskrif vir Landbouwetenskap*] ... 2

SAMMLUNG WISS. ARB. SCHWEIZ INTERN. POLEN *Sammlung wissenschaftlicher Arbeiten der in der Schweiz internierten Polen.* [= *Recueil de Travaux Scientifiques des Polonais Internés en Suisse* = *Zbiór Prac Naukōwych Polaków Internowanych w Szwajcarii.* Eidg. Kommissariat für Internierung und Hospitalisierung. Erster Band 1943. Zweiter Band 1944. LC: AC15.S82] ... 2

SANKHYA *Sankhyā. The Indian Journal of Statistics.* (Calcutta) [Vol. 1-22, 1933-1960. Then SER. A and SER. B; *Sankhyā,* Vol. 4 (1, 2 & 4) duplicated as *Proceedings of the Indian Statistical Conference,* Vol. 1 & 2] ... 114

SANKHYA SER. A  *Sankhyā. Series A.*  (Calcutta)  [Vol.23, 1961-.]  ... 69

SANKHYA SER. B  *Sankhyā. Series B.*  (Calcutta)  [Vol.23, 1961-.]  ... 4

SB. POS. PAM. D.A. GRAVE  *Sbornik Posvjaščeniǐ Pamjati D.A. Grave.*
    (Moscow)  [Memorial Volume dedicated to D.A. Grave, 1940]  ... 2

SCAND. J. PSYCHOL.  *Scandinavian Journal of Psychology.*  (Stockholm)
    ... 1

SCHOOL AND SOCIETY  *School and Society.*  (New York)  ... 1

SCHOOL REV.  *School Review; a journal of secondary education.*  (Ithaca,
    N.Y.)  ... 24

SCHR. MATH. INST. UNIV. BERLIN  *Schriften des Mathematischen Seminars
    und des Instituts für angewandte Mathematik an der Universität Berlin.*
    (Berlin)  ... 1

SCHWEIZ. ARCH. NEUROL. PSYCHIATRIE  *Schweizerische Archiv für Neurologie
    und Psychiatrie.*  (Zürich)  [Became *Schweizerische Archiv für
    Neurologie, Neurochirurgie und Psychiatrie*]  ... 1

SCHWEIZ. Z. PSYCHOL. ANWEND.  *Schweizerische Zeitschrift für Psychologie
    und ihre Anwendungen.*  (Bern)  ... 1

SCHWEIZ. Z. VERMESSUNGS.  *Schweizerische Zeitschrift für Vermessungswesen
    und Kulturtechnik.*  (Winterthur)  [Became *Schweizerische Zeitschrift
    für Vermessung, Kulturtechnik und Photogrammetrie*]  ... 1

SCI. AND CULT.  *Science and Culture; a monthly journal devoted to natural
    and cultural sciences.*  (Calcutta)  ... 18

SCI. REP. KAGOSHIMA UNIV.  *Science Reports of the Kagoshima University.*
    (Kagoshima)  ... 1

SCI. REP. TOKYO KYOIKU DAIGAKU SECT. A  *Science Reports of the Tokyo
    Kyoiku Daigaku. Section A. Mathematics.*  (Tokyo)  ... 2

SCIENCE  *Science.*  (Cambridge, Mass.)  [Vol.103, 1946-. Was SCIENCE NEW
    SER.]  ... 9

SCIENCE NEW SER.  *Science. New Series.*  (Cambridge, Mass.)  [Vol. 1-102,
    1895-1945.  Then SCIENCE]  ... 5

SCIENTIFIC MONTHLY  *The Scientific Monthly.*  (Washington, D.C.)  [Vol.
    1-85, 1915-1957.  Then merged with SCIENCE]  ... 1

SCR. MAT. FILIPPO SIBIRANI *Scritti Matematici in Onore di Filippo Sibirani*. [Cesare Zuffi, Bologna, 1957] ... 1

SELECT. PAPERS NOISE STOCH. PROC. *Selected Papers on Noise and Stochastic Processes*. Edited by Nelson Wax. [Dover Publications, Inc., New York, 1954] ... 3

SELECT. PAPERS STATIST. PROB. WALD *Selected Papers in Statistics and Probability by Abraham Wald*. Edited for the Institute of Mathematical Statistics by T.W. Anderson, H. Cramér, H.A. Freeman, J.L. Hodges, Jr., E.L. Lehmann, A.M. Mood & C.M. Stein. [Reprints of Abraham Wald. McGraw-Hill Book Co., Inc., New York, 1955; Reprinted: Stanford University Press, Stanford, Calif., 1957] ... 24

SELECT. TRANSL. MATH. STATIST. PROB. *Selected Translations in Mathematical Statistics and Probability*. (Providence, R.I.) ... 3

SEMAINE D'ETUDE ROLE ANAL. ECONOMET. PONT. ACAD. SCI. *Semaine d'Étude sur le Rôle de l'Analyse Économétrique dans la Formulation de Plans de Développement, 7-13 Octobre 1963*. [= *Study Week on the Econometric Approach to Development Planning*. Pontificae Academiae Scientiarum Scripta Varia 28. North-Holland Publishing Company, Amsterdam; Rand McNally & Company, Chicago, 1965] ... 1

SHUXUE JINZHAN *Shuxue Jinzhan*. (Peking) [Progress in Mathematics. Chinese Mathematical Society. Academia Sinica] ... 1

SIAM. J. APPL. MATH. *SIAM Journal on Applied Mathematics*. (Philadelphia) [Vol. 14, 1966-. Was J. SOC. INDUST. APPL. MATH.] ... 6

SIAM J. NUMER. ANAL. *SIAM Journal on Numerical Analysis*. (Philadelphia) [Vol. 3, 1966-. Was J. SOC. INDUST. APPL. MATH. SER. B NUMER. ANAL.] ... 1

SIAM NEWSLETTER *SIAM Newsletter*. (Philadelphia) [Vol. 1-7, 1953-1959] ... 1

SIAM REV. *SIAM Review*. (Philadelphia) ... 19

SIMON STEVIN WIS- EN NATUURK. TIJDSCHR. *Simon Stevin. Wis- en natuurkundig Tijdschrift*. (Groningen) [Vol. 25, 1946-.] ... 1

SITZUNGSBER. MATH. NATURWISS. CL. KAIS. AKAD. WISS. WIEN *Sitzungsberichte der Mathematisch-Naturwissenschaftlichen Classe der Kaiserlichen Akademie der Wissenschaften*. (Wien) [Part 2 = *Abtheilung II*, Vol. 43-96 (1861-1888). Part 2A = *Abtheilung IIa*, Vol. 97-142 (1889-1933). *Classe* became *Klasse* and *Abtheilung*, Abteilung circa 1903] ... 4

SITZUNGSBER. MATH. PHYS. CL. KAIS. AKAD. WISS. MUNCHEN *Sitzungsberichte der Matematisch-physikalische Classe der Kaiserlichen Akademie der Wissenschaften*. (München) [Bayerische Akademie der Wissenschaften. *Classe* became *Klasse* in 1903] ... 1

SKAND. AKTUARIETIDSKR. *Skandinavisk Aktuarietidskrift.* (Uppsala)
[Vol. 1, 1918-. Was SVENSKA AKTUARIEFOREN. TIDSKR.] ... 59

SKAND. MAT. KONG. BER. *Skandinaviske Matematiker Kongres. Beretning
udgiven af Kongressens Organisationskomite.* [= *Congrès des Mathé-
maticiens Scandinaves.* 6th Congress, 31 August-4 September, 1925.
10th, 26-30 August, 1946, Pub. 1947. 12th, 1953, Pub. 1954, Jul.
Gjellerups Forlag, København. 13th, 18-23 August, 1957, Pub.
1958, Mercators Tryckeri, Helsinki] ... 5

SMITHSONIAN MISC. COLLECT. *Smithsonian Miscellaneous Collections.*
(Washington, D.C.) [The Smithsonian Institute, Washington, D.C.]
... 2

SOC. ITAL. ECON. DEMOG. STATIST. STUDI MONOG. *Studi e Monografie della
Società Italiana di Economia, Demografia e Statistica.* (Roma)
[Nr. 6: Studi Sulle Relazioni Statistiche] ... 2

SOC. SCI. FENN. COMMENT. PHYS. MATH. *Societas Scientiarum Fennica.
Commentationes Physico-Mathematicae.* (Helsinki) ... 1

SOCIAL FORCES *Social Forces; a scientific medium of social study and
interpretation.* (Chapel Hill, N.C.) ... 1

SOCIOMETRY *Sociometry.* (New York) ... 3

SOCIOMETRY SCI. MAN (MORENO) *Sociometry and the Science of Man.*
Edited by J.L. Moreno. [Beacon House, New York, 1956] ... 1

SOOBSC. AKAD. NAUK GRUZIN. SSR *Soobŝčenija Akademii Nauk Gruzinskoĭ SSR.*
(Tbilisi, Georgia, USSR) ... 3

SOUTHERN. ECON. J. *Southern Economic Journal.* (Chapel Hill, N.C.)
... 1

SOVIET MATH. DOKL. *Soviet Mathematics - Doklady.* (Providence, R.I.)
[Vol. 1, 1960-. Translation by American Mathematical Society of
Vol. 130, 1960-, of DOKL. AKAD. NAUK SSSR] ... 5

SOVIET PHYS. ACOUST. *Soviet Physics - Acoustics.* (New York) [Transla-
tion of AKUST. ZUR.] ... 3

SPRING JOINT COMP. CONF. PROC. *Spring Joint Computer Conference, Proceedings.* [Spartan Books, Washington, D.C., 1962-. Was *Western
Joint Computer Conference, Proceedings*] ... 1

STATIST. ASSOC. METH. MECH. DOC. (STEVENS, GIULIANO, + HEILPRIN)
*Statistical Association Methods for Mechanized Documentation.* Edited
by Mary E. Stevens, Vincent E. Giuliano,& Lawrence B. Heilprin.
[U.S. Govt. Printing Office - N.B.S. Misc. Pub. 269, 1965] ... 1

STATIST. GRAPHIQUE  *Statistique Graphique.*  [Ministère de l'Agriculture, de l'Industrie et du Commerce.  Direction de Statistique.  Pub. Imprimerie Héritiers Botta, Roma, 1880]  ... 1

STATIST. HEFTE  *Statistische Hefte.*  (Saarbücken)  [Title and table of contents also given in English, French and Russian.  Vol. 1, 1960-.]  ... 2

STATIST. MATH. BIOL. (KEMPTHORNE ET AL.)  *Statistics and Mathematics in Biology.*  Edited by Oscar Kempthorne, Theodore A. Bancroft, John W. Growen & Jay L. Lush.  [Iowa State College Press, Ames, Iowa, 1954; Reprinted: Hafner Pub. Co., Inc., New York, 1964]  ... 5

STATIST. METH. RADIO WAVE PROP.  *Statistical Methods in Radio Wave Propagation.  Proceedings of a Symposium held at University of California, Los Angeles, June 18-20, 1958.*  Edited by William C. Hoffman.  [Symposium Publications Division, Pergamon Press Ltd., London, 1960]  ... 5

STATIST. OBZOR  *Statistický Obzor. Revue pro Statistickou Teorii a Praxi.*  (Praha)  [Vol. 1-11 as *Československý Statistický Věstník*]  ... 6

STATIST. PRAX.  *Statistiche Praxis.*  (Berlin)  ... 2

STATIST. RES. MEM. LONDON  *Statistical Research Memoirs.*  Edited by J. Neyman and E.S. Pearson.  [Volume I, June 1936 and II, December 1938. Issued by The Department of Statistics, University College London. Reprinted ca. 1971]  ... 2

STATIST. REV. BEOGRAD  *Statistička Revija.*  (Beograd)  ... 4

STATIST. THEOR. SPECTRA FLUCTUATIONS (PORTER)  *Statistical Theories of Spectra: Fluctuations.*  Edited by Charles E. Porter.  [Academic Press, Inc., New York, 1965]  ... 11

STATIST. VESTNIK  *Statističeskiĭ Vestnik.*  (Moscow)  [Vol. 1-4, 1915-1918.  Became VESTNIK. STATIST.]  ... 2

STATIST. VIERTELJAHRESSCHR.  *Statistische Vierteljahresschrift.*  (Wien)  [Merged with MITT. MATH. STATIST. to form METRIKA]  ... 3

STATISTICA, BOLOGNA  *Statistica.*  (Bologna)  [Vol. 1, 1941-.  Issued under the auspices of the Università di Bologna, Padova & Palermo. Early volumes issued in Ferrara & Milano]  ... 33

STATISTICA, LEIDEN  *Statistica.*  (Leiden)  [Vol. 1-8, 1947-1954.  Some issues published in Rijswijk and some in 's-Gravenhage.  Became STATISTICA NEERLANDICA]  ... 3

STATISTICA NEERLANDICA  *Statistica Neerlandica*.  ('s-Gravenhage)
[Vol. 9, 1955-.  Was STATISTICA, LEIDEN]  ... 9

STATISTICA, WARSZAWA  *Statistica*.  (Warszawa)  [Vol. 1, 1929/1930.
Szkoła Główna Gospodarstwa Weijskiego]  ... 1

STATISTICIAN  *The Statistician*.  (London)  [Vol. 12, 1962-.  Was INC.
STATIST.]  ... 6

STATISZT. SZEMLE  *Statisztikai Szemle*.  (Budapest)  ... 2

STOCKHOLM CONTRIB. GEOL.  *Stockholm Contributions in Geology*.  (Stockholm)
[Acta Universitatis Stockholmiensis]  ... 2

STUD. ITEM ANAL. PREDICT. (SOLOMON)  *Studies in Item Analysis and
Prediction*.  Edited by Herbert Solomon.  [Stanford University Press,
Stanford, Calif., 1961]  ... 19

STUD. MATH. ANAL. RELATED TOPICS (POLYA VOLUME)  *Studies in Mathematical
Analysis and Related Topics. Essays in Honor of George Pólya*.  Edited
by Gabor Szegö, Charles Loewner, Stefan Bergman, Menahem Max Schiffer,
Jerzy Neyman, David Gilbarg & Herbert Solomon.  [Stanford University
Press, Stanford, Calif., 1962]  ... 3

STUD. MATH. ECON. ECONOMET. (SCHULTZ VOLUME)  *Studies in Mathematical
Economics and Econometrics: In Memory of Henry Schultz*.  Edited by
Oscar Lange, Francis McIntyre & Theodore O. Yntema.  [University of
Chicago Press, Chicago, 1942]  ... 2

STUD. MATH. MECH. (VON MISES VOLUME)  *Studies in Mathematics and Mechanics
Presented to Richard von Mises by Friends, Colleagues, and Pupils*.
[Academic Press, Inc., New York, 1954]  ... 1

STUD. PSYCHOL. PSYCHIATRY CATHOLIC UNIV. AMER.  *Studies in Psychology
and Psychiatry. Catholic University of America*.  (Washington, D.C.)
... 1

STUDIA SCI. MATH. HUNGAR.  *Studia Scientiarum Mathematicarum Hungarica*.
(Budapest)  [Was MAGYAR TUD. AKAD. MAT. KUTATO. INT. KOZL.]  ... 2

STUDII CERC. MAT.  *Studii şi Cercetari Matematice*.  (Bucureşti)  ... 3

STUDI MEM. R. BENINI  *Studi in Memoria di Rodolfo Benini*.  [Università
degli Studi di Bari.  Facoltà di Economia e Commercio, Bari, 1956]
... 1

STUDI ONORE C. GINI  *Studi in Onore di Corrado Gini.*  [Istituto di
    Statistica della Facoltà di Scienze Statistiche, Demografiche ed
    Attuariali, Roma, 1960.  Two volumes]  ... 4

SUGAKU  *Sūgaku.*  (Tokyo)  [= *Mathematics*]  ... 6

SUPP. STATIST. NUOVI PROBLEMI POLIT. STORIA ECON.  *Supplemento Statistico
    di Nuovi Problemi di Politica, Storia ed Economia.*  (Ferrara)
    [Vol. 1, 1935-.]  ... 1

SURVEY NUMER. ANAL. (TODD)  *Survey of Numerical Analysis.*  Edited by
    John Todd.  [McGraw-Hill Book Co.,Inc., New York, 1962]  ... 4

SVENSKA AKTUARIEFOREN. TIDSKR.  *Svenska Aktuarieföreningens Tidskrift.*
    (Uppsala) [Vol. 1-4, 1914-1917.  Became SKAND. AKTUARIETIDSKR.]
    ... 5

SYMP. AIR FORCE HUMAN ENGRG. PERS. TRAIN. RES. (FINCH + CAMERON)
    *Symposium on Air Force Human Engineering, Personnel, and Training
    Research, Washington, D.C., November 14-16, 1955.*  Edited by
    Glen Finch & Frank Cameron.  [ARDC Tech. Rep. 56-8, Air Research
    and Development Command, Baltimore, Md., 1956]  ... 4

SYMP. APPL. AUTOCORREL. ANAL. PHYS. PROBLEMS (WOODS HOLE)  *Symposium on
    Applications of Autocorrelation Analysis to Physical Problems, Woods
    Hole, Mass., 13-14 June 1949.*  [Report No. NAVEXOS-P-735, Office of
    Naval Research, Dept. of the Navy, Washington, D.C., 1950]  ... 2

SYSTEMATIC ZOOL.  *Systematic Zoology.*  (Washington, D.C.)  ... 2

T. R. U. MATH.  *T.R.U. Mathematics.  Journal of Mathematics, Tokyo Rika
    University, Tokyo College of Science.*  (Tokyo)  ... 1

TAGUNG WAHRSCHEIN. MATH. STATIST.  *Bericht über die Tagung
    Wahrscheinlichkeitsrechnung und Mathematische Statistik in Berlin,
    Oktober, 1954.*  [Deutscher Verlag der Wissenschaften, Berlin, 1956]
    ... 1

TEC. ORGANIZZ.  *Tecnica ed Organizzazione.*  (Milano) [*Serie Nuova*
    Vol. 1, 1950-.]  ... 1

TECH. STATIST. ANAL. (EISENHART, HASTAY, + WALLIS)  *Techniques of
    Statistical Analysis.*  Edited by C. Eisenhart, M.W. Hastay, & W.A.
    Wallis.  [McGraw-Hill Book Co.,Inc., New York, 1947]  ... 1

TECH. TRANSL.  *Technical Translations.*  [Photoduplication Service,
    Publication Board Project, Library of Congress, Washington, D.C.]
    ... 1

TECHNOMETRICS  *Technometrics.*  (Princeton, N.J.) ... 27

TEOR. FUNKCII FUNKCIONAL. ANAL. I PRILOZEN  *Teorija Funkciĭ, Funkcional'nyĭ Analiz i ih Priloženija.*  (Kharkov (Har'kov), Ukraine, USSR) ... 1

TEOR. VEROJATNOST. I PRIMENEN.  *Teorija Verojatnosteĭ i ee Primenenija. Akademija Nauk SSSR.*  (Moscow) [Translated as THEORY PROB. APPL.] ... 31

TERREST. MAGNET. ATMOSPH. ELEC.  *Terrestrial Magnetism and Atmospheric Electricity.*  (Chicago) [Became J. GEOPHYS. RES.] ... 2

TEST. PROBLEMS PERSPECTIVE (ANASTASI)  *Testing Problems in Perspective, Twenty-Fifth Anniversary Volume of Topical Readings from the Invitational Conference on Testing Problems.*  Edited by Anne Anastasi. [American Council on Education, Washington, D.C., 1966] ... 4

THEOR. MATH. BIOL. (WATERMAN + MOROWITZ)  *Theoretical and Mathematical Biology.*  Edited by Talbot H. Waterman & Harold J. Morowitz. [Blaisdell Pub. Co., Waltham, Mass., 1965] ... 2

THEORY PROB. APPL.  *Theory of Probability and its Applications.*  (Philadelphia) [Translation of TEOR. VEROJATNOST. I PRIMENEN.] ... 31

THEORY PROB. MATH. STATIST. (IZDAT. NAUKA UZBEK. SSR)  *Theory of Probability and Mathematical Statistics. No. I.*  [Academy of Sciences of the Uzbek SSR. V.I. Romanovskiĭ Institute of Mathematics.  Izdat. "Nauka" Uzbek. SSR, Tashkent, 1964] ... 3

TILLOCH'S PHILOS. MAG.  *Tilloch's Philosophical Magazine.*  (London) ... 1

TOHOKU MATH. J.  *Tohoku Mathematical Journal.*  (Sendai) [= *Tōhoku Sūgaku Zasshi*] ... 1

TOWARZ. NAUK. WARSZAWSK. SPRAW. POS. WYD. III NAUK MAT. FIZ.  *Towarzystwo Naukowe Warszawskie. Sprawozdania z Posiedzeń: Wydziału III. Nauk Matematyczno-Fizycznych.*  (Warszawa) ... 1

TRABAJOS ESTADIST.  *Trabajos de Estadística.*  (Madrid) ... 20

TRACTS  COMPUT.  *Tracts for Computers.*  (London) ... 3

TRANS. ACTUAR. SOC. AMER.  *Transactions of the Actuarial Society of America.*  (New York) ... 2

TRANS. AMER. INST. ELEC. ENGRS.  *Transactions of the American Institute of Electrical Engineers.*  (New York) ... 1

TRANS. AMER. MATH. SOC.  *Transactions of the American Mathematical Society.*  (Lancaster, Pa.) ... 15

TRANS. CAMBRIDGE PHILOS. SOC.  *Transactions of the Cambridge Philosophical Society.*  (Cambridge) ... 4

TRANS. CONNECTICUT ACAD. ARTS  SCI.  *Transactions of the Connecticut Academy of Arts and Sciences.*  (New Haven, Conn.) ... 2

TRANS. FAC. ACTUAR. EDINBURGH  *Transactions of the Faculty of Actuaries.*  (Edinburgh) ... 1

TRANS. IRE PROF. GROUP INFORMATION THEORY  *Transactions of the IRE Professional Group on Information Theory.*  (New York) [Vol. PGIT-1 – PGIT-4, 1953-1954.  Then IRE TRANS. INFORMATION THEORY] ... 3

TRANS. ROY. SOC. EDINBURGH  *Transactions of the Royal Society of Edinburgh.*  (Edinburgh) ... 2

TRAVAIL HUMAIN  *Le Travail Humain.*  (Paris) ... 1

TRIMES. SPISANIE GLAV. DIREKT. STATIST.  *Trimesečno Spisanie na Glavnata Direktsija na Statistikata.*  (Sofia) [= *Revue Trimestrielle de la Direction Générale de la Statistique*] ... 1

TRUDY AKAD. NAUK LITOVSK. SSR SER. B  *Trudy.  Akademija Nauk Litovskoĭ SSR.  Serija B.*  (Vilnius, Lithuania, USSR) ... 1

TRUDY KONJUNKTURNYI INST. SER. 2  *Trudy.  Konjunkturnyĭ Institut.  Serija 2.*  (Moscow) [*Trudy:* Vol. 1-4, 1925-1928.  *Trudy, Serija 2:* Vol. 1-2, 1929-1930.  Naučno-issledovatel'skii Konjunkturnyĭ Institut (Institut de Conjoncture Économique)] ... 1

TRUDY MAT. INST. STEKLOVA  *Trudy.  Matematičeskii Institut imeni V.A. Steklova.  Akademija Nauk SSSR.*  (Leningrad) [Translated as PROC. STEKLOV INST. MATH.] ... 4

TRUDY SRED. AZIAT. GOS. UNIV. SER. 5A MAT.  *Trudy.  Sredne-aziatskii Gosudarstvennyi Universitet.  Serija 5a.  Matematika.*  (Tashkent) [Vol. 1, 1929-.] ... 1

TRUDY VYC. CENTRA AKAD. NAUK GRUZIN. SSR  *Trudy Vyčislitel'nogo Centra.*
  *Akademija Nauk Gruzinskoĭ SSR.*  (Tbilisi, Georgia, USSR)  [= *Trans-*
  *actions of the Computer Center of the Academy of Sciences of the*
  *Georgian SSR*] ... 1

TYDSKR. NATUURWETENSK.  *Tydskrif vir Natuurwetenskappe.* (Pretoria) ... 1

UCEN. ZAP. MOSK. GOS. UNIV.  *Učenye Zapiski Moskovskogo Gosudarstvennogo*
  *Universiteta.* (Moscow) ... 2

UCEN. ZAP. STATIST.  *Učenye Zapiski po Statistike. Izdatel'stvo*
  *Akademii Nauk SSR.* (Moscow) ... 3

UKRAIN. MAT. ZUR.  *Ukrainskiĭ Matematičeskiĭ Žurnal. Akademija Nauk*
  *Ukrainskoĭ SSR. Institut Matematiki.* (Kiev) ... 4

UNIV. CALIFORNIA PUBL. GEOL. SCI.  *University of California Publications*
  *in Geological Sciences.* (Berkeley, Calif.) ... 1

UNIV. CALIFORNIA PUBL. MATH.  *University of California Publications in*
  *Mathematics.* (Berkeley, Calif.) ... 1

UNIV. CALIFORNIA PUBL. STATIST.  *University of California Publications*
  *in Statistics.* (Berkeley, Calif.) ... 5

UNIV. REPUB. FAC. INGEN. MONTEVIDEO PUBL. INST. MAT. ESTADIST.
  *Universidad de la República. Facultad de Ingeniería y Agrimensura.*
  *Publicaciones del Instituto de Matemática y Estadística.* (Montevideo)
  ... 1

UPPSALA LANTBRUKSHOGSK. ANN.  *Uppsala Lantbrukshögskolan och Statens*
  *Landbruksförsök Annaler.* (Uppsala) ... 1

USPEHI MAT. NAUK  *Uspehi Matematičeskih Nauk.* (Moscow) [1936-. Ten
  consecutively numbered issues 1936-1944. In 1946 begins new series.
  Bimonthly 1946-1953 and beginning 1956; quarterly 1954-55. Issue
  number always cited; new series also carries volume and part numbers
  (not cited)] ... 13

VERHANDL. KON. AKAD. WETENSCH.  *Verhandelingen der Koninklijke Akademie*
  *van Wetenschappen.* (Amsterdam) ... 1

VERHANDL. NATURFORSCH. GES. BASEL  *Verhandlungen der Naturforschenden*
  *Gesellschaft in Basel.* (Basel) ... 1

VEROFF. FRANKFURTER GES. KONJUNKTURF.  *Veröffentlichungen der Frankfurter*
  *Gesellschaft für Konjunkturforschung.* (Bonn) [Kurt Schröder Verlag]
  ... 2

VERSLAG AFD. NATUURK. KON. AKAD. WETENSCH. *Verslag van de Gewone Vergadering der Afdeeling Natuurkunde. Koninklijke Akademie van Wetenschappen.* (Amsterdam) [Title varies: 1914-1918 *Verslag van de Gewone Vergaderingen der Wis- en Natuurkundige Afdeeling.* In Dutch; English translation (1898-1933) = KON. AKAD. WETENSCH. AMSTERDAM PROC. SECT. SCI.] ... 7

VESTNIK HAR'KOV. GOS. UNIV. *Vestnik Har'kovskogo Gosudarstvennogo Universiteta. Serija Mehaniko-Matematičeskaja. Ministerstvo Vysšego i Srednego Special'nogo Obrazovanija SSSR.* (Kharkov (Har'kov), Ukraine, USSR) ... 1

VESTNIK LENINGRAD. UNIV. *Vestnik Leningradskogo Universiteta.* (Leningrad) ... 3

VESTNIK STATIST. *Vestnik Statistiki.* (Moscow) [1919-1930. Organ Central'nogo Statističeskogo Upravlenija SSSR. Was STATIST. VESTNIK. Became VESTNIK STATIST. SER. 2] ... 4

VESTNIK STATIST. SER. 2 *Vestnik Statistiki.* (Moscow) [Vol. 1, 1949-. Independent resumption of publication of VESTNIK STATIST.] ... 1

VISNIK STATIST. UKRAINI *Visnik Statistiki Ukraini.* (Kharkov (Har'kov), Ukraine, USSR) ... 1

VOPROSY STATIST. IZU. SVJAZI *Voprosy Statističeskogo Izučenija Svjazi.* [Gosplanizdat, Moscow, 1950] ... 1

WISS. Z. HUMBOLDT-UNIV. BERLIN MATH.-NATURWISS. REIHE *Wissenschaftliche Zeitschrift der Humboldt-Universität zu Berlin. Mathematisch-Naturwissenschaftliche Reihe.* (Berlin) ... 2

Z. ANGEW. MATH. MECH. *Zeitschrift für angewandte Mathematik und Mechanik. Ingenieurwissenschaftliche Forschungsarbeiten.* (Berlin) ... 22

Z. ANGEW. PSYCHOL. *Zeitschrift für angewandte Psychologie.* (Leipzig) [Became *Zeitschrift für Psychologie, mit Zeitschrift für angewandte Psychologie und Charakterkunde*] ... 1

Z. GEOPHYS. *Zeitschrift für Geophysik.* (Würzburg) ... 2

Z. INDUKT. ABSTAM. VERERBUNGSL. *Zeitschrift für induktive Abstammungs- und Vererbungslehre.* (Berlin) ... 1

Z. MATH. PHYS. *Zeitschrift für Mathematik und Physik.* (Leipzig) ... 1

Z. NATURFORSCH. PART B *Zeitschrift für Naturforschung. Part B. Anorganische, organische und biologische Chemie, Botanik, Zoologie und verwandte Gebiete.* (Wiesbaden) [Vol. 2, 1947-.] ... 1

Z. PAD. PSYCHOL. EXPER. PAD.  *Zeitschrift für pädagogische Psychologie und experimentelle Pädagogik.*  (Leipzig) [Became *Zeitschrift für pädagogische Psychologie und Jugendkunde*] ... 1

Z. PSYCHOL.  *Zeitschrift für Psychologie.*  (Leipzig) [Became *Zeitschrift für Psychologie, mit Zeitschrift für angewandte Psychologie and Charakterkunde*] ... 1

Z. SCHWEIZ. STATIST. VOLKWIRTSCH.  *Zeitschrift für Schweizerische Statistik und Volkswirtschaft.*  (Basel) [Became *Schweizerische Zeitschrift für Volkswirtschaft und Statistik*] ... 1

Z. VERMESSUNGS.  *Zeitschrift für Vermessungswesen.*  (Stuttgart) ... 1

Z. VERSICHERUNGSW.  *Zeitschrift für Versicherungswissenschaften.*  (Berlin) [Became *Deutsche Versicherungszeitschrift für Sozialversicherung und Privatversicherung*] ... 1

Z. WAHRSCHEIN. VERW. GEBIETE  *Zeitschrift für Wahrscheinlichkeitstheorie und verwandte Gebiete.*  (Berlin) ... 27

ZAP. KIIV. SIL'SKO-GOS. INST.  *Zapiski Kiïvs'kiĭ Sil'sko-gospodars'kiĭ Institut.*  (Kiev) ... 1

ZAP. RUSS. NAUCN. INST. BEOGRAD  *Zapiski Russkiĭ Naučnyĭ Institut.*  (Beograd) ... 1

ZESZYTY NAUK. POLITECH. WARSZAWSK. MAT.  *Politechnika Warszawska. Institut Matematyki. Zeszyty Naukowe. Matematyka i jej Zastosowanie.*  (Warszawa) ... 1

ZUR. GEOFIZ. INST. MOSK.  *Žurnal Geofizičeskogo Instituta, Moskva.*  (Moscow) ... 1

ZUR. INST. MAT. UKRAIN. AKAD. NAUK  *Žurnal Instytutu Matematyky. Ukraïns'koï Akademiï Nauk.*  (Kiev) [= *Journal de l'Institut Mathématique de l'Académie des Sciences d'Ukraine.* In Ukrainian. Vol. 1-4, 1934-1938. LC:Z6655.Am35. Was ZUR. MAT. TSYKLU UKRAIN. AKAD. NAUK] ... 2

ZUR. MAT. TSYKLU UKRAIN. AKAD. NAUK  *Žurnal Matematyčnogo Tsyklu Vseukraïns'koï Akademiï Nauk.*  (Kiev) [= *Journal du Cycle Mathématique de l'Académie des Sciences d'Ukraine.* In Ukrainian. Vol. 1-3, 1931-1934. Then ZUR. INST. MAT. UKRAIN. AKAD. NAUK] ... 1

ZUR. TEH. FIZ. AKAD. NAUK SSSR  *Žurnal Tehničeskoĭ Fiziki. Akademija Nauk SSSR.*  (Moscow) ... 1

636 ANN. MATH. STATIST.
468 BIOMETRIKA
387 PSYCHOMETRIKA
251 J. AMER. STATIST. ASSOC.
114 SANKHYA
113 J. ROY. STATIST. SOC. SER. B
108 J. EDUC. PSYCHOL.
103 ANN. INST. STATIST. MATH. TOKYO
 91 C.R. ACAD. SCI. PARIS
 88 ECONOMETRICA
 72 BRITISH J. PSYCHOL.
 69 BULL. INST. INTERNAT. STATIST.
 69 SANKHYA SER. A
 63 BIOMETRICS
 59 PROC. CAMBRIDGE PHILOS. SOC.
 59 SKAND. AKTUARIETIDSKR.
 51 EDUC. PSYCHOL. MEAS.
 48 METRON
 45 J. ROY. STATIST. SOC.
 42 AMER. MATH. MONTHLY
 38 CALCUTTA STATIST. ASSOC. BULL.
 35 BRITISH J. STATIST. PSYCHOL.
 33 ANN. EUGENICS
 33 BRITISH J. PSYCHOL. STATIST.
     SECT.
 33 STATISTICA, BOLOGNA
 31 TEOR. VEROJATNOST. I PRIMENEN.
 31 THEORY PROB. APPL.
 30 J. INDIAN STATIST. ASSOC.
 30 MULTIVARIATE ANAL. PROC.
     INTERNAT. SYMP. DAYTON
     (KRISHNAIAH)
 30 PROC. INST. STATIST. MATH. TOKYO
 27 J. EXPER. EDUC.
 27 REV. INST. INTERNAT. STATIST.
 27 TECHNOMETRICS
 27 Z. WAHRSCHEIN. VERW. GEBIETE
 26 J. INDIAN SOC. AGRIC. STATIST.
 25 BULL. AMER. MATH. SOC.
 25 RES. MEM. INST. STATIST. MATH.
     TOKYO
 24 PROC. AMER. MATH. SOC.
 24 PSYCHOL. BULL.
 24 SCHOOL REV.
 24 SELECT. PAPERS STATIST. PROB.
     WALD
 23 PSYCHOL. REV.

 22 Z. ANGEW. MATH. MECH.
 20 AUSTRAL. J. STATIST.
 20 METRIKA
 20 PUBL. INST. STATIST. UNIV.
     PARIS
 20 TRABAJOS ESTADIST.
 19 DOKL. AKAD. NAUK SSSR
 19 IRE TRANS. INFORMATION THEORY
 19 PROC. NAT. ACAD. SCI. U.S.A.
 19 SIAM REV.
 19 STUD. ITEM ANAL. PREDICT.
     (SOLOMON)
 18 ANN. MATH. SER. 2
 18 DUKE MATH. J.
 18 MITT. MATH. STATIST.
 18 PROC. ROY. SOC. EDINBURGH
 18 SCI. AND CULT.
 17 APPL. STATIST.
 17 COWLES COMMISS. MONOG.
 17 J. ROY. STATIST. SOC. SER. A.
 17 J. SOC. INDUST. APPL. MATH.
 16 PROC. ROY. SOC. EDINBURGH
     SECT. A
 15 ATTI RIUN. SCI. SOC. ITAL.
     STATIST.
 15 COLLOQ. INTERNAT. CENTRE NAT.
     RECH. SCI.
 15 DOKL. AKAD. NAUK SSSR NOV. SER.
 15 J. APPL. PSYCHOL.
 15 KON. AKAD. WETENSCH. AMSTERDAM
     PROC. SECT. SCI.
 15 PHILOS. MAG. SER. 7
 15 PROC. ROY. SOC. LONDON
 15 PROC. ROY. SOC. LONDON SER. A
 15 TRANS. AMER. MATH. SOC.

### REMAINING JOURNALS AND COLLECTIONS

| | | | | |
|---|---|---|---|---|
| 14 | ... 4 | | 7 | ... 22 |
| 13 | ... 5 | | 6 | ... 23 |
| 12 | ... 7 | | 5 | ... 23 |
| 11 | ... 9 | | 4 | ... 45 |
| 10 | ... 6 | | 3 | ... 80 |
| 9 | ... 10 | | 2 | ... 138 |
| 8 | ... 13 | | 1 | ... 363 |

*Entries in Chapter III including reprints, translations and amendments.
[6093 papers in 819 journals and collections]

## II.4    DUPLICATED PUBLICATIONS

*Econometrica,* Volume 17, Supplement (1949) = BULL. INST. INTERNAT. STATIST., Volume 31, Part 5.

*Journal of Clinical Psychology. Monograph Supplement* is a series of separates from J. CLIN. PSYCHOL.

*Koninklijke Nederlandse Akademie van Wetenschappen. Indagationes Mathematicae ex actis quibus titulus. Proceedings of the Section of Sciences,* Vol. 1-12, 1939-1950 duplicates Vol. 42-53 of KON. NEDERL. AKAD. WETESCH. PROC. SECT. SCI.    ... *Proceedings. Series A. Mathematical Sciences.* Vol. 13, 1951- duplicates Vol. 54- of KON. NEDERL. WETESCH. PROC. SER. A

*Kungl. Fysiografiska Sällskapets i Lund Förhandlingar,* Vol. 1-33, 1931-1963 duplicates Vol. 27-59 of LUNDS UNIV. ARSSKR. N. F. AVD. 2.

*Proceedings of The Indian Statistical Conference.* First Session, Calcutta, January 1938 = SANKHYA, Volume 4, Parts 1 & 2.   Second Session, Lahore, 1939 = SANKHYA, Volume 4, Part 4 (Part 3 includes article by P.C. Mahalanobis "Subhendu Sekhar Bose:  1906-1938", pp. 313-336).

*Psychological Reports, Monograph Supplement* is a series of separates from PSYCHOL. REP.

*Half a Century of Soviet Serials 1917-1968*.  A Bibliography and Union List of Serials Published in the USSR.  Compiled by Rudolf Smits. Pub. The Library of Congress, Washington, D.C., 1968.  2 Volumes.

*Japanese Scientific and Technical Publications in the Collections of The Library of Congress*.  Pub. The Library of Congress, Washington, D.C., 1962.  v + 247 pp.

*List of the Serial Publications of Foreign Governments 1815-1931*. Edited by Winifred Gregory.  Pub. The H.W. Wilson Company, New York, 1932.  [vi] + 720 pp.

*MAST:  Minimum Abbreviations of Serial Titles - Mathematics*.  Edited by Mary L. Tompkins.  Pub. Western Periodicals Company, 13000 Rayner Street, North Hollywood, Calif., 1969.  viii + 427 pp.

*Mathematical Reviews*.  Abbreviations of the Names of Scientific Periodicals reviewed in Mathematical Reviews.  Abbreviations of Names of Journals.  Volumes 9 (1948), 11 (1950), 13 (1952), 15 (1954), 17 (1956), 19 (1958), 21 (1960), 22 (1961), and consecutively from Volume 24 (1962) to date.  Addenda to Abbreviations of Names of Journals.  Volumes 14 (1953), 16 (1955), 18 (1957), 20 (1959).

*New Serial Titles*.  A Union List of Serials Commencing Publication after December 31, 1949.  Supplement to the Union List of Serials, Third Edition.  1950-1960 [Cumulation]:  Pub. The Library of Congress, Washington, D.C., 1961.  2 Volumes.  1961-1965 Cumulation:  Pub. R.R. Bowker Company & Arno Publishing Inc., New York, 1966.  2 Volumes.

*Periodicals in the Social Sciences and Humanities currently received by Canadian Libraries*.  Pub. National Library, Ottawa.  2 Volumes.

*The Times Atlas of the World*.  Comprehensive Edition.  Pub. Times Newspapers Ltd., London and Houghton Mifflin Company, Boston, 1967. xliii + 123 plates + 272 pp.  [Second Edition (revised), 1968]

*Ulrich's International Periodicals Directory*.  A Classified Guide to Current Periodicals, Foreign and Domestic.  Tenth Edition.  Edited by Eileen C. Graves.  Pub. R.R. Bowker Company, New York, 1963. 667 pp.  [Fourteenth Edition / 1971-1972.  Pub. 1971.  2 Volumes]

*Union List of Scientific Serials in Canadian Libraries*.  Third Edition. Pub. National Science Library, Ottawa, 1969.  xi + 1066 pp.

*Union List of Serials in Libraries of The United States and Canada*. Third Edition.  Edited by Edna Brown Titus.  Pub. The H.W. Wilson Company, New York, 1965.  5 Volumes.

*World List of Scientific Periodicals Published in the Years 1900-1960*. Fourth Edition.  Edited by Peter Brown & George Burder Stratton. Pub. Butterworths, London, 1963-1965.  3 Volumes.

CHAPTER III   PAPERS

# III.1 INTRODUCTION

The 6093 entries in §III.2 are references to 5786 research papers, 109 reprints, 72 translations, and 146 amendments published in 819 journals and collections. Each entry is annotated according to subject matter by codes which follow the year of publication. Reference to an abstract or review in *Mathematical Reviews* is often given at the end of the entry. Technical reports and similar unpublished materials as well as abstracts are not included in §III.2.

*Authorship.* The usual form of an author's name is the surname followed by his first given name and middle initial(s). When an author is more readily identified by another form of his name, such as that which usually appears on the title page of his papers, that form is used. In special cases, however, all given names are listed. All entries by a particular author are listed together under the same version of his name; a cross-reference is often provided for other forms. Titles of nobility, office, etc., appear at the end of the author's name, if used frequently. Names not in the Latin alphabet are transliterated; Cyrillic characters are converted following the scheme in *Mathematical Reviews*, Volume 30 (1965), p.1207 (and in the Preface of this *Bibliography*). Cross-references are provided for other transliterations of some names, e.g., KHINCHIN, A. YA. *see* HINCIN, A. JA. All accents and diacritical marks are omitted. When no authorship is mentioned in the paper, the entry is listed an ANONYMOUS; when authorship is supplied from another source, it is listed thereafter in parentheses, with cross-reference. All authors of each paper are cited with appropriate cross-references. Prefixes such as *de* and *von* are treated as part of the surname and alphabetized accordingly with cross-references from other part(s) of the surname. The prefixes *Mac* and *Mc* are alphabetized together as *Mac*. Entries are listed chronologically for each separate authorship; two or more entries with the same authorship and same year of publication are arranged by source of publication. Entries with *n* (*>1*) authors follow all entries by the same first *n-1* authors, and are arranged alphabetically according to the name of the *n*-th author.

*Year of Publication.* The year cited after the authorship is intended to be the year of publication. When a journal or collection was published over a two-year span, the first year is cited when the actual year of publication could not be determined. For conferences and symposia, the year of publication is often at least one year after that of the meeting. Papers, reprints and translations are listed if published in 1966 or earlier. Amendments are listed even if published after 1966.

*Subject-matter Codes.* The codes given in parentheses after the year of publication identify the subject-matter content. These codes are described in Chapter V. Reprints, translations and amendments are coded consistently with their originals.

*Title*.  Whenever possible the title is transcribed from the first page of the article if in a language using the Latin alphabet.  When the title uses another alphabet a translation is given in English, French or German, with the original language in parentheses.  All accents and dia-critical marks are omitted.  Foreign language summaries are cited for papers not in English, French or German.  Reprints, translations and amendments appear as separate entries with cross-references.  Discussion which immediately follows a paper is identified in parentheses; it is considered as part of the paper and is included within its pagination; the discussants are not identified.

*Source of Publication*.  An abbreviated form is used for the journal or collection in which the paper was published.  These abbreviations are identified in Chapter II.  The volume number follows except when none has been allocated; for conferences and symposia, the year or ordinal number of the meeting is cited, in lieu of a volume number.  A part number, in parentheses, is given when the pagination is not consecutive throughout the volume.  For multiple-part numbers only the first number is cited.  A supplement is treated as a part numbered SP.  The first and last page numbers appear in full whenever known.  For an unnumbered page an adjacent page number is quoted instead.

*Mathematical Reviews*.  References to abstracts or reviews appearing in Volumes 1 to 19 are cited by page number, and from Volume 20 onwards by serial number.

   In §§III.3-6 appear various frequency counts of the papers in §III.2 not including reprints, translations and amendments.  §III.3 gives the number of papers per year and §III.4 the number according to language; §III.5 presents the numbers of authors who have written one paper each, two papers each, and so on.  In addition, counts are given of the number of papers which have $k$ authors ($k = 1, \ldots, 6$).  Prominent deceased authors are identified in §III.6 together with their life spans; also given are references to the bibliographies which we searched for their papers.

# III.2 RESEARCH PAPERS, REPRINTS, TRANSLATIONS AND AMENDMENTS ARRANGED ALPHABETICALLY BY AUTHOR

ABBE, CLEVELAND 1871. (8.8)
    A HISTORICAL NOTE ON THE METHOD OF LEAST SQUARES.
  AMER. J. SCI., 101, 411-415.
          1

ABDEL-ATY, S. H. 1954. (2.2/T)
    TABLES OF GENERALIZED K-STATISTICS.
      (REPRINTED AS NO. 3)
  BIOMETRIKA, 41, 253-260. (MR16,P.152)
          2

ABDEL-ATY, S. H. 1954. (2.2/T)
    TABLES OF GENERALIZED K-STATISTICS.
      (REPRINT OF NO. 2)
  NEW STATIST. TABLES, 19, 1-7. (MR16,P.152)
          3

ABDURAHMANOV, T. 1964. (18.4/18.5)
    STATISTICAL ACCEPTANCE CONTROL BY SEVERAL INDICATIONS.
      (IN RUSSIAN)
  THEORY PROB. MATH. STATIST. (IZDAT. NAUKA UZBEK. SSR), 5-12. (MR33-6796)
          4

ABDURAHMANOV, T. SEE ALSO
  5062. SIRAZDINOV, S. H. + ABDURAHMANOV, T. (1964)

ABELSON, ROBERT P. SEE
  3808. MESSICK, SAMUEL J. + ABELSON, ROBERT P. (1956)

ABITA, EMANUELE 1956. (4.9)
    A PROPOSITO DI UN PROBLEMA DELLE CONCORDANZE E DI UNA SUA GENERALIZZAZIONE.
  ARCHIMEDE, 8, 38-40.
          5

ABRAHAM, T. P. + KHOSLA, R. K. 1965. E(11.1/11.2)
    ON THE POSSIBLE USE OF COMPONENT ANALYSIS TECHNIQUE IN PEST AND DISEASE SURVEY DATA.
  J. INDIAN SOC. AGRIC. STATIST., 17, 208-223.
          6

ABRAHAMSON, I. G. 1964. (2.7/T)
    ORTHANT PROBABILITIES FOR THE QUADRIVARIATE NORMAL DISTRIBUTION.
  ANN. MATH. STATIST., 35, 1685-1703. (MR29-4129)
          7

ABRAMSON, LEE R. SEE
  3834. MILLER, KENNETH S. + ABRAMSON, LEE R. (1965)

ABRAMSON, N. M. + BRAVERMAN, D. 1962. (3.1/5.1/E)
    LEARNING TO RECOGNIZE PATTERNS IN A RANDOM ENVIRONMENT.
  IRE TRANS. INFORMATION THEORY, IT-8(5) S58-S63. (MR26-1987)
          8

ABRANISVILI, B. S. 1966. (17.3)
    CONSTRUCTION OF MULTIVARIATE DISTRIBUTION FUNCTIONS AND FUNCTIONS OF
    MUTUAL COVARIANCE IN A CONTROL PROBLEM.
      (IN RUSSIAN, WITH SUMMARY IN GEORGIAN)
  PROBLEMS OPERATIONS RES. (MECNIEREBA), 28-45. (MR36-1226)
          9

ACZEL, J. 1955. (20.5)
    LOSUNG DER VEKTOR-FUNKTIONALGLEICHUNG DER HOMOGENEN UND INHOMOGENEN
    N-DIMENSIONALEN EINPARAMETRICHEN "TRANSLATION" DER ERZEUGENDEN FUNKTION VON
    KETTENREAKTIONEN UND DES STATIONAREN UND NICHT-STATIONAREN BEWEGUNGSINTEGRALS.
  ACTA MATH. ACAD. SCI. HUNGAR., 6, 131-141. (MR17,P.272)
          10

ADAMS, HENRY F. 1931. (15.1)
    THE THEORY OF TWO FACTORS: AN ALTERNATIVE EXPLANATION. PART I.
    GENERAL FACTOR 'G'.
  J. APPL. PSYCHOL., 15, 16-34, 358-377.
          11

ADAMS, HENRY F. 1932. (15.1)
    THE TWO-FACTOR THEORY OF ABILITY: REPLY TO DR. R.H. THOULESS AND REJOINDER.
  J. APPL. PSYCHOL., 16, 572-576.
          12

ADAMS, J. F.: LAX, PETER D. + PHILLIPS, RALPH S. 1965. (20.2)
    ON MATRICES WHOSE REAL LINEAR COMBINATIONS ARE NONSINGULAR.
  PROC. AMER. MATH. SOC., 16, 318-322.
          13

ADAMS, JOHN W. SEE
  3118. KOTZ, SAMUEL + ADAMS, JOHN W. (1964)

  3119. KOTZ, SAMUEL + ADAMS, JOHN W. (1964)

ADCOCK, C. J. 1946. (15.1)                                                    14
    SIMPLIFIED FACTOR ANALYSIS.
    OCCUP. PSYCHOL. LONDON, 20, 188-198.


ADCOCK, C. J. 1952. (6.6)                                                     15
    A NOTE ON CLUSTER-DIRECTED ANALYSIS.
    PSYCHOMETRIKA, 17, 249-253.


ADELMAN, FRANK   SEE
    16. ADELMAN, IRMA • ADELMAN, FRANK (1959)


ADELMAN, IRMA • ADELMAN, FRANK 1959. (16.1)                                   16
    THE DYNAMIC PROPERTIES OF THE KLEIN-GOLDBERGER MODEL.
    ECONOMETRICA, 27, 596-625.


ADHIKARI, BISHWANATH PROSAD 1957. (6.1)                                       17
    ANALYSE DISCRIMINANTE DES MESURES DE PROBABILITE SUR UN ESPACE ABSTRAIT.
    C.R. ACAD. SCI. PARIS, 244, 845-846.   (MR18,P.773)


ADHIKARI, BISHWANATH PROSAD • JOSHI, DEVI DATT 1956. (6.5)                    18
    DISTANCE,DISCRIMINATION ET RESUME EXHAUSTIF.
    PUBL. INST. STATIST. UNIV. PARIS, 5, 57-74.   (MR19,P.329)


ADKE, S. R. 1958. (6.5)                                                       19
    A NOTE ON DISTANCE BETWEEN TWO POPULATIONS.
        (AMENDED BY NO. 20)
    SANKHYA, 19, 195-200.


ADKE, S. R. 1959. (6.5)                                                       20
    CORRIGENDA: A NOTE ON DISTANCE BETWEEN TWO POPULATIONS.
        (AMENDMENT OF NO. 19)
    SANKHYA, 20, 108.


ADKE, S. R. 1964. (16.4)                                                      21
    A MULTI-DIMENSIONAL BIRTH AND DEATH PROCESS.
    BIOMETRICS, 20, 212-216.


ADKINS, DOROTHY C. 1949. (4.1)                                                22
    NOTE ON THE COMPUTATION OF PRODUCT-MOMENT CORRELATION COEFFICIENTS.
    PSYCHOMETRIKA, 14, 69-73.



ADKINS, DOROTHY C. 1950. (15.3)                                               23
    A SUPERIOR ROTATIONAL METHOD IN FACTOR ANALYSIS OF PSYCHOMETRICIANS
    IN GOVERNMENT SERVICE.
    PSYCHOMETRIKA, 15, 331-338.


ADKINS, DOROTHY C. 1964. (1.2)                                                24
    LOUIS LEON THURSTONE:  CREATIVE THINKER, DEDICATED TEACHER, EMINENT
    PSYCHOLOGIST.
    CONTRIB. MATH. PSYCHOL. (FREDERIKSEN • GULLIKSEN), 1-39.


ADKINS, DOROTHY C. • TOOPS, HERBERT A. 1937. (15.4)                           25
    SIMPLIFIED FORMULAS FOR ITEM SELECTION AND CONSTRUCTION.
    PSYCHOMETRIKA, 2, 165-171.


ADKINS, R. M.   SEE
    4508. REMMERS, H. H. • ADKINS, R. M. (1942)


ADYANTHAYA, N. K. 1932. (4.1/4.5/E)                                           26
    A FORMULA FOR THE SPURIOUS CORRELATION BETWEEN TWO FUNCTIONS OF
    MUTUALLY UNCORRELATED VARIABLES.
    J. INDIAN MATH. SOC., 19, 161-164.


AFIFI, A. A. • ELASHOFF, R. M. 1966. (1.2/3.2)                                27
    MISSING OBSERVATIONS IN MULTIVARIATE STATISTICS.  I.  REVIEW OF THE LITERATURE.
    J. AMER. STATIST. ASSOC., 61, 595-604.


AFRIAT, S. N. 1951. (20.1)                                                    28
    BOUNDS FOR THE CHARACTERISTIC VALUES OF MATRIX FUNCTIONS.
    QUART. J. MATH. OXFORD SECOND SER., 2, 81-84.   (MR12,P.793)

[29] AFRIAT, S. N. 1956. (20.2)
ON THE LATENT VECTORS AND CHARACTERISTIC VALUES OF PRODUCTS OF PAIRS
OF SYMMETRIC IDEMPOTENTS.
QUART. J. MATH. OXFORD SECOND SER., 7, 76-78. (MR18,P.371)

[30] AFRIAT, S. N. 1957. (20.1)
ORTHOGONAL AND OBLIQUE PROJECTORS AND THE CHARACTERISTICS OF PAIRS OF
VECTOR SPACES.
PROC. CAMBRIDGE PHILOS. SOC., 53, 800-816. (MR20-1389)

[31] AFRIAT, S. N. 1959. (20.1)
ANALYTIC FUNCTIONS OF FINITE DIMENSIONAL LINEAR TRANSFORMATIONS.
PROC. CAMBRIDGE PHILOS. SOC., 55, 51-61. (MR20-7039)

[32] AGHEVLI, MAHMOUD 1963. (1.2/15.1/15.2/E)
APPLICATION DE L'ANALYSE FACTORIELLE LINEAIRE A DES PHENOMENES ECONOMIQUES.
CAHIERS INST. SCI. ECON. APPL. SER. E, 2, 31-125.

[33] AGNEW, RALPH P. + KAC, MARK 1941. (17.3)
TRANSLATED FUNCTIONS AND STATISTICAL INDEPENDENCE.
BULL. AMER. MATH. SOC., 47, 148-154. (MR2,P.229)

[34] AGRAWAL, H. 1963. (9.1/E)
A NOTE ON OMISSION OF VARIABLES IN MULTIPLE REGRESSION ANALYSIS.
J. INDIAN STATIST. ASSOC., 1, 226-229.

[35] AHMAVAARA, YRJO 1954. (2.4/6.6/15.1)
THE MATHEMATICAL THEORY OF FACTORIAL INVARIANCE UNDER SELECTION.
PSYCHOMETRIKA, 19, 27-38. (MR16,P.940)

[36] AHMAVAARA, YRJO 1957. (15.1/E)
ON THE UNIFIED FACTOR THEORY OF MIND.
ANN. ACAD. SCI. FENN. SER. B, 106, 1-176.

[37] AHMED, MOHAMAD SALAHUDDIN 1961. (18.2/B)
ON A LOCALLY MOST POWERFUL BOUNDARY RANDOMIZED SIMILAR TEST FOR THE
INDEPENDENCE OF TWO POISSON VARIABLES.
ANN. MATH. STATIST., 32, 809-827. (MR24-A1159)

AHUMADA, A.   SEE
1022. COMREY, A. L. + AHUMADA, A. (1964)

[38] AIGNER, DENNIS J. 1966. (17.7/B)
ERRORS OF MEASUREMENT AND LEAST SQUARES ESTIMATION IN A SIMPLE
RECURSIVE MODEL OF DYNAMIC EQUILIBRIUM.
ECONOMETRICA, 34, 424-432. (MR33-6740)

[39] AILAM, GEDALIA 1966. (14.4)
MOMENTS OF COVERAGE AND COVERAGE SPACES.
J. APPL. PROB., 3, 550-555. (MR35-1050)

AIREY, JOHN R.   SEE
5792. WEBB, H. A. + AIREY, JOHN R. (1918)

[40] AITCHISON, JOHN 1965. (8.1/17.5/C)
LIKELIHOOD-RATIO AND CONFIDENCE-REGION TESTS.
J. ROY. STATIST. SOC. SER. B, 27, 245-250. (MR33-6741)

[41] AITCHISON, JOHN + SILVEY, SAMUEL D. 1958. (19.2/A)
MAXIMUM-LIKELIHOOD ESTIMATION OF PARAMETERS SUBJECT TO RESTRAINTS.
ANN. MATH. STATIST., 29, 813-828. (MR20-1382)

[42] AITKEN, A. C. 1928. (20.2)
ON THE LATENT ROOTS OF CERTAIN MATRICES.
PROC. EDINBURGH MATH. SOC. SER. 2, 1, 135-138.

[43] AITKEN, A. C. 1931. (20.3)
NOTE ON THE COMPUTATION OF DETERMINANTS.
TRANS. FAC. ACTUAR. EDINBURGH, 13, 272-275.

AITKEN, A. C. 1932.   (20.3)                                                    44
     ON THE EVALUATION OF DETERMINANTS, THE FORMATION OF THEIR ADJUGATES,
     AND THE PRACTICAL SOLUTIONS OF SIMULTANEOUS LINEAR EQUATIONS.
     PROC. EDINBURGH MATH. SOC. SER. 2, 3, 207-219.

AITKEN, A. C. 1933.   (8.1/20.4/U)                                             45
     ON THE GRADATION OF DATA BY THE ORTHOGONAL POLYNOMIALS OF LEAST SQUARES.
     PROC. ROY. SOC. EDINBURGH, 53, 54-78.

AITKEN, A. C. 1933.   (8.1)                                                    46
     ON FITTING POLYNOMIALS TO DATA WITH WEIGHTED AND CORRELATED ERRORS.
     PROC. ROY. SOC. EDINBURGH, 54, 12-16.

AITKEN, A. C. 1935.   (20.1)                                                   47
     A USEFUL EXPANSION IN APPLICATIONS OF DETERMINANTS.
     MATH. NOTES EDINBURGH, 29, 25-29.

AITKEN, A. C. 1935.   (2.4)                                                    48
     NOTE ON SELECTION FROM A MULTIVARIATE NORMAL POPULATION.
     PROC. EDINBURGH MATH. SOC. SER. 2, 4, 106-110.

AITKEN, A. C. 1935.   (8.1/8.8/20.2)                                           49
     ON LEAST SQUARES AND LINEAR COMBINATION OF OBSERVATIONS.
     PROC. ROY. SOC. EDINBURGH, 55, 42-48.

AITKEN, A. C. 1936.   (2.4)                                                    50
     A FURTHER NOTE ON MULTIVARIATE SELECTION.
     PROC. EDINBURGH MATH. SOC. SER. 2, 5, 37-40.

AITKEN, A. C. 1937.   (20.2)                                                   51
     NOTE ON A SPECIAL DETERMINANT.
     MATH. NOTES EDINBURGH, 30, 27-28.

AITKEN, A. C. 1937.   (20.2/20.3)                                              52
     STUDIES IN PRACTICAL MATHEMATICS. I. THE EVALUATION WITH APPLICATIONS
     OF A CERTAIN TRIPLE PRODUCT MATRIX.
     PROC. ROY. SOC. EDINBURGH, 57, 172-181.

AITKEN, A. C. 1937.   (20.3)                                                   53
     STUDIES IN PRACTICAL MATHEMATICS.  II.   THE EVALUATION OF THE LATENT
     ROOTS AND LATENT VECTORS OF A MATRIX.
     PROC. ROY. SOC. EDINBURGH, 57, 269-304.

AITKEN, A. C. 1938.   (20.5)                                                   54
     STUDIES IN PRACTICAL MATHEMATICS. III. THE APPLICATION OF QUADRATIC
     EXTRAPOLATION TO THE EVALUATION OF DERIVATIVES AND TO INVERSE INTERPOLATION.
     PROC. ROY. SOC. EDINBURGH, 58, 161-175.

AITKEN, A. C. 1940.   (2.5)                                                    55
     ON THE INDEPENDENCE OF LINEAR AND QUADRATIC FORMS IN SAMPLES OF
     NORMALLY DISTRIBUTED VARIATES.
         (AMENDED BY NO. 58)
     PROC. ROY. SOC. EDINBURGH, 60, 40-46.   (MR1,P.346)

AITKEN, A. C. 1945.   (8.1/20.3)                                               56
     STUDIES IN PRACTICAL MATHEMATICS. IV. ON LINEAR APPROXIMATION BY LEAST SQUARES.
     PROC. ROY. SOC. EDINBURGH SECT. A, 62, 138-146.   (MR8,P.54)

AITKEN, A. C. 1948.   (8.1/8.2)                                                57
     ON A PROBLEM IN CORRELATED ERRORS.
     PROC. ROY. SOC. EDINBURGH SECT. A, 62, 273-277.   (MR10,P.312)

AITKEN, A. C. 1948.   (2.5)                                                    58
     CORRIGENDUM:   ON THE INDEPENDENCE OF LINEAR AND QUADRATIC FORMS IN
     SAMPLES OF NORMALLY DISTRIBUTED VARIATES.
         (AMENDMENT OF NO. 55)
     PROC. ROY. SOC. EDINBURGH SECT. A, 62, 277.   (MR1,P.346)

AITKEN, A. C. 1948.   (19.3)                                                   59
     ON THE ESTIMATION OF MANY STATISTICAL PARAMETERS.
     PROC. ROY. SOC. EDINBURGH SECT. A, 62, 369-377.   (MR10,P.201)

AITKEN, A. C. 1949.   (7.1)                                                    60
    ON THE WISHART DISTRIBUTION IN STATISTICS.
    BIOMETRIKA, 36, 59-62.   (MR11,P.528)


AITKEN, A. C. 1950.   (2.5)                                                    61
    ON THE STATISTICAL INDEPENDENCE OF QUADRATIC FORMS IN NORMAL VARIATES.
    BIOMETRIKA, 37, 93-96.   (MR12,P.35)


AITKEN, A. C. 1953.   (20.2)                                                   62
    A SPECIAL DETERMINANT.
    MATH. GAZETTE, 37, 122.


AITKEN, A. C. 1953.   (20.1)                                                   63
    A NOTE ON TRACE-DIFFERENTIATION AND THE   $\Omega$-OPERATOR.
    PROC. EDINBURGH MATH. SOC. SER. 2, 10, 1-4.   (MR14,P.716)


AITKEN, A. C. 1962.   (20.1)                                                   64
    TWO NOTES ON MATRICES.
    PROC. GLASGOW MATH. ASSOC., 5, 109-113.


AITKIN, M. A. 1964.   (2.4/T)                                                  65
    CORRELATION IN A SINGLY TRUNCATED BIVARIATE NORMAL DISTRIBUTION.
    PSYCHOMETRIKA, 29, 263-270.   (MR30-650)


AITKIN, M. A. + HUME, M. W. 1965.   (2.4/4.8/BT)                               66
    CORRELATION IN A SINGLY TRUNCATED BIVARIATE NORMAL DISTRIBUTION.   II.
      RANK CORRELATION.
    BIOMETRIKA, 52, 639-643.   (MR34-5185)


AITKIN, M. A. + HUME, M. W. 1966.   (2.4/4.8/B)                                67
    CORRELATION IN A SINGLY TRUNCATED BIVARIATE NORMAL DISTRIBUTION.   III.
      CORRELATION BETWEEN RANKS AND VARIATE-VALUES.
    BIOMETRIKA, 53, 278-281.   (MR34-5186)


AIZERMAN, M. A.; BRAVERMAN, E. M. + ROZONOER, L. I. 1964.   (6.4/6.6)          68
    THE PROBABILITY PROBLEM OF PATTERN RECOGNITION LEARNING AND THE METHOD
      OF POTENTIAL FUNCTIONS.
        (TRANSLATION OF NO. 69)
    AUTOMAT. REMOTE CONTROL, 25, 1175-1190.


AIZERMAN, M. A.; BRAVERMAN, E. M. + ROZONOER, L. I. 1964.   (6.4/6.6)          69
    PROBABILITY PROBLEM OF PATTERN RECOGNITION LEARNING AND POTENTIAL FUNCTIONS METHOD.
        (IN RUSSIAN, WITH SUMMARY IN ENGLISH, TRANSLATED AS NO. 68)
    AVTOMAT. I TELEMEH., 25, 1307-1323.


AJNE, BJORN 1966.   (2.1/2.6/B)                                                70
    OM EN VALKAND SATS FOR NORMAL FORDELNINGEN.
        (WITH SUMMARY IN ENGLISH)
    NORDISK MAT. TIDSKR., 14, 33-36.   (MR33-4962)


AKAIKE, HIROTUGU 1956.   (20.3)                                               71
    MONTE CARLO METHOD APPLIED TO THE SOLUTION OF SIMULTANEOUS LINEAR EQUATIONS.
    ANN. INST. STATIST. MATH. TOKYO, 7, 107-113.   (MR19,P.1096)


AKAIKE, HIROTUGU 1956.   (18.1/B)                                             72
    ON THE DISTRIBUTION OF THE PRODUCT OF TWO $\Gamma$-DISTRIBUTED VARIABLES.
    ANN. INST. STATIST. MATH. TOKYO, 8, 53-54.   (MR18,P.349)


AKAIKE, HIROTUGU 1958.   (20.3)                                              73
    ON A COMPUTATION METHOD FOR EIGENVALUE PROBLEMS AND ITS APPLICATION
      TO STATISTICAL ANALYSIS.
    ANN. INST. STATIST. MATH. TOKYO, 10, 1-20.   (MR20-4916)


AKAIKE, HIROTUGU 1965.   (8.9/16.5)                                          74
    ON THE STATISTICAL ESTIMATION OF THE FREQUENCY RESPONSE FUNCTION OF A
      SYSTEM HAVING MULTIPLE INPUT.
    ANN. INST. STATIST. MATH. TOKYO, 17, 185-210.


AKAIKE, HIROTUGU 1966.   (16.4)                                              75
    ON THE USE OF NON-GAUSSIAN PROCESS IN THE IDENTIFICATION OF A LINEAR
      DYNAMIC SYSTEM.
    ANN. INST. STATIST. MATH. TOKYO, 18, 269-276.

AKAIKE, HIROTUGU    SEE ALSO
   3753. MATUSITA, KAMEO + AKAIKE, HIROTUGU (1956)

   3754. MATUSITA, KAMEO + AKAIKE, HIROTUGU (1956)


AKESSON, O. A. 1916.  (2.1/17.3)                                              76
   ON THE DISSECTION OF CORRELATION SURFACES.
   ARK. MAT. ASTRONOM. FYS., 11(16) 1-18.


AKUTOWICZ, E. J.   SEE
   5882. WIENER, NORBERT + AKUTOWICZ, E. J. (1959)


ALAM, K. + RIZVI, M. HASEEB 1966.  (6.7)                                      77
   SELECTION FROM MULTIVARIATE NORMAL POPULATIONS.
   ANN. INST. STATIST. MATH. TOKYO, 18, 307-318.


ALAMO, JUAN BEJAR  SEE  BEJAR, JUAN A.


ALBASINY, E. L. 1965.  (20.3)                                                 78
   THE NUMERICAL SOLUTION OF SOME NON-LINEAR EQUATIONS, USEFUL IN THE DESIGN
   OF EXPERIMENTS.
   J. ROY. STATIST. SOC. SER. B, 27, 466-473.


ALBERT, A. ADRIAN 1944.  (15.1/20.2)                                          79
   THE MATRICES OF FACTOR ANALYSIS.
   PROC. NAT. ACAD. SCI. U.S.A., 30, 90-95.  (MR5,P.209)


ALBERT, A. ADRIAN 1944.  (15.1/20.2)                                          80
   THE MINIMUM RANK OF A CORRELATION MATRIX.
   PROC. NAT. ACAD. SCI. U.S.A., 30, 144-146.  (MR6,P.6)


ALBERT, ARTHUR 1963.  (6.4)                                                   81
   A MATHEMATICAL THEORY OF PATTERN RECOGNITION.
   ANN. MATH. STATIST., 34, 284-299.


ALBERT, ARTHUR 1966.  (8.1/U)                                                 82
   FIXED SIZE CONFIDENCE ELLIPSOIDS FOR LINEAR REGRESSION PARAMETERS.
   ANN. MATH. STATIST., 37, 1602-1630.  (MR34-3691)


ALEXAKOS, C. E. 1966.  E(6.2/14.4)                                            83
   PREDICTIVE EFFICIENCY OF TWO MULTIVARIATE SELECTION TECHNIQUES IN COMPARISON
   WITH CLINICAL PREDICTIONS.
   J. EDUC. RES., 57, 297-306.


ALEXANDER, IRVIN E.  SEE
   2933. KARON, BERTRAM P. + ALEXANDER, IRVING E. (1958)


ALEXANDER, LEROY; KLUG, HAROLD P. + KUMMER, ELIZABETH 1948.  E(16.5)          84
   STATISTICAL FACTORS AFFECTING THE INTENSITY OF X-RAYS DIFFRACTED BY
   CRYSTALLINE POWDERS.
   J. APPL. PHYS., 19, 742-753.


ALI, S. M. + SILVEY, SAMUEL D. 1965.  (4.8/12.1/17.4/B)                       85
   ASSOCIATION BETWEEN RANDOM VARIABLES AND THE DISPERSION OF A
   RADON-NIKODYM DERIVATIVE.
      (AMENDED BY NO. 87)
   J. ROY. STATIST. SOC. SER. B, 27, 100-107.


ALI, S. M. + SILVEY, SAMUEL D. 1965.  (4.8)                                   86
   A FURTHER RESULT ON THE RELEVANCE OF THE DISPERSION OF A RADON-NIKODYM
   DERIVATIVE TO THE PROBLEM OF MEASURING ASSOCIATION.
   J. ROY. STATIST. SOC. SER. B, 27, 108-110.


ALI, S. M. + SILVEY, SAMUEL D. 1965.  (4.8/12.1/17.4/B)                       87
   ASSOCIATION BETWEEN RANDOM VARIABLES AND THE DISPERSION OF A RADON-
   NIKODYM DERIVATIVE.
      (AMENDMENT OF NO. 85)
   J. ROY. STATIST. SOC. SER. B, 27, 533.


ALI, S. M. + SILVEY, SAMUEL D. 1966.  (6.5)                                   88
   A GENERAL CLASS OF COEFFICIENTS OF DIVERGENCE OF ONE DISTRIBUTION FROM ANOTHER.
   J. ROY. STATIST. SOC. SER. B, 28, 131-142.

ALLAIS, D. C. 1966.   (3.1/6.2/6.6)                                    89
    THE PROBLEM OF TOO MANY MEASUREMENTS IN PATTERN RECOGNITION AND PREDICTION.
    IEEE INTERNAT. CONV. REC., 14(7) 124–130.


ALLAN, F. E. 1930.   (4.2)                                             90
    A PERCENTILE TABLE OF THE RELATION BETWEEN THE TRUE AND THE OBSERVED
    CORRELATION COEFFICIENT FROM A SAMPLE OF 4.
    PROC. CAMBRIDGE PHILOS. SOC., 26, 536–537.


ALLEN, R. G. D. 1950.   (1.2)                                          91
    THE WORK OF EUGEN SLUTSKY.
    ECONOMETRICA, 18, 209–216.


ALMENDRAS, D. 1946.   (20.4)                                           92
    FUNCIONES ORTOGONALES Y SUS APLICACIONES A LA ESTADÍSTICA.
    ESTADIST. CHILENA SANTIAGO, 19, 681–686.


ALTER, DINSMORE 1931.   (4.6)                                          93
    MULTIPLE CORRELATION FOR PREDICTION PURPOSES.
    J. AMER. STATIST. ASSOC., 26(SP) 258–262.


AMARA, R. C. 1959.   (16.4)                                            94
    APPLICATION OF MATRIX METHODS TO THE LINEAR LEAST SQUARES SYNTHESIS
    OF MULTIVARIABLE SYSTEMS.
    J. FRANKLIN INST., 268, 1–16.   (MR21–4863)


AMATO, VITTORIO 1947.   (8.1)                                          95
    SULLE MATRICI CARATTERISTICHE DI ALCUNI METODI DI INTERPOLAZIONE STATISTICA.
    GIORN. ECON. ANN. ECONOMIA, 6, 646–660.


AMATO, VITTORIO 1949.   (20.2/20.3)                                    96
    SU UN PROCEDIMENTO DI CALCOLO PER L'APPLICAZIONE DEI METODI
    INTERPOLATORI E MATRICI DI VANDERMONDE.
    STATISTICA, BOLOGNA, 9, 205–217.   (MR11,P.618)


AMATO, VITTORIO 1950.   (5.3)                                          97
    SULLE CORRELAZIONI PARZIALI.
    STATISTICA, BOLOGNA, 10, 341–350.   (MR12,P.430)


AMATO, VITTORIO 1954.   (15.1)                                         98
    UNA NUOVA ESPRESSIONE DELL'INDICE DI COGRADUAZIONE $\rho$ DI SPEARMAN.
        (WITH SUMMARY IN FRENCH AND ENGLISH)
    BULL. INST. INTERNAT. STATIST., 34(2) 329–335.


AMATO, VITTORIO 1955.   (4.8)                                          99
    SULLA DISTRIBUZIONE DEL COEFFICIENTE DI CORRELAZIONE PARABOLICA DI ORDINE K
    CALCOLATO PER GRANDI CAMPIONI CASUALI ESTRATTI DA UN DATO UNIVERSO.
    GIORN. ECON. ANN. ECONOMIA, 14, 141–151.


AMATO, VITTORIO 1956.   (4.8/E)                                        100
    ON THE DISTRIBUTION OF GINI'S G COEFFICIENT OF RANK CORRELATION IN
    RANKINGS CONTAINING TIES.
    METRON, 18(1) 83–106.   (MR19,P.1205)


AMATO, VITTORIO 1963.   (20.4)                                         101
    UNA NUOVA TECNICA DI SOLUZIONE DELLE EQUAZIONI ALLE DIFFERENZE.
    LA SCUOLA IN AZIONE, 13, 109–116.


AMBARZUMIAN, G. A. 1946.   (16.4)                                      102
    STOCHASTIC PROCESSES WITH TWO PARAMETERS GIVING IN INFINITY THE
    NORMAL CORRELATION.
        (IN RUSSIAN, WITH SUMMARY IN ARMENIAN AND ENGLISH)
    DOKL. AKAD. NAUK ARMJAN. SSR, 5, 65–70.   (MR8,P.516)


AMBLE, V. N.   SEE
    2751. JAIN, J. P. + AMBLE, V. N. (1962)


AMEMIYA, TAKESHI 1966.   (16.2)                                        103
    SPECIFICATION ANALYSIS IN THE ESTIMATION OF PARAMETERS OF A SIMULTANEOUS
    EQUATION MODEL WITH AUTOREGRESSIVE RESIDUALS.
    ECONOMETRICA, 34, 283–306.   (MR35–6313)

AMES, EDWARD + REITER, STANLEY 1961.  (16.5)                                    104
    DISTRIBUTIONS OF CORRELATION COEFFICIENTS IN ECONOMIC TIME SERIES.
    J. AMER. STATIST. ASSOC., 56, 637-656.  (MR26-3164)

AMOS, D. E. 1964.  (14.5)                                                       105
    REPRESENTATIONS OF THE CENTRAL AND NON-CENTRAL T DISTRIBUTIONS.
    BIOMETRIKA, 51, 451-458.

AMY, LUCIEN 1957.  (17.3)                                                       106
    CONTRIBUTION A L'ETUDE DES RELATIONS ENTRE VARIABLES ALEATOIRES
    FAIBLEMENT LIEES.
    J. SOC. STATIST. PARIS, 98(7) 161-178.

ANASTASI, ANNE 1966.  (15.4)                                                    107
    SOME CURRENT DEVELOPMENTS IN THE MEASUREMENT AND INTERPRETATION OF
    TEST VALIDITY.
    TEST. PROBLEMS PERSPECTIVE (ANASTASI), 307-317.


ANASTASI, ANNE    SEE ALSO
    1856. GARRETT, HENRY E. + ANASTASI, ANNE (1932)

ANDERSEN, ERIK SPARRE 1953.  (17.1)                                            108
    ON SUMS OF SYMMETRICALLY DEPENDENT RANDOM VARIABLES.
    SKAND. AKTUARIETIDSKR., 36, 123-138.  (MR15,P.634)

ANDERSEN, ERIK SPARRE 1954.  (17.3)                                            109
    SOME THEOREMS OF SUMS OF SYMMETRICALLY DEPENDENT RANDOM VARIABLES.
    SKAND. MAT. KONG. BER., 12, 291-296.  (MR16,P.378)

ANDERSON, EDGAR 1960.  (17.9)                                                  110
    A SEMIGRAPHICAL METHOD FOR THE ANALYSIS OF COMPLEX PROBLEMS.
    TECHNOMETRICS, 2, 387-391.  (MR22-5088)

ANDERSON, GEORGE A. 1965.  (3.3/13.2/A)                                        111
    AN ASYMPTOTIC EXPANSION FOR THE DISTRIBUTION OF THE LATENT ROOTS OF
    THE ESTIMATED COVARIANCE MATRIX.
    ANN. MATH. STATIST., 36, 1153-1173.  (MR31-4128)

ANDERSON, HARRY E., JR. 1966.  (1.1)                                           112
    REGRESSION, DISCRIMINANT ANALYSIS, AND A STANDARD NOTATION FOR BASIC
    STATISTICS.
    HANDB. MULTIVARIATE EXPER. PSYCHOL. (CATTELL), 153-173.

ANDERSON, HARRY E., JR. + FRUCHTER, BENJAMIN 1960.  (4.6)                      113
    SOME MULTIPLE CORRELATION AND PREDICTOR SELECTION METHODS.
    PSYCHOMETRIKA, 25, 59-76.  (MR22-3073)

ANDERSON, JOHN E. 1935.  (15.4)                                                114
    THE EFFECT OF ITEM ANALYSIS UPON THE DISCRIMINATIVE POWER OF AN EXAMINATION.
    J. APPL. PSYCHOL., 19, 237-244.

ANDERSON, L. DEWEY + TOOPS, HERBERT A. 1928.  (4.1/20.3)                       115
    A NEW APPARATUS FOR PLOTTING AND A CHECKING METHOD FOR SOLVING LARGE
    NUMBERS OF INTERCORRELATIONS.
    J. EDUC. PSYCHOL., 19, 650-657.

ANDERSON, L. DEWEY + TOOPS, HERBERT A. 1929.  (4.1/20.3)                       116
    A NEW APPARATUS FOR PLOTTING AND A CHECKING METHOD FOR SOLVING LARGE
    NUMBERS OF INTERCORRELATIONS.
    J. EDUC. PSYCHOL., 20, 36-43.

ANDERSON, MONTGOMERY D. 1959.  (4.9/B)                                         117
    UNA TEORIA GENERALIZZATA DI CORRELAZIONE BIVARIATA.
    GIORN. ECON. ANN. ECONOMIA, 18, 427-448.

ANDERSON, OSKAR 1926.  (4.8/6.2)                                               118
    UEBER DIE ANWENDUNG DER DIFFERENZENMETHODE ("VARIATE DIFFERENCE METHOD") BEI
    REIHENAUSGLEICHUNGEN, STABILITATSUNTERSUCHUNGEN UND KORRELATIONSMESSUNGEN.
        (REPRINTED AS NO. 127)
    BIOMETRIKA, 18, 293-320.

ANDERSON, OSKAR 1927.  (4.8)                                                    119
    UEBER DIE ANWENDUNG DER DIFFERENZENMETHODE ("VARIATE DIFFERENCE METHOD") BEI
    REIHENAUSGLEICHUNGEN, STABILITATSUNTERSUCHUNGEN UND KORRELATIONSMESSUNGEN.
        (REPRINTED AS NO. 127)
    BIOMETRIKA, 19, 53-86.


ANDERSON, OSKAR 1927.  (4.1/8.1/16.5)                                           120
    ON THE LOGIC OF THE DECOMPOSITION OF STATISTICAL SERIES INTO SEPARATE
    COMPONENTS.
        (REPRINTED AS NO. 128)
    J. ROY. STATIST. SOC., 90, 548-569.


ANDERSON, OSKAR 1929.  (4.1/4.6/8.1/E)                                          121
    DIE KORRELATIONSRECHNUNG IN DER KONJUNKTURFORSCHUNG.  EIN BEITRAG ZUR
    ANALYSE VON ZEITREIHEN.
        (REPRINTED AS NO. 129)
    VEROFF. FRANKFURTER GES. KONJUNKTURF., 4, 1-141.


ANDERSON, OSKAR 1930.  (4.6/4.8/17.7/E)                                         122
    CORRELATION ET CAUSALITE.   (IN FRENCH AND BULGARIAN)
        (TRANSLATED AS NO. 130)
    TRIMES. SPISANIE GLAV. DIREKT. STATIST., 2, 253-293.


ANDERSON, OSKAR 1954.  (2.6/17.8/A)                                             123
    EIN EXAKTER NICHT-PARAMETRISCHER TEST DER SOGEN. NULL-HYPOTHESE
    IM FALLE VON AUTOKORRELATION UND KORRELATION.
        (REPRINTED AS NO. 131)
    BULL. INST. INTERNAT. STATIST., 34(2) 130-143.  (MR16,P.842)


ANDERSON, OSKAR 1955.  (1.2)                                                    124
    BIBLIOGRAPHIE DER SEIT 1928 IN BUCHFORM ERSCHIENENEN
    DEUTSCHSPRACHIGEN VEROFFENTLICHUNGEN UBER THEORETISCHE STATISTIK UND IHRE
    ANWENDUNGSGEBIETE.
    EINZELSCHR. DEUTSCHEN STATIST. GES., 7, 1-78.


ANDERSON, OSKAR 1955.  (4.5/4.6/4.7)                                            125
    EINE "NICHT-PARAMETRISCHE" (VERTEILUNGSFREIE) ABLEITUNG DER STREUUNG (VARIANCE)
    DES MULTIPLEN ($R_{z.xy}$) UND PARTIELLEN ($R_{xy.z}$) KORRELATIONSKOEFFIZIENTEN IM FALLE
    DER SOGENANNTEN NULL-HYPOTHESE, SOWIE DER DIESER HYPOTHESE ENTSPRECHENDEN
    MITTLEREN QUADRATISCHEN ABWEICHUNGEN (STANDARD DEVIATIONS) DER
    REGRESSIONSKOEFFIZIENTEN.
        (REPRINTED AS NO. 132)
    MITT. MATH. STATIST., 7, 85-112.  (MR17,P.278)


ANDERSON, OSKAR 1955.  (4.8)                                                    126
    WANN IST DER KORRELATIONSINDEX VON FECHNER "GESICHERT" (SIGNIFICANT) ?
        (REPRINTED AS NO. 133)
    MITT. MATH. STATIST., 7, 166-167.


ANDERSON, OSKAR 1963.  (4.8/6.2)                                                127
    UEBER DIE ANWENDUNG DER DIFFERENZENMETHODE ("VARIATE DIFFERENCE METHOD") BEI
    REIHENAUSGLEICHUNGEN, STABILITATSUNTERSUCHUNGEN UND KORRELATIONSMESSUNGEN.
        (REPRINT OF NO. 118, REPRINT OF NO. 119)
    AUSGEWAHLTE SCHR. (ANDERSON REPRINTS), 1, 39-100.


ANDERSON, OSKAR 1963.  (4.1/8.1/16.5)                                           128
    ON THE LOGIC OF THE DECOMPOSITION OF STATISTICAL SERIES INTO SEPARATE
    COMPONENTS.
        (REPRINT OF NO. 120)
    AUSGEWAHLTE SCHR. (ANDERSON REPRINTS), 1, 101-122.


ANDERSON, OSKAR 1963.  (4.1/4.6/8.1/E)                                          129
    DIE KORRELATIONSRECHNUNG IN DER KONJUNKTURFORSCHUNG.  EIN BEITRAG ZUR
    ANALYSE VON ZEITREIHEN.
        (REPRINT OF NO. 121)
    AUSGEWAHLTE SCHR. (ANDERSON REPRINTS), 1, 166-301.


ANDERSON, OSKAR 1963.  (4.6/4.8/17.7/E)                                         130
    KORRELATION UND KAUSALITAT.
        (TRANSLATION OF NO. 122)
    AUSGEWAHLTE SCHR. (ANDERSON REPRINTS), 2, 471-529.

ANDERSON, OSKAR 1963.    (2.6/17.8/A)                                              131
     EIN EXAKTER NICHT-PARPAMETRISCHER TEST DER SOGEN. NULL-HYPOTHESE IM FALLE VON
     AUTOKORRFLATION UND KORRELATION.
          (REPRINT OF NO. 123)
     AUSGEWAHLTE SCHR. (ANDERSON REPRINTS), 2, 864-877.   (MR16,P.842)


ANDERSON, OSKAR 1963.    (4.5/4.6/4.7)                                             132
     EINE "NICHT-PARAMETRISCHE" (VERTEILUNGSFREIE) ABLEITUNG DER STREUUNG (VARIANCE)
     DES MULTIPLEN ($R_{z.xy}$) UND PARTIELLEN ($R_{xy.z}$) KORRELATIONSKOEFFIZIENTEN IM FALLE
     DER SOGENANNTEN NULL-HYPOTHESE, SOWIE DER DIESER HYPOTHESE ENTSPRECHENDEN
     MTTTLEREN QUADRATISCHEN ABWEICHUNGEN (STANDARD DEVIATIONS) DER
     REGRESSIONSKOEFFIZIENTEN.
          (REPRINT OF NO. 125)
     AUSGEWAHLTF SCHR. (ANDERSON REPRINTS), 2, 897-924.


ANDERSON, OSKAR 1963.    (4.8)                                                     133
     WANN IST DER KORRELATIONSINDEX VON FECHNER "GESICHERT" (SIGNIFICANT)?
          (REPRINT OF NO. 126)
     AUSGEWAHLTE SCHR. (ANDERSON REPRINTS), 2, 925-926.


ANDERSON, R. L. 1942.    (2.6/16.5)                                                134
     DISTRIBUTION OF THE SERIAL CORRFLATION COEFFICIENT.
     ANN. MATH. STATIST., 13, 1-13.   (MR4,P.22)


ANDERSON, R. L.    SEE ALSO
     3912. MOTE, V. L. + ANDERSON, R. L. (1965)

     5825. WELLS, W. T.; ANDERSON, R. L. + CELL, JOHN W. (1962)


ANDERSON, T. W. 1946.    (7.1/7.2/8.4)                                             135
     THE NON-CENTRAL WISHART DISTRIBUTION AND CERTAIN PROBLEMS OF
     MULTIVARIATE STATISTICS.
          (AMENDED BY NO. 156)
     ANN. MATH. STATIST., 17, 409-431.   (MR8,P.394)


ANDERSON, T. W. 1948.    (8.7/11.2/13.2/A)                                         136
     THE ASYMPTOTIC DISTRIBUTIONS OF THE ROOTS OF CERTAIN DETERMINANTAL EQUATIONS.
     J. ROY. STATIST. SOC. SER. B, 10, 132-139.   (MR10,P.553)


ANDERSON, T. W. 1948.    (10.2/16.5)                                               137
     ON THE THEORY OF TESTING SERIAL CORRELATION.
     SKAND. AKTUARIETIDSKR., 31, 88-116.   (MR10,P.312)


ANDERSON, T. W. 1950.    (8.7/16.2)                                                138
     ESTIMATION OF THE PARAMETERS OF A SINGLE EQUATION BY THE
     LIMITED-INFORMATION MAXIMUM-LIKELIHOOD METHOD.
     COWLES COMMISS. MONOG., 10, 311-322.


ANDERSON, T. W. 1951.    (8.7/16.2)                                                139
     ESTIMATING LINEAR RESTRICTIONS ON REGRESSION COEFFICIENTS FOR
     MULTIVARIATE NORMAL DISTRIBUTIONS.
     ANN. MATH. STATIST., 22, 327-351.   (MR13,P.144)


ANDERSON, T. W. 1951.    (13.2/A)                                                  140
     THE ASYMPTOTIC DISTRIBUTION OF CERTAIN CHARACTERISTIC ROOTS AND VECTORS.
     PROC. SECOND BERKFLEY SYMP. MATH. STATIST. PROB., 103-130.   (MR13,P.366)


ANDERSON, T. W. 1951.    (6.1/6.2)                                                 141
     CLASSIFICATION BY MULTIVARIATE ANALYSIS.
     PSYCHOMETRIKA, 16, 31-50.   (MR12,P.842)


ANDERSON, T. W. 1954.    (15.5)                                                    142
     ON ESTIMATION OF PARAMETERS IN LATENT STRUCTURE ANALYSIS.
     PSYCHOMETRIKA, 19, 1-10.   (MR17,P.756)


ANDERSON, T. W. 1955.    (8.7/16.2/BE)                                             143
     SOME STATISTICAL PROBLEMS IN RELATING EXPERIMENTAL DATA TO PREDICTING
     PERFORMANCE OF A PRODUCTION PROCESS.
     J. AMER. STATIST. ASSOC., 50, 163-177.

ANDERSON, T. W. 1955.  (5.1/17.2/20.5)                    144
     THE INTEGRAL OF A SYMMETRIC UNIMODAL FUNCTION OVER A SYMMETRIC CONVEX
     SET AND SOME PROBABILITY INEQUALITIES.
     PROC. AMER. MATH. SOC., 6, 170-176.  (MR16,P.1005)

ANDERSON, T. W. 1955.  (15.4/15.5)                    145
     SOME RECENT RESULTS IN LATENT STRUCTURE ANALYSIS.
     PROC. INVIT. CONF. TEST. PROBLEMS. 49-53.

ANDERSON, T. W. 1957.  (3.1)                    146
     MAXIMUM LIKELIHOOD ESTIMATES FOR A MULTIVARIATE NORMAL DISTRIBUTION
     WHEN SOME OBSERVATIONS ARE MISSING.
     J. AMER. STATIST. ASSOC., 52, 200-203.  (MR19,P.332)

ANDERSON, T. W. 1959.  (16.1/A)                    147
     ON ASYMPTOTIC DISTRIBUTIONS OF ESTIMATES OF PARAMETERS OF STOCHASTIC
     DIFFERENCE EQUATIONS.
     ANN. MATH. STATIST., 30, 676-687.  (MR21-6072)

ANDERSON, T. W. 1959.  (15.5)                    148
     SOME SCALING MODELS AND ESTIMATION PROCEDURES IN THE LATENT CLASS MODEL.
     PROB. STATIST. (CRAMER VOLUME), 9-38.  (MR21-7576)

ANDERSON, T. W. 1960.  (15.4)                    149
     SOME STOCHASTIC PROCESS MODELS FOR INTELLIGENCE TEST SCORES.
     MATH. METH. SOC. SCI. (ARROW, KARLIN, + SUPPES), 205-220.  (MR22-9327)

ANDERSON, T. W. 1962.  (8.8/19.1)                    150
     THE CHOICE OF THE DEGREE OF A POLYNOMIAL REGRESSION AS A MULTIPLE
     DECISION PROBLEM.
     ANN. MATH. STATIST., 33, 255-265.  (MR26-5688)

ANDERSON, T. W. 1962.  (8.8/19.3)                    151
     LEAST SQUARES AND BEST UNBIASED ESTIMATES.
     ANN. MATH. STATIST., 33, 266-272.  (MR25-1242)

ANDERSON, T. W. 1963.  (10.2/11.2/13.2/A)                    152
     ASYMPTOTIC THEORY FOR PRINCIPAL COMPONENT ANALYSIS.
     ANN. MATH. STATIST., 34, 122-148.  (MR26-3149)

ANDERSON, T. W. 1963.  (5.2/8.3)                    153
     A TEST FOR EQUALITY OF MEANS WHEN COVARIANCE MATRICES ARE UNEQUAL.
     ANN. MATH. STATIST., 34, 671-672.  (MR26-7090)

ANDERSON, T. W. 1963.  (10.2/16.5/19.1)                    154
     DETERMINATION OF THE ORDER OF DEPENDENCE IN NORMALLY DISTRIBUTED TIME SERIES.
     PROC. SYMP. TIME SERIES ANAL. BROWN UNIV. (ROSENBLATT), 425-446.

ANDERSON, T. W. 1963.  (15.1/16.5)                    155
     THE USE OF FACTOR ANALYSIS IN THE STATISTICAL ANALYSIS OF MULTIPLE TIME SERIES.
     PSYCHOMETRIKA, 28, 1-25.

ANDERSON, T. W. 1964.  (7.1/7.2/8.4)                    156
     CORRECTIONS TO:"SOME EXTENSIONS OF THE WISHART DISTRIBUTION", AND "THE
     NON-CENTRAL WISHART DISTRIBUTION AND CERTAIN PROBLEMS OF MULTIVARIATE
     STATISTICS".
          (AMENDMENT OF NO. 135, AMENDMENT OF NO. 168)
     ANN. MATH. STATIST., 35, 923-924.  (MR6,P.161, MR8,P.394)

ANDERSON, T. W. 1964.  (6.3)                    157
     ON BAYES PROCEDURES FOR A PROBLEM WITH CHOICE OF OBSERVATIONS.
     ANN. MATH. STATIST., 35, 1128-1135.

ANDERSON, T. W. 1965.  (1.2)                    158
     SAMUEL STANLEY WILKS, 1906-1964.
     ANN. MATH. STATIST., 36, 1-27.

ANDERSON, T. W. 1965.  (10.1/11.2/13.2/C)                    159
     SOME OPTIMUM CONFIDENCE BOUNDS FOR ROOTS OF DETERMINANTAL EQUATIONS.
     ANN. MATH. STATIST., 36, 468-488.  (MR30-2616)

ANDERSON, T. W. 1965.  (3.1/10.1/10.2/C)                                        160
    SOME PROPERTIES OF CONFIDENCE REGIONS AND TESTS OF PARAMETERS IN
    MULTIVARIATE DISTRIBUTIONS.
        (WITH DISCUSSION)
    PROC. IBM SCI. COMPUT. SYMP. STATIST., 15-28.


ANDERSON, T. W. 1966.  (6.4/17.8)                                               161
    SOME NONPARAMETRIC MULTIVARIATE PROCEDURES BASED ON STATISTICALLY
    EQUIVALENT BLOCKS.
    MULTIVARIATE ANAL. PROC. INTERNAT. SYMP. DAYTON (KRISHNAIAH), 5-27.  (MR35-5101)


ANDERSON, T. W. + BAHADUR, RAGHU RAJ 1962.  (6.1/6.3/6.5)                       162
    CLASSIFICATION INTO TWO MULTIVARIATE NORMAL DISTRIBUTIONS WITH
    DIFFERERENT COVARIANCE MATRICES.
    ANN. MATH. STATIST., 33, 420-431.  (MR25-4609)


ANDERSON, T. W. + DAS GUPTA, SOMESH 1963.  (20.1)                               163
    SOME INEQUALITIES ON CHARACTERISTIC ROOTS OF MATRICES.
        (AMENDED BY NO. 167)
    BIOMETRIKA, 50, 522-524.  (MR28-5070)


ANDERSON, T. W. + DAS GUPTA, SOMESH 1964.  (9.1)                                164
    MONOTONICITY OF THE POWER FUNCTIONS OF SOME TESTS OF INDEPENDENCE
    BETWEEN TWO SETS OF VARIATES.
    ANN. MATH. STATIST., 35, 206-208.  (MR28-1698)


ANDERSON, T. W. + DAS GUPTA, SOMESH 1964.  (10.1/10.2)                          165
    A MONOTONICITY PROPERTY OF THE POWER FUNCTIONS OF SOME TESTS OF THE
    EQUALITY OF TWO COVARIANCE MATRICES.
        (AMENDED BY NO. 166)
    ANN. MATH. STATIST., 35, 1059-1063.


ANDERSON, T. W. + DAS GUPTA, SOMESH 1965.  (10.1/10.2)                          166
    CORRECTION TO:  A MONOTONICITY PROPERTY OF THE POWER FUNCTIONS OF SOME
    TESTS OF THE EQUALITY OF TWO COVARIANCE MATRICES.
        (AMENDMENT OF NO. 165)
    ANN. MATH. STATIST., 36, 1318.


ANDERSON, T. W. + DAS GUPTA, SOMESH 1965.  (20.1)                               167
    CORRIGENDA:  SOME INEQUALITIES ON CHARACTERISTIC ROOTS OF MATRICES.
        (AMENDMENT OF NO. 163)
    BIOMETRIKA, 52, 669.  (MR28-5070)


ANDERSON, T. W. + GIRSHICK, M. A. 1944.  (7.1)                                  168
    SOME EXTENSIONS OF THE WISHART DISTRIBUTION.
        (AMENDED BY NO. 156)
    ANN. MATH. STATIST., 15, 345-357.  (MR6,P.161)


ANDERSON, T. W. + HURWICZ, LEONID 1947.  (16.2/17.7)                            169
    ERRORS AND SHOCKS IN ECONOMIC RELATIONSHIPS.
    BULL. INST. INTERNAT. STATIST., 31(5) 23-25.


ANDERSON, T. W. + RUBIN, HERMAN 1949.  (8.7/16.2/A)                             170
    ESTIMATION OF THE PARAMETERS OF A SINGLE EQUATION IN A COMPLETE
    SYSTEM OF STOCHASTIC EQUATIONS.
    ANN. MATH. STATIST., 20, 46-63.  (MR10,P.464)


ANDERSON, T. W. + RUBIN, HERMAN 1950.  (8.7/16.2/A)                             171
    THE ASYMPTOTIC PROPERTIES OF ESTIMATES OF THE PARAMETERS OF A SINGLE
    EQUATION IN A COMPLETE SYSTEM OF STOCHASTIC EQUATIONS.
    ANN. MATH. STATIST., 21, 570-582.  (MR12,P.510)


ANDERSON, T. W. + RUBIN, HERMAN 1956.  (15.1/15.2/A)                            172
    STATISTICAL INFERENCE IN FACTOR ANALYSIS.
    PROC. THIRD BERKELEY SYMP. MATH. STATIST. PROB., 5, 111-150.  (MR18,P.954)


ANDERSON, T. W. + WALKER, A.M. 1964.  (16.1)                                    173
    ON THE ASYMPTOTIC DISTRIBUTION OF THE AUTOCORRELATIONS OF A SAMPLE FROM A LINEAR
    STOCHASTIC PROCESS.
    ANN. MATH. STATIST., 35, 1296-1303.

ANDERSON, T. W.    SEE ALSO
    184. ANONYMOUS   (ANDERSON, T. W.) (1952)

    185. ANONYMOUS   (ANDERSON, T. W.) (1955)

    5637. VILLARS, D.S. + ANDERSON, T. W. (1943)

    1219. DAS GUPTA, SOMESH; ANDERSON, T. W. + MUDHOLKAR, GOVIND S. (1964)

ANDERSSON, WALTER 1932.  (8.8)                                                174
    RESEARCHES INTO THE THEORY OF REGRESSION.
    MEDD. LUNDS ASTRONOM. OBS. SER. 2, 64, 1-197.

ANDO, ALBERT + KAUFMAN, G. M. 1965.  (3.1)                                    175
    BAYESIAN ANALYSIS OF THE INDEPENDENT MULTINORMAL PROCESS-NEITHER MEAN
    NOR PRECISION KNOWN.
    J. AMER. STATIST. ASSOC., 60, 347-358.  (MR33-834)

ANDREOLI, GUILIO 1941.  (4.1/4.5)                                             176
    STATISTICA DI CONFIGURAZIONI.  (RICERCHE SU COPPIE DI VARIABILI
    CASUALI IN CORRELAZIONE.)
    REND. ACCAD. SCI. FIS. MAT. NAPOLI SER. 4, 11, 150-158.  (MR8,P.523)

ANDREWS, FRED C.    SEE
    562. BIRNBAUM, Z. W.; PAULSON, EDWARD + ANDREWS, FRED C. (1950)

ANDREWS, T. G. 1943.  E(4.4)                                                  177
    STATISTICAL STUDIES IN ALLERGY.  I. A CORRELATION ANALYSIS.
    J. ALLERGY, 14, 322-328.

ANDREWS, WILLIAM H., JR.    SEE
    3683. MARSCHAK, JACOB + ANDREWS, WILLIAM H., JR. (1944)

    3684. MARSCHAK, JACOB + ANDREWS, WILLIAM H., JR. (1945)

ANGOFF, WILLIAM H. 1956.  (15.4)                                             178
    A NOTE ON THE ESTIMATION OF NONSPURIOUS CORRELATIONS.
    PSYCHOMETRIKA, 21, 295-297.  (MR18,P.343)

ANONYMOUS 1821.  (20.5)                                                      179
    DISSERTATION SUR LA RECHERCHE DU MILIEU LE PLUS PROBABLE, ENTRE LES
    RESULTATS DE PLUSIEURS OBSERVATIONS OU EXPERIENCES.(PAR UN ABONNE.)
    ANN. MATH. PURES APPL., 12, 181-204.

ANONYMOUS 1937.  (1.2)                                                       180
    BIBLIOGRAPHY.  W.F. SHEPPARD'S PUBLISHED WORKS.
    ANN. EUGENICS, 8, 13-14.

ANONYMOUS 1947.  (1.2)                                                       181
    BIBLIOGRAPHIE.
    BIOTYPOLOGIE, 9, 66-80.

ANONYMOUS 1957.  (1.2)                                                       182
    BIBLIOGRAFIE: LUIGI EINAUDI, RONALD AYLMER FISHER, CORRADO GINI.
    STATISTICA, BOLOGNA, 17, 47-105.

ANONYMOUS 1964.  (1.2)                                                       183
    SCIENTIFIC CONTRIBUTIONS OF PROFESSOR P.C.MAHALANOBIS.
    ESSAYS ECONOMET. PLANNING (MAHALANOBIS VOLUME), 321-342.

ANONYMOUS  (ANDERSON, T. W.) 1952.  (1.2)                                    184
    THE PUBLICATIONS OF ABRAHAM WALD.
        (REPRINTED AS NO. 185)
    ANN. MATH. STATIST., 23, 29-33.  (MR13,P.613)

ANONYMOUS  (ANDERSON, T. W.) 1955.  (1.2)                                    185
    THE PUBLICATIONS OF ABRAHAM WALD.
        (REPRINT OF NO. 184)
    SELECT. PAPERS STATIST. PROB. WALD, 20-24.  (MR13,P.613)

ANONYMOUS  (NANDI, HARIKINKAR K.) 1965.  (1.2)                               186
    SAMARENDRA NATH ROY: 1906-1964.
    CALCUTTA STATIST. ASSOC. BULL., 14, 1-8.

ANONYMOUS   (NANDI, HARIKINKAR K.) 1965.   (1.2)                187
    MANINDRA NATH GHOSH:  1918-1965.
   CALCUTTA STATIST. ASSOC. BULL., 14, 89-92.


ANTLE, CHARLES E.   SEE
   1724. FOLKS, JOHN LEROY + ANTLE, CHARLES E. (1965)


AOYAMA, HIROJIRO 1950.   (6.1)                                  188
    A NOTE ON THE CLASSIFICATION OF OBSERVATION DATA.
   ANN. INST. STATIST. MATH. TOKYO, 2, 17-19.   (MR12,P.511)


AOYAMA, HIROJIRO 1952.   (15.5/E)                               189
    A PROBLEM OF QUANTIFICATION.
        (IN JAPANESE)
   RES. MEM. INST. STATIST. MATH. TOKYO, 8, 139-148.


AOYAMA, HIROJIRO 1954.   (4.3)                                  190
    ON THE ESTIMATION OF THE CORRELATION COEFFICIENT IN THE STRATIFIED RANDOM SAMPLING.
        (IN JAPANESE, WITH SUMMARY IN ENGLISH)
   PROC. INST. STATIST. MATH. TOKYO, 1(2) 41-46.


AOYAMA, HIROJIRO 1956.   (17.5)                                 191
    ON THE EVALUATION OF THE SAMPLING ERROR OF A CERTAIN DETERMINANT.
   ANN. INST. STATIST. MATH. TOKYO, 8, 27-33.   (MR18,P.515)


AOYAMA, HIROJIRO 1957.   (15.4)                                 192
    SAMPLING FLUCTUATIONS OF THE TEST RELIABILITY.
   ANN. INST. STATIST. MATH. TOKYO, 8, 129-143.   (MR19,P.991)


AOYAMA, HIROJIRO 1961.   (15.1/E)                               193
    ON SOME STATISTICAL PROBLEMS IN EDUCATION III-APPLICATIONS OF FACTOR ANALYSIS.
        (IN JAPANESE)
   PROC. INST. STATIST. MATH. TOKYO, 8, 143-148.


AOYAMA, HIROJIRO 1962.   (17.6/U)                               194
    NOTE ON ORDERED RANDOM INTERVALS AND ITS APPLICATION.
   ANN. INST. STATIST. MATH. TOKYO, 13, 243-250.


AOYAMA, HIROJIRO 1965.   (6.1/12.1)                             195
    DUMMY VARIABLE AND ITS APPLICATION TO THE QUANTIFICATION METHOD.
        (IN JAPANESE, WITH SUMMARY IN ENGLISH, AMENDED BY NO. 196)
   PROC. INST. STATIST. MATH. TOKYO, 13, 1-12.


AOYAMA, HIROJIRO 1965.   (6.1/12.1)                             196
    DUMMY VARIABLE AND ITS APPLICATION TO THE QUANTIFICATION METHOD. ERRATA.
        (IN JAPANESE, WITH SUMMARY IN ENGLISH, AMENDMENT OF NO. 195)
   PROC. INST. STATIST. MATH. TOKYO, 13, 135-137.


AOYAMA, HIROJIRO + TANAKA, SADAKO 1957.   (20.3)               197
    ON THE PROPAGATION OF THE ERROR TO THE INVERSE OF A CERTAIN MATRIX OF
    20 DEGREES.
        (IN JAPANESE, WITH SUMMARY IN ENGLISH)
   PROC. INST. STATIST. MATH. TOKYO, 5, 49-51.


APARO, ENZO 1955.   (20.2)                                      198
    RISOLUZIONE NUMERICA DI UN PROBLEMA DI MINIMI QUADRATI.
   RICERCA SCI., 25, 3039-3044.   (MR17,P.536)


APARO, ENZO 1963.   (18.5)                                      199
    STRATIFIED RANDOM SAMPLING WITH ALLOCATION FOR MULTIVARIATE POPULATION.
   ANN. INST. STATIST. MATH. TOKYO, 14, 251-258.   (MR28-671)


APPLEBY, R. H. + FREUND, RUDOLF J. 1962.   (5.1/R)             200
    AN EMPIRICAL EVALUATION OF MULTIVARIATE SEQUENTIAL PROCEDURE FOR TESTING MEANS.
   ANN. MATH. STATIST., 33, 1413-1420.   (MR25-4613)


ARACIL, D. JOSE MA. ORTS   SEE   ORTS, JOSE MA.


ARIMA, A. 1956.   (16.5/E)                                      201
    ON CORRELATION BETWEEN TWO TIME-SERIES WITH TREND.
   APPL. STATIST. METEOROL., 7, 6-8.

ARMITAGE, J. V.   SEE
   3142. KRISHNAIAH, PARUCHURI R. + ARMITAGE, J. V. (1965)

   3143. KRISHNAIAH, PARUCHURI R. + ARMITAGE, J. V. (1966)

ARMITAGE, P. 1947.  (5.1/U)                                          202
   SOME SEQUENTIAL TESTS OF STUDENT'S HYPOTHESIS.
   J. ROY. STATIST. SOC. SUPP., 9, 250-263.  (MR9,P.296)

ARMITAGE, P. 1950.  (6.1/19.3)                                       203
   SEQUENTIAL ANALYSIS WITH MORE THAN TWO ALTERNATIVE HYPOTHESES AND ITS
   RELATION TO DISCRIMINANT FUNCTION ANALYSIS.
   J. ROY. STATIST. SOC. SER. B, 12, 137-144.  (MR12,P.429)

ARNAIZ, GONZALO 1958.  (4.1/8.1)                                     204
   PROBLEMAS DE REGRESION Y CORRELACION.
   TRABAJOS ESTADIST., 9, 43-56.

ARNOLD, HARVEY J. 1964.  (5.1/18.4)                                  205
   PERMUTATION SUPPORT FOR MULTIVARIATE TECHNIQUES.
   BIOMETRIKA, 51, 65-70.  (MR30-4333)

ARNOLD, J. NORMAN 1935.  (15.4)                                      206
   NOMOGRAM FOR DETERMINING THE VALIDITY OF TEST ITEMS.
   J. EDUC. PSYCHOL., 26, 151-153.

ARNOLD, J. NORMAN + DUNLAP, JACK W. 1937.  (15.4/T)                  207
   NOMOGRAPHS CONCERNING THE SPEARMAN-BROWN FORMULA AND RELATED FUNCTIONS.
   J. EDUC. PSYCHOL., 27, 371-374.

AROIAN, LEO A. 1941.  (14.5)                                         208
   A STUDY OF R.A.FISHER'S Z-DISTRIBUTION AND THE RELATED F-DISTRIBUTION.
   ANN. MATH. STATIST., 12, 429-448.  (MR3,P.175)

AROIAN, LEO A. 1947.  (2.5/14.5/B)                                   209
   THE PROBABILITY FUNCTION OF THE PRODUCT OF TWO NORMALLY DISTRIBUTED VARIABLES.
   ANN. MATH. STATIST., 18, 265-271.  (MR9,P.48)

AROIAN, LEO A. 1947.  (14.5)                                         210
   NOTE ON THE CUMULANTS OF FISHER'S Z-DISTRIBUTION.
   BIOMETRIKA, 34, 359-360.  (MR9,P.601)

ARRIBAT, PAUL 1964.  (3.1/C)                                         211
   SUR L'ESTIMATION DE LA VALEUR PROBABLE D'UNE VARIABLE ALEATOIRE NORMALE Y AU
   MOYEN D'UNE VARIABLE ALEATOIRE NORMALE AUXILIAIRE X, LIEE A Y, DE VALEUR
   PROBABLE EX CONNUE.
   C.R. ACAD. SCI. PARIS, 258, 4914-4916.  (MR29-2923)

ARROW, K. J. 1960.  (1.2/16.2)                                       212
   THE WORK OF RAGNAR FRISCH, ECONOMETRICIAN.
   ECONOMETRICA, 28, 175-192.

ASAI, AKIRA 1956.  (17.7)                                            213
   EXACT DISTRIBUTION OF LINEAR REGRESSION ESTIMATE WHEN BOTH VARIATES
   ARE STOCHASTIC.
   J. COLL. ARTS SCI. CHIBA UNIV., 2(1) 6-11.

ASANO, CHOOICHIRO 1965.  (18.2)                                      214
   ON ESTIMATING MULTINOMIAL PROBABILITIES BY POOLING INCOMPLETE SAMPLES.
   ANN. INST. STATIST. MATH. TOKYO, 17, 1-13.

ASANO, CHOOICHIRO + SATO, SOKURO 1961.  (3.1/5.2/B)                  215
   A BIVARIATE ANALOGUE OF POOLING OF DATA.
   BULL. MATH. STATIST., 10(3) 39-59.  (MR28-2597)

ASHTON, E. H.; HEALY, M.J.R. + LIPTON, S. 1957.  E(6.2/8.7)          216
   THE DESCRIPTIVE USE OF DISCRIMINANT FUNCTIONS IN PHYSICAL ANTHROPOLOGY.
   PROC. ROY. SOC. LONDON SER. B, 146, 552-572.

ATHANASIADOU, K. A. 1936.  (4.1)                                     217
   REMARQUES SUR L'USAGE DU COEFFICIENT DE CORRELATION.
        (IN GREEK)
   ARCH. OIKON. KOIN. EPIST., 16, 29-38.

ATKINSON, RICHARD C.   SEE
  1575. ESTES, WILLIAM K.; BURKE, C. J.; ATKINSON, RICHARD C. + (1957)
  FRANKMANN, J.P.


AURIC, A. 1924.  (17.9)                                                          218
    SUR LA PROBABILITE DE FAIRE FONCTIONNER N ELEMENTS PAR GROUPES DE N ELEMENTS.
  BULL. SCI. MATH. SER. 2, 48,   292-293.


AUTONNE, LEON 1915.  (20.1/20.2)                                                 219
    SUR LES MATRICES HYPOHERMITIENNES ET SUR LES MATRICES UNITAIRES.
  ANN. UNIV. LYON NOUV. SER. I SCI. MED., 38, 1-77.


AVNER, R. A.   SEE
  1116. CRONBACH, LEE J.; IKEDA, TOYOJI + AVNER, R. A. (1964)


AYANGAR, A. A. K.  SEE  AYYANGAR, A. A. KRISHNASWAMI


AYRES, LEONARD P. 1920.  (4.1)                                                   220
    A SHORTER METHOD FOR COMPUTING THE COEFFICIENT OF CORRELATION.
      (AMENDED BY NO. 222)
  J. EDUC. RES., 1, 216-221.


AYRES, LEONARD P. 1920.  (4.1/T)                                                 221
    THE APPLICATION TO TABLES OF DISTRIBUTION OF A SHORTER METHOD FOR
    COMPUTING COEFFICIENTS OF CORRELATION.
  J. EDUC. RES., 1, 295-298.


AYRES, LEONARD P. 1920.  (4.1)                                                   222
    CORRECTION TO:  A SHORTER METHOD FOR COMPUTING THE COEFFICIENT OF CORRELATION.
      (AMENDMENT OF NO. 220)
  J. EDUC. RES., 1, 298.


AYRES, LEONARD P. 1920.  (4.8)                                                   223
    THE CORRELATION RATIO.
  J. EDUC. RES., 2, 452-456.


AYRES, LEONARD P. 1920.  (4.1)                                                   224
    SUBSTITUTING SMALL NUMBERS FOR LARGE ONES IN THE COMPUTATION OF
    COEFFICIENTS OF CORRELATION.
  J. EDUC. RES., 2, 502-504.


AYYANGAR, A. A. KRISHNASWAMI 1937.  (4.1/15.1)                                   225
    CORRELATION BY THE METHOD OF FACTORS.
  PROC. INDIAN ACAD. SCI. SECT. A, 6, 71-73.


AYYANGAR, A. A. KRISHNASWAMI 1938.  (2.3)                                        226
    ON THE SEMI-INVARIANTS OF TWO VARIATES AND THEIR ADDITIVE PROPERTY.
  J. INDIAN MATH. SOC. NEW SER., 3, 1-7.


AYYANGAR, A. A. KRISHNASWAMI 1938.  (17.3/B)                                     227
    ON THE SEMI-INVARIANTS OF TWO VARIATES AND THEIR ADDITIVE PROPERTY.
  SANKHYA, 4, 85-91.


AZORIN POCH, FRANCISCO 1962.  (4.8/6.5/6.4)                                      228
    NOTAS SOBRE TAXONOMIA Y ESTADISTICA.
  TRABAJOS ESTADIST., 13, 249-263.


AZORIN POCH, FRANCISCO 1963.  (4.8/4.9/6.6)                                      229
    SOBRE LA ASOCIACION EN SUCESIONES DE ATRIBUTOS.
  TRABAJOS ESTADIST., 14, 1-9.


AZZARI, ANTHONY J.   SEE
  2936. KASKEY, GILBERT; KRISHNAIAH, PARUCHURI R. + AZZARI, ANTHONY J. (1962)


BABBAR, M. M. 1955.  (16.2)                                                      230
    DISTRIBUTIONS OF SOLUTIONS OF A SET OF LINEAR EQUATIONS (WITH AN APPLICATION
    TO LINEAR PROGRAMMING).
  J. AMER. STATIST. ASSOC., 50, 854-869.   (MR17,P.380)


BABINGTON SMITH, B.   SEE
  2993. KENDALL, MAURICE G. + BABINGTON SMITH, B. (1950)

  2995. KENDALL, MAURICE G.; KENDALL, SHEILA F. H. + BABINGTON SMITH, B. (1939)

BABITZ, MILTON + KEYS, NOEL 1940.  (4.1)                          231
     A METHOD FOR APPROXIMATING THE AVERAGE INTERCORRELATION COEFFICIENT
     BY CORRELATING THE PARTS WITH THE SUM OF THE PARTS.
     PSYCHOMETRIKA, 5, 283-288.  (MR2,P.110)


BACHELIER, LOUIS 1910.  (17.3)                                   232
     LES PROBABILITES A PLUSIEURS VARIABLES.
     ANN. SCI. ECOLE NORM. SUPER., 27(3) 339-360.


BACHMANN, W. K. 1940.  (14.5)                                    233
     L'ELLIPSOIDE D'ERREUR.
     SCHWEIZ. Z. VERMESSUNGS., 28, 181-197,201-208, 213-216.  (MR3,P.6)


BACON, H. M. 1938.  (4.6)                                        234
     NOTE ON A FORMULA FOR THE MULTIPLE CORRELATION COEFFICIENT.
     ANN. MATH. STATIST., 9, 227-229.


BACON, H. M. 1948.  (2.1/4.1/20.2)                               235
     A MATRIX ARISING IN CORRELATION THEORY.
     ANN. MATH. STATIST., 19, 422-424.  (MR10,P.94)


BACON, RALPH H. 1963.  (2.7/T)                                   236
     APPROXIMATIONS TO MULTIVARIATE NORMAL ORTHANT PROBABILITIES.
     ANN. MATH. STATIST., 34, 191-198.  (MR26-3145)


BADRIKIAN, ALBERT 1959.  (17.1)                                  237
     LES ELEMENTS ALEATOIRES GENERALISES A VALEURS DANS UN ESPACE
     VECTORIEL: DEFINITIONS ET PREMIERS RESULTATS.
     C.R. ACAD. SCI. PARIS, 248, 1603-1605.  (MR20-7335)


BAEDER, HELEN A.   SEE
     392. BATEN, WILLIAM DOWELL: TACK, P.I. + BAEDER, HELEN A. (1958)


BAER, REINHOLD 1939.  (20.5)                                     238
     THE SIGNIFICANCE OF THE SYSTEM OF SUBGROUPS FOR THE STRUCTURE OF THE GROUP.
     AMER. J. MATH., 61, 1-44.


BAETSLE, P.-L. 1961.  (8.1/20.3)                                 239
     UN THEOREME DE LIMITE CONCERNANT LES MATRICES SINGULIERES, ET SON APPLICATION
     A LA TECHNIQUE DES MOINDRES CARRES.
     PUBL. SCI. TECH. MINIST. AIR, 98, 11-29.  (MR24-B2528)


BAGAI, O. P. 1962.  (7.2/13.2)                                   240
     DISTRIBUTION OF THE DETERMINANT OF THE SUM OF PRODUCTS MATRIX IN THE
     NON-CENTRAL LINEAR CASE FOR SOME VALUES OF P.
          (AMENDED BY NO. 243)
     SANKHYA SER. A, 24, 55-62.  (MR26-5667)


BAGAI, O. P. 1962.  (8.4/9.1/10.1)                               241
     STATISTICS PROPOSED FOR VARIOUS TESTS OF HYPOTHESES AND THEIR
     DISTRIBUTIONS IN PARTICULAR CASES.
          (AMENDED BY NO. 242)
     SANKHYA SER. A, 24, 409-418.  (MR30-4334)




BAGAI, O. P. 1963.  (8.4/9.1/10.1)                               242
     ADDENDUM TO: STATISTICS PROPOSED FOR VARIOUS TESTS OF HYPOTHESES AND
     THEIR DISTRIBUTIONS IN PARTICULAR CASES.
          (AMENDMENT OF NO. 241)
     SANKHYA SER. A, 25, 427.  (MR30-4334)


BAGAI, O. P. 1963.  (7.2/13.2)                                   243
     ADDENDUM TO: DISTRIBUTION OF THE DETERMINANT OF THE SUM OF PRODUCTS
     MATRIX IN THE NONCENTRAL LINEAR CASE FOR SOME VALUE OF P.
          (AMENDMENT OF NO. 240)
     SANKHYA SER. A, 25, 428.  (MR30-4335)


BAGAI, O. P. 1964.  (7.2/7.3/8.3)                                244
     DISTRIBUTION OF THE RATIO OF THE GENERALIZED VARIANCE TO ANY OF ITS
     PRINCIPAL MINORS.
     J. INDIAN STATIST. ASSOC., 2, 80-96.  (MR30-2614)

BAGAI, O. P. 1964.  (7.2/8.4/A)                                                      245
    LIMITING DISTRIBUTION OF SOME STATISTICS IN MULTIVARIATE ANALYSIS OF VARIANCE.
    SANKHYA SER. A, 26, 271-278.  (MR32-3211A)

BAGAI, O. P. 1965.  (7.2)                                                            246
    THE DISTRIBUTION OF THE GENERALIZED VARIANCE.
    ANN. MATH. STATIST., 36, 120-130.  (MR30-4336)

BAGAI, O. P. 1965.  (8.4/10.2)                                                       247
    THE DISTRIBUTION OF SOME MULTIVARIATE TEST STATISTICS.
    J. INDIAN STATIST. ASSOC., 3, 116-124.  (MR33-799)

BAGGALEY, ANDREW R. + CATTELL, RAYMOND B. 1956.  (15.2)                              248
    A COMPARISON OF EXACT AND APPROXIMATE LINEAR FUNCTION ESTIMATES OF
    OBLIQUE FACTOR SCORES.
    BRITISH J. STATIST. PSYCHOL., 9, 83-86.

BAHADUR, RAGHU RAJ 1950.  (19.1)                                                     249
    ON A PROBLEM IN THE THEORY OF K POPULATIONS.
    ANN. MATH. STATIST., 21, 362-375.  (MR12,P.117)

BAHADUR, RAGHU RAJ 1954.  (19.1)                                                     250
    SUFFICIENCY AND STATISTICAL DECISION FUNCTIONS.
    ANN. MATH. STATIST., 25, 423-462.  (MR16,P.154)

BAHADUR, RAGHU RAJ 1961.  (18.1/T)                                                   251
    A REPRESENTATION OF THE JOINT DISTRIBUTION OF RESPONSES TO N DICHOTOMOUS ITEMS.
    STUD. ITEM ANAL. PREDICT. (SOLOMON), 158-168.  (MR22-12621B)

BAHADUR, RAGHU RAJ 1961.  (6.4)                                                      252
    ON CLASSIFICATION BASED ON RESPONSES TO N DICHOTOMOUS ITEMS.
    STUD. ITEM ANAL. PREDICT. (SOLOMON), 169-176.  (MR22-12621C)

BAHADUR, RAGHU RAJ 1966.  (17.6/A)                                                  2276
    A NOTE ON QUANTILES IN LARGE SAMPLES.
    ANN. MATH. STATIST., 37, 577-580.

BAHADUR, RAGHU RAJ + ROBBINS, HERBERT E. 1950.  (19.1)                               253
    THE PROBLEM OF THE GREATER MEAN.
        (AMENDED BY NO. 254)
    ANN. MATH. STATIST., 21, 469-487.  (MR12,P.428)

BAHADUR, RAGHU RAJ + ROBBINS, HERBERT E. 1951.  (19.1)                               254
    CORRECTION TO:  THE PROBLEM OF THE GREATER MEAN.
        (AMENDMENT OF NO. 253)
    ANN. MATH. STATIST., 22, 310.  (MR12,P.428)

BAHADUR, RAGHU RAJ    SEE ALSO
    162. ANDERSON, T. W. + BAHADUR, RAGHU RAJ (1962)

BAILEY, D. W. 1956.  E(11.1)                                                         255
    A COMPARISON OF GENETIC AND ENVIRONMENTAL PRINCIPAL COMPONENTS OF
    MORPHOGENESIS IN MICE.
    GROWTH, 20, 63-74.

BAILEY, MARTIN J. 1965.  (16.4)                                                      256
    PREDICTION OF AN AUTORGRESSIVE VARIABLE SUBJECT BOTH TO DISTURBANCES AND TO
    ERRORS OF OBSERVATION.
    J. AMER. STATIST. ASSOC., 60, 164-181.

BAITSCH, HELMUT + BAUER, RAINALD K. 1956.  (6.1/9.3/E)                               257
    ZUM PROBLEM DER MERKMALSAUSWAHL FUR TRENNVERFAHREN (BARNARD-PROBLEM).
    ALLGEMEIN. STATIST. ARCH., 40, 160-167.  (MR18,P.345)

BAKER, GEORGE A. 1930.  (4.3)                                                        258
    THE SIGNIFICANCE OF THE PRODUCT-MOMENT COEFFICIENT OF CORRELATION
    WITH SPECIAL REFERENCE TO THE CHARACTER OF THE MARGINAL DISTRIBUTIONS.
    J. AMER. STATIST. ASSOC., 25, 387-396.

BAKER, GEORGE A. 1937.  (14.2)                                                       259
    CORRELATION SURFACES OF TWO OR MORE INDICES WHEN THE COMPONENTS OF
    THE INDICES ARE NORMALLY DISTRIBUTED.
    ANN. MATH. STATIST., 8, 179-182.

BAKER, GEORGE A. 1941.  (10.2/U)                                                     260
    TEST OF HOMOGENEITY FOR NORMAL POPULATIONS.
    ANN. MATH. STATIST., 12, 233-236.  (MR3,P.7)

BAKER, GEORGE A. 1942.  (4.1)                                      261
    CORRELATIONS BETWEEN FUNCTIONS OF VARIABLES.
    J. AMER. STATIST. ASSOC., 37, 537-539.   (MR4,P.221)


BAKER, GEORGE A. 1954.  (2.4)                                      262
    THE EFFECT OF SELECTION ON LINEAR FUNCTIONS OF NORMALLY DISTRIBUTED
    CORRELATED VARIABLES ON THE DISTRIBUTIONS OF OTHER LINEAR FUNCTIONS.
    ANN. INST. STATIST. MATH. TOKYO, 5, 91-95.


BAKER, GEORGE A. 1956.  (14.2)                                     263
    THE EFFECTS OF WIDE GROUPINGS ON THE DISTRIBUTIONS OF ARRAY MEANS AND
    VARIANCES FOR CORRELATED NORMAL VARIABLES.
    ANN. INST. STATIST. MATH. TOKYO, 7, 103-106.   (MR18,P.79)


BAKER, GEORGE A. 1958.  (14.4/U)                                   264
    EMPIRIC INVESTIGATION OF A TEST OF HOMOGENEITY FOR POPULATIONS
    COMPOSED OF NORMAL DISTRIBUTIONS.
    J. AMER. STATIST. ASSOC., 53, 551-557.


BAKER, KENNETH H. 1939.  (15.4)                                    265
    ITEM VALIDITY BY THE ANALYSIS OF VARIANCE: AN OUTLINE OF METHOD.
    PSYCHOL. REC., 3, 242-248.


BAKER, M. L.   SEE
    2412. HAZEL, L. N.; BAKER, M. L. + REINMILLER, C. F. (1943)



BAKST, AARON 1931.  (4.6/8.1/20.3)                                 266
    A MODIFICATION OF THE COMPILATION OF THE MULTIPLE CORRELATION AND
    REGRESSION COEFFICIENTS BY THE TOLLEY AND EZEKIEL METHOD.
    J. EDUC. PSYCHOL., 22, 629-635.


BALAGANGADHARAN, K. 1947.  (1.2)                                   267
    A CONSOLIDATED LIST OF HINDU MATHEMATICAL WORKS.
    MATH. STUDENT, 15, 55-70.   (MR10,P.667)


BALDESSARI, BRUNO 1961.  (4.9)                                     268
    I CONCETTI DI CONCORDANZA, DISCORDANZA E INDIFFERENZA PROBABILISTICHE.
    GIORN. IST. ITAL. ATTUARI, 24, 218-262.


BALDESSARI, BRUNO 1962.  (4.9)                                     269
    LE PROBABILITA DI CONCORDANZA, DISCORDANZA E "INVARIANZA".
    GIORN. IST. ITAL. ATTUARI, 25, 114-138.   (MR27-2052)


BALDESSARI, BRUNO 1965.  (2.7)                                     270
    REMARQUE SUR LE RAPPORT DE COMBINAISONS LINEAIRES DE $\chi^2$.
    PUBL. INST. STATIST. UNIV. PARIS, 14, 379-391.


BALDESSARI, BRUNO 1965.  (2.5/2.6/16.5)                            271
    REMARQUES SUR LES ECHANTILLONS GAUSSIENS.
    PUBL. INST. STATIST. UNIV. PARIS, 14, 393-405.   (MR34-2102)


BALDESSARI, BRUNO 1966.  (2.5/8.3/8.5)                             272
    ANALYSIS OF VARIANCE OF DEPENDENT DATA.
    STATISTICA, BOLOGNA, 26, 895-903.


BALDWIN, A. L. 1942.  (15.5)                                       273
    PERSONAL STRUCTURE ANALYSIS: A STATISTICAL METHOD FOR INVESTIGATING
    THE SINGLE PERSONALITY.
    J. ABNORM. SOC. PSYCHOL., 37, 163-183.


BALESTRA, PIETRO + NERLOVE, MARC 1966.         (8.1/16.5/E)        274
    POOLING CROSS SECTION AND TIME SERIES DATA IN THE ESTIMATION OF A
    DYNAMIC MODEL:  THE DEMAND FOR NATURAL GAS.
    ECONOMETRICA, 34, 585-612.


BALL, R. J. 1963.  (16.2/E)                                        275
    THE SIGNIFICANCE OF SIMULTANEOUS METHODS OF PARAMETER ESTIMATION IN
    ECONOMETRIC MODELS.
    APPL. STATIST., 12, 14-25.


BALLORE, VICOMTE ROBERT DE MONTESSUS DE  SEE  DE MONTESSUS DE BALLORE, VICOMTE ROBERT

BALOGH, T. 1961.   (16.4)                                                                                      276
    GENERALIZATION OF A THEOREM OF U. GRENANDER.
    DEUX. CONG. MATH. HONGR., 2(4) 1-3.

BANACHIEWICZ, TADEUSZ 1938.   (20.3)                                                                           277
    METODA ROZWIAZYWANIA LICZBOWEGO ROWNAN LINIOWYCH, OBLICZANIA WYZNACZNIKOW I
    ODWROTNOSCI, ORAZ REDUKCJI FORM KWADRATOWYCH. - METHODE DE RESOLUTION
    NUMERIQUE DES EQUATIONS LINEAIRES, DU CALCUL DES DETERMINANTS ET DES INVERSES,
    ET DE REDUCTION DES FORMES QUADRATIQUES.
    BULL. INTERNAT. POLSKA AKAD. UMIEJET. (KRAKOW) SER. A, 393-404.


BANCROFT, T. A.   SEE
    3281. LARSON, HAROLD J. + BANCROFT, T. A. (1963)

BANDYOPADHYAY, SURAJ    SEE
    3923. MUKHERJEE, RAMKRISHNA + BANDYOPADHYAY, SURAJ (1964)

BANERJEE, DURGA PROSAD 1941.   (4.1/8.1/B)                                                                     278
    NOTE ON THE LIMIT OF CORRELATION AND REGRESSION COEFFICIENTS IN MINGLED RECORDS.
    MATH. STUDENT, 9, 155-157.   (MR4,P.104)

BANERJEE, DURGA PROSAD 1951.   (18.1)                                                                          279
    ON SOME NEW RECURRENCE FORMULAE FOR CUMULANTS OF MULTIVARIATE
    MULTINOMIAL DISTRIBUTIONS.
    PROC. INDIAN ACAD. SCI. SECT. A, 34, 20-23.   (MR13,P.665)

BANERJEE, DURGA PROSAD 1952.   (4.7)                                                                           280
    ON THE MOMENTS OF THE MULTIPLE CORRELATION COEFFICIENT IN SAMPLES
    FROM NORMAL POPULATION.
    J. INDIAN SOC. AGRIC. STATIST., 4, 88-90.   (MR14,P.189)

BANERJEE, DURGA PROSAD 1954.   (7.4/B)                                                                         281
    A NOTE ON THE DISTRIBUTION OF THE RATIO OF SAMPLE STANDARD DEVIATIONS IN RANDOM
    SAMPLES OF ANY SIZE FROM A BI-VARIATE CORRELATED NORMAL POPULATION.
    J. INDIAN SOC. AGRIC. STATIST., 6, 93-100.   (MR17,P.639)

BANERJEE, DURGA PROSAD 1958.   (8.4)                                                                           282
    ON THE EXACT DISTRIBUTION OF A TEST IN MULTIVARIATE ANALYSIS.
    J. ROY. STATIST. SOC. SER. B, 20, 108-110.   (MR20-2812)


BANERJEE, DURGA PROSAD 1959.   (18.1)                                                                          283
    ON SOME THEOREMS ON POISSON DISTRIBUTION.
    PROC. NAT. ACAD. SCI. INDIA SECT. A, 28, 30-33.   (MR26-5611)

BANERJEE, DURGA PROSAD 1960.   (7.4/8.4/10.1)                                                                  284
    ON THE MUTUAL INDEPENDENCE OF A SET OF VARIATES IN MULTIVARIATE ANALYSIS.
    PROC. NAT. ACAD. SCI. INDIA SECT. A, 29, 268-272.   (MR26-1966)

BANERJEE, KALISHANKAR 1949.   (20.2)                                                                           285
    ON THE CONSTRUCTION OF HADAMARD MATRICES.
    SCI. AND CULT., 14, 434-435.   (MR10,P.586)

BANERJEE, KALISHANKAR 1964.   (20.2)                                                                           286
    A NOTE ON IDEMPOTENT MATRICES.
    ANN. MATH. STATIST., 35, 880-882.   (MR28-5071)

BANERJEE, KALISHANKAR 1966.   (20.1/20.2)                                                                      287
    SINGULARITY IN HOTELLING'S WEIGHING DESIGNS AND A GENERALIZED INVERSE.
        (AMENDED BY NO. 288)
    ANN. MATH. STATIST., 37, 1021-1032.

BANERJEE, KALISHANKAR 1969.   (20.1/20.2)                                                                      288
    CORRECTION TO: SINGULARITY IN HOTELLING'S WEIGHING DESIGNS AND A GENERALIZED INVERSE.
        (AMENDMENT OF NO. 287)
    ANN. MATH. STATIST., 40, 719.

BANERJEE, KALISHANKAR + MARCUS, L. F. 1965.   (6.3/6.4)                                                        289
    BOUNDS IN A MINIMAX CLASSIFICATION PROCEDURE.
    BIOMETRIKA, 52, 653-654.   (MR34-5236)

BANERJEE, SAIBAL KUMAR 1962.  (8.3/8.4)                                    290
    THE PROBLEM OF TESTING LINEAR HYPOTHESIS ABOUT POPULATION MEANS WHEN
    THE POPULATION VARIANCES ARE NOT EQUAL AND M-TEST.
    SANKHYA SER. A, 24, 363-376.  (MR30-656)


BANERJI, R. B. 1960.  (16.5/E)                                             291
    RADIO-MEASUREMENT OF IONOSPHERIC DRIFT AS A PROBLEM IN PARAMETER ESTIMATION.
    STATIST. METH. RADIO WAVE PROP., 40-48.


BANKS, CHARLOTTE + BURT, CYRIL 1954.  (15.1)                              292
    THE REDUCED CORRELATION MATRIX.
    BRITISH J. STATIST. PSYCHOL., 7, 107-117.


BANTEGUI, CELIA G.   SEE
    4299. PILLAI, K.C.SREEDHARAN + BANTEGUI, CELIA G. (1959)


BAPTIST, JEAN-H. 1945.  (4.9/17.3)                                        293
    ETUDE DE LA DEPENDANCE STOCHASTIQUE.
    ASSOC. ACTUAIR. BELG. BULL., 50, 15-36.  (MR10,P.311)


BARANKIN, EDWARD W. 1945.  (20.1)                                         294
    BOUNDS FOR THE CHARACTERISTIC ROOTS OF A MATRIX.
    BULL. AMER. MATH. SOC., 51, 767-770.  (MR7,P.107)


BARANKIN, EDWARD W. 1949.  (8.3/U)                                        295
    EXTENSION OF THE ROMANOVSKY-BARTLETT-SCHEFFE TEST.
    PROC. BERKELEY SYMP. MATH. STATIST. PROB., 433-449.  (MR10,P.467)


BARANKIN, EDWARD W. 1951.  (19.2/20.3)                                    296
    ON SYSTEMS OF LINEAR EQUATIONS, WITH APPLICATIONS TO LINEAR
    PROGRAMMING AND THE THEORY OF TESTS OF STATISTICAL HYPOTHESES.
    UNIV. CALIFORNIA PUBL. STATIST., 1, 161-214.  (MR14, P.190)


BARANKIN, EDWARD W. 1966.  (19.4)                                         297
    SUFFICIENT STATISTICS IN THE CASE OF NON-CONSTANT CARRIER.
    SANKHYA SER. A, 28, 101-114.


BARANKIN, EDWARD W. 1966.  (19.4)                                        298
    SUFFICIENT STATISTICS IN THE CASE OF NON-CONSTANT CARRIER -II.
    SANKHYA SER. A, 28, 115-122.


BARDWELL, J. + WINKLER, C. A. 1949.  (17.9/E)                            299
    THE FORMATION AND PROPERTIES OF THREE-DIMENSIONAL POLYMERS. I.
    STATISTICS OF NETWORK POLYMERS.
    CANAD. J. RES., 27(B) 116-127.


BARDWELL, J. + WINKLER, C. A. 1949.  (17.9/E)                            300
    THE FORMATION AND PROPERTIES OF THREE-DIMENSIONAL POLYMERS. II.
    NETWORK FORMATION.
    CANAD. J. RES., 27(B) 128-138.


BARDWELL, J. + WINKLER, C. A. 1949.  (17.9/E)                            301
    THE FORMATION AND PROPERTIES OF THREE-DIMENSIONAL POLYMERS. III. THE
    EFFECT OF NETWORK STRUCTURE ON ELASTIC PROPERTIES.
    CANAD. J. RES., 27(B) 139-150.


BARGMANN, ROLF E. 1955.  (15.3/E)                                        302
    UNE EPREUVE STATISTIQUE DE LA STABILITE DE LA STRUCTURE SIMPLE.
        (WITH DISCUSSION)
    COLLOQ. INTERNAT. CENTRE NAT. RECH. SCI., 65, 143-158.


BARGMANN, ROLF E. 1955.  (15.1/15.2/15.3/T)                             303
    SIGNIFIKANZUNTERSUCHUNGEN DER EINFACHEN STRUKTUR IN DER FAKTOREN-ANALYSE.
    MITT. MATH. STATIST., 7, 1-24.


BARGMANN, ROLF E. 1956.  (15.2/E)                                        304
    UBERFUHRUNG DER SCHWERPUNKTS: IN DIE HAUPTACHSENLOSUNG UND WEITERE
    ENTWICKLUNG DER NAHERUNGSVERFAHREN IN DER FAKTORENANALYSE.
    MITT. MATH. STATIST., 8, 1-14.


BARGMANN, ROLF E. 1958.  (10.2/AT)                                       305
    SEPARATION OF RANDOM ERRORS OF SYSTEM AND OF INSTRUMENTATION.
    PROC. STATIST. TECH. MISSILE EVAL. SYMP., 227-240.

BARGMANN, ROLF E. 1961.   (15.1/15.2)                                    306
    MULTIVARIATE STATISTICAL ANALYSIS IN PSYCHOLOGY AND EDUCATION.
    BULL. INST. INTERNAT. STATIST., 38(4) 79-86.


BARGMANN, ROLF E. 1965.   (8.1/8.5/20.3)                                 307
    A STATISTICIAN'S INSTRUCTIONS TO THE COMPUTER:  A REPORT ON A STATISTICAL
        COMPUTER LANGUAGE.
    PROC. IBM SCI. COMPUT. SYMP. STATIST., 301-323.

BARGMANN, ROLF E.    SEE ALSO
    614. BOCK, R. DARRELL + BARGMANN, ROLF E. (1966)

    4341. POSTEN, H. O. + BARGMANN, ROLF E. (1964)

    4681. ROY, S.N. + BARGMANN, ROLF E. (1958)

    5504. TRAWINSKI, IRENE MONAHAN + BARGMANN, ROLF E. (1964)


BARLOW, J.A. + BURT, CYRIL 1954.   (15.2)                                308
    THE IDENTIFICATION OF FACTORS FROM DIFFERENT EXPERIMENTS.
    BRITISH J. STATIST. PSYCHOL., 7, 52-56.


BARLOW, JOHN S. 1959.  E(6.6/16.5)                                       309
    AUTOCORRELATION AND CROSSCORRELATION ANALYSIS IN ELECTROENCEPHALOGRAPHY.
    IRE TRANS. MED. ELECTRONICS, ME-6, 179-183.


BARNARD, M. M. 1935.  E(6.5)                                             310
    THE SECULAR VARIATIONS OF SKULL CHARACTERS IN FOUR SERIES OF EGYPTIAN SKULLS.
    ANN. EUGENICS, 6, 352-371.


BARNDORFF-NIELSEN, O. + SOBEL, MILTON 1966.   (18.6)                     311
    ON THE DISTRIBUTION OF THE NUMBER OF ADMISSIBLE POINTS IN A VECTOR RANDOM SAMPLE.
        (REPRINTED AS NO. 312)
    TEOR. VEROJATNOST. I PRIMENEN., 11, 283-305.   (MR34-6819)


BARNDORFF-NIELSEN, O. + SOBEL, MILTON 1966.   (18.6)                     312
    ON THE DISTRIBUTION OF THE NUMBER OF ADMISSIBLE POINTS IN A VECTOR RANDOM SAMPLE.
        (REPRINT OF NO. 311)
    THEORY PROB. APPL., 11, 249-269.   (MR34-6819)



BARNES, E. W. 1899.  (20.4)                                              313
    THE THEORY OF THE GAMMA FUNCTION.
    MESSENGER  MATH., 29, 64-128.


BARNES, R. B. + SILVERMAN, S. 1934.   (16.4/17.1)                        314
    BROWNIAN MOTION AS A NATURAL LIMIT TO ALL MEASURING PROCESSES.
    REV. MODERN PHYS., 6, 162-192.


BARNETT, V. D. 1962.   (10.2/T)                                          315
    LARGE SAMPLE TABLES OF PERCENTAGE POINTS FOR HARTLEY'S CORRECTION TO BARTLETT'S
    CRITERION FOR TESTING THE HOMOGENEITY OF A SET OF VARIANCES.
    BIOMETRIKA, 49, 487-494.   (MR28-689)


BARNETT, V. D. 1966.   (19.3)                                           316
    EVALUATION OF THE MAXIMUM-LIKELIHOOD ESTIMATOR WHERE THE LIKELIHOOD
        EQUATION HAS MULTIPLE ROOTS.
    BIOMETRIKA, 53, 151-165.


BARR, CAPT. B.R. + RIZVI, M. HASEEB 1966.   (6.7)                       317
    AN INTRODUCTION TO RANKING AND SELECTION PROCEDURES.
    J. AMER. STATIST. ASSOC., 61, 640-646.


BARRA, JEAN-RENE 1960.   (17.1/17.5/17.8)                               318
    EXTENSION DU THEOREME DE VON MISES A CERTAINES CLASSES DE LOIS DE
        PROBABILITE SUR $R^k$ ET PLUS GENERALEMENT SUR UN ESPACE DE RIESZ.
    C.R. ACAD. SCI. PARIS, 250, 52-54.   (MR22-1003)


BARRETT, F. DERMOT + SHEPARD, HERBERT A. 1953.   (1.2)                  319
    A BIBLIOGRAPHY OF CYBERNETICS.
    PROC. AMER. ACAD. ARTS SCI., 80, 204-222.   (MR14,P.887)

BARRETT, J. F. + LAMPARD, D. G. 1955. (16.4)                                  320
    AN EXPANSION FOR SOME SECOND-ORDER PROBABILITY DISTRIBUTIONS AND ITS
    APPLICATION TO NOISE PROBLEMS.
  IRE TRANS. INFORMATION THEORY, IT-1, 10-15.

BARRINGTON, AMY    SEE
  1530. ELDERTON, ETHEL M.; BARRINGTON, AMY; JONES, H. GERTRUDE;   (1913)
  LAMOTTE, EDITH M. M. DE G.; LASKI, H. J. + PEARSON, KARL

BARROW, DAVID F. + COHEN, A. CLIFFORD, JR. 1954.  (2.4/U)                     321
    ON SOME FUNCTIONS INVOLVING MILLS  RATIO.
  ANN. MATH. STATIST., 25, 405-408.  (MR15,P.807)

BARRY, ROBERT F. 1939.  (15.4)                                               322
    AN ANALYSIS OF SOME NEW STATISTICAL METHODS FOR SELECTING TEST ITEMS.
  J. EXPER. EDUC., 7, 221-228.

BARTELME, PHYLLIS    SEE
  730. BROWN, CLARENCE W.; BARTELME, PHYLLIS + COX, GERTRUDE M. (1933)

BARTEN, A. P. 1962.  (4.7/T)                                                 323
    NOTE ON UNBIASED ESTIMATION OF THE SQUARED MULTIPLE CORRELATION COEFFICIENT.
  STATISTICA NEERLANDICA, 16, 151-163.

BARTFAI, P. 1966.  (18.6/A)                                                  324
    LIMITING DISTRIBUTIONS ON THE CIRCUMFERENCE OF A CIRCLE AND IN COMPACT
    ABELIAN GROUPS.
        (IN GERMAN)
  STUDIA SCI. MATH. HUNGAR., 1, 71-86.

BARTHOLOMEW, D. J. 1959.  (5.1/A)                                            325
    A TEST OF HOMOGENEITY FOR ORDERED ALTERNATIVES.
  BIOMETRIKA, 46, 36-48.  (MR21-3067)

BARTLETT, MAURICE S. 1933.  (2.1/8.1)                                        326
    ON THE THEORY OF STATISTICAL REGRESSION.
  PROC. ROY. SOC. EDINBURGH, 53, 260-283.

BARTLETT, MAURICE S. 1934.  (10.2)                                           327
    THE PROBLEM IN STATISTICS OF TESTING SEVERAL VARIANCES.
  PROC. CAMBRIDGE PHILOS. SOC., 30, 164-169.

BARTLETT, MAURICE S. 1934.  (2.1/9.3/17.9/E)                                 328
    THE VECTOR REPRESENTATION OF A SAMPLE.
  PROC. CAMBRIDGE PHILOS. SOC., 30, 327-340.

BARTLETT, MAURICE S. 1935.  (15.2)                                          329
    THE STATISTICAL ESTIMATION OF G.
  BRITISH J. PSYCHOL., 26, 199-206.

BARTLETT, MAURICE S. 1935.  (16.5)                                          330
    SOME ASPECTS OF THE TIME-CORRELATION PROBLEM IN REGARD TO TESTS OF SIGNIFICANCE.
  J. ROY. STATIST. SOC., 98, 536-543.

BARTLETT, MAURICE S. 1935.  (15.2)                                          331
    ESTIMATION OF GENERAL ABILITY.
  NATURE, 135, 71.

BARTLETT, MAURICE S. 1937.  (4.8)                                           332
    NOTE ON THE DEVELOPMENT OF CORRELATIONS AMONG GENETIC COMPONENTS OF ABILITY.
  ANN. EUGENICS, 7, 299-302.

BARTLETT, MAURICE S. 1937.  (15.1)                                          333
    THE STATISTICAL CONCEPTION OF MENTAL FACTORS.
  BRITISH J. PSYCHOL., 28, 97-104.

BARTLETT, MAURICE S. 1938.  (15.2)                                          334
    METHODS OF ESTIMATING MENTAL FACTORS.
  NATURE, 141, 609-610.

BARTLETT, MAURICE S. 1938.  (8.4/11.2/12.2)                                 335
    FURTHER ASPECTS OF THE THEORY OF MULTIPLE REGRESSION.
  PROC. CAMBRIDGE PHILOS. SOC., 34, 33-40.

BARTLETT, MAURICE S. 1939.  (6.3)                                          336
     THE STANDARD ERRORS OF DISCRIMINANT FUNCTION COEFFICIENTS.
     J. ROY. STATIST. SOC. SUPP., 6, 169-173.  (MR1,P.248)


BARTLETT, MAURICE S. 1939.  (6.2/6.3/12.2)                                 337
     A NOTE ON TESTS OF SIGNIFICANCE IN MULTIVARIATE ANALYSIS.
     PROC. CAMBRIDGE PHILOS. SOC., 35, 180-185.


BARTLETT, MAURICE S. 1941.  (12.2)                                         338
     THE STATISTICAL SIGNIFICANCE OF CANONICAL CORRELATIONS.
     BIOMETRIKA, 32, 29-37.  (MR2,P.235)


BARTLETT, MAURICE S. 1947.  (12.2)                                         339
     THE GENERAL CANONICAL CORRELATION DISTRIBUTION.
     ANN. MATH. STATIST., 18, 1-17.  (MR8,P.474)


BARTLETT, MAURICE S. 1947.  (1.1)                                          340
     MULTIVARIATE ANALYSIS.
       (WITH DISCUSSION)
     J. ROY. STATIST. SOC. SUPP., 9, 176-197.  (MR9,P.453)


BARTLETT, MAURICE S. 1948.  (15.1)                                         341
     INTERNAL AND EXTERNAL FACTOR ANALYSIS.
     BRITISH J. PSYCHOL. STATIST. SECT., 1, 73-81.


BARTLETT, MAURICE S. 1948.  (8.7/16.2/E)                                   342
     A NOTE ON THE STATISTICAL ESTIMATION OF DEMAND AND SUPPLY RELATIONS
     FROM TIME SERIES.
     ECONOMETRICA, 16, 323-329.


BARTLETT, MAURICE S. 1949.  (17.7)                                         343
     FITTING A STRAIGHT LINE WHEN BOTH VARIABLES ARE SUBJECT TO ERROR.
     BIOMETRICS, 5, 207-212.  (MR11,P.190)


BARTLETT, MAURICE S. 1950.  (16.3)                                         344
     PERIODOGRAM ANALYSIS AND CONTINUOUS SPECTRA.
     BIOMETRIKA, 37, 1-16.  (MR12,P.35)


BARTLETT, MAURICE S. 1950.  (15.2)                                         345
     TESTS OF SIGNIFICANCE IN FACTOR ANALYSIS.
     BRITISH J. PSYCHOL. STATIST. SECT., 3, 77-85.


BARTLETT, MAURICE S. 1951.  (8.3/8.4/C)                                    346
     THE GOODNESS OF FIT OF A SINGLE HYPOTHETICAL DISCRIMINANT FUNCTION IN
     THE CASE OF SEVERAL GROUPS.
     ANN. EUGENICS, 16, 199-214.  (MR13,P.666)


BARTLETT, MAURICE S. 1951.  (6.2/20.1/20.3)                                347
     AN INVERSE MATRIX ADJUSTMENT ARISING IN DISCRIMINANT ANALYSIS.
     ANN. MATH. STATIST., 22, 107-111.  (MR12,P.639)


BARTLETT, MAURICE S. 1951.  (15.2)                                         348
     THE EFFECT OF STANDARDIZATION ON A $\chi^2$ APPROXIMATION IN FACTOR
     ANALYSIS. (WITH AN APPENDIX BY W. LEDERMAN.)
     BIOMETRIKA, 38, 337-344.  (MR14,P.66)


BARTLETT, MAURICE S. 1951.  (15.2)                                         349
     A FURTHER NOTE ON TESTS OF SIGNIFICANCE IN FACTOR ANALYSIS.
     BRITISH J. PSYCHOL. STATIST. SECT., 4, 1-2.


BARTLETT, MAURICE S. 1953.  (15.1/15.2)                                    350
     FACTOR ANALYSIS IN PSYCHOLOGY AS A STATISTICIAN SEES IT.
     NORDISK PSYKOL. MONOG. NR. 3 (UPPSALA SYMP. PSYCHOL. FACT. ANAL.), 23-34.


BARTLETT, MAURICE S. 1953.  (16.5)                                         351
     THE STATISTICAL APPROACH TO THE ANALYSIS OF TIME-SERIES.
     TRANS. IRE PROF. GROUP INFORMATION THEORY,PGIT-1, 81-101.


BARTLETT, MAURICE S. 1954.  (8.4/9.2/11.2)                                 352
     A NOTE ON THE MULTIPLYING FACTORS FOR VARIOUS $\chi^2$ APPROXIMATIONS.
     J. ROY. STATIST. SOC. SER. B, 16, 296-298.  (MR16,P.1039)

BARTLETT, MAURICE S. 1956.   (1.2)                                    353
     BIBLIOGRAPHY.
   STATISTICA, BOLOGNA, 16, 97-100.


BARTLETT, MAURICE S. 1957.   (8.7/9.2)                                354
     A NOTE ON TESTS OF SIGNIFICANCE FOR LINEAR FUNCTIONAL RELATIONSHIPS.
   BIOMETRIKA, 44, 268-269.


BARTLETT, MAURICE S. 1962.   (1.1)                                    355
     PROBABILITY AND STATISTICS IN THE PHYSICAL SCIENCES.
   BULL. INST. INTERNAT. STATIST., 39(3) 3-21.   (MR28-4596)


BARTLETT, MAURICE S. 1964.   (16.3/B)                                 356
     THE SPECTRAL ANALYSIS OF TWO-DIMENSIONAL POINT PROCESSES.
        (AMENDED BY NO. 357)
   BIOMETRIKA, 51, 299-311.


BARTLETT, MAURICE S. 1965.   (16.3/B)                                 357
     CORRIGENDA: THE SPECTRAL ANALYSIS OF TWO-DIMENSIONAL POINT PROCESSES.
        (AMENDMENT OF NO. 356)
   BIOMETRIKA, 52, 305.


BARTLETT, MAURICE S. 1965.   (1.1/1.2)                                358
     R.A. FISHER AND THE LAST FIFTY YEARS OF STATISTICAL METHODOLOGY.
   J. AMER. STATIST. ASSOC., 60, 395-409.


BARTLETT, MAURICE S. 1965.   (6.5/8.5/8.7/E)                          359
     MULTIVARIATE STATISTICS.
   THEOR. MATH. BIOL. (WATERMAN + MOROWITZ), 201-224.


BARTLETT, MAURICE S. + PLEASE, N.W. 1963.   (5.2/6.3/E)               360
     DISCRIMINATION IN THE CASE OF ZERO MEAN DIFFERENCES.
   BIOMETRIKA, 50, 17-21.   (MR27-6365)


BARTLETT, MAURICE S. + RAJALAKSHMAN, D.V. 1953.   (16.1/A)            361
     GOODNESS OF FIT TESTS FOR SIMULTANEOUS AUTOREGRESSIVE SERIES.
   J. ROY. STATIST. SOC. SER. B, 15, 107-124.   (MR15,P.333)


BARTLETT, MAURICE S.    SEE ALSO
   5966. WISHART, JOHN + BARTLETT, MAURICE S. (1932)

   5967. WISHART, JOHN + BARTLETT, MAURICE S. (1933)

   2692. IRWIN, J. O.; BARTLETT, MAURICE S.; COCHRAN, WILLIAM G. +   (1935)
   FIELLER, E. C.


BARTLETT, NEIL R. 1946.   (4.1/20.3)                                  362
     A PUNCHED-CARD TECHNIQUE FOR COMPUTING MEANS, STANDARD DEVIATIONS AND THE
     PRODUCT MOMENT CORRELATION COEFFICIENT AND FOR LISTING SCATTERGRAMS.
   SCIENCE, 104, 374-375.


BARTON, D. E. + CASLEY, D.J. 1958.   (8.1)                            363
     A QUICK ESTIMATE OF THE REGRESSION COEFFICIENT.
   BIOMETRIKA, 45, 431-435.


BARTON, D. E. + DAVID, F. N. 1960.   (8.8/17.7/E)                     364
     MODELS OF FUNCTIONAL RELATIONSHIP ILLUSTRATED ON ASTRONOMICAL DATA.
   BULL. INST. INTERNAT. STATIST., 37(3) 9-33.   (MR22-9271)


BARTON, D. E. + DAVID, F. N. 1962.   (18.6/BE)                        365
     RANDOMIZATION BASES FOR MULTIVARIATE TESTS. I. THE BIVARIATE CASE.
     RANDOMNESS OF N POINTS IN A PLANE.
   BULL. INST. INTERNAT. STATIST., 39(2) 455-467.   (MR29-1699)


BARTOO, J. B.    SEE
   5192. SRIVASTAVA, OM PRAKASH; HARKNESS, W. L. + BARTOO, J. B. (1964)


BASARINOV, A. E. 1965.  E(6.4)                                        366
     ASYMPTOTICALLY EXTREMAL PROCEDURES FOR THE PARAMETRIC RECOGNITION OF FORMS.
        (IN RUSSIAN)
   RADIOTEHNIKA, 10, 812-816.   (MR31-4646)

BASCH, ALFRED 1914.  (2.1/17.3)                                                          367
    UBER HYPERBELN BZW. HYPERBOLOIDE ALS PRAZISIONSCHARAKTERISTIKA
    EMPIRISCH BESTIMMTER LINEARER FUNKTIONEN.
    SITZUNGSBER. MATH. NATURWISS. CL. KAIS. AKAD. WISS. WIEN, 123(2A) 1659-1678.

BASCH, ALFRED 1928.  (17.3/C)                                                            368
    DIE FEHLERTENSOREN UND DAS FEHLERUBERTRAGUNGSGESETZ DER
    VEKTORALGEBRAISCHEN ELEMENTAROPERATIONEN.
    SITZUNGSBER. MATH. NATURWISS. CL. KAIS. AKAD. WISS. WIEN, 137(2A) 583-598.

BASMANN, R. L. 1957.  (17.7)                                                             369
    A GENERALIZED CLASSICAL METHOD OF LINEAR ESTIMATION OF COEFFICIENTS
    IN A STRUCTURAL EQUATION.
    ECONOMETRICA, 25, 77-83.   (MR19,P.74)

BASMANN, R. L. 1960.  (17.7)                                                             370
    AN EXPOSITORY NOTE ON ESTIMATION OF SIMULTANEOUS STRUCTURAL EQUATIONS.
    BIOMETRICS, 16, 464-480.   (MR22-6047)

BASMANN, R. L. 1960.  (17.7/A)                                                           371
    ON THE ASYMPTOTIC DISTRIBUTION OF GENERALISED LINEAR ESTIMATORS.
    ECONOMETRICA, 28, 97-107.   (MR22-1959)

BASMANN, R. L. 1960.  (8.8/16.2)                                                         372
    ON FINITE SAMPLE DISTRIBUTIONS OF GENERALIZED CLASSICAL LINEAR
    IDENTIFIABILITY TEST STATISTICS.
    J. AMER. STATIST. ASSOC., 55, 650-659.   (MR22-10062)

BASMANN, R. L. 1961.  (16.2)                                                             373
    A NOTE ON THE EXACT FINITE SAMPLE FREQUENCY FUNCTIONS OF GENERALIZED
    CLASSICAL LINEAR ESTIMATORS IN TWO LEADING OVER-IDENTIFIED CASES.
    J. AMER. STATIST. ASSOC., 56, 619-636.   (MR26-5974)

BASMANN, R. L. 1963.  (16.2)                                                             374
    A NOTE ON THE EXACT FINITE SAMPLE FREQUENCY FUNCTIONS OF GENERALIZED
    CLASSICAL LINEAR ESTIMATORS IN A LEADING THREE-EQUATION CASE.
    J. AMER. STATIST. ASSOC., 58, 161-171.   (MR27-6364)

BASMANN, R. L. 1963.  (16.2)                                                             375
    REMARKS CONCERNING THE APPLICATION OF EXACT FINITE SAMPLE DISTRIBUTION FUNCTIONS
    OF GCL ESTIMATORS IN ECONOMETRIC STATISTICAL INFERENCE.
        (AMENDED BY NO. 376)
    J. AMER. STATIST. ASSOC., 58, 943-976.   (MR28-700)

BASMANN, R. L. 1964.  (16.2)                                                             376
    CORRIGENDA:  REMARKS CONCERNING THE APPLICATION OF EXACT FINITE SAMPLE
    DISTRIBUTION FUNCTIONS OF GCL ESTIMATORS IN ECONOMETRIC STATISTICAL INFERENCE.
        (AMENDMENT OF NO. 375)
    J. AMER. STATIST. ASSOC., 59, 1296.   (MR28-700)

BASMANN, R. L. 1965.  (17.2)                                                             377
    A TCHEBYCHEV INEQUALITY FOR THE CONVERGENCE OF A GENERALIZED CLASSICAL
    LINEAR ESTIMATOR, SAMPLE SIZE BEING FIXED.
    ECONOMETRICA, 33, 608-618.   (MR31-6301)

BASMANN, R. L. 1965.  (16.2/16.5)                                                        378
    A NOTE ON THE STATISTICAL TESTABILITY OF 'EXPLICIT CAUSAL CHAINS'
    AGAINST THE CLASS OF 'INTERDEPENDENT' MODELS.
    J. AMER. STATIST. ASSOC., 60, 1080-1093.

BASS, BERNARD M. 1957.  (6.6/15.1/15.3)                                                  379
    ITERATIVE INVERSE FACTOR ANALYSIS--A RAPID METHOD FOR CLUSTERING PERSONS.
    PSYCHOMETRIKA, 22, 105-107.

BASTENAIRE, FRANCOIS 1955.  (17.7)                                                       380
    ESTIMATION D'UNE RELATION STRUCTURALE ET COMPARAISON DE DEUX METHODES
    DE MESURE D'UNE MEME GRANDEUR.
    REV. STATIST. APPL., 3(2) 83-99.

BASU, D. 1956.  (2.2)                                                                    381
    A NOTE ON THE MULTIVARIATE EXTENSION OF SOME THEOREMS RELATED TO THE
    UNIVARIATE NORMAL DISTRIBUTION.
    SANKHYA, 17, 221-224.   (MR19,P.471)

BASU, D. + LAHA, R. G. 1954.   (2.2)                                           382
      ON SOME CHARACTERISATIONS OF THE NORMAL DISTRIBUTION.
           (AMENDED BY NO. 383)
      SANKHYA, 13, 359-362.   (MR16,P.51)

BASU, D. + LAHA, R. G. 1954.   (2.2)                                           383
      ADDENDA:   ON SOME CHARACTERIZATIONS OF THE NORMAL DISTRIBUTION.
           (AMENDMENT OF NO. 382)
      SANKHYA, 14, 180.   (MR16,P.51)

BATEMAN, G. I. 1949.   (2.5/14.5)                                              384
      THE CHARACTERISTIC FUNCTION OF A WEIGHTED SUM OF NON-CENTRAL SQUARES
      OF NORMAL VARIATES SUBJECT TO S LINEAR RESTRAINTS.
      BIOMETRIKA, 36, 460-462.   (MR11,P.608)

BATEN, WILLIAM DOWELL 1931.   (17.3/B)                                         385
      CORRECTION FOR THE MOMENTS OF A FREQUENCY DISTRIBUTION IN TWO VARIABLES.
      ANN. MATH. STATIST., 2, 309-319.

BATEN, WILLIAM DOWELL 1932.   (17.3)                                           386
      FREQUENCY LAWS FOR THE SUM OF N VARIABLES WHICH ARE SUBJECT EACH TO
      GIVEN FREQUENCY LAWS.
      METRON, 10(3) 75-91.

BATEN, WILLIAM DOWELL 1941.   (8.9/E)                                          387
      HOW TO DETERMINE WHICH OF TWO VARIABLES IS BETTER FOR PREDICTING A
      THIRD VARIABLE.
      J. AMER. SOC. AGRON., 33, 695-699.

BATEN, WILLIAM DOWELL 1944.   (6.2/E)                                          388
      THE DISCRIMINANT FUNCTION APPLIED TO SPORE MEASUREMENTS.
      PAPERS MICHIGAN ACAD. SCI. ARTS LETT., 29, 3-7.

BATEN, WILLIAM DOWELL 1945.   E(5.2)                                           389
      THE USE OF DISCRIMINANT FUNCTION IN COMPARING JUDGES' SCORES CONCERNING POTATOES.
      J. AMER. STATIST. ASSOC., 40, 223-228.

BATEN, WILLIAM DOWELL + DEWITT, C.C. 1944.   (6.2/E)                           390
      USE OF THE DISCRIMINANT FUNCTION IN THE COMPARISON OF PROXIMATE COAL ANALYSES.
      J. INDUST. ENGRG. CHEM., 16, 32-34.

BATEN, WILLIAM DOWELL + HATCHER, HAZEL M. 1944.   E(5.2)                       391
      DISTINGUISHING METHOD DIFFERENCES BY USE OF DISCRIMINANT FUNCTIONS.
      J. EXPER. EDUC., 12, 184-186.

BATEN, WILLIAM DOWELL: TACK, P.I. + BAEDER, HELEN A. 1958.   (5.2/5.3/E)       392
      TESTING FOR DIFFERENCES BETWEEN METHODS OF PREPARING FISH BY USE OF A
      DISCRIMINANT FUNCTION.
      INDUST. QUAL. CONTROL, 14(7) 7-10.

BAUER, F. L. 1965.   (20.3)                                                    393
      ESTIMATION WITH WEIGHTED ROW COMBINATIONS FOR SOLVING LINEAR EQUATIONS
      AND LEAST SQUARES PROBLEMS.
      NUMER. MATH., 7, 338-352.   (MR32-3268)

BAUER, F. L. + HOUSEHOLDER, ALSTON S. 1960.   (20.1)                           404
      SOME INEQUALITIES INVOLVING THE EUCLIDEAN CONDITION OF A MATRIX.
      NUMER. MATH., 2, 308-311.   (MR22-6822)

BAUER, F. L.   SEE ALSO
      2613. HOUSEHOLDER, ALSTON S. + BAUER, F. L. (1960)

BAUER, RAINALD K. 1954.   (6.2/6.3)                                            394
      DISKRIMINANZANALYSE.
      ALLGEMEIN. STATIST. ARCH., 38, 205-216.

BAUER, RAINALD K. 1955.   (8.3)                                               395
      DIE LEXISSCHE DISPERSIONSTHEORIE IN IHREN BEZIEHUNGEN ZUR MODERNEN
      STATISTISCHEN METHODENLEHRE INSBESONDERE ZUR STREUUNGSANALYSE (ANALYSIS OF
      VARIANCE).
      MITT. MATH. STATIST., 7, 25-45.   (MR16,P.940)

BAUER, RAINALD K. 1955.   (4.5/4.6)                                                              396
    ZUR NICHTPARAMETRISCHEN ABLEITUNG DER STREUUNGEN DES MULTIPLEN UND
    DES PARTIELLEN KORRELATIONSKOEFFIZIENTEN, SOWIE DES MULTIPLEN
    REGRESSIONSKOEFFIZIENTEN IM FALLE DER NULLHYPOTHESE.
    MITT. MATH. STATIST., 7, 220-223.   (MR17,P.503)

BAUER, RAINALD K.    SEE ALSO
    257. BAITSCH, HELMUT + BAUER, RAINALD K. (1956)

BAUER, THOMAS 1957.   (5.1/20.3)                                                                 397
    THE PRACTICAL CALCULATION OF HOTELLING'S T.
    INDUST. QUAL. CONTROL, 14(1) 7-10.

BAUM, L. E. + KATZ, MELVIN 1965.   (17.1)                                                        398
    CONVERGENCE RATES IN THE LAW OF LARGE NUMBERS.
    TRANS. AMER. MATH. SOC., 120, 108-123.

BAUMERT, L. D. 1966.   (20.2)                                                                    399
    HADAMARD MATRICES OF ORDER 116 AND 232.
    BULL. AMER. MATH. SOC., 72, 237.

BAUMERT, L. D. + HALL, MARSHALL, JR. 1965.   (20.2)                                              400
    A NEW CONSTRUCTION FOR HADAMARD MATRICES.
    BULL. AMER. MATH. SOC., 71, 169-170.

BAUR, F. 1926.   (4.3/4.8/E)                                                                     401
    DIE VERWENDUNG DER KORRELATIONSMETHODE IN DER METEOROLOGIE.
    METEOROL. Z., 43, 386-388.

BAUR, F. 1928.   (4.6)                                                                           402
    PROBLEME DER MEHRFACHKORRELATION.
    Z. ANGEW. MATH. MECH., 8, 438-439.

BAUR, F. 1929.   (4.6/E)                                                                         403
    ZUR THEORIE DER LINEAREN MEHRFACHKORRELATION.
    Z. ANGEW. MATH. MECH., 9, 231-241.

ENTRY NO. 404 APPEARS BETWEEN ENTRY NOS. 393 & 394.

BAYLY, B. DE F. 1938.   (20.5/U)                                                                 405
    GAUSS' QUADRATIC FORMULA WITH TWELVE ORDINATES.
    BIOMETRIKA, 30, 193-194.

BAZU, D.  SEE  BASU, D.

BEALL, GEOFFREY 1945.   (6.2/20.3)                                                               406
    APPROXIMATE METHODS IN CALCULATING DISCRIMINANT FUNCTIONS.
    PSYCHOMETRIKA, 10, 205-217.

BEAN, LOUIS H. 1929.   (4.8)                                                                     407
    A SIMPLIFIED METHOD OF GRAPHIC CURVILINEAR CORRELATION.
    J. AMER. STATIST. ASSOC., 24, 386-397.

BEAN, LOUIS H. 1939.   (4.8/8.1/E)                                                               408
    THE USE OF THE SHORT-CUT GRAPHIC METHOD OF MULTIPLE CORRELATION: COMMENT.
    QUART. J. ECON., 54, 318-331.

BEATTIE, A. W. 1962.   (2.4)                                                                     409
    TRUNCATION IN TWO VARIATES TO MAXIMIZE A FUNCTION OF THE MEANS OF A
    NORMAL MULTIVARIATE DISTRIBUTION.
    AUSTRAL. J. STATIST., 4, 1-3.   (MR25-2665)

BEAUCHAMP, JOHN J. + CORNELL, RICHARD G. 1966.   (8.1/E)                                         410
    SIMULTANEOUS NONLINEAR ESTIMATION.
    TECHNOMETRICS, 8, 319-326.

BECHHOFER, ROBERT E. 1954.   (19.1)                                                              411
    A SINGLE-SAMPLE MULTIPLE DECISION PROCEDURE FOR RANKING MEANS OF
    NORMAL POPULATIONS WITH KNOWN VARIANCES.
    ANN. MATH. STATIST., 25, 16-39.   (MR15,P.638)

BECHHOFER, ROBERT E. 1955.   (19.1)                                                              412
    MULTIPLE DECISION PROCEDURES FOR RANKING MEANS.
    NAT. CONV. TRANS. AMER. SOC. QUAL. CONTROL, 9, 513-519.

BECHHOFER, ROBERT E. + SOBEL, MILTON 1954.  (19.1/U)                    413
    A SINGLE-SAMPLE MULTIPLE DECISION PROCEDURE FOR RANKING VARIANCES OF
    NORMAL POPULATIONS.
    ANN. MATH. STATIST., 25, 273-289.  (MR19,P.1205)

BECHHOFER, ROBERT E.; DUNNETT, CHARLES W. + SOBEL, MILTON 1954.  (19.1)                    414
    A TWO-SAMPLE MULTIPLE DECISION PROCEDURE FOR RANKING MEANS OF NORMAL
    POPULATIONS WITH A COMMON UNKNOWN VARIANCE.
    BIOMETRIKA, 41, 170-176.  (MR15,P.885)

BECHHOFER, ROBERT E.; ELMAGHRABY, SALAH + MORSE, NORMAN 1959.  (18.4/19.1)                    415
    A SINGLE-SAMPLE MULTIPLE-DECISION PROCEDURE FOR SELECTING THE
    MULTINOMIAL EVENT WHICH HAS THE HIGHEST PROBABILITY.
    ANN. MATH. STATIST., 30, 102-119.  (MR21-4515)

BECHTOLDT, H. P. 1961.  (15.1)                    416
    AN EMPIRICAL STUDY OF THE FACTOR ANALYSIS STABILITY HYPOTHESIS.
    PSYCHOMETRIKA, 26, 405-432.

BECK, ANATOLE + SCHWARTZ, JACOB T. 1957.  (16.4)                    417
    A VECTOR-VALUED RANDOM ERGODIC THEOREM.
    PROC. AMER. MATH. SOC., 8, 1049-1059.  (MR20-4624)

BECK, U. P. 1966.  (4.4/B)                    418
    EIN OPTISCHES ANALOGIEGERAT FUR DEN TETRACHORISCHEN KORRELATIONSKOEFFIZIENTEN.
    METRIKA, 10, 46-51.  (MR33-3387)

BECKMAN, F. S. + QUARLES, D.A., JR. 1956.  (8.1/20.3)                    419
    MULTIPLE REGRESSION AND CORRELATION ANALYSIS ON THE IBM TYPE 701 AND
    TYPE 704 ELECTRONIC DATA PROCESSING MACHINES.
    AMER. STATIST., $1_0$(1) 6-8.

BECKMANN, PETR 1962.  (7.4/18.1/B)                    420
    STATISTICAL DISTRIBUTION OF THE AMPLITUDE AND PHASE OF A MULTIPLY
    SCATTERED FIELD.
    J. RES. NAT. BUR. STANDARDS SECT. D, 66, 231-240.

BEDELL, B. J. 1950.  (15.4)                    421
    DETERMINATION OF THE OPTIMUM NUMBER OF ITEMS TO RETAIN IN A TEST
    MEASURING A SINGLE ABILITY.
    PSYCHOMETRIKA, 15, 419-430.

BEECH, DONALD G. 1953.  (4.1/8.8/E)                    422
    EXPERIENCES OF CORRELATION ANALYSIS.
    APPL. STATIST., 2, 73-85.

BEETON, MARY    SEE
    4249. PEARSON, KARL; BEETON, MARY + YULE, G. UDNY (1900)

BEHARI, VINOD    SEE
    1941. GHOSH, MAHINDRA NATH + BEHARI, VINOD (1965)

BEHNKEN, D. W. 1961.  (18.3)                    423
    SAMPLING MOMENTS OF MEANS FROM FINITE MULTIVARIATE POPULATIONS.
    ANN. MATH. STATIST., 32, 406-413.  (MR23-A4213)

BEHRENS, W. U. 1959.  (5.2/B)                    424
    BEITRAG ZUR DISKRIMINANZANALYSE.
    BIOM. Z., 1, 3-14.

BEIGHTLER, CHARLES S. + WILDE, DOUGLASS J. 1966.  (20.3)                    425
    DIAGONALIZATION OF QUADRATIC FORMS BY GAUSS ELIMINATION.
    MANAG. SCI. SER. A SCI., 12, 371-379.

BEJAR, JUAN A. 1956.  (17.8)                    426
    REGRESION EN MEDIANA Y LA PROGRAMACION LINEAL.
        (WITH SUMMARY IN ENGLISH)
    TRABAJOS ESTADIST., 7, 141-158.  (MR18,P.771)

BEJAR, JUAN A. 1957.  (17.8)                    427
    CALCULO PRACTICO DE LA REGRESION EN MEDIANA.
        (WITH SUMMARY IN ENGLISH)
    TRABAJOS ESTADIST., 8, 157-173.

BELL, C. B. 1962. (4.9/18.1)        **428**
    MUTUAL INFORMATION AND MAXIMAL CORRELATION AS MEASURES OF DEPENDENCE.
    ANN. MATH. STATIST., 33, 587-595. (MR26-5690)

BELLMAN, RICHARD E. 1954. (17.1)        **429**
    LIMIT THEOREMS FOR NON-COMMUTATIVE OPERATIONS. I.
    DUKE MATH. J., 21, 491-500. (MR15,P.969)

BELLMAN, RICHARD E. 1955. (17.3)        **430**
    A NOTE ON THE MEAN VALUE OF RANDOM DETERMINANTS.
    QUART. APPL. MATH., 13, 322-324. (MR17,P.274)

BELLMAN, RICHARD E. 1956. (20.4)        **431**
    A GENERALIZATION OF SOME INTEGRAL IDENTITIES DUE TO INGHAM AND SIEGEL.
    DUKE MATH. J., 23, 571-577. (MR18,P.468,P.1118)

BELLMAN, RICHARD E. 1957. (20.1)        **432**
    NOTES ON MATRIX THEORY. IX.
    AMER. MATH. MONTHLY, 64, 189-191. (MR19,P.6)

BENDAT, JULIUS S. 1956. (16.4)        **433**
    A GENERAL THEORY OF LINEAR PREDICTION AND FILTERING.
    J. SOC. INDUST. APPL. MATH., 4, 131-151. (MR18,P.342)

BENEDETTI, CARLO 1958. (4.1)        **434**
    IL COEFFICIENTE DI CORRELAZIONE DEL BRAVAIS COME FUNZIONE NON MOLTIPLICATIVA.
      (WITH SUMMARY IN GERMAN, REPRINT OF NO. 435)
    ATTI RIUN. SCI. SOC. ITAL. STATIST., 17, 97-101.

BENEDETTI, CARLO 1958. (4.1)        **435**
    IL COEFFICIENTE DI CORRELAZIONE DEL BRAVAIS COME FUNZIONE NON MOLTIPLICATIVA.
      (REPRINTED AS NO. 434)
    METRON, 19(1) 193-198.

BENINI, R. 1956. (1.2)        **436**
    BIBLIOGRAPHY.
    STATISTICA, BOLOGNA, 16, 87-92.

BEN-ISRAEL, ADI 1965. (20.3)        **437**
    A MODIFIED NEWTON-RAPHSON METHOD FOR THE SOLUTION OF SYSTEMS OF EQUATIONS.
    ISRAEL J. MATH., 3, 94-98.

BEN-ISRAEL, ADI 1966. (20.3)        **438**
    A NEWTON-RAPHSON METHOD FOR THE SOLUTION OF SYSTEMS OF EQUATIONS.
    J. MATH. ANAL. APPL., 15, 243-252.

BEN-ISRAEL, ADI + CHARNES, ABRAHAM 1963. (20.1)        **439**
    CONTRIBUTIONS TO THE THEORY OF GENERALIZED INVERSES.
    J. SOC. INDUST. APPL. MATH., 11, 667-699.

BENJAMIN, KURT 1945. (20.3)        **440**
    AN I.B.M. TECHNIQUE FOR THE COMPUTATION OF $\sum x^2$ AND $\sum xy$.
    PSYCHOMETRIKA, 10, 61-67. (MR6,P.220,P.334)

BENNETT, B. M. 1951. (5.2)        **441**
    NOTE ON A SOLUTION OF THE GENERALIZED BEHRENS-FISHER PROBLEM.
    ANN. INST. STATIST. MATH. TOKYO, 2, 87-90. (MR12,P.842)

BENNETT, B. M. 1954. (3.1/BC)        **442**
    SOME FURTHER EXTENSIONS OF FIELLER'S THEOREM.
    ANN. INST. STATIST. MATH. TOKYO, 5, 103-106. (MR16,P.54)

BENNETT, B. M. 1955. (18.1/B)        **443**
    ON THE JOINT DISTRIBUTION OF THE MEAN AND STANDARD DEVIATION.
    ANN. INST. STATIST. MATH. TOKYO, 7, 63-66. (MR17,P.639)

BENNETT, B. M. 1955. (1.1/5.3/7.1)        **444**
    CERTAIN MULTIVARIATE DISTRIBUTIONS IN THE PRESENCE OF INTRACLASS CORRELATION.
    J. INDIAN SOC. AGRIC. STATIST., 7, 70-72. (MR19,P.73)

BENNETT, B. M. 1955. (7.2)        **445**
    ON THE CUMULANTS OF THE LOGARITHMIC GENERALIZED VARIANCE AND VARIANCE RATIO.
    SKAND. AKTUARIETIDSKR., 38, 17-21. (MR17,P.638)

BENNETT, B. M. 1956.  (8.1/BC)                                    446
    ON CONFIDENCE LIMITS FOR THE RATIO OF REGRESSION COEFFICIENTS.
    ANN. INST. STATIST. MATH. TOKYO, 8, 41-43.  (MR19,P.74)

BENNETT, B. M. 1957.  (8.3)                                       447
    TESTS FOR LINEARITY OF REGRESSION INVOLVING CORRELATED OBSERVATIONS.
    ANN. INST. STATIST. MATH. TOKYO, 8, 193-195.  (MR19,P.694)

BENNETT, B. M. 1957.  (3.3/17.5/C)                                448
    ON THE PERFORMANCE CHARACTERISTIC OF CERTAIN METHODS OF DETERMINING
    CONFIDENCE LIMITS.
    SANKHYA, 18, 1-12.  (MR20-388)

BENNETT, B. M. 1959.  (3.2/14.3/CE)                               449
    ON A MULTIVARIATE VERSION OF FIELLER'S THEOREM.
    J. ROY. STATIST. SOC. SER. B, 21, 59-62.  (MR21-6659)

BENNETT, B. M. 1961.  (3.1/8.1/14.3/CE)                           450
    CONFIDENCE LIMITS FOR MULTIVARIATE RATIOS.
    J. ROY. STATIST. SOC. SER. B, 23, 108-112.  (MR23-A1428)

BENNETT, B. M. 1961.  (18.3)                                      451
    ON A CERTAIN MULTIVARIATE NON-NORMAL DISTRIBUTION.
    PROC. CAMBRIDGE PHILOS. SOC., 57, 434-436.  (MR22-12637)

BENNETT, B. M. 1962.  (17.8)                                      452
    ON MULTIVARIATE SIGN TESTS.
    J. ROY. STATIST. SOC. SER. B, 24, 159-161.  (MR25-1610)

BENNETT, B. M. 1963.  (3.1)                                       453
    ON COMBINING ESTIMATES OF A RATIO OF MEANS.
    J. ROY. STATIST. SOC. SER. B, 25, 201-205.  (MR29-5329)

BENNETT, B. M. 1963.  (8.3/C)                                     454
    ON TESTS CONCERNING LINES AND PLANES IN HYPERSPACE.
    METRIKA, 7, 41-46.

BENNETT, B. M. 1964.  (17.8/B)                                    455
    A BIVARIATE SIGNED RANK TEST.
    J. ROY. STATIST. SOC. SER. B, 26, 457-461.  (MR30-673)

BENNETT, B. M. 1964.  (3.3)                                       456
    A NOTE ON COMBINING CORRELATED ESTIMATES OF A RATIO OF MULTIVARIATE MEANS.
    TECHNOMETRICS, 6, 463-467.

BENNETT, B. M. 1965.  (17.8)                                      457
    ON MULTIVARIATE SIGNED RANK TESTS.
    ANN. INST. STATIST. MATH. TOKYO, 17, 55-61.  (MR31-1741)

BENNETT, B. M. 1965.  (17.8)                                      458
    CONFIDENCE LIMITS FOR A RATIO USING WILCOXON'S SIGNED RANK TEST.
    BIOMETRICS, 21, 231-234.  (MR31-5292)

BENNETT, B. M. 1965.  (17.8/B)                                    459
    NOTE ON A  $\chi^2$-APPROXIMATION FOR THE MULTIVARIATE SIGN TEST.
    J. ROY. STATIST. SOC. SER. B, 27, 82-85.  (MR32-4786)

BENNETT, B. M. 1966.  (4.8/8.5/8.6/C)                             460
    MULTIVARIATE ANALYSIS OF DISPERSION IN THE PRESENCE OF INTRACLASS CORRELATION.
    METRIKA, 10, 1-5.  (MR33-5043)

BENNETT, B. M. 1966.  (17.8/17.9/BC)                              461
    CONFIDENCE LIMITS FOR A RATIO OF BIVARIATE MEANS.
    METRIKA, 10, 52-54.  (MR33-3408)

BENNETT, B. M. 1966.  (8.3/18.2)                                  462
    MULTIPLE REGRESSION ANALYSIS OF BINARY AND MULTINOMIAL VARIATES.
    SANKHYA SER. A, 28, 301-304.

BENNETT, B. M. 1966.  (18.1)                                      463
    ON ALTERNATIVES TO THE MULTINORMAL DISTRIBUTION.
    TRABAJOS ESTADIST., 17, 45-51.

BENNETT, G. W. + CORNISH, E. A. 1964.  (14.2)                                    464
    A COMPARISON OF THE SIMULTANEOUS FIDUCIAL DISTRIBUTIONS DERIVED FROM
    THE MULTIVARIATE NORMAL DISTRIBUTION.
        (WITH DISCUSSION)
    BULL. INST. INTERNAT. STATIST., 40, 902-939.  (MR31-4110)

BENNETT, JOSEPH F. 1956.  (15.5)                                                 465
    DETERMINATION OF THE NUMBER OF INDEPENDENT PARAMETERS OF A SCORE
    MATRIX FROM THE EXAMINATION OF RANK ORDERS.
    PSYCHOMETRIKA, 21, 383-393.  (MR18,P.427)

BENNETT, W. R. 1953.  (4.1)                                                      466
    THE CORRELATOGRAPH--A MACHINE FOR CONTINUOUS DISPLAY OF SHORT TERM CORRELATION.
    BELL SYSTEM TECH. J., 32, 1173-1185.  (MR15,P.652)

BENTZEL, R. + WOLD, HERMAN O. A. 1946.  (16.2)                                   467
    ON STATISTICAL DEMAND ANALYSIS FROM THE VIEWPOINT OF SIMULTANEOUS EQUATIONS.
    SKAND. AKTUARIETIDSKR., 29, 95-114.  (MR8,P.216)

BENZECRI, J. P. 1964.  (15.2/15.5)                                               468
    SUR L'ANALYSE FACTORIELLE DES PROXIMITES.
    PUBL. INST. STATIST. UNIV. PARIS, 13, 235-281.

BENZECRI, J. P. 1966.  (15.1/15.5)                                               469
    ANALYSE STATISTIQUE ET MODELES PROBABILISTES EN PSYCHOLOGIE.
    REV. INST. INTERNAT. STATIST., 34, 139-155.

BERG, W. F. 1945.  (17.3/E)                                                      470
    AGGREGATES IN ONE- AND TWO-DIMENSIONAL RANDOM DISTRIBUTIONS. (DEVELOPABILITY OF
    SILVER SPECKS OF KNOWN DIMENSIONS AND THE SIZE OF PHOTOGRAPHIC SENSITIVITY SPECKS.)
    PHILOS. MAG. SER. 7, 36, 337-346.  (MR7,P.310)

BERGE, P. O. 1938.  (17.2/B)                                                     471
    A NOTE ON A FORM OF TCHEBYCHEFF'S THEOREM FOR TWO VARIABLES.
    BIOMETRIKA, 29, 405-406.

BERGER, AGNES 1961.  (4.4)                                                       472
    ON COMPARING INTENSITIES OF ASSOCIATION BETWEEN TWO BINARY
    CHARACTERISTICS IN TWO DIFFERENT POPULATIONS.
    J. AMER. STATIST. ASSOC., 56, 889-908.

BERGSTROM, A. R. 1962.  (8.1/8.2/16.5/E)                                         473
    THE EXACT SAMPLING DISTRIBUTIONS OF LEAST SQUARES AND MAXIMUM
    LIKELIHOOD ESTIMATORS OF THE MARGINAL PROPENSITY TO CONSUME.
    ECONOMETRICA, 30, 480-490.  (MR27-5319)

BERGSTROM, HARALD 1945.  (17.1)                                                 474
    ON THE CENTRAL LIMIT THEOREM IN THE SPACE $R_k$, K>1.
    SKAND. AKTUARIETIDSKR., 28, 106-127.  (MR7,P.459)

BERGSTROM, SVERKER 1918.  (2.3)                                                 475
    SUR LES MOMENTS DE LA FONCTION DE CORRELATION NORMALE DE N VARIABLES
    BIOMETRIKA, 12, 177-183.

BERKSON, JOSEPH 1933.  (4.1/4.3)                                                476
    THE COEFFICIENT OF CORRELATION.
    SCIENCE NEW SER., 77, 259.

BERKSON, JOSEPH 1950.  (8.1/17.7)                                               477
    ARE THERE TWO REGRESSIONS?
    J. AMER. STATIST. ASSOC., 45, 164-180.

BERMAN, SIMEON M. 1962.  (14.1/A)                                               478
    CONVERGENCE TO BIVARIATE LIMITING EXTREME VALUE DISTRIBUTIONS.
    ANN. INST. STATIST. MATH. TOKYO, 13, 217-223.  (MR28-1680)

BERMAN, SIMEON M. 1962.  (14.1/A)                                               479
    A LAW OF LARGE NUMBERS FOR THE MAXIMUM IN A STATIONARY GAUSSIAN SEQUENCE.
    ANN. MATH. STATIST., 33, 93-97.

BERMAN, SIMEON M. 1962.  (17.1/17.3)                                            480
    EQUALLY CORRELATED RANDOM VARIABLES.
    SANKHYA SER. A, 24, 155-156.  (MR26-3095)

BERNREUTER, ROBERT G. 1933.  (15.5)                                          481
    THEORY AND CONSTRUCTION PERSONALITY INVENTORY.
    J. SOC. PSYCHOL., 4, 387-405.

BERNREUTER, ROBERT G. 1935.  (15.5)                                          482
    CHANCE AND PERSONALITY INVENTORY SCORES.
    J. EDUC. PSYCHOL., 26, 279-283.

BERNSTEIN, FELIX 1914.  (2.3/8)                                              483
    BERECHNUNG DER KORRELATION ZWISCHEN 2 ARGUMENTEN, FUR DIE NUR DIE
    HAUFIGKEITSKURVE IHRES PRODUKTES GEGEBEN IST.
    NACHR. GES. WISS. GOTTING. MATH. PHYS. KL., 324-333.

BERNSTEIN, FELIX 1932.  (17.3/20.4)                                          484
    DIE MITTLEREN FEHLERQUADRATE UND KORRELATIONEN DER POTENZMOMENTE UND
    IHRE ANWENDUNG AUF FUNKTIONEN  DER POTENZMOMENTE.
    METRON, 10(3) 3-34.

BERNSTEIN, FELIX 1937.  (4.1/8.1)                                            485
    REGRESSION AND CORRELATION EVALUATED BY A METHOD OF PARTIAL SUMS.
    ANN. MATH. STATIST., 8, 77-89.

BERNSTEIN, R. I.    SEE
    3835. MILLER, KENNETH S. + BERNSTEIN, R. I. (1957)

    3837. MILLER, KENNETH S.; BERNSTEIN, R. I. + BLUMENSON, L. E. (1958)

    3838. MILLER, KENNETH S.; BERNSTEIN, R. I. + BLUMENSON, L. E. (1963)

BERNSTEIN, SERGEI N. 1926.  (17.1)                                          486
    SUR LES SOMMES DE QUANTITES DEPENDANTES.
    IZV. AKAD. NAUK SER. 6, 20, 1459-1478.

BERNSTEIN, SERGEI N. 1926.  (17.1)                                          487
    SUR L'EXTENSION DU THEOREME LIMITE DU CALCUL DES PROBABILITES AUX
    SOMMES DE QUANTITES DEPENDANTES.
    MATH. ANN., 97, 1-59.

BERNSTEIN, SERGEI N. 1927.  (4.1/20.5)                                      488
    SUR L'APPLICATION D'UN PRINCIPE GEOMETRIQUE A LA THEORIE DES CORRELATIONS.
        (IN RUSSIAN)
    IN MEM. N. I. LOBATSCHEVSKII, 2, 137-150.

BERNSTEIN, SERGEI N. 1928.  (18.1)                                          489
    SUR LES SOMMES DE QUANTITES DEPENDANTES.
    C. R. ACAD. SCI. URSS SER. A, 55-60.

BERNSTEIN, SERGEI N. 1928.  (4.1/20.5)                                      490
    FONDEMENTS GEOMETRIQUES DE LA THEORIE DES CORRELATIONS.
    METRON, 7(2) 3-27.

BERNSTEIN, SERGEI N. 1928.  (4.9)                                           491
    BEGRIFF DER KORRELATION ZWISCHEN STATISTISCHEN GROSSEN.
        (IN UKRAINIAN)
    VISNIK STATIST. UKRAINI, 1, 111-113.

BERNSTEIN, SERGEI N. 1933.  (17.1)                                          492
    SUR LES LIAISONS ENTRE LES GRANDEURS ALEATOIRES.
    PROC. INTERNAT. CONG. MATH., 1932(1) 288-309.

BERNSTEIN, SERGEI N. 1934.  (16.1)                                          493
    PRINCIPES DE LA THEORIE DES EQUATIONS DIFFERENTIELLES STOCHASTIQUES.  I.
    TRUDY MAT. INST. STEKLOVA, 5, 95-124.

BERNSTEIN, SERGEI N. 1944.  (17.1/17.3)                                     494
    AN EXTENSION OF THE DISTRIBUTION THEOREM OF PROBABILITY THEORY TO THE
    SUM OF DEPENDENT VARIABLES.
        (IN RUSSIAN)
    USPEHI MAT. NAUK, 10, 65-114.

BERNYER, G. 1943.  (15.2)                                            495
    L'ESTIMATION DES FACTEURS PSYCHOLOGIQUES PAR LA REGRESSION.
    ANNEE PSYCHOL., 43, 299-322.

BERNYER, G. 1957.  (15.1)                                            496
    PSYCHOLOGICAL FACTORS: THEIR NUMBER, NATURE, AND IDENTIFICATION.
    BRITISH J. STATIST. PSYCHOL., 10, 17-27.

BERRY, ARTHUR 1898.  (20.3/20.5)                                     497
    ON THE EVALUATION OF A CERTAIN DETERMINANT WHICH OCCURS IN THE MATHEMATICAL
    THEORY OF STATISTICS AND IN THAT OF ELLIPTIC GEOMETRY OF ANY NUMBER OF
    DIMENSIONS.
    PROC. CAMBRIDGE PHILOS. SOC., 10, 2-10.

BERRY, BRIAN J. L. 1961.  (6.6/15.1/15.5/E)                          498
    A METHOD FOR DERIVING MULTI-FACTOR UNIFORM REGIONS.
    PRZEGLAD GEOGRAFICZNY, 33, 263-282.

BERRY, CLIFFORD E. 1945.  (20.3)                                     499
    A CRITERION OF CONVERGENCE FOR THE CLASSICAL ITERATIVE METHOD OF
    SOLVING LINEAR SIMULTANEOUS EQUATIONS.
    ANN. MATH. STATIST., 16, 398-400.  (MR7,P.338)

BERSOT, HENRI 1919.  (4.1/4.8/16.5/BE)                               500
    VARIABILITE ET CORRELATIONS ORGANIQUES. NOUVELLE ETUDE DU REFLEXE PLANTAIRE.
    SCHWEIZ. ARCH. NEUROL. PSYCHIATRIE, 5, 305-324.

BERTRANDIAS, J. P. 1963.  (16.4)                                     501
    DECOMPOSITIONS OF FUNCTIONS POSSESSING AN AUTO-CORRELATION INTO A SUM
    OF THREE FUNCTIONS.
    C.R. ACAD. SCI. PARIS, 256, 1659-1662.

BESICOVITCH, A. S. 1961.  (17.3/19.1)                                502
    ON DIAGONAL VALUES OF PROBABILITY VECTORS OF INFINITELY MANY COMPONENTS.
    PROC. CAMBRIDGE PHILOS. SOC., 57, 759-766.  (MR23-A4154)

BEUTLER, FREDERICK J. 1963.  (16.4)                                  503
    MULTIVARIATE WIDE-SENSE MARKOV PROCESSES AND PREDICTION THEORY.
    ANN. MATH. STATIST., 34, 424-438.  (MR26-7033)

BHAPKAR, V. P. 1959.  (9.1/C)                                        504
    A NOTE ON MULTIPLE INDEPENDENCE UNDER MULTIVARIATE NORMAL LINEAR MODELS.
    ANN. MATH. STATIST., 30, 1248-1251.  (MR22-1024)

BHAPKAR, V. P. 1960.  (8.5/C)                                        505
    CONFIDENCE BOUNDS CONNECTED WITH ANOVA AND MANOVA FOR BALANCED AND
    PARTIALLY BALANCED INCOMPLETE BLOCK DESIGNS.
    ANN. MATH. STATIST., 31, 741-748.  (MR27-2058)

BHAPKAR, V. P. 1966.  (17.8)                                         506
    SOME NONPARAMETRIC TESTS FOR THE MULTIVARIATE SEVERAL SAMPLE LOCATION PROBLEM.
    MULTIVARIATE ANAL. PROC. INTERNAT. SYMP. DAYTON (KRISHNAIAH), 29-41.  (MR35-1147)

BHAPKAR, V. P.   SEE ALSO
    4682. ROY, S.N. + BHAPKAR, V. P. (1960)

BHARGAVA, R. P. 1946.  (4.3/14.3/C)                                  507
    TEST OF SIGNIFICANCE FOR INTRA-CLASS CORRELATION WHEN FAMILY SIZES ARE
      NOT EQUAL.
    SANKHYA, 7, 435-438.  (MR8,P.161)

BHARGAVA, R. P. 1966.  (3.2/4.3/BT)                                  508
    ESTIMATION OF CORRELATION COEFFICIENT WHEN NO SIMULTANEOUS
    OBSERVATIONS ON THE VARIABLE ARE AVAILABLE.
    SANKHYA SER. A, 28, 1-14.

BHAT, B. R. 1959.  (2.5/8.8)                                         509
    ON THE DISTRIBUTION OF VARIOUS SUMS OF SQUARES IN AN ANALYSIS OF
    VARIANCE TABLE FOR DIFFERENT CLASSIFICATIONS WITH CORRELATED AND
    NON-HOMOGENEOUS ERRORS.
    J. ROY. STATIST. SOC. SER. B, 21, 114-119.

BHAT, B. R. 1962.  (2.5)                                                510
    ON THE DISTRIBUTION OF CERTAIN QUADRATIC FORMS IN NORMAL VARIATES.
    J. ROY. STATIST. SOC. SER. B, 24, 148-151.


BHAT, B. R. + KULKARNI, N. V. 1966.  (18.2)                             511
    ON EFFICIENT MULTINOMIAL ESTIMATION.
    J. ROY. STATIST. SOC. SER. B, 28, 45-52.


BHAT, B. R. + KULKARNI, S. R. 1966.  (18.2)                             512
    LAMP TESTS OF LINEAR AND LOGLINEAR HYPOTHESES IN MULTINOMIAL EXPERIMENTS.
        (AMENDED BY NO. 513)
    J. AMER. STATIST. ASSOC., 61, 236-245.


BHAT, B. R. + KULKARNI, S. R. 1966.  (18.2)                             513
    CORRIGENDA:  LAMP TESTS OF LINEAR AND LOGLINEAR HYPOTHESES IN MULTINOMIAL EXPERIMENTS.
        (AMENDMENT OF NO. 512)
    J. AMER. STATIST. ASSOC., 61, 1246.


BHAT, B. R. + NAGNUR, B. N. 1965.  (19.2/A)                             514
    LOCALLY ASYMPTOTICALLY MOST STRINGENT TESTS AND LAGRANGIAN MULTIPLIER
    TESTS OF LINEAR HYPOTHESES.
    BIOMETRIKA, 52, 459-468.


BHATT, N. M. + DAVE, P. H. 1964.  (2.1/14.2)                            515
    A NOTE ON THE CORRELATION BETWEEN POLYNOMIAL TRANSFORMATIONS OF NORMAL VARIATES.
    J. INDIAN STATIST. ASSOC., 2, 177-181.  (MR30-5393)


BHATT, N. M. + DAVE, P. H. 1965.  (4.1)                                 516
    CHANGE IN NORMAL CORRELATION DUE TO EXPONENTIAL TRANSFORMATIONS OF STANDARD
    NORMAL VARIATES.
    J. INDIAN STATIST. ASSOC., 3, 46-54.


BHATT, N. M. + DAVE, P. H. 1965.  (20.2)                                517
    THE-RECIPROCAL OF THE VARIANCE MATRIX OF PARTITION VALUES.
    J. INDIAN STATIST. ASSOC., 3, 170-172.


BHATTACHARYA,N.    SEE
    2724. IYENGAR, N. SREENIVASA + BHATTACHARYA,N. (1965)


BHATTACHARYA, P.K. 1956.  (19.1)                                        518
    COMPARISON OF THE MEANS OF K NORMAL POPULATIONS.
    CALCUTTA STATIST. ASSOC. BULL., 7, 1-16.  (MR19,P.333)


BHATTACHARYA, P.K. 1958.  (19.1/A)                                      519
    ON THE LARGE SAMPLE BEHAVIOUR OF A MULTIPLE-DECISION PROCEDURE.
    CALCUTTA STATIST. ASSOC. BULL., 8, 43-47.


BHATTACHARYA, P.K. 1962.  (8.1)                                         520
    SOME PROPERTIES OF THE LEAST SQUARES ESTIMATOR IN REGRESSION ANALYSIS
    WHEN THE PREDICTOR VARIABLES ARE STOCHASTIC.
    ANN. MATH. STATIST., 33, 1365-1374.  (MR25-4600)


BHATTACHARYA, P.K. 1963.  (17.6/17.8/AB)                                521
    ON AN ANALOG OF REGRESSION ANALYSIS.
    ANN. MATH. STATIST., 34, 1459-1473.  (MR28-672)


BHATTACHARYA, P.K. 1966.  (3.1)                                         522
    ESTIMATING THE MEAN OF A MULTIVARIATE NORMAL POPULATION WITH GENERAL
    QUADRATIC LOSS FUNCTION.
    ANN. MATH. STATIST., 37, 1819-1824.  (MR34-911)


BHATTACHARYA, P.K.    SEE ALSO
    1218. DAS GUPTA, SOMESH + BHATTACHARYA, P.K. (1964)

    4141. PARTHASARATHY,K.R. + BHATTACHARYA, P.K. (1961)


BHATTACHARYA, SAMIR KUMAR + HOLLA, MAHABALESHWARA 1963.  (18.1)         523
    BIVARIATE LIFE-TESTING MODELS FOR TWO COMPONENT SYSTEMS.
    ANN. INST. STATIST. MATH. TOKYO, 15, 37-43.  (MR30-2615)


BHATTACHARYA, SAMIR KUMAR    SEE ALSO
    2515. HOLLA, MAHABALESHWARA + BHATTACHARYA, SAMIR KUMAR (1965)

BHATTACHARYYA, A. 1943.  (6.5/18.4)                                              524
    ON A MEASURE OF DIVERGENCE BETWEEN TWO STATISTICAL POPULATIONS
    DEFINED BY THEIR PROBABILITY DISTRIBUTIONS.
    BULL. CALCUTTA MATH. SOC., 35, 99-109.  (MR6,P.7)

BHATTACHARYYA, A. 1944.  (2.2/B)                                                 525
    ON SOME SETS OF SUFFICIENT CONDITIONS LEADING TO THE NORMAL BIVARIATE
    DISTRIBUTION.
    SANKHYA, 6, 399-406.  (MR6,P.8)

BHATTACHARYYA, A. 1945.  (2.5/14.5)                                              526
    A NOTE ON THE DISTRIBUTION OF THE SUM OF CHI-SQUARES.
    SANKHYA, 7, 27-28.  (MR7,P.131)

BHATTACHARYYA, A. 1946.  (6.5/18.2/18.3/A)                                       527
    ON A MEASURE OF DIVERGENCE BETWEEN TWO MULTINOMIAL POPULATIONS.
    SANKHYA, 7, 401-406.  (MR8,P.282)

BHATTACHARYYA, A. 1950.  (19.3)                                                  528
    UNBIASED STATISTICS WITH MINIMUM VARIANCE.
    PROC. ROY. SOC. EDINBURGH SECT. A, 63, 69-77.  (MR12,P.36)

BHATTACHARYYA, A. 1951.  (17.8/C)                                                529
    THE PROBLEM OF REGRESSION IN A STATISTICAL POPULATION ADMITTING
    LOCATION PARAMETERS.
    BULL. INST. INTERNAT. STATIST., 33(2) 29-54.  (MR16,P.1040)

BHATTACHARYYA, A. 1952.  (4.3/4.7/8.3)                                           530
    ON THE USES OF THE T-DISTRIBUTION IN MULTIVARIATE ANALYSIS.
    SANKHYA, 12, 89-104.  (MR15,P.451)

BHATTACHARYYA, B. C. 1941.  (5.3)                                                531
    ON ALTERNATIVE METHOD OF THE DISTRIBUTION OF MAHALANOBIS'S $D^2$-STATISTIC.
    BULL. CALCUTTA MATH. SOC., 33, 87-92.  (MR4,P.23)

BHATTACHARYYA, D. P. + NARAYAN, RAM DEVA 1941.  (5.3)                            532
    MOMENTS OF THE $D^2$-STATISTIC FOR POPULATIONS WITH UNEQUAL DISPERSIONS
    SANKHYA, 5, 401-412.  (MR4,P.105)

BHATTACHARYYA, G. K. 1966.  (5.1/17.8/AB)                                        533
    A NOTE ON THE ASYMPTOTIC EFFICIENCY OF BENNETT'S BIVARIATE SIGN TEST.
    J. ROY. STATIST. SOC. SER. B, 28, 146-149.  (MR33-5046)

BHATTACHARYYA, M. N. 1954.  (2.4/3.2)                                            534
    ESTIMATION FROM CENSORED BIVARIATE SAMPLES.
    J. INDIAN SOC. AGRIC. STATIST., 6, 83-92.  (MR17,P.639)

BHATTACHARYYA, P. K.  SEE  BHATTACHARYA, P.K.


BHUCHONGKUL, S. 1964.  (17.8/B)                                                  535
    A CLASS OF NONPARAMETRIC TESTS FOR INDEPENDENCE IN BIVARIATE POPULATIONS.
    ANN. MATH. STATIST., 35, 138-149.  (MR28-1709)

BICKEL, PETER J. 1964.  (3.1/17.8)                                              536
    ON SOME ALTERNATIVE ESTIMATES FOR SHIFT IN THE P-VARIATE ONE SAMPLE PROBLEM.
    ANN. MATH. STATIST., 35, 1079-1090.  (MR29-2904)

BICKEL, PETER J. 1965.  (5.1/17.8/A)                                            537
    ON SOME ASYMPTOTICALLY NON-PARAMETRIC COMPETITORS OF HOTELLING'S $T^2$.
        (AMENDED BY NO. 538)
    ANN. MATH. STATIST., 36, 160-173.  (MR31-5281)

BICKEL, PETER J. 1965.  (5.1/17.8/A)                                            538
    CORRECTIONS TO:  ON SOME ASYMPTOTICALLY NON-PARAMETRIC COMPETITORS OF
    HOTELLING'S $T^2$.
        (AMENDMENT OF NO. 537)
    ANN. MATH. STATIST., 36, 1583.  (MR31-5281)

BIDDULPH, R. 1954.  E(16.3)                                                     539
    SHORT-TERM AUTOCORRELATION ANALYSIS AND CORRELATOGRAMS OF SPOKEN DIGITS.
    J. ACOUST. SOC. AMER., 26, 539-541.

BIEHL, KATHERINE    SEE
    1895. GEHLKE, C. E. + BIEHL, KATHERINE (1934)

BIELENSTEIN, U. M.    SEE
    3567. MC GREGOR, J.R. + BIELENSTEIN, U. M. (1965)

BIENAYME, I.-J. 1852.    (2.1/17.3)                               540
    SUR LA PROBABILITE DES ERREURS D'APRES LA METHODE DES MOINDRES CARRES.
    J. MATH. PURES APPL. SER. 1, 17, 33-78.

BIGARELLA, J. J.    SEE
    3814. MICHENER, C. D.; LANGE, R. B.; BIGARELLA, J. J. +    (1958)
    SALAMUNI, R.

BIKELIS, A. 1964.    (17.1)                                       541
    A SHARPENING OF THE REMAINDER TERM IN A HIGHER-DIMENSIONAL CENTRAL
    LIMIT THEOREM.
        (IN RUSSIAN, WITH SUMMARY IN ENGLISH AND LITHUANIAN)
    LITOVSK. MAT. SB., 4, 153-158.    (MR30-2540)

BIKELIS, A. 1964.    (17.1)                                       542
    ON THE IMPROVEMENT OF THE REMAINDER TERM FOR THE MULTIDIMENSIONAL GLOBAL THEOREMS.
        (IN RUSSIAN, WITH SUMMARY IN LITHUANIAN AND ENGLISH)
    LITOVSK. MAT. SB., 4, 159-163.    (MR30-2541)

BIKELIS, A. 1966.    (17.1)                                       543
    THE REMAINDER TERMS IN MULTI-DIMENSIONAL LIMIT THEOREMS.
        (IN RUSSIAN)
    DOKL. AKAD. NAUK SSSR, 168, 731-732.    (MR33-4972)

BIKJALIS, A.    SEE  BIKELIS, A.

BILDIKAR, SHULA    SEE
    4147. PATIL, G. P. + BILDIKAR, SHULA (1966)

BILIMOWIC, ANTON D. 1937.    (4.1)                                544
    ELEMENTARE KORRELATIONSTHEORIE.
    ZAP. RUSS. NAUCN. INST. BEOGRAD, 12, 45-61.

BILLINGSLEY, PATRICK 1956.    (17.1)                              545
    THE INVARIANCE PRINCIPLE FOR DEPENDENT RANDOM VARIABLES.
    TRANS. AMER. MATH. SOC., 83, 250-268.    (MR19,P.891)

BILLINGSLEY, PATRICK 1966.    (17.1)                              546
    CONVERGENCE OF TYPES IN K-SPACE.
    Z. WAHRSCHEIN. VERW. GEBIETE, 5, 175-179.

BINET, F. E. + WATSON, G.S. 1956.    (6.4)                        547
    ALGEBRAIC THEORY OF THE COMPUTING ROUTINE FOR TESTS OF SIGNIFICANCE
    ON THE DIMENSIONALITY OF NORMAL MULTIVARIATE SYSTEMS.
    J. ROY. STATIST. SOC. SER. B, 18, 70-78.    (MR18,P.243)

BINGHAM, M. D. 1941.    (20.3)                                    548
    A NEW METHOD FOR OBTAINING THE INVERSE MATRIX.
    J. AMER. STATIST. ASSOC., 36, 530-534.    (MR3,P.154)

BIRCH, M. W. 1964.    (17.7/B)                                    549
    A NOTE ON THE MAXIMUM LIKELIHOOD ESTIMATION OF A LINEAR STRUCTURAL RELATIONSHIP.
    J. AMER. STATIST. ASSOC., 59, 1175-1178.    (MR31-1732)

BIRCH, M. W. 1964.    (4.9)                                       550
    THE DETECTION OF PARTIAL ASSOCIATION I.   THE 2 X 2 CASE.
    J. ROY. STATIST. SOC. SER. B, 26, 313-324.

BIRCH, M. W. 1965.    (18.2)                                      551
    THE DETECTION OF PARTIAL ASSOCIATION, II. THE GENERAL CASE.
    J. ROY. STATIST. SOC. SER. B, 27, 111-124.

BIRENDRANATH, GHOSH  SEE  GHOSH, BIRENDRANATH

BIRKHOFF, GARRETT 1938.    (17.3)                                 552
    DEPENDENT PROBABILITIES AND SPACES (L).
    PROC. NAT. ACAD. SCI. U.S.A., 24, 154-159.

BIRNBAUM, ALLAN 1955.  (14.3/19.2)                                          553
    CHARACTERIZATIONS OF COMPLETE CLASSES OF TESTS OF SOME
    MULTIPARAMETRIC HYPOTHESES, WITH APPLICATIONS TO LIKELIHOOD RATIO TESTS.
    ANN. MATH. STATIST., 26, 21-36.  (MR16,P.729)

BIRNBAUM, ALLAN 1961.  (8.5/19.1/U)                                         554
    A MULTI-DECISION PROCEDURE RELATED TO THE ANALYSIS OF SINGLE DEGREES OF FREEDOM.
    ANN. INST. STATIST. MATH. TOKYO, 12, 227-236.  (MR23-A1453)

BIRNBAUM, ALLAN + MAXWELL, A.E. 1960.  (6.1)                                555
    CLASSIFICATION PROCEDURES BASED ON BAYES'S FORMULA.
    APPL. STATIST., 9, 152-169.  (MR22-8619)

BIRNBAUM, S. 1965.  (20.3/E)                                                556
    PERTURBATION OF EIGENVALUES--WITH AN ENGINEERING APPLICATION.
    SIAM REV., 7, 13-30.

BIRNBAUM, Z. W. 1942.  (2.7)                                                557
    AN INEQUALITY FOR MILL'S RATIO.
    ANN. MATH. STATIST., 13, 245-246.  (MR4,P.19)

BIRNBAUM, Z. W. 1950.  (2.4)                                                558
    EFFECT OF LINEAR TRUNCATION ON A MULTINORMAL POPULATION.
    ANN. MATH. STATIST., 21, 272-279.  (MR11,P.673)

BIRNBAUM, Z. W. + CHAPMAN, DOUGLAS G. 1950.  (2.4/6.7)                      559
    ON OPTIMUM SELECTIONS FROM MULTINORMAL POPULATIONS.
    ANN. MATH. STATIST., 21, 443-447.  (MR12,P.271)

BIRNBAUM, Z. W. + MARSHALL, ALBERT W. 1961.  (17.2)                         560
    SOME MULTIVARIATE CHEBYSHEV INEQUALITIES WITH EXTENSIONS TO
    CONTINUOUS PARAMETER PROCESSES.
    ANN. MATH. STATIST., 32, 687-703.  (MR26-5615)

BIRNBAUM, Z. W. + MEYER, PAUL L. 1953.  (2.4)                               561
    ON THE EFFECT OF TRUNCATION IN SOME OR ALL CO-ORDINATES OF A
    MULTINORMAL POPULATION.
    J. INDIAN SOC. AGRIC. STATIST., 5, 17-28.  (MR16,P.54)

BIRNBAUM, Z. W.; PAULSON, EDWARD + ANDREWS, FRED C. 1950.  (2.4)            562
    ON THE EFFECT OF SELECTION PERFORMED ON SOME COORDINATES OF A
    MULTI-DIMENSIONAL POPULATION.
    PSYCHOMETRIKA, 15, 191-204.  (MR12,P.36)

BIRNBAUM, Z. W.; RAYMOND, JOSEPH L. + ZUCKERMAN, HERBERT S. 1947.  (17.2/B) 563
    A GENERALIZATION OF TSHEBYSHEV'S INEQUALITY TO TWO DIMENSIONS.
    ANN. MATH. STATIST., 18, 70-79.  (MR8,P.470)

BISHOP, ALBERT B. + CHOPE, HENRY R. 1962.  (8.8/N)                          564
    REGRESSION TECHNIQUES IN MULTIVARIATE ADAPTIVE CONTROL SYSTEMS.
    IRE TRANS. AUTOMAT. CONTROL, AC-7(2) 107-116.

BISHOP, D. J. 1939.  (10.2)                                                 565
    ON A COMPREHENSIVE TEST FOR THE HOMOGENEITY OF VARIANCES AND
    COVARIANCES IN MULTIVARIATE PROBLEMS.
    BIOMETRIKA, 31, 31-55.  (MR1,P.64)

BISPHAM, CAPTAIN J. W., R. E., O.B.E. 1920.  (4.5)                          566
    AN EXPERIMENTAL DETERMINATION OF THE DISTRIBUTION OF THE PARTIAL
    CORRELATION COEFFICIENT IN SAMPLES OF THIRTY.
        (REPRINTED AS NO. 567)
    PROC. ROY. SOC. LONDON SER. A, 97, 218-224.

BISPHAM, CAPTAIN J. W., R. E., O.B.E. 1923.  (4.5)                          567
    AN EXPERIMENTAL DETERMINATION OF THE DISTRIBUTION OF THE PARTIAL
    CORRELATION COEFFICIENT IN SAMPLES OF THIRTY.
        (REPRINT OF NO. 566)
    METRON, 2,   684-696.

BITTNER, REIGN H. + WILDER, CARLTON E. 1946.  (4.1/E)                       568
    EXPECTANCY TABLES: A METHOD OF INTERPRETING CORRELATION COEFFICIENTS.
    J. EXPER. EDUC., 14, 245-252.

BJERHAMMAR, ARNE 1951.  (20.1/E)                                     569
    RECTANGULAR RECIPROCAL MATRICES, WITH SPECIAL REFERENCE TO GEODETIC
    CALCULATIONS.
    BULL. GEODESIQUE, 20, 188-220.  (MR13,P.312)


BJERHAMMAR, ARNE 1951.  (8.1/20.1/EU)                                570
    APPLICATION OF CALCULUS OF MATRICES TO METHOD OF LEAST SQUARES, WITH
    SPECIAL REFERENCE TO GEODETIC CALCULATIONS.
    KUNGL. TEK. HOGSK. HANDL. STOCKHOLM, 49, 1-86.  (MR14,P.1127)


BJERHAMMAR, ARNE 1958.  (20.1)                                       571
    A GENERALIZED MATRIX ALGEBRA.
    KUNGL. TEK. HOGSK. HANDL. STOCKHOLM, 124, 1-32.  (MR20-7038)


BLACK, A. 1898.  (2.1/20.4)                                          572
    REDUCTION OF A CERTAIN MULTIPLE INTEGRAL.
    TRANS. CAMBRIDGE PHILOS. SOC., 16, 219-225.


BLACK, J. D.   SEE
    3629. MALENBAUM, WILFRED + BLACK, J. D. (1937)

    3630. MALENBAUM, WILFRED + BLACK, J. D. (1939)


BLACK, THOMAS P. 1929.  (15.2/E)                                     573
    MENTAL MEASUREMENT: THE PROBABLE ERROR OF SOME BOUNDARY CONDITIONS IN
    DIAGNOSING THE PRESENCE OF GROUP AND GENERAL FACTORS.
    PROC. ROY. SOC. EDINBURGH, 49, 72-77.


BLACKITH, R. E. 1957.  (5.2/6.6/8.5/E)                               574
    POLYMORPHISM IN SOME AUSTRALIAN LOCUSTS AND GRASSHOPPERS.
    BIOMETRICS, 13, 183-196.


BLACKITH, R. E. 1960.  E(11.2)                                       575
    A SYNTHESIS OF MULTIVARIATE TECHNIQUE TO DISTINGUISH PATTERNS OF
    GROWTH IN GRASSHOPPERS.
    BIOMETRICS, 16, 28-40.


BLACKITH, R. E. 1965.  (6.4/17.9)                                    576
    MORPHOMETRICS.
    THEOR. MATH. BIOL. (WATERMAN + MOROWITZ), 225-249.


BLACKWELL, DAVID H. 1956.  (19.1)                                    577
    AN ANALOG OF THE MINIMAX THEOREM FOR VECTOR PAYOFFS.
    PACIFIC J. MATH., 6, 1-8.  (MR18,P.450)


BLACKWELL, DAVID H. + BOWKER, ALBERT H. 1955.  (1.2)                 578
    MEYER ABRAHAM GIRSHICK, 1908-55.
    ANN. MATH. STATIST., 26, 365-367.  (MR17,P.3)


BLACKWELL, DAVID H. + GIRSHICK, M. A. 1946.  (16.4/17.1)             579
    ON FUNCTIONS OF SEQUENCES OF INDEPENDENT CHANCE VECTORS WITH
    APPLICATIONS TO THE PROBLEM OF THE "RANDOM WALK" IN K-DIMENSIONS.
    ANN. MATH. STATIST., 17, 310-317.  (MR8,P.215)


BLAKEMAN, JOHN 1905.  (8.3/U)                                        580
    ON TESTS FOR LINEARITY OF REGRESSION IN FREQUENCY DISTRIBUTIONS.
    BIOMETRIKA, 4, 332-350.


BLAKEY, ROBERT 1940.  (15.2)                                         581
    A RE-ANALYSIS OF A TEST OF THE THEORY OF TWO FACTORS.
    PSYCHOMETRIKA, 5, 121-136.


BLALOCK, HUBERT M. 1961.  (17.7)                                     582
    CORRELATION AND CAUSALITY:  THE MULTIVARIATE CASE.
    SOCIAL FORCES, 39, 246-251.


BLALOCK, HUBERT M. 1962.  (4.5/17.7)                                 583
    FOUR-VARIABLE CAUSAL MODELS AND PARTIAL CORRELATIONS.
    AMER. J. SOCIOL., 68, 182-194.

BLALOCK, HUBERT M. 1962.   (17.7)                                                           584
     FURTHER OBSERVATIONS ON ASYMMETRIC CAUSAL MODELS.
          (AMENDED BY NO. 4586)
     AMER. SOCIOL. REV., 27, 542-545.


BLALOCK, HUBERT M. 1963.   E(17.7)                                                          585
     MAKING CAUSAL INFERENCES FOR UNMEASURED VARIABLES FROM CORRELATIONS
     AMONG INDICATORS.
     AMER. J. SOCIOL., 69, 53-62.


BLANC, CHARLES + LINIGER, WERNER 1956.   (20.3)                                             586
     ERREURS DE CHUTE DANS LA RESOLUTION DE SYSTEMES ALGEBRIQUES LINEAIRES.
     COMMENT. MATH. HELVETIA, 30, 257-264.   (MR17,P.1137)


BLANC-LAPIERRE, ANDRE + FORTET, ROBERT M. 1947.   (16.3)                                    587
     SUR UNE PROPRIETE FONDAMENTALE DES FONCTIONS DE CORRELATION.
     C.R. ACAD. SCI. PARIS, 224, 786-788.   (MR8,P.472)


BLANC-LAPIERRE, ANDRE + FORTET, ROBERT M. 1947.   (16.4)                                    588
     LES FONCTIONS ALEATOIRES STATIONNAIRES DE PLUSIEURS VARIABLES.
     REV. SCI. PARIS, 85, 419-422.   (MR9,P.150)


BLAND, R. P. + OWEN, DONALD B. 1966.   (2.1)                                                589
     A NOTE ON SINGULAR NORMAL DISTRIBUTIONS.
     ANN. INST. STATIST. MATH. TOKYO, 18, 113-116.   (MR33-6666)


BLEJEC, M. 1951.   (4.1)                                                                    590
     IZRACUNAVANJE ARITMETICNE SREDINE, STANDARDNE DEVIACIJE IN
     KORELACIJSKEGA KOEFICIJENTA S POMOCJO KUMULATIVNIH SERIJ.
          (WITH SUMMARY IN FRENCH AND ENGLISH)
     STATIST. REV. BEOGRAD, 1, 266-281.


BLISCHKE, WALLACE R. + HALPIN, ALAN H. 1966.   (2.5/2.6/14.4/B)                             591
     ASYMPTOTIC PROPERTIES OF SOME ESTIMATORS OF QUANTILES OF CIRCULAR ERROR.
     J. AMER. STATIST. ASSOC., 61, 618-632.   (MR34-3687)


BLISS, C.I. 1940.   (8.5/E)                                                                 592
     FACTORIAL DESIGN AND COVARIANCE IN BIOLOGICAL ASSAY OF VITAMIN D.
     J. AMER. STATIST. ASSOC., 35, 498-506.


BLISS, C.I.    SEE ALSO
     1002. COCHRAN, WILLIAM G. + BLISS, C.I. (1948)


BLOMQVIST, NILS 1950.   (4.9)                                                               593
     ON A MEASURE OF DEPENDENCE BETWEEN TWO RANDOM VARIABLES.
     ANN. MATH. STATIST., 21, 593-600.   (MR12,P.510)



BLOOMERS, PAUL 1945.   (1.1)                                                                594
     STATISTICAL THEORY: SOME RECENT DEVELOPMENTS.
     REV. EDUC. RES., 15, 423-440.


BLUM, JULIUS R. 1954.   (17.1/17.9)                                                         595
     MULTIDIMENSIONAL STOCHASTIC APPROXIMATION METHODS.
     ANN. MATH. STATIST., 25, 737-744.   (MR16,P.382)


BLUM, JULIUS R. + ROSENBLATT, JUDAH I. 1966.   (19.3)                                       596
     ON SOME STATISTICAL PROBLEMS REQUIRING PURELY SEQUENTIAL SAMPLING SCHEMES.
     ANN. INST. STATIST. MATH. TOKYO, 18, 351-355.   (MR34-5243)


BLUM, JULIUS R.; HANSON, D. L. + KOOPMANS, LAMBERT H. 1963.   (17.1)                        597
     ON THE STRONG LAW OF LARGE NUMBERS FOR A CLASS OF STOCHASTIC PROCESSES.
     Z. WAHRSCHEIN. VERW. GEBIETE, 2, 1-11.


BLUM, JULIUS R.; KIEFER, J. + ROSENBLATT, MURRAY 1961.   (17.1/17.8)                        598
     DISTRIBUTION FREE TESTS OF INDEPENDENCE BASED ON THE SAMPLE
     DISTRIBUTION FUNCTION.
     ANN. MATH. STATIST., 32, 485-498.   (MR23-A2989)


BLUMEN, ISADORE 1958.   (17.8/B)                                                            599
     A NEW BIVARIATE SIGN TEST.
     J. AMER. STATIST. ASSOC., 53, 448-456.

BLUMENSON, L. E. 1960.   (20.5)                                                          600
    A DERIVATION OF N-DIMENSIONAL SPHERICAL CO-ORDINATES.
    AMER. MATH. MONTHLY, 67, 63-66.

BLUMENSON, L. E. + MILLER, KENNETH S. 1963.   (7.4/14.2/18.1)                             601
    PROPERTIES OF GENERALIZED RAYLEIGH DISTRIBUTIONS.
    ANN. MATH. STATIST., 34, 903-910.   (MR27-846)

BLUMENSON, L. E.    SEE ALSO
    3837. MILLER, KENNETH S.; BERNSTEIN, R. I. + BLUMENSON, L. E. (1958)

    3838. MILLER, KENNETH S.; BERNSTEIN, R. I. + BLUMENSON, L. E. (1963)

BLYTH, COLIN R. 1951.   (19.1)                                                            602
    ON MINIMAX STATISTICAL DECISION PROCEDURES AND THEIR ADMISSIBILITY.
    ANN. MATH. STATIST., 22, 22-42.   (MR12,P.622)

BOAGA, GIOVANNI 1948.   (4.1/20.5)                                                        603
    INTERPRETAZIONE GEOMETRICA DEL COEFFICIENTE DI CORRELAZIONE.
    RIV. ITAL. ECON. DEMOG. STATIST., 2, 471-474.

BOAS, FRANZ 1909.   (4.1)                                                                 604
    DETERMINATION OF THE COEFFICIENT OF CORRELATION.
    SCIENCE NEW SER., 29, 823-824.

BOAS, FRANZ 1921.   (4.1)                                                                 605
    THE COEFFICIENT OF CORRELATION.
    QUART. PUBL. AMER. STATIST. ASSOC., 17, 683-688.

BOCHNER, SALOMON 1954.   (16.4/17.1)                                                      606
    LIMIT THEOREMS FOR HOMOGENEOUS STOCHASTIC PROCESSES.
    PROC. NAT. ACAD. SCI. U.S.A., 40, 699-703.   (MR16,P.379)

BOCHNER, SALOMON 1956.   (16.4)                                                           607
    STATIONARITY, BOUNDEDNESS, ALMOST PERIODICITY OF RANDOM-VALUED FUNCTIONS.
    PROC. THIRD BERKELEY SYMP. MATH. STATIST. PROB., 2, 7-27.   (MR18,P.940)

BOCHNER, SALOMON 1959.   (17.1)                                                           608
    GENERAL ANALYTICAL SETTING FOR THE CENTRAL LIMIT THEORY OF PROBABILITY.
    CALCUTTA MATH. SOC. GOLDEN JUBILEE COMMEM. VOLUME, 1, 111-128.   (MR27-5284)

BOCK, R. DARRELL 1960.   (15.5)                                                           609
    COMPONENTS OF VARIANCE ANALYSIS AS A STRUCTURAL AND DISCRIMINAL ANALYSIS FOR
        PSYCHOLOGICAL TESTS.
    BRITISH J. STATIST. PSYCHOL., 13, 151-163.

BOCK, R. DARRELL 1963.   (8.5/20.2/E)                                                     610
    MULTIVARIATE ANALYSIS OF VARIANCE OF REPEATED MEASUREMENTS.
    PROBLEMS MEAS. CHANGE (HARRIS), 85-103.

BOCK, R. DARRELL 1963.   (8.5/20.3)                                                       611
    PROGRAMMING UNIVARIATE AND MULTIVARIATE ANALYSIS OF VARIANCE.
    TECHNOMETRICS, 5, 95-117.   (MR32-1821)

BOCK, R. DARRELL 1965.   (8.5/20.3)                                                       612
    A COMPUTER PROGRAM FOR UNIVARIATE AND MULTIVARIATE ANALYSIS OF VARIANCE.
    PROC. IBM SCI. COMPUT. SYMP. STATIST., 69-111.

BOCK, R. DARRELL 1966.   (1.1)                                                            613
    CONTRIBUTIONS OF MULTIVARIATE EXPERIMENTAL DESIGNS TO EDUCATIONAL RESEARCH.
    HANDB. MULTIVARIATE EXPER. PSYCHOL. (CATTELL), 820-840.

BOCK, R. DARRELL + BARGMANN, ROLF E. 1966.   (3.2)                                        614
    ANALYSIS OF COVARIANCE STRUCTURES.
    PSYCHOMETRIKA, 31, 507-534.   (MR34-5218)

BOCK, R. DARRELL    SEE ALSO
    1097. CRAMER, ELLIOT M. + BOCK, R. DARRELL (1966)

BODEWIG, E. 1947.   (20.3)                                                                615
    COMPARISON OF SOME DIRECT METHODS FOR COMPUTING DETERMINANTS AND
        INVERSE MATRICES.
    KON. NEDERL. AKAD. WETENSCH. PROC. SECT. SCI., 50, 49-57.

BODIN, N. A. 1965.   (17.5)                                                      616
    ON ROUNDOFF ERRORS IN MULTIDIMENSIONAL MEASUREMENTS.
        (IN RUSSIAN, TRANSLATED AS NO. 617)
    TRUDY MAT. INST. STEKLOVA, 79, 76-105.   (MR34-8516)

BODIN, N. A. 1966.   (17.5)                                                      617
    ON ROUNDOFF ERRORS IN MULTIDIMENSIONAL MEASUREMENTS.
        (TRANSLATION OF NO. 616)
    PROC. STEKLOV INST. MATH., 79, 83-116.   (MR34-8516)

BODIO, LUIGI 1883.   (1.2)                                                       618
    SAGGIO DI BIBLIOGRAFIA STATISTICA ITALIANA.
    ANN. STATIST. SER. 3A, 4, 1-149.

BODIO, LUIGI 1885.   (1.2)                                                       619
    SAGGIO DI BIBLIOGRAFIA STATISTICA ITALIANA. SECONDA EDIZIONE ACCRESCIUTA.
    ANN. STATIST. SER. 3A, 1-179.

BODIO, LUIGI 1890.   (1.2)                                                       620
    SAGGIO DI BIBLIOGRAFIA STATISTICA ITALIANA. TERZA EDIZIONE ACCRESCIUTA.
    ANN. STATIST. SER. 3A, 1-213.

BODIO, LUIGI    SEE ALSO
    3806. MESSEDAGLIA, ANGELO + BODIO, LUIGI (1880)

BOFINGER, EVE + BOFINGER, V.J. 1961.   (17.9)                                    621
    A RUNS TEST FOR SEQUENCES OF RANDOM DIGITS.
    AUSTRAL. J. STATIST., 3, 37-41.   (MR26-1986)

BOFINGER, EVE + BOFINGER, V.J. 1965.   (14.1/17.6/BT)                            622
    THE CORRELATION OF MAXIMA IN SAMPLES DRAWN FROM A BIVARIATE NORMAL
    DISTRIBUTION.
    AUSTRAL. J. STATIST., 7, 57-61.   (MR33-5019)

BOFINGER, V.J.    SEE
    621. BOFINGER, EVE + BOFINGER, V.J. (1961)

    622. BOFINGER, EVE + BOFINGER, V.J. (1965)

BOHM, JOHANNES 1959.   (20.5)                                                    623
    UNTERSUCHUNG DES SIMPLEXINHALTES IN RAUMEN KONSTANTER KRUMMUNG
    BELIEBIGER DIMENSION.
    J. REINE ANGEW. MATH., 202, 16-51.   (MR22-12446)

BOISSEVAIN, C. H. 1939.   (15.1)                                                 624
    DISTRIBUTION OF ABILITIES DEPENDING UPON TWO OR MORE INDEPENDENT FACTORS.
    METRON, 13(4) 49-58.

BOJARSKI, A. JA. 1941.   (4.9/B)                                                 625
    SUR LA CORRELATION GEOMETRIQUE.
        (IN RUSSIAN, WITH SUMMARY IN FRENCH)
    IZV. AKAD. NAUK SSSR SER. MAT., 5, 159-164.   (MR3,P.173)

BOLGER, E. M. + HARKNESS, W. L. 1965.   (17.3)                                   626
    CHARACTERIZATIONS OF SOME DISTRIBUTIONS BY CONDITIONAL MOMENTS.
    ANN. MATH. STATIST., 36, 703-705.

BOLLMAN, DOROTHY A. 1965.   (20.5)                                               627
    SOME PERIODICITY PROPERTIES OF TRANSFORMATIONS ON VECTOR SPACES OVER
    RESIDUE CLASS RINGS.
    J. SOC. INDUST. APPL. MATH., 13, 902-912.

BOL'SEV, L. N. 1955.   (6.3/T)                                                   628
    A NOMOGRAM CONNECTING THE PARAMETERS OF A NORMAL DISTRIBUTION WITH
    THE PROBABILITIES FOR CLASSIFICATION INTO THREE GROUPS.
        (IN RUSSIAN, REPRINTED AS NO. 629, TRANSLATED AS NO. 630)
    INZEN. SB. AKAD. NAUK SSSR, 21, 212-214.   (MR17,P.53)

BOL'SEV, L. N. 1957.   (6.3/T)                                                   629
    A NOMOGRAM CONNECTING THE PARAMETERS OF A NORMAL DISTRIBUTION WITH
    THE PROBABILITIES FOR CLASSIFICATION INTO THREE GROUPS.
        (IN RUSSIAN, WITH SUMMARY IN ENGLISH, TRANSLATED AS NO. 630, REPRINT OF NO. 628)
    TEOR. VEROJATNOST. I PRIMENEN., 2, 124-126.   (MR19,P.691)

BOL'SEV, L. N. 1957.  (6.3/T)                                                    630
    A NOMOGRAM CONNECTING THE PARAMETERS OF A NORMAL DISTRIBUTION WITH
    PROBABILITIES FOR CLASSIFICATION INTO THREE GROUPS.
        (TRANSLATION OF NO. 629, TRANSLATION OF NO. 628)
    THEORY PROB. APPL., 2, 120-122.  (MR17,P.53)


BOLSHEV, L. N.  SEE  BOL'SEV, L. N.


BOMBAY, BARBARA FLORES    SEE
    1845. GARDINER, DONALD A. + BOMBAY, BARBARA FLORES (1965)


BONDARENKO, P. S. 1957.  (20.3)                                                  631
    CONVERGENCE OF AN ALGORITHM OF SUCCESSIVE APPROXIMATIONS AND ERROR ESTIMATES IN
    NUMERICAL SOLUTION OF INFINITE SYSTEMS OF LINEAR ALGEBRAIC EQUATIONS.
        (IN RUSSIAN, REPRINT OF NO. 632)
    KIEV. GOS. UNIV. MAT. SB., 9, 81-89.  (MR20-6187)


BONDARENKO, P. S. 1957.  (20.3)                                                  632
    CONVERGENCE OF AN ALGORITHM OF SUCCESSIVE APPROXIMATIONS AND ERROR ESTIMATES IN
    NUMERICAL SOLUTION OF INFINITE SYSTEMS OF LINEAR ALGEBRAIC EQUATIONS.
        (IN RUSSIAN, REPRINTED AS NO. 631)
    KIIV. DERZ. UNIV. NAUK. ZAP., 16(2) 81-89.  (MR20-6187)


BONFERRONI, CARLO EMILIO 1939.  (4.9)                                            633
    DI UNA ESTENSIONE DEL COEFFICIENTE DI CORRELAZIONE.
    GIORN. ECON. ANN. ECONOMIA, 1, 797-826.


BONFERRONI, CARLO EMILIO 1941.  (4.8)                                            634
    CORRELATION ET INTERPOLATION.
    J. SOC. HONGR. STATIST., 19, 175-185.


BONFERRONI, CARLO EMILIO 1942.  (4.1/4.9/8.1)                                    635
    DI UN COEFFICIENTE DI CORRELAZIONE SIMULTANEA.
    ATTI CONG. UN. MAT. ITAL., 2, 707-714.  (MR8,P.474)


BONFERRONI, CARLO EMILIO 1956.  (4.8/B)                                          636
    INDICI UNILATERALI E BILATERALI DI CONNESSIONE.
    STUDI MEM. R. BENINI, 49-60.


BONIFACIO, GEORGIO 1948.  (4.8)                                                  637
    SULLA CORRELAZIONE FRA REDDITO E TALUNI CONSUMI ALIMENTARI.
    RIV. ITAL. ECON. DEMOG. STATIST., 2, 550-561.


BONNARDEL, R. 1950.  (4.1/4.8/EN)                                                638
    METHODE RAPIDE POUR LE CALCUL DES CORRELATIONS MOYENNES.
    TRAVAIL HUMAIN, 13, 274-282.


BONNER, R. E. 1964.  (6.6/E)                                                     639
    ON SOME CLUSTERING TECHNIQUES.
    IBM SYSTEMS J., 22, 22-32.


BONNER, R. E. 1966.  (4.6)                                                       640
    CLUSTER ANALYSIS.
    ANN. NEW YORK ACAD. SCI., 128(3) 972-983.


BOONSTRA, A. E. H. R. 1943.  (4.1)                                              641
    CORRELATION IN PRACTICE.
    LANDBOUWK. TIJDSCHR. WAGENINGEN, 55, 639-659.


BOOT, J. C. G. 1963.  (20.3)                                                    642
    THE COMPUTATION OF THE GENERALIZED INVERSE OF SINGULAR OR RECTANGULAR MATRICES.
    AMER. MATH. MONTHLY, 70, 302-303.


BOOT, J. C. G.   SEE ALSO
    5394. THEIL, H. + BOOT, J. C. G. (1962)


BORDIN, EDWARD S. 1941.  (15.1)                                                 643
    FACTOR ANALYSIS: ART OR SCIENCE?
    PSYCHOL. BULL., 38, 520-521.

BORGATTA, EDGAR F. + HAYS, DAVID G. 1952.   (15.5)                                    644
   SOME LIMITATION ON THE ARBITRARY CLASSIFICATION OF NON-SCALE RESPONSE
   PATTERNS IN A GUTTMAN SCALE.
   PUBLIC OPINION QUART., 16, 410-416.

BORGATTA, EDGAR F.    SEE ALSO
   2410. HAYS, DAVID G. + BORGATTA, EDGAR F. (1954)

BORGES, R. 1965.   (19.2)                                                            645
   GLEICHMASSIG TRENNSCHÄRFE TESTS.
   Z. WAHRSCHEIN. VERW. GEBIETE, 3, 296-316.

BORGES, R. 1966.   (2.2/U)                                                           646
   A CHARACTERIZATION OF THE NORMAL DISTRIBUTION.
   Z. WAHRSCHEIN. VERW. GEBIETE, 5, 244-246.

BORGES, R. 1966.   (16.4)                                                            647
   ZUR EXISTENZ VON SEPARABLEN STOCHASTISCHEN PROZESSEN.
   Z. WAHRSCHEIN. VERW. GEBIETE, 6, 125-128.

BORGES, R. + PFANZAGL, JOHANN 1963.   (17.3/19.2/U)                                  648
   A CHARACTERIZATION OF THE ONE-PARAMETER EXPONENTIAL FAMILY OF
   DISTRIBUTIONS BY MONOTONICITY OF LIKELIHOOD RATIOS.
   Z. WAHRSCHEIN. VERW. GEBIETE, 2, 111-117.


BOROVKOV, A. A. + ROGOZIN, B. A. 1965.   (17.1)                                      649
   ON THE MULTI-DIMENSIONAL CENTRAL LIMIT THEOREM.
       (IN RUSSIAN, WITH SUMMARY IN ENGLISH, TRANSLATED AS NO. 650)
   TEOR. VEROJATNOST. I PRIMENEN., 10, 61-69.   (MR30-3492)

BOROVKOV, A. A. + ROGOZIN, B. A. 1965.   (17.1)                                      650
   ON THE MULTI-DIMENSIONAL CENTRAL LIMIT THEOREM.
       (TRANSLATION OF NO. 649)
   THEORY PROB. APPL., 10, 55-62.   (MR30-3492)

BOSE, DEB KUMAR + ROY, AMAL KUMAR 1957.   (20.3)                                     651
   INVERSION OF 25x25 MATRIX ON 602A CALCULATING PUNCH.
   SANKHYA, 17, 401-406.

BOSE, MRS. CHAMELI + GAYEN, A. K. 1946.   (17.8)                                     652
   NOTE ON THE EXPECTED DISCREPANCY IN THE ESTIMATION (BY DOUBLE SAMPLING) OF A
   VARIATE IN TERMS OF A CONCOMITANT VARIATE WHEN THERE EXISTS A NON-LINEAR
   REGRESSION BETWEEN THE VARIATES.
   SANKHYA, 8, 73-74.   (MR8,P.476)


BOSE, MRS. CHAMELI    SEE ALSO
   3615. MAHALANOBIS, P.C. + BOSE, MRS. CHAMELI (1941)

BOSE, PURNENDU KUMAR 1941.   (4.7/N)                                                 653
   ON THE REDUCTION FORMULAE FOR THE INCOMPLETE PROBABILITY INTEGRAL OF
   THE MULTIPLE CORRELATION COEFFICIENT OF THE SECOND KIND.
   SCI. AND CULT., 7, 171-172.   (MR5,P.42)

BOSE, PURNENDU KUMAR 1942.   (14.2/B)                                                654
   ON THE EXACT DISTRIBUTION OF THE RATIO OF TWO MEANS BELONGING TO
   SAMPLES DRAWN FROM A GIVEN CORRELATED BIVARIATE NORMAL POPULATION.
   BULL. CALCUTTA MATH. SOC., 34, 139-141.   (MR4,P.103)

BOSE, PURNENDU KUMAR 1942.   (2.3)                                                   655
   CERTAIN MOMENT CALCULATIONS CONNECTED WITH MULTIVARIATE NORMAL POPULATIONS.
   SCI. AND CULT., 7, 411-412.   (MR5,P.42)

BOSE, PURNENDU KUMAR 1944.   (20.4)                                                  656
   ON CONFLUENT HYPERGEOMETRIC SERIES.
   SANKHYA, 6, 407-412.   (MR5,P.245)

BOSE, PURNENDU KUMAR 1947.   (5.3)                                                   657
   PARAMETRIC RELATIONS IN MULTIVARIATE DISTRIBUTIONS.
   SANKHYA, 8, 167-171.   (MR10,P.135)

BOSE, PURNENDU KUMAR 1947. (5.2/20.4/NT)     658
    ON RECURSION FORMULAE, TABLES AND BESSEL FUNCTION POPULATIONS
    ASSOCIATED WITH THE DISTRIBUTION OF CLASSICAL $D^2$-STATISTIC.
    SANKHYA, 8, 235-248. (MR9,P.620)


BOSE, PURNENDU KUMAR 1949. (5.2/NT)     659
    INCOMPLETE PROBABILITY INTEGRAL TABLES CONNECTED WTIH STUDENTISED $D^2$-STATISTIC.
      (AMENDED BY NO. 661)
    CALCUTTA STATIST. ASSOC. BULL., 2, 131-137. (MR11,P.527)


BOSE, PURNENDU KUMAR 1951. (5.3/13.1)     660
    REMARKS ON COMPUTING THE INCOMPLETE PROBABILITY INTEGRAL IN
    MULTIVARIATE DISTRIBUTION FUNCTIONS.
    BULL. INST. INTERNAT. STATIST., 33(2) 55-64. (MR16,P.940)


BOSE, PURNENDU KUMAR 1951. (5.2/NT)     661
    CORRIGENDA: ON THE CONSTRUCTION OF INCOMPLETE PROBABILITY INTEGRAL
    TABLES OF THE CLASSICAL $D^2$-STATISTIC.
      (AMENDMENT OF NO. 659)
    SANKHYA, 11, 96. (MR13,P.52)


BOSE, PURNENDU KUMAR 1957. (18.1/U)     662
    NORMALISATION OF FREQUENCY FUNCTIONS.
    BULL. CALCUTTA MATH. SOC., 48, 109-119. (MR18,P.958)


BOSE, PURNENDU KUMAR 1957. (15.1)     663
    STATISTICAL METHODS IN PSYCHOMETRIC RESEARCH.
    CALCUTTA STATIST. ASSOC. BULL., 7, 150-160.


BOSE, PURNENDU KUMAR + CHAUDHURI, S. B. 1955. (15.4)     664
    SCALING PROCEDURES IN SCHOLASTIC AND VOCATIONAL TESTS.
    SANKHYA, 15, 197-206.


BOSE, PURNENDU KUMAR + CHAUDHURI, S. B. 1957. (15.4)     665
    METHOD OF MATCHING USED FOR THE ESTIMATION OF TEST RELIABILITY.
    SANKHYA, 17, 377-384.


BOSE, PURNENDU KUMAR    SEE ALSO
    4683. ROY, S.N. + BOSE, PURNENDU KUMAR (1939)

    4684. ROY, S.N. + BOSE, PURNENDU KUMAR (1940)

    4685. ROY, S.N. + BOSE, PURNENDU KUMAR (1940)


BOSE, R. C. 1934. (4.6/20.5)     666
    ON THE APPLICATION OF HYPERSPACE GEOMETRY TO THE THEORY OF MULTIPLE CORRELATION.
    SANKHYA, 1, 338-342.


BOSE, R. C. 1935. (6.4)     667
    ON THE EXACT DISTRIBUTION AND MOMENT-COEFFICIENTS OF THE $D^2$-STATISTICS.
    SCI. AND CULT., 1, 205-206.


BOSE, R. C. 1936. (5.2/14.5)     668
    ON THE EXACT DISTRIBUTION AND MOMENT-COEFFICIENTS OF THE $D^2$-STATISTIC.
    SANKHYA, 2, 143-154.


BOSE, R. C. 1936. (5.2/14.5)     669
    A NOTE ON THE DISTRIBUTION OF DIFFERENCES IN MEAN VALUES OF TWO SAMPLES DRAWN
    FROM TWO MULTIVARIATE NORMALLY DISTRIBUTED POPULATIONS, AND THE DEFINITION OF
    THE $D^2$-STATISTIC.
    SANKHYA, 2, 379-384.


BOSE, R. C. 1938. (5.3/18.3/U)     670
    ON THE DISTRIBUTION OF THE MEANS OF SAMPLES DRAWN FROM A BESSEL
    FUNCTION POPULATION.
    SANKHYA, 3, 262-264.


BOSE, R. C. 1950. (17.9)     671
    ON A PROBLEM OF TWO-DIMENSIONAL PROBABILITY.
    SANKHYA, 10, 13-28. (MR12,P.113)

BOSE, R. C. + ROY, S.N. 1935.  (5.2/N)                                          672
    ON THE EVALUATION OF THE PROBABILITY INTEGRAL OF THE $D^2$-STATISTICS.
    SCI. AND CULT., 1, 436–437.

BOSE, R. C. + ROY, S.N. 1937.  (5.3)                                            673
    ON THE DISTRIBUTION OF FISHER'S TAXONOMIC CO-EFFICIENT AND STUDENTISED $D^2$-STATISTIC.
    SCI. AND CULT., 3, 335.

BOSE, R. C. + ROY, S.N. 1938.  (5.3)                                            674
    THE DISTRIBUTION OF THE STUDENTISED $D^2$-STATISTIC.
        (WITH DISCUSSION)
    SANKHYA, 4, 19–38.

BOSE, R. C.   SEE ALSO
    4686. ROY, S.N. + BOSE, R. C. (1940)

    4687. ROY, S.N. + BOSE, R. C. (1953)

    3616. MAHALANOBIS, P.C.; BOSE, R. C. + ROY, S.N. (1937)

BOSE, SATYENDRA NATH 1936.  (5.2/14.5)                                          675
    ON THE COMPLETE MOMENT-COEFFICIENTS OF THE $D^2$-STATISTIC.
    SANKHYA, 2, 385–396.

BOSE, SATYENDRA NATH 1937.  (5.2/14.5/20.4)                                     676
    ON THE MOMENT-COEFFICIENTS OF THE $D^2$-STATISTIC AND CERTAIN INTEGRAL AND
    DIFFERENTIAL EQUATIONS CONNECTED WITH THE MULTIVARIATE NORMAL POPULATION.
    SANKHYA, 3, 105–124.

BOSE, SUBHENDUSEKHAR 1935.  (7.4/B)                                             677
    ON THE DISTRIBUTION OF THE RATIO OF VARIANCES OF TWO SAMPLES DRAWN
    FROM A GIVEN NORMAL BIVARIATE CORRELATED POPULATION.
    SANKHYA, 2, 65–72.

BOSTWICK, ARTHUR E. 1896.  (1.1)                                                678
    THE THEORY OF PROBABILITIES.
    SCIENCE, 3, 66–67.

BOTEZ, M. + COHN, H. 1965.  (17.1)                                              679
    UN THEOREME LIMITE A PLUSIEURS DIMENSIONS POUR LES SYSTEMES ALEATOIRES
    A LIAISONS COMPLETES.
    C. R. ACAD. BULGARE SCI., 18, 703–705.  (MR35-7387)

BOTT, R. + DUFFIN, R. J. 1953.  (20.1)                                          680
    ON THE ALGEBRA OF NETWORKS.
    TRANS. AMER. MATH. SOC., 74, 99–109.

BOTTENBERG, ROBERT A. 1956.  (2.7)                                              681
    A METHOD FOR DETERMINING A CELL PROPORTION IN A MULTIVARIATE NORMAL
    DISTRIBUTION.
    SYMP. AIR FORCE HUMAN ENGRG. PERS. TRAIN. RES. (FINCH + CAMERON), 35–47.

BOUDON, RAYMOND 1965.  (16.2/E)                                                 682
    A METHOD OF LINEAR CAUSAL ANALYSIS:  DEPENDENCE ANALYSIS.
    AMER. SOCIOL. REV., 30, 365–374.

BOULANGER, J. J.   SEE
    1862. GAUCHET, F. + BOULANGER, J. J. (1953)

BOURSIN, JEAN-LOUIS 1965.  (14.4/B)                                             683
    SUR UNE NOTION DE DROITES MOYENNES D'UNE REPARTITION.
    C.R. ACAD. SCI. PARIS, 260, 1842–1844.  (MR31-4059)

BOWKER, ALBERT H. 1946.  (14.4/A)                                               684
    COMPUTATION OF FACTORS FOR TOLERANCE LIMITS ON A NORMAL DISTRIBUTION
    WHEN THE SAMPLE IS LARGE.
    ANN. MATH. STATIST., 17, 238–240.  (MR8,P.524)

BOWKER, ALBERT H. 1947.  (20.1)                                                685
    ON THE NORM OF A MATRIX.
    ANN. MATH. STATIST., 18, 285–288.  (MR9,P.75)

BOWKER, ALBERT H. 1960.   (5.3/6.2)                                      **686**
     A REPRESENTATION OF HOTELLING'S $T^2$ AND ANDERSON'S CLASSIFICATION
     STATISTIC W IN TERMS OF SIMPLE STATISTICS.
          (REPRINTED AS NO. 687)
   CONTRIB. PROB. STATIST. (HOTELLING VOLUME), 142–149.   (MR22–B11450)

BOWKER, ALBERT H. 1961.   (5.3/6.2)                                      **687**
     A REPRESENTATION OF HOTELLING'S $T^2$ AND ANDERSON'S CLASSIFICATION
     STATISTIC W IN TERMS OF SIMPLE STATISTICS.
          (REPRINT OF NO. 686)
   STUD. ITEM ANAL. PREDICT. (SOLOMON), 285–292.   (MR23–A2257)

BOWKER, ALBERT H. + SITGREAVES, ROSEDITH 1961.   (6.2/A)                 **688**
     AN ASYMPTOTIC EXPANSION FOR THE DISTRIBUTION FUNCTION OF THE
     W–CLASSIFICATION STATISTIC.
   STUD. ITEM ANAL. PREDICT. (SOLOMON), 293–310.   (MR23–A4210)

BOWKER, ALBERT H.    SEE ALSO
     578. BLACKWELL, DAVID H. + BOWKER, ALBERT H. (1955)

BOWLEY, A. L. 1928.   (4.2)                                              **689**
     THE STANDARD DEVIATION OF THE CORRELATION COEFFICIENT.
   J. AMER. STATIST. ASSOC., 23, 31–34.

BOWLEY, A. L. 1934.   (1.2)                                              **690**
     FRANCIS YSIDRO EDGEWORTH.
   ECONOMETRICA, 2, 113–124.

BOX, GEORGE F. P. 1949.   (8.4/14.2/14.3/A)                              **691**
     A GENERAL DISTRIBUTION THEORY FOR A CLASS OF LIKELIHOOD CRITERIA.
   BIOMETRIKA, 36, 317–346.   (MR11,P.447)

BOX, GEORGE E. P. 1954.   (2.5/8.8/14.5)                                 **692**
     SOME THEOREMS ON QUADRATIC FORMS APPLIED IN THE STUDY OF ANALYSIS OF VARIANCE
     PROBLEMS. I. EFFECT OF INEQUALITY OF VARIANCE IN THE ONE–WAY CLASSIFICATION.
   ANN. MATH. STATIST., 25, 290–302.   (MR15,P.884)

BOX, GEORGE E. P. 1954.   (2.5/8.8/14.5)                                 **693**
     SOME THEOREMS ON QUADRATIC FORMS APPLIED IN THE STUDY OF ANALYSIS OF VARIANCE
     PROBLEMS. II. EFFECTS OF INEQUALITY OF VARIANCE AND OF CORRELATION BETWEEN
     ERRORS IN THE TWO–WAY CLASSIFICATION.
   ANN. MATH. STATIST., 25, 484–498.   (MR16,P.271)

BOX, GEORGE E. P. 1966.   (8.1/8.9)                                      **694**
     USE AND ABUSE OF REGRESSION.
   TECHNOMETRICS, 8, 625–629.

BOX, GEORGE E. P. + DRAPER, NORMAN R. 1965.   (8.1/17.5)                 **695**
     THE BAYESIAN ESTIMATION OF COMMON PARAMETERS FROM SEVERAL RESPONSES.
   BIOMETRIKA, 52, 355–365.

BOX, GEORGE E. P. + HUNTER, J. STUART 1954.   (14.4/C)                   **696**
     A CONFIDENCE REGION FOR THE SOLUTION OF A SET OF SIMULTANEOUS
     EQUATIONS WITH AN APPLICATION TO EXPERIMENTAL DESIGN.
   BIOMETRIKA, 41, 190–199.   (MR15,P.971)

BOYCE, R.    SEE
     955. CHEW, VICTOR + BOYCE, R. (1962)

BRACKEN, JEROME + SOLAND, RICHARD M. 1966.   (14.4)                      **697**
     STATISTICAL DECISION ANALYSIS OF STOCHASTIC LINEAR PROGRAMMING PROBLEMS.
   NAVAL RES. LOGISTICS QUART., 13, 205–225.

BRADLEY, RALPH A.; MARTIN, DONALD C. + WILCOXON, FRANK 1965.   (17.8/T)  **698**
     SEQUENTIAL RANK TESTS I.   MONTE CARLO STUDIES OF THE TWO–SAMPLE PROCEDURE.
   TECHNOMETRICS, 7, 463–483.

BRADLEY, RALPH A.    SEE ALSO
     2734. JACKSON, J. EDWARD + BRADLEY, RALPH A. (1961)

     2735. JACKSON, J. EDWARD + BRADLEY, RALPH A. (1961)

     2736. JACKSON, J. EDWARD + BRADLEY, RALPH A. (1966)

BRAMBILLA, F. 1954.  (17.7)                                                    699
    ELEMENTI DI ANALISI CONFLUENZIALE.
    TEC. ORGANIZZ., 5(14) 20-23.


BRANDNER, FRED A. 1933.  (4.3/B)                                               700
    A TEST OF THE SIGNIFICANCE OF THE DIFFERENCE OF THE CORRELATION
    COEFFICIENTS IN NORMAL BIVARIATE SAMPLES.
    BIOMETRIKA, 25, 102-109.


BRANDT, A. E. 1928.  (4.6/E)                                                   701
    THE USE OF MACHINE FACTORING IN MULTIPLE CORRELATION.
    J. AMER. STATIST. ASSOC., 23, 291-295.


BRANDWOOD, L.   SEE
    1077. COX, D. R. + BRANDWOOD, L. (1959)


BRANNSTROM, B.   SEE
    4526. REYMENT, R.A. + BRANNSTROM, B. (1962)


BRAUER, ALFRED T. 1946.  (20.1)                                               702
    LIMITS FOR THE CHARACTERISTIC ROOTS OF A MATRIX.
    DUKE MATH. J., 13, 387-395.  (MR8,P.192)


BRAUER, ALFRED T. 1947.  (20.1)                                               703
    LIMITS FOR THE CHARACTERISTIC ROOTS OF A MATRIX.  II.
    DUKE MATH. J., 14, 21-26.  (MR8,P.559)


BRAUER, ALFRED T. 1948.  (20.1)                                               704
    LIMITS FOR THE CHARACTERISTIC ROOTS OF A MATRIX.  III.
    DUKE MATH. J., 15, 871-877.  (MR10,P.231)


BRAVAIS, AUGUSTE 1846.  (2.1/B)                                               705
    ANALYSE MATHEMATIQUE SUR LES PROBABILITES DES ERREURS DE SITUATION D'UN POINT.
    MEM. PRES. ACAD. ROY. SCI. INST. FRANCE, 9, 255-332.


BRAVERMAN, D.   SEE
    8. ABRAMSON, N. M. + BRAVERMAN, D. (1962)


BRAVERMAN, E. M.   SEE
    68. AIZERMAN, M. A.; BRAVERMAN, E. M. + ROZONOER, L. I. (1964)

    69. AIZERMAN, M. A.; BRAVERMAN, E. M. + ROZONOER, L. I. (1964)


BREITENBERGER, ERNST 1963.  (18.1)                                           706
    ANALOGUES OF THE NORMAL DISTRIBUTION ON THE CIRCLE AND THE SPHERE.
    BIOMETRIKA, 50, 81-88.  (MR27-6316)



BRESCIANI, C. 1909.  (4.1)                                                    707
    SUI METODI PER LA MISURA DELLE CORRELAZIONI.
    GIORN. ECON. ANN. ECONOMIA SER. 2, 38, 401-444, 491-516.


BRIDGER, CLYDE A. 1938.  (8.1)                                                708
    NOTE ON REGRESSION FUNCTIONS IN THE CASE OF THREE SECOND-ORDER RANDOM VARIABLES.
    ANN. MATH. STATIST., 9, 309-313.


BRIER, GLENN W.; SCHOOT, R. G. + SIMMONS, V. L. 1940.  E(5.2)                709
    THE DISCRIMINANT FUNCTION APPLIED TO QUALITY RATING IN SHEEP.
    PROC. AMER. SOC. ANIMAL PROD., 33, 153-160.


BRILLINGER, DAVID R. 1965.  (16.3/16.4)                                      710
    AN INTRODUCTION TO POLYSPECTRA.
    ANN. MATH. STATIST., 36, 1351-1374.  (MR31-6333)


BRILLINGER, DAVID R. 1965.  (16.5)                                           711
    A MOVING AVERAGE REPRESENTATION FOR RANDOM VARIABLES COVARIANCE
    STATIONARY ON A FINITE TIME INTERVAL.
    BIOMETRIKA, 52, 295-297.


BRILLINGER, DAVID R. 1966.  (17.3)                                           712
    AN EXTREMAL PROPERTY OF THE CONDITIONAL EXPECTATION.
    BIOMETRIKA, 53, 594-599.

BRODSKII, M. L. 1952.   (20.3)                                    713
    A PROBABILISTIC ESTIMATE OF THE ERROR IN THE DETERMINATION OF THE
    EIGENVALUES AND EIGENVECTORS OF A VARIATION MATRIX.
       (IN RUSSIAN)
  USPEHI MAT. NAUK, 51, 205-214.   (MR14,P.692)

BRODZINSKY, A.    SEE
  4122. PAGE, R. M.; BRODZINSKY, A. + ZIRM, R. R. (1953)

BROGDEN, HUBERT E. 1946.   (15.4)                                  714
    THE EFFECT OF BIAS DUE TO DIFFICULTY FACTORS IN PRODUCT-MOMENT ITEM
    INTERCORRELATIONS ON THE ACCURACY OF ESTIMATION OF RELIABILITY BY THE
    KUDER-RICHARDSON FORMULA NUMBER 20.
  EDUC. PSYCHOL. MEAS., 6, 517-520.

BROGDEN, HUBERT E. 1946.   (4.3/15.4)                              715
    ON THE INTERPRETATION OF THE CORRELATION COEFFICIENT AS A MEASURE OF
    PREDICTIVE EFFICIENCY.
  J. EDUC. PSYCHOL., 37, 65-76.

BROGDEN, HUBERT E. 1946.   (6.6/E)                                 716
    AN APPROACH TO THE PROBLEM OF DIFFERENTIAL PREDICTION.
  PSYCHOMETRIKA, 11, 139-154.

BROGDEN, HUBERT E. 1946.   (15.4)                                  717
    VARIATION IN TEST VALIDITY WITH VARIATION IN THE DISTRIBUTION OF ITEM
    DIFFICULTIES, NUMBER OF ITEMS, AND DEGREE OF THEIR INTERCORRELATION.
  PSYCHOMETRIKA, 11, 197-214.

BROGDEN, HUBERT E. 1949.   (4.4/4.8/15.4)                          718
    A NEW COEFFICIENT:  APPLICATION TO BISERIAL CORRELATION AND TO
    ESTIMATION OF SELECTIVE EFFICIENCY.
  PSYCHOMETRIKA, 14, 169-182.

BROGDEN, HUBERT E. 1951.   (6.7/8.1/15.5)                          719
    INCREASED EFFICIENCY OF SELECTION RESULTING FROM REPLACEMENT OF A
    SINGLE PREDICTOR WITH SEVERAL DIFFERENTIAL PREDICTORS.
  EDUC. PSYCHOL. MEAS., 11, 173-195.

BROGDEN, HUBERT E. 1954.   (6.6)                                   720
    A SIMPLE PROOF OF A PERSONNEL CLASSIFICATION THEOREM.
  PSYCHOMETRIKA, 19, 205-208.

BROGDEN, HUBERT E. 1955.   (6.7/8.1)                               721
    LEAST SQUARES ESTIMATES AND OPTIMAL CLASSIFICATION.
  PSYCHOMETRIKA, 20, 249-252.

BROGDEN, HUBERT E. 1957.   (15.4)                                  722
    THE EXPECTED VARIANCE OF THE SAMPLING ERRORS FOR A SET OF
    ITEM-CRITERION CORRELATIONS.
  PSYCHOMETRIKA, 22, 75-78.

BROGDEN, HUBERT E. 1964.   (6.7/E)                                 723
    SIMPLIFIED REGRESSION PATTERNS FOR CLASSIFICATION.
  PSYCHOMETRIKA, 29, 393-396.

BRONFENBRENNER, JEAN 1953.   (16.1)                                724
    SOURCES AND SIZE OF LEAST-SQUARES BIAS IN A TWO-EQUATION MODEL.
  COWLES COMMISS. MONOG., 14, 221-235.

BRONFIN, H. + NEWHALL, S. M. 1934.   (8.2)                         725
    REGRESSION AND STANDARD ERROR CALCULATIONS WITHOUT THE CORRELATION COEFFICIENT.
  J. EDUC. PSYCHOL., 25, 634-636.

BRONOWSKI, J. + NEYMAN, JERZY 1945.   (17.3)                       726
    THE VARIANCE OF THE MEASURE OF A TWO-DIMENSIONAL RANDOM SET.
  ANN. MATH. STATIST., 16, 330-341.   (MR8,P.389)

BROOK, D. 1966.   (17.2)                                           727
    BOUNDS FOR MOMENT GENERATING FUNCTIONS AND FOR EXTINCTION PROBABILITIES.
  J. APPL. PROB., 3, 171-178.   (MR33-1868)

BROOKNER, RALPH J. 1945.  (19.1)                                              728
    CHOICE OF ONE AMONG SEVERAL STATISTICAL HYPOTHESES.
    ANN. MATH. STATIST., 16, 221-242.  (MR8,P.475)

BROOKNER, RALPH J.   SEE ALSO
    5729. WALD, ABRAHAM + BROOKNER, RALPH J. (1941)

    5730. WALD, ABRAHAM + BROOKNER, RALPH J. (1955)


BROVERMAN, DONALD M. 1961.  (15.1/E)                                          729
    EFFECTS OF SCORE TRANSFORMATIONS IN Q AND R FACTOR ANALYSIS TECHNIQUE.
    PSYCHOL. REV., 68, 68-80.


BROWN, CLARENCE W.; BARTELME, PHYLLIS + COX, GERTRUDE M. 1933.  (15.4)        730
    THE SCORING OF INDIVIDUAL PERFORMANCE ON TESTS SCALED ACCORDING TO
    THE THEORY OF ABSOLUTE SCALING.
    J. EDUC. PSYCHOL., 24, 654-662.


BROWN, GEORGE W. 1939.  (10.2/U)                                             731
    ON THE POWER OF THE $L_1$-TEST FOR EQUALITY OF SEVERAL VARIANCES.
    ANN. MATH. STATIST., 10, 119-128.


BROWN, GEORGE W. 1947.  (8.3)                                                732
    DISCRIMINANT FUNCTIONS.
    ANN. MATH. STATIST., 18, 514-528.  (MR9,P.195)


BROWN, GEORGE W. 1950.  (6.1)                                                733
    BASIC PRINCIPLES FOR CONSTRUCTION AND APPLICATION OF DISCRIMINATORS.
    (WITH DISCUSSION BY JOHN W. TUKEY AND JOHN C. FLANAGAN)
    J. CLIN. PSYCHOL., 6, 58-76.


BROWN, J. F. 1934.  (15.4)                                                   734
    A METHODOLOGICAL CONSIDERATION OF THE PROBLEM OF PSYCHOMETRICS.
    ERKENNTNIS, 4, 46-61.


BROWN, J. F. 1936.  (15.5)                                                   735
    ON THE USE OF MATHEMATICS IN PSYCHOLOGICAL THEORY.
    PSYCHOMETRIKA, 1(1) 77-90; (2) 7-15.


BROWN, J. W.; GREENWOOD, MAJOR + WOOD, FRANCES 1914.  (4.8)                  736
    A STUDY OF INDEX CORRELATIONS.
    J. ROY. STATIST. SOC., 77, 317-346.


BROWN, LAURA M.   SEE
    2382. HARTKEMEIER, HARRY P. + BROWN, LAURA M. (1936)


BROWN, LAWRENCE D. 1966.  (17.5/19.3)                                        737
    ON THE ADMISSIBILITY OF INVARIANT ESTIMATORS OF ONE OR MORE LOCATION
    PARAMETERS.
    ANN. MATH. STATIST., 37, 1087-1136.  (MR35-7476)


BROWN, R. L. 1957.  (17.7/B)                                                 738
    BIVARIATE STRUCTURAL RELATION.
    BIOMETRIKA, 44, 84-96.  (MR19,P.186)


BROWN, R. L. + FEREDAY, F. 1958.  (17.7)                                     739
    MULTIVARIATE LINEAR STRUCTURAL RELATIONS.
    BIOMETRIKA, 45, 136-153.  (MR19,P.1094)


BROWN, T. A. I.   SEE
    3264. LANCASTER, H. O. + BROWN, T. A. I. (1965)


BROWN, VIRGINIA M.   SEE
    2113. GREENE, JOEL E.; MILLER, WILBUR C. + BROWN, VIRGINIA M. (1956)


BROWN, W. R. J.; HOWE, WILLIAM G.; JACKSON, J. EDWARD + 1956.  (2.1/E)       740
    MORRIS, R.H.
    MULTIVARIATE NORMALITY OF THE COLOUR MATCHING PROCESS.
    J. OPT. SOC. AMER., 46, 46-49.

BROWN, WILLIAM 1909.  (4.8/E)                                                          741
    SOME EXPERIMENTAL RESULTS IN CORRELATION.
        (WITH DISCUSSION)
    PROC. INTERNAT. CONG. PSYCHOL., 6, 571-578.


BROWN, WILLIAM 1913.  (15.1/15.2)                                                      742
    THE EFFECTS OF "OBSERVATIONAL ERRORS" AND OTHER FACTORS UPON
    CORRELATION COEFFICIENTS IN PSYCHOLOGY.
    BRITISH J. PSYCHOL., 6, 223-238.

BROWN, WILLIAM 1932.  (15.1/E)                                                         743
    THE MATHEMATICAL AND EXPERIMENTAL EVIDENCE FOR THE EXISTENCE OF A
    CENTRAL INTELLECTIVE FACTOR (G).
    BRITISH J. PSYCHOL., 23, 171-179.

BROWN, WILLIAM 1934.  (15.2)                                                           744
    THE THEORY OF TWO FACTORS VERSUS THE SAMPLING THEORY OF MENTAL ABILITY.
        (AMENDED BY NO. 745)
    NATURE, 133, 724-725.

BROWN, WILLIAM 1935.  (15.2)                                                           745
    A NOTE ON THE THEORY OF TWO FACTORS VERSUS THE SAMPLING THEORY OF
    MENTAL ABILITY.
        (AMENDMENT OF NO. 744)
    BRITISH J. PSYCHOL., 25, 395-398.

BROWN, WILLIAM   SEE ALSO
    5236. STEPHENSON, WILLIAM + BROWN, WILLIAM (1933)

BROWNE-CAVE, F. E. CAVE-  SEE  CAVE-BROWNE-CAVE, F. E.

BROWNELL, WILLIAM H. 1933.  (15.4)                                                     746
    ON THE ACCURACY WITH WHICH RELIABILITY MAY BE MEASURED BY CORRELATING
    TEST HALVES.
    J. EXPER. EDUC., 1, 204-215.

BROWNLEE, JOHN 1910.  (2.1/E)                                                          747
    THE SIGNIFICANCE OF THE CORRELATION COEFFICIENT WHEN APPLIED TO
    MENDELIAN DISTRIBUTIONS.
    PROC. ROY. SOC. EDINBURGH, 30, 473-507.

BROWNLEE, JOHN 1925.  (4.2/E)                                                          748
    ON THE ERROR IN THE CORRELATION DUE TO RANDOM SAMPLING WHEN
    PROPORTIONATE MORTALITIES ARE USED.
    J. ROY. STATIST. SOC., 88, 105-106.

BROYLER, CECIL R. 1932.  (15.2)                                                        749
    A FORMULA FOR A MEAN TETRAD.
    J. GENERAL PSYCHOL., 6, 212-214.

BRUCE, W. J. 1955.  E(4.6)                                                             750
    SOME EVIDENCE ON THE EFFECTS OF THE USE OF A BASIC MATRIX IN MULTIPLE
    CORRELATION.
    EDUC. PSYCHOL. MEAS., 15, 181-185.

BRUCKSHAW, J. MC G.   SEE
    5648. VINCENZ, S.A. + BRUCKSHAW, J. MC G. (1960)

BRUNER, NANCY 1947.  (20.3)                                                            751
    NOTE ON THE DOOLITTLE SOLUTION.
    ECONOMETRICA, 15, 43-44.  (MR8,P.407)


BRUNK, HUGH D. 1960.  (15.5)                                                           752
    MATHEMATICAL MODELS FOR RANKING FROM PAIRED COMPARISONS.
    J. AMER. STATIST. ASSOC., 55, 503-520.  (MR22-6044)

BRYAN, JOSEPH G. 1951.  E(6.3)                                                         753
    THE GENERALIZED DISCRIMINANT FUNCTION:  MATHEMATICAL FOUNDATION AND
    COMPUTATIONAL ROUTINE.
    HARVARD EDUC. REV., 21(2) 90-95.

BRYAN, MIRIAM M.; BURKE, PAUL J. + STEWART, NAOMI 1952. (15.4)        754
    CORRECTION FOR GUESSING IN THE SCORING OF PRETESTS: EFFECT UPON ITEM
    DIFFICULTY AND ITEM VALIDITY INDICES.
    EDUC. PSYCHOL. MEAS., 12, 45–56.

BRYANT, SOPHIE 1893. (2.1/20.3/E)        755
    AN EXAMPLE IN THE "CORRELATION OF AVERAGES" FOR FOUR VARIABLES.
    PHILOS. MAG. SER. 5, 36, 372–377.

BRYSON, MARION R. 1965. (6.4)        756
    ERRORS OF CLASSIFICATION IN A BINOMIAL POPULATION.
    J. AMER. STATIST. ASSOC., 60, 217–224.

BUCK, S. F. 1960. (8.1)        757
    A METHOD OF ESTIMATION OF MISSING VALUES IN MULTIVARIATE DATA SUITABLE
    FOR USE WITH AN ELECTRONIC COMPUTER.
    J. ROY. STATIST. SOC. SER. B, 22, 302–306. (MR22–8606)

BUCKINGHAM, B. R. 1920. (4.1)        758
    DR. AYRES' FORMULA.
    J. EDUC. RES., 2, 505–507.


BUCY, R. S.   SEE
    2913. KALMAN, R. E. + BUCY, R. S. (1961)

BUDDE, C. DONALD LA  SEE  LA BUDDE, C. DONALD

BULA, CLOTILDE A. 1940. (2.3/17.9/B)        759
    THEORY AND EVALUATION OF CENTRAL MOMENTS IN TWO DIMENSIONS.
    SHEPPARD'S CORRECTIONS.  THE SIMPLER METHOD OF MITROPOLSKY.
        (TRANSLATION OF NO. 760)
    PUBL. UN. MAT. ARGENT., 1–97.  (MR2,P.231)

BULA, CLOTILDE A. 1940. (2.3/17.9/B)        760
    THEORY AND EVALUATION OF CENTRAL MOMENTS IN TWO DIMENSIONS.
    SHEPPARD'S CORRECTIONS.  THE SIMPLER METHOD OF MITROPOLSKY.
        (IN SPANISH, TRANSLATED AS NO. 759)
    REV. UN. MAT. ARGENT., 5, 1–97.  (MR2,P.231)

BULLARD, C.   SEE
    1220. DAVENPORT, C. B. + BULLARD, C. (1896)

BUNKE, O. 1964. (6.1/6.2/6.3/A)        761
    UBER OPTIMALE VERFAHREN DER DISKRIMINANZANALYSE.
    ABHANDL. DEUTSCHEN AKAD. WISS. BERLIN KL. MATH. PHYS. TECH., 4, 35–41.  (MR32–6624)

BUNKE, O. 1966. (6.4/17.8)        762
    NICHTPARAMETRISCHE KLASSIFIKATIONSVERFAHREN FUR QUALITATIVE UND
    QUANTITATIVE BEOBACHTUNGEN.
    WISS. Z. HUMBOLDT-UNIV. BERLIN MATH.-NATURWISS. REIHE, 15–18.  (MR36–1031)

BURBURY, S. H. 1895. (18.1)        763
    ON THE LAW OF ERROR IN THE CASE OF CORRELATED VARIATIONS.
    REP. BRITISH ASSOC. ADVANC. SCI., 621–624.

BURBURY, S. H. 1908. (18.1)        764
    ON THE LAW OF EQUIPARTITION OF ENERGY BETWEEN CORRELATED VARIABLES.
    REP. BRITISH ASSOC. ADVANC. SCI., 598–599.

BURBURY, S. H. 1909. (2.1/17.3)        765
    ON THE LAW OF PROBABILITY FOR A SYSTEM OF CORRELATED VARIABLES.
    PHILOS. MAG. SER. 6, 17, 1–28.

BURFORD, THOMAS M. 1955. (4.1)        766
    QUALITATIVE EVALUATION OF CORRELATION COEFFICIENTS FROM SCATTER DIAGRAMS.
    J. APPL. PHYS., 26, 56–57.

BURFORD, THOMAS M.; RIDEOUT, V. C. + SATHER, D. S. 1955. (16.5/E)        767
    THE USE OF CORRELATION TECHNIQUES IN THE STUDY OF SERVOMECHANISMS.
    J. BRITISH INST. RADIO ENGRS., 15, 249–257.

BURKE, C. J. + ESTES, WILLIAM K. 1957.  (15.5/E)                    768
    A COMPONENT MODEL FOR STIMULUS VARIABLES IN DISCRIMINATION LEARNING.
    PSYCHOMETRIKA, 22, 133–145.  (MR19,P.106)

BURKE, C. J.   SEE ALSO
    1572. ESTES, WILLIAM K. + BURKE, C. J. (1955)

    1575. ESTES, WILLIAM K.; BURKE, C. J.; ATKINSON, RICHARD C. +   (1957)
    FRANKMANN, J.P.

BURKE, PAUL J.   SEE
    754. BRYAN, MIRIAM M.; BURKE, PAUL J. + STEWART, NAOMI (1952)

BURKET, GEORGE R. 1964.  (8.1/8.7/E)                    769
    A STUDY OF REDUCED RANK MODELS FOR MULTIPLE PREDICTION.
    PSYCHOMET. MONOG., 12, 1–66.

BURKHARD, JAMES H.   SEE
    4259. PENNY, SAMUEL J. + BURKHARD, JAMES H. (1966)

BURKHARDT, FELIX 1957.  (4.1)                    770
    UBER DIE INVERSION VON KORRELATIONEN.
    BULL. INST. INTERNAT. STATIST., 35(2) 3–5.  (MR23–A728)

BURNABY, T. P. 1966.  (5.3/6.5/8.7)                    771
    GROWTH-INVARIANT DISCRIMINANT FUNCTIONS AND GENERALIZED DISTANCES.
    BIOMETRICS, 22, 96–110.  (MR33–1926)

BURNHAM, PAUL S. + CRAWFORD, ALBERT B. 1935.  (2.1)                    772
    THE VOCATIONAL INTERESTS AND PERSONALITY TEST SCORES OF A PAIR OF DICE.
    J. EDUC. PSYCHOL., 26, 508–512.

BURR, E. J. 1960.  (4.8/T)                    773
    THE DISTRIBUTION OF KENDALL'S SCORE S FOR A PAIR OF TIED RANKINGS.
    BIOMETRIKA, 47, 151–172.

BURR, IRVING W.   SEE
    3468. LONG, W. F. + BURR, IRVING W. (1949)

BURRUS, W. R.   SEE
    4760. RUST, B.; BURRUS, W. R. + SCHNEEBERGER, C. (1966)

BURT, CYRIL 1909.  (15.1/E)                    774
    EXPERIMENTAL TESTS OF GENERAL INTELLIGENCE.
    BRITISH J. PSYCHOL., 3, 94–177.

BURT, CYRIL 1937.  (15.2)                    775
    METHODS OF FACTOR-ANALYSIS WITH AND WITHOUT SUCCESSIVE APPROXIMATION.
    BRITISH J. EDUC. PSYCHOL., 7, 172–195.

BURT, CYRIL 1937.  (15.1)                    776
    CORRELATIONS BETWEEN PERSONS.
    BRITISH J. PSYCHOL., 28, 59–96.

BURT, CYRIL 1938.  (15.2)                    777
    FACTOR ANALYSIS BY SUB-MATRICES.
    J. PSYCHOL., 6, 339–375.

BURT, CYRIL 1938.  (15.1/15.4)                    778
    RECENT DEVELOPMENTS OF STATISTICAL METHODS IN PSYCHOLOGY.  I.
    OCCUP. PSYCHOL. LONDON, 12, 169–177.

BURT, CYRIL 1939.  (15.1)                    779
    THE FACTORIAL ANALYSIS OF ABILITY.  III.  LINES OF POSSIBLE RECONCILEMENT.
    BRITISH J. PSYCHOL., 30, 84–93.

BURT, CYRIL 1943.  (6.7/15.4)                    780
    VALIDATING TESTS FOR PERSONNEL SELECTION.
    BRITISH J. PSYCHOL., 34, 1–19.

BURT, CYRIL 1944.  (15.1)                    781
    MENTAL ABILITIES AND MENTAL FACTORS.
    BRITISH J. EDUC. PSYCHOL., 14, 85–94.

BURT, CYRIL 1944.   (2.4/6.7/15.4/E)                                                   782
    STATISTICAL PROBLEMS IN THE EVALUATION OF ARMY TESTS.
    PSYCHOMETRIKA, 9, 219-235.


BURT, CYRIL 1947.   (15.1)                                                             783
    L'ANALYSE FACTORIELLE DANS LA PSYCHOLOGIE ANGLAISE EN SE REFERANT SPECIALEMENT
    A L'OEUVRE DU PROFESSEUR SPEARMAN.
    BIOTYPOLOGIE, 9, 7-44.


BURT, CYRIL 1947.   (8.5/15.1)                                                         784
    A COMPARISON OF FACTOR ANALYSIS AND ANALYSIS OF VARIANCE.
    BRITISH J. PSYCHOL. STATIST. SECT., 1, 3-26.


BURT, CYRIL 1947.   (15.1)                                                             785
    FACTOR ANALYSIS AND PHYSICAL TYPES.
    PSYCHOMETRIKA, 12, 171-188.


BURT, CYRIL 1948.   (12.1/15.1)                                                        786
    FACTOR ANALYSIS AND CANONICAL CORRELATIONS.
    BRITISH J. PSYCHOL. STATIST. SECT., 1, 95-106.


BURT, CYRIL 1949.   (15.1)                                                             787
    THE STRUCTURE OF THE MIND:  A REVIEW OF THE RESULTS OF FACTOR ANALYSIS.  PART I.
    BRITISH J. EDUC. PSYCHOL., 19, 100-111, 176-199.


BURT, CYRIL 1949.   (15.1)                                                             788
    THE STRUCTURE OF THE MIND.  A REVIEW OF THE RESULTS OF FACTOR ANALYSIS.
    BRITISH J. EDUC. PSYCHOL., 19, 176-199.


BURT, CYRIL 1949.   (15.1)                                                             789
    ALTERNATIVE METHODS OF FACTOR ANALYSIS.
    BRITISH J. PSYCHOL. STATIST. SECT., 2, 98-121.


BURT, CYRIL 1950.   (15.1)                                                             790
    GROUP FACTOR ANALYSIS.
    BRITISH J. PSYCHOL. STATIST. SECT., 3, 40-75.


BURT, CYRIL 1950.   (6.1/6.2/6.3)                                                      791
    APPENDIX:  ON THE DISCRIMINATION BETWEEN MEMBERS OF TWO GROUPS.
    BRITISH J. PSYCHOL. STATIST. SECT., 3, 104.


BURT, CYRIL 1950.   (15.1)                                                             792
    THE FACTORIAL ANALYSIS OF QUALITATIVE DATA.
    BRITISH J. PSYCHOL. STATIST. SECT., 3, 166-185.


BURT, CYRIL 1951.   (20.3)                                                             793
    THE NUMERICAL SOLUTION OF LINEAR EQUATIONS.
    BRITISH J. PSYCHOL. STATIST. SECT., 4, 31-54.


BURT, CYRIL 1951.   (15.4)                                                             794
    TEST-CONSTRUCTION AND THE SCALING OF ITEMS.
    BRITISH J. PSYCHOL. STATIST. SECT., 4, 95-129.


BURT, CYRIL 1952.   (15.2)                                                             795
    TESTS OF SIGNIFICANCE IN FACTOR ANALYSIS.
    BRITISH J. PSYCHOL. STATIST. SECT., 5, 109-133.


BURT, CYRIL 1953.   (15.1/15.5)                                                        796
    SCALE ANALYSIS AND FACTOR ANALYSIS.
    BRITISH J. STATIST. PSYCHOL., 6, 5-23.


BURT, CYRIL 1954.   (15.1)                                                             797
    THE SIGN PATTERN OF FACTOR-MATRICES.
    BRITISH J. STATIST. PSYCHOL., 7, 15-29.


BURT, CYRIL 1955.   (15.4)                                                             798
    TEST RELIABILITY ESTIMATED BY ANALYSIS OF VARIANCE.
    BRITISH J. STATIST. PSYCHOL., 8, 103-118.


BURT, CYRIL 1955.   (8.5/11.1/15.1)                                                    799
    L'ANALYSE FACTORIELLE:  METHODES ET RESULTATS.
        (WITH DISCUSSION)
    COLLOQ. INTERNAT. CENTRE NAT. RECH. SCI., 65, 79-105.

BURT, CYRIL 1962.  (1.1)                                                          800
    FRANCIS GALTON AND HIS CONTRIBUTIONS TO PSYCHOLOGY.
    BRITISH J. STATIST. PSYCHOL., 15, 1–49.

BURT, CYRIL 1966.  (8.5/15.2/E)                                                   801
    THE APPROPRIATE USES OF FACTOR ANALYSIS AND ANALYSIS OF VARIANCE.
    HANDB. MULTIVARIATE EXPER. PSYCHOL. (CATTELL), 267–287.

BURT, CYRIL 1966.  (1.1)                                                          802
    THE EARLY HISTORY OF MULTIVARIATE TECHNIQUES IN PSYCHOLOGICAL RESEARCH.
    MULTIVARIATE BEHAV. RES., 1, 24–42.

BURT, CYRIL + FOLEY, ELIZABETH 1956.  (15.2/BE)                                   803
    THE STATISTICAL ANALYSIS OF THE LEARNING PROCESS. I. TASKS WITH TWO RESPONSES ONLY.
    BRITISH J. STATIST. PSYCHOL., 9, 49–56.

BURT, CYRIL + HOWARD, MARGARET 1956.  (15.1/E)                                    804
    THE MULTIFACTORIAL THEORY OF INHERITANCE AND ITS APPLICATION TO INTELLIGENCE.
    BRITISH J. STATIST. PSYCHOL., 9, 95–131.

BURT, CYRIL + JOHN, ENID 1942.  (2.1)                                             805
    A FACTORIAL ANALYSIS OF TERMAN BINET TEST.  PART I.
    BRITISH J. EDUC. PSYCHOL., 12, 117–127, 156–161.

BURT, CYRIL + STEPHENSON, WILLIAM 1939.  (15.1)                                   806
    ALTERNATIVE VIEWS ON CORRELATIONS BETWEEN PERSONS.
    PSYCHOMETRIKA, 4, 269–281.

BURT, CYRIL    SEE ALSO
    292. BANKS, CHARLOTTE + BURT, CYRIL (1954)

    308. BARLOW, J.A. + BURT, CYRIL (1954)

BUSH, KENNETH A. + OLKIN, INGRAM 1959.  (6.2/12.1/20.1)                           807
    EXTREMA OF QUADRATIC FORMS WITH APPLICATIONS TO STATISTICS.
        (AMENDED BY NO. 808)
    BIOMETRIKA, 46, 483–486.  (MR22-8588)

BUSH, KENNETH A. + OLKIN, INGRAM 1961.  (6.2/12.1/20.1)                           808
    CORRECTION TO:  EXTREMA OF QUADRATIC FORMS WITH APPLICATIONS TO STATISTICS.
        (AMENDMENT OF NO. 807)
    BIOMETRIKA, 48, 474–475.  (MR22-6588)

BUSH, KENNETH A. + OLKIN, INGRAM 1961.  (20.1)                                    809
    EXTREMA OF FUNCTIONS OF A REAL SYMMETRIC MATRIX IN TERMS OF EIGENVALUES.
    DUKE MATH. J., 28, 143–152.

BUSH, ROBERT R.    SEE
    3911. MOSTELLER, FREDERICK + BUSH, ROBERT R. (1954)

BUSINGER, P. + GOLUB, GENE H. 1965.  (20.3)                                       810
    LINEAR LEAST SQUARES SOLUTIONS BY HOUSEHOLDER TRANSFORMATIONS.
    NUMER. MATH., 7, 269–276.

BUSS, ARNOLD H. 1953.  (15.5)                                                     811
    RIGIDITY AS A FUNCTION OF ABSOLUTE AND RELATIONAL SHIFTS IN THE
    LEARNING OF SUCCESSIVE DISCRIMINATIONS.
    J. EXPER. PSYCHOL., 45, 153–156.

BUTLER, JOHN M. 1942.  (15.4)                                                     812
    A RATIO FOR ESTIMATING THE RELIABILITY OF TEST SCORES.
    J. EDUC. PSYCHOL., 33, 391–395.

BUTLER, JOHN M. + HOOK, L. HARMON 1966.  (15.1)                                   813
    MULTIPLE FACTOR ANALYSIS IN TERMS OF WEIGHTED REGRESSION.
    EDUC. PSYCHOL. MEAS., 26, 545–564.

CACOULLOS, THEOPHILOS 1965.  (4.2)                                                814
    A RELATION BETWEEN T AND F-DISTRIBUTIONS.
    J. AMER. STATIST. ASSOC., 60, 528–531.

CACOULLOS, THEOPHILOS 1965.  (6.5/19.1)                                                    815
    COMPARING MAHALANOBIS DISTANCES, I: COMPARING DISTANCES BETWEEN  K  KNOWN
    NORMAL POPULATIONS AND ANOTHER UNKNOWN.
    SANKHYA SER. A, 27, 1-22.  (MR32-4787)

CACOULLOS, THEOPHILOS 1965.  (6.5/19.1)                                                    816
    COMPARING MAHALANOBIS DISTANCES, II: BAYES PROCEDURES WHEN THE MEAN VECTORS ARE UNKNOWN.
    SANKHYA SER. A, 27, 23-32.  (MR32-4788)

CACOULLOS, THEOPHILOS 1966.  (17.8)                                                        817
    ESTIMATION OF A MULTIVARIATE DENSITY.
    ANN. INST. STATIST. MATH. TOKYO, 18, 179-189.  (MR35-1149)

CACOULLOS, THEOPHILOS 1966.  (6.4/19.1/CT)                                                 818
    ON A CLASS OF ADMISSIBLE PARTITIONS.
    ANN. MATH. STATIST., 37, 189-195.

CACOULLOS, THEOPHILOS + OLKIN, INGRAM 1965.  (3.3/17.2/17.3)                               819
    ON THE BIAS OF CHARACTERISTIC ROOTS OF A RANDOM MATRIX.
    BIOMETRIKA, 52, 87-94.  (MR34-6932)

CACOULLOS, THEOPHILOS + SOBEL, MILTON 1966.  (18.2/19.1/A)                                 820
    AN INVERSE-SAMPLING PROCEDURE FOR SELECTING THE MOST PROBABLE EVENT IN
    A MULTINOMIAL DISTRIBUTION.
    MULTIVARIATE ANAL. PROC. INTERNAT. SYMP. DAYTON (KRISHNAIAH), 423-455.

CADWELL, J. H. 1951.  (2.7/B)                                                              821
    THE BIVARIATE NORMAL INTEGRAL.
    BIOMETRIKA, 38, 475-479.  (MR13,P.662)

CAFFREY, JOHN G.   SEE
    2900. KAISER, HENRY F. + CAFFREY, JOHN G. (1965)

CALDER, A. B. 1961.  (6.6/E)                                                               822
    THE USE OF DISCRIMINANT FUNCTIONS IN BIOLOGICAL SAMPLING.
    INC. STATIST., 2, 156-163.

CALDWELL, WILLIAM V.   SEE
    5442. THRALL, ROBERT M.; COOMBS, CLYDE H. + CALDWELL, WILLIAM V. (1958)

CALINSKI, T. 1966.  (8.5/8.6/E)                                                            823
    ON THE DISTRIBUTION OF THE F-TYPE STATISTICS IN THE ANALYSIS OF A
    GROUP OF EXPERIMENTS.
    J. ROY. STATIST. SOC. SER. B, 28, 526-542.

CALVIN, ALLEN D. + SEIBEL, JEAN L. 1954.  (15.5)                                           824
    A FURTHER INVESTIGATION OF RESPONSE SELECTION IN SIMULTANEOUS AND
    SUCCESSIVE DISCRIMINATION.
    J. EXPER. PSYCHOL., 48, 339-342.

CAMACHO, A.   SEE
    1338. DIAZ UNGRIA, A.; CAMACHO, A. + RIOS, S. (1955)

CAMARA, SIXTO 1932.  (4.6)                                                                 825
    PRINCIPIOS DE LA TEORIA DE LA CORRELACION MULTIPLE EN GENERAL.
    REV. MAT. HISP. AMER. SER. 2, 6, 249-262.

CAMARA, SIXTO 1932.  (4.6)                                                                 826
    PRINCIPIOS DE LA TEORIA DE LA CORRELACION MULTIPLE EN GENERAL. (CONTINUACION.)
    REV. MAT. HISP. AMER. SER. 2, 7, 7-21.

CAMARA, SIXTO 1932.  (4.6/8.8)                                                             827
    PRINCIPIOS DE LA TEORIA DE LA CORRELACION MULTIPLE EN GENERAL. (CONTINUACION.)
    CLASIFICACION DE LAS REDES DE REGRESION.
    REV. MAT. HISP. AMER. SER. 2, 7, 71-77.

CAMARA, SIXTO 1932.  (4.6/9.3)                                                             828
    PRINCIPIOS DE LA TEORIA DE LA CORRELACION MULTIPLE EN GENERAL. (CONCLUSION.)
    REV. MAT. HISP. AMER. SER. 2, 7, 97-112.

CAMERANO, LORENZO 1901.  E(4.8)                                                                  829
    LO STUDIO QUANTITATIVO DEGLI ORGANISMI E GLI INDICI DI MANCANZA, DI
    CORRELAZIONE E DI ASIMMETRIA.
    ATTI REALE ACCAD. SCI. TORINO PUBBL. ACCAD. CL. SCI. FIS. MAT. NATUR., 36, 371-376,639-644.

CAMERON, SCOTT H. 1957.  E(16.1)                                                                 830
    MEASUREMENT OF CORRELATION COEFFICIENTS.
    J. APPL. PHYS., 28, 377-378.

CAMP, BURTON H. 1922.  (17.2)                                                                    831
    A NEW GENERALIZATION OF TCHEBYCHEFF'S STATISTICAL INEQUALITY.
    BULL. AMER. MATH. SOC., 28, 427-432.

CAMP, BURTON H. 1925.  (20.4)                                                                    832
    PROBABILITY INTEGRALS FOR A HYPERGEOMETRIC SERIES.
    BIOMETRIKA, 17, 61-67.

CAMP, BURTON H. 1925.  (8.1/17.9)                                                                833
    MUTUALLY CONSISTENT MULTIPLE REGRESSION SURFACES.
    BIOMETRIKA, 17, 443-458.

CAMP, BURTON H. 1929.  (18.1/A)                                                                  834
    THE MULTINOMIAL SOLID AND THE CHI TEST.
        (AMENDED BY NO. 837)
    TRANS. AMER. MATH. SOC., 31, 133-144.

CAMP, BURTON H. 1932.  (15.1)                                                                    835
    THE CONVERSE OF SPEARMAN'S TWO-FACTOR THEOREM.
    BIOMETRIKA, 24, 418-427.

CAMP, BURTON H. 1934.  (15.1)                                                                    836
    SPEARMAN'S GENERAL FACTOR AGAIN.
    BIOMETRIKA, 26, 260-261.

CAMP, BURTON H. 1938.  (18.1/A)                                                                  837
    CORRECTION TO: THE MULTINOMIAL SOLID AND THE CHI TEST.
        (AMENDMENT OF NO. 834)
    TRANS. AMER. MATH. SOC., 44, 151.

CAMP, BURTON H. 1948.  (17.2)                                                                    838
    GENERALIZATION TO N DIMENSIONS OF INEQUALITIES OF THE TCHEBYCHEFF TYPE.
    ANN. MATH. STATIST., 19, 568-574.  (MR10,P.384)

CAMPBELL, DONALD J.    SEE
    1067. COTTON, JOHN W.; CAMPBELL, DONALD J. + MALONE, R. DANIEL (1957)

CAMPBELL, JOEL T. 1934.  (18.1/E)                                                                839
    THE POISSON CORRELATION FUNCTION.
    PROC. EDINBURGH MATH. SOC. SER. 2, 4, 18-26.

CAMPBELL, JOEL T.    SEE ALSO
    5845. WHERRY, ROBERT J.; CAMPBELL, JOEL T. + PERLOFF, ROBERT (1951)

CANELLOS, SPYRIDON  SEE  KANELLOS, SPYRIDON G.

CANSADO MACEDA, ENRIQUE 1947.  (14.5)                                                            840
    CUMULANTES DE LA Z DE FISHER.
    REV. MAT. HISP. AMER. SER. 4, 7, 87-89.  (MR9,P.48,P.735)

CANSADO MACEDA, ENRIQUE 1948.  (18.1)                                                            841
    ON THE COMPOUND AND GENERALIZED POISSON DISTRIBUTIONS.
    ANN. MATH. STATIST., 19, 414-416.  (MR10,P.552)

CANSADO MACEDA, ENRIQUE 1950.  (16.2)                                                            842
    INTERPRETACION VECTORIAL DE LA ECUACION DE SLUTSKY.
        (WITH SUMMARY IN ENGLISH)
    TRABAJOS ESTADIST., 1, 29-36.  (MR13,P.370)

CANSADO MACEDA, ENRIQUE 1951.  (17.3/B)                                                          843
    ESTUDIO DE ALGUNAS DISTRIBUCIONES BIDIMENSIONALES.
        (WITH SUMMARY IN ENGLISH)
    TRABAJOS ESTADIST., 2, 149-178.  (MR13,P.665)

CANSADO MACEDA, ENRIQUE 1951.  (18.1/B)                                          844
    UN EJEMPLO DE DISTRIBUCION BIDIMENSIONAL.
        (WITH SUMMARY IN ENGLISH)
    TRABAJOS ESTADIST., 2, 261-272.  (MR13,P.853)

CAPOCCIA, OTFLLO 1948.  (4.9)                                                    845
    UN INDICE BILATERALE DI CORRELAZIONE.
    STATISTICA, BOLOGNA, 8, 66-67.

CAPON, JACK 1964.  (16.4)                                                        846
    RADON-NIKODYM DERIVATIVES OF STATIONARY GAUSSIAN MEASURES.
    ANN. MATH. STATIST., 35, 517-531.  (MR28-4584)

CAPON, JACK 1965.  (2.5/6.6/19.1)                                                847
    AN ASYMPTOTIC SIMULTANEOUS DIAGONALIZATION PROCEDURE FOR PATTERN RECOGNITION.
    INFORMATION AND CONTROL, 8, 264-281.  (MR32-8452)

CARD, D. G. 1943.  (4.6)                                                         848
    GRAPHIC METHODS OF PRESENTING MULTIPLE CORRELATION ANALYSIS.
    J. FARM ECON., 25, 881-889.

CARLSON, HILDING B. 1945.  (15.3)                                                849
    A SIMPLE ORTHOGONAL MULTIPLE FACTOR APPROXIMATION PROCEDURE.
    PSYCHOMETRIKA, 10, 283-301.

CARLSON, PHILLIP G. 1956.  (4.1/8.1)                                             850
    A LEAST SQUARES INTERPRETATION OF THE BIVARIATE LINE OF ORGANIC CORRELATION.
    SKAND. AKTUARIETIDSKR., 39, 7-10.  (MR18,P.771)

CARMARGO, MANUEL CASTANS  SEE  CASTANS CARMARGO, MANUEL

CAROLIS, LINDA V. DE  SEE  DE CAROLIS, LINDA V.

CARPENTER, A. F. 1940.  (4.1)                                                    851
    SHORT CUTS IN WORKING OUT INTERCORRELATIONS.
    RES. QUART. AMER. ASSOC. HEALTH PHYS. EDUC. REC., 11, 32-37.

CARPENTER, JOHN A.  SEE
    2614. HOUSEHOLDER, ALSTON S. + CARPENTER, JOHN A. (1963)

CARPENTER, OSMER 1950.  (2.5)                                                    852
    NOTE ON THE EXTENSION OF CRAIG'S THEOREM TO NON-CENTRAL VARIATES.
    ANN. MATH. STATIST., 21, 455-457.  (MR12,P.621)

CARPMAEL, C. 1879.  (8.1/20.1)                                                   853
    ON THE VALUES OF THE CONSTANTS IN THE EQUATION
    $_rA_rx^{(r)} + _rA_{r-1}x^{(r-1)} + \ldots + _rA_tx^{(t)} + \ldots + _rA_o - y_x = 0$ OBTAINED BY THE METHOD OF LEAST SQUARES,
    FROM THE $n+1$ VALUES OF $y_x$ WHEN $x = 0, 1, 2 \ldots n$; $n$ BEING GREATER THAN $r$.
    MONTHLY NOTICES ROY. ASTRONOM. SOC., 39, 489-504.

CARROLL, J. DOUGLAS  SEE
    4997. SHEPARD, ROGER N. + CARROLL, J. DOUGLAS (1966)

CARROLL, JOHN B. 1945.  (15.4)                                                   854
    THE EFFECT OF DIFFICULTY AND CHANCE SUCCESS ON CORRELATIONS BETWEEN
    ITEMS OR BETWEEN TESTS.
    PSYCHOMETRIKA, 10, 1-19.

CARROLL, JOHN B. 1953.  (15.3)                                                   855
    AN ANALYTIC SOLUTION FOR APPROXIMATING SIMPLE STRUCTURE IN FACTOR ANALYSIS.
    PSYCHOMETRIKA, 18, 23-38.

CARROLL, JOHN B. 1957.  (15.1/15.3)                                              856
    BIQUARTIMIN CRITERION FOR ROTATION TO OBLIQUE SIMPLE STRUCTURE IN
    FACTOR ANALYSIS.
    SCIENCE, 126, 1114-1115.

CARROLL, JOHN B. + SCHWEIKER, ROBERT F. 1951.  (15.1/E)                          857
    FACTOR ANALYSIS IN EDUCATIONAL RESEARCH.
    REV. EDUC. RES., 21, 368-388.

CARTER, A. H. 1947.  (14.5)                                                      858
    APPROXIMATION TO PERCENTAGE POINTS OF THE Z-DISTRIBUTION.
    BIOMETRIKA, 34, 352-358.  (MR9,P.364)

CARTER, A. H. 1949.  (8.1/8.3/E)                                                 859
    THE ESTIMATION AND COMPARISON OF RESIDUAL REGRESSIONS WHERE THERE ARE
    TWO OR MORE RELATED SETS OF OBSERVATIONS.
    BIOMETRIKA, 36, 26-46.  (MR11,P.673)


CARTWRIGHT, DESMOND S. 1965.  (20.3)                                             860
    A NOTE ON SOME MODIFICATIONS OF LATENT ROOTS AND VECTORS.
    PSYCHOMETRIKA, 30, 319-321.


CASANOVA, TEOBALDO 1939.  (4.2/14.4)                                             861
    A TEST OF THE ASSUMPTIONS OF LINEARITY AND HOMOSCEDASTICITY MADE IN ESTIMATING
    THE CORRELATION IN ONE RANGE FROM THAT OBTAINED IN A DIFFERENT RANGE.
    J. EXPER. EDUC., 7, 245-249.


CASANOVA, TEOBALDO 1940.  (4.2)                                                  862
    CORRECTIONS TO CORRELATION COEFFICIENTS ON ACCOUNT OF HOMOGENEITY IN
    ONE VARIABLE.
    J. EXPER. EDUC., 8, 341-345.


CASESNOVES, DARIO MARAVALL   SEE   MARAVALL CASESNOVES, DARIO


CASLEY, D.J.   SEE
    363. BARTON, D. E. + CASLEY, D.J. (1958)


CASSIE, R. MORRISON 1963.  (6.2/11.1/E)                                          863
    MULTIVARIATE ANALYSIS IN THE INTERPRETATION OF NUMERICAL PLANKTON DATA.
    NEW ZEALAND J. SCI., 6, 36-59.


CASTANS CARMARGO, MANUEL + MEDINA E ISABEL, MARIANO 1956.  (4.8)                 864
    THE LOGARITHMIC CORRELATION.
        (IN SPANISH, WITH SUMMARY IN ENGLISH)
    AN. REAL SOC. ESPAN. FIS. QUIM. SER. A, 52, 117-136.  (MR18,P.79)


CASTELLAN, N. JOHN, JR. 1966.  (4.4)                                             865
    ON THE ESTIMATION OF THE TETRACHORIC CORRELATION COEFFICIENT.
    PSYCHOMETRIKA, 31, 67-73.  (MR35-3785)


CASTELLANI, MARIA 1950.  (18.1)                                                  866
    ON MULTINOMIAL DISTRIBUTIONS WITH LIMITED FREEDOM:  A STOCHASTIC
    GENESIS OF PARETO'S AND PEARSON'S CURVES.
    ANN. MATH. STATIST., 21, 289-293.  (MR11,P.673)


CASTELLANO, VITTORIO 1934.  (4.8)                                               867
    SULLO SCARTO QUADRATICO MEDIO DELLA PROBABILITA DI TRANSVARIAZIONE.
    METRON, 11(4) 19-75.


CASTELLANO, VITTORIO 1957.  (4.9)                                              868
    CONTRIBUTI ALLE TEORIE DELLA CORRELAZIONE E DELLA CONNESSIONE TRA DUE VARIABILI.
    METRON, 18(3) 25-79.


CASTELLANO, VITTORIO 1960.  (17.4/B)                                           869
    SULL'INSIEME DELLE DISTRIBUZIONI DOPPIE E TRIPLE RISULTANTI DALL'ASSOCIAZIONE
    UNA A UNA DELLA UNITA DI DUE O TRE DISTRIBUZIONI SEMPLICI AVENTI LO STESSO
    NUMERO DI UNITA.
    METRON, 20, 251-298.  (MR25-659)


CASTELLANO, VITTORIO 1965.  (1.2)                                              870
    CORRADO GINI: A MEMOIR.
    METRON, 24, 2-84.


CASTELLANO, VITTORIO 1965.  (1.2)                                              871
    CORRADO GINI, 1884-1965.
    REV. INST. INTERNAT. STATIST., 33, 337-344.


CASTELLANO, VITTORIO 1966.  (1.1)                                             872
    SCIENCES, METHOD AND STATISTICS.
    METRON, 25, 1-54.


CASTERMANS, M. 1951.  (4.5/4.7)                                               873
    NOTES SUR L'INTRODUCTION ET LA SIGNIFICATION DES MESURES HABITUELLES DE LA
    CORRELATION ENTRE DEUX OU PLUSIEURS VARIABLES ALEATOIRES EN STATISTIQUE
    MATHEMATIQUE.
    BULL. STATIST. INST. NAT. STATIST. BRUXELLES, 37(1) 3-15.

CASTLE, W. E. 1903.  (2.4)                                                              874
    THE LAWS OF HEREDITY OF GALTON AND MENDEL AND SOME LAW GOVERNING RACE
    IMPROVEMENT BY SELECTION.
    PROC. AMER. ACAD. ARTS SCI., 39, 223–240.

CASTOLDI, LUIGI 1955.  (17.3)                                                           875
    "RIDUCIBILITA" DI OGNI DISTRIBUZIONE STATISTICA MULTIPLA.
    REND. SEM. FAC. SCI. UNIV. CAGLIARI, 25, 137–142.  (MR19,P.70)

CASTORE, GEORGE F. + DYE, WILLIAM S., III 1949.  (20.3)                                 876
    A SIMPLIFIED PUNCH CARD METHOD OF DETERMINING SUMS OF SQUARES AND
    SUMS OF PRODUCTS.
    PSYCHOMETRIKA, 14, 243–250.

CATER, S. 1962.  (20.1)                                                                 877
    AN ELEMENTARY DEVELOPMENT OF THE JORDAN CANONICAL FORM.
    AMER. MATH. MONTHLY, 69, 391–393.

CATTELL, A.K.S.    SEE
    889. CATTELL, RAYMOND B. + CATTELL, A.K.S. (1955)

CATTELL, RAYMOND B. 1944.  (6.6)                                                        878
    A NOTE ON CORRELATION CLUSTERS AND CLUSTER SEARCH METHODS.
    PSYCHOMETRIKA, 9, 169–184.

CATTELL, RAYMOND B. 1944.  (6.6)                                                        879
    "PARALLEL PROPORTIONAL PROFILES" AND OTHER PRINCIPLES FOR DETERMINING
    THE CHOICE OF FACTORS BY ROTATION.
    PSYCHOMETRIKA, 9, 267–283.

CATTELL, RAYMOND B. 1946.  (15.1/E)                                                     880
    SIMPLE STRUCTURE IN RELATION TO SOME ALTERNATIVE FACTORIZATIONS OF
    THE PERSONALITY SPHERE.
    J. GENERAL PSYCHOL., 35, 225–238.

CATTELL, RAYMOND B. 1949.  (15.1)                                                       881
    A NOTE ON FACTOR INVARIANCE AND THE INDENTIFICATION OF FACTORS.
    BRITISH J. PSYCHOL. STATIST. SECT., 2, 134–139.

CATTELL, RAYMOND B. 1949.  (6.6/15.5)                                                   882
    $r_p$ AND OTHER COEFFICIENTS OF PATTERN SIMILARITY.
    PSYCHOMETRIKA, 14, 279–298.

CATTELL, RAYMOND B. 1952.  (15.1)                                                       883
    THE THREE BASIC FACTOR-ANALYTIC RESEARCH DESIGNS--THEIR
    INTERRELATIONS AND DERIVATIVES.
    PSYCHOL. BULL., 49, 499–520.

CATTELL, RAYMOND B. 1955.  (15.1)                                                       884
    GROWING POINTS IN FACTOR ANALYSIS.
    AUSTRAL. J. PSYCHOL., 6, 105–140.

CATTELL, RAYMOND B. 1965.  (15.1)                                                       885
    FACTOR ANALYSIS:  AN INTRODUCTION TO ESSENTIALS.  I.  THE PURPOSE AND
    UNDERLYING MODELS.
    BIOMETRICS, 21, 190–215.  (MR31-5244)

CATTELL, RAYMOND B. 1965.  (15.1)                                                       886
    FACTOR ANALYSIS:  AN INTRODUCTION TO ESSENTIALS.  II.  THE ROLE OF
    FACTOR ANALYSIS IN RESEARCH.
    BIOMETRICS, 21, 405–435.

CATTELL, RAYMOND B. 1966.  (15.1/15.2/15.3)                                             887
    THE MEANING AND STRATEGIC USE OF FACTOR ANALYSIS.
    HANDB. MULTIVARIATE EXPER. PSYCHOL. (CATTELL), 174–243.

CATTELL, RAYMOND B. 1966.  (15.2/E)                                                     888
    THE SCREE TEST FOR THE NUMBER OF FACTORS.
    MULTIVARIATE BEHAV. RES., 1, 245–276.

CATTELL, RAYMOND B. + CATTELL, A.K.S. 1955.  (15.3/E)                                   889
    FACTOR ROTATION FOR PROPORTIONAL PROFILES:  ANALYTICAL SOLUTION AND AN EXAMPLE.
    BRITISH J. STATIST. PSYCHOL., 8, 83–92.

CATTELL, RAYMOND B. + DICKMAN, KERN 1962.  (15.1/E)                    890
     A DYNAMIC MODEL OF PHYSICAL INFLUENCES DEMONSTRATING THE NECESSITY OF
     OBLIQUE SIMPLE STRUCTURE.
   PSYCHOL. BULL., 59, 389-400.

CATTELL, RAYMOND B.; COULTER, MALCOLM A. + TSUJIOKA, BIEN 1966.  (6.6/E)    891
     THE TAXONOMETRIC RECOGNITION OF TYPES AND FUNCTIONAL EMERGENTS.
   HANDB. MULTIVARIATE EXPER. PSYCHOL. (CATTELL), 288-329.

CATTELL, RAYMOND B.   SEE ALSO
     248. BAGGALEY, ANDREW R. + CATTELL, RAYMOND B. (1956)

CAUCHY, AUGUSTIN L. 1853.  (20.3)                                       892
     MEMOIRE SUR L'EVALUATION D'INCONNUES DETERMINEES PAR UN GRAND NOMBRE
     D'EQUATIONS APPROXIMATIVES DU PREMIER DEGRE.
   C.R. ACAD. SCI. PARIS, 36, 1114-1122.

CAUCHY, AUGUSTIN L. 1853.  (20.3)                                       893
     SUR LA PLUS GRANDE ERREUR A CRAINDRE DANS UN RESULTAT MOYEN, ET SUR
     LE SYSTEME DE FACTEURS QUI REND CETTE PLUS GRANDE ERREUR UN MINIMUM.
   C.R. ACAD. SCI. PARIS, 37, 326-334.

CAUSSINUS, HENRI 1966.  (18.1/18.2)                                     894
     SUR LA STRUCTURE DES TABLEAUX DE CORRELATION CARRES.
   C. R. ACAD. SCI. PARIS SER. A, 263, 795-797.

CAUSSINUS, HENRI   SEE ALSO
     4905. SCHEKTMAN, YVES + CAUSSINUS, HENRI (1965)

CAVALLI, LUIGI L. 1949.  (4.8/E)                                        895
     SULLA CORRELAZIONE MEDIA FRA PIU CARATTERI IN RELAZIONE ALLA BIOMETRIA.
          (WITH SUMMARY IN ENGLISH AND GERMAN)
   METRON, 15,   173-188.  (MR11,P.445)

CAVALLI, LUIGI L. 1951.  (2.1)                                          896
     UN METODO RAPIDO DI CALCOLO DELLA CORRELAZIONE MEDIA FRA PIU CARATTERI.
          (WITH SUMMARY IN ENGLISH AND GERMAN)
   METRON, 16(1) 151-167.

CAVALLI-SFORZA, L. L.   SEE
     1505. EDWARDS, A. W. F. + CAVALLI-SFORZA, L. L. (1965)

CAVE, BEATRICE M. + PEARSON, KARL 1914.  (4.8/E)                        897
     NUMERICAL ILLUSTRATIONS OF THE VARIATE DIFFERENCE CORRELATION METHOD.
   BIOMETRIKA, 10, 340-355.

CAVE, BEATRICE M.   SEE ALSO
     5135. SOPER, H.E.; YOUNG, ANDREW W.; CAVE, BEATRICE M.;  (1917)
     LEE, ALICE + PEARSON, KARL

CAVE-BROWNE-CAVE, F. E.   SEE
     4237. PEARSON, KARL + CAVE-BROWNE-CAVE, F. E. (1902)

CELL, JOHN W.   SEE
     5825. WELLS, W. T.; ANDERSON, R. L. + CELL, JOHN W. (1962)

CERNOV, L. A. 1955.  (16.4)                                            898
     CORRELATION OF AMPLITUDE AND PHASE FLUCTUATIONS FOR WAVE PROPAGATION
     IN A MEDIUM WITH RANDOM IRREGULARITIES.
          (IN RUSSIAN, TRANSLATED AS NO. 899)
   AKUST. ZUR., 1, 89-95.  (MR17,P.1252)

CERNOV, L. A. 1955.  (16.4)                                            899
     CORRELATION OF AMPLITUDE AND PHASE FLUCTUATIONS FOR WAVE PROPAGATION
     IN A MEDIUM WITH RANDOM IRREGULARITIES.
          (TRANSLATION OF NO. 898)
   SOVIET PHYS. ACOUST., 1, 94-101.  (MR17,P.1252)

CERVINKA, VLADIMIR 1948.  (15.1/15.2)                                  900
     FACTOR ANALYSIS.
   STATIST. OBZOR, 28, 145-162.

CHAKRABARTI, M. C. 1949.  (2.5/14.5)                                   901
     ON THE MOMENTS OF NON-CENTRAL $\chi^2$.
   BULL. CALCUTTA MATH. SOC., 41, 208-210.  (MR11,P.259)

CHAKRABORTY, P.N.    SEE
    913. CHANDRA SEKAR, C. + CHAKRABORTY, P.N. (1952)

CHAKRAVARTI, I. M. 1954.    (9.3/12.1)                                                    902
    RELATION BETWEEN CANONICAL CORRELATIONS AND PARTIALLY CANONICAL CORRELATIONS.
    CALCUTTA STATIST. ASSOC. BULL., 5, 185-187.    (MR19,P.895)

CHAKRAVARTI, I. M. 1954.    (8.6/17.9)                                                    903
    ON THE PROBLEM OF PLANNING A MULTISTAGE SURVEY FOR MULTIPLE
    CORRELATED CHARACTERS.
    SANKHYA, 14, 211-216.    (MR16,P.730)

CHAKRAVARTI, M. C.  SEE  CHAKRABARTI, M. C.

CHAKRAVARTI, N. C.    SEE
    3950. NAGAR, A.L. + CHAKRAVARTI, N. C. (1965)

CHAKRAVARTI, SUKHARANJAN 1965.    (8.3/U)                                                 904
    THE PROBLEM OF TESTING THE HYPOTHESIS OF EQUALITY OF MEANS OF SEVERAL
    UNIVARIATE NORMAL POPULATIONS WHEN VARIANCES ARE DIFFERENT.
    CALCUTTA STATIST. ASSOC. BULL., 14, 127-150.

CHAKRAVARTI, SUKHARANJAN 1966.    (8.3/8.4)                                               905
    A NOTE ON MULTIVARIATE ANALYSIS OF VARIANCE TEST WHEN DISPERSION
    MATRICES ARE DIFFERENT AND UNKNOWN.
    CALCUTTA STATIST. ASSOC. BULL., 15, 75-92.    (MR34-6933)

CHAMBERLIN, T. C. 1944.    (1.1)                                                          906
    THE METHOD OF MULTIPLE WORKING HYPOTHESES.
    SCIENTIFIC MONTHLY, 59, 357-362.

CHAMBERS, E. G. 1943.    (1.1)                                                            907
    STATISTICS IN PSYCHOLOGY AND LIMITATIONS OF THE TEST METHODS.
    BRITISH J. PSYCHOL., 33, 189-199.

CHAN, MOU TCHEN  SEE  TCHEN, CHAN-MOU

CHAN, NAI NG 1965.    (17.7/B)                                                            908
    ON CIRCULAR FUNCTIONAL RELATIONSHIPS.
    J. ROY. STATIST. SOC. SER. B, 27, 45-56.    (MR32-6590)

CHAND, UTTAM 1951.    (8.2/8.3/B)                                                         909
    TEST CRITERIA FOR HYPOTHESES OF SYMMETRY OF A REGRESSION MATRIX.
    ANN. MATH. STATIST., 22, 513-522.    (MR13,P.367)

CHANDA, K. C. 1954.    (19.3/A)                                                           910
    A NOTE ON THE CONSISTENCY AND MAXIMA OF THE ROOTS OF LIKELIHOOD EQUATIONS.
    BIOMETRIKA. 41, 56-61.

CHANDLER, K. N. 1950.    (4.5/4.6)                                                        911
    ON A THEOREM CONCERNING THE SECONDARY SUBSCRIPTS OF DEVIATIONS IN
    MULTIVARIATE CORRELATION USING YULE'S NOTATION.
    BIOMETRIKA, 37, 451-452.    (MR12,P.347)

CHANDRA SEKAR, C. 1951.    (4.2)                                                          912
    A THEOREM ON THE CORRELATION COEFFICIENT FOR SAMPLES OF THREE WHEN
    THE VARIABLES ARE INDEPENDENT.
    ANN. MATH. STATIST., 22, 132-133.

CHANDRA SEKAR, C. + CHAKRABORTY, P.N. 1952.    (20.4)                                     913
    ON THE CONCEPT AND USE OF ORTHOGONAL SEMI-POLYNOMIALS.
    SANKHYA, 12, 141-150.    (MR14,P.995)

CHANDRASEKHAR, S. 1943.    (16.4/E)                                                       914
    STOCHASTIC PROBLEMS IN PHYSICS AND ASTRONOMY.
        (REPRINTED AS NO. 917)
    REV. MODERN PHYS., 15, 1-89.    (MR4,P.248)

CHANDRASEKHAR, S. 1944.    (16.4)                                                         915
    THE STATISTICS OF THE GRAVITATION FIELD ARISING FROM A RANDOM DISTRIBUTION OF
    STARS.   III. THE CORRELATIONS IN THE FORCES ACTING AT TWO POINTS SEPARATED BY A
    FINITE DISTANCE.
    ASTROPHYS. J., 99, 25-46.    (MR5,P.191)

CHANDRASEKHAR, S. 1949.   (18.1)                                        916
    ON A CLASS OF PROBABILITY DISTRIBUTIONS.
    PROC. CAMBRIDGE PHILOS. SOC., 45, 219-224.   (MR10,P.464)

CHANDRASEKHAR, S. 1954.   (16.4/E)                                      917
    STOCHASTIC PROBLEMS IN PHYSICS AND ASTRONOMY.
        (REPRINT OF NO. 914)
    SELECT. PAPERS NOISE STOCH. PROC., 3-91.   (MR4,P.248)

CHANDRASEKHAR, S. + VON NEUMANN, JOHN 1942.   (16.4/17.6/E)            918
    THE STATISTICS OF THE GRAVITATIONAL FIELD ARISING FROM A RANDOM
    DISTRIBUTION OF STARS.   I.   THE SPEED OF FLUCTUATIONS.
    ASTROPHYS. J., 95, 489-531.   (MR3,P.281)

CHANDRASEKHAR, S. + VON NEUMANN, JOHN 1943.   (16.4/17.6/E)            919
    THE STATISTICS OF THE GRAVITATIONAL FIELD ARISING FROM A RANDOM
    DISTRIBUTION OF STARS.   II.   THE SPEED OF FLUCTUATIONS:   DYNAMICAL
    FRICTION:   SPATIAL CORRELATIONS.
    ASTROPHYS. J., 97, 1-27.   (MR4,P.227)

CHAPANIS, ALPHONSE 1941.   (15.4)                                       920
    NOTES ON THE RAPID CALCULATION OF ITEM VALIDITIES.
    J. EDUC. PSYCHOL., 32, 297-304.

CHAPELIN, JACQUES 1932.   (17.9/18.1/B)                                 921
    ON A METHOD OF PROCEEDING FROM PARTIAL CELL FREQUENCIES TO ORDINATES AND TO
    TOTAL CELL FREQUENCIES IN THE CASE OF A BIVARIATE FREQUENCY SURFACE.
    BIOMETRIKA, 24, 495-497.

CHAPMAN, DOUGLAS G.   SEE
    559. BIRNBAUM, Z. W. + CHAPMAN, DOUGLAS G. (1950)

CHARBONNIER, A.; CYFFERS, B.; SCHWARTZ, D. +   1957.  E(6.3)          922
    VESSEREAU, A.
    DISCRIMINATION ENTRE ICTERES MEDICAUX ET CHIRURGICAUX A PARTIR DES
    RESULTATS DE L'ANALYSE ELECTROPHORETIQUE DES PROTEINES DU SERUM.
    BULL. INST. INTERNAT. STATIST., 35(2) 303-320.

CHARLIER, C. V. L. 1914.   (18.1)                                       923
    CONTRIBUTIONS TO THE MATHEMATICAL THEORY OF STATISTICS 6.   THE
    CORRELATION FUNCTION OF TYPE A.
    ARK. MAT. ASTRONOM. FYS., 9(26) 1-18.

CHARLIER, C. V. L. 1914.   (2.1/4.1/B)                                  924
    OM KORRELATION MELLAN EGENSKAPER INOM DEN HOMOGRADA STATISTIKEN.
    SVENSKA AKTUARIEFOREN. TIDSKR., 1, 21-35.

CHARLIER, C. V. L. 1915.   (2.1/4.1/4.6)                                925
    CONTRIBUTIONS TO MATHEMATICAL STATISTICS.
    SVENSKA AKTUARIEFOREN. TIDSKR., 2, 18-33.

CHARNES, ABRAHAM   SEE
    439. BEN-ISRAEL, ADI + CHARNES, ABRAHAM (1963)

CHARNLEY, F. 1941.   (8.1/18.1/B)                                       926
    SOME PROPERTIES OF A COMPOSITE, BIVARIATE DISTRIBUTION IN WHICH THE
    MEANS OF THE COMPONENT NORMAL DISTRIBUTIONS ARE LINEARLY RELATED.
    CANAD. J. RES. SECT. A, 19, 139-151.   (MR3,P.172)

CHARNLEY, F. 1942.   (2.4/18.3/A)                                       927
    THE VARIANCES OF THE MEANS AND THE VARIANCE OF THE SLOPE OF THE LINE
    OF RELATION OF A LINEAR, COMPOSITE, BIVARIATE DISTRIBUTION.
    CANAD. J. RES. SECT. A, 20, 6-9.   (MR3,P.172)

CHARTRES, B. A. 1963.   (2.1)                                           928
    A GEOMETRICAL PROOF OF A THEOREM DUE TO D. SLEPIAN.
    SIAM REV., 5, 335-341.   (MR28-4559)

CHARTRES, B. A. + MESSEL, HARRY 1954.   (18.1)                          929
    NEW FORMULATION OF A GENERAL THREE-DIMENSIONAL CASCADE THEORY.
    PHYS. REV., 96, 1651-1654.   (MR16,P.496)

CHASE, C. I. 1960.   (4.6)                                                                              930
    COMPUTATION OF VARIANCE ACCOUNTED FOR IN MULTIPLE CORRELATION.
    J. EXPER. EDUC., 28, 265–266.

CHATTERJEE, SHOUTIR KISHORE 1959.   (5.1)                                                               931
    ON AN EXTENSION OF STEIN'S TWO-SAMPLE PROCEDURE TO THE MULTI-NORMAL PROBLEM.
    CALCUTTA STATIST. ASSOC. BULL., 8, 121–148.   (MR21-4501)

CHATTERJEE, SHOUTIR KISHORE 1959.   (5.1)                                                               932
    SOME FURTHER RESULTS ON THE MULTINOMIAL EXTENSION OF STEIN'S TWO-SAMPLE PROCEDURE.
    CALCUTTA STATIST. ASSOC. BULL., 9, 20–28.   (MR22-3064)

CHATTERJEE, SHOUTIR KISHORE 1960.   (8.3/B)                                                             933
    SEQUENTIAL TESTS FOR THE BIVARIATE REGRESSION PARAMATERS WITH KNOWN
    POWER AND RELATED ESTIMATION PROCEDURES.
    CALCUTTA STATIST. ASSOC. BULL., 10, 19–34.   (MR23-A1451)

CHATTERJEE, SHOUTIR KISHORE 1962.   (8.3)                                                               934
    SEQUENTIAL INFERENCE PROCEDURES OF STEIN'S TYPE FOR A CLASS OF
    MULTIVARIATE REGRESSION PROBLEMS.
    ANN. MATH. STATIST., 33, 1039–1064.   (MR25-5575)

CHATTERJEE, SHOUTIR KISHORE 1962.   (3.1/8.1/C)                                                         935
    SIMULTANEOUS CONFIDENCE INTERVALS OF PREDETERMINED LENGTH BASED ON
    SEQUENTIAL SAMPLES.
    CALCUTTA STATIST. ASSOC. BULL., 11, 144–149.   (MR27-2034)

CHATTERJEE, SHOUTIR KISHORE 1966.   (17.8/B)                                                            936
    A BIVARIATE SIGN TEST FOR LOCATION.
    ANN. MATH. STATIST., 37, 1771–1782.   (MR34-902)

CHATTERJEE, SHOUTIR KISHORE + SEN, PRANAB KUMAR 1964.   (17.8/B)                                        937
    NON-PARAMETRIC TESTS FOR THE BIVARIATE TWO-SAMPLE LOCATION PROBLEM.
    CALCUTTA STATIST. ASSOC. BULL., 13, 18–58.   (MR33-831)

CHATTERJEE, SHOUTIR KISHORE + SEN, PRANAB KUMAR 1965.   (17.8/B)                                        938
    SOME NON-PARAMETRIC TESTS FOR THE BIVARIATE TWO-SAMPLE ASSOCIATION PROBLEM.
    CALCUTTA STATIST. ASSOC. BULL., 14, 14–35.   (MR34-901)

CHATTOPADHYAY, K. P. 1941.   (6.5)                                                                      939
    APPLICATION OF STATISTICAL METHODS TO ANTHROPOLOGICAL RESEARCH.
    SANKHYA, 5, 99–104.

CHAUDHURI, S. B. 1954.   (5.2/C)                                                                        940
    THE MOST POWERFUL UNBIASED CRITICAL REGIONS AND THE SHORTEST UNBIASED
    CONFIDENCE INTERVALS ASSOCIATED WITH THE DISTRIBUTION OF CLASSICAL $D^2$-STATISTIC.
    SANKHYA, 14, 71–80.   (MR16,P.383)

CHAUDHURI, S. B. 1956.   (13.1/T)                                                                       941
    STATISTICAL TABLES AND CERTAIN RECURRENCE RELATIONS CONNECTED WITH p-STATISTICS.
    CALCUTTA STATIST. ASSOC. BULL., 6, 181–188.   (MR18,P.517)

CHAUDHURI, S. B.   SEE ALSO
    664. BOSE, PURNENDU KUMAR + CHAUDHURI, S. B. (1955)

    665. BOSE, PURNENDU KUMAR + CHAUDHURI, S. B. (1957)

CHAUNCEY, H. 1939.   (4.6/20.3)                                                                         942
    A METHOD FOR SOLVING SIMULTANEOUS EQUATIONS, WITH PARTICULAR
    REFERENCE TO MULTIPLE CORRELATION PROBLEMS.
    HARVARD EDUC. REV., 9, 63–68.

CHEATHAM, T. P., JR.   SEE
    3350. LEE, Y. W.; CHEATHAM, T. P., JR. + WIESNER, JEROME B. (1950)

CHENG, SHAO-LIEN; TAO, TSUNG-YING + OTHERS (SIC) 1962.   (16.4/16.5/AB)                                 943
    THE ESTIMATION OF REGRESSION COEFFICIENTS OF TIME SERIES WITH MULTIDIMENSIONAL
    STATIONARY RANDOM DISTURBANCES.
        (IN CHINESE, TRANSLATED AS NO. 944)
    ACTA MATH. SINICA, 11, 222–237.   (MR27-6341)

CHENG, SHAO-LIEN; TAO, TSUNG-YING + OTHERS (SIC) 1962.   (16.4/16.5/AB)          944
     THE ESTIMATION OF REGRESSION COEFFICIENTS OF TIME SERIES WITH MULTIDIMENSIONAL
     STATIONARY RANDOM DISTURBANCES.
          (TRANSLATION OF NO. 943)
   CHINESE MATH. ACTA, 2, 248-266.   (MR27-6341)


CHERIAN, K. C. 1941.   (18.1/B)                                                   945
     A BI-VARIATE CORRELATED GAMMA-TYPE DISTRIBUTION.
   J. INDIAN MATH. SOC. NEW SER., 5, 133-144.   (MR3,P.171)


CHERNOFF, HERMAN 1953.   (20.1)                                                   946
     LOCALLY OPTIMAL DESIGNS FOR ESTIMATING PARAMETERS.
     ANN. MATH. STATIST., 24, 586-602.


CHERNOFF, HERMAN 1956.   (17.1/19.2)                                              947
     LARGE SAMPLE THEORY - PARAMETRIC CASE.
     ANN. MATH. STATIST., 27, 1-22.   (MR17,P.869)


CHERNOFF, HERMAN 1962.   (15.4)                                                   948
     THE SCORING OF MULTIPLE CHOICE QUESTIONNAIRES.
     ANN. MATH. STATIST., 33, 375-393.


CHERNOFF, HERMAN 1962.   (15.4)                                                   949
     A NEW APPROACH TO THE EVALUATION OF MULTIPLE CHOICE QUESTIONNAIRES.
     BULL. INST. INTERNAT. STATIST., 39(2) 289-293.


CHERNOFF, HERMAN + RUBIN, HERMAN 1953.   (16.2)                                   950
     ASYMPTOTIC PROPERTIES OF LIMITED-INFORMATION ESTIMATES UNDER GENERALIZED
     CONDITIONS.
     COWLES COMMISS. MONOG., 14, 200-212.


CHERNOFF, HERMAN + TEICHER, HENRY 1965.   (17.1)                                  951
     LIMIT DISTRIBUTIONS OF THE MINIMAX OF INDEPENDENT IDENTICALLY
     DISTRIBUTED RANDOM VARIABLES.
     TRANS. AMER. MATH. SOC., 116, 474-491.


CHERNOV, L. A.   SEE   CERNOV, L. A.


CHERRY, CHARLES N.    SEE
   6033. WRIGLEY, CHARLES F.; CHERRY, CHARLES N.; LEE, MARILYN C. +   (1957)
   MC QUITTY, LOUIS L.


CHESHIRE, LEONE; OLDIS, ELENA + PEARSON, EGON SHARPE 1932.   (4.2)                952
     FURTHER EXPERIMENTS ON THE SAMPLING DISTRIBUTION OF THE CORRELATION COEFFICIENT.
     J. AMER. STATIST. ASSOC., 27, 121-128.


CHESSIN, P. L. 1955.   (1.2)                                                      953
     A BIBLIOGRAPHY ON NOISE.
     IRE TRANS. INFORMATION THEORY, IT-1(2) 15-31.


CHETTY, V. KARUPPAN    SEE
   6073. ZELLNER, ARNOLD + CHETTY, V. KARUPPAN (1965)

   6074. ZELLNER, ARNOLD + CHETTY, V. KARUPPAN (1968)


CHEW, VICTOR 1966.   (5.3/14.4/C)                                                 954
     CONFIDENCE, PREDICTION, AND TOLERANCE REGIONS FOR THE MULTIVARIATE
     NORMAL DISTRIBUTION.
     J. AMER. STATIST. ASSOC., 61, 605-617.   (MR34-3688)


CHEW, VICTOR + BOYCE, R. 1962.   (14.5)                                           955
     DISTRIBUTION OF RADIAL ERROR IN THE BIVARIATE ELLIPTICAL NORMAL DISTRIBUTION.
     TECHNOMETRICS, 4, 138-140.


CHIANG, CHIN LONG 1956.   (17.5)                                                  956
     ON REGULAR BEST ASYMPTOTICALLY NORMAL ESTIMATES.
     ANN. MATH. STATIST., 27, 336-351.   (MR19,P.694)


CHIANG, TSE-PEI  SEE  JIANG, ZE-PEI


CHIARO, A. DEL  SEE  DEL CHIARO, A.

CHIEH-CHIEN P'AN   SEE   PAN, JIE-JIAN

CHING, FONG   SEE   FONG, CHING

CHINO, SADAKO 1963.  (4.8/6.4)                                                      957
    THE RELATION BETWEEN CORRELATION RATIO AND SUCCESS RATE IN THE CLASSIFICATION
    BY THE QUANTIFICATION METHOD.
        (IN JAPANESE, WITH SUMMARY IN ENGLISH)
    PROC. INST. STATIST. MATH. TOKYO, 11, 7-24.

CHINTSCHIN, A. J.  SEE  HINCIN, A. JA.

CHIPMAN, JOHN S. 1964.  (8.1)                                                       958
    ON LEAST SQUARES WITH INSUFFICIENT OBSERVATIONS.
    J. AMER. STATIST. ASSOC., 59, 1078-1111.

CHIPMAN, JOHN S. + RAO, M.M. 1964.  (8.8/20.2)                                      959
    THE TREATMENT OF LINEAR RESTRICTIONS IN REGRESSION ANALYSIS.
    ECONOMETRICA, 32, 198-209.

CHIPMAN, JOHN S. + RAO, M.M. 1964.  (2.5/20.1/20.2)                                 960
    PROJECTIONS, GENERALIZED INVERSES, AND QUADRATIC FORMS.
    J. MATH. ANAL. APPL., 9, 1-11.  (MR29-2272)

CHOI, KEEWHAN 1965.  (2.1/8.3/8.4/T)                                                961
    PROBABILITY BOUNDS FOR A UNION OF HYPERSPHERICAL CONES.
    J. ROY. STATIST. SOC. SER. B, 27, 57-73.

CHOPE, HENRY R.   SEE
    564. BISHOP, ALBERT B. + CHOPE, HENRY R. (1962)

CHOPRA, A. S. 1965.  (8.1/20.4/T)                                                   962
    APPLICATION OF ORTHOGONAL POLYNOMIALS IN FITTING AN ASYMPTOTIC REGRESSION CURVE.
    J. INDIAN SOC. AGRIC. STATIST., 17, 76-82.

CHOUDHURY, S. B.  SEE  CHAUDHURI, S. B.

CHOVER, JOSHUA 1962.  (16.4)                                                        963
    CERTAIN CONVEXITY CONDITIONS ON MATRICES WITH APPLICATIONS TO GAUSSIAN
        PROCESSES.
    DUKE MATH. J., 29, 141-150.  (MR24-A3720)

CHOW, GREGORY C. 1960.  (8.3)                                                       964
    TESTS OF EQUALITY BETWEEN SETS OF COEFFICIENTS IN TWO LINEAR REGRESSIONS.
    ECONOMETRICA, 28, 591-605.  (MR25-4604)

CHOW, GREGORY C. 1964.  (16.2)                                                      965
    A COMPARISON OF ALTERNATIVE ESTIMATORS FOR SIMULTANEOUS EQUATIONS.
    ECONOMETRICA, 32, 532-553.  (MR30-2912)

CHOW, GREGORY C. 1966.  (8.1)                                                       966
    A THEOREM ON LEAST SQUARES AND VECTOR CORRELATION IN MULTIVARIATE
        LINEAR REGRESSION.
    J. AMER. STATIST. ASSOC., 61, 413-414.  (MR34-6934)

CHOWDHURY, S. B.  SEE  CHAUDHURI, S. B.

CHOWN, L. N. + MORAN, P.A.P. 1951.  (4.1)                                           967
    RAPID METHODS FOR ESTIMATING CORRELATION COEFFICIENTS.
    BIOMETRIKA, 38, 464-467.  (MR13,P.667)

CHRIST, CARL F. 1956.  (16.2/E)                                                     968
    AGGREGATE ECONOMETRIC MODELS.
    AMER. ECON. REV., 46, 385-408.

CHRIST, CARL F. 1960.  (16.2)                                                       969
    SIMULTANEOUS EQUATION ESTIMATION:   ANY VERDICT YET ?
    ECONOMETRICA, 28, 835-845.

CHRIST, J. W. 1939.  (4.4/E)                                                        970
    BI-SERIAL R FOR HORTICULTURAL RESEARCH.
    PROC. AMER. SOC. HORT. SCI., 37, 269-271.

CHRIST, J. W. 1940.   (4.8/E)                                                        971
    CORRELATION FROM RANKS FOR HORTICULTURAL RESEARCH.
  PROC. AMER. SOC. HORT. SCI., 38, 593-595.


CHRIST, J. W. 1942.  E(4.4)                                                          972
    TETRACHORIC CORRELATION FOR HORTICULTURAL RESEARCH.
  PROC. AMER. SOC. HORT. SCI., 40, 549-551.


CHRISTENSEN, C. M. 1953.  E(5.2/6.3)                                                 973
    MULTIVARIATE STATISTICAL ANALYSIS OF DIFFERENCES BETWEEN
    PRE-PROFESSIONAL GROUPS OF COLLEGE STUDENTS.
  J. EXPER. EDUC., 21, 221-232.


CHRZASZCZ, ROMAN 1943.   (16.2/E)                                                    974
    EIN PROBLEM DER BESTIMMUNG UND ELIMINIERUNG VON SYSTEMATISCHEN
    BEOBACHTUNGSFEHLERN.
  SAMMLUNG WISS. ARB. SCHWEIZ INTERN. POLEN, 1(4) 23-26.   (MR9,P.296)


CHUNG, J. H. + FRASER, DONALD A. S. 1958.   (17.8)                                   975
    RANDOMIZATION TESTS FOR A MULTIVARIATE TWO-SAMPLE PROBLEM.
  J. AMER. STATIST. ASSOC., 53, 729-735.



CHUNG, KAI LAI 1940.   (17.6)                                                        976
    SUR UN THEOREME DE M. GUMBEL.
  C.R. ACAD. SCI. PARIS, 210, 620-621.   (MR2,P.106)


CHUNG, KAI LAI 1952.   (16.4)                                                        977
    ON THE RENEWAL THEOREM IN HIGHER DIMENSIONS.
  SKAND. AKTUARIETIDSKR., 35, 188-194.   (MR14,P.994)


CHUNG, KAI LAI 1964.   (16.4/U)                                                      978
    THE GENERAL THEORY OF MARKOV PROCESSES ACCORDING TO DOEBLIN.
  Z. WAHRSCHEIN. VERW. GEBIETE, 2, 230-254.


CHUNG, KAI LAI + FUCHS, WOLFGANG H. J. 1951.   (17.1)                                979
    ON THE DISTRIBUTION OF VALUES OF SUMS OF RANDOM VARIABLES.
  MEM. AMER. MATH. SOC., 6, 1-12.   (MR12,P.722)


CHUNG, TEH FAN  SEE  FAN, CHUNG-TEH


CHURCH, A. E. R.   SEE
  1529. ELDERTON, ETHEL M.; MOUL, MARGARET; FIELLER, E. C.;   (1930)
  PRETORIUS, S.J. + CHURCH, A. E. R.


CLAERBOUT, JON F. 1966.   (16.4)                                                     980
    SPECTRAL FACTORIZATION OF MULTIPLE TIME SERIES.
  BIOMETRIKA, 53, 264-267.   (MR33-6688)


CLARINGBOLD, P. J. 1958.   (12.2)                                                    981
    MULTIVARIATE QUANTAL ANALYSIS.
  J. ROY. STATIST. SOC. SER. B, 20, 398-405.


CLARK, CHARLES E. 1961.   (14.1/BT)                                                  982
    THE GREATEST OF A FINITE SET OF RANDOM VARIABLES.
  OPERATIONS RES., 9, 145-162.   (MR23-A2903)


CLARK, E. L. 1949.   (15.4)                                                          983
    METHODS OF SPLITTING VS. SAMPLES AS SOURCES OF INSTABILITY IN TEST-RELIABILITY.
  HARVARD EDUC. REV., 19, 178-182.


CLARK, PHILIP J. 1952.   (6.5)                                                       984
    AN EXTENSION OF THE COEFFICIENT OF DIVERGENCE FOR USE WITH MULTIPLE CHARACTERS.
  COPEIA, 61-64.


CLAY, P.P.F.   SEE
  2547. HOPKINS, J. W. + CLAY, P.P.F. (1963)


CLEMENS, WILLIAM V. 1965.   (15.4/E)                                                 985
    AN ANALYTICAL AND EMPIRICAL EXAMINATION OF SOME PROPERTIES OF IPSATIVE MEASURES.
  PSYCHOMET. MONOG., 14, 1-56.

CLINE, RANDALL E. 1964.  (20.1)                                                    986
    REPRESENTATIONS FOR THE GENERALIZED INVERSE OF A PARTITIONED MATRIX.
    J. SOC. INDUST. APPL. MATH., 12, 588-600.

CLINE, RANDALL E. 1964.  (20.1)                                                    987
    NOTE ON THE GENERALIZED INVERSE OF THE PRODUCT OF MATRICES.
    SIAM REV., 6, 57-58.  (MR28-5076)

CLOUGH, H. W. 1942.  (4.1)                                                         988
    A NOTE ON METHODS OF CORRELATION.
    BULL. AMER. METEOROL. SOC., 23, 410.

CLUNIES-ROSS, CHARLES W. + RIFFENBURGH, ROBERT H. 1960.  (6.1)                     989
    GEOMETRY AND LINEAR DISCRIMINATION.
    BIOMETRIKA, 47, 185-189.  (MR22-1962)

CLUNIES-ROSS, CHARLES W.    SEE ALSO
    4562. RIFFENBURGH, ROBERT H. + CLUNIES-ROSS, CHARLES W. (1960)

    1803. FREUND, RUDOLF J.; VAIL, RICHARD W. + CLUNIES-ROSS, CHARLES W. (1961)

    1804. FREUND, RUDOLF J.; VAIL, RICHARD W. + CLUNIES-ROSS, CHARLES W. (1961)

COBB, J. A. 1908.  (4.2)                                                           990
    THE EFFECT OF ERRORS OF OBSERVATION UPON THE CORRELATION COEFFICIENT.
    BIOMETRIKA, 6, 109.


COBB, WHITFIELD    SEE
    4688. ROY, S.N. + COBB, WHITFIELD (1960)

COCHRAN, WILLIAM G. 1934.  (2.5/8.8)                                               991
    THE DISTRIBUTION OF QUADRATIC FORMS IN A NORMAL SYSTEM, WITH
    APPLICATIONS TO THE ANALYSIS OF VARIANCE.
    PROC. CAMBRIDGE PHILOS. SOC., 30, 178-191.

COCHRAN, WILLIAM G. 1937.  (4.2)                                                   992
    THE EFFICIENCIES OF THE BINOMIAL SERIES TESTS OF SIGNIFICANCE OF A
    MEAN AND OF A CORRELATION COEFFICIENT.
    J. ROY. STATIST. SOC., 100, 69-73.

COCHRAN, WILLIAM G. 1938.  (1.1/1.2)                                               993
    RECENT ADVANCES IN MATHEMATICAL STATISTICS--RECENT WORK ON THE
    ANALYSIS OF VARIANCE.
    J. ROY. STATIST. SOC., 101, 434-449.

COCHRAN, WILLIAM G. 1940.  (14.5)                                                  994
    NOTE ON AN APPROXIMATE FORMULA FOR THE SIGNIFICANCE LEVELS OF Z.
    ANN. MATH. STATIST., 11, 93-95.  (MR1,P.249)

COCHRAN, WILLIAM G. 1941.  (14.2/T)                                                995
    THE DISTRIBUTION OF THE LARGEST OF A SET OF ESTIMATED VARIANCES AS A
    FRACTION OF THEIR TOTAL.
    ANN. EUGENICS, 11, 47-52.  (MR3,P.171)

COCHRAN, WILLIAM G. 1943.  (8.5/8.7/8.9)                                           996
    THE COMPARISON OF DIFFERENT SCALES OF MEASUREMENT FOR EXPERIMENTAL RESULTS.
    ANN. MATH. STATIST., 14, 205-216.  (MR5,P.43)


COCHRAN, WILLIAM G. 1951.  (6.7)                                                   997
    IMPROVEMENT BY MEANS OF SELECTION.
    PROC. SECOND BERKELEY SYMP. MATH. STATIST. PROB., 449-470.  (MR13,P.480)

COCHRAN, WILLIAM G. 1957.  (8.3/8.5/U)                                             998
    ANALYSIS OF COVARIANCE - ITS NATURE AND USES.
    BIOMETRICS, 13, 261-281.  (MR19,P.895)

COCHRAN, WILLIAM G. 1962.  (6.2/6.5/E)                                             999
    ON THE PERFORMANCE OF THE LINEAR DISCRIMINANT FUNCTION.
        (REPRINTED AS NO. 1001)
    BULL. INST. INTERNAT. STATIST., 39(2) 435-447.  (MR29-1690)

COCHRAN, WILLIAM G. 1964. (6.2/8.5)                                                1000
    COMPARISON OF TWO METHODS OF HANDLING COVARIATES IN DISCRIMINATORY ANALYSIS.
    ANN. INST. STATIST. MATH. TOKYO, 16, 43-53. (MR30-4338)

COCHRAN, WILLIAM G. 1964. (6.2/6.5/E)                                              1001
    ON THE PERFORMANCE OF THE LINEAR DISCRIMINANT FUNCTION.
        (REPRINT OF NO. 999)
    TECHNOMETRICS, 6, 179-190. (MR32-1834)

COCHRAN, WILLIAM G. + BLISS, C.I. 1948. (5.2/8.5)                                  1002
    DISCRIMINANT FUNCTIONS WITH COVARIANCE.
    ANN. MATH. STATIST., 19, 151-176. (MR10,P.50)

COCHRAN, WILLIAM G. + HOPKINS, CARL E. 1961. (6.4)                                 1003
    SOME CLASSIFICATION PROBLEMS WITH MULTIVARIATE QUALITATIVE DATA.
    BIOMETRICS, 17, 10-32. (MR22-12636)

COCHRAN, WILLIAM G.   SEE ALSO
    2693. IRWIN, J. O.; COCHRAN, WILLIAM G.; FIELLER, E. C. + (1936)
    STEVENS, W. L.

    2692. IRWIN, J. O.; BARTLETT, MAURICE S.; COCHRAN, WILLIAM G. + (1935)
    FIELLER, E. C.

COCHRANE, D. + ORCUTT, GUY H. 1949. (16.1/E)                                       1004
    APPLICATION OF LEAST SQUARES REGRESSION TO RELATIONSHIPS CONTAINING
    AUTOCORRELATED ERROR TERMS.
    J. AMER. STATIST. ASSOC., 44, 32-61.

COGBURN, ROBERT 1965. (3.1)                                                        1005
    ON THE ESTIMATION OF A MULTIVARIATE LOCATION PARAMETER WITH SQUARED ERROR LOSS.
    BERNOULLI (1713), BAYES (1763), LAPLACE (1813), ANNIV. VOLUME, 24-29. (MR33-6787)

COHEN, A. CLIFFORD, JR. 1955. (2.4/3.2/B)                                          1006
    RESTRICTION AND SELECTION IN SAMPLES FROM BIVARIATE NORMAL DISTRIBUTIONS.
    J. AMER. STATIST. ASSOC., 50, 884-893. (MR17,P.639)

COHEN, A. CLIFFORD, JR. 1955. (2.4/3.1)                                            1007
    MAXIMUM LIKELIHOOD ESTIMATION OF THE DISPERSION PARAMETER OF A CHI-DISTRIBUTED
    RADIAL ERROR FROM TRUNCATED AND CENSORED SAMPLES WITH APPLICATIONS TO TARGET
    ANALYSIS.
    J. AMER. STATIST. ASSOC., 50, 1122-1135. (MR17,P.381)

COHEN, A. CLIFFORD, JR. 1957. (2.4)                                                1008
    RESTRICTION AND SELECTION IN MULTINORMAL DISTRIBUTIONS.
    ANN. MATH. STATIST., 28, 731-741. (MR19,P.895)

COHEN, A. CLIFFORD, JR.   SEE ALSO
    321. BARROW, DAVID F. + COHEN, A. CLIFFORD, JR. (1954)

COHEN, ARTHUR 1965. (3.1)                                                          1009
    ESTIMATES OF LINEAR COMBINATIONS OF THE PARAMETERS IN THE MEAN VECTOR
    OF A MULTIVARIATE DISTRIBUTION.
    ANN. MATH. STATIST., 36, 78-87. (MR30-2618)

COHEN, ARTHUR 1966. (3.1)                                                          1010
    ALL ADMISSIBLE LINEAR ESTIMATES OF THE MEAN VECTOR.
    ANN. MATH. STATIST., 37, 458-463. (MR32-6591)

COHEN, JOZEF   SEE
    2266. GUTTMAN, LOUIS + COHEN, JOZEF (1943)

COHN, H. 1965. (19.2/A)                                                            1011
    THE LIKELIHOOD RATIO TEST FOR RANDOM PROCESSES.
        (IN ROUMAINIAN)
    STUDII CERC. MAT., 17, 413-417.

COHN, H.   SEE ALSO
    679. BOTEZ, M. + COHN, H. (1965)

COHN, J. H. E. 1965. (20.2)                                                        1012
    HADAMARD MATRICES AND SOME GENERALISATIONS.
    AMER. MATH. MONTHLY, 72, 515-518.

COLCORD, CLARENCE G. + DEMING, LOLA S. 1936.   (14.5)                    1013
     THE ONE-TENTH PERCENT LEVEL OF 7.
   SANKHYA, 2, 423-424.

COLE, J. W. L. + GRIZZLE, J. E. 1966.   E(8.5)                           1014
     APPLICATIONS OF MULTIVARIATE ANALYSIS OF VARIANCE TO REPEATED
     MEASUREMENTS EXPERIMENTS.
   BIOMETRICS, 22, 810-828.

COLEMAN, J. R. 1932.   (4.8)                                             1015
     A COEFFICIENT OF LINEAR CORRELATION BASED ON THE METHOD OF LEAST
     SQUARES AND THE LINE OF BEST FIT.
   ANN. MATH. STATIST., 3, 79-85.

COLLIAS, N. E. 1943.   (4.4/E)                                           1016
     STATISTICAL ANALYSIS OF FACTORS WHICH MAKE FOR SUCCESS IN INITIAL
     ENCOUNTERS BETWEEN HENS.
   AMER. NATUR., 77, 519-538.

COLLINS, GWYN 1961.   (15.1)                                             1017
     FACTOR ANALYSIS.
   J. ADV. RES., 1(5) 28-32.

COLOMBO, BERNARDO 1953.   (4.3/T)                                        1018
     ANALISI SEQUENZIALE DELLA CORRELAZIONE NELLE VARIABILI NORMALI.
   INDUST. MILANO, 633-650.

COLOMBO, BERNARDO 1955.   (4.3/10.1/RE)                                  1019
     NUOVI CONTRIBUTI ALL'ANALISI SEQUENZIALE DELLA CORRELAZIONE.
   INDUST. MILANO, 332-343.

COLOMBO, UMBERTO 1965.   (17.3/B)                                        1020
     LE VARIABILI CASUALI DOPPIE EQUIDISTRIBUITE A REGRESSIONE INTERAMENTE LINEARE.
   STATISTICA, BOLOGNA, 25, 457-477.   (MR32-8372)

COLOMBO, UMBERTO 1966.   (6.5/B)                                         1021
     RETTE CENTRALI DISTANZIALI.
        (WITH SUMMARY IN FRENCH AND ENGLISH)
   GIORN. IST. ITAL. ATTUARI, 29, 114-134.

COMREY, A. L. + AHUMADA, A. 1964.   (15.3)                               1022
     AN IMPROVED PROCEDURE AND PROGRAM FOR MINIMUM RESIDUAL FACTOR ANALYSIS.
   PSYCHOL. REP., 15, 91-96.

CONGARD, ROGER 1951.   (8.1)                                             1023
     REGRESSION UNILATERALE ET REGRESSIONS MUTUELLE.
   J. SOC. STATIST. PARIS, 92(10) 284-302.

CONNOLLY, T. W. + MILLER, KENNETH S. 1961.   (16.3/16.4)                 1024
     ANALOG COMPUTATION OF COVARIANCE MATRICES.
   IRE TRANS. ELECTRONIC COMP.,EC-10, 533.

CONRAD, HERBERT S. 1935.   (4.1/15.4)                                    1025
     ON THE CALCULATION OF THE CORRELATION BETWEEN A SINGLE ELEMENT OF A
     COMPOSITE AND THE REMAINDER OF THE COMPOSITE.
   J. EDUC. PSYCHOL., 26, 611-615.

CONSAEL, ROBERT 1949.   (16.4/R)                                         1026
     SUR QUELQUES PROCESSUS STOCHASTIQUES DISCONTINUS A DEUX VARIABLES ALEATOIRES.
   ACAD. ROY. BELG. BULL. CL. SCI. SER. 5, 35, 399-416.   (MR11,P.119)

CONSAEL, ROBERT 1949.   (16.4/B)                                         1027
     SUR QUELQUES PROCESSUS STOCHASTIQUES DISCONTINUS A DEUX VARIABLES ALEATOIRES
     (DEUXIEME COMMUNICATION).
   ACAD. ROY. BELG. BULL. CL. SCI. SER. 5, 35, 743-755.   (MR11,P.256)

CONSAEL, ROBERT 1952.   (16.4/B)                                         1028
     SUR LES PROCESSUS COMPOSES DE POISSON A DEUX VARIABLES ALEATOIRES.
   ACAD. ROY. BELG. CL. SCI. MEM. COLLECT.   IN-8, 27(6) 1-44.   (MR15,P.138)

CONSAEL, ROBERT   SEE ALSO
   3239. LAMENS, A. + CONSAEL, ROBERT (1957)

CONSTANTINE, A. G. 1963.   (7.1/13.2)                                            1029
   SOME NON-CENTRAL DISTRIBUTION PROBLEMS IN MULTIVARIATE ANALYSIS.
   ANN. MATH. STATIST., 34, 1270-1285.   (MR31-5285)

CONSTANTINE, A. G. 1966.   (8.4)                                                 1030
   THE DISTRIBUTION OF HOTELLING'S GENERALISED $T_o^2$.
   ANN. MATH. STATIST., 37, 215-225.   (MR32-6586)

CONSTANTINE, A. G. + JAMES, ALAN T. 1958.   (12.2/13.1/13.2)                     1031
   ON THE GENERAL CANONICAL CORRELATION DISTRIBUTION.
   ANN. MATH. STATIST., 29, 1146-1166.   (MR20-5544)

CONSUL, PREM CHANDRA 1962.   (14.1/U)                                            1032
   ON THE DISTRIBUTION, AND ITS CUMULANTS, OF THE SUM OF SQUARES OF THE
   DISPERSION ABOUT THE MEAN OF N ORDERED VARIATES SUBJECT TO A LINEAR RESTRAINT.
   ACAD. ROY. BELG. BULL. CL. SCI. SER. 5, 48, 715-719.   (MR27-2024)

CONSUL, PREM CHANDRA 1963.   (20.4)                                              1033
   HYPERGEOMETRIC FUNCTION.
   SANKHYA SER. B, 25, 197-214.   (MR32-1871)

CONSUL, PREM CHANDRA 1964.   (6.2/7.2)                                           1034
   DISTRIBUTION OF THE DETERMINANT OF THE SUM OF PRODUCTS MATRIX IN THE
   NON-CENTRAL LINEAR CASE.
   MATH. NACHR., 28, 169-179.   (MR34-6935)

CONSUL, PREM CHANDRA 1964.   (8.4/10.1/13.1)                                     1035
   ON THE LIMITING DISTRIBUTION OF SOME STATISTICS PROPOSED FOR TESTS OF
   HYPOTHESES.
   SANKHYA SER. A, 26, 279-286.

CONSUL, PREM CHANDRA 1965.   (10.2)                                              1036
   THE EXACT DISTRIBUTION OF CERTAIN LIKELIHOOD CRITERIA USEFUL IN
   MULTIVARIATE ANALYSIS.
   ACAD. ROY. BELG. BULL. CL. SCI. SER. 5, 51, 683-691.   (MR32-8453)

CONSUL, PREM CHANDRA 1965.   (6.5/18.1/18.3)                                     1037
   ON A NEW MULTIVARIATE DISTRIBUTION AND ITS PROPERTIES.
   ACAD. ROY. BELG. BULL. CL. SCI. SER. 5, 51, 810-818.   (MR33-3406)

CONSUL, PREM CHANDRA 1966.   (8.4)                                               1038
   ON THE EXACT DISTRIBUTIONS OF THE LIKELIHOOD RATIO CRITERIA FOR
   TESTING LINEAR HYPOTHESES ABOUT REGRESSION COEFFICIENTS.
   ANN. MATH. STATIST., 37, 1319-1330.   (MR34-2104)

CONSUL, PREM CHANDRA 1966.   (6.5/14.2)                                          1039
   ON THE DISTRIBUTION OF THE RATIO OF THE MEASURES OF DIVERGENCE BETWEEN
   TWO MULTIVARIATE POPULATIONS.
   MATH. NACHR., 32, 149-155.

COOK, M. B. 1951.   (2.3/B)                                                      1040
   BI-VARIATE K-STATISTICS AND CUMULANTS OF THEIR JOINT SAMPLING DISTRIBUTION.
   BIOMETRIKA, 38, 179-195.   (MR13,P.142)

COOK, M. B. 1951.   (2.3/B)                                                      1041
   TWO APPLICATIONS OF BIVARIATE K-STATISTICS.
   BIOMETRIKA, 38, 368-376.   (MR13,P.665)

COOLEY, WILLIAM W. + JONES, KENNETH J. 1964.   (15.3)                            1042
   COMPUTER SYSTEMS FOR MULTIVARIATE STATISTICAL ANALYSIS.
   EDUC. PSYCHOL. MEAS., 24, 645-653.

COOLIDGE, JULIAN L. 1923.   (2.1)                                                1043
   THE GAUSSIAN LAW OF ERROR FOR ANY NUMBER OF VARIABLES.
   TRANS. AMER. MATH. SOC., 24, 135-143.

COOMBS, CLYDE H. 1941.   (15.2)                                                  1044
   A CRITERION FOR SIGNIFICANT COMMON FACTOR VARIANCE.
   PSYCHOMETRIKA, 6, 267-272.

COOMBS, CLYDE H. 1948.   (8.6/U)                                                 1045
   THE ROLE OF CORRELATION IN ANALYSIS OF VARIANCE.
   PSYCHOMETRIKA, 13, 233-243.

COOMBS, CLYDE H. 1951.   (15.5)                                                    1046
    MATHEMATICAL MODELS IN PSYCHOLOGICAL SCALING.
  J. AMER. STATIST. ASSOC., 46, 480-489.

COOMBS, CLYDE H.    SEE ALSO
    5442. THRALL, ROBERT M.; COOMBS, CLYDE H. + CALDWELL, WILLIAM V. (1958)

COON, HELEN J.   SEE
    2149. GRUBBS, FRANK E. + COON, HELEN J. (1954)

COOPER, DAVID B. + COOPER, PAUL W. 1964.   (6.6/E)                                 1047
    NONSUPERVISED ADAPTIVE SIGNAL DETECTION AND PATTERN RECOGNITION.
  INFORMATION AND CONTROL, 7, 416-444.   (MR29-6965)

COOPER, PAUL W. 1962.   (15.5)                                                     1048
    THE HYPERPLANE IN PATTERN RECOGNITION.
  CYBERNETICA, 5, 215-238.

COOPER, PAUL W. 1962.   (6.2/6.4)                                                  1049
    THE HYPERSPHERE IN PATTERN RECOGNITION.
  INFORMATION AND CONTROL, 5, 324-346.

COOPER, PAUL W. 1963.   (6.4/18.1/18.4)                                            1050
    STATISTICAL CLASSIFICATION WITH QUADRATIC FORMS.
  BIOMETRIKA, 50, 439-448.   (MR29-1702)

COOPER, PAUL W. 1963.   (6.1/6.2)                                                  1051
    THE MULTIPLE CATEGORY BAYES DECISION PROCEDURE.
  IEEE TRANS. ELECTRONIC COMPUT.,EC-12, 18.

COOPER, PAUL W. 1963.   (18.1)                                                     1052
    MULTIVARIATE EXTENSION OF UNIVARIATE DISTRIBUTIONS.
  IEEE TRANS. ELECTRONIC COMPUT.,EC-12, 572-573.

COOPER, PAUL W. 1964.   (19.1/20.5)                                                1053
    HYPERPLANES, HYPERSPHERES AND HYPERQUADRICS AS DECISION BOUNDARIES.
  COMPUT. INFORMATION SCI. (TOU + WILCOX), 111-138.

COOPER, PAUL W.   SEE ALSO
    1047. COOPER, DAVID B. + COOPER, PAUL W. (1964)

COPELAND, HERMAN A. 1933.   (17.1/17.3)                                            1054
    A MATRIX THEORY OF MEASUREMENT.
  MATH. Z., 37, 542-555.

COPELAND, HERMAN A. 1935.   (15.1)                                                 1055
    A NOTE ON "THE VECTORS OF MIND".
  PSYCHOL. REV., 42, 216-218.

CORNELL, RICHARD G.    SEE
    410. BEAUCHAMP, JOHN J. + CORNELL, RICHARD G. (1966)

CORNFIELD, JEROME 1962.   E(5.2)                                                   1056
    JOINT DEPENDENCE OF RISK OF CORONARY HEART DISEASE ON SERUM CHOLESTEROL AND
    SYSTOLIC BLOOD PRESSURE: A DISCRIMINANT FUNCTION ANALYSIS.
  PROC. FED. AMER. SOC. EXPER. BIOL., 21(2) 58-61.

CORNFIELD, JEROME    SEE ALSO
    1912. GEISSER, SEYMOUR + CORNFIELD, JEROME (1963)

    2219. GURIAN, JOAN M.; CORNFIELD, JEROME + MOSIMANN, JAMES E. (1964)

    2307. HALPERIN, MAX; GREENHOUSE, SAMUEL W.; CORNFIELD, JEROME +   (1955)
  ZALOKAR, JULIA

CORNISH, E. A. 1954.   (14.2/18.1)                                                 1057
    THE MULTIVARIATE T-DISTRIBUTION ASSOCIATED WITH A SET OF NORMAL SAMPLE DEVIATES.
  AUSTRAL. J. PHYS., 7, 531-542.   (MR16,P.602)

CORNISH, E. A. 1955.   (18.3)                                                      1058
    THE SAMPLING DISTRIBUTIONS OF STATISTICS DERIVED FROM THE
    MULTIVARIATE T-DISTRIBUTION.
  AUSTRAL. J. PHYS., 8, 193-199.   (MR19,P.1204)

CORNISH, E. A. 1957.  (8.1/20.1)                                                    1059
    AN APPLICATION OF THE KRONECKER PRODUCT OF MATRICES IN MULTIPLE REGRESSION.
    BIOMETRICS, 13, 19-27.  (MR19,P.895)


CORNISH, E. A. 1961.  (3.1/14.2)                                                    1060
    THE SIMULTANEOUS FIDUCIAL DISTRIBUTION OF THE LOCATION PARAMETERS IN A
    MULTIVARIATE NORMAL DISTRIBUTION.
    CSIRO DIV. MATH. STATIST. TECH. PAPER, 8, 1-12.


CORNISH, E. A. 1962.  (18.1)                                                        1061
    THE MULTIVARIATE T-DISTRIBUTION WITH THE GENERAL MULTIVARIATE NORMAL
    DISTRIBUTION.
    CSIRO DIV. MATH. STATIST. TECH. PAPER, 13, 3-18.


CORNISH, E. A. 1966.  (5.2)                                                         1062
    A MULTIPLE BEHRENS-FISHER DISTRIBUTION.
    MULTIVARIATE ANAL. PROC. INTERNAT. SYMP. DAYTON (KRISHNAIAH), 203-207.


CORNISH, E. A. + EVANS, MARILYN J. 1962.  (20.4)                                    1063
    TABLES FOR GRADUATING BY ORTHOGONAL POLYNOMIALS.
    CSIRO DIV. MATH. STATIST. TECH. PAPER, 12, 1-20.


CORNISH, E. A.   SEE ALSO
    464. BENNETT, G. W. + CORNISH, E. A. (1964)


CORSTEN, L. C. A. 1957.  (4.8/18.1)                                                 1064
    PARTITION OF EXPERIMENTAL VECTORS CONNECTED WITH MULTINOMIAL DISTRIBUTIONS.
    BIOMETRICS, 13, 451-484.  (MR19,P.1095)


CORSTEN, L. C. A. 1958.  (8.1/20.1)                                                 1065
    VECTORS, A TOOL IN STATISTICAL REGRESSION THEORY.
    MEDED. LANDBOUWHOGESCHOOL WAGENINGEN, 58, 1-92.  (MR23-A2993)


COSTNER, HERBERT L. 1965.  (4.9)                                                    1066
    CRITERIA FOR MEASURES OF ASSOCIATION.
    AMER. SOCIOL. REV., 30, 341-353.


COTTON, JOHN W.; CAMPBELL, DONALD J. + MALONE, R. DANIEL 1957.  (15.1/15.4)         1067
    THE RELATIONSHIP BETWEEN FACTORIAL COMPOSITION OF TEST ITEMS AND
    MEASURES OF TEST RELIABILITY.
    PSYCHOMETRIKA, 22, 347-357.  (MR19,P.823)


COULTER, MALCOLM A.   SEE
    891. CATTELL, RAYMOND B.; COULTER, MALCOLM A. + TSUJIOKA, BIEN (1966)


COUNT, EARL W. 1942.  (8.8/E)                                                       1068
    A QUANTITATIVE ANALYSIS OF GROWTH IN CERTAIN HUMAN SKULL DIMENSIONS.
    HUMAN BIOL., 14, 143-165.


COWAN, DONALD R. G. 1932.  (4.8)                                                    1069
    A NOTE ON THE COEFFICIENT OF PART CORRELATION AND OF CORRELATION OF A
    DEPENDENT VARIABLE WITH ALL BUT ONE OF A GROUP OF OTHER VARIABLES.
    J. AMER. STATIST. ASSOC., 27, 177-179.


COWDEN, DUDLEY J. 1943.  (4.1/20.3)                                                 1070
    CORRELATION CONCEPTS AND THE DOOLITTLE METHOD.
    J. AMER. STATIST. ASSOC., 38, 327-334.  (MR5,P.42)



COWDEN, DUDLEY J. 1952.  (4.8)                                                      1071
    THE MULTIPLE PARTIAL CORRELATION COEFFICIENT.
    J. AMER. STATIST. ASSOC., 47, 442-456.


COWDEN, DUDLEY J. 1958.  (8.1/20.3)                                                 1072
    A PROCEDURE FOR COMPUTING REGRESSION COEFFICIENTS.
        (AMENDED BY NO. 1073)
    J. AMER. STATIST. ASSOC., 53, 144-150.  (MR22-286)


COWDEN, DUDLEY J. 1959.  (8.1/20.3)                                                 1073
    CORRIGENDA:  A PROCEDURE FOR COMPUTING REGRESSION COEFFICIENTS.
        (AMENDMENT OF NO. 1072)
    J. AMER. STATIST. ASSOC., 54, 811.  (MR22-286)

COX, D. R. 1957.   (14.4/B)                                                          1074
    NOTE ON GROUPING.
  J. AMER. STATIST. ASSOC., 52, 543-547.

COX, D. R. 1958.   (8.1/18.2)                                                        1075
    THE REGRESSION ANALYSIS OF BINARY SEQUENCES.
        (WITH DISCUSSION)
  J. ROY. STATIST. SOC. SER. B, 20, 215-242.   (MR20-5541)

COX, D. R. 1960.   (17.7)                                                            1076
    REGRESSION ANALYSIS WHEN THERE IS PRIOR INFORMATION ABOUT SUPPLEMENTARY VARIABLES.
  J. ROY. STATIST. SOC. SER. B, 22, 172-176.   (MR22-5094)

COX, D. R. + BRANDWOOD, L. 1959.   (6.4/E)                                           1077
    ON A DISCRIMINATORY PROBLEM CONNECTED WITH THE WORKS OF PLATO.
  J. ROY. STATIST. SOC. SER. B, 21, 195-200.   (MR21-7814)

COX, D. R. + TOWNSEND, M. W. 1948.   E(16.3/U)                                       1078
    THE USE OF CORRELOGRAMS IN MEASURING YARN IRREGULARITY.
  PROC. INTERNAT. WOOL TEXTILE ORG., 2, 28-34.

COX, GERTRUDE M. 1939.   (15.1)                                                      1079
    THE MULTIPLE FACTOR THEORY IN TERMS OF COMMON ELEMENTS.
        (AMENDED BY NO. 1080)
  PSYCHOMETRIKA, 4, 59-68.

COX, GERTRUDE M. 1939.   (15.1)                                                      1080
    ERRATA TO: THE MULTIPLE FACTOR THEORY IN TERMS OF COMMON ELEMENTS.
        (AMENDMENT OF NO. 1079)
  PSYCHOMETRIKA, 4, 178.

COX, GERTRUDE M. + MARTIN, W. P. 1939.   E(8.5)                                      1081
    USE OF A DISCRIMINANT FUNCTION FOR DIFFERENTIATING SOILS WITH
    DIFFERENT AZOTBACTER POPULATIONS.
  IOWA STATE COLL. J. SCI., 11, 323-332.

COX, GERTRUDE M.    SEE ALSO
    730. BROWN, CLARENCE W.; BARTELME, PHYLLIS + COX, GERTRUDE M. (1933)

COXETER, H. S. M. 1935.   (20.4)                                                     1082
    THE FUNCTIONS OF SCHLAFLI AND LOBATSCHEFSKY.
  QUART. J. MATH. OXFORD SER., 6, 13-29.

CRABTREE, DOUGLAS E. 1965.   (20.1)                                                  1083
    ON THE CHARACTERISTIC ROOTS OF MATRICES.
  PROC. AMER. MATH. SOC., 16, 1410-1413.

CRAGG, J. G. 1966.   (16.2)                                                          1084
    ON THE SENSITIVITY OF SIMULTANEOUS-EQUATIONS ESTIMATORS TO THE
    STOCHASTIC ASSUMPTIONS OF THE MODELS.
  J. AMER. STATIST. ASSOC., 61, 136-151.

CRAIG, ALLEN T. 1933.   (4.1)                                                        1085
    ON THE CORRELATION BETWEEN CERTAIN AVERAGES FROM SMALL SAMPLES.
  ANN. MATH. STATIST., 4, 127-142.

CRAIG, ALLEN T. 1933.   (4.1/18.1)                                                   1086
    VARIABLES CORRELATED IN SEQUENCE.
  BULL. AMER. MATH. SOC., 39, 129-136.

CRAIG, ALLEN T. 1936.   (2.5)                                                        1087
    NOTE ON A CERTAIN BILINEAR FORM THAT OCCURS IN STATISTICS.
  AMER. J. MATH., 58, 864-866.

CRAIG, ALLEN T. 1938.   (2.5/8.8)                                                    1088
    ON THE INDEPENDENCE OF CERTAIN ESTIMATES OF VARIANCE.
  ANN. MATH. STATIST., 9, 48-55.

CRAIG, ALLEN T. 1943.   (2.5)                                                        1089
    NOTE ON THE INDEPENDENCE OF CERTAIN QUADRATIC FORMS.
  ANN. MATH. STATIST., 14, 195-197.   (MR5,P.127)

CRAIG, ALLEN T. 1947.   (2.5)                                        1090
    BILINEAR FORMS IN NORMALLY CORRELATED VARIABLES.
    ANN. MATH. STATIST., 18, 565-573.  (MR9,P.294)

CRAIG, ALLEN T.   SEE ALSO
    2511. HOGG, ROBERT V. + CRAIG, ALLEN T. (1958)

    2512. HOGG, ROBERT V. + CRAIG, ALLEN T. (1962)

CRAIG, CECIL C. 1931.  (17.9)                                        1091
    SAMPLING IN THE CASE OF CORRELATED VARIABLES.
    ANN. MATH. STATIST., 2, 324-332.

CRAIG, CECIL C. 1932.  (17.3)                                        1092
    ON THE COMPOSITION  OF DEPENDENT ELEMENTARY ERRORS.
    ANN. MATH. SER. 2, 33, 184-206.

CRAIG, CECIL C. 1933.  (17.2)                                        1093
    ON THE TCHEBYCHEF INEQUALITY OF BERNSTEIN.
    ANN. MATH. STATIST., 4, 94-102.

CRAIG, CECIL C. 1942.  (14.2/17.9/B)                                 1094
    ON THE FREQUENCY DISTRIBUTIONS OF THE QUOTIENT AND OF THE PRODUCT OF
    TWO STATISTICAL VARIABLES.
    AMER. MATH. MONTHLY, 49, 24-32.  (MR3,P.171)

CRAIG, CECIL C. 1942.  (1.1)                                         1095
    RECENT ADVANCES IN MATHEMATICAL STATISTICS, II.
    ANN. MATH. STATIST., 13, 74-85.  (MR4,P.25)

CRAIG, J. I. 1913.  (16.5)                                           1096
    THE PERIODOGRAM AND METHOD OF CORRELATION.
    REP. BRITISH ASSOC. DUNDEE, 12, 416-418.

CRAMER, ELLIOT M. + BOCK, R. DARRELL 1966.  (1.1/1.2)               1097
    MULTIVARIATE ANALYSIS.
    REV. EDUC. RES., 36, 604-617.

CRAMER, HARALD 1924.  (4.1)                                          1098
    REMARKS ON CORRELATION.
    SKAND. AKTUARIETIDSKR., 7, 220-240.

CRAMER, HARALD 1940.  (16.4)                                         1099
    ON THE THEORY OF STATIONARY RANDOM PROCESSES.
    ANN. MATH. SER. 2, 41, 215-230.  (MR1,P.150)

CRAMER, HARALD 1961.  (16.4)                                         1100
    ON THE STRUCTURE OF PURELY NON-DETERMINISTIC PROCESSES.
    ARK. MAT., 4, 249-266.

CRAMER, HARALD 1961.  (16.4)                                         1101
    ON SOME CLASSES OF NON-STATIONARY STOCHASTIC PROCESSES.
    PROC. FOURTH BERKELEY SYMP. MATH. STATIST. PROB., 2, 57-78.   (MR27-815)

CRAMER, HARALD 1962.  (1.2)                                          1102
    A.I. KHINCHIN'S WORK IN MATHEMATICAL PROBABILITY.
    ANN. MATH. STATIST., 33, 1227-1237.  (MR25-2944)

CRAMER, HARALD 1966.  (17.6/A)                                       1103
    ON EXTREME VALUES OF CERTAIN STOCHASTIC PROCESSES.
    RES. PAPERS STATIST. FESTSCHR. NEYMAN (DAVID), 73-78.

CRAMER, HARALD + WOLD, HERMAN O. A. 1936.  (17.1)                    1104
    SOME THEOREMS ON DISTRIBUTION FUNCTIONS.
    J. LONDON MATH. SOC., 11, 290-294.

CRATHORNE, A. R. 1922.  (4.8)                                        1105
    CALCULATION OF THE CORRELATION RATIO.
    J. AMER. STATIST. ASSOC., 18, 394-396.

CRATHORNE, A. R. 1925.  (4.8)                                        1106
    A WEIGHTED RANK CORRELATION PROBLEM.
    METRON, 5(4) 47-52.

CRAWFORD, ALBERT B.     SEE
    772. BURNHAM, PAUL S. + CRAWFORD, ALBERT B. (1935)

CRAWFORD, ISABELLE    SEE
    2298. HALL, D. M.; WELKER, E.L. + CRAWFORD, ISABELLE (1945)

CREAGER, JOHN A. 1958.   (15.1)                                                          1107
    GENERAL RESOLUTION OF CORRELATION MATRICES INTO COMPONENTS AND ITS
    UTILIZATION IN MULTIPLE AND PARTIAL REGRESSION.
    PSYCHOMETRIKA, 23, 1-8.

CREASY, MONICA A. 1956.   (17.7/C)                                                       1108
    CONFIDENCE LIMITS FOR THE GRADIENT IN THE LINEAR FUNCTIONAL RELATIONSHIP.
    J. ROY. STATIST. SOC. SER. B, 18, 65-69.   (MR18,P.426)

CREASY, MONICA A. 1957.   (8.3/15.1/F)                                                   1109
    ANALYSIS OF VARIANCE AS AN ALTERNATIVE TO FACTOR ANALYSIS.
    J. ROY. STATIST. SOC. SER. B, 19, 318-325.   (MR19,P.1095)

CRONBACH, LEE J. 1946.   (15.4)                                                          1110
    A CASE STUDY OF THE SPLIT-HALF RELIABILITY COEFFICIENT.
    J. EDUC. PSYCHOL., 37, 473-480.

CRONBACH, LEE J. 1947.   (15.4)                                                          1111
    TEST "RELIABILITY", ITS MEANING AND DETERMINATION.
    PSYCHOMETRIKA, 12, 1-16.

CRONBACH, LEE J. 1950.   (15.4)                                                          1112
    FURTHER EVIDENCE ON RESPONSE SETS AND TEST DESIGN.
    EDUC. PSYCHOL. MEAS., 10, 3-31.

CRONBACH, LEE J. 1951.   (15.4)                                                          1113
    COEFFICIENT ALPHA AND THE INTERNAL STRUCTURE OF TESTS.
    PSYCHOMETRIKA, 16, 297-334.

CRONBACH, LEE J. + GLESER, GOLDINE C. 1958.   (6.6/15.5)                                 1114
    ASSESSING SIMILARITY BETWEEN PROFILES.
    PSYCHOL. BULL., 50, 456-473.

CRONBACH, LEE J. + WARRINGTON, WILLARD G. 1952.   (15.4)                                 1115
    EFFICIENCY OF MULTIPLE-CHOICE TESTS AS A FUNCTION OF SPREAD OF ITEM
    DIFFICULTIES.
    PSYCHOMETRIKA, 17, 127-147.

CRONBACH, LEE J.; IKEDA, TOYOJI + AVNER, R. A. 1964.   E(4.8/15.4)                       1116
    INTRACLASS CORRELATION AS AN APPROXIMATION TO THE COEFFICIENT OF
    GENERALIZABILITY.
    PSYCHOL. REP., 15, 727-736.

CRONBACH, LEE J.    SEE ALSO
    2022. GLESER, GOLDINE C.; CRONBACH, LEE J. + RAJARATNAM, NAGESWARI (1965)

CROSETTI, ALBERT H.    SEE
    4917. SCHMITT, ROBERT C. + CROSETTI, ALBERT H. (1954)

CROUT, PRESCOTT D. 1941.   (20.3)                                                        1117
    A SHORT METHOD OF EVALUATING DETERMINANTS AND SOLVING SYSTEMS OF
    LINEAR EQUATIONS WITH REAL OR COMPLEX COEFFICIENTS.
    TRANS. AMER. INST. ELEC. ENGRS., 60, 1235-1241.

CRUM, W. L. 1921.   (4.5)                                                                1118
    A SPECIAL APPLICATION OF PARTIAL CORRELATION.
    QUART. PUBL. AMER. STATIST. ASSOC., 17, 949-952.

CRUM, W. L. 1923.   (15.5)                                                               1119
    THE RESEMBLANCE BETWEEN THE ORDINATE OF THE PERIODOGRAM AND THE
    CORRELATION COEFFICIENT.
    J. AMER. STATIST. ASSOC., 18, 889-899.

CRUTCHFIELD, RICHARD S. + TOLMAN, EDWARD C. 1940.   (8.5)                                1120
    MULTIPLE-VARIABLE DESIGN FOR EXPERIMENTS INVOLVING INTERACTION OF BEHAVIOR.
    PSYCHOL. REV., 47, 38-42.

CSAKI, PETER + FISCHER, JANOS 1960.  (4.9/17.3/B)                    1121
    ON BIVARIATE STOCHASTIC CONNECTION.
    MAGYAR TUD. AKAD. MAT. KUTATO. INT. KOZL., 5, 311-323.   (MR23-A4244)


CSAKI, PETER + FISCHER, JANOS 1960.  (4.6/16.4/17.3)                1122
    CONTRIBUTIONS TO THE PROBLEM OF MAXIMUM CORRELATION.
    MAGYAR TUD. AKAD. MAT. KUTATO. INT. KOZL., 5, 325-337.   (MR23-A4245)


CSAKI, PETER + FISCHER, JANOS 1961.  (4.9/16.4/B)                   1123
    ON BIVARIATE STOCHASTIC CONNECTION.
    DEUX. CONG. MATH. HONGR., 2(4) 42-44.


CSAKI, PETER + FISCHER, JANOS 1963.  (4.9/12.1)                     1124
    ON THE GENERAL NOTION OF MAXIMAL CORRELATION.
    MAGYAR TUD. AKAD. MAT. KUTATO. INT. KOZL., 8, 27-51.


CSISZAR, I. 1962.  (6.5/17.1)                                       1125
    INFORMATIONSTHEORETISCHE KONVERGENZBEGRIFFE IM RAUM DER WAHRSCHEINLICHKEITSVERTEILUNGEN.
    MAGYAR TUD. AKAD. MAT. KUTATO. INT. KOZL., 7, 137-158.   (MR32-9135)


CSISZAR, I. 1963.  (6.5/16.4)                                       1126
    EINE INFORMATIONSTHEORETISCHE UNGLEICHUNG UND IHRE ANWENDUNG AUF DEN
    BEWEIS DER ERGODIZITAT VON MARKOFFSCHEN KETTEN.
    MAGYAR TUD. AKAD. MAT. KUTATO. INT. KOZL., 8, 85-107.


CSISZAR, I. 1966.  (17.2)                                           1127
    A NOTE ON JENSEN'S INEQUALITY.
    STUDIA SCI. MATH. HUNGAR., 1, 185-188.


CSISZAR, I. 1966.  (17.1)                                           1128
    ON INFINITE PRODUCTS OF RANDOM ELEMENTS AND INFINITE CONVOLUTIONS OF
    PROBABILITY DISTRIBUTIONS ON LOCALLY COMPACT GROUPS.
    Z. WAHRSCHEIN. VERW. GEBIETE, 5, 279-295.


CSISZAR, I. + FISCHER, JANOS 1962.  (6.5)                           1129
    INFORMATIONSENTFERNUNGEN IM RAUM DER WAHRSCHEINLICHKEITSVERTEILUNGEN.
    MAGYAR TUD. AKAD. MAT. KUTATO. INT. KOZL., 7, 159-180.   (MR32-9136)


CUCCONI, ODOARDO 1963.  (14.3/14.5)                                 1130
    UN CRITERIO PER IL REGETTO DELLE OSSERVAZIONI SPURIE.
    LA SCUOLA IN AZIONE, 21, 92-106.


CUPPENS, ROGER 1966.  (2.3/18.1)                                    1131
    SUR LA DECOMPOSITION DE LA COMPOSITION D'UNE LOI NORMALE ET D'UNE LOI
    DE POISSON.
    C. R. ACAD. SCI. PARIS SER. A, 262, 1113-1116.   (MR33-6667)


CUPPENS, ROGER 1966.  (2.3)                                         1132
    SUR LES FONCTIONS CARACTERISTIQUES ANALYTIQUES.
    C. R. ACAD. SCI. PARIS SER. A, 263, 86-88.   (MR34-838A)


CUPPENS, ROGER 1966.  (2.3)                                         1133
    DECOMPOSITION DES FONCTIONS CARACTERISTIQUES ANALYTIQUES.
    C. R. ACAD. SCI. PARIS SER. A, 263, 133-135.   (MR34-838B)


CUPPENS, ROGER 1966.  (17.3)                                        1134
    DECOMPOSITION DES FONCTIONS CARACTERISTIQUES INDEFINIMENT DIVISIBLES.
    C. R. ACAD. SCI. PARIS SER. A, 263, 616-619.   (MR34-8450)



CUPPENS, ROGER 1966.  (17.3)                                        1135
    EXTENSIONS DE LA NOTION DE FONCTION CARACTERISTIQUE ANALYTIQUE.
    C. R. ACAD. SCI. PARIS SER. A, 263, 682-684.   (MR34-5127)


CUPROV, ALEX. A. 1921.  (2.1/17.3)                                  1136
    UBER DIE KORRELATIONSFLACHE DER ARITHMETISCHEN DURCHSCHNITTE (EIN GRENZTHEOREM).
    METRON, 1(4) 41-48.


CUPROV, ALEX. A. 1923.  (17.3/17.5)                                 1137
    ON THE MATHEMATICAL EXPECTATION OF THE MOMENTS OF FREQUENCY
    DISTRIBUTIONS IN THE CASE OF CORRELATED OBSERVATIONS.
    METRON, 2,   461-493.

CUPROV, ALEX. A. 1923.   (17.3/17.5)                                                    1138
     ON THE MATHEMATICAL EXPECTATION OF THE MOMENTS OF FREQUENCY
     DISTRIBUTIONS IN THE CASE OF CORRELATED OBSERVATIONS.
     METRON, 2,   646-683.


CUPROV, ALEX. A. 1923.   (4.1)                                                          1139
     AUFGABEN UND VORAUSSETZUNGEN DER KORRELATIONSMESSUNG.
     NORDISK STATIST. TIDSKR., 2(1) 24-53.


CUPROV, ALEX. A. 1923.   (4.9)                                                          1140
     UBER NORMAL STABILE KORRELATION.
     SKAND. AKTUARIETIDSKR., 6, 1-17.


CUPROV, ALEX. A. 1925.   (17.3/A)                                                       1141
     ON THE ASYMPTOTIC FREQUENCY DISTRIBUTION OF THE ARITHMETIC MEANS OF N
     CORRELATED OBSERVATIONS FOR VERY GREAT VALUES OF N.
     J. ROY. STATIST. SOC., 88, 91-104.


CUPROV, ALEX. A. 1928.   (4.8/17.3)                                                     1142
     THE MATHEMATICAL THEORY OF THE STATISTICAL METHODS EMPLOYED IN THE
     STUDY OF CORRELATION IN THE CASE OF THREE VARIABLES.
     (TRANSLATED BY L. ISSERLIS)
     TRANS. CAMBRIDGE PHILOS. SOC., 23, 337-382.


CUPROV, ALEX. A. 1931.   (4.9/B)                                                        1143
     THE MATHEMATICAL FOUNDATIONS OF THE METHODS TO BE USED IN STATISTICAL
     INVESTIGATION OF THE DEPENDENCE BETWEEN TWO CHANCE VARIABLES.
     NORDIC STATIST. J., 3, 71-84.


CURETON, EDWARD E. 1929.   (4.1)                                                        1144
     COMPUTATION OF CORRELATION COEFFICIENTS.
     J. EDUC. PSYCHOL., 20, 588-601.


CURETON, EDWARD E. 1931.   (15.1/15.2/15.4)                                             1145
     ERRORS OF MEASUREMENT AND CORRELATION.
     ARCH. OF PSYCHOL. (COLUMBIA), 19(125) 1-63.


CURETON, EDWARD E. 1933.   (4.2/7.3)                                                    1146
     THE STANDARD ERROR OF THE AVERAGE INTERCORRELATION AND OF THE AVERAGE
     CRITERION CORRELATION.
     J. APPL. PSYCHOL., 17, 70-76.


CURETON, EDWARD E. 1935.   (15.2)                                                       1147
     WISHART'S EXACT FORMULA FOR THE STANDARD ERROR OF THE PRODUCT-MOMENT
     TETRAD VERSUS AN APPROXIMATION FORMULA.
     J. EDUC. PSYCHOL., 26, 212-217.


CURETON, EDWARD E. 1936.   (15.2)                                                       1148
     ON CERTAIN ESTIMATED CORRELATION FUNCTIONS AND THEIR STANDARD ERRORS.
     J. EXPER. EDUC., 4, 252-264.


CURETON, EDWARD E. 1951.   (8.1/15.4)                                                   1149
     APPROXIMATE LINEAR RESTRAINTS AND BEST PREDICTOR WEIGHTS.
     EDUC. PSYCHOL. MEAS., 11, 12-15.


CURETON, EDWARD E. 1952.   (15.4)                                                       1150
     NOTE ON THE SCALING OF RATINGS OR RANKINGS WHEN THE NUMBERS PER
     SUBJECT ARE UNEQUAL.
     PSYCHOMETRIKA, 17, 397-399.


CURETON, EDWARD E. 1955.   (15.1/15.2/E)                                                1151
     ON THE USE OF BURT'S FORMULA FOR ESTIMATING FACTOR SIGNIFICANCE.
     BRITISH J. STATIST. PSYCHOL., 8, 28.


CURETON, EDWARD E. 1956.   (4.8)                                                        1152
     RANK-BISERIAL CORRELATION.
     PSYCHOMETRIKA, 21, 287-290.   (MR18,P.343)


CURETON, EDWARD E. 1957.   (15.4)                                                       1153
     THE UPPER AND LOWER TWENTY-SEVEN PER CENT RULE.
     PSYCHOMETRIKA, 22, 293-296.

CURETON, EDWARD E. 1958.   (4.8)                                           1154
    THE AVERAGE SPEARMAN RANK CORRELATION WHEN TIES ARE PRESENT.
        (AMENDED BY NO. 1157)
    PSYCHOMETRIKA, 23, 271-272.


CURETON, EDWARD E. 1959.   (15.1/20.2)                                     1155
    A NOTE ON FACTOR ANALYSIS:   ARBITRARY ORTHOGONAL TRANSFORMATIONS.
    PSYCHOMETRIKA, 24, 169-174.   (MR21-1242)

CURETON, EDWARD E. 1962.   (20.3)                                          1156
    A NOTE ON SIMULTANEOUS EQUATIONS AND MATRIX INVERSION.
    BRITISH J. STATIST. PSYCHOL., 15, 51-58.

CURETON, EDWARD E. 1965.   (4.8)                                           1157
    THE AVERAGE SPEARMAN RANK CORRELATION WHEN TIES ARE PRESENT:   A CORRECTION.
        (AMENDMENT OF NO. 1154)
    PSYCHOMETRIKA, 30, 377.

CURETON, EDWARD E. 1966.   (4.8)                                           1158
    ON CORRELATION COEFFICIENTS.
    PSYCHOMETRIKA, 31, 605-607.

CURETON, EDWARD E. + DUNLAP, JACK W. 1929.   (15.4/T)                      1159
    A NOMOGRAPH FOR ESTIMATING THE RELIABILITY OF A TEST IN ONE RANGE OF
    TALENT WHEN ITS RELIABILITY IS KNOWN IN ANOTHER PLACE.
    J. EDUC. PSYCHOL., 20, 537-538.

CURETON, EDWARD E. + DUNLAP, JACK W. 1930.   (15.1)                        1160
    SOME EFFECTS OF HETEROGENEITY ON THE THEORY OF FACTORS.
    AMER. J. PSYCHOL., 42, 608-620.

CURETON, EDWARD E. + DUNLAP, JACK W. 1930.   (15.4/T)                      1161
    A NOMOGRAPH FOR ESTIMATING A RELIABILITY COEFFICIENT BY THE
    SPEARMAN-BROWN FORMULA AND FOR COMPUTING ITS PROBABLE ERROR.
    J. EDUC. PSYCHOL., 21, 68-69.

CURETON, EDWARD E. + DUNLAP, JACK W. 1938.   (15.4)                        1162
    DEVELOPMENTS IN STATISTICAL METHODS RELATED TO TEST CONSTRUCTION.
    REV. EDUC. RES., 8, 307-362.

CURETON, EDWARD E.    SEE ALSO
    1412. DUNLAP, JACK W. + CURETON, EDWARD E. (1930)

    2943. KATZELL, R. A. + CURETON, EDWARD E. (1947)

    5025. SIEGEL, CARL LUDWIG + CURETON, EDWARD E. (1952)

CURNOW, R. C. 1960.   (17.4)                                               1163
    THE REGRESSION OF TRUE VALUE ON ESTIMATED VALUE.
    BIOMETRIKA, 47, 457-460.

CURNOW, R. N. + DUNNETT, CHARLES W. 1962.   (2.7)                          1164
    THE NUMERICAL EVALUATION OF CERTAIN MULTIVARIATE NORMAL INTEGRALS.
    ANN. MATH. STATIST., 33, 571-579.   (MR25-690)

CURTISS, JOHN H. 1941.   (14.2)                                           1165
    ON THE DISTRIBUTION OF THE QUOTIENT OF TWO CHANCE VARIABLES.
    ANN. MATH. STATIST., 12, 409-421.   (MR4,P.16)

CUSIMANO, GIOVANNI 1961.   (4.1/4.8)                                       1166
    SU UNA NUOVA INTERPRETAZIONE DEL COEFFICIENTE DI CORRELAZIONE SEMPLICE.
    ANN. FAC. ECON. COMM. PALERMO, 15(1) 231-243.

CVETNOV, V. V. 1958.   (16.3)                                              1167
    PHASE CORRELATION PROPERTIES OF SIGNALS AND GAUSSIAN NOISE IN
    TWO-CHANNEL PHASE SYSTEMS.
    RADIOTEHNIKA, 13(4) 53-62.

CYFFERS, B. 1965.   (6.4/E)                                                1168
    DISCRIMINANT ANALYSIS. (IN FRENCH)
    REV. STATIST. APPL., 13(2) 29-46.

CYFFERS, B.    SEE ALSO
    922. CHARBONNIER, A.; CYFFERS, B.; SCHWARTZ, D. +  (1957)
    VESSEREAU, A.

CZUBER, EMANUEL 1881.  (2.1/2.3/B)                                        1169
    ZUR THEORIE DER FEHLERELLIPSE.
    SITZUNGSBER. MATH. NATURWISS. CL. KAIS. AKAD. WISS. WIEN, 82(2) 698-723.

CZUBER, EMANUEL 1920.  (4.1)                                             1170
    UBER FUNKTIONEN VON VARIABLEN, ZWISCHEN WELCHEN KORRELATIONEN BESTEHEN.
    METRON, 1(1) 53-61.

DAGNELIE, P. 1966.  (6.4)                                                1171
    ON DIFFERENT METHODS OF NUMERICAL CLASSIFICATION.
    REV. STATIST. APPL., 14(3) 55-75.

DAGUM, CAMILO 1958.  (17.1/17.3)                                         1172
    VARIANZA E COVARIANZA DEL MOMENTO MISTO DEI CAMPIONI DI UNA VARIABILE
    CASUALE A K-DIMENSIONI.
    ATTI RIUN. SCI. SOC. ITAL. STATIST., 17, 89-96.

DAGUM, CAMILO 1960.  (4.8)                                               1173
    TRANSVARIACION ENTRE MAS DE DOS DISTRIBUCIONES.
    STUDI ONORF C. GINI, 1, 53-92.

DAGUM, CAMILO 1961.  (4.8/9.3/17.9/E)                                    1174
    TRANSVARIACION EN LA HIPOTESIS DE VARIABLES ALEATORIAS NORMALES
    MULTIDIMENSIONALES.
    BULL. INST. INTERNAT. STATIST., 38(4) 473-486.

DAGUM, CAMILO 1966.  (16.2/E)                                            1175
    BASES  Y PRINCIPIOS PARA LA ELABORACION DE MODELOS EN LA CIENCIA ECONOMICA.
    METRON, 25, 401-435.

DAGUM, CAMILO 1966.  (4.8/ET)                                            1176
    PROBABILITY AND MEASURE OF TRANSVARIABILITY IN THE N-DIMENSIONAL SPACE.
        (IN GERMAN)
    STATIST. HEFTE, 7, 3-29.

DALE. M. G.    SEE
    3583. MACNAUGHTON-SMITH, P.; WILLIAMS, W. T.; DALE, M. G. +  (1964)
    MOCKETT, L. G.

DALENIUS, TORE 1953.  (1.1/17.9)                                         1177
    THE MULTI-VARIATE SAMPLING PROBLEM.
    SKAND. AKTUARIETIDSKR., 36, 92-102.  (MR15,P.451)

DALENIUS, TORE + GURNEY, M. 1951.  (18.5)                                1178
    THE PROBLEM OF OPTIMUM STRATIFICATION II.
    SKAND. AKTUARIETIDSKR., 34, 133-148.

DALL'AGLIO, GIORGIO 1960.  (17.4)                                        1179
    LES FONCTIONS EXTREMES DE LA CLASSE DE FRECHET A 3 DIMENSIONS.
    PUBL. INST. STATIST. UNIV. PARIS, 9, 175-188.  (MR23-A1416)

DALL'AGLIO, GIORGIO 1961.  (17.4/B)                                      1180
    SULLE DISTRIBUZIONI DOPPIE CON MARGINI ASSEGNATI SOGGETTE A DELLE LIMITAZIONI.
    GIORN. IST. ITAL. ATTUARI, 24, 94-108.

DALL'AGLIO, GIORGIO 1962.  (8.4/18.1)                                    1181
    THE SIMULTANEOUS DISTRIBUTION OF SEVERAL FISHER FUNCTIONS.
    C.R. ACAD. SCI. PARIS, 254, 412-413.  (MR24-A1154)

DALL'AGLIO, GIORGIO 1966.  (17.8/A)                                      1182
    SU UNA CLASSE DI INDICI CON DISTRIBUZIONE ASINTOTICA DEL TIPO $\chi^2$.
    METRON, 25, 216-227.

DALLAS, A. E. M. M.    SEE
    4285. PIAGGIO, H. T. H. + DALLAS, A. E. M. M. (1934)

DALY, JOSEPH F. 1940.  (9.1)                                              1183
    ON THE UNBIASED CHARACTER OF LIKELIHOOD-RATIO TESTS FOR INDEPENDENCE
    IN NORMAL SYSTEMS.
    ANN. MATH. STATIST., 11, 1-32.  (MR1,P.347)

DALY, JOSEPH F.   SEE ALSO
    1826. FURFEY, PAUL H. + DALY, JOSEPH F. (1935)

    5922. WILKS, S.S. + DALY, JOSEPH F. (1939)

DAMM, H. 1913.  (4.1)                                                     1184
    ZUR EINFUHRUNG IN DIE KORRELATIONSSTATISTIK.
    ARCH. PAD., 1(2) 301-319.

DAMMAN, J. E.   SEE
    3743. MATTSON, R. L. + DAMMAN, J. E. (1965)

DANFORD, MASIL B.; HUGHES, HARRY M. + MC NEE, P. C. 1960.   (8.5)         1185
    ON THE ANALYSIS OF REPEATED-MEASUREMENTS EXPERIMENTS.
    BIOMETRICS, 16, 547-565.  (MR23-A1439)

DANIELL, P. J. 1918.  (20.4)                                              1186
    INTEGRALS IN AN INFINITE NUMBER OF DIMENSIONS.
    ANN. MATH. SER. 2, 20, 281-288.

DANIELL, P. J. 1919.  (20.4)                                              1187
    FUNCTIONS OF LIMITED VARIATION IN AN INFINITE NUMBER OF DIMENSIONS.
    ANN. MATH. SER. 2, 21, 30-38.

DANIELL, P. J. 1929.  (15.1)                                              1188
    BOUNDARY CONDITIONS FOR CORRELATION COEFFICIENTS.
    BRITISH J. PSYCHOL., 20, 190-194.

DANIELS, H. E. 1944.  (4.1)                                               1189
    THE RELATION BETWEEN MEASURES OF CORRELATION IN THE UNIVERSE OF
    SAMPLE PERMUTATIONS.
    BIOMETRIKA, 33, 129-135.  (MR6,P.91)

DANIELS, H. E. 1947.  (2.3/17.6/BT)                                       1190
    GROUPING CORRECTIONS FOR HIGH AUTOCORRELATIONS.
    J. ROY. STATIST. SOC. SUPP., 9, 245-249.  (MR9,P.294)

DANIELS, H. E. 1948.  (4.8)                                               1191
    A PROPERTY OF RANK CORRELATIONS.
    BIOMETRIKA, 35, 416-417.  (MR10,P.386)

DANIELS, H. E. 1950.  (4.8)                                               1192
    RANK CORRELATION AND POPULATION MODELS.
    J. ROY. STATIST. SOC. SER. B, 12, 171-181.  (MR12,P.725)

DANIELS, H. E. 1951.  (4.8)                                               1193
    NOTE ON DURBIN AND STUART'S FORMULA FOR $E(R_S)$.
        (AMENDMENT OF NO. 1431)
    J. ROY. STATIST. SOC. SER. B, 13, 310.  (MR13,P.962)

DANIELS, H. E. 1952.  (14.5/B)                                            1194
    THE COVERING CIRCLE OF A SAMPLE FROM A CIRCULAR NORMAL DISTRIBUTION.
    BIOMETRIKA, 39, 137-143.  (MR13,P.962)

DANIELS, H. E. 1962.  (16.3)                                              1195
    THE ESTIMATION OF SPECTRAL DENSITIES.
    J. ROY. STATIST. SOC. SER. B, 24, 185-198.

DANIELS, H. E. + KENDALL, MAURICE G. 1947.  (4.8)                         1196
    THE SIGNIFICANCE OF RANK CORRELATIONS WHEN PARENTAL CORRELATION EXISTS.
    BIOMETRIKA, 34, 197-208.  (MR9,P.364)

DANIELS, H. E. + KENDALL, MAURICE G. 1958.  (4.2)                         1197
    SHORT PROOF OF MISS HARLEY'S THEOREM ON THE CORRELATION COEFFICIENT.
    BIOMETRIKA, 45, 571-572.

DAR, S. N. 1964.   (8.6/U)                                                                              1198
    COMPARISON OF THE SENSITIVITIES OF DEPENDENT EXPERIMENTS.
  BIOMETRICS, 20, 209-212.

DARBISHIRE, A. D. 1907.   (18.1)                                                                        1199
    SOME TABLES FOR ILLUSTRATING STATISTICAL CORRELATION.
  MEM. PROC. MANCHESTER LIT. PHILOS. SOC., 51(16) 1-21.

DARBISHIRE, A. D. 1909.   E(4.8)                                                                        1200
    AN EXPERIMENTAL ESTIMATION OF THE THEORY OF ANCESTRAL CONTRIBUTIONS IN HEREDITY.
  PROC. ROY. SOC. LONDON SER. B, 81, 61-79.

DARLING, D. A. + SIEGERT, ARNOLD J. F. 1957.   (16.4)                                                   1201
    A SYSTEMATIC APPROACH TO A CLASS OF PROBLEMS IN THE THEORY OF NOISE
    AND OTHER RANDOM PHENOMENA -- PART I.
  IRE TRANS. INFORMATION THEORY, IT-3, 32-37.

DARMOIS, GEORGES 1931.   (2.1)                                                                          1202
    DETERMINATION DE LA MOYENNE ET DE LA DISPERSION DANS LE CAS DES
    EPREUVES DEPENDANTES.
  C. R. ASSOC. FRANC. AVANCE. SCI. CONF., 55, 34-37.

DARMOIS, GEORGES 1934.   (15.1)                                                                         1203
    SUR LA THEORIE DES DEUX FACTEURS DE SPEARMAN.
  C.R. ACAD. SCI. PARIS, 199, 1176-1178, 1358-1360.

DARMOIS, GEORGES 1935.   (15.1)                                                                         1204
    LES METHODES D'ANALYSE FACTORIELLE.
  BIOTYPOLOGIE, 3, 45-57.

DARMOIS, GEORGES 1936.   (15.1)                                                                         1205
    SUR L'INDETERMINATION DU FACTEUR GENERAL DANS LA THEORIE DE SPEARMAN.
  MATHEMATICA, CLUJ, 12, 211-216.

DARMOIS, GEORGES 1940.   (11.1/15.1)                                                                    1206
    LES MATHEMATIQUES DE LA PSYCHOLOGIE.
  MEM. SCI. MATH., 98, 1-51.   (MR3,P.170)

DARMOIS, GEORGES 1953.   (15.1)                                                                         1207
    ANALYSE GENERALE DES LIAISONS STOCHASTIQUES.  ETUDE PARTICULIERE DE L'ANALYSE
    FACTORIELLE LINEAIRE.
  REV. INST. INTERNAT. STATIST., 21, 2-8.   (MR15,P.808)

DARMOIS, GEORGES 1955.   (8.1/15.1/17.5)                                                                1208
    SUR LA REGRESSION.  RESULTATS NOUVEAUX.  PROBLEMES NON RESOLUS.
  COLLOQ. ANAL. STATIST. BRUXELLES, 9-23.   (MR17,P.505)

DARMOIS, GEORGES 1955.   (15.1)                                                                         1209
    OBSERVATIONS THEORIQUES SUR L'ANALYSE FACTORIELLE, LINEAIRE ET GENERALE.
      (WITH DISCUSSION)
  COLLOQ. INTERNAT. CENTRE NAT. RECH. SCI., 65, 295-304.

DARROCH, J. N. 1965.   (11.1)                                                                           1210
    AN OPTIMAL PROPERTY OF PRINCIPAL COMPONENTS.
  ANN. MATH. STATIST., 36, 1579-1582.   (MR32-555)

DARROCH, J. N. 1965.   (15.1)                                                                           1211
    A SET OF INEQUALITIES IN FACTOR ANALYSIS.
  PSYCHOMETRIKA, 30, 449-453.   (MR33-6763)

DAS, A. C. 1948.   (5.2/14.5)                                                                           1212
    A NOTE ON THE $D^2$-STATISTIC WHEN THE VARIANCES AND CO-VARIANCES ARE KNOWN.
  SANKHYA, 8, 372-374.   (MR10,P.134)

DAS, A. C. 1949.   (17.9)                                                                               1213
    TWO-DIMENSTIONAL SYSTEMATIC SAMPLING.
  SCI. AND CULT., 15, 157-158.   (MR11,P.260)

DAS, S. C. 1956.   (20.4)                                                                               1214
    THE NUMERICAL EVALUATION OF A CLASS OF INTEGRALS.  II.
  PROC. CAMBRIDGE PHILOS. SOC., 52, 442-448.   (MR19,P.983)

DAS GUPTA, SOMESH 1960.  (4.4/4.8/15.4)                          1215
    POINT BISERIAL CORRELATION AND ITS GENERALIZATION.
    PSYCHOMETRIKA, 25, 393-408.

DAS GUPTA, SOMESH 1964.  (6.4/17.8/A)                            1216
    NON-PARAMETRIC CLASSIFICATION RULES.
    SANKHYA SER. A, 26, 25-30.

DAS GUPTA, SOMESH 1965.  (6.1)                                   1217
    OPTIMUM CLASSIFICATION RULES FOR CLASSIFICATION INTO TWO MULTIVARIATE
        NORMAL POPULATIONS.
    ANN. MATH. STATIST., 36, 1174-1184.

DAS GUPTA, SOMESH + BHATTACHARYA, P.K. 1964.  (6.2/6.4/19.1/U)   1218
    CLASSIFICATION BETWEEN EXPONENTIAL POPULATIONS.
    SANKHYA SER. A, 26, 17-24.

DAS GUPTA, SOMESH+ ANDERSON, T. W. + MUDHOLKAR, GOVIND S. 1964.  (8.3)   1219
    MONOTONICITY OF THE POWER FUNCTIONS OF SOME TESTS OF THE MULTIVARIATE
        LINEAR HYPOTHESIS.
    ANN. MATH. STATIST., 35, 200-205.  (MR28-1697)

DAS GUPTA, SOMESH   SEE ALSO
    163. ANDERSON, T. W. + DAS GUPTA, SOMESH (1963)

    164. ANDERSON, T. W. + DAS GUPTA, SOMESH (1964)

    165. ANDERSON, T. W. + DAS GUPTA, SOMESH (1964)

    166. ANDERSON, T. W. + DAS GUPTA, SOMESH (1965)

    167. ANDERSON, T. W. + DAS GUPTA, SOMESH (1965)

DAVE, P. H.   SEE
    515. BHATT, N. M. + DAVE, P. H. (1964)

    516. BHATT, N. M. + DAVE, P. H. (1965)

    517. BHATT, N. M. + DAVE, P. H. (1965)

DAVENPORT, C. B. + BULLARD, C. 1896. E(4.8)                      1220
    A CONTRIBUTION TO THE QUANTITATIVE STUDY OF CORRELATED VARIATION AND
    THE COMPARATIVE VARIABILITY OF THE SEXES.
    PROC. AMER. ACAD. ARTS SCI., 32, 87-97.

DAVID, F. N. 1937.  (4.3/C)                                      1221
    A NOTE ON UNBIASED LIMITS FOR THE CORRELATION COEFFICIENT.
    BIOMETRIKA, 29, 157-160.

DAVID, F. N. 1948.  (14.4)                                       1222
    CORRELATIONS BETWEEN $\chi^2$ CELLS.
    BIOMETRIKA, 35, 418-422.  (MR10,P.465)

DAVID, F. N. 1949.  (14.5)                                       1223
    MOMENTS OF THE $z$ AND F DISTRIBUTIONS.
    BIOMETRIKA, 36, 394-403.  (MR11,P.447)

DAVID, F. N. 1953.  (2.7)                                        1224
    A NOTE ON THE EVALUATION OF THE MULTIVARIATE NORMAL INTEGRAL.
    BIOMETRIKA, 40, 458-459.  (MR15,P.354)

DAVID, F. N. + FIX, EVELYN 1966.  (2.7/T)                        1225
    RANDOMIZATION AND THE SERIAL CORRELATION COEFFICIENT.
    RES. PAPERS STATIST. FESTSCHR. NEYMAN (DAVID), 461-468.

DAVID, F. N. + MALLOWS, C.L. 1961.  (4.8)                        1226
    THE VARIANCE OF SPEARMAN'S RHO IN NORMAL SAMPLES.
    BIOMETRIKA, 48, 19-28.  (MR23-A3611)

DAVID, F. N.   SEE ALSO
    364. BARTON, D. E. + DAVID, F. N. (1960)

    365. BARTON, D. E. + DAVID, F. N. (1962)

DAVID, HERBERT T. 1962.   (14.1/17.8/U)                                    1227
     THE SAMPLE MEAN AMONG THE MODERATE ORDER STATISTICS.
     ANN. MATH. STATIST., 33, 1160–1166.   (MR26–860)

DAVID, HERBERT T. 1963.   (14.1/20.5/N)                                    1228
     THE SAMPLE MEAN AMONG THE EXTREME NORMAL ORDER STATISTICS.
     ANN. MATH. STATIST., 34, 33–55.   (MR26–1970)

DAVID, HERBERT T.   SEE ALSO
     1819. FUCHS, CAROL E. + DAVID, HERBERT T. (1965)

     3292. LAWING, WILLIAM D. + DAVID, HERBERT T. (1966)

DAVIDOFF, MELVIN D. 1949.   (4.5/4.6)                                      1229
     A GRAPHIC SOLUTION OF MULTIPLE AND PARTIAL CORRELATION COEFFICIENTS
     FOR THREE VARIABLE PROBLEMS.
     EDUC. PSYCHOL. MEAS., 9, 773–779.

DAVIDOFF, MELVIN D. 1954.   (4.4)                                          1230
     A FURTHER NOTE ON THE RAPID DETERMINATION OF THE TETRACHORIC
     CORRELATION COEFFICIENT.
     PSYCHOMETRIKA, 19, 163–164.

DAVIDOFF, MELVIN D. + GOHEEN, HOWARD W. 1953.   (4.4/T)                    1231
     A TABLE FOR THE RAPID DETERMINATION OF THE TETRACHORIC CORRELATION COEFFICIENT.
     PSYCHOMETRIKA, 18, 115–121.   (MR14,P.995)

DAVIDOFF, MELVIN D.   SEE ALSO
     2039. GOHEEN, HOWARD W. + DAVIDOFF, MELVIN D. (1951)

DAVIES, GEORGE R. 1930.   (4.1)                                           1232
     FIRST MOMENT CORRELATION.
     J. AMER. STATIST. ASSOC., 25, 413–427.

DAVIES, HILDA M. + JOWETT, GEOFFREY H. 1958.   (16.5)                     1233
     THE FITTING OF MARKOFF SERIAL VARIATION CURVES.
     J. ROY. STATIST. SOC. SER. B, 20, 120–142.   (MR20–6183)

DAVIES, O. L. 1933.   (20.4)                                             1234
     ON ASYMPTOTIC FORMULAE FOR THE HYPERGEOMETRIC SERIES.  I.
     HYPERGEOMETRIC SERIES IN WHICH THE FOURTH ELEMENT, X, IS UNITY.
     BIOMETRIKA, 25, 295–322.

DAVIES, O. L. 1934.   (20.4)                                             1235
     ON ASYMPTOTIC FORMULAE FOR THE HYPERGEOMETRIC SERIES.  II.  HYPERGEOMETRIC
     SERIES IN WHICH THE FOURTH ELEMENT, X, IS NOT NECESSARILY UNITY.
     BIOMETRIKA, 26, 59–107.

DAVIS, FREDERICK B. 1944.   (15.4)                                       1236
     A NOTE ON CORRECTING RELIABILITY COEFFICIENTS FOR RANGE.
          (AMENDED BY NO. 3696)
     J. EDUC. PSYCHOL., 35, 500–502.

DAVIS, FREDERICK B. 1947.   (15.1)                                       1237
     THE INTERPRETATION OF PRINCIPAL AXIS FACTORS.
     J. EDUC. PSYCHOL., 38, 471–481.

DAVIS, R. C. 1952.   (16.4)                                              1238
     ON THE THEORY OF PREDICTION OF NONSTATIONARY STOCHASTIC PROCESSES.
     J. APPL. PHYS., 23, 1047–1053.   (MR14,P.295)

DAVIS, R. C. 1953.   (16.4)                                              1239
     ON THE FOURIER EXPANSION OF STATIONARY RANDOM PROCESSES.
     PROC. AMER. MATH. SOC., 4, 564–569.   (MR15,P.45)

DAVIS, R. C. 1957.   (16.4)                                              1240
     OPTIMUM VS. CORRELATION METHODS IN TRACKING RANDOM SIGNALS IN BACKGROUND NOISE.
     QUART. APPL. MATH., 15, 123–138.   (MR20–7372)

DAY, BESSE B. + FISHER, RONALD A. 1937.  E(8.5/B)                         1241
     THE COMPARISON OF VARIABILITY IN POPULATIONS HAVING UNEQUAL MEANS.   AN EXAMPLE
     OF THE ANALYSIS OF COVARIANCE WITH MULTIPLE DEPENDENT AND INDEPENDENT VARIATES.
     ANN. EUGENICS, 7, 333–348.

DAY, BESSE B. + LLOYD, A. 1939.  (17.9/E)                1242
    A THREE-DIMENSIONAL LATTICE DESIGN FOR STUDIES IN FOREST GENETICS.
    J. AGRIC. RES., 54, 101-119.


DAY, BESSE B. + SANDOMIRE, MARION M. 1942.  (6.3/6.4)                1243
    USE OF THE DISCRIMINANT FUNCTION FOR MORE THAN TWO GROUPS.
    J. AMER. STATIST. ASSOC., 37, 461-472.  (MR4,P.104)


DAY, EDMUND E. 1918.  (4.1)                1244
    A NOTE ON KING'S ARTICLE, ON "THE CORRELATION OF HISTORICAL ECONOMIC
    VARIABLES AND THE MISUSE OF COEFFICIENTS IN THIS CONNECTION".
        (AMENDED BY NO. 3046)
    QUART. PUBL. AMER. STATIST. ASSOC., 16, 115-118.


DAYHOFF, EUGENE 1964.  (17.3)                1245
    ON THE EQUIVALENCE OF POLYKAYS OF THE SECOND DEGREE AND $\Sigma$'S.
        (AMENDED BY NO. 1246)
    ANN. MATH. STATIST., 35, 1663-1672.


DAYHOFF, EUGENE 1965.  (17.3)                1246
    CORRECTION TO: ON THE EQUIVALENCE OF POLYKAYS OF THE SECOND DEGREE AND $\Sigma$'S.
        (AMENDMENT OF NO. 1245)
    ANN. MATH. STATIST., 36, 1069.


DAYHOFF, EUGENE 1966.  (17.3)                1247
    GENERALIZED POLYKAYS, AN EXTENSION OF SIMPLE POLYKAYS AND BIPOLYKAYS.
        (AMENDED BY NO. 1248)
    ANN. MATH. STATIST., 37, 226-241.


DAYHOFF, EUGENE 1966.  (17.3)                1248
    CORRECTION TO:  GENERALIZED POLYKAYS, AN EXTENSION OF SIMPLE POLYKAYS
        AND BIPOLYKAYS.
        (AMENDMENT OF NO. 1247)
    ANN. MATH. STATIST., 37, 746.


DE BALLORE, VICOMTE ROBERT DE MONTESSUS   SEE   DE MONTESSUS DE BALLORE, VICOMTE ROBERT


DEBREU, GERARD 1952.  (20.1)                1249
    DEFINITE AND SEMIDEFINITE QUADRATIC FORMS.
    ECONOMETRICA, 20, 295-300.  (MR14,P.125)


DE CAROLIS, LINDA V. 1963.  (8.1/17.3/18.1)                1250
    SU LA REGRESSIONE INTERAMENTE LINEARE.
        (WITH SUMMARY IN FRENCH, ENGLISH, SPANISH, & GERMAN)
    GIORN. IST. ITAL. ATTUARI, 26, 118-139.  (MR29-5340)


DECELL, HENRY P., JR. 1965.  (20.1)                1251
    AN ALTERNATE FORM OF THE GENERALIZED INVERSE OF AN ARBITRARY COMPLEX MATRIX.
    SIAM REV., 7, 356-358.


DEDEBANT, GEORGES + MACHADO, E.A.M. 1954.  (16.1/B)                1252
    EMPLEO DE LA FUNCION GENERATRIZ PARA LA INTEGRACION DE PROCESOS BIDIMENSIONALES.
        (WITH SUMMARY IN FRENCH)
    METEOROS., 4, 194-214.


DEEMER, WALTER L. 1942.  (15.4)                1253
    A METHOD OF ESTIMATING ACCURACY OF TEST SCORING.
    PSYCHOMETRIKA, 7, 65-73.


DEEMER, WALTER L. + OLKIN, INGRAM 1951.  (20.1)                1254
    THE JACOBIANS OF CERTAIN MATRIX TRANSFORMATIONS USEFUL IN MULTIVARIATE
    ANALYSIS.(BASED ON LECTURES OF P. L. HSU, FOREWORD BY HAROLD HOTELLING.)
    BIOMETRIKA, 38, 345-367.  (MR13,P.855)


DEEMER, WALTER L.   SEE ALSO
    5674. VOTAW, DAVID F., JR.; RAFFERTY, J.A. + DEEMER, WALTER L. (1950)


DEEMING, TERENCE J. 1964.  E(11.1)                1255
    STELLAR SPECTRAL CLASSIFICATION.
    MONTHLY NOTICES ROY. ASTRONOM. SOC., 127, 493-516.

DE FERIET, J. KAMPE  SEE  KAMPE DE FERIET, JOSEPH

DE FINETTI, BRUNO 1937.  (4.1)                                                          1256
    A PROPOSITO DI "CORRELAZIONE".
  SUPP. STATIST. NUOVI PROBLEMI POLIT. STORIA ECON., 3, 41-57.

DE FINETTI, BRUNO 1953.  (17.3/17.9)                                                    1257
    SULLA NOZIONE DI "DISPERSIONE" PER DISTRIBUZIONI A PIU DIMENSIONI.
  ATTI CONG. UN. MAT. ITAL., 4, 587-596.  (MR15,P.42)

DE FOREST, E. L. 1881.  (17.3/18.1/B)                                                   1258
    LAW OF FACILITY OF ERRORS IN TWO DIMENSIONS.
  ANALYST, 8, 3-9, 41-48, 73-82.

DE FOREST, E. L. 1882.  (2.1/17.3/18.1)                                                 1259
    LAW OF ERROR IN THE POSITION OF A POINT IN SPACE.
  ANALYST, 9, 33-40, 65-74.

DE FOREST, E. L. 1882.  (18.1)                                                          1260
    ON AN UNSYMMETRICAL LAW OF ERROR IN THE POSITION OF A POINT IN SPACE.
  TRANS. CONNECTICUT ACAD. ARTS SCI., 6, 123-138.

DE FROE, A.; HUIZINGA, J. + VAN GOOL, J. 1947.  (4.1)                                   1261
    VARIATION- AND CORRELATIONCOEFFICIENT.
  KON. NEDERL. AKAD. WETENSCH. PROC. SECT. SCI., 50, 807-815.

DEGAN, JAMES W. 1948.  (15.1)                                                           1262
    A NOTE ON THE EFFECTS OF SELECTION IN FACTOR ANALYSIS.
  PSYCHOMETRIKA, 13, 87-89.

DEGROOT, M. H. + LI, C. C. 1966.  (12.1/12.2/E)                                         1263
    CORRELATIONS BETWEEN SIMILAR SETS OF MEASUREMENTS.
  BIOMETRICS, 22, 781-790.

DEGROOT, M. H. + RAO, M.M. 1966.  (16.4/17.5/19.3)                                      1264
    MULTIDIMENSIONAL INFORMATION INEQUALITIES AND PREDICTION.
  MULTIVARIATE ANAL. PROC. INTERNAT. SYMP. DAYTON (KRISHNAIAH), 287-313.  (MR35-1134)

DEHANEY, K. G.    SEE
    5284. SUMNER, F. C. + DEHANEY, K. G. (1943)

DEHN, MAX 1906.  (20.5)                                                                 1265
    DIE EULERSCHE FORMEL IM ZUSAMMENHANG MIT DEM INHALT IN DER NICHT-EUKLIDISCHEN GEOMETRIE.
  MATH. ANN., 61, 561-586.

DE LAPLACE, P. S.  SEE  LAPLACE, P. S.

DELAPORTE, PIERRE J. 1938.  (4.8/15.1/E)                                                1266
    ANALYSE STATISTIQUE DES INFLUENCES RECIPROQUES DE PLUSIEURS CARACTERES
      BIOLOGIQUES.
  BIOTYPOLOGIE, 6, 160-182.

DELAPORTE, PIERRE J. 1939.  (15.1)                                                      1267
    UNE METHODE D'ANALYSE DES CORRELATIONS.
  C.R. ACAD. SCI. PARIS, 208, 1960-1963.

DELAPORTE, PIERRE J. 1939.  (15.1/E)                                                    1268
    UNE METHODE D'ANALYSE DES CORRELATIONS ET SON APPLICATION.
  C.R. ACAD. SCI. PARIS, 209, 142-145.  (MR1,P.63)

DELAPORTE, PIERRE J. 1945.  (15.1)                                                      1269
    VERIFICATION DE L'EFFICACITE D'UNE METHODE D'ANALYSE FACTORIELLE.
  C.R. ACAD. SCI. PARIS, 220, 212-214.

DELAPORTE, PIERRE J. 1946.  (15.1)                                                      1270
    ESSAI SUR UNE METHODE STATISTIQUE DE RECHERCHE DES TYPES.
  BIOTYPOLOGIE, 8, 14-23.

DELAPORTE, PIERRE J. 1946.  (15.2/A)                                                    1271
    SUR L'ESTIMATION DES CORRELATIONS DES CARACTERES AVEC LE FACTEUR GENERAL ET LES
    FACTEURS DE GROUPE ET SUR L'ECART-TYPE DE CETTE ESTIMATION, EN ANALYSE
    FACTORIELLE.
  C.R. ACAD. SCI. PARIS, 222, 525-527.  (MR7,P.463)

DELAPORTE, PIERRE J. 1947.   (15.1/15.2/E)                                    1272
    PROLONGEMENT DE LA METHODE D'ANALYSE FACTORIELLE DE SPEARMAN EN
    UTILISANT LA STATISTIQUE MATHEMATIQUE.
    BIOTYPOLOGIE, 9, 45-59.

DELAPORTE, PIERRE J. 1947.   (15.1/15.2)                                      1273
    UNE NOUVELLE METHODE D'ANALYSE FACTORIELLE.
    BULL. INST. INTERNAT. STATIST., 31(3A) 241-257.   (MR13,P.668)

DELAPORTE, PIERRE J. 1949.   (15.1)                                           1274
    UNE CONDITION NECESSAIRE QUE LES OBSERVATIONS DOIVENT REMPLIR POUR
    ETRE REPRESENTABLES PAR UN SCHEMA D'ANALYSE FACTORIELLE DE SPEARMAN.
    C.R. ACAD. SCI. PARIS, 229, 973-975.   (MR11,P.448)

DELAPORTE, PIERRE J. 1949.   (15.1/15.2)                                      1275
    SUR UNE UTILISATION SYSTEMATIQUE DE LA STATISTIQUE MATHEMATIQUE EN
    ANALYSE FACTORIELLE.
    COLLOQ. INTERNAT. CENTRE NAT. RECH. SCI., 13, 101-104.   (MR11,P.448)

DELAPORTE, PIERRE J. 1950.   (4.8/E)                                          1276
    ETUDE STATISTIQUE SUR LES PROPRIETES DES FONTES.
    REV. INST. INTERNAT. STATIST., 18, 161-178.


DELAPORTE, PIERRE J. 1955.   (4.3/15.2)                                       1277
    NOUVELLE ESTIMATION DU COEFFICIENT DE CORRELATION D'UN CARACTERE AVEC LE
    FACTEUR GENERAL OU UN FACTEUR DE GROUPE ET SON ECART-TYPE EN ANALYSE
    FACTORIELLE.
    C.R. ACAD. SCI. PARIS, 240, 1398-1400.   (MR16,P.731)

DELAPORTE, PIERRE J. 1955.   (15.2/T)                                         1278
    NOUVELLE METHODE DE STATISTIQUE MATHEMATIQUE POUR L'ESTIMATION DES
    FACTEURS ET DE LEUR ECART-TYPE EN ANALYSE FACTORIELLE.
        (WITH DISCUSSION)
    COLLOQ. INTERNAT. CENTRE NAT. RECH. SCI., 65, 305-317.

DELAPORTE, PIERRE J. 1961.   (17.7)                                           1279
    ETUDE STATISTIQUE DES ERREURS DE MESURE.
    BULL. INST. INTERNAT. STATIST., 38(4) 387-411.   (MR27-859)

DEL CHIARO, A. 1936.   (18.1)                                                 1280
    SUI MOMENTI DELLE LEGGI DI DISTRIBUZIONE DEL POLYA A PIU VARIABILI.
    GIORN. IST. ITAL. ATTUARI, 7, 151-159.

DE LEVE, G. 1957.   (15.1)                                                    1281
    ENIGE STATISTISCHE ASPECTEN VAN DE FACTORANALYSE.
        (WITH SUMMARY IN ENGLISH)
    STATISTICA NEERLANDICA, 11, 201-209.

D'ELIA, EUGENIO 1947.   (4.8)                                                 1282
    ALCUNI METODI PER LA MISURA DELLE CORRELAZIONI STATISTICHE.
    RIV. ITAL. ECON. DEMOG. STATIST., 1, 175-199.

D'ELIA, EUGENIO 1954.   (4.9)                                                 1283
    SUR LA CORRELATION ENTRE FONCTIONS ANALYTIQUES NON LINEAIRES.
    BULL. INST. INTERNAT. STATIST., 34(2) 163-169.

DELPORTE, JEAN 1965.   (16.4/17.1/B)                                          1284
    REPRESENTATION DES FONCTIONS CONTINUES DE DEUX VARIABLES ET DES COVARIANCES
    CONTINUES PAR DES DEVELOPEMENTS EN SERIES DE HAAR UNIFORMEMENT CONVERGENTS.
    C.R. ACAD. SCI. PARIS, 260, 2701-2704.   (MR32-494)

DELPORTE, JEAN 1965.   (17.1/B)                                               1285
    FONCTIONS ALEATOIRES NORMALES DE DEUX VARIABLES PRESQUE SUREMENT A ECHANTILLONS
    CONTINUMENT DERIVABLES.
    C.R. ACAD. SCI. PARIS, 260, 3544-3557.   (MR32-495)

DELPORTE, JEAN 1966.   (16.4/B)                                               1286
    FONCTIONS ALEATOIRES DE DEUX VARIABLES PRESQUE SUREMENT A ECHANTILLONS
    CONTINUS SUR UN DOMAINE RECTANGULAIRE BORNE.
    Z. WAHRSCHEIN. VERW. GEBIETE, 6, 181-205.   (MR35-3723)

DE LUCIA, LUIGI 1954.  (4.8)                                                                      1287
     TRANSVARIAZIONE TRA CARATTERI CONNESSI.
   ATTI RIUN. SCI. SOC. ITAL. STATIST., 12, 289-364.


DE LUCIA, LUIGI 1955.  (17.9/B)                                                                   1288
     VARIABILITA A DUE DIMENSIONI.
         (REPRINTED AS NO. 1289)
   RIV. ITAL. ECON. DEMOG. STATIST., 9(1)   145-155.  (MR20-2050)


DE LUCIA, LUIGI 1956.  (17.9/B)                                                                   1289
     VARIABILITA A DUE DIMENSIONI.
         (REPRINT OF NO. 1288)
   FAC. SCI. STATIST. DEMOG. ATTUAR. IST. STATIST. CALC. PROB., 2, 1-13.  (MR20-2050)


DE LUCIA, LUIGI 1965.  (4.8/17.4/B)                                                               1290
     VARIABILITA SUPERFICIALE E DISSOMIGLIANZA TRA DISTRIBUZIONI SEMPLICI.
         (WITH SUMMARY IN FRENCH AND ENGLISH)
   METRON, 24, 225-331.  (MR33-5015)


DELURY, DANIEL B. 1938.  (4.1)                                                                    1291
     NOTE ON CORRELATIONS.
   ANN. MATH. STATIST., 9, 149-151.


DE MEERSMAN, R. + SCHOTSMANS, L. 1964.  (20.3)                                                    1292
     NOTE ON THE INVERSION OF SYMMETRIC MATRICES BY THE GAUSS-JORDAN METHOD.
   ICC BULL., 3, 152-155.


DEMING, LOLA S. 1960.  (1.2)                                                                      1293
     SELECTED BIBLIOGRAPHY OF STATISTICAL LITERATURE, 1930-1957: I. CORRELATION AND
     REGRESSION THEORY.
   J. RES. NAT. BUR. STANDARDS SECT. B, 64, 55-68.  (MR22-260)


DEMING, LOLA S. 1960.  (1.2)                                                                      1294
     SELECTED BIBLIOGRAPHY OF STATISTICAL LITERATURE, 1930-1957  II. TIME SERIES.
   J. RES. NAT. BUR. STANDARDS SECT. B, 64, 69-76.  (MR22-261)


DEMING, LOLA S. 1960.  (1.2)                                                                      1295
     SELECTED BIBLIOGRAPHY OF STATISTICAL LITERATURE, 1930-1957  III. LIMIT THEOREMS.
   J. RES. NAT. BUR. STANDARDS SECT. B, 64, 175-192.  (MR23-A1396)


DEMING, LOLA S. 1962.  (1.2)                                                                      1296
     SELECTED BIBLIOGRAPHY OF STATISTICAL LITERATURE, 1930-1957: VI.
     THEORY OF ESTIMATION AND TESTING OF HYPOTHESES, SAMPLING
     DISTRIBUTIONS, AND THEORY OF SAMPLE SURVEYS.
   J. RES. NAT. BUR. STANDARDS SECT. B, 66, 109-151.


DEMING, LOLA S. 1963.  (1.2)                                                                      1297
     SELECTED BIBLIOGRAPHY OF STATISTICAL LITERATURE:   SUPPLEMENT, 1958-1960.
   J. RES. NAT. BUR. STANDARDS SECT. B, 67, 91-133.


DEMING, LOLA S.    SEE ALSO
   1013. COLCORD, CLARENCE G. + DEMING, LOLA S. (1936)


DEMING, W. EDWARDS 1934.  (8.8/E)                                                                 1298
     ON THE APPLICATION OF LEAST SQUARES.  II.
   PHILOS. MAG. SER. 7, 17, 804-829.


DEMING, W. EDWARDS 1935.  (8.8)                                                                   1299
     ON THE APPLICATION OF LEAST SQUARES.  III. A NEW PROPERTY OF LEAST SQUARES.
   PHILOS. MAG. SER. 7, 19, 389-402.


DEMING, W. EDWARDS 1937.  (4.1/8.1/20.3)                                                          1300
     ON THE SIGNIFICANT FIGURES OF LEAST SQUARES AND CORRELATIONS.
         (WITH DISCUSSION)
   SCIENCE, 85, 451-454.


DEMING, W. EDWARDS    SEE ALSO
   3810. MEYER, H. ARTHUR + DEMING, W. EDWARDS (1935)


DE MONTESSUS DE BALLORE, VICOMTE ROBERT 1926.  (4.1/8.8)                                          1301
     CORRELATION ET MOINDRES CARRES.
   C. R. ASSOC. FRANC. AVANCE. SCI. CONF., 49, 133-135.

DE MONTESSUS DE BALLORE, VICOMTE ROBERT 1926.   (4.1)                         1302
    LA METHODE DE LA CORRELATION.
    REV. GENERALF SCI. PURES APPL., 37, 207-213.

DE MONTESSUS DE BALLORE, VICOMTE ROBERT 1933.   (4.1/E)                       1303
    LA  CORRELATION COMPAREE APPLIQUEE A L'ETUDE DES STATISTIQUES
    LINEAIRES ...PREVISION D'EVENEMENTS.
    BAROMETRO ECON. ITAL., 5, 188-190.

DEMPSTER, A. P. 1958.   (5.2)                                                 1304
    A HIGH DIMENSIONAL TWO SAMPLE SIGNIFICANCE TEST.
    ANN. MATH. STATIST., 29, 995-1010.   (MR22-3062A)

DEMPSTER, A. P. 1960.   (5.2)                                                 1305
    A SIGNIFICANCE TEST FOR THE SEPARATION OF TWO HIGHLY MULTIVARIATE SMALL SAMPLES.
    BIOMETRICS, 16, 41-50.   (MR22-3062B)

DEMPSTER, A. P. 1963.   (8.5/10.1/19.2)                                       1306
    MULTIVARIATE THEORY FOR GENERAL STEPWISE METHODS.
    ANN. MATH. STATIST., 34, 873-883.   (MR27-2038)

DEMPSTER, A. P. 1963.   (19.3)                                               1307
    FURTHER EXAMPLES OF INCONSISTENCIES IN THE FIDUCIAL ARGUMENT.
    ANN. MATH. STATIST., 34, 884-891.   (MR27-851)

DEMPSTER, A. P. 1963.   (3.1/7.4/14.4)                                       1308
    ON A PARADOX CONCERNING INFERENCE ABOUT A COVARIANCE MATRIX.
    ANN. MATH. STATIST., 34, 1414-1418.   (MR27-6322)

DEMPSTER, A. P. 1963.   (8.2/10.1)                                           1309
    STEPWISE MULTIVARIATE ANALYSIS OF VARIANCE BASED ON PRINCIPAL VARIABLES.
    BIOMETRICS, 19, 478-490.   (MR28-2609)

DEMPSTER, A. P. 1964.   (10.1)                                               1310
    TESTS FOR THE EQUALITY OF TWO COVARIANCE MATRICES IN RELATION TO A
    BEST LINEAR DISCRIMINATOR ANALYSIS.
    ANN. MATH. STATIST., 35, 190-199.   (MR28-4626)

DEMPSTER, A. P. 1966.   (3.1/12.2/19.3)                                      1311
    ESTIMATION IN MULTIVARIATE ANALYSIS.
    MULTIVARIATE ANAL. PROC. INTERNAT. SYMP. DAYTON (KRISHNAIAH), 315-334.   (MR35-2414)

DENNEY, H. R. + REMMERS, H. H. 1940.   (15.4)                                1312
    RELIABILITY OF MULTIPLE-CHOICE MEASURING INSTRUMENTS AS A FUNCTION OF
    THE SPEARMAN-BROWN PROPHECY FORMULA.   II.
    J. EDUC. PSYCHOL., 31, 699-704.

DENNY, J.L. 1964.   (17.3)                                                   1313
    ON CONTINUOUS SUFFICIENT STATISTICS.
    ANN. MATH. STATIST., 35, 1229-1233.

DENT, B. 1935.   (8.1)                                                       1314
    ON OBSERVATIONS OF POINTS CONNECTED BY A LINEAR RELATION.
    PROC. PHYS. SOC., 47, 92-106.

DENTON, F. T. 1963.   (16.5/A)                                               1315
    SOME TECHNIQUES FOR ANALYSING A SET OF TIME SERIES SUBJECT TO A LINEAR
    RESTRICTION.
    J. AMER. STATIST. ASSOC., 58, 513-518.

DEO, MAHADEORA CHANDRAKANT 1965.   (16.4)                                    1316
    PREDICTION THEORY OF NON-STATIONARY PROCESSES.
    SANKHYA SER. A, 27, 113-132.

DE OLIVEIRA, J. TIAGO  SEE  TIAGO DE OLIVEIRA, J.

DE PANNE, C. VAN  SEE  VAN DE PANNE, C.

DERKSEN, Y. B. D. 1935.   (4.5/4.6)                                          1317
    NOTE ON A REMARK OF D.J. STRUIK ON CORRELATION COEFFICIENTS.
    BULL. AMER. MATH. SOC., 41, 394-398.

DERKSEN, Y. R. D. 1938.   (4.1/E)                                                    1318
    DE CORRELATIEREKENING EN HAAR TOEPASSIN, VOORNAMELIJK BIJ HET
    CONJUNCTUURONDERZOEK. (LE CALCUL DE CORRELATION ET SON APPLICATION
    PRINCIPALEMENT DANS LES RECHERCHES SUR LA CONJUNCTURE).
    DE ECONOMIST, 87, 665-684.

DERRYBERRY, MAHEW   SEE
    1760. FRANZEN, RAYMOND + DERRYBERRY, MAHEW (1931)

DER VAART, H. R. VAN  SEE   VAN DER VAART, H. ROBERT

DES RAJ 1952.   (2.4/3.2)                                                            1319
    ON ESTIMATING THE PARAMETERS OF NORMAL POPULATIONS FROM SINGLY
    TRUNCATED SAMPLES.
    GANITA, 3, 41-57.   (MR14,P.569)

DES RAJ 1953.   (20.4)                                                              1320
    ON A GENERALISED BESSEL FUNCTION POPULATION.
    GANITA, 3, 111-115.   (MR14,P.775)

DES RAJ 1953.   (2.4/3.2/B)                                                         1321
    ON ESTIMATING THE PARAMETERS OF BINORMAL POPULATIONS FROM LINEARLY
    TRUNCATED SAMPLES.
    GANITA, 4, 147-154.   (MR16,P.498)

DES RAJ 1953.   (2.4/3.2/B)                                                         1322
    ON ESTIMATING THE PARAMETERS OF BIVARIATE NORMAL POPULATIONS FROM
    DOUBLY AND SINGLY LINEARLY TRUNCATED SAMPLES.
    SANKHYA, 12, 277-290.   (MR15,P.241)

DES RAJ 1955.   (2.4/6.7)                                                           1323
    ON OPTIMUM SELECTIONS FROM MULTIVARIATE POPULATIONS.
    SANKHYA, 14, 363-366.   (MR17,P.279)

DES RAJ 1965.   (18.5)                                                              1324
    ON A METHOD OF USING MULTI-AUXILIARY INFORMATION IN SAMPLE SURVEYS.
    J. AMER. STATIST. ASSOC., 60, 270-277.

DETAMBEL, MARVIN H. 1955.   (15.5)                                                  1325
    A TEST OF A MODEL FOR MULTI-CHOICE BEHAVIOUR.
    J. EXPER. PSYCHOL., 49, 97-104.

DE TROCONIZ, ANTONIA FZ.   SEE   FZ. DE TROCONIZ, ANTONIA

DEUCHLER, GUSTAV 1914.   (4.1)                                                      1326
    UBER DIE METHODEN DER KORRELATIONSRECHNUNG IN DER PADAGOGIK UND PSYCHOLOGIE.
    Z. PAD. PSYCHOL. EXPER. PAD., 15, 114-131, 145-159, 229-242.

DEUCHLER, GUSTAV 1917.   (4.8)                                                      1327
    UBER DIE BESTIMMUNG VON RANGKORRELATIONEN AUS ZEUGNISNOTEN.
    Z. ANGEW. PSYCHOL., 12, 395-439.

DEUEL, P. D. 1956.   (15.3)                                                         1328
    A NOMOGRAM FOR FACTOR ANALYSIS.
    PSYCHOMETRIKA, 21, 291-294.   (MR18,P.516)

DEUKER, ERNST-A. 1954.   (17.1)                                                     1329
    UBER DIE VERTEILUNGSFUNKTIONEN VON VEKTORSUMMEN.
    Z. ANGEW. MATH. MECH., 34, 162-174.   (MR15,P.969)

DE VERGOTTINI, MARIO 1941.   (4.8)                                                  1330
    SULLA RELAZIONE TRA IL RAPPORTO DI CORRELAZIONE E IL COEFFICIENTE DI
    CORRELAZIONE.
    GIORN. ECON. ANN. ECONOMIA, 3, 308-318.

DE VERGOTTINI, MARIO 1952.   (4.8/E)                                                1331
    SUL SIGNIFICATO DEGLI INDICI DI RELAZIONE.
    RIV. ITAL. ECON. DEMOG. STATIST., 6(1) 32-39.

DE VERGOTTINI, MARIO 1953.   (4.8)                                                  1332
    SULL'INDICE DI CORRELAZIONE INTRA CLASSE.
    SOC. ITAL. ECON. DEMOG. STATIST. STUDI MONOG., 6, 5-14.

DEWITT, C.C.    SEE
    390. BATEN, WILLIAM DOWELL + DEWITT, C.C. (1944)

DHONDT, ANDRE 1960.   (17.7)                                                          1333
    SUR UNE GENERALISATION D'UN THEOREME DE R. FRISCH EN ANALYSE DE LA CONFLUENCE.
    CAHIERS CENTRE ETUDES RECH. OPERAT., 2, 37-46.   (MR22-6041)

DIANANDA, P. H. 1953.   (17.1)                                                        1334
    SOME PROBABILITY LIMIT THEOREMS WITH STATISTICAL APPLICATIONS.
    PROC. CAMBRIDGE PHILOS. SOC., 49. 239-246.   (MR14,P.771)

DIANANDA, P. H. 1954.   (17.1)                                                        1335
    THE CENTRAL LIMIT THEOREM FOR M-DEPENDENT VARIABLES ASYMPTOTICALLY
    STATIONARY TO SECOND ORDER.
    PROC. CAMBRIDGE PHILOS. SOC., 50, 287-292.   (MR15,P.635)

DIANANDA, P. H. 1955.   (17.1)                                                        1336
    THE CENTRAL LIMIT THEOREM FOR M-DEPENDENT VARIABLES.
    PROC. CAMBRIDGE PHILOS. SOC., 51, 92-95.   (MR16,P.724)

DIAZ, EMILIO L. 1955.   (4.1)                                                         1337
    METODO ABREVIADO PARA CALCULAR CORRELACIONES Y OTROS PROCEDIMIENTOS PRACTICOS.
        (WITH SUMMARY IN ENGLISH)
    METEOROS., 5, 238-242.

DIAZ UNGRIA, A.; CAMACHO, A. + RIOS, S. 1955.   E(6.5/B)                              1338
    ANALISIS DISCRIMINANTE DE DOS MUESTRAS DE INDIOS VENEZOLANOS.
    TRABAJOS ESTADIST., 6, 237-242.

DICK, I. D. 1947.   (4.1)                                                             1339
    A NOTE ON THE CORRELATION OF PRODUCTS.
    NEW ZEALAND J. SCI. TECH. SECT. A, 29, 75.

DICK, I. D. 1949.   E(17.7)                                                           1340
    THE APPLICATION OF CONFLUENCE ANALYSIS TO AGRO-ECONOMIC SURVEYS.
    NEW ZEALAND J. SCI. TECH. SECT. A, 31(6) 1-10.

DICK, I. D. 1951.   (15.1/17.7)                                                       1341
    THE EQUIVALENCE OF FACTOR ANALYSIS AND CONFLUENCE ANALYSIS IN
    PROBLEMS OF MULTICOLLINEARITY.
    NEW ZEALAND J. SCI. TECH. SECT. A. 33, 345-347.

DICKMAN, KERN    SEE
    890. CATTELL, RAYMOND B. + DICKMAN, KERN (1962)

DI DONATO, A. R. + JARNAGIN, M.P. 1961.   (2.7/B)                                     1342
    INTEGRATION OF THE GENERAL BIVARIATE GAUSSIAN DISTRIBUTION OVER AN
    OFFSET CIRCLE.
    MATH. COMPUT., 15. 375-382.   (MR23-B2153)

DI DONATO, A. R. + JARNAGIN, M.P. 1962.   (2.7)                                       1343
    A METHOD FOR COMPUTING THE CIRCULAR COVERAGE FUNCTION.
    MATH. COMPUT., 16, 347-355.   (MR26-5669)

DI DONATO, A. R.    SEE ALSO
    5804. WEINGARTEN, HARRY + DI DONATO, A. R. (1961)

DIEDERICH, GERTRUDE W.; MESSICK, SAMUEL J. + TUCKER, LEDYARD R. 1957.   (15.4)        1344
    A GENERAL LEAST SQUARES SOLUTION FOR SUCCESSIVE INTERVALS.
    PSYCHOMETRIKA, 22, 159-173.   (MR19,P.589)

DIETZIUS, ROBERT 1916.   (4.1/8.8)                                                    1345
    AUSDEHNUNG DER KORRELATIONSMETHODE UND DIE METHODE DER KLEINSTEN
    QUADRATE AUF VEKTOREN.
    SITZUNGSBER. MATH. NATURWISS. CL. KAIS. AKAD. WISS. WIEN, 125(2A) 3-20.

DIEULEFAIT, CARLOS F. 1933.   (4.1/17.3/B)                                            1346
    FUNDAMENTALES PARA UNA TEORIA GENERAL DE LA CORRELACION.
    AN. SOC. CI. ARGENT., 115, 20-22.

DIEULEFAIT, CARLOS F. 1934.   (4.8)                                                   1347
    GENERALIZACION DEL COEFICIENTE DE RELACION DE LA CORRELACION.
    AN. SOC. CI. ARGENT., 117, 260-262.

DIEULEFAIT, CARLOS F. 1934.   (8.1/20.4)                                                    1348
    CONTRIBUTION A L'ETUDE DE LA THEORIE DE LA CORRELATION.
    BIOMETRIKA, 26, 379-403.


DIEULEFAIT, CARLOS F. 1934.   (4.8/8.1/18.1)                                                1349
    LA CORRELACION EN EL SENTIDO MODAL.
    REV. UNIV. NAC. LITORAL FAC. CI. ECON. COM. POLIT. (ROSARIO) SER. 3, 4, 140-147.


DIEULEFAIT, CARLOS E. 1935.   (4.8/18.1)                                                    1350
    SUR LA CORRELATION AU SENS DES MODES.
    C.R. ACAD. SCI. PARIS, 200, 1511-1513.


DIEULEFAIT, CARLOS E. 1940.   (17.4)                                                        1351
    SOBRE UN RESULTADO DEL PROF. BEPPO LEVI Y SU RELACION CON EL PROBLEMA
    DE LAS SUPERFICIES DE FRECUENCIAS.
    AN. SOC. CI. ARGENT., 129, 249-253.   (MR3,P.171)


DIEULEFAIT, CARLOS E. 1943.   (2.1/18.1/20.4)                                               1352
    LA LEY DE GAUSS MULTIDIMENSIONAL Y SU GENERALIZACION.
    AN. SOC. CI. ARGENT., 136, 193-215.   (MR6,P.159)


DIEULEFAIT, CARLOS F. 1951.   (2.5)                                                         1353
    SOBRE LAS FORMAS CUADRIATICAS A VARIABLES ALEATORIAS.
    AN. SOC. CI. ARGENT., 151, 167-172.   (MR12,P.838)


DIGMAN, JOHN M. 1966.   (8.5/15.2)                                                          1354
    INTERACTION AND NON-LINEARITY IN MULTIVARIATE EXPERIMENT.
    HANDB. MULTIVARIATE EXPER. PSYCHOL. (CATTELL), 459-475.


DIN, M. ZIA UD-  SEE  ZIA UD-DIN, M.


DINGMAN, HARVEY F. 1954.   (4.4/T)                                                          1355
    A COMPUTING CHART FOR THE POINT BISERIAL CORRELATION COEFFICIENT.
    PSYCHOMETRIKA, 19, 257-259.


DINGMAN, HARVEY F. 1958.   (15.1/15.4)                                                      1356
    THE RELATION BETWEEN COEFFICIENTS OF CORRELATION AND DIFFICULTY FACTORS.
    BRITISH J. STATIST. PSYCHOL., 11, 13-17.


DINGMAN, HARVEY F.   SEE ALSO
    3829. MILLER, C. R.; SABAGH, G. + DINGMAN, HARVEY F. (1962)


DIPAOLA, PETER P. 1945.   (8.8/C)                                                           1357
    USE OF CORRELATION IN QUALITY CONTROL.
    INDUST. QUAL. CONTROL, 2(1) 10-14.


DIVISIA, F. J. 1957.   (4.1/8.1/E)                                                          1358
    CORRELATION OU REGRESSION?  UN EXEMPLE INSTRUCTIF DE STATISTIQUE
    INDUCTIVE LA HAUSSE DES PRIX 1913-1953 EN FRANCE.
    BULL. INST. INTERNAT. STATIST., 35(4) 231-239.


DIXON, WILFRID J. 1944.   (2.6/16.5)                                                        1359
    FURTHER CONTRIBUTIONS TO THE PROBLEM OF SERIAL CORRELATION.
    ANN. MATH. STATIST., 15, 119-144.   (MR6,P.6)


D'JACENKO, Z. N. 1961.   (18.1)                                                             1360
    THE MOMENTS OF A TWO-VARIATE GAMMA-DISTRIBUTION.
        (IN RUSSIAN)
    IZV. VYSS. UCEBN. ZAVED. MAT., 1(20) 55-65.   (MR25-2659)


D'OCAGNE, MAURICE 1893.   (18.6)                                                            1361
    SUR LA DETERMINATION GEOMETRIQUE DU POINT LE PLUS PROBABLE DONNE PAR
    UN SYSTEME DE DROITES NON CONVERGENTES.
    J. ECOLE POLYTECH., 63, 1-25.


D'OCAGNE, MAURICE 1894.   (2.1/B)                                                           1362
    DEMONSTRATION DES FORMULES RELATIVES A LA COMPOSITION DES LOIS D'ERREURS DE
    SITUATION D'UN POINT PUBLIEES DANS LES COMPTES RENDUS DE L'ACADEMIE DES
    SCIENCES DE PARIS.
    ANN. SOC. SCI. BRUXELLES SECT. A, 18, 86-90.

D'OCAGNE, MAURICE 1897.   (20.5)                                     1363
     SUR LA REPRESENTATION PAR DES DROITES ET PAR DES CERCLES DES
     EQUATIONS DU SECOND DEGRE A TROIS VARIABLES.
     IN MEM. N. I. LOBATSCHEVSKII, 1, 93-97.

DODD, EDWARD L. 1938.   (17.7/B)                                     1364
     CERTAIN COEFFICIENTS OF REGRESSION OR TREND ASSOCIATED WITH LARGEST LIKELIHOOD.
     ACTUAL. SCI. INDUST., 740(7) 5-14.

DODD, STUART C. 1926.   (4.1)                                        1365
     THE APPLICATIONS AND MECHANICAL CALCULATION OF CORRELATION COEFFICIENTS.
     J. FRANKLIN INST., 201, 337-350.

DODD, STUART C. 1927.   (15.1)                                       1366
     ON CRITERIA FOR FACTORISING CORRELATED VARIABLES.
     BIOMETRIKA, 19, 45-52.

DODD, STUART C. 1928.   (15.2)                                       1367
     THE THEORY OF FACTORS, II.
     PSYCHOL. REV., 35, 261-279.

DODD, STUART C. 1929.   (15.2)                                       1368
     ON THE SAMPLING THEORY OF INTELLIGENCE.
     BRITISH J. PSYCHOL., 19, 306-327.

DODD, STUART C. 1940.   (15.5)                                       1369
     ANALYSES OF THE INTERRELATION MATRIX BY ITS SURFACE AND STRUCTURE.
     SOCIOMETRY, 3, 133-143.

DOEBLIN, WOLFGANG 1939.   (17.3)                                     1370
     SUR LES SOMMES D'UN GRAND NOMBRE DE VARIABLES ALEATOIRES INDEPENDANTES.
     BULL. SCI. MATH. SER. 2, 63, 22-32, 35-64.

DOERING, CARL R.    SEE
     3464. LOMBARD, HERBERT L. + DOERING, CARL R. (1947)

DOKSUM, K. 1966.   (17.8/TU)                                         1371
     DISTRIBUTION-FREE STATISTICS BASED ON NORMAL DEVIATES IN ANALYSIS OF VARIANCE.
     REV. INST. INTERNAT. STATIST., 34, 376-388.

DOLEZAL, E. 1928.   (1.2)                                            1372
     EMANUEL CZUBER.
     JAHRESBER. DEUTSCHEN MATH. VEREIN., 37, 287-297.

DONATO, A. R. DI  SEE  DI DONATO, A. R.

DONEY, R. A. 1965.   (16.4)                                         1373
     RECURRENT AND TRANSIENT SETS FOR 3-DIMENSIONAL RANDOM WALKS.
     Z. WAHRSCHEIN. VERW. GEBIETE, 4, 253-259.

DONEY, R. A. 1966.   (16.4)                                         1374
     AN ANALOGUE OF THE RENEWAL THEOREM IN HIGHER DIMENSIONS.
     PROC. LONDON MATH. SOC. SER. 2, 16, 669-684.

DOOB, J. L. 1936.   (19.3)                                           1375
     STATISTICAL ESTIMATION.
     TRANS. AMER. MATH. SOC., 39, 410-421.

DOOB, J. L. 1942.   (16.4)                                           1376
     THE BROWNIAN MOVEMENT AND STOCHASTIC EQUATIONS.
          (REPRINTED AS NO. 1379)
     ANN. MATH. SER. 2, 43, 351-369.   (MR4,P.17)

DOOB, J. L. 1944.   (16.4)                                           1377
     THE ELEMENTARY GAUSSIAN PROCESSES.
     ANN. MATH. STATIST., 15, 229-282.   (MR6,P.89)

DOOB, J. L. 1949.   (16.1)                                           1378
     THE TRANSITION FROM STOCHASTIC DIFFERENCE TO STOCHASTIC DIFFERENTIAL EQUATIONS.
     ECONOMETRICA, 17, 68-70.

DOOB, J. L. 1954. (16.4)          1379
    THE BROWNIAN MOVEMENT AND STOCHASTIC EQUATIONS.
       (REPRINT OF NO. 1376)
  SELECT. PAPERS NOISE STOCH. PROC., 319–337. (MR4,P.17)

DOR, LEOPOLD 1944. (2.1/14.4/B)        1380
    QUELQUES REMARQUES SUR LES VARIABLES ALEATOIRES COMBINEES $xy, \sqrt{\alpha^2 x^2 + \beta^2 y^2}$.
  BULL. SOC. ROY. SCI. LIEGE, 13, 203–209. (MR7,P.18)

DORFF, M. + GURLAND, JOHN 1961. (17.7)      1381
    ESTIMATION OF THE PARAMETERS OF A LINEAR FUNCTIONAL RELATION.
  J. ROY. STATIST. SOC. SER. B, 23, 160–170.

DOROGOVCEV, A. JA. 1964. (16.4)        1382
    PROBLEMS OF LINEAR EXTRAPOLATION FOR A CLASS OF VECTOR PROCESSES.
       (IN RUSSIAN)
  UKRAIN. MAT. ZUR., 16, 830–834. (MR30-2549)

DOSS, S. A. D. C. 1962. (18.2)        1383
    A NOTE ON CONSISTENCY AND ASYMPTOTIC EFFICIENCY OF MAXIMUM LIKELIHOOD
    ESTIMATES IN MULTI-PARAMETRIC PROBLEMS.
  CALCUTTA STATIST. ASSOC. BULL., 11, 85–93.

DOSS, S. A. D. C. 1963. (18.2)        1384
    ON CONSISTENCY AND ASYMPTOTIC EFFICIENCY OF MAXIMUM LIKELIHOOD ESTIMATES.
  J. INDIAN SOC. AGRIC. STATIST., 15, 232–241. (MR32-8435)

DOUGLASS, HARL R. 1934. (4.2)        1385
    SOME OBSERVATIONS AND DATA ON CERTAIN METHODS OF MEASURING THE PREDICTIVE
    SIGNIFICANCE OF THE PEARSON PRODUCT MOMENT COEFFICIENT OF CORRELATION.
  J. EDUC. PSYCHOL., 25, 225–231.

DOUST, A. + PRICE, V. E. 1964. (20.1)      1386
    THE LATENT ROOTS AND VECTORS OF A SINGULAR MATRIX.
  COMPUT. J., 7, 222–227.

DOWNS, THOMAS D. 1966. (17.3)        1387
    SOME RELATIONSHIPS AMONG THE VON MISES DISTRIBUTIONS OF DIFFERENT DIMENSIONS.
  BIOMETRIKA, 53, 269–272. (MR33-6668)

DRAPER, NORMAN R. + HUNTER, WILLIAM G. 1966. (8.1)    1388
    DESIGN OF EXPERIMENTS FOR PARAMETER ESTIMATION IN MULTIRESPONSE SITUATIONS.
  BIOMETRIKA, 53, 525–533.

DRAPER, NORMAN R.    SEE ALSO
    695. BOX, GEORGE E. P. + DRAPER, NORMAN R. (1965)

DRAZIN, M.P. 1951. (20.2)        1389
    ON DIAGONAL AND NORMAL MATRICES.
  QUART. J. MATH. OXFORD SECOND SER., 2, 189–198. (MR13,P.200)

DRAZIN, M.P. 1958. (20.1)        1390
    PSEUDO-INVERSES IN ASSOCIATIVE RINGS AND SEMIGROUPS.
  AMER. MATH. MONTHLY, 65, 506–514.

DRESSEL, PAUL L. 1940. (15.4)        1391
    SOME REMARKS ON THE KUDER-RICHARDSON RELIABILITY COEFFICIENT.
  PSYCHOMETRIKA, 5, 305–310. (MR2,P.110)

DRION, E. F. 1951. (17.7)        1392
    ESTIMATION OF THE PARAMETERS OF A STRAIGHT LINE AND OF THE VARIANCES
    OF THE VARIABLES, IF THEY ARE BOTH SUBJECT TO ERROR.
  KON. NEDERL. WETENSCH. PROC. SER. A, 54, 256–260. (MR13,P.144)

DUBLIN, LOUIS I.; LOTKA, ALFRED J. + SPIEGELMAN, MORTIMER 1935. (17.9)   1393
    THE CONSTRUCTION OF LIFE TABLE BY CORRELATION.
  METRON, 12(2) 121–131.

DU BOIS, PHILIP H. 1939. (4.8/T)       1394
    FORMULAS AND TABLES FOR RANK CORRELATION.
  PSYCHOL. REC., 3, 46–56.

DU BOIS, PHILIP H. 1942.   (4.4/15.4)                                      1395
    A NOTE ON THE COMPUTATION OF BISERIAL R IN ITEM VALIDATION.
    PSYCHOMETRIKA, 7, 143-146.

DU BOIS, PHILIP H. 1960.   (15.1/15.2)                                     1396
    AN ANALYSIS OF GUTTMAN'S SIMPLEX.
    PSYCHOMETRIKA, 25, 173-182.   (MR22-3613)

DU BOIS, PHILIP H.   SEE ALSO
    3461. LOEVINGER, JANE; GLESER, GOLDINE C. + DU BOIS, PHILIP H. (1953)

    6020. WRIGHT, E. M. J.; MANNING, W. H. + DU BOIS, PHILIP H. (1959)

DUDEK, F. J. 1952.   (4.4)                                                 1397
    A COMPARISON OF BISERIAL R WITH PEARSON R.
    EDUC. PSYCHOL. MEAS., 12, 759-766.

DUDLEY, R. M. 1965.   (16.4)                                              1398
    GAUSSIAN PROCESSES ON SEVERAL PARAMETERS.
    ANN. MATH. STATIST., 36, 771-788.

DUFFELL, J. H. 1909.   (20.4/T)                                           1399
    TABLES OF THE $\Gamma$-FUNCTION.
    BIOMETRIKA, 7, 43-47.

DUFFIN, R. J.   SEE
    680. BOTT, R. + DUFFIN, R. J. (1953)

DUGUE, DANIEL 1959.   (8.4/8.5)                                           1400
    SUR L'ANALYSE DE LA VARIANCE A PLUSIEURS DIMENSIONS (EXTENSION DE LA
    LOI D'HOTELLING).
    COLLOQ. INTERNAT. CENTRE NAT. RECH. SCI., 87, 81-87.   (MR22-3072)

DUGUE, DANIEL 1960.   (2.3/4.8)                                           1401
    SUR UNE CERTAINE FORME DE CORRELATION ET SES APPLICATIONS A LA THEORIE
    DE L'INFORMATION. THEOREMES DE MAJORATION SUR LES FONCTIONS CARACTERISTIQUES.
    STUDI ONORE C. GINI, 1, 93-101.

DUMAS, F. M. 1949.   (15.5)                                               1402
    THE COEFFICIENT OF PROFILE SIMILARITY.
    J. CLIN. PSYCHOL., 5, 123-131.

DUMON, A. G. 1940.   (4.8/ET)                                             1403
    CORRELATIESTUDIE EN GENETISCH ONDERZOEK BIJ KIPPEN.
    NATUURWETENSCH. TIJDSCHR. (ANTWERP), 21, 393-399.

DUNCAN, DAVID B. + JONES, RICHARD H. 1966.   (8.1/E)                      1404
    MULTIPLE REGRESSION WITH STATIONARY ERRORS.
    J. AMER. STATIST. ASSOC., 61, 917-928.   (MR34-6936)

DUNCAN, W. J. 1944.   (20.3)                                              1405
    SOME DEVICES FOR THE SOLUTION OF LARGE SETS OF SIMULTANEOUS LINEAR
    EQUATIONS.   (WITH AN APPENDIX ON THE RECIPROCATION OF PARTITIONED MATRICES.)
    PHILOS. MAG. SER. 7, 35, 660-670.   (MR7,P.84)

DUNLAP, HILDA F. 1931.   (18.3)                                           1406
    AN EMPIRICAL DETERMINATION OF THE DISTRIBUTION OF MEANS, STANDARD DEVIATIONS,
    AND CORRELATION COEFFICIENTS DRAWN FROM RECTANGULAR POPULATIONS.
    ANN. MATH. STATIST., 2, 66-81.

DUNLAP, JACK W. 1933.   (15.4)                                            1407
    COMPARABLE TESTS AND RELIABILITY.
    J. EDUC. PSYCHOL., 24, 442-453.

DUNLAP, JACK W. 1936.   (4.4/15.4)                                        1408
    NOTE ON COMPUTATION OF BI-SERIAL CORRELATIONS IN ITEM EVALUATION.
    PSYCHOMETRIKA, 1(2) 51-58.

DUNLAP, JACK W. 1936.   (4.4/T)                                           1409
    NOMOGRAPH FOR COMPUTING BI-SERIAL CORRELATIONS.
    PSYCHOMETRIKA, 1(2) 59-60.

DUNLAP, JACK W. 1938.  (1.1)                                                    1410
    RECENT ADVANCES IN STATISTICAL THEORY AND APPLICATIONS.
    AMER. J. PSYCHOL., 51, 558-571.

DUNLAP, JACK W. 1940.  (4.4)                                                    1411
    NOTE ON THE COMPUTATION OF TETRACHORIC CORRELATION.
    PSYCHOMETRIKA, 5, 137-140.  (MR2,P.110)

DUNLAP, JACK W. + CURETON, EDWARD E. 1930.  (4.1)                               1412
    ON THE ANALYSIS OF CAUSATION.
    J. EDUC. PSYCHOL., 21, 657-680.

DUNLAP, JACK W.   SEE ALSO
    207. ARNOLD, J. NORMAN + DUNLAP, JACK W. (1937)

    1159. CURETON, EDWARD E. + DUNLAP, JACK W. (1929)

    1160. CURETON, EDWARD E. + DUNLAP, JACK W. (1930)

    1161. CURETON, EDWARD E. + DUNLAP, JACK W. (1930)

    1162. CURETON, EDWARD E. + DUNLAP, JACK W. (1938)

DUNN, OLIVE JEAN 1958.  (3.1/3.2/C)                                             1413
    ESTIMATION OF THE MEANS OF DEPENDENT VARIABLES.
    ANN. MATH. STATIST., 29, 1095-1111.  (MR21-399)

DUNN, OLIVE JEAN 1959.  (3.1/3.2/C)                                             1414
    CONFIDENCE INTERVALS FOR THE MEANS OF DEPENDENT, NORMALLY DISTRIBUTED VARIABLES.
    J. AMER. STATIST. ASSOC., 54, 613-621.  (MR22-283)

DUNN, OLIVE JEAN 1965.  (18.1)                                                  1415
    A PROPERTY OF THE MULTIVARIATE T DISTRIBUTION.
    ANN. MATH. STATIST., 36, 712-714.  (MR30-2591)

DUNN, OLIVE JEAN + MASSEY, FRANK J., JR. 1965.  (3.1/5.1/CT)                    1416
    ESTIMATION OF MULTIPLE CONTRASTS USING T-DISTRIBUTIONS.
    J. AMER. STATIST. ASSOC., 60, 573-583.  (MR33-804)

DUNN, OLIVE JEAN + VARADY, PAUL V. 1966.  (6.2/T)                               1417
    PROBABILITIES OF CORRECT CLASSIFICATION IN DISCRIMINANT ANALYSIS.
    BIOMETRICS, 22, 908-924.

DUNN, OLIVE JEAN   SEE ALSO
    5803. WEINER, JOHN M. + DUNN, OLIVE JEAN (1966)

DUNNETT, CHARLES W. 1955.  (19.1/U)                                             1418
    A MULTIPLE COMPARISON PROCEDURE FOR COMPARING SEVERAL TREATMENTS WITH A CONTROL.
    J. AMER. STATIST. ASSOC., 50, 1096-1121.

DUNNETT, CHARLES W. 1960.  (19.1/U)                                             1419
    ON SELECTING THE LARGEST OF K NORMAL POPULATION MEANS.
        (WITH DISCUSSION)
    J. ROY. STATIST. SOC. SER. B, 22, 1-40.  (MR22-7202)

DUNNETT, CHARLES W. + SOBEL, MILTON 1954.  (14.2/18.1/T)                        1420
    A BIVARIATE GENERALIZATION OF STUDENT'S T-DISTRIBUTION, WITH TABLES
    FOR CERTAIN SPECIAL CASES.
    BIOMETRIKA, 41, 153-169.  (MR15,P.885)

DUNNETT, CHARLES W. + SOBEL, MILTON 1955.  (18.1/T)                             1421
    APPROXIMATIONS TO THE PROBABILITY INTEGRAL AND CERTAIN PERCENTAGE
    POINTS OF A MULTIVARIATE ANALOGUE OF STUDENT'S T-DISTRIBUTION.
    BIOMETRIKA, 42, 258-260.  (MR16,P.840)

DUNNETT, CHARLES W.   SEE ALSO
    1164. CURNOW, R. N. + DUNNETT, CHARLES W. (1962)

    414. BECHHOFER, ROBERT E.; DUNNETT, CHARLES W. + SOBEL, MILTON (1954)

DUNSMORE, I. R. 1966. (6.1/6.2/U)                                          1422
    A BAYESIAN APPROACH TO CLASSIFICATION.
    J. ROY. STATIST. SOC. SER. B, 28, 568-577.

DURAIN, GENEVIEVE 1956. (1.2/15.1)                                         1423
    L'ANALYSE FACTORIELLE: LE COLLOQUE INTERNATIONAL DE 1955.
    BULL. CENTRE ETUDES RECH. PSYCHOTECH., 5, 79-89.

DURAND, DAVID 1954. (8.8/C)                                                1424
    JOINT CONFIDENCE REGIONS FOR MULTIPLE REGRESSION COEFFICIENTS.
    J. AMER. STATIST. ASSOC., 49, 130-146.

DURAND, DAVID 1956. (20.3)                                                 1425
    A NOTE ON MATRIX INVERSION BY THE SQUARE ROOT METHOD.
    J. AMER. STATIST. ASSOC., 51, 288-292.   (MR18,P.72)

DURAND, DAVID + GREENWOOD, J. ARTHUR 1957. (18.3/20.4)                     1426
    RANDOM UNIT VECTORS II: USEFULNESS OF GRAM-CHARLIER AND RELATED
    SERIES IN APPROXIMATING DISTRIBUTIONS.
    ANN. MATH. STATIST., 28, 978-986.   (MR20-325)

DURAND, DAVID    SEE ALSO
    2115. GREENWOOD, J. ARTHUR + DURAND, DAVID (1955)

    2208. GUMBEL, EMILE J.; GREENWOOD, J. ARTHUR + DURAND, DAVID (1953)

DURBIN, JAMES 1953. (8.1)                                                  1427
    A NOTE ON REGRESSION WHEN THERE IS EXTRANEOUS INFORMATION ABOUT ONE
    OF THE COEFFICIENTS.
    J. AMER. STATIST. ASSOC., 48, 799-808.

DURBIN, JAMES 1954. (17.7)                                                 1428
    ERRORS IN VARIABLES.
    REV. INST. INTERNAT. STATIST., 22, 23-32.   (MR17,P.52)

DURBIN, JAMES 1957. (16.2)                                                 1429
    TESTING FOR SERIAL CORRELATION IN SYSTEMS OF SIMULTANEOUS REGRESSION EQUATIONS.
    BIOMETRIKA, 44, 370-377.   (MR19,P.991)

DURBIN, JAMES 1961. (16.5)                                                 1430
    EFFICIENT FITTING OF LINEAR MODELS FOR CONTINUOUS STATIONARY TIME
    SERIES FROM DISCRETE DATA.
    BULL. INST. INTERNAT. STATIST., 38(4) 273-282.   (MR26-4463)

DURBIN, JAMES + STUART, ALAN 1951. (4.8)                                   1431
    INVERSIONS AND RANK CORRELATION COEFFICIENTS.
        (AMENDED BY NO. 1193)
    J. ROY. STATIST. SOC. SER. B, 13, 303-309.   (MR13,P.962)

DURBIN, JAMES    SEE ALSO
    5774. WATSON, G.S. + DURBIN, JAMES (1951)

DUTTA, M. 1966. (19.3)                                                     1432
    ON MAXIMUM (INFORMATION-THEORETIC) ENTROPY ESTIMATION.
    SANKHYA SER. A, 28, 319-328.

DUTTON, A. M. 1954. E(8.5)                                                 1433
    APPLICATION OF SOME MULTIVARIATE ANALYSIS TECHNIQUES TO DATA FROM
    RADIATION EXPERIMENTS.
    STATIST. MATH. BIOL. (KEMPTHORNE ET AL.), 81-91.

DVORETZKY, ARYEH; KIEFER, J. + WOLFOWITZ, J. 1953. (19.2)                  1434
    SEQUENTIAL DECISION PROBLEMS FOR PROCESSES WITH CONTINUOUS TIME
    PARAMETER. I. TESTING HYPOTHESES.
    ANN. MATH. STATIST., 24, 254-264.   (MR14,P.997,P.1279)

DVORETZKY, ARYEH; KIEFER, J. + WOLFOWITZ, J. 1953. (19.1)                  1435
    SEQUENTIAL DECISION PROBLEMS FOR PROCESSES WITH CONTINUOUS TIME
    PARAMETER. II. PROBLEMS OF ESTIMATION.
    ANN. MATH. STATIST., 24, 403-415.   (MR15,P.242)

DVORETZKY, ARYEH; WALD, ABRAHAM + WOLFOWITZ, J. 1951.   (17.3)                    **1436**
    RELATIONS AMONG CERTAIN RANGES OF VECTOR MEASURES.
        (REPRINTED AS NO. 1437)
    PACIFIC J. MATH., 1, 59-74.   (MR13,P.331)

DVORETZKY, ARYEH; WALD, ABRAHAM + WOLFOWITZ, J. 1955.   (17.3)                    **1437**
    RELATIONS AMONG CERTAIN RANGES OF VECTOR MEASURES.
        (REPRINT OF NO. 1436)
    SELECT. PAPERS STATIST. PROB. WALD, 586-601.   (MR13,P.331)

DVORETZKY, ARYEH; ERDOS, PAUL; KAKUTANI, SHIZUO + 1957.   (16.4)                    **1438**
    TAYLOR, S. J.
        TRIPLE POINTS OF BROWNIAN PATHS IN 3-SPACE.
    PROC. CAMBRIDGE PHILOS. SOC., 53, 856-862.   (MR20-1364)

DWASS, MEYER 1955.   (3.2/C)                    **1439**
    A NOTE ON SIMULTANEOUS CONFIDENCE INTERVALS.
    ANN. MATH. STATIST., 26, 146-147.   (MR16,P.728)

DWASS, MEYER 1959.   (19.3/C)                    **1440**
    MULTIPLE CONFIDENCE PROCEDURES.
    ANN. INST. STATIST. MATH. TOKYO, 10, 277-282.   (MR21-6055)

DWASS, MEYER 1961.   (18.5)                    **1441**
    THE DISTRIBUTION OF LINEAR COMBINATIONS OF RANDOM DIVISIONS OF AN INTERVAL.
    TRABAJOS ESTADIST., 12, 11-17.   (MR26-790)

DWASS, MEYER + TEICHER, HENRY 1957.   (17.1/17.3)                    **1442**
    ON INFINITELY DIVISIBLE RANDOM VECTORS.
    ANN. MATH. STATIST., 28, 461-470.   (MR19,P.986)

DWINAS, S. 1948.   (4.8/20.2)                    **1443**
    SOBRE CIERTOS DETERMINANTES QUE SE UTILIZAN EN ESTADISTICA MATEMATICA.
    MAT. ELEM. SER. 4, 8, 79-80.

DWYER, PAUL S. 1937.   (4.6/8.1)                    **1444**
    THE SIMULTANEOUS COMPUTATION OF GROUPS OF REGRESSION EQUATIONS AND
    ASSOCIATED MULTIPLE CORRELATION COEFFICIENTS.
    ANN. MATH. STATIST., 8, 224-231.

DWYER, PAUL S. 1937.   (4.1/E)                    **1445**
    THE USE OF SUBCORRELATION IN THE ANALYSIS OF NONLINEAR OR
    NON-HOMOSCEDASTIC CORRELATION CHARTS.
    J. EDUC. PSYCHOL., 28, 541-547.

DWYER, PAUL S. 1937.   (15.2/15.3)                    **1446**
    THE DETERMINATION OF THE FACTOR LOADINGS OF A GIVEN TEST FROM THE
    KNOWN FACTOR LOADINGS OF OTHER TESTS.
        (AMENDED BY NO. 3899)
    PSYCHOMETRIKA, 2, 173-178.

DWYER, PAUL S. 1939.   (4.6/15.1)                    **1447**
    THE CONTRIBUTION OF AN ORTHOGONAL MULTIPLE-FACTOR SOLUTION TO
    MULTIPLE CORRELATION.
    PSYCHOMETRIKA, 4, 163-171.

DWYER, PAUL S. 1940.   (4.1)                    **1448**
    THE CALCULATION OF CORRELATION COEFFICIENTS FROM UNGROUPED DATA WITH
    MODERN CALCULATING MACHINES.
    J. AMER. STATIST. ASSOC., 35, 671-673.

DWYER, PAUL S. 1940.   (4.5/4.6/15.3)                    **1449**
    THE EVALUATION OF MULTIPLE AND PARTIAL CORRELATION COEFFICIENTS FROM
    THE FACTORIAL MATRIX.
    PSYCHOMETRIKA, 5, 211-232.   (MR2,P.234)

DWYER, PAUL S. 1941.   (8.9/U)                    **1450**
    THE SKEWNESS OF THE RESIDUALS IN LINEAR REGRESSION THEORY.
    ANN. MATH. STATIST., 12, 104-110.   (MR2,P.233)

DWYER, PAUL S. 1941.   (20.3)                    **1451**
    THE DOOLITTLE TECHNIQUE.
    ANN. MATH. STATIST., 12, 449-458.   (MR3,P.276)

DWYER, PAUL S. 1941.  (20.3)                                            1452
     THE SOLUTION OF SIMULTANEOUS EQUATIONS.
   PSYCHOMETRIKA, 6, 101-129.  (MR2,P.367)

DWYER, PAUL S. 1941.  (20.3)                                            1453
     THE EVALUATION OF DETERMINANTS.
   PSYCHOMETRIKA, 6, 191-204.  (MR2,P.367)

DWYER, PAUL S. 1941.  (20.3)                                            1454
     THE IMPLICIT EVALUATION OF LINEAR FORMS AND THE SOLUTION OF SIMPLE
     MATRIX EQUATIONS.
   PSYCHOMETRIKA, 6, 355-365.  (MR3,P.154)

DWYER, PAUL S. 1942.  (4.1)                                             1455
     RECENT DEVELOPMENTS IN CORRELATION TECHNIQUE.
   J. AMER. STATIST. ASSOC., 37, 441-460.  (MR4,P.164)

DWYER, PAUL S. 1944.  (20.3)                                            1456
     A MATRIX PRESENTATION OF LEAST SQUARES AND CORRELATION THEORY WITH
     MATRIX JUSTIFICATION OF IMPROVED METHODS OF SOLUTION.
   ANN. MATH. STATIST., 15, 82-89.  (MR5,P.245)

DWYER, PAUL S. 1945.  (4.1/8.8/20.3)                                    1457
     THE SQUARE ROOT METHOD AND ITS USE IN CORRELATION AND REGRESSION.
   J. AMER. STATIST. ASSOC., 40, 493-503.  (MR7,P.338)

DWYER, PAUL S. 1947.  (4.1/20.3/T)                                      1458
     SIMULTANEOUS COMPUTATION OF CORRELATION COEFFICIENTS WITH MISSING VARIATES.
     (THE PROCEDURE DESCRIBED WAS DEVELOPED IN COOPERATION WITH ALAN D. MEACHAM)
     (WITH DISCUSSION)
   PROC. RES. FORUM, 20-27.

DWYER, PAUL S. 1949.  (4.1)                                             1459
     PEARSONIAN CORRELATION COEFFICIENTS ASSOCIATED WITH LEAST SQUARES THEORY.
   ANN. MATH. STATIST., 20, 404-416.  (MR11,P.191)

DWYER, PAUL S. 1954.  (6.7/15.5)                                        1460
     SOLUTION OF THE PERSONNEL CLASSIFICATION PROBLEM WITH THE METHOD OF
     OPTIMAL REGIONS.
   PSYCHOMETRIKA, 19, 11-26.

DWYER, PAUL S. 1956.  (6.7/E)                                           1461
     THE PROBLEM OF OPTIMUM GROUP ASSEMBLY.
   SYMP. AIR FORCE HUMAN ENGRG. PERS. TRAIN. RES. (FINCH + CAMERON), 104-114.

DWYER, PAUL S. 1957.  (6.7/15.5)                                        1462
     THE DETAILED METHOD OF OPTIMAL REGIONS.
   PSYCHOMETRIKA, 22, 43-52.

DWYER, PAUL S. 1958.  (8.1/20.2)                                        1463
     GENERALIZATIONS OF A GAUSSIAN THEOREM.
         (AMENDED BY NO. 1464)
   ANN. MATH. STATIST., 29, 106-117.  (MR20-360)

DWYER, PAUL S. 1960.  (8.1/20.2)                                        1464
     CORRECTION TO: GENERALIZATIONS OF A GAUSSIAN THEOREM.
         (AMENDMENT OF NO. 1463)
   ANN. MATH. STATIST., 31, 227.  (MR20-360)

DWYER, PAUL S. 1964.  (6.7/T)                                           1465
     THE MEAN AND STANDARD DEVIATION OF THE DISTRIBUTION OF GROUP ASSEMBLY SUMS.
   PSYCHOMETRIKA, 29, 397-408.

DWYER, PAUL S. 1964.  (20.3)                                            1466
     MATRIX INVERSION WITH THE SQUARE ROOT METHOD.
   TECHNOMETRICS, 6, 197-213.

DWYER, PAUL S. + MACPHAIL, M.S. 1948.  (20.1)                           1467
     SYMBOLIC MATRIX DERIVATIVES.
   ANN. MATH. STATIST., 19, 517-534.  (MR10,P.278)

DWYER, PAUL S. + MEACHAM, ALAN D. 1937.  (4.1/20.3)                     1468
     THE PREPARATION OF CORRELATION TABLES ON A TABULATOR EQUIPPED WITH
     DIGIT SELECTION.
   J. AMER. STATIST. ASSOC., 32, 654-662.

DWYER, PAUL S. + WAUGH, FREDERICK V. 1953.   (20.3)                                        1469
    ON ERRORS IN MATRIX INVERSION.
        (AMENDED BY NO. 1470)
  J. AMER. STATIST. ASSOC., 48, 289-319.   (MR15,P.66)

DWYER, PAUL S. + WAUGH, FREDERICK V. 1953.   (20.3)                                        1470
    ERRATA:   ON ERRORS IN MATRIX INVERSION.
        (AMENDMENT OF NO. 1469)
  J. AMER. STATIST. ASSOC., 48, 911-912.   (MR15,P.66)

DWYER, PAUL S.    SEE ALSO
    1838. GALLER, BERNARD A. + DWYER, PAUL S. (1957)

    4892. SCHAEFFER, ESTHER + DWYER, PAUL S. (1963)

    5788. WAUGH, FREDERICK V. + DWYER, PAUL S. (1945)

DYE, WILLIAM S., III   SEE
    876. CASTORE, GEORGE F. + DYE, WILLIAM S., III (1949)

DYEN, I. 1962.  E(6.6)                                                                     1471
    THE LEXICOSTATISTICAL CLASSIFICATION OF THE MALAYOPOLYNESIAN LANGUAGES.
  LANGUAGE, 38, 38-46.

DYKSTRA, OTTO, JR. 1966.   (8.1/8.2/E)                                                     1472
    THE ORTHOGONALIZATION OF UNDESIGNED EXPERIMENTS.
  TECHNOMETRICS, 8, 279-289.

DYSON, FREEMAN J. 1962.   (13.1)                                                           1473
    A BROWNIAN-MOTION MODEL FOR THE EIGENVALUES OF A RANDOM MATRIX.
        (REPRINTED AS NO. 1474)
  J. MATH. PHYS., 3, 1191-1198.   (MR26-5904)

DYSON, FREEMAN J. 1965.   (13.1)                                                           1474
    A BROWNIAN-MOTION MODEL FOR THE EIGENVALUES OF A RANDOM MATRIX.
        (REPRINT OF NO. 1473)
  STATIST. THEOR. SPECTRA FLUCTUATIONS (PORTER), 421-428.   (MR26-5904)

EAGLE, ALBERT R. 1954.  E(15.4)                                                            1475
    A METHOD FOR HANDLING ERRORS IN TESTING AND MEASURING.
  INDUST. QUAL. CONTROL, 10(5) 10-15.

EAGLESON, G. K. 1964.   (17.3/B)                                                           1476
    POLYNOMIAL EXPANSIONS OF BIVARIATE DISTRIBUTIONS.
  ANN. MATH. STATIST., 35, 1208-1215.

EAST, D. A. + OCHINSKY, L. 1958.  E(6.5/6.6)                                               1477
    A COMPARISON OF SEROLOGICAL AND SOMATOMETRICAL METHODS USED IN DIFFERENTIATING
    BETWEEN CERTAIN EAST AFRICAN RACIAL GROUPS, WITH SPECIAL REFERENCE TO
    $D^2$-ANALYSIS.
  SANKHYA, 20, 31-68.

EATON, M. L. 1966.   (2.2/17.3)                                                            1478
    CHARACTERIZATION OF DISTRIBUTIONS BY THE IDENTICAL DISTRIBUTION OF
    LINEAR FORMS.
  J. APPL. PROB., 3, 481-494.   (MR34-2038)

EBEL, ROBERT L. 1947.   (6.4/15.5)                                                         1479
    THE FREQUENCY OF ERRORS IN THE CLASSIFICATION OF INDIVIDUALS ON THE
    BASIS OF FALLIBLE TEST SCORES.
  EDUC. PSYCHOL. MEAS., 7, 725-734.

ECKART, CARL + YOUNG, GALE 1936.   (15.1/20.1)                                             1480
    THE APPROXIMATION OF ONE MATRIX BY ANOTHER OF LOWER RANK.
  PSYCHOMETRIKA, 1, 211-218.

ECKART, CARL + YOUNG, GALE 1939.   (20.1)                                                  1481
    A PRINCIPAL AXIS TRANSFORMATION FOR NON-HERMITIAN MATRICES.
  BULL. AMER. MATH. SOC., 45, 118-121.

EDDINGTON, ARTHUR S. 1933.   (8.1/18.2/U)                                                  1482
    NOTES ON THE METHOD OF LEAST SQUARES.
  PROC. PHYS. SOC., 45, 271-287.

EDEN, MURRAY 1961.   (16.4/18.1/B)                                         1483
    A TWO-DIMENSIONAL GROWTH PROCESS.
  PROC. FOURTH BERKELEY SYMP. MATH. STATIST. PROB., 4, 223-239.   (MR24-B2493)


EDGERTON, HAROLD A. 1935.   (4.1)                                          1484
    A FORMULA FOR FINDING THE AVERAGE CORRELATION OF ANY ONE VARIABLE WITH THE (N-1)
    OTHER VARIABLES WITHOUT SOLVING ANY OF THE INDIVIDUAL CORRELATIONS.
  J. EDUC. PSYCHOL., 26, 373-376.


EDGERTON, HAROLD A. + THOMSON, KENNETH F. 1942.   (15.4)                   1485
    TEST SCORES EXAMINED WITH LEXIS RATIO.
  PSYCHOMETRIKA, 7, 281-288.


EDGERTON, HAROLD A. + TOOPS, HERBERT A. 1928.   (4.1)                      1486
    A FORMULA FOR FINDING THE AVERAGE INTER-CORRELATION COEFFICIENT FOR UNRANKED
    RAW SCORES WITHOUT SOLVING ANY OF THE INDIVIDUAL INTER-CORRELATIONS.
  J. EDUC. PSYCHOL., 19, 131-138.


EDGETT, GEORGE L. 1956.   (8.8)                                            1487
    MULTIPLE REGRESSION WITH MISSING OBSERVATIONS AMONG THE INDEPENDENT VARIABLES.
  J. AMER. STATIST. ASSOC., 51, 122-131.   (MR17,P.981)


EDGEWORTH, F. Y. 1883.   (8.8)                                             1488
    THE METHOD OF LEAST SQUARES.
  PHILOS. MAG. SER. 5, 16, 360-375.


EDGEWORTH, F. Y. 1884.   (8.8/17.9)                                        1489
    ON THE REDUCTION OF OBSERVATIONS.
  PHILOS. MAG. SER. 5, 17, 135-141.


EDGEWORTH, F. Y. 1886.   (8.1)                                             1490
    THE MATHEMATICAL METHOD OF STATISTICS.
  J. STATIST. SOC. LONDON, 49, 649-654.


EDGEWORTH, F. Y. 1887.   (8.8)                                             1491
    THE CHOICE OF MEANS.
  PHILOS. MAG. SER. 5, 24, 268-271.


EDGEWORTH, F. Y. 1888.   (8.8)                                             1492
    ON OBSERVATIONS RELATING TO SEVERAL QUANTITIES.
  HERMATHENA, 6, 279-283.


EDGEWORTH, F. Y. 1888.   (8.8/17.9/E)                                      1493
    ON A NEW METHOD OF REDUCING OBSERVATIONS RELATING TO SEVERAL QUANTITIES.
  PHILOS. MAG. SER. 5, 25, 184-191.


EDGEWORTH, F. Y. 1892.   (3.1/E)                                           1494
    CORRELATED AVERAGES.
  PHILOS. MAG. SER. 5, 34, 190-204.


EDGEWORTH, F. Y. 1893.   (4.1)                                             1495
    STATISTICAL CORRELATION BETWEEN SOCIAL PHENOMENA.
  J. ROY. STATIST. SOC., 56, 670-675.


EDGEWORTH, F. Y. 1893.   (3.1)                                             1496
    A NEW METHOD OF TREATING CORRELATED AVERAGES.
  PHILOS. MAG. SER. 5, 35, 63-64.


EDGEWORTH, F. Y. 1893.   (3.1/7.3/E)                                       1497
    EXERCISES IN THE CALCULATION OF ERRORS.
  PHILOS. MAG. SER. 5, 36, 98-111.


EDGEWORTH, F. Y. 1893.   (2.1/20.3)                                        1498
    NOTE ON THE CALCULATION OF CORRELATION BETWEEN ORGANS.
  PHILOS. MAG. SER. 5, 36, 350-351.


EDGEWORTH, F. Y. 1894.   (18.1/BE)                                         1499
    ASYMMETRICAL CORRELATION BETWEEN SOCIAL PHENOMENA.
  J. ROY. STATIST. SOC., 57, 563-568.


EDGEWORTH, F. Y. 1895.   (2.1/U)                                           1500
    ON SOME RECENT CONTRIBUTIONS TO THE THEORY OF STATISTICS.
  J. ROY. STATIST. SOC., 506-515.

EDGEWORTH, F. Y. 1905.   (17.3/U)                                                          1501
    THE LAW OF ERROR.
    TRANS. CAMBRIDGE PHILOS. SOC., 20(1) 36-65.

EDGEWORTH, F. Y. 1909.   (1.1)                                                             1502
    ON THE APPLICATION OF THE CALCULUS OF PROBABILITIES TO STATISTICS.
    BULL. INST. INTERNAT. STATIST., 18(1) 505-536.

EDMUNDSON, H. P. 1965.   (4.8)                                                             1503
    A CORRELATION COEFFICIENT FOR ATTRIBUTES OR EVENTS.
    STATIST. ASSOC. METH. MECH. DOC. (STEVENS, GIULIANO, + HEILPRIN), 41-46.

EDWARDS, A. W. F. 1963.   (4.8)                                                            1504
    THE MEASURE OF ASSOCIATION IN A 2 X 2 TABLE.
    J. ROY. STATIST. SOC. SER. A, 126, 109-114.

EDWARDS, A. W. F. + CAVALLI-SFORZA, L. L. 1965.   (6.5)                                    1505
    A METHOD FOR CLUSTER ANALYSIS.
    BIOMETRICS, 21, 362-375.

EDWARDS, ALLEN L. 1948.   (18.2)                                                           1506
    NOTE ON THE "CORRECTION FOR CONTINUITY" IN TESTING THE SIGNIFICANCE
    OF THE DIFFERENCE BETWEEN CORRELATED PROPORTIONS.
    PSYCHOMETRIKA, 13, 185-187.

EDWARDS, ALLEN L. + KENNEY, KATHRYN C. 1946.   (15.5)                                      1507
    A COMPARISON OF THE THURSTONE AND LIKERT TECHNIQUES OF ATTITUDE
    SCALE CONSTRUCTION.
    J. APPL. PSYCHOL., 30, 72-83.

EDWARDS, ALLEN L. + KILPATRICK, FRANKLIN P. 1948.   (15.5)                                 1508
    SCALE ANALYSIS AND THE MEASUREMENT OF SOCIAL ATTITUDES.
    PSYCHOMETRIKA, 13, 99-114.

EDWARDS, D. A. + MOYAL, J. E. 1955.   (16.4)                                               1509
    STOCHASTIC DIFFERENTIAL EQUATIONS.
    PROC. CAMBRIDGE PHILOS. SOC., 51, 663-677.   (MR17,P.276)

EELLS, WALTER CROSBY 1929.   (4.2)                                                         1510
    FORMULAS FOR PROBABLE ERRORS OF COEFFICIENTS OF CORRELATION.
    J. AMER. STATIST. ASSOC., 24, 170-173.

EFRON, BRADLEY 1965.   (20.5)                                                              1511
    THE CONVEX HULL OF A RANDOM SET OF POINTS.
    BIOMETRIKA, 52, 331-343.   (MR34-6820)

EGERSDORFER, L. 1932.   (4.5)                                                              1512
    BEMERKUNGEN ZUR THEORIE DER PARTIELLEN KORRELATIONSKOEFFIZIENTEN.
    ERFAHRUNGSBER. DEUTSCHEN FLUGWETTERD., 41-51.

EHRENBERG, A. S. C. 1962.   (15.1/15.2/E)                                                  1513
    SOME QUESTIONS ABOUT FACTOR ANALYSIS.
    STATISTICIAN, 12, 191-208.

EHRENBERG, A. S. C. 1963.   (8.1/17.7/B)                                                   1514
    BIVARIATE REGRESSION ANALYSIS IS USELESS.
    APPL. STATIST., 12, 161-179.

EHRENBERG, A. S. C. 1963.   (15.1/15.2)                                                    1515
    SOME QUERIES TO FACTOR ANALYSTS.
    STATISTICIAN, 13, 257-262.

EHRENFELD, SYLVAIN 1955.   (8.5/U)                                                         1516
    ON THE EFFICIENCY OF EXPERIMENTAL DESIGNS.
    ANN. MATH. STATIST., 26, 247-255.   (MR17,P.56)

EICKER, F. 1965.   (16.4/17.5)                                                             1517
    MULTILINEAR MAPPING OF TOPOLOGICAL VECTOR SPACES AND A PROBABILISTIC
    APPLICATION.
    MATH. Z., 88, 295-300.

EICKER, F. 1966.   (17.1)                                              1518
    A MULTIVARIATE CENTRAL LIMIT THEOREM FOR RANDOM LINEAR VECTOR FORMS.
    ANN. MATH. STATIST., 37, 1825-1828.   (MR34-2048)


EIDEMILLER, R. L.   SEE
    3445. LIPOW, M. + EIDEMILLER, R. L. (1964)


EISENHART, CHURCHILL 1947.   (8.5/U)                                   1519
    THE ASSUMPTIONS UNDERLYING THE ANALYSIS OF VARIANCE.
    BIOMETRICS, 3, 1-21.   (MR8,P.593)


EISENPRESS, HARRY 1956.   (16.5)                                       1520
    REGRESSION TECHNIQUES APPLIED TO SEASONAL CORRECTIONS AND ADJUSTMENTS
    FOR CALENDAR SHIFTS.
    J. AMER. STATIST. ASSOC., 51, 615-620.


EISENPRESS, HARRY 1960.   (16.2)                                       1521
    SIMULTANEOUS LINEAR STOCHASTIC EQUATIONS:   METHODS OF ESTIMATION.
    PROC. FIRST IBM CONF. STATIST., 79-99.


EISENPRESS, HARRY 1962.   (16.2)                                       1522
    NOTE ON THE COMPUTATION OF FULL-INFORMATION MAXIMUM-LIKELIHOOD
    ESTIMATES OF COEFFICIENTS OF A SIMULTANEOUS SYSTEM.
    ECONOMETRICA, 30, 343-348.


EISENPRESS, HARRY + GREENSTADT, JOHN 1966.   (16.2/E)                  1523
    THE ESTIMATION OF NONLINEAR ECONOMETRIC SYSTEMS.
    ECONOMETRICA, 34, 851-861.


EISENSCHITZ, R. 1944.   (16.4)                                         1524
    MATRIX THEORY OF CORRELATIONS IN A LATTICE. I.
    PROC. ROY. SOC. LONDON SER. A, 182, 244-259.   (MR5,P.280)


EISENSCHITZ, R. 1944.   (16.4)                                         1525
    MATRIX THEORY OF CORRELATIONS IN A LATTICE.  II.
    PROC. ROY. SOC. LONDON SER. A, 182, 260-269.   (MR5,P.280)


ELANDT-JOHNSON, REGINA C. 1966.   (20.5)                               1526
    MULTI-DIMENSIONAL ORTHOGONAL POLYNOMIALS FOR CERTAIN MODELS IN
    MULTIVARIATE ANALYSIS.
    SANKHYA SER. B, 28, 191-198.   (MR35-5100)


ELASHOFF, R. M.   SEE
    27. AFIFI, A. A. + ELASHOFF, R. M. (1966)


ELDERTON, ETHEL M. 1929.   (4.2/T)                                     1527
    TABLE OF THE PRODUCT MOMENT $T_m$ FUNCTION.
    BIOMETRIKA, 21, 194-201.


ELDERTON, ETHEL M. + PEARSON, KARL 1915.   (4.1/17.9/E)               1528
    FURTHER EVIDENCE OF NATURAL SELECTION IN MAN.
    BIOMETRIKA, 10, 488-506.


ELDERTON, ETHEL M.; MOUL, MARGARET; FIELLER, E. C.;   1930.   (2.7/BT)  1529
    PRETORIUS, S.J. + CHURCH, A. E. R.
    TABLES OF THE VOLUMES OF THE NORMAL SURFACE.(TABLES FOR DETERMINING THE VOLUMES
    OF A BI-VARIATE NORMAL SURFACE. (II), TABLES FOR R=.00 TO  -.75.)
    BIOMETRIKA, 22, 12-35.


ELDERTON, ETHEL M.; BARRINGTON, AMY; JONES, H. GERTRUDE;  1913.   (4.1/4.5/8.8/E)  1530
    LAMOTTE, EDITH M. M. DE G.; LASKI, H. J. + PEARSON, KARL
    ON THE CORRELATION OF FERTILITY WITH SOCIAL VALUE.  A COOPERATIVE STUDY.
    EUGENICS LAB. MEM., 18, 1-72.


ELDERTON, ETHEL M.   SEE ALSO
    3080. KONDO, TAKAYUKI + ELDERTON, ETHEL M. (1931)

    4238. PEARSON, KARL + ELDERTON, ETHEL M. (1923)

    4239. PEARSON, KARL + ELDERTON, ETHEL M. (1928)

    4265. PEROTT, B. + ELDERTON, ETHEL M. (1927)

    4936. SCHUSTER, EDGAR + ELDERTON, ETHEL M. (1907)

    4250. PEARSON, KARL; JEFFERY, G.B. + ELDERTON, ETHEL M. (1929)

ELDERTON, WILLIAM PALIN 1908.   (20.5)                                                    1531
    SOME NOTES ON INTERPOLATION IN N-DIMENSIONAL SPACE.
    BIOMETRIKA, 6, 94-103.

ELDERTON, WILLIAM PALIN 1945.   (18.1)                                                    1532
    CRICKET SCORES AND SOME SKEW CORRELATION DISTRIBUTIONS (AN ARITHMETICAL STUDY).
    J. ROY. STATIST. SOC., 108, 1-11, 22-40.

ELFVING, GUSTAV 1947.   (1.1)                                                             1533
    A SIMPLE METHOD OF DEDUCING CERTAIN DISTRIBUTIONS CONNECTED WITH
    MULTIVARIATE SAMPLING.
    SKAND. AKTUARIETIDSKR., 30, 56-74.   (MR9,P.48)

ELFVING, GUSTAV 1952.   (8.9)                                                             1534
    OPTIMUM ALLOCATION IN LINEAR REGRESSION THEORY.
    ANN. MATH. STATIST., 23, 255-262.   (MR13,P.963)

ELFVING, GUSTAV 1954.   (8.8/8.9)                                                         1535
    GEOMETRIC ALLOCATION THEORY.
    SKAND. AKTUARIETIDSKR., 37, 170-190.   (MR17,P.640)

ELFVING, GUSTAV 1954.   (20.5)                                                            1536
    CONVEX SETS IN STATISTICS.
    SKAND. MAT. KONG. BER., 12, 34-39.   (MR16,P.499)

ELFVING, GUSTAV 1956.   (8.1/15.4/U)                                                      1537
    SELECTION OF NON-REPEATABLE OBSERVATIONS FOR ESTIMATION.
        (REPRINTED AS NO. 1542)
    PROC. THIRD BERKELEY SYMP. MATH. STATIST. PROB., 1, 69-75.   (MR18,P.946)

ELFVING, GUSTAV 1956.   (8.8/8.9)                                                         1538
    UBER OPTIMALE ALLOKATION.
    TAGUNG WAHRSCHEIN. MATH. STATIST., 1954, 89-95.   (MR18,P.425)

ELFVING, GUSTAV 1957.   (8.8/8.9)                                                         1539
    MINIMAX CHARACTER OF BALANCED EXPERIMENTAL DESIGNS.
    SKAND. MAT. KONG. BER., 13, 69-76.   (MR21-4514)

ELFVING, GUSTAV 1959.   (8.8/8.9)                                                         1540
    DESIGN OF LINEAR EXPERIMENTS.
    PROB. STATIST. (CRAMER VOLUME), 58-74.   (MR22-1970)

ELFVING, GUSTAV 1961.   (15.4)                                                            1541
    THE ITEM-SELECTION PROBLEM AND EXPERIMENTAL DESIGN.
    STUD. ITEM ANAL. PREDICT. (SOLOMON), 81-87.   (MR23-A739)

ELFVING, GUSTAV 1961.   (8.1/15.4/U)                                                      1542
    ITEM SELECTION AND CHOICE OF NONREPEATABLE OBSERVATIONS FOR ESTIMATION.
        (REPRINT OF NO. 1537)
    STUD. ITEM ANAL. PREDICT. (SOLOMON), 88-95.   (MR23-A737)

ELFVING, GUSTAV 1961.   (15.4)                                                            1543
    CONTRIBUTIONS TO A TECHNIQUE FOR ITEM SELECTION.
    STUD. ITEM ANAL. PREDICT. (SOLOMON), 96-108.   (MR23-A738)

ELFVING, GUSTAV 1961.   (6.2/14.5/U)                                                      1544
    AN EXPANSION PRINCIPLE FOR DISTRIBUTION FUNCTIONS WTIH APPLICATIONS TO STUDENT'S
    STATISTIC AND THE ONE-DIMENSIONAL CLASSIFICATION STATISTIC.
    STUD. ITEM ANAL. PREDICT. (SOLOMON), 276-284.   (MR23-A4209)

ELFVING, GUSTAV; SITGREAVES, ROSEDITH + SOLOMON, HERBERT 1959.   (15.1/15.4/E)            1545
    ITEM SELECTION PROCEDURES FOR ITEM VARIABLES WITH A KNOWN FACTOR STRUCTURE.
        (REPRINTED AS NO. 1546)
    PSYCHOMETRIKA, 24, 189-205.   (MR21-6054)

ELFVING, GUSTAV; SITGREAVES, ROSEDITH + SOLOMON, HERBERT 1961.   (15.1/15.4/E)            1546
    ITEM-SELECTION PROCEDURES FOR ITEM VARIABLES WITH A KNOWN FACTOR STRUCTURE.
        (REPRINT OF NO. 1545)
    STUD. ITEM ANAL. PREDICT. (SOLOMON), 64-80.   (MR22-12638)

ELLIS, MAX E. + RIOPELLE, ARTHUR J. 1948.   (20.3)                                        1547
    AN EFFICIENT PUNCHED-CARD METHOD OF COMPUTING $\Sigma x$, $\Sigma x^2$, $\Sigma xy$ AND HIGHER MOMENTS.
    PSYCHOMETRIKA, 13, 79-85.   (MR9,P.622)

ELLISON, BOB E. 1962. (6.1/6.2/6.3)      1548
     A CLASSIFICATION PROBLEM IN WHICH INFORMATION ABOUT ALTERNATIVE
     DISTRIBUTIONS IS BASED ON SAMPLES.
     ANN. MATH. STATIST., 33, 213-223. (MR25-689)

ELLISON, BOB E. 1964. (14.4/AU)      1549
     ON TWO-SIDED TOLERANCE INTERVALS FOR A NORMAL DISTRIBUTION.
     ANN. MATH. STATIST., 35, 762-772. (MR28-4609)

ELLISON, BOB E. 1965. (6.3)      1550
     MULTIVARIATE-NORMAL CLASSIFICATION WITH COVARIANCES KNOWN.
     ANN. MATH. STATIST., 36, 1787-1793. (MR32-3212)

ELMAGHRABY, SALAH   SEE
     415. BECHHOFER, ROBERT E.; ELMAGHRABY, SALAH + MORSE, NORMAN (1959)

EL SHANAWANY, M. R. 1936. (18.1/AT)      1551
     AN ILLUSTRATION OF THE ACCURACY OF THE $\chi^2$ APPROXIMATION.
     BIOMETRIKA, 28, 179-187.

ELSTON, R. C. + GRIZZLE, J. E. 1962. (8.1/8.6/CE)      1552
     ESTIMATION OF TIME-RESPONSE CURVES AND THEIR CONFIDENCE BANDS.
     BIOMETRICS, 18, 148-159.

EMMETT, W. G. 1936. (15.2)      1553
     SAMPLING ERROR AND THE TWO-FACTOR THEORY.
     BRITISH J. PSYCHOL., 26, 362-387.

EMMETT, W. G. 1949. (15.2)      1554
     FACTOR ANALYSIS BY LAWLEY'S METHOD OF MAXIMUM LIKELIHOOD.
       (AMENDED BY NO. 3302)
     BRITISH J. PSYCHOL. STATIST. SECT., 2, 90-97.

EMMETT, W. G. 1951. (4.3/15.4)      1555
     THE ESTIMATION OF CORRELATION COEFFICIENTS FROM THE VARIANCE OF SCORE
     DIFFERENCES.
     BRITISH J. PSYCHOL. STATIST. SECT., 4, 55-59.

EMMONS, ARTHUR BREWSTER 1913. E(17.9)      1556
     A STUDY OF VARIATIONS IN THE FEMALE PELVIS BASED ON OBSERVATIONS MADE
     ON 217 SPECIMENS OF THE AMERICAN INDIAN SQUAW.
     BIOMETRIKA, 9, 34-57.

ENGELHART, MAX D. 1936. (15.5)      1557
     THE TECHNIQUE OF PATH COEFFICIENTS.
     PSYCHOMETRIKA, 1, 287-293.

ENGLEFIELD, M. J. 1966. (20.1)      1558
     THE COMMUTING INVERSES OF A SQUARE MATRIX.
     PROC. CAMBRIDGE PHILOS. SOC., 62, 667-671.

ENGLMAN, ROBERT 1958. (17.3)      1559
     THE EIGENVALUES OF A RANDOMLY DISTRIBUTED MATRIX.
     NUOVO CIMENTO, 10, 615-621. (MR21-3040)

EPSTEIN, BENJAMIN 1948. (16.2/U)      1560
     SOME APPLICATIONS OF THE MELLIN TRANSFORM IN STATISTICS.
     ANN. MATH. STATIST., 19, 370-379. (MR10,P.552)

EPSTEIN, MARVIN + FLANDERS, HARLEY 1955. (20.1)      1561
     ON THE REDUCTION OF A MATRIX TO DIAGONAL FORM.
     AMER. MATH. MONTHLY, 62, 168-171. (MR16,P.784)

ERDELYI, ARTUR 1940. (20.4)      1562
     SOME CONFLUENT HYPERGEOMETRIC FUNCTIONS OF TWO VARIABLES.
     PROC. ROY. SOC. EDINBURGH, 60, 344-361. (MR2,P.287)

ERDELYI, ARTUR 1948. (20.4)      1563
     TRANSFORMATIONS OF HYPERGEOMETRIC FUNCTIONS OF TWO VARIABLES.
     PROC. ROY. SOC. EDINBURGH SECT. A, 62, 378-385. (MR10,P.115)

ERDELYI, IVAN 1966.  (20.1)                                                        1564
    ON THE "REVERSE ORDER LAW" RELATED TO THE GENERALIZED INVERSE OF
    MATRIX PRODUCTS.
    J. ASSOC. COMPUT. MACH., 13, 439-443.


ERDOS, PAUL    SEE
    1438. DVORETZKY, ARYEH; ERDOS, PAUL; KAKUTANI, SHIZUO +  (1957)
    TAYLOR, S. J.


ERLANDER, S. + GUSTAVSON, J. 1965.  (8.2/8.3/CE)                                   1565
    SIMULTANEOUS CONFIDENCE REGIONS IN NORMAL REGRESSION ANALYSIS WITH AN
    APPLICATION TO ROAD ACCIDENTS.
    REV. INST. INTERNAT. STATIST., 33, 364-377.


ERTEL, HANS 1930.  (17.3)                                                          1566
    ZUR THEORIE DER MAXWELLSCHEN GESCHWINDIGKEITSVERTEILUNG IN
    TURBULENTEN STROMUNGEN.
    Z. GEOPHYS., 6, 329-333.


ESSCHER, FREDRIK F. 1921.  (4.6)                                                   1567
    SOME GENERAL FORMULAE IN THE THEORY OF MULTIPLE CORRELATION.
    ARK. MAT. ASTRONOM. FYS., 15(19) 1-14.


ESSCHER, FREDRIK F. 1924.  (4.8)                                                   1568
    ON A METHOD OF DETERMINING CORRELATION FROM THE RANKS OF THE VARIATES.
    SKAND. AKTUARIETIDSKR., 7, 201-219.


ESSEEN, C. G. 1966.  (17.2)                                                        1569
    ON THE KOLMOGOROV-ROGOZIN INEQUALITY FOR THE CONCENTRATION FUNCTION.
    Z. WAHRSCHEIN. VERW. GEBIETE, 5, 210-216.   (MR34-5128)


ESTES, WILLIAM K. 1950.  (15.5)                                                    1570
    TOWARD A STATISTICAL THEORY OF LEARNING.
        (REPRINTED AS NO. 1571)
    PSYCHOL. REV., 57, 94-107.


ESTES, WILLIAM K. 1963.  (15.5)                                                    1571
    TOWARD A STATISTICAL THEORY OF LEARNING.
        (REPRINT OF NO. 1570)
    READINGS MATH. PSYCHOL. (LUCE, BUSH, + GALANTER), 1, 308-321.


ESTES, WILLIAM K. + BURKE, C. J. 1955.  (15.5)                                     1572
    APPLICATION OF A STATISTICAL MODEL TO SIMPLE DISCRIMINATION LEARNING
    IN HUMAN SUBJECTS.
    J. EXPER. PSYCHOL., 50, 81-88.


ESTES, WILLIAM K. + STRAUGHAN, J. H. 1954.  (15.5)                                 1573
    ANALYSIS OF A VERBAL CONDITIONING SITUATION IN TERMS OF STATISTICAL
    LEARNING THEORY.
        (REPRINTED AS NO. 1574)
    J. EXPER. PSYCHOL., 47, 225-234.


ESTES, WILLIAM K. + STRAUGHAN, J. H. 1963.  (15.5)                                 1574
    ANALYSIS OF A VERBAL CONDITIONING SITUATION IN TERMS OF STATISTICAL
    LEARNING THEORY.
        (REPRINT OF NO. 1573)
    READINGS MATH. PSYCHOL. (LUCE, BUSH, + GALANTER), 1, 343-352.


ESTES, WILLIAM K.; BURKE, C. J.; ATKINSON, RICHARD C. +  1957.  (15.5)             1575
    FRANKMANN, J.P.
    PROBABILISTIC DISCRIMINATION LEARNING.
    J. EXPER. PSYCHOL., 54, 233-239.



ESTES, WILLIAM K.    SEE ALSO
    768. BURKE, C. J. + ESTES, WILLIAM K. (1957)


ETHERINGTON, I. M. H. 1932.  (20.3)                                                1576
    ON ERRORS IN DETERMINANTS.
    PROC. EDINBURGH MATH. SOC. SER. 2, 3, 107-117.

ETTINGER, PIERRE 1966.   (17.1/17.6)                                                    1577
    CONDITIONS NECESSAIRES ET SUFFISANTES POUR QU'UNE VARIABLE ALEATOIRE X
    SOIT ATTIREE PAR L'UNE DES TROIS CLASSES DE LA THEORIE DES VALEURS EXTREMES
    QUAND X EST A VALEURS DANS UN ESPACE EUCLIDIEN A P DIMENSIONS;   CLASSE
    QUI ATTIRE LA SOMME DE DEUX VARIABLES ALEATOIRES INDEPENDANTES A VALEURS DANS
    UN ESPACE EUCLIDIEN A P DIMENSIONS.
    C. R. ACAD. SCI. PARIS SER. A, 263, 620-623.


EUGENE, S. A. 1943.   (15.1)                                                            1578
    NEW LIGHT ON FACTOR ANALYSIS.
    J. FARM ECON., 25, 477-486.


EVANS, E. A. 1965.   (14.4/17.9)                                                        1579
    ON THE NUMBER OF REAL ROOTS OF A RANDOM ALGEBRAIC EQUATION.
    PROC. LONDON MATH. SOC. SER. 3, 15, 731-749.


EVANS, I. G. 1965.   (3.1)                                                              1580
    BAYESIAN ESTIMATION OF PARAMETERS OF A MULTIVARIATE NORMAL DISTRIBUTION.
    J. ROY. STATIST. SOC. SER. B, 27, 279-283.   (MR33-3394)


EVANS, MARILYN J. + HILL, G.W. 1961.   (4.2/T)                                          1581
    TABLES OF Z = ARC TANH(R).
    CSIRO DIV. MATH. STATIST. TECH. PAPER, 7, 1-13.


EVANS, MARILYN J.    SEE ALSO
    1063. CORNISH, E. A. + EVANS, MARILYN J. (1962)


EVERITT, P. F. 1910.   (4.4/T)                                                          1582
    TABLES OF THE TETRACHORIC FUNCTIONS FOR FOURFOLD CORRELATION TABLES.
    BIOMETRIKA, 7, 437-451.


EVERITT, P. F. 1912.   (4.4/T)                                                          1583
    SUPPLEMENTARY TABLES FOR FINDING THE CORRELATION COEFFICIENT FROM
    TETRACHORIC GROUPINGS.
    BIOMETRIKA, 8, 385-395.


EWART, EDWIN    SEE
    4509. REMMERS, H. H. + EWART, EDWIN (1941)


EXNER, F. M. 1913.   (15.1)                                                             1584
    DIE KORRELATIONSMETHODE UND IHRE VERWENDUNG IN DER STATISTIK.
    NATURWISS., 1, 206-208.


EXNER, F. M. 1913.   (15.1)                                                             1585
    UBER DIE KORRELATIONSMETHODE.   NACH EINIGEN VORTRAGEN.
    NATURWISS. WOCHENSCHR. JENA, 28, 705-710.


EYRAUD, HENRI 1934.   (18.1)                                                            1586
    SUR UNE REPRESENTATION NOUVELLE DES CORRELATIONS CONTINUES.
    C.R. ACAD. SCI. PARIS, 199, 1356-1358.


EYRAUD, HENRI 1935.   (4.1/4.8)                                                         1587
    CORRELAZIONE E CAUSALITA.
    GIORN. IST. ITAL. ATTUARI, 6, 57-68.


EYRAUD, HENRI 1936.   (4.1/4.9)                                                         1588
    LES PRINCIPES DE LA MESURE DES CORRELATIONS.
    ANN. UNIV. LYON TROISIEME SER. SCI. SECT. A, 1, 30-47.


EYRAUD, HENRI 1938.   (18.1/B)                                                          1589
    SUR QUELQUES LOIS D'ERREURS A DEUX DIMENSIONS.
    C.R. ACAD. SCI. PARIS, 206, 402-404.


EYRAUD, HENRI 1939.   (2.2/17.3/18.1)                                                   1590
    LES LOIS D'ERREURS DANS DEUX DIMENSIONS.
    ANN. UNIV. LYON TROISIEME SER. SCI. SECT. A, 2, 19-23.   (MR8,P.282)


EYSENCK, HANS J. 1950.   (15.1/15.5)                                                    1591
    CRITERION ANALYSIS--AN APPLICATION OF THE HYPOTHETICO-DEDUCTIVE METHOD
    TO FACTOR ANALYSIS.
    PSYCHOL. REV., 57, 38-53.

EYSENCK, HANS J. 1952.   (15.1)                                                1592
     USES AND ABUSES OF FACTOR ANALYSIS.
     APPL. STATIST., 1, 45-49.

EZEKIEL, MORDECAI J. B. 1924.   (4.8)                                          1593
     A METHOD OF HANDLING CURVILINEAR CORRELATION FOR ANY NUMBER OF VARIABLES.
     J. AMER. STATIST. ASSOC., 19, 431-453.

EZEKIEL, MORDECAI J. B. 1926.   (8.1)                                          1594
     DETERMINATION OF CURVILINEAR REGRESSION "SURFACES" IN THE PRESENCE OF
     OTHER VARIABLES.
     J. AMER. STATIST. ASSOC., 21, 310-320.

EZEKIEL, MORDECAI J. B. 1929.   (4.9)                                          1595
     MEANING AND SIGNIFICANCE OF CORRELATION.
     AMER. ECON. REV., 19, 246-250.

EZEKIEL, MORDECAI J. B. 1929.   (4.6/4.8)                                      1596
     THE APPLICATION OF THE THEORY OF ERROR TO MULTIPLE AND CURVILINEAR CORRELATION.
          (WITH DISCUSSION)
     J. AMER. STATIST. ASSOC., 24(SP) 99-107.

EZEKIEL, MORDECAI J. B. 1932.   (8.8)                                          1597
     FURTHER REMARKS ON THE GRAPHIC METHOD OF CORRELATION: I. A REPLY TO "SOME
     CHARACTERISTICS OF THE GRAPHIC METHOD OF CORRELATION".
          (AMENDMENT OF NO. 5685)
     J. AMER. STATIST. ASSOC., 27, 183-185.

EZEKIEL, MORDECAI J. B. 1939.   (4.8/8.1/E)                                    1598
     FURTHER COMMENT.
     QUART. J. ECON., 54, 331-346.

EZEKIEL, MORDECAI J. B. 1943.   (8.8)                                          1599
     CHOICE OF THE DEPENDENT VARIABLE IN REGRESSION ANALYSIS--COMMENTS.
     J. AMER. STATIST. ASSOC., 38, 214-216.

EZEKIEL, MORDECAI J. B.     SEE ALSO
     5491. TOLLEY, H. R. + EZEKIEL, MORDECAI J. B. (1923)

     5492. TOLLEY, H. R. + EZEKIEL, MORDECAI J. B. (1927)

FABIAN, VACLAV 1954.   (17.7)                                                  1600
     STRUCTURAL RELATION.
     CZECH. MATH. J., 4, 354-363.   (MR16,P.842)

FABIAN, VACLAV 1957.   (17.9)                                                  1601
     ZUFALLIGES ABRUNDEN UND DIE KONVERGENZ DES LINEAREN (SEIDELSCHEN)
     ITERATIONSVERFAHRENS.
     MATH. NACHR., 16, 265-270.   (MR20-1424)

FABIAN, VACLAV 1958.   (20.3)                                                  1602
     ODHAD CHYBY ZAOKROUHLOVANI PRI LINEARNICH ITERACNICH PROCESECH, ZEJMENA PRI
     SEIDLOVE RESENI DIRICHLETOVA PROBLEMU PRO CTVEREC 10 X 10.
     APL. MAT., 3, 22-44.   (MR23-B1098)

FABIAN, VACLAV 1958.   (20.3)                                                  1603
     L'INFLUENCE DE L'ARRONDISSEMENT SUR LES EVALUATIONS NUMERIQUES LINEAIRES.
     CZECH. MATH. J., 8, 203-221.   (MR24-B1280)

FABIAN, VACLAV 1962.   (19.1)                                                  1604
     ON MULTIPLE DECISION METHODS FOR RANKING POPULATION MEANS.
     ANN. MATH. STATIST., 33, 248-254.   (MR26-5699)

FABIAN, VACLAV 1963.   (17.7)                                                  1605
     NOTE ON HALPERIN'S METHOD OF FITTING STRAIGHT LINES WHEN BOTH
     VARIABLES ARE SUBJECT TO ERROR.
          (IN CZECH)
     APL. MAT., 8, 197-200.

FAIRTHORNE, D. 1964.   (17.9/B)                                                1606
     THE DISTANCES BETWEEN RANDOM POINTS IN TWO CONCENTRIC CIRCLES.
     BIOMETRIKA, 51, 275-276.

FALLIS, ROBERT F.    SEE
   5846. WHERRY, ROBERT J.; NAYLOR, JAMES C.; WHERRY, ROBERT J., JR. + (1965)
  FALLIS, ROBERT F.

FAN, CHUNG-TEH 1954. (15.4)                                                          1607
    NOTE ON CONSTRUCTION OF AN ITEM ANALYSIS TABLE FOR THE HIGH-LOW 27
    PER CENT GROUP METHOD.
  PSYCHOMETRIKA, 19, 231-237.

FAN, CHUNG-TEH 1957. (15.5)                                                          1608
    ON THE APPLICATIONS OF THE METHOD OF ABSOLUTE SCALING.
  PSYCHOMETRIKA, 22, 175-183.  (MR19,P.333)

FAN, CHUNG-TEH   SEE ALSO
   5305. SWINEFORD, FRANCES + FAN, CHUNG-TEH (1957)

FARLIE, D.J.G. 1960. (4.9/B)                                                         1609
    THE PERFORMANCE OF SOME CORRELATION COEFFICIENTS FOR A GENERAL
    BIVARIATE DISTRIBUTION.
  BIOMETRIKA, 47, 307-323.  (MR22-10078)

FARLIE, D.J.G. 1961. (4.8/A)                                                         1610
    THE ASYMPTOTIC EFFICIENCY OF DANIELS'S GENERALIZED CORRELATION COEFFICIENTS.
  J. ROY. STATIST. SOC. SER. B, 23, 128-142.  (MR23-A1425)

FARLIE, D.J.G. 1963. (4.8/17.4)                                                      1611
    THE ASYMPTOTIC EFFICIENCY OF DANIELS'S GENERALISED CORRELATION.
  BIOMETRIKA, 50, 499-504.

FARNELL, A. B. 1944. (20.1)                                                          1612
    LIMITS FOR THE CHARACTERISTIC ROOTS OF A MATRIX.
  BULL. AMER. MATH. SOC., 50, 789-794.  (MR6,P.113)

FARNELL, A. B. 1945. (20.1)                                                          1613
    LIMITS FOR THE FIELD OF VALUES OF A MATRIX.
  AMER. MATH. MONTHLY, 52, 488-493.  (MR7,P.108)

FARRELL, R. H. 1962. (17.5/17.8/C)                                                   1614
    BOUNDED LENGTH CONFIDENCE INTERVALS FOR THE ZERO OF A REGRESSION FUNCTION.
  ANN. MATH. STATIST., 33, 237-247.  (MR25-687)

FAVA, LINDO 1958. (18.5)                                                             1615
    UBER DIE VARIANZ DES SCHATZWERTES DER REGRESSION BEI ZUFALLIGER
    STICHPROBENNATURE OHNE ZURUCKLEGEN AUS EINER ENDLICHEN POPULATION.
    (IN PORTUGUESE, WITH SUMMARY IN ENGLISH)
  BOL. SOC. MAT. SAO PAULO, 10, 121-128.  (MR22-4109)

FAVERGE, J. M. 1955. (15.1/E)                                                        1616
    UTILISATION DU SCHEMA DE SPEARMAN DANS LE CALCUL DES IMAGES.
    (WITH DISCUSSION)
  COLLOQ. INTERNAT. CENTRE NAT. RECH. SCI., 65, 221-236.

FAXER, P.   SEE
   5996. WOLD, HERMAN O. A. + FAXER, P. (1957)

FEDER, DANIEL D., LT. COMM. (U.S.N.R.) 1947. (15.4)                                  1617
    ESTIMATION OF TEST ITEM DIFFICULTY BY AVERAGING HIGHEST AND LOWEST
    QUARTER PERFORMANCES COMPARED WITH TOTAL POPULATION COUNT.
  EDUC. PSYCHOL. MEAS., 7, 133-134.

FEDERER, WALTER T. 1951. (10.2)                                                      1618
    TESTING PROPORTIONALITY OF COVARIANCE MATRICES.
  ANN. MATH. STATIST., 22, 102-106.  (MR12,P.622)

FEINBERG, E. L.   SEE
   1924. GERSMAN, S. G. + FEINBERG, E. L. (1955)

   1925. GERSMAN, S. G. + FEINBERG, E. L. (1955)

FEIVESON, A. H.   SEE
   4035. ODELL, P. L. + FEIVESON, A. H. (1966)

   4036. ODELL, P. L. + FEIVESON, A. H. (1966)

FELDHEIM, ERVIN 1936.   (17.3/B)                                                                      1619
    SUR LA STABILITE DES LOIS DE PROBABILITE A DEUX VARIABLES.
    BULL. SOC. MATH. FRANCE, 64, 209-212.


FELDHEIM, ERVIN 1937.   (18.1/B)                                                                      1620
    SULLE LEGGI DI PROBABILITA STABILI A DUE VARIABILI.
    GIORN. IST. ITAL. ATTUARI, 8, 146-158.


FELDMAN, HYMAN M. 1935.   (17.3)                                                                      1621
    MATHEMATICAL EXPECTATION OF PRODUCT MOMENTS OF SAMPLES DRAWN FROM A
    SET OF INFINITE POPULATIONS.
    ANN. MATH. STATIST., 6, 30-52.


FELDT, LEONARD S. 1961.   (4.3/B)                                                                     1622
    THE USE OF EXTREME GROUPS TO TEST FOR THE PRESENCE OF A RELATIONSHIP.
    PSYCHOMETRIKA, 26, 307-316.   (MR26-878)


FENSTAD, GRETI USTERUD 1966.   (19.3/A)                                                               1623
    ON ASYMPTOTICALLY NORMAL ESTIMATORS.
    SKAND. AKTUARIETIDSKR., 49, 48-60.


FERAUD, LUCIEN 1942.   (1.1)                                                                          1624
    PROBLEME D'ANALYSE STATISTIQUE A PLUSIEURS VARIABLES.
    ANN. UNIV. LYON TROISIEME SER. SCI. SECT. A, 5, 42-53.   (MR8,P.282)


FERAUD, LUCIEN 1943.   (7.4/B)                                                                        1625
    STATISTIQUE MATHEMATIQUE:   DISTRIBUTIONS DE PRODUITS INTERIEURS.
    C.R. SEANCES SOC. PHYS. HIST. NATUR. GENEVE, 60, 196-200, 296.   (MR7,P.212)


FERBER, ROBERT 1956.   (4.1)                                                                          1626
    ARE CORRELATIONS ANY GUIDE TO PREDICTIVE VALUE?
    APPL. STATIST., 5, 113-121.


FEREDAY, F.   SEE
    739. BROWN, R. L. + FEREDAY, F. (1958)


FERGUSON, GEORGE A. 1940.   (15.4)                                                                    1627
    A BI-FACTOR ANALYSIS OF RELIABILITY COEFFICIENTS.
    BRITISH J. PSYCHOL., 31, 172-182.


FERGUSON, GEORGE A. 1941.   (15.4)                                                                    1628
    THE FACTORIAL INTERPRETATION OF TEST DIFFICULTY.
    PSYCHOMETRIKA, 6, 323-329.


FERGUSON, GEORGE A. 1942.   (15.4)                                                                    1629
    ITEM SELECTION BY THE CONSTANT PROCESS.
    PSYCHOMETRIKA, 7, 19-29.


FERGUSON, GEORGE A. 1949.   (15.4)                                                                    1630
    ON THE THEORY OF TEST DISCRIMINATION.
    PSYCHOMETRIKA, 14, 61-68.


FERGUSON, GEORGE A. 1954.   (15.1)                                                                    1631
    THE CONCEPT OF PARSIMONY IN FACTOR ANALYSIS.
    PSYCHOMETRIKA, 19, 281-290.


FERGUSON, GEORGE A.   SEE ALSO
    2741. JACKSON, ROBERT W. B. + FERGUSON, GEORGE A. (1941)

    2742. JACKSON, ROBERT W. B. + FERGUSON, GEORGE A. (1943)


FERGUSON, JAMES 1964.   (20.5)                                                                        1632
    MULTIVARIATE CURVE INTERPOLATION.
    J. ASSOC. COMPUT. MACH., 11, 221-228.


FERGUSON, LEONARD W. 1941.   (15.5)                                                                   1633
    A STUDY OF THE LIKERT TECHNIQUE OF ATTITUDE SCALE CONSTRUCTION.
    J. SOC. PSYCHOL., 13, 51-57.


FERGUSON, LEONARD W. + LAWRENCE, WARREN R. 1942.   (15.1/E)                                           1634
    AN APPRAISAL OF THE VALIDITY OF THE FACTOR LOADINGS EMPLOYED IN THE
    CONSTRUCTION OF THE PRIMARY SOCIAL ATTITUDE SCALES.
    PSYCHOMETRIKA, 7, 135-138.

FERGUSON, THOMAS S. 1955.   (2.2/17.3/17.7)                                    1635
   ON THE EXISTENCE OF LINEAR REGRESSION IN LINEAR STRUCTURAL RELATIONS.
  UNIV. CALIFORNIA PUBL. STATIST., 2, 143-165.   (MR17,P.634)

FERGUSON, THOMAS S. 1962.   (18.1)                                             1636
   A REPRESENTATION OF THE SYMMETRIC BIVARIATE CAUCHY DISTRIBUTION.
  ANN. MATH. STATIST., 33, 1256-1266.   (MR26-840)

FERIBERGER, W. + GRENANDER, ULF 1965.   (11.1/16.3/16.4)                       1637
   ON THE FORMULATION OF STATISTICAL METEOROLOGY.
  REV. INST. INTERNAT. STATIST., 33, 59-86.

FERIET, J. KAMPE DE   SEE   KAMPE DE FERIET, JOSEPH

FERNANDEZ BANOS, OLEGARIO 1930.   (4.5)                                        1638
   CONTRIBUCION AL ESTUDIO DE LA CORRELACION. NUEVO METODO PARA HALLAR LA
    LINEA DE REGRESION.
  REV. MAT. HISP. AMER. SER. 2, 5, 48-53.

FERNANDEZ BANOS, OLEGARIO 1930.   (4.5/4.6/8.8)                                1639
   SOBRE LA CORRELACION Y ECUACION DE REGRESION.
  REV. MAT. HISP. AMER. SER. 2, 5, 161-177.

FERNIQUE, X. 1963.   (16.4)                                                    1640
   STOCHASTIC PROCESSES AND DISTRIBUTION THEORY.
  C.R. ACAD. SCI. PARIS, 256, 5274-5275.

FERON, ROBERT 1947.   (4.9)                                                    1641
   MERITES COMPARES DES DIVERS INDICES DE CORRELATION.
    (WITH DISCUSSION)
  J. SOC. STATIST. PARIS, 88, 328-352.   (MR10,P.722)

FERON, ROBERT 1952.   (4.9/17.3)                                               1642
   INFORMATION ET CORRELATION.
  C.R. ACAD. SCI. PARIS, 234, 1343-1345.   (MR13,P.761)

FERON, ROBERT 1952.   (8.1)                                                    1643
   DE LA REGRESSION.
  C.R. ACAD. SCI. PARIS, 234, 2143-2145.   (MR13,P.853)

FERON, ROBERT 1956.   (17.4)                                                   1644
   SUR LES TABLEAUX DE CORRELATION DONT LES MARGES SONT DONNEES.  CAS DE
   L'ESPACE A TROIS DIMENSIONS.
  PUBL. INST. STATIST. UNIV. PARIS, 5, 3-12.   (MR18,P.522)

FERON, ROBERT 1956.   (4.1/8.1/14.4)                                           1645
   INFORMATION, REGRESSION, CORRELATION.
  PUBL. INST. STATIST. UNIV. PARIS, 5, 111-215.   (MR19,P.1027)

FERON, ROBERT + FOURGEAUD, CLAUDE 1951.   (8.1/14.4)                           1646
   INFORMATION ET REGRESSION.
  C.R. ACAD. SCI. PARIS, 232, 1636-1638.   (MR12,P.843)

FERON, ROBERT + FOURGEAUD, CLAUDE 1952.   (2.2)                                1647
   QUELQUES PROPRIETES CARACTERISTIQUES DE LA LOI DE LAPLACE-GAUSS.
  PUBL. INST. STATIST. UNIV. PARIS, 1(2) 44-49.   (MR15,P.805)

FERON, ROBERT + FOURGEAUD, CLAUDE 1952.   (17.3/B)                             1648
   SUR LE RAPPORT DE DEUX VARIABLES ALEATOIRES.
  PUBL. INST. STATIST. UNIV. PARIS, 1(2) 50-52.   (MR15,P.805)

FERREIRA MURTEIRA, BENTO JOSE   SEE   MURTEIRA, BENTO J.

FERRERI, CARLO 1958.   (4.8)                                                   1649
   UN NUOVO IMPORTANTE SCHEMA DI DIPENDENZA.
  ANN. FAC. ECON. COMM. PALERMO, 12(2) 221-238.

FERSCHL, FRANZ 1956.   (17.7)                                                  1650
   DIE IDENTIFIKATION STRUKTURELLER BEZIEHUNGEN.
  STATIST. VIERTELJAHRESSCHR., 9, 141-151.

FERTIG, JOHN W. 1936.   (14.3)                                                      1651
    ON A METHOD OF TESTING THE HYPOTHESIS THAN AN OBSERVED SAMPLE OF n VARIABLES
    AND OF SIZE N HAS BEEN DRAWN FROM A SPECIFIED POPULATION OF THE SAME NUMBER OF
    VARIABLES.
    ANN. MATH. STATIST., 7, 113-121.

FERTIG, JOHN W. 1936.   (4.5)                                                       1652
    THE USE OF INTERACTION IN THE REMOVAL OF CORRELATED VARIATION.
    BIOM. BULL. (NEURO-ENDOCRINE RES.), 1(1) 1-14.

FESTINGER, LEON 1947.   (15.5)                                                      1653
    THE TREATMENT OF QUALITATIVE DATA BY "SCALE ANALYSIS".
    PSYCHOL. BULL., 44, 149-161.

FESTINGER, LEON 1949.   (15.5)                                                      1654
    THE ANALYSIS OF SOCIOGRAMS USING MATRIX ALGEBRA.
    HUMAN RELATIONS, 2, 153-158.

FETTIS, HENRY E. 1965.   (20.3)                                                     1655
    NOTE ON THE MATRIX EQUATION  AX = $\lambda$ BX.
    COMPUT. J., 8, 279.

FIEDLER, MIROSLAV 1957.   (20.1)                                                    1656
    O NEKTERYCH VLASTNOSTECH HERMITOVSKYCH MATIC.
    MAT.-FYZ. CASOPIS SLOVEN. AKAD. VIED, 7, 168-176.   (MR21-58)

FIEDLER, MIROSLAV 1961.   (20.1)                                                    1657
    UBER EINE UNGLEICHUNG FUR POSITIV DEFINITE MATRIZEN.
    MATH. NACHR., 23, 197-199.   (MR25-3049)

FIEDLER, MIROSLAV 1965.   (20.1)                                                    1658
    SOME ESTIMATES OF THE PROPER VALUES OF MATRICES.
    J. SOC. INDUST. APPL. MATH., 13, 1-5.

FIELDS, JERRY L. + WIMP, JET 1961.   (20.4)                                         1659
    EXPANSIONS OF HYPERGEOMETRIC FUNCTIONS IN HYPERGEOMETRIC FUNCTIONS.
    MATH. COMPUT., 15, 390-395.   (MR23-A3289)

FIELLER, E. C. 1932.   (14.2/B)                                                     1660
    THE DISTRIBUTION OF THE INDEX IN A NORMAL BIVARIATE DISTRIBUTION.
    BIOMETRIKA, 24, 428-440.

FIELLER, E. C. 1954.   (8.8/C)                                                      1661
    SYMPOSIUM ON INTERVAL ESTIMATION:  SOME PROBLEMS IN INTERVAL ESTIMATION.
    J. ROY. STATIST. SOC. SER. B, 16, 175-185.   (MR19,P.1204)

FIELLER, E. C. + SMITH, CEDRIC A.B. 1951.   (4.8/8.8)                               1662
    NOTE ON THE ANALYSIS OF VARIANCE AND INTRACLASS CORRELATION.
    ANN. EUGENICS, 16, 97-104.   (MR13,P.261)

FIELLER, E. C.; LEWIS, T. + PEARSON, EGON SHARPE 1955.   (2.1/14.4/NT)              1663
    CORRELATED RANDOM NORMAL DEVIATES.
        (AMENDED BY NO. 1664)
    TRACTS COMPUT., 26, 1-60.   (MR17,P.638)

FIELLER, E. C.; LEWIS, T. + PEARSON, EGON SHARPE 1956.   (2.1/14.4/T)               1664
    CORRIGENDA TO:  CORRELATED RANDOM NORMAL DEVIATES.
        (AMENDMENT OF NO. 1663)
    BIOMETRIKA, 43, 496-497.   (MR18,P.422)

FIELLER, E. C.    SEE ALSO
    2693. IRWIN, J. O.; COCHRAN, WILLIAM G.; FIELLER, E. C. +   (1936)
    STEVENS, W. L.

    1529. ELDERTON, ETHEL M.; MOUL, MARGARET; FIELLER, E. C.;   (1930)
    PRETORIUS, S.J. + CHURCH, A. E. R.

    2692. IRWIN, J. O.; BARTLETT, MAURICE S.; COCHRAN, WILLIAM G. +   (1935)
    FIELLER, E. C.

FIERING, MYRON B. 1962.   (4.1)                                                     1665
    ON THE USE OF CORRELATION TO AUGMENT DATA.
    J. AMER. STATIST. ASSOC., 57, 20-32.   (MR26-7088)

FILON, L.N.G.   SEE
    4240. PEARSON, KARL + FILON, L.N.G. (1898)

    4241. PEARSON, KARL + FILON, L.N.G. (1956)

FINCH, J. H.   SEE
    3364. LENTZ, THEODORE F., JR.; HIRSCHSTEIN, BERTHA + FINCH, J. H. (1932)

FINCH, P. D. 1960.   (16.5/A)                                          1666
    ON THE COVARIANCE DETERMINANTS OF MOVING AVERAGE AND AUTOREGRESSION MODELS.
    BIOMETRIKA, 47, 194–196.   (MR22–10137)

FINDLEY, WARREN 1934.   (15.3)                                         1667
    A MULTIPLE FACTOR METHOD YIELDING ONLY POSITIVE WEIGHTS.
    PSYCHOL. BULL., 31, 676–677.

FINE, NATHAN J. + NIVEN, IVAN 1944.   (20.1)                           1668
    THE PROBABILITY THAT A DETERMINANT BE CONGRUENT TO A(MOD M).
    BULL. AMER. MATH. SOC., 50, 89–93.   (MR5,P.169)

FINETTI, BRUNO DE   SEE   DE FINETTI, BRUNO

FINKEL'STEIN, B. V. 1953.   (17.1)                                     1669
    ON THE LIMITING DISTRIBUTIONS OF THE EXTREME TERMS OF A VARIATIONAL
    SERIES OF A TWO-DIMENSIONAL RANDOM QUANTITY.
        (IN RUSSIAN)
    DOKL. AKAD. NAUK SSSR NOV. SER., 91, 209–211.   (MR15,P.444)

FINNEY, D. J. 1938.   (7.4)                                            1670
    THE DISTRIBUTION OF THE RATIO OF ESTIMATES OF THE TWO VARIANCES IN A
    SAMPLE FROM A NORMAL BI-VARIATE POPULATION.
    BIOMETRIKA, 30, 190–192.

FINNEY, D. J. 1941.   (7.4/T)                                          1671
    THE JOINT DISTRIBUTION OF VARIANCE RATIOS BASED ON A COMMON ERROR MEAN SQUARE.
    ANN. EUGENICS, 11, 136–140.   (MR3,P.172)

FINNEY, D. J. 1946.   (8.4/14.2)                                       1672
    THE FREQUENCY DISTRIBUTION OF DEVIATES FROM MEANS AND REGRESSION
    LINES IN SAMPLES FROM A MULTIVARIATE NORMAL POPULATION.
    ANN. MATH. STATIST., 17, 344–349.   (MR8,P.161)

FINNEY, D. J. 1956.   (1.1)                                            1673
    MULTIVARIATE ANALYSIS AND AGRICULTURAL EXPERIMENTS.
    BIOMETRICS, 12, 67–71.

FINNEY, D. J. 1962.   (2.3/2.4)                                        1674
    CUMULANTS OF TRUNCATED MULTINORMAL DISTRIBUTIONS.
    J. ROY. STATIST. SOC. SER. B, 24, 535–536.   (MR26–7077)

FIRSCHEIN, O. + FISCHLER, M. 1963.   (6.6)                             1675
    AUTOMATIC SUBCLASS DETERMINATION FOR PATTERN RECOGNITION APPLICATIONS.
    IEEE TRANS. ELECTRONIC COMPUT.,EC-12, 137–141.

FISCHER, CARL H. 1933.   (2.1/17.3)                                    1676
    ON CORRELATION SURFACES OF SUMS WITH A CERTAIN NUMBER OF RANDOM
    ELEMENTS IN COMMON.
    ANN. MATH. STATIST., 4, 103–126.

FISCHER, CARL H. 1933.   (4.5/4.6)                                     1677
    ON MULTIPLE AND PARTIAL CORRELATION COEFFICIENTS OF A CERTAIN SEQUENCE OF SUMS.
    ANN. MATH. STATIST., 4, 278–284.

FISCHER, CARL H. 1942.   (4.5/4.6/18.1/B)                              1678
    A SEQUENCE OF DISCRETE VARIABLES EXHIBITING CORRELATION DUE TO COMMON ELEMENTS.
    ANN. MATH. STATIST., 13, 97–101.   (MR4,P.23)

FISCHER, ERNST 1905.   (20.1)                                          1679
    UBER QUADRATISCHE FORMEN MIT REELLEN KOEFFIZIENTEN.
    MONATSHEFTE MATH. PHYS., 16, 234–249.

FISCHER, G. + ROPPERT, JOSEF 1966.  (20.1)                                    1680
    UBER EIN THEOREM VON ECKART UND YOUNG: "EINE VERALLGEMEINERUNG EINES
    TRANSFORMATIONSVERFAHRENS VON JACOBI".
  METRIKA, 10, 161-170.

FISCHER, JANOS   SEE
  1121. CSAKI, PETER + FISCHER, JANOS (1960)

  1122. CSAKI, PETER + FISCHER, JANOS (1960)

  1123. CSAKI, PETER + FISCHER, JANOS (1961)

  1124. CSAKI, PETER + FISCHER, JANOS (1963)

  1129. CSISZAR, I. + FISCHER, JANOS (1962)

FISCHER, OTTO F. 1949.  (6.2/E)                                               1681
    DISKRIMINACNI ANALYSA A HODNOCENI ZKOUSEK SCHOPNOSTI. (DISCRIMINATORY
    ANALYSIS AND THE WEIGHTING OF THE RESULTS OF PSYCHOLOGICAL MEASUREMENTS.)
      (WITH SUMMARY IN ENGLISH)
  STATIST. OBZOR, 29, 106-129.

FISCHLER, M.   SEE
  1675. FIRSCHEIN, O. + FISCHLER, M. (1963)

FISHER, FRANKLIN M. 1961.  (16.2)                                             1682
    ON THE COST OF APPROXIMATE SPECIFICATION IN SIMULTANEOUS EQUATION ESTIMATION.
  ECONOMETRICA, 29, 139-170.

FISHER, FRANKLIN M. 1962.  (1.1)                                              1683
    THE PLACE OF LEAST SQUARES IN ECONOMETRICS:  A COMMMENT.
  ECONOMETRICA, 30, 565-567.

FISHER, FRANKLIN M. 1965.  (16.2)                                            1684
    NEAR-IDENTIFIABILITY AND THE VARIANCES OF THE DISTURBANCE TERMS.
  ECONOMETRICA, 33, 409-419.

FISHER, FRANKLIN M. 1966.  (16.2)                                            1685
    THE RELATIVE SENSITIVITY TO SPECIFICATION ERROR OF DIFFERENT K-CLASS
    ESTIMATORS.
  J. AMER. STATIST. ASSOC., 61, 345-356.  (MR35-2416)

FISHER, MICHAEL E. + FULLER, A. T. 1958.  (20.3/20.5)                        1686
    ON THE STABILIZATION OF MATRICES AND THE CONVERGENCE OF LINEAR ITERATIVE PROCESSES.
  PROC. CAMBRIDGE PHILOS. SOC., 54, 417-425.  (MR20-2086)

FISHER, RONALD A. 1915.  (4.2)                                               1687
    FREQUENCY DISTRIBUTION OF THE VALUES OF THE CORRELATION COEFFICIENT
    IN SAMPLES FROM AN INDEFINITELY LARGE POPULATION.
  BIOMETRIKA, 10, 507-521.

FISHER, RONALD A. 1918.  (4.1/4.8/E)                                         1688
    THE CORRELATION BETWEEN RELATIVES ON THE SUPPOSITION OF MENDELIAN INHERITANCE.
  TRANS. ROY. SOC. EDINBURGH, 52, 399-433.

FISHER, RONALD A. 1920.  (4.3/8.8/E)                                         1689
    STUDIES IN CROP VARIATION I.  AN EXAMINATION OF THE YIELD OF DRESSED
    GRAINS FROM BROADBALK.
  J. AGRIC. SCI., 11, 107-135.

FISHER, RONALD A. 1921.  (4.2)                                               1690
    ON THE "PROBABLE ERROR" OF A COEFFICIENT OF CORRELATION DEDUCED FROM
    A SMALL SAMPLE.
  METRON, 1(4) 3-32.

FISHER, RONALD A. 1922.  (8.2/8.3)                                           1691
    THE GOODNESS OF FIT OF REGRESSION FORMULAE, AND THE DISTRIBUTION OF
    REGRESSION COEFFICIENTS.
  J. ROY. STATIST. SOC., 85, 597-612.

FISHER, RONALD A. 1924.  (4.5)                                               1692
    THE DISTRIBUTION OF THE PARTIAL CORRELATION COEFFICIENT.
  METRON, 3,   329-333.

FISHER, RONALD A. 1924.   (4.6/8.8/E)                                    1693
     THE INFLUENCE OF RAINFALL ON THE YIELD OF WHEAT AT ROTHAMSTED.
     PHILOS. TRANS. ROY. SOC. LONDON SER. B, 213, 89-142.

FISHER, RONALD A. 1925.   E(4.8/10.1/17.9/B)                             1694
     THE RESEMBLANCE BETWEEN TWINS, A STATISTICAL EXAMINATION OF
     LAUTERBACH'S MEASUREMENTS.
     GENETICS, 10, 569-579.

FISHER, RONALD A. 1926.   (4.1/17.8/U)                                   1695
     ON THE RANDOM SEQUENCE.
     QUART. J. ROY. METEOROL. SOC., 52, 250.

FISHER, RONALD A. 1928.   (4.7)                                          1696
     THE GENERAL SAMPLING DISTRIBUTION OF THE MULTIPLE CORRELATION COEFFICIENT.
         (REPRINTED AS NO. 1704)
     PROC. ROY. SOC. LONDON SER. A, 121, 654-673.   (MR12,P.427)

FISHER, RONALD A. 1929.   (16.5/U)                                       1697
     TESTS OF SIGNIFICANCE IN HARMONIC ANALYSIS.
     PROC. ROY. SOC. LONDON SER. A, 125, 54-59.

FISHER, RONALD A. 1936.   (5.2/6.2/8.5/E)                                1698
     THE USE OF MULTIPLE MEASUREMENTS IN TAXONOMIC PROBLEMS.
     ANN. EUGENICS, 7, 179-188.

FISHER, RONALD A. 1936.   (6.5)                                          1699
     "THE COEFFICIENT OF RACIAL LIKENESS" AND THE FUTURE OF CRANIOMETRY.
     J. ROY. ANTHROP. INST., 66, 57-63.

FISHER, RONALD A. 1938.   (5.2/6.5/8.5)                                  1700
     THE STATISTICAL UTILIZATION OF MULTIPLE MEASUREMENTS.
         (REPRINTED AS NO. 1705)
     ANN. EUGENICS, 8, 376-386.   (MR12.P.427)

FISHER, RONALD A. 1939.   (8.4/13.1)                                     1701
     THE SAMPLING DISTRIBUTION OF SOME STATISTICS OBTAINED FROM NON-LINEAR EQUATIONS.
     ANN. EUGENICS, 9, 238-249.   (MR1,P.248)

FISHER, RONALD A. 1940.   (5.2/6.2/6.5/E)                                1702
     THE PRECISION OF DISCRIMINANT FUNCTIONS.
         (REPRINTED AS NO. 1706)
     ANN. EUGENICS, 10, 422-429.   (MR2.P.235)

FISHER, RONALD A. 1947.   E(8.5)                                         1703
     THE ANALYSIS OF COVARIANCE METHOD FOR THE RELATION BETWEEN A PART AND THE WHOLE.
     BIOMETRICS, 3, 65-68.

FISHER, RONALD A. 1950.   (4.7)                                          1704
     THE GENERAL SAMPLING DISTRIBUTION OF THE MULTIPLE CORRELATION
     COEFFICIENT (REPRINT, WITH AUTHOR'S NOTE).
         (REPRINT OF NO. 1696)
     CONTRIB. MATH. STATIST. (FISHER REPRINTS), 14, 653A-673.   (MR12,P.427)

FISHER, RONALD A. 1950.   (5.2/6.5/8.5)                                  1705
     THE STATISTICAL UTILIZATION OF MULTIPLE MEASUREMENTS.
         (REPRINT OF NO. 1700)
     CONTRIB. MATH. STATIST. (FISHER REPRINTS), 33, 375A-386.   (MR12,P.427)

FISHER, RONALD A. 1950.   (5.2/6.2/6.5/E)                                1706
     THE PRECISION OF DISCRIMINANT FUNCTIONS.
         (REPRINT OF NO. 1702)
     CONTRIB. MATH. STATIST. (FISHER REPRINTS), 34, 421A-429.   (MR12,P.427)

FISHER, RONALD A. 1962.   (7.1)                                          1707
     THE SIMULTANEOUS DISTRIBUTION OF CORRELATION COEFFICIENTS.
     SANKHYA SER. A, 24, 1-8.   (MR26-3135)

FISHER, RONALD A. + MACKENZIE, W.A. 1922.   (4.3/8.8/E)                  1708
     THE CORRELATION OF WEEKLY RAINFALL.
         (WITH DISCUSSION)
     QUART. J. ROY. METEOROL. SOC., 48, 234-245.

FISHER, RONALD A.    SEE ALSO
   1241. DAY, BESSE B. + FISHER, RONALD A. (1937)


FISHER, WALTER D. 1958.    (6.6/20.3)                                          1709
     ON GROUPING FOR MAXIMUM HOMOGENEITY.
   J. AMER. STATIST. ASSOC., 53, 789-798. (MR21-387)


FISHER, WALTER D. 1962.    (8.1)                                               1710
     ESTIMATION IN THE LINEAR DECISION MODEL.
   INTERNAT. ECON. REV., 3, 1-29.


FISHMAN, JOSHUA A. 1956.    (4.4)                                              1711
     A NOTE ON JENKINS' "IMPROVED METHOD FOR TETRACHORIC R".
        (AMENDMENT OF NO. 2781)
   PSYCHOMETRIKA, 21, 305.


FISZ, MAREK 1953.    (17.1/U)                                                  1712
     THE LIMITING DISTRIBUTIONS OF SUMS OF ARBITRARY INDEPENDENT AND
     EQUALLY DISTRIBUTED R-POINT (R $\geq$ 2) RANDOM VARIABLES.
   BULL. ACAD. POLON. SCI. CL. 3, 1, 235-238.    (MR15,P.635)


FISZ, MAREK 1953.    (17.1)                                                    1713
     THE LIMITING DISTRIBUTIONS OF SUMS OF ARBITRARY INDEPENDENT AND
     EQUALLY DISTRIBUTED R-POINT RANDOM VARIABLES.
   POLSKA AKAD. NAUK STUD. MATH., 14, 111-123.    (MR15,P.882)


FISZ, MAREK 1954.    (18.4/A)                                                  1714
     THE LIMITING DISTRIBUTIONS OF THE MULTINOMIAL DISTRIBUTION.
   POLSKA AKAD. NAUK STUD. MATH., 14, 272-275.    (MR16,P.839)


FISZ, MAREK 1954.    (17.1)                                                    1715
     A GENERALIZATION OF A THEOREM OF KHINTCHIN.
   POLSKA AKAD. NAUK STUD. MATH., 14, 310-313.    (MR16,P.839)


FIX, EVELYN    SEE
   1225. DAVID, F. N. + FIX, EVELYN (1966)


FLANAGAN, JOHN C. 1939.    (15.4)                                             1716
     NOTE ON CALCULATING THE STANDARD ERROR OF MEASUREMENT AND RELIABILITY
     COEFFICIENTS WITH THE TEST-SCORING MACHINE.
   J. APPL. PSYCHOL., 23, 529.


FLANAGAN, JOHN C. 1939.    (4.2/15.4)                                         1717
     GENERAL CONSIDERATIONS IN THE SELECTION OF TEST ITEMS AND A SHORT METHOD OF
     ESTIMATING THE PRODUCT-MOMENT COEFFICIENTS FROM DATA AT THE TAILS OF THE
     DISTRIBUTION.
   J. EDUC. PSYCHOL., 30, 674-680.


FLANAGAN, JOHN C. 1952.    (4.2)                                              1718
     THE EFFECTIVENESS OF SHORT METHODS FOR CALCULATING CORRELATION COEFFICIENTS.
   PSYCHOL. BULL., 49, 342-348.


FLANDERS, HARLEY    SEE
   1561. EPSTEIN, MARVIN + FLANDERS, HARLEY (1955)


FLATTO, L. + KONHEIM, A. G. 1962.    (18.6)                                   1719
     THE RANDOM DIVISION OF AN INTERVAL AND THE RANDOM COVERING OF A CIRCLE.
   SIAM REV., 4, 211-222.


FLEISS, JOSEPH L. 1966.    (8.1/8.5/E)                                        1720
     ASSESSING THE ACCURACY OF MULTIVARIATE OBSERVATIONS.
   J. AMER. STATIST. ASSOC., 61, 403-412.    (MR34-8552)


FLEMING, H.    SEE
   4594. ROGERS, DAVID J. + FLEMING, H. (1965)


FLOOD, MERRILL M. 1940.    (11.1)                                             1721
     A COMPUTATIONAL PROCEDURE FOR THE METHOD OF PRINCIPAL COMPONENTS.
   PSYCHOMETRIKA, 5, 169-172.


FOG, DAVID 1948.    (7.1)                                                     1722
     THE GEOMETRICAL METHOD IN THE THEORY OF SAMPLING.
   BIOMETRIKA, 35, 46-54.    (MR9,P.600)

FOGELSON, S. 1934.  (4.8)                                           1723
     O INTERPRETACJI I ZAKRESIE STOSOWALNOSCI MIAR KORELACJI.
          (WITH SUMMARY IN FRENCH)
     KWART. STATYST. GLOWNY URZAD STATYST., 11, 1-38.


FOLEY, ELIZABETH   SEE
     803. BURT, CYRIL + FOLEY, ELIZABETH (1956)


FOLKS, JOHN LEROY + ANTLE, CHARLES F. 1965.  (17.9)                 1724
     OPTIMUM ALLOCATION OF SAMPLING UNITS TO STRATA WHEN THERE ARE  R  RESPONSES OF INTEREST.
     J. AMER. STATIST. ASSOC., 60, 225-233.  (MR31-844)


FOLLIN, JAMES W., JR.   SEE
     1801. FRENKIEL, FRANCOIS N. + FOLLIN, JAMES W., JR. (1954)


FONG, CHING   SEE
     5370. TAYLOR, WILSON L. + FONG, CHING (1963)


FOOTE, RICHARD J. 1953.  (4.6)                                      1725
     THE MATHEMATICAL BASIS FOR THE BEAN METHOD OF GRAPHIC MULTIPLE CORRELATION.
     J. AMER. STATIST. ASSOC., 48, 778-788.


FOOTE, RICHARD J. 1958.  (4.5/4.6/20.3)                             1726
     A MODIFIED DOOLITTLE APPROACH FOR MULTIPLE AND PARTIAL CORRELATION
     AND REGRESSION.
     J. AMER. STATIST. ASSOC., 53, 133-143.  (MR22-287)


FORD, ADELBERT 1931.  (4.1)                                         1727
     THE CORRELATOR.
     J. EXPER. PSYCHOL., 14, 155-163.


FORD, F. A. J. 1959.  (7.4/16.4/U)                                  1728
     A NOTE ON THE PAPER OF MILLER, BERNSTEIN AND BLUMENSON.
          (AMENDMENT OF NO. 3837)
     QUART. APPL. MATH., 17, 446.  (MR22-256)


FOREST, E. L. DE  SEE  DE FOREST, E. L.


FORLANO, GEORGE  SEE
     4306. PINTNER, RUDOLF + FORLANO, GEORGE (1937)


FORMAN, VIRGINIA P.   SEE
     2536. HOLZMAN, MATHILDA S. + FORMAN, VIRGINIA P. (1966)


FORSYTHE, GEORGE E. + GOLUB, GENE H. 1965.  (20.5)                  1729
     ON THE STATIONARY VALUES OF A SECOND-DEGREE POLYNOMIAL ON THE UNIT SPHERE.
     J. SOC. INDUST. APPL. MATH., 13, 1050-1068.


FORSYTHE, GEORGE E. + HENRICI, P. 1960.  (20.3)                     1730
     THE CYCLIC JACOBI METHOD FOR COMPUTING THE PRINCIPAL VALUES OF A COMPLEX MATRIX.
     TRANS. AMER. MATH. SOC., 94, 1-23.


FORSYTHE, GEORGE E. + LEIBLER, RICHARD A. 1950.  (20.3)             1731
     MATRIX INVERSION BY A MONTE CARLO METHOD.
     MATH. TABLES AIDS COMPUT., 4, 127-129.  (MR12,P.361)


FORSYTHE, GEORGE E. + STRAUS, ERNST G. 1955.  (20.3)               1732
     ON BEST CONDITIONED MATRICES.
     PROC. AMER. MATH. SOC., 6, 340-345.  (MR16,P.1054)


FORTET, ROBERT M. 1951.  (17.1/17.3)                                1733
     RANDOM DETERMINANTS.
     J. RES. NAT. BUR. STANDARDS, 47, 465-470.  (MR13,P.852)


FORTET, ROBERT M.   SEE ALSO
     587. BLANC-LAPIERRE, ANDRE + FORTET, ROBERT M. (1947)

     588. BLANC-LAPIERRE, ANDRE + FORTET, ROBERT M. (1947)


FORTIER, JEAN J. 1966.  (17.5)                                      1734
     SIMULTANEOUS NON-LINEAR PREDICTION.
     PSYCHOMETRIKA, 31, 447-455.  (MR34-5566)

FORTIER, JEAN J. + SOLOMON, HERBERT 1966.   (6.6/E)                              1735
    CLUSTERING PROCEDURES.
   MULTIVARIATE ANAL. PROC. INTERNAT. SYMP. DAYTON (KRISHNAIAH), 493–506.


FORTUNATI, P. 1961.   (17.3/TU)                                                  1736
    RELATIONSHIPS AMONG THE VARIABILITY MEASURES.
        (IN ITALIAN)
   STATISTICA. BOLOGNA, 21, 239–291.


FORTUNATI, P. 1962.   (17.3/TU)                                                  1737
    ABOUT THE VARIABILITY MEASURES OF SOME PARTICULAR DISTRIBUTIONS.   (IN ITALIAN)
   STATISTICA, BOLOGNA, 22, 383–395.


FORTUNATI, P. 1963.   (17.3/T)                                                   1738
    SOME REMARKS ON THE CONCENTRATION RATIO.
        (IN ITALIAN)
   STATISTICA. BOLOGNA, 23, 427–445.


FOSTER, F. G. 1957.   (8.4/18.1/20.4/T)                                          1739
    UPPER PERCENTAGE POINTS OF THE GENERALIZED BETA DISTRIBUTION.   II.
        (REPRINTED AS NO. 1742)
   BIOMETRIKA, 44, 441–453.   (MR19,P.781)


FOSTER, F. G. 1958.   (8.4/18.1/20.4/T)                                          1740
    UPPER PERCENTAGE POINTS OF THE GENERALIZED BETA DISTRIBUTION.   III.
        (REPRINTED AS NO. 1742)
   BIOMETRIKA, 45, 492–503.   (MR20-6799)


FOSTER, F. G. + REES, D.H. 1957.   (8.4/18.1/20.4/T)                             1741
    UPPER PERCENTAGE POINTS OF THE GENERALIZED BETA DISTRIBUTION.   I.
        (REPRINTED AS NO. 1742)
   BIOMETRIKA, 44, 237–247.   (MR19,P.188)


FOSTER, F. G. + REES, D.H. 1958.   (8.4/18.1/20.4/T)                             1742
    TABLES OF THE UPPER PERCENTAGE POINTS OF THE GENERALIZED BETA DISTRIBUTION.
        (REPRINT OF NO. 1741, REPRINT OF NO. 1739, REPRINT OF NO. 1740)
   NEW STATIST. TABLES, 26, 1–30.   (MR19,P.188,781; MR20-6799)


FOURGEAUD, CLAUDE 1951.   (16.2/E)                                               1743
    RECHERCHE DE RELATIONS A FORME LINEAIRE DANS UN SYSTEME ECONOMIQUE.
    APPLICATION AU MARCHE DU TEXTILE EN FRANCE.
   CAHIERS SEM. ECONOMET., 1, 43–55.


FOURGEAUD, CLAUDE    SEE ALSO
   1646. FERON, ROBERT + FOURGEAUD, CLAUDE (1951)

   1647. FERON, ROBERT + FOURGEAUD, CLAUDE (1952)

   1648. FERON, ROBERT + FOURGEAUD, CLAUDE (1952)


FOX, A. J. + JOHNSON, F. A. 1966.   (20.3)                                       1744
    ON FINDING THE EIGENVALUES OF REAL SYMMETRIC TRIDIAGONAL MATRICES.
   COMPUT. J., 9, 98–105.


FOX, CHARLES 1957.   (17.3/B)                                                    1745
    SOME APPLICATIONS OF MELLIN TRANSFORMS TO THE THEORY OF BIVARIATE
    STATISTICAL DISTRIBUTIONS.
   PROC. CAMBRIDGE PHILOS. SOC., 53, 620–628.   (MR19,P.781)


FOX, CHARLES 1965.   (17.3/B)                                                    1746
    A FAMILY OF DISTRIBUTIONS WITH THE SAME RATIO PROPERTY AS NORMAL DISTRIBUTION.
   CANAD. MATH. BULL., 8, 631–636.   (MR32-6507)


FOX, KARL A.    SEE
   4938. SCHWARTZ, F. R. + FOX, KARL A. (1942)

   5789. WAUGH, FREDERICK V. + FOX, KARL A. (1957)

   5790. WAUGH, FREDERICK V. + FOX, KARL A. (1958)

FOX, LESLIE 1950.  (20.3)                                          1747
    PRACTICAL METHODS FOR THE SOLUTION OF LINEAR EQUATIONS AND THE
    INVERSION OF MATRICES.
    J. ROY. STATIST. SOC. SER. B, 12, 120-136.  (MR12,P.538)

FOX, LESLIE + HAYES, J.G. 1951.  (20.3)                            1748
    PRACTICAL METHODS FOR THE INVERSION OF MATRICES.
    J. ROY. STATIST. SOC. SER. B, 13, 83-91.  (MR13,P.990)

FRANCIS, J. G. F. 1961.  (20.3)                                    1749
    THE QR TRANSFORMATION: A UNITARY ANALOGUE TO THE LR TRANSFORMATION. I.
    COMPUT. J., 4, 265-271.  (MR23-B3143)

FRANCIS, J. G. F. 1962.  (20.3)                                    1750
    THE QR TRANSFORMATION, II.
    COMPUT. J., 4, 332-345.  (MR25-744)

FRANCKX, EDOUARD D. 1937.  (2.3/17.3)                              1751
    SUR LES FONCTIONS DE FREQUENCE DE N VARIABLES.  RELATION GENERALE
    ENTRE LES MOMENTS ET LES SEMI-INVARIANTS.
    AKTUAR. VEDY, 6, 163-167.

FRANK, OVE 1966.  (3.1/19.3/C)                                     1752
    SIMULTANEOUS CONFIDENCE INTERVALS.
    SKAND. AKTUARIETIDSKR., 49, 78-84.

FRANK, RONALD E. 1966.  (20.3)                                     1753
    USE OF TRANSFORMATIONS.
    J. MARKET. RES., 3, 247-253.

FRANK, RONALD E.; MASSY, WILLIAM F. + MORRISON, DONALD G. 1965.  (6.3/E)   1754
    BIAS IN MULTIPLE DISCRIMINANT ANALYSIS.
    J. MARKET. RES., 2, 250-258.

FRANK, WILLIAM M. 1965.  (20.1)                                    1755
    A BOUND ON DETERMINANTS.
    PROC. AMER. MATH. SOC., 16, 360-363.

FRANKEL, EUGEN T. 1948.  (4.8)                                     1756
    NOVY SPOSOB VYPOCITANIA PARABOLICKEJ TRENDOVEJ CIARY A KORELACIE.(A NEW METHOD
    OF MATHEMATICAL TREATMENT OF PARABOLIC TREND LINES AND CORRELATION.)
        (WITH SUMMARY IN ENGLISH)
    STATIST. OBZOR, 28, 391-444.

FRANKEL, LESTER R.   SEE
    2607. HOTELLING, HAROLD + FRANKEL, LESTER R. (1938)

FRANKLIN, JOEL N. 1963.  (16.4)                                    1757
    THE COVARIANCE MATRIX OF A CONTINUOUS AUTOREGRESSIVE VECTOR TIME-SERIES.
    ANN. MATH. STATIST., 34, 1259-1264.  (MR27-5297)

FRANKLIN, JOEL N. 1965.  (16.4)                                    1758
    NUMERICAL SIMULATION OF STATIONARY AND NON-STATIONARY GAUSSIAN RANDOM PROCESSES.
    SIAM REV., 7, 68-80.

FRANKMANN, J.P.   SEE
    1575. ESTES, WILLIAM K.; BURKE, C. J.; ATKINSON, RICHARD C. +  (1957)
    FRANKMANN, J.P.

FRANZEN, RAYMOND 1928.  (4.5)                                      1759
    A COMMENT ON PARTIAL CORRELATION.
    J. EDUC. PSYCHOL., 19, 194-197.

FRANZEN, RAYMOND + DERRYBERRY, MAHEW 1931.  (4.5/4.6)              1760
    THE ROUTINE COMPUTATION OF PARTIAL AND MULTIPLE CORRELATION.
    J. EDUC. PSYCHOL., 22, 641-651.

FRASER, A. R. + KOVATS, M. 1966.  (8.1/17.9/F)                     1761
    STEREOSCOPIC MODELS OF MULTIVARIATE STATISTICAL DATA.
    BIOMETRICS, 22, 358-367.

FRASER, DONALD A. S. 1951.   (2.1/14.4)                                    1762
    NORMAL SAMPLES WITH LINEAR CONSTRAINTS AND GIVEN VARIANCES.
    CANAD. J. MATH., 3, 363-366.   (MR13,P.53)

FRASER, DONALD A. S. 1956.   (17.1/17.8)                                   1763
    A VECTOR FORM OF THE WALD-WOLFOWITZ-HOEFFDING THEOREM.
    ANN. MATH. STATIST., 27, 540-543.   (MR17,P.1219)

FRASER, DONALD A. S. 1957.   (8.3/19.2)                                    1764
    A REGRESSION ANALYSIS USING THE INVARIANCE METHOD.
    ANN. MATH. STATIST., 28, 517-520.   (MR19,P.474)

FRASER, DONALD A. S. 1964.   (14.3/17.8)                                   1765
    FIDUCIAL INFERENCE FOR LOCATION AND SCALE PARAMETERS.
    BIOMETRIKA, 51, 17-24.   (MR30-4339)

FRASER, DONALD A. S. 1966.   (19.4)                                        1766
    SUFFICIENCY FOR REGULAR MODELS.
    SANKHYA SER. A, 28, 137-144.

FRASER, DONALD A. S. 1966.   (19.4)                                        1767
    ON SUFFICIENCY AND CONDITIONAL SUFFICIENCY.
    SANKHYA SER. A, 28, 145-150.

FRASER, DONALD A. S. + GUTTMAN, IRWIN 1956.   (19.3)                       1768
    TOLERANCE REGIONS.
    ANN. MATH. STATIST., 27, 162-179.   (MR17,P.871)

FRASER, DONALD A. S.    SEE ALSO
    975. CHUNG, J. H. + FRASER, DONALD A. S. (1958)

FREAR, D. E. H. 1945.   (4.1/20.3)                                         1769
    PUNCH CARDS IN CORRELATION STUDIES.
    CHEM. AND ENGRG. NEWS, 23, 2077.

FRECHET, MAURICE 1915.   (20.1)                                           1770
    SUR LES FONCTIONNELLES BILINEAIRES.
    TRANS. AMER. MATH. SOC., 16, 215-234.

FRECHET, MAURICE 1929.   (6.5)                                            1771
    SUR LA DISTANCE DE DEUX VARIABLES ALEATOIRES.
    C.R. ACAD. SCI. PARIS, 188, 368-370.

FRECHET, MAURICE 1931.   (6.5/B)                                          1772
    NOUVELLES EXPRESSIONS DE LA "DISTANCE" DE DEUX VARIABLES ALEATOIRES
    ET DE LA "DISTANCE" DE DEUX FONCTIONS MESURABLES.
    ANN. SOC. POLON. MATH., 9, 45-48.

FRECHET, MAURICE 1933.   (4.1/4.9)                                        1773
    SUR LE COEFFICIENT, DIT DE CORRELATION.
    C.R. ACAD. SCI. PARIS, 197, 1268-1269.

FRECHET, MAURICE 1933.   (4.8/4.9)                                        1774
    SUR LE COEFFICIENT, DIT DE CORRELATION ET SUR LA CORRELATION EN GENERAL.
    REV. INST. INTERNAT. STATIST., 1(3) 16-23.

FRECHET, MAURICE 1935.   (4.1/4.3/4.8)                                    1775
    COMMUNICATION SUR L'USAGE DU SOI-DISANT COEFFICIENT DE CORRELATION.
        (WITH DISCUSSION)
    BULL. INST. INTERNAT. STATIST., 28(2) 25-52.

FRECHET, MAURICE 1935.   (4.1/4.8)                                        1776
    SUR DEUX RELATIONS SIMPLES ENTRE LE COEFFICIENT DE CORRELATION ET LE
    "RAPPORT" DE CORRELATION.
    CASOPIS PEST. MAT. FYS., 64, 209-210.

FRECHET, MAURICE 1935.   (4.1/4.9)                                        1777
    SUR LE COEFFICIENT DE LINEARITE, DIT DE CORRELATION.
    REV. INST. INTERNAT. STATIST., 3, 365-379.

FRECHET, MAURICE 1936.   (4.1)                                            1778
    ON THE SO-CALLED CORRELATION COFFFICIENT.
    J. EDUC. PSYCHOL., 27, 304-306.

FRECHET, MAURICE 1936.   (4.1/4.9)                                          1779
    RAPPORT DE LA COMMISSION D'ETUDE DE L'USAGE DU COEFFICIENT DE CORRELATION.
       (REPRINTED AS NO. 1780)
  REV. INST. INTERNAT. STATIST., 4, 238-242.

FRECHET, MAURICE 1937.   (4.1/4.9)                                          1780
    RAPPORT DE LA COMMISSION D'ETUDE DE L'USAGE DU COEFFICIENT DE CORRELATION.
       (REPRINT OF NO. 1779)
  BULL. INST. INTERNAT. STATIST., 29(3) 318-322.

FRECHET, MAURICE 1946.   (4.9)                                             1781
    A GENERAL METHOD OF CONSTRUCTING CORRELATION INDICES.
  PROC. MATH. PHYS. SOC. EGYPT, 3, 13-20.   (MR8,P.592; MR11, P.870)

FRECHET, MAURICE 1947.   (4.9/6.5/B)                                       1782
    ANCIENS ET NOUVEAUX INDICES DE CORRELATION.  LEUR APPLICATION AU
    CALCUL DES RETARDS ECONOMIQUES.
       (AMENDED BY NO. 1783)
  ECONOMETRICA, 15, 1-30.   (MR8,P.393)

FRECHET, MAURICE 1947.   (4.9/6.5/B)                                       1783
    ERRATA TO: ANCIENS ET NOUVEAUX INDICES DE CORRELATION.  LEUR
    APPLICATION AU CALCUL DES RETARDS ECONOMIQUES.
       (AMENDMENT OF NO. 1782)
  ECONOMETRICA, 15, 374-375.   (MR9,P.194)

FRECHET, MAURICE 1948.   (4.8)                                             1784
    LE COEFFICIENT DE CONNEXION STATISTIQUE DE GINI-SALVEMINI.
  MATHEMATICA, TIMISOARA, 23, 46-51.   (MR10,P.50)

FRECHET, MAURICE 1948.   (4.9)                                             1785
    ADDITIONAL NOTE ON A GENERAL METHOD OF CONSTRUCTING CORRELATION INDICES.
  PROC. MATH. PHYS. SOC. EGYPT, 3, 73-74.   (MR11,P.191, P.871)

FRECHET, MAURICE 1950.   (4.1)                                             1786
    SUR UN ESSAI INFONDE DE SAUVER LE COEFFICIENT CLASSIQUE DIT DE CORRELATION.
  REV. INST. INTERNAT. STATIST., 18, 157-160.   (MR12,P.840)

FRECHET, MAURICE 1951.   (2.1/2.2/17.3)                                    1787
    GENERALISATIONS DE LA LOI DE PROBABILITE DE LAPLACE.
  ANN. INST. H. POINCARE, 12, 1-29.   (MR12,P.839)

FRECHET, MAURICE 1951.   (17.4)                                           1788
    SUR LES TABLEAUX DE CORRELATION DONT LES MARGES SONT DONNEES.
  ANN. UNIV. LYON TROISIEME SER. SCI. SECT. A, 14, 53-77.   (MR14,P.189)

FRECHET, MAURICE 1954.   (17.4)                                           1789
    SUR LES TABLEAUX DE CORRELATION DONT LES MARGES SONT DONNEES.
  BULL. INST. INTERNAT. STATIST., 34(2) 3-4.   (MR14,P.189)

FRECHET, MAURICE 1956.   (17.4)                                           1790
    SUR LES TABLEAUX DE CORRELATION DONT LES MARGES SONT DONNEES.
  C.R. ACAD. SCI. PARIS, 242, 2426-2428.   (MR17,P.1217)

FRECHET, MAURICE 1957.   (17.4)                                           1791
    LES TABLEAUX DE CORRELATION DONT LES MARGES ET DES BORNES SONT DONNEES.
  ANN. UNIV. LYON TROISIEME SER. SCI. SECT. A, 20, 13-31.   (MR20-6168)

FRECHET, MAURICE 1957.   (6.5)                                            1792
    SUR LA DISTANCE DE DEUX LOIS DE PROBABILITE.
  C.R. ACAD. SCI. PARIS, 244, 689-692.   (MR18,P.679)

FRECHET, MAURICE 1957.   (4.1/17.4/E)                                     1793
    LES TABLEAUX DE CORRELATION ET LES PROGRAMMES LINEAIRES.
  REV. INST. INTERNAT. STATIST., 25, 23-40.

FRECHET, MAURICE 1958.   (17.4)                                          1794
    SUR LES TABLEAUX DE CORRELATION DONT LES MARGES ET DES BORNES SONT DONNEES.II.
  ANN. UNIV. LYON TROISIEME SER. SCI. SECT. A, 21, 19-32.   (MR22-1010)

FRECHET, MAURICE 1959.   (6.5/17.3)                                       1795
    LES DEFINITIONS DE LA SOMME ET DU PRODUIT SCALAIRES EN TERMES DE
    DISTANCE DANS UN ESPACE ABSTRAIT.  (AVEC SUPPLEMENT)
  CALCUTTA MATH. SOC. GOLDEN JUBILEE COMMEM. VOLUME, 1, 151-157, 159-160.

FRECHET, MAURICE 1959. (4.1)
    A NOTE ON SIMPLE CORRELATION.
    MATH. MAG., 32, 265-268. (MR21-2335)
        1796

FREIBERGER, WALTER F. 1963. (16.3)
    APPROXIMATE DISTRIBUTIONS OF CROSS-SPECTRAL ESTIMATES FOR GAUSSIAN PROCESSES.
    PROC. SYMP. TIME SERIES ANAL. BROWN UNIV. (ROSENBLATT), 244-259. (MR27-849)
        1797

FREIBERGER, WALTER F. + JONES, RICHARD H. 1960. (2.5/14.5/N)
    COMPUTATION OF THE FREQUENCY FUNCTION OF A QUADRATIC FORM IN RANDOM
    NORMAL VARIABLES.
    J. ASSOC. COMPUT. MACH., 7, 245-250. (MR23-A1459)
        1798

FREIBERGER, WALTER F.; ROSENBLATT, MURRAY + VAN NESS, J. 1962. (16.4/16.5)
    REGRESSION ANALYSIS OF VECTOR-VALUED RANDOM PROCESSES.
    J. SOC. INDUST. APPL. MATH., 10, 89-102. (MR25-722)
        1799

FRENCH, JOHN W. 1966. (15.4)
    THE LOGIC OF AND ASSUMPTIONS UNDERLYING DIFFERENTIAL TESTING.
    TEST. PROBLEMS PERSPECTIVE (ANASTASI), 321-330.
        1800

FRENKIEL, FRANCOIS N. + FOLLIN, JAMES W., JR. 1954. (2.1)
    ON MULTIVARIATE NORMAL PROBABILITY DISTRIBUTIONS.
    STUD. MATH. MECH. (VON MISES VOLUME), 295-300. (MR16,P.377)
        1801

FRENKIEL, FRANCOIS N. SEE ALSO
    2922. KAMPE DE FERIET, JOSEPH + FRENKIEL, FRANCOIS N. (1959)

FREUND, JOHN E. 1961. (18.1/B)
    A BIVARIATE EXTENSION OF THE EXPONENTIAL DISTRIBUTION.
    J. AMER. STATIST. ASSOC., 56, 971-977. (MR24-A2465)
        1802

FREUND, JOHN E. SEE ALSO
    5942. WINE, R.L. + FREUND, JOHN E. (1957)

FREUND, RUDOLF J.; VAIL, RICHARD W. + CLUNIES-ROSS, CHARLES W. 1961. (8.1/8.9)
    RESIDUAL ANALYSIS.
        (AMENDED BY NO. 1804)
    J. AMER. STATIST. ASSOC., 56, 98-104. (MR22-12639)
        1803

FREUND, RUDOLF J.; VAIL, RICHARD W. + CLUNIES-ROSS, CHARLES W. 1961. (8.9)
    CORRIGENDA: RESIDUAL ANALYSIS.
        (AMENDMENT OF NO. 1803)
    J. AMER. STATIST. ASSOC., 56, 1005. (MR22-12639)
        1804

FREUND, RUDOLF J. SEE ALSO
    200. APPLEBY, R. H. + FREUND, RUDOLF J. (1962)

FRIEDE, GEORG + MUNZNER, HANS-FRIEDRICH 1948. (4.9)
    ZUR MAXIMALKORRELATION.
    Z. ANGEW. MATH. MECH., 28, 158-160. (MR10,P.50)
        1805

FRIEDLAND, BERNARD 1958. (16.3)
    LEAST SQUARES FILTERING AND PREDICTION OF NON-STATIONARY SAMPLED DATA.
    INFORMATION AND CONTROL, 1, 297-313. (MR21-382)
        1806

FRIEDMAN, MORRIS D. 1950. (20.3/20.4)
    DETERMINATION OF EIGENVALUES USING A GENERALIZED LAPLACE TRANSFORM.
    J. APPL. PHYS., 21, 1333-1337.
        1807

FRISCH, RAGNAR 1929. (8.1/16.2/17.7)
    CORRELATION AND SCATTER IN STATISTICAL VARIABLES.
    NORDIC STATIST. J., 1, 36-102.
        1808

FRISCH, RAGNAR 1933. (17.7)
    PROPAGATION PROBLEMS AND IMPULSE PROBLEMS IN DYNAMIC ECONOMICS.
    ECON. ESSAYS (CASSEL VOLUME), 171-205.
        1809

FRISCH, RAGNAR 1937. (17.7)
    NOTE ON THE PHASE DIAGRAM OF TWO VARIATES.
    ECONOMETRICA, 5, 326-328.
        1810

FRISCH, RAGNAR + MUDGETT, BRUCE D. 1931.  (17.7)                    1811
     STATISTICAL CORRELATION AND THE THEORY OF CLUSTER TYPES.
          (AMENDED BY NO. 1812)
     J. AMER. STATIST. ASSOC., 26, 375-392.

FRISCH, RAGNAR + MUDGETT, BRUCE D. 1932.  (17.7)                    1812
     ERRATA TO: STATISTICAL CORRELATION AND THE THEORY OF CLUSTER TYPES.
          (AMENDMENT OF NO. 1811)
     J. AMER. STATIST. ASSOC., 27, 187.

FROE, A. DE  SEE  DE FROE, A.

FRUCHTER, BENJAMIN 1949.  (20.2/20.3)                    1813
     NOTE ON THE COMPUTATION OF THE INVERSE OF A TRIANGULAR MATRIX.
     PSYCHOMETRIKA, 14, 89-93.  (MR11,P.403)

FRUCHTER, BENJAMIN 1966.  (15.1/15.2/E)                    1814
     MANIPULATIVE AND HYPOTHESIS-TESTING FACTOR-ANALYTIC EXPERIMENTAL DESIGNS.
     HANDB. MULTIVARIATE EXPER. PSYCHOL. (CATTELL), 330-354.

FRUCHTER, BENJAMIN + NOVAK, EDWIN 1958.  (15.3)                    1815
     A COMPARATIVE STUDY OF THREE METHODS OF ROTATION.
     PSYCHOMETRIKA, 23, 211-221.

FRUCHTER, BENJAMIN   SEE ALSO
     113. ANDERSON, HARRY E., JR. + FRUCHTER, BENJAMIN (1960)

FU, YUMIN; SRINATH, MANDYAM D. + YEN, ANDREW T. 1965.  (16.4)                    1816
     A NOTE ON MULTIVARIATE GENERATING FUNCTIONS AND APPLICATIONS TO
     DISCRETE STOCHASTIC PROCESSES.
     SIAM REV., 7, 31-41.

FUCHS, AIME 1957.  (17.1)                    1817
     SOME LIMIT THEOREMS FOR NONHOMOGENEOUS MARKHOFF PROCESSES.
     TRANS. AMER. MATH. SOC., 86, 511-531.  (MR20-1357)

FUCHS, AIME 1963.  (2.2)                    1818
     LOIS DE PROBABILITE LIEES A LA LOI DE LAPLACE-GAUSS A K DIMENSIONS.
     PUBL. INST. STATIST. UNIV. PARIS, 12, 117-129.  (MR30-4277)

FUCHS, CAROL E. + DAVID, HERBERT T. 1965.  (17.8/A)                    1819
     POISSON LIMITS OF MULTIVARIATE RUN DISTRIBUTIONS.
     ANN. MATH. STATIST., 36, 215-225.

FUCHS, KLAUS 1908.  (2.1)                    1820
     DAS NORMALELLIPSOID.
     Z. VERMESSUNGS., 37, 721-731, 753-763.

FUCHS, WOLFGANG H. J.   SEE
     979. CHUNG, KAI LAI + FUCHS, WOLFGANG H. J. (1951)

FUHRICH, J. 1929.  (2.1)                    1821
     UBER DIE ALLGEMEINE FORM DES KORRELATIONSMASSES.
     Z. ANGEW. MATH. MECH., 9, 77-79.

FUJISAWA, H.   SEE
     3190. KUDO, AKIO + FUJISAWA, H. (1966)

FUKUDA, YOICHIRO   SEE
     2345. HARMAN, HARRY H. + FUKUDA, YOICHIRO (1966)

FULCHER, JOHN S. + ZUBIN, JOSEPH 1942.  (15.4)                    1822
     THE ITEM ANALYZER:  A MECHANICAL DEVICE FOR TREATING THE FOUR FOLD
     TABLE IN LARGE SAMPLES.
     J. APPL. PSYCHOL., 26, 511-522.

FULLER, A. T.   SEE
     1686. FISHER, MICHAEL E. + FULLER, A. T. (1958)

FULLER, E. L. + HEMMERLE, W. J. 1966.  (15.2)                    1823
     ROBUSTNESS OF THE MAXIMUM-LIKELIHOOD ESTIMATION PROCEDURE IN FACTOR ANALYSIS.
     PSYCHOMETRIKA, 31, 255-266.  (MR33-1927)

FULLER, WILBUR N.    SEE
   4156. PEARL, RAYMOND + FULLER, WILBUR N. (1905)


FULTON, CURTIS M. 1965.   (20.5)                                                    1824
    PARALLEL VECTOR FIELDS.
  PROC. AMER. MATH. SOC., 16, 136-137.


FUNKHOUSER, H. G. 1937.   (1.1)                                                     1825
    HISTORICAL DEVELOPMENT OF THE GRAPHICAL REPRESENTATION OF STATISTICAL DATA.
  OSIRIS, 3, 269-404.


FURFEY, PAUL H. + DALY, JOSEPH F. 1935.   (4.1/E)                                   1826
    PRODUCT-MOMENT CORRELATION AS A RESEARCH TECHNIQUE IN EDUCATION.
  J. EDUC. PSYCHOL., 26, 206-211.


FURST, DARIO 1955.   (8.1/20.3)                                                     1827
    NOTA ALLA TABELLA PER IL CALCOLO DEI COEFFICIENTI DI REGRESSIONE QUADRATICA.
  STATISTICA, BOLOGNA, 15, 492-495.   (MR17,P.665)


FURSTENBERG, HARRY + KESTEN, HARRY 1960.   (16.4/A)                                 1828
    PRODUCTS OF RANDOM MATRICES.
  ANN. MATH. STATIST., 31, 457-469.   (MR22-12558)


FURUKAWA, NAGATA 1962.   (2.1/3.2/8.6)                                              1829
    THE INFERENCE THEORY IN MULTIVARIATE RANDOM EFFECT MODEL.   I.
  KUMAMOTO J. SCI. SER. A, 5, 158-170.   (MR26-4448)


FZ. DE TROCONIZ, ANTONIA 1965.   (2.1/2.3)                                          1830
    DISTRIBUCION NORMAL MULTIVARIANTE.
       (WITH SUMMARY IN ENGLISH)
  TRABAJOS ESTADIST., 16, 33-41.   (MR31-6258)


GABRIEL, K.R. 1954.   (15.5)                                                        1831
    THE SIMPLEX STRUCTURE OF THE PROGRESSIVE MATRICES TEST.
  BRITISH J. STATIST. PSYCHOL., 7, 9-14.


GABRIEL, K.R. 1962.   (9.3/10.2/16.1)                                              1832
    ANTE-DEPENDENCE ANALYSIS OF AN ORDERED SET OF VARIABLES.
  ANN. MATH. STATIST., 33, 201-212.   (MR26-3141)


GABRIEL, K.R. 1962.   (10.2/E)                                                      1833
    THE MODEL OF ANTE-DEPENDENCE FOR DATA OF BIOLOGICAL GROWTH.
  BULL. INST. INTERNAT. STATIST., 39(2) 253-264.


GABRIEL, K.R. 1966.   (17.8/T)                                                      1834
    SIMULTANEOUS TEST PROCEDURES FOR MULTIPLE COMPARISONS ON CATEGORICAL DATA.
  J. AMER. STATIST. ASSOC., 61, 1081-1096.


GADDIS, L. WESLEY    SEE
   3813. MICHAEL, WILLIAM B.; JONES, ROBERT A.; GADDIS, L. WESLEY +   (1962)
   KAISER, HENRY F.


GAGE, N. L.    SEE
   4512. REMMERS, H. H.; KARSLAKE, R. + GAGE, N. L. (1940)


GAIER, EUGENE L. + LEE, MARILYN C. 1953.   (15.5)                                   1835
    PATTERN ANALYSIS:  THE CONFIGURAL APPROACH TO PREDICTIVE MEASUREMENT.
  PSYCHOL. BULL., 50, 140-148.


GALE, DAVID 1951.   (20.5)                                                          1836
    CONVEX POLYHEDRAL CONES AND LINEAR INEQUALITIES.
  COWLES COMMISS. MONOG., 13, 287-297.   (MR13,P.60)


GALES, KATHLEEN 1957.   E(6.3)                                                      1837
    DISCRIMINANT FUNCTIONS OF SOCIO-ECONOMIC CLASS.
  APPL. STATIST., 6, 123-132.


GALLER, BERNARD A. + DWYER, PAUL S. 1957.   (20.3)                                  1838
    TRANSLATING THE METHOD OF REDUCED MATRICES TO MACHINES.
  NAVAL RES. LOGISTICS QUART., 4, 55-71.   (MR19,P.515)

GALLOT, S. 1966.  (17.2)                                                    1839
    A BOUND FOR THE MAXIMUM OF A NUMBER OF RANDOM VARIABLES.
    J. APPL. PROB., 3, 556–558.  (MR34-3607)


GALTON, FRANCIS 1886.  (4.1/E)                                              1840
    FAMILY LIKENESS IN STATURE (WITH AN APPENDIX BY J.D. HAMILTON DICKSON).
    PROC. ROY. SOC. LONDON, 40, 42–73.


GALTON, FRANCIS 1888.  (4.1/15.1/E)                                         1841
    CO-RELATIONS AND THEIR MEASUREMENT, FROM ANTHROPOMETRIC DATA.
    PROC. ROY. SOC. LONDON, 45, 135–145.


GALTON, FRANCIS 1897.  (4.1)                                                1842
    NOTE TO THE MEMOIR BY PROFESSOR KARL PEARSON ON SPURIOUS CORRELATION.
    PROC. ROY. SOC. LONDON, 60, 498–502.


GALVENIUS, I. + WOLD, HERMAN O. A. 1947.  E(16.5)                           1843
    STATISTICAL TESTS OF H. ALFVEN'S THEORY OF SUNSPOTS.
    ARK. MAT. ASTRONOM. FYS., 34A(24) 1–9.


GARCIA TRANQUE, TOMAS 1950.  (4.9)                                          1844
    SOBRE EL COEFICIENTE DE CORRELACION LINEAL.
    GACETA MAT., 2, 184–190.  (MR12,P.512)


GARDINER, DONALD A. + BOMBAY, BARBARA FLORES 1965.  (14.5)                  1845
    AN APPROXIMATION TO STUDENT'S T.
    TECHNOMETRICS, 7, 71–72.


GARDING, LARS 1941.  (4.5/4.7/14.2)                                         1846
    THE DISTRIBUTIONS OF THE FIRST AND SECOND ORDER MOMENTS, THE PARTIAL
    CORRELATION COEFFICIENTS AND THE MULTIPLE CORRELATION COEFFICIENT IN SAMPLES
    FROM A NORMAL MULTIVARIATE POPULATION.
    SKAND. AKTUARIETIDSKR., 24, 185–202.  (MR7,P.212)


GARDNER, ROBERT S.   SEE
    3763. MAXFIELD, JOHN E. + GARDNER, ROBERT S. (1955)


GARNER, F. H.; GRANTHAM, J. + SANDERS, H. G. 1934.  (8.5/U)                 1847
    THE VALUE OF COVARIANCE IN ANALYSING FIELD EXPERIMENTAL DATA.
    J. AGRIC. SCI., 24, 250–259.


GARNER, W. R. + MC GILL, WILLIAM J. 1956.  (8.5/17.7)                       1848
    THE RELATION BETWEEN INFORMATION AND VARIANCE ANALYSES.
    PSYCHOMETRIKA, 21, 219–228.  (MR18,P.367)


GARNER, W. R.   SEE ALSO
    3830. MILLER, GEORGE A. + GARNER, W. R. (1944)


GARNETT, J. C. MAXWELL 1919.  (15.1)                                        1849
    ON CERTAIN INDEPENDENT FACTORS IN MENTAL MEASUREMENTS.
    PROC. ROY. SOC. LONDON, 96, 91–111.


GARNETT, J. C. MAXWELL 1934.  (15.1)                                        1850
    THE SINGLE GENERAL FACTOR.  A NOTE ON LINEAR TRANSFORMATION OF
    HIERARCHICAL SYSTEMS.
    BRITISH J. PSYCHOL., 25, 100–105.


GARNETT, J. C. MAXWELL + THOMSON, GODFREY H. 1919.  (15.1)                  1851
    JOINT NOTE ON "THE HIERARCHY OF ABILITIES".
    BRITISH J. PSYCHOL., 9, 367–368.


GARRETT, HENRY E. 1928.  (4.6)                                             1852
    A MODIFICATION OF TOLLEY AND EZEKIEL'S METHOD OF HANDLING MULTIPLE
    CORRELATION PROBLEMS.
    J. EDUC. PSYCHOL., 19, 45–49.


GARRETT, HENRY E. 1933.  (15.2)                                            1853
    THE SAMPLING DISTRIBUTION OF THE TETRAD EQUATION.
    J. EDUC. PSYCHOL., 24, 536–542.


GARRETT, HENRY E. 1935.  (15.1)                                            1854
    THE TWO-FACTOR THEORY AND ITS CRITICISM.
    PSYCHOL. REV., 42, 293–301.

GARRETT, HENRY E. 1936.  (15.1)                                              1855
    THURSTONE'S "VECTORS OF MIND".
  PSYCHOL. BULL., 33, 395-404.


GARRETT, HENRY E. + ANASTASI, ANNE 1932.  (15.2/E)                           1856
    THE TETRAD-DIFFERENCE CRITERION AND THE MEASUREMENT OF MENTAL TRAITS.
  ANN. NEW YORK ACAD. SCI., 33, 235-282.

GARRIGUE, V. ROUQUET LA   SEE  LA GARRIGUE, V. ROUQUET

GARSIDE, M.J. 1965.  (8.3/19.1/20.2)                                         1857
    THE BEST SUB-SET IN MULTIPLE REGRESSION ANALYSIS.
  APPL. STATIST., 14, 196-200.

GARTI, Y. 1940.  (17.3)                                                      1858
    LES LOIS DE PROBABILITE POUR LES FONCTIONS STATISTIQUES (CAS DE
    COLLECTIFS A PLUSIEURS DIMENSIONS).
  REV. MATH. UN. INTERBALK., 3, 21-39.  (MR2,P.106)

GARWOOD, F. 1933.  (4.2)                                                     1859
    THE PROBABILITY INTEGRAL OF THE CORRELATION COEFFICIENT IN SAMPLES
    FROM A NORMAL BIVARIATE POPULATION.
  BIOMETRIKA, 25, 71-78.

GARWOOD, F. 1947.  (14.4)                                                    1860
    THE VARIANCE OF THE OVERLAP OF GEOMETRICAL FIGURES WITH REFERENCE TO
    A BOMBING PROBLEM.
  BIOMETRIKA, 34, 1-17.  (MR8,P.389)

GATTY, RONALD 1966.  (1.1)                                                   1861
    MULTIVARIATE ANALYSIS FOR MARKETING RESEARCH:  AN EVALUATION.
  APPL. STATIST., 15, 157-172.

GAUCHET, F. + BOULANGER, J. J. 1953.  E(17.9)                                1862
    LA RECHERCHE DES FACTEURS INDEPENDANTS ET SON ROLE DANS LA SCHEMATISATION ET
    L'EXPLICATION DES PHENOMENES A CARACTERES MULTIPLES.
  J. SOC. STATIST. PARIS, 94, 277-286.

GAUDIN, MICHEL 1961.  (13.1/20.1/A)                                          1863
    SUR LA LOI LIMITE DE L'ESPACEMENT DES VALEURS PROPRES D'UNE MATRICE ALEATOIRE.
         (REPRINTED AS NO. 1864)
  NUCLEAR PHYS., 25, 447-458.

GAUDIN, MICHEL 1965.  (13.1/20.1/A)                                          1864
    SUR LA LOI LIMITE DE L'ESPACEMENT DES VALEURS PROPRES D'UNE MATRICE ALEATOIRE.
         (REPRINT OF NO. 1863)
  STATIST. THEOR. SPECTRA FLUCTUATIONS (PORTER), 356-367.

GAUDIN, MICHEL    SEE ALSO
   3784. MEHTA, MADAN LAL + GAUDIN, MICHEL (1960)

   3785. MEHTA, MADAN LAL + GAUDIN, MICHEL (1961)

   3786. MEHTA, MADAN LAL + GAUDIN, MICHEL (1965)

   3787. MEHTA, MADAN LAL + GAUDIN, MICHEL (1965)

GAUSS, CAROLO FRIDERICO  SEE  GAUSS, KARL FRIEDRICH


GAUSS, KARL FRIEDRICH 1823.  (8.8)                                           1865
    THEORIA COMBINATIONIS OBSERVATIONUM ERRORIBUS MINIMIS OBNOXIAE.  PARS PRIOR AND
    PARS POSTERIOR. (SOCIETATI REGIAE EXHIBITA, FEB. 15, 1821 AND FEB. 2, 1823.)
         (REPRINTED AS NO. 1866)
  COMMENT. SOC. REG. SCI. GOTTING. REC. CL. MATH., 5, 33-62, 63-90.

GAUSS, KARL FRIEDRICH 1823.  (8.8)                                           1866
    THEORIA COMBINATIONIS OBSERVATIONUM ERRORIBUS MINIMIS OBNOXIAE.  PARS PRIOR AND
    PARS POSTERIOR. (SOCIETATI REGIAE EXHIBITA, FEB. 15, 1821 - FEB. 2, 1823.)
         (REPRINT OF NO. 1865)
  MONOG. GOTTING. APUD HEN. DIET., 1-30, 31-80.

GAUSS, KARL FRIEDRICH 1828.   (8.8)                                          **1867**
    THEORIA RESIDUORUM BIQUADRATICORUM.  COMMENTATIO PRIMA.
    COMMENT. SOC. REG. SCI. GOTTING. REC. CL. MATH., 6, 27-56.

GAUSS, KARL FRIEDRICH 1828.   (8.8)                                          **1868**
    SUPPLEMENTUM THEORIAE COMBINATIONIS OBSERVATIONUM ERRORIBUS MINIMIS
    ORNOXIAE.  (SOCIETATI REGIAE EXHIBITUM, SEPT. 16, 1826.)
        (REPRINTED AS NO. 1869)
    COMMENT. SOC. REG. SCI. GOTTING. REC. CL. MATH., 6, 57-98.

GAUSS, KARL FRIEDRICH 1828.   (8.8)                                          **1869**
    SUPPLEMENTUM THEORIAE COMBINATIONIS OBSERVATIONUM ERRORIBUS MINIMIS
    ORNOXIAE.  (SOCIETATI REGIAE EXHIBITUM, SEPT. 16, 1826.)
        (REPRINT OF NO. 1868)
    MONOG. GOTTING. APUD HEN. DIET., 1-40.

GAUSS, KARL FRIEDRICH 1832.   (8.8)                                          **1870**
    THEORIA RESIDUORUM BIQUADRATICORUM.  COMMENTATIO SECUNDA.
    COMMENT. SOC. REG. SCI. GOTTING. REC. CL. MATH., 7, 89-148.

GAYEN, A. K. 1951.  (4.2/18.3)                                               **1871**
    THE FREQUENCY DISTRIBUTION OF THE PRODUCT-MOMENT CORRELATION COEFFICIENT IN
    RANDOM SAMPLES OF ANY SIZE DRAWN FROM NON-NORMAL UNIVERSES.
    BIOMETRIKA, 38, 219-247.  (MR13,P.53)

GAYEN, A. K.   SEE ALSO
    652. BOSE, MRS. CHAMELI + GAYEN, A. K. (1946)

GAYLORD, RICHARD H.   SEE
    5841. WHERRY, ROBERT J. + GAYLORD, RICHARD H. (1943)

    5842. WHERRY, ROBERT J. + GAYLORD, RICHARD H. (1944)

GEARY, ROBERT C. 1927.  (4.1/8.1/18.5)                                       **1872**
    SOME PROPERTIES OF CORRELATION AND REGRESSION IN A LIMITED UNIVERSE.
    METRON, 7(1) 83-119.

GEARY, ROBERT C. 1930.  (14.2/U)                                             **1873**
    THE FREQUENCY DISTRIBUTIONS OF THE QUOTIENT OF TWO NORMAL VARIATES.
    J. ROY. STATIST. SOC., 93, 442-446.

GEARY, ROBERT C. 1935.  (14.2/U)                                             **1874**
    I. THE RATIO OF THE MEAN DEVIATION TO THE STANDARD DEVIATION AS A TEST OF NORMALITY.
    BIOMETRIKA, 27, 310-332.

GEARY, ROBERT C. 1935.  (14.2/U)                                             **1875**
    III. NOTE ON THE CORRELATION BETWEEN $\beta_2$ AND $\omega'$.
    BIOMETRIKA, 27, 353-355.

GEARY, ROBERT C. 1936.  (14.2/U)                                             **1876**
    MOMENTS OF THE RATIO OF THE MEAN DEVIATION TO THE STANDARD DEVIATION FOR
    NORMAL SAMPLES.
    BIOMETRIKA, 28, 295-305.

GEARY, ROBERT C. 1942.  (19.3)                                              2331
    THE ESTIMATION OF MANY PARAMETERS.
    J. ROY. STATIST. SOC., 105, 213-217.   (MR4,P.165)

GEARY, ROBERT C. 1942.  (17.7)                                               **1877**
    INHERENT RELATIONS BETWEEN RANDOM VARIABLES.
    PROC. ROY. IRISH ACAD. A, 47, 63-76.   (MR4,P.21)

GEARY, ROBERT C. 1944.  (17.3)                                               **1878**
    EXTENSION OF A THEOREM BY HARALD CRAMER ON THE FREQUENCY DISTRIBUTION
    OF THE QUOTIENT OF TWO VARIABLES.
    J. ROY. STATIST. SOC., 107, 56-57.   (MR6,P.159)

GEARY, ROBERT C. 1948.  (16.5)                                               **1879**
    STUDIES IN RELATIONS BETWEEN ECONOMIC TIME SERIES.
    J. ROY. STATIST. SOC. SER. B, 10, 140-158.   (MR10,P.553)

GEARY, ROBERT C. 1949.  (17.7)                                               **1880**
    DETERMINATION OF LINEAR RELATIONS BETWEEN SYSTEMATIC PARTS OF VARIABLES WITH
    ERRORS OF OBSERVATION THE VARIANCES OF WHICH ARE UNKNOWN.
    ECONOMETRICA, 17, 30-58.   (MR10,P.465, P. 856)

GEARY, ROBERT C. 1953.  (8.1)                                                    1881
    NON-LINEAR FUNCTIONAL RELATIONSHIP BETWEEN TWO VARIABLES WHEN ONE
    VARIABLE IS CONTROLLED.
  J. AMER. STATIST. ASSOC., 48, 94-103.  (MR14,P.776)


GEARY, ROBERT C. 1955.  (17.7)                                                   1882
    ESTIMATIONS DES RELATIONS ENTRE DES VARIABLES ALEATOIRES.
  BULL. SOC. BELG. STATIST., 25-60.



GEARY, ROBERT C. 1963.  (4.9)                                                    1883
    SOME REMARKS ABOUT RELATIONS BETWEEN STOCHASTIC VARIABLES: A
    DISCUSSION DOCUMENT.
  REV. INST. INTERNAT. STATIST., 31, 163-181.  (MR31-6324)


GEARY, ROBERT C. 1966.  (8.8)                                                    1884
    A NOTE ON RESIDUAL HETEROVARIANCE AND ESTIMATION EFFICIENCY IN REGRESSION.
  AMER. STATIST., 20(4) 30-31.


GEARY, ROBERT C. + LINEHAN, THOMAS P. 1960.  (4.1/6.6)                           1885
    PARADOXES IN STATISTICAL CLASSIFICATION.
  STUDI ONORE C. GINI, 1, 177-199.


GEBELEIN, HANS 1941.  (4.1/4.9)                                                  1886
    DAS STATISTISCHE PROBLEM DER KORRELATION ALS VARIATIONS- UND EIGENWERTPROBLEM
    UND SEIN ZUSAMMENHANG MIT DER AUSGLEICHSRECHNUNG.
  Z. ANGEW. MATH. MECH., 21, 364-379.  (MR4,P.104)


GEBELEIN, HANS 1942.  (4.8)                                                      1887
    BEMERKUNG UBER EIN VON W. HOEFFDING VORGESCHLAGENES,
    MASZSTABSINVARIANTES KORRELATIONSMASS.
  Z. ANGEW. MATH. MECH., 22, 171-173.  (MR4,P.279)



GEBELEIN, HANS 1942.  (4.9)                                                      1888
    VERFAHREN ZUR BEURTEILUNG EINER SEHR GERINGEN KORRELATION ZWISCHEN
    STATISTISCHEN MERKMALSREIHEN.
  Z. ANGEW. MATH. MECH., 22, 286-298, 553-592.  (MR6,P.6)


GEBELEIN, HANS 1952.  (4.9/20.4)                                                 1889
    MAXIMALKORRELATION UND KORRELATIONSSPEKTRUM.
  Z. ANGEW. MATH. MECH., 32, 9-19.  (MR13,P.669)


GEFFROY, JEAN 1957.  (17.1/17.3/17.6/B)                                          1890
    SUR LA NOTION D'INDEPENDANCE LIMITE DE DEUX VARIABLES ALEATOIRES.
    APPLICATION DE L'ETENDUE ET AU MILIEU D'UN ECHANTILLON.
  C.R. ACAD. SCI. PARIS, 245, 1291-1293.  (MR19,P.690)


GEFFROY, JEAN 1961.  (14.2/17.1)                                                 1891
    LOCALISATION ASYMPTOTIQUE DU POLYEDRE D'APPUI D'UN ECHANTILLON
    LAPLACIEN A K DIMENSIONS.
  PUBL. INST. STATIST. UNIV. PARIS, 10, 213-228.  (MR25-4559)


GEFFROY, JEAN 1964.  (18.6)                                                      1892
    SUR UN PROBLEME D'ESTIMATION GEOMETRIQUE.
  PUBL. INST. STATIST. UNIV. PARIS, 13, 191-210.  (MR34-2110)


GEFFROY, JEAN 1965.  (17.6/AU)                                                   1893
    SUR UNE CONDITION NECESSAIRE ET SUFFISANTE DE STABILITE PRESQUE SURE
    DES VALEURS EXTREMES D'UN ECHANTILLON.
  C.R. ACAD. SCI. PARIS, 260, 3271-3273.


GEHAN, EDMUND A. 1962.  E(8.1/14.4)                                              1894
    AN APPLICATION OF MULTIVARIATE REGRESSION ANALYSIS TO THE PROBLEM OF
    PREDICTING SURVIVAL IN PATIENTS WITH SEVERE LEUKEMIA.
  BULL. INST. INTERNAT. STATIST., 39(3) 173-179.


GEHLKE, C. E. + BIEHL, KATHERINE 1934.  (4.1)                                    1895
    CERTAIN EFFECTS OF GROUPING UPON THE SIZE OF THE CORRELATION
    COEFFICIENT IN CENSUS TRACT MATERIAL.
  J. AMER. STATIST. ASSOC., 29(SP) 169-170.

GEIDAROV, T. G. 1966.  (17.1)                                                            1896
EXPANSION OF N-DIMENSIONAL DISTRIBUTION LAWS IN A SERIES OF EDGEWORTH TYPE.
     (IN RUSSIAN, WITH SUMMARY IN AZERBAIJAN)
IZV. AKAD. NAUK AZERBAIDZAN. SSR SER. FIZ.-TEH. MAT. NAUK, 4, 54-61.   (MR35-7365)

GEIRINGER, HILDA 1933.  (4.8)                                                            1897
BEMERKUNGEN ZUR KORRELATIONSTHEORIE.
PROC. INTERNAT. CONG. MATH., 1932(2) 229-230.

GEIRINGER, HILDA 1933.  (4.1/B)                                                          1898
KORRELATIONSMESSUNG AUF GRUND DER SUMMENFUNKTION.
Z. ANGEW. MATH. MECH., 13, 121-124.

GEIRINGER, HILDA 1934.  (4.8/18.1)                                                       1899
KORRELATIONSMODELLE.
Z. ANGEW. MATH. MECH., 14, 19-35.

GEIRINGER, HILDA 1935.  (17.5/18.1/B)                                                    1900
UNE NOUVELLE METHODE DE STATISTIQUE THEORIQUE (PROBLEMES A DEUX DIMENSIONS).
ACAD. ROY. BELG. BULL. CL. SCI. SER. 5, 21, 157-165.

GEIRINGER, HILDA 1935.  (18.1/18.3/B)                                                    1901
UNE NOUVELLE METHODE DE STATISTIQUE THEORIQUE (PROBLEME A DEUX DIMENSIONS)
     (DEUXIEME COMMUNICATION).
ACAD. ROY. BELG. BULL. CL. SCI. SER. 5, 21, 307-324.

GEIRINGER, HILDA 1936.  (2.1/8.8)                                                        1902
ZUR VERWENDUNG DER MEHRDIMENSIONALEN NORMALVERTEILUNG IN DER
     STATISTIK.(1. MITTEILUNG.)
MONATSHEFTE MATH. PHYS., 43, 425-434.

GEIRINGER, HILDA 1936.  (2.1/2.3/4.3)                                                    1903
ZUR VERWENDUNG DER MEHRDIMENSIONALEN NORMALVERTEILUNG IN DER
     STATISTIK. (2. MITTEILUNG.)
MONATSHEFTE MATH. PHYS., 44, 97-112.

GEISSER, SEYMOUR 1957.  (2.6/14.5/16.5)                                                  1904
THE DISTRIBUTION OF THE RATIOS OF CERTAIN QUADRATIC FORMS IN TIME SERIES.
ANN. MATH. STATIST., 28, 724-730.   (MR19,P.897)

GEISSER, SEYMOUR 1963.  (8.3/8.4)                                                        1905
MULTIVARIATE ANALYSIS OF VARIANCE FOR A SPECIAL COVARIANCE CASE.
     (AMENDED BY NO. 1906)
J. AMER. STATIST. ASSOC., 58, 660-669.   (MR27-6343)

GEISSER, SEYMOUR 1964.  (8.3/8.4)                                                        1906
CORRIGENDA:  MULTIVARIATE ANALYSIS OF VARIANCE FOR A SPECIAL COVARIANCE CASE.
     (AMENDMENT OF NO. 1905)
J. AMER. STATIST. ASSOC., 59, 1296.   (MR27-6343)

GEISSER, SEYMOUR 1964.  (6.3)                                                            1907
POSTERIOR ODDS FOR MULTIVARIATE NORMAL CLASSIFICATIONS.
J. ROY. STATIST. SOC. SER. B, 26, 69-76.   (MR30-4340)

GEISSER, SEYMOUR 1964.  (3.1/4.3)                                                        1908
ESTIMATION IN THE UNIFORM COVARIANCE CASE.
J. ROY. STATIST. SOC. SER. B, 26, 477-483.   (MR31-4114)

GEISSER, SEYMOUR 1965.  (3.1)                                                            1909
BAYESIAN ESTIMATES IN MULTIVARIATE ANALYSIS.
ANN. MATH. STATIST., 36, 150-159.   (MR30-4341)

GEISSER, SEYMOUR 1965.  (3.2)                                                            1910
A BAYES APPROACH FOR COMBINING CORRELATED ESTIMATES.
J. AMER. STATIST. ASSOC., 60, 602-607.   (MR32-532)

GEISSER, SEYMOUR 1966.  (6.1/6.3)                                                        1911
PREDICTIVE DISCRIMINATION.
MULTIVARIATE ANAL. PROC. INTERNAT. SYMP. DAYTON (KRISHNAIAH), 149-163.   (MR35-2417)

GEISSER, SEYMOUR + CORNFIELD, JEROME 1963.  (14.2)                                       1912
POSTERIOR DISTRIBUTIONS FOR MULTIVARIATE NORMAL PARAMETERS.
J. ROY. STATIST. SOC. SER. B, 25, 368-376.   (MR30-1585)

GEISSER, SEYMOUR + GREENHOUSE, SAMUEL W. 1958.   (8.4)                    1913
    AN EXTENSION OF BOX'S RESULTS ON THE USE OF THE F DISTRIBUTION IN
    MULTIVARIATE ANALYSIS.
    ANN. MATH. STATIST., 29, 885-891.   (MR20-2815)

GEISSER, SEYMOUR + MANTEL, NATHAN 1962.   (7.4/8.9/18.1)                  1914
    PAIRWISE INDEPENDENCE OF JOINTLY DEPENDENT VARIABLES.
    ANN. MATH. STATIST., 33, 290-291.   (MR25-644)

GEISSER, SEYMOUR    SEE ALSO
    2114. GREENHOUSE, SAMUEL W. + GEISSER, SEYMOUR (1959)

GEL'FANDBEIN, JA. A. 1964.   (16.4)                                       1915
    DETERMINATION OF THE DYNAMICAL CHARACTERISTICS OF MULTI-VARIATE
    NON-STATIONARY SYSTEMS BY A MATRIX METHOD DURING NORMAL OPERATION.
        (IN RUSSIAN, WITH SUMMARY IN LATVIAN AND ENGLISH)
    LATVIJAS PSR ZINATNU AKAD. VESTIS FIZ. TEH. ZINATNU SER. 6, 6, 101-106.   (MR30-666)

GELFOND, A. O. + HINCIN, A. JA. 1933.   (20.2)                            1916
    GRAMSCHE DETERMINANTEN FUR STAZIONARE REIHEN.
        (IN RUSSIAN, WITH SUMMARY IN GERMAN)
    UCEN. ZAP. MOSK. GOS. UNIV., 1, 3-5.

GENGERELLI, J. A. 1948.   (8.1)                                           1917
    A SIMPLIFIED METHOD FOR APPROXIMATING MULTIPLE REGRESSION COEFFICIENTS.
    PSYCHOMETRIKA, 13, 135-146.   (MR10,P.134)

GEORGE, ALEYAMMA 1957.   (7.3)                                            1918
    THE DISTRIBUTION OF THE NON-CENTRAL RECTANGULAR CO-ORDINATES.
    BULL. RES. INST. UNIV. TRAVANCORE SER. B, 4, 1-32.

GEORGESCU-ROEGEN, NICULAE 1947.   (16.5)                                  1919
    FURTHER CONTRIBUTIONS TO THE SCATTER ANALYSIS.
    BULL. INST. INTERNAT. STATIST., 31(5) 39-43.

GEORGIOS, PAULOS  SEE  GEORGIOU, PAUL

GEORGIOU, PAUL 1966.   (16.4/17.9)                                        1920
    VEKTORWERTIGE ZUFALLSVARIABLEN UND IHRE WAHRSCHEINLICHKEITSVERTEILUNGEN.
    ARCH. MATH., 17, 70-77.   (MR34-5115)

GEPPERT, MARIA PIA 1939.   (18.1/B)                                       1921
    SU UNA CLASSE DI DISTRIBUZIONI IN DUE VARIABILI CASUALI.
    GIORN. IST. ITAL. ATTUARI, 10, 225-228.   (MR1,P.246)

GEPPERT, MARIA PIA 1940.   (18.1/B)                                       1922
    UBER EINE KLASSE VON ZWEIDIMENSIONALEN VERTEILUNGEN.
    Z. ANGEW. MATH. MECH., 20, 45-49.   (MR1,P.340)

GERSMAN, S. G. 1953.   E(16.4/T)                                          1923
    AN INTERFERENCE METHOD FOR THE MEASUREMENT OF THE CORRELATION
    COEFFICIENT OF STATIONARY NOISE.
        (IN RUSSIAN)
    DOKL. AKAD. NAUK SSSR NOV. SER., 92, 33-35.

GERSMAN, S. G. + FEINBERG, E. L. 1955.   (4.1/4.6)                        1924
    ON THE MEASUREMENT OF THE CORRELATION COEFFICIENT.
        (TRANSLATED AS NO. 1925)
    AKUST. ZUR., 1, 327-338.

GERSMAN, S. G. + FEINBERG, E. L. 1955.   (4.1/6.6)                        1925
    ON THE MEASUREMENT OF THE CORRELATION COEFFICIENT.
        (TRANSLATION OF NO. 1924)
    SOVIET PHYS. ACOUST., 1, 340-352.

GERSTENHABER, MURRAY 1951.   (20.5)                                       1926
    THEORY OF CONVEX POLYHEDRAL CONES.
    COWLES COMMISS. MONOG., 13, 298-316.   (MR13,P.60)

GHISELLI, EDWIN E. 1942.   (15.4)                                        1927
    ESTIMATING THE MINIMAL RELIABILITY OF A TOTAL TEST FROM THE INTER-CORRELATIONS
    AMONG, AND THE STANDARD DEVIATIONS OF, THE COMPONENT PARTS.
    J. APPL. PSYCHOL., 26, 332-337.

GHISELLI, EDWIN E. + KUZNETS, GEORGE 1937.  (15.4)                    1928
    SHORT-CUT METHODS FOR CALCULATING RAW AND CORRECTED CORRELATIONS
    BETWEEN A COMPOSITE VARIABLE AND ITS COMPONENTS.
    J. EDUC. PSYCHOL., 28, 237-240.

GHOSH, BHASKAR K. 1966.  (4.2/AB)                    1929
    ASYMPTOTIC EXPANSIONS FOR THE MOMENTS OF THE DISTRIBUTION OF
    CORRELATION COEFFICIENT.
    BIOMETRIKA, 53, 258-262.  (MR33-3388)

GHOSH, BIRENDRANATH 1943.  (18.3/B)                    1930
    ON THE DISTRIBUTION OF RANDOM DISTANCES IN A RECTANGLE.
    SCI. AND CULT., 8, 388.  (MR5,P.40)

GHOSH, BIRENDRANATH 1943.  (18.3/B)                    1931
    ON RANDOM DISTANCES BETWEEN TWO RECTANGLES.
    SCI. AND CULT., 8, 464.  (MR5,P.40)

GHOSH, BIRENDRANATH 1947.  (17.9)                    1932
    DOUBLE SAMPLING WITH MANY AUXILIARY VARIATES.
    CALCUTTA STATIST. ASSOC. BULL., 1, 91-93.

GHOSH, BIRENDRANATH 1951.  (16.4)                    1933
    ON MULTI-DIMENSIONAL STOCHASTIC FIELDS.
    CALCUTTA STATIST. ASSOC. BULL., 4, 29-32.

GHOSH, BIRENDRANATH 1951.  (16.4)                    1934
    SOME EXPONENTAIL FORMS FOR TOPOGRAPHIC CORRELATION.
    SANKHYA, 11, 29-36.  (MR13,P.141)

GHOSH, J. K. 1963.  (18.5)                    1935
    A GAME THEORY APPROACH TO THE PROBLEM OF OPTIMUM ALLOCATION IN
    STRATIFIED SAMPLING WITH MULTIPLE CHARACTERS.
    CALCUTTA STATIST. ASSOC. BULL., 12, 4-12.

GHOSH, MAHINDRA NATH 1948.  (16.4/17.1/17.8/B)                    1936
    A TEST FOR FIELD UNIFORMITY BASED ON THE SPACE CORRELATION METHOD.
    SANKHYA, 9, 39-46.  (MR10,P.389)

GHOSH, MAHINDRA NATH 1955.  (8.3)                    1937
    SIMULTANEOUS TESTS OF LINEAR HYPOTHESES.
    BIOMETRIKA, 42, 441-449.  (MR17,P.640)

GHOSH, MAHINDRA NATH 1961.  (8.8)                    1938
    LINEAR ESTIMATION OF PARAMETERS IN THE CASE OF UNEQUAL VARIANCES.
    CALCUTTA STATIST. ASSOC. BULL., 10, 117-130.  (MR23-A3608)

GHOSH, MAHINDRA NATH 1963.  (8.3/8.4/13.2/C)                    1939
    HOTELLING'S GENERALIZED $T^2$ IN THE MULTIVARIATE ANALYSIS OF VARIANCE.
    J. ROY. STATIST. SOC. SER. B, 25, 358-367.  (MR30-3534)

GHOSH, MAHINDRA NATH 1964.  (8.3)                    1940
    ON THE ADMISSIBILITY OF SOME TESTS OF MANOVA.
    ANN. MATH. STATIST., 35, 789-794.  (MR29-701)

GHOSH, MAHINDRA NATH + BEHARI, VINOD 1965.  (8.5)                    1941
    ANALYSIS OF RANDOMISED BLOCK EXPERIMENTS WHEN TREATMENTS HAVE DIFFERENT ERRORS.
    CALCUTTA STATIST. ASSOC. BULL., 14, 93-105.

GHOSH, MAHINDRA NATH    SEE ALSO
    5631. VERMA, M. C. + GHOSH, MAHINDRA NATH (1963)

GHOSH, S. P. 1963.  (18.5)                    1942
    OPTIMUM STRATIFICATION WITH TWO CHARACTERS.
    ANN. MATH. STATIST., 34, 866-872.  (MR27-4330)

GHURYE, S. G. 1954.  (16.1)                    1943
    RANDOM FUNCTIONS SATISFYING CERTAIN LINEAR RELATIONS.
    ANN. MATH. STATIST., 25, 543-554.  (MR16,P.150)

GHURYE, S. G. 1955.  (16.1)                    1944
    RANDOM FUNCTIONS SATISFYING CERTAIN LINEAR RELATIONS.  II.
    ANN. MATH. STATIST., 26, 105-111.  (MR16,P.723)

GHURYE, S. G. 1955.  (16.1)                                                                1945
    NOTE ON ASYMPTOTIC ESTIMATION OF PARAMETERS OF AN AUTOREGRESSIVE PROCESS.
    GANITA, 6, 1-7.  (MR18,P.771)

GHURYE, S. G. + OLKIN, INGRAM 1962.  (2.2/7.4)                                              1946
    A CHARACTERIZATION OF THE MULTIVARIATE NORMAL DISTRIBUTION.
    ANN. MATH. STATIST., 33, 533-541.  (MR25-657)

GIACCARDI, FERNANDO 1950.  (4.8)                                                            1947
    COEFFICIENTI DI CORRELAZIONE QUADRATICA.
    RIV. ITAL. ECON. DEMOG. STATIST., 3(1) 79-86.

GIBSON, GORDON D. 1947.  E(4.8)                                                             1948
    ON GLADWIN'S METHODS OF CORRELATION IN TREE-RING ANALYSIS.
    AMER. ANTHROP., 49, 337-340.

GIBSON, W. A. 1952.  (15.3)                                                                 1949
    ORTHOGONAL AND OBLIQUE SIMPLE STRUCTURES.
    PSYCHOMETRIKA, 17, 317-323.

GIBSON, W. A. 1955.  (15.5)                                                                 1950
    AN EXTENSION OF ANDERSON'S SOLUTION FOR THE LATENT STRUCTURE EQUATIONS.
    PSYCHOMETRIKA, 20, 69-73.  (MR17,P.756)

GIBSON, W. A. 1956.  (15.5)                                                                 1951
    PROPORTIONAL PROFILES AND LATENT STRUCTURE.
    PSYCHOMETRIKA, 21, 135-144.

GIBSON, W. A. 1959.  (15.1/15.5/E)                                                          1952
    THREE MULTIVARIATE MODELS:  FACTOR ANALYSIS, LATENT STRUCTURE
    ANALYSIS, AND LATENT PROFILE ANALYSIS.
        (REPRINTED AS NO. 1956)
    PSYCHOMETRIKA, 24, 229-252.  (MR21-7809)

GIBSON, W. A. 1960.  (15.2)                                                                 1953
    REMARKS ON TUCKER'S INTER-BATTERY METHOD OF FACTOR ANALYSIS.
    PSYCHOMETRIKA, 25, 19-25.  (MR22-2501)

GIBSON, W. A. 1960.  (15.1)                                                                 1954
    NONLINEAR FACTORS IN TWO DIMENSIONS.
    PSYCHOMETRIKA, 25, 381-392.  (MR24-B424)

GIBSON, W. A. 1963.  (15.1/E)                                                               1955
    FACTORING THE CIRCUMPLEX.
    PSYCHOMETRIKA, 28, 87-92.

GIBSON, W. A. 1966.  (15.1/15.5/E)                                                          1956
    THREE MULTIVARIATE MODELS:  FACTOR ANALYSIS, LATENT STRUCTURE
    ANALYSIS, AND LATENT PROFILE ANALYSIS.
        (REPRINT OF NO. 1952)
    READINGS MATH. SOCIAL SCI. (LAZARSFELD + HENRY), 54-77.  (MR21-7809)

GIBSON, WENDY M. + JOWETT, GEOFFREY H. 1957.  (8.1/8.2)                                     1957
    "THREE-GROUP" REGRESSION ANALYSIS.  1. SIMPLE REGRESSION ANALYSIS.
    2. MULTIPLE REGRESSION ANALYSIS.
    APPL. STATIST., 6, 114-122, 189-197.

GIBSON, WINIFRED + PEARSON, KARL 1908.  (4.1/17.9/E)                                        1958
    FURTHER CONSIDERATIONS ON THE CORRELATION OF STELLAR CHARACTERS.
    MONTHLY NOTICES ROY. ASTRONOM. SOC., 69, 128-151.

GILBERT, E. N. 1965.  (18.6)                                                                1959
    THE PROBABILITY OF COVERING A SPHERE WITH N CIRCULAR CAPS.
    BIOMETRIKA, 52, 323-330.  (MR34-6821)

GILES, EUGENE 1960.  E(5.2)                                                                 1960
    MULTIVARIATE ANALYSIS OF PLEISTOCENE AND RECENT COYOTES (CANIS LATRANS) FROM
    CALIFORNIA.
    UNIV. CALIFORNIA PUBL. GEOL. SCI., 36, 369-390.

GILI, ADOLFO 1966.  (4.1/4.6/17.3/E)                                                        1961
    SULLA CORRELAZIONE LINEARE MULTIPLA.
    STATISTICA, BOLOGNA, 26, 73-102.

GILI, ADOLFO 1966.  (17.3)                                                              **1962**
     PRIMI APPUNTI PER UNA IMPOSTAZIONE VETTORIALE DELLA VARIABILITA DELLE
     VARIABLI MULTIPLE.
     STATISTICA, BOLOGNA, 26, 746-749.

GILI, ADOLFO 1966.  (4.6/4.8/17.3)                                                      **1963**
     AU SUJET DE LA REPRESENTATION VECTORIELLE DES VARIABLES STATISTIQUES.
     STATISTICA, BOLOGNA, 26, 937-942.

GILLILAND, DENNIS C. 1962.  (14.5/B)                                                    **1964**
     INTEGRAL OF THE BIVARIATE NORMAL DISTRIBUTION OVER AN OFFSET CIRCLE.
     J. AMER. STATIST. ASSOC., 57, 758-768.  (MR25-5564)

GILLILAND, DENNIS C. 1964.  (14.5)                                                      **1965**
     A NOTE ON THE MAXIMIZATION OF A NON-CENTRAL CHI-SQUARE PROBABILITY.
     ANN. MATH. STATIST., 35, 441-442.  (MR28-1677)

GILLILAND, DENNIS C. 1966.  (14.4)                                                      **1966**
     SOME BOMBING PROBLEMS.
     AMER. MATH. MONTHLY, 73, 713-716.  (MR34-5126)

GILLIS, J. 1955.  (16.4)                                                                **1967**
     CORRELATED RANDOM WALK.
     PROC. CAMBRIDGE PHILOS. SOC., 51, 639-651.  (MR17,P.275)

GILLMAN, LEONARD + GOODE, HARRY H. 1946.  (4.4)                                         **1968**
     AN ESTIMATE OF THE CORRELATION COEFFICIENT OF A BIVARIATE NORMAL
     POPULATION WHEN X IS TRUNCATED AND Y IS DICHOTOMIZED.
     HARVARD EDUC. REV., 16, 52-55.

GILLOIS, M. 1965.  (4.8)                                                                **1969**
     RELATION D'IDENTITE EN GENETIQUE.  I. POSTULATS ET AXIOMES MENDELIENS.
     ANN. INST. H. POINCARE SECT. B, 2, 1-38.  (MR32-9073A)

GILLOIS, M. 1965.  (4.8)                                                                **1970**
     RELATION D'IDENTITE EN GENETIQUE.  II. CORRELATION GENETIQUE DANS LE
     CAS DE DOMINANCE.
     ANN. INST. H. POINCARE SECT. B, 2, 39-94.  (MR32-9073B)

GINI, CORRADO 1910.  (4.8)                                                              **1971**
     INDICI DI CONCENTRAZIONE E DI DIPENDENZA.
     ATTI SOC. ITAL. PROG. SCI., 3, 453-469.

GINI, CORRADO 1914.  (4.8)                                                              **1972**
     SULLA MISURA DELLA CONCENTRAZIONE E DELLA VARIABILITA DEI CARATTERI.
     ATTI REALE IST. VENETO SCI. LETT. ARTI, 73(2) 1203-1248.

GINI, CORRADO 1914.  (4.8)                                                              **1973**
     DI UNA MISURA DELLA DISSOMIGLIANZA TRA DUE GRUPPI DI QUANTITA E DELLE
     SUE APPLICAZIONI ALLO STUDIO DELLE RELAZIONI STATISTICHE.
     ATTI REALE IST. VENETO SCI. LETT. ARTI, 74(2) 185-213.

GINI, CORRADO 1914.  (4.8)                                                              **1974**
     INDICI DI OMOFILIA E DI RASSOMIGLIANZA E LORO RELAZIONI COL
     COEFFICIENTE DI CORRELAZIONE E CON GLI INDICI DI ATTRAZIONE.
     ATTI REALE IST. VENETO SCI. LETT. ARTI, 74(2) 583-610.

GINI, CORRADO 1915.  (4.8)                                                              **1975**
     NUOVI CONTRIBUTI ALLA TEORIA DELLE RELAZIONI STATISTICHE.
     ATTI REALE IST. VENETO SCI. LETT. ARTI, 74(2) 1903-1942.

GINI, CORRADO 1915.  (4.8/B)                                                            **1976**
     SUL CRITERIO DI CONCORDANZA TRA DUE CARATTERI.
     ATTI REALE IST. VENETO SCI. LETT. ARTI, 75(2) 309-331.

GINI, CORRADO 1916.  (4.8)                                                              **1977**
     INDICI DI CONCORDANZA.
     ATTI REALE IST. VENETO SCI. LETT. ARTI, 75(2) 1419-1461.

GINI, CORRADO 1917.  (4.8)                                                              **1978**
     DELLE RELAZIONI TRA LE INTENSITA COGRADUATE DI DUE CARATTERI.
     ATTI REALE IST. VENETO SCI. LETT. ARTI, 76(2) 1147-1185.

GINI, CORRADO 1936.  (4.9)                                                                      1979
    SUR LE COEFFICIENT DE CORRELATION COMME CARACTERISTIQUE DU DEGRE DE
    LA CONCORDANCE.
    REV. INST. INTERNAT. STATIST., 4, 355-366.

GINI, CORRADO 1937.  (4.9)                                                                      1980
    METHODS OF ELIMINATING THE INFLUENCE OF SEVERAL GROUPS OF FACTORS.
    ECONOMETRICA, 5, 56-73.

GINI, CORRADO 1941.  (4.8)                                                                      1981
    DEGLI INDICI SINTETICI DI CORRELAZIONE E DELLE LORO RELAZIONI CON L'INDICE
    INTERNO DI CORRELAZIONE (INTRACLASS CORRELATION COEFFICIENT) E CON GLI INDICI
    DI CORRELAZIONE TRA SERIE DI GRUPPI.
        (WITH SUMMARY IN GERMAN, FRENCH AND ENGLISH)
    METRON, 14(2) 241-261.  (MR8,P.474)

GINI, CORRADO 1942.  (4.9)                                                                      1982
    INDICI SINTETICI DI CORRELAZIONE.
    ATTI RIUN. SCI. SOC. ITAL. STATIST., 3, 60-61.

GINI, CORRADO 1952.  (7.2/17.3)                                                                 1983
    ESTENSIONE DELLA TEORIA DELLA DISPERSIONE E DELLA CONNESSIONE A SERIE
    DI GRANDEZZE ASSOLUTE.
    GIORN. IST. ITAL. ATTUARI, 15, 4-24.  (MR16,P.54)

GINI, CORRADO 1953.  (4.8/E)                                                                    1984
    THE MEASUREMENT OF THE DIFFERENCES BETWEEN TWO QUANTITY GROUPS AND IN
    PARTICULAR BETWEEN THE CHARACTERISTICS OF TWO POPULATIONS.
    ACTA GENET. STATIST. MED., 4, 175-191.

GINI, CORRADO 1957.  (1.1)                                                                      1985
    POSTILLA SUGLI INDICI DI CONNESSIONE.
    METRON, 18(3) 80-86.

GINI, CORRADO 1965.  (1.1)                                                                      1986
    ON THE CHARACTERISTICS OF ITALIAN STATISTICS.
    J. ROY. STATIST. SOC. SER. A, 128, 89-109.

GINI, CORRADO 1965.  (6.5/17.3/E)                                                               1987
    LA DISSOMIGLIANZA.
    METRON, 24, 85-215.

GINI, CORRADO + LIVADA, GREGORIO 1945.  (4.8/T)                                                 1988
    TRANSVARIAZIONE A PIU DIMENSIONI.
    ATTI RIUN. SCI. SOC. ITAL. STATIST., 6, 25-62.

GINZBURG, G. M. 1955.  (17.1)                                                                   1989
    ON LIMIT DISTRIBUTIONS DETERMINED BY STOCHASTIC EQUATIONS WITH AN
    INFINITE SET OF ZEROS OF THE DISPERSION FUNCTION.
        (IN RUSSIAN)
    DOKL. AKAD. NAUK SSSR, 102, 441-444.  (MR17,P.166)

GIRAULT, MAURICE 1948.  (15.1)                                                                  1990
    SUR LA NOTION DE FACTEUR COMMUN EN ANALYSE FACTORIELLE GENERALE.
    C.R. ACAD. SCI. PARIS, 227, 499-500.  (MR10,P.136)

GIRAULT, MAURICE 1954.  (8.1/18.1)                                                              1991
    DROITES DE REGRESSION CONFONDUES.
    C.R. ACAD. SCI. PARIS, 239, 1266-1267.

GIRI, N. C. 1962.  (5.1)                                                                        1992
    ON A MULTIVARIATE TESTING PROBLEM.
    CALCUTTA STATIST. ASSOC. BULL., 11, 55-60.

GIRI, N. C. 1964.  (6.2)                                                                        1993
    ON THE LIKELIHOOD RATIO TEST OF A NORMAL MULTIVARIATE TESTING PROBLEM.
        (AMENDED BY NO. 1994)
    ANN. MATH. STATIST., 35, 181-189.  (MR31-6318)

GIRI, N. C. 1964.   (6.2)                                                          1994
    CORRECTION TO: ON THE LIKELIHOOD RATIO TEST OF A NORMAL MULTIVARIATE
    TESTING PROBLEM.
        (AMENDMENT OF NO. 1993)
  ANN. MATH. STATIST., 35, 1388.   (MR31-6318)


GIRI, N. C. 1965.   (4.7/5.2)                                                      1995
    ON THE COMPLEX ANALOGUES OF $T^2$-AND $R^2$-TESTS.
  ANN. MATH. STATIST., 36, 664-670.   (MR31-6312)


GIRI, N. C. 1965.   (6.2/6.3)                                                      1996
    ON THE LIKELIHOOD RATIO TEST OF A NORMAL MULTIVARIATE TESTING PROBLEM. II.
  ANN. MATH. STATIST., 36, 1061-1065.   (MR33-815)


GIRI, N. C. + KIEFER, J. 1964.   (5.1/8.3/19.2)                                    1997
    LOCAL AND ASYMPTOTIC MINIMAX PROPERTIES OF MULTIVARIATE TESTS.
  ANN. MATH. STATIST., 35, 21-35.   (MR28-2605)


GIRI, N. C. + KIEFER, J. 1964.   (4.7)                                             1998
    MINIMAX CHARACTER OF THE $R^2$-TEST IN THE SIMPLEST CASE.
  ANN. MATH. STATIST., 35, 1475-1490.


GIRI, N. C.+ KIEFER, J. + STEIN, CHARLES M. 1963.   (5.2/8.4/13.2/A)               1999
    MINIMAX CHARACTER OF HOTELLING'S $T^2$-TEST IN THE SIMPLEST CASE.
  ANN. MATH. STATIST., 34, 1524-1535.   (MR27-6331)


GIRSANOV, I. V. 1958.   (16.4)                                                     2000
    SPECTRA OF DYNAMICAL SYSTEMS GENERATED BY STATIONARY GAUSSIAN PROCESSES.
        (IN RUSSIAN)
  DOKL. AKAD. NAUK SSSR, 119, 851-853.   (MR22-5075)


GIRSHICK, M. A. 1936.   (11.1)                                                     2001
    PRINCIPAL COMPONENTS.
  J. AMER. STATIST. ASSOC., 31, 519-528.


GIRSHICK, M. A. 1939.   (13.1)                                                     2002
    ON THE SAMPLING THEORY OF ROOTS OF DETERMINANTAL EQUATIONS.
  ANN. MATH. STATIST., 10, 203-224.   (MR1,P.22)


GIRSHICK, M. A. 1941.   (10.2)                                                     2003
    THE DISTRIBUTION OF THE ELLIPTICITY STATISTIC $L_e$ WHEN THE HYPOTHESIS IS FALSE.
        (REPRINTED AS NO. 2004)
  STATISTICA, BOLOGNA, 2, 157-162.


GIRSHICK, M. A. 1941.   (10.2)                                                     2004
    THE DISTRIBUTION OF THE ELLIPTICITY STATISTIC $L_e$ WHEN THE HYPOTHESIS IS FALSE.
        (REPRINT OF NO. 2003)
  TERREST. MAGNET. ATMOSPH. ELEC., 46, 455-457.


GIRSHICK, M. A. 1942.   (17.9/20.5)                                                2005
    NOTE ON THE DISTRIBUTION OF ROOTS OF A POLYNOMIAL WITH RANDOM COMPLEX
    COEFFICIENTS.
        (AMENDED BY NO. 2006)
  ANN. MATH. STATIST., 13, 235-238.   (MR4,P.21)


GIRSHICK, M. A. 1942.   (17.9/20.5)                                                2006
    CORRECTION TO: NOTE ON THE DISTRIBUTION OF ROOTS OF A POLYNOMIAL WITH
    RANDOM COMPLEX COEFFICIENTS.
        (AMENDMENT OF NO. 2005)
  ANN. MATH. STATIST., 13, 447.   (MR4,P.164)


GIRSHICK, M. A. + HAAVELMO, TRYGVE 1947.   (17.7/E)                                2007
    STATISTICAL ANALYSIS OF THE DEMAND FOR FOOD:  EXAMPLES OF
    SIMULTANEOUS ESTIMATION OF STRUCTURAL EQUATIONS.
        (REPRINTED AS NO. 2008)
  ECONOMETRICA, 15, 79-110.


GIRSHICK, M. A. + HAAVELMO, TRYGVE 1953.   (17.7/E)                                2008
    STATISTICAL ANALYSIS OF THE DEMAND FOR FOOD:  EXAMPLES OF
    SIMULTANEOUS ESTIMATION OF STRUCTURAL EQUATIONS.
        (REPRINT OF NO. 2007)
  COWLES COMMISS. MONOG., 14, 92-111.

GIRSHICK, M. A.; KARLIN, SAMUEL + ROYDEN, H.L. 1957.   (19.1)                    2009
    MULTISTAGE STATISTICAL DECISION PROCEDURES.
  ANN. MATH. STATIST., 28, 111-125.   (MR18,P.832)

GIRSHICK, M. A.    SEE ALSO
    168. ANDERSON, T. W. + GIRSHICK, M. A. (1944)

    579. BLACKWELL, DAVID H. + GIRSHICK, M. A. (1946)

GJACJAUSKAS, E. 1966.   (14.4)                                                    2010
    DISTRIBUTION FUNCTION OF THE DISTANCE BETWEEN TWO POINTS IN AN OVAL.
        (IN RUSSIAN, WITH SUMMARY IN LITHUANIAN)
  LITOVSK. MAT. SB., 6, 245-248.   (MR35-1051)

GLADYSEV, E. G. 1958.   (16.4)                                                    2011
    ON MULTI-DIMENSIONAL STATIONARY RANDOM PROCESSES.
        (IN RUSSIAN, WITH SUMMARY IN ENGLISH, TRANSLATED AS NO. 2012)
  TEOR. VEROJATNOST. I PRIMENEN., 3, 458-462.   (MR21-927)

GLADYSEV, E. G. 1958.   (16.4)                                                    2012
    ON MULTI-DIMENSIONAL STATIONARY RANDOM PROCESSES.
        (TRANSLATION OF NO. 2011)
  THEORY PROB. APPL., 3, 425-428.   (MR21-927)

GLADYSEV, E. G. 1960.   (16.5)                                                    2013
    AN INTERPOLATIONAL PROBLEM FOR MULTIDIMENSIONAL STATIONARY SEQUENCES.
        (IN RUSSIAN)
  PROC. SIXTH ALL-UNION CONF. THEORY PROB. MATH. STATIST. (VILNIUS), 203-208.   (MR32-1761)

GLAHN, HARRY R. 1965.   (6.4/11.1/E)                                              2014
    OBJECTIVE WEATHER FORECASTING BY STATISTICAL METHODS.
  STATISTICIAN, 15, 111-142.

GLAISHER, J. W. L. 1872.   (8.1)                                                  2015
    ON THE LAW OF FACILITY OF ERRORS OF OBSERVATIONS AND ON THE METHOD OF
    LEAST SQUARES.
        (AMENDED BY NO. 2016)
  MEM. ROY. ASTRONOM. SOC., 39(2) 75-124.

GLAISHER, J. W. L. 1872.   (8.1)                                                  2016
    ERRATA TO:  ON THE LAW OF FACILITY OF ERRORS OF OBSERVATIONS AND ON THE METHOD
    OF LEAST SQUARES.
        (AMENDMENT OF NO. 2015)
  MONTHLY NOTICES ROY. ASTRONOM. SOC., 32, 241-242.

GLAISHER, J. W. L. 1872.   (8.8/17.1)                                             2017
    REMARKS ON CERTAIN PORTIONS OF LAPLACE'S PROOF OF THE METHOD OF LEAST SQUARES.
  PHILOS. MAG. SER. 4, 43, 194-201.

GLAISHER, J. W. L. 1874.   (8.8)                                                  2018
    ON THE SOLUTION OF THE EQUATIONS IN THE METHOD OF LEAST SQUARES.
  MONTHLY NOTICES ROY. ASTRONOM. SOC., 34, 311-334.

GLASGOW, MARK O.    SEE
    2119. GREENWOOD, ROBERT E., JR. + GLASGOW, MARK O. (1950)

GLASS, GENE V. 1966.   (15.1/E)                                                   2019
    ALPHA FACTOR ANALYSIS OF INFALLIBLE VARIABLES.
  PSYCHOMETRIKA, 31, 545-561.

GLASSER, G. J. 1962.   (17.8)                                                     2020
    A DISTRIBUTION-FREE TEST OF INDEPENDENCE WITH A SAMPLE OF PAIRED OBSERVATIONS.
  J. AMER. STATIST. ASSOC., 57, 116-133.   (MR25-666)

GLAUERT, H. 1925.   (17.7)                                                        2021
    THE DETERMINATION OF THE BEST LINEAR RELATIONSHIP CONNECTING ANY
    NUMBER OF VARIABLES.
  PHILOS. MAG. SER. 6, 50, 205-207.

GLESER, GOLDINE C.; CRONBACH, LEE J. + RAJARATNAM, NAGESWARI 1965.   (8.6/E)      2022
    GENERALIZABILITY OF SCORES INFLUENCED BY MULTIPLE SOURCES OF VARIANCE.
  PSYCHOMETRIKA, 30, 395-418.

GLESER, GOLDINE C.    SEE ALSO
  1114. CRONBACH, LEE J. + GLESER, GOLDINE C. (1958)

  3461. LOEVINGER, JANE; GLESER, GOLDINE C. + DU BOIS, PHILIP H. (1953)

GLESER, LEON J. 1965.  (8.8/C)                                                2023
    ON THE ASYMPTOTIC THEORY OF FIXED-SIZE SEQUENTIAL CONFIDENCE BOUNDS
    FOR LINEAR REGRESSION PARAMETERS.
    ANN. MATH. STATIST., 36, 463-467.

GLESER, LEON J. 1966.  (10.2)                                                 2024
    A NOTE ON THE SPHERICITY TEST.
        (AMENDED BY NO. 2026)
    ANN. MATH. STATIST., 37, 464-467.  (MR32-4781)

GLESER, LEON J. 1966.  (5.1/5.2/8.3/A)                                        2025
    THE COMPARISON OF MULTIVARIATE TESTS OF HYPOTHESIS BY MEANS OF BAHADUR
      EFFICIENCY.
    SANKHYA SER. A, 28, 157-174.  (MR35-2419)

GLESER, LEON J. 1968.  (10.2)                                                 2026
    CORRECTION TO: A NOTE ON THE SPHERICITY TEST.
        (AMENDMENT OF NO. 2024)
    ANN. MATH. STATIST., 39, 684-685.  (MR32-4781)

GLESER, LEON J. + OLKIN, INGRAM 1966.  (8.1/8.3/8.4/AC)                       2027
    A K-SAMPLE REGRESSION MODEL WITH COVARIANCE.
    MULTIVARIATE ANAL. PROC. INTERNAT. SYMP. DAYTON (KRISHNAIAH), 59-72.  (MR35-2418)

GLOCK, WALDO S. 1942.  (4.8/16.5/U)                                           2028
    A RAPID METHOD OF CORRELATION FOR CONTINUOUS TIME SERIES.
    AMER. J. SCI., 240, 437-442.  (MR4,P.27)

GNANADESIKAN, R. 1959.  (10.1)                                                2029
    EQUALITY OF MORE THAN TWO VARIANCES AND OF MORE THAN TWO DISPERSION
    MATRICES AGAINST CERTAIN ALTERNATIVES.
        (AMENDED BY NO. 2030)
    ANN. MATH. STATIST., 30, 177-184.  (MR21-2326)

GNANADESIKAN, R. 1960.  (10.1)                                                2030
    CORRECTION TO AND COMMENT ON: EQUALITY OF MORE THAN TWO VARIANCES AND
    OF MORE THAN TWO DISPERSION MATRICES AGAINST CERTAIN ALTERNATIVES.
        (AMENDMENT OF NO. 2029)
    ANN. MATH. STATIST., 31, 227-228.  (MR31-227)

GNANADESIKAN, R. 1963.  (17.9/E)                                              2031
    SOME REMARKS ON MULTIVARIATE STATISTICAL METHODS FOR ANALYSIS OF EXPERIMENTAL DATA.
    INDUST. QUAL. CONTROL, 19(9) 22-26, 32.

GNANADESIKAN, R.    SEE ALSO
  4689. ROY, S.N. + GNANADESIKAN, R. (1957)

  4690. ROY, S.N. + GNANADESIKAN, R. (1958)

  4691. ROY, S.N. + GNANADESIKAN, R. (1959)

  4692. ROY, S.N. + GNANADESIKAN, R. (1959)

  4693. ROY, S.N. + GNANADESIKAN, R. (1961)

  4694. ROY, S.N. + GNANADESIKAN, R. (1962)

  5904. WILK, MARTIN B. + GNANADESIKAN, R. (1961)

  5905. WILK, MARTIN B. + GNANADESIKAN, R. (1964)

  5102. SMITH, HARRY, JR.; GNANADESIKAN, R. + HUGHES, J.B. (1962)

GNEDENKO, BORIS V. 1944.  (17.3)                                              2032
    ELEMENTS OF THE THEORY OF DISTRIBUTION FUNCTIONS OF RANDOM VECTORS.
        (IN RUSSIAN)
    USPEHI MAT. NAUK, 10, 230-244.  (MR7,P.19)

GNEDENKO, BORIS V. 1948.  (2.2)                                                    2033
    ON A THEOREM OF S. N. BERNSTEIN.
        (IN RUSSIAN)
    IZV. AKAD. NAUK SSSR SER. MAT., 12, 97-100.  (MR9,P.450)

GODDARD, L. S. 1945.  (17.1/U)                                                     2034
    THE ACCUMULATION OF CHANCE EFFECTS AND THE GAUSSIAN FREQUENCY DISTRIBUTION.
    PHILOS. MAG. SER. 7, 36, 428-433.  (MR7,P.311)

GODDARD, L. S. 1955.  (20.1)                                                       2035
    NOTE ON A MATRIX THEOREM OF A. BRAUER AND ITS EXTENSION.
    CANAD. J. MATH., 7, 527-530.  (MR17,P.228)

GODWIN, H. J. 1955.  (17.2)                                                        2036
    ON GENERALIZATIONS OF TCHEBYCHEF'S INEQUALITY.
    J. AMER. STATIST. ASSOC., 50, 923-945.  (MR17,P.165)

GODWIN, H. J. + ZAREMBA, S. K. 1961.  (4.9/17.1)                                   2037
    A CENTRAL LIMIT THEOREM FOR PARTLY DEPENDENT VARIABLES.
    ANN. MATH. STATIST., 32, 677-686.  (MR23-A5388)

GOFFMAN, CASPER 1946.  (8.8/C)                                                     2038
    THE EFFECT OF A NUMBER OF VARIABLES ON A QUALITY CHARACTERISTIC.
    INDUST. QUAL. CONTROL, 2(4) 3-5, 10.

GOGGANS, JAMES F.   SEE
    4929. SCHULTZ, E. FRED, JR. + GOGGANS, JAMES F. (1961)

GOHEEN, HOWARD W. + DAVIDOFF, MELVIN D. 1951.  (4.4)                               2039
    A GRAPHICAL METHOD FOR THE RAPID CALCULATION OF BISERIAL AND POINT
    BISERIAL CORRELATION IN TEST RESEARCH.
    PSYCHOMETRIKA, 16, 239-242.

GOHEEN, HOWARD W. + KAVRUCK, SAMUEL 1948.  (4.4)                                   2040
    A WORK-SHEET FOR TETRACHORIC R AND STANDARD ERROR OF TETRACHORIC R
    USING HAYES DIAGRAMS AND TABLES.
    PSYCHOMETRIKA, 13, 279-280.

GOHEEN, HOWARD W.   SEE ALSO
    1231. DAVIDOFF, MELVIN D. + GOHEEN, HOWARD W. (1953)

GOLDBERG, K. 1966.  (20.2)                                                         2041
    HADAMARD MATRICES OF ORDER CUBE PLUS ONE.
    PROC. AMER. MATH. SOC., 17, 744-746.

GOLDBERGER, ARTHUR S. 1961.  (8.9)                                                 2042
    STEPWISE LEAST SQUARES:  RESIDUAL ANALYSIS AND SPECIFICATION ERROR.
    J. AMER. STATIST. ASSOC., 56, 998-1000.  (MR23-A4221)

GOLDBERGER, ARTHUR S. 1962.  (8.1)                                                 2043
    BEST LINEAR UNBIASED PREDICTION IN THE GENERALIZED LINEAR REGRESSION MODEL.
    J. AMER. STATIST. ASSOC., 57, 369-375.  (MR26-854)

GOLDBERGER, ARTHUR S. + JOCHEMS, D. B. 1961.  (8.1)                                2044
    NOTE ON STEPWISE LEAST SQUARES.
    J. AMER. STATIST. ASSOC., 56, 105-110.  (MR22-12640)

GOLDBERGER, ARTHUR S.; NAGAR, A.L. + ODEH, H.S. 1961.  (16.2/16.5)                 2045
    THE COVARIANCE MATRICES OF REDUCED-FORM COEFFICIENTS AND OF FORECASTS
    FOR A STRUCTURAL ECONOMETRIC MODEL.
    ECONOMETRICA, 29, 556-573.  (MR30-5800)

GOLDIE, N. 1956.  (4.1/17.9/E)                                                     2046
    FREQUENCIES AND CORRELATION IN UPPER AIR DATA.
    METEOROL. MAG. LONDON, 85, 299-311.

GOLDMAN, A. J. + NEWMAN, MORRIS 1966.  (20.1)                                      2047
    FINDING A RANK-MAXIMIZING MATRIX BLOCK.
    J. RES. NAT. BUR. STANDARDS SECT. B, 70, 219-220.

GOLDMAN, A. J. + ZELEN, MARVIN 1964.  (8.8/20.1)                                   2048
    WEAK GENERALIZED INVERSES AND MINIMUM VARIANCE LINEAR UNBIASED ESTIMATION.
    J. RES. NAT. BUR. STANDARDS SECT. B, 68, 151-172.  (MR30-3525)

GOLDSTEIN, NEIL    SEE
   2207. GUMBEL, EMILE J. + GOLDSTEIN, NEIL (1964)


GOLDSTINE, HERMAN H. + VON NEUMANN, JOHN 1951.  (20.3)                 2049
     NUMERICAL INVERTING OF MATRICES OF HIGH ORDER. II.
   PROC. AMER. MATH. SOC., 2, 188-202.  (MR12,P.861)


GOLDSTINE, HERMAN H.    SEE ALSO
   5662. VON NEUMANN, JOHN + GOLDSTINE, HERMAN H. (1947)


GOLUB, GENE H. 1962.  (20.3)                                          2050
     BOUNDS FOR EIGENVALUES OF TRIDIAGONAL SYMMETRIC MATRICES COMPUTED BY
     THE LR METHOD.
   MATH. COMPUT., 16, 438-445.


GOLUB, GENE H. 1962.  (17.2/20.3)                                     2051
     BOUNDS FOR THE ROUND-OFF ERRORS IN THE RICHARDSON SECOND ORDER METHOD.
   NORDISK TIDSSKR. INFORMATIONS-BEHANDLING, 2, 212-223.


GOLUB, GENE H. 1963.  (8.2)                                           2052
     COMPARISON OF THE VARIANCE OF MINIMUM VARIANCE AND WEIGHTED LEAST
     SQUARES REGRESSION COEFFICIENTS.
   ANN. MATH. STATIST., 34, 984-991.  (MR27-5336)


GOLUB, GENE H. 1963.  (20.1)                                          2053
     ON THE LOWER BOUND FOR THE RANK OF A PARTITIONED SQUARE MATRIX.
   MATH. COMPUT., 17, 186-188.


GOLUB, GENE H. 1965.  (20.3)                                          2054
     NUMERICAL METHODS FOR SOLVING LINEAR LEAST SQUARES PROBLEMS.
   NUMER. MATH., 7, 206-216.


GOLUB, GENE H. + KAHAN, WILLIAM 1965.  (20.3)                         2055
     CALCULATING THE SINGULAR VALUES AND PSEUDO-INVERSE OF A MATRIX.
   J. SOC. INDUST. APPL. MATH. SER. B NUMER. ANAL., 2, 205-224.


GOLUB, GENE H. + WILKINSON, J. H. 1966.  (20.3)                       2056
     NOTE ON ITERATIVE REFINEMENT OF LEAST SQUARES SOLUTION.
   NUMER. MATH., 9, 139-148.


GOLUB, GENE H.    SEE ALSO
    810. BUSINGER, P. + GOLUB, GENE H. (1965)

   1729. FORSYTHE, GEORGE E. + GOLUB, GENE H. (1965)


GONIN, H. T. 1936.  (18.1/18.2/18.3)                                  2057
     THE USE OF FACTORIAL MOMENTS IN THE TREATMENT OF THE HYPERGEOMETRIC
     DISTRIBUTION AND IN TESTS FOR REGRESSION.
   PHILOS. MAG. SER. 7, 21, 215-226.


GONIN, H. T. 1966.  (18.1/B)                                          2058
     POISSON AND BINOMIAL FREQUENCY SURFACES.
   BIOMETRIKA, 53, 617-619.


GOOD, I. J. 1956.  (2.1)                                              2059
     THE SURPRISE INDEX FOR THE MULTIVARIATE NORMAL DISTRIBUTION.
        (AMENDED BY NO. 2061)
   ANN. MATH. STATIST., 27, 1130-1135.  (MR19,P.70)


GOOD, I. J. 1957.  (18.2/18.3/20.4)                                   2060
     SADDLE-POINT METHODS FOR THE MULTINOMIAL DISTRIBUTION.
        (AMENDED BY NO. 2064)
   ANN. MATH. STATIST., 28, 861-881.  (MR20-386)


GOOD, I. J. 1957.  (2.1)                                              2061
     CORRECTION TO:  THE SURPRISE INDEX FOR THE MULTIVARIATE NORMAL DISTRIBUTION.
        (AMENDMENT OF NO. 2059)
   ANN. MATH. STATIST., 28, 1055.  (MR19,P.70)


GOOD, I. J. 1960.  (16.4)                                            2062
     GENERALIZATIONS TO SEVERAL VARIABLES OF LAGRANGE'S EXPANSION, WITH APPLICATIONS
     TO STOCHASTIC PROCESSES.
   PROC. CAMBRIDGE PHILOS. SOC., 56, 367-380.  (MR23-A352)

GOOD, I. J. 1961.   (18.3/20.4)                                                    2063
    THE MULTIVARIATE SADDLE POINT METHOD AND CHI-SQUARED FOR THE
    MULTINOMIAL DISTRIBUTION.
    ANN. MATH. STATIST., 32, 535-548.   (MR25-685)

GOOD, I. J. 1961.   (18.2/18.3/20.4)                                               2064
    CORRECTIONS TO "THE MULTIVARIATE SADDLE POINT METHOD AND CHI-SQUARED
    FOR THE MULTINOMIAL DISTRIBUTION".
        (AMENDMENT OF NO. 2060)
    ANN. MATH. STATIST., 32, 619.

GOOD, I. J. 1962.   (16.5)                                                         2065
    WEIGHTED COVARIANCE FOR ESTIMATING THE DIRECTION OF A GAUSSIAN SOURCE.
    PROC. SYMP. TIME SERIES ANAL. BROWN UNIV. (ROSENBLATT), 447-470.   (MR26-7460)

GOOD, I. J. 1963.   (4.8/17.9/18.2)                                                2066
    MAXIMUM ENTROPY FOR HYPOTHESIS FORMULATION, ESPECIALLY FOR
    MULTIDIMENSIONAL CONTINGENCY TABLES.
    ANN. MATH. STATIST., 34, 911-934.   (MR27-866)

GOOD, I. J. 1963.   (2.5)                                                          2067
    ON THE INDEPENDENCE OF QUADRATIC EXPRESSIONS (WITH AN APPENDIX BY L.R. WELCH).
        (AMENDED BY NO. 2069)
    J. ROY. STATIST. SOC. SER. B, 25, 377-381.   (MR29-5341)

GOOD, I. J. 1965.   (6.1)                                                          2068
    CATEGORIZATION OF CLASSIFICATION.
        (WITH DISCUSSION)
    MATH. COMPUT. SCI. BIOL. MED., 115-128.

GOOD, I. J. 1966.   (2.5)                                                          2069
    ON THE INDEPENDENCE OF QUADRATIC EXPRESSIONS:  CORRIGENDA.
        (AMENDMENT OF NO. 2067)
    J. ROY. STATIST. SOC. SER. B, 28, 584.   (MR34-8532)

GOODALL, D. W. 1953.  E(6.5)                                                       2070
    OBJECTIVE METHODS FOR THE CLASSIFICATION OF VEGETATION. I: THE USE OF
    POSITIVE INTERSPECIFIC CORRELATION.
    AUSTRAL. J. BOT., 1, 39-63.

GOODALL, D. W. 1954.  E(6.5)                                                       2071
    OBJECTIVE METHODS FOR THE CLASSIFICATION OF VEGETATION. III: AN ESSAY
    IN THE USE OF FACTOR ANALYSIS.
    AUSTRAL. J. BOT., 2, 304-324.

GOODE, HARRY H.    SEE
    1968. GILLMAN, LEONARD + GOODE, HARRY H. (1946)

GOODMAN, LEO A. 1963.   (11.1)                                                     2072
    ON METHODS OF COMPARING CONTINGENCY TABLES.
    J. ROY. STATIST. SOC. SER. A, 126, 94-108.

GOODMAN, LEO A. 1964.   (4.8/17.8)                                                 2073
    SIMULTANEOUS CONFIDENCE INTERVALS FOR CONTRASTS AMONG MULTINOMIAL POPULATIONS.
    ANN. MATH. STATIST., 35, 716-725.   (MR28-5515)

GOODMAN, LEO A. 1965.   (18.2/C)                                                   2074
    ON SIMULTANEOUS CONFIDENCE INTERVALS FOR MULTINOMIAL PROPORTIONS.
    TECHNOMETRICS, 7, 247-254.

GOODMAN, LEO A. + KRUSKAL, WILLIAM H. 1954.   (4.8/4.9)                            2075
    MEASURES OF ASSOCIATION FOR CROSS CLASSIFICATIONS.
        (AMENDED BY NO. 2076, AMENDED BY NO. 2077)
    J. AMER. STATIST. ASSOC., 49, 732-764.

GOODMAN, LEO A. + KRUSKAL, WILLIAM H. 1957.   (4.8/4.9)                            2076
    CORRECTION TO: MEASURES OF ASSOCIATION FOR CROSS CLASSIFICATIONS.
        (AMENDMENT OF NO. 2075)
    J. AMER. STATIST. ASSOC., 52, 578.

GOODMAN, LEO A. + KRUSKAL, WILLIAM H. 1958.  (4.9)                    2077
    CORRIGENDA:  MEASURES OF ASSOCIATION FOR CROSS CLASSIFICATIONS.
        (AMENDMENT OF NO. 2075)
    J. AMER. STATIST. ASSOC., 53, 1031.

GOODMAN, LEO A. + KRUSKAL, WILLIAM H. 1963.  (4.9/A)                    2078
    MEASURES OF ASSOCIATION FOR CROSS CLASSIFICATIONS. III.  APPROXIMATE
    SAMPLING THEORY.
    J. AMER. STATIST. ASSOC., 58, 310-364.

GOODMAN, N. R. 1961.  (16.3)                    2079
    SOME COMMENTS ON SPECTRAL ANALYSIS OF TIME SERIES.
    TECHNOMETRICS, 3, 221-228.  (MR23-A3033A)

GOODMAN, N. R. 1963.  (14.4)                    2080
    STATISTICAL ANALYSIS BASED ON A CERTAIN MULTIVARIATE COMPLEX GAUSSIAN
    DISTRIBUTION.
    ANN. MATH. STATIST., 34, 152-177.  (MR26-3148A)

GOODMAN, N. R. 1963.  (7.2)                    2081
    THE DISTRIBUTION OF THE DETERMINANT OF A COMPLEX WISHART DISTRIBUTED MATRIX.
    ANN. MATH. STATIST., 34, 178-180.  (MR26-3148B)

GOODMAN, N. R. 1963.  (16.3)                    2082
    SPECTRAL ANALYSIS OF MULTIPLE TIME SERIES.
    PROC. SYMP. TIME SERIES ANAL. BROWN UNIV. (ROSENBLATT), 260-266.  (MR27-902)

GOODMAN, THOMAS P. 1956.  (4.3)                    2083
    TECHNIQUE FOR APPROXIMATE MEASUREMENT OF CORRELATION COEFFICIENTS.
    J. APPL. PHYS., 27, 773-775.  (MR17,P.1219)

GORDON, M. A. 1954. E(4.6)                    2084
    EMPIRICAL COMPARISON OF THREE MULTIPLE CORRELATION TECHNIQUES.
    EDUC. PSYCHOL. MEAS., 14, 133-137.

GORDON, R. A. 1937.  (1.2)                    2085
    A SELECTED BIBLIOGRAPHY OF THE LITERATURE OF ECONOMIC FLUCTUATIONS.
    REV. ECON. STATIST., 19, 37-68.

GORDON, ROBERT D. 1941.  (14.4)                    2086
    THE ESTIMATION OF A QUOTIENT WHEN THE DENOMINATOR IS NORMALLY DISTRIBUTED.
    ANN. MATH. STATIST., 12, 115-118.  (MR3,P.8)

GOSSET, W. S.  SEE "STUDENT"

GOSWAMI, J. N. + SUKHATME, B. V. 1965.  (17.7)                    2087
    RATIO METHOD OF ESTIMATION IN MULTI-PHASE SAMPLING WITH SEVERAL AUXILIARY VARIABLES.
    J. INDIAN SOC. AGRIC. STATIST., 17, 83-103.

GOTAAS, PER 1936.  (2.3/17.3)                    2088
    FORMULES DE RECURRENCE POUR LES SEMI-INVARIANTS A QUELQUES LOIS DE
    DISTRIBUTION A PLUSIEURS VARIABLES.
    C.R. ACAD. SCI. PARIS, 202, 619-621.

GOTAAS, PER 1936.  (17.3)                    2089
    FORMULAE OF RECURRENCE FOR THE SEMIINVARIANTS AT A CLASS OF FREQUENCY
    FUNCTIONS OF MORE VARIABLES.
    SKAND. AKTUARIETIDSKR., 19, 200-211.

GOTTSCHICK, J. 1936.  (4.1/E)                    2090
    DER QUOTIENT DER ZUFALLSABWEICHUNGEN EIN MASS DER KORRELATION.
    BEITRAG ZUM MATHEMATISCHEN RUSTZEUG DER BIOLOGIE.
    ARCH. RASS. GES. BIOL., 30(3) 237-261.

GOUARNE, RENE 1957.  (20.3)                    2091
    CALCUL AUTOMATIQUE DES DETERMINANTS.
    C.R. ACAD. SCI. PARIS, 245, 824-826.  (MR19,P.1085)

GOUARNE, RENE 1957.  (20.3)                    2092
    REMARQUES SUR LE CALCUL AUTOMATIQUE DES DETERMINANTS ET POLYNOMES
    CARACTERISTIQUES PAR LA METHODE DES CYCLES.
    C.R. ACAD. SCI. PARIS, 245, 1998-2000.  (MR19,P.1081)

GOUDSMIT, S. A. + SAUNDERSON, J. L. 1940.   (16.4/E)                                    2093
    MULTIPLE SCATTERING OF ELECTRONS.
  PHYS. REV., 57, 24–29.


GOUYET, JEAN-FRANCOIS 1965.   (16.4/A)                                                   2094
    AUTOCORRELATION D'UNE ONDE ELECTROMAGNETIQUE STATIONNAIRE.
  C.R. ACAD. SCI. PARIS, 261, 1941–1944.


GOVINDARAJULU, ZAKKULA 1956.   (18.1)                                                    2095
    A NOTE ON THE CORRELATION COEFFICIENT OF THE BIVARIATE GAMMA TYPE DISTRIBUTION.
  J. MADRAS UNIV. SECT. B, 26, 639–642.   (MR20–2814)


GOVINDARAJULU, ZAKKULA 1965.   (1.2/18.1/A)                                              2096
    NORMAL APPROXIMATIONS TO THE CLASSICAL DISCRETE DISTRIBUTIONS.
  SANKHYA SER. A, 27, 143–172.


GOVINDARAJULU, ZAKKULA 1966.   (17.3/17.6/U)                                             2097
    CHARACTERIZATION OF THE EXPONENTIAL AND POWER DISTRIBUTIONS.
  SKAND. AKTUARIETIDSKR., 49, 132–136.


GOWER, J. C. 1966.   (6.5/10.1/14.1)                                                     2098
    SOME DISTANCE PROPERTIES OF LATENT ROOT AND VECTOR METHODS USED IN
    MULTIVARIATE ANALYSIS.
  BIOMETRIKA, 53, 325–338.   (MR35–5075)


GOWER, J. C. 1966.   (6.3/8.5/12.1)                                                      2099
    A Q-TECHNIQUE FOR THE CALCULATION OF CANONICAL VARIATES.
  BIOMETRIKA, 53, 588–590.   (MR35–7484)


GRAD, ARTHUR + SOLOMON, HERBERT 1955.   (2.5/14.5)                                       2100
    DISTRIBUTION OF QUADRATIC FORMS AND SOME APPLICATIONS.
  ANN. MATH. STATIST., 26, 464–477.   (MR17,P.634)


GRANTHAM, J.   SEE
  1847. GARNER, F. H.; GRANTHAM, J. + SANDERS, H. G. (1934)


GRAVELIUS, H. 1921.   (4.1/4.5/8.1/E)                                                    2101
    UBER DIE KORRELATIONSMETHODE.
  Z. ANGEW. MATH. MECH., 1, 199–205.


GRAYBILL, FRANKLIN A. + HULTQUIST, ROBERT A. 1961.   (8.6/U)                             2102
    THEOREMS CONCERNING EISENHART'S MODEL II.
  ANN. MATH. STATIST., 32, 261–269. (MR22–11489)


GRAYBILL, FRANKLIN A. + MARSAGLIA, GEORGE 1957.   (2.5/8.8/20.2)                         2103
    IDEMPOTENT MATRICES AND QUADRATIC FORMS IN THE GENERAL LINEAR HYPOTHESIS.
  ANN. MATH. STATIST., 28, 678–686.   (MR19,P.1095)


GRAYBILL, FRANKLIN A.; MEYER, C. D. + PAINTER, R. J. 1966.   (20.3)                      2104
    NOTE ON THE COMPUTATION OF THE GENERALIZED INVERSE OF A MATRIX.
  SIAM REV., 8, 522–524.


GREEN, BERT F., JR. 1951.   (15.5)                                                       2105
    A GENERAL SOLUTION FOR THE LATENT CLASS MODEL OF LATENT STRUCTURE ANALYSIS.
  PSYCHOMETRIKA, 16, 151–166.


GREEN, BERT F., JR. 1952.   (15.1/15.2/15.5)                                             2106
    LATENT STRUCTURE ANALYSIS AND ITS RELATION TO FACTOR ANALYSIS.
  J. AMER. STATIST. ASSOC., 47, 71–76.


GREEN, BERT F., JR. 1952.   (15.2)                                                       2107
    THE ORTHOGONAL APPROXIMATION OF AN OBLIQUE STRUCTURE IN FACTOR ANALYSIS.
  PSYCHOMETRIKA, 17, 429–440.   (MR14,P.715)


GREEN, PAUL E. 1956.   (1.2)                                                             2108
    A BIBLIOGRAPHY OF SOVIET LITERATURE ON NOISE, CORRELATION AND
    INFORMATION THEORY.
  IRE TRANS. INFORMATION THEORY, IT-2, 91–94.


GREEN, PAUL E.; HALBERT, MICHAEL H. + ROBINSON, PATRICK J. 1966.  E(12.1/12.2)           2109
    CANONICAL ANALYSIS: AN EXPOSITION AND ILLUSTRATIVE APPLICATION.
  J. MARKET. RES., 3, 32–39.

GREENBERG, BERNARD G. + SARHAN, AHMED E. 1959.  (20.2/E)                    2110
    MATRIX INVERSION, ITS INTEREST AND APPLICATION IN ANALYSIS OF DATA.
        (REPRINTED AS NO. 2111)
  J. AMER. STATIST. ASSOC., 54, 755-766.  (MR21-7578)

GREENBERG, BERNARD G. + SARHAN, AHMED E. 1960.  (20.2/E)                    2111
    MATRIX INVERSION, ITS INTEREST AND APPLICATION IN ANALYSIS OF DATA.
        (REPRINT OF NO. 2110)
  BULL. INST. INTERNAT. STATIST., 37(3) 107-120.  (MR22-10165)

GREENBERG, BERNARD G. + SARHAN, AHMED E. 1960.  (20.1/20.2)                 2112
    GENERALIZATION OF SOME RESULTS FOR INVERSION OF PARTITIONED MATRICES.
  CONTRIB. PROB. STATIST. (HOTELLING VOLUME), 216-223.  (MR22-11509)

GREENBERG, BERNARD G.    SEE ALSO
    4703. ROY, S.N.; GREENBERG, BERNARD G. + SARHAN, AHMED E. (1960)

    4842. SARHAN, AHMED E.; GREENBERG, BERNARD G. + ROBERTS, ELEANOR (1962)

GREENE, JOEL E.; MILLER, WILBUR C. + BROWN, VIRGINIA M. 1956.  (15.4)       2113
    RELIABILITY AND INTERCORRELATION CRITERIA FOR ITEM SELECTION.
  SYMP. AIR FORCE HUMAN ENGRG. PERS. TRAIN. RES. (FINCH + CAMERON), 166-172.

GREENHOUSE, SAMUEL W. + GEISSER, SEYMOUR 1959.  (4.8/15.5)                  2114
    ON METHODS IN THE ANALYSIS OF PROFILE DATA.
  PSYCHOMETRIKA, 24, 95-112.

GREENHOUSE, SAMUEL W.    SEE ALSO
    1913. GEISSER, SEYMOUR + GREENHOUSE, SAMUEL W. (1958)

    2307. HALPERIN, MAX; GREENHOUSE, SAMUEL W.; CORNFIELD, JEROME + (1955)
    ZALOKAR, JULIA

GREENSTADT, JOHN    SEE
    1523. EISENPRESS, HARRY + GREENSTADT, JOHN (1966)

GREENWOOD, J. ARTHUR + DURAND, DAVID 1955.  (18.1)                          2115
    THE DISTRIBUTION OF LENGTH AND COMPONENTS OF THE SUM OF N RANDOM UNIT VECTORS.
  ANN. MATH. STATIST., 26, 233-246.  (MR16,P.1034)

GREENWOOD, J. ARTHUR    SEE ALSO
    1426. DURAND, DAVID + GREENWOOD, J. ARTHUR (1957)

    2208. GUMBEL, EMILE J.; GREENWOOD, J. ARTHUR + DURAND, DAVID (1953)

GREENWOOD, MAJOR 1904.  (4.1/ET)                                            2116
    A FIRST STUDY OF THE WEIGHT, VARIABILITY, AND CORRELATION OF THE HUMAN VISCERA,
    WITH SPECIAL REFERENCE TO THE HEALTHY AND DISEASED HEART.
  BIOMETRIKA, 3, 63-83.

GREENWOOD, MAJOR 1926.  (4.1)                                               2117
    PROFESSOR TSCHUPROW ON THE THEORY OF CORRELATION.
  J. ROY. STATIST. SOC., 89, 320-325.

GREENWOOD, MAJOR + YULE, G. UDNY 1920.  (18.1/E)                            2118
    AN INQUIRY INTO THE NATURE OF FREQUENCY DISTRIBUTIONS REPRESENTATIVE OF
    MULTIPLE HAPPENINGS WITH PARTICULAR REFERENCE TO THE OCCURRENCE OF MULTIPLE
    ATTACKS OF DISEASE OR OF REPEATED ACCIDENTS.
  J. ROY. STATIST. SOC., 83, 255-279.


GREENWOOD, MAJOR    SEE ALSO
    736. BROWN, J. W.; GREENWOOD, MAJOR + WOOD, FRANCES (1914)

GREENWOOD, ROBERT E., JR. + GLASGOW, MARK O. 1950.  (18.3)                  2119
    DISTRIBUTION OF MAXIMUM AND MINIMUM FREQUENCIES IN A SAMPLE DRAWN
    FROM A MULTINOMIAL DISTRIBUTION.
  ANN. MATH. STATIST., 21, 416-424.  (MR12,P.428)

GREGSON, R. A. M. 1966.  (15.5/E)                                           2120
    THEORETICAL AND EMPIRICAL MULTIDIMENSIONAL SCALINGS OF TASTE MIXTURES MATCHINGS.
  BRITISH J. MATH. STATIST. PSYCHOL., 19, 59-75.

GRENANDER, ULF 1952.  (16.3)                                                          2121
    ON EMPIRICAL SPECTRAL ANALYSIS OF STOCHASTIC PROCESSES.
  ARK. MAT., 1, 503-531.  (MR14,P.187)

GRENANDER, ULF 1952.  (16.4)                                                          2122
    ON TOEPLITZ FORMS AND STATIONARY PROCESSES.
  ARK. MAT., 1, 555-571.  (MR14,P.187)

GRENANDER, ULF 1957.  (16.5)                                                          2123
    MODERN TRENDS IN TIME SERIES ANALYSIS.
  SANKHYA, 18, 149-158.  (MR19,P.1206)

GRENANDER, ULF 1959.  (16.5/17.9)                                                     2124
    SOME NON LINEAR PROBLEMS IN PROBABILITY THEORY.
  PROB. STATIST. (CRAMER VOLUME), 108-129.

GRENANDER, ULF + ROSENBLATT, MURRAY 1952.  (16.3)                                     2125
    ON SPECTRAL ANALYSIS OF STATIONARY TIME SERIES.
  PROC. NAT. ACAD. SCI. U.S.A., 38, 519-521.  (MR14,P.61)

GRENANDER, ULF + ROSENBLATT, MURRAY 1953.  (16.3)                                     2126
    STATISTICAL SPECTRAL ANALYSIS OF TIME SERIES ARISING FROM STATIONARY
    STOCHASTIC PROCESS.
  ANN. MATH. STATIST., 24, 537-558.  (MR15,P.448)

GRENANDER, ULF + ROSENBLATT, MURRAY 1953.  (16.3)                                     2127
    COMMENTS ON STATISTICAL SPECTRAL ANALYSIS.
  SKAND. AKTUARIETIDSKR., 36, 182-202.  (MR15,P.728)

GRENANDER, ULF + ROSENBLATT, MURRAY 1956.  (16.3)                                     2128
    SOME PROBLEMS IN ESTIMATING THE SPECTRUM OF A TIME SERIES.
  PROC. THIRD BERKELEY SYMP. MATH. STATIST. PROB., 1, 77-93.  (MR18,P.946)

GRENANDER, ULF: POLLAK, H.O. + SLEPIAN, DAVID 1959.  (2.5/14.5/16.3)                  2129
    THE DISTRIBUTION OF QUADRATIC FORMS IN NORMAL VARIATES:  A SMALL
    SAMPLE THEORY WITH APPLICATIONS TO SPECTRAL ANALYSIS.
  J. SOC. INDUST. APPL. MATH., 7, 374-401.  (MR22-1978)

GRENANDER, ULF   SEE ALSO
    1637. FREIBERGER, W. + GRENANDER, ULF (1965)

GRESSENS, O. + MOUZON, E. D., JR. 1927.  (15.5)                                       2130
    THE VALIDITY OF CORRELATION IN TIME SEQUENCES AND A NEW COEFFICIENT
    OF SIMILARITY.
  J. AMER. STATIST. ASSOC., 22, 483-492.

GREVILLE, T. N. E. 1957.  (20.3)                                                      2131
    THE PSEUDOINVERSE OF A RECTANGULAR OR SINGULAR MATRIX AND ITS
    APPLICATION TO THE SOLUTION OF SYSTEMS OF LINEAR EQUATIONS.
        (REPRINTED AS NO. 2133)
  SIAM NEWSLETTER, 5(2) 3-6.  (MR19,P.243)

GREVILLE, T. N. E. 1958.  (20.1)                                                      2132
    THE PSEUDOINVERSE OF A RECTANGULAR MATRIX AND ITS STATISTICAL APPLICATIONS.
  PROC. SOCIAL STATIST. SECT. AMER. STATIST. ASSOC., 116-121.

GREVILLE, T. N. E. 1959.  (20.3)                                                      2133
    THE PSEUDOINVERSE OF A RECTANGULAR OR SINGULAR MATRIX AND ITS
    APPLICATION TO THE SOLUTION OF SYSTEMS OF LINEAR EQUATIONS.
        (REPRINT OF NO. 2131)
  SIAM REV., 1, 38-43.  (MR21-424)

GREVILLE, T. N. E. 1960.  (20.1)                                                      2134
    SOME APPLICATIONS OF THE PSEUDOINVERSE OF A MATRIX.
  SIAM REV., 2, 15-22.  (MR22-1067)

GREVILLE, T. N. E. 1966.  (20.1)                                                      2135
    NOTE ON THE GENERALIZED INVERSE OF A MATRIX PRODUCT.
  SIAM REV., 8, 518-521.

GRIFFIN, HAROLD D. 1930.  (4.1/4.6/T)                                                 2136
    NOMOGRAMS FOR CORRECTING SIMPLE AND MULTIPLE CORRELATION COEFFICIENTS.
  J. AMER. STATIST. ASSOC., 25, 316-319.

GRIFFIN, HAROLD D. 1931.  (20.3)                                         2137
    FUNDAMENTAL FORMULAS FOR THE DOOLITTLE METHOD, USING ZERO-ORDER
    CORRELATION COEFFICIENTS.
    ANN. MATH. STATIST., 2, 150-153.


GRIFFIN, HAROLD D. 1931.  (8.1)                                          2138
    ON PARTIAL CORRELATION VERSUS PARTIAL REGRESSION FOR OBTAINING THE
    MULTIPLE REGRESSION EQUATIONS.
    J. EDUC. PSYCHOL., 22, 35-44.


GRIFFIN, HAROLD D. 1932.  (4.8)                                          2139
    ON THE COEFFICIENT OF PART CORRELATION.
    J. AMER. STATIST. ASSOC., 27, 298-301.


GRIFFIN, HAROLD D. 1933.  (4.6)                                          2140
    SIMPLIFIED SCHEMES FOR MULTIPLE LINEAR CORRELATION.
    J. EXPER. EDUC., 1, 239-254.


GRIFFIN, HAROLD D. 1936.  (4.5/4.6)                                      2141
    A FURTHER SIMPLIFICATION OF THE MULTIPLE AND PARTIAL CORRELATION PROCESS.
    PSYCHOMETRIKA, 1, 219-228.


GRIFFIN, JOHN S., JR.: KING, J. H., JR. + TUNIS, C. J. 1963.  (6.4/19.1)  2142
    A PATTERN IDENTIFICATION SYSTEM USING LINEAR DECISION FUNCTIONS.
    IBM SYSTEMS J., 2, 248-267.


GRIMZE, L. B. 1966.  (7.2/20.1)                                          2143
    AN ESTIMATE OF THE GRAM DETERMINANT.
         (IN RUSSIAN)
    MET. VYC. (LENINGRAD), 3, 30-33.  (MR34-8624)


GRIZZLE, J. E.    SEE
    1014. COLE, J. W. L. + GRIZZLE, J. E. (1966)

    1552. ELSTON, R. C. + GRIZZLE, J. E. (1962)


GROEBNER, WOLFGANG 1948.  (20.4)                                         2144
    UBER DIE KONSTRUKTION VON SYSTEMEN ORTHOGONALER POLYNOME IN EIN- UND
    ZWEI-DIMENSIONALEN BEREICHEN.
    MONATSHEFTE MATH., 52, 38-54.  (MR9,P.430)


GROSJEAN, CARL C. 1953.  (17.3/17.9)                                     2145
    SOLUTION OF THE NON-ISOTROPIC RANDOM FLIGHT PROBLEM IN THE K-DIMENSIONAL SPACE.
    PHYSICA, 19, 29-45.  (MR14,P.772)


GROSSMAN, DAVID P., SGT. 1944.  (15.4)                                   2146
    TECHNIQUE FOR WEIGHTING OF CHOICES AND ITEMS ON I.B.M. SCORING MACHINES.
    PSYCHOMETRIKA, 9, 101-104.


GROSSNICKLE, LOUISE T. 1942.  (15.4)                                     2147
    THE SCALING OF TEST SCORES BY THE METHOD OF PAIRED COMPARISONS.
    PSYCHOMETRIKA, 7, 43-64.


GRUBBS, FRANK E. 1944.  (14.5)                                           2148
    ON THE DISTRIBUTION OF THE RADIAL STANDARD DEVIATION.
    ANN. MATH. STATIST., 15, 75-81.  (MR5,P.209)


GRUBBS, FRANK E. + COON, HELEN J. 1954.  (14.4)                          2149
    ON SETTING TEST LIMITS RELATIVE TO SPECIFICATION LIMITS.
    INDUST. QUAL. CONTROL, 10(5) 15-20.


GRUNEBERG, HANS-JOACHIM 1952.  (15.1)                                    2150
    DIE MULTIPLE FAKTORANALYSE.
    MITT. MATH. STATIST., 4, 9-31.  (MR13,P.963)


GRUNERT, JOHANN AUGUST 1858.  (8.8/17.8/20.5)                            2151
    DREI GROSSEN X, Y, Z, DEREN SUMME DIE GEGEBENE GROSSE S  IST, SIND DURCH
    MESSUNG BESTIMMT WORDEN, UND MAN HABE DADURCH FUR DIESE DREI GROSSEN RESPECTIVE
    DIE WERTHE A, B, C ERHALTEN.  DA DIESE WERTHE MIT BEOBACHTUNGSFEHLERN
    BEHAFTET SIND, UND IHRE SUMME ALSO IM ALLGEMEINEN NICHT GENAU S IST,
    SO SOLL MAN DIESELBEN SO VERBESSEREN, DASS DIE VERBESSERTEN WERTHE GENAU DIE
    SUMME  S  GEBEN, UND DIE SUMME DER QUADRATE DER VERBESSERUNGEN EIN MINIMUM IST.
    ARCH. MATH. PHYS., 31, 479-481.

GRUNWALD, H. 1965. (16.4)
THE CORRELATION THEORY FOR STATIONARY STOCHASTIC PROCESSES APPLIED TO
EXPONENTIAL SMOOTHING.
STATISTICA NEERLANDICA, 19, 129-138.

**2152**

GRUZEWSKA, H. MILICER SEE MILICER-GRUZEWSKA, HALINA

GUENTHER, WILLIAM C. 1961. (2.6)
CIRCULAR PROBABILITY PROBLEMS.
AMER. MATH. MONTHLY, 68, 541-544.

**2153**

GUENTHER, WILLIAM C. 1961. (14.4/T)
ON THE PROBABILITY OF CAPTURING A RANDOMLY SELECTED POINT IN THREE DIMENSIONS.
SIAM REV., 3, 247-251. (MR24-A566)

**2154**

GUENTHER, WILLIAM C. 1964. (14.5)
A GENERALIZATION OF THE INTEGRAL OF THE CIRCULAR COVERAGE FUNCTION.
AMER. MATH. MONTHLY, 71, 278-283. (MR28-4612)

**2155**

GUENTHER, WILLIAM C. + TERRAGNO, PAUL J. 1964. (1.2/2.6/14.4)
A REVIEW OF THE LITERATURE ON A CLASS OF COVERAGE PROBLEMS.
ANN. MATH. STATIST., 35, 232-260. (MR28-2623)

**2156**

GUEST, P. G. 1950. (8.1/20.4/U)
ORTHOGONAL POLYNOMIALS IN THE LEAST SQUARES FITTING OF OBSERVATIONS.
PHILOS. MAG. SER. 7, 41, 124-137. (MR11,P.692)

**2157**

GUEST, P. G. 1953. (8.8/20.3)
THE DOOLITTLE METHOD AND THE FITTING OF POLYNOMIALS TO WEIGHTED DATA.
BIOMETRIKA, 40, 229-231. (MR14,P.1127)

**2158**

GUEST, P. G. 1956. (8.8)
GROUPING METHODS IN THE FITTING OF POLYNOMIALS TO UNEQUALLY SPACED OBSERVATIONS.
BIOMETRIKA, 43, 149-160. (MR17,P.1219)

**2159**

GUEST, P. G. 1958. (8.8)
THE SPACING OF OBSERVATIONS IN POLYNOMIAL REGRESSION.
ANN. MATH. STATIST., 29, 294-299. (MR20-1392)

**2160**

GUILDBAUD, G. TH. 1951. (16.2)
L'ETUDE STATISTIQUE DES OSCILLATIONS ECONOMIQUES.
CAHIERS SEM. ECONOMET., 1, 5-42.

**2161**

GUILFORD, J. P. 1936. (15.1)
UNITARY TRAITS OF PERSONALITY AND FACTOR THEORY.
AMER. J. PSYCHOL., 48, 673-680.

**2162**

GUILFORD, J. P. 1936. (15.4)
THE DETERMINATION OF ITEM DIFFICULTY WHEN CHANCE SUCCESS IS A FACTOR.
PSYCHOMETRIKA, 1, 259-264.

**2163**

GUILFORD, J. P. 1941. (15.4)
THE PHI COEFFICIENT AND CHI SQUARE AS INDICES OF ITEM VALIDITY.
PSYCHOMETRIKA, 6, 11-19.

**2164**

GUILFORD, J. P. 1946. (15.4)
NEW STANDARDS FOR TEST EVALUATION.
(AMENDED BY NO. 2166)
EDUC. PSYCHOL. MEAS., 6, 427-438.

**2165**

GUILFORD, J. P. 1947. (15.4)
CORRECTION TO: NEW STANDARDS FOR TEST EVALUATION.
(AMENDMENT OF NO. 2165)
EDUC. PSYCHOL. MEAS., 7, 2.

**2166**

GUILFORD, J. P. 1948. (15.4)
FACTOR ANALYSIS IN A TEST-DEVELOPMENT PROGRAM.
PSYCHOL. REV., 55, 79-94.

**2167**

GUILFORD, J. P. 1957. (1.2)
LOUIS LEON THURSTONE, 1887-1955.
BIOGRAPH. MEM. NAT. ACAD. SCI. U.S.A., 30, 349-382.

**2168**

GUILFORD, J. P. 1963.  (15.1/15.4/E)                                              2169
    PREPARATION OF ITEM SCORES FOR CORRELATION BETWEEN PERSONS IN Q FACTOR ANALYSIS.
    EDUC. PSYCHOL. MEAS., 23, 13-22.

GUILFORD, J. P. 1965.  (4.8)                                                      2170
    THE MINIMAL PHI COEFFICIENT AND THE MAXIMAL PHI.
    EDUC. PSYCHOL. MEAS., 25, 3-8.

GUILFORD, J. P. + LYONS, THOBURN C. 1942.  (4.4/15.4)                             2171
    ON DETERMINING THE RELIABILITY AND SIGNIFICANCE OF A TETRACHORIC
    COEFFICIENT OF CORRELATION.
    PSYCHOMETRIKA, 7, 243-249.  (MR4,P.165)

GUILFORD, J. P. + MICHAEL, WILLIAM B. 1948.  (15.1)                               2172
    APPROACHES TO UNIVOCAL FACTOR SCORES.
    PSYCHOMETRIKA, 13, 1-22.

GUILFORD, J. P. + MICHAEL, WILLIAM B. 1950.  (15.1)                               2173
    CHANGES IN COMMON-FACTOR LOADINGS AS TESTS ARE ALTERED HOMOGENEOUSLY IN LENGTH.
    PSYCHOMETRIKA, 15, 237-249.

GUILFORD, J. P. + PERRY, NORMAN C. 1951.  (4.8)                                   2174
    ESTIMATION OF OTHER COEFFICIENTS OF CORRELATION FROM THE PHI COEFFICIENT.
    PSYCHOMETRIKA, 16. 335-346.

GUILFORD, J. P.   SEE ALSO
    2517. HOLLEY, J. W. + GUILFORD, J. P. (1964)

    2518. HOLLEY, J. W. + GUILFORD, J. P. (1966)

    3812. MICHAEL, WILLIAM B.; PERRY, NORMAN C. + GUILFORD, J. P. (1952)

GUILLAUME, EDOUARD 1915.  (2.1)                                                   2175
    SUR L'IMPOSSIBILITE DE RAMENER A UNE PROBABILITE COMPOSEE LA LOI DES
    ECARTS A PLUSIEURS VARIABLES.
    ENSEIGNEMENT MATH., 17, 358-359.

GUIRAUM MARTIN, ALFONSO 1953.  (15.2)                                            2176
    MODIFICACION DE TEST DE LAWLEY PARA PEQUENA MUESTRA.
    LAS CIENCIAS (MADRID), 18, 753-756.  (MR17,P.983)

GULATI, R. L. 1956.  (4.3)                                                        2177
    SEQUENTIELLE TESTS FUR DEN KORRELATIONSKOEFFIZIENTEN.
    MITT. MATH. STATIST., 8, 202-233.  (MR18,P.957)

GULDBERG, ALF. 1919.  (2.1/17.1/B)                                                2178
    A REMARK ON CORRELATION.
    SKAND. AKTUARIETIDSKR., 2, 197-203.

GULDBERG, ALF. 1920.  (2.1)                                                       2179
    ON THE LAW OF ERRORS IN THE SPACE OF P DIMENSIONS.
    TOHOKU MATH. J., 17, 18-23.

GULDBERG, ALF. 1921.  (4.1)                                                       2180
    ON THE CORRELATION OF SUCCESSIVE OBSERVATIONS.
    SKAND. AKTUARIETIDSKR., 4, 145-151.

GULDBERG, ALF. 1923.  (4.1/8.1)                                                   2181
    ZUR THEORIE DER KORRELATION.
    METRON, 2, 637-645.

GULDBERG, ALF. 1933.  (18.1/B)                                                    2182
    SUR LA LOI DE BERNOULLI A DEUX VARIABLES ET LA CORRELATION.
    C.R. ACAD. SCI. PARIS, 196, 1634-1636.

GULDBERG, ALF. 1934.  (18.1)                                                      2183
    ON DISCONTINUOUS FUNCTIONS OF TWO VARIABLES.
    SKAND. AKTUARIETIDSKR., 17, 89-117.

GULDBERG, ALF. 1935.  (4.8)                                                       2184
    SUR LES LOIS DE PROBABILITES ET LA CORRELATION.
    ANN. INST. H. POINCARE, 5, 159-176.

GULDBERG, SVEN 1935.  (18.1)                                                      2185
    SUR LES FORMULES DE RECURRENCE DES SEMI-VARIANTS DE LA LOI DE
    BERNOULLI ET DE LA LOI DE PASCAL A N VARIABLES.
    C.R. ACAD. SCI. PARIS, 201, 376-378.


GULDBERG, SVEN 1935.  (18.1)                                                      2186
    RECURRENCES FORMULAE FOR THE SEMI-INVARIANTS OF SOME DISCONTINUOUS
    FREQUENCY FUNCTIONS IN N VARIABLES.
    SKAND. AKTUARIETIDSKR., 18, 270-278.


GULLIKSEN, HAROLD 1936.  (15.4)                                                   2187
    THE CONTENT RELIABILITY OF A TEST.
    PSYCHOMETRIKA, 1, 189-194.


GULLIKSEN, HAROLD 1945.  (15.4)                                                   2188
    THE RELATION OF ITEM DIFFICULTY AND INTER-ITEM CORRELATION TO TEST
    VARIANCE AND RELIABILITY.
    PSYCHOMETRIKA, 10, 79-91.


GULLIKSEN, HAROLD 1953.  (15.5)                                                   2189
    A GENERALIZATION OF THURSTONE'S LEARNING FUNCTION.
    PSYCHOMETRIKA, 18, 297-307.  (MR15,P.813)


GULLIKSEN, HAROLD 1954.  (15.5)                                                   2190
    A LEAST SQUARES SOLUTION FOR SUCCESSIVE INTERVALS ASSUMING UNEQUAL
    STANDARD DEVIATIONS.
    PSYCHOMETRIKA, 19, 117-139.


GULLIKSEN, HAROLD 1961.  (15.5)                                                   2191
    LINEAR AND MULTIDIMENSIONAL SCALING.
    PSYCHOMETRIKA, 26, 9-25.  (MR27-4698)


GULLIKSEN, HAROLD + TUCKER, LEDYARD R. 1951.  (15.1/E)                            2192
    A MECHANICAL MODEL ILLUSTRATING THE SCATTER DIAGRAM WITH OBLIQUE TEST VECTORS.
    PSYCHOMETRIKA, 16, 233-238.


GULLIKSEN, HAROLD + WILKS, S.S. 1950.  (8.3)                                      2193
    REGRESSION TESTS FOR SEVERAL SAMPLES.
    PSYCHOMETRIKA, 15, 91-114.  (MR12,P.193)


GUMBEL, EMILE J. 1923.  (4.1)                                                     2194
    UEBER DIE BEI FUNKTIONEN VON VARIABELN AUFTRETENDE KORRELATION.
    Z. ANGEW. MATH. MECH., 3, 396-398.


GUMBEL, EMILE J. 1926.  (4.1)                                                     2195
    SPURIOUS CORRELATION AND ITS SIGNIFICANCE TO PHYSIOLOGY.
    J. AMER. STATIST. ASSOC., 21, 179-194.


GUMBEL, EMILE J. 1926.  (4.1)                                                     2196
    UBER SCHEINBARE KORRELATIONEN UND IHR AUFTRETEN IN DER PHYSIOLOGISCHEN
    STATISTIK.
    Z. ANGEW. MATH. MECH., 6, 401-410.


GUMBEL, EMILE J. 1954.  (18.1)                                                    2197
    APPLICATIONS OF THE CIRCULAR NORMAL DISTRIBUTION.
    J. AMER. STATIST. ASSOC., 49, 267-297.


GUMBEL, EMILE J. 1958.  (17.4)                                                    2198
    DISTRIBUTIONS A PLUSIEURS VARIABLES DONT LES MARGES SONT DONNEES.
    C.R. ACAD. SCI. PARIS, 246, 2717-2720.  (MR20-6166)


GUMBEL, EMILE J. 1959.  (17.4)                                                    2199
    MULTIVARIATE DISTRIBUTIONS WITH GIVEN MARGINS.
    REV. FAC. CI. UNIV. LISBOA SER. 2A CI. MAT., 7, 179-218.


GUMBEL, EMILE J. 1960.  (17.4/18.1)                                               2200
    MULTIVARIATE DISTRIBUTIONS WITH GIVEN MARGINS AND ANALYTICAL EXAMPLES.
    BULL. INST. INTERNAT. STATIST., 37(3) 363-373.  (MR22-10070)


GUMBEL, EMILE J. 1960.  (18.1/B)                                                  2201
    BIVARIATE EXPONENTIAL DISTRIBUTIONS.
    J. AMER. STATIST. ASSOC., 55, 698-707.  (MR22-7191)

GUMBEL, EMILE J. 1960.   (17.6)                                                      2202
    DISTRIBUTIONS DES VALEURS EXTREMES EN PLUSIEURS DIMENSIONS.
    PUBL. INST. STATIST. UNIV. PARIS, 9, 171-173.   (MR22-10045)

GUMBEL, EMILE J. 1961.   (18.1/B)                                                    2203
    BIVARIATE LOGISTIC DISTRIBUTIONS.
    J. AMER. STATIST. ASSOC., 56, 335-349.   (MR28-1674)

GUMBEL, EMILE J. 1961.   (17.6/18.1/B)                                               2204
    UNE DISTRIBUTION A DEUX VARIABLES ET GENERALISATIONS DE SES VALEURS EXTREMES.
    REV. STATIST. APPL., 10(1) 97-98.

GUMBEL, EMILE J. 1962.   (14.1/17.6)                                                 2205
    MULTIVARIATE EXTREMAL DISTRIBUTIONS.
    BULL. INST. INTERNAT. STATIST., 39(2) 471-475.   (MR31-5261)

GUMBEL, EMILE J. 1966.   (14.1/17.6/B)                                               2206
    TWO SYSTEMS OF BIVARIATE EXTREMAL DISTRIBUTIONS.
    BULL. INST. INTERNAT. STATIST., 41, 749-763.   (MR35-6262)

GUMBEL, EMILE J. + GOLDSTEIN, NEIL 1964.   (17.6)                                    2207
    ANALYSIS OF EMPIRICAL BIVARIATE EXTREMAL DISTRIBUTIONS.
    J. AMER. STATIST. ASSOC., 59, 794-816.   (MR29-4134)

GUMBEL, EMILE J.; GREENWOOD, J. ARTHUR + DURAND, DAVID 1953.   (18.1/T)              2208
    THE CIRCULAR NORMAL DISTRIBUTION:   THEORY AND TABLES.
    J. AMER. STATIST. ASSOC., 48, 131-152.   (MR14,P.664)

GUNSTAD, BORGHILD    SEE
    2369. HARRIS, J. ARTHUR + GUNSTAD, BORGHILD (1930)

    2370. HARRIS, J. ARTHUR + GUNSTAD, BORGHILD (1931)

GUPTA, R. P. 1964.   (7.1/18.1)                                                      2209
    SOME EXTENSIONS OF THE WISHART AND MULTIVARIATE T-DISTRIBUTIONS IN
    THE COMPLEX CASE.
    J. INDIAN STATIST. ASSOC., 2, 131-136.   (MR33-822)

GUPTA, R. P. 1965.   (11.1/13.2/A)                                                   2210
    ASYMPTOTIC THEORY FOR PRINCIPAL COMPONENT ANALYSIS IN THE COMPLEX CASE.
    J. INDIAN STATIST. ASSOC., 3, 97-106.   (MR32-8424)

GUPTA. R. P.   SEE ALSO
    3176. KSHIRSAGAR, A. M. + GUPTA, R. P. (1965)

    3177. KSHIRSAGAR, A. M. + GUPTA, R. P. (1965)

GUPTA, SHANTI S. 1963.   (2.7/14.1/18.1/T)                                           2211
    PROBABILITY INTEGRALS OF MULTIVARIATE NORMAL AND MULTIVARIATE T.
    ANN. MATH. STATIST., 34, 792-828.   (MR27-2048)

GUPTA, SHANTI S. 1963.   (1.2/2.7)                                                   2212
    BIBLIOGRAPHY ON THE MULTIVARIATE NORMAL INTEGRALS AND RELATED TOPICS.
    ANN. MATH. STATIST., 34, 829-838.   (MR27-2049)

GUPTA, SHANTI S. 1965.   (6.7)                                                       2213
    ON SOME MULTIPLE DECISION (SELECTION AND RANKING) RULES.
    TECHNOMETRICS, 7, 225-245.

GUPTA, SHANTI S. 1966.   (6.5/6.7/ACT)                                               2214
    ON SOME SELECTION AND RANKING PROCEDURES FOR MULTIVARIATE NORMAL POPULATIONS
    USING DISTANCE FUNCTIONS.
    MULTIVARIATE ANAL. PROC. INTERNAT. SYMP. DAYTON (KRISHNAIAH), 457-475.

GUPTA, SHANTI S. + PILLAI, K.C.SREEDHARAN 1965.   (14.1/TU)                          2215
    ON LINEAR FUNCTIONS OF ORDERED CORRELATED NORMAL RANDOM VARIABLES.
    BIOMETRIKA, 52, 367-379.   (MR34-6942)

GUPTA, SHANTI S. + SOBEL, MILTON 1957.   (14.1/14.2/NT)                              2216
    ON A STATISTIC WHICH ARISES IN SELECTION AND RANKING PROBLEMS.
    ANN. MATH. STATIST., 28, 957-967.   (MR20-366)

GUPTA, SHANTI S. + SOBEL, MILTON 1962.   (17.6/18.1/18.3/U)                    2217
    ON THE SMALLEST OF SEVERAL CORRELATED F STATISTICS.
    BIOMETRIKA. 49, 509-523.  (MR28-3514)


GUPTA, SHANTI S.; PILLAI, K.C.SREEDHARAN + STECK, GEORGE P. 1964.  (14.1)      2218
    ON THE DISTRIBUTION OF LINEAR FUNCTIONS AND RATIOS OF LINEAR FUNCTIONS OF
    ORDERED CORRELATED NORMAL RANDOM VARIABLES WITH EMPHASIS ON RANGE.
    BIOMETRIKA. 51, 143-151.  (MR30-3517)

GUPTA, SOMESH DAS  SEE  DAS GUPTA, SOMESH


GURIAN, JOAN M.; CORNFIELD, JEROME + MOSIMANN, JAMES E. 1964.  (18.2/18.3/T)   2219
    COMPARISONS OF POWER FOR SOME EXACT MULTINOMIAL SIGNIFICANCE TESTS.
    PSYCHOMETRIKA. 29, 409-419.


GURLAND, JOHN 1948.  (2.5/14.5/17.9)                                            2220
    INVERSION FORMULAE FOR THE DISTRIBUTION OF RATIOS.
    ANN. MATH. STATIST., 19, 228-237.  (MR9,P.582)


GURLAND, JOHN 1953.  (2.5/2.6/14.5)                                             2221
    DISTRIBUTION OF QUADRATIC FORMS AND RATIOS OF QUADRATIC FORMS.
    ANN. MATH. STATIST., 24, 416-427.  (MR15,P.885)


GURLAND, JOHN 1955.  (2.5/14.5)                                                 2222
    DISTRIBUTION OF DEFINITE AND OF INDEFINITE QUADRATIC FORMS.
        (AMENDED BY NO. 2226)
    ANN. MATH. STATIST., 26, 122-127.  (MR16,P.727)


GURLAND, JOHN 1955.  (18.1/18.3)                                                2223
    DISTRIBUTION OF THE MAXIMUM OF THE ARITHMETIC MEAN OF CORRELATED
    RANDOM VARIABLES.
        (AMENDED BY NO. 2225)
    ANN. MATH. STATIST., 26, 294-300.  (MR19,P.1204)


GURLAND, JOHN 1956.  (2.5/14.5)                                                 2224
    QUADRATIC FORMS IN NORMALLY DISTRIBUTED RANDOM VARIABLES.
    SANKHYA, 17, 37-50.  (MR18,P.607)


GURLAND, JOHN 1959.  (18.1/18.3)                                                2225
    CORRECTIONS TO: DISTRIBUTION OF THE MAXIMUM OF THE ARITHMETIC MEAN
    OF CORRELATED RANDOM VARIABLES.
        (AMENDMENT OF NO. 2223)
    ANN. MATH. STATIST., 30, 1265-1266.  (MR19,P.1204)


GURLAND, JOHN 1962.  (2.5/14.5)                                                 2226
    ERRATA:  DISTRIBUTION OF DEFINITE AND OF INDEFINITE QUADRATIC FORMS.
        (AMENDMENT OF NO. 2222)
    ANN. MATH. STATIST., 33, 813.  (MR16,P.727)


GURLAND, JOHN   SEE ALSO
    1381. DORFF, M. + GURLAND, JOHN (1961)


GURNEY, M.   SEE
    1178. DALENIUS, TORE + GURNEY, M. (1951)


GUSTAFSON, ROBERT L. 1961.  (4.5)                                               2227
    PARTIAL CORRELATIONS IN REGRESSION COMPUTATIONS.
    J. AMER. STATIST. ASSOC., 56, 363-367.  (MR23-A2997)


GUSTAVSON, J.   SEE
    1565. ERLANDER, S. + GUSTAVSON, J. (1965)


GUTIERREZ CABRIA, S. 1966.  (4.3)                                               2228
    USE AND ABUSE OF REGRESSION AND CORRELATION.
        (IN SPANISH)
    ESTADIST. ESPAN., 30, 24-39.


GUTTMAN, IRWIN 1961.  (19.1)                                                    2229
    BEST POPULATIONS AND TOLERANCE REGIONS.
    ANN. INST. STATIST. MATH. TOKYO, 13, 9-26.  (MR25-656)

GUTTMAN, IRWIN + MEETER, DUANE A. 1965.   (8.8/14.3/17.9/E)                2230
    ON BEALE'S MEASURES OF NON-LINEARITY.
    TECHNOMETRICS, 7, 623-637.

GUTTMAN, IRWIN    SEE ALSO
    1768. FRASER, DONALD A. S. + GUTTMAN, IRWIN (1956)

    5466. TIAO, GEORGE C. + GUTTMAN, IRWIN (1965)

GUTTMAN, LESTER    SEE
    5095. SMITH, CYRIL STANLEY + GUTTMAN, LESTER (1953)

GUTTMAN, LOUIS 1938.   (4.5/4.6)                                            2231
    A NOTE ON THE DERIVATION OF FORMULAE FOR MULTIPLE AND PARTIAL CORRELATION.
    ANN. MATH. STATIST., 9, 305-308.

GUTTMAN, LOUIS 1940.   (8.1/15.1)                                           2232
    MULTIPLE RECTILINEAR PREDICTION AND THE RESOLUTION INTO COMPONENTS.
    PSYCHOMETRIKA, 5, 75-99.   (MR2,P.234)

GUTTMAN, LOUIS 1944.   (15.5)                                               2233
    A BASIS FOR SCALING QUALITATIVE DATA.
    AMER. SOCIOL. REV., 9, 139-150.

GUTTMAN, LOUIS 1944.   (15.1/20.1)                                          2234
    GENERAL THEORY AND METHODS FOR MATRIC FACTORING.
    PSYCHOMETRIKA, 9, 1-16.   (MR6,P.6)

GUTTMAN, LOUIS 1945.   (15.4)                                               2235
    A BASIS FOR ANALYZING TEST-RETEST RELIABILITY.
    PSYCHOMETRIKA, 10, 255-282.   (MR7,P.318)

GUTTMAN, LOUIS 1946.   (20.3)                                               2236
    ENLARGEMENT METHODS FOR COMPUTING THE INVERSE MATRIX.
    ANN. MATH. STATIST., 17, 336-343.   (MR8,P.171)

GUTTMAN, LOUIS 1946.   (15.4)                                              2237
    THE TEST-RETEST RELIABILITY OF QUALITATIVE DATA.
    PSYCHOMETRIKA, 11, 81-95.   (MR8,P.41)

GUTTMAN, LOUIS 1947.   (15.5)                                               2238
    THE CORNELL TECHNIQUE FOR SCALE AND INTENSITY ANALYSIS.
    EDUC. PSYCHOL. MEAS., 7, 247-279.

GUTTMAN, LOUIS 1947.   (15.5)                                               2239
    ON FESTINGER'S EVALUATION OF SCALE ANALYSIS.
    PSYCHOL. BULL., 44, 451-465.

GUTTMAN, LOUIS 1950.   (11.1)                                               2240
    THE PRINCIPAL COMPONENTS OF SCALE ANALYSIS.
    MEAS. PREDICT. (STOUFFER ET AL.), 312-361.

GUTTMAN, LOUIS 1951.   (15.1/15.2/15.5)                                     2241
    SCALE ANALYSIS, FACTOR ANALYSIS, AND DR. EYSENCK.
    INTERNAT. J. OPINION ATTITUDE RES., 5, 103-120.

GUTTMAN, LOUIS 1952.   (15.1/15.3)                                          2242
    MULTIPLE GROUP METHODS FOR COMMON FACTOR ANALYSIS:   THEIR BASIS,
    COMPUTATION AND INTERPRETATION.
    PSYCHOMETRIKA, 17, 209-223.

GUTTMAN, LOUIS 1953.   (15.1)                                               2243
    A NOTE ON SIR CYRIL BURT'S "FACTORIAL ANALYSIS OF QUALITATIVE DATA"
    BRITISH J. STATIST. PSYCHOL., 6, 1-4.

GUTTMAN, LOUIS 1953.   (15.4)                                               2244
    ON SMITH'S PAPER ON "RANDOMNESS OF ERROR" IN REPRODUCIBLE SCALES.
    EDUC. PSYCHOL. MEAS., 13, 505-511.

GUTTMAN, LOUIS 1953.   (8.1/15.1/17.3)                                      2245
    IMAGE THEORY FOR THE STRUCTURE OF QUANTITATIVE VARIATES.
    PSYCHOMETRIKA, 18, 277-296.   (MR15,P.543)

GUTTMAN, LOUIS 1954.  (11.1/15.5/E)
   THE PRINCIPAL COMPONENTS OF SCALABLE ATTITUDES.
  MATH. THINK. SOC. SCI. (LAZARSFELD), 216-257, 427-429.                    2246

GUTTMAN, LOUIS 1954.  (15.1/15.5)
   A NEW APPROACH TO FACTOR ANALYSIS:  THE RADEX.
  MATH. THINK. SOC. SCI. (LAZARSFELD), 258-348, 430-433.   (MR19, P.932)      2247

GUTTMAN, LOUIS 1954.  (15.1)
   SOME NECESSARY CONDITIONS FOR COMMON-FACTOR ANALYSIS.
  PSYCHOMETRIKA, 19, 149-161.  (MR19,P.932)                              2248

GUTTMAN, LOUIS 1954.  (15.5)
   AN OUTLINE OF SOME NEW METHODOLOGY FOR SOCIAL RESEARCH.
  PUBLIC OPINION QUART., 18, 395-404.                                    2249

GUTTMAN, LOUIS 1955.  (11.1/15.5)
   AN ADDITIVE METRIC FROM ALL THE PRINCIPAL COMPONENTS OF A PERFECT SCALE.
  BRITISH J. STATIST. PSYCHOL., 8, 17-24.                                2250

GUTTMAN, LOUIS 1955.  (15.1)
   THE DETERMINACY OF FACTOR SCORE MATRICES WITH IMPLICATIONS FOR FIVE
   OTHER BASIC PROBLEMS OF COMMON-FACTOR THEORY.
  BRITISH J. STATIST. PSYCHOL., 8, 65-81.                                2251

GUTTMAN, LOUIS 1955.  (15.1/E)
   LA METHODE RADEX D'ANALYSE FACTORIELLE.
  COLLOQ. INTERNAT. CENTRE NAT. RECH. SCI., 65, 209-219.                 2252

GUTTMAN, LOUIS 1955.  (15.1)
   A GENERALIZED SIMPLEX FOR FACTOR ANALYSIS.
  PSYCHOMETRIKA, 20, 173-192.  (MR19,P.1205)                             2253

GUTTMAN, LOUIS 1956.  (15.2)
   UNE SOLUTION AU PROBLEME DES COMMUNAUTES.
  BULL. CENTRE ETUDES RECH. PSYCHOTECH., 5, 123-128.                     2254

GUTTMAN, LOUIS 1956.  (15.2)
   "BEST POSSIBLE" SYSTEMATIC ESTIMATES OF COMMUNALITIES.
  PSYCHOMETRIKA, 21, 273-285.  (MR18,P.343)                              2255

GUTTMAN, LOUIS 1957.  (15.5/E)
   EMPIRICAL VERIFICATION OF THE RADEX STRUCTURE OF MENTAL ABILITIES AND
   PERSONALITY TRAITS.
  EDUC. PSYCHOL. MEAS., 17, 391-407.                                     2256

GUTTMAN, LOUIS 1957.  (20.1)
   SOME INEQUALITIES BETWEEN LATENT ROOTS AND MINIMAX (MAXIMIN) ELEMENTS
   OF REAL MATRICES.
  PACIFIC J. MATH., 7, 897-902.  (MR19,P.242)                            2257

GUTTMAN, LOUIS 1957.  (15.1/20.1)
   A NECESSARY AND SUFFICIENT FORMULA FOR MATRIX FACTORING.
  PSYCHOMETRIKA, 22, 79-81.  (MR19,P.114)                                2258

GUTTMAN, LOUIS 1957.  (4.6/15.1)
   SIMPLE PROOFS OF RELATIONS BETWEEN THE COMMUNALITY PROBLEM AND
   MULTIPLE CORRELATION.
  PSYCHOMETRIKA, 22, 147-157.  (MR19,P.476)                              2259

GUTTMAN, LOUIS 1958.  (15.1)
   WHAT LIES AHEAD FOR FACTOR ANALYSIS?
  EDUC. PSYCHOL. MEAS., 18, 497-515.                                     2260

GUTTMAN, LOUIS 1958.  (15.1/20.2)
   TO WHAT EXTENT CAN COMMUNALITIES REDUCE RANK?
  PSYCHOMETRIKA, 23, 297-308.                                            2261

GUTTMAN, LOUIS 1959.  (15.5/E)
   A STRUCTURAL THEORY FOR INTER GROUP BELIEFS AND ACTION.
  AMER. SOCIOL. REV., 24, 318-328.                                       2262

GUTTMAN, LOUIS 1959.  (15.1/15.5)                              2263
    METRICIZING RANK-ORDERED OR UNORDERED DATA FOR A LINEAR FACTOR ANALYSIS.
    SANKHYA, 21, 257-268.  (MR22-1969)


GUTTMAN, LOUIS 1960.  (8.8/15.1/20.2)                          2264
    THE MATRICES OF LINEAR LEAST-SQUARES IMAGE ANALYSIS.
    BRITISH J. STATIST. PSYCHOL., 13, 109-118.

GUTTMAN, LOUIS 1966.  (15.1/15.5/E)                            2265
    ORDER ANALYSIS OF CORRELATION MATRICES.
    HANDB. MULTIVARIATE EXPER. PSYCHOL. (CATTELL), 438-458.

GUTTMAN, LOUIS + COHEN, JOZEF 1943.  (8.1/15.1)               2266
    MULTIPLE RECTILINEAR PREDICTION AND THE RESOLUTION INTO COMPONENTS.  II.
    PSYCHOMETRIKA, 8, 169-183.  (MR5,P.127)

GUTTMAN, LOUIS + GUTTMAN, RUTH 1965.  (15.2/15.5/E)          2267
    A NEW APPROACH TO THE ANALYSIS OF GROWTH PATTERNS:  THE SIMPLEX
    STRUCTURE OF INTERCORRELATIONS OF MEASUREMENTS.
    GROWTH, 29, 219-232.

GUTTMAN, RUTH    SEE
    2267. GUTTMAN, LOUIS + GUTTMAN, RUTH (1965)

GUYOU. E. 1888.  (8.8)                                         2268
    NOTE RELATIVE A L'EXPRESSION DE L'ERREUR PROBABLE D'UN SYSTEME D'OBSERVATIONS.
    C.R. ACAD. SCI. PARIS, 106, 1282-1285.

HAAVELMO, TRYGVE 1938.  E(17.7)                               2269
    THE METHOD OF SUPPLEMENTARY CONFLUENT RELATIONS, ILLUSTRATED BY A
    STUDY OF STOCK PRICES.
    ECONOMETRICA, 6, 203-218.

HAAVELMO, TRYGVE 1943.  (16.2)                                2270
    THE STATISTICAL IMPLICATIONS OF A SYSTEM OF SIMULTANEOUS EQUATIONS.
    ECONOMETRICA, 11, 1-12.  (MR4,P.220)

HAAVELMO, TRYGVE 1944.  (16.2/E)                              2271
    THE PROBABILITY APPROACH IN ECONOMETRICS.
    ECONOMETRICA, 12(SP) 1-118.  (MR6,P.93)

HAAVELMO, TRYGVE 1950.  (17.7)                                2272
    REMARKS ON FRISCH'S CONFLUENCE ANALYSIS AND ITS USE IN ECONOMETRICS.
    COWLES COMMISS. MONOG., 10, 258-265.

HAAVELMO, TRYGVE    SEE ALSO
    2007. GIRSHICK, M. A. + HAAVELMO, TRYGVE (1947)

    2008. GIRSHICK, M. A. + HAAVELMO, TRYGVE (1953)

HABA, J. 1937.  (17.1)                                        2273
    SUR LES PROBABILITES DES EVENEMENTS DEPENDANTS.
    PUBL. FAC. SCI. UNIV. CHARLES, 154, 21-24.

HABERMAN, SOL 1955.  (17.8)                                   2274
    DISTRIBUTIONS OF KENDALL'S TAU BASED ON PARTIALLY ORDERED SYSTEMS.
    BIOMETRIKA, 42, 417-424.  (MR17,P.278)

HADAMARD, JACQUES 1893.  (20.1)                               2275
    RESOLUTION D'UNE QUESTION RELATIVE AUX DETERMINANTS.
    BULL. SCI. MATH. SER. 2, 17, 240-246.

ENTRY NO. 2276 APPEARS BETWEEN ENTRY NOS. 252 & 253.


HAGIS, PETER,JR.    SEE
    3147. KRISHNAIAH, PARUCHURI R.; HAGIS, PETER,JR. + STEINBERG, LEON (1963)

    3148. KRISHNAIAH, PARUCHURI R.; HAGIS, PETER,JR. + STEINBERG, LEON (1965)

HAGSTROEM, K.-G. 1929.   (17.9)                                              2277
    LES POTENTIELS A N DIMENSIONS ET LA FONCTION DE PROBABILITE.
    SKAND. AKTUARIETIDSKR., 12, 218-238.

HAGSTROEM, K.-G. 1930.   (4.8/ET)                                            2278
    CORRELATION ONCE MORE.
    SKAND. AKTUARIETIDSKR., 13, 193-204.

HAGSTROEM, K.-G. 1954.   (2.1/18.6)                                          2279
    HEXAEDERNS DIAGONALSNITT OCH GAUSS' FORDELNINGSKLOCKA.
        (WITH SUMMARY IN ENGLISH)
    NORDISK MAT. TIDSKR., 2, 97-100, 135-136. (MR16,P.602)

HAHURIJA, C. G. 1965.   (7.2/18.1/18.2)                                      2280
    A LEMMA ON RANDOM DETERMINANTS AND ITS APPLICATIONS TO CHARACTERIZING
    MULTIVARIATE DISTRIBUTIONS.
        (IN RUSSIAN, WITH SUMMARY IN ENGLISH, TRANSLATED AS NO. 2281)
    TEOR. VEROJATNOST. I PRIMENEN., 10, 755-758.   (MR32-8373)

HAHURIJA, C. G. 1965.   (7.2/18.1/18.2)                                      2281
    A LEMMA ON RANDOM DETERMINANTS AND ITS APPLICATION TO THE
    CHARACTERIZATION OF MULTIVARIATE DISTRIBUTIONS.
        (TRANSLATION OF NO. 2280)
    THEORY PROB. APPL., 10, 685-689.   (MR32-8373)

HAIGHT, FRANK A. 1961.   (1.2)                                               2282
    INDEX TO THE DISTRIBUTIONS OF MATHEMATICAL STATISTICS.
    J. RES. NAT. BUR. STANDARDS SECT. B, 65, 23-60.   (MR23-A3606)

HAITOVSKY, YOEL 1966.   (11.1)                                               2283
    A NOTE ON REGRESSION ON PRINCIPAL COMPONENTS.
    AMER. STATIST., 20(4) 28-29.

HAITOVSKY, YOEL 1966.   (8.1/E)                                              2284
    UNBIASED MULTIPLE REGRESSION COEFFICIENTS ESTIMATED FROM ONE-WAY-CLASSIFICATION
    TABLES WHEN THE CROSS CLASSIFICATIONS ARE UNKNOWN.
    J. AMER. STATIST. ASSOC., 61, 720-728.

HAJEK, JAROSLAV 1958.   (3.2/4.8/10.2)                                       2285
    O ROZDELENI NEKTERYCH STATISTIK ZA PRITOMNOSTI VNITROTRIDNI KORELACE.
        (WITH SUMMARY IN RUSSIAN AND ENGLISH, TRANSLATED AS NO. 2289)
    CASOPIS PEST. MAT., 83, 327-329.   (MR21-389)

HAJEK, JAROSLAV 1958.   (16.4)                                               2286
    PREDICTING A STATIONARY PROCESS WHEN THE CORRELATION FUNCTION IS CONVEX.
    CZECH. MATH. J., 8, 150-154.   (MR20-2042)

HAJEK, JAROSLAV 1958.   (16.4)                                               2287
    ON A PROPERTY OF NORMAL DISTRIBUTIONS OF ANY STOCHASTIC PROCESS.
        (IN RUSSIAN, WITH SUMMARY IN ENGLISH)
    CZECH. MATH. J., 8, 610-618.   (MR22-1428)

HAJEK, JAROSLAV 1961.   (2.5/14.5/17.1)                                      2288
    NECESSARY AND SUFFICIENT CONDITIONS FOR ASYMPTOTIC NORMALITY OF A
    SCALAR PRODUCT OF TWO NORMALLY PERMUTATED VECTORS.
    DEUX. CONG. MATH. HONGR., 2(4) 7-9.

HAJEK, JAROSLAV 1962.   (3.2/4.8/10.2)                                       2289
    ON THE DISTRIBUTION OF SOME STATISTICS IN THE PRESENCE OF INTRACLASS
    CORRELATION.
        (TRANSLATION OF NO. 2285)
    SELECT. TRANSL. MATH. STATIST. PROB., 2, 75-77.   (MR27-855)

HAKE, HAROLD W. 1966.   (15.5/E)                                             2290
    THE STUDY OF PERCEPTION IN THE LIGHT OF MULTIVARIATE METHODS.
    HANDB. MULTIVARIATE EXPER. PSYCHOL. (CATTELL), 502-534.

HALBERT, MICHAEL H.   SEE
    2109. GREEN, PAUL E.; HALBERT, MICHAEL H. + ROBINSON, PATRICK J. (1966)

HALD, A. HJORTH + RASCH, G. 1943.   (14.4)                                   2291
    NOGLE ANVENDELSER AF TRANSFORMATIONSMETODEN I DEN NORMALE FORDELINGS TEORI.
    FESTSK. STEFFENSEN, 52-65.   (MR8,P.42)

HALDANE, J. B. S. 1948.   (17.3)                                              2292
    NOTE ON THE MEDIAN OF A MULTIVARIATE DISTRIBUTION.
    BIOMETRIKA, 35, 414–415.

HALDANE, J. B. S. 1949.   (4.1/4.2/18.1)                                      2293
    A NOTE ON NON-NORMAL CORRELATION.
    BIOMETRIKA, 36, 467–468.   (MR11,P.733)

HALDANE, J. B. S. 1959.   (1.1)                                               2294
    THE MATHEMATICAL THEORY OF NATURAL SELECTION.
    CALCUTTA MATH. SOC. GOLDEN JUBILEE COMMEM. VOLUME, 1, 99–103.   (MR28-2927)

HALDANE, J. B. S. 1960.   (18.1/18.3)                                         2295
    THE ADDITION OF RANDOM VECTORS.
    SANKHYA, 22, 213–220.   (MR24-A565)

HALDANE, J. B. S.    SEE ALSO
    2998. KERMACK, K. A. + HALDANE, J. B. S. (1950)

HALIKOV, M. K. 1958.   (17.1)                                                 2296
    A LOCAL THEOREM FOR SUMS OF INDEPENDENT RANDOM VECTORS.
        (IN RUSSIAN, WITH SUMMARY IN UZBEKISH)
    IZV. AKAD. NAUK UZSSR. SER. FIZ.-MAT. NAUK, 1958(2) 95–105.   (MR22-10001)

HALIKOV, M. K. 1959.   (17.1)                                                 2297
    AN EXACT ESTIMATE FOR A MULTIDIMENSIONAL LOCAL THEOREM.
        (IN RUSSIAN, WITH SUMMARY IN UZBEKISH)
    IZV. AKAD. NAUK UZSSR. SER. FIZ.-MAT. NAUK, 1959(2) 3–6.   (MR24-A571)

HALIKOV, M. K.    SEE ALSO
    3634. MAMATOV, M. + HALIKOV, M. K. (1964)

HALL, D. M.; WELKER, E.L. + CRAWFORD, ISABELLE 1945.   (15.3)                 2298
    FACTOR ANALYSIS CALCULATIONS BY TABULATING MACHINES.
    PSYCHOMETRIKA, 10, 93–125.   (MR7,P.87)

HALL, MARSHALL, JR.    SEE
    400. BAUMERT, L. D. + HALL, MARSHALL, JR. (1965)

HALL, P. 1927.   (4.5/4.6)                                                    2299
    MULTIPLE AND PARTIAL CORRELATION COEFFICIENTS IN THE CASE OF AN
    M-FOLD VARIATE SYSTEM.
    BIOMETRIKA, 19, 100–109.

HALL, WM. JACKSON 1958.   (19.1)                                              2300
    MOST ECONOMICAL MULTIPLE-DECISION RULES.
    ANN. MATH. STATIST., 29, 1079–1094.   (MR20-6761)

HALMOS, PAUL R. 1948.   (17.3)                                               2301
    THE RANGE OF A VECTOR MEASURE.
    BULL. AMER. MATH. SOC., 54, 416–421.   (MR9,P.574)

HALPERIN, MAX 1951.   (8.1/8.2/16.5)                                          2302
    NORMAL REGRESSION THEORY IN THE PRESENCE OF INTRA-CLASS CORRELATION.
    ANN. MATH. STATIST., 22, 573–580.   (MR13,P.261)

HALPERIN, MAX 1961.   (17.7)                                                  2303
    FITTING OF STRAIGHT LINES AND PREDICTION WHEN BOTH VARIABLES ARE
    SUBJECT TO ERROR.
    J. AMER. STATIST. ASSOC., 56, 657–669.   (MR23-A2260)

HALPERIN, MAX 1964.   (17.7/C)                                                2304
    INTERVAL ESTIMATION IN LINEAR REGRESSION WHEN BOTH VARIABLES ARE
    SUBJECT TO ERROR.
    J. AMER. STATIST. ASSOC., 59, 1112–1120.

HALPERIN, MAX + MANTEL, NATHAN 1963.   (3.1/3.2/BC)                           2305
    INTERVAL ESTIMATION OF NON-LINEAR PARAMETRIC FUNCTIONS.
    J. AMER. STATIST. ASSOC., 58, 611–627.   (MR27-4304)

HALPERIN, MAX; HARTLEY, H. O. + HOEL, PAUL G. 1965.   (1.1)                   2306
    RECOMMENDED STANDARDS FOR STATISTICAL SYMBOLS AND NOTATION.
    AMER. STATIST., 19(3) 12–14.

HALPERIN, MAX; GREENHOUSE, SAMUEL W.; CORNFIELD, JEROME + 1955. (14.2/TU)    2307
  ZALOKAR, JULIA
    TABLES OF PERCENTAGE POINTS FOR THE STUDENTIZED MAXIMUM ABSOLUTE
    DEVIATE IN NORMAL SAMPLES.
  J. AMER. STATIST. ASSOC., 50, 185-195.  (MR16,P.726)


HALPHEN, ETIENNE 1939.  (14.4)                                              2308
    SUR LA COVARIATION.
  C.R. ACAD. SCI. PARIS, 208, 780-781.


HALPIN, ALAN H.   SEE
    591. BLISCHKE, WALLACE R. + HALPIN, ALAN H. (1966)


HAMAKER, H. C. 1962.  (8.1/19.1/20.2)                                       2309
    DE L'ANALYSE A REGRESSION MULTIPLE.
  REV. STATIST. APPL., 10(1) 23-48.


HAMBURGER, HANS 1951.  (20.1/20.5)                                          2310
    NON-SYMMETRIC OPERATORS IN HILBERT SPACE.
  PROC. SYMP. SPECTRAL THEORY DIFFER. PROBLEMS, 67-112.


HAMDAN, H. A.   SEE
    3265. LANCASTER, H. O. + HAMDAN, H. A. (1964)


HAMILTON, MAX 1949.  (4.8)                                                  2311
    A SIMPLE DIAGRAM FOR OBTAINING TETRACHORIC CORRELATION COEFFICIENTS.
  BRITISH J. PSYCHOL., 39, 168-171.


HAMMERSLEY, J. M. 1950.  (18.3)                                             2312
    THE DISTRIBUTION OF DISTANCE IN A HYPERSPHERE.
  ANN. MATH. STATIST., 21, 447-452.  (MR12,P.268)


HAMMERSLEY, J. M. 1960.  (20.3)                                             2313
    MONTE CARLO METHODS FOR SOLVING MULTIVARIABLE PROBLEMS.
  ANN. NEW YORK ACAD. SCI., 86, 844-874.  (MR22-8644)


HAMMING, R. W. + PINKHAM, ROGER S. 1966.  (20.4)                            2314
    A CLASS OF INTEGRATION FORMULAS.
  J. ASSOC. COMPUT. MACH., 13, 430-438.


HAMON, B. V. + HANNAN, E. J. 1963.  (16.5)                                  2315
    ESTIMATING RELATIONS BETWEEN TIME SERIES.
  J. GEOPHYS. RES., 68, 6033-6041.


HANAJIRI, AKIRA 1966.  (4.5/8.1/17.9/E)                                     2316
    AN APPLICATION OF MULTIPLE REGRESSION ANALYSIS TO CUPOLA MELTING OPERATIONS.
  REP. STATIST. APPL. RES. UN. JAPAN. SCI. ENGRS., 13, 58-68.


HANDY, UVAN + LENTZ, THEODORE F., JR. 1934.  (15.4)                         2317
    ITEM VALUE AND TEST RELIABILITY.
  J. EDUC. PSYCHOL., 25, 703-708.


HANNAN, E. J. 1955.  (17.1/U)                                               2318
    EXACT TESTS FOR SERIAL CORRELATION.
  BIOMETRIKA, 42, 133-142.  (MR16,P.1040)


HANNAN, E. J. 1955.  (16.5)                                                 2319
    AN EXACT TEST FOR CORRELATION BETWEEN TIME SERIES.
  BIOMETRIKA, 42, 316-326.  (MR17,P.381)


HANNAN, E. J. 1956.  (8.3/16.5)                                             2320
    THE ASYMPTOTIC POWERS OF CERTAIN TESTS BASED ON MULTIPLE CORRELATIONS.
  J. ROY. STATIST. SOC. SER. B, 18, 227-233.  (MR18,P.682)


HANNAN, E. J. 1956.  (16.4/16.5)                                            2321
    EXACT TESTS FOR SERIAL CORRELATION IN VECTOR PROCESSES.
  PROC. CAMBRIDGE PHILOS. SOC., 52, 482-487.  (MR18,P.520)

HANNAN, E. J. 1958.  (16.5/U)                                                    2322
     THE ASYMPTOTIC POWERS OF CERTAIN TESTS OF GOODNESS OF FIT FOR TIME SERIES.
     J. ROY. STATIST. SOC. SER. B, 20, 143-151.   (MR20-6180)

HANNAN, E. J. 1960.  (16.5)                                                      2323
     THE ESTIMATION OF SEASONAL VARIATION.
     AUSTRAL. J. STATIST., 2, 1-15.   (MR22-3087)

HANNAN, E. J. 1961.  (9.3)                                                       2324
     THE GENERAL THEORY OF CANONICAL CORRELATION AND ITS RELATION TO
     FUNCTIONAL ANALYSIS.
     J. AUSTRAL. MATH. SOC., 2, 229-242.   (MR29-4142)

HANNAN, E. J. 1961.  (8.2/16.5/A)                                                2325
     A CENTRAL LIMIT THEOREM FOR SYSTEMS OF REGRESSIONS.
     PROC. CAMBRIDGE PHILOS. SOC., 57, 583-588.   (MR23-A724)

HANNAN, E. J. 1963.  (16.5)                                                      2326
     REGRESSION FOR TIME SERIES WITH ERRORS OF MEASUREMENT.
     BIOMETRIKA, 50, 293-302.

HANNAN, E. J. 1963.  (16.5/U)                                                    2327
     THE ESTIMATION OF SEASONAL VARIATION IN ECONOMIC TIME SERIES.
          (AMENDED BY NO. 2328)
     J. AMER. STATIST. ASSOC., 58, 31-44.

HANNAN, E. J. 1963.  (16.5/U)                                                    2328
     CORRIGENDA:  THE ESTIMATION OF SEASONAL VARIATION IN ECONOMETRIC TIME SERIES.
          (AMENDMENT OF NO. 2327)
     J. AMER. STATIST. ASSOC., 58, 1162.

HANNAN, E. J. 1963.  (8.1/8.2/16.5/U)                                            2329
     REGRESSION FOR TIME SERIES.
     PROC. SYMP. TIME SERIES ANAL. BROWN UNIV. (ROSENBLATT), 17-37.   (MR27-3062)

HANNAN, E. J. 1965.  (1.1)                                                       2330
     GROUP REPRESENTATIONS AND APPLIED PROBABILITY.
     J. APPL. PROB., 2, 1-68.  (SEE ALSO CHAPTER I)


HANNAN, E. J.    SEE ALSO
     2315. HAMON, B. V. + HANNAN, E. J. (1963)

ENTRY NO. 2331 APPEARS BETWEEN ENTRY NOS. 1876 & 1877.


HANNAN, J. F. 1960.  (19.3/A)                                                    2332
     CONSISTENCY OF MAXIMUM LIKELIHOOD ESTIMATION OF DISCRETE DISTRIBUTIONS.
     CONTRIB. PROB. STATIST. (HOTELLING VOLUME), 249-257.

HANNAN, J. F. + TATE, ROBERT F. 1965.  (3.2)                                     2333
     ESTIMATION OF THE PARAMETERS FOR A MULTIVARIATE NORMAL DISTRIBUTION
     WHEN ONE VARIABLE IS DICHOTOMIZED.
     BIOMETRIKA, 52, 664-668.   (MR34-5199)

HANSON, D. L.    SEE
     597. BLUM, JULIUS R.; HANSON, D. L. + KOOPMANS, LAMBERT H. (1963)

HANSON, M. A. 1965.  (19.3/A)                                                    2334
     INEQUALITY CONSTRAINED MAXIMUM LIKELIHOOD ESTIMATION.
     ANN. INST. STATIST. MATH. TOKYO, 17, 311-321.

HANSON, R. M.    SEE
     3935. MUTALIPASSI, L. R. + HANSON, R. M. (1964)

HARKNESS, W. L.    SEE
     626. BOLGER, E. M. + HARKNESS, W. L. (1965)

     5192. SRIVASTAVA, OM PRAKASH; HARKNESS, W. L. + BARTOO, J. B. (1964)

HARLEY, B. I. 1954.  (4.2)                                                       2335
     A NOTE ON THE PROBABILITY INTEGRAL OF THE CORRELATION COEFFICIENT.
     BIOMETRIKA, 41, 278-280.   (MR15,P.971)

HARLEY, B. I. 1956.  (4.2)                                                    **2336**
    SOME PROPERTIES OF AN ANGULAR TRANSFORMATION FOR THE CORRELATION COEFFICIENT.
    BIOMETRIKA, 43, 219-224.  (MR17,P.981)

HARLEY, B. I. 1957.  (4.2)                                                    **2337**
    RELATION BETWEEN THE DISTRIBUTIONS OF NON-CENTRAL T AND OF A
    TRANSFORMED CORRELATION COEFFICIENT.
    BIOMETRIKA, 44, 219-224.  (MR19,P.472)

HARLEY, B. I. 1957.  (4.2)                                                    **2338**
    FURTHER PROPERTIES OF AN ANGULAR TRANSFORMATION OF THE CORRELATION COEFFICIENT.
    BIOMETRIKA, 44, 273-275.

HARMAN, HARRY H. 1938.  (15.2)                                               **2339**
    SYSTEMS OF REGRESSION EQUATIONS FOR THE ESTIMATION OF FACTORS.
    J. EDUC. PSYCHOL., 29, 431-441.

HARMAN, HARRY H. 1938.  (15.2)                                               **2340**
    EXTENSIONS OF FACTORIAL SOLUTIONS.
    PSYCHOMETRIKA, 3, 75-84.

HARMAN, HARRY H. 1941.  (15.1)                                               **2341**
    FOUR ASPECTS OF FACTOR ANALYSIS.
    PROC. EDUC. RES. FORUM, 60-67.

HARMAN, HARRY H. 1941.  (15.2)                                               **2342**
    ON THE RECTILINEAR PREDICTION OF OBLIQUE FACTORS.
    PSYCHOMETRIKA, 6, 29-35.  (MR2,P.234)

HARMAN, HARRY H. 1954.  (15.3/20.3)                                          **2343**
    THE SQUARE ROOT METHOD AND MULTIPLE GROUP METHODS OF FACTOR ANALYSIS.
    PSYCHOMETRIKA, 19, 39-55.  (MR16,P.177)

HARMAN, HARRY H. 1960.  (15.3)                                               **2344**
    FACTOR ANALYSIS.
    MATH. METH. DIGIT. COMPUT. (RALSTON + WILF), 204-212.  (MR22-8698)

HARMAN, HARRY H. + FUKUDA, YOICHIRO 1966.  (15.2)                            **2345**
    RESOLUTION OF THE HEYWOOD CASE IN THE MINRES SOLUTION.
    PSYCHOMETRIKA, 31, 563-571.  (MR35-2420)

HARMAN, HARRY H. + JONES, WAYNE H. 1966.  (15.2)                             **2346**
    FACTOR ANALYSIS BY MINIMIZING RESIDUALS (MINRES).
    PSYCHOMETRIKA, 31, 351-368.  (MR34-893)

HARMAN, HARRY H.    SEE ALSO
    2531. HOLZINGER, KARL J. + HARMAN, HARRY H. (1937)

    2532. HOLZINGER, KARL J. + HARMAN, HARRY H. (1938)

    2533. HOLZINGER, KARL J. + HARMAN, HARRY H. (1939)

HARRIS, A. J. 1957.  (2.1/17.9)                                             **2347**
    A MAXIMUM-MINIMUM PROBLEM RELATED TO STATISTICAL DISTRIBUTIONS IN TWO
    DIMENSIONS.
    BIOMETRIKA, 44, 384-398.  (MR19,P.691)

HARRIS, CHESTER W. 1951.  (15.1/20.2)                                        **2348**
    THE SYMMETRICAL IDEMPOTENT MATRIX IN FACTOR ANALYSIS.
    J. EXPER. EDUC., 19, 237-246.

HARRIS, CHESTER W. 1954.  (5.3)                                              **2349**
    NOTE ON DISPERSION ANALYSIS.
    J. EXPER. EDUC., 22, 289-291.

HARRIS, CHESTER W. 1955.  (15.1)                                             **2350**
    SEPARATION OF DATA AS A PRINCIPLE IN FACTOR ANALYSIS.
    PSYCHOMETRIKA, 20, 23-28.

HARRIS, CHESTER W. 1955.  (15.5)                                             **2351**
    CHARACTERISTICS OF TWO MEASURES OF PROFILE SIMILARITY.
    PSYCHOMETRIKA, 20, 289-297.

HARRIS, CHESTER W. 1958.   (15.1)                                          2352
    RELATIONSHIP BETWEEN TWO SYSTEMS OF FACTOR ANALYSIS.
    PSYCHOMETRIKA, 21, 185-190.

HARRIS, CHESTER W. 1962.   (4.6/12.1/15.1)                                 2353
    SOME RAO-GUTTMAN RELATIONSHIPS.
    PSYCHOMETRIKA, 27, 247-263.   (MR26-3146)

HARRIS, CHESTER W. 1963.   (15.1/15.5/E)                                   2354
    CANONICAL FACTOR MODELS FOR THE DESCRIPTION OF CHANGE.
    PROBLEMS MEAS. CHANGE (HARRIS), 138-155.

HARRIS, CHESTER W. + KNOELL, DOROTHY M. 1948.   (15.3)                     2355
    THE OBLIQUE SOLUTION IN FACTOR ANALYSIS.
    J. EDUC. PSYCHOL., 39, 385-403.

HARRIS, CHESTER W. + SCHMID, JOHN, JR. 1950.  E(15.3)                      2356
    FURTHER APPLICATION OF THE PRINCIPLES OF DIRECT ROTATION IN FACTOR ANALYSIS.
    J. EXPER. EDUC., 18, 175-193.

HARRIS, D.L.   SEE
    2389. HARTLEY, H. O. + HARRIS, D.L. (1963)

HARRIS, J. ARTHUR 1909.   (8.9/U)                                          2357
    THE CORRELATION BETWEEN A VARIABLE AND THE DEVIATION OF A DEPENDENT
    VARIABLE FROM ITS PROBABLE VALUE.
    BIOMETRIKA, 6, 438-443.

HARRIS, J. ARTHUR 1909.   (4.1/20.4)                                       2358
    A SHORT METHOD OF CALCULATING THE COEFFICIENT OF CORRELATION IN THE
    CASE OF INTEGRAL VARIATES.
    BIOMETRIKA, 7, 214-218.

HARRIS, J. ARTHUR 1910.   (4.1)                                            2359
    THE ARITHMETIC OF THE PRODUCT MOMENT METHOD OF CALCULATING THE
    COEFFICIENT OF CORRELATION.
    AMER. NATUR., 44, 693-699.

HARRIS, J. ARTHUR 1911.   (4.1)                                           2360
    ON THE FORMATION OF CORRELATION AND CONTINGENCY TABLES WHEN THE
    NUMBER OF COMBINATIONS IS LARGE.
    AMER. NATUR., 45, 566-571.

HARRIS, J. ARTHUR 1912.   (4.1)                                            2361
    THE FORMATION OF CONDENSED CORRELATION TABLES WHEN THE NUMBER OF
    COMBINATIONS IS LARGE.
    AMER. NATUR., 46, 477-486.

HARRIS, J. ARTHUR 1913.   (4.8)                                            2362
    ON THE CALCULATION OF INTRA-CLASS AND INTER-CLASS COEFFICIENTS OF
    CORRELATION FROM CLASS MOMENTS WHEN THE NUMBER OF POSSIBLE
    COMBINATIONS IS LARGE.
    BIOMETRIKA, 9, 446-472.

HARRIS, J. ARTHUR 1914.   (4.8)                                            2363
    ON SPURIOUS VALUES OF INTRA-CLASS CORRELATION COEFFICIENTS ARISING
    FROM DISORDERLY DIFFERENTIATION WITHIN THE CLASSES.
    BIOMETRIKA, 10, 412-416.

HARRIS, J. ARTHUR 1917.   (4.1/8.1/EU)                                     2364
    NOTE ON "THE COEFFICIENT OF CORRELATION (BY W.G. REED)."
       (AMENDMENT OF NO. 4496)
    QUART. PUBL. AMER. STATIST. ASSOC., 15, 803-805.

HARRIS, J. ARTHUR 1917.   (4.1)                                           2365
    THE CORRELATION BETWEEN A COMPONENT, AND BETWEEN THE SUM OF TWO OR MORE
    COMPONENTS, AND THE SUM OF THE REMAINING COMPONENTS OF A VARIABLE.
    QUART. PUBL. AMER. STATIST. ASSOC., 15, 854-859.

HARRIS, J. ARTHUR 1918.   (4.1/4.8/E)                                      2366
    FURTHER ILLUSTRATIONS OF THE APPLICABILITY OF A COEFFICIENT MEASURING THE
    CORRELATION BETWEEN A VARIABLE AND THE DEVIATION OF A DEPENDENT VARIABLE FROM
    ITS PROBABLE VALUE.
    GENETICS, 3, 328-352.

HARRIS, J. ARTHUR 1922. (4.1)                                                    2367
    ON THE CORRELATION BETWEEN A VARIABLE AND THE DEVIATION OF AN
    ASSOCIATED BUT NOT DEPENDENT VARIABLE FROM ITS PROBABLE VALUE.
    J. AMER. STATIST. ASSOC., 18, 393-394.


HARRIS, J. ARTHUR 1930. (4.8)                                                    2368
    THE DETERMINATION OF INTRA-CLASS AND INTER-CLASS EQUIVALENT
    PROBABILITY COEFFICIENTS OF CORRELATION.
    AMER. NATUR., 64, 115-141.

HARRIS, J. ARTHUR + GUNSTAD, BORGHILD 1930. E(4.8/T)                             2369
    THE CORRELATION BETWEEN THE SEX OF HUMAN SIBLINGS.I. THE CORRELATION IN THE
    GENERAL POPULATION.
    GENETICS, 15, 445-461.

HARRIS, J. ARTHUR + GUNSTAD, BORGHILD 1931. (4.4/4.8/E)                          2370
    EXTENSION OF PEARSON'S CORRELATION METHOD TO INTRA-CLASS AND
    INTER-CLASS RELATIONSHIPS.
    J. AGRIC. RES., 42, 279-292.

HARRIS, T. E. 1965. (16.4)                                                       2371
    DIFFUSION WITH "COLLISIONS" BETWEEN PARTICLES.
    J. APPL. PROB., 2, 323-328.

HARRIS, W. A., JR. + HELVIG, T. N. 1966. (17.3)                                  2372
    MARGINAL AND CONDITIONAL DISTRIBUTIONS OF SINGULAR DISTRIBUTIONS.
    PUBL. RES. INST. MATH. SCI. KYOTO UNIV. SER. A, 1, 199-204. (MR33-4965)

HARRISON, P. J. 1965. (16.5/E)                                                   2373
    SHORT-TERM SALES FORECASTING.
    APPL. STATIST., 14, 102-139.

HARSH, CHARLES M. 1937. (15.3)                                                   2374
    A NOTE ON ROTATION OF AXES.
    PSYCHOMETRIKA, 2, 211-214.

HARSH, CHARLES M. 1940. (15.1)                                                   2375
    CONSTANCY AND VARIATION IN PATTERNS OF FACTOR LOADINGS.
    J. EDUC. PSYCHOL., 31, 335-359.

HART, B. I. 1942. (2.6/T)                                                        2376
    SIGNIFICANCE LEVELS FOR THE RATIO OF THE MEAN SQUARE SUCCESSIVE DIFFERENCE TO
    THE VARIANCE.
    ANN. MATH. STATIST., 13, 445-447. (MR4,P.165)

HART, B. I. + VON NEUMANN, JOHN 1942. (2.6/T)                                    2377
    TABULATION OF THE PROBABILITIES FOR THE RATIO OF THE MEAN SQUARE SUCCESSIVE
    DIFFERENCE TO THE VARIANCE.
    ANN. MATH. STATIST., 13, 207-214. (MR4,P.22)


HARTE, CORNELIA 1948. (4.8)                                                      2378
    DIE AUFSTELLUNG VON KOPPELUNGSGRUPPEN UNTER VERWENDUNG DER
    KORRELATIONSRECHNUNG.
    Z. NATURFORSCH. PART B, 3, 99-105.

HARTER, H. LEON 1951. (6.2)                                                      2379
    ON THE DISTRIBUTION OF WALD'S CLASSIFICATION STATISTIC.
    ANN. MATH. STATIST., 22, 58-67. (MR12,P.620)

HARTER, H. LEON 1960. (14.5/BT)                                                  2380
    CIRCULAR ERROR PROBABILITIES.
    J. AMER. STATIST. ASSOC., 55, 723-731. (MR26-1948)

HARTIGAN, J. A. 1966. (19.2/19.3)                                                2381
    ESTIMATION BY RANKING PARAMETERS.
    J. ROY. STATIST. SOC. SER. B, 28, 32-44.

HARTKEMEIER, HARRY P. + BROWN, LAURA M. 1936. (4.6/15.1)                         2382
    MULTIPLE CORRELATION AND THE MULTIPLE FACTOR METHOD.
    J. APPL. PSYCHOL., 20, 396-415.

HARTLEY, H. O. 1939.  (1.2)                                                     2383
    RECENT ADVANCES IN MATHEMATICAL STATISTICS.  BIBLIOGRAPHY OF
    MATHEMATICAL STATISTICS (1937, SECOND HALF, AND 1938).
    J. ROY. STATIST. SOC., 102, 406-444.  (MR1,P.22)

HARTLEY, H. O. 1940.  (10.2/U)                                                  2384
    TESTING THE HOMOGENEITY OF A SET OF VARIANCES.
    BIOMETRIKA, 31, 249-255.  (MR1,P.346)

HARTLEY, H. O. 1940.  (1.2)                                                     2385
    BIBLIOGRAPHY OF MATHEMATICAL STATISTICS (1939).
    J. ROY. STATIST. SOC., 103, 534-560.  (MR2,P.231)

HARTLEY, H. O. 1941.  (20.4/T)                                                  2386
    TABLES OF PERCENTAGE POINTS OF THE INCOMPLETE BETA-FUNCTION:  METHODS
    OF INTERPOLATION.
    BIOMETRIKA, 32, 161-167.

HARTLEY, H. O. 1944.  (14.4/U)                                                  2387
    STUDENTIZATION OR THE ELIMINATION OF THE STANDARD DEVIATION OF THE
    PARENT POPULATION FROM THE RANDOM-SAMPLE DISTRIBUTION OF STATISTICS.
    BIOMETRIKA, 33, 173-180.  (MR6,P.10)

HARTLEY, H. O. 1950.  (10.1/U)                                                  2388
    THE MAXIMUM F-RATIO AS A SHORT-CUT TEST FOR HETEROGENEITY OF VARIANCES.
    BIOMETRIKA, 37, 308-312.  (MR12,P.345)

HARTLEY, H. O. + HARRIS, D.L. 1963.  (8.3/E)                                    2389
    MONTE CARLO COMPUTATIONS IN NORMAL CORRELATION PROBLEMS.
    J. ASSOC. COMPUT. MACH., 10, 302-306.  (MR27-2040)

HARTLEY, H. O.   SEE ALSO
    2306. HALPERIN, MAX; HARTLEY, H. O. + HOEL, PAUL G. (1965)

HARTLEY, RAYMOND E. 1954.  (15.1)                                               2390
    TWO KINDS OF FACTOR ANALYSIS?
    PSYCHOMETRIKA, 19, 195-203.

HARTMAN, PHILIP + WINTNER, AUREL 1940.  (2.1)                                   2391
    ON THE SPHERICAL APPROACH TO THE NORMAL DISTRIBUTION LAW.
    AMER. J. MATH., 62, 759-779.  (MR2,P.107)

HARVEY, J. R.   SEE
    4598. ROHDE, CHARLES A. + HARVEY, J. R. (1965)

HATCHER, HAZEL M.   SEE
    391. BATEN, WILLIAM DOWELL + HATCHER, HAZEL M. (1944)

HAUPTMAN, HERBERT + KARLE, J. 1952.  (16.4/E)                                   2392
    CRYSTAL-STRUCTURE DETERMINATION BY MEANS OF A STATISTICAL
    DISTRIBUTION OF INTERATOMIC VECTORS.
    ACTA CRYST., 5, 48-59.

HAUPTMAN, HERBERT + KARLE, J. 1953.  (16.4/17.3/18.1/E)                         2393
    THE PROBABILITY DISTRIBUTION OF THE MAGNITUDE OF A STRUCTURE
    FACTOR.II. THE NON-CENTROSYMMETRIC CRYSTAL.
    ACTA CRYST., 6, 136-141.

HAUPTMAN, HERBERT   SEE ALSO
    2931. KARLE, J. + HAUPTMAN, HERBERT (1953)

HAVILAND, E. K. 1935.  (17.3)                                                   2394
    ON THE INVERSION FORMULA FOR FOURIER-STIELTJES TRANSFORMS IN MORE
    THAN ONE DIMENSION.
    AMER. J. MATH., 57, 94-101, 382, 569.

HAVILAND, E. K. 1935.  (17.3/B)                                                 2395
    ON THE MOMENTUM PROBLEM FOR DISTRIBUTION FUNCTIONS IN MORE THAN ONE DIMENSION.
    AMER. J. MATH., 57, 562-568.

HAYAKAWA, TAKESI 1965.  (20.1/20.5)                                             2396
    ON THE POSITIVE DEFINITENESS AND CONVEXITY OF A CERTAIN MATRIX.
        (IN JAPANESE, WITH SUMMARY IN ENGLISH)
    PROC. INST. STATIST. MATH. TOKYO, 13, 87-93.

HAYAKAWA, TAKESI 1966.   (2.5/7.4/14.5)                                          2397
    ON THE DISTRIBUTION OF A QUADRATIC FORM IN A MULTIVARIATE NORMAL SAMPLE.
    ANN. INST. STATIST. MATH. TOKYO, 18, 191-201.


HAYAKAWA, TAKESI + KABE, D. G. 1965.   (8.3/8.4)                                 2398
    ON TESTING THE HYPOTHESIS THAT SUBMATRICES OF THE MULTIVARIATE
    REGRESSION MATRICES OF  K  POPULATIONS ARE EQUAL.
    ANN. INST. STATIST. MATH. TOKYO, 17, 67-73.   (MR31-2796)

HAYAKAWA, TAKESI   SEE ALSO
    5060. SIOTANI, MINORU + HAYAKAWA, TAKESI (1964)

HAYASHI, CHIKIO 1946.   (16.4)                                                   2399
    ON THE PHENOMENA OF THE CORRELATED CHAIN.
        (IN JAPANESE)
    RES. MEM. INST. STATIST. MATH. TOKYO, 2, 439-456.

HAYASHI, CHIKIO 1950.   (15.5/E)                                                 2400
    ON THE QUANTIFICATION OF QUALITATIVE DATA FROM THE MATHEMATICO-STATISTICAL POINT OF VIEW.
        (AMENDED BY NO. 2402)
    ANN. INST. STATIST. MATH. TOKYO, 2, 35-47.   (MR12,P.511)


HAYASHI, CHIKIO 1954.   (15.5)                                                   2401
    MULTIDIMENSIONAL QUANTIFICATION--WITH THE APPLICATIONS TO ANALYSIS OF
    SOCIAL PHENOMENA.
    ANN. INST. STATIST. MATH. TOKYO, 5, 121-143.

HAYASHI, CHIKIO 1954.   (15.5/E)                                                 2402
    ERRATA TO: ON THE QUANTIFICATION OF QUALITATIVE DATA FROM THE
    MATHEMATICO-STATISTICAL POINT OF VIEW.
        (AMENDMENT OF NO. 2400)
    ANN. INST. STATIST. MATH. TOKYO, 5, 144.   (MR12,P.511)

HAYASHI, CHIKIO 1954.   (15.5)                                                   2403
    MULTIDIMENSIONAL QUANTIFICATION.  I, II.
    PROC. JAPAN ACAD., 30, 61-65, 165-169.  (MR16,P.381)

HAYASHI, CHIKIO 1961.   (15.5)                                                   2404
    THEORY OF QUANTIFICATION AND ITS EXAMPLES V.
        (IN JAPANESE, WITH SUMMARY IN ENGLISH)
    PROC. INST. STATIST. MATH. TOKYO, 8, 149-151.

HAYASHI, CHIKIO 1961.   (15.1/15.5)                                              2405
    THEORY OF QUANTIFICATION AND ITS EXAMPLES VI.
        (IN JAPANESE, WITH SUMMARY IN ENGLISH)
    PROC. INST. STATIST. MATH. TOKYO, 9, 29-35.

HAYASHI, CHIKIO 1964.   (15.5/17.8)                                              2406
    MULTIDIMENSIONAL QUANTIFICATION OF THE DATA OBTAINED BY THE METHOD OF
    PAIRED COMPARISON.
    ANN. INST. STATIST. MATH. TOKYO, 16, 231-245.   (MR31-5297)

HAYES, J.G.   SEE
    1748. FOX, LESLIE + HAYES, J.G. (1951)

HAYES, SAMUEL P., JR. 1939.   (4.8)                                              2407
    CONVERTING PERCENTAGE DIFFERENCES INTO TETRACHORIC CORRELATION COEFFICIENTS.
    J. EDUC. PSYCHOL., 30, 391-396.

HAYES, SAMUEL P., JR. 1943.   (4.4/T)                                            2408
    TABLES OF THE STANDARD ERROR OF TETRACHORIC CORRELATION COEFFICIENT.
    PSYCHOMETRIKA, 8, 193-203.   (MR5,P.42)

HAYES, SAMUEL P., JR. 1946.   (4.4/T)                                            2409
    DIAGRAMS FOR COMPUTING TETRACHORIC CORRELATION COEFFICIENTS FROM
    PERCENTAGE DIFFERENCES.
    PSYCHOMETRIKA, 11, 163-172.   (MR7,P.41)

HAYS, DAVID G. + BORGATTA, EDGAR F. 1954.   (15.5)                               2410
    AN EMPIRICAL COMPARISON OF RESTRICTED AND GENERAL LATENT DISTANCE ANALYSIS.
    PSYCHOMETRIKA, 19, 271-279.

HAYS, DAVID G.    SEE ALSO
    644. BORGATTA, EDGAR F. + HAYS, DAVID G. (1952)


HAZEL, L. N. + LUSH, J. L. 1942.  (6.7/14.4/E)                                        2411
    THE EFFICIENCY OF THREE METHODS OF SELECTION.
    J. HEREDITY, 33, 393-399.


HAZEL, L. N.; BAKER, M. L. + REINMILLER, C. F. 1943.  E(8.5)                          2412
    GENETIC AND ENVIRONMENTAL CORRELATIONS BETWEEN THE GROWTH RATES OF
    PIGS AT DIFFERENT AGES.
    J. ANIMAL SCI., 2, 118-128.


H'DOUBLER, F. 1910.  (8.1/U)                                                          2413
    A FORMULA FOR DRAWING TWO CORRELATED CURVES SO AS TO MAKE THE
    RESEMBLANCE AS CLOSE AS POSSIBLE.
    QUART. PUBL. AMER. STATIST. ASSOC., 12, 165-167.


HEALY, M.J.R. 1957.  (12.1/20.3)                                                      2414
    A ROTATION METHOD FOR COMPUTING CANONICAL CORRELATIONS.
    MATH. TABLES AIDS COMPUT., 11, 83-86.  (MR19,P.64)


HEALY, M.J.R. 1963.  (1.2)                                                            2415
    BIBLIOGRAPHY OF THE WORKS OF SIR RONALD FISHER.
    J. ROY. STATIST. SOC. SER. A, 126, 170-178.


HEALY, M.J.R. 1963.  (1.2)                                                            2416
    A SUBJECT INDEX TO THE KENDALL-DOIG BIBLIOGRAPHY OF STATISTICAL LITERATURE.
    J. ROY. STATIST. SOC. SER. A, 126, 270-275.  (MR30-5389)


HEALY, M.J.R. 1965.  (5.2)                                                            2417
    COMPUTING A DISCRIMINANT FUNCTION FROM WITHIN-SAMPLE DISPERSIONS.
    BIOMETRICS, 21, 1011-1012.


HEALY, M.J.R. 1965.  E(6.2)                                                           2418
    DESCRIPTIVE USES OF DISCRIMINANT FUNCTIONS.
    MATH. COMPUT. SCI. BIOL. MED., 93-102.


HEALY, M.J.R.    SEE ALSO
    216. ASHTON, E. H.; HEALY, M.J.R. + LIPTON, S. (1957)


HECK, D. L. 1960.  (8.4/13.1/20.4/T)                                                  2419
    CHARTS OF SOME UPPER PERCENTAGE POINTS OF THE DISTRIBUTION OF THE
    LARGEST CHARACTERISTIC ROOT.
    ANN. MATH. STATIST., 31, 625-642.  (MR22-10067)


HECKENDORFF, HARTMUT 1964.  (17.1)                                                    2420
    A HIGHER-DIMENSIONAL LIMIT THEOREM FOR DISTRIBUTION FUNCTIONS.
        (IN RUSSIAN, WITH SUMMARY IN GERMAN)
    UKRAIN. MAT. ZUR., 16, 365-373.  (MR29-643)


HECKENDORFF, HARTMUT 1965.  (17.1)                                                    2421
    A MULTIDIMENSIONAL LOCAL LIMIT THEOREM FOR DISTRIBUTIONS OF MIXED TYPE.
        (IN RUSSIAN)
    FIRST REPUBL. MATH. CONF. OF YOUNG RES. (AKAD. NAUK UKRAIN. SSR, KIEV), 2, 690-697.  (MR33-758)


HEDMAN, HATTIE B. 1938.  (15.1)                                                       2422
    A CRITICAL COMPARISON BETWEEN THE SOLUTIONS OF THE FACTOR PROBLEM
    OFFERED BY SPEARMAN AND THURSTONE.
    J. EDUC. PSYCHOL., 29, 671-685.


HEDMAN, HATTIE B.    SEE ALSO
    3423. LINE, W. + HEDMAN, HATTIE B. (1933)


HEERMAN, EMIL F. 1965.  (15.5)                                                        2423
    COMMENTS ON OVERALL'S "MULTIVARIATE METHODS FOR PROFILE ANALYSIS".
        (AMENDMENT OF NO. 4113)
    PSYCHOL. BULL., 63, 128.


HEERMAN, EMIL F. 1966.  (15.1)                                                        2424
    THE ALGEBRA OF FACTORIAL INDETERMINANCY.
    PSYCHOMETRIKA, 31, 539-543.  (MR35-1136)

HEILMAN, J. D. 1937.  (4.1/E)                                                    2425
    THE K AND G METHODS OF INTERPRETING THE COEFFICIENT OF CORRELATION.
    J. EDUC. PSYCHOL., 28, 232-236.


HEINE, VOLKER 1955.  (16.4/B)                                                    2426
    MODELS FOR TWO-DIMENSIONAL STATIONARY STOCHASTIC PROCESSES.
    BIOMETRIKA, 42, 170-178.  (MR17,P.167)


HEKKER, TH. 1947.  (2.1/4.1)                                                     2427
    CONSTRUCTIE VAN KANSELLISPEN IN EEN CORRELATIEDIAGRAM.
        (WITH SUMMARY IN ENGLISH)
    STATISTICA, LEIDEN, 1, 203-208.  (MR11,P.259)


HELLINGER, E. D. 1909.  (20.1)                                                   2428
    NEUE BEGRUENDUNG DER THEORIE QUADRATISCHER FORMEN VON UNENDLICH VIELEN
    VERAENDERLICHEN.
    J. REINE ANGEW. MATH., 136, 210-271.


HELM, C. E.; MESSICK, SAMUEL J. + TUCKER, LEDYARD R. 1961.  (15.5)              2429
    PSYCHOLOGICAL MODELS FOR RELATING DISCRIMINATION AND MAGNITUDE
    ESTIMATION SCALES.
    PSYCHOL. REV., 68, 167-177.


HELMERT, F. R. 1876.  (2.1/2.3/B)                                                2430
    UBER DIE WAHRSCHEINLICHKEIT DER POTENZSUMMEN DER BEOBACHTUNGSFEHLER
    UND UBER EINIGE DAMIT IM ZUSAMMENHANGE STEHENDE FRAGEN.
    Z. MATH. PHYS., 21, 192-218.


HELSON, HENRY + LOWDENSLAGER, DAVID B. 1958.  (16.4)                            2431
    PREDICTION THEORY AND FOURIER SERIES IN SEVERAL VARIABLES.
    ACTA MATH., 99, 165-202.  (MR20-4155)


HELSON, HENRY + LOWDENSLAGER, DAVID B. 1961.  (16.4)                            2432
    VECTOR-VALUED PROCESSES.
    PROC. FOURTH BERKELEY SYMP. MATH. STATIST. PROB., 2, 203-212.  (MR25-4577)


HELVIG, T. N.   SEE
    2372. HARRIS, W. A., JR. + HELVIG, T. N. (1966)


HEMMERLE, W. J. 1965.  (15.3)                                                    2433
    OBTAINING MAXIMUM-LIKELIHOOD ESTIMATES OF FACTOR LOADINGS AND
    COMMUNALITIES USING AN EASILY IMPLEMENTED ITERATIVE COMPUTER PROCEDURE.
    PSYCHOMETRIKA, 30, 291-302.  (MR34-3715)


HEMMERLE, W. J.   SEE ALSO
    1823. FULLER, E. L. + HEMMERLE, W. J. (1966)


HENDERSON, JAMES 1922.  (4.4)                                                    2434
    ON EXPANSIONS IN TETRACHORIC FUNCTIONS.
    BIOMETRIKA, 14, 157-185.


HENGARTNER, OTTO 1965.  (2.1/18.1)                                               2435
    AUSGEWAHLTE ZWEI- UND DREIDIMENSIONALE PRUFVERTEILUNGEN.
    METRIKA, 9, 1-37, 105-148.  (MR31-1729)


HENRICI, P.   SEE
    1730. FORSYTHE, GEORGE E. + HENRICI, P. (1960)


HENRYSSON, STEN 1950.  (15.2)                                                    2436
    THE SIGNIFICANCE OF FACTOR LOADINGS.
    BRITISH J. PSYCHOL. STATIST. SECT., 3, 159-165.


HENRYSSON, STEN 1962.  (4.4/15.2)                                                2437
    THE RELATION BETWEEN FACTOR LOADINGS AND BISERIAL CORRELATIONS IN ITEM ANALYSIS.
    PSYCHOMETRIKA, 27, 419-424.


HENSHAW, RICHARD C., JR. 1966.  (16.5/E)                                         2438
    TESTING SINGLE-EQUATION LEAST-SQUARES REGRESSION MODELS FOR AUTOCORRELATED
    DISTURBANCES.
    ECONOMETRICA, 34, 646-660.

HEPPES, ALADAR 1956.  (17.4)                                                    2439
    ON THE DETERMINATION OF PROBABILITY DISTRIBUTIONS OF MORE DIMENSIONS
    BY THEIR PROJECTIONS.
    ACTA MATH. ACAD. SCI. HUNGAR., 7, 403-410.  (MR19,P.70)

HERBACH, LEON H. 1959.  (8.6/U)                                                 2440
    PROPERTIES OF MODEL II-TYPE ANALYSIS OF VARIANCE TESTS, A:  OPTIMUM
    NATURE OF THE F-TEST FOR MODEL II IN THE BALANCED CASE.
    ANN. MATH. STATIST., 30, 939-959.  (MR22-1040)

HERON, DAVID 1910.  (4.2/T)                                                     2441
    BIOMETRIC NOTES.  (I) AN ABAC TO DETERMINE THE PROBABLE ERRORS OF
    CORRELATION COEFFICIENTS.
    BIOMETRIKA. 7, 416-411.

HERON, DAVID 1910.  (4.5)                                                       2442
    BIOMETRIC NOTES.  (II) ON THE PROBABLE ERROR OF A PARTIAL CORRELATION
    COEFFICIENT.
    BIOMETRIKA. 7, 411-412.

HERON, DAVID 1911.  (4.8)                                                       2443
    THE DANGER OF CERTAIN FORMULAE SUGGESTED AS SUBSTITUTES FOR THE
    CORRELATION COEFFICIENT.
    BIOMETRIKA. 8, 109-122.

HERON, DAVID   SEE ALSO
    4242. PEARSON, KARL + HERON, DAVID (1913)

HERRICK, ALLYN 1944.  E(4.6)                                                    2444
    MULTIPLE CORRELATION IN PREDICTING THE GROWTH OF MANY-AGED OAK-HICKORY STANDS.
    J. FORESTRY, 42, 812-817.

HERZ, CARL S. 1955.  (20.4)                                                     2445
    BESSEL FUNCTIONS OF MATRIX ARGUMENT.
    ANN. MATH. SER. 2, 61, 474-523.  (MR16,P.1107)

HERZBERG, AGNES M. 1966.  (17.9)                                                2446
    CYLINDRICALLY ROTATABLE DESIGNS.
    ANN. MATH. STATIST., 37, 242-247.

HERZEL, A. 1961.  (17.9)                                                        2447
    FREQUENCY DISTRIBUTION OF MULTIPLE STATISTICAL VARIABLES WITH
    GRADUATED AND CONTRAGRADUATED COMPONENTS.
        (IN ITALIAN)
    ATTI RIUN. SCI. SOC. ITAL. STATIST., 21, 67-90.

HESS, F. G. 1952.  (4.9/16.3/E)                                                 2448
    SOME DIRECTIONAL CORRELATION FUNCTIONS FOR SUCCESSIVE NUCLEAR RADIATIONS.
    CANAD. J. PHYS., 30, 130-146.  (MR13,P.853)

HESTENES, MAGNUS R. 1958.  (20.1)                                               2449
    INVERSION OF MATRICES BY BIORTHOGONALIZATION AND RELATED RESULTS.
    J. SOC. INDUST. APPL. MATH., 6, 51-90.

HEXT, G. R. 1963.  (11.2/E)                                                     2450
    THE ESTIMATION OF SECOND ORDER TENSORS, WITH RELATED TESTS AND DESIGNS.
    BIOMETRIKA, 50, 353-374.

HICKS, C. 1955.  E(5.2/C)                                                       2451
    SOME APPLICATIONS OF HOTELLING'S T.
    INDUST. QUAL. CONTROL, 11(9) 23-26.

HIDA, TAKEYUKI 1958.  (16.4)                                                    2452
    ON THE UNIFORM CONTINUITY OF WIENER PROCESS WITH A MULTI-DIMENSIONAL PARAMETER.
    NAGOYA MATH. J., 13, 53-61.  (MR20-2806)

HIGGINS, M. S. LONGUET-  SEE  LONGUET-HIGGINS, M. S.

HIGGINS, T. J. 1944.  (1.2)                                                     2453
    BIOGRAPHIES AND COLLECTED WORKS OF MATHEMATICIANS.
        (AMENDED BY NO. 2454)
    AMER. MATH. MONTHLY, 51, 433-445.

HIGGINS, T. J. 1949.  (1.2)
    BIOGRAPHIES AND COLLECTED WORKS OF MATHEMATICIANS- ADDENDA.
        (AMENDMENT OF NO. 2453)
    AMER. MATH. MONTHLY, 56, 310-312.                                    2454

HIGHLEYMAN, W. H. 1962.  (6.6/15.5)
    LINEAR DECISION FUNCTIONS, WITH APPLICATION TO PATTERN RECOGNITION.
    PROC. INST. RADIO ENGRS., 50, 1501-1514.   (MR25-1032)              2455

HIGUTI, ISAO 1950.  (8.4/B)
    ON SOME DISCRIMINATION (AN EXAMPLE FOR WHICH THE USUAL LEAST SQUARE
    METHOD IS NOT APPLICABLE).
        (IN JAPANESE)
    RES. MEM. INST. STATIST. MATH. TOKYO, 6, 374-395.                    2456


HIGUTI, ISAO 1951.  (14.2/14.3/U)
    A NOTE ON "R-DISTRIBUTIONS".
    RES. MEM. INST. STATIST. MATH. TOKYO, 7, 312-322.                    2457

HIGUTI, ISAO 1952.  (14.1/14.5/A)
    ON PROPERTIES OF THE SOLUTIONS OF A CERTAIN SYSTEM OF EQUATIONS.
        (IN JAPANESE)
    RES. MEM. INST. STATIST. MATH. TOKYO, 8, 235-292.                    2458

HIGUTI, ISAO 1953.  (7.2/B)
    THE DISTRIBUTION OF A STATISTIC WHICH REPRESENTS TWO DIMENSIONAL DISPERSION.
        (IN JAPANESE, WITH SUMMARY IN ENGLISH)
    PROC. INST. STATIST. MATH. TOKYO, 1(1) 51-53.                        2459

HIGUTI, ISAO 1954.  (14.4/20.3/U)
    ON THE SOLUTIONS OF CERTAIN SIMULTANEOUS EQUATIONS IN THE THEORY OF
    SYSTEMATIC STATISTICS.
    ANN. INST. STATIST. MATH. TOKYO, 5, 77-90.                           2460

HIGUTI, ISAO + OZAWA, MASARU 1958.  (1.1)
    ANALYSIS OF FACTORS WHICH HAVE INFLUENCE ON THE ACCURACIES OF
    DIMENSIONS OF MACHINE PARTS.
        (IN JAPANESE, WITH SUMMARY IN ENGLISH)
    PROC. INST. STATIST. MATH. TOKYO, 6, 95-99.   (MR24-A1790)           2461

HIGUTI, ISAO + SIBUYA, MASAAKI 1965.  (18.5/18.6)
    SAMPLING OF SCATTERED SPHERES.
    REP. STATIST. APPL. RES. UN. JAPAN. SCI. ENGRS., 12, 112-114.        2462

HILDRETH, CLIFFORD 1960.  (17.7)
    SIMULTANEOUS EQUATIONS:  ANY VERDICT YET?
    ECONOMETRICA, 28, 846-854.                                           2463

HILL, BRUCE M. 1960.  (17.8/B)
    A RELATIONSHIP BETWEEN HODGES' BIVARIATE SIGN TEST AND A
    NON-PARAMETRIC TEST OF DANIELS.
        (AMENDED BY NO. 2465)
    ANN. MATH. STATIST., 31, 1190-1192.   (MR22-7216)                    2464

HILL, BRUCE M. 1961.  (17.8/B)
    CORRECTIONS TO: A RELATIONSHIP BETWEEN HODGES' BIVARIATE SIGN TEST
    AND A NON-PARAMETRIC TEST OF DANIELS.
        (AMENDMENT OF NO. 2464)
    ANN. MATH. STATIST., 32, 619.   (MR22-7216)                          2465

HILL, G.W.   SEE
    1581. EVANS, MARILYN J. + HILL, G.W. (1961)

HILL, I. D. 1954.  (18.3)
    THE DISTRIBUTION OF THE REGRESSION COEFFICIENT IN SAMPLES FROM A
    NON-NORMAL POPULATION.
    BIOMETRIKA, 41, 548-552.   (MR16,P.381)                              2466


HILL, L. E.   SEE
    2467. HILL, TERRELL L. + HILL, L. E. (1945)

HILL, TERRELL L. + HILL, L. E. 1945.   (15.5)                          2467
    CONTRIBUTION TO THE THEORY OF DISCRIMINATION LEARNING.
    BULL. MATH. BIOPHYS., 7, 107-114.


HILLS, G. F. S. 1921.   (2.7/20.4)                                    2468
    A MULTIPLE INTEGRAL OF IMPORTANCE IN THE THEORY OF STATISTICS.
    PROC. LONDON MATH. SOC. SER. 2, 19, 281-291.

HILLS, M. 1966.   (6.1/6.2/6.4/T)                                     2469
    ALLOCATION RULES AND THEIR ERROR RATES.
    J. ROY. STATIST. SOC. SER. B, 28, 1-31.   (MR33-5063)


HINCHEN, J. D. 1956.   (8.1/CE)                                       2470
    CORRELATION ANALYSIS IN BATCH PROCESS CONTROL.
    INDUST. QUAL. CONTROL, 12(11) 54-59.


HINCIN, A. JA. 1928.   (4.1)                                          2471
    BERGRUNDUNG DER NORMALKORRELATION NACH DER LINDEBERGSCHEN METHODE.
    IZV. ASSOTS. NAUC. INST. UNIV., 1, 37-45.


HINCIN, A. JA. 1928.   (2.1/B)                                        2472
    URER DIE STABILTTAT ZWEIDIMENSIONALER VERTEILUNGSGESETZE.
    MAT. SB., 35, 19-23.


HINCIN, A. JA. 1934.   (16.4)                                         2473
    KORRELATIONSTHEORIE DER STATIONAREN STOCHASTISCHEN PROZESSE.
        (REPRINTED AS NO. 2474)
    MATH. ANN., 109, 604-615.


HINCIN, A. JA. 1938.   (16.4)                                         2474
    KORRELATIONSTHEORIE DER STATIONAREN STOCHASTISCHEN PROZESSE.
        (IN RUSSIAN, REPRINT OF NO. 2473)
    USPEHI MAT. NAUK, 5, 42-51.


HINCIN, A. JA. 1941.   (4.8)                                          2475
    SUR LA CORRELATION INTERMOLECULAIRE.
    C. R. ACAD. SCI. URSS NOUV. SER., 33, 482-484.   (MR5,P.168)


HINCIN, A. JA. 1943.   (17.3)                                         2476
    SUR UN CAS DE CORRELATION A POSTERIORI.
    MAT. SB. NOV. SER., 12(54) 185-196.   (MR5,P.168)


HINCIN, A. JA. 1955.   (20.5)                                         2477
    SYMMETRIC FUNCTIONS ON MULTIDIMENSIONAL SURFACES.
        (IN RUSSIAN)
    PAM. ALEKS. ALEKS. ANDRONOVA, 541-574.   (MR17,P.567)


HINCIN, A. JA.   SEF ALSO
    1916. GELFOND, A. O. + HINCIN, A. JA. (1933)


HIRSCHFELD, H. O. 1935.   (4.8)                                       2478
    A CONNECTION BETWEEN CORRELATION AND CONTINGENCY.
    PROC. CAMBRIDGE PHILOS. SOC., 31, 520-524.


HIRSCHFELD, H. O. 1937.   (14.2)                                      2479
    THE DISTRIBUTION OF THE RATIO OF COVARIANCE ESTIMATES IN TWO SAMPLES
    DRAWN FROM NORMAL BIVARIATE POPULATIONS.
    BIOMETRIKA, 29, 65-79.

HIRSCHSTEIN, BERTHA    SEE
    3364. LENTZ, THEODORE F., JR.; HIRSCHSTEIN, BERTHA + FINCH, J. H. (1932)


HLAWKA, EDMUND 1963.   (17.9)                                         2480
    GEORDNETE SCHAT7FUNKTIONEN UND DISKREPANZ.
    MATH. ANN., 150, 259-267.


HOBBY, CHARLES R. + PYKE, RONALD 1963.   (2.1/7.3/16.4)              2481
    COMBINATORIAL RFSULTS IN MULTI-DIMENSIONAL FLUCTUATION THEORY.
    ANN. MATH. STATIST., 34, 402-404.   (MR26-5602)


HOCH, JOHANNES 1958.   (15.2)                                        2482
    SIMULTANEOUS EQUATION BIAS IN THE CONTEXT OF THE COBB-DOUGLAS
    PRODUCTION FUNCTION.
    ECONOMETRICA, 26, 566-578.

HOCH, JOHANNES 1962. (4.8/17.7/20.3)
   BERECHNUNG VON LINEAREN MEHRFACHKORRELATIONEN MIT HILFE VON
   ADJUNKTIONSMATRIZEN NACH FRISCH.
   METRIKA, 5, 17-30.

2483

HOCKING, R. R. 1965. (8.8)
   THE DISTRIBUTION OF A PROJECTED LEAST SQUARES ESTIMATOR.
   ANN. INST. STATIST. MATH. TOKYO, 17, 357-362.

2484

HODGES, JOSEPH L., JR. 1955. (17.8/B)
   A BIVARIATE SIGN TEST.
   ANN. MATH. STATIST., 26, 523-527. (MR17,P.56)

2485

HOEFFDING, WASSILY 1940. (2.1/4.9/17.3)
   MASZSTABINVARIANTE KORRELATIONSTHEORIE.
   SCHR. MATH. INST. UNIV. BERLIN, 5, 181-233. (MR3,P.5)

2486

HOEFFDING, WASSILY 1941. (17.3)
   MASZSTABINVARIANTE KORRELATIONSMASSE FUR DISKONTINUIERLICHE VERTEILUNGEN.
   ARCH. MATH. WIRTSCH., 7, 49-70.

2487

HOEFFDING, WASSILY 1942. (4.9)
   STOCHASTISCHE ABHANGIGKEIT UND FUNKTIONALER ZUSAMMENHANG.
   SKAND. AKTUARIETIDSKR., 25, 200-227. (MR7,P.212)

2488

HOEFFDING, WASSILY 1947. (4.8/A)
   ON THE DISTRIBUTION OF THE RANK CORRELATION COEFFICIENT $\tau$ , WHEN THE
   VARIATES ARE NOT INDEPENDENT.
   BIOMETRIKA, 34, 1a3-196. (MR9,P.364)

2489

HOEFFDING, WASSILY 1960. (4.8)
   AN UPPER BOUND FOR THE VARIANCE OF KENDALL'S "TAU" AND OF RELATED STATISTICS.
   CONTRIB. PROB. STATIST. (HOTELLING VOLUME), 258-264.

2490

HOEFFDING, WASSILY 1961. (17.1)
   ON SEQUENCES OF SUMS OF INDEPENDENT RANDOM VECTORS.
   PROC. FOURTH BERKELEY SYMP. MATH. STATIST. PROB., 2, 213-226. (MR25-1563)

2491

HOEFFDING, WASSILY + ROBBINS, HERBERT E. 1948. (17.1)
   THE CENTRAL LIMIT THEOREM FOR DEPENDENT RANDOM VARIABLES.
   DUKE MATH. J., 15, 773-780. (MR10,P.200)

2492

HOEL, PAUL G. 1937. (11.2/A)
   A SIGNIFICANCE TEST FOR COMPONENT ANALYSIS.
   ANN. MATH. STATIST., 8, 149-158.

2493

HOEL, PAUL G. 1939. (15.2)
   A SIGNIFICANCE TEST FOR MINIMUM RANK IN FACTOR ANALYSIS.
   PSYCHOMETRIKA, 4, 245-253.

2494

HOEL, PAUL G. 1940. (20.2/20.3)
   THE ERRORS INVOLVED IN EVALUATING CORRELATION DETERMINANTS.
   ANN. MATH. STATIST., 11, 58-65. (MR1,P.248)

2495

HOEL, PAUL G. 1944. (6.5/E)
   ON STATISTICAL COEFFICIENTS OF LIKENESS.
   UNIV. CALIFORNIA PUBL. MATH., 2(1) 1-8. (MR6,P.6)

2496

HOEL, PAUL G. 1951. (8.1/C)
   CONFIDENCE REGIONS FOR LINEAR REGRESSION.
   PROC. SECOND BERKELEY SYMP. MATH. STATIST. PROB., 75-81. (MR13,P.481)

2497

HOEL, PAUL G. 1954. (8.1/C)
   CONFIDENCE BANDS FOR POLYNOMIAL CURVES.
   ANN. MATH. STATIST., 25, 534-542. (MR16,P.155)

2498

HOEL, PAUL G. 1955. (8.3/U)
   ON A SEQUENTIAL TEST FOR THE GENERAL LINEAR HYPOTHESIS.
   ANN. MATH. STATIST., 26, 136-139. (MR16,P.842)

2499

HOEL, PAUL G. 1958. (7.2/8.1)
   EFFICIENCY PROBLEMS IN POLYNOMIAL ESTIMATION.
   ANN. MATH. STATIST., 29, 1134-1145. (MR20-7359)

2500

HOEL, PAUL G. 1965.  (7.2/8.5/B)                                          2501
    MINIMAX DESIGNS IN TWO DIMENSIONAL REGRESSION.
    ANN. MATH. STATIST., 36, 1097-1106.

HOEL, PAUL G. 1966.  (7.2/8.5)                                           2502
    A SIMPLE SOLUTION FOR OPTIMAL CHEBYSHEV REGRESSION EXTRAPOLATION.
    ANN. MATH. STATIST., 37, 720-725.

HOEL, PAUL G. + PETERSON, RAYMOND P. 1949.  (6.1/A)                       2503
    A SOLUTION TO THE PROBLEM OF OPTIMUM CLASSIFICATION.
    ANN. MATH. STATIST., 20, 433-438.  (MR11,P.191)

HOEL, PAUL G. + SCHEUER, ERNEST M. 1961.  (17.8/C)                        2504
    CONFIDENCE SETS FOR MULTIVARIATE MEDIANS.
    ANN. MATH. STATIST., 32, 477-484.  (MR23-A2992)

HOEL, PAUL G.   SEE ALSO
    2306. HALPERIN, MAX; HARTLEY, H. O. + HOEL, PAUL G. (1965)

HOEFFDING, WASSILIJ  SEE  HOEFFDING, WASSILY

HOFFMAN, WILLIAM C. 1960.  (7.4/14.1/E)                                   2505
    SOME STATISTICAL METHODS OF POTENTIAL VALUE IN RADIO WAVE PROPOGATION
    INVESTIGATIONS.
    STATIST. METH. RADIO WAVE PROP., 117-135.

HOFLUND, OLLE 1963.  (4.8/T)                                             2506
    SIMULATED DISTRIBUTIONS FOR SMALL N OF KENDALL'S PARTIAL RANK
    CORRELATION COEFFICIENT.
    BIOMETRIKA, 50, 520-522.

HOFSTATTER, PETER R. 1938.  (15.1/E)                                      2507
    UBER FAKTOREN-ANALYSE.
    ARCH. GES. PSYCHOL., 100, 223-279.

HOGG, ROBERT V. 1951.  (2.6)                                             2508
    ON RATIOS OF CERTAIN ALGEBRAIC FORMS.
    ANN. MATH. STATIST., 22, 567-572.  (MR13,P.366)

HOGG, ROBERT V. 1960.  (4.9/17.3)                                        2509
    CERTAIN UNCORRELATED STATISTICS.
    J. AMER. STATIST. ASSOC., 55, 265-267.  (MR22-3051)

HOGG, ROBERT V. 1963.  (7.4)                                            2510
    ON THE INDEPENDENCE OF CERTAIN WISHART VARIABLES.
    ANN. MATH. STATIST., 34, 935-939.  (MR27-2043)

HOGG, ROBERT V. + CRAIG, ALLEN T. 1958.  (2.5)                           2511
    ON THE DECOMPOSITION OF CERTAIN $\chi^2$ VARIABLES.
    ANN. MATH. STATIST., 29, 608-610.

HOGG, ROBERT V. + CRAIG, ALLEN T. 1962.  (19.3)                          2512
    SOME RESULTS ON UNBIASED ESTIMATION.
    SANKHYA SER. A, 24, 333-338.

HOHLE, RAYMOND H. 1966.  (15.5)                                          2513
    AN EMPIRICAL EVALUATION AND COMPARISON OF TWO MODELS FOR DISCRIMINABILITY SCALES.
    J. MATH. PSYCHOL., 3, 174-183.

HOLGATE, P. 1964.  (18.2)                                               2514
    ESTIMATION FOR THE BIVARIATE POISSON DISTRIBUTION.
    BIOMETRIKA, 51, 241-245.  (MR30-2593)

HOLLA, MAHABALESHWARA + BHATTACHARYA, SAMIR KUMAR 1965.   (18.1/B)        2515
    A BIVARIATE GAMMA DISTRIBUTION IN LIFE TESTING.
    DEFENCE SCI. J., 15, 65-74.  (MR33-851)

HOLLA, MAHABALESHWARA   SEE ALSO
    523. BHATTACHARYA, SAMIR KUMAR + HOLLA, MAHABALESHWARA (1963)

HOLLAND, J. G.   SEE
    3063. KNOWLES, W. B.; HOLLAND, J. G. + NEWLIN, E. P. (1957)

HOLLEY, J. W. 1947. (15.3)                                                    2516
    A NOTE ON THE REFLECTION OF SIGNS IN THE EXTRACTION OF CENTROID FACTORS.
    PSYCHOMETRIKA, 12, 263-265.

HOLLEY, J. W. + GUILFORD, J. P. 1964. (4.8)                                   2517
    A NOTE ON THE G INDEX OF AGREEMENT.
    EDUC. PSYCHOL. MEAS., 24, 749-753.

HOLLEY, J. W. + GUILFORD, J. P. 1966. (20.3)                                  2518
    NOTE ON THE DOUBLE CENTERING OF DICHOTOMIZED MATRICES.
    SCAND. J. PSYCHOL., 7, 97-101.

HOLLISTER, L. E.    SEE
    4114. OVERALL, JOHN E. + HOLLISTER, L. E. (1964)

HOLWERDA, A. O. 1916. (4.1)                                                   2519
    CORRELATIE-COEFFICIENTEN.
    ARCH. VERZEKERINGS-WETENSCH. AANVERW. VAKKEN, 15, 109-116.

HOLZINGER, KARL J. 1923. (4.1/4.8)                                            2520
    A COMBINATION FORM FOR CALCULATING THE CORRELATION COEFFICIENT AND RATIOS.
    J. AMER. STATIST. ASSOC., 18, 623-628.

HOLZINGER, KARL J. 1925. (4.2/T)                                              2521
    TABLES OF THE PROBABLE ERROR OF THE COEFFICIENT OF CORRELATION AS
    FOUND BY THE PRODUCT MOMENT METHOD.
    TRACTS COMPUT., 12, 1-36.

HOLZINGER, KARL J. 1929. (15.2)                                               2552
    ON TETRAD DIFFERENCES WITH OVERLAPPING VARIABLES.
    J. EDUC. PSYCHOL., 20, 91-97.

HOLZINGER, KARL J. 1932. (15.4)                                              2522
    THE RELIABILITY OF A SINGLE TEST ITEM.
    J. EDUC. PSYCHOL., 23, 411-417.

HOLZINGER, KARL J. 1938. (15.1)                                              2523
    RELATIONSHIP BETWEEN THREE MULTIPLE ORTHOGONAL FACTORS AND FOUR
    BI-FACTORS.
    J. EDUC. PSYCHOL., 29, 513-519.

HOLZINGER, KARL J. 1940. (15.1)                                              2524
    A SYNTHETIC APPROACH TO FACTOR ANALYSIS.
    PSYCHOMETRIKA, 5, 235-250.

HOLZINGER, KARL J. 1942. (15.1)                                              2525
    WHY DO PEOPLE FACTOR?
    PSYCHOMETRIKA, 7, 147-156.

HOLZINGER, KARL J. 1944. (15.1)                                              2526
    FACTORING TEST SCORES AND IMPLICATIONS FOR THE METHOD OF AVERAGES.
    PSYCHOMETRIKA, 9, 155-167. (MR6,P.92)

HOLZINGER, KARL J. 1944. (15.2/15.3)                                         2527
    A SIMPLE METHOD OF FACTOR ANALYSIS.
    PSYCHOMETRIKA, 9, 257-262. (MR6,P.162)

HOLZINGER, KARL J. 1945. (15.1)                                             2528
    INTERPRETATION OF SECOND ORDER FACTORS.
    PSYCHOMETRIKA, 10, 21-25.

HOLZINGER, KARL J. 1945. (1.2)                                              2529
    SPEARMAN AS I KNEW HIM.
    PSYCHOMETRIKA, 10, 231-235.

HOLZINGER, KARL J. 1946. (15.1)                                             2530
    A COMPARISON OF THE PRINCIPAL AXIS AND CENTROID FACTORS.
    J. EDUC. PSYCHOL., 37, 449-472.

HOLZINGER, KARL J. + HARMAN, HARRY H. 1937. (15.1/E)                        2531
    RELATIONSHIPS BETWEEN FACTORS OBTAINED FROM CERTAIN ANALYSES.
    J. EDUC. PSYCHOL., 28, 321-345.

HOLZINGER, KARL J. + HARMAN, HARRY H. 1938. (15.2)                          2532
    COMPARISON OF TWO FACTORIAL ANALYSES.
    PSYCHOMETRIKA, 3, 45-60.

HOLZINGER, KARL J. + HARMAN, HARRY H. 1939.   (11.1/15.1/15.2)                    2533
    FACTOR ANALYSIS.
    REV. EDUC. RES., 9, 528-531, 619-621.

HOLZINGER, KARL J. + SWINEFORD, FRANCES 1932.   (15.1/E)                    2534
    UNIQUENESS OF FACTOR PATTERNS.
    J. EDUC. PSYCHOL., 23, 247-258.

HOLZINGER, KARL J. + SWINEFORD, FRANCES 1937.   (15.1)                    2535
    THE BI-FACTOR METHOD.
    PSYCHOMETRIKA, 2, 41-54.

HOLZINGER, KARL J.    SEE ALSO
    5164. SPEARMAN, CHARLES + HOLZINGER, KARL J. (1924)

    5165. SPEARMAN, CHARLES + HOLZINGER, KARL J. (1925)

    5166. SPEARMAN, CHARLES + HOLZINGER, KARL J. (1930)

    5306. SWINEFORD, FRANCES + HOLZINGER, KARL J. (1933)

    5307. SWINEFORD, FRANCES + HOLZINGER, KARL J. (1934)

    5308. SWINEFORD, FRANCES + HOLZINGER, KARL J. (1935)

    5309. SWINEFORD, FRANCES + HOLZINGER, KARL J. (1936)

    5310. SWINEFORD, FRANCES + HOLZINGER, KARL J. (1937)

    5311. SWINEFORD, FRANCES + HOLZINGER, KARL J. (1938)

    5312. SWINEFORD, FRANCES + HOLZINGER, KARL J. (1939)

    5313. SWINEFORD, FRANCES + HOLZINGER, KARL J. (1940)

    5314. SWINEFORD, FRANCES + HOLZINGER, KARL J. (1941)

    5315. SWINEFORD, FRANCES + HOLZINGER, KARL J. (1942)

    5316. SWINEFORD, FRANCES + HOLZINGER, KARL J. (1943)

    5317. SWINEFORD, FRANCES + HOLZINGER, KARL J. (1944)

    5318. SWINEFORD, FRANCES + HOLZINGER, KARL J. (1945)

    5319. SWINEFORD, FRANCES + HOLZINGER, KARL J. (1946)

    5320. SWINEFORD, FRANCES + HOLZINGER, KARL J. (1947)

    5321. SWINEFORD, FRANCES + HOLZINGER, KARL J. (1948)

    5322. SWINEFORD, FRANCES + HOLZINGER, KARL J. (1949)

    5323. SWINEFORD, FRANCES + HOLZINGER, KARL J. (1950)

    5324. SWINEFORD, FRANCES + HOLZINGER, KARL J. (1951)

HOLZMAN, MATHILDA S. + FORMAN, VIRGINIA P. 1966.   (15.5/E)                    2536
    A MULTIDIMENSIONAL CONTENT-ANALYSIS SYSTEM APPLIED TO THE ANALYSIS OF
    THE THERAPEUTIC TECHNIQUE IN PSYCHOTHERAPY WITH SCHIZOPHRENIC PATIENTS.
    PSYCHOL. BULL., 66, 263-281.

HOMER, L. D.   SEE
    5562. TURNER, MALCOLM E.; MONROE, ROBERT J. + HOMER, L. D. (1963)

HOOD, WILLIAM C.   SEE
    3099. KOOPMANS, TJALLING C. + HOOD, WILLIAM C. (1953)

HOOK, L. HARMON   SEE
    813. BUTLER, JOHN M. + HOOK, L. HARMON (1966)

HOOKER, R. H. 1901.   (16.5/E)                    2537
    CORRELATION OF THE MARRIAGE-RATE WITH TRADE.
    J. ROY. STATIST. SOC., 64, 485-492.

HOOKER, R. H. 1907.  E(4.5/4.6)                                                      2538
    CORRELATION OF THE WEATHER AND CROPS.
       (WITH DISCUSSION)
  J. ROY. STATIST. SOC., 70, 1-51.


HOOKER, R. H. + YULE, G. UDNY 1906.  E(4.5/4.6)                                       2539
    NOTE ON ESTIMATING THE RELATIVE INFLUENCE OF TWO VARIABLES UPON A THIRD.
  J. ROY. STATIST. SOC., 69, 197-200.



HOOPER, JOHN W. 1958.  (4.2)                                                          2540
    THE SAMPLING VARIANCE OF CORRELATION COEFFICIENTS UNDER ASSUMPTIONS
    OF FIXED AND MIXED VARIATES.
    BIOMETRIKA, 45, 471-477.  (MR21-945)


HOOPER, JOHN W. 1959.  (12.1/16.2)                                                    2541
    SIMULTANEOUS EQUATIONS AND CANONICAL CORRELATION THEORY.
    ECONOMETRICA, 27, 245-256.  (MR21-4505)


HOOPER, JOHN W. 1962.  (9.3/12.1/12.2/A)                                              2542
    PARTIAL TRACE CORRELATIONS.
    ECONOMETRICA, 30, 324-331.


HOOPER, JOHN W. + THEIL, H. 1958.  (8.1/17.7)                                         2543
    THE EXTENSION OF WALD'S METHOD OF FITTING STRAIGHT LINES TO MULTIPLE REGRESSION.
    REV. INST. INTERNAT. STATIST., 26, 37-47.  (MR21-6065)


HOOPER, JOHN W. + ZELLNER, ARNOLD 1961.  (8.1/8.9/12.1/C)                             2544
    THE ERROR OF FORECAST FOR MULTIVARIATE REGRESSION MODELS.
    ECONOMETRICA, 29, 544-555.  (MR30-5427)


HOPE, A. E. 1933.  (4.1)                                                              2545
    A FORMULA IN THE THEORY OF CORRELATION.
    MATH. NOTES EDINBURGH, 28, 14-16.



HOPKINS, CARL E.   SEE
    1003. COCHRAN, WILLIAM G. + HOPKINS, CARL E. (1961)


HOPKINS, J. W. 1966.  (15.2/E)                                                        2546
    SOME CONSIDERATIONS IN MULTIVARIATE ALLOMETRY.
    BIOMETRICS, 22, 747-760.


HOPKINS, J. W. + CLAY, P.P.F. 1963.  (5.3/10.2/17.5/B)                                2547
    SOME EMPIRICAL DISTRIBUTIONS OF BIVARIATE $T^2$ AND HOMOSCEDASTICITY
    CRITERION M UNDER UNEQUAL VARIANCE AND LEPTOKURTOSIS.
  J. AMER. STATIST. ASSOC., 58, 1048-1053.  (MR27-6344)


HORN, DANIEL 1942.  (4.8)                                                             2548
    A CORRECTION FOR THE EFFECT OF TIED RANKS ON THE VALUE OF THE RANK
    DIFFERENCE CORRELATION COEFFICIENT.
  J. EDUC. PSYCHOL., 33, 686-690.


HORN, JOHN L. 1964.  (15.2)                                                           2549
    A NOTE ON THE ESTIMATION OF FACTOR SCORES.
    EDUC. PSYCHOL. MEAS., 24, 525-527.


HORN, JOHN L. 1965.  (15.2/E)                                                         2550
    A RATIONALE AND TEST FOR THE NUMBER OF FACTORS IN FACTOR ANALYSIS.
    PSYCHOMETRIKA, 30, 179-185.


HORN, JOHN L. 1966.  (15.2)                                                           2551
    A NOTE ON THE USE OF RANDOM VARIABLES IN FACTOR ANALYSIS.
    BRITISH J. MATH. STATIST. PSYCHOL., 19, 127-129.


ENTRY NO. 2552 APPEARS BETWEEN ENTRY NOS. 2521 & 2522.


HORST, PAUL 1932.  (4.6)                                                              2553
    A SHORT METHOD FOR SOLVING FOR A COEFFICIENT OF MULTIPLE CORRELATION.
    ANN. MATH. STATIST., 3, 40-44.

HORST, PAUL 1932.  (15.4)                                                                    2554
    THE CHANCE ELEMENT IN THE MULTIPLE CHOICE TEST ITEM.
  J. GENERAL PSYCHOL., 6, 209-211.

HORST, PAUL 1932.  (4.1)                                                                      2555
    MEASUREMENT RELATIONSHIP AND CORRELATION.
  J. PHILOS., 29, 631-637.

HORST, PAUL 1934.  (15.4)                                                                     2556
    ITEM ANALYSIS BY THE METHOD OF SUCCESSIVE RESIDUALS.
  J. EXPER. EDUC., 2, 254-263.

HORST, PAUL 1935.  (20.1)                                                                     2557
    A METHOD FOR DETERMINING THE COEFFICIENTS OF A CHARACTERISTIC EQUATION.
  ANN. MATH. STATIST., 6, 83-84.

HORST, PAUL 1936.  (15.4)                                                                     2558
    ITEM SELECTION BY MEANS OF A MAXIMIZING FUNCTION.
  PSYCHOMETRIKA, 1, 229-244.

HORST, PAUL 1937.  (15.2)                                                                     2559
    A METHOD OF FACTOR ANALYSIS BY MEANS OF WHICH ALL COORDINATES OF THE
    FACTOR MATRIX ARE GIVEN SIMULTANEOUSLY.
  PSYCHOMETRIKA, 2, 225-236.

HORST, PAUL 1941.  (15.3)                                                                     2560
    A NON-GRAPHICAL METHOD FOR TRANSFORMING AN ARBITRARY FACTOR MATRIX
    INTO A SIMPLE STRUCTURE FACTOR MATRIX.
  PSYCHOMETRIKA, 6, 79-99.   (MR2,P.234)

HORST, PAUL 1949.  (4.6/15.4)                                                                 2561
    DETERMINATION OF THE OPTIMAL TEST LENGTH TO MAXIMIZE THE MULTIPLE CORRELATION.
  PSYCHOMETRIKA, 14, 79-88.

HORST, PAUL 1954.  (15.4)                                                                     2562
    THE MAXIMUM EXPECTED CORRELATION BETWEEN TWO MULTIPLE-CHOICE TESTS.
  PSYCHOMETRIKA, 19, 291-296.

HORST, PAUL 1956.  (15.3)                                                                     2563
    SIMPLIFIED COMPUTATIONS FOR THE MULTIPLE GROUP METHOD OF FACTOR ANALYSIS.
  EDUC. PSYCHOL. MEAS., 16, 101-109.

HORST, PAUL 1956.  F(6.3)                                                                     2564
    MULTIPLE CLASSIFICATION BY THE METHOD OF LEAST SQUARES.
  J. CLIN. PSYCHOL., 12, 3-16.

HORST, PAUL 1956.  F(6.3)                                                                      2565
    LEAST SQUARE MULTIPLE CLASSIFICATION FOR UNEQUAL SUBGROUPS.
  J. CLIN. PSYCHOL., 12, 309-315.

HORST, PAUL 1956.  (15.3)                                                                      2566
    A SIMPLE METHOD OF ROTATING A CENTROID FACTOR MATRIX TO A SIMPLE
    STRUCTURE HYPOTHESIS.
  J. EXPER. EDUC., 24, 251-258.

HORST, PAUL 1960.  (6.7/8.1/E)                                                                 2567
    OPTIMAL ESTIMATES OF MULTIPLE CRITERIA WITH RESTRICTIONS ON THE
    COVARIANCE MATRIX OF ESTIMATED CRITERIA.
  PSYCHOL. REP., 6, 427-444.

HORST, PAUL 1961.  (12.1/12.2/E)                                                              2568
    GENERALIZED CANONICAL CORRELATIONS AND THEIR APPLICATIONS TO EXPERIMENTAL DATA.
  J. CLIN. PSYCHOL., 17, 331-347.

HORST, PAUL 1961.  (12.1/15.4)                                                                2569
    RELATIONS AMONG M SETS OF MEASURES.
  PSYCHOMETRIKA, 26, 129-149.   (MR24-A2468)

HORST, PAUL 1962.  (6.7)                                                                      2570
    THE LOGIC OF PERSONNEL SELECTION AND CLASSIFICATION.
  PSYCHOL. PRINCIPLES SYSTEM DEVELOP., 231-272.

HORST, PAUL 1962.   (15.3/20.1)                                                    2571
    MATRIX REDUCTION AND APPROXIMATION TO PRINCIPAL AXES.
    PSYCHOMETRIKA, 27, 169–179.

HORST, PAUL 1963.   (15.5)                                                         2572
    MULTIVARIATE MODELS FOR EVALUATING CHANGE.
    PROBLEMS MEAS. CHANGE (HARRIS), 104–121.

HORST, PAUL 1964.   (15.1/15.4)                                                    2573
    MATRIX FACTORING AND TEST THEORY.
    CONTRIB. MATH. PSYCHOL. (FREDERIKSEN + GULLIKSEN), 129–139.

HORST, PAUL 1966.   (1.1)                                                          2574
    AN OVERVIEW OF THE ESSENTIALS OF MULTIVARIATE ANALYSIS METHODS.
    HANDB. MULTIVARIATE EXPER. PSYCHOL. (CATTELL), 129–152.

HORST, PAUL + MAC EWAN, CHARLOTTE 1957.   (15.4)                                   2575
    OPTIMAL TEST LENGTH FOR MULTIPLE PREDICTION:  THE GENERAL CASE.
    PSYCHOMETRIKA, 22, 311–324.

HORST, PAUL + MAC EWAN, CHARLOTTE 1960.   (12.1/15.4)                              2576
    PREDICTOR ELIMINATION TECHNIQUES FOR DETERMINING MULTIPLE PREDICTION BATTERIES.
    PSYCHOL. REP., 7, 19–50.

HORST, PAUL + SCHAIE, K. WARNER 1956.   (15.1/15.3)                                2577
    THE MULTIPLE GROUP METHOD OF FACTOR ANALYSIS AND ROTATION TO A SIMPLE
    STRUCTURE HYPOTHESIS.
    J. EXPER. EDUC., 24, 231–237.

HORST, PAUL + SMITH, STEVENSON 1950.   (5.2/E)                                     2578
    THE DISCRIMINATION OF TWO RACIAL SAMPLES.
    PSYCHOMETRIKA, 15, 271–290.

HOSTINSKY, BOHUSLAV 1929.   (18.6)                                                 2579
    PROBABILITES RELATIVES A LA POSITION D'UNE SPHERE A CENTRE FIXE.
    J. MATH. PURES APPL. SER. 9, 8, 35–43.

HOSTINSKY, BOHUSLAV 1931.   (20.4)                                                 2580
    SUR L'INTEGRATION DES TRANSFORMATIONS FONCTIONELLES LINEAIRES. NOTA (I).
    ATTI REALE ACCAD. NAZ. LINCEI SER. 6, 13, 921–923.

HOSTINSKY, BOHUSLAV 1931.   (20.4)                                                 2581
    SUR L'INTEGRATION DES TRANSFORMATIONS FONCTIONELLES LINEAIRES. NOTA II.
    ATTI REALE ACCAD. NAZ. LINCEI SER. 6, 14, 326–331.

HOSTINSKY, BOHUSLAV 1932.   (20.4)                                                 2582
    SUR L'INTEGRATION DES TRANSFORMATIONS FONCTIONELLES LINEAIRES. NOTA (III).
    ATTI REALE ACCAD. NAZ. LINCEI SER. 6, 16, 25–27.

HOSTINSKY, BOHUSLAV 1940.   (4.1/18.1)                                             2583
    SUR LE COEFFICIENT DE CORRELATION.
    SB. POS. PAM. D. A. GRAVE, 48–51.   (MR2,P.229)

HOTELLING, HAROLD 1925.   (4.2)                                                    2584
    THE DISTRIBUTION OF CORRELATION RATIOS CALCULATED FROM RANDOM DATA.
    PROC. NAT. ACAD. SCI. U.S.A., 11, 657–662.

HOTELLING, HAROLD 1931.   (5.2/5.3)                                                2585
    THE GENERALIZATION OF STUDENT'S RATIO.
    ANN. MATH. STATIST., 2, 360–378.

HOTELLING, HAROLD 1931.   (1.1)                                                    2586
    RECENT IMPROVEMENTS IN STATISTICAL INFERENCE.
        (WITH DISCUSSION)
    J. AMER. STATIST. ASSOC., 26(SP) 79–89.

HOTELLING, HAROLD 1933.   (11.1)                                                   2587
    ANALYSIS OF A COMPLEX OF STATISTICAL VARIABLES INTO PRINCIPAL COMPONENTS.
    J. EDUC. PSYCHOL., 24, 417–441, 498–520.   (SEE ALSO CHAPTER 1)

HOTELLING, HAROLD 1935.   (12.1)                                                   2588
    THE MOST PREDICTABLE CRITERION.
    J. EDUC. PSYCHOL., 26, 139–142.

HOTELLING, HAROLD 1936.  (9.1/12.1)                                    2589
    RELATIONS BETWEEN TWO SETS OF VARIATES.
    BIOMETRIKA, 28, 321–377.


HOTELLING, HAROLD 1936.  (11.1/20.2)                                   2590
    SIMPLIFIED CALCULATION OF PRINCIPAL COMPONENTS.
    PSYCHOMETRIKA, 1(1) 27–35.


HOTELLING, HAROLD 1939.  (20.5)                                        2591
    TUBES AND SPHERES IN N-SPACES, AND A CLASS OF STATISTICAL PROBLEMS.
    AMER. J. MATH., 61, 440–460.



HOTELLING, HAROLD 1940.  (4.6)                                         2592
    THE SELECTION OF VARIATES FOR USE IN PREDICTION WITH SOME COMMENTS ON
    THE GENERAL PROBLEM OF NUISANCE PARAMETERS.
    ANN. MATH. STATIST., 11, 271–283.  (MR2,P.111)


HOTELLING, HAROLD 1942.  (15.1)                                        2593
    ROTATIONS IN PSYCHOLOGY AND THE STATISTICAL REVOLUTION.
    SCIENCE, 95, 504–507.


HOTELLING, HAROLD 1943.  (20.3)                                        2594
    SOME NEW METHODS IN MATRIX CALCULATION.
    ANN. MATH. STATIST., 14, 1–34.  (MR4,P.202)


HOTELLING, HAROLD 1943.  (20.3)                                        2595
    FURTHER POINTS ON MATRIX CALCULATION AND SIMULTANEOUS EQUATIONS.
    ANN. MATH. STATIST., 14, 440–441.  (MR5,P.245)


HOTELLING, HAROLD 1944.  (2.5/20.1)                                    2596
    NOTE ON A MATRIC THEOREM OF A. T. CRAIG.
    ANN. MATH. STATIST., 15, 427–429.  (MR6,P.160)


HOTELLING, HAROLD 1947.  (8.3/8.4/E)                                   2597
    MULTIVARIATE QUALITY CONTROL, ILLUSTRATED BY THE AIR TESTING OF
    SAMPLE BOMBSIGHTS.
    TECH. STATIST. ANAL. (EISENHART, HASTAY, + WALLIS), 111–184.


HOTELLING, HAROLD 1949.  (1.1)                                         2598
    THE PLACE OF STATISTICS IN THE UNIVERSITY.
        (WITH DISCUSSION)
    PROC. BERKELEY SYMP. MATH. STATIST. PROB., 21–49.


HOTELLING, HAROLD 1949.  (20.3)                                        2599
    PRACTICAL PROBLEMS OF MATRIX CALCULATION.
    PROC. BERKELEY SYMP. MATH. STATIST. PROB., 275–293.  (MR10,P.574)


HOTELLING, HAROLD 1951.  (8.3)                                         2600
    A GENERALIZED T TEST AND MEASURE OF MULTIVARIATE DISPERSION.
    PROC. SECOND BERKELEY SYMP. MATH. STATIST. PROB., 23–41.  (MR13,P.479)


HOTELLING, HAROLD 1953.  (4.2/A)                                       2601
    NEW LIGHT ON THE CORRELATION COEFFICIENT AND ITS TRANSFORMS.
        (WITH DISCUSSION)
    J. ROY. STATIST. SOC. SER. B, 15, 193–232.  (MR15,P.728)


HOTELLING, HAROLD 1954.  (1.1)                                         2602
    MULTIVARIATE ANALYSIS.
    STATIST. MATH. BIOL. (KEMPTHORNE ET AL.), 67–80.


HOTELLING, HAROLD 1955.  (8.5/15.1)                                    2603
    LES RAPPORTS ENTRE LES METHODES STATISTIQUES RECENTES PORTANT SUR DES
    VARIABLES MULTIPLES ET L'ANALYSE FACTORIELLE.
        (WITH DISCUSSION, TRANSLATED AS NO. 2605)
    COLLOQ. INTERNAT. CENTRE NAT. RECH. SCI., 65, 107–125.


HOTELLING, HAROLD 1955.  (8.3/8.4)                                     2604
    MULTIVARIATE METHODS IN TESTING COMPLEX EQUIPMENT.
    PAPERS PRES. COLLOQ. STATIST. DESIGN LAB. EXPER., 4028, 84–91.

HOTELLING, HAROLD 1957.   (8.5/15.1)                                                            2605
    THE RELATIONS OF THE NEWER MULTIVARIATE STATISTICAL METHODS TO FACTOR ANALYSIS.
        (TRANSLATION OF NO. 2603)
    BRITISH J. STATIST. PSYCHOL., 10, 69-79.

HOTELLING, HAROLD 1960.   (1.2)                                                                 2606
    BIBLIOGRAPHY OF HAROLD HOTELLING.
    CONTRIB. PROB. STATIST. (HOTELLING VOLUME), 25-32.

HOTELLING, HAROLD + FRANKEL, LESTER R. 1938.   (18.1/U)                                         2607
    THE TRANSFORMATION OF STATISTICS TO SIMPLIFY THEIR DISTRIBUTION.
    ANN. MATH. STATIST., 9, 87-96.

HOTELLING, HAROLD + HOWELLS, W. W. 1936.   (4.1/E)                                              2608
    MEASUREMENTS AND CORRELATIONS ON PELVES OF INDIANS OF THE SOUTHWEST.
    AMER. J. PHYS. ANTHROP., 21, 91-106.

HOTELLING, HAROLD + PABST, MARGARET R. 1936.   (4.8/17.8)                                       2609
    RANK CORRELATION AND TESTS OF SIGNIFICANCE INVOLVING NO ASSUMPTION OF NORMALITY.
    ANN. MATH. STATIST., 7, 29-43.


HOUSE, J. MILTON   SEE
    4510. REMMERS, H. H. + HOUSE, J. MILTON (1941)

HOUSEHOLDER, ALSTON S. 1956.   (1.2)                                                            2610
    BIBLIOGRAPHY ON NUMERICAL ANALYSIS.
    J. ASSOC. COMPUT. MACH., 3, 85-100.


HOUSEHOLDER, ALSTON S. 1958.   (20.3)                                                           2611
    UNITARY TRIANGULARIZATION OF A NONSYMMETRIC MATRIX.
    J. ASSOC. COMPUT. MACH., 5, 339-342.   (MR22-1992)

HOUSEHOLDER, ALSTON S. 1964.   (20.1)                                                           2612
    LOCALIZATION OF THE CHARACTERISTIC ROOTS OF MATRICES.
    RECENT ADVANC. MATRIX THEORY (SCHNEIDER), 39-60.

HOUSEHOLDER, ALSTON S. + BAUER, F. L. 1960.   (20.3)                                            2613
    ON CERTAIN ITERATIVE METHODS FOR SOLVING LINEAR SYSTEMS.
    NUMER. MATH., 2, 55-59.   (MR22-7251)

HOUSEHOLDER, ALSTON S. + CARPENTER, JOHN A. 1963.   (20.2)                                      2614
    THE SINGULAR VALUES OF INVOLUTORY AND OF IDEMPOTENT MATRICES.
    NUMER. MATH., 5, 234-237.

HOUSEHOLDER, ALSTON S. + YOUNG, GALE 1938.   (20.1)                                             2615
    MATRIX APPROXIMATION AND LATENT ROOTS.
    AMER. MATH. MONTHLY, 45, 165-171.

HOUSEHOLDER, ALSTON S.   SEE ALSO
    404. BAUER, F. L. + HOUSEHOLDER, ALSTON S. (1960)

    6050. YOUNG, GALE + HOUSEHOLDER, ALSTON S. (1938)

    6051. YOUNG, GALE + HOUSEHOLDER, ALSTON S. (1938)

    6052. YOUNG, GALE + HOUSEHOLDER, ALSTON S. (1940)

    6053. YOUNG, GALE + HOUSEHOLDER, ALSTON S. (1941)

HOWARD, MARGARET   SEE
    804. BURT, CYRIL + HOWARD, MARGARET (1956)

HOWE, WILLIAM G.   SEE
    740. BROWN, W. R. J.; HOWE, WILLIAM G.; JACKSON, J. EDWARD +   (1956)
    MORRIS, R.H.

HOWELLS, W. W.   SEE
    2608. HOTELLING, HAROLD + HOWELLS, W. W. (1936)

HOYT, CYRIL J. 1941. (15.4)                                                    **2616**
    NOTE ON A SIMPLIFIED METHOD OF COMPUTING TEST RELIABILITY.
    EDUC. PSYCHOL. MEAS., 1, 93-95.

HOYT, CYRIL J. 1941. (15.4)                                                    **2617**
    TEST RELIABILITY ESTIMATED BY ANALYSIS OF VARIANCE.
    PSYCHOMETRIKA, 6, 153-160.

HOYT, CYRIL J.   SEE ALSO
    2823. JOHNSON, PALMER O. + HOYT, CYRIL J. (1947)

    2824. JOHNSON, PALMER O. + HOYT, CYRIL J. (1947)

HSU, CHUNG-TSI 1940. (4.3/5.1/10.2/BT)                                         **2618**
    ON SAMPLES FROM A NORMAL BIVARIATE POPULATION.
    ANN. MATH. STATIST., 11, 410-426.   (MR2,P.236)

HSU, CHUNG-TSI 1941. (4.3/10.2/14.3)                                           **2619**
    SAMPLES FROM TWO BIVARIATE NORMAL POPULATIONS.
    ANN. MATH. STATIST., 12, 279-292.   (MR3,P.174)

HSU, F. H. 1951. (4.2/T)                                                       **2620**
    SIMPLE GRAPHICAL DETERMINATION OF STANDARD ERROR OF DIFFERENCE AND
    CORRELATION COEFFICIENT.
    EDUC. PSYCHOL. MEAS., 11, 516-523.

HSU, LEETZE CHING 1951. (20.4)                                                 **2621**
    THE ASYMPTOTIC BEHAVIOUR OF A KIND OF MULTIPLE INTEGRALS INVOLVING A PARAMETER.
    QUART. J. MATH. OXFORD SECOND SER., 2, 175-188.   (MR13,P.340)

HSU, LEETZE CHING 1959. (20.4)                                                 **2622**
    CONCERNING THE NUMERICAL INTEGRATION OF PERIODIC FUNCTIONS OF SEVERAL VARIABLES.
    ACTA UNIV. SZEGED. ACTA SCI. MATH., 20, 230-233.   (MR22-2005)

HSU, P. L. 1938. (5.2/5.3)                                                     **2623**
    NOTES ON HOTELLING'S GENERALIZED T.
    ANN. MATH. STATIST., 9, 231-243.

HSU, P. L. 1939. (8.4/10.1/13.1)                                              **2624**
    ON THE DISTRIBUTION OF ROOTS OF CERTAIN DETERMINANTAL EQUATIONS.
    ANN. EUGENICS, 9, 250-258.   (MR1,P.248)

HSU, P. L. 1939. (7.1)                                                         **2625**
    A NEW PROOF OF THE JOINT PRODUCT MOMENT DISTRIBUTION.
    PROC. CAMBRIDGE PHILOS. SOC., 35, 336-338.

HSU, P. L. 1940. (8.4/8.5)                                                     **2626**
    ON GENERALIZED ANALYSIS OF VARIANCE (I).
    BIOMETRIKA, 31, 221-237.   (MR2,P.111)

HSU, P. L. 1940. (7.3)                                                         **2627**
    AN ALGEBRAIC DERIVATION OF THE DISTRIBUTION OF RECTANGULAR COORDINATES.
    PROC. EDINBURGH MATH. SOC. SER. 2, 6, 185-189.   (MR2,P.109)

HSU, P. L. 1941. (8.7/13.2/A)                                                  **2628**
    ON THE PROBLEM OF RANK AND THE LIMITING DISTRIBUTION OF FISHER'S TEST FUNCTION.
    ANN. EUGENICS, 11, 39-41.   (MR3,P.8)

HSU, P. L. 1941. (8.2/8.3)                                                     **2629**
    CANONICAL REDUCTION OF THE GENERAL REGRESSION PROBLEM.
    ANN. EUGENICS, 11, 42-46.   (MR3,P.8)

HSU, P. L. 1941. (9.2/12.2/13.1/A)                                             **2630**
    ON THE LIMITING DISTRIBUTION OF THE CANONICAL CORRELATIONS.
    BIOMETRIKA, 32, 38-45.   (MR2,P.234)

HSU, P. L. 1941. (8.4/10.1/13.1/A)                                             **2631**
    ON THE LIMITING DISTRIBUTION OF ROOTS OF A DETERMINANTAL EQUATION.
    J. LONDON MATH. SOC., 16, 183-194.   (MR3,P.174)

HSU, P. L. 1942. (17.1)                                                        **2632**
    THE LIMITING DISTRIBUTION OF A GENERAL CLASS OF STATISTICS.
    ACAD. SINICA SCI. REC., 1, 37-41.   (MR5,P.42)

HSU, P. L. 1945.  (4.7/5.1)                                                        2633
    ON THE POWER FUNCTIONS FOR THE $E^2$-TEST AND THE $T^2$-TEST.
    ANN. MATH. STATIST., 16, 278-286.  (MR7,P.212)


HSU, P. L. 1946.  (9.2/14.2/A)                                                     2634
    ON THE ASYMPTOTIC DISTRIBUTIONS OF CERTAIN STATISTICS USED IN TESTING
    THE INDEPENDENCE BETWEEN SUCCESSIVE OBSERVATIONS FROM A NORMAL POPULATION.
    ANN. MATH. STATIST., 17, 350-354.  (MR8,P.161)


HSU, P. L. 1946.  (20.2)                                                           2635
    ON A FACTORIZATION OF PSEUDO-ORTHOGONAL MATRICES.
    QUART. J. MATH. OXFORD SER., 17, 162-165.  (MR8,P.129)


HSU, P. L. 1949.  (17.5/A)                                                         2636
    THE LIMITING DISTRIBUTION OF FUNCTIONS OF SAMPLE MEANS AND
    APPLICATIONS TO TESTING HYPOTHESES.
    PROC. BERKELEY SYMP. MATH. STATIST. PROB., 359-402.  (MR10,P.387)


HSU, P. L. 1953.  (20.2)                                                           2637
    ON SYMMETRIC, ORTHOGONAL AND SKEW-SYMMETRIC MATRICES.
    PROC. EDINBURGH MATH. SOC. SER. 2, 10, 37-44.  (MR15,P.671)


HSU, P. L. 1955.  (20.1)                                                           2638
    ON A KIND OF TRANSFORMATIONS OF MATRICES.
        (IN CHINESE, WITH SUMMARY IN ENGLISH)
    ACTA MATH. SINICA, 5, 333-346.  (MR17,P.339)


HSU, P. L. 1955.  (20.3)                                                           2639
    ON A KIND OF TRANSFORMATION OF MATRIC PAIRS.
        (IN CHINESE, WITH SUMMARY IN ENGLISH)
    PEKING UNIV. J., 1, 1-16.


HSU, P. L. 1957.  (20.2)                                                           2640
    ON THE SIMULTANEOUS TRANSFORMATION OF A HERMITIAN MATRIX AND A
    SYMMETRIC OR SKEW-SYMMETRIC MATRIX.
        (IN CHINESE, WITH SUMMARY IN ENGLISH)
    PEKING UNIV. J., 3, 167-209.



HUANG, DAVID S.   SEE
    6075. ZELLNER, ARNOLD + HUANG, DAVID S. (1962)


HUBER, H. 1956.  (5.2/E)                                                           2641
    UBER DIE ANWENDUNG STATISTISCHER TRENNFUNKTIONEN ZUR UNTERSCHEIDUNG NAHE
    VERWANDTER ARTEN.
    VERHANDL. NATURFORSCH. GES. BASEL, 67, 149-175.  (MR18,P.452)


HUDDLESTON, HAROLD F. 1955.  (8.1/20.3/E)                                          2642
    AN INVERTED MATRIX APPROACH FOR DETERMINING CROPWEATHER REGRESSION EQUATIONS.
    BIOMETRICS, 11, 231-236.


HUDIMOTO, HIROSI 1956.  (17.7/E)                                                   2643
    NOTE ON FITTING A STRAIGHT LINE WHEN BOTH VARIABLES ARE SUBJECT TO
    ERROR AND SOME APPLICATIONS.
    ANN. INST. STATIST. MATH. TOKYO, 7, 159-167.  (MR18,P.242)


HUDIMOTO, HIROSI 1956.  (6.4)                                                      2644
    ON THE DISTRIBUTION-FREE CLASSIFICATION OF AN INDIVIDUAL INTO ONE OF TWO GROUPS.
    ANN. INST. STATIST. MATH. TOKYO, 8, 105-118.  (MR19,P.472)


HUDIMOTO, HIROSI 1957.  (6.4)                                                      2645
    A NOTE ON THE PROBABILITY OF THE CORRECT CLASSIFICATION WHEN THE
    DISTRIBUTIONS ARE NOT SPECIFIED.
    ANN. INST. STATIST. MATH. TOKYO, 9, 31-36.  (MR19,P.1094)


HUDIMOTO, HIROSI 1963.  (6.4/B)                                                    2646
    ON THE CLASSIFICATION-I. THE CASE OF TWO POPULATIONS.
        (IN JAPANESE, WITH SUMMARY IN ENGLISH)
    PROC. INST. STATIST. MATH. TOKYO, 11, 31-38.

HUDIMOTO, HIROSI 1965.  (6.4/17.8)                                          2647
    ON THE CLASSIFICATION II.
        (IN JAPANESE, WITH SUMMARY IN ENGLISH)
    PROC. INST. STATIST. MATH. TOKYO, 12, 274-276.  (MR32-3220)

HUGHES, HARRY M. 1949.  (3.1/3.3/B)                                         2648
    ESTIMATION OF THE VARIANCE OF THE BIVARIATE NORMAL DISTRIBUTION.
    UNIV. CALIFORNIA PUBL. STATIST., 1, 37-51.  (MR12,P.346)

HUGHES, HARRY M.   SEE ALSO
    1185. DANFORD, MASIL B.; HUGHES, HARRY M. + MC NEE, P. C. (1960)

HUGHES, J.B.   SEE
    5102. SMITH, HARRY, JR.; GNANADESIKAN, R. + HUGHES, J.B. (1962)

HUGHES, R. E. + LINDLEY, D. V. 1955.  E(6.1)                                2649
    APPLICATION OF BIOMETRIC METHODS TO PROBLEMS OF CLASSIFICATION IN ECOLOGY.
    NATURE, 175, 806-807.

HUIZINGA, J.   SEE
    1261. DE FROE, A.; HUIZINGA, J. + VAN GOOL, J. (1947)

HULL, CLARK L. 1922.  (15.4)                                                2650
    THE CONVERSION OF TEST SCORES INTO SERIES WHICH SHALL HAVE ANY
    ASSIGNED MEAN AND DEGREE OF DISPERSION.
    J. APPL. PSYCHOL., 6, 298-300.

HULL, CLARK L. 1925.  (4.1)                                                 2651
    AN AUTOMATIC CORRELATION CALCULATING MACHINE.
    J. AMER. STATIST. ASSOC., 20, 522-531.

HULTQUIST, ROBERT A.   SEE
    2102. GRAYBILL, FRANKLIN A. + HULTQUIST, ROBERT A. (1961)

HUME, M. W.   SEE
        66. AITKIN, M. A. + HUME, M. W. (1965)

        67. AITKIN, M. A. + HUME, M. W. (1966)

HUMM, DONCASTER G. 1956.  (4.9/B)                                           2652
    A SUGGESTED MEASURE OF BIVARIATE RELATIONSHIP BASED ON THE ELLIPSE OF
    CONCENTRATION.
    J. PSYCHOL., 41, 395-403.

HUNTER, J. STUART   SEE
    696. BOX, GEORGE E. P. + HUNTER, J. STUART (1954)

HUNTER, WILLIAM G.   SEE
    1388. DRAPER, NORMAN R. + HUNTER, WILLIAM G. (1966)

HUNTINGTON, EDWARD V. 1919.  (4.1/4.8)                                      2653
    MATHEMATICS AND STATISTICS, WITH AN ELEMENTARY ACCOUNT OF THE
    CORRELATION COEFFICIENT AND THE CORRELATION RATIO.
    AMER. MATH. MONTHLY, 26, 421-435.

HUNTINGTON, EDWARD V. 1932.  (2.1)                                          2654
    AN IMPROVED EQUAL-FREQUENCY MAP OF THE NORMAL CORRELATION SURFACE,
    USING CIRCLES INSTEAD OF ELLIPSES.
        (REPRINT OF NO. 2656, AMENDED BY NO. 2655)
    J. AMER. STATIST. ASSOC., 27, 251-255.

HUNTINGTON, EDWARD V. 1932.  (2.1)                                          2655
    ERRATA: AN IMPROVED EQUAL-FREQUENCY MAP OF THE NORMAL CORRELATION SURFACE,
    USING CIRCLES INSTEAD OF ELLIPSES.
        (AMENDMENT OF NO. 2656, AMENDMENT OF NO. 2654)
    J. AMER. STATIST. ASSOC., 27, 431.

HUNTINGTON, EDWARD V. 1932.  (2.1)                                          2656
    AN IMPROVED EQUAL-FREQUENCY MAP OF THE NORMAL CORRELATION SURFACE,
    USING CIRCLES INSTEAD OF ELLIPSES.
        (REPRINTED AS NO. 2654, AMENDED BY NO. 2655)
    PROC. NAT. ACAD. SCI. U.S.A., 18, 465-467.

HURON, ROGER 1955.   (18.1)                                                                    2657
    LOI MULTINOMIALE ET TEST DU  $\chi^2$.
    C.R. ACAD. SCI. PARIS, 240, 2047-2048.   (MR17,P.56)

HURON, ROGER 1959.   (4.1/18.1)                                                                2658
    ETUDE DE LA CORRELATION ENTRE CERTAINES VARIABLES ALEATOIRES LIEES A
    LA LOI MULTINOMIALE.
    C.R. ACAD. SCI. PARIS, 249, 2268-2269.   (MR22-1015)

HURWICZ, LEONID 1950.   (16.2)                                                                 2659
    GENERALIZATION OF THE CONCEPT OF IDENTIFICATION.
    COWLES COMMISS. MONOG., 10, 245-257.

HURWICZ, LEONID 1950.   (16.2)                                                                 2660
    PREDICTION AND LEAST SQUARES.
    COWLES COMMISS. MONOG., 10, 266-300.

HURWICZ, LEONID 1962.   (17.7)                                                                 2661
    ON THE STRUCTURAL FORM OF INTERDEPENDENT SYSTEMS.
    PROC. INTERNAT. CONG. LOGIC METH. PHILOS. SCI., 232-239.

HURWICZ, LEONID    SEE ALSO
    169. ANDERSON, T. W. + HURWICZ, LEONID (1947)

HUSSON, R. 1934.   (1.1)                                                                       2662
    STATISTIQUE, PSYCHOTECHNIQUE ET SELECTION.
    J. SOC. STATIST. PARIS, 75, 83-85.

HUTTLY, N. A. 1956.   (8.1/16.5)                                                               2663
    THE FITTING OF REGRESSION CURVES WITH AUTOCORRELATED DATA.
    BIOMETRIKA, 43, 468-474.   (MR18,P.522)

HUTTLY, N. A. 1960.   (4.4/16.5/E)                                                             2664
    USE OF THE TETRACHORIC CROSS-CORRELATION IN HYPOTHESES CONCERNING
    AUTO-CORRELATED FADING SIGNALS.
    STATIST. METH. RADIO WAVE PROP., 154-175.

HUZURBAZAR, V. S. 1948.   (19.3/A)                                                             2665
    THE LIKELIHOOD EQUATION, CONSISTENCY AND THE MAXIMA OF THE LIKELIHOOD FUNCTION.
    ANN. EUGENICS, 14, 185-200.

HUZURBAZAR, V. S. 1950.   (19.3/20.2)                                                          2666
    PROBABILITY DISTRIBUTIONS AND ORTHOGONAL PARAMETERS.
    PROC. CAMBRIDGE PHILOS. SOC., 46, 281-284.   (MR11,P.608)

HYRENIUS, HANNES 1952.   (18.3)                                                                2667
    SAMPLING FROM BIVARIATE NON-NORMAL UNIVERSES BY MEANS OF COMPOUND
    NORMAL DISTRIBUTIONS.
    BIOMETRIKA, 39, 238-246.   (MR14,P.487)

IGNAT'EV, N. K. 1958.   (17.9)                                                                 2668
    STATISTICAL CHARACTERISTICS OF MANY-DIMENSIONAL COMMUNICATIONS.
        (IN RUSSIAN)
    NAUCN. DOKL. VYSS. SKOLY, 3, 3-12.

IHM, PETER 1954.   (5.2)                                                                       2669
    ANWENDUNG VON HOTELLINGS VERALLGEMEINERTEN T-TEST ZUR PRUFUNG DER
    DIFFERENZ ZWEIER MITTELWERTEPAARE.
    Z. INDUKT. ABSTAM. VERERBUNGSL., 86, 143-156.

IHM, PETER 1955.   (10.2/B)                                                                    2670
    EIN KRITERIUM FUR ZWEI TYPEN ZWEIDIMENSIONALER NORMALVERTEILUNGEN.
    MITT. MATH. STATIST., 7, 46-52.   (MR16,P.1039)

IHM, PETER 1957.   (18.1/20.4)                                                                 2671
    BERECHNUNG VON INTEGRALEN DER N-DIMENSIONALEN STUDENT-VERTEILUNG
    MITTELS STIELTJESINTEGRALPAPIER.
    MITT. MATH. STATIST., 9, 143-146.   (MR19,P.780)

IHM, PETER 1959.  (2.7)                                                    2672
    NUMERICAL EVALUATION OF CERTAIN MULTIVARIATE NORMAL INTEGRALS.
    SANKHYA, 21, 363-366.  (MR22-2004)


IHM, PETER 1960.  (2.7)                                                    2673
    ZUR NUMERISCHEN INTEGRATION N-DIMENSIONALER NORMALVERTEILUNGEN.
    METRIKA, 3, 74-78.  (MR22-1077)


IHM, PETER 1961.  (2.7)                                                    2674
    A FURTHER CONTRIBUTION TO THE NUMERICAL EVALUATION OF CERTAIN
    MULTIVARIATE NORMAL INTEGRALS.
    SANKHYA SER. A, 23, 205-206.  (MR25-1613)



IKEDA, S. 1962.  (6.5/17.1)                                                2675
    NECESSARY CONDITIONS FOR THE CONVERGENCE OF KULLBACK-LEIBLER'S MEAN
    INFORMATION.
    ANN. INST. STATIST. MATH. TOKYO, 14, 107-118.


IKEDA, TOYOJI 1950.  (14.4)                                                2676
    ON THE INEQUALITIES BASED ON THE K-VARIATE NORMAL DISTRIBUTION.
        (IN JAPANESE)
    RES. MEM. INST. STATIST. MATH. TOKYO, 5, 519-522.


IKEDA, TOYOJI   SEE ALSO
    1116. CRONBACH, LEE J.; IKEDA, TOYOJI + AVNER, R. A. (1964)


IMHOF, J. P. 1961.  (2.5/14.5)                                             2677
    COMPUTING THE DISTRIBUTION OF QUADRATIC FORMS IN NORMAL VARIABLES.
    BIOMETRIKA, 48, 419-426.  (MR25-655)


IMMER, F. R.; TYSDAL, H. M. + STEECE, H. M. 1939.  (1.2)                   2678
    BIBLIOGRAPHY OF FIELD EXPERIMENTS.
    J. AMER. SOC. AGRON., 31, 1049-1052.


INDO, TARO 1950.  (15.1/15.2)                                              2679
    FACTOR ANALYSIS OF HOLZINGER AND HARMAN.
        (IN JAPANESE)
    JAPAN. J. PSYCHOL., 20(3) 38-46.


INGHAM, A. E. 1933.  (7.2/20.4)                                            2680
    AN INTEGRAL THAT OCCURS IN STATISTICS.
    PROC. CAMBRIDGE PHILOS. SOC., 29, 271-276.


IRWIN, J. O. 1929.  (14.2)                                                 2681
    ON THE FREQUENCY DISTRIBUTION OF ANY NUMBER OF DEVIATES FROM THE MEAN OF A
    SAMPLE FROM A NORMAL POPULATION, AND THE PARTIAL CORRELATIONS BETWEEN THEM.
    J. ROY. STATIST. SOC., 92, 580-584.


IRWIN, J. O. 1931.  (1.1/1.2)                                              2682
    RECENT ADVANCES IN MATHEMATICAL STATISTICS.
    J. ROY. STATIST. SOC., 94, 568-578.


IRWIN, J. O. 1932.  (15.1)                                                 2683
    ON THE UNIQUENESS OF THE FACTOR G FOR GENERAL INTELLIGENCE.
    BRITISH J. PSYCHOL., 22, 359-363.


IRWIN, J. O. 1932.  (1.1/1.2)                                              2684
    RECENT ADVANCES IN MATHEMATICAL STATISTICS (1931).
    J. ROY. STATIST. SOC., 95, 498-530.


IRWIN, J. O. 1933.  (15.1)                                                 2685
    A CRITICAL DISCUSSION OF THE SINGLE-FACTOR THEORY.
    BRITISH J. PSYCHOL., 23, 371-381.


IRWIN, J. O. 1934.  (4.1/4.8)                                              2686
    CORRELATION METHODS IN PSYCHOLOGY.
    BRITISH J. PSYCHOL., 25, 86-91.


IRWIN, J. O. 1934.  (1.1/1.2)                                              2687
    RECENT ADVANCES IN MATHEMATICAL STATISTICS (1932).
    J. ROY. STATIST. SOC., 97, 114-154.

IRWIN, J. O. 1935.  (15.2)                                                        **2688**
    ON THE INDETERMINACY IN THE ESTIMATE OF G.
  BRITISH J. PSYCHOL., 25, 393-394.

IRWIN, J. O. 1935.  (15.1/15.2)                                                   **2689**
    STATISTICAL METHODS IN PSYCHOLOGY-THE PRESENT POSITION OF THE THEORY
    OF TWO FACTORS.
  BULL. INST. INTERNAT. STATIST., 28(2) 192-204.

IRWIN, J. O. 1938.  (1.2)                                                         **2690**
    RECENT ADVANCES IN MATHEMATICAL STATISTICS.  BIBLIOGRAPHY OF
    MATHEMATICAL STATISTICS (1935, 1936, AND FIRST HALF OF 1937).
  J. ROY. STATIST. SOC., 101, 394-433.

IRWIN, J. O. 1966.  (1.2)                                                         **2691**
    LEON ISSERLIS, M.A., D.SC. (1881-1966).
  J. ROY. STATIST. SOC. SER. A, 129, 612-616.

IRWIN, J. O.; BARTLETT, MAURICE S.; COCHRAN, WILLIAM G. +  1935.  (1.1/1.2)       **2692**
  FIELLER, E. C.
    RECENT ADVANCES IN MATHEMATICAL STATISTICS (1933).
  J. ROY. STATIST. SOC., 98, 83-127.

IRWIN, J. O.; COCHRAN, WILLIAM G.; FIELLER, E. C. +  1936.  (1.1/1.2)             **2693**
  STEVENS, W. L.
    RECENT ADVANCES IN MATHEMATICAL STATISTICS (1934).
  J. ROY. STATIST. SOC., 99, 714-769.

IRWIN, LARRY 1966.  (15.2/15.3)                                                   **2694**
    A METHOD FOR CLUSTERING EIGENVALUES.
  PSYCHOMETRIKA, 31, 11-16.  (MR35-7489)

ISAACSON, STANLEY L. 1954.  (6.1/6.2/6.3)                                         **2695**
    PROBLEMS IN CLASSIFYING POPULATIONS.
  STATIST. MATH. BIOL. (KEMPTHORNE ET AL.), 107-117.

ISABEL, MARIANO MEDINA E  SEE  MEDINA E ISABEL, MARIANO

ISII, KEIITI 1959.  (1.2/17.2)                                                    **2696**
    ON TCHEBYCHEFF TYPE INEQUALITIES.
      (IN JAPANESE, WITH SUMMARY IN ENGLISH)
  PROC. INST. STATIST. MATH. TOKYO, 7, 123-143.

ISII, KEIITI 1962.  (17.2)                                                        **2697**
    ON SHARPNESS OF TCHEBYCHEFF-TYPE INEQUALITIES.
  ANN. INST. STATIST. MATH. TOKYO, 14, 185-197.  (MR31-6260)

ISRAEL, ADI BEN-  SEE  BEN-ISRAEL, ADI

ISSERLIS, L. 1914.  (4.5/4.8)                                                     **2698**
    ON THE PARTIAL CORRELATION RATIO.  PART I.  THEORETICAL.
  BIOMETRIKA, 10, 391-411.

ISSERLIS, L. 1914.  (4.1/18.1/18.2/E)                                             **2699**
    THE APPLICATION OF SOLID HYPERGEOMETRICAL SERIES TO FREQUENCY
    DISTRIBUTIONS IN SPACE.
  PHILOS. MAG. SER. 6, 28, 379-403.

ISSERLIS, L. 1915.  (4.5/4.8)                                                     **2700**
    ON THE PARTIAL CORRELATION-RATIO.  PART II.  NUMERICAL.
  BIOMETRIKA, 11, 50-66.  (MR35-5091)

ISSERLIS, L. 1916.  (4.7/14.2)                                                    **2701**
    ON CERTAIN PROBABLE ERRORS AND CORRELATION COEFFICIENTS OF MULTIPLE FREQUENCY
    DISTRIBUTIONS WITH SKEW REGRESSION.
  BIOMETRIKA, 11, 185-190.

ISSERLIS, L. 1917.  (4.2/4.5/4.7)                                                 **2702**
    THE VARIATION OF THE MULTIPLE CORRELATION COEFFICIENT IN SAMPLES
    DRAWN FROM AN INFINITE POPULATION WITH NORMAL DISTRIBUTION.
  PHILOS. MAG. SER. 6, 34, 205-220.

ISSERLIS, L. 1918.  (4.1/4.5)                                                     **2703**
    ON A FORMULA FOR THE PRODUCT-MOMENT COEFFICIENT OF ANY ORDER OF A
    NORMAL FREQUENCY DISTRIBUTION IN ANY NUMBER OF VARIABLES.
  BIOMETRIKA, 12, 134-139.

ISSERLIS, L. 1918.   (17.3/17.9)                2704
    FORMULAE FOR DETERMINING THE MEAN VALUES OF PRODUCTS OF DEVIATIONS OF
    MIXED MOMENT COEFFICIENTS IN TWO TO EIGHT VARIABLES IN SAMPLES TAKEN
    FROM A LIMITED POPULATION.
    BIOMETRIKA, 12, 183-184.

ITO, KIYOSI 1951.   (20.4)                2705
    MULTIPLE WIENER INTEGRAL.
    J. MATH. SOC. JAPAN, 3, 157-169.   (MR13,P.364)

ITO, KIYOSI 1951.   (16.4)                2706
    ON STOCHASTIC DIFFERENTIAL EQUATIONS.
    MEM. AMER. MATH. SOC., 4, 1-51.   (MR12,P.724)

ITO, KIYOSI 1952.   (20.4)                2707
    COMPLEX MULTIPLE WIENER INTEGRAL.
    JAPAN. J. MATH., 22, 63-86.   (MR16,P.151)

ITO, KIYOSI 1953.   (16.4/U)                2708
    STOCHASTIC DIFFERENTIAL EQUATIONS IN A DIFFERENTIABLE MANIFOLD.   II.
    MEM. COLL. SCI. UNIV. KYOTO SER. A, 28, 81-85.   (MR15,P.636)

ITO, KOICHI 1956.   (8.4/A)                2709
    ASYMPTOTIC FORMULAE FOR THE DISTRIBUTION OF HOTELLING'S GENERALIZED $T_0^2$ STATISTIC.
    ANN. MATH. STATIST., 27, 1091-1105.   (MR18,P.958)

ITO, KOICHI 1960.   (8.4/A)                2710
    ASYMPTOTIC FORMULAE FOR THE DISTRIBUTION OF HOTELLING'S GENERALIZED $T_0^2$ STATISTIC. II.
    ANN. MATH. STATIST., 31, 1148-1153.   (MR22-7189)

ITO, KOICHI 1961.   (8.4/8.5/A)                2711
    ON MULTIVARIATE ANALYSIS OF VARIANCE TESTS.
    BULL. INST. INTERNAT. STATIST., 38(4) 87-98.   (MR27-2039)

ITO, KOICHI 1962.   (8.3/8.4/AT)                2712
    A COMPARISON OF THE POWERS OF TWO MULTIVARIATE ANALYSIS OF VARIANCE TESTS.
        (AMENDED BY NO. 2713)
    BIOMETRIKA, 49, 455-462.   (MR27-6330)

ITO, KOICHI 1963.   (8.3/8.4/AT)                2713
    CORRIGENDA:  A COMPARISON OF THE POWERS OF TWO MULTIVARIATE ANALYSIS OF VARIANCE TESTS.
        (AMENDMENT OF NO. 2712)
    BIOMETRIKA, 50, 546.   (MR27-6330)

ITO, KOICHI 1966.   (8.3/8.4/T)                2714
    ON THE HETEROSCEDASTICITY IN THE LINEAR NORMAL REGRESSION MODEL.
    RES. PAPERS STATIST. FESTSCHR. NEYMAN (DAVID), 147-155.

ITO, KOICHI + SCHULL, WILLIAM J. 1964.   (8.3/8.4)                2715
    ON THE ROBUSTNESS OF THE $T_0^2$ TEST IN MULTIVARIATE ANALYSIS OF VARIANCE
    WHEN VARIANCE-COVARIANCE MATRICES ARE NOT EQUAL.
    BIOMETRIKA, 51, 71-82.   (MR30-4342)

IVANOVIC, BRANISLAV V. 1954.   (6.1/6.5/17.8)                2716
    SUR LA DISCRIMINATION DES ENSEMBLES STATISTIQUES.
    PUBL. INST. STATIST. UNIV. PARIS, 3, 207-269.   (MR16,P.1038)

IVANOVIC, BRANISLAV V. 1955.   (4.9)                2717
    KOLEKTIVNI KOEFICIJENT KORELACIJE.(COEFFICIENT COLLECTIF DE CORRELATION.)
        (WITH SUMMARY IN FRENCH)
    STATIST. REV. BEOGRAD, 5, 7-16.

IVANOVIC, BRANISLAV V. 1957.   (6.1/6.5/E)                2718
    NOV NACIN ODREDIVANJA OTSTOJANJA IZMEDU VISEDIMENZIONALNIH STATISTICKIH SKUPOVA
    SA PRIMENOM U PROBLEMU KLASIFIKACIJE SREZOVA FNRJ PREMA STEPENU EKONOMSKE
    RAZVIJENOSTI. (NOUVELLE METHODE DE DETERMINATION DE LA DISTANCE ENTRE
    LES ENSEMBLES STATISTIQUES A PLUSIEURS DIMENSIONS AVEC L'APPLICATION DANS
    LE PROBLEME DE LA CLASSIFICATION DES DEPARTEMENTS DE LA RFPY D'APRES
    LEUR DEGRE DU DEVELOPPEMENT ECONOMIQUE.)
        (WITH SUMMARY IN FRENCH)
    STATIST. REV. BEOGRAD, 7, 125-154.

IVANOVIC, BRANISLAV V. 1963.   (17.4/B)                                            2719
    THE ESTIMATION OF THE TWO-DIMENSIONAL DISTRIBUTION OF A STATISTICAL SET, BASED
    ON THE MARGINAL DISTRIBUTIONS OF ITS STRATA.
    STATISTICA NEERLANDICA, 17, 13-23.

IVANOVIC, BRANISLAV V. + KARAMATA, JOVAN 1955.   (4.9/20.5)                        2720
    DOKAZ JEDNOG STAVA O KOLEKTIVNOM KOEFICIENTU KORELACIJE.
        (WITH SUMMARY IN FRENCH)
    STATIST. REV. BEOGRAD, 5, 309-313.

IVCHENKO, G. I. + MEDVEDEV, YU. I. 1965.   (17.1)                                  2721
    SOME MULTIDIMENSIONAL THEOREMS CONCERNING ONE CLASSICAL PROBLEM ON PERMUTATION.
        (IN RUSSIAN, WITH SUMMARY IN ENGLISH, TRANSLATED AS NO. 2722)
    TEOR. VEROJATNOST. I PRIMENEN., 10, 156-162.

IVCHENKO, G. I. + MEDVEDEV, YU. I. 1965.   (17.1)                                  2722
    SOME MULTIDIMENSIONAL THEOREMS ON A CLASSICAL PROBLEM OF PERMUTATIONS.
        (TRANSLATION OF NO. 2721)
    THEORY PROB. APPL., 10, 144-149.

IWASZKIEWICZ, C. 1933.   (4.5)                                                     2723
    SUR LA GENERALISATION DE LA METHODE DE CORRELATION PARTIELLE POUR LE CAS OU LA
    VARIABLE ELIMINEE N'EST PAS MESURABLE.
        (IN POLISH, WITH SUMMARY IN FRENCH)
    STATISTICA, WARSZAWA, 3, .

IYENGAR, N. SREENIVASA + BHATTACHARYA,N. 1965.   (17.8/B)                          2724
    SOME OBSERVATIONS ON FRACTILE GRAPHICAL ANALYSIS.
    ECONOMETRICA, 33, 644-645.

JACK, HENRY 1965.   (20.1/20.2)                                                    2725
    JACOBIANS OF TRANSFORMATIONS INVOLVING ORTHOGONAL MATRICES.
    PROC. ROY. SOC. EDINBURGH SECT. A, 67, 81-103.   (MR33-8063)

JACKSON, DUNHAM 1924.   (4.1/4.5/4.8)                                              2726
    THE ALGEBRA OF CORRELATION.
    AMER. MATH. MONTHLY, 31, 110-121.

JACKSON, DUNHAM 1924.   (4.1)                                                      2727
    THE TRIGONOMETRY OF CORRELATION.
    AMER. MATH. MONTHLY, 31, 275-280.

JACKSON, DUNHAM 1924.   (8.8/20.3)                                                 2728
    ON THE METHOD OF LEAST $m^{th}$ POWERS FOR A SET OF SIMULTANEOUS EQUATIONS.
    ANN. MATH. SER. 2, 25, 185-192.

JACKSON, DUNHAM 1938.   (20.4)                                                     2729
    ORTHOGONAL POLYNOMIALS IN THREE VARIABLES.
    DUKE MATH. J., 4, 441-454.

JACKSON, J. EDWARD 1959.  E(5.2/8.5)                                               2730
    SOME MULTIVARIATE STATISTICAL TECHNIQUES USED IN COLOR MATCHING DATA.
        (AMENDED BY NO. 2733)
    J. OPT. SOC. AMER., 49, 585-592.

JACKSON, J. EDWARD 1959.   (3.1/5.2/11.1/C)                                        2731
    QUALITY CONTROL METHODS FOR SEVERAL RELATED VARIABLES.
    TECHNOMETRICS, 1, 359-377.   (MR22-1053)

JACKSON, J. EDWARD 1960.   (20.4)                                                  2732
    A SHORT TABLE OF THE HYPERGEOMETRIC FUNCTION $_0F_1(C;X)$.
    SANKHYA, 22, 351-356.   (MR23-A3861)

JACKSON, J. EDWARD 1962.  E(5.2/8.5)                                               2733
    SOME MULTIVARIATE STATISTICAL TECHNIQUES USED IN COLOR MATCHING DATA - ADDENDA
    AND ERRATA.
        (AMENDMENT OF NO. 2730)
    J. OPT. SOC. AMER., 52, 835-836.

JACKSON, J. EDWARD + BRADLEY, RALPH A. 1961.   (5.1)                               2734
    SEQUENTIAL $\chi^2$ - AND $T^2$-TESTS.
    ANN. MATH. STATIST., 32, 1063-1077.   (MR24-A631)

JACKSON, J. EDWARD + BRADLEY, RALPH A. 1961.   (5.1)                    2735
   SEQUENTIAL $\chi^2$- AND $T^2$-TESTS AND THEIR APPLICATION TO AN ACCEPTANCE
   SAMPLING PROBLEM.
   TECHNOMETRICS, 3, 519-534.   (MR24-A632)

JACKSON, J. EDWARD + BRADLEY, RALPH A. 1966.   (5.1/5.2/E)              2736
   SEQUENTIAL MULTIVARIATE PROCEDURES FOR MEANS WITH QUALITY CONTROL
   APPLICATIONS.
   MULTIVARIATE ANAL. PROC. INTERNAT. SYMP. DAYTON (KRISHNAIAH), 507-519.

JACKSON, J. EDWARD + MORRIS, R.H. 1957.   E(5.1/11.1)                   2737
   AN APPLICATION OF MULTIVARIATE QUALITY CONTROL TO PHOTOGRAPHIC PROCESSING.
   J. AMER. STATIST. ASSOC., 52, 186-199.

JACKSON, J. EDWARD    SEE ALSO
   740. BROWN, W. R. J.; HOWE, WILLIAM G.; JACKSON, J. EDWARD +  (1956)
   MORRIS, R.H.

JACKSON, ROBERT W. B. 1939.   (15.4)                                    2738
   RELIABILITY OF MENTAL TESTS.
   BRITISH J. PSYCHOL., 29, 267-287.

JACKSON, ROBERT W. B. 1942.   (15.4)                                    2739
   NOTE ON THE RELATIONSHIP BETWEEN INTERNAL CONSISTENCY AND TEST-RETEST
   ESTIMATES OF THE RELIABILITY OF A TEST.
   PSYCHOMETRIKA, 7, 157-164.

JACKSON, ROBERT W. B. 1950.   E(5.2/8.5)                                2740
   THE SELECTION OF STUDENTS FOR FRESHMAN CHEMISTRY BY MEANS OF
   DISCRIMINANT FUNCTION.
   J. EXPER. EDUC., 18, 209-214.

JACKSON, ROBERT W. B. + FERGUSON, GEORGE A. 1941.   (15.4/E)            2741
   STUDIES ON THE RELIABILITY OF TESTS.
   BULL. DEPT. EDUC. RES. UNIV. TORONTO, 12, 1-132.

JACKSON, ROBERT W. B. + FERGUSON, GEORGE A. 1943.   (15.4)             2742
   A FUNCTIONAL APPROACH IN TEST CONSTRUCTION.
   EDUC. PSYCHOL. MEAS., 3, 23-28.

JACOBSON, BERNARD + WISNER, ROBERT J. 1965.   (20.1)                   2743
   MATRIX NUMBER THEORY:  AN EXAMPLE OF NONUNIQUE FACTORIZATION.
   AMER. MATH. MONTHLY, 72, 399-402.

JAGLOM, A. M. 1949.   (16.4)                                           2744
   ON THE STATISTICAL REVERSIBILITY OF BROWNIAN MOTION.
      (IN RUSSIAN)
   MAT. SB. NOV. SER., 24(66) 457-492.   (MR11,P.41)

JAGLOM, A. M. 1949.   (16.4)                                           2745
   ON PROBLEMS ABOUT THE LINEAR INTERPOLATION OF STATIONARY RANDOM
   SEQUENCES AND PROCESSES.
      (IN RUSSIAN)
   USPEHI MAT. NAUK, 32, 173-178.   (MR11,P.119)

JAGLOM, A. M. 1955.   (16.4)                                           2746
   CORRELATION THEORY OF PROCESSES WITH RANDOM STATIONARY nTH INCREMENTS.
      (IN RUSSIAN)
   MAT. SB. NOV. SER., 37(79) 141-196.   (MR17,P.167)

JAGLOM, A. M. 1957.   (16.4)                                           2747
   CERTAIN TYPES OF RANDOM FIELDS IN N-DIMENSIONAL SPACE SIMILAR TO
   STATIONARY STOCHASTIC PROCESSES.
      (IN RUSSIAN, WITH SUMMARY IN ENGLISH, TRANSLATED AS NO. 2748)
   TEOR. VEROJATNOST. I PRIMENEN., 2, 292-338.   (MR20-1353)

JAGLOM, A. M. 1957.   (16.4)                                           2748
   SOME CLASSES OF RANDOM FIELDS IN N-DIMENSIONAL SPACE, RELATED TO
   STATIONARY RANDOM PROCESSES.
      (TRANSLATION OF NO. 2747)
   THEORY PROB. APPL., 2, 273-320.   (MR20-1353)

JAGLOM, A. M. 1960.  (16.4)                                                              2749
    EFFECTIVE SOLUTIONS OF LINEAR APPROXIMATION PROBLEMS FOR MULTIVARIATE
    STATIONARY PROCESSES WITH A RATIONAL SPECTRUM.
        (IN RUSSIAN, WITH SUMMARY IN ENGLISH. TRANSLATED AS NO. 2750)
    TEOR. VEROJATNOST. I PRIMENEN., 5, 265-292.  (MR26-1934)

JAGLOM, A. M. 1960.  (16.4)                                                              2750
    EFFECTIVE SOLUTIONS OF LINEAR APPROXIMATION PROBLEMS FOR MULTIVARIATE
    STATIONARY PROCESSES WITH A RATIONAL SPECTRUM.
        (TRANSLATION OF NO. 2749)
    THEORY PROB. APPL., 5, 239-264.  (MR26-1934)

JAIN, J. P. + AMBLE, V. N. 1962.  (6.7/E)                                                2751
    IMPROVEMENT THROUGH SELECTION AT SUCCESSIVE STAGES.
    J. INDIAN SOC. AGRIC. STATIST., 14, 88-109.

JAISWAL, M.C.  SEE
    3032. KHATRI, C. G. + JAISWAL, M.C. (1963)

JAMBUNATHAN, M. V. 1966.  (8.1/20.1)                                                     2752
    SOME MATRIX SIMPLIFICATIONS IN MULTI-VARIATE AND REGRESSION ANALYSIS.
    J. KARNATAK UNIV., 11, 269-274.  (MR35-7490)

JAMBUNATHAN, M. V. 1966.  (7.1)                                                          2753
    METHODS OF DERIVING WISHART'S DISTRIBUTION.
    J. KARNATAK UNIV., 11, 275-279.  (MR35-7491)

JAMES, A. N.  SEE
    2951. KEEN, JOAN + JAMES, A. N. (1953)

JAMES, ALAN T. 1954.  (13.1/20.4)                                                        2754
    NORMAL MULTIVARIATE ANALYSIS AND THE ORTHOGONAL GROUP.
    ANN. MATH. STATIST., 25, 40-75.  (MR15,P.726)

JAMES, ALAN T. 1955.  (7.1)                                                              2755
    THE NON-CENTRAL WISHART DISTRIBUTION.
    PROC. ROY. SOC. LONDON SER. A, 229, 364-366.  (MR17,P.53)

JAMES, ALAN T. 1955.  (20.1)                                                             2756
    A GENERATING FUNCTION FOR AVERAGES OVER THE ORTHOGONAL GROUP.
    PROC. ROY. SOC. LONDON SER. A, 229, 367-375.  (MR17,P.53)

JAMES, ALAN T. 1960.  (13.2)                                                             2757
    THE DISTRIBUTION OF THE LATENT ROOTS OF THE COVARIANCE MATRIX.
    ANN. MATH. STATIST., 31, 151-158.  (MR23-A4195)

JAMES, ALAN T. 1961.  (20.1/20.4)                                                        2758
    ZONAL POLYNOMIALS OF THE REAL POSITIVE DEFINITE SYMMETRIC MATRICES.
    ANN. MATH. SER. 2, 74, 456-469.  (MR25-4155)

JAMES, ALAN T. 1961.  (7.1)                                                              2759
    THE DISTRIBUTION OF NONCENTRAL MEANS WITH KNOWN COVARIANCE.
    ANN. MATH. STATIST., 32, 874-882.  (MR24-A607)

JAMES, ALAN T. 1964.  (4.7/13.2/20.4)                                                    2760
    DISTRIBUTIONS OF MATRIX VARIATES AND LATENT ROOTS DERIVED FROM NORMAL SAMPLES.
    ANN. MATH. STATIST., 35, 475-501.  (MR31-5286)

JAMES, ALAN T. 1966.  (13.2/20.4/ET)                                                     2761
    INFERENCE ON LATENT ROOTS BY CALCULATION OF HYPERGEOMETRIC FUNCTIONS
    OF MATRIX ARGUMENT.
    MULTIVARIATE ANAL. PROC. INTERNAT. SYMP. DAYTON (KRISHNAIAH), 209-235.  (MR35-2421)

JAMES, ALAN T.  SEE ALSO
    1031. CONSTANTINE, A. G. + JAMES, ALAN T. (1958)

JAMES, G. S. 1951.  (8.3/8.5)                                                            2762
    THE COMPARISON OF SEVERAL GROUPS OF OBSERVATIONS WHEN THE RATIOS OF
    THE POPULATION VARIANCES ARE UNKNOWN.
    BIOMETRIKA, 38, 324-329.  (MR13,P.762)

JAMES, G. S. 1952.   (2.5/8.8)                                  2763
     NOTES ON A THEOREM OF COCHRAN.
     PROC. CAMBRIDGE PHILOS. SOC., 48, 443-446.   (MR13,P.962)


JAMES, G. S. 1954.   (8.3/8.4)                                  2764
     TESTS OF LINEAR HYPOTHESES IN UNIVARIATE AND MULTIVARIATE ANALYSIS
     WHEN THE RATIOS OF THE POPULATION VARIANCES ARE UNKNOWN.
     BIOMETRIKA, 41, 19-43.   (MR16,P.842)


JAMES, S. F.    SEE
     4093. ORCUTT, GUY H. + JAMES, S. F. (1948)


JAMES, W. + STEIN, CHARLES M. 1961.   (3.1)                     2765
     ESTIMATION WITH QUADRATIC LOSS.
     PROC. FOURTH BERKELEY SYMP. MATH. STATIST. PROB., 1, 361-379.   (MR24-A3025)


JANG, ZE-PEI  SEE  JIANG, ZE-PEI


JANKO, JAROSLAV 1926.   (4.6)                                   2766
     KOEFICIENT KORELACE V HOMOGRADNI STATISTICE.
          (WITH SUMMARY IN FRENCH)
     CASOPIS PEST. MAT. FYS., 55, 160-164.


JANOS, WILLIAM A. 1960.   (20.2/20.3)                           2767
     ANALYTIC INVERSION OF A CLASS OF COVARIANCE MATRICES.
     IRE TRANS. INFORMATION THEORY, IT-6, 477-484.   (MR22-8631)


JANSSON, BIRGER 1964.   (2.7/R)                                 2768
     GENERATION OF RANDOM BIVARIATE NORMAL DEVIATES AND COMPUTATION OF
     RELATED INTEGRALS.
     NORDISK TIDSSKR. INFORMATIONS-BEHANDLING, 4, 205-212.   (MR31-2817)


JARNAGIN, M.P.    SEE
     1342. DI DONATO, A. R. + JARNAGIN, M.P. (1961)


     1343. DI DONATO, A. R. + JARNAGIN, M.P. (1962)


JASTREMSKII, B. S. 1924.   (4.1)                                2769
     VARIABLE CORRELATION.
          (IN RUSSIAN)
     VESTNIK STATIST., 17, 39-58.


JEEVES, T. A. 1954.   (17.7)                                    2770
     IDENTIFICATION AND ESTIMATION OF LINEAR MANIFOLDS IN N-DIMENSIONS.
     ANN. MATH. STATIST., 25, 714-723.   (MR16,P.604)


JEEVES, T. A. 1954.   (4.8/15.4)                                2771
     TESTING HOMOGENEITY OF RELIABILITY COEFFICIENTS WHEN SPLIT-HALF AND
     TEST-RETEST R'S ARE TO BE COMBINED.
     J. GENET. PSYCHOL., 85, 75-77.


JEFFERS, J. N. R. 1962.  E(11.1)                                2772
     PRINCIPAL COMPONENT ANALYSIS OF DESIGNED EXPERIMENT.
     STATISTICIAN, 12, 230-242.


JEFFERY, G.B.    SEE
     4250. PEARSON, KARL; JEFFERY, G.B. + ELDERTON, ETHEL M. (1929)


JEFFREYS, HAROLD 1938.   (4.2/4.8/14.2)                         2773
     THE POSTERIOR PROBABILITY DISTRIBUTIONS OF THE ORDINARY AND
     INTRACLASS CORRELATION COEFFICIENTS.
     PROC. ROY. SOC. LONDON SER. A, 167, 464-483.


JENKINS, G. M. 1954.   (16.1/U)                                 2774
     TESTS OF HYPOTHESES IN THE LINEAR AUTOREGRESSIVE MODEL.  I.  NULL
     HYPOTHESIS DISTRIBUTIONS IN THE YULE SCHEME.
     BIOMETRIKA, 41, 405-419.   (MR16,P.605)


JENKINS, G. M. 1956.   (16.1/U)                                 2775
     TESTS OF HYPOTHESES IN THE LINEAR AUTOREGRESSIVE MODEL.  II.  NULL
     DISTRIBUTIONS FOR HIGHER ORDER SCHEMES:  NON-NULL DISTRIBUTIONS.
     BIOMETRIKA, 43, 186-199.   (MR18,P.79)

JENKINS, G. M. 1961.  (16.3)                                                      2776
   GENERAL CONSIDERATIONS IN THE ANALYSIS OF SPECTRA.
   TECHNOMETRICS, 3, 133-166.  (MR23-A3030)

JENKINS, G. M. 1963.  (16.3)                                                      2777
   CROSS-SPECTRAL ANALYSIS AND THE ESTIMATION OF LINEAR OPEN LOOP
   TRANSFER FUNCTIONS.
   PROC. SYMP. TIME SERIES ANAL. BROWN UNIV. (ROSENBLATT), 267-276.  (MR27-2069)

JENKINS, WILLIAM LEROY 1946.  (4.5/4.6)                                           2778
   A QUICK METHOD FOR MULTIPLE R AND PARTIAL r'S.
   EDUC. PSYCHOL. MEAS., 6, 273-286.

JENKINS, WILLIAM LEROY 1946.  (4.6/4.7)                                           2779
   A SHORT-CUT METHOD FOR σ AND R.
   EDUC. PSYCHOL. MEAS., 6, 533-536.

JENKINS, WILLIAM LEROY 1950.  (4.6)                                               2780
   QUICK ESTIMATION OF MULTIPLE R.
   EDUC. PSYCHOL. MEAS., 10, 346-348.

JENKINS, WILLIAM LEROY 1955.  (4.4)                                               2781
   AN IMPROVED METHOD FOR TETRACHORIC R.
       (AMENDED BY NO. 1711)
   PSYCHOMETRIKA, 20, 253-258.

JENKINS, WILLIAM LEROY 1956.  (4.4)                                               2782
   TRISERIAL R--A NEGLECTED STATISTIC.
   J. APPL. PSYCHOL., 40, 63-64.

JENKINSON, J. W. 1912.  E(4.3)                                                    2783
   GROWTH, VARIABILITY AND CORRELATION IN YOUNG TROUT.
   BIOMETRIKA, 8, 444-455.

JENNINGS, EARL 1965.  (4.5)                                                       2784
   MATRIX FORMULAS FOR PART AND PARTIAL CORRELATION.
   PSYCHOMETRIKA, 30, 353-356.

JENNISON, R. F.; PENFOLD, J. B. + ROBERTS, J. A. FRASER 1948.  E(6.3)             2785
   AN APPLICATION TO A LABORATORY PROBLEM OF DISCRIMINANT FUNCTION
   ANALYSIS INVOLVING MORE THAN TWO GROUPS.
   BRITISH J. SOC. MED., 2, 139-148.

JESSEN, BORGE 1934.  (20.5)                                                       2786
   THE THEORY OF INTEGRATION IN A SPACE OF AN INFINITE NUMBER OF DIMENSIONS.
   ACTA MATH., 63, 249-323.

JIANG, ZE-PEI 1963.  (16.4)                                                       2787
   THE PREDICTION THEORY OF MULTIVARIATE STATIONARY PROCESSES. I.
       (IN CHINESE, AMENDED BY NO. 2788, TRANSLATED AS NO. 2789, AMENDED BY NO. 2790)
   ACTA MATH. SINICA, 13, 269-298.  (MR30-667)

JIANG, ZE-PEI 1963.  (16.4)                                                       2788
   CORRECTIONS TO: THE PREDICTION THEORY OF MULTIVARIATE STATIONARY PROCESSES. I.
       (IN CHINESE, AMENDMENT OF NO. 2787, TRANSLATED AS NO. 2790, AMENDMENT OF NO. 2789)
   ACTA MATH. SINICA, 13, 653-656.  (MR30-668)

JIANG, ZE-PEI 1963.  (16.4)                                                       2789
   THE PREDICTION THEORY OF MULTIVARIATE STATIONARY PROCESSES. I.
       (TRANSLATION OF NO. 2787, AMENDED BY NO. 2790, AMENDED BY NO. 2788)
   CHINESE MATH. ACTA, 4, 291-321.  (MR30-667)

JIANG, ZE-PEI 1963.  (16.4)                                                       2790
   CORRECTIONS TO: THE PREDICTION THEORY OF MULTIVARIATE STATIONARY PROCESSES. I.
       (TRANSLATION OF NO. 2788, AMENDMENT OF NO. 2789, AMENDMENT OF NO. 2787)
   CHINESE MATH. ACTA, 4, 709-711.  (MR30-668)

JIANG, ZE-PEI 1964.  (16.4)                                                       2791
   THE PREDICTION THEORY OF MULTIVARIATE STATIONARY PROCESSES. II.
       (IN CHINESE, TRANSLATED AS NO. 2792)
   ACTA MATH. SINICA, 14, 438-450.  (MR30-5369)

JIANG, ZE-PEI 1964.  (16.4)                                                  2792
     THE PREDICTION THEORY OF MULTIVARIATE STATIONARY PROCESSES.  II.
          (TRANSLATION OF NO. 2791)
     CHINESE MATH. ACTA, 5, 471-484.  (MR30-5369)


JIE-JIAN PAN  SEE  PAN, JIE-JIAN


JOCHEMS, D. B.   SEE
     2044. GOLDBERGER, ARTHUR S. + JOCHEMS, D. B. (1961)


JOFFE, ANATOLE + KLOTZ, JEROME 1962.  (17.8/A)                                2793
     NULL DISTRIBUTION AND BAHADUR EFFICIENCY OF THE HODGES BIVARIATE SIGN TEST.
     ANN. MATH. STATIST., 33, 803-807.  (MR25-696)


JOHLER, J. RALPH + WALTERS, LILLIE C. 1959.  (18.1)                           2794
     MEAN ABSOLUTE VALUE AND STANDARD DEVIATION OF THE PHASE OF A CONSTANT
     VECTOR PLUS A RAYLEIGH-DISTRIBUTED VECTOR.
     J. RES. NAT. BUR. STANDARDS, 62, 183-186.  (MR21-2321)


JOHN, ENID   SEE
     805. BURT, CYRIL + JOHN, ENID (1942)


JOHN, FRITZ 1966.  (20.2)                                                     2795
     ON SYMMETRIC MATRICES WHOSE EIGENVALUES SATISFY LINEAR INEQUALITIES.
     PROC. AMER. MATH. SOC., 17, 1140-1145.


JOHN, PETER W. M. 1964.  (20.1)                                               2796
     PSEUDO-INVERSES IN THE ANALYSIS OF VARIANCE.
     ANN. MATH. STATIST., 35, 895-896.  (MR28-5514)


JOHN, S. 1959.  (2.7)                                                         2797
     ON THE EVALUATION OF THE PROBABILITY INTEGRAL OF A MULTIVARIATE
     NORMAL DISTRIBUTION.
     SANKHYA, 21, 367-370.  (MR22-1034)


JOHN, S. 1959.  (6.2)                                                         2798
     THE DISTRIBUTION OF WALD'S CLASSIFICATION STATISTIC WHEN THE
     DISPERSION MATRIX IS KNOWN.
     SANKHYA, 21, 371-376.  (MR22-1958)


JOHN, S. 1960.  (6.2/6.3)                                                     2799
     ON SOME CLASSIFICATION PROBLEMS.  I, II.
          (AMENDED BY NO. 2802)
     SANKHYA, 22, 301-308, 309-316.  (MR23-A4212)


JOHN, S. 1961.  (6.2)                                                         2800
     ERRORS IN DISCRIMINATION.
     ANN. MATH. STATIST., 32, 1125-1144.  (MR24-A2487)


JOHN, S. 1961.  (18.1/N)                                                      2801
     ON THE EVALUATION OF THE PROBABILITY INTEGRAL OF THE MULTIVARIATE
     T-DISTRIBUTION.
     BIOMETRIKA, 48, 409-417.  (MR26-1951)


JOHN, S. 1961.  (6.2/6.3)                                                     2802
     CORRIGENDA:  ON SOME CLASSIFICATION STATISTICS.
          (AMENDMENT OF NO. 2799)
     SANKHYA SER. A, 23, 308.


JOHN, S. 1963.  (6.2)                                                         2803
     ON CLASSIFICATION BY THE STATISTICS R AND Z.
          (AMENDED BY NO. 2807)
     ANN. INST. STATIST. MATH. TOKYO, 14, 237-246.  (MR28-670)


JOHN, S. 1963.  (14.4)                                                        2804
     A TOLERANCE REGION FOR MULTIVARIATE NORMAL DISTRIBUTIONS.
     SANKHYA SER. A, 25, 363-368.  (MR30--5428)


JOHN, S. 1964.  (6.2/A)                                                       2805
     FURTHER RESULTS ON CLASSIFICATION BY W.
     SANKHYA SER. A, 26, 39-46.

JOHN, S. 1964.   (18.1/BN)                                                                    2806
    METHODS FOR THE EVALUATION OF PROBABILITIES OF POLYGONAL AND ANGULAR
    REGIONS WHEN THE DISTRIBUTION IS BIVARIATE T.
    SANKHYA SER. A, 26, 47–54.

JOHN, S. 1965.   (6.2)                                                                        2807
    CORRECTIONS TO:  ON CLASSIFICATION BY THE STATISTICS  R  AND  Z.
    (AMENDMENT OF NO. 2803)
    ANN. INST. STATIST. MATH. TOKYO, 17, 113.   (MR28-670)

JOHN, S. 1966.   (2.1/18.1/N)                                                                 2808
    ON THE EVALUATION OF PROBABILITIES OF CONVEX POLYHEDRA UNDER
    MULTIVARIATE NORMAL AND  T   DISTRIBUTIONS.
    J. ROY. STATIST. SOC. SER. B, 28, 366–369.   (MR35-6237)

JOHNS, MILTON V., JR. 1961.   (6.4/A)                                                         2809
    AN EMPIRICAL BAYES APPROACH TO NON-PARAMETRIC TWO-WAY CLASSIFICATION.
    STUD. ITEM ANAL. PREDICT. (SOLOMON), 221–232.   (MR22-12648)

JOHNSON, A. P. 1947.   (15.4)                                                                 2810
    AN INDEX OF ITEM VALIDITY PROVIDING A CORRECTION FOR CHANCE SUCCESS.
    PSYCHOMETRIKA, 12, 51–58.

JOHNSON, F. A.   SEE
    1744. FOX, A. J. + JOHNSON, F. A. (1966)

JOHNSON, H. M. 1942.   (15.4)                                                                 2811
    GENERAL RULES FOR PREDICTING THE SELECTIVITY OF A TEST (WHEN THE STANDARDIZING
    POPULATION AND THE PARENT POPULATION ARE NOT NECESSARILY HOMOGENEOUS).
    AMER. J. PSYCHOL., 55, 436–442.

JOHNSON, H. M. 1944.   (4.8)                                                                  2812
    A USEFUL INTERPRETATION OF PEARSONIAN R IN 2X2 CONTINGENCY TABLES.
    AMER. J. PSYCHOL., 57, 236–242.

JOHNSON, H. M. 1945.   (4.8/15.4)                                                             2813
    MAXIMAL SELECTIVITY, CORRECTIVITY AND CORRELATION OBTAINABLE IN 2X2
    CONTINGENCY TABLES.
    AMER. J. PSYCHOL., 58, 65–68.

JOHNSON, HELMER G. 1944.   (17.7/E)                                                           2814
    AN EMPIRICAL STUDY OF THE INFLUENCE OF ERRORS OF MEASUREMENT UPON CORRELATION.
    AMER. J. PSYCHOL., 57, 521–536.

JOHNSON, HELMER G. 1946.   (4.1/4.8/F)                                                        2815
    CERTAIN PROPERTIES OF THE CORRELATION COEFFICIENT.
    J. EXPER. EDUC., 14, 263–266.

JOHNSON, M.C. 1955.   E(6.2/6.3)                                                              2816
    CLASSIFICATION BY MULTIVARIATE ANALYSIS WITH OBJECTIVES OF MINIMIZING
    RISK, MINIMIZING MAXIMUM RISK, AND MINIMIZING PROBABILITY OF MISCLASSIFICATION.
    J. EXPER. EDUC., 23, 259–264.

JOHNSON, N. L. 1949.   (18.1/B)                                                               2817
    BIVARIATE DISTRIBUTIONS BASED ON SIMPLE TRANSLATION SYSTEMS.
    BIOMETRIKA, 36, 297–304.   (MR11,P.527)

JOHNSON, N. L. 1960.   (18.1)                                                                 2818
    AN APPROXIMATION TO THE MULTINOMIAL DISTRIBUTION:  SOME PROPERTIES
    AND APPLICATIONS.
    BIOMETRIKA, 47, 93–102.   (MR22-1963)

JOHNSON, N. L. 1962.   (10.1/U)                                                               2819
    SOME NOTES ON THE INVESTIGATION OF HETEROGENEITY IN INTERACTIONS.
    TRABAJOS ESTADIST., 13, 183–199.   (MR32-1822)

JOHNSON, N. L. + WELCH, B.L. 1940.   (14.4/T)                                                 2820
    APPLICATIONS OF THE NON-CENTRAL T-DISTRIBUTION.
    BIOMETRIKA, 31, 362–389.   (MR1,P.346)

JOHNSON, N. L. + YOUNG, D. H. 1960.   (18.1/A)                                                2821
    SOME APPLICATIONS OF TWO APPROXIMATIONS TO THE MULTINOMIAL DISTRIBUTION.
    BIOMETRIKA, 47, 463–469.

JOHNSON, PALMER O. 1950.  (6.4/E)                                                          2822
     THE QUANTIFICATION OF QUALITATIVE DATA IN DISCRIMINANT ANALYSIS.
     J. AMER. STATIST. ASSOC., 45, 65-76.

JOHNSON, PALMER O. + HOYT, CYRIL J. 1947.  (8.3/ET)                                        2823
     ON DETERMINING THREE DIMENSIONAL REGIONS OF SIGNIFICANCE.
          (REPRINTED AS NO. 2824)
     J. EXPER. EDUC., 15, 203-212.

JOHNSON, PALMER O. + HOYT, CYRIL J. 1947.  (8.3/ET)                                        2824
     ON DETERMINING THREE DIMENSIONAL REGIONS OF SIGNIFICANCE.
          (REPRINT OF NO. 2823)
     J. EXPER. EDUC., 15, 342-353.

JOHNSON, R. C. ELANDT-  SEE  ELANDT-JOHNSON, REGINA C.

JOHNSON, RICHARD M. 1963.  (15.1/20.1)                                                     2825
     ON A THEOREM STATED BY ECKART AND YOUNG.
     PSYCHOMETRIKA, 28, 259-263.  (MR27-6376)

JOHNSON, RICHARD M. 1966.  (4.8/E)                                                         2826
     NOTE ON THE USE OF PHI AS A SIMPLIFIED PARTIAL RANK CORRELATION COEFFICIENT.
     PSYCHOL. REP., 18, 973-974.

JOHNSON, RICHARD M. 1966.  (8.1/20.2)                                                      2827
     THE MINIMAL TRANSFORMATION TO ORTHONORMALITY.
     PSYCHOMETRIKA, 31, 61-66.  (MR35-5076)

JOHNSON, WHITNEY L.   SEE
     4992. SHENTON, L. R. + JOHNSON, WHITNEY L. (1965)

JOLICOEUR, PIERRE 1959.  E(8.5/C)                                                          2828
     MULTIVARIATE GEOGRAPHICAL VARIATION IN THE WOLF CANIS LUPUS L.
     EVOLUTION, 13, 283-299.

JOLICOEUR, PIERRE 1962.  (11.1/F)                                                          2829
     THE DEGREE OF GENERALITY OF ROBUSTNESS IN MARTES AMERICANA.
     GROWTH, 27, 1-27.

JOLICOEUR, PIERRE 1963.  E(11.1/17.9)                                                      2830
     BILATERAL SYMMETRY AND ASYMMETRY IN LIMB BONES OF MARTES AMERICANA AND MAN.
     REV. CANAD. BIOL., 22, 409-432.

JOLICOEUR, PIERRE + MOSIMANN, JAMES E. 1960.  E(11.1)                                      2831
     SIZE AND SHAPE VARIATION IN THE PAINTED TURTLE.  A PRINCIPAL COMPONENT ANALYSIS.
     GROWTH, 24, 339-354.

JONES, H. GERTRUDE   SEE
     1530. ELDERTON, ETHEL M.; BARRINGTON, AMY; JONES, H. GERTRUDE;  (1913)
     LAMOTTE, EDITH M. M. DE G.; LASKI, H. J. + PEARSON, KARL

JONES, HERBERT E. 1937.  (8.7/16.5)                                                        2832
     THE NATURE OF REGRESSION FUNCTIONS IN THE CORRELATION ANALYSIS OF TIME SERIES.
     ECONOMETRICA, 5, 305-325.

JONES, KENNETH J.   SEE
     1042. COOLEY, WILLIAM W. + JONES, KENNETH J. (1964)

JONES, LL. WYNN 1933.  (15.1)                                                              2833
     PRESENT DAY THEORIES OF INTELLECTUAL FACTORS (GENERAL, GROUP AND SPECIFIC).
     BRITISH J. EDUC. PSYCHOL., 3, 1-12.

JONES, LYLE V. 1966.  (8.5/E)                                                              2834
     ANALYSIS OF VARIANCE IN ITS MULTIVARIATE DEVELOPMENTS.
     HANDB. MULTIVARIATE EXPER. PSYCHOL. (CATTELL), 244-266.

JONES, RICHARD H. 1963.  (16.4)                                                           2835
     STOCHASTIC PROCESSES ON A SPHERE.
     ANN. MATH. STATIST., 34, 213-218.

JONES, RICHARD H. 1966.  (16.5)                                                           2836
     EXPONENTIAL SMOOTHING FOR MULTIVARIATE TIME SERIES.
     J. ROY. STATIST. SOC. SER. B, 28, 241-251.  (MR33-6802)

JONES, RICHARD H.    SEE ALSO
   1404. DUNCAN, DAVID B. + JONES, RICHARD H. (1966)

   1798. FREIBERGER, WALTER F. + JONES, RICHARD H. (1960)

JONES, ROBERT A.    SEE
   3813. MICHAEL, WILLIAM B.; JONES, ROBERT A.; GADDIS, L. WESLEY +   (1962)
   KAISER, HENRY F.

JONES, WAYNE H.    SEE
   2346. HARMAN, HARRY H. + JONES, WAYNE H. (1966)

JORDAN, CAMILLE 1875.  (20.5)                                                      2837
     ESSAI SUR LA GEOMETRIE A N DIMENSIONS.
   BULL. SOC. MATH. FRANCE, 3, 103-174.

JORDAN, CHARLES  SEE  JORDAN, KAROLY

JORDAN, KAROLY 1927.  (4.2/4.3/AE)                                                 2838
     A KORRELACIOS MODSZEREK ALKALMAZASA A METEOROLOGIABAN.
   IDOJARAS, 31, 65-70, 93-94.

JORDAN, KAROLY 1932.  (8.8/20.4)                                                   2839
     APPROXIMATION AND GRADUATION ACCORDING TO THE PRINCIPLE OF LEAST
     SQUARES BY ORTHOGONAL POLYNOMIALS.
        (AMENDED BY NO. 2845)
   ANN. MATH. STATIST., 3, 257-357.

JORDAN, KAROLY 1937.  (20.5)                                                       2840
     SUR L'APPROXIMATION D'UNE FONCTION A PLUSIEURS VARIABLES.
   ACTA LITT. SCI. REG. UNIV. HUNGAR. SECT. SCI. MATH., 8, 205-225.

JORDAN, KAROLY 1937.  (4.2/4.3/AE)                                                 2841
     A KORRELACIO SZAMITAS ALKALMAZASA A METEOROLOGIABAN.
        (WITH SUMMARY IN FRENCH)
   IDOJARAS, 41, 93-110, 136-140.

JORDAN, KAROLY 1938.  (2.1/4.1/4.4)                                                2842
     CRITIQUE DE LA CORRELATION AU POINT DE VUE DES PROBABILITES.
   ACTUAL. SCI. INDUST., 740(7) 15-33.

JORDAN, KAROLY 1941.  (4.8)                                                        2843
     A KORRELACIO SZAMITASA.  I.
   STATISZT. SZEMLE, 19, 1-47.

JORDAN, KAROLY 1953.  (17.3)                                                       2844
     ESZLELESEK TORVENYSZERUSEGENEK MEGHATAROZASA TOBB VALTOZO ESETEN.
     (DETERMINATION OF THE LAW IN OBSERVATIONS IN THE CASE OF MANY VARIABLES.)
   MAGYAR TUD. AKAD. MAT. OSZT. KOZL., 3, 459-466.

JORDAN, KAROLY 1959.  (8.8/20.4)                                                   2845
     CORRECTIONS TO: APPROXIMATIONS AND GRADUATION ACCORDING TO THE PRINCIPLE OF
     LEAST SQUARES BY ORTHOGONAL POLYNOMIALS.
        (AMENDMENT OF NO. 2839)
   ANN. MATH. STATIST., 30, 1266.

JORESKOG, K. G. 1962.  (15.2)                                                      2846
     ON THE STATISTICAL TREATMENT OF RESIDUALS IN FACTOR ANALYSIS.
   PSYCHOMETRIKA, 27, 335-354.  (MR29-2910)

JORESKOG, K. G. 1966.  (15.2)                                                      2847
     TESTING A SIMPLE STRUCTURE HYPOTHESIS IN FACTOR ANALYSIS.
   PSYCHOMETRIKA, 31, 165-178.  (MR33-6767)

JORGENSON, DALE W. 1961.  (16.4)                                                   2848
     MULTIPLE REGRESSION ANALYSIS OF A POISSON PROCESS.
   J. AMER. STATIST. ASSOC., 56, 235-245.  (MR23-A1466)

JOSEPH, A. W. 1947.  (18.3)                                                        2849
     THIRD-ORDER MOMENTS OF A BIVARIATE FREQUENCY DISTRIBUTION.
   J. INST. ACTUAR., 73, 427-429.

JOSEPH, A. W. 1948.  (20.5)                                                          2850
   A COMMENT ON INTERPOLATION IN TWO VARIABLES.
   J. INST. ACTUAR., 74, 82-85.  (MR11,P.136)


JOSEPHSON, NORA S.   SEE
   4758. RUSSELL, ALLAN M. + JOSEPHSON, NORA S. (1965)

JOSHI, DEVI DATT   SEE
   18. ADHIKARI, BISHWANATH PROSAD + JOSHI, DEVI DATT (1956)

JOSIFKO, MARCEL 1958.  (4.8)                                                         2851
   CHARAKTERISTICKA FUNKCE KENDALLOVA KOEFICIENTU KORELACE PORADI.
      (WITH SUMMARY IN RUSSIAN AND ENGLISH)
   CASOPIS PEST. MAT., 83, 56-59.  (MR20-2820)

JOWETT, GEOFFREY H. 1955.  (8.8/16.5)                                                2852
   LEAST SQUARES REGRESSION ANALYSIS FOR TREND-REDUCED TIME SERIES.
   J. ROY. STATIST. SOC. SER. B, 17, 91-104.  (MR17,P.505)

JOWETT, GEOFFREY H. 1958.  (15.1)                                                    2853
   FACTOR ANALYSIS.
   APPL. STATIST., 7, 114-125.

JOWETT, GEOFFREY H. 1963.  (4.6/5.1/20.3)                                            2854
   APPLICATIONS OF JORDAN'S PROCEDURE FOR MATRIX INVERSION IN MULTIPLE
   REGRESSION AND MULTIVARIATE DISTANCE ANALYSIS.
   J. ROY. STATIST. SOC. SER. B, 25, 352-357.  (MR30-669)

JOWETT, GEOFFREY H.   SEE ALSO
   1233. DAVIES, HILDA M. + JOWETT, GEOFFREY H. (1958)

   1957. GIBSON, WENDY M. + JOWETT, GEOFFREY H. (1957)

JUDGE, G. G. + TAKAYAMA, T. 1966.  (8.1)                                             2855
   INEQUALITY RESTRICTIONS IN REGRESSION ANALYSIS.
   J. AMER. STATIST. ASSOC., 61, 166-181.

JULIN, A. 1935.  (4.1/4.8/17.8)                                                      2856
   DE QUELQUES METHODES DE DETERMINATION DE LA COVARIATION.
      (AMENDED BY NO. 2857)
   REV. INST. INTERNAT. STATIST., 3, 251-277.  (MR35-6265)

JULIN, A. 1935.  (4.1/4.8/17.8)                                                      2857
   ERRATUM: DE QUELQUES METHODES DE DETERMINATION DE LA COVARIATION.
      (AMENDMENT OF NO. 2856)
   REV. INST. INTERNAT. STATIST., 3, 399.  (MR35-6265)

JUNCOSA, MARIO L. 1945.  (20.4)                                                      2858
   AN INTEGRAL EQUATION RELATED TO BESSEL FUNCTIONS.
   DUKE MATH. J., 12, 465-471.  (MR7,P.156)

JURGENSEN, C. E. 1947.  (4.8/T)                                                      2859
   TABLE FOR DETERMINING PHI COEFFICIENTS.
   PSYCHOMETRIKA, 12, 17-29.  (MR8,P.477)

K. N. M. (SIC) 1952.  (19.3/U)                                                       2860
   A NOTE ON CORRELATION BETWEEN TWO UNBIASED ESTIMATORS.
   CALCUTTA STATIST. ASSOC. BULL., 4, 72-73.  (MR14,P.190)

KAARSEMAKER, L. + VAN WIJNGAARDEN, A. 1953.  (4.8/T)                                 2861
   TABLES FOR USE IN RANK CORRELATION.
   STATISTICA, LEIDEN, 7, 41-54.  (MR15,P.331)


KABE, D. G. 1958.  (20.4)                                                            2862
   APPLICATIONS OF MEIJER-G FUNCTIONS TO DISTRIBUTION PROBLEMS IN STATISTICS.
   BIOMETRIKA, 45, 578-580.

KABE, D. G. 1962.  (8.4)                                                             2863
   ON THE EXACT DISTRIBUTION OF A CLASS OF MULTIVARIATE TEST CRITERIA.
   ANN. MATH. STATIST., 33, 1197-1200.  (MR27-870)

KABE, D. G. 1963. (8.1)                                                                    2864
    ESTIMATION OF A SET OF FIXED VARIATES FOR OBSERVED VALUES OF DEPENDENT VARIATES
    WITH NORMAL MULTIVARIATE REGRESSION MODELS SUBJECTED TO LINEAR RESTRICTIONS.
    ANN. INST. STATIST. MATH. TOKYO, 15, 51-59.   (MR30-4343)

KABE, D. G. 1963. (6.2/7.3)                                                                 2865
    SOME RESULTS ON THE DISTRIBUTION OF TWO RANDOM MATRICES USED IN
    CLASSIFICATION PROCEDURES.
        (AMENDED BY NO. 2869)
    ANN. MATH. STATIST., 34, 181-185.   (MR26-833)

KABE, D. G. 1963. (16.2)                                                                    2866
    A NOTE ON THE EXACT DISTRIBUTIONS OF THE GCL ESTIMATORS IN TWO
    LEADING OVER IDENTIFIED CASES.
    J. AMER. STATIST. ASSOC., 58, 535-537.   (MR27-858)

KABE, D. G. 1963. (8.1/8.2/8.3)                                                             2867
    STEPWISE MULTIVARIATE LINEAR REGRESSION.
    J. AMER. STATIST. ASSOC., 58, 770-773.   (MR27-2045)

KABE, D. G. 1963. (8.1/8.2/8.3)                                                             2868
    MULTIVARIATE LINEAR HYPOTHESIS WITH LINEAR RESTRICTIONS.
    J. ROY. STATIST. SOC. SER. B, 25, 348-351.

KABE, D. G. 1964. (6.2/7.3)                                                                 2869
    CORRECTIONS TO: SOME RESULTS ON THE DISTRIBUTION OF TWO RANDOM
    MATRICES USED IN CLASSIFICATION PROCEDURES.
        (AMENDMENT OF NO. 2865)
    ANN. MATH. STATIST., 35, 924.   (MR26-833)

KABE, D. G. 1964. (7.1/7.3)                                                                 2870
    DECOMPOSITION OF WISHART DISTRIBUTION.
    BIOMETRIKA, 51, 267.   (MR30-4316)

KABE, D. G. 1964. (16.2)                                                                    2871
    ON THE EXACT DISTRIBUTIONS OF THE GCL ESTIMATORS IN A LEADING
    THREE-EQUATION CASE.
    J. AMER. STATIST. ASSOC., 59, 881-894.   (MR30-2653)

KABE, D. G. 1964. (7.3)                                                                     2872
    A NOTE ON THE BARTLETT DECOMPOSITION OF A WISHART MATRIX.
    J. ROY. STATIST. SOC. SER. B, 26, 270-273.   (MR30-1589)

KABE, D. G. 1964. (8.3/8.4)                                                                 2873
    ON TESTS CONCERNING COEFFICIENT MATRICES OF LINEAR RESTRICTION WITH
    NORMAL REGRESSION MODELS.
    METRIKA, 8, 231-234.   (MR30-4344)

KABE, D. G. 1965. (5.3)                                                                     2874
    ON THE NONCENTRAL DISTRIBUTION OF RAO'S U  STATISTIC.
    ANN. INST. STATIST. MATH. TOKYO, 17, 75-80.   (MR31-2778)

KABE, D. G. 1965. (7.1)                                                                     2875
    GENERALIZATION OF SVERDRUP'S LEMMA AND ITS APPLICATIONS TO
    MULTIVARIATE DISTRIBUTION THEORY.
    ANN. MATH. STATIST., 36, 671-676.   (MR30-4317)

KABE, D. G. 1966. (18.1/20.4)                                                               2876
    DIRICHLET'S TRANSFORMATION AND DISTRIBUTIONS OF LINEAR FUNCTIONS OF ORDERED
    GAMMA VARIATES.
    ANN. INST. STATIST. MATH. TOKYO, 18, 367-374.   (MR35-6238)

KABE, D. G. 1966. (8.1/8.2)                                                                 2877
    SOME RESULTS FOR THE NORMAL MULTIVARIATE REGRESSION MODELS.
    AUSTRAL. J. STATIST., 8, 22-27.   (MR32-572)

KABE, D. G. 1966. (1.1)                                                                     2878
    COMPLEX ANALOGUES OF SOME RESULTS IN CLASSICAL NORMAL MULTIVARIATE
    REGRESSION THEORY.
    AUSTRAL. J. STATIST., 8, 87-91.   (MR34-6937)

KABE, D. G. 1966.  (4.5/4.7/5.3)                                                    **2879**
    COMPLEX ANALOGUES OF SOME CLASSICAL NON-CENTRAL MULTIVARIATE DISTRIBUTIONS.
    AUSTRAL. J. STATIST., 8, 99-103.

KABE, D. G. 1966.  (5.1/8.4)                                                        **2880**
    ON THE DISTRIBUTION OF COMPLEX ANALOG OF RAO'S U STATISTIC.
    J. INDIAN STATIST. ASSOC., 4, 189-194.

KABE, D. G.   SEE ALSO
    2398. HAYAKAWA, TAKESI + KABE, D. G. (1965)

    5472. TIKKIWAL, B.D. + KABE, D. G. (1958)

KAC, MARK 1939.  (2.2)                                                              **2881**
    ON A CHARACTERIZATION OF THE NORMAL DISTRIBUTION.
    AMER. J. MATH., 61, 726-728.  (MR1,P.62)

KAC, MARK 1943.  (14.4)                                                             **2882**
    ON THE AVERAGE NUMBER OF REAL ROOTS OF A RANDOM ALGEBRAIC EQUATION.
        (AMENDED BY NO. 2883)
    BULL. AMER. MATH. SOC., 49, 314-320.  (MR4,P.196)

KAC, MARK 1943.  (14.4)                                                             **2883**
    A CORRECTION TO:  ON THE AVERAGE NUMBER OF REAL ROOTS OF A RANDOM
    ALGEBRAIC EQUATION.
        (AMENDMENT OF NO. 2882)
    BULL. AMER. MATH. SOC., 49, 938.  (MR5,P.179)

KAC, MARK 1945.  (2.5)                                                              **2884**
    A REMARK ON INDEPENDENCE OF LINEAR AND QUADRATIC FORMS INVOLVING
    INDEPENDENT GAUSSIAN VARIABLES.
    ANN. MATH. STATIST., 16, 400-401.  (MR7,P.310)

KAC, MARK 1947.  (16.4)                                                             **2885**
    RANDOM WALK AND THE THEORY OF BROWNIAN MOTION.
    AMER. MATH. MONTHLY, 54, 369-391.  (MR9,P.46)

KAC, MARK 1949.  (17.9)                                                             **2886**
    ON THE AVERAGE NUMBER OF REAL ROOTS OF A RANDOM ALGEBRAIC EQUATION. (II).
    PROC. LONDON MATH. SOC. SER. 2, 50, 390-408.  (MR11,P.40)

KAC, MARK 1955.  (16.4/A)                                                           **2887**
    DISTRIBUTION OF EIGENVALUES OF CERTAIN INTEGRAL OPERATORS.
    MICHIGAN MATH. J., 3, 141-148.  (MR19,P.70)

KAC, MARK + SLEPIAN, DAVID 1959.  (16.4/A)                                          **2888**
    LARGE EXCURSIONS OF GAUSSIAN PROCESSES.
    ANN. MATH. STATIST., 30, 1215-1227.  (MR22-6024)

KAC, MARK + VAN KAMPEN, E. R. 1939.  (18.1)                                         **2889**
    CIRCULAR EQUIDISTRIBUTIONS AND STATISTICAL INDEPENDENCE.
    AMER. J. MATH., 61, 677-682.  (MR1,P.21)

KAC, MARK; KIEFER, J. + WOLFOWITZ, J. 1955.  (14.3/17.8/AU)                         **2890**
    ON TESTS OF NORMALITY AND OTHER TESTS OF GOODNESS OF FIT BASED ON
    DISTANCE METHODS.
    ANN. MATH. STATIST., 26, 189-211.  (MR17,P.55)

KAC, MARK   SEE ALSO
    33. AGNEW, RALPH P. + KAC, MARK (1941)

KAERSNA, ALFRED 1935.  (2.1/4.1/B)                                                  **2891**
    VEREINFACHTE METHODEN ZUR BERECHNUNG DES KORRELATIONSKOEFFIZIENTEN
    BEI NORMALER KORRELATION.
    ACTA COMMENT. SER. A, 29(2) 1-16.

KAGAN, A. M.; LINNIK, YU. V. + RAO, C. RADHAKRISHNA 1965.  (2.2/U)                  **2892**
    ON A CHARACTERIZATION OF THE NORMAL LAW BASED ON A PROPERTY OF SAMPLE VARIANCE.
    SANKHYA SER. A, 27, 405-406.  (MR34-884)

KAHAN, WILLIAM   SEE
    2055. GOLUB, GENE H. + KAHAN, WILLIAM (1965)

KAISER, HENRY F. 1956.  (15.3)                                                    2893
    NOTE ON CARROLL'S ANALYTIC SIMPLE STRUCTURE.
    PSYCHOMETRIKA, 21, 89-92.

KAISER, HENRY F. 1958.  (15.2)                                                    2894
    THE VARIMAX CRITERION FOR ANALYTIC ROTATION IN FACTOR ANALYSIS.
    PSYCHOMETRIKA, 23, 187-200.

KAISER, HENRY F. 1959.  (15.3)                                                    2895
    COMPUTER PROGRAM FOR VARIMAX ROTATION IN FACTOR ANALYSIS.
    EDUC. PSYCHOL. MEAS., 19, 413-420.

KAISER, HENRY F. 1962.  (15.5)                                                    2896
    SCALING A SIMPLEX.
    PSYCHOMETRIKA, 27, 155-162.   (MR25-4617)

KAISER, HENRY F. 1963.  (15.5)                                                    2897
    IMAGE ANALYSIS.
    PROBLEMS MEAS. CHANGE (HARRIS), 156-166.

KAISER, HENRY F. 1964.  (20.3)                                                    2898
    A METHOD FOR DETERMINING EIGENVALUES.
    J. SOC. INDUST. APPL. MATH., 12, 238-248.

KAISER, HENRY F. 1966.  (15.1/15.2)                                               2899
    PSYCHOMETRIC APPROACHES TO FACTOR ANALYSIS.
    TEST. PROBLEMS PERSPECTIVE (ANASTASI), 360-368.

KAISER, HENRY F. + CAFFREY, JOHN G. 1965.  (15.1)                                 2900
    ALPHA FACTOR ANALYSIS.
    PSYCHOMETRIKA, 30, 1-14.

KAISER, HENRY F.   SEE ALSO
    3813. MICHAEL, WILLIAM B.; JONES, ROBERT A.; GADDIS, L. WESLEY +  (1962)
    KAISER, HENRY F.


KAITZ, HYMAN B. 1945.  (15.4)                                                     2901
    A COMMENT ON THE CORRECTION OF RELIABILITY COEFFICIENTS FOR
    RESTRICTION OF RANGE.
        (AMENDED BY NO. 3696)
    J. EDUC. PSYCHOL., 36, 510-512.

KAITZ, HYMAN B. 1945.  (15.4)                                                     2902
    A NOTE ON RELIABILITY.
    PSYCHOMETRIKA, 10, 127-131.   (MR7,P.22)

KAKUTANI, SHIZUO 1944.  (16.4)                                                    2903
    ON BROWNIAN MOTIONS IN N-SPACE.
    PROC. IMP. ACAD. TOKYO, 20, 648-652.   (MR7,P.315)

KAKUTANI, SHIZUO 1944.  (16.4)                                                    2904
    TWO-DIMENSIONAL BROWNIAN MOTION AND HARMONIC FUNCTIONS.
    PROC. IMP. ACAD. TOKYO, 20, 706-714.   (MR7,P.315)

KAKUTANI, SHIZUO 1949.  (16.4/B)                                                  2905
    TWO-DIMENSIONAL BROWNIAN MOTION AND THE TYPE PROBLEM OF RIEMANN SURFACES.
    PROC. JAPAN ACAD. 21, 138-140.   (MR11,P.257)

KAKUTANI, SHIZUO   SEE ALSO
    1438. DVORETZKY, ARYEH; ERDOS, PAUL; KAKUTANI, SHIZUO +  (1957)
    TAYLOR, S. J.

KAKWANI, N. C. 1965.  (8.9)                                                       2906
    NOTE ON THE UNBIASED ESTIMATION OF THE THIRD MOMENT OF THE RESIDUAL IN
    REGRESSION ANALYSIS.
    ECONOMETRICA, 33, 434-436.

KAKWANI, N. C. 1965.  (8.1/14.4)                                                  2907
    NOTE ON THE USE OF PRIOR INFORMATION IN FORECASTING WITH LINEAR REGRESSION.
    SANKHYA SEP. A, 27, 101-104.

KALE, B. K. 1962.  (19.3)                                                          2908
    ON THE SOLUTION OF LIKELIHOOD EQUATIONS BY ITERATION PROCESSES.  THE
    MULTIPARAMETRIC CASE.
    BIOMETRIKA, 49, 479-486.  (MR27-6326)


KALLIANPUR, GOPINATH 1955.  (17.1)                                                 2909
    ON A LIMIT THEOREM FOR DEPENDENT RANDOM VARIABLES.
         (IN RUSSIAN)
    DOKL. AKAD. NAUK SSSR, 101, 13-16.  (MR16,P.1035)


KALLIANPUR, GOPINATH 1964.  (19.3/A)                                               2910
    VON MISES FUNCTIONALS AND MAXIMUM LIKELIHOOD ESTIMATION.
    CONTRIB. STATIST. (MAHALANOBIS VOLUME), 137-146.


KALMAN, R. E. 1960.  (16.5)                                                        2911
    A NEW APPROACH TO LINEAR FILTERING AND PREDICTION PROBLEMS.
    J. BASIC ENGRG., 82, 35-45.


KALMAN, R. E. 1965.  (20.2)                                                        2912
    IRREDUCIBLE REALIZATIONS AND THE DEGREE OF A RATIONAL MATRIX.
    J. SOC. INDUST. APPL. MATH., 13, 520-544.


KALMAN, R. E. + BUCY, R. S. 1961.  (16.5)                                          2913
    NEW RESULTS IN LINEAR FILTERING AND PREDICTION THEORY.
    J. BASIC ENGRG., 83, 95-108.


KALYANASUNDARAM, G.   SEE
    4656. ROY, J. + KALYANASUNDARAM, G. (1964)


KAMALOV, M. K. 1953.  (2.5)                                                        2914
    CONDITION FOR INDEPENDENCE OF RATIOS OF QUADRATIC FORMS.
         (IN RUSSIAN)
    AKAD. NAUK UZBEK. SSR TRUDY INST. MAT. MEH., 11, 88-91.  (MR17,P.871)


KAMALOV, M. K. 1955.  (2.5)                                                        2915
    MOMENTS AND CUMULANTS OF QUADRATIC AND BILINEAR FORMS OF NORMALLY
    DISTRIBUTED VARIABLES.
         (IN RUSSIAN)
    AKAD. NAUK UZBEK. SSR TRUDY INST. MAT. MEH., 15, 69-77.  (MR17,P.862)


KAMAT, A. R. 1953.  (2.3/T)                                                        2916
    INCOMPLETE AND ABSOLUTE MOMENTS OF THE MULTIVARIATE NORMAL
    DISTRIBUTION WITH SOME APPLICATIONS.
    BIOMETRIKA, 40, 20-34.  (MR15,P.451)


KAMAT, A. R. 1958.  (2.4)                                                          2917
    HYPERGEOMETRIC EXPANSIONS FOR INCOMPLETE MOMENTS OF THE BIVARIATE
    NORMAL DISTRIBUTION.
    SANKHYA, 20, 317-320.  (MR21-6050)


KAMAT, A. R. 1958.  (2.4)                                                          2918
    INCOMPLETE MOMENTS OF THE TRIVARIATE NORMAL DISTRIBUTION.
    SANKHYA, 20, 321-322.  (MR21-6052)


KAMAT, A. R. 1962.  (2.6)                                                          2919
    SOME MORE ESTIMATES OF CIRCULAR PROBABLE ERROR.
    J. AMER. STATIST. ASSOC., 57, 191-195.  (MR24-A3739)


KAMAT, A. R. + SATHE, Y. S. 1962.  (2.6/AT)                                        2920
    ASYMPTOTIC POWER OF CERTAIN TEST CRITERIA (BASED ON FIRST AND SECOND
    DIFFERENCES) FOR SERIAL CORRELATION BETWEEN SUCCESSIVE OBSERVATIONS.
    ANN. MATH. STATIST., 33, 186-200.


KAMENSKY, L. A. + LIU, C. N. 1964.  (6.1/6.6)                                      2921
    A THEORETICAL AND EXPERIMENTAL STUDY OF A MODEL FOR PATTERN RECOGNITION.
    COMPUT. INFORMATION SCI. (TOU + WILCOX), 194-218.


KAMPE DE FERIET, JOSEPH + FRENKIEL, FRANCOIS N. 1959.  (16.3)                      2922
    ESTIMATION DE LA CORRELATION D'UNE FONCTION ALEATOIRE NON STATIONNAIRE.
    C.R. ACAD. SCI. PARIS, 249, 348-351.  (MR21-6025)


KANEHISA, M.   SEE
    3957. NAKAGAMI, MINORU; TANAKA, K. + KANEHISA, M. (1957)

KANELLOS, SPYRIDON G. 1953.   (4.2/17.1/B)                                      2923
     ON THE COMPARATIVE FREQUENCY OF AN EVENT.
          (IN GREEK, WITH SUMMARY IN ENGLISH)
     BULL. SOC. MATH. GRECE, 27, 25-67.


KANO, SEIGO 1964.   (16.5)                                                      2924
     THE APPROXIMATE DISTRIBUTION OF THE CROSSCORRELATION COEFFICIENT
     BETWEEN TWO STATIONARY MARKOV SERIES.
     SCI. REP. KAGOSHIMA UNIV., 13, 1-10.   (MR30-1569)

KANTOROWICZ, I. 1930.   (17.3)                                                  2925
     SUR L'INDEPENDANCE DES VARIABLES ALEATOIRES.
     ENSEIGNEMENT MATH., 29, 18-26.


KANTOROWITZ, SHMUEL + KOTZ, SAMUEL 1957.   (17.4)                               2926
     TWO-DIMENSIONAL DISTRIBUTIONS WITH GIVEN MARGINALS.
          (IN HEBREW, WITH SUMMARY IN ENGLISH)
     RIVEON LEMATEMATIKA, 11, 32-38, 91. (MR20-4306)

KAPLAN, EDWARD L. 1952.   (17.3)                                                2927
     TENSOR NOTATION AND THE SAMPLING CUMULANTS OF K-STATISTICS.
     BIOMETRIKA, 39, 319-323.   (MR14,P.486)


KAPROCKI, STANISLAW 1943.   (4.1/E)                                             2928
     LE PROBLEME DE LA CORRELATION.
     SAMMLUNG WISS. ARB. SCHWEIZ INTERN. POLEN, 1(1) 54-70.


KAPTEYN, J. C. 1912.   (2.1/4.1/B)                                              2929
     DEFINITION OF THE CORRELATION COEFFICIENT.
     MONTHLY NOTICES ROY. ASTRONOM. SOC., 72, 518-525.


KARAMATA, JOVAN    SEE
     2720. IVANOVIC, BRANISLAV V. + KARAMATA, JOVAN (1955)


KARHUNEN, KARI 1947.   (16.4)                                                   2930
     LINEARE TRANSFORMATIONEN STATIONARER STOCHASTISCHER PROZESSE.
     SKAND. MAT. KONG. BER., 10, 320-324.   (MR8,P.391)


KARLE, J. + HAUPTMAN, HERBERT 1953.   (16.4/17.3/18.1/E)                        2931
     THE PROBABILITY DISTRIBUTION OF THE MAGNITUDE OF A STRUCTURE
     FACTOR.I. THE CENTROSYMMETRIC CRYSTAL.
     ACTA CRYST., 6, 131-135.

KARLE, J.    SEE ALSO
     2392. HAUPTMAN, HERBERT + KARLE, J. (1952)

     2393. HAUPTMAN, HERBERT + KARLE, J. (1953)


KARLIN, SAMUEL + TRUAX, DONALD R. 1960.   (19.1)                                2932
     SLIPPAGE PROBLEMS.
     ANN. MATH. STATIST., 31, 296-324.   (MR23-A1450)


KARLIN, SAMUEL    SEE ALSO
     2009. GIRSHICK, M. A.; KARLIN, SAMUEL + ROYDEN, H.L. (1957)


KARON, BERTRAM P. + ALEXANDER, IRVING E. 1958.   (4.8)                          2933
     A MODIFICATION OF KENDALL'S TAU FOR MEASURING ASSOCIATION IN CONTINGENCY TABLES.
     PSYCHOMETRIKA, 23, 379-383.


KARSLAKE, R.    SEE
     4512. REMMERS, H. H.; KARSLAKE, R. + GAGE, N. L. (1940)


KARSNA, ALFRED  SEE  KAERSNA, ALFRED


KARTASOV, G. D.  SEE  KHARTASHOV, G. D.


KASANIN, RADIVOJE 1947.   (8.8/17.2)                                            2934
     LE COEFFICIENT D'APPROXIMATION MOYENNE ET LE COEFFICIENT DE CORRELATION.
     ACAD. SERBE  SCI. PUBL. INST. MATH., 1, 71-87.   (MR10,P.552)


KASHIWAGI, SHIGEO 1965.   (15.3)                                                2935
     GEOMETRIC VECTOR ORTHOGONAL ROTATION METHOD IN MULTIPLE-FACTOR ANALYSIS.
     PSYCHOMETRIKA, 30, 515-530.

KASKEY, GILBERT; KRISHNAIAH, PARUCHURI R. + AZZARI, ANTHONY J. 1962.  E(6.6)                2936
    CLUSTER FORMATION AND DIAGNOSTIC SIGNIFICANCE IN PSYCHIATRIC SYMPTOM EVALUATION.
    FALL JOINT COMP. CONF. PROC., 22, 285-303.


KASTENBAUM, MARVIN A.   SEE
    4695. ROY, S.N. + KASTENBAUM, MARVIN A. (1956)


KATTI, S. K. 1961.  (8.4)                2937
    DISTRIBUTION OF THE LIKELIHOOD RATIO FOR TESTING MULTIVARIATE LINEAR HYPOTHESES.
    ANN. MATH. STATIST., 32, 333-335.  (MR22-10061)

KATTI, S. K. 1965.  (8.3/8.5/E)                2938
    MULTIPLE COVARIATE ANALYSIS.
        (AMENDED BY NO. 2939)
    BIOMETRICS, 21, 957-974.  (MR32-8461)

KATTI, S. K. 1966.  (8.3/8.5/E)                2939
    CORRECTION: MULTIPLE COVARIATE ANALYSIS.
        (AMENDMENT OF NO. 2938)
    BIOMETRICS, 22, 962.  (MR32-8461)

KATZ, I. J. + PEARL, M. H. 1966.  (20.2)                2940
    SOLUTIONS OF THE MATRIX EQUATIONS A* = XA = AX.
    J. LONDON MATH. SOC., 41, 443-452.

KATZ, LEO + OLKIN, INGRAM 1953.  (20.2)                2941
    PROPERTIES AND FACTORIZATIONS OF MATRICES DEFINED BY THE OPERATION OF
    PSEUDO-TRANSPOSITION.
    DUKE MATH. J., 20, 331-337.


KATZ, LEO + WILSON, THURLOW R. 1956.  (15.5)                2942
    THE VARIANCE OF THE NUMBER OF MUTUAL CHOICES IN SOCIOMETRY.
    PSYCHOMETRIKA, 21, 299-304.  (MR18,P.367)

KATZ, MELVIN   SEE
    398. BAUM, L. E. + KATZ, MELVIN (1965)

KATZELL, R. A. + CURETON, EDWARD E. 1947.  (4.4)                2943
    BISERIAL CORRELATION AND PREDICTION.
    J. PSYCHOL., 24, 273-278.

KAUFMAN, G. M.   SEE
    175. ANDO, ALBERT + KAUFMAN, G. M. (1965)

KAUFMAN, S. 1966.  (19.3/A)                2944
    ASYMPTOTIC EFFICIENCY OF THE MAXIMUM LIKELIHOOD ESTIMATOR.
    ANN. INST. STATIST. MATH. TOKYO, 18, 155-178.

KAVANAGH, ARTHUR J. 1941.  (8.1)                2945
    NOTE ON THE ADJUSTMENT OF OBSERVATIONS.
    ANN. MATH. STATIST., 12, 111-114.  (MR3,P.6)

KAVRUCK, SAMUEL   SEE
    2040. GOHEEN, HOWARD W. + KAVRUCK, SAMUEL (1948)

KAWADA, YUKIYOSI 1949.  (2.2)                2946
    ON A CHARACTERIZATION OF MULTIPLE NORMAL DISTRIBUTIONS.
    KODAI MATH. SEM. REP., 3, 1-2.  (MR11,P.118)

KAWADA, YUKIYOSI 1950.  (2.5)                2947
    INDEPENDENCE OF QUADRATIC FORMS IN NORMALLY CORRELATED VARIABLES.
    ANN. MATH. STATIST., 21, 614-615.  (MR12,P.346)

KAWAKAMI, HISAKO   SEE
    5061. SIOTANI, MINORU + KAWAKAMI, HISAKO (1963)

KAZINEC, L. S. 1955.  (4.9/E)                2948
    COEFFICIENT OF CORRELATION AND QUESTIONS OF THE THEORY OF INDICES.
        (IN RUSSIAN)
    UCEN. ZAP. STATIST., 1, 125-138.

KEATS, JOHN A. 1957.  (15.4)                                                    2949
    ESTIMATION OF ERROR VARIANCES OF TEST SCORES.
  PSYCHOMETRIKA, 22, 29-41.  (MR19,P.329)


KEEHN, DANIEL G. 1965.  (6.3)                                                   2950
    A NOTE ON LEARNING FOR GAUSSIAN PROPERTIES.
  IEEE TRANS. INFORMATION THEORY,IT-11, 126-132.  (MR32-5441)


KEEN, JOAN + JAMES, A. N. 1953.  (4.1/E)                                        2951
    CORRELATION FOR THE NON-MATHEMATICIAN. PART II. 3-D GRAPHS.
  INC. STATIST., 4, 128-130, 134-135.


KEEN, JOAN + PAGE, D.J. 1953.  (4.1/E)                                          2952
    CORRELATION FOR THE NON-MATHEMATICIAN. PART I. SIMPLE CORRELATION OF
    TWO VARIABLES.
  INC. STATIST., 4, 55-65.



KEEPING, E. S. 1956.  (17.7)                                                    2953
    NOTE ON WALD'S METHOD OF FITTING A STRAIGHT LINE WHEN BOTH VARIABLES
    ARE SUBJECT TO ERROR.
  BIOMETRICS, 12, 445-448.  (MR18,P.602)


KELLEHER, ROGER T. 1956.  (15.5)                                                2954
    DISCRIMINATION LEARNING AS A FUNCTION OF REVERSAL AND NONREVERSAL SHIFTS.
  J. EXPER. PSYCHOL., 51, 379-384.


KELLER, JOSEPH B. 1962.  (15.2/20.1)                                            2955
    FACTORIZATION OF MATRICES BY LEAST-SQUARES.
  BIOMETRIKA, 49, 239-242.  (MR25-3944)


KELLERER, HANS G. 1960.  (17.3)                                                 2956
    LAGEPARAMETER N-DIMENSIONALER VERTEILUNGSFUNKTIONEN.
  MATH. Z., 73, 197-218.  (MR22-3050)


KELLERER, HANS G. 1963.  (19.4)                                                 2957
    ZUR EXISTENZ ANALOGER BEREICHE.
  Z. WAHRSCHEIN. VERW. GEBIETE, 1, 240-246.


KELLERER, HANS G. 1964.  (17.4)                                                 2958
    MASSTHEORETISCHE MARGINALPROBLEME.
  MATH. ANN., 153, 168-198.


KELLERER, HANS G. 1964.  (17.4)                                                 2959
    MARGINALPROBLEME FUR FUNKTIONEN.
  MATH. ANN., 154, 147-156.


KELLERER, HANS G. 1964.  (17.3/17.4)                                            2960
    VERTEILUNGSFUNKTIONEN MIT GEGEBENEN MARGINALVERTEILUNGEN.
  Z. WAHRSCHEIN. VERW. GEBIETE, 3, 247-270.


KELLEY, TRUMAN L. 1916.  (4.5/8.8/T)                                            2961
    TABLES:  TO FACILITATE THE CALCULATION OF PARTIAL COEFFICIENTS OF
    CORRELATION AND REGRESSION EQUATIONS.
  BULL. UNIV. TEXAS.1916(27) 1-53.


KELLEY, TRUMAN L. 1921.  (15.4)                                                 2962
    THE RELIABILITY OF TEST SCORES.
  J. EDUC. RES., 3, 370-379.


KELLEY, TRUMAN L. 1923.  (15.4/E)                                               2963
    A NEW METHOD FOR DETERMINING THE SIGNIFICANCE OF DIFFERENCES IN
    INTELLIGENCE AND ACHIEVEMENT TEST SCORES.
  J. EDUC. PSYCHOL., 14, 321-333.


KELLEY, TRUMAN L. 1925.  (15.4)                                                 2964
    MEASURES OF CORRELATION DETERMINED FROM GROUPS OF VARYING HOMOGENEITY.
  J. AMER. STATIST. ASSOC., 20, 512-521.


KELLEY, TRUMAN L. 1935.  (4.8)                                                  2965
    AN UNBIASED CORRELATION RATIO MEASURE.
  PROC. NAT. ACAD. SCI. U.S.A., 21, 554-559.

KELLEY, TRUMAN L. 1935.  (15.1)                                                 2966
    THE ANALYSIS OF A COMPLEX OF MENTAL MEASUREMENTS INTO COMPONENTS.
    PSYCHOL. BULL., 32, 718-719.


KELLEY, TRUMAN L. 1939.  (4.3/15.4)                                             2967
    THE SELECTION OF UPPER AND LOWER GROUPS FOR THE VALIDATION OF TEST ITEMS.
    J. EDUC. PSYCHOL., 30, 17-24.


KELLEY, TRUMAN L. 1940.  (15.1/15.2/E)                                          2968
    COMMENT ON WILSON AND WORCESTER'S "NOTE ON FACTOR ANALYSIS".
        (AMENDMENT OF NO. 5940)
    PSYCHOMETRIKA, 5, 117-120.


KELLEY, TRUMAN L. 1942.  (15.1)                                                 2969
    THE RELIABILITY COEFFICIENT.
    PSYCHOMETRIKA, 7, 75-83.


KELLEY, TRUMAN L. 1944.  (15.2)                                                 2970
    A VARIANCE-RATIO TEST OF THE UNIQUENESS OF PRINCIPAL-AXIS COMPONENTS AS THEY
    EXIST AT ANY STAGE OF THE KELLEY ITERATIVE PROCESS FOR THEIR DETERMINATION.
    PSYCHOMETRIKA, 9, 199-200.


KELLEY, TRUMAN L. + SALISBURY, FRANK S. 1926.  (4.6)                            2971
    AN ITERATION METHOD FOR DETERMINING MULTIPLE CORRELATION CONSTANTS.
    J. AMER. STATIST. ASSOC., 21, 282-292.


KELLOGG, CHESTER E. 1936.  (11.1/15.3)                                          2972
    THE PROBLEM OF PRINCIPAL COMPONENTS:  DERIVATION OF HOTELLING'S
    METHOD FROM THURSTONE'S.
    J. EDUC. PSYCHOL., 27, 512-520.


KELLOGG, CHESTER E. 1936.  (11.1/15.1)                                          2973
    THE PROBLEM OF PRINCIPAL COMPONENTS.  II.  THE ARGUMENT FOR COMMUNALITIES.
    J. EDUC. PSYCHOL., 27, 581-590.


KEMENY, J. G. + SNELL, J. L. 1966.  (15.5/E)                                    2974
    A MARKOV CHAIN MODEL IN SOCIOLOGY.
    READINGS MATH. SOCIAL SCI. (LAZARSFELD + HENRY), 148-158.


KEMPERMAN, J. H. B. 1965.  (17.2)                                               2975
    ON THE SHARPNESS OF TCHEBYCHEFF TYPE INEQUALITIES. I-III.
    KON. NEDERL. WETENSCH. PROC. SER. A, 68, 554-601.


KEMPTHORNE, OSCAR 1955.  (4.8/8.8/E)                                            2976
    THE THEORETICAL VALUES OF CORRELATIONS BETWEEN RELATIVES IN RANDOM POPULATIONS.
    GENETICS, 40, 153-167.


KEMPTHORNE, OSCAR 1966.  (1.1)                                                  2977
    MULTIVARIATE RESPONSES IN COMPARATIVE EXPERIMENTS.
    MULTIVARIATE ANAL. PROC. INTERNAT. SYMP. DAYTON (KRISHNAIAH), 521-540.


KENDALL, DAVID G. 1960.  (1.2)                                                  2978
    CHARLES JORDAN.
    J. LONDON MATH. SOC., 35, 380-383.  (MR23-A776)


KENDALL, DAVID G. + MOYAL, J. E. 1957.  (20.5)                                  2979
    ON THE CONTINUITY PROPERTIES OF VECTOR-VALUED FUNCTIONS OF BOUNDED VARIATION.
    QUART. J. MATH. OXFORD SECOND SER., 8, 54-57.  (MR19,P.295)


KENDALL, MAURICE G. 1938.  (4.8)                                                2980
    A NEW MEASURE OF RANK CORRELATION.
    BIOMETRIKA, 30, 81-93.


KENDALL, MAURICE G. 1940.  (17.3)                                              2981
    THE DERIVATION OF MULTIVARIATE SAMPLING FORMULAE FROM UNIVARIATE
    FORMULAE BY SYMBOLIC OPERATION.
    ANN. EUGENICS, 10, 392-402.  (MR2,P.235)


KENDALL, MAURICE G. 1941.  (2.1/2.7)                                           2982
    PROOF OF RELATIONS CONNECTED WITH THE TETRACHORIC SERIES AND ITS GENERALIZATION.
    BIOMETRIKA, 32, 196-198.  (MR3,P.173)

KENDALL, MAURICE G. 1941.   (4.5/4.6/20.4)                                         2983
    THE RELATIONSHIP BETWEEN CORRELATION FORMULAE AND ELLIPTIC FUNCTIONS.
    J. ROY. STATIST. SOC., 104, 281-283.   (MR3,P.173)


KENDALL, MAURICE G. 1942.   (4.8)                                                  2984
    PARTIAL RANK CORRELATION.
    BIOMETRIKA, 32, 277-283.   (MR4,P.22)


KENDALL, MAURICE G. 1947.   (16.1)                                                 2985
    THE ESTIMATION OF PARAMETERS IN LINEAR AUTOREGRESSIVE TIME SERIES.
    BULL. INST. INTERNAT. STATIST., 31(5) 44-57.   (MR12,P.512)


KENDALL, MAURICE G. 1949.   (4.1/4.8)                                              2986
    RANK AND PRODUCT-MOMENT CORRELATION.
    BIOMETRIKA, 36, 177-193.   (MR11,P.673)


KENDALL, MAURICE G. 1951.   (8.1/8.7/17.7)                                         2987
    REGRESSION, STRUCTURE AND FUNCTIONAL RELATIONSHIP.  I.
    BIOMETRIKA, 38, 11-25.   (MR13,P.144)


KENDALL, MAURICE G. 1952.   (8.1/8.7/17.7)                                         2988
    REGRESSION, STRUCTURE AND FUNCTIONAL RELATIONSHIP.  II.
    BIOMETRIKA, 39, 96-108.   (MR14,P.66)


KENDALL, MAURICE G. 1952.   (1.2)                                                  2989
    GEORGE UDNY YULE, C.B.E., F.R.S.
    J. ROY. STATIST. SOC. SER. A, 115, 156-161.


KENDALL, MAURICE G. 1960.   (4.2)                                                  2990
    THE EVERGREEN CORRELATION COEFFICIENT.
    CONTRIB. PROB. STATIST. (HOTELLING VOLUME), 274-277.   (MR22-11457)


KENDALL, MAURICE G. 1963.   (1.2)                                                  2991
    RONALD AYLMER FISHER, 1890-1962.
    BIOMETRIKA, 50, 1-15.   (MR27-4730)


KENDALL, MAURICE G. 1966.   (6.4/6.6/E)                                            2992
    DISCRIMINATION AND CLASSIFICATION.
    MULTIVARIATE ANAL. PROC. INTERNAT. SYMP. DAYTON (KRISHNAIAH), 165-185.


KENDALL, MAURICE G. + BABINGTON SMITH, B. 1950.   (15.1/15.2)                      2993
    FACTOR ANALYSIS.   (PART 1:  FACTOR ANALYSIS AS A STATISTICAL TECHNIQUE, BY M.G.
    KENDALL. PART 2:   AN EVALUATION OF FACTOR ANALYSIS FROM THE POINT OF VIEW OF A
    PSYCHOLOGIST, BY B. BABINGTON SMITH.)
        (WITH DISCUSSION)
    J. ROY. STATIST. SOC. SER. B, 12, 60-94.   (MR12,P.621)


KENDALL, MAURICE G. + LAWLEY, D. N. 1956.   (15.1)                                 2994
    THE PRINCIPLES OF FACTOR ANALYSIS.
    J. ROY. STATIST. SOC. SER. A, 119, 83-84.   (MR18,P.79)


KENDALL, MAURICE G.; KENDALL, SHEILA F. H. + BABINGTON SMITH, B. 1939.   (4.8)     2995
    THE DISTRIBUTION OF SPEARMAN'S COEFFICIENT OF RANK CORRELATION IN A
    UNIVERSE IN WHICH ALL RANKINGS OCCUR AN EQUAL NUMBER OF TIMES.
    BIOMETRIKA, 30, 251-273.

KENDALL, MAURICE G.   SEE ALSO
    1196. DANIELS, H. E. + KENDALL, MAURICE G. (1947)

    1197. DANIELS, H. E. + KENDALL, MAURICE G. (1958)


KENDALL, SHEILA F. H.   SEE
    2995. KENDALL, MAURICE G.; KENDALL, SHEILA F. H. + BABINGTON SMITH, B. (1939)


KENNEY, J. F. 1939.   (1.1)                                                        2996
    SOME TOPICS IN MATHEMATICAL STATISTICS.
    AMER. MATH. MONTHLY, 46, 59-74.

KENNEY, KATHRYN C.   SEE
    1507. EDWARDS, ALLEN L. + KENNEY, KATHRYN C. (1946)

KENNY, D. T. 1953.  (10.2)                                                                    2997
    TESTING OF DIFFERENCES BETWEEN VARIANCES BASED ON CORRELATED VARIATES.
    CANAD. J. PSYCHOL., 7, 25-28.

KERMACK, K. A. + HALDANE, J. B. S. 1950.   (11.2/E)                                           2998
    ORGANIC CORRELATION AND ALLOMETRY.
    BIOMETRIKA, 37, 30-41.  (MR12,P.430)

KESTELMAN, H. 1952.  (15.1)                                                                   2999
    THE FUNDAMENTAL EQUATION OF FACTOR ANALYSIS.
    BRITISH J. PSYCHOL. STATIST. SECT., 5, 1-6.

KESTEN, HARRY + MORSE, NORMAN 1959.  (18.1)                                                   3000
    A PROPERTY OF THE MULTINOMIAL DISTRIBUTION.
    ANN. MATH. STATIST., 30, 120-127.  (MR21-4516)

KESTEN, HARRY + RUNNENBURG, JOHANNES THEODORUS 1956.   (17.8)                                 3001
    ENIGE OPMERKINGEN OVER DE BEREKENING VAN DE VERWACHTING EN DE SPREIDING VAN HET
    AANTAL INCONSISTENTIES IN EEN BEPAALD RANGCORRELATIESCHEMA.
        (WITH SUMMARY IN ENGLISH)
    STATISTICA NEERLANDICA, 10, 197-204.  (MR19,P.187)

KESTEN, HARRY + STIGUM, B. P. 1966.  (17.1)                                                   3002
    A LIMIT THEOREM FOR MULTIDIMENSIONAL GALTON-WATSON PROCESSES.
    ANN. MATH. STATIST., 37, 1211-1223.

KESTEN, HARRY + STIGUM, B. P. 1966.  (17.1)                                                   3003
    ADDITIONAL LIMIT THEOREMS FOR INDECOMPOSABLE MULTIDIMENSIONAL GALTON-WATSON PROCESSES.
    ANN. MATH. STATIST., 37, 1463-1481.

KESTEN, HARRY    SEE ALSO
    1828. FURSTENBERG, HARRY + KESTEN, HARRY (1960)

KEULS, M. + VERDOOREN, L. R. 1964.  (1.2)                                                     3004
    OVERZICHT VAN DE PUBLIKATIES DOOR PROF. DR. M. J. VAN UVEN.
    STATISTICA NEERLANDICA, 18, 19-25.

KEYFITZ, NATHAN 1938.  (2.4/TU)                                                               3005
    GRADUATION BY A TRUNCATED NORMAL.
    ANN. MATH. STATIST., 9, 66-67.

KEYS, NOEL    SEE
    231. BABITZ, MILTON + KEYS, NOEL (1940)

KHAKHUBIYA, TS. G.  SEE  HAHUBIJA, C. G.

KHAN, F. 1953.  (4.4/15.4)                                                                    3006
    SIMPLIFIED BISERIAL R FOR ITEM VALIDATION.
    PAKISTAN J. SCI., 5, 73-75.

KHAN, N. A.   SEE
    3659. MARCUS, MARVIN + KHAN, N. A. (1959)

KHARTASHOV, G. D. 1965.  (17.3)                                                               3007
    ON FINDING A FUNCTIONAL DEPENDENCE BETWEEN RANDOM VARIABLES.
        (IN RUSSIAN, WITH SUMMARY IN ENGLISH, TRANSLATED AS NO. 3008)
    TEOR. VEROJATNOST. I PRIMENEN., 10, 584-593.  (MR34-5117)

KHARTASHOV, G. D. 1965.  (17.3)                                                               3008
    ON DETERMINING A FUNCTIONAL RELATION BETWEEN RANDOM VARIABLES.
        (TRANSLATION OF NO. 3007)
    THEORY PROB. APPL., 10, 528-535.  (MR34-5117)

KHATRI, C. G. 1959.  (7.4/14.2)                                                               3009
    ON THE MUTUAL INDEPENDENCE OF CERTAIN STATISTICS.
        (AMENDED BY NO. 3013)
    ANN. MATH. STATIST., 30, 1258-1262.  (MR22-1017)

KHATRI, C. G. 1959.  (18.1)                                                                   3010
    ON CERTAIN PROPERTIES OF POWER-SERIES DISTRIBUTIONS.
    BIOMETRIKA, 46, 486-490.  (MR22-267)

KHATRI, C. G. 1959.   (7.4)                                                          3011
    ON CONDITIONS FOR THE FORMS OF THE TYPE XAX' TO BE DISTRIBUTED
    INDEPENDENTLY OR TO OBEY WISHART DISTRIBUTION.
    CALCUTTA STATIST. ASSOC. BULL., 8, 162-168.   (MR21-4490)


KHATRI, C. G. 1961.   (8.1/C)                                                        3012
    A NOTE ON THE INTERVAL ESTIMATION RELATED TO THE REGRESSION MATRIX.
    ANN. INST. STATIST. MATH. TOKYO, 13, 145-146.   (MR24-A3733)


KHATRI, C. G. 1961.   (7.4/14.2)                                                     3013
    CORRECTION TO: ON THE MUTUAL INDEPENDENCE OF CERTAIN STATISTICS.
        (AMENDMENT OF NO. 3009)
    ANN. MATH. STATIST., 32, 1344.   (MR22-1017)


KHATRI, C. G. 1961.   (2.5)                                                          3014
    CUMULANTS AND HIGHER ORDER UNCORRELATION OF CERTAIN FUNCTIONS OF
    NORMAL VARIATES.
        (AMENDED BY NO. 3019)
    CALCUTTA STATIST. ASSOC. BULL., 10, 93-98.   (MR24-A2466)



KHATRI, C. G. 1961.   (20.1)                                                         3015
    A SIMPLIFIED APPROACH TO THE DERIVATION OF THE THEOREMS ON THE RANK OF A MATRIX.
    J. MAHARAJA SAYAJIRAO UNIV. BARODA, 10, 1-5.   (MR26-6189)


KHATRI, C. G. 1961.   (8.5/C)                                                        3016
    SIMULTANEOUS CONFIDENCE BOUNDS CONNECTED WITH A GENERAL LINEAR HYPOTHESIS.
    J. MAHARAJA SAYAJIRAO UNIV. BARODA, 10, 11-13.   (MR25-4606)


KHATRI, C. G. 1962.   (8.1/8.7/C)                                                    3017
    SIMULTANEOUS CONFIDENCE BOUNDS ON THE DEPARTURES FROM A PARTICULAR
    KIND OF MULTICOLLINEARITY.
    ANN. INST. STATIST. MATH. TOKYO, 13, 239-242.   (MR27-6337)


KHATRI, C. G. 1962.   (7.1/7.4)                                                      3018
    CONDITIONS FOR WISHARTNESS AND INDEPENDENCE OF SECOND DEGREE
    POLYNOMIALS IN A NORMAL VECTOR.
    ANN. MATH. STATIST., 33, 1002-1007.   (MR25-4594)


KHATRI, C. G. 1963.   (2.5)                                                          3019
    ERRATA:  CUMULANTS AND HIGHER ORDER UNCORRELATION OF CERTAIN
    FUNCTIONS OF NORMAL VARIATES.
        (AMENDMENT OF NO. 3014)
    CALCUTTA STATIST. ASSOC. BULL., 12, 67.   (MR24-A2466)


KHATRI, C. G. 1963.   (14.5)                                                         3020
    SOME MORE ESTIMATES OF CIRCULAR PROBABLE ERROR.
    J. INDIAN STATIST. ASSOC., 1, 40-47.   (MR29-4143)


KHATRI, C. G. 1963.   (7.1)                                                          3021
    FURTHER CONTRIBUTIONS TO WISHARTNESS AND INDEPENDENCE OF SECOND
    DEGREE POLYNOMIALS IN NORMAL VECTORS.
    J. INDIAN STATIST. ASSOC., 1, 61-70.   (MR29-5342)


KHATRI, C. G. 1963.   (3.2)                                                          3022
    JOINT ESTIMATION OF THE PARAMETERS OF MULTIVARIATE NORMAL POPULATIONS.
    J. INDIAN STATIST. ASSOC., 1, 125-133.   (MR29-5343)


KHATRI, C. G. 1964.   (8.4/8.7/14.2)                                                 3023
    DISTRIBUTION OF THE "GENERALISED" MULTIPLE CORRELATION MATRIX IN THE DUAL CASE.
    ANN. MATH. STATIST., 35, 1801-1806.   (MR30-4345)



KHATRI, C. G. 1964.   (13.1)                                                         3024
    DISTRIBUTION OF THE LARGEST OR THE SMALLEST CHARACTERISTIC ROOT UNDER
    NULL HYPOTHESIS CONCERNING COMPLEX MULTIVARIATE NORMAL POPULATIONS.
    ANN. MATH. STATIST., 35, 1807-1810.   (MR30-5429)


KHATRI, C. G. 1965.   (10.1/11.2/14.3)                                               3025
    A NOTE ON THE CONFIDENCE BOUNDS FOR THE CHARACTERISTIC ROOTS OF
    DISPERSION MATRICES OF NORMAL VARIATES.
    ANN. INST. STATIST. MATH. TOKYO, 17, 175-183.   (MR36-2264)

KHATRI, C. G. 1965.  (7.1/8.3)                                3026
    CLASSICAL STATISTICAL ANALYSIS BASED ON A CERTAIN MULTIVARIATE COMPLEX
    GAUSSIAN DISTRIBUTION.
    ANN. MATH. STATIST., 36, 98-114.  (MR33-823)


KHATRI, C. G. 1965.  (9.2/10.2)                              3027
    A TEST FOR REALITY OF A COVARIANCE MATRIX IN A CERTAIN COMPLEX
    GAUSSIAN DISTRIBUTION.
    ANN. MATH. STATIST., 36, 115-119.  (MR33-824)


KHATRI, C. G. 1965.  (8.7)                                   3028
    NON-NULL DISTRIBUTION OF THE LIKELIHOOD RATIO STATISTIC FOR A
    PARTICULAR KIND OF MULTICOLLINEARITY.
    J. INDIAN STATIST. ASSOC., 3, 212-219.  (MR34-6922)


KHATRI, C. G. 1966.  (8.3)                                   3029
    A NOTE ON A MANOVA MODEL APPLIED TO PROBLEMS IN GROWTH CURVE.
    ANN. INST. STATIST. MATH. TOKYO, 18, 75-86.


KHATRI, C. G. 1966.  (4.7/A)                                 3030
    A NOTE ON A LARGE SAMPLE DISTRIBUTION OF A TRANSFORMED MULTIPLE
    CORRELATION COEFFICIENT.
    ANN. INST. STATIST. MATH. TOKYO, 18, 375-380.  (MR34-6893)


KHATRI, C. G. 1966.  (7.1/7.3/7.4)                           3031
    ON CERTAIN DISTRIBUTION PROBLEMS BASED ON POSITIVE DEFINITE QUADRATIC
    FUNCTIONS IN NORMAL VECTORS.
    ANN. MATH. STATIST., 37, 468-479.  (MR32-8375)


KHATRI, C. G. + JAISWAL, M.C. 1963.  (3.2/B)                 3032
    ESTIMATION OF PARAMETERS OF A TRUNCATED BIVARIATE NORMAL DISTRIBUTION.
    J. AMER. STATIST. ASSOC., 58, 519-526.  (MR27-2022)


KHATRI, C. G. + PILLAI, K.C.SREEDHARAN 1965.  (7.3/8.4)      3033
    SOME RESULTS ON THE NON-CENTRAL MULTIVARIATE BETA DISTRIBUTION AND
    MOMENTS OF TRACES OF TWO MATRICES.
    ANN. MATH. STATIST., 36, 1511-1520.  (MR32-1790)


KHATRI, C. G. + PILLAI, K.C.SREEDHARAN 1966.  (8.4/13.1/13.2/T)   3034
    ON THE MOMENTS OF THE TRACE OF A MATRIX AND APPROXIMATIONS TO ITS
    NON-CENTRAL DISTRIBUTION.
    ANN. MATH. STATIST., 37, 1312-1318.  (MR34-2118)


KHATRI, C. G.    SEE ALSO
    4398. RAMACHANDRAN, K.V. + KHATRI, C. G. (1957)

    4983. SHAH, B.K. + KHATRI, C. G. (1961)

    4984. SHAH, B.K. + KHATRI, C. G. (1963)

KHINCHIN, A. YA.  SEE  HINCIN, A. JA.

KHOSLA, R. K.    SEE
    6. ABRAHAM, T. P. + KHOSLA, R. K. (1965)

KIBBLE, W. F. 1941.  (18.1)                                  3035
    A TWO-VARIATE GAMMA TYPE DISTRIBUTION.
    SANKHYA, 5, 137-150.  (MR4,P.103)


KIBBLE, W. F. 1945.  (2.7/20.4)                              3036
    AN EXTENSION OF A THEOREM OF MEHLER'S ON HERMITE POLYNOMIALS.
    PROC. CAMBRIDGE PHILOS. SOC., 41, 12-15.  (MR7,P.65)


KIEFER, J. 1961.  (17.1)                                     3037
    ON LARGE DEVIATIONS OF THE EMPIRIC D. F. OF VECTOR CHANCE VARIABLES
    AND A LAW OF THE ITERATED LOGARITHM.
    PACIFIC J. MATH., 11, 649-660.  (MR24-A1732)


KIEFER, J. 1966.  (3.1/8.3/19.2)                            3038
    MULTIVARIATE OPTIMALITY RESULTS.
    MULTIVARIATE ANAL. PROC. INTERNAT. SYMP. DAYTON (KRISHNAIAH), 255-274.

KIEFER, J. + SCHWARTZ, R. 1965.   (6.3/8.3/10.1)                                    3039
    ADMISSIBLE BAYES CHARACTER OF $T^2$-, $R^2$-, AND OTHER FULLY INVARIANT
    TESTS FOR CLASSICAL MULTIVARIATE NORMAL PROBLEMS.
    ANN. MATH. STATIST., 36, 747-770.  (MR30-5430)


KIEFER, J. + WOLFOWITZ, J. 1952.   (17.5)                                           3040
    STOCHASTIC ESTIMATION OF THE MAXIMUM OF A REGRESSION FUNCTION.
    ANN. MATH. STATIST., 23, 462-466.  (MR14,P.299, P.1278)

KIEFER, J. + WOLFOWITZ, J. 1956.   (17.5)                                           3041
    CONSISTENCY OF THE MAXIMUM LIKELIHOOD ESTIMATOR IN THE PRESENCE OF INFINITELY
    MANY INCIDENTAL PARAMETERS.
    ANN. MATH. STATIST., 27, 887-906.  (MR19,P.189)

KIEFER, J. + WOLFOWITZ, J. 1959.   (17.1/17.8/19.3)                                 3042
    ASYMPTOTIC MINIMAX CHARACTER OF THE SAMPLE DISTRIBUTION FUNCTION FOR
    VECTOR CHANCE VARIABLES.
    ANN. MATH. STATIST., 30, 463-489.  (MR21-6642)

KIEFER, J.   SEE ALSO
    1997. GIRI, N. C. + KIEFER, J. (1964)

    1998. GIRI, N. C. + KIEFER, J. (1964)

     598. BLUM, JULIUS R.; KIEFER, J. + ROSENBLATT, MURRAY (1961)

    1434. DVORETZKY, ARYEH; KIEFER, J. + WOLFOWITZ, J. (1953)

    1435. DVORETZKY, ARYEH; KIEFER, J. + WOLFOWITZ, J. (1953)

    1999. GIRI, N. C.; KIEFER, J. + STEIN, CHARLES M. (1963)

    2890. KAC, MARK; KIEFER, J. + WOLFOWITZ, J. (1955)

KIERKEGAARD-HANSEN, PETER 1963.   (20.1)                                            3043
    MATRIX-SYMBOLIK ANVENDT PA DIFFERENTIALREGNING.
    NORDISK MAT. TIDSKR., 11, 13-21, 40. (MR26-6312)

KILPATRICK, FRANKLIN P.   SEE
    1508. EDWARDS, ALLEN L. + KILPATRICK, FRANKLIN P. (1948)

KIMBALL, ALLYN W.   SEE
    5673. VOTAW, DAVID F., JR.; KIMBALL, ALLYN W. + RAFFERTY, J.A. (1950)

KIMURA, HITOSI 1951.   (4.3/17.8)                                                   3044
    ON A DISTRIBUTION-FREE TEST OF CORRELATION COEFFICIENT.
    BULL. JAPAN STATIST. SOC., 48-49.


KIMURA, MOTOO   SEE
    5806. WEISS, GEORGE H. + KIMURA, MOTOO (1966)

KINCAID, W. M. 1948.   (20.5)                                                       3045
    SOLUTION OF EQUATIONS BY INTERPOLATION.
    ANN. MATH. STATIST., 19, 207-219.  (MR9,P.621)

KING, J. H., JR.   SEE
    2142. GRIFFIN, JOHN S., JR.; KING, J. H., JR. + TUNIS, C. J. (1963)

KING, WILLFORD I. 1918.   (4.1)                                                     3046
    REPLY TO DR. DAY'S CRITICISM OF MY ARTICLE ON CORRELATION COEFFICIENTS.
        (AMENDMENT OF NO. 1244)
    QUART. PUBL. AMER. STATIST. ASSOC., 16, 171-173.

KINGMAN, J. F. C. 1963.   (17.2)                                                    3047
    ON INEQUALITIES OF THE TCHEBYCHEV TYPE.
    PROC. CAMBRIDGE PHILOS. SOC., 59, 135-146.  (MR29-5285)

KINNEY, J. R. + PITCHER, T. S. 1966.   (17.3/B)                                     3048
    DIMENSIONAL PROPERTIES OF A RANDOM DISTRIBUTION FUNCTION ON THE SQUARE.
    ANN. MATH. STATIST., 37, 849-854.  (MR34-841)

KIRKHAM, WILLIAM J. 1937.  (4.6)                                                                3049
   NOTE ON THE DERIVATION OF THE MULTIPLE CORRELATION COEFFICIENT.
   ANN. MATH. STATIST., 8, 68-71.


KITAGAWA, TOSIO 1959.  (8.1/8.3)                                                                3050
   SUCCESSIVE PROCESS OF STATISTICAL INFERENCES APPLIED TO LINEAR REGRESSION
   ANALYSIS AND ITS SPECIALISATIONS TO RESPONSE SURFACE ANALYSIS.
   BULL. MATH. STATIST., 8, 80-114.  (MR22-11490)

KITAGAWA, TOSIO 1960.  (8.1/14.1/AB)                                                            3051
   SAMPLING DISTRIBUTIONS OF STATISTICS ASSOCIATED WITH A FRACTILE GRAPHIC METHOD.
   BULL. MATH. STATIST., 9(1) 10-42.  (MR23-A721)

KIYOSHI, ITO  SEE  ITO, KIYOSI

KLAUBER, L. M. 1945.  (4.1/4.6/E)                                                               3052
   HERPETOLOGICAL CORRELATIONS.  I.  CORRELATIONS IN HOMOGENEOUS POPULATIONS.
   BULL. ZOOL. SOC. SAN DIEGO, 21, 1-101.

KLEIBER, JOSEPH 1887.  (18.6/E)                                                                 3053
   ON "RANDOM SCATTERING" OF POINTS ON A SURFACE.
   PHILOS. MAG. SER. 5, 24, 439-445.

KLEIN, LAWRENCE R. 1960.  (17.7)                                                                3054
   THE EFFICIENCY OF ESTIMATION IN ECONOMETRIC MODELS.
   ESSAYS ECON. ECONOMET. (HOTELLING VOLUME), 216-232.  (MR23-B1585)

KLEMM, P. G. 1964.  (4.4)                                                                       3055
   NEUE DIAGRAMME FUR DIE BERECHNUNG VON VIERFELDERKORRELATIONEN.
   BIOM. Z., 6, 103-109.


KLEPIKOV, N. P. + SOKOLOV, S.N. 1957.  (17.7)                                                   3056
   NON-LINEAR CONFLUENCE ANALYSIS.
       (IN RUSSIAN, WITH SUMMARY IN ENGLISH, TRANSLATED AS NO. 3057)
   TEOR. VEROJATNOST. I PRIMENEN., 2, 473-475.  (MR20-377)

KLEPIKOV, N. P. + SOKOLOV, S.N. 1957.  (17.7)                                                   3057
   NON-LINEAR CONFLUENCE ANALYSIS.
       (TRANSLATION OF NO. 3056)
   THEORY PROB. APPL., 2, 465-468.  (MR20-377)

KLOEK, T. + MENNES, L. B. M. 1960.  (16.2)                                                      3058
   SIMULTANEOUS EQUATIONS ESTIMATION BASED ON PRINCIPAL COMPONENTS OF
   PREDETERMINED VARIABLES.
   ECONOMETRICA, 28, 45-61.  (MR23-B2039)

KLOEK, T.  SEE ALSO
   5395. THEIL, H. + KLOEK, T. (1959)

KLOOT, N. H.  SEE
   5930. WILLIAMS, F.J. + KLOOT, N. H. (1953)

KLOTZ, JEROME 1964.  (17.8/B)                                                                   3059
   SMALL SAMPLE POWER OF THE BIVARIATE SIGN TESTS OF BLUMEN AND HODGES.
   ANN. MATH. STATIST., 35, 1576-1582.  (MR30-1586)

KLOTZ, JEROME  SEE ALSO
   2793. JOFFE, ANATOLE + KLOTZ, JEROME (1962)

KLUG, A. 1958.  (16.4/17.3/18.1/E)                                                              3060
   JOINT PROBABILITY DISTRIBUTIONS OF STRUCTURE FACTORS AND THE PHASE PROBLEM.
   ACTA CRYST., 11, 515-543.

KLUG, HAROLD P.  SEE
   84. ALEXANDER, LEROY; KLUG, HAROLD P. + KUMMER, ELIZABETH (1948)

KLUYVER, J. C. 1906.  (17.1)                                                                    3061
   A LOCAL PROBABILITY PROBLEM.
   KON. AKAD. WETENSCH. AMSTERDAM PROC. SECT. SCI., 8, 341-350.

KNIGHT, WILLIAM 1965. (19.1)                                           3062
    A LEMMA FOR MULTIPLE INFERENCE.
    ANN. MATH. STATIST., 36, 1873-1874.

KNOELL, DOROTHY M.   SEE
    2355. HARRIS, CHESTER W. + KNOELL, DOROTHY M. (1948)

KNOTT, VIRGINIA   SEE
    3549. MC CLOY, C. H.; METHENY, ELEANOR + KNOTT, VIRGINIA (1938)

KNOWLES, W. B.; HOLLAND, J. G. + NEWLIN, E. P. 1957.  (4.6/8.1/E)      3063
    A CORRELATIONAL ANALYSIS OF TRACKING BEHAVIOUR.
    PSYCHOMETRIKA, 22, 275-287.

KOEPPLER, HANS 1913.  (18.1)                                          3064
    NOCH EINMAL DAS BERNOULLISCHE THEOREM FUR MEHR ALS ZWEI EREIGNISSE.
    ANN. VERSICHERUNGSW., 44, 749-750.

KOEPPLER, HANS 1935.  (4.2)                                           3065
    DAS FEHLERGESETZ DES KORRELATIONSKOEFFIZIENTEN UND ANDERE
    WAHRSCHEINLICHKEITSGESETZE DER KORRELATIONSTHEORIE.
    METRON, 12(2) 105-120.

KOEPPLER, HANS 1938.  (17.3/B)                                        3066
    DIE ANWENDUNG DER THEORIE DER ELEMENTARWAHRSCHEINLICHKEIT ZWEIER ABWEICHUNGEN
    AUF DIE DARSTELLUNG DER DIFFERENTIALGLEICHUNG DER FREQUENZFUNKTION ZWEIER
    VARIABLEN.
    AKTUAR. VEDY, 7, 72-85.

KOGA, Y.   SEE
    3880. MORANT, G. M. + KOGA, Y. (1923)

KOGO, KUSUNORI   SEE
    4058. OGAWA, JUNJIRO + KOGO, KUSUNORI (1955)

KOKAN, A. R. 1963.  (18.5/20.2)                                       3067
    OPTIMUM ALLOCATION IN MULTIVARIATE SURVEYS.
        (AMENDED BY NO. 3068)
    J. ROY. STATIST. SOC. SER. A, 126, 557-565.  (MR28-3507)

KOKAN, A. R. 1965.  (18.5/20.2)                                       3068
    ERRATA TO:  OPTIMUM ALLOCATION IN MULTIVARIATE SURVEYS.
        (AMENDMENT OF NO. 3067)
    J. ROY. STATIST. SOC. SER. A, 128, 468.  (MR28-3507)

KOLLER, DIETER 1963.  (17.8/A)                                        3069
    PRUFUNG DER NORMALITAT VERTEILUNG.
    Z. WAHRSCHEIN. VERW. GEBIETE, 2, 147-166.

KOLLER, SIEGFRIED 1936.  (2.1/B)                                      3070
    DIE ANALYSE DER ABHANGIGKEITSVERHALTNISSE IN ZWEI KORRELATIONSSYSTEMEN.
    METRON, 12(4) 73-105.

KOLLER, SIEGFRIED 1963.  (4.9)                                        3071
    TYPISIERUNG KORRELATIVER ZUSAMMENHANGE.
    METRIKA, 6, 65-75.

KOLMOGOROV, A. N. 1946.  (8.1/8.2/U)                                  3072
    ON THE PROOF OF THE METHOD OF LEAST SQUARES.
        (IN RUSSIAN)
    USPEHI MAT. NAUK, 11, 57-70.  (MR8,P.523)

KOLMOGOROV, A. N. 1948.  (1.2)                                        3073
    EVGENII EVGENEVICH SLUTSKII, (1880-1948).
        (IN RUSSIAN)
    USPEHI MAT. NAUK, 26, 143-151.

KOLMOGOROV, A. N. + SARMANOV, O. V. 1960.  (1.2)                      3074
    THE WORK OF S. N. BERNSHTEIN ON THE THEORY OF PROBABILITY (ON HIS EIGHTIETH BIRTHDAY).
        (IN RUSSIAN. TRANSLATED AS NO. 3075)
    TEOR. VEROJATNOST. I PRIMENEN., 5, 215-221.  (MR24-A1200)

KOLMOGOROV, A. N. + SARMANOV, O. V. 1960.   (1.2)                    3075
     THE WORK OF S. N. BERNSHTEIN ON THE THEORY OF PROBABILITY (ON HIS
     EIGHTIETH BIRTHDAY).
         (TRANSLATION OF NO. 3074)
   THEORY PROB. APPL., 5, 197-203.   (MR24-A1200)


KOMATU, YUSAKU 1955.   (2.1/2.7/U)                    3076
     ELEMENTARY INEQUALITIES FOR MILLS' RATIO.
   REP. STATIST. APPL. RES. UN. JAPAN. SCI. ENGRS., 4, 69-70.   (MR18,P.155)


KOMAZAWA, TSUTOMU 1965.   (20.3)                    3077
     A REVISED METHOD FOR COMPUTING EIGENVALUES AND EIGENVECTORS OF A REAL MATRIX.
         (IN JAPANESE, WITH SUMMARY IN ENGLISH)
   PROC. INST. STATIST. MATH. TOKYO, 12, 263-272.


KOMAZAWA, TSUTOMU    SEE ALSO
     5330. TAGA, YASUSHI + KOMAZAWA, TSUTOMU (1960)

     5335. TAKAHASI, KOITI + KOMAZAWA, TSUTOMU (1966)

     3744. MATUFUJI, HAJIME; YAMAZAKI, KAZUHIDE + KOMAZAWA, TSUTOMU (1966)


KONDO, H. + TAKEDA, M. 1956.   (16.5/E)                    3078
     CORRELOGRAM AND SPECTRUM.
   APPL. STATIST. METEOROL., 6, 72-73.


KONDO, JIRO 1964.   (16.4)                    3079
     A THEORY OF MULTIPLE PREDICTION.
   J. OPERATIONS RES. SOC. JAPAN, 6, 63-81.


KONDO, TAKAYUKI + ELDERTON, ETHEL M. 1931.   (2.7/TU)                    3080
     TABLE OF NORMAL CURVE FUNCTIONS TO EACH PERMILLE OF FREQUENCY.
   BIOMETRIKA, 22, 368-376.


KONHEIM, A. G.    SEE
     1719. FLATTO, L. + KONHEIM, A. G. (1962)


KONIJN, H. S. 1956.   (17.8/B)                    3081
     ON THE POWER OF CERTAIN TESTS FOR INDEPENDENCE IN BIVARIATE POPULATIONS.
         (AMENDED BY NO. 3085)
   ANN. MATH. STATIST., 27, 300-323.   (MR18,P.78)


KONIJN, H. S. 1957.   (15.1)                    3082
     A PROBLEM ARISING IN FACTOR ANALYSIS.
   AUSTRAL. J. SCI., 19, 202-203.


KONIJN, H. S. 1957.   (8.1/17.7)                    3083
     A NOTE ON COMPUTATIONS IN REGRESSION AND CONFLUENCE ANALYSIS.
   BULL. STATIST. SOC. N. S. W., 6-17.


KONIJN, H. S. 1957.   (18.1/B)                    3084
     A CLASS OF TWO-DIMENSIONAL RANDOM VARIABLES AND DISTRIBUTION FUNCTIONS.
   SANKHYA, 18, 167-172.   (MR19,P.1087)


KONIJN, H. S. 1958.   (17.8/B)                    3085
     CORRECTION TO: "ON THE POWER OF CERTAIN TESTS FOR INDEPENDENCE IN
     BIVARIATE POPULATIONS".
         (AMENDMENT OF NO. 3081)
   ANN. MATH. STATIST., 29, 935-936.   (MR18,P.78)


KONIJN, H. S. 1958.   (16.2)                    3086
     A RESTATEMENT OF THE CONDITIONS FOR IDENTIFIABILITY IN COMPLETE
     SYSTEMS OF LINEAR DIFFERENCE EQUATIONS.
   METROECONOMICA, 10, 182-190.   (MR21-4054)


KONIJN, H. S. 1962.   (16.2)                    3087
     IDENTIFICATION AND ESTIMATION IN A SIMULTANEOUS EQUATIONS MODEL WITH
     ERRORS IN THE VARIABLES.
   ECONOMETRICA, 30, 79-87.   (MR25-664)


KONIJN, H. S. 1963.   (3.2)                    3088
     NOTE ON THE NON-EXISTENCE OF A MAXIMUM LIKELIHOOD ESTIMATE.
   AUSTRAL. J. STATIST., 5, 143-146.

KONIJN, H. S. 1966. (19.3/C)
SOME REMARKS ON THE CHOICE OF SYSTEMS OF CONFIDENCE INTERVALS.
SANKHYA SER. A, 28, 305-308.

3089

KONO, KAZUMASA 1952. (4.3/B)
ON INEFFICIENT STATISTICS FOR MEASUREMENT OF DEPENDENCY OF NORMAL BIVARIATES.
MEM. FAC. SCI. KYUSHU UNIV. SER. A, 7, 1-12. (MR14,P.1102)

3090

KONUS. A. A. 1966. (8.8/17.7)
THE SINGLE REGRESSION LINE.
METRON, 25, 108-115.

3091

KOOPMANS, LAMBERT H. 1964. (16.4)
ON THE COEFFICIENT OF COHERENCE FOR WEAKLY STATIONARY STOCHASTIC PROCESSES.
ANN. MATH. STATIST., 35, 532-549. (MR28-4611)

3092

KOOPMANS, LAMBERT H. 1964. (16.4)
ON THE MULTIVARIATE ANALYSIS OF WEAKLY STATIONARY STOCHASTIC PROCESSES.
ANN. MATH. STATIST., 35, 1765-1780. (MR30-618)

3093

KOOPMANS, LAMBERT H. SEE ALSO
597. BLUM, JULIUS R.; HANSON, D. L. + KOOPMANS, LAMBERT H. (1963)

KOOPMANS, TJALLING C. 1942. (2.5/14.5)
SERIAL CORRELATION AND QUADRATIC FORMS IN NORMAL VARIABLES.
ANN. MATH. STATIST., 13, 14-33. (MR4,P.22)

3094

KOOPMANS, TJALLING C. 1945. (16.2)
STATISTICAL ESTIMATION OF SIMULTANEOUS ECONOMIC RELATIONS.
J. AMER. STATIST. ASSOC., 40, 448-466. (MR7,P.215)

3095

KOOPMANS, TJALLING C. 1949. (16.2)
IDENTIFICATION PROBLEMS IN ECONOMIC MODEL CONSTRUCTION.
    (REPRINTED AS NO. 3098)
ECONOMETRICA, 17, 125-144. (MR11,P.192)

3096

KOOPMANS, TJALLING C. 1950. (16.2)
WHEN IS AN EQUATION SYSTEM COMPLETE FOR STATISTICAL PURPOSES ?
COWLES COMMISS. MONOG., 10, 393-409.

3097

KOOPMANS, TJALLING C. 1953. (16.2)
IDENTIFICATION PROBLEMS IN ECONOMIC MODEL CONSTRUCTION.
    (REPRINT OF NO. 3096)
COWLES COMMISS. MONOG., 14, 27-48. (MR11,P.192)

3098

KOOPMANS, TJALLING C. + HOOD, WILLIAM C. 1953. (16.2)
THE ESTIMATION OF SIMULTANEOUS ECONOMIC RELATIONSHIPS.
COWLES COMMISS. MONOG., 14, 112-199.

3099

KOOPMANS, TJALLING C. + REIERSOL, OLAV 1950. (17.7)
THE IDENTIFICATION OF STRUCTURAL CHARACTERISTICS.
ANN. MATH. STATIST., 21, 165-181. (MR12,P.622)

3100

KOOPMANS, TJALLING C.; RUBIN, HERMAN + LEIPNIK, ROY B. 1950. (16.2)
MEASURING THE EQUATION SYSTEMS OF DYNAMIC ECONOMICS.
COWLES COMMISS. MONOG., 10, 53-237.

3101

KOPOCINSKI, B. 1966. (16.4)
ON THE LOCAL OPTIMALITY OF SOME REGULAR SAMPLING PATTERNS IN THE PLANE (II).
    (IN POLISH)
POLSKA AKAD. NAUK ZASTOS. MAT., 9, 67-73.

3102

KOPOCINSKI, B. + ZUBRZYCKA, L. 1964. (12.1/17.3)
REMARK ON THE DIVISION OF A SET OF CHARACTERISTICS INTO CONSISTENT SUBSETS.
    (IN POLISH)
POLSKA AKAD. NAUK ZASTOS. MAT., 7, 317-321.

3103

KORNFELD, PETER SEE
5997. WOLF, ROBERT L.; MENDLOWITZ, MILTON; ROBOZ, JULIA; (1966)
STYAN, GEORGE P. H.; KORNFELD, PETER + WEIGL, ALFRED

KORZHIK, V. I. 1966.   (17.5)                                                          3104
    A METHOD FOR OBTAINING ESTIMATES OF THE EXPECTATION OF SOME FUNCTION
    OF N-DIMENSIONAL RANDOM VECTOR.
        (IN RUSSIAN, WITH SUMMARY IN ENGLISH, TRANSLATED AS NO. 3105)
    TEOR. VEROJATNOST. I PRIMENEN., 11, 537-541.

KORZHIK, V. I. 1966.   (17.5)                                                          3105
    A METHOD OF OBTAINING ESTIMATES OF THE MATHEMATICAL EXPECTATION OF
    SOME FUNCTION OF THE COMPONENTS OF AN N-DIMENSIONAL RANDOM VECTOR.
        (TRANSLATION OF NO. 3104)
    THEORY PROB. APPL., 11, 476-480.

KOSAMBI, DAMODAR 1941.   (7.2/AB)                                                      3106
    A BIVARIATE EXTENSION OF FISHER'S Z-TEST.
    CURRENT SCI., 10, 191-192.   (MR3,P.175)

KOSAMBI, DAMODAR 1941.   (4.8)                                                         3107
    CORRELATION AND TIME SERIES.
    CURRENT SCI., 10, 372-374.   (MR3,P.175)

KOSAMBI, DAMODAR 1942.   (8.3)                                                         3108
    A TEST OF SIGNIFICANCE FOR MULTIPLE OBSERVATIONS.
    CURRENT SCI., 11, 271-274.   (MR4,P.107)

KOSAMBI, DAMODAR 1959.   (20.1)                                                        3109
    THE METHOD OF LEAST SQUARES.
    J. INDIAN SOC. AGRIC. STATIST., 11, 49-57.   (MR22-5089)

KOSSACK, CARL F. 1945.   (6.1)                                                         3110
    ON THE MECHANICS OF CLASSIFICATION.
    ANN. MATH. STATIST., 16, 95-98.

KOSSACK, CARL F. 1948.   (4.1)                                                         3111
    ON THE COMPUTATION OF ZERO-ORDER CORRELATION COEFFICIENTS.
    PSYCHOMETRIKA, 13, 91-93.

KOSSACK, CARL F. 1949.   (6.2/T)                                                       3112
    SOME TECHNIQUES FOR SIMPLE CLASSIFICATION.
    PROC. BERKELEY SYMP. MATH. STATIST. PROB., 345-352.

KOSSACK, CARL F. 1963.   (6.1/6.4)                                                     3113
    STATISTICAL CLASSIFICATION TECHNIQUES.
    IBM SYSTEMS J., 2, 136-151.

KOTLARSKI, IGNACY 1964.   (17.3/18.1/B)                                                3114
    ON BIVARIATE RANDOM VARIABLES WHERE THE QUOTIENT OF THEIR COORDINATES
    FOLLOWS SOME KNOWN DISTRIBUTION.
    ANN. MATH. STATIST., 35, 1673-1684.   (MR29-5326)

KOTLARSKI, IGNACY 1964.   (17.3/B)                                                     3115
    ON BIVARIATE RANDOM VECTORS FOR WHICH THE QUOTIENT OF THEIR COORDINATES HAS
    STUDENT'S DISTRIBUTION.
    ZESZYTY NAUK. POLITECH. WARSZAWSK. MAT., 3, 207-220.   (MR32-8377)

KOTLARSKI, IGNACY 1966.   (2.2/U)                                                      3116
    ON CHARACTERIZING THE NORMAL DISTRIBUTION BY STUDENT'S LAW.
    BIOMETRIKA, 53, 603-606.

KOTZ, SAMUEL 1965.   (20.5)                                                            3117
    SOME INEQUALITIES FOR CONVEX FUNCTIONS USEFUL IN INFORMATION THEORY.
    SIAM REV., 7, 395-402.

KOTZ, SAMUEL + ADAMS, JOHN W. 1964.   (18.1)                                           3118
    DISTRIBUTION OF SUM OF IDENTICALLY DISTRIBUTED EXPONENTIALLY
    CORRELATED GAMMA-VARIABLES.
        (AMENDED BY NO. 3119)
    ANN. MATH. STATIST., 35, 277-283.   (MR28-1682)

KOTZ, SAMUEL + ADAMS, JOHN W. 1964.   (18.1)                                           3119
    ACKNOWLEDGEMENT OF PRIORITY.
        (AMENDMENT OF NO. 3118)
    ANN. MATH. STATIST., 35, 925.   (MR28-1682)

KOTZ, SAMUEL    SEE ALSO
    2926. KANTOROWITZ, SHMUEL + KOTZ, SAMUEL (1957)


KOVALENKO, A. D. 1960.  (16.4)                                              3120
    INQUIRY INTO THE MULTILINEAR SERVICING SYSTEM WITH SUCCESSION AND
    LIMITED TIME OF REMAINING IN THE SYSTEM.
        (IN UKRAINIAN)
    UKRAIN. MAT. ZUR., 12, 471-476.


KOVATS, M.   SEE
    1761. FRASER, A. R. + KOVATS, M. (1966)


KOZELKA, ROBERT M. 1956.  (18.3)                                           3121
    APROXIMATE UPPER PERCENTAGE POINTS FOR EXTREME VALUES IN MULTINOMIAL SAMPLING.
    ANN. MATH. STATIST., 27, 507-512.  (MR17,P.1222)


KOZESNIK, JAROSLAV 1965.  (16.4)                                           3122
    STOCHASTIC THEORY OF BIOLOGICAL AND SECONOMIC CONFIGURATIONS.
    SANKHYA SER. A, 27, 213-230.


KRAFT, L. G. 1950.  (16.3/E)                                               3123
    CORRELATION FUNCTION ANALYSIS.
    J. ACOUST. SOC. AMER., 22, 762-764.


KRAMER, AMIHUD 1956.  (17.8)                                               3124
    A QUICK RANK TEST FOR SIGNIFICANCE OF DIFFERENCES IN MULTIPLE COMPARISONS.
    FOOD TECH., 10, 391-392.


KRAMER, CLYDE YOUNG 1957.  (19.1)                                          3125
    EXTENSION OF MULTIPLE RANGE TESTS TO GROUP CORRELATED ADJUSTED MEANS.
    BIOMETRICS, 13, 13-18.  (MR19,P.331)


KRAMER, CLYDE YOUNG 1957.  (8.1)                                           3126
    SIMPLIFIED COMPUTATIONS FOR MULTIPLE REGRESSION.
    INDUST. QUAL. CONTROL, 13(8) 8-11.


KRAMER, CLYDE YOUNG 1966.  (14.5/N)                                        3127
    APPROXIMATION TO THE CUMULATIVE T-DISTRIBUTION.
    TECHNOMETRICS, 8, 358-359.


KRAMER, G. 1966.  (19.1)                                                   3128
    ENTSCHEIDUNGSPROBLEM, ENTSCHEIDUNGSKRITERIEN BEI VOLLIGER UNGEWISSHEIT
    UND CHERNOFFSCHES AXIOMENSYSTEM.
    METRIKA, 11, 15-38.


KRAMER, H. P. + MATHEWS, M. V. 1956.  (6.6)                                3129
    A LINEAR CODING FOR TRANSMITTING A SET OF CORRELATED SIGNALS.
    IRE TRANS. INFORMATION THEORY, IT-2, 41-46.


KRAMER, K. H. 1963.  (4.7)                                                 3130
    TABLES FOR CONSTRUCTING CONFIDENCE LIMITS ON THE MULTIPLE CORRELATION
    COEFFICIENT.
    J. AMER. STATIST. ASSOC., 58, 1082-1085.


KRAMER, M. 1940.  (2.1/18.1)                                               3131
    FREQUENCY SURFACES IN TWO VARIABLES EACH OF WHICH IS UNIFORMLY DISTRIBUTED.
    AMER. J. EPIDEM., 32, 45-65.


KRANTZ, DAVID H. 1964.  (15.5)                                             3132
    CONJOINT MEASUREMENT:  THE LUCE-TUKEY AXIOMATIZATION AND SOME EXTENSIONS.
    J. MATH. PSYCHOL., 1, 248-277.


KRAUSCH, F. 1964.  (4.1/4.9)                                               3133
    GRAPHISCHE SCHATZUNG VON KORRELATIONSKOEFFIZIENTEN.
    BIOM. Z., 6, 246-250.


KRAVCHUK, M. 1935.  (4.5/B)                                                3134
    SUR LES ECARTS MOYENS DES COEFFICIENTS DE CORRELATION ET DE REGRESSION.
        (IN UKRAINIAN, WITH SUMMARY IN FRENCH)
    ZUR. INST. MAT. UKRAIN. AKAD. NAUK, 3, 121-131.

KRAVCHUK, M. + OKONENKO, A. 1926.  (4.1)                                          3135
    UBER DIE NORMALE KORRELATION.
          (IN UKRAINIAN, WITH SUMMARY IN GERMAN)
    ZAP. KIIV. SIL SKO-GOS. INST., 1, 95-99.


KREWERAS, G. 1966.  (16.4)                                                        3136
    SPECTRE DES PROCESSUS MULTINOMIAUX ET APPLICATION A CERTAINS
    COMPORTEMENTS COLLECTIFS.
    PUBL. INST. STATIST. UNIV. PARIS, 15, 91-112.

KREWERAS, G. 1966.  (15.5/E)                                                      3137
    A MODEL FOR OPINION CHANGE DURING REPEATED BALLOTING.
    READINGS MATH. SOCIAL SCI. (LAZARSFELD + HENRY), 174-190.

KRISHNAIAH, PARUCHURI R. 1965.  (8.5/C)                                           3138
    ON THE SIMULTANEOUS ANOVA AND MANOVA TESTS.
    ANN. INST. STATIST. MATH. TOKYO, 17, 35-53.  (MR33-817)

KRISHNAIAH, PARUCHURI R. 1965.  (8.3/8.5/C)                                       3139
    ON A MULTIVARIATE GENERALIZATION OF THE SIMULTANEOUS ANALYSIS OF VARIANCE TEST.
    ANN. INST. STATIST. MATH. TOKYO, 17, 167-173.  (MR32-3215)

KRISHNAIAH, PARUCHURI R. 1965.  (10.1/U)                                          3140
    SIMULTANEOUS TESTS FOR THE EQUALITY OF VARIANCES AGAINST CERTAIN ALTERNATIVES.
    AUSTRAL. J. STATIST., 7, 105-109.  (MR33-6756)

KRISHNAIAH, PARUCHURI R. 1965.  (8.3/8.4/18.1)                                    3141
    MULTIPLE COMPARISON TESTS IN MULTI-RESPONSE EXPERIMENTS.
    SANKHYA SER. A, 27, 65-72.


KRISHNAIAH, PARUCHURI R. + ARMITAGE, J. V. 1965.  (14.1/18.1/T)                   3142
    TABLES FOR THE DISTRIBUTION OF THE MAXIMUM OF CORRELATED CHI-SQUARE
    VARIATES WITH ONE DEGREE OF FREEDOM.
    TRABAJOS ESTADIST., 16, 91-115.  (MR33-807)

KRISHNAIAH, PARUCHURI R. + ARMITAGE, J. V. 1966.  (18.1/T)                        3143
    TABLES FOR MULTIVARIATE T DISTRIBUTION.
    SANKHYA SER. B, 28, 31-56.  (MR35-2391)

KRISHNAIAH, PARUCHURI R. + MURTHY, V. KRISHNA 1966.  (16.1/16.5)                  3144
    SIMULTANEOUS TESTS FOR TREND AND SERIAL CORRELATIONS FOR GAUSSIAN
    MARKOV RESIDUALS.
    ECONOMETRICA, 34, 472-480.

KRISHNAIAH, PARUCHURI R. + RAO, M.M. 1961.  (18.1)                                3145
    REMARKS ON A MULTIVARIATE GAMMA DISTRIBUTION.
    AMER. MATH. MONTHLY, 68, 342-346.  (MR24-A587)

KRISHNAIAH, PARUCHURI R. + RIZVI, M. HASEEB 1966.  (6.5/6.7/19.1)                 3146
    SOME PROCEDURES FOR SELECTION OF MULTIVARIATE NORMAL
    POPULATIONS BETTER THAN A CONTROL.
    MULTIVARIATE ANAL. PROC. INTERNAT. SYMP. DAYTON (KRISHNAIAH), 477-490.  (MR36-2261)

KRISHNAIAH, PARUCHURI R.; HAGIS, PETER,JR. + STEINBERG, LEON 1963.  (7.4/18.1/B)  3147
    A NOTE ON THE BIVARIATE CHI DISTRIBUTION.
    SIAM REV., 5, 140-144.  (MR27-3038)

KRISHNAIAH, PARUCHURI R.; HAGIS, PETER,JR. + STEINBERG, LEON 1965.  (10.2/BT)     3148
    TESTS FOR THE EQUALITY OF STANDARD DEVIATIONS IN A BIVARIATE NORMAL POPULATION.
    TRABAJOS ESTADIST., 16, 3-15.  (MR31-6314)


KRISHNAIAH, PARUCHURI R.   SEE ALSO
    2936. KASKEY, GILBERT; KRISHNAIAH, PARUCHURI R. + AZZARI, ANTHONY J. (1962)

KRISHNAMOORTHY, A. S. 1951.  (18.1)                                               3149
    MULTIVARIATE BINOMIAL AND POISSON DISTRIBUTIONS.
    SANKHYA, 11, 117-124.  (MR13,P.478)

KRISHNAMOORTHY, A. S. + PARTHASARATHY, M. 1951.   (18.1)                    3150
    A MULTIVARIATE GAMMA-TYPE DISTRIBUTION.
        (AMENDED BY NO. 3151)
    ANN. MATH. STATIST., 22, 549-557.   (MR13,P.478)

KRISHNAMOORTHY, A. S. + PARTHASARATHY, M. 1960.   (18.1)                    3151
    CORRECTION TO:  A MULTIVARIATE GAMMA-TYPE DISTRIBUTION.
        (AMENDMENT OF NO. 3150)
    ANN. MATH. STATIST., 31, 229.   (MR31-229)

KRISHNAN, MARAKATHA 1966.   (19.2)                                          3152
    LOCALLY UNBIASED TYPE M TEST.
    J. ROY. STATIST. SOC. SER. B, 28, 298-309.

KRISHNA SASTRY, K. V.   SEE  SASTRY, K.V. KRISHNA

KRISHNASWAMI, G. V. + VENKATACHARI, S. 1934.   (4.5/4.6)                    3153
    ON A REPRESENTATION OF THE COEFFICIENTS OF CORRELATION, BY THE
    CONSTANTS OF A SPHERICAL TRIANGLE.
    J. ANNAMALAI UNIV., 3, 189-193.

KRISHNASWAMI AYYANGAR, A. A.   SEE  AYYANGAR, A. A. KRISHNASWAMI

KRISTOF, WALTER 1963.   (15.4)                                             3154
    THE STATISTICAL THEORY OF STEPPED-UP RELIABILITY COEFFICIENTS WHEN A
    TEST HAS BEEN DIVIDED INTO SEVERAL EQUIVALENT PARTS.
    PSYCHOMETRIKA, 28, 221-238.   (MR27-5339)

KROLL, ABRAHAM 1940.   (15.4)                                              3155
    ITEM VALIDITY AS A FACTOR IN TEST VALIDITY.
    J. EDUC. PSYCHOL., 31, 425-436.

KRUEGER, F. + SPEARMAN, CHARLES 1906.   (4.1/E)                            3156
    DIE KORRELATION ZWISCHEN VERSCHIEDENEN GEISTIGEN LEISTUNGSFAHIGKEITEN.
    Z. PSYCHOL., 44, 50-114.

KRULL, WOLFGANG 1951.   (4.8/B)                                            3157
    ZUR KORRELATIONSTHEORIE ZWEIDIMENSIONALER MERKMALE.
    MITT. MATH. STATIST., 3, 15-29.   (MR13,P.762)

KRULL, WOLFGANG 1951.   (4.8)                                              3158
    KORRELATIONSTHEORIE MEHRDIMENSIONALER MERKMALE.
    MITT. MATH. STATIST., 3, 185-200.   (MR13,P.762)

KRUSKAL, J. B. 1964.   (15.4/15.5/E)                                       3159
    MULTIDIMENSIONAL SCALING BY OPTIMIZING GOODNESS OF FIT TO A NON-METRIC
        HYPOTHESIS.
    PSYCHOMETRIKA, 29, 1-27.

KRUSKAL, J. B. 1964.   (15.5/N)                                            3160
    NONMETRIC MULTIDIMENSIONAL SCALING:  A NUMERICAL METHOD.
    PSYCHOMETRIKA, 29, 115-129.

KRUSKAL, WILLIAM H. 1953.   (2.1/4.9)                                      3161
    ON THE UNIQUENESS OF THE LINE OF ORGANIC CORRELATION.
    BIOMETRICS, 9, 47-58.   (MR14,P.890)

KRUSKAL, WILLIAM H. 1958.   (4.9)                                          3162
    ORDINAL MEASURES OF ASSOCIATION.
    J. AMER. STATIST. ASSOC., 53, 814-861.   (MR20-7366)

KRUSKAL, WILLIAM H.   SEE ALSO
    2075. GOODMAN, LEO A. + KRUSKAL, WILLIAM H. (1954)

    2076. GOODMAN, LEO A. + KRUSKAL, WILLIAM H. (1957)

    2077. GOODMAN, LEO A. + KRUSKAL, WILLIAM H. (1958)

    2078. GOODMAN, LEO A. + KRUSKAL, WILLIAM H. (1963)

KSHIRSAGAR, A. M. 1959.   (7.1/7.3)                                        3163
    BARTLETT DECOMPOSITION AND WISHART DISTRIBUTION.
    ANN. MATH. STATIST., 30, 239-241.   (MR21-2320)

KSHIRSAGAR, A. M. 1960.  (8.7/13.1)                                          3164
    A NOTE ON THE DERIVATION OF SOME EXACT MULTIVARIATE TESTS.
    BIOMETRIKA, 47, 480-482.

KSHIRSAGAR, A. M. 1961.  (18.1)                                              3165
    THE NON-CENTRAL MULTIVARIATE BETA DISTRIBUTION.
    ANN. MATH. STATIST., 32, 104-111.  (MR22-8607)

KSHIRSAGAR, A. M. 1961.  (10.2/11.2)                                         3166
    THE GOODNESS-OF-FIT OF A SINGLE (NON-ISOTROPIC) HYPOTHETICAL PRINCIPAL
        COMPONENT.
    BIOMETRIKA, 48, 397-407.  (MR26-1919)

KSHIRSAGAR, A. M. 1961.  (18.1)                                              3167
    SOME EXTENSIONS OF THE MULTIVARIATE T-DISTRIBUTION AND THE MULTIVARIATE
    GENERALIZATION OF THE DISTRIBUTION OF THE REGRESSION COEFFICIENT.
    PROC. CAMBRIDGE PHILOS. SOC., 57, 80-85.  (MR22-8601)

KSHIRSAGAR, A. M. 1962.  (8.7)                                               3168
    A NOTE ON DIRECTION AND COLLINEARITY FACTORS IN CANONICAL ANALYSIS.
    BIOMETRIKA, 49, 255-259.

KSHIRSAGAR, A. M. 1962.  (16.2)                                             3169
    PREDICTION FROM SIMULTANEOUS EQUATION SYSTEMS AND WOLD'S IMPLICIT
    CAUSAL CHAIN MODEL.
    ECONOMETRICA, 30, 801-811.  (MR28-2621)

KSHIRSAGAR, A. M. 1963.  (7.1)                                              3170
    EFFECT OF NON-CENTRALITY ON THE BARTLETT DECOMPOSITION OF A WISHART MATRIX.
    ANN. INST. STATIST. MATH. TOKYO, 14, 217-228.  (MR28-1683)

KSHIRSAGAR, A. M. 1963.  (6.1/8.7/CF)                                       3171
    CONFIDENCE INTERVALS FOR DISCRIMINANT FUNCTION COEFFICIENTS.
    J. INDIAN STATIST. ASSOC., 1, 1-7.  (MR29-1701)

KSHIRSAGAR, A. M. 1964.  (13.1)                                             3172
    DISTRIBUTION OF THE RESIDUAL ROOTS IN PRINCIPAL COMPONENTS ANALYSIS.
    DEFENCE SCI. J., 14, 288-294.

KSHIRSAGAR, A. M. 1964.  (1.1)                                              3173
    WILKS' Λ CRITERION.
    J. INDIAN STATIST. ASSOC., 2, 1-20.  (MR28-5517)

KSHIRSAGAR, A. M. 1964.  (8.7)                                              3174
    DISTRIBUTIONS OF THE DIRECTION AND COLLINEARITY FACTORS IN
    DISCRIMINANT ANALYSIS.
    PROC. CAMBRIDGE PHILOS. SOC., 60, 217-225.  (MR29-2903)

KSHIRSAGAR, A. M. 1966.  (10.1/11.2)                                        3175
    THE NON-NULL DISTRIBUTION OF A STATISTIC IN PRINCIPAL COMPONENT ANALYSIS.
    BIOMETRIKA, 53, 590-594.  (MR35-6272)

KSHIRSAGAR, A. M. + GUPTA, R. P. 1965.  (11.2)                              3176
    THE GOODNESS OF FIT OF TWO (OR MORE) HYPOTHETICAL PRINCIPAL COMPONENTS.
    ANN. INST. STATIST. MATH. TOKYO, 17, 347-356.  (MR33-6768)

KSHIRSAGAR, A. M. + GUPTA, R. P. 1965.  (12.1/15.2)                         3177
    A NOTE ON THE USE OF CANONICAL ANALYSIS IN FACTORIAL EXPERIMENTS.
    J. INDIAN STATIST. ASSOC., 3, 165-169.

KUBOTA, ISAO 1963.  (4.1/4.5)                                               3178
    AN ELUCIDATION ON SOME CORRELATION COEFFICIENTS.
    J. NARA GAKUGEI UNIV. NATUR. SCI., 11, 1-2.  (MR27-2046)

KUDER, G. F. 1937.  (4.4/4.8/T)                                             3179
    NOMOGRAPH FOR POINT BISERIAL R, BISERIAL R, AND FOURFOLD CORRELATIONS.
    PSYCHOMETRIKA, 2, 135-138.

KUDER, G. F. + RICHARDSON, MOSES W. 1937.  (15.4)                           3180
    THE THEORY OF THE ESTIMATION OF TEST RELIABILITY.
    PSYCHOMETRIKA, 2, 151-160.

KUDER, G. F.    SEE ALSO
   4541. RICHARDSON, MOSES W. + KUDER, G. F. (1939)


KUDO, AKIO 1956.   (19.1)                                                            3181
     ON THE INVARIANT MULTIPLE DECISION PROCEDURES.
   BULL. MATH. STATIST., 6, 57-68.   (MR18,P.520)


KUDO, AKIO 1956.   (19.3)                                                            3182
     ON SUFFICIENCY AND COMPLETENESS OF STATISTICS.
   SUGAKU, 8, 129-138.


KUDO, AKIO 1957.   (14.1)                                                            3183
     THE EXTREME VALUE IN A MULTIVARIATE NORMAL SAMPLE.
   MEM. FAC. SCI. KYUSHU UNIV. SER. A, 11, 143-156.   (MR20-2813)


KUDO, AKIO 1958.   (14.1)                                                            3184
     ON THE DISTRIBUTION OF THE MAXIMUM VALUE OF AN EQUALLY CORRELATED
     SAMPLE FROM A NORMAL POPULATION.
   SANKHYA, 20, 309-316.   (MR21-5247)


KUDO, AKIO 1959.   (6.1/6.2)                                                         3185
     THE CLASSIFICATORY PROBLEM VIEWED AS A TWO-DECISION PROBLEM.
   MEM. FAC. SCI. KYUSHU UNIV. SER. A, 13, 96-125.   (MR22-11495)


KUDO, AKIO 1960.   (6.1/6.2)                                                         3186
     THE CLASSIFICATORY PROBLEM VIEWED AS A TWO-DECISION PROBLEM, II.
   MEM. FAC. SCI. KYUSHU UNIV. SER. A, 14, 63-83.   (MR22-11496)


KUDO, AKIO 1960.   (19.1)                                                            3187
     THE SYMMETRIC MULTIPLE DECISION PROBLEMS.
   MEM. FAC. SCI. KYUSHU UNIV. SER. A, 14, 179-206.   (MR23-A741)


KUDO, AKIO 1961.   (6.1/17.6/19.1)                                                   3188
     SOME PROBLEMS OF SYMMETRIC MULTIPLE DECISIONS IN MULTIVARIATE ANALYSIS.
   BULL. INST. INTERNAT. STATIST., 38(4) 165-171.   (MR27-2041)


KUDO, AKIO 1963.   (5.1)                                                             3189
     A MULTIVARIATE ANALOGUE OF THE ONE-SIDED TEST.
   BIOMETRIKA, 50, 403-418.   (MR29-689)


KUDO, AKIO + FUJISAWA, H. 1966.   (5.1/T)                                            3190
     SOME MULTIVARIATE TESTS WITH RESTRICTED ALTERNATIVE HYPOTHESES.
   MULTIVARIATE ANAL. PROC. INTERNAT. SYMP. DAYTON (KRISHNAIAH), 73-85.


KUDO, AKIO    SEE ALSO
   4928. SCHULL, WILLIAM J. + KUDO, AKIO (1962)


KUDO, HIROKICHI 1961.   (19.2)                                                       3191
     LOCALLY COMPLETE CLASS OF TESTS.
   BULL. INST. INTERNAT. STATIST., 38(4) 173-180.


KUIPER, NICOLAAS H. 1961.   (1.1)                                                    3192
     RANDOM VARIABLES RATHER THAN DISTRIBUTION FUNCTIONS, A DIFFERENT AND
     USEFUL APPROACH IN STATISTICS.
   MEDED. LANDBOUWHOGESCHOOL WAGENINGEN, 61, 1-19.


KULKARNI, N. V.    SEE
   511. BHAT, B. R. + KULKARNI, N. V. (1966)


KULKARNI, S. R.    SEE
   512. BHAT, B. R. + KULKARNI, S. R. (1966)

   513. BHAT, B. R. + KULKARNI, S. R. (1966)


KULLBACK, SOLOMON 1934.   (5.3/7.1/7.2)                                              3193
     AN APPLICATION OF CHARACTERISTIC FUNCTIONS TO THE DISTRIBUTION PROBLEM
     OF STATISTICS.
   ANN. MATH. STATIST., 5, 263-307.


KULLBACK, SOLOMON 1935.   (4.7/7.4/14.2)                                             3194
     ON SAMPLES FROM A MULTIVARIATE NORMAL POPULATION.
   ANN. MATH. STATIST., 6, 202-213.

KULLBACK, SOLOMON 1935.    (4.5)                                                    3195
    A NOTE ON THE DISTRIBUTION OF A CERTAIN PARTIAL BELONGING COEFFICIENT.
    METRON, 12(3) 65-70.


KULLBACK, SOLOMON 1936.    (1.1)                                                    3196
    ON CERTAIN DISTRIBUTION THEOREMS OF STATISTICS.
    BULL. AMER. MATH. SOC., 42, 407-410.


KULLBACK, SOLOMON 1936.    (4.7)                                                    3197
    A NOTE ON THE MULTIPLE CORRELATION COEFFICIENT.
    METRON, 12(4) 67-72.


KULLBACK, SOLOMON 1937.    (18.3)                                                   3198
    ON CERTAIN DISTRIBUTIONS DERIVED FROM THE MULTINOMIAL DISTRIBUTION.
    ANN. MATH. STATIST., 8, 127-144.


KULLBACK, SOLOMON 1952.    (6.2/11.1/12.1)                                          3199
    AN APPLICATION OF INFORMATION THEORY TO MULTIVARIATE ANALYSIS, I.
    ANN. MATH. STATIST., 23, 88-102.   (MR13,P.854)


KULLBACK, SOLOMON 1954.    (19.3)                                                   3200
    CERTAIN INEQUALITIES IN INFORMATION THEORY AND THE CRAMER-RAO INEQUALITY.
    ANN. MATH. STATIST., 25, 745-751.   (MR16,P.495)


KULLBACK, SOLOMON 1956.    (8.3/8.4/10.2)                                           3201
    AN APPLICATION OF INFORMATION THEORY TO MULTIVARIATE ANALYSIS, II.
        (AMENDED BY NO. 3202)
    ANN. MATH. STATIST., 27, 122-146.   (MR17,P.982)


KULLBACK, SOLOMON 1956.    (8.3/8.4/10.2)                                           3202
    CORRECTION TO: AN APPLICATION OF INFORMATION THEORY TO MULTIVARIATE ANALYSIS,II.
        (AMENDMENT OF NO. 3201)
    ANN. MATH. STATIST., 27, 860.   (MR17,P.982)


KULLBACK, SOLOMON + ROSENBLATT, HARRY M. 1957.   (8.2/8.7)                          3203
    ON THE ANALYSIS OF MULTIPLE REGRESSION IN K CATEGORIES.
    BIOMETRIKA, 44, 67-83.   (MR19,P.186)


KULLDORFF, GUNNAR 1957.    (18.2)                                                   3204
    ON THE CONDITIONS FOR CONSISTENCY AND ASYMPTOTIC EFFICIENCY OF MAXIMUM
        LIKELIHOOD ESTIMATES.
    SKAND. AKTUARIETIDSKR., 40, 129-144.   (MR20-4336)


KUMMELL, CHAS. H. 1879.    (8.8/17.9)                                               3205
    REDUCTION OF OBSERVATION EQUATIONS WHICH CONTAIN MORE THAN ONE
        OBSERVED QUANTITY.
    ANALYST, 6, 97-105.


KUMMELL, CHAS. H. 1882.    (8.9/20.5)                                               3206
    ON THE COMPOSITION OF ERRORS FROM SINGLE CAUSES OF ERROR.
    ASTRONOM. NACHR., 103, 177-206.


KUMMELL, CHAS. H. 1888.    (20.4)                                                   3207
    THE QUADRIC TRANSFORMATION OF ELLIPTIC INTEGRALS, COMBINED WITH THE
        ALGORITHM OF THE ARITHMETICO-GEOMETRIC MEAN.
    SMITHSONIAN MISC. COLLECT., 33(2) 102-121.


KUMMER, ELIZABETH    SEE
    84. ALEXANDER, LEROY; KLUG, HAROLD P. + KUMMER, ELIZABETH (1948)


KUMPEL, P. G., JR. 1965.    (20.5)                                                  3208
    LIE GROUPS AND PRODUCTS OF SPHERES.
    PROC. AMER. MATH. SOC., 16, 1350-1356.


KUNETZ, GEZA 1935.    (15.1)                                                        3209
    SUR LA CONSERVATION DU FACTEUR COMMUN DE SPEARMAN DANS UNE
        SUBSTITUTION LINEAIRE.
    C.R. ACAD. SCI. PARIS, 201, 864-867.


KUNREUTHER, HOWARD 1966.    (4.1)                                                   3210
    THE USE OF THE PEARSONIAN APPROXIMATION IN COMPARING DEFLATED AND UNDEFLATED
        REGRESSION ESTIMATES.
    ECONOMETRICA, 34, 232-234.   (MR33-1929)

KURTZ, ALBERT K. 1929.  (4.6)                                                    3211
   A SPECIAL CASE OF THE MULTIPLE CORRELATION COEFFICIENT.
   J. EDUC. PSYCHOL., 20, 378–385.

KURTZ, ALBERT K. 1936.  (4.6)                                                    3212
   THE USE OF THE DOOLITTLE METHOD IN OBTAINING RELATED MULTIPLE
   CORRELATION COEFFICIENTS.
   PSYCHOMETRIKA. 1(1) 45–51.

KURTZ, ALBERT K. 1937.  (15.2)                                                   3213
   THE SIMULTANEOUS PREDICTION OF ANY NUMBER OF CRITERIA BY THE USE OF A
   UNIQUE SET OF WEIGHTS.
   PSYCHOMETRIKA, 2, 95–101.

KURTZ, T. E.; LINK,R.F.; TUKEY, JOHN W. + 1966.  (14.1/B)                        3214
   WALLACE, D. L.
      CORRELATION OF RANGES OF CORRELATED DEVIATES. (WITH EDITORIAL NOTE).
   BIOMETRIKA. 53, 191–197.  (MR33-8052)

KURUP, R. S. 1965.  (17.7)                                                       3215
   CERTAIN DISTRIBUTION-FREE TESTS OF REGRESSION.
   J. INDIAN SOC. AGRIC. STATIST., 17, 104–110.

KUSAMA, TOKITAKE 1966.  (19.2)                                                   3216
   REMARKS ON ADMISSIBILITY OF DECISION FUNCTIONS.
   ANN. INST. STATIST. MATH. TOKYO, 18, 141–148.

KUTIKOV, L. M. 1965.  (20.3)                                                     3217
   INVERSION OF CORRELATION MATRICES.
      (TRANSLATION OF NO. 3218)
   ENGRG. CYBERNETICS, 5, 35–39.  (MR34-3718)

KUTIKOV, L. M. 1965.  (20.3)                                                     3218
   INVERSION OF CORRELATION MATRICES.
      (IN RUSSIAN, TRANSLATED AS NO. 3217)
   IZV. AKAD. NAUK SSSR TEH. KIBERNET., 5, 42–47.  (MR34-3718)

KUZMIN, R.O. 1939.  (2.1/4.2/B)                                                  3219
   SUR LA LOI DE DISTRIBUTION DU COEFFICIENT DE CORRELATION DANS LES
   TIRAGES D'UN ENSEMBLE NORMAL.
   C. R. ACAD. SCI. URSS NOUV. SER., 22, 298–301.

KUZNECOV, E. 1935.  (14.2/14.5/B)                                                3220
   LA LOI DE PROBABILITE D'UN VECTEUR ALEATOIRE.
      (IN RUSSIAN AND FRENCH)
   C. R. ACAD. SCI. URSS NOUV. SER., 7, 187–193.

KUZNETS, GEORGE   SEE
   1928. GHISELLI, EDWIN E. + KUZNETS, GEORGE (1937)

KUZNETS, SIMON 1928.  (16.5/E)                                                   3221
   ON MOVING CORRELATION OF TIME SEQUENCES.
   J. AMER. STATIST. ASSOC., 23, 121–136.

LA BUDDE, C. DONALD 1964.  (20.3)                                               3222
   TWO NEW CLASSES OF ALGORITHMS FOR FINDING THE EIGENVALUES AND
   EIGENVECTORS OF REAL SYMMETRIC MATRICES.
   J. ASSOC. COMPUT. MACH., 11, 53–58.

LACHENBRUCH, P. A. 1966.  (6.1/6.2/T)                                            3223
   DISCRIMINANT ANALYSIS WHEN THE INITIAL SAMPLES ARE MISCLASSIFIED.
   TECHNOMETRICS, 8, 657–662.  (MR34-3719)

LADD, GEORGE W. 1966.  (5.2/6.2)                                                 3224
   LINEAR PROBABILITY FUNCTIONS AND DISCRIMINANT FUNCTIONS.
   ECONOMETRICA. 34, 873–885.

LADERMAN, JACK 1939.  (17.5/U)                                                   3225
   THE DISTRIBUTION OF "STUDENT'S" RATIO FOR SAMPLES OF TWO ITEMS DRAWN
      FROM NON-NORMAL UNIVERSES.
   ANN. MATH. STATIST., 10, 376–379.  (MR1,P.153)

LA GARRIGUE, V. ROUQUET 1963.  (1.1)                              3226
    AVANT-PROPOS.
  CAHIERS INST. SCI. ECON. APPL. SER. E, 2, 3-13.


LAHA, R. G. 1954.  (14.5)                                        3227
    ON SOME PROPERTIES OF THE BESSEL FUNCTION DISTRIBUTIONS.
  BULL. CALCUTTA MATH. SOC., 46, 59-71.  (MR16,P.152)


LAHA, R. G. 1954.  (12.1)                                        3228
    ON SOME PROBLEMS IN CANONICAL CORRELATIONS.
  SANKHYA, 14, 61-66.  (MR16,P.1038)


LAHA, R. G. 1955.  (2.2)                                         3229
    ON A CHARACTERIZATION OF THE MULTIVARIATE NORMAL DISTRIBUTION.
  SANKHYA, 14, 367-368.  (MR16,P.940)


LAHA, R. G. 1956.  (2.5)                                         3230
    ON THE STOCHASTIC INDEPENDENCE OF TWO SECOND-DEGREE POLYNOMIAL
    STATISTICS IN NORMALLY DISTRIBUTED VARIATES.
  ANN. MATH. STATIST., 27, 790-796.  (MR18,P.160)


LAHA, R. G. 1956.  (2.2)                                         3231
    ON STOCHASTIC INDEPENDENCE OF A HOMOGENEOUS QUADRATIC STATISTIC AND OF THE MEAN.
      (IN RUSSIAN)
  VESTNIK LENINGRAD. UNIV., 11(1) 25-32.  (MR17,P.871)


LAHA, R. G. 1957.  (2.2/15.1)                                    3232
    ON SOME CHARACTERIZATION PROBLEMS CONNECTED WITH LINEAR STRUCTURAL RELATIONS.
  ANN. MATH. STATIST., 28, 405-414.  (MR19,P.471)


LAHA, R. G. + LUKACS, EUGENE 1960.  (2.5/4.8)                    3233
    ON CERTAIN FUNCTIONS OF NORMAL VARIATES WHICH ARE UNCORRELATED OF A
    HIGHER ORDER.
  BIOMETRIKA, 47, 175-176.


LAHA, R. G. + LUKACS, EUGENE 1960.  (17.3)                       3234
    ON A PROBLEM CONNECTED WITH QUADRATIC REGRESSION.
  BIOMETRIKA, 47, 335-343.  (MR22-12649)


LAHA, R. G.   SEE ALSO
    382. BASU, D. + LAHA, R. G. (1954)

    383. BASU, D. + LAHA, R. G. (1954)

ENTRY NO. 3235 APPEARS BETWEEN ENTRY NOS. 6067 & 6068.

LAKSHMANAMURTI, M. 1945.  (4.8)                                  3236
    COEFFICIENT OF ASSOCIATION BETWEEN TWO ATTRIBUTES IN STATISTICS.
  PROC. INDIAN ACAD. SCI. SECT. A, 22, 123-133.  (MR7,P.212)


LAL, D. N. 1955.  (17.2)                                         3237
    A NOTE ON A FORM OF TCHEBYCHEFF'S INEQUALITY FOR TWO OR MORE VARIABLES.
  SANKHYA, 15, 317-320.  (MR18,P.155)


LAMENS, A. 1957.  (16.4/B)                                       3238
    SUR LE PROCESSUS NON HOMOGENE DE NAISSANCE ET DE MORT A DEUX
    VARIABLES ALEATOIRES.
  ACAD. ROY. BELG. BULL. CL. SCI. SER. 5, 43, 711-719.  (MR20-358)


LAMENS, A. + CONSAEL, ROBERT 1957.  (16.4/B)                     3239
    SUR LE PROCESSUS NON HOMOGENE DE NAISSANCE ET DE MORT A DEUX VARIABLES
    ALEATOIRES.
  ACAD. ROY. BELG. BULL. CL. SCI. SER. 5, 43, 597-605.  (MR19,P.1203)

LAMOTTE, EDITH M. M. DE G.   SEE
    1530. ELDERTON, ETHEL M.; BARRINGTON, AMY; JONES, H. GERTRUDE;  (1913)
    LAMOTTE, EDITH M. M. DE G.; LASKI, H. J. + PEARSON, KARL

LAMPARD, D. G. 1955. (16.4)
    A NEW METHOD OF DETERMINING CORRELATION FUNCTIONS OF STATIONARY TIME SERIES.
  PROC. INST. ELEC. ENGRS. LONDON, 102(C) 35-41. (MR16,P.1134)
                            **3240**

LAMPARD, D. G. 1956. (16.4)
    THE PROBABILITY DISTRIBUTION FOR THE FILTERED OUTPUT OF A MULTIPLIER
    WHOSE INPUTS ARE CORRELATED, STATIONARY, GAUSSIAN TIME-SERIES.
  IRE TRANS. INFORMATION THEORY, IT-2, 4-11.
                            **3241**

LAMPARD, D. G.    SEE ALSO
  320. BARRETT, J. F. + LAMPARD, D. G. (1955)


LANCASTER, H. O. 1949. (18.2/18.3)
    THE DERIVATION AND PARTITION OF $\chi^2$ IN CERTAIN DISCRETE DISTRIBUTIONS.
       (AMENDED BY NO. 3244)
  BIOMETRIKA, 36, 117-129. (MR12,P.191)
                            **3242**

LANCASTER, H. O. 1950. (18.2)
    THE EXACT PARTITION OF $\chi^2$ AND ITS APPLICATION TO THE PROBLEM OF THE
    POOLING OF SMALL EXPECTATIONS.
  BIOMETRIKA, 37, 267-270. (MR12,P.429)
                            **3243**

LANCASTER, H. O. 1950. (18.2/18.3)
    CORRIGENDA.
       (AMENDMENT OF NO. 3242)
  BIOMETRIKA, 37, 452. (MR12,P.191)
                            **3244**

LANCASTER, H. O. 1951. (18.2/18.3)
    COMPLEX CONTINGENCY TABLES TREATED BY THE PARTITION OF $\chi^2$.
  J. ROY. STATIST. SOC. SER. B, 13, 242-249. (MR14,P.486)
                            **3245**

LANCASTER, H. O. 1953. (4.8/E)
    A RECONCILIATION OF $\chi^2$ CONSIDERED FROM METRICAL AND ENUMERATIVE ASPECTS.
  SANKHYA, 13, 1-10. (MR15,P.972)
                            **3246**

LANCASTER, H. O. 1954. (2.5/20.2)
    TRACES AND CUMULANTS OF QUADRATIC FORMS IN NORMAL VARIABLES.
  J. ROY. STATIST. SOC. SER. B, 16, 247-254. (MR16,P.1128)
                            **3247**

LANCASTER, H. O. 1957. (2.1/B)
    SOME PROPERTIES OF THE BIVARIATE NORMAL DISTRIBUTION CONSIDERED IN
    THE FORM OF A CONTINGENCY TABLE.
  BIOMETRIKA, 44, 289-292.
                            **3248**

LANCASTER, H. O. 1958. (17.3/17.8/AB)
    THE STRUCTURE OF BIVARIATE DISTRIBUTIONS.
       (AMENDED BY NO. 3258)
  ANN. MATH. STATIST., 29, 719-736. (MR21-944)
                            **3249**

LANCASTER, H. O. 1959. (17.3)
    ZERO CORRELATION AND INDEPENDENCE.
  AUSTRAL. J. STATIST., 1, 53-56. (MR21-6611)
                            **3250**

LANCASTER, H. O. 1960. (17.5)
    ON TESTS OF INDEPENDENCE IN SEVERAL DIMENSIONS.
       (AMENDED BY NO. 3254)
  J. AUSTRAL. MATH. SOC., 1, 241-254.
                            **3251**

LANCASTER, H. O. 1960. (2.2)
    THE CHARACTERISATION OF THE NORMAL DISTRIBUTION.
  J. AUSTRAL. MATH. SOC., 1, 368-383. (MR24-A3023)
                            **3252**

LANCASTER, H. O. 1960. (4.9)
    ON STATISTICAL INDEPENDENCE AND ZERO CORRELATION IN SEVERAL DIMENSIONS.
  J. AUSTRAL. MATH. SOC., 1, 492-496. (MR24-A3024)
                            **3253**

LANCASTER, H. O. 1960. (17.5)
    CORRECTION TO: ON TESTS OF INDEPENDENCE IN SEVERAL DIMENSIONS.
       (AMENDMENT OF NO. 3251)
  J. AUSTRAL. MATH. SOC., 1, 496.
                            **3254**

LANCASTER, H. O. 1963.  (4.9/18.1)                                    3255
    CORRELATIONS AND CANONICAL FORMS OF BIVARIATE DISTRIBUTIONS.
    ANN. MATH. STATIST., 34, 532-538.  (MR26-4431)

LANCASTER, H. O. 1963.  (4.9/17.3)                                    3256
    CORRELATION AND COMPLETE DEPENDENCE OF RANDOM VARIABLES.
    ANN. MATH. STATIST., 34, 1315-1321.  (MR27-4325)

LANCASTER, H. O. 1963.  (12.1/18.1/18.2/A)                            3257
    CANONICAL CORRELATIONS AND PARTITIONS OF $\chi^2$.
    QUART. J. MATH. OXFORD SECOND SER., 14, 220-224.  (MR27-3035)

LANCASTER, H. O. 1964.  (17.3/17.8/AB)                                3258
    CORRECTION TO: "THE STRUCTURE OF BIVARIATE DISTRIBUTIONS".
        (AMENDMENT OF NO. 3249)
    ANN. MATH. STATIST., 35, 1388.  (MR30-5431)

LANCASTER, H. O. 1965.  (20.2)                                        3259
    THE HELMERT MATRICES.
    AMER. MATH. MONTHLY, 72, 4-12.

LANCASTER, H. O. 1965.  (17.3)                                        3260
    PAIRWISE STATISTICAL INDEPENDENCE.
    ANN. MATH. STATIST., 36, 1313-1317.

LANCASTER, H. O. 1965.  (4.8/17.3/17.8)                               3261
    SYMMETRY IN MULTIVARIATE DISTRIBUTIONS.
    AUSTRAL. J. STATIST., 7, 115-126.

LANCASTER, H. O. 1966.  (8.8/14.5)                                    3262
    FORERUNNERS OF THE PEARSON  $\chi^2$.
    AUSTRAL. J. STATIST., 8, 117-126.

LANCASTER, H. O. 1966.  (12.1)                                        3263
    KOLMOGOROV'S REMARK ON THE HOTELLING CANONICAL CORRELATIONS.
    BIOMETRIKA, 53, 585-588.  (MR36-2266)

LANCASTER, H. O. + BROWN, T. A. I. 1965.  (18.2/18.3)                 3264
    SIZES OF THE $\chi^2$ TEST IN THE SYMMETRICAL MULTINOMIALS.
    AUSTRAL. J. STATIST., 7, 40-44.

LANCASTER, H. O. + HAMDAN, H. A. 1964.  (4.8)                         3265
    ESTIMATION OF THE CORRELATION COEFFICIENT IN CONTINGENCY TABLES WITH
    POSSIBLY NONMETRICAL CHARACTERS.
    PSYCHOMETRIKA, 29, 383-391.  (MR31-858)

LANDAHL, HERBERT D. 1938.  (15.3)                                     3266
    CENTROID ORTHOGONAL TRANSFORMATIONS.
    PSYCHOMETRIKA, 3, 219-223.

LANDAHL, HERBERT D. 1939.  (15.5)                                     3267
    A CONTRIBUTION TO THE MATHEMATICAL BIOPHYSICS OF PSYCHOPHYSICAL
    DISCRIMINATION.  II, III.
    BULL. MATH. BIOPHYS., 1, 159-176.

LANDAHL, HERBERT D. 1940.  (15.1)                                     3268
    TIME SCORES AND FACTOR ANALYSIS.
    PSYCHOMETRIKA, 5, 67-74.

LANDAHL, HERBERT D. 1941.  (15.5)                                     3269
    STUDIES IN THE MATHEMATICAL BIOPHYSICS OF DISCRIMINATION AND CONDITIONING.  II.
    SPECIAL CASE.  ERRORS, TRIALS, AND NUMBER OF POSSIBLE RESPONSES.
    BULL. MATH. BIOPHYS., 3, 71-77.

LANDENNA, GIAMPIERO 1957.  (4.9)                                      3270
    OSSERVAZIONI SULLA CONNESSIONE.
    STATISTICA. BOLOGNA, 17, 351-392.  (MR19,P.780)

LANG, E.D.   SEE
    4757. RUSHTON, S. + LANG, E.D. (1954)

LANGE, R. B.   SEE
    3814. MICHENER, C. D.; LANGE, R. B.; BIGARELLA, J. J. +  (1958)
    SALAMUNI, R.

LANGER, RUDOLPH E. 1929. (20.5)                                              3271
    THE ASYMPTOTIC LOCATION OF THE ROOTS OF A CERTAIN TRANSCENDENTAL EQUATION.
    TRANS. AMER. MATH. SOC., 31, 837-844.

LANGMACK, P. V. F. P. 1920.  (4.8)                                          3272
    EN KORRELATIONSFORMEL.
    MAT. TIDSSKR. A, 89-103.

LAPIERRE, ANDRE BLANC-  SEE  BLANC-LAPIERRE, ANDRE

LAPLACE, P. S. 1811.  (2.1/20.4/B)                                          3273
    MEMOIRE SUR LES INTEGRALES DEFINIES ET LEUR APPLICATION AUX PROBABILITES.
    MEM. INST. IMP. FRANCE (ANNEE 1810), 279-347.

LAPLACE, P. S. 1812.  (2.1/20.4)                                            3274
    MEMOIRE SUR LES INTEGRALES DEFINIES, ET LEUR APPLICATION AUX PROBABILITES, ET
    SPECIALEMENT A LA RECHERCHE DU MILIEU QU'IL FAUT CHOISIR ENTRE LES RESULTATS
    DES OBSERVATIONS.  (IN ENGLISH)
    TILLOCH'S PHILOS. MAG., 39, 240-244.

LAREMBA, S. K. 1965.  (16.1/16.3/AE)                                        3275
    KURTOSIS AND DETERMINATE COMPONENTS IN LINEAR PROCESSES.
    REV. INST. INTERNAT. STATIST., 33, 378-413.

LARGUIER, EVERETT H., S.J. 1935.  (20.2)                                    3276
    ON CERTAIN COEFFICIENTS USED IN MATHEMATICAL STATISTICS.
    ANN. MATH. STATIST., 6, 220-226.

LARGUIER, EVERETT H., S.J. 1939.  (17.3/20.1)                               3277
    A MATRIX THEORY OF N-DIMENSIONAL MEASUREMENT.
    DUKE MATH. J., 5, 729-739.  (MR1,P.149)

LARMAN, D. G. + WARD, D. J. 1966.  (20.5)                                   3278
    ON CONVEX SETS AND MEASURES.
    PROC. CAMBRIDGE PHILOS. SOC., 62, 33-41.

LARSEN, HAROLD D. 1939.  (18.1)                                             3279
    ON THE MOMENTS ABOUT THE ARITHMETIC MEAN OF A HYPERGEOMETRIC
    FREQUENCY DISTRIBUTION FUNCTION.
    ANN. MATH. STATIST., 10, 198-201.

LARSON, HAROLD J. 1966.  (8.7)                                              3280
    LEAST SQUARES ESTIMATION OF THE COMPONENTS OF A SYMMETRIC MATRIX.
    TECHNOMETRICS, 8, 360-362.

LARSON, HAROLD J. + BANCROFT, T. A. 1963.  (8.3/19.1)                       3281
    SEQUENTIAL MODEL BUILDING FOR PREDICTION IN REGRESSION ANALYSIS, I.
    ANN. MATH. STATIST., 34, 462-479.

LARSON, SELMER C. 1931.  (4.6)                                              3282
    THE SHRINKAGE OF THE COEFFICIENT OF MULTIPLE CORRELATION.
    J. EDUC. PSYCHOL., 22, 45-55.

LASKA, V. 1913.  (4.1)                                                      3283
    ZUR KORRELATION.
    METEOROL. Z., 30, 558-560.

LASKI, H. J.    SEE
    1530. ELDERTON, ETHEL M.; BARRINGTON, AMY; JONES, H. GERTRUDE;  (1913)
    LAMOTTE, EDITH M. M. DE G.; LASKI, H. J. + PEARSON, KARL

LATICHEVA, K. JA. 1934.  (2.1/4.1/B)                                        3284
    SUR QUELQUES SCHEMAS D'URNES POUR L'EXPLICATION D'UNE CORRELATION NORMALE.
        (IN UKRAINIAN)
    ZUR. MAT. TSYKLU UKRAIN. AKAD. NAUK, 3, 71-75.

LATICHEVA, K. JA. 1935.  (2.1/B)                                            3285
    DE LA LOI DE DISTRIBUTION DE GAUSS POUR LA FONCTION DE DEUX VARIABLES.
    ZUR. INST. MAT. UKRAIN. AKAD. NAUK URSR, 2, 113-119.

LATIMER, H. B. 1947.  E(4.3)                                                3286
    CORRELATIONS OF ORGAN WEIGHTS WITH BODY WEIGHT, BODY LENGTH AND WITH
    OTHER WEIGHTS IN THE ADULT CAT.
    GROWTH, 11, 61-75.

LAUBSCHER, NICO F. 1959.   (4.2)                                                3287
     A NOTE ON FISHER'S TRANSFORMATION ON CORRELATION COEFFICIENT.
     J. ROY. STATIST. SOC. SER. B, 21, 409-410.

LAUENSTEIN, G. 1966.   (8.1/E)                                                  3288
     A RECURSION FORMULA FOR CALCULATING MULTIPLE REGRESSION COEFFICIENTS.
          (IN GERMAN)
     STATIST. PRAX., 21, 113-116.

LAURENT, ANDRE G. 1957.   (14.4/14.5/B)                                         3289
     BOMBING PROBLEMS - A STATISTICAL APPROACH.
          (AMENDED BY NO. 3290)
     OPERATIONS RES., 5, 75-89.   (MR19,P.333)

LAURENT, ANDRE G. 1958.   (14.4/14.5/B)                                         3290
     ERRATA TO: BOMBING PROBLEMS - A STATISTICAL APPROACH.
          (AMENDMENT OF NO. 3289)
     OPERATIONS RES., 6, 297.   (MR19,P.333)

LAURENT, ANDRE G. 1962.   (8.6/14.4/B)                                          3291
     BOMBING PROBLEMS - A STATISTICAL APPROACH, II.
     OPERATIONS RES., 10, 380-387.   (MR25-1052)

LAWING, WILLIAM D. + DAVID, HERBERT T. 1966.   (19.1/19.2)                      3292
     LIKELIHOOD RATIO COMPUTATIONS OF OPERATING CHARACTERISTICS.
     ANN. MATH. STATIST., 37, 1704-1716.

LAWLEY, D. N. 1938.   (8.3/8.4)                                                 3293
     A GENERALIZATION OF FISHER'S Z TEST.
          (AMENDED BY NO. 3294)
     BIOMETRIKA, 30, 180-187.

LAWLEY, D. N. 1939.   (8.3/8.4)                                                 3294
     A CORRECTION TO "A GENERALIZATION OF FISHER'S Z TEST".
          (AMENDMENT OF NO. 3293)
     BIOMETRIKA, 30, 467-469.

LAWLEY, D. N. 1940.   (15.2)                                                    3295
     THE ESTIMATION OF FACTOR LOADINGS BY THE METHOD OF MAXIMUM LIKELIHOOD.
     PROC. ROY. SOC. EDINBURGH, 60, 64-82.   (MR2,P.110)

LAWLEY, D. N. 1942.   (15.2)                                                    3296
     FURTHER INVESTIGATION IN FACTOR ESTIMATION.
     PROC. ROY. SOC. EDINBURGH SECT. A, 61, 176-185.   (MR3,P.174)

LAWLEY, D. N. 1943.   (15.2)                                                    3297
     THE APPLICATION OF THE MAXIMUM LIKELIHOOD METHOD TO FACTOR ANALYSIS.
     BRITISH J. PSYCHOL., 33, 172-175.

LAWLEY, D. N. 1943.   (15.4)                                                    3298
     ON PROBLEMS CONNECTED WITH ITEM SELECTION AND TEST CONSTRUCTION.
     PROC. ROY. SOC. EDINBURGH SECT. A, 61, 273-287.   (MR4,P.222)

LAWLEY, D. N. 1943.   (2.4)                                                     3299
     A NOTE ON KARL PEARSON'S SELECTION FORMULAE.
     PROC. ROY. SOC. EDINBURGH SECT. A, 62, 28-30.   (MR5,P.127)

LAWLEY, D. N. 1944.   (15.1/15.4)                                               3300
     THE FACTORIAL ANALYSIS OF MULTIPLE ITEM TESTS.
     PROC. ROY. SOC. EDINBURGH SECT. A, 62, 74-82.   (MR6,P.92)

LAWLEY, D. N. 1949.   (15.2/A)                                                  3301
     PROBLEMS IN FACTOR ANALYSIS.
     PROC. ROY. SOC. EDINBURGH SECT. A, 62, 394-399.   (MR11,P.192)

LAWLEY, D. N. 1950.   (15.2)                                                    3302
     FACTOR ANALYSIS BY MAXIMUM LIKELIHOOD:   A CORRECTION.
          (AMENDMENT OF NO. 1554)
     BRITISH J. PSYCHOL. STATIST. SECT., 3, 76.

LAWLEY, D. N. 1950.   (15.2)                                                    3303
     A FURTHER NOTE ON A PROBLEM IN FACTOR ANALYSIS.
     PROC. ROY. SOC. EDINBURGH SECT. A, 63, 93-94.   (MR11,P.733)

LAWLEY, D. N. 1953.   (15.2/A)                                                    3304
   A MODIFIED METHOD OF ESTIMATION IN FACTOR ANALYSIS AND SOME LARGE
   SAMPLE RESULTS.
   NORDISK PSYKOL. MONOG. NR. 3 (UPPSALA SYMP. PSYCHOL. FACT. ANAL.), 35–42.   (MR15,P.972)


LAWLEY, D. N. 1956.   (10.2/13.2/15.2)                                            3305
   TESTS OF SIGNIFICANCE FOR THE LATENT ROOTS OF COVARIANCE AND
   CORRELATION MATRICES.
   BIOMETRIKA. 43, 128–136.   (MR17,P.1220)


LAWLEY, D. N. 1956.   (4.3/13.2)                                                  3306
   A GENERAL METHOD FOR APPROXIMATING TO THE DISTRIBUTION OF LIKELIHOOD
   RATIO CRITERIA.
   BIOMETRIKA. 43, 295–303.   (MR18,P.521)


LAWLEY, D. N. 1956.   (15.2)                                                      3307
   A STATISTICAL EXAMINATION OF THE CENTROID METHOD.
   PROC. ROY. SOC. EDINBURGH SECT. A. 64, 175–189.   (MR17,P.984)


LAWLEY, D. N. 1958.   (15.2)                                                      3308
   ESTIMATION IN FACTOR ANALYSIS UNDER VARIOUS INITIAL ASSUMPTIONS.
   BRITISH J. STATIST. PSYCHOL., 11, 1–12.


LAWLEY, D. N. 1959.   (12.2)                                                      3309
   TESTS OF SIGNIFICANCE IN CANONICAL ANALYSIS.
   BIOMETRIKA, 46, 59–66.   (MR21–1667)


LAWLEY, D. N. 1960.   (15.2)                                                      3310
   APPROXIMATE METHODS IN FACTOR ANALYSIS.
   BRITISH J. STATIST. PSYCHOL., 13, 11–17.


LAWLEY, D. N. 1963.   (10.2/A)                                                    3311
   ON TESTING A SET OF CORRELATION COEFFICIENTS FOR EQUALITY.
   ANN. MATH. STATIST., 34, 149–151.   (MR26–3147)


LAWLEY, D. N. + MAXWELL, A.E. 1962.   (15.2/E)                                    3312
   FACTOR ANALYSIS AS A STATISTICAL METHOD.
   STATISTICIAN, 12, 209–229.


LAWLEY, D. N. + MAXWELL, A.E. 1964.   (15.3)                                      3313
   FACTOR TRANSFORMATION METHODS.
   BRITISH J. STATIST. PSYCHOL., 17, 97–103.


LAWLEY, D. N. + SWANSON, Z. 1954.   (15.2/E)                                      3314
   TESTS OF SIGNIFICANCE IN A FACTOR ANALYSIS OF ARTIFICIAL DATA.
   BRITISH J. STATIST. PSYCHOL., 7, 75–79.


LAWLEY, D. N.     SEE ALSO
   2994. KENDALL, MAURICE G. + LAWLEY, D. N. (1956)


LAWRENCE, WARREN R.     SEE
   1634. FERGUSON, LEONARD W. + LAWRENCE, WARREN R. (1942)


LAWSHE, C. H., JR. 1942.   (15.4/T)                                               3315
   A NOMOGRAPH FOR ESTIMATING THE VALIDITY OF TEST ITEMS.
   J. APPL. PSYCHOL., 26, 846–849.


LAX, PETER D.     SEE
   13. ADAMS, J. F.; LAX, PETER D. + PHILLIPS, RALPH S. (1965)


LAZARSFELD, PAUL F. 1950.   (15.5)                                                3316
   THE LOGICAL AND MATHEMATICAL FOUNDATION OF LATENT STRUCTURE ANALYSIS.
   MEAS. PREDICT. (STOUFFER ET AL.), 362–412.


LAZARSFELD, PAUL F. 1950.   (15.5)                                                3317
   THE INTERPRETATION AND COMPUTATION OF SOME LATENT STRUCTURES.
   MEAS. PREDICT. (STOUFFER ET AL.), 413–472.


LAZARSFELD, PAUL F. 1954.   E(15.5)                                               3318
   A CONCEPTUAL INTRODUCTION TO LATENT STRUCTURE ANALYSIS.
   MATH. THINK. SOC. SCI. (LAZARSFELD), 349–387, 433–435.

LAZARSFELD, PAUL F. 1955.   (15.5)                                              **3319**
     RECENT DEVELOPMENTS IN LATENT STRUCTURE ANALYSIS.
          (REPRINTED AS NO. 3320)
     SOCIOMETRY, 18, 391-403.

LAZARSFELD, PAUL F. 1956.   (15.5)                                              **3320**
     RECENT DEVELOPMENTS IN LATENT STRUCTURE ANALYSIS.
          (REPRINT OF NO. 3319)
     SOCIOMETRY SCI. MAN (MORENO), 647-659.

LAZARSFELD, PAUL F. 1959.   (15.5)                                              **3321**
     LATENT STRUCTURE ANALYSIS.
     PSYCHOL. STUDY SCI. (KOCH), 3, 476-543.

LAZARSFELD, PAUL F. 1960.   (15.4/15.5/E)                                       **3322**
     LATENT STRUCTURE ANALYSIS AND TEST THEORY.
          (REPRINTED AS NO. 3324)
     PSYCHOL. SCALING THEORY APPL. (GULLIKSEN + MESSICK), 83-95.

LAZARSFELD, PAUL F. 1961.   (15.4/18.1/E)                                       **3323**
     THE ALGEBRA OF DICHOTOMOUS SYSTEMS.
     STUD. ITEM ANAL. PREDICT. (SOLOMON), 111-157.

LAZARSFELD, PAUL F. 1966.   (15.4/15.5/E).                                      **3324**
     LATENT STRUCTURE ANALYSIS AND TEST THEORY.
          (REPRINT OF NO. 3322)
     READINGS MATH. SOCIAL SCI. (LAZARSFELD + HENRY), 78-88.

LBOV, G. S. 1964.   (6.3/E)                                                     **3325**
     ERRORS IN THE CLASSIFICATION OF PATTERNS FOR UNEQUAL COVARIANCE MATRICES.
          (IN RUSSIAN)
     AKAD. NAUK SSSR  SIBIRSK. OTDEL. INST. MAT. VYC. SISTEMY  SB. TRUDOV, 14, 31-38. (MR31-5724)

LEBEDINCEVA, E. K. 1957.   (17.1)                                               **3326**
     ON LIMIT DISTRIBUTIONS FOR NORMALIZED SUMS OF INDEPENDENT RANDOM VECTORS.
          (IN UZBEKISH, TRANSLATION OF NO. 3327)
     DOKL. AKAD. NAUK UZBEK. SSR, 3, 219-223.   (MR19,P.1201)

LEBEDINCEVA, E. K. 1957.   (17.1)                                               **3327**
     ON THE LIMITING DISTRIBUTIONS FOR NORMED SUMS OF INDEPENDENT RANDOM VECTORS.
          (IN UKRAINIAN, WITH SUMMARY IN RUSSIAN, TRANSLATED AS NO. 3326)
     DOPOVIDI AKAD. NAUK UKRAIN. RSR, 219-221.   (MR19,P.1201)

LECAM, LUCIEN 1951.   (19.3/A)                                                  **3328**
     ON SOME ASYMPTOTIC PROPERTIES OF MAXIMUM LIKELIHOOD ESTIMATES AND
     RELATED BAYES' ESTIMATES.
     UNIV. CALIFORNIA PUBL. STATIST., 1, 277-329.   (MR14,P.998)

LECAM, LUCIEN 1956.   (19.2/19.3/A)                                             **3329**
     ON THE ASYMPTOTIC THEORY OF ESTIMATION AND TESTING HYPOTHESES.
     PROC. THIRD BERKELEY SYMP. MATH. STATIST. PROB., 1, 129-156.   (MR18,P.947)

LECAM, LUCIEN 1959.   (17.3)                                                    **3330**
     REMARQUES SUR LES VARIABLES ALEATOIRES DANS LES ESPACE VECTORIELS NON
     SEPARABLES.
     COLLOQ. INTERNAT. CENTRE NAT. RECH. SCI., 87, 39-53.   (MR21-4463)

LECAM, LUCIEN 1964.   (19.4)                                                    **3331**
     SUFFICIENCY AND APPROXIMATE SUFFICIENCY.
     ANN. MATH. STATIST., 35, 1419-1455.

LECAM, LUCIEN 1965.   (17.1)                                                    **3332**
     A REMARK ON THE CENTRAL LIMIT THEOREM.
     PROC. NAT. ACAD. SCI. U.S.A., 54, 354-359.

LECAM, LUCIEN 1966.   (17.3/19.2/19.3/A)                                        **3333**
     LIKELIHOOD FUNCTIONS FOR LARGE NUMBERS OF INDEPENDENT OBSERVATIONS.
     RES. PAPERS STATIST. FESTSCHR. NEYMAN (DAVID), 167-187.

LEDERMAN, WALTER  SEE  LEDERMANN, WALTER

LEDERMANN, WALTER 1936.   (15.1)                                              **3334**
   SOME MATHEMATICAL REMARKS CONCERNING BOUNDARY CONDITIONS IN THE
   FACTORIAL ANALYSIS OF ABILITY.
  PSYCHOMETRIKA, 1, 165-174.

LEDERMANN, WALTER 1937.   (20.3)                                              **3335**
   ON AN UPPER LIMIT FOR THE LATENT ROOTS OF A CERTAIN CLASS OF MATRICES.
  J. LONDON MATH. SOC., 12, 12-18.

LEDERMANN, WALTER 1937.   (15.1)                                              **3336**
   ON THE RANK OF THE REDUCED CORRELATIONAL MATRIX IN MULTIPLE-FACTOR ANALYSIS.
  PSYCHOMETRIKA, 2, 85-93.

LEDERMANN, WALTER 1938.   (2.4/15.1)                                          **3337**
   NOTE OF PROF. G.H. THOMSON'S ARTICLE "THE INFLUENCE OF UNIVARIATE
   SELECTION ON FACTORIAL ANALYSIS OF ABILITY".
  BRITISH J. PSYCHOL., 29, 69-73.

LEDERMANN, WALTER 1938.   (15.3)                                              **3338**
   THE ORTHOGONAL TRANSFORMATIONS OF A FACTORIAL MATRIX INTO ITSELF.
  PSYCHOMETRIKA, 3, 181-187.

LEDERMANN, WALTER 1939.   (2.4/7.1)                                           **3339**
   SAMPLING DISTRIBUTION AND SELECTION IN A NORMAL POPULATION.
  BIOMETRIKA, 30, 295-304.

LEDERMANN, WALTER 1939.   (2.4/15.1)                                          **3340**
   THE INFLUENCE OF MULTIVARIATE SELECTION ON THE FACTORIAL ANALYSIS OF
   ABILITY.   SECTIONS VI-VIII.
  BRITISH J. PSYCHOL., 29, 297-306.

LEDERMANN, WALTER 1940.   (15.1/20.2)                                         **3341**
   ON A PROBLEM CONCERNING MATRICES WITH VARIABLE DIAGONAL ELEMENTS.
  PROC. ROY. SOC. EDINBURGH, 60, 1-17.   (MR1,P.195)

LEE, ALICE 1917.   (4.8/T)                                                    **3342**
   FURTHER SUPPLEMENTARY TABLES FOR DETERMINING HIGH CORRELATIONS FROM
   TETRACHORIC GROUPINGS.
  BIOMETRIKA, 11, 284-291.

LEE, ALICE 1925.   (4.4/T)                                                    **3343**
   TABLE OF THE FIRST TWENTY TETRACHORIC FUNCTIONS TO SEVEN DECIMAL PLACES.
  BIOMETRIKA, 17, 343-354.

LEE, ALICE 1927.   (4.4/T)                                                    **3344**
   SUPPLEMENTARY TABLES FOR DETERMINING CORRELATION FROM TETRACHORIC GROUPINGS.
  BIOMETRIKA, 19, 354-404.

LEE, ALICE + PEARSON, KARL 1897.  E(17.9)                                     **3345**
   MATHEMATICAL CONTRIBUTIONS TO THE THEORY OF EVOLUTION.  ON THE RELATIVE
   VARIATION AND CORRELATION IN CIVILISED AND UNCIVILISED RACES.
  PROC. ROY. SOC. LONDON, 61, 343-357.

LEE, ALICE + PEARSON, KARL 1901.   (4.1/8.8/E)                                **3346**
   DATA FOR THE PROBLEM OF EVOLUTION IN MAN.   VI.  A FIRST STUDY OF THE
   CORRELATION OF THE HUMAN SKULL.
  PHILOS. TRANS. ROY. SOC. LONDON SER. A, 196, 225-264.

LEE, ALICE; LEWENZ, MARIE A. + PEARSON, KARL 1902.  E(4.8)                    **3347**
   ON THE CORRELATION OF THE MENTAL AND PHYSICAL CHARACTERS IN MAN.   PART II.
  PROC. ROY. SOC. LONDON, 71, 106-114.

LEE, ALICE   SEE ALSO
   4243. PEARSON, KARL + LEE, ALICE (1897)

   4244. PEARSON, KARL + LEE, ALICE (1908)

   4251. PEARSON, KARL; LEE, ALICE + ELDERTON, ETHEL M. (1910)

   5135. SOPER, H.E.; YOUNG, ANDREW W.; CAVE, BEATRICE M.;   (1917)
  LEE, ALICE + PEARSON, KARL

LEE, MARILYN C.   SEE
  1835. GAIFR, EUGENE L. + LEE, MARILYN C. (1953)

    6033. WRIGLEY, CHARLES F.; CHERRY, CHARLES N.; LEE, MARILYN C. +   (1957)
  MC QUITTY, LOUIS L.

LEE, Y. W. 1950.   (16.3/B)                                                          3348
    COMMUNICATION APPLICATIONS OF CORRELATION ANALYSIS.
  SYMP. APPL. AUTOCORREL. ANAL. PHYS. PROBLEMS (WOODS HOLE), 4-23.

LEE, Y. W. + WIESNER, JEROME B. 1950.   E(16.3)                                      3349
    CORRELATION FUNCTIONS AND COMMUNICATIONS APPLICATIONS.
  ELECTRONICS, 23, 86-92.

LEE, Y. W.; CHEATHAM, T. P., JR. + WIESNER, JEROME B. 1950.   (16.3/E)               3350
    APPLICATION OF CORRELATION ANALYSIS TO THE DETECTION OF PERIODIC
    SIGNALS IN NOISE.
  PROC. INST. RADIO ENGRS., 38, 1165-1171.   (MR12,P.191)

LEENDERS, C. T.   SEE
  4643. ROTHENBERG, T. J. + LEENDERS, C. T. (1964)

LEES, RUTH W. + LORD, FREDERIC M. 1961.   (4.5/T)                                    3351
    A NOMOGRAPH FOR COMPUTING PARTIAL CORRELATION COEFFICIENTS.
    (AMENDED BY NO. 3352)
  J. AMER. STATIST. ASSOC., 56, 995-997.   (MR23-B2611)

LEES, RUTH W. + LORD, FREDERIC M. 1962.   (4.5/T)                                    3352
    CORRIGENDA:  A NOMOGRAPH FOR COMPUTING PARTIAL CORRELATION COEFFICIENTS.
    (AMENDMENT OF NO. 3351)
  J. AMER. STATIST. ASSOC., 57, 917-918.   (MR23-B2611)

LEE-TSCH, C.   SEE  HSU, LEETZE CHING

LEGENDRE, ROBERT 1965.   (16.4/A)                                                    3353
    SPECTRE DU COURANT DANS UN TUBE A VIDE.
  C.R. ACAD. SCI. PARIS, 260, 1845-1847.

LEHMANN, E. L. 1947.   (19.2)                                                        3354
    ON FAMILIES OF ADMISSIBLE TESTS.
  ANN. MATH. STATIST., 18, 97-104.   (MR9,P.151)

LEHMANN, E. L. 1952.   (19.2)                                                        3355
    TESTING MULTIPARAMETER HYPOTHESES.
  ANN. MATH. STATIST., 23, 541-552.   (MR14,P.666)

LEHMANN, E. L. 1957.   (8.1/19.1)                                                    3356
    A THEORY OF SOME MULTIPLE DECISION PROBLEMS, I.
  ANN. MATH. STATIST., 28, 1-25.   (MR18,P.955)

LEHMANN, E. L. 1957.   (19.1)                                                        3357
    A THEORY OF SOME MULTIPLE DECISION PROBLEMS.  II.
  ANN. MATH. STATIST., 28, 547-572.   (MR20-2822)

LEHMANN, E. L. 1961.   (19.1)                                                        3358
    SOME MODEL I PROBLEMS OF SELECTION.
  ANN. MATH. STATIST., 32, 990-1012.   (MR26-4442)

LEHMANN, E. L. 1966.   (4.9/B)                                                       3359
    SOME CONCEPTS OF DEPENDENCE.
  ANN. MATH. STATIST., 37, 1137-1153.   (MR34-2101)

LEHMER, EMMA 1944.   (8.4/T)                                                         3360
    INVERSE TABLES OF PROBABILITIES OF ERRORS OF THE SECOND KIND.
  ANN. MATH. STATIST., 15, 388-398.   (MR6,P.161)

LEIBLER, RICHARD A.   SEE
  1731. FORSYTHE, GEORGE E. + LEIBLER, RICHARD A. (1950)

LEIGH-DUGMORE, C. H. 1953.   (4.3)                                                   3361
    A RAPID METHOD FOR ESTIMATING THE CORRELATION COEFFICIENT FROM THE
    RANGE OF THE DEVIATIONS ABOUT THE REDUCED MAJOR AXIS.
  BIOMETRIKA, 40, 218-219.

LEIMAN, JOHN M.   SEE
    4914. SCHMID, JOHN, JR. + LEIMAN, JOHN M. (1957)


LEIPNIK, ROY B. 1947.  (2.6)                                                    3362
    DISTRIBUTION OF THE SERIAL CORRELATION COEFFICIENT IN A CIRCULARLY
    CORRELATED UNIVERSE.
    ANN. MATH. STATIST., 18, 80-87.  (MR8,P.476)

LEIPNIK, ROY B. 1958.  (2.5/14.5)                                               3363
    MOMENT GENERATING FUNCTIONS OF QUADRATIC FORMS IN SERIALLY CORRELATED
    NORMAL VARIABLES.
    BIOMETRIKA, 45, 198-210.  (MR19,P.1095)

LEIPNIK, ROY B.   SEE ALSO
    3101. KOOPMANS, TJALLING C.; RUBIN, HERMAN + LEIPNIK, ROY B. (1950)

LENTZ, THEODORE F., JR.; HIRSCHSTEIN, BERTHA + FINCH, J. H. 1932.  (15.4/E)     3364
    EVALUATION OF METHODS OF EVALUATING TEST ITEMS.
    J. EDUC. PSYCHOL., 23, 344-350.

LENTZ, THEODORE F., JR.   SEE ALSO
    2317. HANDY, UVAN + LENTZ, THEODORE F., JR. (1934)

LERWICK, TRYGVE R. 1965.  (16.1)                                               3365
    MAXIMUM LIKELIHOOD ESTIMATORS OF REGRESSION COEFFICIENTS FOR THE CASE
    OF AUTOCORRELATED RESIDUALS.
    TECHNOMETRICS, 7, 51-58.

LESER, C. E. V. 1942.  (17.2)                                                  3366
    INEQUALITIES FOR MULTIVARIATE FREQUENCY DISTRIBUTIONS.
    BIOMETRIKA, 32, 284-293.  (MR4,P.16)

LESLIE, P. H. 1948.  (20.1)                                                    3367
    SOME FURTHER NOTES ON THE USE OF MATRICES IN POPULATION MATHEMATICS.
    BIOMETRIKA, 35, 213-245.  (MR10,P.386)

LETESTU, SERGE 1948.  (8.5)                                                    3368
    NOTE SUR L'ANALYSE DISCRIMINATOIRE.
    EXPERIENTIA, 4, 22-23.

LETI, GIUSEPPE 1965.  (17.9/E)                                                 3369
    SULL'ENTROPOIA, SU UN INDICE DEL GINI E SU ALTRE MISURE DELL'ETEROGENEITA DI UN
    COLLETTIVO.
    METRON, 24, 332-378.

LETI, GIUSEPPE 1966.  (4.9/6.5/E)                                              3370
    ON THE INDEPENDENCE TABLE FOR TWO DISTRIBUTIONS AND THE MEAN DIFFERENCE
    BETWEEN THEM.
    METRON, 25, 150-171.

LEV, JOSEPH 1936.  (15.3)                                                      3371
    A NOTE ON FACTOR ANALYSIS BY THE METHOD OF PRINCIPAL AXES.
    PSYCHOMETRIKA, 1, 283-286.

LEV, JOSEPH 1949.  (4.4)                                                       3372
    THE POINT BISERIAL COEFFICIENT OF CORRELATION.
    ANN. MATH. STATIST., 20, 125-126.  (MR10,P.465)

LEV, JOSEPH 1956.  (8.1/15.4)                                                  3373
    MAXIMIZING TEST BATTERY PREDICTION WHEN THE WEIGHTS ARE REQUIRED TO
    BE NON-NEGATIVE.
    PSYCHOMETRIKA, 21, 245-252.  (MR18,P.452)

LEV, JOSEPH 1965.  (15.1)                                                      3374
    A GENERALIZED FORMULATION OF FACTOR THEORY.
    NEW YORK STATIST., 17(4) 1-2.

LEVENE, HOWARD + WOLFOWITZ, J. 1944.  (17.8)                                   3375
    THE COVARIANCE MATRIX OF RUNS UP AND DOWN.
    ANN. MATH. STATIST., 15, 58-69.  (MR5,P.208)

LEVERT, C. 1964.    (16.1/E)                                                        3376
    DEUX MODELES STATISTIQUES DE PERSISTANCE DES SERIES CHRONOLOGIQUES CLIMATOLOGIQUES.
    BIOM. PRAX., 5, 63-79.


LEVI, BEPPO 1950.    (17.1)                                                         3377
    ON A LIMITING PROPERTY OF THE SPHERE IN N DIMENSIONS.
    MATH. NOTAE, 10, 36-40.    (MR12,P.722)


LEVIN, JOSEPH 1965.    (15.1)                                                       3378
    THREE-MODE FACTOR ANALYSIS.
    PSYCHOL. BULL., 64, 442-452.


LEVIN, JOSEPH 1966.    (15.1/15.2)                                                  3379
    SIMULTANEOUS FACTOR ANALYSIS OF SEVERAL GRAMIAN MATRICES.
    PSYCHOMETRIKA, 31, 413-419.    (MR33-6769)


LEVITT, EUGENE L.    SEE
    4893. SCHAEFFER, MAURICE S. + LEVITT, EUGENE L. (1956)


LEVY, PAUL 1935.    (17.1)                                                          3380
    LA LOI FORTE DES GRANDS NOMBRES POUR LES VARIABLES ALEATOIRES ENCHAINEES.
        (REPRINTED AS NO. 3383)
    C.R. ACAD. SCI. PARIS, 201, 493-495, 800.


LEVY, PAUL 1935.    (17.1)                                                          3381
    PROPRIETES ASYMPTOTIQUES DES SOMMES DE VARIABLES ALEATOIRES
    INDEPENDANTES OU ENCHAINEES.
    J. MATH. PURES APPL. SER. 9, 14, 347-402.


LEVY, PAUL 1936.    (17.3)                                                          3382
    INTEGRALES A ELEMENTS ALEATOIRES INDEPENDANTS ET LOIS STABLES A N VARIABLES.
    C.R. ACAD. SCI. PARIS, 202, 543-545.


LEVY, PAUL 1936.    (17.1)                                                          3383
    LA LOI FORTE DES GRANDS NOMBRES POUR LES VARIABLES ALEATOIRES ENCHAINEES.
        (REPRINT OF NO. 3380)
    J. MATH. PURES APPL. SER. 9, 15, 11-24.


LEVY, PAUL 1938.    (4.1/17.3)                                                      3384
    SUR LA DEFINITION DES LOIS DE PROBABILITES PAR LEURS PROJECTIONS.
    C.R. ACAD. SCI. PARIS, 206, 1240-1242.


LEVY, PAUL 1939.    (17.3)                                                          3385
    SUR LES PROJECTIONS D'UNE LOI DE PROBABILITE A N VARIABLES.
    BULL. SCI. MATH. SER. 2, 63, 148-160.


LEVY, PAUL 1939.    (18.6)                                                          3386
    L'ADDITION DES VARIABLES ALEATOIRES DEFINIES SUR UNE CIRCONFERENCE.
    BULL. SOC. MATH. FRANCE, 67, 1-41.    (MR1,P.62)


LEVY, PAUL 1944.    (16.4)                                                          3387
    UNE PROPRIETE D'INVARIANCE PROJECTIVE DANS LE MOUVEMENT BROWNIEN.
    C.R. ACAD. SCI. PARIS, 219, 377-379.    (MR7,P.314)


LEVY, PAUL 1944.    (16.4)                                                          3388
    UN THEOREME D'INVARIANCE PROJECTIVE RELATIF AU MOUVEMENT BROWNIEN.
    COMMENT. MATH. HELVETIA, 16, 242-248.    (MR6,P.5)


LEVY, PAUL 1945.    (16.4)                                                          3389
    SUR LE MOUVEMENT BROWNIEN DEPENDANT DE PLUSIEURS PARAMETRES.
    C.R. ACAD. SCI. PARIS, 220, 420-422.    (MR7,P.130)


LEVY, PAUL 1948.    (16.4)                                                          3390
    CHAINES DOUBLES DE MARKOFF ET FONCTIONS ALEATOIRES DE DEUX VARIABLES.
    C.R. ACAD. SCI. PARIS, 226, 53-55.    (MR9,P.361)


LEVY, PAUL 1948.    (16.4)                                                          3391
    EXEMPLES DE PROCESSUS DOUBLES DE MARKOFF.
    C.R. ACAD. SCI. PARIS, 226, 307-308.    (MR9,P.361)


LEVY, PAUL 1948.    (7.1)                                                           3392
    THE ARITHMETICAL CHARACTER OF THE WISHART DISTRIBUTION.
    PROC. CAMBRIDGE PHILOS. SOC., 44, 295-297.    (MR10,P.131)

LEVY, PAUL 1949.   (16.4)                                                                    3393
    PROCESSUS DOUBLES DE MARKOFF.
  COLLOQ. INTERNAT. CENTRE NAT. RECH. SCI., 13, 53-59.   (MR12,P.114)


LEVY, PAUL 1950.   (16.4)                                                                    3394
    WIENER'S RANDOM FUNCTION, AND OTHER LAPLACIAN RANDOM FUNCTIONS.
  PROC. SECOND BERKELEY SYMP. MATH. STATIST. PROB., 171-187.   (MR13,P.476)


LEVY, PAUL 1951.   (16.4)                                                                    3395
    LA MESURE DE HAUSDORFF DE LA COURBE DU MOUVEMENT BROWNIEN A N DIMENSIONS.
  C.R. ACAD. SCI. PARIS, 233, 600-602.   (MR13,P.363)


LEVY, PAUL 1953.   (17.3)                                                                    3396
    RANDOM FUNCTIONS: GENERAL THEORY WITH SPECIAL REFERENCE TO LAPLACIAN
    RANDOM FUNCTIONS.
  UNIV. CALIFORNIA PUBL. STATIST., 1, 331-390.   (MR14,P.1099)


LEVY, PAUL 1954.   (16.4)                                                                    3397
    RECTIFICATION A UN THEOREME SUR LE MOUVEMENT BROWNIEN A P PARAMETRES.
  C.R. ACAD. SCI. PARIS, 238, 2140-2141.   (MR16,P.150)


LEVY, PAUL 1955.   (16.4/A)                                                                  3398
    PROPRIETES ASYMPTOTIQUES DE LA COURBE DU MOUVEMENT BROWNIEN A N DIMENSIONS.
  C.R. ACAD. SCI. PARIS, 241, 689-690.   (MR17,P.275)


LEVY, PAUL 1956.   (16.4)                                                                    3399
    FONCTIONS ALEATOIRES A CORRELATION LINEAIRE.
      (REPRINTED AS NO. 3401)
  C.R. ACAD. SCI. PARIS, 242, 2095-2097.   (MR17,P.978)


LEVY, PAUL 1957.   (17.1)                                                                    3400
    SUR QUELQUES PROBLEMES DE LA THEORIE DES LIAISONS STOCHASTIQUES.
  C.R. ACAD. SCI. PARIS, 244, 1313-1316.   (MR19,P.185)


LEVY, PAUL 1957.   (16.4)                                                                    3401
    FONCTIONS ALEATOIRES A CORRELATION LINEAIRE.
      (REPRINT OF NO. 3399)
  ILLINOIS J. MATH., 1, 217-258.   (MR20-7339)


LEVY, PAUL 1957.   (16.4/17.3)                                                               3402
    FONCTIONS LINEAIREMENT MARKOVIENNES D'ORDRE N.
  MATH. JAPON., 4, 113-121.   (MR20-7340)


LEVY, PAUL 1960.   (17.4)                                                                    3403
    SUR LES CONDITIONS DE COMPATIBILITE DES DONNEES MARGINALES RELATIVES AUX LOIS DE
    PROBABILITE.
  C.R. ACAD. SCI. PARIS, 250, 2507-2509.   (MR22-9985)


LEVY, PAUL 1966.   (16.4)                                                                    3404
    FONCTIONS BROWNIENNES DANS L'ESPACE EUCLIDIEN ET DANS L'ESPACE DE HILBERT.
  RES. PAPERS STATIST. FESTSCHR. NEYMAN (DAVID), 189-223.


LEWENZ, MARIE A.   SEE
  3347. LEE, ALICE; LEWENZ, MARIE A. + PEARSON, KARL (1902)


LEWIS, PHILIP M., II 1962.   (6.7)                                                           3405
    THE CHARACTERISTIC SELECTION PROBLEM IN RECOGNITION SYSTEMS.
  IRE TRANS. INFORMATION THEORY, IT-8, 171-178.


LEWIS, T. + ODELL, P. L. 1966.   (8.7)                                                       3406
    A GENERALIZATION OF THE GAUSS-MARKOV THEOREM.
  J. AMER. STATIST. ASSOC., 61, 1063-1066.


LEWIS, T.   SEE ALSO
  1663. FIELLER, E. C.; LEWIS, T. + PEARSON, EGON SHARPE (1955)

  1664. FIELLER, E. C.; LEWIS, T. + PEARSON, EGON SHARPE (1956)


LI, C. C.   SEE
  1263. DEGROOT, M. H. + LI, C. C. (1966)

LI, CHEN-NAN 1934.   (4.1/4.8)                                                    3407
     SUMMATION METHOD OF FITTING PARABOLIC CURVES AND CALCULATING LINEAR
     AND CURVILINEAR CORRELATION COEFFICIENTS ON A SCATTER-DIAGRAM.
     J. AMER. STATIST. ASSOC., 29, 405-409.

LIAN, CHENG SHAW   SEE  CHENG, SHAO-LIEN

LIDDELL, F.D.K. 1957.   (4.5/T)                                                   3408
     DIAGRAMS FOR RAPID ESTIMATION OF PARTIAL CORRELATION COEFFICIENTS.
     INC. STATIST., 7, 167-169.

LIENAU, C. C. 1941.   (4.8/18.1/E)                                                3409
     DISCRETE BIVARIATE DISTRIBUTION IN CERTAIN PROBLEMS OF STATISTICAL ORDER.
     AMER. J. EPIDEM., 33, 65-85.

LIENERT, G.A. 1962.   (15.5)                                                      3410
     UBER DIE ANWENDUNG VON VARIABLEN-TRANSFORMATIONEN IN DER PYSCHOLOGIE.
     BIOM. Z., 4, 145-181.

LINCOLN, EDWARD A. 1930.   (4.1)                                                  3411
     THE INTERPRETATION OF CORRELATION COEFFICIENTS IN TERMS OF DEPARTURE
     FROM PERFECT CORRELATION.
     J. EDUC. PSYCHOL., 21, 284-285.

LINDBLAD, TORD 1937.   (1.1)                                                      3412
     ZUR THEORIE DER KORRELATION BEI MEHRDIMENSIONALEN ZUFALLIGEN VARIABLEN.
     ACTA  SOC. SCI. FENN. NOVA SER. A. 2(10) 1-81.

LINDBLOM, SVEN G. 1946.   (4.3/5.1)                                               3413
     ON THE CONNECTION BETWEEN TESTS OF SIGNIFICANCE FOR CORRELATION
     COEFFICIENTS AND FOR DIFFERENCES BETWEEN MEANS.
     SKAND. AKTUARIETIDSKR., 29, 12-29.   (MR8,P.42)

LINDEBERG, J. W. 1925.   (4.1/E)                                                  3414
     UBER DIE KORRELATION.
     SKAND. MAT. KONG. BER., 6, 437-446.

LINDEBERG, J. W. 1929.   (4.8)                                                    3415
     SOME REMARKS ON THE MEAN ERROR OF THE PERCENTAGE OF CORRELATION.
          (REPRINT OF NO. 3416)
     NORDIC STATIST. J., 1, 137-141.

LINDEBERG, J. W. 1929.   (4.8)                                                    3416
     SOME REMARKS ON THE MEAN ERROR OF THE PERCENTAGE OF CORRELATION.
          (REPRINTED AS NO. 3415)
     NORDISK STATIST. TIDSKR., 8, 138-142.

LINDER, ARTHUR 1963.   (6.4)                                                      3417
     TRENNVERFAHREN BEI QUALITATIVEN MERKMALEN.
     METRIKA, 6, 76-83.   (MR28-4629)

LINDERS, F. J. 1932.   (2.3)                                                      3418
     A SIMPLE DEMONSTRATION OF A GENERAL FORMULA FOR THE MOMENTS ABOUT THE
     MEAN OF THE NORMAL CORRELATION OF 2 VARIABLES.
     NORDIC STATIST. J., 4, 73-74.

LINDLEY, D. V. 1947.   (8.1/17.7)                                                 3419
     REGRESSION LINES AND THE LINEAR FUNCTIONAL RELATIONSHIP.
     J. ROY. STATIST. SOC. SUPP., 9, 218-244.   (MR9,P.363)

LINDLEY, D. V. 1951.   (8.1/8.4)                                                  3420
     A REGRESSION PROBLEM.
     PROC. CAMBRIDGE PHILOS. SOC., 47, 337-346.   (MR12,P.842)

LINDLEY, D. V. 1953.   (17.7)                                                     3421
     ESTIMATION OF A FUNCTIONAL RELATIONSHIP.
     BIOMETRIKA, 40, 47-49.   (MR14,P.1104)

LINDLEY, D. V.   SEE ALSO
     2649. HUGHES, R. E. + LINDLEY, D. V. (1955)

LINE, W. 1933. (15.1)                                                                    3422
    FACTORIAL ANALYSIS AND ITS RELATIONSHIP TO PSYCHOLOGICAL METHOD.
    BRITISH J. PSYCHOL., 24, 187-198.

LINE, W. + HEDMAN, HATTIE B. 1933. (15.1)                                                3423
    A SIMPLIFIED STATEMENT OF THE TWO FACTOR THEORY.
    J. EDUC. PSYCHOL., 24, 195-220.

LINEHAN, THOMAS P.   SEE
    1885. GEARY, ROBERT C. + LINEHAN, THOMAS P. (1960)

LINFOOT, E. H. 1957. (4.8)                                                               3424
    AN INFORMATIONAL MEASURE OF ASSOCIATION.
    INFORMATION AND CONTROL, 1, 85-89.  (MR19,P.1148)

LINGOES, JAMES C. 1963. (15.5/E)                                                         3425
    MULTIPLE SCALOGRAM ANALYSIS.
    EDUC. PSYCHOL. MEAS., 23, 501-524.

LINGOES, JAMES C. 1966. (15.5/20.3)                                                      3426
    AN IBM-7090 PROGRAM FOR GUTTMAN-LINGOES MULTIDIMENSIONAL SCALOGRAM
    ANALYSIS--I.
    BEHAV. SCI., 11, 76-78.

LINHART, H. 1959. (6.4)                                                                  3427
    TECHNIQUES FOR DISCRIMINANT ANALYSIS WITH DISCRETE VARIABLES.
    METRIKA, 2, 138-149.  (MR21-6067)

LINHART, H. 1960. (17.7)                                                                 3428
    A MEASURE OF PREDICTIVE PRECISION IN REGRESSION ANALYSIS.
    ANN. MATH. STATIST., 31, 399-404.  (MR22-8605)

LINHART, H. 1960. (8.9/19.1/20.2/CU)                                                     3429
    A CRITERION FOR SELECTING VARIABLES IN A REGRESSION ANALYSIS.
    PSYCHOMETRIKA, 25, 45-58.  (MR22-3069)

LINHART, H. 1961. (6.2/T)                                                                3430
    ZUR WAHL VON VARIABLEN IN DER TRENNANALYSE.
    METRIKA, 4, 126-139.

LINHART, H. 1962. (14.3/BT)                                                              3431
    KONFIDENZINTERVALLE FUR DEN UBERLAPPENDEN TEIL ZWEIER NORMALVERTEILUNGEN.
    METRIKA, 5, 31-48.  (MR26-839)

LINIGER, WERNER   SEE
    586. BLANC, CHARLES + LINIGER, WERNER (1956)

LINK,P.F.   SEE
    3214. KURTZ, T. E.; LINK,R.F.; TUKEY, JOHN W. +   (1966)
    WALLACE, D. L.

LINN, HANS-JOCHEN + POSCHL, KLAUS 1956.  (4.8)                                           3432
    EINE EINFACHE DEUTUNG DES KOMPLEXEN KORRELATIONSKOEFFIZIENTEN.
    ARCH. ELEK. UBERTR., 10, 105-106.  (MR17,P.984)

LINNIK, YU. V. 1957. (8.1/CE)                                                            3433
    SOME REMARKS ON LEAST SQUARES IN CONNECTION WITH DIRECT AND INVERSE
    LOCATION PROBLEMS.
        (IN RUSSIAN, WITH SUMMARY IN ENGLISH. TRANSLATED AS NO. 3434)
    TEOR. VEROJATNOST. I PRIMENEN., 2, 349-359.  (MR19,P.1096)

LINNIK, YU. V. 1957. (8.1/CE)                                                            3434
    SOME REMARKS ON THE METHOD OF LEAST SQUARES, WITH APPLICATION TO
    DIRECT AND INVERSE LOCATION PROBLEMS.
        (TRANSLATION OF NO. 3433)
    THEORY PROB. APPL., 2, 332-344.  (MR19,P.1096)

LINNIK, YU. V. 1959. (17.3)                                                              3435
    POLYNOMIAL STATISTICS AND POLYNOMIAL IDEALS.
    CALCUTTA MATH. SOC. GOLDEN JUBILEE COMMEM. VOLUME, 1, 95-98.

LINNIK, YU. V. 1966.  (5.1)                                                    3436
      APPROXIMATELY MINIMAX DETECTION OF VECTOR SIGNAL ON A GAUSSIAN BACKGROUND.
            (IN RUSSIAN)
      DOKL. AKAD. NAUK SSSR, 169, 523–524.  (MR34–2117)


LINNIK, YU. V. 1966.  (5.2/U)                                                  3437
      LATEST INVESTIGATIONS ON BEHRENS–FISHER PROBLEM.
      SANKHYA SER. A, 28, 15–24.


LINNIK, YU. V. 1966.  (5.3/6.6)                                                3438
      APPROXIMATELY MINIMAX DETECTING OF A VECTOR SIGNAL IN GAUSSIAN NOISE.
            (IN RUSSIAN, WITH SUMMARY IN ENGLISH, TRANSLATED AS NO. 3439)
      TEOR. VEROJATNOST. I PRIMENEN., 11, 561–578.  (MR34–8582)


LINNIK, YU. V. 1966.  (5.3/6.6)                                                3439
      APPROXIMATELY MINIMAX DETECTION OF A VECTOR SIGNAL IN GAUSSIAN NOISE.
            (TRANSLATION OF NO. 3438)
      THEORY PROB. APPL., 11, 497–512.  (MR34–8582)


LINNIK, YU. V. + MITROFANOVA, N. M. 1964.  (19.3/A)                            3440
      SOME ASYMPTOTIC EXPANSIONS FOR THE DISTRIBUTION OF THE MAXIMUM
      LIKELIHOOD ESTIMATE.
            (REPRINTED AS NO. 3441)
      CONTRIB. STATIST. (MAHALANOBIS VOLUME), 229–238.


LINNIK, YU. V. + MITROFANOVA, N. M. 1965.  (19.3/A)                            3441
      SOME ASYMPTOTIC EXPANSIONS FOR THE DISTRIBUTION OF THE MAXIMUM
      LIKELIHOOD ESTIMATE.
            (REPRINT OF NO. 3440)
      SANKHYA SER. A, 27, 73–82.


LINNIK, YU. V. + PLISS, V. A. 1966.  (6.2)                                     3442
      ON THE THEORY OF THE HOTELLING TEST.
            (IN RUSSIAN)
      DOKL. AKAD. NAUK SSSR, 168, 743–746.  (MR33–3403)


LINNIK, YU. V. + SAPOGOV, N.A. 1949.  (16.4/17.1)                              3443
      ON AN INTEGRAL AND A LOCAL LAW FOR A MULTI-DIMENSIONAL
      NON-HOMOGENEOUS MARKOV CHAIN.
      DOKL. AKAD. NAUK UZBEK. SSR, 6, 7–10.


LINNIK, YU. V. + SAPOGOV, N.A. 1949.  (16.4/17.1)                              3444
      MULTI-DIMENSIONAL INTEGRAL LAWS AND A MULTI-DIMENSIONAL LOCAL LAW FOR
      NON-HOMOGENEOUS MARKOV CHAINS.
            (IN RUSSIAN)
      IZV. AKAD. NAUK SSSR SER. MAT., 13, 533–566.  (MR11,P.606)


LINNIK, YU. V.   SEE ALSO
      2892. KAGAN, A. M.; LINNIK, YU. V. + RAO, C. RADHAKRISHNA (1965)


LINS MARTINS, OCTAVIO A.   SEE   MARTINS, OCTAVIO A. LINS


LIPOW, M. + EIDEMILLER, R. L. 1964.  E(2.4/2.7)                                3445
      APPLICATION  OF THE BIVARIATE NORMAL DISTRIBUTION TO A STRESS VS
      STRENGTH PROBLEM IN RELIABILITY ANALYSIS.
      TECHNOMETRICS, 6, 325–328.


LIPTON, S.   SEE
      216. ASHTON, E. H.; HEALY, M.J.R. + LIPTON, S. (1957)


LITTLEJOHN, F. S. 1921.  (18.1)                                                3446
      ON AN ELEMENTARY METHOD OF FINDING THE MOMENTS OF THE TERMS OF A
      MULTIPLE HYPERGEOMETRIC SERIES.
      METRON, 1(4) 49–56.


LITTLEWOOD, J. E. + OFFORD, A. C. 1938.  (14.4/17.9)                           3447
      ON THE NUMBER OF REAL ROOTS OF A RANDOM ALGEBRAIC EQUATION.
      J. LONDON MATH. SOC., 13, 288–295.


LITTLEWOOD, J. E. + OFFORD, A. C. 1939.  (17.9)                                3448
      ON THE NUMBER OF REAL ROOTS OF A RANDOM ALGEBRAIC EQUATION.  II.
      PROC. CAMBRIDGE PHILOS. SOC., 35, 133–148.  (MR5,P.179)

LITTLEWOOD, J. E. + OFFORD, A. C. 1943.  (17.9)                                    3449
    ON THE NUMBER OF REAL ROOTS OF A RANDOM ALGEBRAIC EQUATION.  III.
    MAT. SB. NOV. SER., 12(54) 277-286.  (MR5,P.179)

LIU, C. N.   SEE
    2921. KAMENSKY, L. A. + LIU, C. N. (1964)

LIU, TA-CHUNG 1960.  (17.7)                                                        3450
    UNDER-IDENTIFICATION, STRUCTURAL ESTIMATIONS, AND FORECASTING.
    ECONOMETRICA, 28, 855-865.

LIVADA, GREGORIO 1945.  (4.8/NT)                                                   3451
    PROCEDIMENTO PER IL CALCOLO DELLA INTENSITA DI TRANSVARIAZIONE.
    ATTI RIUN. SCI. SOC. ITAL. STATIST., 6, 63-73.

LIVADA, GREGORIO   SEE ALSO
    1988. GINI, CORRADO + LIVADA, GREGORIO (1945)

LIVINGSTONE, D.  SEE
    4495. RAYNER, A. A. + LIVINGSTONE, D. (1965)

LLOYD, A.   SEE
    1242. DAY, BESSE B. + LLOYD, A. (1939)

LLOYD, E.H. 1952.  (3.1)                                                           3452
    ON THE ESTIMATION OF VARIANCE AND COVARIANCE.
    PROC. ROY. SOC. EDINBURGH SECT. A, 63, 280-289.  (MR14,P.64)

LLOYD, E.H. 1962.  (8.8)                                                           3453
    GENERALIZED LEAST-SQUARES THEOREM.
    CONTRIB. ORDER STATIST. (SARHAN + GREENBERG), 20-27.

LLOYD, S. P. 1962.  (4.8)                                                          3454
    ON A MEASURE OF STOCHASTIC DEPENDENCE.
        (IN RUSSIAN, WITH SUMMARY IN ENGLISH, REPRINTED AS NO. 3455)
    TEOR. VEROJATNOST. I PRIMENEN., 7, 312-322.

LLOYD, S. P. 1962.  (4.8)                                                          3455
    ON A MEASURE OF STOCHASTIC DEPENDENCE.
        (REPRINT OF NO. 3454)
    THEORY PROB. APPL., 7, 301-312.

LOCKHEAD, G.R. 1966.  (15.5/E)                                                     3456
    EFFECTS OF DIMENSIONAL REDUNDANCY ON VISUAL DISCRIMINATION.
    J. EXPER. PSYCHOL., 72, 95-104.

LOEHLIN, JOHN C. 1965.  (15.1)                                                     3457
    SOME METHODOLOGICAL PROBLEMS IN CATTELL'S MULTIPLE ABSTRACT VARIANCE ANALYSIS.
    PSYCHOL. REV., 72, 156-161.

LOEVE, MICHEL 1945.  (17.1/17.3)                                                   3458
    SUR LA COVARIANCE D'UNE FONCTION ALEATOIRE.
    C.R. ACAD. SCI. PARIS, 220, 295-296.  (MR7,P.129)

LOEVE, MICHEL 1946.  (16.4)                                                        3459
    SUR LES FONCTIONS ALEATOIRES VECTORIELLES DE SECOND ORDRE.
    C.R. ACAD. SCI. PARIS, 222, 942-944.  (MR7,P.458)

LOEVINGER, JANE 1948.  (15.1/15.5)                                                 3460
    THE TECHNIC OF HOMOGENEOUS TESTS COMPARED WITH SOME ASPECTS OF "SCALE
    ANALYSIS" AND FACTOR ANALYSIS.
    PSYCHOL. BULL., 45, 507-529.

LOEVINGER, JANE; GLESER, GOLDINE C. + DU BOIS, PHILIP H. 1953.  (15.4)             3461
    MAXIMIZING THE DISCRIMINATING POWER OF A MULTIPLE SCORE TEST.
    PSYCHOMETRIKA, 18, 309-317.

LOFTSGAARDEN, D. O. + QUESENBERRY, C. P. 1965.  (17.8/17.9)                        3462
    A NONPARAMETRIC ESTIMATE OF A MULTIVARIATE DENSITY FUNCTION.
    ANN. MATH. STATIST., 36, 1049-1051.  (MR31-839)

LOHNES, PAUL R. 1961.   (6.1/15.4)                                3463
    TEST SPACE AND DISCRIMINANT SPACE CLASSIFICATION MODELS AND RELATED
    SIGNIFICANCE TESTS.
    EDUC. PSYCHOL. MEAS., 21, 559-574.

LOMBARD, HERBERT L. + DOERING, CARL R. 1947.   (4.8/E)           3464
    TREATMENT OF THE FOURFOLD TABLE BY PARTIAL CORRELATION AS IT RELATES
    TO PUBLIC HEALTH PROBLEMS.
    BIOMETRICS, 3, 123-128.

LOMNICKI, Z. ANTONIO 1930.   (4.8)                                3465
    SULLA NECESSITA DI DISTINGUERE DUE GENERI DI DIPENDENZA NELLA
    STATISTICA A DUE VARIABILI.
    GIORN. IST. ITAL. ATTUARI, 1, 83-94.

LONDON, DAVID 1966.   (20.1)                                      3466
    INEQUALITIES IN QUADRATIC FORMS.
    DUKE MATH. J., 33, 511-522.

LONG, JOHN A. 1934.   (15.4)                                      3467
    IMPROVED OVERLAPPING METHODS FOR DETERMINING THE VALIDITIES OF TEST  ITEMS.
    J. EXPER. EDUC., 2, 264-268.

LONG, W. F. + BURR, IRVING W. 1949.   (4.6/15.4)                 3468
    DEVELOPMENT OF A METHOD FOR INCREASING THE UTILITY OF MULTIPLE
    CORRELATIONS BY CONSIDERING BOTH TESTING TIME AND TEST VALIDITY.
    PSYCHOMETRIKA, 14, 137-161.

LONGUET-HIGGINS, M. S. 1958.   (16.4/U)                           3469
    THE STATISTICAL DISTRIBUTION OF THE CURVATURE OF A RANDOM GAUSSIAN SURFACE.
    PROC. CAMBRIDGE PHILOS. SOC., 54, 439-453.   (MR20-2048)

LONGUET-HIGGINS, M. S. 1962.   (18.6)                             3470
    THE STATISTICAL GEOMETRY OF RANDOM SURFACES.
    PROC. SIXTH ALL-UNION CONF. THEORY PROB. MATH. STATIST. (VILNIUS), 13, 105-143.   (MR25-3597)

LONSETH, ARVID T. 1942.   (17.9/20.3)                             3471
    SYSTEMS OF LINEAR EQUATIONS WITH COEFFICIENTS SUBJECT TO ERROR.
    ANN. MATH. STATIST., 13, 332-337.   (MR4,P.90)

LONSETH, ARVID T. 1944.   (20.3)                                  3472
    ON RELATIVE ERRORS IN SYSTEMS OF LINEAR EQUATIONS.
    ANN. MATH. STATIST., 15, 323-325.   (MR6,P.51)

LONSETH, ARVID T. 1949.   (20.1)                                  3473
    AN EXTENSION OF AN ALGORITHM OF HOTELLING.
    PROC. BERKELEY SYMP. MATH. STATIST. PROB., 353-358.   (MR10,P.627)

LOPES QUEIROZ, A. + SERRA, J. A. 1944.   (4.1/4.8/E)            3474
    CORRELACOES ENTRE A ESTATURA E ALGUNS CARACTERES OSTEOMETRICOS.
        (WITH SUMMARY IN FRENCH)
    QUESTOES MET. INST. ANTROP. COIMBRA, 4, 317-357.

LORD, FREDERIC M. 1944.   (15.4)                                  3475
    RELIABILITY OF MULTIPLE-CHOICE TESTS AS A FUNCTION OF NUMBER OF
    CHOICES PER ITEM.
    J. EDUC. PSYCHOL., 35, 175-180.

LORD, FREDERIC M. 1944.   (4.8/T)                                 3476
    ASIGNMENT CHART FOR CALCULATING THE FOURFOLD POINT CORRELATION COEFFICIENT.
    PSYCHOMETRIKA, 9, 41-42.   (MR5,P.208)

LORD, FREDERIC M. 1950.   (8.9)                                   3477
    EFFICIENCY OF PREDICTION WHEN A REGRESSION EQUATION FROM ONE SAMPLE
    IS USED IN A NEW SAMPLE.
    RES. BULL. EDUC. TEST. SERVICE, RB-50-40, 1-6.

LORD, FREDERIC M. 1952.   (15.4)                                  3478
    A THEORY OF TEST SCORES.
    PSYCHOMET. MONOG., 7, 1-84.

LORD, FREDERIC M. 1952.   (15.4)                                                                    3479
     THE RELATION OF THE RELIABILITY OF MULTIPLE CHOICE TESTS TO THE
     DISTRIBUTION OF ITEM DIFFICULTIES.
     PSYCHOMETRIKA, 17, 181-194.

LORD, FREDERIC M. 1952.   (6.7)                                                                     3480
     NOTES ON A PROBLEM OF MULTIPLE CLASSIFICATION.
     PSYCHOMETRIKA, 17, 297-304.

LORD, FREDERIC M. 1953.   (15.4/T)                                                                  3481
     THE RELATION OF TEST SCORE TO THE TRAIT UNDERLYING THE TEST.
          (REPRINTED AS NO. 3504)
     EDUC. PSYCHOL. MEAS., 13, 517-548.

LORD, FREDERIC M. 1953.   (15.4/C)                                                                  3482
     AN APPLICATION OF CONFIDENCE INTERVALS AND OF MAXIMUM LIKELIHOOD TO
     THE ESTIMATION OF AN EXAMINEE'S ABILITY.
     PSYCHOMETRIKA, 18, 57-76.

LORD, FREDERIC M. 1955.   (3.2/E)                                                                   3483
     ESTIMATION OF PARAMETERS FROM INCOMPLETE DATA.
     J. AMER. STATIST. ASSOC., 50, 870-876.   (MR17,P.169)

LORD, FREDERIC M. 1955.   (4.6/T)                                                                   3484
     NOMOGRAPH FOR COMPUTING MULTIPLE CORRELATION COEFFICIENTS.
     J. AMER. STATIST. ASSOC., 50, 1073-1077.

LORD, FREDERIC M. 1955.   (15.4)                                                                    3485
     SAMPLING FLUCTUATIONS RESULTING FROM THE SAMPLING OF TEST ITEMS.
     PSYCHOMETRIKA, 20, 1-22.   (MR16,P.841)

LORD, FREDERIC M. 1955.   (15.4)                                                                    3486
     EQUATING TEST SCORES--A MAXIMUM LIKELIHOOD SOLUTION.
     PSYCHOMETRIKA, 20, 193-200.

LORD, FREDERIC M. 1956.   (15.4)                                                                    3487
     THE MEASUREMENT OF GROWTH.
     EDUC. PSYCHOL. MEAS., 16, 421-437.

LORD, FREDERIC M. 1957.   (10.2/15.4)                                                               3488
     A SIGNIFICANCE TEST FOR THE HYPOTHESIS THAT TWO VARIABLES MEASURE THE
     SAME TRAIT EXCEPT FOR ERRORS OF MEASUREMENT.
     PSYCHOMETRIKA, 22, 207-220.   (MR19,P.1095)

LORD, FREDERIC M. 1958.   (15.4/E)                                                                  3489
     FURTHER PROBLEMS IN THE MEASUREMENT OF GROWTH.
     EDUC. PSYCHOL. MEAS., 18, 437-451.

LORD, FREDERIC M. 1958.   (15.5)                                                                    3490
     SOME RELATIONS BETWEEN GUTTMAN'S PRINCIPAL COMPONENTS OF SCALE
     ANALYSIS AND OTHER PSYCHOMETRIC THEORY.
          (AMENDED BY NO. 3502)
     PSYCHOMETRIKA, 23, 291-296.

LORD, FREDERIC M. 1959.   (17.3)                                                                    3491
     THE JOINT CUMULANTS OF TRUE VALUES AND ERRORS OF MEASUREMENT.
     ANN. MATH. STATIST., 30, 1000-1004.   (MR22-1018)

LORD, FREDERIC M. 1959.   (15.4)                                                                    3492
     PROBLEMS IN MENTAL TEST THEORY ARISING FROM ERRORS OF MEASUREMENT.
     J. AMER. STATIST. ASSOC., 54, 472-479.

LORD, FREDERIC M. 1959.   (15.4)                                                                    3493
     STATISTICAL INFERENCES ABOUT TRUE SCORES.
     PSYCHOMETRIKA, 24, 1-17.

LORD, FREDERIC M. 1959.   (15.4)                                                                    3494
     AN APPROACH TO MENTAL TEST THEORY.
     PSYCHOMETRIKA, 24, 283-302.

LORD, FREDERIC M. 1960.   (17.7)                                                                    3495
     LARGE-SAMPLE COVARIANCE ANALYSIS WHEN THE CONTROL VARIABLE IS FALLIBLE.
     J. AMER. STATIST. ASSOC., 55, 307-321.   (MR22-3074)

LORD, FREDERIC M. 1960.   (15.4)                                              3496
    AN EMPIRICAL STUDY OF THE NORMALITY AND INDEPENDENCE OF ERRORS OF
    MEASUREMENT IN TEST SCORES.
    PSYCHOMETRIKA, 25, 91-104.

LORD, FREDERIC M. 1960.   (15.4)                                              3497
    USE OF TRUE-SCORE THEORY TO PREDICT MOMENTS OF UNIVARIATE AND
    BIVARIATE OBSERVED-SCORE DISTRIBUTIONS.
    PSYCHOMETRIKA, 25, 325-342.

LORD, FREDERIC M. 1962.   (15.4)                                              3498
    CUTTING SCORES AND ERRORS OF MEASUREMENT.
    PSYCHOMETRIKA, 27, 19-30.

LORD, FREDERIC M. 1963.   (15.4)                                              3499
    CUTTING SCORES AND ERRORS OF MEASUREMENT--A SECOND CASE.
    EDUC. PSYCHOL. MEAS., 23, 63-68.

LORD, FREDERIC M. 1963.   (4.5/8.1/15.4)                                      3500
    ELEMENTARY MODELS FOR MEASURING CHANGE.
    PROBLEMS MEAS. CHANGE (HARRIS), 21-38.

LORD, FREDERIC M. 1963.   (4.4)                                              3501
    BISERIAL ESTIMATES OF CORRELATION.
    PSYCHOMETRIKA, 28, 81-85.   (MR29-2886)

LORD, FREDERIC M. 1963.   (15.5)                                             3502
    ERRATA TO:  SOME RELATIONS BETWEEN GUTTMAN'S PRINCIPAL COMPONENTS OF SCALE
    ANALYSIS AND OTHER PSYCHOMETRIC THEORY.
        (AMENDMENT OF NO. 3490)
    PSYCHOMETRIKA, 28, 129.

LORD, FREDERIC M. 1965.   (15.4/E)                                           3503
    A STRONG TRUE-SCORE THEORY, WITH APPLICATIONS.
    PSYCHOMETRIKA, 30, 239-270.

LORD, FREDERIC M. 1966.   (15.4/T)                                           3504
    THE RELATION OF TEST SCORE TO THE TRAIT UNDERLYING THE TEST.
        (REPRINT OF NO. 3481)
    READINGS MATH. SOCIAL SCI. (LAZARSFELD + HENRY), 21-53.

LORD, FREDERIC M.    SEE ALSO
    3351. LEES, RUTH W. + LORD, FREDERIC M. (1961)

    3352. LEES, RUTH W. + LORD, FREDERIC M. (1962)

    6010. WOODBURY, MAX A. + LORD, FREDERIC M. (1956)

LORD, R.D. 1948.   (18.3)                                                    3505
    A PROBLEM ON RANDOM VECTORS.
    PHILOS. MAG. SER. 7, 39, 66-71.   (MR9,P.360)

LORD, R.D. 1954.   (18.3)                                                    3506
    THE DISTRIBUTION OF DISTANCE IN A HYPERSPHERE.
    ANN. MATH. STATIST., 25, 794-798.   (MR16,P.377)

LORD, R.D. 1954.   (18.1/20.4)                                               3507
    THE USE OF THE HANKEL TRANSFORM IN STATISTICS.   I. GENERAL THEORY AND EXAMPLES.
    BIOMETRIKA, 41, 44-55.   (MR15,P.885)

LOREY, W. 1926.   (1.2)                                                      3508
    NACHRUF AUF EMANUEL CZUBER.
    Z. VERSICHERUNGSW., 26, 117-124.

LORGE, IRVING 1940.   (6.2)                                                  3509
    TWO-GROUP COMPARISONS BY MULTIVARIATE ANALYSIS.
    OFF. REP. 1940 MEET. AMER. EDUC. RES. ASSOC., 117-121.

LORGE, IRVING 1940.   (12.1)                                                 3510
    THE COMPUTATION OF THE HOTELLING CANONICAL CORRELATION.
        (WITH DISCUSSION)
    PROC. EDUC. RES. FORUM, 68-74.

LORGE, IRVING + MORRISON, N. 1938.  (11.2)                                                          3511
     THE RELIABILITY OF PRINCIPAL COMPONENTS.
     SCIENCE, 87, 491-492.


LOTKA, ALFRED J.    SEE
     1393. DUBLIN, LOUIS I.; LOTKA, ALFRED J. + SPIEGELMAN, MORTIMER (1935)


LOTKIN, MARK M. + REMAGE, RUSSELL 1953.  (20.1/20.3)                                                3512
     SCALING AND ERROR ANALYSIS FOR MATRIX INVERSION BY PARTITIONING.
     ANN. MATH. STATIST., 24, 428-439.  (MR15,P.66)


LOUCHARD, G. 1965.  (16.4)                                                                          3513
     POTENTIAL THEORY AND BROWNIAN MOTION ON THE SPHERE.
     CAHIERS CENTRE ETUDES RECH. OPERAT., 7, 248-257.


LOVERA, GIUSEPPE 1942.  (4.8/E)                                                                     3514
     UN'APPLICAZIONE DEL COEFFICIENTE DI CORRELAZIONE ALLE MEDIE STATISTICHE.
     ATTI REALE ACCAD. SCI. TORINO PUBBL. ACCAD. CL. SCI. FIS. MAT. NATUR., 77, 341-346.  (MR7,P.4


LOVERA, GIUSEPPE 1945.  (4.6)                                                                       3515
     METODO ABBREVIATO DI CALCOLO DELLE CARATTERISTICHE DI UNA CORRELAZIONE MULTIPLA.
     ATTI REALE ACCAD. SCI. TORINO PUBBL. ACCAD. CL. SCI. FIS. MAT. NATUR., 80, 194-198.  (MR8,P.4


LOWDENSLAGER, DAVID B.    SEE
     2431. HELSON, HENRY + LOWDENSLAGER, DAVID B. (1958)

     2432. HELSON, HENRY + LOWDENSLAGER, DAVID B. (1961)


LOWE, J.R. 1960.  (2.1/BT)                                                                          3516
     A TABLE OF THE INTEGRAL OF THE BIVARIATE NORMAL DISTRIBUTION OVER AN
     OFFSET CIRCLE.
     J. ROY. STATIST. SOC. SER. B, 22, 177-187.  (MR22-8600)


LOYNES, R. M. 1966.  (20.2)                                                                         3517
     ON IDEMPOTENT MATRICES.
     ANN. MATH. STATIST., 37, 295-296.


LOZANO LOPEZ, VICENTE 1965.  (17.7)                                                                 3518
     MODELOS ECONOMETRICOS CON ERRORES EN LAS VARIABLES.
     ESTADIST. ESPAN., 28, 5-44.


LOZANO LOPEZ, VICENTE 1965.  (17.7)                                                                 3519
     MODELOS UNIECUACIONALES CON ERRORES EN LAS ECUACIONES Y EN LAS VARIABLES
     Y CON REGRESORES ESTOCASTICOS.
     ESTADIST. ESPAN., 29, 5-17.


LUBENOW, KLAUS 1958.  (17.1)                                                                        3520
     VERALLGEMEINERUNG EINES SATZES VON PROCHOROV AUF MEHRERE DIMENSIONEN.
     WISS. Z. HUMBOLDT-UNIV. BERLIN MATH.-NATURWISS. REIHE, 8, 535-547.  (MR22-8551)


LUBIN, ARDIE 1950.  (6.1/6.2/6.3)                                                                   3521
     LINEAR AND NON-LINEAR DISCRIMINATING FUNCTIONS.
     BRITISH J. PSYCHOL. STATIST. SECT., 3, 90-103.


LUBIN, ARDIE 1950.  (15.5)                                                                          3522
     A NOTE ON "CRITERION ANALYSIS".
     PSYCHOL. REV., 57, 54-57.


LUBIN, ARDIE + OSBURN, HOBART G. 1957.  (15.5)                                                      3523
     A THEORY OF PATTERN ANALYSIS FOR THE PREDICTION OF A QUANTITATIVE CRITERION.
     PSYCHOMETRIKA, 22, 63-73.


LUBIN, ARDIE    SEE ALSO
     5283. SUMMERFIELD, A. + LUBIN, ARDIE (1951)


LUBISCHEW, ALEXANDER A. 1962.  (6.2/6.3/6.5/E)                                                      3524
     ON THE USE OF DISCRIMINANT FUNCTIONS IN TAXONOMY.
     BIOMETRICS, 18, 455-477.


LUCAS, HENRY L., JR.    SEE
     5563. TURNER, MALCOLM E.; MONROE, ROBERT J. + LUCAS, HENRY L., JR. (1961)

LUCE, R. DUNCAN 1950.   (15.5)                                               3525
    CONNECTIVITY AND GENERALIZED CLIQUES IN SOCIOMETRIC GROUP STRUCTURE.
  PSYCHOMETRIKA, 15, 169-190.   (MR12,P.39)

LUCE, R. DUNCAN 1963.   (15.5)                                               3526
    DISCRIMINATION.
  HANDB. MATH. PSYCHOL. (LUCE, BUSH. + GALANTER), 1, 191-243.

LUCE, R. DUNCAN 1964.   (15.5)                                               3527
    A GENERALIZATION OF A THEOREM OF DIMENSIONAL ANALYSIS.
  J. MATH. PSYCHOL., 1, 278-284.

LUCE, R. DUNCAN + PERRY, ALBERT D. 1949.   (15.5/E)                          3528
    A METHOD OF MATRIX ANALYSIS OF GROUP STRUCTURE.
       (REPRINTED AS NO. 3529)
  PSYCHOMETRIKA, 14, 95-116.   (MR12,P.39)

LUCE, R. DUNCAN + PERRY, ALBERT D. 1966.   (15.5/E)                          3529
    A METHOD OF MATRIX ANALYSIS OF GROUP STRUCTURE.
       (REPRINT OF NO. 3528)
  READINGS MATH. SOCIAL SCI. (LAZARSFELD + HENRY), 111-130.   (MR12,P.39)

LUCE, R. DUNCAN + TUKEY, JOHN W. 1964.   (15.4)                              3530
    SIMULTANEOUS CONJOINT MEASUREMENT:  A NEW TYPE OF FUNDAMENTAL MEASUREMENT.
  J. MATH. PSYCHOL. 1, 1-27.

LUCE, R. DUNCAN: MACY, JOSIAH, JR. + TAGIURI, RENATO 1955.   (15.5)          3531
    A STATISTICAL MODEL FOR RELATIONAL ANALYSIS.
  PSYCHOMETRIKA, 20, 319-327.

LUCIA, LUIGI DE  SEE  DE LUCIA, LUIGI

LUDWIG, ROLF 1965.   (4.1)                                                   3532
    NOMOGRAMM ZUR PRUFUNG DES PRODUKT-MOMENT-KORRELATIONSKOEFFIZIENTEN R.
  BIOM. Z., 7, 94-95.

LUKACS, EUGENE 1952.   (2.2)                                                 3533
    THE STOCHASTIC INDEPENDENCE OF SYMMETRIC AND HOMOGENEOUS LINEAR AND
    QUADRATIC STATISTICS.
  ANN. MATH. STATIST., 23, 442-449.   (MR14,P.297)

LUKACS, EUGENE 1962.   (2.2/U)                                               3534
    ON TUBE STATISTICS AND CHARACTERIZATION PROBLEMS.
  Z. WAHRSCHEIN. VERW. GEBIETE, 1, 116-125.

LUKACS, EUGENE   SEE ALSO
    3233. LAHA, R. G. + LUKACS, EUGENE (1960)

    3234. LAHA, R. G. + LUKACS, EUGENE (1960)

LUKOMSKII, JA. I. 1939.   (17.3)                                             3535
    ON SOME PROPERTIES OF MULTIDIMENSIONAL DISTRIBUTIONS.
  ANN. MATH. STATIST., 10, 236-246.   (MR1,P.23)

LUKOMSKII, JA. I. 1955.   (4.9/AE)                                           3536
    PROPOSALS OF THE THEORY OF LARGE NUMBERS TO THAT OF INDICES.
       (IN RUSSIAN)
  UCEN. ZAP. STATIST., 1, 193-220.

LURQUIN, CONSTANT 1937.   (18.1/B)                                           3537
    SUR LA LOI DE BERNOULLI A DEUX VARIABLES.
  ACAD. ROY. BELG. BULL. CL. SCI. SER. 5, 23, 857-860.

LUSH, J. L. 1947.  E(17.9)                                                   3538
    FAMILY MERIT AND INDIVIDUAL MERIT AS BASES FOR SELECTION.
  AMER. NATUR., 81, 241-261, 362-379.

LUSH, J. L.   SEE ALSO
    2411. HAZEL, L. N. + LUSH, J. L. (1942)

LUTHER, NORMAN Y. 1965.   (2.5)                                             3539
    DECOMPOSITION OF SYMMETRIC MATRICES AND DISTRIBUTIONS OF QUADRATIC FORMS.
  ANN. MATH. STATIST., 36, 683-690.   (MR30-2592)

LUU-MAU-THANH 1963.   (12.1/15.1/15.2/F)                                               3540
    ANALYSE CANONIQUE ET ANALYSE FACTORIELLE. ESSAI DE RAPPROCHEMENT.
    CAHIERS INST. SCI. ECON. APPL. SER. E, 2, 127-164.


LYERLY, SAMUEL B. 1952.   (4.8)                                                        3541
    THE AVERAGE SPEARMAN RANK CORRELATION COEFFICIENT.
    PSYCHOMETRIKA, 17, 421-428.   (MR14,P.665)


LYONS, THOBURN C.   SEE
    2171. GUILFORD, J. P. + LYONS, THOBURN C. (1942)


LYTTKENS, EJNAR 1950.   (17.3)                                                         3542
    ON A CLASS OF MULTIDIMENSIONAL CONDITIONAL CHARACTERISTIC FUNCTIONS
    AND SEMI-INVARIANTS.
    ARK. ASTRONOM., 1, 27-45.   (MR12,P.115)


LYTTKENS, EJNAR 1950.   (17.3)                                                         3543
    A GENERALISATION OF THE MULTIDIMENSIONAL A-SERIES.
    ARK. ASTRONOM., 1, 47-57.   (MR12,P.115)


LYTTKENS, EJNAR 1950.   (17.3/17.4)                                                    3544
    DETERMINATION OF UNKNOWN DISTRIBUTIONS BY MEANS OF CONDITIONAL
    FREQUENCY FUNCTIONS.
    ARK. ASTRONOM., 1, 69-75.   (MR12,P.115)


LYTTKENS, EJNAR 1964.   (8.2/16.5/A)                                                   3545
    STANDARD ERRORS OF REGRESSION COEFFICIENTS BY AUTOCORRELATED RESIDUALS.
    ECONOMET. MODEL BUILDING (WOLD), 169-228.


LYTTKENS, EJNAR 1964.   (8.3/16.5/A)                                                   3546
    A LARGE-SAMPLE $\chi^2$-DIFFERENCE TEST FOR REGRESSION COEFFICIENTS.
    ECONOMET. MODEL BUILDING (WOLD), 236-278.


LYTTKENS, EJNAR 1964.   (16.2/16.3)                                                    3547
    SOME NOTES ON ECONOMETRIC MODELS.
    ECONOMET. MODEL BUILDING (WOLD), 322-340.


LYTTKENS, EJNAR 1966.   (11.2)                                                         3548
    ON THE FIX-POINT PROPERTY OF WOLD'S ITERATIVE ESTIMATION METHOD FOR
    PRINCIPAL COMPONENTS.
    MULTIVARIATE ANAL. PROC. INTERNAT. SYMP. DAYTON (KRISHNAIAH), 335-350.   (MR35-2422)


MC CALLAN, S. F. A.   SEE
    5824. WELLMAN, R. H. + MC CALLAN, S. E. A. (1943)


MC CLOY, C. H.; METHENY, ELEANOR + KNOTT, VIRGINIA 1938.   (11.1/15.2)                 3549
    A COMPARISON OF THE THURSTONE METHOD OF MULTIPLE FACTORS WITH THE
    HOTELLING METHOD OF PRINCIPAL COMPONENTS.
    PSYCHOMETRIKA, 3, 61-67.


MC CORMICK, THOMAS C. 1934.   (4.8/E)                                                  3550
    A COEFFICIENT OF INDEPENDENT DETERMINATION.
        (AMENDED BY NO. 3551)
    J. AMER. STATIST. ASSOC., 29, 76-78.


MC CORMICK, THOMAS C. 1934.   (4.8/E)                                                  3551
    CORRECTION TO:  A COEFFICIENT OF INDEPENDENT DETERMINATION.
        (AMENDMENT OF NO. 3550)
    J. AMER. STATIST. ASSOC., 29, 319.


MC CREA, WILLIAM H. + WHIPPLE, F. J. W. 1940.   (18.1)                                 3552
    RANDOM PATHS IN TWO AND THREE DIMENSIONS.
    PROC. ROY. SOC. EDINBURGH, 60, 281-298.   (MR2,P.107)


MC DONALD, RODERICK P. 1962.   (15.5)                                                  3553
    A NOTE ON THE DERIVATION OF THE GENERAL LATENT CLASS MODEL.
    PSYCHOMETRIKA, 27, 203-206.


MC DONALD, RODERICK P. 1962.   (15.1)                                                  3554
    A GENERAL APPROACH TO NONLINEAR FACTOR ANALYSIS.
    PSYCHOMETRIKA, 27, 397-415.   (MR29-2905)

MC DONALD, RODERICK P. 1965.   (15.1)                        3555
    DIFFICULTY FACTORS AND NON-LINEAR FACTOR ANALYSIS.
    BRITISH J. MATH. STATIST. PSYCHOL., 18, 11-23.


MC DONOUGH, M. ROSA 1929.   (15.2)                           3556
    GENERAL FACTORS IN A TABLE ON INTERCORRELATIONS.
    SCIENCE NEW SER., 69, 402.


MACEDA, ENRIQUE CANSADO   SEE   CANSADO MACEDA, ENRIQUE


MAC EWAN, CHARLOTTE    SEE
    2575, HORST, PAUL + MAC EWAN, CHARLOTTE (1957)

    2576. HORST, PAUL + MAC EWAN, CHARLOTTE (1960)


MC EWEN, G. F. + MICHAEL, E. L. 1919.   (8.1/E)             3557
    THE FUNCTIONAL RELATION OF ONE VARIABLE TO EACH OF A NUMBER OF CORRELATED
    VARIABLES DETERMINED BY A METHOD OF SUCCESSIVE APPROXIMATION TO GROUP AVERAGES:
    A CONTRIBUTION TO STATISTICAL METHODS.
    PROC. AMER. ACAD. ARTS SCI., 55, 95-133.


MC FADDEN, J.A. 1955.   (2.1)                                3558
    URN MODELS OF CORRELATION AND A COMPARISON WITH THE MULTIVARIATE
    NORMAL INTEGRAL.
    ANN. MATH. STATIST., 26, 478-489.   (MR17,P.47)


MC FADDEN, J.A. 1956.   (2.7)                                3559
    AN APPROXIMATION FOR THE SYMMETRIC, QUADRIVARIATE NORMAL INTEGRAL.
    BIOMETRIKA, 43, 206-207.   (MR17,P.983)


MC FADDEN, J.A. 1956.   (16.3/B)                             3560
    THE CORRELATION FUNCTION OF A SINE WAVE PLUS NOISE AFTER EXTREME CLIPPINGS.
    IRE TRANS. INFORMATION THEORY, IT-2, 82-83.


MC FADDEN, J.A. 1960.   (2.7)                                3561
    TWO EXPANSIONS FOR THE QUADRI-VARIATE NORMAL INTEGRAL.
    BIOMETRIKA, 47, 325-333.   (MR22-9987)


MC GEE, VICTOR E. 1966.   (15.4)                             3562
    THE MULTIDIMENSIONAL ANALYSIS OF "ELASTIC" DISTANCES.
    BRITISH J. MATH. STATIST. PSYCHOL., 19, 181-196.


MC GILL, WILLIAM J. 1954.   (8.5/17.7)                       3563
    MULTIVARIATE INFORMATION TRANSMISSION.
        (REPRINTED AS NO. 3564, REPRINTED AS NO. 3565)
    PSYCHOMETRIKA, 19, 97-116.   (MR19,P.476)


MC GILL, WILLIAM J. 1954.   (8.5/17.7)                       3564
    MULTIVARIATE INFORMATION TRANSMISSION.
        (REPRINT OF NO. 3563, REPRINTED AS NO. 3565)
    TRANS. IRE PROF. GROUP INFORMATION THEORY,PGIT-4, 93-111.   (MR19,P.476)


MC GILL, WILLIAM J. 1963.   (8.5/17.7)                       3565
    MULTIVARIATE INFORMATION TRANSMISSION.
        (REPRINT OF NO. 3564, REPRINT OF NO. 3563)
    READINGS MATH. PSYCHOL. (LUCE, BUSH, + GALANTER), 1, 84-103.   (MR19,P.476)


MC GILL, WILLIAM J.   SEE ALSO
    1848. GARNER, W. R. + MC GILL, WILLIAM J. (1956)


MC GREGOR, J.R. 1962.   (16.5/A)                             3566
    THE APPROXIMATE DISTRIBUTION OF THE CORRELATION BETWEEN TWO
    STATIONARY LINEAR MARKOV SERIES.
    BIOMETRIKA, 49, 379-388.   (MR27-6325)


MC GREGOR, J.R. + BIELENSTEIN, U. M. 1965.   (16.5/A)       3567
    THE APPROXIMATE DISTRIBUTION OF THE CORRELATION BETWEEN TWO STATIONARY
    LINEAR MARKOV SERIES.  II.
    BIOMETRIKA, 52, 301-303.


MACHADO, E.A.M.   SEE
    1252. DEDEBANT, GEORGES + MACHADO, E.A.M. (1954)

MACHOVER, MAURICE 1965.   (20.4)                                                    3568
    A GENERALIZED EIGENFUNCTION EXPANSION OF THE GREEN'S FUNCTION.
    PROC. AMER. MATH. SOC., 16, 348-352.


MC HUGH, RICHARD B. 1956.   (15.5)                                                  3569
    EFFICIENT ESTIMATION AND LOCAL IDENTIFICATION IN LATENT CLASS ANALYSIS.
    PSYCHOMETRIKA, 21, 331-347.   (MR18,P.548)


MC HUGH, RICHARD B. + WALL, F. J. 1962.   (16.5)                                    3570
    ESTIMATING THE PRECISION OF TIME PERIOD EFFECTS IN LONGITUDINAL MODELS
    WITH SERIALLY CORRELATED AND HETEROGENEOUS ERRORS.
    BIOMETRICS, 18, 520-528.


MC INTYRE, FRANCIS 1937.   (4.1)                                                    3571
    AUTOMATIC CHECKS IN CORRELATION ANALYSIS.
    J. AMER. STATIST. ASSOC., 32, 119-123.


MACK, C. 1950.   (18.6)                                                             3572
    THE EXPECTED NUMBER OF AGGREGATES IN A RANDOM DISTRIBUTION OF N POINTS.
    PROC. CAMBRIDGE PHILOS. SOC., 46, 285-292.   (MR11,P.605)


MC KAY, A. T. 1932.   (20.4)                                                        3573
    A BESSEL FUNCTION DISTRIBUTION.
    BIOMETRIKA, 24, 39-44.


MACKENZIE, J. K. 1957.   (8.8)                                                      3574
    A LEAST SQUARES SOLUTION OF LINEAR EQUATIONS WITH COEFFICIENTS
    SUBJECT TO A SPECIAL TYPE OF ERROR.
    AUSTRAL. J. PHYS., 10, 103-109.   (MR18,P.937)


MACKENZIE, W.A.   SEE
    1708. FISHER, RONALD A. + MACKENZIE, W.A. (1922)


MC KEON, JAMES J. 1965.   (12.1/12.2/E)                                             3575
    CANONICAL ANALYSIS: SOME RELATIONS BETWEEN CANONICAL CORRELATION, FACTOR
    ANALYSIS, DISCRIMINANT FUNCTION ANALYSIS, AND SCALING THEORY.
    PSYCHOMET. MONOG., 13, 1-43.


MACKIE, JOHN 1928.   (15.2)                                                         3576
    THE PROBABLE VALUE OF THE TETRAD DIFFERENCE ON THE SAMPLING THEORY.
    BRITISH J. PSYCHOL., 19, 65-76.


MACKIE, JOHN 1928.   (15.2)                                                         3577
    THE SAMPLING THEORY AS A VARIANT OF THE TWO FACTOR THEORY.
    J. EDUC. PSYCHOL., 19, 614-621.


MACKIE, JOHN 1929.   (15.1)                                                         3578
    MATHEMATICAL CONSEQUENCES OF CERTAIN THEORIES OF MENTAL ABILITY.
    PROC. ROY. SOC. EDINBURGH, 49, 16-37.


MC LAIN, RICHARD E.   SEE
    5031. SILVERSTEIN, A. B. + MC LAIN, RICHARD E. (1964)


MC LEOD, ROBERT M. 1965.   (20.5)                                                   3579
    MEAN VALUE THEOREMS FOR VECTOR VALUED FUNCTIONS.
    PROC. EDINBURGH MATH. SOC. SER. 2, 14, 197-209.


MC MAHON, JAMES 1923.   (4.9/20.5)                                                  3580
    HYPERSPHERICAL GONIOMETRY; AND ITS APPLICATION TO CORRELATION THEORY
    FOR N VARIABLES.
    BIOMETRIKA, 15, 173-208.


MC MILLAN, J. R. A. 1938.   (4.8)                                                   3581
    THE USE OF THE COEFFICIENT OF CORRELATION OF QUANTITATIVE CHARACTERS
    AS A MEASURE OF GENE LINKAGE.
    AUSTRAL. COUNC. SCI. INDUST. RES. J., 11, 311-316.


MACNAUGHTON-SMITH, P. 1963.   (6.4)                                                 3582
    THE CLASSIFICATION OF INDIVIDUALS BY THE POSSESSION OF ATTRIBUTES
    ASSOCIATED WITH A CRITERION.
    BIOMETRICS, 19, 364-366.

MACNAUGHTON-SMITH, P.; WILLIAMS, W. T.; DALE, M. G. + 1964. (17.9)          3583
  MOCKETT, L. G.
    DISSIMILARITY ANALYSIS: A NEW TECHNIQUE OF HIERARCHICAL SUB-DIVISION.
  NATURE, 202, 1034-1035.

MC NEE, P. C.   SEE
  1185. DANFORD, MASIL B.; HUGHES, HARRY M. + MC NEE, P. C. (1960)


MC NEMAR, QUINN 1941. (15.2)                                                3584
    ON THE SAMPLING ERRORS OF FACTOR LOADINGS.
  PSYCHOMETRIKA, 6, 141-152.

MC NEMAR, QUINN 1942. (15.1)                                                3585
    ON THE NUMBER OF FACTORS.
  PSYCHOMETRIKA, 7, 9-18.

MC NEMAR, QUINN 1947. (18.3)                                                3586
    NOTE ON THE SAMPLING ERROR OF THE DIFFERENCE BETWEEN CORRELATED
    PROPORTIONS OR PERCENTAGES.
  PSYCHOMETRIKA, 12, 153-157.

MC NOLTY, FRANK W. 1961. (14.4)                                             3587
    PELLET-EFFECTIVENESS ANALYSIS.
  OPERATIONS RES., 9, 522-534. (MR26-1206)

MC NOLTY, FRANK W. 1962. (14.4)                                             3588
    KILL PROBABILITY WHEN THE WEAPON BIAS IS RANDOMLY DISTRIBUTED.
  OPERATIONS RES., 10, 693-701.

MACPHAIL, M.S.   SEE
  1467. DWYER, PAUL S. + MACPHAIL, M.S. (1948)

MACQUEEN. J. 1965. (6.6/17.1/19.1)                                          3589
    SOME METHODS FOR CLASSIFICATION AND ANALYSIS OF MULTIVARIATE OBSERVATIONS.
  PROC. FIFTH BERKELEY SYMP. MATH. STATIST. PROB., 1, 281-297.

MC QUITTY, J. V.   SEE
  3906. MOSIER, CHARLES I. + MC QUITTY, J. V. (1940)

MC QUITTY, LOUIS L. 1956. (6.1/15.5)                                        3590
    AGREEMENT ANALYSIS: CLASSIFYING PERSONS BY PREDOMINANT PATTERNS OF RESPONSES.
  BRITISH J. STATIST. PSYCHOL., 9, 5-16.

MC QUITTY, LOUIS L. 1957. (15.1/E)                                          3591
    ELEMENTARY LINKAGE ANALYSIS FOR ISOLATING ORTHOGONAL AND OBLIQUE TYPES
    AND TYPAL RELEVANCIES.
  EDUC. PSYCHOL. MEAS., 17, 207-229.

MC QUITTY, LOUIS L.   SEE ALSO
  6033. WRIGLEY, CHARLES F.; CHERRY, CHARLES N.; LEE, MARILYN C. + (1957)
  MC QUITTY, LOUIS L.

MACY, JOSIAH, JR.   SEE
  3531. LUCE, R. DUNCAN; MACY, JOSIAH, JR. + TAGIURI, RENATO (1955)


MADANSKY, ALBERT 1959. (17.2)                                              3592
    BOUNDS ON THE EXPECTATION OF A CONVEX FUNCTION OF A MULTIVARIATE
    RANDOM VARIABLE.
  ANN. MATH. STATIST., 30, 743-746. (MR21-6062)

MADANSKY, ALBERT 1959. (8.1/17.7)                                          3593
    THE FITTING OF STRAIGHT LINES WHEN BOTH VARIABLES ARE SUBJECT TO ERROR.
  J. AMER. STATIST. ASSOC., 54, 173-205. (MR21-1661)

MADANSKY, ALBERT 1960. (15.5)                                             3594
    DETERMINANTAL METHODS IN LATENT CLASS ANALYSIS.
  PSYCHOMETRIKA, 25, 183-198. (MR22-3614)

MADANSKY, ALBERT 1964. (16.2)                                             3595
    ON THE EFFICIENCY OF THREE-STAGE LEAST-SQUARES ESTIMATION.
  ECONOMETRICA, 32, 51-56. (MR30-4324)

MADANSKY, ALBERT 1964.  (2.4)                                                              3596
   SPURIOUS CORRELATION DUE TO DEFLATING VARIABLES.
   ECONOMETRICA, 32, 652–655.  (MR30–5432)

MADANSKY, ALBERT 1964.  (15.1/17.7)                                                        3597
   INSTRUMENTAL VARIABLES IN FACTOR ANALYSIS.
   PSYCHOMETRIKA, 29, 105–113.  (MR29–2925)

MADANSKY, ALBERT 1965.  (15.1)                                                             3598
   ON ADMISSIBLE COMMUNALITIES IN FACTOR ANALYSIS.
   PSYCHOMETRIKA, 30, 455–458.  (MR33–6764)

MADOW, WILLIAM G. 1937.  (17.9)                                                            3599
   CONTRIBUTIONS TO THE THEORY OF COMPARATIVE STATISTICAL ANALYSIS.  I.
   FUNDAMENTAL THEOREMS OF COMPARATIVE ANALYSIS.
   ANN. MATH. STATIST., 8, 159–176.

MADOW, WILLIAM G. 1938.  (1.1)                                                             3600
   CONTRIBUTIONS TO THE THEORY OF MULTIVARIATE STATISTICAL ANALYSIS.
   TRANS. AMER. MATH. SOC., 44, 454–495.

MADOW, WILLIAM G. 1940.  (2.5)                                                             3601
   THE DISTRIBUTION OF QUADRATIC FORMS IN NON-CENTRAL NORMAL RANDOM VARIABLES.
   ANN. MATH. STATIST., 11, 100–103.  (MR1,P.248)

MADOW, WILLIAM G. 1940.  (17.1)                                                            3602
   LIMITING DISTRIBUTIONS OF QUADRATIC AND BILINEAR FORMS.
   ANN. MATH. STATIST., 11, 125–146.  (MR1,P.341)

MADOW, WILLIAM G. 1945.  (2.6)                                                             3603
   NOTE ON THE DISTRIBUTION OF THE SERIAL CORRELATION COEFFICIENT.
   ANN. MATH. STATIST., 16, 308–310.  (MR7,P.131)

MA, ER-CHIEH 1966.  (20.3)                                                                 3604
   A FINITE SERIES SOLUTION OF THE MATRIX EQUATION AX - XB = C$^*$.
   SIAM J. APPL. MATH., 14, 490–495.

MAHALANOBIS, P.C. 1922.  (4.1/6.5/17.9/E)                                                  3605
   ANTHROPOLOGICAL OBSERVATIONS ON THE ANGLO-INDIANS OF CALCUTTA. PART
   I, ANALYSIS OF MALE STATURE.
   REC. INDIAN MUSEUM, 23, 1–96.

MAHALANOBIS, P.C. 1927.  (6.5)                                                             3606
   ANALYSIS OF RACE MIXTURE IN BENGAL.
   J. AND PROC. ASIAT. SOC. BENGAL, 23, 301–333.

MAHALANOBIS, P.C. 1928.  (6.5/E)                                                           3607
   A STATISTICAL STUDY OF THE CHINESE HEAD.
   MAN IN INDIA, 8, 107–122.

MAHALANOBIS, P.C. 1930.  E(5.2/6.4)                                                        3608
   A STATISTICAL STUDY OF CERTAIN ANTHROPOMETRIC MEASUREMENTS FROM SWEDEN.
   BIOMETRIKA, 22, 94–108.

MAHALANOBIS, P.C. 1930.  (6.5)                                                             3609
   ON TESTS AND MEASURES OF GROUP DIVERGENCE.  PART I.  THEORETICAL FORMULAE.
   J. AND PROC. ASIAT. SOC. BENGAL, 26, 541–588.

MAHALANOBIS, P.C. 1931.  (4.1/6.5/17.9/E)                                                  3610
   ANTHROPOLOGICAL OBSERVATIONS ON THE ANGLO-INDIANS OF CALCUTTA. PART
   II, ANALYSIS OF ANGLO-INDIAN HEAD LENGTH.
   REC. INDIAN MUSEUM, 23, 97–149.

MAHALANOBIS, P.C. 1932.  (14.5/T)                                                          3611
   AUXILIARY TABLES FOR FISHER'S Z-TEST FOR USE IN THE ANALYSIS OF VARIANCE.
   INDIAN J. AGRIC. SCI., 2, 679–693.

MAHALANOBIS, P.C. 1936.  (6.5/6.2)                                                         3612
   ON THE GENERALIZED DISTANCE IN STATISTICS.
   PROC. NAT. INST. SCI. INDIA, 2, 49–55.

MAHALANOBIS, P.C. 1940.   (4.1/6.5/17.9/E)                                3613
    ANTHROPOLOGICAL OBSERVATIONS ON THE ANGLO-INDIANS OF CALCUTTA. PART
    III. STATISTICAL ANALYSIS OF MEASUREMENTS OF SEVEN CHARACTERS.
    REC. INDIAN MUSEUM, 23, 151-187.

MAHALANOBIS, P.C. 1940.   (5.2/6.5)                                       3614
    THE APPLICATION OF STATISTICAL METHODS IN PHYSICAL ANTHROPOMETRY.
    SANKHYA, 4, 594-598.

MAHALANOBIS, P.C. + ROSE, MRS. CHAMELI 1941.   (4.1/4.8/8.1/E)            3615
    CORRELATION BETWEEN ANTHROPOMETRIC CHARACTERS IN SOME BENGAL CASTES AND TRIBES.
    SANKHYA, 5, 249-260.

MAHALANOBIS, P.C.; ROSE, R. C. + ROY, S.N. 1937.   (7.1/7.3)             3616
    NORMALISATION OF STATISTICAL VARIATES AND THE USE OF RECTANGULAR
    CO-ORDINATES IN THE THEORY OF SAMPLING DISTRIBUTIONS.
    SANKHYA, 3, 1-40.

MAHALANOBIS, P.C.; MAJUMDAR, D.N. + RAO, C. RADHAKRISHNA 1949.   (6.3/6.5/8.5/T)   3617
    ANTHROPOMETRIC SURVEY OF THE UNITED PROVINCES, 1941:  A STATISTICAL STUDY.
    SANKHYA, 9, 89-324.

MAHLMANN, HENRY 1935.   (2.1/18.1/B)                                      3618
    EIN BEITRAG ZU UNTERSUCHUNGEN UBER ZWEIDIMENSIONALE VERTEILUNGEN VON
    MASSENPUNKTEN BEI ZUFALLSARTIG BEDINGTEN BEWEGUNGEN.
    BIOMETRIKA, 27, 191-226.


MAHMOUD, A.F. 1955.   (15.1/15.4)                                         3619
    TEST RELIABILITY IN TERMS OF FACTOR THEORY.
    BRITISH J. STATIST. PSYCHOL., 8, 119-135.

MAHMOUD, M. W.   SEE
    3915. MOUSTAFA, M.D. + MAHMOUD, M. W. (1964)

MAILLET, P. 1955.   (16.2)                                               3620
    INTRODUCTION A L'ETUDE DES MODELES ECONOMETRIQUES.
    CAHIERS SEM. ECONOMET., 3, 7-30.

MAITRA, ASKOK 1966.   (17.3)                                             3621
    ON STABLE TRANSFORMATIONS.
    SANKHYA SER. A, 28, 25-34.

MAJINDAR, KULENDRA N. 1963.   (20.1)                                     3622
    ON A FACTORISATION OF POSITIVE DEFINITE MATRICES.
    CANAD. MATH. BULL., 6, 405-407.   (MR28-1206)

MAJUMDAR, D.N.   SEE
    4461. RAO, C. RADHAKRISHNA + MAJUMDAR, D.N. (1958)

    3617. MAHALANOBIS, P.C.; MAJUMDAR, D.N. + RAO, C. RADHAKRISHNA (1949)


MAKELAINEN, TIMO 1966.   (2.5)                                           3623
    ON QUADRATIC FORMS IN NORMAL VARIABLES.
    SOC. SCI. FENN. COMMENT. PHYS. MATH., 31(12) 1-6.   (MR33-8053)

MAKOVER, S.G. 1956.   (20.3)                                             3624
    SOLUTION OF A SYSTEM OF NORMAL EQUATIONS WITH THE AID OF MATRICES.
    ASTRONOM. ZUR., 33, 423-439.   (MR18,P.676)

MALECOT, GUSTAVE 1938.   (4.8)                                           3625
    SUR LES ALEATOIRES MENDELIENNES ET LES CORRELATIONS DE L'HEREDITE.
    C.R. ACAD. SCI. PARIS, 206, 404-406.

MALECOT, GUSTAVE 1939.   (4.8)                                           3626
    LES CORRELATIONS ENTRE INDIVIDUS APPARENTES, DANS L'HYPOTHESE D'HOMOGAMIE.
    C.R. ACAD. SCI. PARIS, 208, 552-554.

MALECOT, GUSTAVE 1966.   (17.9)                                          3627
    LES COVARIANCES DANS UN MILIEU EN EQUILIBRE STATISTIQUE.
    BULL. INST. INTERNAT. STATIST., 41, 811-821.

MALENBAUM, WILFRED 1939.  (4.8/8.1/E)                                          3628
     CONCLUDING REMARKS.
     QUART. J. ECON., 54, 358-364.

MALENBAUM, WILFRED + BLACK, J. D. 1937.  (4.6)                                 3629
     USE OF THE SHORT-CUT GRAPHIC METHOD OF MULTIPLE CORRELATION.
          (AMENDED BY NO. 3630)
     QUART. J. ECON., 52, 66-112.

MALENBAUM, WILFRED + BLACK, J. D. 1939.  (4.6)                                 3630
     REJOINDER.
     QUART. J. ECON., 54, 346-358.


MALLIK, A. K.   SEE
     4647. ROY, A. R. + MALLIK, A. K. (1966)

MALLOWS, C.L. 1953.  (6.1/6.2)                                                 3631
     SEQUENTIAL DISCRIMINATION.
     SANKHYA, 12, 321-338.  (MR15,P.453)

MALLOWS, C.L. 1956.  (17.2)                                                    3632
     GENERALIZATIONS OF TCHEBYCHEFF'S INEQUALITIES.
          (WITH DISCUSSION)
     J. ROY. STATIST. SOC. SER. B, 18, 139-176.  (MR19,P.326)

MALLOWS, C.L. 1961.  (8.7/13.2/17.9/C)                                         3633
     LATENT VECTORS OF RANDOM SYMMETRIC MATRICES.
     BIOMETRIKA, 48, 133-149.  (MR24-A1164)

MALLOWS, C.L.   SEE ALSO
     1226. DAVID, F. N. + MALLOWS, C.L. (1961)


MALONE, R. DANIEL   SEE
     1067. COTTON, JOHN W.; CAMPBELL, DONALD J. + MALONE, R. DANIEL (1957)

MAMATOV, M. + HALIKOV, M. K. 1964.  (17.1)                                     3634
     GLOBAL LIMIT THEOREMS FOR DISTRIBUTION FUNCTIONS IN THE HIGHER-DIMENSIONAL CASE.
          (IN RUSSIAN, WITH SUMMARY IN UZBEKISH)
     IZV. AKAD. NAUK UZSSR. SER. FIZ.-MAT. NAUK, 1964(1) 13-21.  (MR29-647)

MANE, CESARO VILLEGAS  SEE  VILLEGAS, CESAREO

MANIYA, G.M. 1958.  (3.3)                                                      3635
     QUADRATIC ERROR OF ESTIMATION OF DENSITY OF A NORMAL TWO-DIMENSIONAL
     DISTRIBUTION IN TERMS OF SAMPLING DATA.
     SOOBSC. AKAD. NAUK GRUZIN. SSR, 20, 655-658.  (MR20-6169)

MANIYA, G.M. 1962.  (3.1/14.4/B)                                               3636
     A QUADRATIC ESTIMATE OF THE DEVIATION OF DENSITIES OF A
     TWO-DIMENSIONAL NORMAL DISTRIBUTION BASED ON A GIVEN SAMPLE.
          (IN RUSSIAN)
     TRUDY VYC. CENTRA AKAD. NAUK GRUZIN. SSR, 2, 153-211.  (MR32-3193)

MANN, HENRY B. 1943.  (20.1)                                                   3637
     QUADRATIC FORMS WITH LINEAR CONSTRAINTS.
     AMER. MATH. MONTHLY, 50, 430-433.  (MR5,P.30)

MANN, HENRY B. 1960.  (8.3/20.1)                                               3638
     THE ALGEBRA OF A LINEAR HYPOTHESIS.
     ANN. MATH. STATIST., 31, 1-15.  (MR24-A1773)

MANN, HENRY B. + WALD, ABRAHAM 1942.  (17.8/A)                                 3639
     ON THE CHOICE OF THE NUMBER OF CLASS INTERVALS IN THE APPLICATION OF
     THE CHI SQUARE TEST.
          (REPRINTED AS NO. 3642)
     ANN. MATH. STATIST., 13, 306-317.  (MR4,P.105)

MANN, HENRY B. + WALD, ABRAHAM 1943.  (17.1)                                   3640
     ON STOCHASTIC LIMIT AND ORDER RELATIONSHIPS.
          (REPRINTED AS NO. 3643)
     ANN. MATH. STATIST., 14, 217-226.  (MR5,P.125)

MANN, HENRY B. + WALD, ABRAHAM 1943.  (16.1/16.2)                3641
    ON THE STATISTICAL TREATMENT OF LINEAR STOCHASTIC DIFFERENCE EQUATIONS.
        (REPRINTED AS NO. 3644)
    ECONOMETRICA, 11, 173-220.  (MR5,P.129)

MANN, HENRY B. + WALD, ABRAHAM 1955.  (17.8/A)                3642
    ON THE CHOICE OF THE NUMBER OF CLASS INTERVALS IN THE APPLICATION OF
    THE CHI SQUARE TEST.
        (REPRINT OF NO. 3639)
    SELECT. PAPERS STATIST. PROB. WALD, 218-229.  (MR4,P.105)

MANN, HENRY B. + WALD, ABRAHAM 1955.  (17.1)                3643
    ON STOCHASTIC LIMIT AND ORDER RELATIONSHIPS.
        (REPRINT OF NO. 3640)
    SELFCT. PAPERS STATIST. PROB. WALD, 265-274.  (MR5,P.125)

MANN, HENRY B. + WALD, ABRAHAM 1955.  (16.1/16.2)                3644
    ON THE STATISTICAL TREATMENT OF LINEAR STOCHASTIC DIFFERENCE EQUATIONS.
        (REPRINT OF NO. 3641)
    SELFCT. PAPERS STATIST. PROB. WALD, 275-322.  (MR5,P.129)


MANNING, W. H.   SEE
    6020. WRIGHT, E. M. J.; MANNING, W. H. + DU BOIS, PHILIP H. (1959)

MANTEL, NATHAN 1966.  (4.3)                3645
    CORRECTED CORRELATION COEFFICIENTS WHEN OBSERVATION ON ONE VARIABLE IS
        RESTRICTED.
    BIOMETRICS. 22, 182-187.

MANTEL, NATHAN   SEE ALSO
    1914. GEISSER, SEYMOUR + MANTEL, NATHAN (1962)

    2305. HALPERIN, MAX + MANTEL, NATHAN (1963)

MARAIS, HENRI 1930.  (2.2/4.1/4.8)                3646
    REMARQUES SUR LA CORRELATION NORMALE.
    BULL. TRIMEST. INST. ACTUAIR. FRANC., 36, 21-34.

MARAVALL CASESNOVES, DARIO 1958.  (17.3)                3647
    LA ADICION DE VECTORES ALEATORIOS ISOTROPOS EN UN ESPACIO DE DIMENSION N.
        (WITH SUMMARY IN ENGLISH)
    TRABAJOS ESTADIST., 9, 183-202.  (MR21-2297)

MARCANTONI, ALESSANDRO 1942.  (4.1/8.8)                3648
    PESI E CORRELAZIONI PER MISURE INDIRETTE CONDIZIONATE.
        (REPRINT OF NO. 3649)
    ATTI REALE ACCAD. NAZ. LINCEI SER. 7, 3, 23-32.  (MR8,P.284)

MARCANTONI, ALESSANDRO 1942.  (4.1/8.8)                3649
    PESI E CORRELAZIONI PER MISURE INDIRETTE E CONDIZIONATE.
        (REPRINTED AS NO. 3648)
    REND. IST. LOMBARDO SCI. LETT. CL. SCI. MAT. NATUR., 6(75) 37-46.  (MR8,P.284)

MARCANTONI, ALESSANDRO 1946.  (8.8/20.1/20.3)                3650
    SAGGIO DI UN'APPLICAZIONE DEL CALCOLO DELLE MATRICI ALLA TEORIA DEGLI ERRORI.
        (REPRINT OF NO. 3651)
    PONT. ACAD. SCI. ACTA, 10, 301-320.  (MR10,P.68)


MARCANTONI, ALESSANDRO 1946.  (8.8/20.1/20.3)                3651
    SAGGIO DI UN'APPLICAZIONE DEL CALCOLO DELLE MATRICI ALLA TEORIA DEGLI ERRORI.
        (REPRINTED AS NO. 3650)
    REND. MAT. E APPL. REG. UNIV. ROMA REALE IST. NAZ. ALTA MAT., 5, 252-270.  (MR9,P.49)

MARCH, M. LUCIEN 1910.  (4.9)                3652
    ESSAI SUR UN MODE D'EXPOSER LES PRINCIPAUX ELEMENTS DE LA THEORIE STATISTIQUE.
    J. SOC. STATIST. PARIS, 51, 447-486.

MARCH, M. LUCIEN 1928.  (4.1)                3653
    DIFFERENCES ET CORRELATION EN STATISTIQUE.
    J. SOC. STATIST. PARIS, 69, 38-63.

MARCH, M. LUCIEN 1932.  (4.1)                                                    3654
    NOTE SUR LA CORRELATION.
        (WITH DISCUSSION)
    PROC. INTERNAT. CONG. MATH., 1928(6) 133-147.


MARCH, M. LUCIEN 1933.  (4.3/4.9)                                               3655
    PARELLELISME. CORRELATION, CAUSALITE.
    REV. INST. INTERNAT. STATIST., 1(1) 9-22.


MARCUS, L. F. 1964.  (17.9/E)                                                    3656
    MEASUREMENT OF NATURAL SELECTION IN NATURAL POPULATIONS.
    NATURE, 202, 1033-1034.


MARCUS, L. F.   SEE ALSO
    289. BANERJEE, KALISHANKAR + MARCUS, L. F. (1965)


MARCUS, MARVIN 1962.  (20.1)                                                    3657
    HERMITIAN FORMS AND EIGENVALUES.
    SURVEY NUMER. ANAL. (TODD), 298-313.


MARCUS, MARVIN 1964.  (20.1)                                                    3658
    THE USE OF MULTILINEAR ALGEBRA FOR PROVING MATRIX INEQUALITIES.
    RECENT ADVANC. MATRIX THEORY (SCHNEIDER), 61-80.


MARCUS, MARVIN + KHAN, N. A. 1959.  (20.1)                                      3659
    A NOTE ON THE HADAMARD PRODUCT.
    CANAD. MATH. BULL., 2, 81-83.  (MR21-4166)


MARCUS, MARVIN + NEWMAN, MORRIS 1965.  (20.1)                                   3660
    GENERALIZED FUNTIONS OF SYMMETRIC MATRICES.
    PROC. AMER. MATH. SOC., 16, 826-830.


MARCUS, MARVIN + THOMPSON, ROBERT C. 1963.  (20.1)                             3661
    THE FIELD OF VALUES OF THE HADAMARD PRODUCT.
    ARCH. MATH., 14, 283-288.  (MR27-3648)


MARCUS, MARVIN: MOYLS, V.N. + WESTWICK, ROY 1957.  (20.1)                      3662
    SOME EXTREME VALUE RESULTS FOR INDEFINITE HERMITIAN MATRICES.
    ILLINOIS J. MATH., 1, 449-457.  (MR19,P.523)


MARCUS, MARVIN   SEE ALSO
    5364. TAUSSKY, OLGA + MARCUS, MARVIN (1962)



MARDIA, K.V. 1962.  (18.1)                                                      3663
    MULTIVARIATE PARETO DISTRIBUTIONS.
        (AMENDED BY NO. 3664)
    ANN. MATH. STATIST., 33, 1008-1015.  (MR27-871)


MARDIA, K.V. 1963.  (18.1)                                                      3664
    CORRECTION TO:  MULTIVARIATE PARETO DISTRIBUTIONS.
        (AMENDMENT OF NO. 3663)
    ANN. MATH. STATIST., 34, 1603.  (MR27-871)


MARDIA, K.V. 1964.  (14.1/18.3)                                                 3665
    SOME RESULTS ON THE ORDER STATISTICS OF THE MULTIVARIATE NORMAL AND
    PARETO TYPE 1 POPULATIONS.
    ANN. MATH. STATIST., 35, 1815-1818.


MARDIA, K.V. 1964.  (17.6)                                                      3666
    ASYMPTOTIC INDEPENDENCE OF BIVARIATE EXTREMES.
    CALCUTTA STATIST. ASSOC. BULL., 13, 172-178.  (MR31-837)


MARDIA, K.V. 1964.  (17.6)                                                      3667
    EXACT DISTRIBUTIONS OF EXTREMES, RANGES AND MID-RANGES IN SAMPLES
    FROM ANY MULTIVARIATE POPULATION.
    J. INDIAN STATIST. ASSOC., 2, 126-130.  (MR31-838)


MARINESCU, G. 1951.  (17.6/U)                                                   3668
    LA FONCTION DE DISTRIBUTION DU MAXIMUM DU MODULE DE N VARIABLES STATISTIQUES.
        (IN ROUMAINIAN, WITH SUMMARY IN RUSSIAN AND FRENCH)
    COMUN. ACAD. REPUR. POP. ROMINE, 1(4) 309-313.  (MR17,P.52)

MARITZ, J.S. 1953.  (4.3/B)                                                    3669
     ESTIMATION OF THE CORRELATION COEFFICIENT IN THE CASE OF A BIVARIATE
     NORMAL POPULATION WHEN ONE OF THE VARIABLES IS DICHOTOMIZED.
     PSYCHOMETRIKA, 18, 97-110.  (MR14,P.996)

MARITZ, J.S. 1962.  (8.1/C)                                                    3670
     CONFIDENCE REGIONS FOR REGRESSION PARAMETERS.
     AUSTRAL. J. STATIST., 4, 4-10.  (MR25-1617)

MARKOFF, ANDRE  SEE  MARKOV, A. A.


MARKOV, A. A. 1907.  (17.1/B)                                                  3671
     RECHERCHES SUR UN CAS REMARQUABLE D'EPREUVES DEPENDANTES.
          (IN RUSSIAN. TRANSLATED AS NO. 3672)
     IZV. AKAD. NAUK SER. 6, 1, 61-80.

MARKOV, A. A. 1909.  (17.2)                                                    3672
     RECHERCHES SUR UN CAS REMARQUABLE D'EPREUVES DEPENDANTES.
          (TRANSLATION OF NO. 3671)
     ACTA MATH., 33, 87-104.

MARKOV, A. A. 1924.  (8.8)                                                     3673
     UBER DIE ELLIPSOIDE UND DIE ELLIPSE DER DISPERSION UND KORRELATION.
     IZV. AKAD. NAUK SER. 6, 18, 117-126.

MARRIOTT, F.H.C. 1952.  (12.2)                                                3674
     TESTS OF SIGNIFICANCE IN CANONICAL ANALYSIS.
     BIOMETRIKA. 39, 58-64.  (MR13,P.963)


MARRIOTT, F.H.C. + POPE, J. A. 1954.  (16.1/U)                                3675
     BIAS IN THE ESTIMATION OF AUTOCORRELATIONS.
     BIOMETRIKA, 41, 390-402.  (MR16,P.385)

MARSAGLIA, GEORGE 1954.  (17.1)                                               3676
     ITERATED LIMITS AND THE CENTRAL LIMIT THEOREM FOR DEPENDENT RANDOM VARIABLES.
     PROC. AMER. MATH. SOC., 5, 987-991.  (MR16,P.494)

MARSAGLIA, GEORGE 1957.  (2.1/R)                                              3677
     A NOTE ON THE CONSTRUCTION OF A MULTIVARIATE NORMAL SAMPLE.
     IRE TRANS. INFORMATION THEORY, IT-3, 149.

MARSAGLIA, GEORGE 1963.  (2.1/2.7)                                            3678
     EXPRESSING THE NORMAL DISTRIBUTION WITH COVARIANCE MATRIX A+B IN
     TERMS OF ONE WITH COVARIANCE MATRIX A.
     BIOMETRIKA. 50, 535-538.  (MR31-5290)

MARSAGLIA, GEORGE 1964.  (2.1)                                                3679
     CONDITIONAL MEANS AND COVARIANCES OF NORMAL VARIABLES WITH SINGULAR
     COVARIANCE MATRIX.
     J. AMER. STATIST. ASSOC., 59, 1203-1204.

MARSAGLIA, GEORGE 1965.  (20.2)                                              3680
     SHORT PROOF OF A RESULT ON DETERMINANTS.
     AMER. MATH. MONTHLY, 72, 173.

MARSAGLIA, GEORGE 1965.  (3.3/17.5)                                          3681
     RATIOS OF NORMAL VARIABLES AND RATIOS OF SUMS OF UNIFORM VARIABLES.
     J. AMER. STATIST. ASSOC., 60, 193-204.

MARSAGLIA, GEORGE   SEE ALSO
     2103. GRAYBILL, FRANKLIN A. + MARSAGLIA, GEORGE (1957)

MARSCHAK, JACOB 1942.  (16.2/17.7)                                           3682
     ECONOMIC INTERDEPENDENCE AND STATISTICAL ANALYSIS.
     STUD. MATH. ECON. ECONOMET. (SCHULTZ VOLUME), 135-150.

MARSCHAK, JACOB + ANDREWS, WILLIAM H., JR. 1944.  (16.2/E)                   3683
     RANDOM SIMULTANEOUS EQUATIONS AND THE THEORY OF PRODUCTION.
          (AMENDED BY NO. 3684)
     ECONOMETRICA, 12, 143-205.  (MR6,P.238)

MARSCHAK, JACOB + ANDREWS, WILLIAM H., JR. 1945.  (16.2/E)                    3684
    ERRATA:  RANDOM SIMULTANEOUS EQUATIONS AND THE THEORY OF PRODUCTION.
        (AMENDMENT OF NO. 3683)
    ECONOMETRICA, 13, 91.  (MR6,P.238)


MARSHALL, ALBERT W. 1960.  (17.2)                                            3685
    A ONE-SIDED ANALOG OF KOLMOGOROV'S INEQUALITY.
    ANN. MATH. STATIST., 31, 483-487.  (MR22-9995)


MARSHALL, ALBERT W. + OLKIN, INGRAM 1960.  (17.2)                            3686
    A ONE-SIDED INEQUALITY OF THE CHEBYSHEV TYPE.
    ANN. MATH. STATIST., 31, 488-491.  (MR22-9996)


MARSHALL, ALBERT W. + OLKIN, INGRAM 1960.  (17.2)                            3687
    MULTIVARIATE CHEBYSHEV INEQUALITIES.
    ANN. MATH. STATIST., 31, 1001-1014.  (MR22-10000)


MARSHALL, ALBERT W. + OLKIN, INGRAM 1960.  (17.2/B)                          3688
    A BIVARIATE CHEBYSHEV INEQUALITY FOR SYMMETRIC CONVEX POLYGONS.
    CONTRIB. PROB. STATIST. (HOTELLING VOLUME), 299-308.  (MR22-11418)


MARSHALL, ALBERT W. + OLKIN, INGRAM 1961.  (17.2)                            3689
    GAME THEORETIC PROOF THAT CHEBYSHEV INEQUALITIES ARE SHARP.
    PACIFIC J. MATH., 11, 1421-1429.  (MR25-1592)


MARSHALL, ALBERT W. + OLKIN, INGRAM 1964.  (20.1/20.5)                       3690
    REVERSAL OF THE LYAPUNOV, HOLDER, AND MINKOWSKI INEQUALITIES AND
    OTHER EXTENSIONS OF THE KANTOROVICH INEQUALITY.
    J. MATH. ANAL. APPL., 8, 503-514.


MARSHALL, ALBERT W. + OLKIN, INGRAM 1964.  (17.2/20.1)                       3691
    INCLUSION THEOREMS FOR EIGENVALUES FROM PROBABILITY INEQUALITIES.
    NUMER. MATH., 7, 98-102.  (MR29-2824)


MARSHALL, ALBERT W.   SEE ALSO
    560. BIRNBAUM, Z. W. + MARSHALL, ALBERT W. (1961)


MARTIN, ALFONSO GUIRAUM  SEE  GUIRAUM MARTIN, ALFONSO


MARTIN, D. 1946.  (14.5)                                                     3692
    ON THE RADIAL ERROR IN A GAUSSIAN ELLIPTICAL SCATTER.
    PHILOS. MAG. SER. 7, 37, 636-639.  (MR8,P.523)


MARTIN, DONALD C.   SEE
    698. BRADLEY, RALPH A.; MARTIN, DONALD C. + WILCOXON, FRANK (1965)


MARTIN, E.S. 1936.  (6.1/8.3/E)                                              3693
    A STUDY OF THE EGYPTIAN SERIES OF MANDIBLES WITH SPECIAL REFERENCE TO
    MATHEMATICAL METHODS OF SEXING.
    BIOMETRIKA, 28, 149-178.


MARTIN, LEOPOLD J. 1960.  (8.8/17.7)                                         3694
    AJUSTEMENT D'UN FAISCEAU DE REGRESSIONS CURVILIGNES AU MOYEN D'UN SYSTEME DE
    POLYNOMES ORTHOGONAUX.
    BIOM. PRAX., 1, 35-52.


MARTIN, W. P.   SEE
    1081. COX, GERTRUDE M. + MARTIN, W. P. (1939)


MARTINDALE, J. G. 1941.  (16.3/U)                                            3695
    A CORRELATION PERIODOGRAPH FOR THE MEASUREMENT OF PERIODS IN
    DISTURBED WAVE-FORMS.
    J. TEXTILE INST., 32, T71-T82.


MARTINEZ SALAS, J.  SEE  SALAS, J. MARTINEZ


MARTINS, OCTAVIO A. LINS 1946.  (15.4)                                       3696
    NOTE ON A COMMENT ON THE "CORRECTION" OF RELIABILITY COEFFICIENTS FOR
    RESTRICTION OF RANGE.
        (AMENDMENT OF NO. 2901, AMENDMENT OF NO. 1236)
    J. EDUC. PSYCHOL., 37, 182-183.

MARTINS, OCTAVIO A. LINS 1946.   (4.4)                                         3697
    SOBRE O COEFFICIENTE DE CORRELACAO BISSERIAL: ESTUDO EXPERIMENTAL DA
    DISTRIBUCAO DE SEUS VALORES OBTIDOS EM PEQUENAS AMOSTRAS.
    REV. BRASIL ESTATIST., 7, 713-762.

MARTINS, OCTAVIO A. LINS 1947.   (15.1/15.2)                                   3698
    O METODO FACTORIAL DE INVESTIGACAO DAS FACULDADES MENTAIS: ANALISE DE
    RESULTADOS EXPERIMENTAIS OBTIDOS EM SAO PAULO EM 1944.
    REV. BRASIL ESTATIST., 8, 303-338.

MARTYNIHINA, T.F. 1959.   (18.2/B)                                             3699
    AN ANALYTICAL METHOD FOR ESTABLISHING CORRELATION EQUATIONS.
        (IN RUSSIAN)
    IZV. VYSS. UCEBN. ZAVED. MAT., 2(9) 138-143.   (MR24-A613)

MARUYAMA, GISHIRO 1947.   (16.4/16.5/20.5)                                     3700
    REPRESENTATIONS OF THE INTEGRATION OF NORMAL PROBABILITY PROCESS.
        (IN JAPANESE)
    RES. MEM. INST. STATIST. MATH. TOKYO, 3, 55-60.

MARUYAMA, GISHIRO + TANAKA, HIROSHI 1959.   (16.4)                             3701
    ERGODIC PROPERTY OF N-DIMENSIONAL RECURRENT MARKOV PROCESSES.
    MEM. FAC. SCI. KYUSHU UNIV. SER. A, 13, 157-172.   (MR22-3030)

MASANI, PESI R. 1958.   (16.4/A)                                               3702
    SUR LA PREVISION LINEAIRE D'UN PROCESSUS VECTORIEL A DENSITE
    SPECTRALE NON BORNEE.
    C.R. ACAD. SCI. PARIS, 246, 2337-2339.   (MR20-4326B)

MASANI, PESI R. 1959.   (16.4)                                                 3703
    SUR LA FONCTION GENERATRICE D'UN PROCESSUS STOCHASTIQUE VECTORIEL.
    C.R. ACAD. SCI. PARIS, 249, 360-362.   (MR21-6024)

MASANI, PESI R. 1959.   (16.4)                                                 3704
    ISOMORPHIE ENTRE LES DOMAINES TEMPOREL ET SPECTRAL D'UN PROCESSUS
    VECTORIEL, REGULIER.
    C.R. ACAD. SCI. PARIS, 249, 496-498.   (MR23-A2234)

MASANI, PESI R. 1959.   (16.4)                                                 3705
    CRAMER'S THEOREM ON MONOTONE MATRIX-VALUED FUNCTIONS AND THE WOLD DECOMPOSITION.
    PROB. STATIST. (CRAMER VOLUME), 175-189.   (MR23-A2236)

MASANI, PESI R. 1960.   (16.4)                                                 3706
    THE PREDICTION THEORY OF MULTIVARIATE STOCHASTIC PROCESSES.  III.
        UNBOUNDED SPECTRAL DENSITIES.
    ACTA MATH., 104, 141-162.   (MR22-12679)

MASANI, PESI R. 1962.   (16.4)                                                 3707
    SHIFT INVARIANT SPACES AND PREDICTION THEORY.
    ACTA MATH., 107, 275-290.   (MR25-4344)

MASANI, PESI R. 1966.   (16.4)                                                 3708
    WIENER'S CONTRIBUTIONS TO GENERALIZED HARMONIC ANALYSIS, PREDICTION THEORY AND
    FILTER THEORY.
    BULL. AMER. MATH. SOC., 72, 73-125.

MASANI, PESI R. 1966.   (16.4)                                                 3709
    RECENT TRENDS IN MULTIVARIATE PREDICTION THEORY.
    MULTIVARIATE ANAL. PROC. INTERNAT. SYMP. DAYTON (KRISHNAIAH), 351-382.   (MR35-5079)

MASANI, PESI R. + WIENER, NORBERT 1959.   (16.4)                               3710
    NON-LINEAR PREDICTION.
    PROB. STATIST. (CRAMER VOLUME), 190-212.

MASANI, PESI R. + WIENER, NORBERT 1959.   (16.4)                               3711
    ON BIVARIATE STATIONARY PROCESSES AND THE FACTORIZATION OF MATRIX-VALUED FUNCTIONS.
        (REPRINTED AS NO. 3712)
    TEOR. VEROJATNOST. I PRIMENEN., 4, 322-331.   (MR22-5074)

MASANI, PESI R. + WIENER, NORBERT 1959.   (16.4)                               3712
    ON BIVARIATE STATIONARY PROCESSES AND THE FACTORIZATION OF MATRIX-
    VALUED FUNCTIONS.
        (REPRINT OF NO. 3711)
    THEORY PROB. APPL., 4, 300-308.   (MR23-A695)

MASANI, PESI R.    SEE ALSO
   5883. WIENER, NORBERT + MASANI, PESI R. (1957)

   5884. WIENER, NORBERT + MASANI, PESI R. (1958)

   5885. WIENER, NORBERT + MASANI, PESI R. (1958)

   5886. WIENER, NORBERT + MASANI, PESI R. (1958)

MASLOV, P. 1950.   (4.8)                                                  3713
   SERIES OF CORRELATION COEFFICIENTS.
   VOPROSY STATIST. IZU. SVJAZI, 62-64.

MASLOV, P. 1955.   (8.8/16.5)                                             3714
   SMOOTHING DYNAMIC SERIES BY MEANS OF LEAST SQUARES.
       (IN RUSSIAN)
   VESTNIK STATIST. SER. 2, 1, 68-72.

MASSEY, FRANK J., JR.    SEE
   1416. DUNN, OLIVE JEAN + MASSEY, FRANK J., JR. (1965)

MASSY, WILLIAM F. 1965.   E(6.3)                                          3715
   DISCRIMINATION ANALYSIS OF AUDIENCE CHARACTERISTICS.
   J. ADV. RES., 5(1) 39-48.

MASSY, WILLIAM F. 1965.   (11.2)                                          3716
   PRINCIPAL COMPONENTS REGRESSION IN EXPLORATORY STATISTICAL RESEARCH.
   J. AMER. STATIST. ASSOC., 60, 234-256.

MASSY, WILLIAM F.    SEE ALSO
   1754. FRANK, RONALD E.; MASSY, WILLIAM F. + MORRISON, DONALD G. (1965)

MASTERS, J.I. 1955.   (14.5/ET)                                          3717
   SOME APPLICATIONS ON PHYSICS OF THE P FUNCTIONS.
   J. CHEM. PHYS., 23, 1865-1874.

MASUYAMA, MOTOSABURO 1939.   (9.3)                                        3718
   CORRELATION BETWEEN TENSOR QUANTITIES.
   PROC. PHYS. MATH. SOC. JAPAN SER. 3, 21, 638-647.   (MR1,P.151)

MASUYAMA, MOTOSABURO 1939.   (20.1)                                       3719
   TENSOR CHARACTERISTIC OF VECTOR SET AND ITS APPLICATION TO GEOPHYSICS.
   PROC. PHYS. MATH. SOC. JAPAN SER. 3, 21, 647-655.

MASUYAMA, MOTOSABURO 1940.   (4.9)                                        3720
   ON THE MEANING OF THE SYMMETRIC CORRELATION COEFFICIENT BETWEEN VECTOR SETS.
   PROC. PHYS. MATH. SOC. JAPAN SER. 3, 22, 579-585.   (MR2,P.110)

MASUYAMA, MOTOSABURO 1940.   (4.9)                                        3721
   ON THE SUBDEPENDENCY.
   PROC. PHYS. MATH. SOC. JAPAN SER. 3, 22, 855-858.   (MR2,P.233)

MASUYAMA, MOTOSABURO 1940.   (4.9)                                        3722
   THE VARIANCE TENSOR OF VECTOR SET AND A NATURE OF THE SYMMETRIC
   CORRELATION COEFFICIENT.
   PROC. PHYS. MATH. SOC. JAPAN SER. 3, 22, 858-861.   (MR2,P.234)

MASUYAMA, MOTOSABURO 1941.   (3.1/17.5)                                   3723
   THE STANDARD ERROR OF THE MEAN VECTOR.
   PROC. PHYS. MATH. SOC. JAPAN SER. 3, 23, 194-195.   (MR3,P.6)

MASUYAMA, MOTOSABURO 1941.   (2.1)                                        3724
   THE NORMAL LAW OF FREQUENCY FOR VECTOR QUANTITIES.
   PROC. PHYS. MATH. SOC. JAPAN SER. 3, 23, 196-199.   (MR3,P.6)

MASUYAMA, MOTOSABURO 1941.   (4.9)                                        3725
   ON THE CHARACTERISTIC VALUES OF THE CORRELATION TENSOR AND A NEW
   MEASURE OF CORRELATION BETWEEN VECTOR QUANTITIES.
   PROC. PHYS. MATH. SOC. JAPAN SER. 3, 23, 199-204.   (MR3,P.6)

MASUYAMA, MOTOSABURO 1941.   (4.9)                                        3726
   THE TOTALLY ORTHONORMALISED VECTOR SET AND THE NORMAL FORM OF
   CORRELATION TENSOR.
   PROC. PHYS. MATH. SOC. JAPAN SER. 3, 23, 346-351.   (MR3,P.6)

MASUYAMA, MOTOSABURO 1941.  (9.3/20.5)                    3727
    THE MEAN ANGLE BETWEEN TWO VECTOR SETS.
    PROC. PHYS. MATH. SOC. JAPAN SER. 3, 23, 351-355.  (MR3,P.6)

MASUYAMA, MOTOSABURO 1941.  (9.3)                         3728
    CORRELATION COEFFICIENT BETWEEN TWO SETS OF COMPLEX VECTORS.
    PROC. PHYS. MATH. SOC. JAPAN SER. 3, 23, 918-924.  (MR7,P.316)

MASUYAMA, MOTOSABURO 1941.  (4.9)                         3729
    ON THE SIGNIFICANCE TEST OF THE ADDITIVE CORRELATION COEFFICIENT.
    PROC. PHYS. MATH. SOC. JAPAN SER. 3, 23, 1016-1019.  (MR7,P.316)

MASUYAMA, MOTOSABURO 1942.  (17.2)                        3730
    THE BIENAYME-TCHEBYCHEFF INEQUALITY FOR HERMITIC TENSORS.
    PROC. PHYS. MATH. SOC. JAPAN SER. 3, 24, 409-411.  (MR7,P.310)

MASUYAMA, MOTOSABURO 1948.  (18.5)                        3731
    SOME MEASURES OF VARIABILITIES IN SAMPLING FROM FINITE MULTIVARIATE POPULATIONS.
        (IN JAPANESE, WITH SUMMARY IN ENGLISH)
    BULL. MATH. STATIST., 2(3) 53-54, 61-62.


MASUYAMA, MOTOSABURO 1952.  (6.4)                         3732
    THE MISCLASSIFICATION IN THE SAMPLING INSPECTION.
    REP. STATIST. APPL. RES. UN. JAPAN. SCI. ENGRS., 1(4) 7-9.  (MR14,P.665)

MASUYAMA, MOTOSABURO 1952.  (14.2)                        3733
    THE EXACT DISTRIBUTION OF GEARY'S STATISTIC AND ITS GENERALIZATION.
    REP. STATIST. APPL. RES. UN. JAPAN. SCI. ENGRS., 2(1) 1-3.  (MR14,P.775)

MASUYAMA, MOTOSABURO 1953.  (18.6)                        3734
    A RAPID METHOD OF ESTIMATING BASAL AREA IN TIMBER SURVEY--AN
    APPLICATION OF INTEGRAL GEOMETRY TO AREAL SAMPLING PROBLEMS.
    SANKHYA, 12, 291-302.  (MR15,P.332)

MATERN, BERTIL 1949.  (2.5)                               3735
    INDEPENDENCE OF NON-NEGATIVE QUADRATIC FORMS IN NORMALLY CORRELATED VARIABLES.
    ANN. MATH. STATIST., 20, 119-120.  (MR10,P.553)

MATERN, BERTIL 1962.  (3.1/3.2/17.5)                      3736
    ON THE USE OF INFORMATION ON SUPPLEMENT VARIABLES IN ESTIMATING A DISTRIBUTION.
    REV. INST. INTERNAT. STATIST., 30, 121-135.  (MR28-1686)

MATHAI, A. M. 1963.  (2.2/17.3/18.1)                      3737
    ON MULTIVARIATE EXPONENTIAL TYPE DISTRIBUTIONS.
    J. INDIAN STATIST. ASSOC., 4, 143-154.

MATHAI, A. M. 1966.  (3.2)                                3738
    SAMPLING DISTRIBUTIONS UNDER MISSING VALUES.
    TRABAJOS ESTADIST., 17(2) 59-83.  (MR34-8528)

MATHAI, A. M. + SAXENA, ASHOK K. 1966.  (20.4)           3739
    ON A GENERALIZED HYPERGEOMETRIC DISTRIBUTION.
    METRIKA, 11, 127-132.

MATHER, K.   SEE
    6041. YATES, FRANK + MATHER, K. (1963)

MATHEWS, M. V.   SEE
    3129. KRAMER, H. P. + MATHEWS, M. V. (1956)

MATHISEN, HAROLD C. 1943.  (17.8/U)                       3740
    A METHOD OF TESTING THE HYPOTHESIS THAT TWO SAMPLES ARE FROM THE SAME
    POPULATION.
    ANN. MATH. STATIST., 14, 188-194.  (MR5,P.128)

MATSUDA, K. + ROHLF, F. J. 1961.  E(11.1)                 3741
    STUDIES OF RELATIVE GROWTH IN CERRIDAO. (5) COMPARISON OF TWO POPULATIONS.
    GROWTH, 25, 211-217.

MATSUMOTO, HIROSHI   SEE
    5526. TSUDA, TAKAO + MATSUMOTO, HIROSHI (1966)

MATSUMURA, SOJI 1951. (4.1)
BEMERKUNG ZU KORRELATIONS THEORIE.
J. OSAKA INST. SCI. TECH. PART I, 3, 33–34. (MR15,P.45)

3742

MATTSON, R. L. + DAMMAN. J. E. 1965. (6.6)
A TECHNIQUE FOR DETERMINING AND CODING SUBCLASSES IN PATTERN
RECOGNITION PROBLEMS.
IBM SYSTEMS J., 23, 294–302.

3743

MATTSON, R. L.   SEE ALSO
4278. PETERSON, D. W. + MATTSON, R. L. (1966)

MATUFUJI, HAJIME; YAMAZAKI, KAZUHIDE + KOMAZAWA, TSUTOMU 1966. (15.1/E)
FACTOR ANALYSIS OF CONTRACTION OF APPENDICITIS.
(IN JAPANESE)
MED. AND BIOL. (IGAKU TO SEIBUTSUGAKU), 72(5) 282–286.

3744

MATUSITA, KAMEO 1949. (2.5)
NOTE ON THE INDEPENDENCE OF CERTAIN STATISTICS.
ANN. INST. STATIST. MATH. TOKYO, 1, 79–82. (MR11,P.260)

3745

MATUSITA, KAMEO 1951. (19.1)
ON THE THEORY OF STATISTICAL DECISION FUNCTIONS.
(AMENDED BY NO. 3747)
ANN. INST. STATIST. MATH. TOKYO, 3, 17–35. (MR13,P.668)

3746

MATUSITA, KAMEO 1952. (19.1)
CORRECTION TO THE PAPER "ON THE THEORY OF STATISTICAL DECISION FUNCTIONS".
(AMENDMENT OF NO. 3746)
ANN. INST. STATIST. MATH. TOKYO, 4, 51–53. (MR14,P.488)

3747

MATUSITA, KAMEO 1956. (6.4/19.1)
DECISION RULE, BASED ON THE DISTANCE, FOR THE CLASSIFICATION PROBLEM.
ANN. INST. STATIST. MATH. TOKYO, 8, 67–77. (MR19,P.186)

3748

MATUSITA, KAMEO 1958. (1.2)
KINSAKU TAKANO: 1915–1958.
ANN. INST. STATIST. MATH. TOKYO, 10, I–II.

3749

MATUSITA, KAMEO 1961. (19.3)
INTERVAL ESTIMATION BASED ON THE NOTION OF AFFINITY.
BULL. INST. INTERNAT. STATIST., 38(4) 241–244. (MR26-7085)

3750

MATUSITA, KAMEO 1964. (6.5/19.1)
DISTANCE AND DECISION RULES.
ANN. INST. STATIST. MATH. TOKYO, 16, 305–315. (MR30-2638)

3751

MATUSITA, KAMEO 1966. (5.2/6.5/10.2)
A DISTANCE AND RELATED STATISTICS IN MULTIVARIATE ANALYSIS.
MULTIVARIATE ANAL. PROC. INTERNAT. SYMP. DAYTON (KRISHNAIAH), 187–200. (MR35-1139)

3752

MATUSITA, KAMEO + AKAIKE, HIROTUGU 1956. (19.1)
DECISION RULES, BASED ON THE DISTANCE, FOR THE PROBLEMS OF
INDEPENDENCE, INVARIANCE AND TWO SAMPLES.
(AMENDED BY NO. 3754)
ANN. INST. STATIST. MATH. TOKYO, 7, 67–80. (MR18,P.158)

3753

MATUSITA, KAMEO + AKAIKE, HIROTUGU 1956. (19.1)
ERRATA TO: DECISION RULES, BASED ON THE DISTANCE, FOR THE PROBLEMS OF
INDEPENDENCE, INVARIANCE AND TWO SAMPLES.
(AMENDMENT OF NO. 3753)
ANN. INST. STATIST. MATH. TOKYO, 7, 221. (MR18,P.158)

3754

MATUSITA, KAMEO + MOTOO, MINORU 1956. (19.1)
ON THE FUNDAMENTAL THEOREM FOR THE DECISION RULE BASED ON DISTANCE.
ANN. INST. STATIST. MATH. TOKYO, 7, 137–142. (MR18,P.158)

3755

MATVEEV, R.F. 1959. (16.4)
ON THE REGULARITY OF MULTIDIMENSIONAL STATIONARY RANDOM PROCESSES WITH
DISCRETE TIME.
(IN RUSSIAN)
DOKL. AKAD. NAUK SSSR, 126, 713–715. (MR22-6017)

3756

MATYAS, JOSEF 1961.  (16.4)                                        3757
     SHAPING FILTERS FOR GENERATING AN ARBITRARY NUMBER OF RANDOM
     PROCESSES WITH PRESCRIBED STATISTICAL PROPERTIES.
   APL. MAT., 6, 274-287.  (MR24-A2518)

MAUCHLY, JOHN W. 1940.  (10.2)                                     3758
     SIGNIFICANCE TEST FOR SPHERICITY OF A NORMAL N-VARIATE DISTRIBUTION.
   ANN. MATH. STATIST., 11, 204-209.  (MR1,P.348)

MAUCHLY, JOHN W. 1940.  (10.2)                                     3759
     A SIGNIFICANCE TEST FOR ELLIPTICITY IN THE HARMONIC DIAL.
   TERREST. MAGNET. ATMOSPH. ELEC., 45, 145-148.

MAULDON, J.G. 1955.  (7.1/C)                                       3760
     PIVOTAL QUANTITIES FOR WISHART'S AND RELATED DISTRIBUTIONS AND A
     PARADOX IN FIDUCIAL THEORY.
   J. ROY. STATIST. SOC. SER. B, 17, 79-85.  (MR17,P.380)

MAUNG, KHINT 1941.  E(6.2/T)                                       3761
     DISCRIMINANT ANALYSIS OF TOCHER'S EYE-COLOUR DATA FOR SCOTTISH SCHOOL CHILDREN.
   ANN. EUGENICS, 11, 64-76.

MAUNG, KHINT 1942.  (4.8/9.3/E)                                    3762
     MEASUREMENT OF ASSOCIATION IN A CONTINGENCY TABLE WITH SPECIAL REFERENCE TO THE
     PIGMENTATION OF HAIR AND EYE COLOURS OF SCOTTISH SCHOOL CHILDREN.
   ANN. EUGENICS, 11, 189-223.

MAU-THANH, LUU-   SEE   LUU-MAU-THANH

MAXFIELD, JOHN E. + GARDNER, ROBERT S. 1955.  (8.8/20.1)          3763
     NOTE ON LINEAR HYPOTHESES WITH PRESCRIBED MATRIX OF NORMAL EQUATIONS.
   ANN. MATH. STATIST., 26, 149-150.  (MR16,P.665)

MAXIMON, LEONARD C. 1956.  (20.4)                                  3764
     ON THE REPRESENTATION OF INDEFINITE INTEGRALS CONTAINING BESSEL
     FUNCTIONS BY SIMPLE NEUMANN SERIES.
   PROC. AMER. MATH. SOC., 7, 1054-1062.  (MR18,P.650)

MAXWELL, A.E. 1956.  (15.1)                                        3765
     FACTOR MODELS.
   J. EDUC. PSYCHOL., 47, 129-132.

MAXWELL, A.E. 1959.  (15.2)                                        3766
     STATISTICAL METHODS IN FACTOR ANALYSIS.
   PSYCHOL. BULL., 56, 228-235.

MAXWELL, A.E. 1961.  (15.1)                                        3767
     RECENT TRENDS IN FACTOR ANALYSIS.
   J. ROY. STATIST. SOC. SER. A, 124, 49-59.  (MR26-4439)

MAXWELL, A.E. 1964.  (15.3/E)                                      3768
     CALCULATING MAXIMUM-LIKELIHOOD FACTOR LOADINGS.
   J. ROY. STATIST. SOC. SER. A, 127, 238-241.

MAXWELL, A.E.   SEE ALSO
     555. BIRNBAUM, ALLAN + MAXWELL, A.E. (1960)

   3312. LAWLEY, D. N. + MAXWELL, A.E. (1962)

   3313. LAWLEY, D. N. + MAXWELL, A.E. (1964)

MAXWELL, JAMES C. 1862.  (20.5)                                    3769
     ON THE GEOMETRICAL MEAN DISTANCE OF TWO FIGURES IN A PLANE.
   TRANS. ROY. SOC. EDINBURGH, 26, 729-733.

MAY, MARK A. 1929.  (4.2)                                          3770
     A METHOD FOR CORRECTING COEFFICIENTS OF CORRELATION FOR HETEROGENEITY
     IN THE DATA.
   J. EDUC. PSYCHOL., 20, 417-423.

MAY, MARK A. 1931.  (15.4)                                         3771
     STATISTICAL METHODS IN PERSONALITY STUDIES:  RELIABILITY.
   J. AMER. STATIST. ASSOC., 26(SP) 168-174.

MAYBU, JOHN S. 1966.  (20.2)
   NEW GENERALIZATIONS OF JACOBI MATRICES.
   SIAM J. APPL. MATH., 14, 1032–1037.
                                                                                3772

MAYNE, DAVID 1966.  (20.3)
   AN ALGORITHM FOR THE CALCULATION OF THE PSEUDO-INVERSE OF A SINGULAR MATRIX.
   COMPUT. J., 9, 312–317.
                                                                                3773

MAYOH, B. H. 1965.  (20.3)
   A GRAPH TECHNIQUE FOR INVERTING CERTAIN MATRICES.
   MATH. COMPUT., 19, 644–646.  (MR33-5108)
                                                                                3774

MEACHAM, ALAN D. 1941.  (20.3)
   THE VALUE OF THE COLLATOR IN USING PREPUNCHED CARDS FOR OBTAINING
   MOMENTS AND PRODUCT MOMENTS.
        (WITH DISCUSSION)
   PROC. EDUC. RES. FORUM, 9–15.
                                                                                3775

MEACHAM, ALAN D.   SEE ALSO
   1468. DWYER, PAUL S. + MEACHAM, ALAN D. (1937)

MEDINA E ISABEL, MARIANO   SEE
   864. CASTANS CARMARGO, MANUEL + MEDINA E ISABEL, MARIANO (1956)

MEDLAND, FRANCIS F. 1947.  (15.1/15.2/E)
   AN EMPIRICAL COMPARISON OF METHODS OF COMMUNALITY ESTIMATION.
   PSYCHOMETRIKA, 12, 101–109.
                                                                                3776

MEDOLAGHI, PAOLO 1933.  (16.5/B)
   SULLE CORRELAZIONI TRA SERIE STATISTICHE RAPPRESENTATIVE DI FENOMENI ECONOMICI.
   BAROMETRO ECON. ITAL., 5, 501–503.
                                                                                3777

MEDVEDEV, YU. I.   SEE
   2721. IVCHENKO, G. I. + MEDVEDEV, YU. I. (1965)

   2722. IVCHENKO, G. I. + MEDVEDEV, YU. I. (1965)

MEEHL, PAUL E. 1945.  (8.1)
   A SIMPLE ALGEBRAIC DEVELOPMENT OF HORST'S SUPPRESSOR VARIABLES.
   AMER. J. PSYCHOL., 58, 550–554.
                                                                                3778

MEERSMAN, R. DE   SEE   DE MEERSMAN, R.

MEETER, DUANE A. 1966.  (17.5)
   ON A THEOREM USED IN NONLINEAR LEAST SQUARES.
   SIAM J. APPL. MATH., 14, 1176–1179.
                                                                                3779

MEETER, DUANE A.   SEE ALSO
   2230. GUTTMAN, IRWIN + MEETER, DUANE A. (1965)

MEGEE, MARY 1965.  (15.1)
   ON ECONOMIC GROWTH AND THE FACTOR ANALYSIS METHOD.
   SOUTHERN. ECON. J., 31, 215–228.
                                                                                3780

MEHLER, F.G. 1866.  (17.1)
   UEBER DIE ENTWICKLUNG EINER FUNCTION VON BELIEBIG VIELEN VARIABLEN
   NACH LAPLACESCHEN FUNCTIONEN HOHERER ORDNUNG.
   J. REINE ANGEW. MATH., 66, 161–176.
                                                                                3781

MEHMKE, RUDOLF 1892.  (20.3)
   UBER DAS SEIDEL'SCHE VERFAHREN, UND LINEARE GLEICHUNGEN BEI EINER SEHR GROSSEN
   ANZAHL DER UNBEKANNTEN DURCH SUCCESSIVE ANNAHERUNG AUFZULOSEN.
   MAT. SB., 16, 342–345.
                                                                                3782

MEHMKE, RUDOLF + NEKRASSOV, P. A. 1892.  (20.3)
   AUFLOSUNG EINES LINEARES SYSTEMS VON GLEICHUNGEN DURCH SUCCESSIVE ANNAHERUNG.
        (IN GERMAN AND RUSSIAN)
   MAT. SB., 16, 437–459.
                                                                                3783

MEHTA, MADAN LAL + GAUDIN, MICHEL 1960.  (13.1)
   ON THE DENSITY OF EIGENVALUES OF A RANDOM MATRIX.
        (AMENDED BY NO. 3785, REPRINTED AS NO. 3786, AMENDED BY NO. 3787)
   NUCLEAR PHYS., 18. 420–427.  (MR22-3741)
                                                                                3784

MEHTA, MADAN LAL + GAUDIN, MICHEL 1961.   (13.1)                           3785
     ADDENDUM AND ERRATUM.
          (AMENDMENT OF NO. 3784, REPRINTED AS NO. 3787, AMENDMENT OF NO. 3786)
     NUCLEAR PHYS., 22, 340.   (MR22-P.2546)

MEHTA, MADAN LAL + GAUDIN, MICHEL 1965.   (13.1)                           3786
     ON THE DENSITY OF EIGENVALUES OF A RANDOM MATRIX.
          (AMENDED BY NO. 3787, REPRINT OF NO. 3784, AMENDED BY NO. 3785)
     STATIST. THEOR. SPECTRA FLUCTUATIONS (PORTER), 342-349.   (MR22-3741)

MEHTA, MADAN LAL + GAUDIN, MICHEL 1965.   (13.1)                           3787
     ADDENDUM AND ERRATUM.
          (AMENDMENT OF NO. 3786, REPRINT OF NO. 3785, AMENDMENT OF NO. 3784)
     STATIST. THEOR. SPECTRA FLUCTUATIONS (PORTER), 350.   (MR22,P.2546)


MEILI, RICHARD 1930.   (15.2)                                             3788
     A PROPOS DE LA THEORIE DES FACTEURS.   REPONSE A MONSIEUR C. SPEARMAN.
     ARCH. DE PSYCHOL. GENEVA, 22, 328-332.

MEILI, RICHARD 1947.   (15.1)                                             3789
     DIE FAKTORENTHEORIE VON CHARLES EDWARD SPEARMAN.
     SCHWEIZ. Z. PSYCHOL. ANWEND., 6, 137-140.

MEIZLER, D.G.; PARASJUK, O. S. + RVACEVA, E. L. 1948.   (17.1)            3790
     A MULTIDIMENSIONAL, LOCAL LIMIT THEOREM OF THE THEORY OF PROBABILITY.
          (IN RUSSIAN)
     DOKL. AKAD. NAUK SSSR NOV. SER., 60, 1127-1128.   (MR10.P.132)

MEIZLER, D.G.; PARASJUK, O. S. + RVACEVA, E. L. 1949.   (17.1)            3791
     ON A MANY DIMENSIONAL LOCAL LIMIT THEOREM OF THE THEORY OF PROBABILITY.
     UKRAIN. MAT. ZUR., 1(1) 9-20.   (MR14,P.61)

MELTON, RICHARD S. 1963.   (6.3)                                          3792
     SOME REMARKS ON FAILURE TO MEET ASSUMPTIONS IN DISCRIMINANT ANALYSES.
     PSYCHOMETRIKA, 28, 49-53.

MENDERSHAUSEN, HORST 1939.   (17.7)                                       3793
     CLEARING VARIATES IN CONFLUENCE ANALYSIS.
     J. AMER. STATIST. ASSOC., 34, 93-105.

MENDLOWITZ, MILTON   SEE
     5997. WOLF, ROBERT L.; MENDLOWITZ, MILTON; ROBOZ, JULIA;   (1966)
     STYAN, GEORGE P. H.; KORNFELD, PETER + WEIGL, ALFRED

MENNES, L. B. M.   SEE
     3058. KLOEK, T. + MENNES, L. B. M. (1960)

MENON, M.V. 1966.   (17.3)                                               3794
     CHARACTERIZATION THEOREMS FOR SOME UNIVARIATE PROBABILITY DISTRIBUTIONS.
     J. ROY. STATIST. SOC. SER. B, 28, 143-145.

MENTASTI, F. 1965.   (16.5)                                              3795
     THE USE OF SPECTRAL ANALYSIS IN ECONOMIC RESEARCH.
          (IN ITALIAN)
     INDUST. MILANO, 4, 537-559.


MENZEL, HERBERT 1950.   (4.8/E)                                          3796
     COMMENT ON ROBINSON'S ECOLOGICAL CORRELATION AND THE BEHAVIOUR  OF INDIVIDUALS.
     AMER. SOCIOL. REV., 15, 674.

MEREDITH, WILLIAM 1964.   (9.3/12.1)                                     3797
     CANONICAL CORRELATIONS WITH FALLIBLE DATE.
     PSYCHOMETRIKA, 29, 55-65.   (MR29-4146)

MEREDITH, WILLIAM 1964.   (15.1)                                         3798
     NOTES ON FACTORIAL INVARIANCE.
     PSYCHOMETRIKA, 29, 177-185.   (MR29-5344)

MEREDITH, WILLIAM 1964.   (15.1)                                         3799
     ROTATION TO ACHIEVE FACTORIAL INVARIANCE.
     PSYCHOMETRIKA, 29, 187-206.   (MR29-5345)

MEREDITH, WILLIAM 1965.   (2.4/3.2/14.3/E)                                          3800
   A METHOD FOR STUDYING DIFFERENCES BETWEEN GROUPS.
   PSYCHOMETRIKA, 30, 15-29.   (MR32-6612)

MEREDITH, WILLIAM 1965.   (15.4/E)                                                  3801
   SOME RESULTS BASED ON A GENERAL STOCHASTIC MODEL FOR MENTAL TESTS.
   PSYCHOMETRIKA, 30, 419-440.

MEREDITH, WILLIAM   SEE ALSO
   5652. VISONHALER, JOHN F. + MEREDITH, WILLIAM (1966)

MERRIL, WALTER W., JR. 1937.   (15.4)                                               3802
   SAMPLING THEORY IN ITEM ANALYSIS.
   PSYCHOMETRIKA, 2, 215-223.

MERRILL, A. S. 1928.   (14.2/BT)                                                    3803
   FREQUENCY DISTRIBUTION OF AN INDEX WHEN BOTH COMPONENTS FOLLOW THE NORMAL LAW.
   BIOMETRIKA, 20(A) 53-63.

MERRIMAN, M. 1877.   (1.2)                                                          3804
   A LIST OF WRITINGS RELATING TO THE METHOD OF LEAST SQUARES WITH
   HISTORICAL AND CRITICAL NOTES.
   TRANS. CONNECTICUT ACAD. ARTS SCI., 4, 151-232.

MERZRATH, E. 1933.   (2.1/2.3/17.3/B)                                               3805
   ANPASSUNG VON FLACHEN AN ZWEIDIMENSIONALE KOLLEKTIVGEGENSTANDE UND
   IHRE AUSWERTUNG FUR DIE KORRELATIONSTHEORIE.
   METRON. 11(2) 103-136.

MESSEDAGLIA, ANGELO + BODIO, LUIGI 1880.   (17.9)                                   3806
   PRESENTAZIONE DEI DIAGRAMMI A TRE DIMENSIONI, O STEREOGRAMMI.
   ESEGUITI DALLA DIREZIONE DI STATISTICA.   (WITH DISCUSSION)
   ANN. STATIST. SER. 2A, 15, 53-60.

MESSEL, HARRY   SEE
   929. CHARTRES, R. A. + MESSEL, HARRY (1954)

MESSICK, SAMUEL J. 1956.   (15.5)                                                   3807
   SOME RECENT THEORETICAL DEVELOPMENTS IN MULTI-DIMENSIONAL SCALING.
   EDUC. PSYCHOL. MEAS., 16, 82-100.

MESSICK, SAMUEL J. + ABELSON, ROBERT P. 1956.   (15.5)                              3808
   THE ADDITIVE CONSTANT PROBLEM IN MULTIDIMENSIONAL SCALING.
   PSYCHOMETRIKA, 21, 1-15.

MESSICK, SAMUEL J.   SEE ALSO
   5546. TUCKER, LEDYARD R. + MESSICK, SAMUEL J. (1963)

   1344. DIEDERICH, GERTRUDE W.; MESSICK, SAMUEL J. + TUCKER, LEDYARD R. (1957)

   2429. HELM, C. E.; MESSICK, SAMUEL J. + TUCKER, LEDYARD R. (1961)

METAKIDES, THEOCHARIS A. 1953.   (6.2)                                              3809
   CALCULATION AND TESTING OF DISCRIMINANT FUNCTIONS.
   TRABAJOS ESTADIST., 4, 339-368.   (MR15,P.728)

METHENY, ELEANOR   SEE
   3549. MC CLOY, C. H.; METHENY, ELEANOR + KNOTT, VIRGINIA (1938)

MEYER, C. D.   SEE
   2104. GRAYBILL, FRANKLIN A.; MEYER, C. D. + PAINTER, R. J. (1966)

MEYER, H. ARTHUR + DEMING, W. EDWARDS 1935.   (6.2)                                 3810
   ON THE INFLUENCE OF CLASSIFICATION ON THE DETERMINATION OF A
   MEASURABLE CHARACTERISTIC.
   J. AMER. STATIST. ASSOC., 30, 671-677.

MEYER, PAUL L.   SEE
   561. BIRNBAUM, Z. W. + MEYER, PAUL L. (1953)

MICHAEL, E. L.   SEE
   3557. MC EWEN, G. F. + MICHAEL, E. L. (1919)

MICHAEL, WILLIAM B. 1956.   (4.5/T)                                              3811
    TABLES TO FACILITATE COMPUTATION OF PARTIAL CORRELATION COEFFICIENTS.
  EDUC. PSYCHOL. MEAS., 16, 232-236.

MICHAEL, WILLIAM B.; PERRY, NORMAN C. + GUILFORD, J. P. 1952.   (4.4/T)          3812
    THE ESTIMATION OF A POINT BISERIAL COEFFICIENT OF CORRELATION FROM A
    PHI COEFFICIENT.
  BRITISH J. PSYCHOL. STATIST. SECT., 5, 139-150.

MICHAEL, WILLIAM B.; JONES, ROBERT A.; GADDIS, L. WESLEY + 1962.   (4.1/T)       3813
  KAISER, HENRY F.
    ABACS FOR DETERMINATION OF A CORRELATION COEFFICIENT CORRECTED FOR
    RESTRICTION OF RANGE.
  PSYCHOMETRIKA, 27, 197-202.

MICHAEL, WILLIAM B.   SEE ALSO
    2172. GUILFORD, J. P. + MICHAEL, WILLIAM B. (1948)

    2173. GUILFORD, J. P. + MICHAEL, WILLIAM B. (1950)

    4268. PERRY, NORMAN C. + MICHAEL, WILLIAM B. (1954)

    4269. PERRY, NORMAN C. + MICHAEL, WILLIAM B. (1958)

MICHENER, C. D.; LANGE, R. B.; BIGARELLA, J. J. + 1958. E(17.9)                  3814
  SALAMUNI, P.
    FACTORS INFLUENCING THE DISTRIBUTION OF BEE'S NESTS IN EARTH BANKS.
  ECOLOGY, 39, 207-217.

MICKEY, RAY 1959.   (13.1)                                                       3815
    SOME BOUNDS ON THE DISTRIBUTION FUNCTION OF THE LARGEST AND SMALLEST
    ROOTS OF NORMAL DETERMINANTAL EQUATIONS.
  ANN. MATH. STATIST., 30, 242-243.   (MR21-3068)

MIDDLETON, DAVID 1948.   (16.4)                                                  3816
    SOME GENERAL RESULTS IN THE THEORY OF NOISE THROUGH NON-LINEAR DEVICES.
  QUART. APPL. MATH., 5, 445-498.   (MR9,P.362)

MIDDLETON, DAVID 1950.   (6.6/B)                                                 3817
    NOISE AND NONLINEAR COMMUNICATION PROBLEMS.
  SYMP. APPL. AUTOCORREL. ANAL. PHYS. PROBLEMS (WOODS HOLE), 24-44.

MIDDLETON, DAVID   SEE ALSO
    4277. PETERSEN, DANIEL P. + MIDDLETON, DAVID (1964)

MIHOC, GH. 1963.   (16.4/17.1)                                                   3818
    LEGEA LIMITA SUMELE DE VARIABILE VECTORIALE ALE UNUI LANT CU
    LEGATURI COMPLETE STATIONAR MULTIPLU.
  COMUN. ACAD. REPUB. POP. ROMINE, 13, 5-9.

MIHOC, GH. 1963.   (16.4/17.1)                                                   3819
    LA LOI LIMITE POUR LES SOMMES DES VARIABLES D'UNE CHAINE A LIAISON
    COMPLETE, STATIONNAIRE ET MULTIPLE.
  REV. MATH. PURES APPL., 8, 217-226.

MIHOC, GH. + THEILER, G. 1961.   (17.9)                                          3820
    THE LAWS OF PROBABILITY FOR CERTAIN STATISTICAL INDICES.
        (IN ROUMAINIAN, WITH SUMMARY IN RUSSIAN AND FRENCH)
  AN. UNIV. BUCURESTI SER. STI. NATUR. MAT. FIZ., 10(29) 233-253.   (MR30-5420)

MIJARES, TITO A. 1961.   (8.4/13.1)                                              3821
    THE MOMENTS OF ELEMENTARY SYMMETRIC FUNCTIONS OF THE ROOTS OF A
    MATRIX IN MULTIVARIATE ANALYSIS.
  ANN. MATH. STATIST., 32, 1152-1160.   (MR24-A610)

MIJARES, TITO A. 1964.   (8.4)                                                   3822
    ON ELEMENTARY SYMMETRIC FUNCTIONS OF THE ROOTS OF A MULTIVARIATE
    MATRIX: DISTRIBUTIONS.
  ANN. MATH. STATIST., 35, 1186-1198.   (MR29-2906)

MIJARES, TITO A.   SEE ALSO
    4300. PILLAI, K.C.SREEDHARAN + MIJARES, TITO A. (1959)

MIKHAIL, N. N. 1965.   (8.4/13.2/T)                                              3823
    A COMPARISON OF TESTS OF THE WILKS- LAWLEY HYPOTHESIS IN MULTIVARIATE ANALYSIS.
    BIOMETRIKA, 52, 149-156.   (MR35-3810)

MIKHAIL, WADIE F. 1962.   (10.1)                                                 3824
    ON A PROPERTY OF A TEST FOR THE EQUALITY OF TWO NORMAL DISPERSION
    MATRICES AGAINST ONE-SIDED ALTERNATIVES.
    ANN. MATH. STATIST., 33, 1463-1465.   (MR25-4602)

MIKHAIL, WADIE F.   SEE ALSO
    4696. ROY, S.N. + MIKHAIL, WADIE F. (1961)

MIKIEWICZ, J. 1963.   (6.1/6.6)                                                  3825
    ON LEVELS OF CONFIDENCE IN WROCLAW TAXONOMY.
        (IN POLISH)
    POLSKA AKAD. NAUK ZASTOS. MAT., 7, 1-40.

MILES, R. E. 1965.   (18.6)                                                      3826
    ON RANDOM ROTATIONS IN $R^3$.
    BIOMETRIKA, 52, 636-639.   (MR35-1052)

MILICER-GRUZEWSKA, HALINA 1946.   (17.3/A)                                       3827
    THE COEFFICIENT OF CORRELATION A POSTERIORI OF EQUIVALENT VARIABLES.
    TOWARZ. NAUK. WARSZAWSK. SPRAW. POS. WYD. III NAUK MAT. FIZ., 39, 3-17.   (MR11,P.374)

MILICER-GRUZEWSKA, HALINA 1951.   (17.3/B)                                       3828
    SUR LA DISTRIBUANTE DE DEUX VARIABLES DEPENDANTES.
    C.R. ACAD. SCI. PARIS, 233, 1256-1258.   (MR13,P.664)

MILLER, C. R.; SABAGH, G. + DINGMAN, HARVEY F. 1962.   E(15.5)                   3829
    LATENT CLASS ANALYSIS AND DIFFERENTIAL MORTALITY.
    J. AMER. STATIST. ASSOC., 57, 430-438.

MILLER, GEORGE A. + GARNER, W. R. 1944.   (15.5/E)                               3830
    EFFECT OF RANDOM PRESENTATION ON THE PSYCHOMETRIC FUNCTION
        IMPLICATIONS FOR A QUANTAL THEORY OF DISCRIMINATION.
    AMER. J. PSYCHOL., 57, 451-467.

MILLER, KENNETH S. 1963.   (16.3/16.5)                                           3831
    A NOTE ON INPUT-OUTPUT SPECTRAL DENSITIES.
    QUART. APPL. MATH., 21, 249-252.   (MR27-6636)

MILLER, KENNETH S. 1964.   (7.4)                                                 3832
    DISTRIBUTIONS INVOLVING NORMS OF CORRELATED GAUSSIAN VECTORS.
    QUART. APPL. MATH., 22, 235-243.

MILLER, KENNETH S. 1965.   (14.2)                                               3833
    SOME MULTIVARIATE DENSITY FUNCTIONS OF PRODUCTS OF GAUSSIAN VARIATES.
    BIOMETRIKA, 52, 645-646.   (MR34-6830)

MILLER, KENNETH S. + ABRAMSON, LEE R. 1965.   (16.4)                            3834
    A REPRESENTATION OF GAUSSIAN PROCESSES.
    INTERNAT. J. ENGRG. SCI., 2, 431-439.

MILLER, KENNETH S. + BERNSTEIN, R. I. 1957.   (16.3/16.4/E)                     3835
    AN ANALYSIS OF COHERENT INTEGRATION AND ITS APPLICATION TO SIGNAL DETECTION.
    IRE TRANS. INFORMATION THEORY, IT-3, 237-248.

MILLER, KENNETH S. + SACKROWITZ, HAROLD 1965.   (2.1/14.2)                      3836
    DISTRIBUTIONS ASSOCIATED WITH THE QUADRIVARIATE NORMAL.
    INDUST. MATH., 15(2) 1-14.

MILLER, KENNETH S.; BERNSTEIN, R. I. + BLUMENSON, L. E. 1958.   (7.4/16.4/U)    3837
    GENERALIZED RAYLEIGH PROCESSES.
        (AMENDED BY NO. 1728, AMENDED BY NO. 3838)
    QUART. APPL. MATH., 16, 137-145.   (MR20-1371)

MILLER, KENNETH S.; BERNSTEIN, R. I. + BLUMENSON, L. E. 1963.   (7.4/16.4/U)    3838
    A NOTE ON THE PAPER: GENERALIZED RAYLEIGH PROCESSES.
        (AMENDMENT OF NO. 3837)
    QUART. APPL. MATH., 20, 395.   (MR20-1371)

MILLER, KENNETH S.    SEE ALSO
     601. BLUMENSON, L. E. + MILLER, KENNETH S. (1963)

    1024. CONNOLLY, T. W. + MILLER, KENNETH S. (1961)


MILLER, ROBERT G. 1962.   (6.3/6.4/8.5/E)                                3839
     STATISTICAL PREDICTION BY DISCRIMINANT ANALYSIS.
    METEOROL. MONOG., 4(25) 1-54.

MILLER, WILBUR C.    SEE
    2113. GREENE, JOEL E.; MILLER, WILBUR C. + BROWN, VIRGINIA M. (1956)


MILLS, FREDERICK C. 1924.   (4.6/4.8)                                    3840
     THE MEASUREMENT OF CORRELATION AND THE PROBLEM OF ESTIMATION.
    J. AMER. STATIST. ASSOC., 19, 273-300.


MINE, AKIKO 1955.   (16.5)                                               3841
     ESTIMATION OF LINEAR REGRESSION COEFFICIENTS IN TIME SERIES.   (A NOTE
    ON THE GENERALIZATION OF MINIMUM DISTANCE METHOD.)
    ANN. INST. STATIST. MATH. TOKYO, 6, 181-189.   (MR17,P.505)


MIRSKY, L. 1964.   (20.1)                                                3842
     INEQUALITIES AND EXISTENCE THEOREMS IN THE THEORY OF MATRICES.
    J. MATH. ANAL. APPL., 9, 99-118.


MIRSKY, L. 1966.   (20.1)                                                3843
     AN INEQUALITY FOR CHARACTERISTIC ROOTS AND SINGULAR VALUES OF COMPLEX MATRICES.
    MONATSHEFTE MATH., 70, 357-359.


MISRA, R. K. 1966.   E(8.5/11.1)                                         3844
     VECTORIAL ANALYSIS FOR GENETIC CLINES IN BODY DIMENSIONS IN
    POPULATIONS OF DROSOPHILIA SUBOBSCURA COLL. AND A COMPARISON WITH
    THOSE OF D. ROBUSTA STURT.
    BIOMETRICS, 22, 469-487.


MITALAUSKAS, A. A. 1960.   (17.1)                                        3845
     A LOCAL LIMIT THEOREM IN HIGHER DIMENSIONS FOR LATTICE DISTRIBUTIONS.
    TRUDY AKAD. NAUK LITOVSK. SSR SER. B, 2(22) 3-14.   (MR24-A1736)


MITRA, SAMARENDA K. 1955.   (20.3)                                       3846
     ELECTRICAL ANALOG COMPUTING MACHINE FOR SOLVING LINEAR EQUATIONS AND
    RELATED PROBLEMS.
    REV. SCI. INSTRUMENTS, 26, 453-457.   (MR17,P.903)


MITRA, SUJIT KUMAR    SEE
    4657. ROY, J. + MITRA, SUJIT KUMAR (1954)

    4697. ROY, S.N. + MITRA, SUJIT KUMAR (1956)


MITROFANOVA, N. M.    SEE
    3440. LINNIK, YU. V. + MITROFANOVA, N. M. (1964)

    3441. LINNIK, YU. V. + MITROFANOVA, N. M. (1965)


MITROPOLSKY, A. K. 1937.   (4.1/4.6/4.9)                                 3847
     ON ESTABLISHING THE CORRELATION EQUATIONS BY TCHEBYCHEFF'S METHOD.
          (IN RUSSIAN, WITH SUMMARY IN ENGLISH)
    IZV. AKAD. NAUK SSSR SER. MAT., 1, 125-134.


MITROPOLSKY, A. K. 1939.   (8.1/20.4)                                    3848
     ON THE MULTIPLE NON-LINEAR CORRELATION EQUATIONS.
          (IN RUSSIAN, WITH SUMMARY IN ENGLISH)
    IZV. AKAD. NAUK SSSR SER. MAT., 3, 399-406.   (MR1,P.345)


MITROPOLSKY, A. K. 1940.   (4.8/N)                                       3849
     ON THE CALCULATION OF ORDINARY CORRELATION EQUATIONS.
          (IN RUSSIAN)
    ZUR. TEH. FIZ. AKAD. NAUK SSSR, 10, 1227-1241.


MITROPOLSKY, A. K. 1949.   (8.1)                                         3850
     ORDINARY CORRELATION EQUATIONS.
          (IN RUSSIAN)
    USPEHI MAT. NAUK, 33, 142-175.   (MR11,P.259)

MITTMANN, OTFRID M. J. 1960.   (4.5)                                          3851
    EINE METHODE ZUR AUFFINDUNG URSACHLICHER ZUSAMMENHANGE MITTELS
    PARTIELLER KORRELATIONSKOEFFIZIENTEN.
    METRON, 20,   3-44.   (MR27-3064)

MIYAKAWA, HIROSHI 1959.   (16.4)                                             3852
    SAMPLING THEOREM OF STATIONARY STOCHASTIC VARIABLES IN MULTI-DIMENSIONAL SPACE.
        (IN JAPANESE, WITH SUMMARY IN ENGLISH)
    J. INST. ELEC. COMMUN. ENGRS. JAPAN, 42, 421-427.

MOCKETT, L. G.   SEE
    3583. MACNAUGHTON-SMITH, P.; WILLIAMS, W. T.; DALE, M. G. +  (1964)
    MOCKETT, L. G.

MODE, C. J. 1962.   (16.4)                                                   3853
    SOME MULTI-DIMENSIONAL BIRTH AND DEATH PROCESSES AND THEIR
    APPLICATIONS IN POPULATION GENETICS.
        (AMENDED BY NO. 3854)
    BIOMETRICS, 18, 543-567.

MODE, C. J. 1963.   (16.4)                                                   3854
    SOME MULTI-DIMENSIONAL BIRTH AND DEATH PROCESSES AND THEIR
    APPLICATIONS IN POPULATION GENETICS.
        (AMENDMENT OF NO. 3853)
    BIOMETRICS, 19, 667.

MODER, JOSEPH J., JR. 1956.   (1.1)                                          3855
    A TEACHING AID FOR REGRESSION, CORRELATION, ANALYSIS OF VARIANCE AND
    OTHER STATISTICAL TECHNIQUES.
    INDUST. QUAL. CONTROL, 13(4) 16-21.

MOGNO, ROBERTO 1952.   (4.8)                                                 3856
    SUL CALCOLO DI NUOVI COEFFICIENTI DI CORRELAZIONE.
    SOC. ITAL. ECON. DEMOG. STATIST. STUDI MONOG., 6, 43-53.

MONROE, ROBERT J.   SEE
    5562. TURNER, MALCOLM E.; MONROE, ROBERT J. + HOMER, L. D. (1963)

    5563. TURNER, MALCOLM E.; MONROE, ROBERT J. + LUCAS, HENRY L., JR. (1961)

MONROE, WALTER S. 1922.   (4.1/4.9)                                          3857
    USE AND INTERPRETATION OF COEFFICIENTS OF CORRELATION.
    SCHOOL AND SOCIETY, 16, 288-292.

MONROE, WALTER S. 1936.   (4.1)                                              3858
    NOTE ON THE INTERPRETATION OF COEFFICIENTS OF CORRELATION.
    J. EDUC. PSYCHOL., 27, 551-553.

MONROE, WALTER S. + STUIT, D. B. 1935.   (4.1/4.5/8.1/E)                     3859
    CORRELATION ANALYSIS AS A MEANS OF STUDYING CONTRIBUTIONS OF CAUSES.
    J. EXPER. EDUC., 3, 155-165.

MONTELLO, JESSE 1959.   (17.1)                                               3860
    ON EXTENSION OF THE CONCEPT OF ASYMPTOTICALLY NORMAL DISTRIBUTION TO
    MULTIDIMENSIONAL RANDOM VARIABLES.
    ESTADISTICA, 17, 457-475.   (MR22-1964)

MONTESSUS DE BALLORE, VICOMTE ROBERT DE   SEE   DE MONTESSUS DE BALLORE, VICOMTE ROBERT

MONTZINGO, LLOYD J., JR.   SEE
    4979. SEVERO, NORMAN C.; MONTZINGO, LLOYD J., JR. + SCHILLO, PAUL J. (1965)

MOOD, A. M. 1941.   (17.8/17.9)                                              3861
    ON THE JOINT DISTRIBUTION OF THE MEDIANS IN SAMPLES FROM A
    MULTIVARIATE POPULATION.
    ANN. MATH. STATIST., 12, 268-278.   (MR3,P.172)

MOOD, A. M. 1951.   (13.1)                                                   3862
    ON THE DISTRIBUTION OF THE CHARACTERISTIC ROOTS OF NORMAL
    SECOND-MOMENT MATRICES.
    ANN. MATH. STATIST., 22, 266-273.   (MR13,P.52)

3863
MOONAN, WILLIAM J. 1954. (5.2)
    ON THE PROBLEM OF SAMPLE SIZE FOR MULTIVARIATE SIMPLE RANDOM SAMPLING.
    J. EXPER. EDUC., 22, 285-288.

3864
MOONAN, WILLIAM J. 1957. (2.1/20.1)
    LINEAR TRANSFORMATION TO A SET OF STOCHASTICALLY DEPENDENT NORMAL VARIABLES.
    J. AMER. STATIST. ASSOC., 52, 247-252. (MR19,P.70)

3865
MOONAN, WILLIAM J. 1957. (4.6)
    A QUICK AND DIRTY METHOD FOR ESTIMATING MULTIPLE CORRELATIONS.
    J. EXPER. EDUC., 25, 339-343.

3866
MOORE, ELIAKIM HASTINGS 1935. (20.1)
    GENERAL ANALYSIS. PART I.
    MEM. AMER. PHILOS. SOC., 1, 1-231.

3867
MOORE, P.G. 1956. (2.4/3.2)
    THE ESTIMATION OF THE MEAN OF A CENSORED NORMAL DISTRIBUTION BY
    ORDERED VARIABLES.
    BIOMETRIKA, 43, 482-485. (MR18,P.772)

3868
MOORE, THOMAS V. 1931. (4.6/15.1/E)
    MULTIPLE CORRELATION AND THE CORRELATION BETWEEN GENERAL FACTORS.
    STUD. PSYCHOL. PSYCHIATRY CATHOLIC UNIV. AMER., 3(1) 1-32.

3869
MOOS, H. 1958. (4.6/U)
    BERECHNUNG MEHRFACHER KORRELATIONEN AUF ELEKTRONISCHEN RECHENMASCHINEN.
    METRIKA, 1, 148-153.

3870
MORAN, P.A.P. 1947. (16.4/U)
    SOME THEOREMS ON TIME SERIES. I.
    BIOMETRIKA, 34, 281-291. (MR9,P.361,P.735)

3871
MORAN, P.A.P. 1948. (4.1/4.8)
    RANK CORRELATION AND PRODUCT-MOMENT CORRELATION.
    BIOMETRIKA, 35, 203-206. (MR9,P.601)

3872
MORAN, P.A.P. 1948. (4.8)
    RANK CORRELATION AND PERMUTATION DISTRIBUTIONS.
    PROC. CAMBRIDGE PHILOS. SOC., 44, 142-144. (MR9,P.263)

3873
MORAN, P.A.P. 1950. (4.7)
    THE DISTRIBUTION OF THE MULTIPLE CORRELATION COEFFICIENT.
    PROC. CAMBRIDGE PHILOS. SOC., 46, 521-522. (MR11,P.732)

3874
MORAN, P.A.P. 1951. (4.8)
    PARTIAL AND MULTIPLE RANK CORRELATION.
    BIOMETRIKA, 38, 26-32. (MR13,P.141)

3875
MORAN, P.A.P. 1956. (20.4)
    THE NUMERICAL EVALUATION OF A CLASS OF INTEGRALS.
    PROC. CAMBRIDGE PHILOS. SOC., 52, 230-233. (MR17,P.901)

3876
MORAN, P.A.P. 1966. (18.6)
    A NOTE ON RECENT RESEARCH IN GEOMETRICAL PROBABILITY.
    J. APPL. PROB., 3, 453-463.

MORAN, P.A.P.   SEE ALSO
    967. CHOWN. L. N. + MORAN, P.A.P. (1951)

3877
MORANDA, PAUL B. 1960. (14.4/B)
    EFFECT OF BIAS ON ESTIMATES OF THE CIRCULAR PROBABLE ERROR.
    J. AMER. STATIST. ASSOC., 55, 732-735. (MR22-10049)

3878
MORANT, G. M. 1928. F(6.5/6.6)
    A PRELIMINARY CLASSIFICATION OF EUROPEAN RACES BASED ON CRANIAL MEASUREMENTS.
    BIOMETRIKA, 20(B) 301-375.

3879
MORANT, G. M. 1939. E(6.4)
    THE USE OF STATISTICAL METHODS IN THE INVESTIGATION OF PROBLEMS OF
    CLASSIFICATION IN ANTHROPOLOGY. 1. THE GENERAL NATURE OF THE MATERIAL AND THE
    FORMS OF INTRARACIAL DISTRIBUTIONS OF METRICAL CHARACTERS.
    BIOMETRIKA, 31, 72-98.

MORANT, G. M. + KOGA, Y. 1923.   (4.1/8.8/E)                    3880
   ON THE DEGREE OF ASSOCIATION BETWEEN REACTION TIMES IN THE CASE OF
   DIFFERENT SENSES.
   BIOMETRIKA, 15, 346-372.

MORANT, G. M.   SEE ALSO
   6008. WOO, T.L. + MORANT, G. M. (1932)

MORDELL, LOUIS J. 1958.   (20.5)                               3881
   ON THE EVALUATION OF SOME MULTIPLE SERIES.
   J. LONDON MATH. SOC., 33, 368-371.   (MR20-6615)

MORGAN, JAMES N. + SONQUIST, JOHN A. 1963.   (1.1/E)          3882
   PROBLEMS IN THE ANALYSIS OF SURVEY DATA, AND A PROPOSAL.
   J. AMER. STATIST. ASSOC., 58, 415-434.

MORGAN, R.W. + WELSH, D.J.A. 1965.   (16.4)                   3883
   A TWO-DIMENSIONAL POISSON GROWTH PROCESS.
   J. ROY. STATIST. SOC. SER. B, 27, 497-504.

MORGAN, W.A. 1939.   (10.2/B)                                 3884
   A TEST FOR THE SIGNIFICANCE OF THE DIFFERENCE BETWEEN THE VARIANCES
   IN A SAMPLE FROM A NORMAL BIVARIATE POPULATION.
   BIOMETRIKA, 31, 13-19.   (MR1,P.64)

MORGENSTERN, DIETRICH 1956.   (18.1/B)                        3885
   EINFACHE BEISPIELE ZWEIDIMENSIONALER VERTEILUNGEN.
   MITT. MATH. STATIST., 8, 234-235.   (MR18,P.423)

MORGENTHALER, GEORGE W. 1961.   (14.4/B)                      3886
   SOME CIRCULAR COVERAGE PROBLEMS.
   BIOMETRIKA, 48, 313-324.   (MR24-B2015)

MORI, HAZIME 1933.   (4.5/T)                                  3887
   TABLES OF VALUES OF LOG $(1 - R^2)^{\frac{1}{2}}$ FOR THE CALCULATION OF COEFFICIENTS
   OF PARTIAL CORRELATION.
       (IN JAPANESE)
   JAPAN. J. GENET., 8, 131-136.

MORIGUTI, SIGEITI 1960.   (20.3)                              3888
   NOTES ON THE NUMERICAL CONVERGENCE OF ITERATIVE PROCESSES.
   CONTRIB. PROB. STATIST. (HOTELLING VOLUME), 309-321.

MORISHIMA, MICHIO 1965.   (16.2)                              3889
   ON THE TWO THEOREMS OF GROWTH ECONOMICS:  A MATHEMATICAL EXERCISE.
   ECONOMETRICA, 33, 829-840.   (MR33-7108)

MORRIS, R.H.   SEE
   2737. JACKSON, J. EDWARD + MORRIS, R.H. (1957)

   740. BROWN, W. R. J.; HOWE, WILLIAM G.; JACKSON, J. EDWARD +   (1956)
   MORRIS, R.H.

MORRISON, ALEXANDER W. 1942.   (15.4/T)                       3890
   A GRAPHICAL DEVICE FOR COMPARING THE FORM OF A GIVEN DISTRIBUTION OF
   TEST SCORES WITH THE FORM OF THE STANDARD DISTRIBUTION FOR THE TEST.
   J. EDUC. RES., 36, 218-220.

MORRISON, DONALD F. 1962.   (7.1)                             3891
   ON THE DISTRIBUTION OF SUMS OF SQUARES AND CROSS PRODUCTS OF NORMAL
   VARIATES IN THE PRESENCE OF INTRA-CLASS CORRELATION.
   ANN. MATH. STATIST., 33, 1461-1463.   (MR25-5566)

MORRISON, DONALD F. 1963.   (16.5)                            3892
   EXPECTATIONS AND COVARIANCES OF SERIAL AND CROSS-CORRELATION
   COEFFICIENTS IN A COMPLEX STATIONARY TIME SERIES.
   BIOMETRIKA, 50, 213-216.   (MR27-5347)

MORRISON, DONALD G.   SEE
   1754. FRANK, RONALD E.; MASSY, WILLIAM F. + MORRISON, DONALD G. (1965)

MORRISON, J.T. 1934.  (16.5)
    NOTE ON THE CORRELATION OF TIME SERIES.
  PHILOS. MAG. SER. 7, 18, 545-554.
3893

MORRISON, J.T. 1937.  (4.6)
    NOTE ON THE MAXIMUM CORRELATION COEFFICIENT BETWEEN TWO SERIES WHOSE
    VALUES HAVE BEEN DETERMINED AT EQUAL INTERVALS.
  PHILOS. MAG. SER. 7, 24, 240-245.
3894

MORRISON, N.   SEE
  3511. LORGE, IRVING + MORRISON, N. (1938)

MORRISON, WINIFRED J.   SEE
  5005. SHERMAN, JACK + MORRISON, WINIFRED J. (1950)

MORSE, NORMAN   SEE
  3000. KESTEN, HARRY + MORSE, NORMAN (1959)

   415. BECHHOFER, ROBERT E.; ELMAGHRABY, SALAH + MORSE, NORMAN (1959)

MORTARA, GIORGIO 1914.  (4.1/E)
    SULL'IMPIEGO DEL COEFFICIENTE DI CORRELAZIONE NELLA SEMIOLOGIA ECONOMICA.
  GIORN. ECON. RIV. STATIST. SER. 3, 48, 296-305.
3895

MORTARA, GIORGIO 1927.  (4.9)
    INDICI DI DIPENDENZA IN USO NELLA STATISTICA E LORO SIGNIFICATO.
  REND. SEM. MAT. FIS. MILANO, 1, 119-128.
3896

MORTON, R. R. A. 1966.  (18.6)
    THE EXPECTED NUMBER AND ANGLE OF INTERSECTIONS BETWEEN RANDOM CURVES
    IN A PLANE.
  J. APPL. PROB., 3, 559-562.
3897

MOSIER, CHARLES I. 1936.  (15.4)
    A NOTE ON ITEM ANALYSIS AND THE CRITERION OF INTERNAL CONSISTENCY.
  PSYCHOMETRIKA, 1, 275-282.
3898

MOSIER, CHARLES I. 1938.  (15.2/15.3)
    A NOTE ON DWYER: THE DETERMINATION OF THE FACTOR LOADINGS OF A GIVEN TEST.
      (AMENDMENT OF NO. 1446)
  PSYCHOMETRIKA, 3, 297-299.
3899

MOSIER, CHARLES I. 1939.  E(15.2)
    INFLUENCE OF CHANCE ERROR IN SIMPLE STRUCTURE:  AN EMPIRICAL INVESTIGATION OF
    THE EFFECT OF CHANCE ERROR AND ESTIMATED COMMUNALITIES ON SIMPLE STRUCTURE IN
    FACTORIAL ANALYSIS.
  PSYCHOMETRIKA, 4, 33-44.
3900

MOSIER, CHARLES I. 1939.  (15.3)
    DETERMINING A SIMPLE STRUCTURE WHEN LOADINGS FOR CERTAIN TESTS ARE KNOWN.
  PSYCHOMETRIKA, 4, 149-162.
3901

MOSIER, CHARLES I. 1940.  (15.5)
    PSYCHOPHYSICS AND MENTAL TEST THEORY:  FUNDAMENTAL POSTULATES AND
    ELEMENTARY THEOREMS.
  PSYCHOL. REV., 47, 355-366.
3902

MOSIER, CHARLES I. 1941.  (15.4)
    A SHORT CUT IN THE ESTIMATION OF THE SPLIT-HALVES COEFFICIENTS.
  EDUC. PSYCHOL. MEAS., 1, 407-408.
3903

MOSIER, CHARLES I. 1941.  (15.4/15.5)
    PSYCHOPHYSICS AND MENTAL TEST THEORY   II.  THE CONSTANT PROCESS.
  PSYCHOL. REV., 48, 235-249.
3904

MOSIER, CHARLES I. 1943.  (15.4)
    ON THE RELIABILITY OF A WEIGHTED COMPOSITE.
  PSYCHOMETRIKA, 8, 161-168.
3905

MOSIER, CHARLES I. + MC QUITTY, J. V. 1940.  (15.4/T)
    METHODS OF ITEM VALIDATION AND THE ABACS FOR ITEM-TEST CORRELATION
    AND CRITICAL RATIO OF UPPER-LOWER DIFFERENCE.
  PSYCHOMETRIKA, 5, 57-65.
3906

MOSIMANN, JAMES E. 1962. (18.1/18.3)                                    3907
    ON THE COMPOUND MULTINOMIAL DISTRIBUTION, THE MULTIVARIATE
    $\beta$-DISTRIBUTION, AND CORRELATIONS AMONG PROPORTIONS.
    BIOMETRIKA. 49, 65–82. (MR26-858)

MOSIMANN, JAMES E.   SEE ALSO
    2831. JOLICOEUR, PIERRE + MOSIMANN, JAMES E. (1960)

    2219. GURLAND, JOAN M.; CORNFIELD, JEROME + MOSIMANN, JAMES E. (1964)

MOSTELLER, FREDERICK 1951. (3.2)                                         3908
    REMARKS ON THE METHOD OF PAIRED COMPARISONS: I. THE LEAST SQUARES SOLUTION
    ASSUMING EQUAL STANDARD DEVIATIONS AND EQUAL CORRELATIONS.
        (REPRINTED AS NO. 3910)
    PSYCHOMETRIKA. 16, 3–9.

MOSTELLER, FREDERICK 1956. (15.5)                                        3909
    STOCHASTIC LEARNING MODELS.
    PROC. THIRD BERKELEY SYMP. MATH. STATIST. PROB., 5, 151–167.  (MR18,P.955)

MOSTELLER, FREDERICK 1963. (3.2)                                         3910
    REMARKS ON THE METHOD OF PAIRED COMPARISONS: I. THE LEAST SQUARES SOLUTION
    ASSUMING EQUAL STANDARD DEVIATIONS AND EQUAL CORRELATIONS.
        (REPRINT OF NO. 3908)
    READINGS MATH. PSYCHOL. (LUCE, BUSH, + GALANTER), 1, 152–158.

MOSTELLER, FREDERICK + BUSH, ROBERT R. 1954. (15.5)                      3911
    SELECTED QUANTITATIVE TECHNIQUES.
    HANDB. SOC. PSYCHOL. (LINDZEY), 1, 289–334.

MOTE, V. L. + ANDERSON, R. L. 1965. (17.8/18.2/18.3/AT)                  3912
    AN INVESTIGATION OF THE EFFECT OF MISCLASSIFICATION ON THE PROPERTIES
    OF $\chi^2$-TESTS IN THE ANALYSIS OF CATEGORIAL DATA.
    BIOMETRIKA, 52, 95–109.

MOTOO, MINORU   SEE
    3755. MATUSITA, KAMEO + MOTOO, MINORU (1956)

MOU, TCHEN CHAN  SEE  TCHEN, CHAN-MOU

MOUL, MARGARET   SEE
    4245. PEARSON, KARL + MOUL, MARGARET (1927)

    1529. ELDERTON, ETHEL M.; MOUL, MARGARET; FIELLER, E. C.;   (1930)
    PRETORIUS, S.J. + CHURCH, A. E. R.

MOUSTAFA, M.D. 1955. (4.5)                                               3913
    A NOTE ON PARTIAL CORRELATION.
    EGYPTIAN STATIST. J., 1, 1–5.

MOUSTAFA, M.D. 1958. (18.3)                                              3914
    TESTING OF HYPOTHESES ON A MULTIVARIATE POPULATION; SOME OF THE
    VARIATES BEING CONTINUOUS AND THE REST CATEGORICAL.
    EGYPTIAN STATIST. J., 2, 73–96.

MOUSTAFA, M.D. + MAHMOUD, M. W. 1964. (18.2/B)                           3915
    ON THE PROBLEM OF ESTIMATION FOR THE BIVARIATE LOGNORMAL DISTRIBUTION.
    BIOMETRIKA. 51, 522–527. (MR31-5274)

MOUZON, E. D., JR.  SEE
    2130. GRESSENS, O. + MOUZON, E. D., JR. (1927)

MOWBRAY, A. H. 1941. (4.6/E)                                             3916
    OBSERVATION ON CORRELATION ANALYSIS.
    J. AMER. STATIST. ASSOC., 36, 248–252.

MOYAL, J. E.   SEE
    1509. EDWARDS, D. A. + MOYAL, J. E. (1955)

    2979. KENDALL, DAVID G. + MOYAL, J. E. (1957)

MOYLS, V.N.   SEE
    3662. MARCUS, MARVIN; MOYLS, V.N. + WESTWICK, ROY (1957)

MUDGETT, BRUCE D. 1929.  (16.5)                                      3917
     THE APPLICATION OF THE THEORY OF SAMPLING TO SUCCESSIVE OBSERVATIONS
     NOT INDEPENDENT OF EACH OTHER.  (WITH DISCUSSION)
     J. AMER. STATIST. ASSOC., 24(SP) 108-117.

MUDGETT, BRUCE D.    SEE ALSO
     1811. FRISCH, RAGNAR + MUDGETT, BRUCE D. (1931)

     1812. FRISCH, RAGNAR + MUDGETT, BRUCE D. (1932)

MUDHOLKAR, GOVIND S. 1965.  (8.3/9.1)                                 3918
     A CLASS OF TESTS WITH MONOTONE POWER FUNCTIONS FOR TWO PROBLEMS IN
       MULTIVARIATE STATISTICAL ANALYSIS.
     ANN. MATH. STATIST., 36, 1794-1801.  (MR33-5044)

MUDHOLKAR, GOVIND S. 1966.  (8.3/9.1/C)                               3919
     ON CONFIDENCE BOUNDS ASSOCIATED WITH MULTIVARIATE ANALYSIS OF VARIANCE
       AND NON-INDEPENDENCE BETWEEN TWO SETS OF VARIATES.
     ANN. MATH. STATIST., 37, 1736-1746.  (MR35-5055)

MUDHOLKAR, GOVIND S. 1966.  (17.2)                                    3920
     THE INTEGRAL OF AN INVARIANT UNIMODAL FUNCTION OVER AN INVARIANT CONVEX
       SET - AN INEQUALITY AND APPLICATIONS.
     PROC. AMER. MATH. SOC., 17, 1327-1333.   (MR34-7741)

MUDHOLKAR, GOVIND S.    SEE ALSO
     1219. DAS GUPTA, SOMESH; ANDERSON, T. W. + MUDHOLKAR, GOVIND S. (1964)

MUIR, SIR THOMAS 1924.  (20.2)                                        3921
     A SECOND BUDGET OF EXERCISES ON DETERMINANTS.
     AMER. MATH. MONTHLY, 31, 264-274.

MUKERJI, H. K.    SEE
     3953. NAIR, K. RAGHAVAN + MUKERJI, H. K. (1960)

MUKHERJEE, BISHWA NATH 1966.  (10.2/20.2/E)                          3922
     DERIVATION OF LIKELIHOOD-RATIO TESTS FOR GUTTMAN QUASI-SIMPLEX
       COVARIANCE STRUCTURES.
     PSYCHOMETRIKA, 31, 97-123.  (MR35-6265)

MUKHERJEE, RAMKRISHNA + BANDYOPADHYAY, SURAJ 1964.   E(5.2/6.2/6.5)  3923
     SOCIAL RESEARCH AND MAHALANOBIS'S D$^2$.
     CONTRIB. STATIST. (MAHALANOBIS VOLUME), 259-282.

MULLER, J. H. 1931.  (20.4)                                           3924
     ON THE APPLICATION OF CONTINUED FRACTIONS TO THE EVALUATION OF
       CERTAIN INTEGRALS, WITH SPECIAL REFERENCE TO THE INCOMPLETE BETA-FUNCTION.
     BIOMETRIKA, 22, 284-297.

MUNSINGER, HARRY 1966.  (15.5/E)                                      3925
     MULTIVARIATE ANALYSIS OF PREFERENCE FOR VARIABILITY.
     J. EXPER. PSYCHOL., 71, 889-895.

MUNZNER, HANS-FRIEDRICH 1936.  (4.1/4.2/4.8)                         3926
     GRUNDBEGRIFFE UND PROBLEME DER KORRELATIONSRECHNUNG.
     DEUTSCHE MATH., 1, 290-307.

MUNZNER, HANS-FRIEDRICH    SEE ALSO
     1805. FRIEDE, GEORG + MUNZNER, HANS-FRIEDRICH (1948)

MURTEIRA, BENTO J. 1954.  (17.7)                                      3927
     ANALISE DO CONFLUENCIA.
     AN. INST. SUPER. CI. ECON. FINAN., 22, 40-52.

MURTHY, V. KRISHNA 1963.  (16.3/16.4)                                3928
     ESTIMATION OF THE CROSS-SPECTRUM.
     ANN. MATH. STATIST., 34, 1012-1021.   (MR27-900)

MURTHY, V. KRISHNA 1966.  (17.5/17.8/A)                              3929
     NONPARAMETRIC ESTIMATION OF MULTIVARIATE DENSITIES WITH APPLICATIONS.
     MULTIVARIATE ANAL. PROC. INTERNAT. SYMP. DAYTON (KRISHNAIAH), 43-56.  (MR35-1152)

MURTHY, V. KRISHNA    SEE ALSO
  3144. KRISHNAIAH, PARUCHURI R. + MURTHY, V. KRISHNA (1966)

  4658. ROY, J. + MURTHY, V. KRISHNA (1960)

MURTY, V. N. 1952.  (2.1/B)                                                     3930
  ON A RESULT OF BIRNBAUM REGARDING THE SKEWNESS OF X IN A BIVARIATE
  NORMAL POPULATION.
  J. INDIAN SOC. AGRIC. STATIST., 4, 85-87.  (MR14,P.389)

MUSHAM, M. V. 1947.  (4.1)                                                      3931
  SUR L'INTERPRETATION DU COEFFICIENT DE CORRELATION.
  J. SOC. STATIST. PARIS, 88, 134-138.

MUSHAM, M. V. 1954.  (4.8)                                                      3932
  A PROBABILITY APPROACH TO TIES IN RANK CORRELATION.
  BULL. RES. COUNC. ISRAEL, 3, 321-327.  (MR16,P.731)

MUSSELMAN, JOHN ROGERS 1923.  (4.5/15.1)                                        3933
  SPURIOUS CORRELATION APPLIED TO URN SCHEMATA.
  J. AMER. STATIST. ASSOC., 18, 908-911.

MUSSELMAN, JOHN ROGERS 1926.  (4.8)                                             3934
  ON THE LINEAR CORRELATION RATIO IN THE CASE OF CERTAIN SYMMETRICAL
  FREQUENCY DISTRIBUTIONS.
  BIOMETRIKA, 18, 228-231.

MUTALIPASSI, L. R. + HANSON, R. M. 1964.  E(15.5)                               3935
  THE NONMETRIC MULTIDIMENSIONAL APPROACH APPLIED TO RANK-ORDER SIMILARITY DATA.
  PSYCHOL. REP., 15, 399-403.

NABEYA, SEIJI 1948.  (4.7)                                                      3936
  ON THE DISTRIBUTION OF MULTIPLE CORRELATION COEFFICIENTS.
       (IN JAPANESE)
  RES. MEM. INST. STATIST. MATH. TOKYO, 4, 381-385.

NABEYA, SEIJI 1949.  (15.1)                                                     3937
  ON FACTOR ANALYSIS.
       (IN JAPANESE)
  RES. MEM. INST. STATIST. MATH. TOKYO, 4, 460-499.

NABEYA, SEIJI 1949.  (4.9)                                                      3938
  ON THE INDEPENDENCE OF STATISTICS.
  SUGAKU, 2, 69-73.

NABEYA, SEIJI 1951.  (4.2)                                                      3939
  NOTE ON THE MOMENTS OF THE TRANSFORMED CORRELATION.
  ANN. INST. STATIST. MATH. TOKYO, 3, 1.  (MR13,P.478)

NABEYA, SEIJI 1951.  (2.3/B)                                                    3940
  ABSOLUTE MOMENTS IN 2-DIMENSIONAL NORMAL DISTRIBUTION.
  ANN. INST. STATIST. MATH. TOKYO, 3, 2-6.  (MR13,P.570)

NABEYA, SEIJI 1951.  (4.8)                                                      3941
  ON THE SAMPLING DISTRIBUTION OF INTRACLASS CORRELATION COEFFICIENT.
       (IN JAPANESE)
  RES. MEM. INST. STATIST. MATH. TOKYO, 7, 355-359.

NABEYA, SEIJI 1952.  (2.3)                                                      3942
  ABSOLUTE MOMENTS IN 3-DIMENSIONAL NORMAL DISTRIBUTION.
  ANN. INST. STATIST. MATH. TOKYO, 4, 15-30.  (MR14,P.569)

NABEYA, SEIJI 1961.  (2.3)                                                      3943
  ABSOLUTE AND INCOMPLETE MOMENTS OF THE MULTIVARIATE NORMAL DISTRIBUTION.
  BIOMETRIKA, 48, 77-84.  (MR23-A4211)

NADARAJA, E. A. 1963.  (17.8/A)                                                 3944
  ON ESTIMATING THE DENSITY OF RANDOM VARIABLES.
       (IN RUSSIAN, WITH SUMMARY IN GEORGIAN)
  SOOBSC. AKAD. NAUK GRUZIN. SSR, 32, 277-280.  (MR29-2840)

NADARAJA, E. A. 1964.  (17.8/AB)                                                     3945
    AN ESTIMATE FOR THE DENSITY OF A TWO-DIMENSIONAL DISTRIBUTION.
        (IN RUSSIAN, WITH SUMMARY IN GEORGIAN)
    SOOBSC. AKAD. NAUK GRUZIN. SSR, 36, 267-268.   (MR30-601)


NAGABHUSHANAM,K. 1951.  (20.1)                                                       3946
    LINEAR TRANSFORMATIONS AND THE PRODUCT-MOMENT MATRIX.
    ANN. MATH. STATIST., 22, 302-304.  (MR12,P.841)


NAGAEV, A. V. 1966.  (17.1/B)                                                        3947
    LIMIT THEOREMS FOR SUMS OF INDEPENDENT TWO-DIMENSIONAL RANDOM VECTORS.
        (IN RUSSIAN)
    LIMIT THEOR. STATIST. INFER. (IZDAT FAN, TASHKENT), 67-82.   (MR34-3628)


NAGAR, A.L. 1959.  (16.2)                                                            3948
    THE BIAS AND MOMENT MATRIX OF THE GENERAL K-CLASS ESTIMATORS OF THE
    PARAMETERS IN SIMULTANEOUS EQUATIONS.
    ECONOMETRICA, 27, 575-595.  (MR22-11479)


NAGAR, A.L. 1961.  (16.2)                                                            3949
    A NOTE ON THE RESIDUAL VARIANCE ESTIMATION IN SIMULTANEOUS EQUATIONS.
    ECONOMETRICA, 29, 238-243.  (MR26-7080)


NAGAR, A.L. + CHAKRAVARTI, N. C. 1965.  (16.1)                                       3950
    NOTE ON THE USE OF PRIOR INFORMATION IN STATISTICAL ESTIMATION OF
    ECONOMIC RELATIONS.
    SANKHYA SER. A, 27, 105-112.  (MR33-825)


NAGAR, A.L.   SEE ALSO
    2045. GOLDBERGER, ARTHUR S.; NAGAR, A.L. + ODEH, H.S. (1961)


NAGNUR, B. N.   SEE
    514. BHAT, B. R. + NAGNUR, B. N. (1965)


NAIDIN, D. P.   SEE
    4527. REYMENT, R.A. + NAIDIN, D. P. (1962)


NAIR, A. N. KRISHNAN 1941.  (4.3/18.4)                                               3951
    DISTRIBUTION OF STUDENT'S "T" AND THE CORRELATION COEFFICIENT IN
    SAMPLES FROM NON-NORMAL POPULATIONS.
    SANKHYA, 5, 383-400.  (MR4,P.164)


NAIR, K. RAGHAVAN 1952.  E(6.5)                                                      3952
    USE OF MEASUREMENTS OF MORE THAN ONE BIOMETRIC CHARACTER REGARDING SIZE OF
    BODY-PARTS FOR DISCRIMINATING BETWEEN PHASES IN THE CASE OF SIX-EYE-STRIPED
    SPECIMENS OF THE DESERT LOCUST.(PART OF LONGER PAPER BY S.D.MISRA, K.R.NAIR
    AND M.L.ROONWAL. STUDIES IN INTRASPECIFIC VARIATION, PART VI., P.95-152.)
    INDIAN J. ENT., 14, 126-136.


NAIR, K. RAGHAVAN + MUKERJI, H. K. 1960.  E(6.6/8.5)                                 3953
    CLASSIFICATION OF NATURAL AND PLANTATION TECK (TECTONA GRANDIS) GROWN AT
    DIFFERENT LOCALITIES OF INDIA AND BURMA WITH RESPECT TO PHYSICAL AND MECHANICAL
    PROPERTIES.
    SANKHYA, 22, 1-20.


NAIR, U. SIVARAMAN 1939.  (10.1)                                                     3954
    THE APPLICATION OF THE MOMENT FUNCTION IN THE STUDY OF DISTRIBUTION
    LAWS IN STATISTICS.
    BIOMETRIKA, 30, 274-294.


NAIR, U. SIVARAMAN 1941.  (3.1/4.3/BC)                                               3955
    PROBABILITY STATEMENTS REGARDING THE RATIO OF STANDARD DEVIATIONS AND
    CORRELATION COEFFICIENT IN A BIVARIATE NORMAL POPULATION.
    SANKHYA, 5, 151-156.  (MR4,P.164)


NAKAGAMI, MINORU 1960.  (7.4/16.4/E)                                                 3956
    THE M-DISTRIBUTION - A GENERAL FORMULA OF INTENSITY DISTRIBUTION OF
    RAPID FADING.
    STATIST. METH. RADIO WAVE PROP., 3-36.


NAKAGAMI, MINORU; TANAKA, K. + KANEHISA, M. 1957.  (18.1/BE)                         3957
    THE M-DISTRIBUTION AS THE GENERAL FORMULA OF INTENSITY DISTRIBUTION
    OF RAPID FADING.
    MEM. FAC. ENGRG. KOBE UNIV., 78-128.

NANDA, D.N. 1948.   (13.1)                                                                      3958
    DISTRIBUTION OF A ROOT OF A DETERMINANTAL EQUATION.
    ANN. MATH. STATIST., 19, 47–57.   (MR9,P.453)

NANDA, D.N. 1948.   (13.1/A)                                                                    3959
    LIMITING DISTRIBUTION OF A ROOT OF A DETERMINANTAL EQUATION.
    ANN. MATH. STATIST., 19, 340–350.  ·(MR10,P.135)

NANDA, D.N. 1949.   (6.2/E)                                                                     3960
    EFFICIENCY OF THE APPLICATION OF DISCRIMINANT FUNCTION IN PLANT-SELECTION.
    J. INDIAN SOC. AGRIC. STATIST., 2, 8–19.   (MR11,P.674)

NANDA, D.N. 1949.   (6.3/E)                                                                     3961
    THE STANDARD ERRORS OF DISCRIMINANT FUNCTION COEFFICIENTS IN
    PLANT-BREEDING EXPERIMENTS.
    J. ROY. STATIST. SOC. SER. B, 11, 283–290.   (MR11,P.674)

NANDA, D.N. 1950.   (13.1)                                                                      3962
    DISTRIBUTION OF THE SUM OF ROOTS OF A DETERMINANTAL EQUATION UNDER A
    CERTAIN CONDITION.
    ANN. MATH. STATIST., 21, 432–439.   (MR12,P.192)

NANDA, D.N. 1951.   (13.1/T)                                                                    3963
    PROBABILITY DISTRIBUTION TABLES OF THE LERGER (SIC) ROOT OF A
    DETERMINANTAL EQUATION WITH TWO ROOTS.
    J. INDIAN SOC. AGRIC. STATIST., 3, 175–177.   (MR13,P.478)

NANDI, HARIKINKAR K. 1946.   (5.3)                                                              3964
    ON THE POWER FUNCTION OF STUDENTIZED $D^2$-STATISTIC.
    BULL. CALCUTTA MATH. SOC., 38, 79–84.   (MR8,P.394)

NANDI, HARIKINKAR K. 1946.   (4.3/10.2/B)                                                       3965
    NOTE ON TESTS APPLIED TO SAMPLES FROM NORMAL BIVARIATE POPULATION.
    SCI. AND CULT., 12, 249.   (MR8,P.283)

NANDI, HARIKINKAR K. 1947.   (8.3)                                                              3966
    TESTS OF SIGNIFICANCE ON MULTIVARIATE SAMPLES.
    CALCUTTA STATIST. ASSOC. BULL., 1, 42–43.

NANDI, HARIKINKAR K. 1963.   (5.3/10.1/19.2)                                                    3967
    ON THE ADMISSIBILITY OF A CLASS OF TESTS.
    CALCUTTA STATIST. ASSOC. BULL., 12, 13–18.   (MR27-4317)

NANDI, HARIKINKAR K. 1965.   (8.3)                                                              3968
    ON SOME PROPERTIES OF ROY'S UNION-INTERSECTION TESTS.
    CALCUTTA STATIST. ASSOC. BULL., 14, 9–13.

NANDI, HARIKINKAR K.    SEE ALSO
    186. ANONYMOUS + NANDI, HARIKINKAR K. (1965)

    187. ANONYMOUS + NANDI, HARIKINKAR K. (1965)

NARAIN, R. D.   SEE   NARAYAN, RAM DEVA

NARAYAN, RAM DEVA 1948.   (4.7/5.3/8.4)                                                         3969
    A NEW APPROACH TO SAMPLING DISTRIBUTIONS OF THE MULTIVARIATE NORMAL
    THEORY.  PART I.
    J. INDIAN SOC. AGRIC. STATIST., 1, 59–69.   (MR10,P.387)

NARAYAN, RAM DEVA 1948.   (7.1/7.4/8.4)                                                         3970
    A NEW APPROACH TO SAMPLING DISTRIBUTIONS OF THE MULTIVARIATE NORMAL
    THEORY, PART II.
    J. INDIAN SOC. AGRIC. STATIST., 1, 137–146.   (MR11,P.607)

NARAYAN, RAM DEVA 1949.   (6.2)                                                                 3971
    SOME RESULTS ON DISCRIMINANT FUNCTIONS.
    J. INDIAN SOC. AGRIC. STATIST., 2, 49–59.   (MR12,P.192)

NARAYAN, RAM DEVA 1950.   (9.1)                                                                 3972
    ON THE COMPLETELY UNBIASED CHARACTER OF TESTS OF INDEPENDENCE IN
    MULTIVARIATE NORMAL SYSTEMS.
    ANN. MATH. STATIST., 21, 293–298.   (MR12,P.37)

NARAYAN, RAM DEVA    SEE ALSO
    532. BHATTACHARYYA, D. P. + NARAYAN, RAM DEVA (1941)

NARAYANAN, R.    SEE
    5489. TINTNER, GERHARD + NARAYANAN, R. (1966)

NARUMI, SEIMATSU 1923.   (17.3/18.1)                                          3973
    ON THE GENERAL FORMS OF BIVARIATE FREQUENCY DISTRIBUTIONS WHICH ARE
    MATHEMATICALLY POSSIBLE WHEN REGRESSION AND VARIATION ARE SUBJECTED TO LIMITING
    CONDITIONS. PART I.
    BIOMETRIKA, 15, 77-88.

NARUMI, SEIMATSU 1923.   (17.4/B)                                             3974
    ON THE GENERAL FORMS OF BIVARIATE FREQUENCY DISTRIBUTIONS WHICH ARE
    MATHEMATICALLY POSSIBLE WHEN REGRESSION AND VARIATION ARE SUBJECTED TO
    LIMITING CONDITIONS. II.
    BIOMETRIKA, 15, 209-221.

NATAF, ANDRE 1962.   (17.4)                                                   3975
    DETERMINATION DES DISTRIBUTIONS DE PROBABILITES DONT LES MARGES SONT DONNEES.
    C.R. ACAD. SCI. PARIS, 255, 42-43.   (MR25-2623)

NATAF, ANDRE 1962.   (17.4)                                                   3976
    ETUDE GRAPHIQUE DE DETERMINATION DE DISTRIBUTIONS DE PROBABILITES
    PLANES DONT LES MARGES SONT DONNEES.
    PUBL. INST. STATIST. UNIV. PARIS, 11, 257-260.   (MR27-785)

NATIONAL PHYSICAL LABORATORY 1951.   (1.2)                                    3977
    RECENT ADVANCES IN MATHEMATICAL STATISTICS.  BIBLIOGRAPHY, 1943-47.
    J. ROY. STATIST. SOC. SER. A, 114, 497-558.   (MR13,P.478)

NAUS, J. I. 1965.   (18.6/B)                                                  3978
    CLUSTERING OF RANDOM POINTS IN TWO DIMENSIONS.
    BIOMETRIKA, 52, 263-267.

NAYLOR, JAMES C.    SEE
    5846. WHERRY, ROBERT J.; NAYLOR, JAMES C.; WHERRY, ROBERT J., JR. +   (1965)
    FALLIS, ROBERT F.

NEELY, PETER M. 1966.   (20.3)                                                3979
    COMPARISON OF SEVERAL ALGORITHMS FOR COMPUTATION OF MEANS, STANDARD
    DEVIATIONS AND CORRELATION COEFFICIENTS.
    COMMUN. ASSOC. COMP. MACH., 9, 497-499.

NEIFELD, M. R. 1927.   (4.1/4.5/15.1)                                         3980
    A STUDY OF SPURIOUS CORRELATION.
    J. AMER. STATIST. ASSOC., 22, 331-338.

NEKRASSOV, P. A.    SEE
    3783. MEHMKE, RUDOLF + NEKRASSOV, P. A. (1892)

NELDER, J.A. 1951.   (2.5/8.8)                                                3981
    A NOTE ON THE STATISTICAL INDEPENDENCE OF QUADRATIC FORMS IN THE
    ANALYSIS OF VARIANCE.
    BIOMETRIKA, 38, 482-483.   (MR13,P.668)

NELDER, J.A. 1966.   (20.4)                                                   3982
    INVERSE POLYNOMIALS, A USEFUL GROUP OF MULTIFACTOR RESPONSE FUNCTIONS.
    BIOMETRICS, 22, 128-141.

NEMCINOV, V. S. 1955.   (4.6/17.2/20.5/E)                                     3983
    CONTEMPORARY MATHEMATICAL STATISTICS AND THE CEBYSEV SERIES (ON THE
    100-TH ANNIVERSARY OF THE INTERPOLATIONAL SERIES OF CEBYSEV).
       (IN RUSSIAN)
    UCEN. ZAP. STATIST., 1, 240-252.

NEPRASH, JERRY A. 1934.   (4.1)                                               3984
    SOME PROBLEMS IN THE CORRELATION OF SPATIALLY DISTRIBUTED VARIABLES.
    J. AMER. STATIST. ASSOC., 29(SP) 167-168.

NERLOVE, MARC    SEE
    274. BALESTRA, PIETRO + NERLOVE, MARC (1966)

NEUDECKER, H. + VAN DE PANNE, C. 1966.   (16.2)                              3985
     NOTE ON THE ASYMPTOTIC STANDARD ERRORS OF LATENT ROOTS OF ECONOMETRIC
     EQUATION SYSTEMS.
     REV. INST. INTERNAT. STATIST., 34, 43-47.   (MR33-3762)

NEUHAUS, JACK O. + WRIGLEY, CHARLES F. 1954.   (15.3)                        3986
     THE QUARTIMAX METHOD. AN ANALYTIC APPROACH TO ORTHOGONAL SIMPLE STRUCTURE.
     BRITISH J. STATIST. PSYCHOL., 7, 81-91.


NEUHAUS, JACK O.    SEE ALSO
     6030. WRIGLEY, CHARLES F. + NEUHAUS, JACK O. (1952)

     6031. WRIGLEY, CHARLES F. + NEUHAUS, JACK O. (1955)

     6032. WRIGLEY, CHARLES F.; SAUNDERS, DAVID R. + NEUHAUS, JACK O. (1958)

NEWBOLD, ETHEL M. 1925.   (4.5)                                              3987
     NOTES ON AN EXPERIMENTAL TEST OF ERRORS IN PARTIAL CORRELATION
     COEFFICIENTS, DERIVED FROM FOUR-FOLD AND BISERIAL TOTAL COEFFICIENTS.
     BIOMETRIKA, 17, 251-265.

NEWCOMB, ROBERT W. 1961.   (20.1)                                           3988
     ON THE SIMULTANEOUS DIAGONALIZATION OF TWO SEMI-DEFINITE MATRICES.
     QUART. APPL. MATH., 19, 144-146.   (MR23-A1650)

NEWHALL, S. M.   SEE
     725. BRONFIN, H. + NEWHALL, S. M. (1934)

NEWLIN, E. P.   SEE
     3063. KNOWLES, W. B.; HOLLAND, J. G. + NEWLIN, E. P. (1957)

NEWMAN, MORRIS 1962.   (20.3)                                               3989
     MATRIX COMPUTATIONS.
     SURVEY NUMER. ANAL. (TODD), 222-254.   (MR26-7139)

NEWMAN, MORRIS   SEE ALSO
     2047. GOLDMAN, A. J. + NEWMAN, MORRIS (1966)

     3660. MARCUS, MARVIN + NEWMAN, MORRIS (1965)

NEWTON, D. 1966.   (16.4)                                                   3990
     ON GAUSSIAN PROCESSES WITH SIMPLE SPECTRUM.
     Z. WAHRSCHEIN. VERW. GEBIETE, 5, 207-209.

NEYMAN, JERZY 1926.   (17.5/U)                                              3991
     ON THE CORRELATION OF THE MEAN AND THE VARIANCE IN SAMPLES DRAWN FROM
     AN "INFINITE" POPULATION.
     BIOMETRIKA, 18, 401-413.

NEYMAN, JERZY 1930.   (4.3/17.8/18.3/A)                                     3992
     CONTRIBUTION TO THE THEORY OF CERTAIN TEST CRITERIA.
     BULL. INST. INTERNAT. STATIST., 24(2) 44-88.


NEYMAN, JERZY 1941.   (10.2)                                                3993
     ON A STATISTICAL PROBLEM ARISING IN ROUTINE ANALYSES AND IN SAMPLING
     INSPECTIONS OF MASS PRODUCTION.
     ANN. MATH. STATIST., 12, 46-76.   (MR3,P.9)

NEYMAN, JERZY 1951.   (17.7)                                                3994
     EXISTENCE OF CONSISTENT ESTIMATES OF THE DIRECTIONAL PARAMETER IN A
     LINEAR STRUCTURAL RELATION BETWEEN TWO VARIABLES.
     ANN. MATH. STATIST., 22, 497-512.   (MR13,P.481)

NEYMAN, JERZY 1951.   (19.3)                                                3995
     FOUNDATION OF THE GENERAL THEORY OF ESTIMATION.
     CONG. INTERNAT. PHILOS. SCI., 4, 83-95.   (MR13,P.762)

NEYMAN, JERZY 1959.   (19.2/A)                                              3996
     OPTIMAL ASYMPTOTIC TESTS OF COMPOSITE STATISTICAL HYPOTHESES.
     PROB. STATIST. (CRAMER VOLUME), 213-234.

NEYMAN, JERZY 1963.   (6.6)                                                        3997
    ON FINITENESS OF THE PROCESS OF CLUSTERING.
    SANKHYA SER. A, 25, 69-74.

NEYMAN, JERZY 1965.   (18.1/E)                                                     3998
    CERTAIN CHANCE MECHANISMS INVOLVING DISCRETE DISTRIBUTIONS.
    SANKHYA SER. A, 27, 249-258.

NEYMAN, JERZY + PEARSON, EGON SHARPE 1931.   (10.1/U)                              3999
    ON THE PROBLEM OF K SAMPLES.
        (TRANSLATED AS NO. 4000)
    BULL. INTERNAT. POLSKA AKAD. UMIEJET. (KRAKOW) SER. A, 460-481.

NEYMAN, JERZY + PEARSON, EGON SHARPE 1932.   (10.1/U)                              4000
    O ZAGADNIENIU K PROB. ON THE PROBLEM OF K SAMPLES.
        (IN POLISH, TRANSLATION OF NO. 3999)
    ROCZNIK TOWARZ. NAUK. WARSZAWSK., 24(3) 122-126.

NEYMAN, JERZY + PEARSON, EGON SHARPE 1936.   (19.2)                                4001
    CONTRIBUTIONS TO THE THEORY OF TESTING STATISTICAL HYPOTHESES.
    STATIST. RES. MEM. LONDON, 1, 1-37.

NEYMAN, JERZY + SCOTT, ELIZABETH L. 1948.   (19.3/A)                               4002
    CONSISTENT ESTIMATES BASED ON PARTIALLY CONSISTENT OBSERVATIONS.
    ECONOMETRICA, 16, 1-32.

NEYMAN, JERZY + SCOTT, ELIZABETH L. 1951.   (17.7)                                 4003
    ON CERTAIN METHODS OF ESTIMATING THE LINEAR STRUCTURAL RELATION.
        (AMENDED BY NO. 4004)
    ANN. MATH. STATIST., 22, 352-361.   (MR13,P.259)

NEYMAN, JERZY + SCOTT, ELIZABETH L. 1952.   (17.7)                                 4004
    ERRATA TO: ON CERTAIN METHODS OF ESTIMATING THE LINEAR STRUCTURAL RELATION.
        (AMENDMENT OF NO. 4003)
    ANN. MATH. STATIST., 23, 135.   (MR13,P.259)

NEYMAN, JERZY + SCOTT, ELIZABETH L. 1962.   (6.6)                                  4005
    CONTRIBUTION TO THE STUDY OF THE ABUNDANCE OF MULTIPLE GALAXIES.
    STUD. MATH. ANAL. RELATED TOPICS (POLYA VOLUME), 262-269.

NEYMAN, JERZY + SCOTT, ELIZABETH L. 1965.   (19.2)                                 4006
    ASYMPTOTICALLY OPTIMAL TESTS OF COMPOSITE HYPOTHESES FOR RANDOMIZED EXPERIMENTS
    WITH NONCONTROLLED PREDICTOR VARIABLES.
    J. AMER. STATIST. ASSOC., 60, 699-721.

NEYMAN, JERZY + SCOTT, ELIZABETH L. 1966.   (16.2/19.2/E)                          4007
    ON THE USE OF C(X) OPTIMAL TESTS OF COMPOSITE HYPOTHESES.
    BULL. INST. INTERNAT. STATIST., 41, 477-497.

NEYMAN, JERZY    SEE ALSO
    726. BRONOWSKI, J. + NEYMAN, JERZY (1945)

NICHOLSON, C. 1941.   (14.2)                                                       4008
    A GEOMETRICAL ANALYSIS OF THE FREQUENCY DISTRIBUTION OF THE RATIO
    BETWEEN TWO VARIABLES.
    BIOMETRIKA, 32, 16-28.   (MR2,P.231)

NICHOLSON, C. 1943.   (2.7/T)                                                      4009
    THE PROBABILITY INTEGRAL FOR TWO VARIABLES.
    BIOMETRIKA, 33, 59-72.   (MR6,P.161)

NICHOLSON, GEORGE E., JR. 1957.   (3.1)                                            4010
    ESTIMATION OF PARAMETERS FROM INCOMPLETE MULTIVARIATE SAMPLES.
    J. AMER. STATIST. ASSOC., 52, 523-526.   (MR19,P.783)

NICHOLSON, GEORGE E., JR. 1960.   (8.9/E)                                          4011
    PREDICTION IN FUTURE SAMPLES.
    CONTRIB. PROB. STATIST. (HOTELLING VOLUME), 322-330.

NICHOLSON, W.L. 1958.   (7.2)                                                      4012
    ON THE DISTRIBUTION OF 2X2 RANDOM NORMAL DETERMINANTS.
    ANN. MATH. STATIST., 29, 575-580.   (MR20-364)

NIELSEN, O. BARNDORFF-  SEE  BARNDORFF-NIELSEN, O.

NIEVERGELT, F. 1957.  (4.8/T)                                                    4013
    DIE RANGKORRELATION U. ERSTER TEIL.  DER FALL DER UNABHANGIGKEIT.
  MITT. MATH. STATIST., 9, 196-232.

NILES, HENRY E. 1922.  E(17.7)                                                   4014
    CORRELATION, CAUSATION AND WRIGHT'S THEORY OF PATH COEFFICIENTS.
  GENETICS, 7, 258-273.

NILES, HENRY E. 1923.  (17.7)                                                    4015
    THE METHOD OF PATH COEFFICIENTS. AN ANSWER TO WRIGHT.
  GENETICS, 8, 256-260.

NISIDA, T. 1954.  (16.4/U)                                                       4016
    NOTE ON BROWNIAN MOTION WITH PARAMETER SPACE R".
  MATH. JAPON., 3, 85-91.  (MR16,P.1130)

NIVEN, IVAN  SEE
  1668. FINE, NATHAN J. + NIVEN, IVAN (1944)

NOBLE, BEN 1966.  (20.3)                                                         4017
    A METHOD FOR COMPUTING THE GENERALIZED INVERSE OF A MATRIX.
  SIAM J. NUMER. ANAL., 3, 582-584.

NOMACHI, YUKIO 1955.  (18.1/T)                                                   4018
    AUXILIARY TABLES FOR THE APPLICATIONS OF N-DIMENSIONAL
    T-DISTRIBUTIONS TO CERTAIN CLASS OF EMPIRICAL FUNCTIONS.
  BULL. MATH. STATIST., 6, 25-47.  (MR17,P.984)

NOMOKONOV, M. K. 1950.  (17.9)                                                   4019
    ON THE SIMPLENESS OF THE SECOND CHARACTERISTIC VALUE OF CORRELATION
    INTEGRAL EQUATIONS.
        (IN RUSSIAN)
  DOKL. AKAD. NAUK SSSR NOV. SER., 72, 1021-1024.  (MR13,P.561)

NORTON, H. W. 1946.  (4.3/E)                                                     4020
    ESTIMATING THE CORRELATION COEFFICIENT.
  BULL. AMER. METEOROL. SOC., 27, 589-590.

NORTON, KENNETH A.; SHULTZ, EDNA L. + YARBROUGH, HELEN 1952.  (18.3)             4021
    THE PROBABILITY DISTRIBUTION OF THE PHASE OF THE RESULTANT VECTOR SUM
    OF A CONSTANT VECTOR PLUS A RAYLEIGH DISTRIBUTED VECTOR.
  J. APPL. PHYS., 23, 137-141.  (MR13,P.761)

NOVAK, EDWIN  SEE
  1815. FRUCHTER, BENJAMIN + NOVAK, EDWIN (1958)

NOVELLIS, A. 1880.  (20.4)                                                       4022
    DI UN METODO D'INTERPOLAZIONE PER PASSARE DALLE CLASSI QUINQUENNALI
    DI POPOLAZIONE ALLE CLASSI ANNUALI.
  ANN. STATIST. SER. 2A, 15, 17-29.

NOVICK, MELVIN R. 1966.  (15.5)                                                  4023
    THE AXIOMS AND PRINCIPAL RESULTS OF CLASSICAL TEST THEORY.
  J. MATH. PSYCHOL., 3, 1-18.

NUESCH, PETER E. 1966.  (5.1)                                                    4024
    ON THE PROBLEM OF TESTING LOCATION IN MULTIVARIATE POPULATIONS FOR
    RESTRICTED ALTERNATIVES.
  ANN. MATH. STATIST., 37, 113-119.  (MR32-6613)

NUNNALLY, J. 1962.  (6.6/15.1/E)                                                 4025
    THE ANALYSIS OF PROFILE DATA.
  PSYCHOL. BULL., 59, 311-319.

NYBOLLE, H. CL. 1936.  (2.1/5.1/BE)                                              4026
    ON THE STATISTICAL DISTINCTION BETWEEN SETS OF TWO-DIMENSIONAL OBSERVATIONS.
  SKAND. AKTUARIETIDSKR., 19, 1-26.

NYDELL, STURE 1924.  (18.1/18.2/E)                                               4027
    THE CONSTRUCTION OF CURVES OF EQUAL FREQUENCY IN CASE OF LOGARITHMIC
    A-CORRELATION, WITH APPLICATIONS TO THE DISTRIBUTION OF AGES AT FIRST MARRIAGE.
  SKAND. AKTUARIETIDSKR., 7, 36-53.

NYGAARD, P. H. 1932.  (4.1/15.1)                                        4028
     INTERPRETATION OF CORRELATION ON THE BASIS OF COMMON ELEMENTS.
     J. EDUC. PSYCHOL., 23, 578-585.

NYQUIST, H.; RICE, S.O. + RIORDAN, JOHN 1954.  (7.2)                    4029
     THE DISTRIBUTION OF RANDOM DETERMINANTS.
     QUART. APPL. MATH., 12, 97-104.  (MR16,P.148)

OBERG, EDWIN N. 1947.  (2.6/B)                                          4030
     APPROXIMATE FORMULAS FOR THE RADII OF CIRCLES WHICH INCLUDE A
     SPECIFIED FRACTION OF A NORMAL BIVARIATE DISTRIBUTION.
     ANN. MATH. STATIST., 18, 442-447.  (MR9,P.47)

OBOUKHOFF, A. M.  SEE  OBUHOV, A. M.

OBRECHKOV, N. 1937.  (20.3)                                             4031
     SUR LES SOLUTIONS D'UN SYSTEME D'EQUATIONS LINEAIRES AUX DIFFERENCES
     FINIES DU PREMIER ORDRE.
     C.R. ACAD. SCI. PARIS, 205, 265-268.

OBUHOV, A. M. 1938.  (9.3)                                              4032
     SUR LA CORRELATION NORMALE DES VECTEURS.
         (IN RUSSIAN, WITH SUMMARY IN FRENCH)
     IZV. AKAD. NAUK SSSR SER. MAT., 2, 339-370.

OBUHOV, A. M. 1940.  (12.1)                                             4033
     EINE KORRELATIONSTHEORIE DER VEKTOREN.
         (IN RUSSIAN)
     UCEN. ZAP. MOSK. GOS. UNIV., 45, 73-92.  (MR3,P.6)

OCHINSKY, L.  SEE
     1477. EAST, D. A. + OCHINSKY, L. (1958)

ODANAKA, TOSHIO 1961.  (16.4/E)                                         4034
     PREDICTION THEORY AND DYNAMIC PROGRAMMING.
     BULL. INST. INTERNAT. STATIST., 38(4) 333-341.  (MR23-A3221)

ODEH, H.S.  SEE
     2045. GOLDBERGER, ARTHUR S.; NAGAR, A.L. + ODEH, H.S. (1961)

ODELL, P. L. + FEIVESON, A. H. 1966.  (2.5/2.7/7.1/R)                   4035
     A NUMERICAL PROCEDURE TO GENERATE A SAMPLE COVARIANCE MATRIX.
         (AMENDED BY NO. 4036)
     J. AMER. STATIST. ASSOC., 61, 199-203.  (MR33-860)

ODELL, P. L. + FEIVESON, A. H. 1966.  (2.5/2.6/7.1/R)                   4036
     CORRIGENDA TO: A NUMERICAL PROCEDURE TO GENERATE A SAMPLE COVARIANCE MATRIX.
         (AMENDMENT OF NO. 4035)
     J. AMER. STATIST. ASSOC., 61, 1248-1249.  (MR33-860)

ODELL, P. L.  SEE ALSO
     3406. LEWIS, T. + ODELL, P. L. (1966)

     4950. SCROGGS, JAMES E. + ODELL, P. L. (1966)

ODERFELD, J. 1958.  (3.1/14.4/B)                                        4037
     STATISTICAL TESTING FOR CORRELATED CHARACTERISTICS.
     POLSKA AKAD. NAUK ZASTOS. MAT., 4, 255-264.  (MR23-A2983)

OFFORD, A. C.  SEE
     3447. LITTLEWOOD, J. E. + OFFORD, A. C. (1938)

     3448. LITTLEWOOD, J. E. + OFFORD, A. C. (1939)

     3449. LITTLEWOOD, J. E. + OFFORD, A. C. (1943)

OGASAWARA, TOZIRO + TAKAHASHI, MASAYUKI 1951.  (2.5/20.1)               4038
     INDEPENDENCE OF QUADRATIC QUANTITIES IN A NORMAL SYSTEM.
     J. SCI. HIROSHIMA UNIV. SER. A, 15, 1-9.  (MR13,P.142)

OGASAWARA, TOZIRO + TAKAHASHI, MASAYUKI 1953.  (8.8/20.5)               4039
     ORTHOGONALITY RELATION IN THE ANALYSIS OF VARIANCE. I.
     J. SCI. HIROSHIMA UNIV. SER. A, 16, 457-470.

OGASAWARA, TOZIRO + TAKAHASHI, MASAYUKI 1953.  (8.8/20.5)                          4040
    ORTHOGONALITY RELATION IN THE ANALYSIS OF VARIANCE.  II.
    J. SCI. HIROSHIMA UNIV. SER. A, 17, 27-41.

OGAWA, JUNJIRO 1946.  (2.5/20.1)                                                   4041
    ON A PROOF OF THE FUNDAMENTAL LEMMA IN H. SAKAMOTO'S PAPER ON THE
    INDEPENDENCE OF STATISTICS.
        (IN JAPANESE)
    RES. MEM. INST. STATIST. MATH. TOKYO, 1(15) 25-28.

OGAWA, JUNJIRO 1946.  (2.5/20.1)                                                   4042
    ON THE INDEPENDENCE OF STATISTICS OF QUADRATIC FORMS.
        (IN JAPANESE, TRANSLATED AS NO. 4043)
    RES. MEM. INST. STATIST. MATH. TOKYO, 2, 98-111.

OGAWA, JUNJIRO 1947.  (2.5/20.1)                                                   4043
    ON THE INDEPENDENCE OF STATISTICS OF QUADRATIC FORMS.
        (TRANSLATION OF NO. 4042)
    RES. MEM. INST. STATIST. MATH. TOKYO, 3, 137-151.

OGAWA, JUNJIRO 1948.  (2.5/20.1)                                                   4044
    ON THE INDEPENDENCE OF STATISTICS BETWEEN LINEAR FORMS, QUADRATIC
    FORMS AND BILINEAR FORMS FROM NORMAL POPULATIONS.
        (IN JAPANESE, TRANSLATED AS NO. 4045)
    RES. MEM. INST. STATIST. MATH. TOKYO, 4, 1-40.  (MR11,P.260)

OGAWA, JUNJIRO 1949.  (2.5/20.1)                                                   4045
    ON THE INDEPENDENCE OF BILINEAR AND QUADRATIC FORMS OF A RANDOM
    SAMPLE FROM A NORMAL POPULATION.
        (TRANSLATION OF NO. 4044)
    ANN. INST. STATIST. MATH. TOKYO, 1, 83-108.  (MR11,P.260)

OGAWA, JUNJIRO 1949.  (8.1/8.2)                                                    4046
    REGRESSION THEORY OF MULTIVARIATE NORMAL DISTRIBUTION.
        (IN JAPANESE)
    RES. MEM. INST. STATIST. MATH. TOKYO, 5, 9-16.

OGAWA, JUNJIRO 1949.  (2.5)                                                        4047
    INDEPENDENCE OF STATISTICS OF QUADRATIC FORMS FROM NON-CENTRAL NORMAL
    DISTRIBUTION.
        (IN JAPANESE, TRANSLATED AS NO. 4048)
    RES. MEM. INST. STATIST. MATH. TOKYO, 5, 51-55.  (MR12,P.509)

OGAWA, JUNJIRO 1950.  (2.5)                                                        4048
    ON THE INDEPENDENCE OF QUADRATIC FORMS IN A NON-CENTRAL NORMAL SYSTEM.
        (TRANSLATION OF NO. 4047)
    OSAKA MATH. J., 2, 151-159.  (MR12,P.509)

OGAWA, JUNJIRO 1951.  (3.1/C)                                                      4049
    ON A CONFIDENCE INTERVAL OF THE RATIO OF POPULATION MEANS OF A
    BIVARIATE NORMAL DISTRIBUTION.
    PROC. JAPAN ACAD., 27, 313-316.  (MR13,P.962)

OGAWA, JUNJIRO 1951.  (2.5)                                                        4050
    ON THE ANALYTICAL DERIVATION OF INTER-CLASS CORRELATION COEFFICIENTS
    IN THE SAMPLE DISTRIBUTION.
        (IN JAPANESE)
    RES. MEM. INST. STATIST. MATH. TOKYO, 7(5), 1-?.

OGAWA, JUNJIRO 1952.  (4.8)                                                        4051
    ANALYTICAL DERIVATION OF SAMPLING DISTRIBUTION OF INTRACLASS
    CORRELATION COEFFICIENT.
    OSAKA MATH. J., 4, 69-76.  (MR14,P.389)

OGAWA, JUNJIRO 1952.  (7.1)                                                        4052
    DERIVATION OF THE WISHART DISTRIBUTION.
        (IN JAPANESE)
    RES. MEM. INST. STATIST. MATH. TOKYO, 8, 149-152.

OGAWA, JUNJIRO 1952.  (4.7)                                                        4053
    SAMPLING DISTRIBUTION OF A MULTIPLE CORRELATION COEFFICIENT.
        (IN JAPANESE)
    RES. MEM. INST. STATIST. MATH. TOKYO, 8, 153-158.

OGAWA, JUNJIRO 1953.  (4.5/4.7/5.3)                                              4054
    ON THE SAMPLING DISTRIBUTIONS OF CLASSICAL STATISTICS IN MULTIVARIATE ANALYSIS.
    OSAKA MATH. J., 5, 13-52.  (MR15,P.141)

OGAWA, JUNJIRO 1956.  (9.3)                                                      4055
    ON MASUYAMA'S CORRELATION COEFFICIENT OF VECTORS.
        (IN JAPANESE, WITH SUMMARY IN ENGLISH)
    PROC. INST. STATIST. MATH. TOKYO, 4(2) 53-60.

OGAWA, JUNJIRO 1960.  (6.3)                                                      4056
    A REMARK ON WALD'S PAPER:"ON A STATISTICAL PROBLEM ARISING IN THE
    CLASSIFICATION OF AN INDIVIDUAL INTO ONE OF TWO GROUPS".
    COLLECTED PAPERS 70-TH ANNIV. NIHON UNIV. NATUR. SCI., 3, 11-20.

OGAWA, JUNJIRO 1962.  (4.3/14.1)                                                 4057
    ESTIMATION OF CORRELATION COEFFICIENT BY ORDER STATISTICS.
    CONTRIB. ORDER STATIST. (SARHAN + GREENBERG), 283-291.

OGAWA, JUNJIRO + KOGO, KUSUNORI 1955.  (19.3)                                    4058
    ON THE CORRELATION OF EFFICIENT ESTIMATES OF UNKNOWN PARAMETERS.
    OSAKA MATH. J., 7, 15-22.  (MR17,P.54)

OGAWARA, MASAMI 1948.  (9.3)                                                     4059
    ON CORRELATION FUNCTIONS OF RANDOM VECTOR IN DIFFERENT DIMENSIONS.
        (IN JAPANESE)
    SUGAKU, 1, 216-219.

OGAWARA, MASAMI 1958.  (16.3/16.4/U)                                             4060
    TIME SERIES ANALYSIS AND STOCHASTIC PREDICTION. (I)
    BULL. MATH. STATIST., 8, 8-53.  (MR22-5106)

OGAWARA, MASAMI 1959.  (16.3/16.4/U)                                             4061
    TIME SERIES ANALYSIS AND STOCHASTIC PREDICTION. (II)
    BULL. MATH. STATIST., 8, 55-72.  (MR22-5106)

OGAWARA, MASAMI 1960.  (16.4)                                                    4062
    TIME SERIES ANALYSIS AND STOCHASTIC PREDICTION. (III)
    BULL. MATH. STATIST., 9(1) 1-9.  (MR23-A3614)

OGAWARA, MASAMI + YAMAZAKI, H. 1952.  (8.1/8.9/E)                                4063
    MULTIVARIATE REGRESSION THEORY AND ITS APPLICATION TO THE PROBLEM OF
    CLIMATIC CHANGE.
    J. METEOROL. SOC. JAPAN, 30, 158-166.

OKAMOTO, MASASHI 1961.  (6.3)                                                    4064
    DISCRIMINATION FOR VARIANCE MATRICES.
    OSAKA MATH. J., 13, 1-39.  (MR25-1615)

OKAMOTO, MASASHI 1963.  (6.2/A)                                                  4065
    AN ASYMPTOTIC EXPANSION FOR THE DISTRIBUTION OF LINEAR DISCRIMINANT FUNCTION.
        (AMENDED BY NO. 4066)
    ANN. MATH. STATIST., 34, 1286-1301.  (MR27-6342)

OKAMOTO, MASASHI 1968.  (6.2/A)                                                  4066
    CORRECTION TO: AN ASYMPTOTIC EXPANSION FOR THE DISTRIBUTION OF THE LINEAR
    DISCRIMINANT FUNCTION.
        (AMENDMENT OF NO. 4065)
    ANN. MATH. STATIST., 39, 1358-1380.  (MR27-6342)

OKONENKO, A.    SEE
    3135. KRAVCHUK, M. + OKONENKO, A. (1926)

OLDIS, ELENA    SEE
    952. CHESHIRE, LEONE; OLDIS, ELENA + PEARSON, EGON SHARPE (1932)

OLIVEIRA, J. TIAGO DE  SEE  TIAGO DE OLIVEIRA, J.

OLKIN, INGRAM 1953.  (20.1)                                                      4067
    NOTE ON: THE JACOBIANS OF CERTAIN MATRIX TRANSFORMATIONS USEFUL IN MULTIVARIATE ANALYSIS.
    BIOMETRIKA, 40, 43-46.  (MR15,P.94)

OLKIN, INGRAM 1958.   (18.5)                                                    4068
    MULTIVARIATE RATIO ESTIMATION FOR FINITE POPULATIONS.
    BIOMETRIKA, 45, 154–165.   (MR19,P.1097)

OLKIN, INGRAM 1958.   (18.1/U)                                                  4069
    AN INEQUALITY SATISFIED BY THE GAMMA FUNCTION.
    SKAND. AKTUARIETIDSKR., 41, 37–39.   (MR22-9988)

OLKIN, INGRAM 1959.   (20.4)                                                    4070
    A CLASS OF INTEGRAL IDENTITIES WITH MATRIX ARGUMENT.
    DUKE MATH. J., 26, 207–213.   (MR21-36)

OLKIN, INGRAM 1959.   (20.1)                                                    4071
    ON INEQUALITIES OF SZEGO AND BELLMAN.
    PROC. NAT. ACAD. SCI. U.S.A., 45, 230–231.

OLKIN, INGRAM + PRATT, JOHN W. 1958.   (4.3)                                    4072
    UNBIASED ESTIMATION OF CERTAIN CORRELATION COEFFICIENTS.
    ANN. MATH. STATIST., 29, 201–211.   (MR20-374)

OLKIN, INGRAM + PRATT, JOHN W. 1958.   (17.2)                                   4073
    A MULTIVARIATE TCHEBYCHEFF INEQUALITY.
    ANN. MATH. STATIST., 29, 226–234.   (MR20-385)

OLKIN, INGRAM + ROY, S.N. 1954.   (7.1/13.1)                                    4074
    ON MULTIVARIATE DISTRIBUTION THEORY.
    ANN. MATH. STATIST., 25, 329–339.   (MR15,P.885)

OLKIN, INGRAM + RUBIN, HERMAN 1962.   (7.1)                                     4075
    A CHARACTERIZATION OF THE WISHART DISTRIBUTION.
    ANN. MATH. STATIST., 33, 1272–1280.   (MR25-4597)

OLKIN, INGRAM + RUBIN, HERMAN 1964.   (7.1/18.1)                                4076
    MULTIVARIATE BETA DISTRIBUTIONS AND INDEPENDENCE PROPERTIES OF THE
    WISHART DISTRIBUTION.
        (AMENDED BY NO. 4077)
    ANN. MATH. STATIST., 35, 261–269.   (MR28-3511)

OLKIN, INGRAM + RUBIN, HERMAN 1966.   (7.1/18.1)                                4077
    CORRECTION TO: MULTIVARIATE BETA DISTRIBUTIONS AND INDEPENDENCE PROPERTIES OF
    THE WISHART DISTRIBUTION.
        (AMENDMENT OF NO. 4076)
    ANN. MATH. STATIST., 37, 297.   (MR32-1792)

OLKIN, INGRAM + SOBEL, MILTON 1965.   (18.1/N)                                  4078
    INTEGRAL EXPRESSIONS FOR TAIL PROBABILITIES OF THE MULTINOMIAL
    AND NEGATIVE MULTINOMIAL DISTRIBUTIONS.
    BIOMETRIKA, 52, 167–179.

OLKIN, INGRAM + TATE, ROBERT F. 1961.   (18.1/18.3)                             4079
    MULTIVARIATE CORRELATION MODELS WITH MIXED DISCRETE AND CONTINUOUS VARIABLES.
    ANN. MATH. STATIST., 32, 448–465.   (MR27-2042)

OLKIN, INGRAM + TATE, ROBERT F. 1965.   (18.1/18.3)                             4080
    CORRECTION TO: MULTIVARIATE CORRELATION MODELS WITH MIXED DISCRETE AND
    CONTINUOUS VARIABLES.
    ANN. MATH. STATIST., 36, 343–344.   (MR30-3545)

OLKIN, INGRAM    SEE ALSO
    807. BUSH, KENNETH A. + OLKIN, INGRAM (1959)

    808. BUSH, KENNETH A. + OLKIN, INGRAM (1961)

    809. BUSH, KENNETH A. + OLKIN, INGRAM (1961)

    819. CACOULLOS, THEOPHILOS + OLKIN, INGRAM (1965)

    1946. GHURYE, S. G. + OLKIN, INGRAM (1962)

    2027. GLESER, LEON J. + OLKIN, INGRAM (1966)

    2941. KATZ, LEO + OLKIN, INGRAM (1953)                        CONTINUED...

OLKIN, INGRAM  SEE ALSO  (CONTINUED)

3686. MARSHALL, ALBERT W. + OLKIN, INGRAM (1960)

3687. MARSHALL, ALBERT W. + OLKIN, INGRAM (1960)

3688. MARSHALL, ALBERT W. + OLKIN, INGRAM (1960)

3689. MARSHALL, ALBERT W. + OLKIN, INGRAM (1961)

3690. MARSHALL, ALBERT.W. + OLKIN, INGRAM (1964)

3691. MARSHALL, ALBERT W. + OLKIN, INGRAM (1964)

1254. DEEMER, WALTER L.; OLKIN, INGRAM; HSU, P. L. +  (1951)
HOTELLING, HAROLD

OLMSTEAD, P. S. + TUKEY, JOHN W. 1947.  (4.8)                     4081
    A CORNER TEST FOR ASSOCIATION.
    ANN. MATH. STATIST., 18, 495-513.  (MR9,P.294)

ONICESCU, OCTAV 1933.  (4.1)                                      4082
    FUNCTION AND CORRELATION.
    MATHEMATICA, CLUJ, 7, 51-60.

ONICESCU, OCTAV 1958.  (17.1)                                     4083
    SUR LES CHAMPS DE VECTEURS-SOMME.
    C.R. ACAD. SCI. PARIS, 246, 3574-3576.  (MR20-320)

ONICESCU, OCTAV 1961.  (4.9)                                      4084
    LFS FONCTIONS-ENTROPIE, CORRELATION ET QUANTITE D'INFORMATION.
    DEUX. CONG. MATH. HONGR., 2(4) 37-41.

ONICESCU, OCTAV + SACUIU, I. 1964.  (4.8)                         4085
    EXTENSION OF THE NOTIONS OF MOMENT AND COEFFICIENT OF CORRELATION TO
    ARBITRARY VARIABLES.
    STUDII CERC. MAT., 15, 325-330.  (MR32-1741)

ONO, SUMINOSUKE 1931.  (20.1)                                     4086
    ON VECTOR QUANTITY.  II.  VECTOR QUANTITY REDUCIBLE FROM A KIND OF PROBABILITY.
        (REPRINTED AS NO. 4087)
    PROC. PHYS. MATH. SOC. JAPAN SER. 3, 13, 157-165.

ONO, SUMINOSUKE 1931.  (20.1)                                     4087
    ON VECTOR QUANTITY.  II.  VECTOR QUANTITY IS REDUCIBLE FROM A KIND
    OF PROBABILITY.
        (REPRINT OF NO. 4086)
    SCI. REP. TOKYO KYOIKU DAIGAKU SECT. A, 1, 85-95.

ONO, SUMINOSUKE 1935.  (17.3)                                     4088
    GENERALIZED THEORY OF VECTOR PROBABILITY.
    SCI. REP. TOKYO KYOIKU DAIGAKU SECT. A, 2, 279-303.

ONSAGER, LARS 1944.  (16.4)                                       4089
    CRYSTAL STATISTICS.  I.  A TWO-DIMENSIONAL MODEL WITH AN
    ORDER-DISORDER TRANSITION.
    PHYS. REV., 65, 117-149.  (MR5,P.280)

OPARIN, D. I. 1958.  (4.8)                                        4090
    A KORRELACIOSZAMITAS LENYEGE ES JELENTOSEGE.   (THE ESSENCE AND
    IMPORTANCE OF THE CORRELATION ANALYSIS.)
    STATISZT. SZEMLE, 36, 562-571.

OPPENHEIM, A. 1930.  (20.1)                                       4091
    INEQUALITIES CONNECTED WITH DEFINITE HERMITIAN FORMS.
    J. LONDON MATH. SOC., 5, 114-119.

OPPENHEIM, A. 1954.  (20.1)                                       4092
    INEQUALITIES CONNECTED WITH DEFINITE HERMITIAN FORMS, II.
    AMER. MATH. MONTHLY, 61, 463-466.

ORCUTT, GUY H. + JAMES, S. F. 1948.  (16.5)                                    4093
    TESTING THE SIGNIFICANCE OF CORRELATION BETWEEN TIME SERIES.
    BIOMETRIKA, 35, 397-413.

ORCUTT, GUY H.   SEE ALSO
    1004. COCHRANE, D. + ORCUTT, GUY H. (1949)


ORDEN, ALEX 1960.  (20.3)                                                      4094
    MATRIX INVERSION AND RELATED TOPICS BY DIRECT METHODS.
    MATH. METH. DIGIT. COMPUT. (RALSTON + WILF), 39-55.  (MR22-8682)

ORDEN, ALEX 1964.  (20.5)                                                      4095
    STATIONARY POINTS OF QUADRATIC FUNCTIONS UNDER LINEAR CONSTRAINTS.
    COMPUT. J., 7, 238-242.

ORLOCI, L. 1966.  (11.1/17.9/E)                                                4096
    GEOMETRIC MODELS IN ECOLOGY.  I.
    J. ECOL., 54, 193-215.

ORTS, JOSE MA. 1941.  (2.1/B)                                                  4097
    ESTABILIDAD DE LA LEY NORMAL DE PROBABILIDAD DEPENDIENTE DE DOS
    VARIABLES ALEATORIAS.
    REV. MAT. HISP. AMER. SER. 4, 1, 34-36.  (MR3,P.2)

ORTS ARACIL, D. JOSE MA.  SEE  ORTS, JOSE MA.

OSBORNE, E. F. 1961.  (20.3)                                                   4098
    ON LEAST SQUARES SOLUTIONS OF LINEAR EQUATIONS.
    J. ASSOC. COMPUT. MACH., 8, 628-636.

OSBORNE, E. E. 1965.  (20.3)                                                   4099
    SMALLEST LEAST SQUARES SOLUTION OF LINEAR EQUATIONS.
    J. SOC. INDUST. APPL. MATH. SER. B NUMER. ANAL., 2, 300-307.

OSBURN, HOBART G.   SEE
    3523. LUBIN, ARDIE + OSBURN, HOBART G. (1957)

OSTROVSKII, I. V. 1965.  (17.3)                                               4100
    A MULTI-DIMENSIONAL ANALOG OF JU. V. LINNIK'S THEOREM ON
    DECOMPOSITIONS OF A COMPOSITION OF GAUSSIAN AND POISSON LAWS.
        (IN RUSSIAN, WITH SUMMARY IN ENGLISH, TRANSLATED AS NO. 4101)
    TEOR. VEROJATNOST. I PRIMENEN., 10, 742-745.  (MR35-3715)


OSTROVSKII, I. V. 1965.  (17.3)                                               4101
    THE MULTIDIMENSIONAL ANALOGUE OF YU. V. LINNIK'S THEOREM ON
    DECOMPOSITIONS OF A CONVOLUTION OF GAUSSIAN AND POISSON LAWS.
        (TRANSLATION OF NO. 4100)
    THEORY PROB. APPL., 10, 673-677.  (MR35-3715)

OSTROVSKII, I. V. 1966.  (17.3)                                               4102
    DECOMPOSITION OF MULTI-DIMENSIONAL PROBABILISTIC LAWS.
        (IN RUSSIAN)
    DOKL. AKAD. NAUK SSSR, 169, 1017-1019.  (MR34-3619)

OSTROVSKII, I. V. 1966.  (17.3)                                               4103
    CERTAIN PROPERTIES OF HOLOMORPHIC CHARACTERISTIC FUNCTIONS OF MULTI-DIMENSIONAL
    PROBABILISTIC LAWS.
        (IN RUSSIAN)
    TEOR. FUNKCII FUNKCIONAL. ANAL. I PRILOZEN, 2, 169-177.  (MR34-5132)

OSTROVSKII, I. V. 1966.  (17.1)                                               4104
    DECOMPOSITION OF MULTI-DIMENSIONAL INFINITELY DIVISIBLE LAWS WITHOUT
    GAUSSIAN COMPONENT.
        (IN RUSSIAN)
    VESTNIK HAR'KOV. GOS. UNIV., 14, 51-72.  (MR35-6178)

OSTROWSKI, ALEXANDRE M. 1937.  (20.2)                                         4105
    SUR LA DETERMINATION DES BORNES INFERIEURES POUR UNE CLASSE DES DETERMINANTS.
    BULL. SCI. MATH. SER. 2, 61,   19-32.

OSTROWSKI, ALEXANDRE M. 1940.  (20.3)                                                **4106**
    SUR LA CONVERGENCE ET L'ESTIMATION DES ERREURS DANS QUELQUES PROCEDES
    DE RESOLUTION DES EQUATIONS NUMERIQUES.
    SB. POS. PAM. D. A. GRAVE, 213-234.  (MR2,P.367)

OSTROWSKI, ALEXANDRE M. 1958.  (20.3)                                                **4107**
    ON GAUSS' SPEEDING UP DEVICE IN THE THEORY OF SINGLE STEP ITERATION.
    MATH. TABLES AIDS COMPUT., 12, 116-132.  (MR20-6185)

OTHERS (SIC)    SEE
    943. CHENG, SHAO-LIEN; TAO, TSUNG-YING + OTHERS (SIC) (1962)

    944. CHENG, SHAO-LIEN; TAO, TSUNG-YING + OTHERS (SIC) (1962)

O'TOOLE, A.L. 1932.  (17.3)                                                          **4108**
    ON SYMMETRIC FUNCTIONS OF MORE THAN ONE VARIABLE AND OF FREQUENCY FUNCTIONS.
    ANN. MATH. STATIST., 3, 56-63.

O'TOOLE, A.L. 1934.  (4.1/18.5)                                                      **4109**
    ON A BEST VALUE OF R IN SAMPLES OF R FROM A FINITE POPULATION OF N.
    ANN. MATH. STATIST., 5, 146-152.

OTTAVIANI, GIUSEPPE 1939.  (18.1)                                                    **4110**
    SULLA PROBABILITA CHE UNA PROVA SU DUE VARIABILI CASUALI X E Y VERIFICHI LA
    DISUGUAGLIANZA x < Y E SUL CORRISPONDENTE SCARTO QUADRATICO MEDIO.
    GIORN. IST. ITAL. ATTUARI, 10, 185-192.  (MR1,P.340)

OTTAVIANI, GIUSEPPE 1959.  (4.8)                                                     **4111**
    SU ALCUNI PROBLEMI RIGUARDANTI LA TRANSVARIAZIONE.
    ATTI RIUN. SCI. SOC. ITAL. STATIST., 19, 11-25.

OTTAVIANI, GIUSEPPE 1960.  (4.8)                                                     **4112**
    ALCUNE CONSIDERAZIONI SUL CONCETTO DI TRANSVARIAZIONE.
    STUDI ONORE C. GINI, 1, 313-314.

OVERALL, JOHN E. 1964.  (15.5)                                                       **4113**
    NOTE ON MULTIVARIATE METHODS FOR PROFILE ANALYSIS.
        (AMENDED BY NO. 2423)
    PSYCHOL. BULL., 61, 195-198.

OVERALL, JOHN E. + HOLLISTER, L. E. 1964.  (6.4/6.6/20.3/E)                          **4114**
    COMPUTER PROCEDURES FOR PSYCHIATRIC CLASSIFICATION.
    J. AMER. MED. ASSOC., 187, 115-120.

OVERALL, JOHN E. + WILLIAMS, CLYDE M. 1961.  (15.1/E)                                **4115**
    MODELS FOR MEDICAL DIAGNOSIS:  FACTOR ANALYSIS.  PART ONE, THEORETICAL.
    MEDIZIN. DOKUMENT., 5, 51-56.

OWEN, DONALD B. 1953.  (5.1)                                                         **4116**
    A DOUBLE SAMPLE TEST PROCEDURE.
    ANN. MATH. STATIST., 24, 449-457.  (MR15,P.46)

OWEN, DONALD B. 1956.  (2.7/B)                                                       **4117**
    TABLES FOR COMPUTING BIVARIATE NORMAL PROBABILITIES.
    ANN. MATH. STATIST., 27, 1075-1090.  (MR23-B607)

OWEN, DONALD B. 1965.  (18.1/B)                                                      **4118**
    A SPECIAL CASE OF A BIVARIATE NONCENTRAL T-DISTRIBUTION.
    BIOMETRIKA, 52, 437-446.  (MR34-5190)

OWEN, DONALD B. + STECK, GEORGE P. 1962.  (14.1)                                     **4119**
    MOMENTS OF ORDER STATISTICS FROM THE EQUICORRELATED MULTIVARIATE
    NORMAL DISTRIBUTION.
    ANN. MATH. STATIST., 33, 1286-1291.  (MR25-4589)

OWEN, DONALD B. + WIESEN, J.M. 1959.  (2.7/BE)                                       **4120**
    A METHOD OF COMPUTING BIVARIATE NORMAL PROBABILITIES WITH AN
    APPLICATION TO HANDLING ERRORS IN TESTING AND MEASURING.
    BELL SYSTEM TECH. J., 38, 553-572.  (MR21-954)

OWEN, DONALD B.　SEE ALSO
　　589. BLAND, R. P. + OWEN, DONALD B. (1966)

　　5206. STECK, GEORGE P. + OWEN, DONALD B. (1962)

OZAWA, MASARU　SEE
　　2461. HIGUTI, ISAO + OZAWA, MASARU (1958)

PABST, MARGARET R.　SEE
　　2609. HOTELLING, HAROLD + PABST, MARGARET R. (1936)

PACHARES, JAMES 1955.　(2.5/14.5/17.3)　　　　　　　　　　　　　　　　　4121
　　NOTE ON THE DISTRIBUTION OF A DEFINITE QUADRATIC FORM.
　ANN. MATH. STATIST., 26, 128–131.　(MR16,P.727)

PAGE, D.J.　SEE
　　2952. KEEN, JOAN + PAGE, D.J. (1953)

PAGE, R. M.; BRODZINSKY, A. + ZIRM, R. R. 1953.　(6.6/16.3)　　　　　　4122
　　A MICROWAVE CORRELATOR.
　PROC. INST. RADIO ENGRS., 41, 128–131.

PAINTER, R. J.　SEE
　　2104. GRAYBILL, FRANKLIN A.; MEYER, C. D. + PAINTER, R. J. (1966)

PALEY, R. E. A. C. 1933.　(20.2)　　　　　　　　　　　　　　　　　　　4123
　　ON ORTHOGONAL MATRICES.
　J. MATH. AND PHYS., 12, 311–320.

PALL, GORDON 1945.　(20.1)　　　　　　　　　　　　　　　　　　　　　4124
　　THE ARITHMETICAL INVARIANTS OF QUADRATIC FORMS.
　BULL. AMER. MATH. SOC., 51, 185–197.　(MR7,P.50)

PAN, JIE-JIAN 1964.　(2.6)　　　　　　　　　　　　　　　　　　　　　　4125
　　DISTRIBUTION OF THE SERIAL CORRELATION COEFFICIENTS WITH NON-CIRCULAR
　　STATISTICS.
　　　　(IN CHINESE)
　SHUXUE JINZHAN, 7, 328–337.　(MR32-8426)

PANNE, C. VAN DE　SEE　VAN DE PANNE, C.

PANOFSKY, H. 1949.　(4.3/E)　　　　　　　　　　　　　　　　　　　　4126
　　SIGNIFICANCE OF METEOROLOGICAL CORRELATION COEFFICIENTS.
　BULL. AMER. METEOROL. SOC., 30, 326–327.

PANSE, V. G. 1946.　E(6.2/6.4/8.5)　　　　　　　　　　　　　　　　　4127
　　AN APPLICATION OF THE DISCRIMINANT FUNCTION FOR SELECTION IN POULTRY.
　J. GENET., 47, 242–248.

PAPAMICHAIL, D. 1961.　(4.2)　　　　　　　　　　　　　　　　　　　4128
　　QUELQUES SERIES CONVERGENTES DONT LES TERMES SONT FONCTIONS-Γ, UNE
　　METHODE PROBABILISTE POUR LES SOMMER.
　PUBL. INST. STATIST. UNIV. PARIS, 10, 287–296.　(MR26-7067)

PARASJUK, O. S.　SEE
　　3790. MEIZLER, D.G.; PARASJUK, O. S. + RVACEVA, E. L. (1948)

　　3791. MEIZLER, D.G.; PARASJUK, O. S. + RVACEVA, E. L. (1949)

PARENTI, GIUSEPPE 1941.　(4.1/4.8)　　　　　　　　　　　　　　　　　4129
　　ESTENSIONE DI UNA RELAZIONE DEL FRECHET AL COEFFICIENTE DI
　　CORRELAZIONE PARABOLICA DI ORDINE K.
　GIORN. ECON. ANN. ECONOMIA, 3, 319–324.

PARENTI, GIUSEPPE 1941.　(4.8)　　　　　　　　　　　　　　　　　　4130
　　SU UN'ESTENSIONE DEL COEFFICIENTE DI CORRELAZIONE.
　STATISTICA, BOLOGNA, 1, 382–400.

PARENTI, GIUSEPPE 1943.　(4.8)　　　　　　　　　　　　　　　　　　4131
　　SUL RAPPORTO DI CORRELAZIONE DEL PEARSON E SULLA MISURA DELLA
　　"STRETTEZZA" DELLE RELAZIONI STATISTICHE.
　STATISTICA, BOLOGNA, 3, 217–242.

PARENTI, GIUSEPPE 1946.  (4.4)                                              4132
    SULL'ERRORE MEDIO DEL COEFFICIENTE DI VARIABILITA RELATIVA.
    ARCH. ANTROP. ETNOLOGIA, 76, 69-71.


PARENTI, GIUSEPPE 1954.  (4.8/E)                                            4133
    SU DI UN INDICE DI CONCORDANZA FRA DUE CARATTERI.
         (WITH SUMMARY IN FRENCH)
    STATISTICA, BOLOGNA, 14, 594-603.


PARK, JOHN H., JR. 1961.  (7.4)                                            4134
    MOMENTS OF THE GENERALIZED RAYLEIGH DISTRIBUTION.
    QUART. APPL. MATH., 19, 45-49.  (MR22-9988)


PARK, R. E. 1966.  (8.8)                                                   4135
    ESTIMATION WITH HETEROSCEDASTIC ERROR TERMS.
    ECONOMETRICA, 34, 888.


PARKER, W. V. 1937.  (20.1)                                                4136
    THE CHARACTERISTIC ROOTS OF A MATRIX.
    DUKE MATH. J., 3, 484-487.


PARKER, W. V. 1943.  (20.1)                                                4137
    LIMITS TO THE CHARACTERISTIC ROOTS OF A MATRIX.
    DUKE MATH. J., 10, 479-482.  (MR5,P.30)


PARKER, W. V. 1948.  (20.1)                                                4138
    CHARACTERISTIC ROOTS AND THE FIELD OF VALUES OF A MATRIX.
    DUKE MATH. J., 15, 439-442.  (MR10,P.4)


PARKER, W. V. 1951.  (20.1)                                                4139
    CHARACTERISTIC ROOTS AND FIELD OF VALUES OF A MATRIX.
    BULL. AMER. MATH. SOC., 57, 103-108.  (MR12,P.581)


PARLETT, BERESFORD 1965.  (20.1)                                           4140
    MATRIX EIGENVALUE PROBLEMS.
    AMER. MATH. MONTHLY, 72(2) 59-66.


PARTHASARATHY,K.R. + BHATTACHARYA, P.K. 1961.   (8.1/17.1/17.8/B)          4141
    SOME LIMIT THEOREMS IN REGRESSION THEORY.
    SANKHYA SER. A, 23, 91-102.  (MR24-A3748)


PARTHASARATHY, M.    SEE
    3150. KRISHNAMOORTHY, A. S. + PARTHASARATHY, M. (1951)

    3151. KRISHNAMOORTHY, A. S. + PARTHASARATHY, M. (1960)


PARTHASARATHY, T. 1965.  (19.1)                                           4142
    A NOTE ON A MINIMAX THEOREM OF T. T. TIE.
    SANKHYA SER. A, 27, 407-408.


PATHAK, P. K. 1966.  (18.5)                                               4143
    AN ESTIMATOR IN PPS SAMPLING FOR MULTIPLE CHARACTERISTICS.
    SANKHYA SER. A, 28, 35-40.


PATIL, G. P. 1965.  (18.1/18.2)                                           4144
    ON MULTIVARIATE GENERALIZED POWER SERIES DISTRIBUTION AND ITS
    APPLICATION TO THE MULTINOMIAL AND NEGATIVE MULTINOMIAL.
         (REPRINTED AS NO. 4146)
    PROC. INTERNAT. SYMP. CLASS. CONTAG. DISCRETE DISTNS. (PATIL), 183-194.  (MR35-2393)


PATIL, G. P. 1965.  (18.1)                                                4145
    CERTAIN CHARACTERISTIC PROPERTIES OF MULTIVARIATE DISCRETE
    PROBABILITY DISTRIBUTIONS AKIN TO THE BATES-NEYMAN MODEL IN THE THEORY
       OF ACCIDENT PRONENESS.
    SANKHYA SER. A, 27, 259-270.  (MR34-881)


PATIL, G. P. 1966.  (18.1/18.2)                                           4146
    ON MULTIVARIATE GENERALIZED POWER SERIES DISTRIBUTION AND ITS
    APPLICATION TO THE MULTINOMIAL AND NEGATIVE MULTINOMIAL.
         (REPRINT OF NO. 4144)
    SANKHYA SER. A, 28, 225-238.  (MR34-8543)

PATIL, G. P. + BILDIKAR, SHULA 1966.  (17.3/18.1/20.1)                    4147
    IDENTIFIABILITY OF COUNTABLE MIXTURES OF DISCRETE PROBABILITY
    DISTRIBUTIONS USING METHODS OF INFINITE MATRICES.
    PROC. CAMBRIDGE PHILOS. SOC., 62, 485-494.

PATIL, G. P.   SEE ALSO
    4973. SESHADRI, VANAMAMALAI + PATIL, G. P. (1964)

PATNAIK, P. B. 1949.  (14.5)                                              4148
    THE NON-CENTRAL $\chi^2$ AND F-DISTRIBUTIONS AND THEIR APPLICATIONS.
    BIOMETRIKA, 36, 202-232.  (MR11,P.608)

PAULSON, EDWARD 1942.  (3.1)                                              4149
    A NOTE ON THE ESTIMATION OF SOME MEAN VALUES FOR A BIVARIATE DISTRIBUTION.
    ANN. MATH. STATIST., 13, 440-445.  (MR4,P.280)

PAULSON, EDWARD 1949.  (19.1)                                            4150
    A MULTIPLE DECISION PROCEDURE FOR CERTAIN PROBLEMS IN THE ANALYSIS OF VARIANCE.
    ANN. MATH. STATIST., 20, 95-98.  (MR10,P.467)

PAULSON, EDWARD 1952.  (19.1)                                            4151
    ON THE COMPARISON OF SEVERAL EXPERIMENTAL CATEGORIES WITH A CONTROL.
    ANN. MATH. STATIST., 23, 239-246.  (MR14,P.299)

PAULSON, EDWARD 1961.  (17.8/19.1/U)                                     4152
    A NON-PARAMETRIC SOLUTION FOR THE K-SAMPLE SLIPPAGE PROBLEM.
    STUD. ITEM ANAL. PREDICT. (SOLOMON), 233-238.   (MR22-12647)

PAULSON, EDWARD   SEE ALSO
    562. BIRNBAUM, Z. W.; PAULSON, EDWARD + ANDREWS, FRED C. (1950)

PAVLOV, V. V. 1965.  (16.4)                                              4153
    CERTAIN QUESTIONS CONCERNING THE REALIZATION OF INVARIANCE CONDITIONS
    IN MANY-DIMENSIONAL NONLINEAR AUTOMATIC SYSTEMS.
        (IN RUSSIAN)
    COMPLEX SYSTEMS CONTROL (NAUKOVA DUMKA, KIEV), 113-124.   (MR32-9086)

PAYEN, RAYMOND 1966.  (16.4)                                             4154
    SUITES STATIONNAIRES DE VECTEURS ALEATOIRES A VALEURS DANS UN ESPACE DE HILBERT.
    C. R. ACAD. SCI. PARIS SER. A, 262, 579-582.

PEARL, M. H. 1966.  (20.1)                                              4155
    ON GENERALIZED INVERSES OF MATRICES.
    PROC. CAMBRIDGE PHILOS. SOC., 62, 673-677.

PEARL, M. H.   SEE ALSO
    2940. KATZ, I. J. + PEARL, M. H. (1966)

PEARL, RAYMOND + FULLER, WILBUR N. 1905. E(4.8)                         4156
    VARIATION AND CORRELATION IN THE EARTHWORM.
    BIOMETRIKA, 4, 213-229.

PEARSON, EGON SHARPE 1922.  (4.1/E)                                     4157
    ON THE VARIATIONS IN PERSONAL EQUATIONS AND THE CORRELATION OF
    SUCCESSIVE JUDGEMENTS.
    BIOMETRIKA, 14, 23-102.

PEARSON, EGON SHARPE 1923.  (4.8)                                       4158
    THE PROBABLE ERROR OF A CLASS-INDEX CORRELATION.
    BIOMETRIKA, 14, 261-280.

PEARSON, EGON SHARPE 1924.  (4.2)                                       4159
    NOTE ON THE APPROXIMATIONS TO THE PROBABLE ERROR OF A COEFFICIENT OF
    CORRELATION.
    BIOMETRIKA, 16, 196-198.

PEARSON, EGON SHARPE 1927.  (4.8)                                       4160
    FURTHER NOTE ON THE "LINEAR CORRELATION RATIO".
    BIOMETRIKA, 19, 223-224.

PEARSON, EGON SHARPE 1929.  (4.8/8.3)                                   4161
    SOME NOTES ON SAMPLING TESTS WITH TWO VARIABLES.
    BIOMETRIKA, 21, 337-360.

PEARSON, EGON SHARPE 1931.   (4.3)                                          4162
    THE TEST OF SIGNIFICANCE FOR THE CORRELATION COEFFICIENT.
    J. AMER. STATIST. ASSOC., 26, 128-134.


PEARSON, EGON SHARPE 1932.   (4.2/18.3)                                     4163
    THE TEST OF SIGNIFICANCE FOR THE CORRELATION COEFFICIENT:   SOME FURTHER RESULTS.
    J. AMER. STATIST. ASSOC., 27, 424-426.


PEARSON, EGON SHARPE 1934.   (8.2)                                          4164
    APPENDIX I.
    J. ROY. STATIST. SOC. SUPP., 1, 178-181.


PEARSON, EGON SHARPE 1935.   (14.2/U)                                       4165
    IT. A COMPARISON OF $\beta_2$ AND MR GEARY'S $\omega_n$ CRITERIA.
    BIOMETRIKA, 27, 333-352.


PEARSON, EGON SHARPE 1957.   (1.2)                                          4166
    JOHN WISHART 1898-1956.
    BIOMETRIKA, 44, 1-8.   (MR18,P.892)


PEARSON, EGON SHARPE + SNOW, BARBARA 1962.   (4.8/A)                        4167
    TESTS FOR RANK CORRELATION COEFFICIENTS.
    BIOMETRIKA, 49, 185-191.


PEARSON, EGON SHARPE + WILKS, S.S. 1933.   (10.1/B)                         4168
    METHODS OF STATISTICAL ANALYSIS APPROPRIATE FOR K SAMPLES OF TWO VARIABLES.
    BIOMETRIKA, 25, 353-378.


PEARSON, EGON SHARPE    SEE ALSO
    3999. NEYMAN, JERZY + PEARSON, EGON SHARPE (1931)

    4000. NEYMAN, JERZY + PEARSON, EGON SHARPE (1932)

    4001. NEYMAN, JERZY + PEARSON, EGON SHARPE (1936)

    4246. PEARSON, KARL + PEARSON, EGON SHARPE (1922)

    4247. PEARSON, KARL + PEARSON, EGON SHARPE (1931)

     952. CHESHIRE, LEONE; OLDIS, ELENA + PEARSON, EGON SHARPE (1932)

    1663. FIELLER, E. C.; LEWIS, T. + PEARSON, EGON SHARPE (1955)

    1664. FIELLER, E. C.; LEWIS, T. + PEARSON, EGON SHARPE (1956)


PEARSON, KARL 1896.   (2.4/4.1/8.1/E)                                       4169
    MATHEMATICAL CONTRIBUTIONS TO THE THEORY OF EVOLUTION.   III.
    REGRESSION, HEREDITY AND PANMIXIA.
        (REPRINTED AS NO. 4233)
    PHILOS. TRANS. ROY. SOC. LONDON SER. A, 187, 253-318.


PEARSON, KARL 1897.   (17.3/AE)                                             4170
    MATHEMATICAL CONTRIBUTIONS TO THE THEORY OF EVOLUTION.   ON A FORM OF SPURIOUS
    CORRELATION WHICH MAY ARISE WHEN INDICES ARE USED IN THE MEASUREMENT OF ORGANS.
    PROC. ROY. SOC. LONDON, 60, 489-498.


PEARSON, KARL 1899.   (20.4)                                                4171
    ON CERTAIN PROPERTIES OF THE HYPERGEOMETRICAL SERIES, AND ON THE FITTING OF
    SUCH SERIES TO OBSERVATION POLYGONS IN THE THEORY OF CHANCE.
    PHILOS. MAG. SER. 5, 47, 236-246.


PEARSON, KARL 1899.   (4.1/E)                                               4172
    DATA FOR THE PROBLEM OF EVOLUTION IN MAN.   III.   ON THE MAGNITUDE OF
    CERTAIN COEFFICIENTS OF CORRELATION IN MAN, ETC.
    PROC. ROY. SOC. LONDON, 66, 23-32.


PEARSON, KARL 1900.   (5.1/17.5/18.2/E)                                     4173
    ON THE CRITERION THAT A GIVEN SYSTEM OF DEVIATIONS FROM THE PROBABLE IN THE
    CASE OF A CORRELATED SYSTEM OF VARIABLES IS SUCH THAT IT CAN BE REASONABLY
    SUPPOSED TO HAVE ARISEN FROM RANDOM SAMPLING.
        (REPRINTED AS NO. 4234)
    PHILOS. MAG. SER. 5, 50, 157-175.

PEARSON, KARL 1900.  (4.8)                                                                          4174
    MATHEMATICAL CONTRIBUTIONS TO THE THEORY OF EVOLUTION.  VII.  ON THE
    CORRELATION OF CHARACTERS NOT QUANTITATIVELY MEASURABLE.
    PHILOS. TRANS. ROY. SOC. LONDON SER. A, 195, 1-47.

PEARSON, KARL 1901.  (11.1/17.9/E)                                                                  4175
    ON LINES AND PLANES OF CLOSEST FIT TO SYSTEMS OF POINTS IN SPACE.
    PHILOS. MAG. SER. 6, 2, 559-572.

PEARSON, KARL 1902.  (2.4)                                                                          4176
    MATHEMATICAL CONTRIBUTIONS TO THE THEORY OF EVOLUTION.  XI.  ON THE INFLUENCE
    OF NATURAL SELECTION ON THE VARIABILITY AND CORRELATION OF ORGANS.
    PHILOS. TRANS. ROY. SOC. LONDON SER. A, 200, 1-66.

PEARSON, KARL 1902.  E(4.8)                                                                         4177
    ON THE CORRELATION OF INTELLECTUAL ABILITY WITH THE SIZE AND SHAPE OF
    THE HEAD.  PRELIMINARY NOTICE.  (A NEW YEAR'S GREETING TO FRANCIS GALTON, 1902)
    PROC. ROY. SOC. LONDON, 69, 333-342.

PEARSON, KARL 1903.  (17.3)                                                                         4178
    ON THE PROBABLE ERRORS OF FREQUENCY CONSTANTS.  EDITORIAL.  I.
    BIOMETRIKA, 2, 273-281.

PEARSON, KARL 1904.  (4.1/4.3/E)                                                                    4179
    ON THE RANDOM INCREASE AND DECREASE OF SEGMENTS AND ON THE
    CORRELATIONS BETWEEN THE THREE VERTEBRAL REGIONS OF SPINAX NIGER.
    BIOMETRIKA, 3, 338-341.

PEARSON, KARL 1904.  (2.1/4.8/8.8/ET)                                                               4180
    MATHEMATICAL CONTRIBUTIONS TO THE THEORY OF EVOLUTION. —XIII.  ON THE THEORY OF
    CONTINGENCY AND ITS RELATION TO ASSOCIATION AND NORMAL CORRELATION.
        (REPRINTED AS NO. 4235)
    DRAPERS CO. RES. MEM. BIOM. SER., 1, 1-35.

PEARSON, KARL 1905.  (4.1/4.8/8.1/E)                                                                4181
    MATHEMATICAL CONTRIBUTIONS TO THE THEORY OF EVOLUTION. —XIV.  ON THE
    GENERAL THEORY OF SKEW CORRELATION AND NON-LINEAR REGRESSION.
        (REPRINTED AS NO. 4236)
    DRAPERS CO. RES. MEM. BIOM. SER., 2, 1-54.

PEARSON, KARL 1906.  E(4.8)                                                                         4182
    ON THE RELATIONSHIP OF INTELLIGENCE TO SIZE AND SHAPE OF HEAD, AND TO
    OTHER PHYSICAL AND MENTAL CHARACTERS.
    BIOMETRIKA, 5, 105-146.

PEARSON, KARL 1906.  (4.1/8.8)                                                                      4183
    ON CERTAIN POINTS CONNECTED WITH SCALE ORDER IN THE CASE OF CORRELATION OF TWO
    CHARACTERS  WHICH FOR SOME ARRANGEMENT GIVE A LINEAR REGRESSION LINE.
    BIOMETRIKA, 5, 176-178.

PEARSON, KARL 1907.  (4.4/17.3/17.5/E)                                                              4184
    REPLY TO CERTAIN CRITICISMS OF MR G.U. YULE.
    BIOMETRIKA, 5, 470-476.

PEARSON, KARL 1907.  (4.1/4.4/E)                                                                    4185
    MATHEMATICAL CONTRIBUTIONS TO THE THEORY OF EVOLUTION.  XVI.  ON
    FURTHER METHODS OF DETERMINING CORRELATION.
    DRAPERS CO. RES. MEM. BIOM. SER., 4, 1-39.

PEARSON, KARL 1907.  (2.4/4.8/6.7)                                                                  4186
    APPENDIX: INFLUENCE OF ACADEMIC SELECTION ON CORRELATION COEFFICIENTS.
    EUGENICS LAB. MEM., 1, 41-42.

PEARSON, KARL 1907.  (4.1/4.8/E)                                                                    4187
    ON CORRELATION AND THE METHODS OF MODERN STATISTICS.
    NATURE, 76, 517-518, 613-615, 662.

PEARSON, KARL 1908.  (4.2/4.3/E)                                                                    4188
    ON HEREDITY IN SEX.  REMARKS ON MR COBB'S NOTE.
    BIOMETRIKA, 6, 109-111.

PEARSON, KARL 1908.  (2.4)                                                    4189
     ON THE INFLUENCE OF DOUBLE SELECTION ON THE VARIATION AND CORRELATION
     OF TWO CHARACTERS.
     BIOMETRIKA, 6, 111-112.

PEARSON, KARL 1909.  (4.8)                                                    4190
     ON A NEW METHOD OF DETERMINING CORRELATION BETWEEN A MEASURED CHARACTER A, AND
     A CHARACTER B, OF WHICH ONLY THE PERCENTAGE OF CASES WHEREIN B EXCEEDS (OR
     FALLS SHORT OF) A GIVEN INTENSITY IS RECORDED FOR EACH GRADE OF A.
     BIOMETRIKA, 7, 96-105.

PEARSON, KARL 1909.  E(4.8)                                                   4191
     THE THEORY OF ANCESTRAL CONTRIBUTIONS IN HEREDITY.
     PROC. ROY. SOC. LONDON SER. B, 81, 219-224.

PEARSON, KARL 1909.  E(4.8)                                                   4192
     ON THE ANCESTRAL GAMETIC CORRELATIONS OF A MENDELIAN POPULATION
     MATING AT RANDOM.
     PROC. ROY. SOC. LONDON SER. B, 81, 225-229.

PEARSON, KARL 1909.  (4.1)                                                    4193
     DETERMINATION OF THE COEFFICIENT OF CORRELATION.
     SCIENCE NEW SER., 30, 23-25.

PEARSON, KARL 1910.  (4.8)                                                    4194
     ON A NEW METHOD OF DETERMINING CORRELATION  WHEN ONE VARIABLE IS
     GIVEN BY ALTERNATIVE AND THE OTHER BY MULTIPLE CATEGORIES.
     BIOMETRIKA, 7, 248-257.

PEARSON, KARL 1911.  (4.8)                                                    4195
     ON A CORRECTION TO BE MADE TO THE CORRELATION RATIO $\eta$.
     BIOMETRIKA, 8, 254-256.

PEARSON, KARL 1912.  (2.4)                                                    4196
     ON THE GENERAL THEORY OF THE INFLUENCE OF SELECTION ON CORRELATION AND
     VARIATION.
     BIOMETRIKA, 8, 437-443.

PEARSON, KARL 1913.  (4.2/17.9)                                              4197
     ON THE PROBABLE ERRORS OF FREQUENCY CONSTANTS, PART II.
     BIOMETRIKA, 9, 1-10.

PEARSON, KARL 1913.  (4.4)                                                    4198
     ON THE PROBABLE ERROR OF A COEFFICIENT OF CORRELATION AS FOUND FROM A
     FOURFOLD TABLE.
     BIOMETRIKA, 9, 22-27.

PEARSON, KARL 1913.  (4.1)                                                    4199
     ON THE MEASUREMENT OF THE INFLUENCE OF 'BROAD CATEGORIES' ON CORRELATION.
     BIOMETRIKA, 9, 116-139.

PEARSON, KARL 1914.  (4.6)                                                    4200
     ON CERTAIN ERRORS WITH REGARD TO MULTIPLE CORRELATION OCCASIONALLY
     MADE BY THOSE WHO HAVE NOT ADEQUATELY STUDIED THE SUBJECT.
     BIOMETRIKA, 10, 181-187.

PEARSON, KARL 1914.  (4.4/4.8)                                               4201
     ON AN EXTENSION OF THE METHOD OF CORRELATION BY GRADES OR RANKS.
     BIOMETRIKA, 10, 416-418.

PEARSON, KARL 1915.  (4.8)                                                    4202
     ON THE PARTIAL CORRELATION RATIO.
     PROC. ROY. SOC. LONDON SER. A, 91, 492-498.

PEARSON, KARL 1916.  (4.5/4.6)                                               4203
     ON SOME NOVEL PROPERTIES OF PARTIAL AND MULTIPLE CORRELATION
     COEFFICIENTS IN A UNIVERSE OF MANIFOLD CHARACTERISTICS.
     BIOMETRIKA, 11, 231-238.

PEARSON, KARL 1917.  (4.4)                                                    4204
     THE PROBABLE ERROR OF BISERIAL $\eta$.
     BIOMETRIKA, 11, 292-302.

PEARSON, KARL 1919.  (17.2/B)                                                                          4205
    ON GENERALISED TCHEBYCHEFF THEOREMS IN THE MATHEMATICAL THEORY OF STATISTICS.
    BIOMETRIKA, 12, 284–296.

PEARSON, KARL 1920.  (4.1)                                                                             4206
    NOTES ON THE HISTORY OF CORRELATION.
    BIOMETRIKA, 13, 25–45.

PEARSON, KARL 1920.  (2.7/20.5/BT)                                                                     4207
    ON THE CONSTRUCTION OF TABLES AND ON INTERPOLATION.  PART II.  BIVARIATE TABLES.
    TRACTS COMPUT., 3, 1–54.

PEARSON, KARL 1921.  E(4.4/4.8)                                                                        4208
    SECOND NOTE ON THE COEFFICIENT OF CORRELATION AS DETERMINED FROM THE
    QUANTITATIVE MEASUREMENT OF ONE VARIATE AND THE RANKING OF A SECOND VARIATE.
    BIOMETRIKA, 13, 302–305.

PEARSON, KARL 1923.  (4.8)                                                                             4209
    ON THE CORRECTION NECESSARY FOR THE CORRELATION RATIO $\eta$.
    BIOMETRIKA, 14, 412–417.

PEARSON, KARL 1924.  (18.1)                                                                            4210
    ON THE MOMENTS OF THE HYPERGEOMETRICAL SERIES.
    BIOMETRIKA, 16, 157–162.

PEARSON, KARL 1924.  (18.1)                                                                            4211
    ON A CERTAIN DOUBLE HYPERGEOMETRICAL SERIES AND ITS REPRESENTATION BY
    CONTINUOUS FREQUENCY SURFACES.
    BIOMETRIKA, 16, 172–188.

PEARSON, KARL 1924.  (20.4)                                                                            4212
    NOTE ON THE RELATIONSHIP OF THE INCOMPLETE B-FUNCTION TO THE SUM OF
    THE FIRST P TERMS OF THE BINOMIAL $(A + B)^n$.
    BIOMETRIKA, 16, 202–203.

PEARSON, KARL 1925.  (4.1)                                                                             4213
    ON THE CORRELATION BETWEEN TWO VARIATES X AND $Y = KX^s$.
        (AMENDED BY NO. 4214)
    AMER. MATH. MONTHLY, 32, 70–73.

PEARSON, KARL 1925.  (4.1)                                                                             4214
    ERRATA TO:  ON THE CORRELATION BETWEEN TWO VARIATES  X  AND  $Y=KX^s$.
        (AMENDMENT OF NO. 4213)
    AMER. MATH. MONTHLY, 32, 538.

PEARSON, KARL 1925.  (4.6/E)                                                                           4215
    ON THE MULTIPLE CORRELATION OF BROTHERS, BEING A NOTE ON MR J.O. IRWIN'S
    MEMOIR, AND ON MY STATEMENT OF THE APPLICATION OF GALTON'S DIFFERENCE PROBLEM
    TO THE DETERMINATION OF THE DEGREE OF RELATIONSHIP OF BROTHERS, MADE
    IN AUGUST 1902.
    BIOMETRIKA, 17, 129–141.

PEARSON, KARL 1925.  (7.1/8.2/18.1/BT)                                                                 4216
    FURTHER CONTRIBUTIONS TO THE THEORY OF SMALL SAMPLES.
        (AMENDED BY NO. 4222)
    BIOMETRIKA, 17, 176–199.

PEARSON, KARL 1925.  (4.1/4.8)                                                                         4217
    NOTE BY THE EDITOR (OF BIOMETRIKA) ON "NOTES ON AN EXPERIMENTAL TEST OF ERRORS
    IN PARTIAL CORRELATION COEFFICIENTS, DERIVED FROM FOUR-FOLD AND BISERIAL TOTAL
    COEFFICIENTS".
    BIOMETRIKA, 17, 266–267.

PEARSON, KARL 1925.  (18.1)                                                                            4218
    THE FIFTEEN CONSTANT BIVARIATE FREQUENCY SURFACE.
    BIOMETRIKA, 17, 268–313.

PEARSON, KARL 1925.  (4.1/T)                                                                           4219
    ON FIRST POWER METHODS OF FINDING CORRELATION.
    BIOMETRIKA, 17, 459–469.

PEARSON, KARL 1926.  (6.5)                                                                             4220
    ON THE COEFFICIENT OF RACIAL LIKENESS.
    BIOMETRIKA, 18, 105–117.

PEARSON, KARL 1926.  (8.2/14.2)                                                    4221
    RESEARCHES ON THE MODE OF DISTRIBUTION OF THE CONSTANTS OF SAMPLES
    TAKEN AT RANDOM FROM A BIVARIATE NORMAL POPULATION.
    PROC. ROY. SOC. LONDON SER. A, 112, 1-14.

PEARSON, KARL 1927.  (7.1/8.2/18.1/BT)                                             4222
    ERRATA TO: FURTHER CONTRIBUTIONS TO THE THEORY OF SMALL SAMPLES.
        (AMENDMENT OF NO. 4216)
    BIOMETRIKA, 19, 441-442.

PEARSON, KARL 1928.  (2.1/B)                                                       4223
    THE CONTRIBUTION OF GIOVANNI PLANA TO THE NORMAL BIVARIATE FREQUENCY SURFACE.
    BIOMETRIKA, 20(A) 295-298.

PEARSON, KARL 1928.  E(6.4)                                                        4224
    THE APPLICATION OF THE COEFFICIENT OF RACIAL LIKENESS TO TEST THE
    CHARACTER OF SAMPLES.
    BIOMETRIKA, 20(B) 294-300.

PEARSON, KARL 1928.  (6.5)                                                         4225
    NOTE ON STANDARDISATION OF METHOD OF USING THE COEFFICIENT OF RACIAL LIKENESS.
    BIOMETRIKA, 20(B) 376-378.

PEARSON, KARL 1930.  (2.7/BT)                                                      4226
    ON THE REMAINING TABLES FOR DETERMINING THE VOLUMES OF A BI-VARIATE NORMAL
    SURFACE.  EDITORIAL.  (TABLES FOR DETERMINING THE VOLUMES OF A BI-VARIATE
    NORMAL SURFACE.  (I) INTRODUCTION TO THE TABLES.)
    BIOMETRIKA, 22, 1-11.

PEARSON, KARL 1931.  (14.2/B)                                                      4227
    ON THE NATURE OF THE RELATIONSHIP BETWEEN TWO OF "STUDENT'S" VARIATES ($Z_1$ AND $Z_2$)
        WHEN SAMPLES ARE TAKEN FROM A  BIVARIATE NORMAL POPULATION.
    BIOMETRIKA, 22, 405-422.

PEARSON, KARL 1931.  (14.2)                                                        4228
    SOME PROPERTIES OF "STUDENT'S" Z:  CORRELATION, REGRESSION AND
    SCEDASTICITY OF Z WITH THE MEAN AND STANDARD DEVIATION OF THE SAMPLE.
    BIOMETRIKA, 23, 1-9.

PEARSON, KARL 1931.  (14.2)                                                        4229
    FURTHER REMARKS ON THE "Z" TEST.
    BIOMETRIKA, 23, 408-415.

PEARSON, KARL 1932.  (4.1)                                                         4230
    PROFESSOR RIETZ'S PROBLEM.
    BIOMETRIKA, 24, 290-291.

PEARSON, KARL 1933.  (20.4)                                                        4231
    ON THE APPLICATIONS OF THE DOUBLE BESSEL FUNCTION $K_{\tau_1 \tau_2}(x)$   TO
    STATISTICAL PROBLEMS.  PART I. THEORETICAL.
    BIOMETRIKA, 25, 158-178.

PEARSON, KARL 1934.  (4.8)                                                         4232
    REMARKS ON PROFESSOR STEFFENSEN'S MEASURE OF CONTINGENCY (EDITORIAL).
    BIOMETRIKA, 26, 255-260.

PEARSON, KARL 1956.  (2.4/4.1/8.1/E)                                               4233
    MATHEMATICAL CONTRIBUTIONS TO THE THEORY OF EVOLUTION. III. REGRESSION,HEREDITY,
    AND PANMIXIA.
        (REPRINT OF NO. 4169)
    EARLY STATIST. PAPERS (PEARSON REPRINTS), 3,113-178.

PEARSON, KARL 1956.  (5.1/17.5/18.2/E)                                             4234
    ON THE CRITERION THAT A GIVEN SYSTEM OF DEVIATIONS FROM THE PROBABLE IN THE
    CASE OF A CORRELATED SYSTEM OF VARIABLES IS SUCH THAT IT CAN BE REASONABLY
    SUPPOSED TO HAVE ARISEN FROM RANDOM SAMPLING.
        (REPRINT OF NO. 4173)
    EARLY STATIST. PAPERS (PEARSON REPRINTS), 6, 339-357.

PEARSON, KARL 1956.  (2.1/4.8/8.8/ET)                                             4235
    MATHEMATICAL CONTRIBUTIONS TO THE THEORY OF EVOLUTION. —XIII.  ON THE THEORY OF
    CONTINGENCY AND ITS RELATION TO ASSOCIATION AND NORMAL CORRELATION.
        (REPRINT OF NO. 4180)
    EARLY STATIST. PAPERS (PEARSON REPRINTS), 9, 443-475.

PEARSON, KARL 1956.    (4.1/4.8/8.1/E)                                              4236
    MATHEMATICAL CONTRIBUTIONS TO THE THEORY OF EVOLUTION. - XIV. ON THE GENERAL
    THEORY OF SKEW CORRELATION AND NON-LINEAR REGRESSION.
        (REPRINT OF NO. 4181)
  EARLY STATIST. PAPERS (PEARSON REPRINTS), 10, 477-528.


PEARSON, KARL + CAVE-BROWNE-CAVE, F. E. 1902.    (4.1/E)                            4237
    ON THE CORRELATION BETWEEN THE BAROMETRIC HEIGHT AT STATIONS ON THE
    EASTERN SIDE OF THE ATLANTIC.
  PROC. ROY. SOC. LONDON, 70, 465-470.


PEARSON, KARL + ELDERTON, ETHEL M. 1923.    (16.5/E)                                4238
    ON THE VARIATE DIFFERENCE METHOD.  BEING A PAPER READ BEFORE THE
    SOCIETY OF BIOMETRICIANS AND STATISTICIANS.
  BIOMETRIKA, 14, 281-310.


PEARSON, KARL + ELDERTON, ETHEL M. 1928.    (4.6)                                   4239
    ON THE CORRELATION OF TWO FIRST ORDER MULTIPLE CORRELATION COEFFICIENTS.
  BIOMETRIKA, 20(A) 310-313.


PEARSON, KARL + FILON, L.N.G. 1898.    (4.2/14.2/17.5)                              4240
    MATHEMATICAL CONTRIBUTIONS TO THE THEORY OF EVOLUTION.  IV.  ON THE PROBABLE
    ERRORS OF FREQUENCY CONSTANTS AND ON THE INFLUENCE OF RANDOM SELECTION ON
    VARIATION AND CORRELATION.
        (REPRINTED AS NO. 4241)
  PHILOS. TRANS. ROY. SOC. LONDON SER. A, 191, 229-311.


PEARSON, KARL + FILON, L.N.G. 1956.    (4.2/14.2/17.5)                              4241
    MATHEMATICAL CONTRIBUTIONS TO THE THEORY OF EVOLUTION.  IV.  ON THE PROBABLE
    ERRORS OF FREQUENCY CONSTANTS AND ON THE INFLUENCE OF RANDOM SELECTION ON
    VARIATION AND CORRELATION.
        (REPRINT OF NO. 4240)
  EARLY STATIST. PAPERS (PEARSON REPRINTS), 4, 179-261.


PEARSON, KARL + HERON, DAVID 1913.    (4.9)                                         4242
    ON THEORIES OF ASSOCIATION.
  BIOMETRIKA, 9, 159-315.


PEARSON, KARL + LEE, ALICE 1897.    (4.1/E)                                         4243
    ON THE DISTRIBUTION OF FREQUENCY (VARIATION AND CORRELATION) OF THE
    BAROMETRIC HEIGHT AT DIVERS STATIONS.
  PHILOS. TRANS. ROY. SOC. LONDON SER. A, 190, 423-470.


PEARSON, KARL + LEE, ALICE 1908.    (4.7)                                           4244
    ON THE GENERALISED PROBABLE ERROR IN MULTIPLE NORMAL CORRELATION.
  BIOMETRIKA, 6, 59-68.


PEARSON, KARL + MOUL, MARGARET 1927.    (15.1/15.2)                                 4245
    THE MATHEMATICS OF INTELLIGENCE.  I. THE SAMPLING ERRORS IN THE
    THEORY OF A GENERALISED FACTOR.
  BIOMETRIKA, 19, 246-291.


PEARSON, KARL + PEARSON, EGON SHARPF 1922.    (4.8)                                 4246
    ON POLYCHORIC COEFFICIENTS OF CORRELATION.
  BIOMETRIKA, 14, 127-156.


PEARSON, KARL + PEARSON, EGON SHARPF 1931.    (4.7)                                 4247
    APPENDIX TO A PAPER BY DR WISHART.  TABLES OF THE MEAN VALUE AND SQUARED
    STANDARD DEVIATION OF THE SQUARE OF A MULTIPLE CORRELATION COEFFICIENT.
  BIOMETRIKA, 22, 362-367.


PEARSON, KARL + YOUNG, ANDREW W. 1918.    (2.1/4.1/BT)                              4248
    ON THE PRODUCT-MOMENTS OF VARIOUS ORDERS OF THE NORMAL CORRELATION
    SURFACE OF TWO VARIATES.
  BIOMETRIKA, 12, 86-92.


PEARSON, KARL; BEETON, MARY + YULE, G. UDNY 1900.    (4.1/E)                        4249
    DATA FOR THE PROBLEM OF EVOLUTION IN MAN.  V.  ON THE CORRELATION
    BETWEEN DURATION OF LIFE AND THE NUMBER OF OFFSPRING.
  PROC. ROY. SOC. LONDON, 67, 159-179.

PEARSON, KARL; JEFFERY, G.B. + ELDERTON, ETHEL M. 1929. (4.2/T)                    **4250**
    ON THE DISTRIBUTION OF THE FIRST PRODUCT MOMENT-COEFFICIENT, IN
    SAMPLES DRAWN FROM AN INDEFINITELY LARGE NORMAL POPULATION.
    BIOMETRIKA, 21, 164-193.

PEARSON, KARL; LEE, ALICE + ELDERTON, ETHEL M. 1910.  E(4.5)                       **4251**
    ON THE CORRELATION OF DEATH-RATES.
    J. ROY. STATIST. SOC., 73, 534-539.

PEARSON, KARL    SEE ALSO
    897. CAVE, BEATRICE M. + PEARSON, KARL (1914)

    1528. ELDERTON, ETHEL M. + PEARSON, KARL (1915)

    1958. GIBSON, WINIFRED+ PEARSON, KARL (1908)

    3345. LEE, ALICE + PEARSON, KARL (1897)

    3346. LEE, ALICE + PEARSON, KARL (1901)

    5569. UCHIDA, GINZO + PEARSON, KARL (1904)

    5850. WHITELEY, M. A. + PEARSON, KARL (1899)

    3347. LEE, ALICE; LEWENZ, MARIE A. + PEARSON, KARL (1902)

    5135. SOPER, H.E.; YOUNG, ANDREW W.; CAVE, BEATRICE M.;  (1917)
    LEE, ALICE + PEARSON, KARL

    1530. ELDERTON, ETHEL M.; BARRINGTON, AMY; JONES, H. GERTRUDE;  (1913)
    LAMOTTE, EDITH M. M. DE G.; LASKI, H. J. + PEARSON, KARL

PEEL, E. A. 1948.  (15.4)                                                          **4252**
    PREDICTION OF A COMPLEX CRITERION AND BATTERY RELIABILITY.
    BRITISH J. PSYCHOL. STATIST. SECT., 1, 84-94.

PEEL, E. A. 1949.  (15.4)                                                          **4253**
    ITEM DIFFICULTY AS THE MEASURING DEVICE IN OBJECTIVE MENTAL TESTS.
    BRITISH J. PSYCHOL. STATIST. SECT., 2, 69-75.

PEEL, E. A. 1953.  (15.1/E)                                                        **4254**
    FACTORIAL ANALYSIS AS A PSYCHOLOGICAL TECHNIQUE.
    NORDISK PSYKOL. MONOG. NR. 3 (UPPSALA SYMP. PSYCHOL. FACT. ANAL.), 7-22.

PEEL, E. A. 1955.  (15.1)                                                          **4255**
    RESUME DU SYMPOSIUM D'UPPSALA "L'ANALYSE FACTORIELLE EN PSYCHOLOGIE" (1953).
        (WITH DISCUSSION)
    COLLOQ. INTERNAT. CENTRE NAT. RECH. SCI., 65, 45-54.

PEEL, E. A. 1955.  (15.1)                                                          **4256**
    ANALYSE FACTORIELLE DE CORRELATIONS ENTRE PERSONNES ET UTILISATION DE
    QUALITES DETERMINANTES INDEPENDANTES POUR IDENTIFIER LES FACTEURS.
        (WITH DISCUSSION)
    COLLOQ. INTERNAT. CENTRE NAT. RECH. SCI., 65, 377-392.

PEINE, W. 1930.  (4.1)                                                             **4257**
    DER KORRELATIONSFAKTOR FUR VEKTOREN UND SEIN ZUSAMMENHANG MIT DEN
    NUMERISCHEN ELEMENTEN DER BRUNS'SCHEN $\Phi$-REIHE.
    METEOROL. Z., 47, 151-154.

PENA TRAPERO, JESUS BERNARDO 1965.  (16.1)                                         **4258**
    CONSIDERACIONES ECONOMETRICAS SOBRE AUTOCORRELACION.
    ESTADIST. ESPAN., 26, 5-24.

PENFOLD, J. B.    SEE
    2785. JENNISON, R. F.; PENFOLD, J. B. + ROBERTS, J. A. FRASER (1948)

PENNY, SAMUEL J. + BURKHARD, JAMES H. 1966.  (6.6)                                 **4259**
    MULTIDIMENSIONAL CORRELATION LATTICES AS AN AID TO
    THREE DIMENSIONAL PATTERN RECOGNITION.
    SPRING JOINT COMP. CONF. PROC., 449-456.

PENROSE, L.S. 1947.  (5.2/E)                                                          4260
    SOME NOTES ON DISCRIMINATION.
    ANN. EUGENICS, 13, 228-237.  (MR8,P.592)


PENROSE, L.S. 1954.  (6.5/E)                                                          4261
    DISTANCE, SIZE AND SHAPE.
    ANN. EUGENICS, 18, 337-343.


PENROSE, R. A. 1955.  (20.1)                                                          4262
    A GENERALIZED INVERSE FOR MATRICES.
    PROC. CAMBRIDGE PHILOS. SOC., 51, 406-413.  (MR16,P.1082)


PENROSE, R. A. 1956.  (20.1)                                                          4263
    ON BEST APPROXIMATE SOLUTIONS OF LINEAR MATRIX EQUATIONS.
    PROC. CAMBRIDGE PHILOS. SOC., 52, 17-19.  (MR17,P.536)


PERLOFF, ROBERT 1953.  (4.6)                                                          4264
    MULTIPLE CORRELATION FOR FOUR PREDICTORS USING ZERO-ORDER COEFFICIENTS ALONE.
    EDUC. PSYCHOL. MEAS., 13, 655-659.


PERLOFF, ROBERT   SEE ALSO
    5845. WHERRY, ROBERT J.; CAMPBELL, JOEL T. + PERLOFF, ROBERT (1951)


PEROTT, R. + ELDERTON, ETHEL M. 1927.  E(4.8)                                         4265
    CORRELATION BETWEEN PROGNOSIS BASED ON CONDITION OF THE TUBERCULOUS PATIENT
    AT ENTRY TO A SANATORIUM AND THE ISSUE.
    ANN. EUGENICS, 2, 63-75.


PEROZZO, LUIGI 1880.  (17.9/E)                                                        4266
    DELLA RAPPRESENTAZIONE GRAFICA DI UNA COLLETTIVITA DI INDIVIDUI NELLA
    SUCCESSIONE DEL TEMPO, E IN PARTICOLARE DEI DIAGRAMMI A TRE COORDINATE.
        (TRANSLATED AS NO. 4267)
    ANN. STATIST. SER. 2A, 12, 1-16.


PEROZZO, LUIGI 1880.  (17.9/E)                                                        4267
    SUR LA REPRESENTATION GRAPHIQUE D'UNE COLLECTIVITE D'INDIVIDUS DANS
    LA SUCCESSION DU TEMPS ET EN PARTICULIER SUR LES DIAGRAMMES A TROIS COORDONNEES.
        (TRANSLATION OF NO. 4266)
    STATIST. GRAPHIQUE, 1-19.


PERRY, ALBERT D.   SEE
    3528. LUCE, R. DUNCAN + PERRY, ALBERT D. (1949)

    3529. LUCE, R. DUNCAN + PERRY, ALBERT D. (1966)


PERRY, NORMAN C. + MICHAEL, WILLIAM B. 1954.  (4.4)                                   4268
    THE RELIABILITY OF A POINT BISERIAL COEFFICIENT OF CORRELATION.
    PSYCHOMETRIKA, 19, 313-325.


PERRY, NORMAN C. + MICHAEL, WILLIAM B. 1958.  (4.4)                                   4269
    A NOTE CONCERNING THE RELIABILITY OF A POINT BISERIAL COEFFICIENT FOR
    LARGE SAMPLING.
    EDUC. PSYCHOL. MEAS., 18, 139-143.


PERRY, NORMAN C.   SEE ALSO
    2174. GUILFORD, J. P. + PERRY, NORMAN C. (1951)

    3812. MICHAEL, WILLIAM B.; PERRY, NORMAN C. + GUILFORD, J. P. (1952)


PERSONS, WARREN M. 1910.  (4.1/4.6/8.1/E)                                             4270
    THE CORRELATION OF ECONOMIC STATISTICS.
    QUART. PUBL. AMER. STATIST. ASSOC., 12, 287-323.


PERSONS, WARREN M. 1917.  (4.8/E)                                                     4271
    ON THE VARIATE DIFFERENCE CORRELATION METHOD AND CURVE-FITTING.
    QUART. PUBL. AMER. STATIST. ASSOC., 15, 602-642.


PERSONS, WARREN M. 1923.  (16.5/E)                                                    4272
    CORRELATION OF TIME SERIES.
    J. AMER. STATIST. ASSOC., 18, 713-726.

PERSSON, OLLE 1958.  (16.2/17.7)                                           4273
    OM LOSNINGEN AV ETT OVERBESTAMT EKVATIONSSYSTEM ENLIGT MINSTA KVADRATMETODEN.
        (WITH SUMMARY IN ENGLISH)
    NORDISK MAT. TIDSKR., 6, 69-77, 95.  (MR20-5543)

PETERS, CHARLES C. 1946.  (4.8)                                            4274
    A NEW DESCRIPTIVE STATISTIC:  THE PARABOLIC CORRELATION COEFFICIENT.
    PSYCHOMETRIKA, 11, 57-69.  (MR7,P.462)

PETERS, CHARLES C. + WYKES, ELIZABETH CROSSLEY 1931.  (4.5/4.6/8.8)        4275
    SIMPLIFIED METHODS FOR COMPUTING REGRESSION COEFFICIENTS AND PARTIAL
    AND MULTIPLE CORRELATIONS.
    J. EDUC. RES., 23, 383-393.

PETERS, CHARLES C. + WYKES, ELIZABETH CROSSLEY 1931.  (4.5/4.6/8.8)        4276
    SIMPLIFIED METHODS FOR COMPUTING REGRESSION COEFFICIENTS AND MULTIPLE
    AND PARTIAL CORRELATIONS.
    J. EDUC. RES., 24, 44-52.

PETERSEN, DANIEL P. + MIDDLETON, DAVID 1964.  (16.4)                       4277
    RECONSTRUCTION OF MULTIDIMENSIONAL STOCHASTIC FIELDS FROM DISCRETE
    MEASUREMENTS OF AMPLITUDE AND GRADIENT.
    INFORMATION AND CONTROL, 7, 445-476.

PETERSON, D. W. + MATTSON, R. L. 1966.  (6.3)                              4278
    A METHOD OF FINDING LINEAR DISCRIMINANT FUNCTIONS FOR A CLASS OF PERFORMANCE CRITERIA.
    IEEE TRANS. INFORMATION THEORY,IT-12, 380-387.  (MR34-895)

PETERSON, RAYMOND P.   SEE
    2503. HOEL, PAUL G. + PETERSON, RAYMOND P. (1949)

PETRYSHYN, W. V. 1965.  (20.3)                                             4279
    ON THE INVERSION OF MATRICES AND LINEAR OPERATORS.
    PROC. AMER. MATH. SOC., 16, 893-901.

PFANZAGL, JOHANN 1962.  (19.2)                                             4280
    UBERALL TRENNSCHARFE TESTS UND MONOTONE DICHTEQUOTIENTEN.
    Z. WAHRSCHEIN. VERW. GEBIETE, 1, 109-115.  (MR26-4443)

PFANZAGL, JOHANN   SEE ALSO
    648. BORGES, R. + PFANZAGL, JOHANN (1963)

PHILIP, J.R. 1966.  (18.6)                                                 4281
    SOME INTEGRAL EQUATIONS IN GEOMETRICAL PROBABILITY.
    BIOMETRIKA, 53, 365-374.

PHILLIPS, F. M. 1917.  (4.1/E)                                             4282
    SHORT METHOD OF OBTAINING A PEARSON COEFFICIENT OF CORRELATION AND
    OTHER SHORT STATISTICAL PROCESSES.
    MONTHLY WEATHER REV., 50, 854-859.

PHILLIPS, RALPH S.   SEE
    13. ADAMS, J. F.; LAX, PETER D. + PHILLIPS, RALPH S. (1965)

PIAGGIO, H. T. H. 1933.  (15.1)                                            4283
    THREE SETS OF CONDITIONS NECESSARY FOR THE EXISTENCE OF A G THAT IS
    REAL AND UNIQUE EXCEPT IN SIGN.
    BRITISH J. PSYCHOL., 24, 88-105.

PIAGGIO, H. T. H. 1935.  (15.1)                                            4284
    APPROXIMATE GENERAL AND SPECIFIC FACTORS WITHOUT INDETERMINATE PARTS.
    BRITISH J. PSYCHOL., 25, 485-489.

PIAGGIO, H. T. H. + DALLAS, A. E. M. M. 1934.  (15.2)                      4285
    AN ANALYSIS OF RECENT TESTS OF THE TWO-FACTOR THEORY.
    BRITISH J. PSYCHOL., 25, 217-220.

PICARD, H.C. 1951.  (12.1)                                                 4286
    EINE ALLGEMEINE THEORIE DER MEHRDIMENSIONALEN KORRELATION.
    MITT. MATH. STATIST., 3, 103-112.  (MR13,P.665)

PICKREL, E. W. 1958.  (6.1)                                                4287
    CLASSIFICATION THEORY AND TECHNIQUES.
    EDUC. PSYCHOL. MEAS., 18, 37-46.

PIETRA, GAETANO 1936.   (4.1/4.8/4.9)                                                                    4288
    DE LA CONCORDANCE ET DU COEFFICIENT DE CORRELATION.
    REV. INST. INTERNAT. STATIST., 4, 500-515.

PIETRA, GAETANO 1942.   (4.1/4.8/E)                                                                       4289
    INTORNO ALLA MISURA DELLA CONNESSIONE E DELLA CONCORDANZA.
    STATISTICA. BOLOGNA, 2, 59-64.

PILLAI, K.C.SREEDHARAN 1946.   (4.3/C)                                                                    4290
    CONFIDENCE INTERVAL FOR THE CORRELATION COEFFICIENT.
    SANKHYA, 7, 415-422.   (MR8,P.283)

PILLAI, K.C.SREEDHARAN 1955.   (8.4/9.2/10.1)                                                             4291
    SOME NEW TEST CRITERIA IN MULTIVARIATE ANALYSIS.
    ANN. MATH. STATIST., 26, 117-121.   (MR16,P.728)

PILLAI, K.C.SREEDHARAN 1956.   (13.1/20.1)                                                                4292
    SOME RESULTS USEFUL IN MULTIVARIATE ANALYSIS.
    ANN. MATH. STATIST., 27, 1106-1114.   (MR18,P.456)

PILLAI, K.C.SREEDHARAN 1956.   (13.1)                                                                     4293
    ON THE DISTRIBUTION OF THE LARGEST OR THE SMALLEST ROOT OF A MATRIX
    IN MULTIVARIATE ANALYSIS.
    BIOMETRIKA, 43, 122-127.   (MR17,P.983)

PILLAI, K.C.SREEDHARAN 1964.   (13.1)                                                                     4294
    ON THE MOMENTS OF ELEMENTARY SYMMETRIC FUNCTIONS OF THE ROOTS OF TWO MATRICES.
    ANN. MATH. STATIST., 35, 1704-1712.   (MR29-5327)

PILLAI, K.C.SREEDHARAN 1964.   (13.1)                                                                     4295
    ON THE DISTRIBUTION OF THE LARGEST OF SEVEN ROOTS OF A MATRIX IN
    MULTIVARIATE ANALYSIS.
    BIOMETRIKA, 51, 270-275.   (MR30-4347)

PILLAI, K.C.SREEDHARAN 1965.   (13.1/T)                                                                   4296
    ON THE DISTRIBUTION OF THE LARGEST CHARACTERISTIC ROOT OF A MATRIX IN
    MULTIVARIATE ANALYSIS.
    BIOMETRIKA, 52, 405-414.

PILLAI, K.C.SREEDHARAN 1965.   (13.1)                                                                     4297
    ON ELEMENTARY SYMMETRIC FUNCTIONS OF THE ROOTS OF TWO MATRICES IN
    MULTIVARIATE ANALYSIS.
    BIOMETRIKA, 52, 499-506.   (MR35-1140)

PILLAI, K.C.SREEDHARAN 1966.   (8.4/13.2/18.1/T)                                                          4298
    NONCENTRAL MULTIVARIATE BETA DISTRIBUTION AND THE MOMENTS OF TRACES OF
      SOME MATRICES.
    MULTIVARIATE ANAL. PROC. INTERNAT. SYMP. DAYTON (KRISHNAIAH), 237-251.   (MR35-1141)

PILLAI, K.C.SREEDHARAN + BANTEGUI, CELIA G. 1959.   (13.1)                                                4299
    ON THE DISTRIBUTION OF THE LARGEST OF SIX ROOTS OF A MATRIX IN
    MULTIVARIATE ANALYSIS.
    BIOMETRIKA, 46, 237-240.   (MR21-946)

PILLAI, K.C.SREEDHARAN + MIJARES, TITO A. 1959.   (8.4/13.1/A)                                            4300
    ON THE MOMENTS OF THE TRACE OF A MATRIX AND APPROXIMATIONS TO ITS DISTRIBUTION.
    ANN. MATH. STATIST., 30, 1135-1140.   (MR21-7579)

PILLAI, K.C.SREEDHARAN + RAMACHANDRAN, K.V. 1954.   (14.2/U)                                              4301
    ON THE DISTRIBUTION OF THE RATIO OF THE $i$TH OBSERVATION IN AN ORDERED SAMPLE
    FROM A NORMAL POPULATION TO AN INDEPENDENT ESTIMATE OF THE STANDARD DEVIATION.
    ANN. MATH. STATIST., 25, 565-572.   (MR16,P.270)

PILLAI, K.C.SREEDHARAN + SAMSON, PABLO, JR. 1959.   (8.3/8.4)                                             4302
    ON HOTELLING'S GENERALIZATION OF $T^2$.
    BIOMETRIKA, 46, 160-168.   (MR21-940)

PILLAI, K.C.SREEDHARAN    SEE ALSO
    2215. GUPTA, SHANTI S. + PILLAI, K.C.SREEDHARAN (1965)

    3033. KHATRI, C. G. + PILLAI, K.C.SREEDHARAN (1965)

    3034. KHATRI, C. G. + PILLAI, K.C.SREEDHARAN (1966)

    2218. GUPTA, SHANTI S.; PILLAI, K.C.SREEDHARAN + STECK, GEORGE P. (1964)

PINEAU, H. 1955.  (11.1/15.1/E)                                    4303
    REMARQUES SUR L'ANALYSE FACTORIELLE DE HOTELLING.  COMPARAISON AVEC
    LES METHODES CENTROIDES.
        (WITH DISCUSSION)
    COLLOQ. INTERNAT. CENTRE NAT. RECH. SCI., 65, 127-141.


PINKHAM, ROGER S.   SEE
    2314. HAMMING, R. W. + PINKHAM, ROGER S. (1966)


PINSKER, M. S. 1954.  (16.4/U)                                     4304
    THE QUANTITY OF INFORMATION ABOUT A GAUSSIAN RANDOM STATIONARY PROCESS,
    CONTAINED IN A SECOND PROCESS CONNECTED WITH IT IN A STATIONARY MANNER.
        (IN RUSSIAN)
    DOKL. AKAD. NAUK SSSR NOV. SER., 99, 213-216.  (MR16,P.495)


PINSKER, M. S. 1958.  (16.4)                                       4305
    EXTRAPOLATION OF RANDOM VECTOR PROCESSES AND THE QUANTITY OF
    INFORMATION CONTAINED IN A STATIONARY RANDOM VECTOR PROCESS RELATIVE
    TO ANOTHER ONE STATIONARILY CONNECTED WITH IT.
        (IN RUSSIAN)
    DOKL. AKAD. NAUK SSSR, 121, 49-51.  (MR22-9336)


PINTNER, RUDOLF + FORLANO, GEORGE 1937.  E(15.4)                   4306
    A COMPARISON OF METHODS OF ITEM SELECTION FOR A PERSONALITY TEST.
    J. APPL. PSYCHOL., 21, 643-652.


PITCHER, T. S.   SEE
    3048. KINNEY, J. R. + PITCHER, T. S. (1966)


PITMAN, A. E. N. T.   SEE
    4490. RAY, W.D. + PITMAN, A. E. N. T. (1961)


PITMAN, E.J.G. 1937.  (17.8/B)                                     4307
    SIGNIFICANCE TESTS WHICH MAY BE APPLIED TO SAMPLES FROM ANY
    POPULATIONS.  I.
    J. ROY. STATIST. SOC. SUPP., 4, 119-130.


PITMAN, E.J.G. 1937.  (17.8)                                       4308
    SIGNIFICANCE TESTS WHICH MAY BE APPLIED TO SAMPLES FROM ANY
    POPULATIONS.  II. THE CORRELATION COEFFICIENT TEST.
    J. ROY. STATIST. SOC. SUPP., 4, 225-232.


PITMAN, E.J.G. 1939.  (10.2)                                       4309
    A NOTE ON NORMAL CORRELATION.
    BIOMETRIKA, 31, 9-12.  (MR1,P.63)


PITMAN, E.J.G. 1939.  (10.1/19.2/U)                                4310
    TESTS OF HYPOTHESES CONCERNING LOCATION AND SCALE PARAMETERS.
    BIOMETRIKA, 31, 200-215.  (MR1,P.63)


PITMAN, E.J.G.   SEE ALSO
    4577. ROBBINS, HERBERT E. + PITMAN, E.J.G. (1949)


PLACKETT, R.L. 1946.  (4.3/10.1/10.2)                              4311
    LITERATURE ON TESTING THE EQUALITY OF VARIANCES AND COVARIANCES IN
    NORMAL POPULATIONS.
    J. ROY. STATIST. SOC., 109, 457-468.


PLACKETT, R.L. 1947.  (10.2)                                       4312
    AN EXACT TEST FOR THE EQUALITY OF VARIANCES.
    BIOMETRIKA, 34, 311-319.  (MR9,P.453)


PLACKETT, R.L. 1949.  (8.8)                                        4313
    A HISTORICAL NOTE ON THE METHOD OF LEAST SQUARES.
    BIOMETRIKA, 36, 458-460.


PLACKETT, R.L. 1950.  (8.1)                                        4314
    SOME THEOREMS IN LEAST SQUARES.
    BIOMETRIKA, 37, 149-157.  (MR12,P.194)


PLACKETT, R.L. 1954.  (2.7/20.4)                                   4315
    A REDUCTION FORMULA FOR NORMAL MULTIVARIATE INTEGRALS.
    BIOMETRIKA, 41, 351-360.  (MR16,P.377)

PLACKETT, R.L. 1965.   (17.6/18.1/18.2/B)                           4316
    A CLASS OF BIVARIATE DISTRIBUTIONS.
    J. AMER. STATIST. ASSOC., 60, 516-522.   (MR32-524)


PLANA, G.A.A. 1813.   (2.1/B)                                       4317
    MEMOIRE SUR DIVERS PROBLEMES DE PROBABILITE.
    MEM. REALE ACCAD. SCI. TORINO, 20, 355-408.


PLATT, JOHN R. 1943.   (4.1)                                        4318
    A MECHANICAL DETERMINATION OF CORRELATION COEFFICIENTS AND STANDARD DEVIATIONS.
    J. AMER. STATIST. ASSOC., 38, 311-318.   (MR5,P.42)


PLEASE, N.W.   SEE
    360. BARTLETT, MAURICE S. + PLEASE, N.W. (1963)


PLESS, VERA 1965.   (20.5)                                          4319
    ON THE INVARIANTS OF A VECTOR SUBSPACE OF A VECTOR SPACE OVER A FIELD
    OF CHARACTERISTIC TWO.
    PROC. AMER. MATH. SOC., 16, 1062-1067.


PLESZCZYNSKA, ELZBIETA 1960.   (14.4/B)                             4320
    SCREENING IN STATISTICAL TESTING FOR CORRELATED CHARACTERISTICS.
        (IN POLISH, WITH SUMMARY IN RUSSIAN)
    POLSKA AKAD. NAUK ZASTOS. MAT., 5, 47-59.   (MR23-A2266)


PLESZCZYNSKA, ELZBIETA 1966.   (17.5/A)                             4321
    SAMPLE SIZE WARRANTING THE DIVERGENCE OF HYPOTHESES.
        (IN POLISH)
    POLSKA AKAD. NAUK ZASTOS. MAT., 8, 309-324.


PLISS, V. A.   SEE
    3442. LINNIK, YU. V. + PLISS, V. A. (1966)


POCH, FRANCISCO AZORIN  SEE  AZORIN POCH, FRANCISCO


PODTJAGIN, M. E. 1916.   (4.9)                                      4322
    ON METHODS OF STUDYING INTERRELATIONS BETWEEN PHENOMENA.
        (IN RUSSIAN)
    STATIST. VESTNIK, 2, 15-32.


POINCARE, H. 1905.   (20.5)                                         4323
    SUR LA GENERALISATION D'UN THEOREME ELEMENTAIRE DE GEOMETRIE.
    C.R. ACAD. SCI. PARIS, 140, 113-117.


POLK, KENNETH 1962.   (17.7)                                        4324
    A NOTE ON ASYMMETRIC CAUSAL MODELS.
        (AMENDED BY NO. 4586)
    AMER. SOCIOL. REV., 27, 539-542.


POLLACZEK-GEIRINGER, HILDA   SEE  GEIRINGER, HILDA


POLLAK, H.O. 1956.   (2.1/2.7)                                      4325
    A REMARK ON "ELEMENTARY INEQUALITIES FOR MILLS' RATIO" BY YUSAKU KOMATU.
    REP. STATIST. APPL. RES. UN. JAPAN. SCI. ENGRS., 4, 110.   (MR18,P.722)


POLLAK, H.O.   SEE ALSO
    2129. GRENANDER, ULF; POLLAK, H.O. + SLEPIAN, DAVID (1959)


POLYA, GEORGE 1949.   (2.7/17.2)                                    4326
    REMARKS ON COMPUTING THE PROBABILITY INTEGRAL IN ONE AND TWO DIMENSIONS.
    PROC. BERKELEY SYMP. MATH. STATIST. PROB., 63-78.   (MR10,P.384)


POLYA, GEORGE 1962.   (1.2)                                         4327
    PUBLICATIONS OF GEORGE POLYA.
    STUD. MATH. ANAL. RELATED TOPICS (POLYA VOLUME), XIII-XXI.


POMPILJ, GIUSEPPE 1946.   (8.1)                                     4328
    SULLA REGRESSIONE.
    REND. MAT. E APPL. REG. UNIV. ROMA REALE IST. NAZ. ALTA MAT., 5, 186-219.   (MR9,P.150)


POMPILJ, GIUSEPPE 1956.   (1.1)                                     4329
    TEORIA AFFINE DELLE VARIABILI CASUALI.
    FAC. SCI. STATIST. DEMOG. ATTUAR. IST. STATIST. CALC. PROB., 15, 1-23.   (MR20-3586)

POMPILJ, GIUSEPPE 1957. (2.1/17.3)                                 4330
    ON INTRINSIC INDEPENDENCE.
    BULL. INST. INTERNAT. STATIST., 35(2) 91-97. (MR23-A672)


POMPILJ, GIUSEPPE 1957. (8.1/17.3)                                 4331
    SULLA REGRESSIONE INTERAMENTE PSEUDO-LINEARE.
    FAC. SCI. STATIST. DEMOG. ATTUAR. IST. STATIST. CALC. PROB., 30, 1-18. (MR21-955)

POMPILJ, GIUSEPPE 1958. (17.7)                                     4332
    ON ENTIRELY PSUEDO-LINEAR REGRESSION.
    BULL. INST. INTERNAT. STATIST., 36(3) 60-63. (MR22-11486)

POMPILJ, GIUSEPPE 1960. (8.1/17.3)                                 4333
    LA REGRESSIONE.
    STATISTICA, BOLOGNA, 20, 585-603. (MR23-A4222)

POMPILJ, GIUSEPPE 1966. (17.3/17.9)                                4334
    ON THE AVERAGES OF MULTIPLE RANDOM VARIABLES.
        (WITH DISCUSSION)
    BULL. INST. INTERNAT. STATIST., 41, 741-749.

POPE, J. A.   SEE
    3675. MARRIOTT, F.H.C. + POPE, J. A. (1954)

POREBSKI, OLGIERD R. 1966. (4.7/5.1/8.3)                           4335
    ON THE INTERRELATED NATURE OF THE MULTIVARIATE STATISTICS USED IN
    DISCRIMINATORY ANALYSIS.
    BRITISH J. MATH. STATIST. PSYCHOL., 19, 197-214.

POREBSKI, OLGIERD R. 1966. (5.2/8.3/8.5/E)                         4336
    DISCRIMINATORY AND CANONICAL ANALYSIS OF TECHNICAL COLLEGE DATA.
    BRITISH J. MATH. STATIST. PSYCHOL., 19, 215-236.

PORTER, CHARLES E. 1963. (13.1/20.3)                               4337
    RANDOM MATRIX DIAGONALIZATION - SOME NUMERICAL COMPUTATIONS.
        (REPRINTED AS NO. 4339)
    J. MATH. PHYS., 4, 1039-1042. (MR27-2094)

PORTER, CHARLES E. 1965. (13.1/20.1)                               4338
    FLUCTUATIONS OF QUANTAL SPECTRA.
    STATIST. THEOR. SPECTRA FLUCTUATIONS (PORTER), 3-87.

PORTER, CHARLES E. 1965. (13.1/20.3)                               4339
    RANDOM MATRIX DIAGONALIZATION - SOME NUMERICAL COMPUTATIONS.
        (REPRINT OF NO. 4337)
    STATIST. THEOR. SPECTRA FLUCTUATIONS (PORTER), 511-514. (MR27-2094)

PORTIG, W. 1936. (4.1)                                             4340
    MESSGENAUIGKEIT UND KORRELATIONS-KOEFFIZIENT.
    ANN. HYDROGRAPHIE, 64, 32-37.

POSCHL, KLAUS   SEE
    3432. LINN, HANS-JOCHEN + POSCHL, KLAUS (1956)

POSNER, EDWARD C.   SEE
    4755. RUMSEY, HOWARD, JR. + POSNER, EDWARD C. (1965)

POSTEN, H. O. + BARGMANN, ROLF E. 1964. (8.4)                      4341
    POWER OF THE LIKELIHOOD-RATIO TEST OF THE GENERAL LINEAR HYPOTHESIS
    IN MULTIVARIATE ANALYSIS.
    BIOMETRIKA, 51, 467-480. (MR30-3546)


POTTHOFF, RICHARD F. 1965. (8.3)                                   4342
    SOME SCHEFFE-TYPE TESTS FOR SOME BEHRENS-FISHER-TYPE REGRESSION PROBLEMS.
    J. AMER. STATIST. ASSOC., 60, 1163-1190.

POTTHOFF, RICHARD F. 1966. (15.4)                                 4343
    EQUATING OF GRADES OR SCORES ON THE BASIS OF A COMMON BATTERY OF
    MEASUREMENTS.
    MULTIVARIATE ANAL. PROC. INTERNAT. SYMP. DAYTON (KRISHNAIAH), 541-559.

POTTHOFF, RICHARD F. + ROY, S.N. 1964.  (8.3)                              **4344**
    A GENERALIZED MULTIVARIATE ANALYSIS OF VARIANCE MODEL USEFUL
    ESPECIALLY FOR GROWTH CURVE PROBLEMS.
    BIOMETRIKA, 51, 313-326.  (MR31-5291)


POTTHOFF, RICHARD F. + WHITTINGHILL, MAURICE 1966.  (18.2)                 **4345**
    TESTING FOR HOMOGENEITY.  I.  THE BINOMIAL AND MULTINOMIAL DISTRIBUTIONS.
    BIOMETRIKA, 53, 167-182.


POTTHOFF, RICHARD F.   SEE ALSO
    4698. ROY, S.N. + POTTHOFF, RICHARD F. (1958)

PRAK, J. L. 1931.  F(4.2)                                                  **4346**
    AN EMPIRICAL RESEARCH ON THE RELIABILITY OF CORRELATION COEFFICIENTS.
    BRITISH J. PSYCHOL., 21, 394-403.

PRAKASH SRIVASTAVA, OM  SEE  SRIVASTAVA, OM PRAKASH

PRATT, JOHN W.  SEE
    4072. OLKIN, INGRAM + PRATT, JOHN W. (1958)

    4073. OLKIN, INGRAM + PRATT, JOHN W. (1958)


PREDETTI, ALDO 1960.  (6.1/6.2)                                            **4347**
    IN TEMA DI ANALISI DISCRIMINATORIA.
    GIORN. ECON. ANN. ECONOMIA, 19, 223-258.


PRESS, SHELDON JAMES 1966.  (2.5)                                          **4348**
    LINEAR COMBINATIONS OF NON-CENTRAL CHI-SQUARE VARIATES.
    ANN. MATH. STATIST., 37, 480-487.


PRETORIUS, S.J. 1930.  (18.1/B)                                            **4349**
    SKEW BIVARIATE FREQUENCY SURFACES, EXAMINED IN THE LIGHT OF NUMERICAL
    ILLUSTRATIONS.
    BIOMETRIKA, 22, 109-223.


PRETORIUS, S.J.  SEE ALSO
    1529. ELDERTON, ETHEL M.; MOUL, MARGARET; FIELLER, E. C.;  (1930)
    PRETORIUS, S.J. + CHURCH, A. E. R.


PRICE, BRONSON 1935.  (4.9)                                                **4350**
    PROPOSED METHOD FOR THE DIRECT MEASUREMENT OF CORRELATION.
    SCIENCE, 82, 497-498.


PRICE, BRONSON 1936.  E(4.8)                                               **4351**
    HOMOGAMY AND THE INTERCORRELATION OF CAPACITY TRAITS.
    ANN. EUGENICS, 7, 22-27.


PRICE, BRONSON 1937.  (15.1)                                               **4352**
    A NOTE ON PROFESSOR TRYON'S THEORY OF INTERCORRELATION.
    PSYCHOL. REV., 44, 183-187.


PRICE, CHARLES M. 1964.  (8.8/20.1)                                        **4353**
    THE MATRIX PSEUDO-INVERSE AND MINIMAL VARIANCE ESTIMATES.
    SIAM REV., 6, 115-120.  (MR29-6619)


PRICE, G. BALEY 1946.  (18.1)                                             **4354**
    DISTRIBUTIONS DERIVED FROM THE MULTINOMIAL EXPANSION.
    AMER. MATH. MONTHLY, 53, 59-74.  (MR7,P.309)


PRICE, G. BALEY 1947.  (20.1)                                             **4355**
    SOME IDENTITIES IN THE THEORY OF DETERMINANTS.
    AMER. MATH. MONTHLY, 54, 75-90.  (MR8,P.366)


PRICE, ROBERT 1964.  (14.5)                                               **4356**
    SOME NON-CENTRAL F-DISTRIBUTIONS EXPRESSED IN CLOSED FORM.
    BIOMETRIKA, 51, 107-122.


PRICE, V. E.  SEE
    1386. DOUST, A. + PRICE, V. E. (1964)

PRIESTLEY, M. B. 1962.   (16.3)                                            4357
    BASIC CONSIDERATIONS IN THE ESTIMATION OF SPECTRA.
    TECHNOMETRICS, 4, 551-564.

PRIESTLEY, M. B. 1964.   (16.3)                                            4358
    ESTIMATION OF THE SPECTRAL DENSITY FUNCTION IN THE PRESENCE OF HARMONIC
    COMPONENTS.
    J. ROY. STATIST. SOC. SER. B, 26, 123-132.

PRIESTLEY, M. B. 1965.   (16.3)                                            4359
    THE ROLE OF BANDWIDTH IN SPECTRAL ANALYSIS.
    APPL. STATIST., 14, 33-47.

PROKHOROV, YU. V. 1965.   (1.2)                                            4360
    ON THE WORKS OF ACADEMICIAN JU. V. LINNIK IN THE THEORY OF PROBABILITY
    AND MATHEMATICAL STATISTICS  (ON HIS FIFTIETH BIRTHDAY).
        (IN RUSSIAN, TRANSLATED AS NO. 4361)
    TEOR. VEROJATNOST. I PRIMENEN., 10, 117-129.   (MR30-3486)

PROKHOROV, YU. V. 1965.   (1.2)                                            4361
    YU. V. LINNIK'S WORK IN PROBABILITY THEORY AND MATHEMATICAL STATISTICS
    (ON THE OCCASION OF HIS FIFTIETH BIRTHDAY).
        (TRANSLATION OF NO. 4360)
    THEORY PROB. APPL., 10, 107-119.   (MR30-3486)

PROKOPOVIC, S. N. 1935.   (4.1/4.5)                                        4362
    KORELACE KVANTITIVNICH RAD.
        (WITH SUMMARY IN FRENCH)
    STATIST. OBZOR, 16, 64-92.

PROSCHAN, F. 1961.   (8.8)                                                 4363
    PRECISION OF LEAST SQUARES POLYNOMIAL ESTIMATES.
    SIAM REV., 3, 230-235.

PRYTZ, KNUD STEENSTRUP 1948.   (4.1/4.5/4.6/AE)                            4364
    LANDBRUGSMETEOROLOGISKE KORRELATIONSUNDERSOGELSER.
        (WITH SUMMARY IN ENGLISH)
    INGEN. SKR., 4, 1-127.

PUGACEV, V.S. 1953.   (16.4)                                               4365
    THE GENERAL THEORY OF CORRELATION OF RANDOM FUNCTIONS.
        (IN RUSSIAN)
    IZV. AKAD. NAUK SSSR SER. MAT., 17, 401-420.   (MR15,P.238)

PURI, MADAN LAL + SEN, PRANAB KUMAR 1966.   (17.8/A)                       4366
    ON A CLASS OF MULTIVARIATE MULTISAMPLE RANK-ORDER TESTS.
    SANKHYA SER. A, 28, 353-376.

PYKE, RONALD    SEE
    2481. HOBBY, CHARLES R. + PYKE, RONALD (1963)

QUADE, DANA 1966.   (17.8/AU)                                             4367
    ON ANALYSIS OF VARIANCE FOR THE K-SAMPLE PROBLEM.
    ANN. MATH. STATIST., 37, 1747-1758.

QUARLES, D.A., JR.    SEE
    419. BECKMAN, F. S. + QUARLES, D.A., JR. (1956)

QUARTEY, JAMES 1955.   (8.1/20.3/T)                                       4368
    TABLE OF INVERTED MATRICES FOR THE SOLUTION OF QUADRATIC REGRESSION
    COEFFICIENTS.
    STATISTICA, BOLOGNA, 15, 491.   (MR17,P.665)

QUEIROZ, A. LOPES   SEE   LOPES QUEIROZ, A.

QUENOUILLE, MAURICE H. 1947.   (16.5)                                     4369
    NOTES ON THE CALCULATION OF AUTO-CORRELATIONS OF LINEAR AUTOREGRESSIVE SCHEMES.
    BIOMETRIKA, 34, 365-367.   (MR9,P.361)

QUENOUILLE, MAURICE H. 1949.   (6.3/8.3)                                  4370
    NOTE ON THE ELIMINATION OF INSIGNIFICANT VARIATES IN DISCRIMINATORY ANALYSIS.
    ANN. EUGENICS, 14, 305-308.   (MR11,P.259)

QUENOUILLE, MAURICE H. 1949.  (2.6/16.5)                                          4371
    THE JOINT DISTRIBUTION OF SERIAL CORRELATION COEFFICIENTS.
    ANN. MATH. STATIST., 20, 561-571.   (MR12,P.118)

QUENOUILLE, MAURICE H. 1949.  (16.5)                                              4372
    APPROXIMATE TESTS OF CORRELATION IN TIME-SERIES.
    J. ROY. STATIST. SOC. SER. B, 11, 68-84.   (MR11,P.262)

QUENOUILLE, MAURICE H. 1949.  (14.5)                                              4373
    THE EVALUATION OF PROBABILITIES IN A NORMAL MULTIVARIATE
    DISTRIBUTION, WITH SPECIAL REFERENCE TO THE CORRELATION RATIO.
    PROC. EDINBURGH MATH. SOC. SER. 2, 8, 95-100.   (MR11,P.673)

QUENOUILLE, MAURICE H. 1950.  (6.2/6.3)                                           4374
    A FURTHER NOTE ON DISCRIMINATORY ANALYSIS.
    ANN. EUGENICS, 15, 11-14.   (MR11,P.743)

QUENOUILLE, MAURICE H. 1950.  (8.5/F)                                             4375
    MULTIVARIATE EXPERIMENTATION.
    BIOMETRICS, 6, 303-316.

QUENOUILLE, MAURICE H. 1950.  (8.8/20.3)                                          4376
    COMPUTATIONAL DEVICES IN THE APPLICATION OF LEAST SQUARES.
    J. ROY. STATIST. SOC. SER. B, 12, 256-272.   (MR13,P.54)

QUENOUILLE, MAURICE H. 1957.  (4.2)                                               4377
    THE EFFECT OF TRANSFORMATIONS OF VARIABLES UPON THEIR CORRELATION COEFFICIENTS.
    BIOMETRIKA, 44, 272-273.

QUENOUILLE, MAURICE H. 1958.  (16.5)                                              4378
    THE COMPARISON OF CORRELATIONS IN TIME-SERIES.
    J. ROY. STATIST. SOC. SER. B, 20, 158-164.   (MR20-2830)

QUENSEL, CARL-ERIK 1936.  (17.4/18.3)                                             4379
    A METHOD OF DETERMINING THE REGRESSION CURVE WHEN THE MARGINAL
    DISTRIBUTION IS OF THE NORMAL LOGARITHMIC TYPE.
    ANN. MATH. STATIST., 7, 196-201.

QUENSEL, CARL-ERIK 1938.  (18.3)                                                  4380
    THE DISTRIBUTIONS OF THE SECOND MOMENT AND OF THE CORRELATION
    COEFFICIENT IN SAMPLES FROM POPULATIONS OF TYPE A.
    LUNDS UNIV. ARSSKR. N. F. AVD. 2, 34(4) 1-111.

QUENSEL, CARL-ERIK 1940.  (2.4)                                                   4381
    TRUNCATED NORMAL CURVES AND CORRELATION DISTRIBUTIONS.
    LUNDS UNIV. ARSSKR. N. F. AVD. 2, 36(15) 1-17.   (MR2,P.231)

QUENSEL, CARL-ERIK 1952.  (18.3)                                                  4382
    THE DISTRIBUTION OF THE SECOND ORDER MOMENTS IN RANDOM SAMPLES FROM
    NON-NORMAL MULTIVARIATE UNIVERSES.
    LUNDS UNIV. ARSSKR. N. F. AVD. 2, 48(4) 1-11.   (MR14,P.486)

QUENSEL, CARL-ERIK 1953.  (4.5/18.3)                                              4383
    THE DISTRIBUTION OF THE PARTIAL CORRELATION COEFFICIENT IN SAMPLES FROM
    MULTIVARIATE UNIVERSES IN A SPECIAL CASE OF NON-NORMALLY DISTRIBUTED RANDOM
    VARIABLES.
    SKAND. AKTUARIETIDSKR., 36, 16-23.   (MR15,P.331)

QUENSEL, CARL-ERIK 1958.  (8.1/18.3)                                              4384
    A CONTRIBUTION TO THE THEORY OF CORRELATION AND REGRESSION IN
    NON-RANDOM SAMPLES.
    LUNDS UNIV. ARSSKR. N. F. AVD. 2, 54(7) 1-31.   (MR20-404)

QUENSEL, CARL-ERIK 1966.  (2.1/4.8/17.8/BE)                                       4385
    ON TESTS OF INDEPENDENCE AND EVALUATION OF THE MAGNITUDE OF THE DEPENDENCE
    BETWEEN TWO CATEGORIZED DATA.
    SKAND. AKTUARIETIDSKR., 49, 199-217.

QUESENBERRY, C. P.   SEE
    3462. LOFTSGAARDEN, D. O. + QUESENBERRY, C. P. (1965)

RAATH, M. J.   SEE
    4519. REYBURN, H. A. + RAATH, M. J. (1949)

RADCLIFFE, J. 1964.  (9.2/15.2)                                                4386
    THE CONSTRUCTION OF A MATRIX USED IN DERIVING TESTS OF SIGNIFICANCE
    IN MULTIVARIATE ANALYSIS.
    BIOMETRIKA, 51, 503-504.  (MR30-4348)

RADCLIFFE, J. 1966.  (8.4/8.7/AE)                                              4387
    FACTORIZATIONS OF THE RESIDUAL LIKELIHOOD CRITERION IN DISCRIMINANT ANALYSIS.
    PROC. CAMBRIDGE PHILOS. SOC., 62, 743-752.  (MR36-1032)

RADHAKRISHNA, S. 1964.  (6.6/E)                                                4388
    DISCRIMINATION ANALYSIS IN MEDICINE.
    STATISTICIAN, 14, 147-167.

RADNER, ROY 1958.  (8.1)                                                       4389
    MINIMAX ESTIMATION FOR LINEAR REGRESSIONS.
    ANN. MATH. STATIST., 29, 1244-1250.  (MR20-6750)

RADOK, JENS RAINER MARIA 1955.  (20.5)                                         4390
    THE SOLUTION OF EIGENVALUE PROBLEMS OF INTEGRAL EQUATIONS BY POWER SERIES.
    QUART. APPL. MATH., 12, 413-417.  (MR16,P.630)

RAFFERTY, J.A.   SEE
    5674. VOTAW, DAVID F., JR.; RAFFERTY, J.A. + DEEMER, WALTER L. (1950)

    5673. VOTAW, DAVID F., JR.; KIMBALL, ALLYN W. + RAFFERTY, J.A. (1950)

RAGHAVAN NAIR, K.  SEE  NAIR, K. RAGHAVAN

RAHMAN, N. A. 1963.  (5.2/6.5/14.2/F)                                          4391
    ON THE SAMPLING DISTRIBUTION OF THE STUDENTIZED PENROSE MEASURE OF DISTANCE.
    ANN. HUMAN GENET., 26, 97-106.

RAIFFA, HOWARD 1961.  (15.4/E)                                                 4392
    STATISTICAL DECISION THEORY APPROACH TO ITEM SELECTION FOR
    DICHOTOMOUS TEST AND CRITERION VARIABLES.
    STUD. ITEM ANAL. PREDICT. (SOLOMON), 187-220.  (MR22-12661)

RAJ, DES  SEE  DES RAJ

RAJALAKSHMAN, D.V.   SEE
    361. BARTLETT, MAURICE S. + RAJALAKSHMAN, D.V. (1953)

RAJARATNAM, NAGESWARI   SEE
    2022. GLESER, GOLDINE C.; CRONBACH, LEE J. + RAJARATNAM, NAGESWARI (1965)

RAM, SIYA 1954.  (18.1)                                                        4393
    A NOTE ON THE CALCULATION OF MOMENTS OF THE TWO-DIMENSIONAL
    HYPERGEOMETRIC DISTRIBUTION.
    GANITA, 5, 97-101.

RAM, SIYA 1955.  (18.1)                                                        4394
    MULTIDIMENSIONAL HYPERGEOMETRIC DISTRIBUTION.
    SANKHYA, 15, 391-398.  (MR17,P.753)

RAMABHADRAN, V.K. 1951.  (18.1)                                                4395
    A MULTIVARIATE GAMMA-TYPE DISTRIBUTION.
    SANKHYA, 11, 45-46.  (MR13,P.142)

RAMACHANDRAN, K.V. 1956.  (8.3)                                                4396
    ON THE SIMULTANEOUS ANALYSIS OF VARIANCE TEST.
    ANN. MATH. STATIST., 27, 521-528.  (MR18,P.77)

RAMACHANDRAN, K.V. 1956.  (3.1/8.5/C)                                          4397
    CONTRIBUTIONS TO SIMULTANEOUS CONFIDENCE INTERVAL ESTIMATION.
    BIOMETRICS, 12, 51-56.  (MR17,P.1102)

RAMACHANDRAN, K.V. + KHATRI, C. G. 1957.  (8.3/19.1)                           4398
    ON A DECISION PROCEDURE BASED ON THE TUKEY STATISTIC.
    ANN. MATH. STATIST., 28, 802-806.  (MR19,P.991)

RAMACHANDRAN, K.V.    SEE ALSO
    4301. PILLAI, K.C.SREEDHARAN + RAMACHANDRAN, K.V. (1954)

RAMAKRISHNAN, ALLADI + SRINIVASAN, S. K. 1956.    (16.4)                    4399
    CORRELATION PROBLEMS IN THE STUDY OF THE BRIGHTNESS OF THE MILKY WAY.
    ASTROPHYS. J., 123, 479–485.    (MR19,P.616)


RAND, GERTRUDE 1925.    (15.4)                                              4400
    DISCUSSION OF THE QUOTIENT METHOD FOR SPECIFYING TEST RESULTS.
    J. EDUC. PSYCHOL., 16, 599–618.

RANGA RAO, R. 1961.    (17.1)                                               4401
    ON THE CENTRAL LIMIT THEOREM IN $R_k$.
    BULL. AMER. MATH. SOC., 67, 359–361.    (MR24–A2984)

RANGA RAO, R. + VARADARAJAN, V.S. 1960.    (17.1)                           4402
    A LIMIT THEOREM FOR DENSITIES.
    SANKHYA, 22, 261–266.    (MR24–A1734)

RANKE, K.E. 1906.    (4.1)                                                  4403
    DIE THEORIE DER KORRELATION.  NACH DEN GRUNDLEGENDEN ARBEITEN VON
    FRANCIS GALTON, KARL PEARSON UND UDNY YULE.
    ARCH. ANTHROP., 32, 168–202.

RAO, C. RADHAKRISHNA 1942.    (18.1/20.5)                                   4404
    ON THE VOLUME OF A PRISMOID IN N-SPACE AND SOME PROBLEMS IN
    CONTINUOUS PROBABILITY.
    MATH. STUDENT, 10, 68–74.    (MR4,P.248)

RAO, C. RADHAKRISHNA 1942.    (2.1/18.1/B)                                  4405
    ON BIVARIATE CORRELATION SURFACES.
    SCI. AND CULT., 8, 236–237.    (MR5,P.126)

RAO, C. RADHAKRISHNA 1945.    (8.1/8.3/U)                                   4406
    GENERALIZATION OF MARKOFF'S THEOREM AND TESTS OF LINEAR HYPOTHESES.
    SANKHYA, 7, 9–16.    (MR7,P.132)

RAO, C. RADHAKRISHNA 1945.    (8.1/U)                                       4407
    MARKOFF'S THEOREM WITH LINEAR RESTRICTIONS ON PARAMETERS.
    SANKHYA, 7, 16–19.    (MR7,P.132)

RAO, C. RADHAKRISHNA 1945.    (8.3/U)                                       4408
    STUDENTISED TESTS OF LINEAR HYPOTHESES.
    SCI. AND CULT., 11, 202–203.    (MR7,P.213)

RAO, C. RADHAKRISHNA 1946.    (8.1)                                         4409
    ON THE LINEAR COMBINATION OF OBSERVATIONS AND THE GENERAL THEORY OF
    LEAST SQUARES.
    SANKHYA, 7, 237–256.    (MR8,P.41)

RAO, C. RADHAKRISHNA 1946.    (5.3/8.4)                                     4410
    TESTS WITH DISCRIMINANT FUNCTIONS IN MULTIVARIATE ANALYSIS.
    SANKHYA, 7, 407–414.    (MR8,P.162)

RAO, C. RADHAKRISHNA 1947.    (8.1/17.7)                                    4411
    NOTE ON A PROBLEM OF RAGNAR FRISCH.
        (AMENDED BY NO. 4412)
    ECONOMETRICA, 15, 245–249.    (MR8,P.592)

RAO, C. RADHAKRISHNA 1947.    (8.1/17.7)                                    4412
    A CORRECTION TO NOTE ON A PROBLEM OF RAGNAR FRISCH.
        (AMENDMENT OF NO. 4411)
    ECONOMETRICA, 17, 212.    (MR11,P.259)

RAO, C. RADHAKRISHNA 1947.    (6.1/6.2)                                     4413
    A STATISTICAL CRITERION TO DETERMINE THE GROUP TO WHICH AN INDIVIDUAL BELONGS.
    NATURE, 160, 835–836.

RAO, C. RADHAKRISHNA 1947.    (19.3)                                        4414
    MINIMUM VARIANCE AND THE ESTIMATION OF SEVERAL PARAMETERS.
    PROC. CAMBRIDGE PHILOS. SOC., 43, 280–283.    (MR8,P.478)

RAO, C. RADHAKRISHNA 1948.   (5.2/8.3/8.4/E)                                     4415
    TESTS OF SIGNIFICANCE IN MULTIVARIATE ANALYSIS.
    BIOMETRIKA, 35, 58–79.   (MR9,P.602)


RAO, C. RADHAKRISHNA 1948.   (6.1/6.6/E)                                         4416
    THE UTILIZATION OF MULTIPLE MEASUREMENTS IN PROBLEMS OF BIOLOGICAL
    CLASSIFICATION.
        (WITH DISCUSSION)
    J. ROY. STATIST. SOC. SER. B, 10, 159–203.   (MR11,P.191)


RAO, C. RADHAKRISHNA 1948.   (19.2/A)                                            4417
    LARGE SAMPLE TESTS OF STATISTICAL HYPOTHESES CONCERNING SEVERAL PARAMETERS WITH
    APPLICATIONS TO PROBLEMS OF ESTIMATION.
    PROC. CAMBRIDGE PHILOS. SOC., 44, 50–57.   (MR9,P.454)


RAO, C. RADHAKRISHNA 1949.   (19.3)                                              4418
    SUFFICIENT STATISTICS AND MINIMUM VARIANCE ESTIMATES.
    PROC. CAMBRIDGE PHILOS. SOC., 45, 215–218.   (MR10,P.446)


RAO, C. RADHAKRISHNA 1949.   (8.4/8.5)                                           4419
    ON SOME PROBLEMS ARISING OUT OF DISCRIMINATION WITH MULTIPLE CHARACTERS.
    SANKHYA, 9, 343–366.   (MR11,P.448)


RAO, C. RADHAKRISHNA 1950.   (6.1/6.5/19.2)                                      4420
    STATISTICAL INFERENCE APPLIED TO CLASSIFICATORY PROBLEMS.
    SANKHYA, 10, 229–256.   (MR12,P.511)


RAO, C. RADHAKRISHNA 1950.   (5.2/8.4)                                           4421
    A NOTE ON THE DISTRIBUTION OF $D_{p+q}^2 - D_p^2$ AND SOME COMPUTATIONAL
    ASPECTS OF $D^2$ STATISTIC AND DISCRIMINANT FUNCTION.
    SANKHYA, 10, 257–268.   (MR12,P.428)


RAO, C. RADHAKRISHNA 1951.   (8.4/A)                                             4422
    AN ASYMPTOTIC EXPANSION OF THE DISTRIBUTION OF WILKS' $\Lambda$-CRITERION.
    BULL. INST. INTERNAT. STATIST., 33(2) 177–180.   (MR16,P.841)


RAO, C. RADHAKRISHNA 1951.   (8.1)                                              4423
    A THEOREM IN LEAST SQUARES.
    SANKHYA, 11, 9–12.   (MR13,P.54)


RAO, C. RADHAKRISHNA 1951.   (6.7)                                              4424
    STATISTICAL INFERENCE APPLIED TO CLASSIFICATORY PROBLEMS.  II.   THE PROBLEM OF
    SELECTING INDIVIDUALS FOR VARIOUS DUTIES IN A SPECIFIED RATIO.
    SANKHYA, 11, 107–116.   (MR13,P.480)


RAO, C. RADHAKRISHNA 1951.   (16.5)                                             4425
    THE APPLICABILITY OF LARGE SAMPLE TESTS FOR MOVING AVERAGE AND AUTOREGRESSIVE
    SCHEMES TO SERIES OF SHORT LENGTH--AN EXPERIMENTAL STUDY.   PART 3:   THE
    DISCRIMINANT FUNCTION APPROACH IN THE CLASSIFICATION OF TIME SERIES (PART III
    OF STATISTICAL INFERENCE APPLIED TO CLASSIFICATORY PROBLEMS).
    SANKHYA, 11, 257–272.   (MR14,P.391)


RAO, C. RADHAKRISHNA 1953.   (6.6)                                              4426
    DISCRIMINANT FUNCTIONS FOR GENETIC DIFFERENTIATION AND SELECTION.
    SANKHYA, 12, 229–246.   (MR15,P.543)


RAO, C. RADHAKRISHNA 1953.   (8.4)                                              4427
    ON TRANSFORMATIONS USEFUL IN THE DISTRIBUTION PROBLEMS OF LEAST SQUARES.
    SANKHYA, 12, 339–346.   (MR15,P.451)


RAO, C. RADHAKRISHNA 1954.   (6.1/6.3)                                          4428
    A GENERAL THEORY OF DISCRIMINATION WHEN THE INFORMATION ABOUT
    ALTERNATIVE POPULATION DISTRIBUTIONS IS BASED ON SAMPLES.
    ANN. MATH. STATIST., 25, 651–670.   (MR16,P.380)


RAO, C. RADHAKRISHNA 1954.   (8.3/8.5/14.2/CE)                                  4429
    ESTIMATION OF RELATIVE POTENCY FROM MULTIPLE RESPONSE DATA.
    BIOMETRICS, 10, 208–220.


RAO, C. RADHAKRISHNA 1954.   (6.5)                                              4430
    ON THE USE AND INTERPRETATION OF DISTANCE FUNCTIONS IN STATISTICS.
    BULL. INST. INTERNAT. STATIST., 34(2) 90–97.   (MR16,P.1037)

RAO, C. RADHAKRISHNA 1955.  (15.2)                                                    4431
    ESTIMATION AND TESTS OF SIGNIFICANCE IN FACTOR ANALYSIS.
  PSYCHOMETRIKA, 20, 93–111.  (MR17,P.55)

RAO, C. RADHAKRISHNA 1955.  (8.4/8.5)                                                 4432
    ANALYSIS OF DISPERSION FOR MULTIPLY CLASSIFIED DATA WITH UNEQUAL
    NUMBERS IN CELLS.
  SANKHYA, 15, 253–280.  (MR17,P.277)

RAO, C. RADHAKRISHNA 1956.  (8.4/8.5)                                                 4433
    ANALYSIS OF DISPERSION WITH INCOMPLETE OBSERVATIONS ON ONE OF THE CHARACTERS.
  J. ROY. STATIST. SOC. SER. B, 18, 259–264.  (MR18,P.608)

RAO, C. RADHAKRISHNA 1957.  (18.2/U)                                                  4434
    MAXIMUM LIKELIHOOD ESTIMATION FOR THE MULTINOMIAL DISTRIBUTION.
  SANKHYA, 18, 139–148.  (MR21-3926)

RAO, C. RADHAKRISHNA 1958.  (8.3/11.1/16.5/E)                                         4435
    SOME STATISTICAL METHODS FOR COMPARISON OF GROWTH CURVES.
  BIOMETRICS, 14, 1–17.

RAO, C. RADHAKRISHNA 1958.  (18.2/U)                                                  4436
    MAXIMUM LIKELIHOOD ESTIMATION FOR THE MULTINOMIAL DISTRIBUTION WITH
    INFINITE NUMBER OF CELLS.
  SANKHYA, 20, 211–218.  (MR21-6059)

RAO, C. RADHAKRISHNA 1959.  (8.1/8.3/8.4/C)                                           4437
    SOME PROBLEMS INVOLVING LINEAR HYPOTHESES IN MULTIVARIATE ANALYSIS.
  BIOMETRIKA, 46, 49–58.  (MR20-7365)

RAO, C. RADHAKRISHNA 1959.  (2.2)                                                     4438
    SUR UNE CARACTERISATION DE LA DISTRIBUTION NORMALE ETABLIE D'APRES
    UNE PROPRIETE OPTIMUM DES ESTIMATIONS LINEAIRES.
  COLLOQ. INTERNAT. CENTRE NAT. RECH. SCI., 87, 165–172.  (MR21-4500)

RAO, C. RADHAKRISHNA 1960.  (1.1)                                                     4439
    MULTIVARIATE ANALYSIS:  AN INDISPENSABLE STATISTICAL AID IN APPLIED RESEARCH.
  SANKHYA, 22, 317–338.  (MR23-A2995)

RAO, C. RADHAKRISHNA 1961.  (1.1/E)                                                   4440
    SOME OBSERVATIONS ON MULTIVARIATE STATISTICAL METHODS IN
    ANTHROPOLOGICAL RESEARCH.
  BULL. INST. INTERNAT. STATIST., 38(4) 99–109.

RAO, C. RADHAKRISHNA 1961.  (19.3)                                                    4441
    APPARENT ANOMALIES AND IRREGULARITIES IN MAXIMUM LIKELIHOOD ESTIMATION.
      (REPRINTED AS NO. 4449)
  BULL. INST. INTERNAT. STATIST., 38(4) 439–453.  (MR26-7081)

RAO, C. RADHAKRISHNA 1961.  (19.3)                                                    4442
    ASYMPTOTIC EFFICIENCY AND LIMITING INFORMATION.
  PROC. FOURTH BERKELEY SYMP. MATH. STATIST. PROB., 1, 531–554.  (MR24-A3026)

RAO, C. RADHAKRISHNA 1961.  (19.2/19.3)                                               4443
    A STUDY OF LARGE SAMPLE TEST CRITERIA THROUGH PROPERTIES OF EFFICIENT ESTIMATES.
  SANKHYA SER. A, 23, 25–40.  (MR25-718)

RAO, C. RADHAKRISHNA 1962.  (6.1/6.4/E)                                               4444
    SOME OBSERVATIONS IN ANTHROPOMETRIC SURVEYS.
  INDIAN ANTHROP. ESSAYS MEM. MAJUMDAR, 135–149.

RAO, C. RADHAKRISHNA 1962.  (19.2/19.3/A)                                             4445
    EFFICIENT ESTIMATES AND OPTIMUM INFERENCE PROCEDURES IN LARGE SAMPLES.
      (WITH DISCUSSION)
  J. ROY. STATIST. SOC. SER. B, 24, 46–72.

RAO, C. RADHAKRISHNA 1962.  (2.5/20.1)                                                4446
    A NOTE ON A GENERALIZED INVERSE OF A MATRIX WITH APPLICATIONS TO
    PROBLEMS IN MATHEMATICAL STATISTICS.
  J. ROY. STATIST. SOC. SER. B, 24, 152–158.  (MR25-1596)

RAO, C. RADHAKRISHNA 1962.    (8.1)                                                   4447
    PROBLEMS OF SELECTION WITH RESTRICTIONS.
 J. ROY. STATIST. SOC. SER. B, 24, 401-405.   (MR26-5677)


RAO, C. RADHAKRISHNA 1962.    (6.1/6.2)                                               4448
    USE OF DISCRIMINANT AND ALLIED FUNCTIONS IN MULTIVARIATE ANALYSIS.
 SANKHYA SER. A, 24, 149-154.   (MR26-1967)


RAO, C. RADHAKRISHNA 1962.    (19.3)                                                  4449
    APPARENT ANOMALIES AND IRREGULARITIES IN MAXIMUM LIKELIHOOD ESTIMATION.
       (WITH DISCUSSION, REPRINT OF NO. 4441)
    SANKHYA SER. B, 24, 73-102.   (MR26-5676)


RAO, C. RADHAKRISHNA 1964.    (1.1/1.2)                                               4450
    SIR RONALD AYLMER FISHER--THE ARCHITECT OF MULTIVARIATE ANALYSIS.
 BIOMETRICS, 20, 286-300.


RAO, C. RADHAKRISHNA 1964.    (19.3/A)                                                4451
    CRITERIA OF ESTIMATION IN LARGE SAMPLES.
 CONTRIB. STATIST. (MAHALANOBIS VOLUME), 345-362.


RAO, C. RADHAKRISHNA 1964.    (1.2)                                                   4452
    SCIENTIFIC CONTRIBUTIONS OF PROFESSOR P. C. MAHALANOBIS.
 CONTRIB. STATIST. (MAHALANOBIS VOLUME), 495-516.


RAO, C. RADHAKRISHNA 1964.    (11.1/11.2/15.1/E)                                      4453
    THE USE AND INTERPRETATION OF PRINCIPAL COMPONENT ANALYSIS IN APPLIED RESEARCH.
 SANKHYA SER. A, 26, 329-358.   (MR32-1848)


RAO, C. RADHAKRISHNA 1965.    (8.1/8.3/8.6/CE)                                        4454
    THE THEORY OF LEAST SQUARES WHEN THE PARAMETERS ARE STOCHASTIC AND ITS
    APPLICATIONS TO THE ANALYSIS OF GROWTH CURVES.
 BIOMETRIKA, 52, 447-458.   (MR34-5221)


RAO, C. RADHAKRISHNA 1965.    (6.7)                                                   4455
    PROBLEMS OF SELECTION INVOLVING PROGRAMMING TECHNIQUES.
 PROC. IBM SCI. COMPUT. SYMP. STATIST., 29-51.


RAO, C. RADHAKRISHNA 1966.    (5.2/6.1/6.3)                                           4456
    DISCRIMINANT FUNCTION BETWEEN COMPOSITE HYPOTHESES AND RELATED PROBLEMS.
 BIOMETRIKA, 53, 339-345.   (MR35-2424)


RAO, C. RADHAKRISHNA 1966.    (19.3/A)                                                4457
    EFFICIENCY OF AN ESTIMATOR AND FISHER'S LOWER BOUND TO ASYMPTOTIC VARIANCE.
 BULL. INST. INTERNAT. STATIST., 41, 55-63.


RAO, C. RADHAKRISHNA 1966.    (5.2/6.2/8.3/E)                                         4458
    COVARIANCE ADJUSTMENT AND RELATED PROBLEMS IN MULTIVARIATE ANALYSIS.
 MULTIVARIATE ANAL. PROC. INTERNAT. SYMP. DAYTON (KRISHNAIAH), 87-103.


RAO, C. RADHAKRISHNA 1966.    (2.5/8.8/20.1)                                          4459
    GENERALIZED INVERSE FOR MATRICES AND ITS APPLICATIONS IN MATHEMATICAL
    STATISTICS.
 RES. PAPERS STATIST. FESTSCHR. NEYMAN (DAVID), 263-279.


RAO, C. RADHAKRISHNA 1966.    (2.2/15.1/17.3)                                         4460
    CHARACTERISATION OF THE DISTRIBUTION OF RANDOM VARIABLES IN LINEAR
    STRUCTURAL RELATIONS.
 SANKHYA SER. A, 28, 251-260.   (MR35-1142)


RAO, C. RADHAKRISHNA + MAJUMDAR, D.N. 1958.    (6.5/6.6/E)                            4461
    BENGAL ANTHROPOMETRIC SURVEY, 1945:  A STATISTICAL STUDY.
 SANKHYA, 19, 201-408.


RAO, C. RADHAKRISHNA + SHAH, D. C. 1948.    (8.1/8.3/E)                               4462
    ON A FORMULA FOR THE PREDICTION OF CRANIAL CAPACITY.
 BIOMETRICS, 4, 247-253.


RAO, C. RADHAKRISHNA + SLATER, PATRICK 1949.    (5.2/E)                               4463
    MULTIVARIATE ANALYSIS APPLIED TO DIFFERENCES BETWEEN NEUROTIC GROUPS.
 BRITISH J. PSYCHOL. STATIST. SECT., 2, 17-29.

RAO, C. RADHAKRISHNA + VARADARAJAN, V.S. 1963.   (6.5/16.4)                    4464
    DISCRIMINATION OF GAUSSIAN PROCESSES.
        (REPRINTED AS NO. 4465)
    SANKHYA SER. A, 25, 303-330.   (MR32-572)

RAO, C. RADHAKRISHNA + VARADARAJAN, V.S. 1964.   (6.5/16.4)                    4465
    DISCRIMINATION OF GAUSSIAN PROCESSES.
        (REPRINT OF NO. 4464)
    CONTRIB. STATIST. (MAHALANOBIS VOLUME), 363-390.   (MR32-572)

RAO, C. RADHAKRISHNA   SEE ALSO
    2892. KAGAN, A. M.; LINNIK, YU. V. + RAO, C. RADHAKRISHNA (1965)

    3617. MAHALANOBIS, P.C.; MAJUMDAR, D.N. + RAO, C. RADHAKRISHNA (1949)

RAO, J.N.K. 1958.   (2.2)                                                      4466
    A CHARACTERIZATION OF THE NORMAL DISTRIBUTION.
        (AMENDED BY NO. 4467)
    ANN. MATH. STATIST., 29, 914-919.   (MR20-328)

RAO, J.N.K. 1959.   (2.2)                                                      4467
    ACKNOWLEDGMENT OF PRIORITY.
        (AMENDMENT OF NO. 4466)
    ANN. MATH. STATIST., 30, 610.

RAO, J.N.K. 1964.   (16.2)                                                     4468
    ESTIMATION OF SIMULTANEOUS EQUATIONS.
    J. INDIAN STATIST. ASSOC., 2, 210-228.   (MR32-1803)

RAO, J.N.K. 1966.   (18.5)                                                     4469
    ALTERNATIVE ESTIMATORS IN PPS SAMPLING FOR MULTIPLE CHARACTERISTICS.
    SANKHYA SER. A, 28, 47-60.

RAO, J.N.K. 1966.   (18.5)                                                     4470
    ON THE RELATIVE EFFICIENCY OF SOME ESTIMATORS IN PPS SAMPLING FOR
    MULTIPLE CHARACTERISTICS.
    SANKHYA SER. A, 28, 61-70.

RAO, K. SAMBASIVA 1951.   (5.3/14.2)                                           4471
    ON THE MUTUAL INDEPENDENCE OF A SET OF HOTELLING'S $T^2$ DERIVABLE FROM
    A SAMPLE OF SIZE N FROM A K-VARIATE NORMAL POPULATION.
    BULL. INST. INTERNAT. STATIST., 33(2) 171-176.   (MR17,P.278)

RAO, K. SAMBASIVA 1954.   (16.5)                                              4472
    TESTING FOR SERIAL CORRELATION IN A STATIONARY MULTIDIMENSIONAL
    DISCRETE STOCHASTIC PROCESS.
    BULL. INST. INTERNAT. STATIST., 34(2) 185-194.   (MR16,P.1134)

RAO, M. BHASKAR 1965.   (8.5/20.1/20.2/U)                                      4473
    APPLICATION OF GREENBERG AND SARHAN'S METHOD OF INVERSION OF
    PARTITIONED MATRICES IN THE ANALYSIS OF NON-ORTHOGONAL DATA.
    J. AMER. STATIST. ASSOC., 60, 1200-1202.   (MR32-8466)

RAO, M.M. 1961.   (16.1/A)                                                     4474
    CONSISTENCY AND LIMIT DISTRIBUTIONS OF ESTIMATORS OF PARAMETERS IN
    EXPLOSIVE STOCHASTIC DIFFERENCE EQUATIONS.
    ANN. MATH. STATIST., 32, 195-218.   (MR27-5313)

RAO, M.M. 1963.   (6.3/E)                                                      4475
    DISCRIMINANT ANALYSIS.
    ANN. INST. STATIST. MATH. TOKYO, 15, 11-24.

RAO, M.M. 1966.   (16.1/A)                                                     4476
    INFERENCE IN STOCHASTIC PROCESSES. II.
    Z. WAHRSCHEIN. VERW. GEBIETE, 5, 317-335.

RAO, M.M.   SEE ALSO
    959. CHIPMAN, JOHN S. + RAO, M.M. (1964)

    960. CHIPMAN, JOHN S. + RAO, M.M. (1964)

    1264. DEGROOT, M. H. + RAO, M.M. (1966)

    3145. KRISHNAIAH, PARUCHURI R. + RAO, M.M. (1961)

RAO, M. SUDHAKARA 1965.  (17.3)                                          4477
    A NOTE ON THE EXISTENCE OF MOMENTS OF INFINITELY DIVISIBLE DISTRIBUTIONS IN
    N-DIMENSIONS.
    J. INDIAN STATIST. ASSOC., 3, 220-222.  (MR33-6672)

RAO, R. RANGA  SEE  RANGA RAO, R.

RAO, V. V. NARAYAN 1966.  (8.4/10.2)                                     4478
    DISTRIBUTION OF SOME MULTIVARIATE TEST STATISTICS.
    J. INDIAN STATIST. ASSOC., 4, 195-201.

RASCH, D. 1962.  (15.2/ET)                                              4479
    DIE FAKTORANALYSE UND IHRE ANWENDUNG IN DER TIERZUCHT.
    BIOM. 7., 4, 15-39.

RASCH, D. 1964.  (15.2)                                                 4480
    DER VERGLEICH DER ERGEBNISSE DER FAKTORANALYSE IN MEHREREN POPULATIONEN.
    ABHANDL. DEUTSCHEN AKAD. WISS. BERLIN KL. MATH. PHYS. TECH., 4, 97-100.  (MR33-1933)

RASCH, G. 1948.  (7.1)                                                  4481
    A FUNCTIONAL EQUATION FOR WISHART'S DISTRIBUTION.
    ANN. MATH. STATIST., 19, 262-266.  (MR9,P.600)

RASCH, G. 1950.  (7.3/8.4)                                              4482
    A VECTORIAL T-TEST IN THE THEORY OF MULTIVARIATE NORMAL DISTRIBUTIONS.
    MAT. TIDSSKR. B, 76-81.  (MR12,P.345)

RASCH, G. 1953.  (15.1)                                                 4483
    ON SIMULTANEOUS FACTOR ANALYSIS IN SEVERAL POPULATIONS.
    NORDISK PSYKOL. MONOG. NR. 3 (UPPSALA SYMP. PSYCHOL. FACT. ANAL.), 65-71.  (MR16,P.385)

RASCH, G. 1961.  (15.4)                                                 4484
    ON GENERAL LAWS AND THE MEANING OF MEASUREMENT IN PSYCHOLOGY.
    PROC. FOURTH BERKELEY SYMP. MATH. STATIST. PROB., 4, 321-333.  (MR24-B1246)

RASCH, G. 1966.  (15.4)                                                 4485
    AN ITEM ANALYSIS WHICH TAKES INDIVIDUAL DIFFERENCES INTO ACCOUNT.
    BRITISH J. MATH. STATIST. PSYCHOL., 19, 49-57.

RASCH, G. 1966.  (15.4)                                                 4486
    AN INDIVIDUALISTIC APPROACH TO ITEM ANALYSIS.
    READINGS MATH. SOCIAL SCI. (LAZARSFELD + HENRY), 89-107.

RASCH, G.   SEE ALSO
    2291. HALD, A. HJORTH + RASCH, G. (1943)

RASMUSSEN, P. NORREGAARD 1948.  (16.2/E)                                4487
    SOME REMARKS ON THE JOINT EFFECTS OF SIMULTANEOUS RELATIONS BETWEEN
    ECONOMIC VARIABLES.
    NORDISK TIDSSKR. TEK. OKON., 12, 215-222.

RAUDELJUNAS, A.K. 1961.  (16.4/17.1)                                    4488
    LIMIT THEOREMS FOR SUMS OF RANDOM VECTORS FORMING A NON-HOMOGENEOUS
    MARKOV CHAIN. I. (IN RUSSIAN, WITH SUMMARY IN LITHUANIAN AND ENGLISH)
    LITOVSK. MAT. SB., 1, 203-230.  (MR26-7037)

RAUDELJUNAS, A.K. 1962.  (16.4/17.1)                                    4489
    LIMIT THEOREMS OR SUMS OF RANDOM VECTORS FORMING A NON-HOMOGENEOUS
    MARKOV CHAIN. II. (IN RUSSIAN, WITH SUMMARY IN LITHUANIAN AND ENGLISH)
    LITOVSK. MAT. SB., 1, 115-124.  (MR26-7038)

RAY, W.D. + PITMAN, A. E. N. T. 1961.  (5.2/U)                          4490
    AN EXACT DISTRIBUTION OF THE FISHER-BEHRENS-WELCH STATISTIC FOR TESTING THE
    DIFFERENCE BETWEEN THE MEANS OF TWO NORMAL POPULATIONS WITH UNKNOWN VARIANCES.
    J. ROY. STATIST. SOC. SER. B, 23, 377-384.  (MR25-2660)

RAY, W.D. + WYLD, C. 1965.  (16.5)                                      4491
    POLYNOMIAL PROJECTING PROPERTIES OF MULTITERM PREDICTORS/CONTROLLERS
    IN  NON-STATIONARY TIME SERIES.
    J. ROY. STATIST. SOC. SER. B, 27, 144-158.

RAYLEIGH, JOHN W. S., BARON 1880.  (2.7/14.4/14.5/BE)                     4492
    ON THE RESULTANT OF A LARGE NUMBER OF VIBRATIONS OF THE SAME PITCH AND
      OF ARBITRARY PHASE.
    PHILOS. MAG. SER. 5, 10, 73-78.


RAYLEIGH, LORD 1919.  (14.4/20.4)                                        4493
    ON THE PROBLEM OF RANDOM VIBRATIONS, AND OF RANDOM FLIGHTS IN ONE,
    TWO OR THREE DIMENSIONS.
    PHILOS. MAG. SER. 6, 37, 331-347.


RAYMOND, JOSEPH L.   SEE
    563. BIRNBAUM, Z. W.; RAYMOND, JOSEPH L. + ZUCKERMAN, HERBERT S. (1947)


RAYNER, A. A. 1966.  (2.5/14.5)                                          4494
    ON THE DISTRIBUTION OF PEARSON'S CHI-SQUARED CRITERION.
    S. AFRICAN J. AGRIC. SCI., 9, 1029-1031.


RAYNER, A. A. + LIVINGSTONE, D. 1965.  (2.5)                             4495
    ON THE DISTRIBUTION OF QUADRATIC FORMS IN SINGULAR NORMAL VARIATES.
    S. AFRICAN J. AGRIC. SCI., 8, 357-369.


REED, W. G. 1917.  (4.1/8.1/EU)                                          4496
    THE COEFFICIENT OF CORRELATION.
        (AMENDED BY NO. 2364)
    QUART. PUBL. AMER. STATIST. ASSOC., 15, 670-684.


REES, D.H.   SEE
    1741. FOSTER, F. G. + REES, D.H. (1957)

    1742. FOSTER, F. G. + REES, D.H. (1958)


REEVE, E. C. R. 1955.  (4.8/15.4)                                        4497
    THE VARIANCE OF THE GENETIC CORRELATION COEFFICIENT.
    BIOMETRICS, 11, 357-374.


REGAN, MARY C. 1965.  (1.1)                                              4498
    DEVELOPMENT AND CLASSIFICATION OF MODELS FOR MULTIVARIATE ANALYSIS.
    EDUC. PSYCHOL. MEAS., 25, 997-1010.


REIERSOL, OLAV 1940.  (17.7/20.3)                                        4499
    A METHOD FOR RECURRENT COMPUTATION OF ALL THE PRINCIPAL MINORS OF A
    DETERMINANT, AND ITS APPLICATION IN CONFLUENCE ANALYSIS.
    ANN. MATH. STATIST., 11, 193-198.  (MR2,P.61)


REIERSOL, OLAV 1941.  (17.7)                                             4500
    CONFLUENCE ANALYSIS BY MEANS OF LAG MOMENTS AND OTHER METHODS OF
      CONFLUENCE ANALYSIS.
    ECONOMETRICA, 9, 1-24.  (MR2,P.237)


REIERSOL, OLAV 1945.  (17.7)                                             4501
    CONFLUENCE ANALYSIS BY MEANS OF INSTRUMENTAL SETS OF VARIABLES.
    ARK. MAT. ASTRONOM. FYS., 32A(4) 1-119.  (MR7,P.317)   (SEE ALSO CHAPTER I)


REIERSOL, OLAV 1945.  (17.7)                                             4502
    RESIDUAL VARIABLES IN REGRESSION AND CONFLUENCE ANALYSIS.
    SKAND. AKTUARIETIDSKR., 28, 201-217.  (MR7,P.318)


REIERSOL, OLAV 1950.  (17.7)                                             4503
    IDENTIFIABILITY OF A LINEAR RELATION BETWEEN VARIABLES WHICH ARE
      SUBJECT TO ERROR.
    ECONOMETRICA, 18, 375-389.  (MR12,P.347)


REIERSOL, OLAV 1950.  (15.1)                                             4504
    ON THE IDENTIFIABILITY OF PARAMETERS IN THURSTONE'S MULTIPLE FACTOR ANALYSIS.
    PSYCHOMETRIKA, 15, 121-149.  (MR12,P.38)


REIERSOL, OLAV 1956.  (4.5)                                              4505
    A NOTE ON THE SIGNS OF GROSS CORRELATION COEFFICIENTS AND PARTIAL
      CORRELATION COEFFICIENTS.
    BIOMETRIKA, 43, 480-482.  (MR18,P.344)


REIERSOL, OLAV   SEE ALSO
    3100. KOOPMANS, TJALLING C. + REIERSOL, OLAV (1950)

REINMILLER, C. F.    SEE
    2412. HAZFL, L. N.; BAKER, M. L. + REINMILLER, C. F. (1943)

REINSCHKE, K. 1966.  (17.1)                                                    4506
    ZUM ZENTRALEN GRENZWERTSATZ FUR ZUFALLIGE ELEMENTE MIT WERTEN AUS
    EINEM HILBERTRAUM.
    Z. WAHRSCHEIN. VERW. GEBIETE, 6, 161-169.


REITER, STANLEY    SEE
    104. AMES, EDWARD + REITER, STANLEY (1961)

REMAGE, RUSSELL    SEE
    3512. LOTKIN, MARK M. + REMAGE, RUSSELL (1953)

REMMERS, H. H. 1931.  (15.4)                                                   4507
    THE EQUIVALENCE OF JUDGMENTS TO TEST ITEMS IN THE SENSE OF THE
    SPEARMAN-BROWN FORMULA.
    J. EDUC. PSYCHOL., 22, 66-71.

REMMERS, H. H. + ADKINS, R. M. 1942.  (15.4)                                   4508
    RELIABILITY OF MULTIPLE-CHOICE MEASURING INSTRUMENTS AS A FUNCTION OF
    THE SPEARMAN-BROWN PROPHECY FORMULA, VI.
    J. EDUC. PSYCHOL., 33, 385-390.

REMMERS, H. H. + EWART, EDWIN 1941.  (15.4)                                    4509
    RELIABILITY OF MULTIPLE-CHOICE MEASURING INSTRUMENTS AS A FUNCTION OF
    THE SPEARMAN-BROWN PROPHECY FORMULA.  III.
    J. EDUC. PSYCHOL., 32, 61-66.

REMMERS, H. H. + HOUSE, J. MILTON 1941.  (15.4)                                4510
    RELIABILITY OF MULTIPLE-CHOICE MEASURING INSTRUMENTS AS A FUNCTION OF
    THE SPEARMAN-BROWN PROPHECY FORMULA.  IV.
    J. EDUC. PSYCHOL., 32, 372-376.

REMMERS, H. H. + SAGESER, H. W. 1941.  (15.4)                                  4511
    RELIABILITY OF MULTIPLE-CHOICE MEASURING INSTRUMENTS AS A FUNCTION OF
    THE SPEARMAN-BROWN PROPHECY FORMULA.  V.
    J. EDUC. PSYCHOL., 32, 445-451.

REMMERS, H. H.; KARSLAKE, R. + GAGE, N. L. 1940.  (15.4)                       4512
    RELIABILITY OF MULTIPLE-CHOICE MEASURING INSTRUMENTS AS A FUNCTION OF
    THE SPEARMAN-BROWN PROPHECY FORMULA.  I.
    J. EDUC. PSYCHOL., 31, 583-590.


REMMERS, H. H.    SEE ALSO
    1312. DENNEY, H. R. + REMMERS, H. H. (1940)

RENSHAW, T. 1952.  (15.3)                                                      4513
    FACTOR ROTATION BY THE METHOD OF EXTENDED VECTORS.
    BRITISH J. PSYCHOL. STATIST. SECT., 5, 7-18.

RENYI, ALFRED 1952.  (17.3)                                                    4514
    ON PROJECTIONS OF PROBABILITY DISTRIBUTIONS.
    ACTA MATH. ACAD. SCI. HUNGAR., 3, 131-142.  (MR14,P.771)

RENYI, ALFRED 1959.  (4.9)                                                     4515
    ON MEASURES OF DEPENDENCE.
    ACTA MATH. ACAD. SCI. HUNGAR., 10, 441-451.  (MR22-6005)

RENYI, ALFRED 1961.  (17.2)                                                    4516
    ON KOLMOGOROFF'S INEQUALITY.
    MAGYAR TUD. AKAD. MAT. KUTATO. INT. KOZL., 6, 411-415.  (MR26-5623)

RENYI, ALFRED + SULANKE, ROLF 1963.  (17.3/18.6/20.5/A)                        4517
    UBER DIE KONVEXE HULLE VON N ZUFALLIG GEWAHLTEN PRUNKTEN.
    Z. WAHRSCHEIN. VERW. GEBIETE, 2, 75-84.

RENYI, ALFRED + SULANKE, ROLF 1964.  (17.3/18.6/A)                             4518
    UBER DIE KONVEXE HULLE VON  N  ZUFALLIG GEWAHLTEN PUNKTEN.  II.
    Z. WAHRSCHEIN. VERW. GEBIETE, 3, 138-147.

REYBURN, H. A. + RAATH, M. J. 1949.  (15.1)                                    4519
    SIMPLE STRUCTURE:  A CRITICAL EXAMINATION.
  BRITISH J. PSYCHOL. STATIST. SECT., 2, 125–133.

REYBURN, H. A. + TAYLOR, JOHN G. 1943.  (15.1)                                 4520
    ON THE INTERPRETATION OF COMMON FACTORS:  A CRITICISM AND A STATEMENT.
  PSYCHOMETRIKA, 8, 53–64.

REYMENT, R.A. 1961.  E(5.2/10.1)                                               4521
    A NOTE ON GEOGRAPHICAL VARIATION IN EUROPEAN RANA.
  GROWTH, 25, 219–227.

REYMENT, R.A. 1961.  (11.2/E)                                                  4522
    QUADRIVARIATE PRINCIPAL COMPONENT ANALYSIS OF GLOBIGERINA YEGUAENSIS.
  STOCKHOLM CONTRIB. GEOL., 8, 17–26.

REYMENT, R.A. 1962.  (10.1/E)                                                  4523
    OBSERVATIONS ON HOMOGENEITY OF COVARIANCE MATRICES IN PALEONTOLOGIC BIOMETRY.
  BIOMETRICS, 18, 1–11.

REYMENT, R.A. 1963.  (1.1/E)                                                   4524
    PALEONTOLOGICAL APPLICABILITY OF CERTAIN RECENT ADVANCES IN
    MULTIVARIATE STATISTICAL ANALYSIS.
  GEOL. FOREN. I STOCKHOLM FORHANDL., 85, 236–265.

REYMENT, R.A. 1963.  (11.1/E)                                                  4525
    MULTIVARIATE ANALYTICAL TREATMENT OF QUANTITATIVE SPECIES
    ASSOCIATIONS:  AN EXAMPLE FROM PALAEOECOLOGY.
  J. ANIMAL ECOL., 32, 535–547.

REYMENT, R.A. + BRANNSTROM, B. 1962.  (10.1/11.1/12.1/E)                       4526
    CERTAIN ASPECTS OF THE PHYSIOLOGY OF CYPRIDOPSIS (OSTRACODA, CRUSTACEA).
  GEOL. FOREN. I STOCKHOLM FORHANDL., 9, 207–242.

REYMENT, R.A. + NAIDIN, D. P. 1962.  E(11.2/18.6)                             4527
    BIOMETRIC STUDY OF ACTINOCAMAX VERSUS S.L. FROM THE UPPER CRETACEOUS
    OF THE RUSSIAN PLATFORM.
  STOCKHOLM CONTRIB. GEOL., 9, 147–206.

REZNY, ZDENEK 1965.  (16.1)                                                    4528
    VYHLAZOVANI V DISKRETNIM NAHODNEM PROCESU AUTOREGRESNIHO TYPU.
  PROBLEMY KYBERNETIKY (PRAGUE), 96–106.

RHODES, E.C. 1923.  (18.1)                                                     4529
    ON A CERTAIN SKEW CORRELATION SURFACE.
  BIOMETRIKA, 14, 355–377.

RHODES, E.C. 1925.  (18.1)                                                     4530
    ON A SKEW CORRELATION SURFACE.
  BIOMETRIKA, 17, 314–326.

RHODES, E.C. 1927.  (8.8)                                                      4531
    ON LINES AND PLANES OF CLOSEST FIT.
  PHILOS. MAG. SER. 7, 3, 357–364.

RHODES, E.C. 1928.  (18.5)                                                     4532
    ON THE NORMAL CORRELATION FUNCTION AS AN APPROXIMATION TO THE
    DISTRIBUTION OF PAIRED DRAWINGS.
  J. ROY. STATIST. SOC., 91, 548–550.

RICE, S.O. 1944.  (16.4)                                                       4533
    MATHEMATICAL ANALYSIS OF RANDOM NOISE.
        (REPRINTED AS NO. 4535)
  BELL SYSTEM TECH. J., 23, 282–332.  (MR6,P.89)

RICE, S.O. 1945.  (16.4)                                                       4534
    MATHEMATICAL ANALYSIS OF RANDOM NOISE (CONCLUDED).
        (REPRINTED AS NO. 4535)
  BELL SYSTEM TECH. J., 24, 46–156.  (MR6,P.233)

RICE, S.O. 1954.  (16.4)                                                       4535
    MATHEMATICAL ANALYSIS OF RANDOM NOISE.
        (REPRINT OF NO. 4533, REPRINT OF NO. 4534)
  SELECT. PAPERS NOISE STOCH. PROC., 133–294.  (MR6,P.89, P. 233)

RICE, S.O.   SEE ALSO
   4029. NYQUIST, H.; RICE, S.O. + RIORDAN, JOHN (1954)

RICHARDS, HENRY I. 1931.  (4.1/8.1)                              4536
   ANALYSIS OF THE SPURIOUS EFFECT OF HIGH INTERCORRELATION OF
   INDEPENDENT VARIABLES ON REGRESSION AND CORRELATION COEFFICIENTS.
   J. AMER. STATIST. ASSOC., 26, 21-29.

RICHARDSON, LEWIS F. 1950.  (11.1)                              4537
   A METHOD FOR COMPUTING PRINCIPAL AXES.
   BRITISH J. PSYCHOL. STATIST. SECT., 3, 16-20.

RICHARDSON, MOSES W. 1936.  (15.4)                             4538
   NOTES ON THE RATIONALE OF ITEM ANALYSIS.
   PSYCHOMETRIKA, 1(1) 69-76.

RICHARDSON, MOSES W. 1936.  (15.4)                             4539
   THE RELATION BETWEEN THE DIFFICULTY AND THE DIFFERENTIAL VALIDITY OF A TEST.
   PSYCHOMETRIKA, 1(2) 33-49.

RICHARDSON, MOSES W. 1938.  (15.5)                             4540
   MULTIDIMENSIONAL PSYCHOPHYSICS.
   PSYCHOL. BULL., 35, 659-660.

RICHARDSON, MOSES W. + KUDER, G. F. 1939.  (15.4)             4541
   THE CALCULATION OF TEST RELIABILITY COEFFICIENTS BASED ON THE METHOD
   OF RATIONAL EQUIVALENCE.
   J. EDUC. PSYCHOL., 30, 681-687.

RICHARDSON, MOSES W. + STALNAKER, JOHN M. 1933.  (4.4/15.4)   4542
   A NOTE ON THE USE OF BI-SERIAL R IN TEST RESEARCH.
   J. GENERAL PSYCHOL., 8, 463-465.

RICHARDSON, MOSES W.   SEE ALSO
   3180. KUDER, G. F. + RICHARDSON, MOSES W. (1937)

RICHTER, HANS 1949.  (4.6)                                     4543
   ZUR MAXIMALKORRELATION.
   Z. ANGEW. MATH. MECH., 29, 127.   (MR11,P.42)

RICHTER, HANS 1949.  (2.1/17.3)                                4544
   ZUR GAUSSISCHEN VERTEILUNG IM N-DIMENSIONALEN RAUME.
   Z. ANGEW. MATH. MECH., 29, 161-164.   (MR11,P.258)

RICHTER, HANS 1963.  (19.2)                                    4545
   SUBJECTIVE WAHRSCHEINLICHKEIT UND MULTISUBJECTIVE TESTS.
   Z. WAHRSCHEIN. VERW. GEBIETE, 1, 271-277.

RICHTER, K. J. 1966.  (4.8/E)                                  4546
   CYBERNETIC ASPECTS OF CORRELATION AND REGRESSION ANALYSIS.
   STATIST. PRAX., 21, 69-73.

RICHTER, WOLFGANG 1958.  (17.1)                                4547
   MEHRDIMENSIONALE LOKALE GRENZWERTSATZE FUR GROSSE ABWEICHUNGEN.
   (IN RUSSIAN, WITH SUMMARY IN GERMAN, TRANSLATED AS NO. 4548)
   TEOR. VEROJATNOST. I PRIMENEN., 3, 107-114.   (MR20-334)

RICHTER, WOLFGANG 1958.  (17.1)                                4548
   MULTI-DIMENSIONAL LOCAL LIMIT THEOREMS FOR LARGE DEVIATIONS.
   (TRANSLATION OF NO. 4547)
   THEORY PROB. APPL., 3, 100-106.   (MR20-334)

RIDEOUT, V. C.   SEE
   767. BURFORD, THOMAS M.; RIDEOUT, V. C. + SATHER, D. S. (1955)

RIDER, PAUL R. 1924.  (2.1/4.1/4.4)                            4549
   THE CORRELATION BETWEEN TWO VARIATES, ONE OF WHICH IS NORMALLY DISTRIBUTED.
   (AMENDED BY NO. 4550)
   AMER. MATH. MONTHLY, 31, 227-231.

RIDER, PAUL R. 1924.  (2.1/4.1/4.4)                            4550
   ERRATA TO:  THE CORRELATION BETWEEN TWO VARIATES, ONE OF WHICH IS NORMALLY DISTRIBUTED.
   (AMENDMENT OF NO. 4549)
   BULL. AMER. MATH. SOC., 30, 397-398.

RIDER, PAUL R. 1932.  (4.2)                                                              4551
    ON THE DISTRIBUTION OF THE CORRELATION COEFFICIENT IN SMALL SAMPLES.
    BIOMETRIKA, 24, 382-403.

RIDER, PAUL R. 1936.  (5.1/17.7)                                                         4552
    ANNUAL SURVEY OF STATISTICAL TECHNIQUE.   DEVELOPMENTS IN THE ANALYSIS
    OF MULTIVARIATE DATA.  PART I.
    ECONOMETRICA, 4, 264-268.

RIDLER-ROWE, C. J. 1966.  (16.4/A)                                                       4553
    ON FIRST HITTING OF SOME RECURRENT TWO-DIMENSIONAL RANDOM WALKS.
    Z. WAHRSCHEIN. VERW. GEBIETE, 5, 187-201.

RIEBESELL, PAUL 1922.  (4.1/4.6/4.8)                                                     4554
    UBER DIE KORRELATIONSMETHODE.
    Z. ANGEW. MATH. MECH., 2, 195-199.

RIETZ, H.L. 1912.  (2.1/4.1)                                                             4555
    ON THE THEORY OF CORRELATION WITH SPECIAL REFERENCE TO CERTAIN SIGNIFICANT LOCI
    ON THE PLANE OF DISTRIBUTION IN THE CASE OF NORMAL CORRELATION.
    ANN. MATH. SER. 2, 13, 187-199.

RIETZ, H.L. 1918.  (4.1/4.8)                                                             4556
    ON FUNCTIONAL RELATIONS FOR WHICH THE COEFFICIENT OF CORRELATION IS ZERO.
    QUART. PUBL. AMER. STATIST. ASSOC., 16, 472-476.

RIETZ, H.L. 1920.  (4.1/18.5)                                                            4557
    URN SCHEMATA AS A BASIS FOR THE DEVELOPMENT OF CORRELATION THEORY.
    ANN. MATH. SER. 2, 21, 306-322.

RIETZ, H.L. 1923.  (2.1/U)                                                               4558
    FREQUENCY DISTRIBUTIONS OBTAINED BY CERTAIN TRANSFORMATIONS OF
    NORMALLY DISTRIBUTED VARIATES.
    ANN. MATH. SER. 2, 23, 292-300.

RIETZ, H.L. 1932.  (18.1)                                                                4559
    A SIMPLE NON-NORMAL CORRELATION SURFACE.
    BIOMETRIKA, 24, 288-290.

RIETZ, H.L. 1938.  (1.1)                                                                 4560
    ON A RECENT ADVANCE IN STATISTICAL INFERENCE.
    AMER. MATH. MONTHLY, 45, 149-158.

RIFFENBURGH, ROBERT H. 1966.  (15.5)                                                     4561
    A METHOD OF SOCIOMETRIC IDENTIFICATION ON THE BASIS OF MULTIPLE MEASUREMENTS.
    SOCIOMETRY, 29, 280-290.

RIFFENBURGH, ROBERT H. + CLUNIES-ROSS, CHARLES W. 1960.  (6.1/6.2/6.3)                   4562
    LINEAR DISCRIMINANT ANALYSIS.
    PACIFIC SCI., 14, 251-256.

RIFFENBURGH, ROBERT H.    SEE ALSO
    989. CLUNIES-ROSS, CHARLES W. + RIFFENBURGH, ROBERT H. (1960)

RIGBY, CONSTANCE M. 1933.  (20.4)                                                        4563
    ON A RECURRENCE RELATION CONNECTED WITH THE DOUBLE BESSEL FUNCTIONS
    $K_{\tau_1,\tau_2}(x)$ AND $T_{\tau_1,\tau_2}(x)$.
    BIOMETRIKA, 25, 420-421.

RIGG, MISS F. A. 1946.  (1.2)                                                            4564
    RECENT ADVANCES IN MATHEMATICAL STATISTICS: BIBLIOGRAPHY OF
    MATHEMATICAL STATISTICS (1940-'42).
    J. ROY. STATIST. SOC., 109, 395-450.  (MR9,P.194)

RIHTER, VOL≠FGANG  SEE  RICHTER, WOLFGANG

RIOPELLE, ARTHUR J.   SEE
    1547. ELLIS, MAX E. + RIOPELLE, ARTHUR J. (1948)

RIORDAN, JOHN 1937.  (18.1)                                                              4565
    MOMENT RECURRENCE RELATIONS FOR BINOMIAL, POISSON AND HYPERGEOMETRIC
    FREQUENCY DISTRIBUTIONS.
    ANN. MATH. STATIST., 8, 103-111.

RIORDAN, JOHN    SEE ALSO
    4029. NYQUIST, H.; RICE, S.O. + RIORDAN, JOHN (1954)

RIOS, S.    SEE
    1338. DIAZ UNGRIA, A.; CAMACHO, A. + RIOS, S. (1955)

RIPPE, DAYLE D. 1953.  (15.2)                                              4566
    APPLICATION OF A LARGE SAMPLING CRITERION TO SOME SAMPLING PROBLEMS
    IN FACTOR ANALYSIS.
    PSYCHOMETRIKA, 18, 191-205.  (MR15,P.726)

RISSER, RENE 1930.  (2.1/4.1/B)                                            4567
    INDICATIONS SUR LA CORRELATION.
    BULL. TRIMEST. INST. ACTUAIR. FRANC., 36, 87-116.

RITCHIE-SCOTT, A. 1915.  (4.2)                                             4568
    NOTE ON THE PROBABLE ERROR OF THE COEFFICIENT OF CORRELATION IN THE
    VARIATE DIFFERENCE CORRELATION METHOD.
    BIOMETRIKA, 11, 136-138.

RITCHIE-SCOTT, A. 1918.  (4.8)                                             4569
    THE CORRELATION COEFFICIENT OF A POLYCHORIC TABLE.
    BIOMETRIKA, 12, 93-133.

RITCHIE-SCOTT, A. 1921.  (2.7/B)                                           4570
    THE INCOMPLETE MOMENTS OF A NORMAL SOLID.
    BIOMETRIKA, 13, 401-425.

RIZVI, M. HASEEB    SEE
      77. ALAM, K. + RIZVI, M. HASEEB (1966)

     317. BARR, CAPT. B.R. + RIZVI, M. HASEEB (1966)

    3146. KRISHNAIAH, PARUCHURI R. + RIZVI, M. HASEEB (1966)

RIZZI, A. 1961.  (4.1/18.1/BE)                                             4571
    ON THE BIVARIATE DISTRIBUTIONS OF THE FRECHEL'S CLASS HAVING THE SAME
    CORRELATION COEFFICIENT.
          (IN ITALIAN)
    STATISTICA, BOLOGNA, 21, 812-817.

ROBBINS, HERBERT E. 1944.  (17.3)                                         4572
    ON THE MEASURE OF A RANDOM SET. I.
          (AMENDED BY NO. 4574)
    ANN. MATH. STATIST., 15, 70-74.  (MR6,P.5)

ROBBINS, HERBERT E. 1945.  (17.3)                                        4573
    ON THE MEASURE OF A RANDOM SET. II.
          (AMENDED BY NO. 4574)
    ANN. MATH. STATIST., 16, 342-347.  (MR8,P.389)

ROBBINS, HERBERT E. 1947.  (17.3)                                        4574
    ACKNOWLEDGEMENT OF PRIORITY.
          (AMENDMENT OF NO. 4572, AMENDMENT OF NO. 4573)
    ANN. MATH. STATIST., 18, 297.  (MR8,P.592)

ROBBINS, HERBERT E. 1948.  (2.5/14.5)                                    4575
    THE DISTRIBUTION OF A DEFINITE QUADRATIC FORM.
    ANN. MATH. STATIST., 19, 266-270.  (MR9,P.601)

ROBBINS, HERBERT E. 1951.  (19.1)                                        4576
    ASYMPTOTICALLY SUBMINIMAX SOLUTIONS OF COMPOUND STATISTICAL DECISION PROBLEMS.
    PROC. SECOND BERKELEY SYMP. MATH. STATIST. PROB., 131-148.  (MR13,P.480)

ROBBINS, HERBERT E. + PITMAN, E.J.G. 1949.  (2.5/14.5)                   4577
    APPLICATION OF THE METHOD OF MIXTURES TO QUADRATIC FORMS IN NORMAL VARIATES.
    ANN. MATH. STATIST., 20, 552-560.  (MR11,P.259)

ROBBINS, HERBERT E.    SEE ALSO
     253. BAHADUR, RAGHU RAJ + ROBBINS, HERBERT E. (1950)

     254. BAHADUR, RAGHU RAJ + ROBBINS, HERBERT E. (1951)

    2492. HOEFFDING, WASSILY + ROBBINS, HERBERT E. (1948)

ROBERTS, CHARLES 1966.  (4.3/14.1/14.2/B)                                    4578
    A CORRELATION MODEL USEFUL IN THE STUDY OF TWINS.
    J. AMER. STATIST. ASSOC., 61, 1184-1190.  (MR34-5203)

ROBERTS, ELEANOR    SEE
    4842. SARHAN, AHMED E.; GREENBERG, BERNARD G. + ROBERTS, ELEANOR (1962)

ROBERTS, J. A. FRASER    SEE
    2785. JENNISON, R. F.; PENFOLD, J. B. + ROBERTS, J. A. FRASER (1948)

ROBERTS, JOHN L. 1937.  E(4.4)                                               4579
    A COEFFICIENT OF CORRELATION BETWEEN SCHOLARSHIP AND SALARIES.
    ANN. MATH. STATIST., 8, 66-68.

ROBINSON, ENDERS A. 1963.  (16.4)                                           4580
    STRUCTURAL PROPERTIES OF STATIONARY STOCHASTIC PROCESSES WITH APPLICATIONS.
    PROC. SYMP. TIME SERIES ANAL. BROWN UNIV. (ROSENBLATT), 170-192.  (MR27-3050)

ROBINSON, ENDERS A. 1964.  (16.4/16.5)                                      4581
    WAVELET COMPOSITION OF TIME-SERIES.
    ECONOMET. MODEL BUILDING (WOLD), 37-106.

ROBINSON, ENDERS A. 1964.  (16.3/16.4/16.5)                                 4582
    RECURSIVE DECOMPOSITION OF STOCHASTIC PROCESSES.
    ECONOMET. MODEL BUILDING (WOLD), 111-168.

ROBINSON, ENDERS A. + WOLD, HERMAN O. A. 1963.  (16.1)                      4583
    APPENDIX.  MINIMUM-DELAY STRUCTURE OF LEAST-SQUARES AND EO-IPSO
    PREDICTING SYSTEMS FOR STATIONARY STOCHASTIC PROCESSES.
    PROC. SYMP. TIME SERIES ANAL. BROWN UNIV. (ROSENBLATT), 192-196.  (MR27-3051)

ROBINSON, ENDERS A.    SEE ALSO
    5889. WIGGINS, RALPH A. + ROBINSON, ENDERS A. (1965)

ROBINSON, J. 1965.  (2.5/14.5)                                              4584
    THE DISTRIBUTION OF A GENERAL QUADRATIC FORM IN NORMAL VARIATES.
    AUSTRAL. J. STATIST., 7, 110-114.  (MR33-6739)

ROBINSON, PATRICK J.    SEE
    2109. GREEN, PAUL E.; HALBERT, MICHAEL H. + ROBINSON, PATRICK J. (1966)

ROBINSON, W. S. 1950.  (4.8/E)                                              4585
    ECOLOGICAL CORRELATIONS AND THE BEHAVIOUR OF INDIVIDUALS.
    AMER. SOCIOL. REV., 15, 351-357.

ROBINSON, W. S. 1962.  (17.7)                                              4586
    ASYMMETRIC CAUSAL MODELS:  COMMENT ON POLK AND BLALOCK.
        (AMENDMENT OF NO. 584, AMENDMENT OF NO. 4324)
    AMER. SOCIOL. REV., 27, 545-548.

ROBOZ, JULIA    SEE
    5997. WOLF, ROBERT L.; MENDLOWITZ, MILTON; ROBOZ, JULIA;   (1966)
    STYAN, GEORGE P. H.; KORNFELD, PETER + WEIGL, ALFRED

ROBSON, D. S. 1957.  (18.5)                                                4587
    APPLICATIONS OF MULTIVARIATE POLYKAYS TO THE THEORY OF UNBIASED
    RATIO-TYPE ESTIMATION.
    J. AMER. STATIST. ASSOC., 52, 511-522.  (MR19,P.1097)

RODGERS, DAVID A. 1957.  (15.3)                                            4588
    A FAST APPROXIMATE ALGEBRAIC FACTOR ROTATION METHOD TO MAXIMIZE
    AGREEMENT BETWEEN LOADINGS AND PREDETERMINED WEIGHTS.
    PSYCHOMETRIKA, 22, 199-205.  (MR19,P.234)

ROEGEN, NICHOLAS GEORGESCU    SEE   GEORGESCU-ROEGEN, NICULAE

ROESER, H. M; 1924.  (4.1)                                                 4589
    NOTE ON THE NATURE OF THE CORRELATION COEFFICIENT.
    AMER. MATH. MONTHLY, 31, 346.

ROFF, MERRILL 1936.  (15.1)                                                4590
    SOME PROPERTIES OF THE COMMUNALITY IN MULTIPLE FACTOR THEORY.
    PSYCHOMETRIKA, 1(2) 1-6.

ROFF, MERRILL 1937.   (15.2)                                                    4591
    THE RELATION BETWEEN RESULTS OBTAINABLE WITH RAW AND CORRECTED
    CORRELATION COEFFICIENTS IN MULTIPLE FACTOR ANALYSIS.
    PSYCHOMETRIKA, 2, 35-39.

ROFF, MERRILL 1940.   (4.5/4.6)                                                 4592
    LINEAR DEPENDENCE IN MULTIPLE CORRELATION WORK.
    PSYCHOMETRIKA, 5, 295-298.

ROGERS, CHARLES A. 1961.   (2.7/20.4)                                           4593
    AN ASYMPTOTIC EXPANSION FOR CERTAIN SCHLAFLI FUNCTIONS.
    J. LONDON MATH. SOC., 36, 78-80.   (MR26-2863)

ROGERS, DAVID J. + FLEMING, H. 1965.   (6.6/20.3/E)                            4594
    A COMPUTER PROGRAM FOR CLASSIFYING PLANTS:  II.   A NUMERICAL
    HANDLING OF NON-NUMERICAL DATA.
    BIOSCIENCE, 14, 15-28.

ROGERS, DAVID J. + TANIMOTO, TAFFEE T. 1960.   (20.3)                          4595
    A COMPUTER PROGRAM FOR CLASSIFYING PLANTS.
    SCIENCE, 132, 1115-1118.

ROGOZIN, B. A.    SEE
    649. BOROVKOV, A. A. + ROGOZIN, B. A. (1965)

    650. BOROVKOV, A. A. + ROGOZIN, B. A. (1965)

ROHDE, CHARLES A. 1965.   (20.1)                                               4596
    GENERALIZED INVERSES OF PARTITIONED MATRICES.
    J. SOC. INDUST. APPL. MATH., 13, 1033-1035.

ROHDE, CHARLES A. 1966.   (20.1)                                               4597
    SOME RESULTS ON GENERALIZED INVERSES.
    SIAM REV., 8, 201-205.

ROHDE, CHARLES A. + HARVEY, J. R. 1965.   (8.8)                                4598
    UNIFIED LEAST SQUARES ANALYSIS.
    J. AMER. STATIST. ASSOC., 60, 523-527.

ROHLF, F. J. 1965.   (6.1/6.6)                                                 4599
    MULTIVARIATE METHODS IN TAXONOMY.
    PROC. IBM SCI. COMPUT. SYMP. STATIST., 3-14.

ROHLF, F. J. + SOKAL, ROBERT R. 1962.   (15.1)                                 4600
    THE DESCRIPTION OF TAXONOMIC RELATIONSHIPS BY FACTOR ANALYSIS.
    SYSTEMATIC ZOOL., 11, 1-16.

ROHLF, F. J.    SEE ALSO
    3741. MATSUDA, K. + ROHLF, F. J. (1961)

ROMANOVSKII, VSEVOLOD I. 1922.   (4.8)                                         4601
    ABOUT THE LINEAR CORRELATION OF TWO QUANTITIES.
         (IN RUSSIAN)
    VESTNIK STATIST., 12, 23-41.

ROMANOVSKII, VSEVOLOD I. 1925.   (18.1)                                        4602
    ON THE MOMENTS OF THE HYPERGEOMETRICAL SERIES.
    BIOMETRIKA, 17, 57-60.

ROMANOVSKII, VSEVOLOD I. 1925.   (4.8/17.2)                                    4603
    GENERALISATION OF AN INEQUALITY OF MARKOFF AND ITS APPLICATION IN THE
    THEORY OF CORRELATION.
         (IN RUSSIAN, WITH SUMMARY IN ENGLISH)
    BJULL. SREDNE-AZIAT. GOS. UNIV., 8, 107-111.

ROMANOVSKII, VSEVOLOD I. 1925.   (4.2)                                         4604
    SUR CERTAINES ESPERANCES MATHEMATIQUES ET SUR L'ERREUR MOYENNE DU
    COEFFICIENT DE CORRELATION.
    C.R. ACAD. SCI. PARIS, 180, 1897-1899.

ROMANOVSKII, VSEVOLOD I. 1925.   (4.2)                                         4605
    ON THE MOMENTS OF STANDARD DEVIATIONS AND OF CORRELATION COEFFICIENT
    IN SAMPLES FROM NORMAL POPULATION.
    METRON, 5(4) 3-46.

ROMANOVSKII, VSEVOLOD I. 1925.   (4.1/4.8)                                        4606
    THE CURRENT STATUS OF THE THEORY OF CORRELATION.
        (IN RUSSIAN)
    VESTNIK STATIST., 20, 1-46.


ROMANOVSKII, VSEVOLOD I. 1926.   (20.4)                                           4607
    AN ABRIDGED DEDUCTION OF THE K. PEARSON'S FORMULA FOR THE MOMENTS OF THE
    HYPERGEOMETRICAL SERIES.
        (IN RUSSIAN, WITH SUMMARY IN ENGLISH)
    BJULL. SREDNE-AZIAT. GOS. UNIV., 12, 127-129.


ROMANOVSKII, VSEVOLOD I. 1928.   (6.5/14.3)                                       4608
    ON THE CRITERIA THAT TWO GIVEN SAMPLES BELONG TO THE SAME NORMAL
    POPULATION (ON THE DIFFERENT COEFFICIENTS OF RACIAL LIKENESS).
    METRON, 7(3) 3-46.


ROMANOVSKII, VSEVOLOD I. 1929.   (17.3/17.9)                                      4609
    ON THE MOMENTS OF MEANS OF FUNCTIONS OF ONE AND MORE RANDOM VARIABLES.
    METRON, 8(1) 251-289.


ROMANOVSKII, VSEVOLOD I. 1932.   (2.1/B)                                          4610
    SUR LA COMPARAISON DES MOYENNES DES VARIABLES ALEATOIRES DEPENDANTES
    POUR LES EPREUVES INDEPENDANTES.
    PROC. INTERNAT. CONG. MATH., 1928(6) 101-102.


ROMANOVSKII, VSEVOLOD I. 1932.   (2.1/17.3/B)                                     4611
    SUR LE CALCUL DES MOMENTS DES MOYENNES DES FONCTIONS DES VARIABLES ALEATOIRES
    POUR LES EPREUVES INDEPENDANTES.
    PROC. INTERNAT. CONG. MATH., 1928(6) 103-105.


ROMANOVSKII, VSEVOLOD I. 1934.   (17.3/B)                                         4612
    ON THE TCHEBYCHEFF'S INEQUALITY FOR THE TWO-DIMENSIONAL CASE.
    TRUDY SRED. AZIAT. GOS. UNIV. SER. 5A MAT., 11-13.


ROMANOVSKII, VSEVOLOD I. 1935.   (20.5)                                           4613
    SUI MOMENTI DELLA DISTRIBUZIONE IPERGEOMETRICA.
    GIORN. IST. ITAL. ATTUARI, 6, 170-177.


ROMANOVSKII, VSEVOLOD I. 1945.   (17.1)                                           4614
    ON A RANDOM SERIES OF CORRELATED EVENTS.
        (IN RUSSIAN)
    BJULL. SREDNE-AZIAT. GOS. UNIV., 23, 17-25.


ROMANOVSKII, VSEVOLOD I. 1953.   (20.1)                                           4615
    ON APPLICATIONS OF INFINITE MATRICES TO THE THEORY OF PROBABILITY.
        (IN RUSSIAN, WITH SUMMARY IN UZBEKISH)
    DOKL. AKAD. NAUK UZBEK. SSR, 9, 3-6.   (MR17,P.980)


ROMANOFSKY, V. I.   SEE   ROMANOVSKII, VSEVOLOD I.


ROOS, CHARLES F. 1936.   (16.5/U)                                                 4616
    ANNUAL SURVEY OF STATISTICAL TECHNIQUES.  II.  THE CORRELATION AND
    ANALYSIS OF TIME SERIES.
    ECONOMETRICA, 4, 368-381.


ROOS, CHARLES F. 1937.   (17.7/E)                                                 4617
    A GENERAL INVARIANT CRITERION OF FIT FOR LINES AND PLANES, WHERE ALL
    VARIATES ARE SUBJECT TO ERROR.
    METRON, 13(1) 3-20.


ROOT, WILLIAM L. 1966.   (1.2)                                                    4618
    BIBLIOGRAPHY OF NORBERT WIENER.
    BULL. AMER. MATH. SOC., 72, 135-145.


ROPPERT, JOSEF 1959.   (14.5/20.1/20.2)                                           4619
    EINIGE ANWENDUNGEN DES MATRIZENKALKULS BEI DER UNTERSUCHUNG VON STATISTISCHEN
    VERTEILUNGEN.
    ANWEND. MATRIZ. WIRTSCH. STATIST. PROBLEME (ADAM ET AL.), 216-234.


ROPPERT, JOSEF 1964.   (20.3)                                                     4620
    EIN SCHNELL KONVERGIERENDES ITERATIONSVERFAHREN ZUR MATRIXINVERSION.
    METRIKA, 8, 152-154.

ROPPERT, JOSEF    SEE ALSO
   1680. FISCHER, G. + ROPPERT, JOSEF (1966)


ROSANDER, A. C. 1936.  (4.8)                                            4621
      A NOTE ON CORRELATION BY RANKS.
   J. EDUC. PSYCHOL., 27, 145-148.


ROSE, M. J. 1964.  (6.4/20.2)                                          4622
      CLASSIFICATION OF A SET OF ELEMENTS.
   COMPUT. J., 7, 208-211.


 ROSE, NICHOLAS J. 1965.  (20.2)                                       4623
      ON THE EIGENVALUES OF A MATRIX WHICH COMMUTES WITH ITS DERIVATIVE.
   PROC. AMER. MATH. SOC., 16, 752-754.


ROSEN, J. B. 1964.  (20.3)                                             4624
      MINIMUM AND BASIC SOLUTIONS TO SINGULAR LINEAR SYSTEMS.
   J. SOC. INDUST. APPL. MATH., 12, 156-162.


ROSENBAUM, S. 1961.  (2.4/B)                                           4625
      MOMENTS OF A TRUNCATED BIVARIATE NORMAL DISTRIBUTION.
   J. ROY. STATIST. SOC. SER. B, 23, 405-408.   (MR24-A2470)


ROSENBERG, LLOYD 1965.  (2.2/17.4/B)                                   4626
      NONNORMALITY OF LINEAR COMBINATIONS OF NORMALLY DISTRIBUTED RANDOM VARIABLES.
   AMER. MATH. MONTHLY, 72, 888-890.  (MR32-1742)


ROSENBERG, LLOYD 1966.  (16.5)                                         4627
      THE ITERATION OF MEANS.
   MATH. MAG., 39, 58-62.  (MR33-1950)


ROSENBERG, MILTON 1964.  (20.4)                                        4628
      THE SQUARE-INTEGRABILITY OF MATRIX-VALUED FUNCTIONS WITH RESPECT TO A
      NON-NEGATIVE HERMITIAN MEASURE.
   DUKE MATH. J., 31, 291-298.


ROSENBERG, S.    SEE
   4645. ROUANET, H. + ROSENBERG, S. (1964)


ROSENBLATT, HARRY M.    SEE
   3203. KULLBACK, SOLOMON + ROSENBLATT, HARRY M. (1957)


ROSENBLATT, JUDAH I. 1962.  (17.8/A)                                   4629
      NOTE ON MULTIVARIATE GOODNESS-OF-FIT TESTS.
   ANN. MATH. STATIST., 33, 807-810.  (MR25-691)


ROSENBLATT, JUDAH I.    SEE ALSO
   596. BLUM, JULIUS R. + ROSENBLATT, JUDAH I. (1966)


ROSENBLATT, MURRAY 1952.  (20.3)                                       4630
      REMARKS ON A MULTIVARIATE TRANSFORMATION.
   ANN. MATH. STATIST., 23, 470-472.  (MR14,P.189)


ROSENBLATT, MURRAY 1956.  (8.1/16.5/B)                                 4631
      ON THE ESTIMATION OF REGRESSION COEFFICIENTS OF A VECTOR-VALUED TIME
      SERIES WITH A STATIONARY RESIDUAL.
   ANN. MATH. STATIST., 27, 99-121.  (MR17,P.871)


ROSENBLATT, MURRAY 1956.  (16.5)                                       4632
      SOME REGRESSION PROBLEMS IN TIME SERIES ANALYSIS.
   PROC. THIRD BERKELEY SYMP. MATH. STATIST. PROB., 1, 165-186.   (MR18,P.948)


ROSENBLATT, MURRAY 1957.  (16.4/16.5/E)                                4633
      THE MULTIDIMENSIONAL PREDICTION PROBLEM.
   PROC. NAT. ACAD. SCI. U.S.A., 43, 989-992.  (MR20-4322)


ROSENBLATT, MURRAY 1958.  (16.4)                                       4634
      A MULTI-DIMENSIONAL PREDICTION PROBLEM.
   ARK. MAT., 3, 407-424.  (MR19,P.1098)


ROSENBLATT, MURRAY 1959.  (16.3/16.4/16.5)                             4635
      STATISTICAL ANALYSIS OF STOCHASTIC PROCESSES WITH STATIONARY RESIDUALS.
   PROB. STATIST. (CRAMER VOLUME), 246-275.

ROSENBLATT, MURRAY 1960.   (16.4/16.5/E)                                    4636
    THE MULTIDIMENSIONAL PREDICTION PROBLEM.
    STATIST. METH. RADIO WAVE PROP., 99–111.

ROSENBLATT, MURRAY 1966.   (16.3)                                          4637
    REMARKS ON HIGHER ORDER SPECTRA.
    MULTIVARIATE ANAL. PROC. INTERNAT. SYMP. DAYTON (KRISHNAIAH), 383–389.   (MR35-2446)

ROSENBLATT, MURRAY   SEE ALSO
    2125. GRENANDER, ULF + ROSENBLATT, MURRAY (1952)

    2126. GRENANDER, ULF + ROSENBLATT, MURRAY (1953)

    2127. GRENANDER, ULF + ROSENBLATT, MURRAY (1953)

    2128. GRENANDER, ULF + ROSENBLATT, MURRAY (1956)

    1799. FREIBERGER, WALTER F.; ROSENBLATT, MURRAY + VAN NESS, J. (1962)

    598. BLUM, JULIUS R.; KIEFER, J. + ROSENBLATT, MURRAY (1961)

ROSENSTIEHL, PIERRE 1958.   (17.3)                                          4638
    SUR L'INTEGRATION DE CERTAINES FONCTIONS CARACTERISTIQUES SIMPLES OU COMPOSEES.
    C.R. ACAD. SCI. PARIS. 246, 2213–2215.   (MR20-384)

ROSENTHALL, IRENE 1966.   (4.8/17.8/E)                                      4639
    DISTRIBUTION OF THE SAMPLE VERSION OF THE MEASURE OF ASSOCIATION, GAMMA.
    J. AMER. STATIST. ASSOC., 61, 440–453.

ROSNER, BENJAMIN   SEE
    5122. SOLOMON, HERBERT + ROSNER, BENJAMIN (1954)

ROSNER, BURTON S. 1948.   (15.3)                                           4640
    AN ALGEBRAIC SOLUTION FOR THE COMMUNALITIES.
    PSYCHOMETRIKA, 13, 181–184.

ROSS, C. W. CLUNIES-  SEE  CLUNIES-ROSS, CHARLES W.

ROSS, JOHN 1962.   (17.9)                                                  4641
    INFORMATIONAL COVERAGE AND CORRELATIONAL ANALYSIS.
    PSYCHOMETRIKA, 27, 297–306.   (MR26-2334)

ROSS, JOHN + WEITZMAN, R. A. 1964.   (4.3/4.4/15.4/T)                      4642
    THE TWENTY-SEVEN PER CENT RULE.
    ANN. MATH. STATIST., 35, 214–221.   (MR28-3502)

ROTHENBERG, T. J. + LEENDERS, C. T. 1964.   (16.2)                         4643
    EFFICIENT ESTIMATION OF SIMULTANEOUS EQUATION SYSTEMS.
    ECONOMETRICA, 32, 57–76.

ROTHER, KARL 1950.   (4.8)                                                 4644
    KONNUPTIALINDEX UND KORRELATIONSKOEFFIZIENT.
    MITT. MATH. STATIST., 2, 184–189.

ROUANET, H. + ROSENBERG, S. 1964.   (16.4)                                4645
    STOCHASTIC MODELS FOR THE RESPONSE CONTINUUM IN A DETERMINATE SITUATION:
    COMPARISONS AND EXTENSIONS.
    J. MATH. PSYCHOL., 1, 215–232.

ROUQUET LA GARRIGUE, V.  SEE  LA GARRIGUE, V. ROUQUET

ROUVIER, R. 1966.   (4.8/11.1/E)                                           4646
    L'ANALYSE EN COMPOSANTES PRINCIPALES:  SON UTILISATION EN GENETIQUE ET
    SES RAPPORTS AVEC L'ANALYSE DISCRIMINATOIRE.
    BIOMETRICS, 22, 343–357.

ROY, A. R. + MALLIK, A. K. 1966.   (19.3)                                  4647
    SOME USES OF BHATTACHARYA BOUNDS IN FINDING ADMISSIBLE ESTIMATES.
    CALCUTTA STATIST. ASSOC. BULL., 15, 1–13.

ROY, A. R. + SRIVASTAVA, V. K. 1965.   (16.5)                              4648
    ON THE METHOD OF MIXED ESTIMATION WITH AUTOCORRELATED DISTURBANCES.
    J. INDIAN STATIST. ASSOC., 3, 107–115.

ROY, AMAL KUMAR    SEE
    651. BOSE, DEB KUMAR + ROY, AMAL KUMAR (1957)

ROY, J. 1951.   (10.1/A)                                                          4649
    THE DISTRIBUTION OF CERTAIN LIKELIHOOD CRITERIA USEFUL IN MULTIVARIATE ANALYSIS.
    BULL. INST. INTERNAT. STATIST., 33(2) 219-230.   (MR16,P.841)

ROY, J. 1954.   (10.2)                                                           4650
    ON SOME TESTS OF SIGNIFICANCE IN SAMPLES FROM BIPOLAR NORMAL DISTRIBUTIONS.
    SANKHYA, 14, 203-210.   (MR19,P.896)

ROY, J. 1956.   (14.3)                                                           4651
    ON SOME QUICK DECISION METHODS IN MULTIVARIATE AND UNIVARIATE ANALYSIS.
    SANKHYA, 17, 77-88.   (MR18,P.521)

ROY, J. 1958.   (8.5/10.1/C)                                                     4652
    STEP-DOWN PROCEDURE IN MULTIVARIATE ANALYSIS.
    ANN. MATH. STATIST., 29, 1177-1187.   (MR20-7363)

ROY, J. 1960.   (9.1/9.2/10.2/A)                                                 4653
    TESTS FOR INDEPENDENCE AND SYMMETRY IN MULTIVARIATE NORMAL POPULATIONS.
    SANKHYA, 22, 267-278.   (MR24-A2483)

ROY, J. 1960.   (8.4)                                                            4654
    NON-NULL DISTRIBUTION OF THE LIKELIHOOD-RATIO IN ANALYSIS OF DISPERSION.
    SANKHYA, 22, 289-294.   (MR23-A3610)

ROY, J. 1966.   (8.4/NT)                                                         4655
    POWER OF THE LIKELIHOOD-RATIO TEST USED IN ANALYSIS OF DISPERSION.
    MULTIVARIATE ANAL. PROC. INTERNAT. SYMP. DAYTON (KRISHNAIAH), 105-127.   (MR35-1143)

ROY, J. + KALYANASUNDARAM, G. 1964.   (17.8/20.3/B)                              4656
    PUNCHED CARD PROCESSING OF SAMPLE SURVEY DATA FOR FRACTILE GRAPHICAL ANALYSIS.
    CONTRIB. STATIST. (MAHALANOBIS VOLUME), 411-418.

ROY, J. + MITRA, SUJIT KUMAR 1954.   (6.7)                                       4657
    A METHOD OF SELECTION FOR IMPROVEMENT.
    CALCUTTA STATIST. ASSOC. BULL., 5, 82-86.

ROY, J. + MURTHY, V. KRISHNA 1960.   (10.2/T)                                    4658
    PERCENTAGE POINTS OF WILKS' $L_{mvc}$ AND $L_{vc}$ CRITERIA.
    PSYCHOMETRIKA, 25, 243-250.   (MR24-A605)

ROY, J.   SEE ALSO
    4699. ROY, S.N. + ROY, J. (1956)

    4700. ROY, S.N. + ROY, J. (1959)

ROY, S.N. 1938.   (7.3)                                                          4659
    A GEOMETRICAL NOTE ON THE USE OF RECTANGULAR COORDINATES IN THE THEORY OF
    SAMPLING DISTRIBUTIONS CONNECTED WITH A MULTIVARIATE NORMAL POPULATION.
    SANKHYA, 3, 273-284.

ROY, S.N. 1939.   (5.3)                                                          4660
    A NOTE ON THE DISTRIBUTION OF THE STUDENTIZED $D^2$-STATISTIC.
    SANKHYA, 4, 373-380.

ROY, S.N. 1939.   (8.3/8.4/13.1)                                                 4661
    P-STATISTICS OR SOME GENERALISATIONS IN ANALYSIS OF VARIANCE
    APPROPRIATE TO MULTIVARIATE PROBLEMS.
    SANKHYA, 4, 381-396.

ROY, S.N. 1940.   (17.5)                                                         4662
    DISTRIBUTION OF P-STATISTICS ON THE NON-NULL HYPOTHESIS.
    SCI. AND CULT., 5, 562-563.

ROY, S.N. 1940.   (17.5)                                                         4663
    ON THE DISTRIBUTION OF CERTAIN SYMMETRIC FUNCTIONS OF P-STATISTICS ON THE NULL HYPOTHESIS.
    SCI. AND CULT., 5, 563.

ROY, S.N. 1942.  (8.4/13.2)                                                    4664
    THE SAMPLING DISTRIBUTION OF P-STATISTICS AND CERTAIN ALLIED
    STATISTICS ON THE NON-NULL HYPOTHESIS.
    SANKHYA, 6, 15-34.  (MR4,P.106)

ROY, S.N. 1942.  (8.4/13.1/13.2)                                              4665
    ANALYSIS OF VARIANCE FOR MULTIVARIATE NORMAL POPULATIONS:  THE SAMPLING
    DISTRIBUTION OF THE REQUISITE P-STATISTICS ON THE NULL AND NON-NULL HYPOTHESES.
    SANKHYA, 6, 35-50.  (MR4,P.106)

ROY, S.N. 1945.  (20.4)                                                       4666
    ON CERTAIN CLASS OF MULTIPLE INTEGRALS.
    BULL. CALCUTTA MATH. SOC., 37, 69-77.  (MR8,P.16)

ROY, S.N. 1945.  (8.4/13.1)                                                   4667
    THE INDIVIDUAL SAMPLING DISTRIBUTION OF THE MAXIMUM, THE MINIMUM AND
    ANY INTERMEDIATE OF THE P-STATISTICS ON THE NULL HYPOTHESIS.
    SANKHYA, 7, 133-158.  (MR7,P.317)

ROY, S.N. 1946.  (8.4/13.2)                                                   4668
    MULTIVARIATE ANALYSIS OF VARIANCE:  THE SAMPLING DISTRIBUTION OF THE
    NUMERICALLY LARGEST OF THE P-STATISTICS ON THE NON-NULL HYPOTHESES.
    SANKHYA, 8, 15-52.  (MR8,P.475)

ROY, S.N. 1946.  (8.3)                                                        4669
    A NOTE ON MULTIVARIATE ANALYSIS OF VARIANCE WHEN THE NUMBER OF VARIATES IS
    GREATER THAN THE NUMBER OF LINEAR HYPOTHESES PER CHARACTER.
    SANKHYA, 8, 53-66.  (MR8,P.475)

ROY, S.N. 1947.  (12.1/20.5)                                                  4670
    A NOTE ON CRITICAL ANGLES BETWEEN TWO FLATS IN HYPERSPACE WITH
    CERTAIN STATISTICAL APPLICATIONS.
    SANKHYA, 8, 177-194.  (MR10,P.134)

ROY, S.N. 1950.  (19.2)                                                       4671
    UNIVARIATE AND MULTIVARIATE ANALYSIS AS PROBLEMS IN TESTING OF
    COMPOSITE HYPOTHESES.
    SANKHYA, 10, 29-80.  (MR12,P.37)

ROY, S.N. 1952.  (20.1)                                                       4672
    SOME USEFUL RESULTS IN JACOBIANS.
    CALCUTTA STATIST. ASSOC. BULL., 4, 117-122.  (MR14,P.959)

ROY, S.N. 1952.  (8.5/10.1)                                                   4673
    ON SOME ASPECTS OF STATISTICAL INFERENCE.
    PROC. INTERNAT. CONG. MATH., 1950(1) 555-564.  (MR13,P.366)

ROY, S.N. 1953.  (8.3/10.1/19.2)                                              4674
    ON A HEURISTIC METHOD OF TEST CONSTRUCTION AND ITS USE IN MULTIVARIATE ANALYSIS.
    ANN. MATH. STATIST., 24, 220-238.  (MR15,P.241)

ROY, S.N. 1954.  (8.3/9.1/10.1/C)                                             4675
    SOME FURTHER RESULTS IN SIMULTANEOUS CONFIDENCE INTERVAL ESTIMATION.
        (AMENDED BY NO. 4677)
    ANN. MATH. STATIST., 25, 752-761.  (MR16,P.382)

ROY, S.N. 1954.  (20.1)                                                       4676
    A USEFUL THEOREM IN MATRIX THEORY.
    PROC. AMER. MATH. SOC., 5, 635-638.  (MR16,P.4)

ROY, S.N. 1956.  (8.3/9.1/10.1/C)                                             4677
    A NOTE ON "SOME FURTHER RESULTS IN SIMULTANEOUS CONFIDENCE INTERVAL ESTIMATION".
        (AMENDMENT OF NO. 4675)
    ANN. MATH. STATIST., 27, 856-858.  (MR18,P.772)

ROY, S.N. 1959.  (10.1/C)                                                     4678
    A NOTE ON CONFIDENCE BOUNDS CONNECTED WITH THE HYPOTHESIS OF EQUALITY
    OF TWO DISPERSION MATRICES.
    CALCUTTA MATH. SOC. GOLDEN JUBILEE COMMEM. VOLUME, 2, 329-332.  (MR27-6336)

ROY, S.N. 1961.  (1.1)                                                        4679
    ON THE PLANNING AND INTERPRETATION OF MULTIFACTOR MULTIRESPONSE EXPERIMENTS.
    BULL. INST. INTERNAT. STATIST., 38(4) 59-72.  (MR26-7108)

ROY, S.N. 1962.   (8.4/9.1/10.1/C)                                                              4680
    A SURVEY OF SOME RECENT RESULTS IN NORMAL MULTIVARIATE CONFIDENCE BOUNDS.
    BULL. INST. INTERNAT. STATIST., 39(2) 405-422.   (MR28-4631)

ROY, S.N. + BARGMANN, ROLF E. 1958.   (9.1/9.2/C)                                               4681
    TESTS OF MULTIPLE INDEPENDENCE AND THE ASSOCIATED CONFIDENCE BOUNDS.
    ANN. MATH. STATIST., 29, 491-503.   (MR20-382)

ROY, S.N. + BHAPKAR, V. P. 1960.   (17.8)                                                       4682
    SOME NON-PARAMETRIC ANALOGS OF "NORMAL" ANOVA, MANOVA, AND OF STUDIES
    IN "NORMAL" ASSOCIATION.
    CONTRIB. PROB. STATIST. (HOTELLING VOLUME), 371-387.   (MR22-11464)

ROY, S.N. + BOSE, PURNENDU KUMAR 1939.   (4.7)                                                  4683
    THE DISTRIBUTION OF THE ROOT-MEAN SQUARE OF THE MULTIPLE CORRELATION
    CO-EFFICIENT ON THE MULTIVARIATE NORMAL HYPOTHESIS.
    SCI. AND CULT., 5, 714-715.

ROY, S.N. + BOSE, PURNENDU KUMAR 1940.   (6.5/N)                                                4684
    ON THE REDUCTION FORMULAE FOR THE INCOMPLETE PROBABILITY INTEGRAL OF
    THE STUDENTIZED $D^2$-STATISTIC.
    SCI. AND CULT., 5, 773-775.

ROY, S.N. + BOSE, PURNENDU KUMAR 1940.   (4.7)                                                  4685
    THE DISTRIBUTION OF THE ROOT-MEAN-SQUARE OF THE SECOND TYPE OF THE
    MULTIPLE CORRELATION CO-EFFICIENT.
    SCI. AND CULT., 6, 59.   (MR5,P.126)

ROY, S.N. + BOSE, R. C. 1940.   (5.3)                                                           4686
    THE USE AND DISTRIBUTION OF THE STUDENTIZED $D^2$-STATISTIC WHEN THE
    VARIANCES AND COVARIANCES ARE BASED ON k SAMPLES.
    SANKHYA, 4, 535-542.   (MR4,P.105)

ROY, S.N. + BOSE, R. C. 1953.   (8.3/9.1/10.1/C)                                                4687
    SIMULTANEOUS CONFIDENCE INTERVAL ESTIMATION.
    ANN. MATH. STATIST., 24, 513-536.   (MR15,P.726)

ROY, S.N. + COBB, WHITFIELD 1960.   (8.6)                                                       4688
    MIXED MODEL VARIANCE ANALYSIS WITH NORMAL ERROR AND POSSIBLY
    NON-NORMAL OTHER RANDOM EFFECTS. II. THE MULTIVARIATE CASE.
    ANN. MATH. STATIST., 31, 958-968.   (MR25-4610B)

ROY, S.N. + GNANADESIKAN, R. 1957.   (10.1/C)                                                   4689
    FURTHER CONTRIBUTIONS TO MULTIVARIATE CONFIDENCE BOUNDS.
        (AMENDED BY NO. 4693)
    BIOMETRIKA, 44, 399-410.   (MR19,P.895)

ROY, S.N. + GNANADESIKAN, R. 1958.   (10.1/C)                                                   4690
    A NOTE ON "FURTHER CONTRIBUTIONS TO MULTIVARIATE CONFIDENCE BOUNDS".
    BIOMETRIKA, 45, 581.

ROY, S.N. + GNANADESIKAN, R. 1959.   (8.3/8.5/8.6/U)                                            4691
    SOME CONTRIBUTIONS TO ANOVA IN ONE OR MORE DIMENSIONS.  I.
    ANN. MATH. STATIST., 30, 304-317.   (MR21-6664)

ROY, S.N. + GNANADESIKAN, R. 1959.   (8.3/8.5/8.6/C)                                            4692
    SOME CONTRIBUTIONS TO ANOVA IN ONE OR MORE DIMENSIONS.  II.
    ANN. MATH. STATIST., 30, 318-340.   (MR21-6665)

ROY, S.N. + GNANADESIKAN, R. 1961.   (10.1/C)                                                   4693
    CORRIGENDA:  FURTHER CONTRIBUTIONS TO MULTIVARIATE CONFIDENCE BOUNDS.
        (AMENDMENT OF NO. 4689)
    BIOMETRIKA, 48, 474.   (MR19,P.895)

ROY, S.N. + GNANADESIKAN, R. 1962.   (10.1)                                                     4694
    TWO-SAMPLE COMPARISONS OF DISPERSION MATRICES FOR ALTERNATIVES OF
    INTERMEDIATE SPECIFICITY.
    ANN. MATH. STATIST., 33, 432-437.   (MR24-A3743)

ROY, S.N. + KASTENBAUM, MARVIN A. 1956.   (17.8)                                                4695
    ON THE HYPOTHESIS OF NO "INTERACTION" IN A MULTI-WAY CONTINGENCY TABLE.
    ANN. MATH. STATIST., 27, 749-757.   (MR18,P.160)

ROY, S.N. + MIKHAIL, WADIE F. 1961.  (8.3/10.1)                                    4696
    ON THE MONOTONIC CHARACTER OF THE POWER FUNCTIONS OF TWO MULTIVARIATE TESTS.
    ANN. MATH. STATIST., 32, 1145-1151.  (MR24-A604)


ROY, S.N. + MITRA, SUJIT KUMAR 1956.  (17.8)                                       4697
    AN INTRODUCTION TO SOME NONPARAMETRIC GENERALIZATIONS OF ANALYSIS OF
    VARIANCE AND MULTIVARIATE ANALYSIS.
    BIOMETRIKA, 43, 361-376.  (MR18,P.522)


ROY, S.N. + POTTHOFF, RICHARD F. 1958.  (10.2/BC)                                  4698
    CONFIDENCE BOUNDS ON VECTOR ANALOGUES OF THE "RATIO OF MEANS" AND THE "RATIO OF
    VARIANCES" FOR TWO CORRELATED NORMAL VARIATES AND SOME ASSOCIATED TESTS.
    ANN. MATH. STATIST., 29, 829-841.  (MR20-7364)


ROY, S.N. + ROY, J. 1956.  (8.3)                                                   4699
    A NOTE ON TESTABILITY IN NORMAL MODEL I ANOVA AND MANOVA.
    CALCUTTA MATH. SOC. GOLDEN JUBILEE COMMEM. VOLUME, 2, 333-339.  (MR28-1700)


ROY, S.N. + ROY, J. 1959.  (8.3)                                                   4700
    A NOTE ON A CLASS OF PROBLEMS IN "NORMAL" MULTIVARIATE ANALYSIS OF VARIANCE.
    ANN. MATH. STATIST., 30, 577-581.  (MR22-1035)


ROY, S.N. + SARHAN, AHMED E. 1956.  (20.2/20.3)                                    4701
    ON INVERTING A CLASS OF PATTERNED MATRICES.
    BIOMETRIKA, 43, 227-231.  (MR17,P.936)


ROY, S.N. + SRIVASTAVA, J. N. 1964.  (8.3/8.5)                                     4702
    HIERARCHICAL AND P-BLOCK MULTIRESPONSE DESIGNS AND THEIR ANALYSIS.
    CONTRIB. STATIST. (MAHALANOBIS VOLUME), 419-428.  (MR35-6287)


ROY, S.N.; GREENBERG, BERNARD G. + SARHAN, AHMED E. 1960.  (20.2/20.3)             4703
    EVALUATION OF DETERMINANTS, CHARACTERISTIC EQUATIONS AND THEIR ROOTS
    FOR A CLASS OF PATTERNED MATRICES.
    J. ROY. STATIST. SOC. SER. B, 22, 348-359.  (MR22-6071)


ROY, S.N.    SEE ALSO
    672. BOSE, R. C. + ROY, S.N. (1935)

    673. BOSE, R. C. + ROY, S.N. (1937)

    674. BOSE, R. C. + ROY, S.N. (1938)

    4074. OLKIN, INGRAM + ROY, S.N. (1954)

    4344. POTTHOFF, RICHARD F. + ROY, S.N. (1964)

    3616. MAHALANOBIS, P.C.; BOSE, R. C. + ROY, S.N. (1937)


ROYDEN, H.L.    SEE
2009. GIRSHICK, M. A.; KARLIN, SAMUEL + ROYDEN, H.L. (1957)


ROYER, ELMER B. 1941.  (4.4)                                                       4704
    A MACHINE METHOD FOR COMPUTING THE BISERIAL CORRELATION COEFFICIENT
    IN ITEM VALIDATION.
    PSYCHOMETRIKA, 6, 55-59.


ROZANOV, JU. A. 1958.  (16.3/16.4)                                                 4705
    SPECTRAL THEORY OF N-DIMENSIONAL STATIONARY STOCHASTIC PROCESSES WITH
    DISCRETE TIME.
        (IN RUSSIAN, TRANSLATED AS NO. 4711)
    USPEHI MAT. NAUK, 80, 93-142.  (MR22-5076)


ROZANOV, JU. A. 1959.  (16.4)                                                      4706
    LINEAR EXTRAPOLATION OF MULTI-DIMENSIONAL STATIONARY PROCESSES OF THE
    FIRST RANK WITH DISCRETE TIME.
        (IN RUSSIAN)
    DOKL. AKAD. NAUK SSSR, 125, 277-280.  (MR25-1580)


ROZANOV, JU. A. 1959.  (16.4)                                                      4707
    ON THE EXTRAPOLATION OF GENERALIZED STATIONARY RANDOM PROCESSES.
        (IN RUSSIAN, WITH SUMMARY IN ENGLISH, TRANSLATED AS NO. 4708)
    TEOR. VEROJATNOST. I PRIMENEN., 4, 465-471.  (MR22-6018)

**4708**
ROZANOV, JU. A. 1959.  (16.4)
    ON THE EXTRAPOLATION OF GENERALIZED STATIONARY RANDOM PROCESSES.
        (TRANSLATION OF NO. 4707)
    THEORY PROB. APPL., 4, 426–431.  (MR22-6018)

**4709**
ROZANOV, JU. A. 1960.  (16.4)
    SPECTRAL PROPERTIES OF MULTIVARIATE STATIONARY PROCESSES AND BOUNDARY
    PROPERTIES OF ANALYTIC MATRICES.
        (IN RUSSIAN, WITH SUMMARY IN ENGLISH, TRANSLATED AS NO. 4710)
    TEOR. VEROJATNOST. I PRIMENEN., 5, 399–414.  (MR24-A2432)

**4710**
ROZANOV, JU. A. 1960.  (16.4)
    SPECTRAL PROPERTIES OF MULTIVARIATE STATIONARY PROCESSES AND BOUNDARY
    PROPERTIES OF ANALYTIC MATRICES.
        (TRANSLATION OF NO. 4709)
    THEORY PROB. APPL., 5, 362–376.  (MR24-A2432)

**4711**
ROZANOV, JU. A. 1961.  (16.3/16.4)
    SPECTRAL THEORY OF MULTI-DIMENSIONAL STATIONARY RANDOM PROCESSES WITH
    DISCRETE TIME.
        (TRANSLATION OF NO. 4705)
    SELECT. TRANSL. MATH. STATIST. PROB., 1, 253–306.  (MR22-7168)

**4712**
ROZEBOOM, WILLIAM W. 1965.  (12.1/E)
    LINEAR CORRELATIONS BETWEEN SETS OF VARIABLES.
    PSYCHOMETRIKA, 30, 57–71.  (MR30-5390)

ROZONOER, L. I.    SEE
    68. AIZERMAN, M. A.; BRAVERMAN, E. M. + ROZONOER, L. I. (1964)

    69. AIZERMAN, M. A.; BRAVERMAN, E. M. + ROZONOER, L. I. (1964)

**4713**
RUBEN, HAROLD 1954.  (14.1/U)
    ON THE MOMENTS OF ORDER STATISTICS IN SAMPLES FROM NORMAL POPULATIONS.
        (AMENDED BY NO. 4714)
    BIOMETRIKA, 41, 200–227.  (MR16,P.153)

**4714**
RUBEN, HAROLD 1954.  (14.1/U)
    CORRIGENDA.
        (AMENDMENT OF NO. 4713)
    BIOMETRIKA, 41, 268A(568).  (MR16,P.153)

**4715**
RUBEN, HAROLD 1956.  (14.1/U)
    ON THE SUM OF SQUARES OF NORMAL SCORES.
    BIOMETRIKA, 43, 456–458.  (MR18,P.426)

**4716**
RUBEN, HAROLD 1956.  (14.1)
    ON THE MOMENTS OF THE RANGE AND PRODUCT MOMENTS OF EXTREME ORDER STATISTICS
    IN NORMAL SAMPLES.
    BIOMETRIKA, 43, 458–460.  (MR18,P.607)

**4717**
RUBEN, HAROLD 1960.  (17.6/20.5)
    ON THE GEOMETRICAL MOMENTS OF SKEW-REGULAR SIMPLICES IN HYPERSPHERICAL SPACE,
    WITH SOME APPLICATIONS IN GEOMETRY AND MATHEMATICAL STATISTICS.
    ACTA MATH., 103, 1–23.  (MR22-A12447)

**4718**
RUBEN, HAROLD 1960.  (2.7)
    PROBABILITY CONTENT OF REGIONS UNDER SPHERICAL NORMAL DISTRIBUTIONS.I.
    ANN. MATH. STATIST., 31, 598–618.  (MR22-8602)

**4719**
RUBEN, HAROLD 1960.  (2.7)
    PROBABILITY CONTENT OF REGIONS UNDER SPHERICAL NORMAL
    DISTRIBUTIONS.II. THE DISTRIBUTION OF RANGE IN NORMAL SAMPLES.
        (AMENDED BY NO. 4722)
    ANN. MATH. STATIST., 31, 1113–1121.  (MR22-8603)

**4720**
RUBEN, HAROLD 1960.  (14.1/U)
    ON THE GEOMETRICAL SIGNIFICANCE OF THE MOMENTS OF ORDER STATISTICS AND OF
    DEVIATIONS OF ORDER STATISTICS FROM THE MEAN IN SAMPLES FROM GAUSSIAN
    POPULATIONS.
    J. MATH. AND MECH., 9, 631–638.  (MR22-10073)

RUBEN, HAROLD 1961.  (2.7/B)                                                          4721
    PROBABILITY CONTENT OF REGIONS UNDER SPHERICAL NORMAL
    DISTRIBUTIONS.III:   THE BIVARIATE NORMAL INTEGRAL.
    ANN. MATH. STATIST., 32, 171-186.  (MR22-8604)

RUBEN, HAROLD 1961.  (2.7)                                                            4722
    CORRECTION TO: PROBABILITY CONTENT OF REGIONS UNDER SPHERICAL NORMAL
    DISTRIBUTIONS.  II. THE DISTRIBUTION OF THE RANGE IN NORMAL SAMPLES.
        (AMENDMENT OF NO. 4719)
    ANN. MATH. STATIST., 32, 620.  (MR23-A2969)

RUBEN, HAROLD 1961.  (2.7/A)                                                          4723
    AN ASYMPTOTIC EXPANSION FOR A CLASS OF MULTIVARIATE NORMAL INTEGRALS.
    J. AUSTRAL. MATH. SOC., 2, 253-264.  (MR25-4607)

RUBEN, HAROLD 1961.  (2.7/20.4)                                                       4724
    A POWER SERIES EXPANSION FOR A CLASS OF SCHLAFLI FUNCTIONS.
    J. LONDON MATH. SOC., 36, 69-77.  (MR26-1510)

RUBEN, HAROLD 1961.  (2.7)                                                            4725
    ON THE NUMERICAL EVALUATION OF A CLASS OF MULTIVARIATE NORMAL INTEGRALS.
    PROC. ROY. SOC. EDINBURGH SECT. A, 65, 272-281.  (MR23-B3133)

RUBEN, HAROLD 1961.  (20.4)                                                           4726
    A MULTIDIMENSIONAL GENERALIZATION OF THE INVERSE SINE FUNCTION.
    QUART. J. MATH. OXFORD SECOND SER., 12, 257-264.  (MR25-2170)

RUBEN, HAROLD 1962.  (16.4)                                                           4727
    SOME ASPECTS OF THE EMIGRATION-IMMIGRATION PROCESS.
    ANN. MATH. STATIST., 33, 119-129.  (MR25-1577)

RUBEN, HAROLD 1962.  (2.5/2.7/14.5)                                                   4728
    PROBABILITY CONTENT OF REGIONS UNDER SPHERICAL NORMAL DISTRIBUTIONS, IV:   THE
    DISTRIBUTION OF HOMOGENEOUS AND NON-HOMOGENEOUS QUADRATIC FUNCTIONS OF NORMAL
    VARIABLES.
    ANN. MATH. STATIST., 33, 542-570.  (MR25-645)

RUBEN, HAROLD 1962.  (14.1/NT)                                                        4729
    THE MOMENTS OF THE ORDER STATISTICS AND OF THE RANGE IN SAMPLES FROM
    NORMAL POPULATIONS.
    CONTRIB. ORDER STATIST. (SARHAN + GREENBERG), 165-190.

RUBEN, HAROLD 1963.  (16.5)                                                          4730
    THE ESTIMATION OF A FUNDAMENTAL INTERACTION PARAMETER IN AN
    EMIGRATION-IMMIGRATION PROCESS.
    ANN. MATH. STATIST., 34, 238-259.  (MR26-1959)

RUBEN, HAROLD 1963.  (2.5/14.5)                                                      4731
    A NEW RESULT ON THE DISTRIBUTION OF QUADRATIC FORMS.
    ANN. MATH. STATIST., 34, 1582-1584.  (MR27-5309)

RUBEN, HAROLD 1964.  (16.4/E)                                                        4732
    GENERALISED CONCENTRATION FLUCTUATIONS UNDER DIFFUSION EQUILIBRIUM.
    J. APPL. PROB., 1, 47-68.

RUBEN, HAROLD 1964.  (2.7)                                                           4733
    AN ASYMPTOTIC EXPANSION FOR THE MULTIVARIATE NORMAL DISTRIBUTION AND
    MILLS' RATIO.
    J. RES. NAT. BUR. STANDARDS SECT. B, 68, 3-11.  (MR29-2902)

RUBEN, HAROLD 1966.  (7.1/A)                                                         4734
    ON THE SIMULTANEOUS STABILIZATION OF VARIANCES AND COVARIANCES.
    ANN. INST. STATIST. MATH. TOKYO, 18, 203-210.

RUBEN, HAROLD 1966.  (4.2)                                                           4735
    SOME NEW RESULTS ON THE DISTRIBUTION OF THE SAMPLE CORRELATION COEFFICIENT.
    J. ROY. STATIST. SOC. SER. B, 28, 513-525.  (MR35-2394)

RUBIN, HERMAN 1945.  (2.6)                                                           4736
    ON THE DISTRIBUTION OF THE SERIAL CORRELATION COEFFICIENT.
    ANN. MATH. STATIST., 16, 211-215.  (MR7,P.20)

RUBIN, HERMAN 1950.   (16.1/U)                                                    4737
     CONSISTENCY OF MAXIMUM-LIKELIHOOD ESTIMATES IN THE EXPLOSIVE CASE.
   COWLES COMMISS. MONOG., 10, 356-364.

RUBIN, HERMAN 1950.   (16.2)                                                      4738
     NOTE ON RANDOM COEFFICIENTS.
   COWLES COMMISS. MONOG., 10, 419-421.

RUBIN, HERMAN 1956.   (17.1)                                                      4739
     UNIFORM CONVERGENCE OF RANDOM FUNCTIONS WITH APPLICATIONS TO STATISTICS.
   ANN. MATH. STATIST., 27, 200-203.   (MR17,P.860)

RUBIN, HERMAN 1958.   (17.7)                                                      4740
     ESTIMATION OF A REGRESSION LINE WITH BOTH VARIABLES SUBJECT TO ERROR
     UNDER AN UNUSUAL IDENTIFICATION CONDITION.
   ANN. MATH. STATIST., 29, 606-608.   (MR20-1386)

RUBIN, HERMAN 1961.   (16.4/17.8)                                                 4741
     THE ESTIMATION OF DISCONTINUITIES IN MULTIVARIATE DENSITIES, AND
     RELATED PROBLEMS IN STOCHASTIC PROCESSES.
   PROC. FOURTH BERKELEY SYMP. MATH. STATIST. PROB., 1, 563-574.   (MR25-4605)

RUBIN, HERMAN + SETHURAMAN, J. 1965.   (17.1)                                     4742
     PROBABILITIES OF MODERATE DEVIATIONS.
   SANKHYA SER. A, 27, 325-346.

RUBIN, HERMAN + SETHURAMAN, J. 1965.   (19.2/A)                                   4743
     BAYES RISK EFFICIENCY.
   SANKHYA SER. A, 27, 347-356.

RUBIN, HERMAN + TUCKER, HOWARD G. 1959.   (16.4)                                  4744
     ESTIMATING THE PARAMETERS OF A DIFFERENTIAL PROCESS.
   ANN. MATH. STATIST., 30, 641-658.

RUBIN, HERMAN + WESLER, OSCAR 1958.   (20.5)                                      4745
     A NOTE ON CONVEXITY IN EUCLIDEAN N-SPACE.
   PROC. AMER. MATH. SOC., 9, 522-523.   (MR20-4239)

RUBIN, HERMAN    SEE ALSO
     170. ANDERSON, T. W. + RUBIN, HERMAN (1949)

     171. ANDERSON, T. W. + RUBIN, HERMAN (1950)

     172. ANDERSON, T. W. + RUBIN, HERMAN (1956)

     950. CHERNOFF, HERMAN + RUBIN, HERMAN (1953)

    4075. OLKIN, INGRAM + RUBIN, HERMAN (1962)

    4076. OLKIN, INGRAM + RUBIN, HERMAN (1964)

    4077. OLKIN, INGRAM + RUBIN, HERMAN (1966)

    3101. KOOPMANS, TJALLING C.; RUBIN, HERMAN + LEIPNIK, ROY B. (1950)

RUDEMO, MATS 1964.   (16.4)                                                       4746
     DIMENSION AND ENTROPY FOR A CLASS OF STOCHASTIC PROCESSES.
   MAGYAR TUD. AKAD. MAT. KUTATO. INT. KOZL., 9, 73-88.   (MR31-6279)

RUDRA, A. 1952.   (16.5)                                                          4747
     DISCRIMINATION IN TIME-SERIES ANALYSIS.
   BIOMETRIKA, 39, 434-439.

RUDRA, A. 1954.   (16.5)                                                          4748
     A METHOD OF DISCRIMINATION IN TIME SERIES ANALYSIS.
   CALCUTTA STATIST. ASSOC. BULL., 5, 59-72.

RUDRA, A. 1955.   (16.5)                                                          4749
     A METHOD OF DISCRIMINATION IN TIME-SERIES ANALYSIS.  I.
   SANKHYA, 15, 9-34.   (MR17,P.170)

RUDRA, A. 1956.   (16.5)                                                          4750
     A METHOD OF DISCRIMINATION IN TIME-SERIES ANALYSIS.  II.
   SANKHYA, 17, 51-66.   (MR18,P.683)

RUIST, ERIK 1946.  (17.7)                                                                4751
    STANDARD ERRORS OF THE TILLING COEFFICIENTS USED IN CONFLUENCE ANALYSIS.
    ECONOMETRICA, 14, 235–241.   (MR8,P.163)

RULON, PHILLIP J. 1930.  (15.4/T)                                                         4752
    A GRAPH FOR ESTIMATING RELIABILITY IN ONE RANGE, KNOWING IT IN ANOTHER.
    J. EDUC. PSYCHOL., 21, 140-142.

RULON, PHILLIP J. 1939.  (15.4)                                                           4753
    A SIMPLIFIED PROCEDURE FOR DETERMINING THE RELIABILITY OF A TEST BY
    SPLIT-HALVES.
    HARVARD EDUC. REV., 9, 99-103.

RULON, PHILLIP J. 1951.  E(6.2/6.3)                                                       4754
    DISTINCTIONS BETWEEN DISCRIMINANT AND REGRESSION ANALYSIS AND A
    GEOMETRIC INTERPRETATION OF THE DISCRIMINANT FUNCTION.
    HARVARD EDUC. REV., 21, 80-90.

RUMSEY, HOWARD, JR. + POSNER, EDWARD C. 1965.  (17.4)                                     4755
    JOINT DISTRIBUTIONS WITH PRESCRIBED MOMENTS.
    ANN. MATH. STATIST., 36, 286-298.   (MR30-662)

RUNNENBURG, JOHANNES THEODORUS   SEE
    3001. KESTEN, HARRY + RUNNENBURG, JOHANNES THEODORUS (1956)

RUSHTON, S. 1954.  (20.4)                                                                 4756
    ON THE CONFLUENT HYPERGEOMETRIC FUNCTION $M(\alpha,\gamma,x)$.
    SANKHYA, 13, 369-376.   (MR16,P.129)

RUSHTON, S. + LANG, E.D. 1954.  (20.4/T)                                                  4757
    TABLES OF THE CONFLUENT HYPERGEOMETRIC FUNCTION.
    SANKHYA, 13, 377-411.   (MR16,P.129)

RUSSELL, ALLAN M. + JOSEPHSON, NORA S. 1965.  (18.6)                                      4758
    MEASUREMENT OF AREA BY COUNTING.
    J. APPL. PROB., 2, 339-351.

RUSSELL, J. T.   SEE
    5367. TAYLOR, H. C. + RUSSELL, J. T. (1939)

RUSSELL, OLIVE R. 1935.  (15.1)                                                           4759
    SOME OBSERVATIONS ON MULTIPLE-FACTOR ANALYSIS.
    J. EDUC. PSYCHOL., 26, 284-290.

RUST, B.; BURRUS, W. R. + SCHNEEBERGER, C. 1966.  (20.3)                                  4760
    A SIMPLE ALGORITHM FOR COMPUTING THE GENERALIZED INVERSE OF A MATRIX.
    COMMUN. ASSOC. COMP. MACH., 9, 381-385.

RUST, HANSPETIR 1965.  (18.3/A)                                                           4761
    DIE MOMENTE DER TESTGROSSE DES $\chi^2$-TESTS.
    Z. WAHRSCHEIN. VERW. GEBIETE, 4, 222-231.

RUTOVITZ, D. 1966.  (6.6)                                                                 4762
    PATTERN RECOGNITION.
        (WITH DISCUSSION)
    J. ROY. STATIST. SOC. SER. A, 129, 504-530.

RVACEVA, E. L. 1950.  (17.1)                                                              4763
    DOMAINS OF ATTRACTION OF MANY DIMENSIONAL STABLE DISTRIBUTIONS.
        (IN UKRAINIAN, WITH SUMMARY IN RUSSIAN)
    DOPOVIDI AKAD. NAUK UKRAIN. RSR, 179-181.   (MR13,P.663)

RVACEVA, E. L. 1950.  (17.1)                                                              4764
    A MANY DIMENSIONAL LOCAL THEOREM FOR STABLE LIMIT DISTRIBUTIONS.
        (IN UKRAINIAN, WITH SUMMARY IN RUSSIAN, TRANSLATED AS NO. 4765)
    DOPOVIDI AKAD. NAUK UKRAIN. RSR, 183-189.   (MR13,P.663)

RVACEVA, E. L. 1953.  (17.1)                                                              4765
    A MANY DIMENSIONAL LOCAL THEOREM FOR STABLE LIMIT DISTRIBUTIONS.
        (IN RUSSIAN, TRANSLATION OF NO. 4764)
    AKAD. NAUK UZBEK. SSR TRUDY INST. MAT. MEH., 10(1) 106-121.   (MR13,P.663)

RVACEVA, E. L. 1954.  (17.1)                                                      4766
    ON DOMAINS OF ATTRACTION OF MULTIDIMENSIONAL DISTRIBUTIONS.
        (IN RUSSIAN)
    L'VOV GOS. UNIV. UCEN. ZAP. SER. MEH. MAT., 29(6) 5-44.   (MR17,P.864)

RVACEVA, E. L.   SEE ALSO
    3790. MEIZLER, D.G.; PARASJUK, O. S. + RVACEVA, E. L. (1948)

    3791. MEIZLER, D.G.; PARASJUK, O. S. + RVACEVA, E. L. (1949)

RYDBERG, SVEN 1962.  (4.1)                                                        4767
    METHODS OF CORRECTING CORRELATIONS FOR INDIRECT RESTRICTION OF RANGE
    WITH NON-INTERVAL DATA.
    PSYCHOMETRIKA, 27, 49-58.   (MR26-852)

RYSER, HERBERT J. 1956.  (20.1)                                                   4768
    MAXIMAL DETERMINANTS IN COMBINATORIAL INVESTIGATIONS.
    CANAD. J. MATH., 8, 245-249.   (MR18,P.105)

SABAGH, G.   SEE
    3829. MILLER, C. R.; SABAGH, G. + DINGMAN, HARVEY F. (1962)

SACKROWITZ, HAROLD   SEE
    3836. MILLER, KENNETH S. + SACKROWITZ, HAROLD (1965)

SACUIU, I.   SEE
    4085. ONICESCU, OCTAV + SACUIU, I. (1964)

SADA, MASATO 1956.  (8.1/17.7)                                                    4769
    TWO-STAGE OPTIMUM ALLOCATION IN REGRESSION WHEN BOTH VARIABLES ARE
    SUBJECT TO ERRORS.
    BULL. FUKUOKA GAKUGEI UNIV. SECT. III, 5, 1-4.   (MR24-A3749)

SADIKOVA, S. M. 1966.  (17.1/B)                                                   4770
    ON TWO-DIMENSIONAL ANALOGS OF AN INEQUALITY OF K. G. ESSEEN AND THEIR
    APPLICATION TO THE CENTRAL LIMIT THEOREM.
        (IN RUSSIAN, WITH SUMMARY IN ENGLISH, TRANSLATED AS NO. 4772)
    TEOR. VEROJATNOST. I PRIMENEN., 11, 369-380.   (MR34-6832)

SADIKOVA, S. M. 1966.  (17.2)                                                     4771
    CERTAIN INEQUALITIES FOR CHARACTERISTIC FUNCTIONS.
        (IN RUSSIAN, TRANSLATED AS NO. 4773)
    TEOR. VEROJATNOST. I PRIMENEN., 11, 500-506.   (MR34-6833)

SADIKOVA, S. M. 1966.  (17.1/B)                                                   4772
    TWO-DIMENSIONAL ANALOGUES OF AN INEQUALITY OF ESSEEN WITH APPLICATIONS
    TO THE CENTRAL LIMIT THEOREM.
        (TRANSLATION OF NO. 4770)
    THEORY PROB. APPL., 11, 325-335.   (MR34-6832)

SADIKOVA, S. M. 1966.  (17.2)                                                     4773
    CERTAIN INEQUALITIES FOR CHARACTERISTIC FUNCTIONS.
        (TRANSLATION OF NO. 4771)
    THEORY PROB. APPL., 11, 441-447.   (MR34-6833)

SAGESER, H. W.   SEE
    4511. REMMERS, H. H. + SAGESER, H. W. (1941)

SAGRISTA, SEBASTIAN N. 1952.  (18.1/B)                                            4774
    SOBRE UNA GENERALIZACION DE LAS CURVAS DE PEARSON AL CASO
    BIDIMENSIONAL.
        (WITH SUMMARY IN ENGLISH)
    TRABAJOS ESTADIST., 3, 273-314.   (MR14,P.665)

SAHAIDAROVA, N. 1965.  (17.1/B)                                                   4775
    ON THE MULTIDIMENSIONAL LIMIT THEOREM.
        (IN RUSSIAN, WITH SUMMARY IN UZBEKISH)
    IZV. AKAD. NAUK UZSSR. SER. FIZ.-MAT. NAUK, 9(4) 11-16.   (MR32-6532)

SAIBEL, EDWARD 1944.  (20.3)                                                      4776
    A RAPID METHOD OF INVERSION OF CERTAIN TYPES OF MATRICES.
    J. FRANKLIN INST., 237, 197-201.   (MR5,P.245)

SAKAGUCHI,MINORU 1952.   (17.1)                                                                                    4777
    ON A CERTAIN LIMIT DISTRIBUTION.
  REP. STATIST. APPL. RES. UN. JAPAN. SCI. ENGRS., 1(4) 10-14.   (MR14,P.567)

SAKAGUCHI,MINORU 1957.   (4.9/8.1/14.4)                                                                             4778
    NOTES ON INFORMATION TRANSMISSION IN MULTIVARIATE PROBABILITY DISTRIBUTIONS.
  REP. UNIV. ELECTRO-COMMUN. PHYS. SCI. ENGRG., 9, 25-31.   (MR23-B3063)

SAKAGUCHI,MINORU 1959.   (19.1)                                                                                     4779
    ON A CERTAIN MULTI-STAGE GAME.
  REP. STATIST. APPL. RES. UN. JAPAN. SCI. ENGRS., 6, 1-4.   (MR21-2540)

SAKAMOTO, HEIHACHI 1944.   (2.5)                                                                                    4780
    PROBLEMS ON THE INDEPENDENCE OF STATISTICS AND DEGREES OF FREEDOM.
        (IN JAPANESE)
  RES. MEM. INST. STATIST. MATH. TOKYO, 1(3) 65-74.

SAKAMOTO, HEIHACHI 1944.   (2.5)                                                                                    4781
    ON THE INDEPENDENCE OF TWO STATISTICS.
        (IN JAPANESE)
  RES. MEM. INST. STATIST. MATH. TOKYO, 1(9) 1-25.

SAKAMOTO, HEIHACHI 1946.   (2.5/8.8)                                                                                4782
    ON THE INDEPENDENCE OF STATISTICS AND THE JUDGEMENT OF DEGREE OF
    FREEDOM IN THE THEORY OF ANALYSIS OF VARIANCE.
        (IN JAPANESE, TRANSLATED AS NO. 4784)
  RES. MEM. INST. STATIST. MATH. TOKYO, 2, 155-162.   (MR11,P.260)

SAKAMOTO, HEIHACHI 1948.   (2.5)                                                                                    4783
    ON THE INDEPENDENCE AND DEGREES OF FREEDOM OF THE MAXIMUM LIKELIHOOD ESTIMATES.
        (IN JAPANESE)
  BULL. MATH. STATIST., 2(1) 36-37, (2) 24-27.

SAKAMOTO, HEIHACHI 1949.   (2.5/8.8)                                                                                4784
    ON THE CRITERIA OF THE INDEPENDENCE AND THE DEGREES OF FREEDOM OF
    STATISTICS AND THEIR APPLICATIONS TO THE ANALYSIS OF VARIANCE.
        (TRANSLATION OF NO. 4782)
  ANN. INST. STATIST. MATH. TOKYO, 1, 109-122.   (MR11,P.260)

SAKAMOTO, HEIHACHI 1949.   (2.5)                                                                                    4785
    ON INDEPENDENCE OF STATISTICAL QUANTITIES.
        (IN JAPANESE)
  SUGAKU, 1, 263-274.   (MR13,P.366)

SAKODA, JAMES M. 1954.   (6.6/15.1)                                                                                 4786
    OSGOOD AND SUCI'S MEASURE OF PATTERN SIMILARITY AND Q-TECHNIQUE FACTOR ANALYSIS.
  PSYCHOMETRIKA, 19, 253-256.

SALAEVSKII, O.V. 1960.   (8.8)                                                                                      4787
    SOME REMARKS ON THE LEVELLING  OF OBSERVATIONS WITH UNKNOWN WEIGHTS.
        (IN RUSSIAN, TRANSLATED AS NO. 4788)
  DOKL. AKAD. NAUK SSSR, 130, 37-40.   (MR22-11473)

SALAEVSKII, O.V. 1960.   (8.8)                                                                                      4788
    SOME REMARKS ON THE ADJUSTMENT OF OBSERVATIONS WITH UNKNOWN WEIGHTS.
        (TRANSLATION OF NO. 4787)
  SOVIET MATH. DOKL., 1, 27-30.   (MR22-11473)

SALAEVSKII, O.V. 1963.   (5.2/U)                                                                                    4789
    ON NON-EXISTING REGULARLY VARYING TESTS FOR BEHRENS-FISHER PROBLEM.
        (IN RUSSIAN, TRANSLATED AS NO. 4790)
  DOKL. AKAD. NAUK SSSR, 151, 509-510.   (MR27-2037)

SALAEVSKII, O.V. 1963.   (5.2/U)                                                                                    4790
    ON THE NONEXISTENCE OF REGULARLY VARYING TESTS FOR THE BEHRENS-FISHER PROBLEM.
        (TRANSLATION OF NO. 4789)
  SOVIET MATH. DOKL., 4, 1043-1045.   (MR27-2037)

SALAEVSKII, O.V. 1963.   (10.1)                                                                                     4791
    TESTING OF THE FUNDAMENTAL HYPOTHESES IN MULTIVARIATE ANALYSIS.
        (IN RUSSIAN, WITH SUMMARY IN ENGLISH)
  VESTNIK LENINGRAD. UNIV., 18(3) 150-152.   (MR27-5335)

SALAMUNI, R.    SEE
    3814. MICHENER, C. D.; LANGE, R. B.; BIGARELLA, J. J. +  (1958)
    SALAMUNI, R.

SALAS, J. MARTINEZ 1946.  (17.2)                                            4792
    LA DERIVACION DE FUNCIONES DE  N  VARIABLES REALES DE VARIACION ACOTADA.
    REV. MAT. HISP. AMER. SER. 4, 6, 249-259.

SALISBURY, FRANK S. 1929.  (4.6)                                            4793
    A SIMPLIFIED METHOD OF COMPUTING MULTIPLE CORRELATION CONSTANTS.
    J. EDUC. PSYCHOL., 20, 44-52.

SALISBURY, FRANK S.    SEE ALSO
    2971. KELLEY, TRUMAN L. + SALISBURY, FRANK S. (1926)

SALVEMINI, TOMMASO 1937.  (17.3)                                            4794
    LEGGE DI FREQUENZA DI UNA VARIABILE CASUALE SOMMA DI VARIABILI DIPENDENTI.
    ATTI CONG. UN. MAT. ITAL., 1, 421-425.

SALVEMINI, TOMMASO 1938.  (4.8)                                             4795
    VALORI MEDI DI PIU VARIABILI CASUALI DIPENDENTI TRA LORO.
    ATTI SOC. ITAL. PROG. SCI., 12, 46-49.

SALVEMINI, TOMMASO 1942.  (4.8/4.9)                                         4796
    GLI INDICI SEMPLICI DI OMOFILIA E DI CORRELAZIONE TRA CLASSI DI
    VALORI EQUISPAZIATI.
    ATTI RIUN. SCI. SOC. ITAL. STATIST., 3, 62-75.

SALVEMINI, TOMMASO 1947.  (4.8/E)                                           4797
    SU ALCUNI INDICI USATI PER LA MISURE DELLE RELAZIONI STATISTICHE.
    STATISTICA, BOLOGNA, 7, 200-218.

SALVEMINI, TOMMASO 1947.  (4.8)                                            4798
    SU UN NUOVO INDICE PROPOSTO DA FRECHET PER LA MISURA DELLA DIPENDENZA
    STATISTICA.
    STATISTICA, BOLOGNA, 7, 219-226.

SALVEMINI, TOMMASO 1951.  (4.9)                                            4799
    ASPETTI DELLA CORRELAZIONE.
    ATTI CONG. UN. MAT. ITAL., 3, 216-220.

SALVEMINI, TOMMASO 1951.  (4.8/E)                                          4800
    SUI VARI INDICI DI COGRADUAZIONE.
    STATISTICA, BOLOGNA, 11, 133-154.

SALVEMINI, TOMMASO 1952.  (4.8/4.9)                                        4801
    INDIPENDENZA STATISTICA TRA DUE O PIU VARIABILI.
    ATTI RIUN. SCI. SOC. ITAL. STATIST., 9, 145-181.

SALVEMINI, TOMMASO 1952.  (4.8)                                           4802
    INDICI DI CONNESSIONE E INDICI DI CONCORDANZA.
    GENUS, 9, 283-292.

SALVEMINI, TOMMASO 1952.  (2.1/EU)                                         4803
    L'INDICE DI DISSOMIGLIANZA FRA DISTRIBUZIONI CONTINUE.
    METRON, 16(3) 75-100.

SALVEMINI, TOMMASO 1953.  (4.8/E)                                          4804
    SU ALCUNI ASPETTI DELLA DISSOMIGLIANZA E DELLA CONCORDANZA CON
    APPLICAZIONE ALLE DISTRIBUZIONI DEGLI SPOSI SECONDO L'ETA.
        (REPRINTED AS NO. 4807)
    RIV. ITAL. ECON. DEMOG. STATIST., 7(2) 49-69.

SALVEMINI, TOMMASO 1954.  (4.8)                                           4805
    DISSOMIGLIANZA A PIU DIMENSIONI.
    ATTI RIUN. SCI. SOC. ITAL. STATIST., 11, 67-94.

SALVEMINI, TOMMASO 1954.  (4.8/B)                                         4806
    CORREZIONE DEI MOMENTI DELLE DISTRIBUZIONI IN DUE VARIABILI E
    INFLUENZA DEL RAGGRUPPAMENTO IN CLASSI SUGLI INDICI DI CONCORDANZA.
    ATTI RIUN. SCI. SOC. ITAL. STATIST., 12, 239-287.

SALVEMINI, TOMMASO 1954.  (4.8/E)                                                    4807
     SU ALCUNI ASPETTI DELLA DISSOMIGLIANZA E DELLA CONCORDANZA CON
     APPLICAZIONE ALLE DISTRIBUZIONI DEGLI SPOSI SECONDO L'ETA.
     (WITH SUMMARY IN FRENCH AND ENGLISH, REPRINT OF NO. 4804)
     BULL. INST. INTERNAT. STATIST., 34(2) 283-300.


SALVEMINI, TOMMASO 1955.  (4.8/B)                                                    4808
     L'INDICE QUADRATICO DI DISSOMIGLIANZA TRA DISTRIBUZIONI GAUSSIANE.
     ATTI RIUN. SCI. SOC. ITAL. STATIST., 14, 197-201.


SALVEMINI, TOMMASO 1955.  (4.1/4.8/4.9/E)                                            4809
     GLI INDICI DI CONNESSIONE NEL CASO DI VARIABILI CASUALI NORMALI E
     CONSIDERAZIONI SULLA GRADUATORIA TRA INDICI DI CONNESSIONE E INDICI DI
     CONCORDANZA.
     STATISTICA, BOLOGNA, 15, 77-90.  (MR16,P.1037)


SALVEMINI, TOMMASO 1956.  (4.8/4.9/B)                                                4810
     FONDAMENTI RAZIONALI, SCHEMI TEORICI E MODERNI ASPETTI DELLE RELAZIONI TRA DUE
     O PIU VARIABILI.
     ATTI RIUN. SCI. SOC. ITAL. STATIST., 15, 11-47.


SALVEMINI, TOMMASO 1960.  (2.1/4.8/17.3/T)                                           4811
     TRANSVARIAZIONE TRA MEDIE DI K VALORI DI DUE VARIABILI STATISTICHE.
     STATISTICA, BOLOGNA, 20, 135-143.  (MR22-5082)


SALVEMINI, TOMMASO 1961.  (6.2/6.3/B)                                                4812
     SULLA DISCRIMINAZIONE TRA DUE VARIABILI STATISTICHE SEMPLICI E MULTIPLE.
     STATISTICA, BOLOGNA, 21, 121-144.


SALVEMINI, TOMMASO 1963.  (4.8/E)                                                    4813
     SULLA DIPENDENZA IN MEDIA TRA LE COMPONENTI DI UNA VARIABILE TRIPLA.
     ATTI RIUN. SCI. SOC. ITAL. STATIST., 23, 59-125.


SALVEMINI, TOMMASO 1966.  (4.9/6.5/F)                                                4814·
     INDICI DI DISSOMIGLIANZA TRA MUTABILI STATISTICHE CICLICHE.
     METRON, 25, 116-149.


SALZER, HERBERT E. 1957.  (20.5)                                                     4815
     NOTE ON MULTIVARIATE INTERPOLATION FOR EQUALLY SPACED ARGUMENTS, WITH
     AN APPLICATION TO DOUBLE SUMMATION.
     J. SOC. INDUST. APPL. MATH., 5, 254-262.


SAMANTA, M. 1965.  (2.4/18.5/B)                                                      4816
     A NOTE ON THE PROBLEM OF OPTIMUM TRUNCATION OF A  BIVARIATE POPULATION
     IN STRATIFIED RANDOM SAMPLING.
     ANN. INST. STATIST. MATH. TOKYO, 17, 363-375.  (MR33-6800)


SAMPFORD, M. R. 1953.  (2.7)                                                         4817
     SOME INEQUALITIES IN MILL'S RATIO AND RELATED FUNCTIONS.
     ANN. MATH. STATIST., 24, 130-132.  (MR14,P.995)


SAMSON, PABLO, JR.   SEE
     4302. PILLAI, K.C.SREEDHARAN + SAMSON, PABLO, JR. (1959)


SAMUEL, ESTER 1963.  (6.1)                                                           4818
     NOTE ON A SEQUENTIAL CLASSIFICATION PROBLEM.
     ANN. MATH. STATIST., 34, 1095-1097.


SAMUEL, ISAAC 1965.  (20.2)                                                          4819
     RELATIONS ENTRE LES VECTEURS PROPRES DE DEUX MATRICES SINGULIERES, TRANSPOSEES
     L'UNE DE L'AUTRE.
     C.R. ACAD. SCI. PARIS, 261, 2431-2434.


SAMUELSON, P. ANTHONY 1942.  (20.2)                                                  4820
     A METHOD OF DETERMINING EXPLICITLY THE COEFFICIENTS OF THE
     CHARACTERISTIC EQUATION.
     ANN. MATH. STATIST., 13, 424-429.  (MR4,P.148)


SANDERS, H. G.   SEE
     1847. GARNER, F. H.; GRANTHAM, J. + SANDERS, H. G. (1934)

SANDLER, JOSEPH 1949.    (15.1)                                          4821
    THE RECIPROCITY PRINCIPLE AS AN AID TO FACTOR ANALYSIS.
  BRITISH J. PSYCHOL. STATIST. SECT., 2, 180-187.


SANDLER, JOSEPH 1952.    (15.3)                                          4822
    A TECHNIQUE FOR FACILITATING THE ROTATION OF FACTOR AXES, BASED ON AN
    EQUIVALENCE BETWEEN PERSONS AND TESTS.
  PSYCHOMETRIKA, 17, 223-229.


SANDOMIRE, MARION M.    SEE
  1243. DAY, BESSE B. + SANDOMIRE, MARION M. (1942)


SANKARAN, MUNUSWAMY 1958.    (4.2)                                       4823
    ON NAIR'S TRANSFORMATION OF THE CORRELATION COEFFICIENT.
  BIOMETRIKA, 45, 567-571.


SANKARAN, MUNUSWAMY 1963.    (14.5)                                      4824
    APPROXIMATIONS TO THE NON-CENTRAL CHI-SQUARE DISTRIBUTION.
  BIOMETRIKA, 50, 199-204.


SANOV, I. N. 1965.    (18.3/A)                                           4825
    A METHOD FOR OBTAINING ASYMPTOTIC EXPRESSIONS FOR PROBABILITIES OF
    LARGE DEVIATIONS OF NONLINEAR STATISTICS FROM THE MULTINOMIAL
    DISTRIBUTION AND ITS APPLICATIONS.
        (IN RUSSIAN, TRANSLATED AS NO. 4826)
  TEOR. VEROJATNOST. I PRIMENEN., 10, 761-763.


SANOV, I. N. 1965.    (18.3/A)                                           4826
    A METHOD FOR OBTAINING ASYMPTOTIC EXPRESSIONS FOR PROBABILITIES OF
    LARGE DEVIATIONS OF NONLINEAR STATISTICS FROM THE MULTINOMIAL
    DISTRIBUTION AND ITS APPLICATIONS.
        (TRANSLATION OF NO. 4825)
  THEORY PROB. APPL., 10, 692-694.


SANTALO, LUIS A. 1947.    (17.3)                                         4827
    ON THE FIRST TWO MOMENTS OF THE MEASURE OF A RANDOM SET.
  ANN. MATH. STATIST., 18, 37-49.    (MR8,P.389)


SAPOGOV, N.A. 1948.    (17.1)                                            4828
    ON SUMS OF DEPENDENT RANDOM VARIABLES.
        (IN RUSSIAN)
  DOKL. AKAD. NAUK SSSR NOV. SER., 63, 353-356.    (MR10,P.310)


SAPOGOV, N.A. 1948.    (17.1)                                            4829
    ON THE LAW OF THE ITERATED LOGARITHM FOR DEPENDENT VARIABLES.
        (IN RUSSIAN)
  DOKL. AKAD. NAUK SSSR NOV. SER., 63, 487-490.    (MR10,P.384)


SAPOGOV, N.A. 1949.    (17.1)                                            4830
    ON A LIMIT THEOREM.
        (IN RUSSIAN)
  DOKL. AKAD. NAUK SSSR NOV. SER., 69, 15-18.    (MR11,P.256)


SAPOGOV, N.A. 1949.    (16.4)                                            4831
    ON MULTIDIMENSIONAL INHOMOGENEOUS MARKOV CHAINS.
        (IN RUSSIAN)
  DOKL. AKAD. NAUK SSSR NOV. SER., 69, 133-135.    (MR11,P.256)


SAPOGOV, N.A. 1949.    (16.4/17.1/B)                                     4832
    A TWO-DIMENSIONAL LIMIT THEOREM FOR TWO-DIMENSIONAL CHAINS.
        (IN RUSSIAN)
  IZV. AKAD. NAUK SSSR SER. MAT., 13, 301-314.    (MR11,P.40)


SAPOGOV, N.A. 1949.    (16.4/17.1/B)                                     4833
    AN INTEGRAL LIMIT THEOREM FOR MULTIDIMENSIONAL MARKOV CHAINS.
        (IN RUSSIAN)
  USPEHI MAT. NAUK, 32, 190-192.    (MR11,P.189)


SAPOGOV, N.A. 1950.    (17.1)                                            4834
    GENERAL FORM OF A LIMIT THEOREM FOR INDEPENDENT RANDOM VECTORS.
        (IN RUSSIAN)
  DOKL. AKAD. NAUK SSSR NOV. SER., 70, 765-768.    (MR11,P.444)

SAPOGOV, N.A. 1950.  (17.1)                                                          4835
    ON THE LAW OF LARGE NUMBERS FOR DEPENDENT RANDOM VARIABLES.
        (IN RUSSIAN)
    IZV. AKAD. NAUK SSSR SER. MAT., 14, 145-154.  (MR11,P.606)


SAPOGOV, N.A. 1950.  (17.1)                                                          4836
    ON A MULTIDIMENSIONAL LIMIT THEOREM OF THE THEORY OF PROBABILITY.
        (IN RUSSIAN)
    USPEHI MAT. NAUK, 37, 137-151.  (MR12,P.34)


SAPOGOV, N.A. 1959.  (17.1/17.3)                                                     4837
    ON INDEPENDENT TERMS OF A SUM OF RANDOM VARIABLES WHICH IS
    DISTRIBUTED ALMOST NORMALLY.
    VESTNIK LENINGRAD. UNIV., 14(19) 78-105.  (MR22-5056)


SAPOGOV, N.A.   SEE ALSO
    3443. LINNIK, YU. V. + SAPOGOV, N.A. (1949)

    3444. LINNIK, YU. V. + SAPOGOV, N.A. (1949)


SAPOZKOV, M. A. 1956.  (16.4)                                                        4838
    CORRELATION METHOD FOR MEASURING THE DISTORTION COEFFICIENT OF TRANSMISSION.
        (IN RUSSIAN, TRANSLATED AS NO. 4839)
    AKUST. ZUR., 2, 279-284.  (MR18,P.368)


SAPOZKOV, M. A. 1956.  (16.4)                                                        4839
    CORRELATION METHOD FOR MEASURING THE DISTORTION COEFFICIENT OF TRANSMISSION.
        (TRANSLATION OF NO. 4838)
    SOVIET PHYS. ACOUST., 2, 294-300.  (MR18,P.368)


SARGAN, JOHN D. 1961.  (16.1)                                                        4840
    THE MAXIMUM LIKELIHOOD ESTIMATION OF ECONOMIC RELATIONSHIPS WITH
    AUTOREGRESSIVE RESIDUALS.
    ECONOMETRICA, 29, 414-426.  (MR24-B2471)


SARGAN, JOHN D. 1964.  (16.1/16.2)                                                   4841
    THREE-STAGE LEAST-SQUARES AND FULL MAXIMUM LIKELIHOOD ESTIMATES.
    ECONOMETRICA, 32, 77-81.  (MR31-5279)


SARHAN, AHMED E.; GREENBERG, BERNARD G. + ROBERTS, ELEANOR 1962.  (20.3)             4842
    MODIFIED SQUARE ROOT METHOD OF MATRIX INVERSION.
    TECHNOMETRICS, 4, 282-287.


SARHAN, AHMED E.   SEE ALSO
    2110. GREENBERG, BERNARD G. + SARHAN, AHMED E. (1959)

    2111. GREENBERG, BERNARD G. + SARHAN, AHMED E. (1960)

    2112. GREENBERG, BERNARD G. + SARHAN, AHMED E. (1960)

    4701. ROY, S.N. + SARHAN, AHMED E. (1956)

    4703. ROY, S.N.; GREENBERG, BERNARD G. + SARHAN, AHMED E. (1960)


SARKADI, KAROLY; SCHNELL, EDIT + VINCZE, ISTVAN 1962.  (17.6/U)                      4843
    ON THE POSITION OF THE SAMPLE MEAN AMONG THE ORDERED SAMPLE ELEMENTS.
    MAGYAR TUD. AKAD. MAT. KUTATO. INT. KOZL., 7, 239-254.  (MR27-3055)


SARMA, Y. RAMAKRISHNA 1965.  (6.5/19.2)                                              4844
    UNE PROPRIETE DU LEMME DE NEYMAN ET PEARSON.
    C.R. ACAD. SCI. PARIS, 260, 2402-2404.


SARMANOFF, O. W.  SEE  SARMANOV, O. V.


SARMANOV, O. V. 1941.  (18.1)                                                        4845
    SUR LA CORRELATION ISOGENE.
    C. R. ACAD. SCI. URSS NOUV. SER., 32, 28-30.  (MR3,P.173)


SARMANOV, O. V. 1945.  (18.1/B)                                                      4846
    ON ISOGENEOUS CORRELATION.
        (IN RUSSIAN, WITH SUMMARY IN ENGLISH)
    IZV. AKAD. NAUK SSSR SER. MAT., 9, 169-200.  (MR7,P.130)

SARMANOV, O. V. 1946.  (17.9)                                                           4847
     SUR LES SOLUTIONS MONOTONES DES EQUATIONS INTEGRALES DE CORRELATION.
     C. R. ACAD. SCI. URSS NOUV. SER., 53, 773-776.  (MR8,P.467)

SARMANOV, O. V. 1947.  (4.1/17.3/B)                                                     4848
     ON THE RECTIFICATION OF A SYMMETRIC CORRELATION.
          (IN RUSSIAN)
     DOKL. AKAD. NAUK SSSR NOV. SER., 58, 745-747.  (MR10,P.45)

SARMANOV, O. V. 1948.  (4.1/17.3/B)                                                     4849
     ON THE RECTIFICATION OF AN ASYMMETRIC CORRELATION.
          (IN RUSSIAN)
     DOKL. AKAD. NAUK SSSR NOV. SER., 59, 861-863.  (MR9,P.442)

SARMANOV, O. V. 1948.  (17.3/B)                                                         4850
     ON THE RECTIFICATION OF CORRELATION.
          (IN RUSSIAN)
     USPEHI MAT. NAUK, 27, 190-192.  (MR10,P.305)

SARMANOV, O. V. 1952.  (4.9/17.3/B)                                                     4851
     ON FUNCTIONAL MOMENTS OF A SYMMETRIC CORRELATION.
          (IN RUSSIAN)
     DOKL. AKAD. NAUK SSSR NOV. SER., 84, 887-890.  (MR14,P.64)

SARMANOV, O. V. 1952.  (4.9/17.3/B)                                                     4852
     ON FUNCTIONAL MOMENTS OF AN ASYMMETRIC CORRELATION.
          (IN RUSSIAN)
     DOKL. AKAD. NAUK SSSR NOV. SER., 84, 1139-1142.  (MR14,P.64)

SARMANOV, O. V. 1958.  (4.6)                                                            4853
     THE MAXIMUM CORRELATION COEFFICIENT (SYMMETRICAL CASE).
          (IN RUSSIAN)
     DOKL. AKAD. NAUK SSSR, 120, 715-718.  (MR21-2334)

SARMANOV, O. V. 1958.  (4.6/18.1/B)                                                     4854
     MAXIMUM CORRELATION COEFFICIENT (NON-SYMMETRICAL CASE).
          (IN RUSSIAN)
     DOKL. AKAD. NAUK SSSR, 121, 52-55.  (MR20-5539)

SARMANOV, O. V. 1960.  (4.9/18.1)                                                       4855
     PSEUDONORMAL CORRELATION AND ITS VARIOUS GENERALIZATIONS.
          (IN RUSSIAN. TRANSLATED AS NO. 4857)
     DOKL. AKAD. NAUK SSSR, 132, 299-302.  (MR22-12576)

SARMANOV, O. V. 1960.  (16.4)                                                           4856
     CHARACTERISTIC CORRELATION FUNCTIONS AND THEIR USE IN THE THEORY OF
     STATIONARY MARKOFF PROCESSES.
          (IN RUSSIAN. TRANSLATED AS NO. 4858)
     DOKL. AKAD. NAUK SSSR, 132, 769-772.  (MR22-12575)

SARMANOV, O. V. 1960.  (4.9/18.1)                                                       4857
     PSEUDONORMAL CORRELATION AND ITS VARIOUS GENERALIZATIONS.
          (TRANSLATION OF NO. 4855)
     SOVIET MATH. DOKL., 1, 564-567.  (MR22-12576)

SARMANOV, O. V. 1960.  (16.4)                                                           4858
     CHARACTERISTIC CORRELATION FUNCTIONS AND THEIR APPLICATIONS IN THE
     THEORY OF STATIONARY MARKOV PROCESSES.
          (TRANSLATION OF NO. 4856)
     SOVIET MATH. DOKL., 1, 651-654.  (MR22-12575)

SARMANOV, O. V. 1961.  (16.4)                                                           4859
     INVESTIGATION OF STATIONARY MARKOV PROCESSES BY THE METHOD OF EIGENFUNCTION EXPANSION.
          (IN RUSSIAN, TRANSLATED AS NO. 4860)
     TRUDY MAT. INST. STEKLOVA, 60, 238-261.  (MR26-1935)

SARMANOV, O. V. 1963.  (16.4)                                                           4860
     INVESTIGATION OF STATIONARY MARKOV PROCESSES BY THE METHOD OF EIGENFUNCTION EXPANSION.
          (TRANSLATION OF NO. 4859)
     SELECT. TRANSL. MATH. STATIST. PROB., 4, 245-269.  (MR26-1935)

SARMANOV, O. V. 1966.   (17.3/17.4/B)                                                          4861
    GENERALIZED NORMAL CORRELATION AND TWO-DIMENSIONAL FRECHET CLASSES.
        (IN RUSSIAN)
  DOKL. AKAD. NAUK SSSR, 168, 32–35.   (MR33–4969)

SARMANOV, O. V. + ZAKHAROV, V. K. 1960.   (4.6)                                                4862
    MAXIMUM COEFFICIENTS OF MULTIPLE CORRELATION.
        (IN RUSSIAN, TRANSLATED AS NO. 4863, TRANSLATED AS NO. 4864)
  DOKL. AKAD. NAUK SSSR, 130, 269–271.   (MR22–8548)

SARMANOV, O. V. + ZAKHAROV, V. K. 1960.   (4.6)                                                4863
    MAXIMUM COEFFICIENTS OF MULTIPLE CORRELATION.
        (TRANSLATION OF NO. 4862)
  SOVIET MATH. DOKL., 1, 51–53.   (MR22–8548)


SARMANOV, O. V. + ZAKHAROV, V. K. 1961.   (4.6)                                                4864
    MAXIMUM COEFFICIENTS OF MULTIPLE CORRELATION.
        (TRANSLATION OF NO. 4862)
  TECH. TRANSL., 1–3.   (MR22–8548)

SARMANOV, O. V.   SEE ALSO
    3074. KOLMOGOROV, A. N. + SARMANOV, O. V. (1960)

    3075. KOLMOGOROV, A. N. + SARMANOV, O. V. (1960)

SARNDAL, CARL-ERIK 1965.   (17.9)                                                              4865
    DERIVATION OF A CLASS OF FREQUENCY DISTRIBUTIONS VIA BAYES'S THEOREM.
  J. ROY. STATIST. SOC. SER. B, 27, 290–300.   (MR34–5191)

SARYMSAKOV, T. A. 1955.   (1.2)                                                                4866
    VSEVOLOD IVANOVIC ROMANOVSKII.   OBITUARY.
        (IN RUSSIAN)
  USPEHI MAT. NAUK, 63, 79–88.   (MR16,P.782)

SASAKI, TATSUJIRO 1955.   (17.9)                                                               4867
    ON THE THEORY OF STATISTICS WITH SCALAR AS A VARIABLE.
        (IN JAPANESE, WITH SUMMARY IN ENGLISH)
  PROC. INST. STATIST. MATH. TOKYO, 2(2) 1.

SASAKI, TATSUJIRO 1955.   (17.9)                                                               4868
    ON THE THEORY OF STATISTICS CONCERNING THE QUANTITIES CHARACTERIZED AS TENSORS.
        (IN JAPANESE, WITH SUMMARY IN ENGLISH)
  PROC. INST. STATIST. MATH. TOKYO, 3(1) 1–2.

SASTRI, A. V. K. 1958.   (4.8/8.8)                                                             4869
    A RESEARCH NOTE ON "THE VARIANCE OF THE COEFFICIENT OF GENETIC
    CORRELATION ESTIMATED IN PLANT BREEDING EXPERIMENTS".
  J. INDIAN SOC. AGRIC. STATIST., 10, 156–159.

SASTRY, K.V. KRISHNA 1948.   (8.4/20.4)                                                        4870
    ON A BESSEL FUNCTION OF THE SECOND KIND AND WILKS' Z-DISTRIBUTION.
  PROC. INDIAN ACAD. SCI. SECT. A, 28, 532–536.   (MR10,P.387)


SASTRY, K.V.R. 1965.   (17.7)                                                                  4871
    UNBIASED RATIO ESTIMATORS.
  J. INDIAN SOC. AGRIC. STATIST., 17, 20–29.

SASTRY, M. V. R. 1961.   (18.3/BE)                                                             4872
    USE OF BIVARIATE K-STATISTICS IN FIXING QUALITY STANDARDS FOR MARKET MILK.
  J. INDIAN SOC. AGRIC. STATIST., 13, 195–210.

SASULY, MAX 1930.   (16.5)                                                                     4873
    GENERALIZED MULTIPLE CORRELATION ANALYSIS OF ECONOMIC STATISTICAL SERIES.
  J. AMER. STATIST. ASSOC., 25(SP) 146–152.

SATHE, Y. S.   SEE
    2920. KAMAT, A. R. + SATHE, Y. S. (1962)

SATHER, D. S.   SEE
    767. BURFORD, THOMAS M.; RIDEOUT, V. C. + SATHER, D. S. (1955)

SATO, RYOICHIRO 1951.   (14.2/14.3/U)                                          4874
    'R' DISTRIBUTIONS AND 'R' TESTS.
          (AMENDED BY NO. 4876)
    ANN. INST. STATIST. MATH. TOKYO, 2, 91-124.   (MR13,P.52)

SATO, RYOICHIRO 1951.   (8.2/8.3/U)                                            4875
    THE 'R' TESTS RELATING TO THE REGRESSION.
          (AMENDED BY NO. 4876)
    ANN. INST. STATIST. MATH. TOKYO, 3, 45-56.   (MR13,P.571)

SATO, RYOICHIRO 1952.   (8.2/8.3/14.2/U)                                       4876
    ERRATA TO: 'R' DISTRIBUTIONS AND 'R' TESTS, AND THE 'R' TESTS RELATING TO THE REGRESSION.
          (AMENDMENT OF NO. 4874, AMENDMENT OF NO. 4875)
    ANN. INST. STATIST. MATH. TOKYO, 3, 127-128.   (MR14,P.189)

SATO, SOKURO 1961.   (3.2)                                                     4877
    A MULTIVARIATE ANALOGUE OF POOLING OF DATA.
    BULL. MATH. STATIST., 10(3) 61-76.   (MR28-2598)

SATO, SOKURO   SEE ALSO
    215. ASANO, CHOOICHIRO + SATO, SOKURO (1961)

SATTERTHWAITE, FRANKLIN E. 1941.   (20.1)                                      4878
    A CONCISE ANALYSIS OF CERTAIN ALGEBRAIC FORMS.
    ANN. MATH. STATIST., 12, 77-83.   (MR2,P.244)

SATTERTHWAITE, FRANKLIN E. 1944.   (20.3)                                      4879
    ERROR CONTROL IN MATRIX CALCULATION.
    ANN. MATH. STATIST., 15, 373-387.   (MR6,P.218)

SAUNDERS, DAVID R. 1948.   (15.2)                                             4880
    FACTOR ANALYSIS. I.   SOME EFFECTS OF CHANCE ERROR.
    PSYCHOMETRIKA, 13, 251-257.

SAUNDERS, DAVID R. 1949.   (15.3)                                             4881
    FACTOR ANALYSIS. II.   A NOTE CONCERNING ROTATION OF AXES TO SIMPLE STRUCTURE.
    EDUC. PSYCHOL. MEAS., 9, 753-756.

SAUNDERS, DAVID R. 1960.   (15.3)                                             4882
    A COMPUTER PROGRAM TO FIND THE BEST-FITTING ORTHOGONAL FACTORS FOR A
       GIVEN HYPOTHESIS.
    PSYCHOMETRIKA, 25, 199-205.   (MR22-6098)

SAUNDERS, DAVID R.   SEE ALSO
    6032. WRIGLEY, CHARLES F.; SAUNDERS, DAVID R. + NEUHAUS, JACK O. (1958)

SAUNDERSON, J. L.   SEE
    2093. GOUDSMIT, S. A. + SAUNDERSON, J. L. (1940)

SAVAGE, I. RICHARD 1953.   (1.2/17.8)                                         4883
    BIBLIOGRAPHY OF NONPARAMETRIC STATISTICS AND RELATED TOPICS.
    J. AMER. STATIST. ASSOC., 48, 844-906.   (MR15,P.450)

SAVAGE, I. RICHARD 1962.   (2.1/2.7)                                          4884
    MILL'S RATIO FOR MULTIVARIATE NORMAL DISTRIBUTIONS.
    J. RES. NAT. BUR. STANDARDS SECT. B, 66, 93-96.

SAW, J. G. 1966.   (14.2/14.3)                                                4885
    A CONSERVATIVE TEST FOR THE CONCURRENCE OF SEVERAL REGRESSION LINES
       AND RELATED PROBLEMS.
    BIOMETRIKA, 53, 272-275.

SAWKINS, D. T. 1944.   (4.1/8.1)                                              4886
    SIMPLE REGRESSION AND CORRELATION.
    PROC. ROY. SOC. N. S. W., 77, 85-95.   (MR6,P.5)

SAXENA, ASHOK K. 1966.   (5.2)                                                4887
    ON THE COMPLEX ANALOGUE OF $T^2$ FOR TWO POPULATIONS.
    J. INDIAN STATIST. ASSOC., 4, 99-102.

SAXENA, ASHOK K.   SEE ALSO
    3739. MATHAI, A. M. + SAXENA, ASHOK K. (1966)

SAZONOV, VJACHESLAV V. 1963.   (17.1)                                                4888
     ON A HIGHER-DIMENSIONAL CENTRAL LIMIT THEOREM.
          (IN RUSSIAN, WITH SUMMARY IN ENGLISH AND LATVIAN)
     LITOVSK. MAT. SB., 3, 219-224.   (MR28-5454)


SAZONOV, VJACHESLAV V. 1966.   (17.3/A)                                              4889
     ON MULTIDIMENSIONAL CONCENTRATION FUNCTIONS.
          (IN RUSSIAN, WITH SUMMARY IN ENGLISH, TRANSLATED AS NO. 4890)
     TEOR. VEROJATNOST. I PRIMENEN., 11, 683-690.   (MR35-1062)


SAZONOV, VJACHESLAV V. 1966.   (17.3/A)                                              4890
     ON MULTI-DIMENSIONAL CONCENTRATION FUNCTIONS.
          (TRANSLATION OF NO. 4889)
     THEORY PROB. APPL., 11, 603-609.   (MR35-1062)


SCARDOVI, ITALO 1966.   (4.9)                                                        4891
     OSSERVAZIONI CRITICHE SOPRA UN TEMA DI CONNESSIONE STATISTICA.
     STATISTICA, BOLOGNA, 26, 217-230.


SCHAEFFER, ESTHER + DWYER, PAUL S. 1963.   (8.1/17.3/20.3)                           4892
     COMPUTATION WITH MULTIPLE K-STATISTICS.
     J. AMER. STATIST. ASSOC., 58, 120-151.   (MR26-3150)


SCHAEFFER, MAURICE S. + LEVITT, EUGENE E. 1956.   (4.8)                              4893
     CONCERNING KENDALL'S TAU, A NONPARAMETRIC CORRELATION COEFFICIENT.
     PSYCHOL. BULL., 53, 338-346.


SCHAIE, K. WARNER 1958.   (10.2/E)                                                   4894
     TESTS OF HYPOTHESES ABOUT DIFFERENCES BETWEEN TWO INTERCORRELATION MATRICES.
     J. EXPER. EDUC., 26, 241-245.


SCHAIE, K. WARNER   SEE ALSO
     2577. HORST, PAUL + SCHAIE, K. WARNER (1956)


SCHATZOFF, MARTIN 1966.   (8.4)                                                      4895
     EXACT DISTRIBUTIONS OF WILKS'S LIKELIHOOD RATIO CRITERION.
          (AMENDED BY NO. 4897)
     BIOMETRIKA, 53, 347-358.   (MR35-3812)


SCHATZOFF, MARTIN 1966.   (8.3/T)                                                    4896
     SENSITIVITY COMPARISONS AMONG TESTS OF THE GENERAL LINEAR HYPOTHESIS.
     J. AMER. STATIST. ASSOC., 61, 415-435.


SCHATZOFF, MARTIN 1967.   (8.4)                                                      4897
     CORRECTION TO: EXACT DISTRIBUTIONS OF WILKS'S LIKELIHOOD-RATIO CRITERION.
          (AMENDMENT OF NO. 4895)
     BIOMETRIKA, 54, 347-348.   (MR35-3812)


SCHEFFE, HENRY 1942.   (17.9/20.2)                                                   4898
     AN INVERSE PROBLEM IN CORRELATION THEORY.
     AMER. MATH. MONTHLY, 49, 99-104.   (MR4,P.23)


SCHEFFE, HENRY 1942.   (10.2/U)                                                      4899
     ON THE RATIO OF THE VARIANCES OF TWO NORMAL POPULATIONS.
     ANN. MATH. STATIST., 13, 371-388.   (MR4,P.164)


SCHEFFE, HENRY 1943.   (5.2/CU)                                                      4900
     ON SOLUTIONS OF THE BEHRENS-FISHER PROBLEM BASED ON THE T-DISTRIBUTION.
     ANN. MATH. STATIST., 14, 35-44.   (MR4,P.221)


SCHEFFE, HENRY 1961.   (19.3/C)                                                      4901
     SIMULTANEOUS INTERVAL ESTIMATES OF LINEAR FUNCTIONS OF PARAMETERS.
     BULL. INST. INTERNAT. STATIST., 38(4) 245-253.   (MR26-7086)


SCHEID, FRANCIS 1958.   (18.6/R)                                                     4902
     RADIAL DISTRIBUTION OF THE CENTER OF GRAVITY OF RANDOM POINTS ON A UNIT CIRCLE.
     J. RES. NAT. BUR. STANDARDS, 60, 307-308.


SCHEINBERG, ELIYAHU 1966.   (4.2)                                                    4903
     THE SAMPLING VARIANCE OF THE CORRELATION COEFFICIENTS ESTIMATED IN
          GENETIC EXPERIMENTS.
     BIOMETRICS, 22, 187-191.

SCHEINOK, PERRY A.    SEE
    5505. TRENCH, WILLIAM F. + SCHEINOK, PERRY A. (1966)

SCHEKTMAN, YVES 1965.    (15.5)                                                    4904
    SUR L'ESTIMATION DES MOMENTS DE LA VARIABLE LATENTE POUR UN MODELE CONTINU
    EN ANALYSE DE STRUCTURE LATENTE.
    C.R. ACAD. SCI. PARIS, 260, 6513-6516.

SCHEKTMAN, YVES + CAUSSINUS, HENRI 1965.    (15.5)                                 4905
    ETUDE D'UNE CLASSE DE MODELES CONTINUS NON-LINEAIRES EN ANALYSE DE STRUCTURE LATENTE.
    C.R. ACAD. SCI. PARIS, 261, 2805-2808.

SCHELLING, HERMANN VON   SEE   VON SCHELLING, HERMANN

SCHENCK, A.M.    SEE
    5354. TARVER, M.G. + SCHENCK, A.M. (1957)

SCHEUER, ERNEST M. 1962.    (14.5)                                                 4906
    MOMENTS OF THE RADIAL ERROR.
    J. AMER. STATIST. ASSOC., 57, 187-190.    (MR25-652)

SCHEUER, ERNEST M. + STOLLER, DAVID S. 1962.    (2.1/R)                            4907
    ON THE GENERATION OF NORMAL RANDOM VECTORS.
    TECHNOMETRICS, 4, 278-281.

SCHEUER, ERNEST M.    SEE ALSO
    2504. HOEL, PAUL G. + SCHEUER, ERNEST M. (1961)

SCHILLO, PAUL J.    SEE
    4979. SEVERO, NORMAN C.; MONTZINGO, LLOYD J., JR. + SCHILLO, PAUL J. (1965)

SCHLAFLI, LUDWIG 1858.    (20.4)                                                   4908
    ON THE MULTIPLE INTEGRAL WHOSE LIMITS ARE $p_1 = a_1 x + b_1 y + ... + c_1 z > 0, p_2 > 0, ... p_n > 0,$
    AND $x^2 + y^2 + ... + z^2 < 1$.
    QUART. J. PURE APPL. MATH., 2, 269-301.

SCHLAFLI, LUDWIG 1860.    (20.4)                                                   4909
    ON THE MULTIPLE INTEGRAL WHOSE LIMITS ARE $p_1 = a_1 x + b_1 y + ... + c_1 z > 0, p_2 > 0, ... p_n > 0,$
    AND $x^2 + y^2 + ... + z^2 < 1$.
    QUART. J. PURE APPL. MATH., 3, 54-68, 97-107.

SCHMERKOTTE, H. 1966.    (19.2/19.3/A)                                             4910
    ENTSCHEIDBARE BEHAUPTUNGEN, KONSISTENTE SCHATZER-UND TESTFOLGEN.
    Z. WAHRSCHEIN. VERW. GEBIETE, 5, 139-145.

SCHMETTERER, L. 1966.    (17.1/19.3)                                               4911
    ON THE ASYMPTOTIC EFFICIENCY OF ESTIMATES.
    RES. PAPERS STATIST. FESTSCHR. NEYMAN (DAVID), 301-317.

SCHMID, JOHN, JR. 1947.    (4.1)                                                   4912
    THE RELATIONSHIP BETWEEN THE COEFFICIENT OF CORRELATION AND THE ANGLE
    INCLUDED BETWEEN REGRESSION LINES.
    J. EDUC. RES., 41, 311-313.

SCHMID, JOHN, JR. 1950.    (6.2/20.3)                                              4913
    A COMPARISON OF TWO PROCEDURES FOR CALCULATING DISCRIMINANT FUNCTION
    COEFFICIENTS.
    PSYCHOMETRIKA, 15, 431-434.

SCHMID, JOHN, JR. + LEIMAN, JOHN M. 1957.    (15.1)                                4914
    THE DEVELOPMENT OF HIERARCHICAL FACTOR SOLUTIONS.
    PSYCHOMETRIKA, 22, 53-61.

SCHMID, JOHN, JR.    SEE ALSO
    2356. HARRIS, CHESTER W. + SCHMID, JOHN, JR. (1950)

    5567. TWERY, RAYMOND J.; SCHMID, JOHN, JR. + WRIGLEY, CHARLES F. (1958)

SCHMIDT, A. 1926.    (4.1)                                                         4915
    ZUR KRITIK DES KORRELATIONSFAKTORS.
    METEOROL. Z., 43, 329-334.

SCHMIDT, R. J. 1941.    (20.3)                                                     4916
    ON THE NUMERICAL SOLUTION OF LINEAR SIMULTANEOUS EQUATIONS BY AN
    ITERATIVE METHOD.
    PHILOS. MAG. SER. 7, 32, 369-383.    (MR3,P.276)

SCHMITT, ROBERT C. + CROSETTI, ALBERT H. 1954.   E(4.8)                                    4917
    ACCURACY OF THE RATIO-CORRELATION METHOD FOR ESTIMATING POSTCENSAL POPULATION.
    LAND ECON., 30, 279-281.

SCHNEEBERGER, C.   SEE
    4760. RUST, B.; BURRUS, W. R. + SCHNEEBERGER, C. (1966)

SCHNEEBERGER, H. 1966.   (8.3/8.4)                                                          4918
    IDENTITATSTEST VON REGRESSIONKOEFFIZIENTEN.
    METRIKA, 11, 39-45.

SCHNEIDER, OTTO 1942.   (18.1/B)                                                            4919
    SOBRE UN PARAMETRO USADO PARA LA CARACTERIZACION DE DISTRIBUCIONES
    ESTADISTICAS BIDIMENSIONALES.
    AN. SOC. CI. ARGENT., 133, 397-401.   (MR4,P.280)

SCHNELL, EDIT   SEE
    4843. SARKADI, KAROLY; SCHNELL, EDIT + VINCZE, ISTVAN (1962)

SCHOLS, CH. M. 1875.   (2.3/2.7/17.3/BT)                                                    4920
    OVER DE THEORIE DER FOUTEN IN DE RUIMTE EN IN HET PLATTE VLAK.
        (TRANSLATED AS NO. 4921)
    VERHANDL. KON. AKAD. WETENSCH., 15(6) 1-67.

SCHOLS, CH. M. 1886.   (2.3/2.7/17.3/BT)                                                    4921
    THEORIE DES ERREURS DANS LE PLAN ET DANS L'ESPACE.
        (TRANSLATION OF NO. 4920)
    ANN. ECOLE POLYTECH. DELFT, 2, 123-175.

SCHOLS, CH. M. 1887.   (2.3/17.3)                                                           4922
    LA LOI DE L'ERREUR RESULTANTE.
    ANN. ECOLE POLYTECH. DELFT, 3, 140-150.

SCHOLZ, H. 1959.   (20.3)                                                                   4923
    PRAKTISCHE LOSUNG VON AUFGABEN DER MATRIZENRECHNUNG.
    ANWEND. MATRIZ. WIRTSCH. STATIST. PROBLEME (ADAM ET AL.), 235-262.

SCHONEMAN, PETER H. 1966.   (20.2)                                                          4924
    A GENERALIZED SOLUTION OF THE ORTHOGONAL PROCRUSTES PROBLEM.
    PSYCHOMETRIKA, 31, 1-10.

SCHONEMAN, PETER H. 1966.   (15.3)                                                          4925
    VARISIM:   A NEW MACHINE METHOD FOR ORTHOGONAL ROTATION.
    PSYCHOMETRIKA, 31, 235-248.   (MR34-896)

SCHOOT, R. G.   SEE
    709. BRIER, GLENN W.; SCHOOT, R. G. + SIMMONS, V. L. (1940)

SCHORER, EDGAR 1941.   (4.9)                                                                4926
    ZUR FRAGE DER KORRELATION.
    Z. SCHWEIZ. STATIST. VOLKWIRTSCH., 77, 376-397.

SCHOTSMANS, L.   SEE
    1292. DE MEERSMAN, R. + SCHOTSMANS, L. (1964)

SCHREK, ROBERT 1942.   (4.1/4.8/8.8/E)                                                      4927
    LOGARITHMIC CORRELATION COEFFICIENTS AND REGRESSION EQUATIONS.
    HUMAN BIOL., 14, 95-103.

SCHULL, WILLIAM J. + KUDO, AKIO 1962.   (8.5)                                               4928
    CERTAIN MULTIVARIATE PROBLEMS ARISING IN HUMAN GENETICS.
    BULL. MATH. STATIST., 10(3) 77-88.

SCHULL, WILLIAM J.   SEE ALSO
    2715. ITO, KOICHI + SCHULL, WILLIAM J. (1964)

SCHULTZ, E. FRED, JR. + GOGGANS, JAMES F. 1961.   (8.9)                                     4929
    A SYSTEMATIC PROCEDURE FOR DETERMINING POTENT INDEPENDENT VARIABLES
    IN MULTIPLE REGRESSION AND DISCRIMINANT ANALYSIS.
    AGRIC. EXPER. STATION AUBURN UNIV. BULL., 336, 1-75.

SCHULTZ, EDNA L.  SEE  SHULTZ, EDNA L.

SCHULTZ, H. 1932.  (4.6)                                                      4930
    HOHE KORRELATIONSKOEFFIZIENTEN UND IHRE BEDEUTUNG FUR DAS STUDIUM DER
    NACHFRAGEKURVEN.
    ALLGEMEIN. STATIST. ARCH., 22, 293-298.

SCHULZ, GUNTHER 1933.  (20.3)                                                 4931
    ITERATIVE BERECHNUNG DER REZIPROKEN MATRIX.
    Z. ANGEW. MATH. MECH., 13, 57-59.

SCHULZ, GUNTHER 1947.  (17.1)                                                 4932
    DAS SUMMENPROBLEM BEI MEHRDIMENSIONALEN ARITHMETISCHEN
    WAHRSCHEINLICHKEITSVERTEILUNGEN.
    BER. MATH. TAGUNG TUBINGEN, 131-134.  (MR9,P.46)

SCHUMANN, T. E. W. 1940.  (8.1/20.3)                                          4933
    THE PRINCIPLES OF A MECHANICAL METHOD FOR CALCULATING REGRESSION EQUATIONS AND
    MULTIPLE CORRELATION COEFFICIENTS AND FOR THE SOLUTION OF SIMULTANEOUS LINEAR
    EQUATIONS.
    PHILOS. MAG. SER. 7, 29, 258-273.  (MR1,P.253)

SCHUR, J. 1911.  (20.1)                                                       4934
    BERMERKUNGEN ZUR THEORIE DER BESCHRANKTEN BILINEARFORMEN MIT UNENDLICH
    VIELEN VERANDERLICHEN.
    J. REINE ANGEW. MATH., 140, 1-28.

SCHUR, J. 1917.  (20.1)                                                       4935
    UBER POTENZREIHEN, DIE IM INNERN DES EINHEITSKREISES BESCHRANKT SIND.
    J. REINE ANGEW. MATH., 147, 205-232.

SCHUSTER, EDGAR + ELDERTON, ETHEL M. 1907.  E(4.8/6.7/T)                      4936
    THE INHERITANCE OF ABILITY,BEING A STATISTICAL STUDY OF THE OXFORD CLASS LISTS
    AND OF THE SCHOOL LISTS OF HARROW AND CHARTERHOUSE.
    EUGENICS LAB. MEM., 1, 1-40.

SCHUTZENBERGER, MARCEL-PAUL 1948.  (4.8)                                      4937
    VALEURS CARACTERISTIQUES DU COEFFICIENT DE CORRELATION PAR RANG DE
    KENDALL DANS LE CAS GENERAL.
    C.R. ACAD. SCI. PARIS, 226, 2122-2123.  (MR10,P.134)

SCHWARTZ, D.  SEE
    922. CHARBONNIER, A.; CYFFERS, B.; SCHWARTZ, D. +  (1957)
    VESSEREAU, A.

SCHWARTZ, E. R. + FOX, KARL A. 1942.  (4.8/E)                                 4938
    APPLICATION OF RANK CORRELATION TO THE DEVELOPMENT OF TESTING METHODS.
    BULL. AMER. SOC. TEST. MATER., 119, 21-24.

SCHWARTZ, JACOB T.  SEE
    417. BECK, ANATOLE + SCHWARTZ, JACOB T. (1957)

SCHWARTZ, R. 1966.  (8.3/19.2)                                                4939
    FULLY INVARIANT PROPER BAYES TESTS.
    MULTIVARIATE ANAL. PROC. INTERNAT. SYMP. DAYTON (KRISHNAIAH), 275-284.  (MR35-2425)

SCHWARTZ, R.  SEE ALSO
    3039. KIEFER, J. + SCHWARTZ, R. (1965)

SCHWEIGER, B. 1966.  (6.5/17.1)                                               4940
    ON THE UNIFORM CONTINUITY OF THE PROBABILITY DISTANCE.
    Z. WAHRSCHEIN. VERW. GEBIETE, 5, 357-360.

SCHWEIKER, ROBERT F.  SEE
    857. CARROLL, JOHN B. + SCHWEIKER, ROBERT F. (1951)

SCHWEITZER, BERTHOLD + SKLAR, A. 1960.  (17.3)                                4941
    STATISTICAL METRIC SPACES.
    PACIFIC J. MATH., 10, 313-334.  (MR22-5955)

SCIGOLEV, B. M. 1925.  (2.3)                                                          4942
    A DISSECTION OF THE DISTRIBUTION FUNCTION OF THREE VARIABLES INTO TWO
    SPHERICAL DISTRIBUTIONS.
        (IN RUSSIAN, WITH SUMMARY IN ENGLISH)
    RUSS. ASTRONOM. ZUR., 2(1) 1-6.


SCIGOLEV, B. M. 1926.  (2.3)                                                          4943
    ZERLEGUNG DER VERTEILUNG DER DREI VERANDERLICHEN IN ZWEI NORMALE
    REDUZIERTE HAUFIGKEITSFUNKTIONEN.
        (IN RUSSIAN, WITH SUMMARY IN ENGLISH)
    RUSS. ASTRONOM. ZUR., 3, 145-156.


SCIGOLEV, B. M. 1928.  (4.1/4.8)                                                      4944
    ON THE COEFFICIENT OF CORRELATION.
        (IN RUSSIAN, WITH SUMMARY IN ENGLISH)
    RUSS. ASTRONOM. ZUR., 5, 173-182.



SCIGOLEV, B. M. 1931.  E(18.6)                                                        4945
    GEOMETRICAL PROBABILITY OF DISCOVERY OF ECLIPSING BINARIES.
        (IN RUSSIAN, WITH SUMMARY IN ENGLISH)
    ASTRONOM. ZUR., 8, 214-222.


SCOTT, A. D. 1951.  (1.2/16.2)                                                        4946
    BIBLIOGRAPHY OF APPLICATIONS OF MATHEMATICAL STATISTICS TO ECONOMICS, 1943-1949.
    J. ROY. STATIST. SOC. SER. A, 114, 372-393.  (MR13,P.370)


SCOTT, A. D. 1953.  (1.2/16.2)                                                        4947
    BIBLIOGRAPHY OF APPLICATIONS OF MATHEMATICAL STATISTICS TO ECONOMICS,
    SUPPLEMENT FOR 1950.
    J. ROY. STATIST. SOC. SER. A, 116, 177-185.  (MR15,P.50)


SCOTT, A. RITCHIE-  SEE  RITCHIE-SCOTT, A.


SCOTT, ELIZABETH L. 1950.  (17.7)                                                     4948
    NOTE ON CONSISTENT ESTIMATES OF THE LINEAR STRUCTURAL RELATION
    BETWEEN TWO VARIABLES.
    ANN. MATH. STATIST., 21, 284-288.  (MR11,P.733)


SCOTT, ELIZABETH L.    SEE ALSO
    4002. NEYMAN, JERZY + SCOTT, ELIZABETH L. (1948)

    4003. NEYMAN, JERZY + SCOTT, ELIZABETH L. (1951)

    4004. NEYMAN, JERZY + SCOTT, ELIZABETH L. (1952)

    4005. NEYMAN, JERZY + SCOTT, ELIZABETH L. (1962)

    4006. NEYMAN, JERZY + SCOTT, ELIZABETH L. (1965)

    4007. NEYMAN, JERZY + SCOTT, ELIZABETH L. (1966)


SCOTT, JOHN T., JR. 1966.  (11.2/15.2)                                                4949
    FACTOR ANALYSIS AND REGRESSION.
    ECONOMETRICA, 34, 552-562.  (MR35-6283)



SCROGGS, JAMES E. + ODELL, P. L. 1966.  (20.1)                                        4950
    AN ALTERNATE DEFINITION OF A PSEUDOINVERSE OF A MATRIX.
    SIAM J. APPL. MATH., 14, 796-810.


SEAL, K. C. 1955.  (19.1/U)                                                           4951
    ON A CLASS OF DECISION PROCEDURES FOR RANKING MEANS OF NORMAL POPULATIONS.
    ANN. MATH. STATIST., 26, 387-398.  (MR19,P.1205)


SEAL, K. C. 1956.  (14.1/U)                                                           4952
    A NOTE ON SUMS OF COVARIANCES OF ORDER STATISTICS FROM NORMAL POPULATIONS.
    CALCUTTA STATIST. ASSOC. BULL., 7, 33-34.  (MR19,P.74)


SEAL, K. C. 1957.  (14.1/AU)                                                          4953
    APPROXIMATE DISTRIBUTION OF CERTAIN LINEAR FUNCTION OF ORDER STATISTICS.
    SANKHYA, 17, 345-348.  (MR19,P.1095)

SEAL, K. C. 1959.  (14.4)                                                        4954
     A SINGLE SAMPLING PLAN FOR CORRELATED VARIABLES WITH A SINGLE-SIDED
     SPECIFICATION LIMIT.
     J. AMER. STATIST. ASSOC., 54, 248-259.  (MR21-1678)


SEARLE, S.R. 1956.  (8.5/8.6/20.2/U)                                             4955
     MATRIX METHODS IN COMPONENTS OF VARIANCE AND COVARIANCE ANALYSIS.
     ANN. MATH. STATIST., 27, 737-748.  (MR18,P.346)


SEARLE, S.R. 1965.  (8.1/8.3)                                                    4956
     ADDITIONAL RESULTS CONCERNING ESTIMABLE FUNCTIONS AND GENERALIZED
     INVERSE MATRICES.
     J. ROY. STATIST. SOC. SER. B, 27, 486-490.


SEBER, G.A.F. 1964.  (19.2)                                                      4957
     ORTHOGONALITY IN ANALYSIS OF VARIANCE.
     ANN. MATH. STATIST., 35, 705-710.  (MR28-4643)


SEBER, G.A.F. 1964.  (2.5/8.3/20.2/A)                                            4958
     THE LINEAR HYPOTHESIS AND IDEMPOTENT MATRICES.
     J. ROY. STATIST. SOC. SER. B, 26, 261-266.


SEBESTYEN, GEORGE S. 1961.  (6.6)                                                4959
     RECOGNITION OF MEMBERSHIP IN CLASSES.
     IRE TRANS. INFORMATION THEORY, IT-7, 44-50.


SEBESTYEN, GEORGE S. 1962.  (6.6)                                                4960
     PATTERN RECOGNITION BY AN ADAPTIVE PROCESS OF SAMPLE SET CONSTRUCTION.
     IRE TRANS. INFORMATION THEORY, IT-8(5) S82-S91.


SEBESTYEN, GEORGE S. 1966.  (6.6/17.9)                                           4961
     AUTOMATIC OFF-LINE MULTIVARIATE DATA ANALYSIS.
     FALL JOINT COMP. CONF. PROC., 29, 685-694.


SEGAL, IRVING E. 1938.  (14.2)                                                   4962
     FIDUCIAL DISTRIBUTION OF SEVERAL PARAMETERS WITH APPLICATION TO A NORMAL SYSTEM.
     PROC. CAMBRIDGE PHILOS. SOC., 34, 41-47.


SEIBEL, JEAN L.    SEE
     824. CALVIN, ALLEN D. + SEIBEL, JEAN L. (1954)


SEIDEN, ESTHER + ZEMACH, RITA 1966.  (20.2)                                      4963
     ON ORTHOGONAL ARRAYS.
     ANN. MATH. STATIST., 37, 1355-1370.


SEIDEN, JOSEPH 1955.  (16.4)                                                     4964
     RELAXATION TRANSVERSALE ET FONCTION DE CORRELATION D'UNE VARIABLE QUANTIQUE.
     C.R. ACAD. SCI. PARIS, 241, 1450-1452.


SELLS, SAUL B. 1966.  (6.7)                                                      4965
     MULTIVARIATE TECHNOLOGY IN INDUSTRIAL AND MILITARY PERSONNEL PSYCHOLOGY.
     HANDB. MULTIVARIATE EXPER. PSYCHOL. (CATTELL), 841-855.


SELTZER, CARL C. 1937.  (6.5)                                                    4966
     A CRITIQUE OF THE COEFFICIENT OF RACIAL LIKENESS.
     AMER. J. PHYS. ANTHROP., 23, 101-109.


SEN, PRANAB KUMAR 1963.  (17.8/A)                                                4967
     ON THE PROPERTIES OF U-STATISTICS WHEN THE OBSERVATIONS ARE NOT
     INDEPENDENT I.  ESTIMATION OF NON-SERIAL PARAMETERS IN SOME STATIONARY
     STOCHASTIC PROCESS.
     CALCUTTA STATIST. ASSOC. BULL., 12, 69-92.


SEN, PRANAB KUMAR 1965.  (17.8/A)                                                4968
     ON SOME ASYMPTOTIC PROPERTIES OF A CLASS OF NONPARAMETRIC TESTS BASED
     ON THE NUMBER OF RARE EXCEEDANCES.
     ANN. INST. STATIST. MATH. TOKYO, 17, 233-255.


SEN, PRANAB KUMAR 1965.  (16.5/17.8)                                             4969
     SOME NON-PARAMETRIC TESTS FOR M-DEPENDENT TIME SERIES.
     J. AMER. STATIST. ASSOC., 60, 134-147.

SEN, PRANAB KUMAR 1966.   (17.8)                                                          4970
    ON SOME NONPARAMETRIC TESTS FOR SYMMETRY IN TWOWAY TABLES.
    J. INDIAN STATIST. ASSOC., 4, 125–142.

SEN, PRANAB KUMAR    SEE ALSO
    937. CHATTERJEE, SHOUTIR KISHORE + SEN, PRANAB KUMAR (1964)

    4366. PURI, MADAN LAL + SEN, PRANAB KUMAR (1966)

    938. CHATTERJEE, SHOUTIR KISHORE; SEN, PRANAB KUMAR + KISHORE, SHOUTIR (1965)


SENGSTAKE, CORD B. 1965.   (4.7/E)                                                        4971
    PERCEPTION OF DEVIATIONS IN REPETITIVE PATTERNS.
    J. EXPER. PSYCHOL., 70, 210–217.

SERRA, J. A.    SEE
    3474. LOPES QUEIROZ, A. + SERRA, J. A. (1944)

SESHADRI, VANAMAMALAI 1966.   (2.2)                                                       4972
    A CHARACTERISTIC PROPERTY OF THE MULTIVARIATE NORMAL DISTRIBUTION.
    ANN. MATH. STATIST., 37, 1829–1831.   (MR33-8013)

SESHADRI, VANAMAMALAI + PATIL, G. P. 1964.   (17.4/18.1)                                  4973
    A CHARACTERIZATION OF A BIVARIATE DISTRIBUTION BY THE MARGINAL AND
    THE CONDITIONAL DISTRIBUTIONS OF THE SAME COMPONENT.
    ANN. INST. STATIST. MATH. TOKYO, 15, 215–221.   (MR30-4318)

SETHURAMAN, J. 1961.   (17.1)                                                             4974
    SOME LIMIT THEOREMS FOR JOINT DISTRIBUTIONS.
    SANKHYA SER. A, 23, 379–386.   (MR25-2626)

SETHURAMAN, J. 1963.   (17.1/18.5)                                                        4975
    SOME LIMIT DISTRIBUTIONS CONNECTED WITH FIXED INTERVAL ANALYSIS.
    SANKHYA SER. A, 25, 395–398.

SETHURAMAN, J. 1964.   (17.1)                                                             4976
    ON THE PROBABILITY OF LARGE DEVIATIONS OF FAMILIES OF SAMPLE MEANS.
    ANN. MATH. STATIST., 35, 1304–1316.

SETHURAMAN, J. 1964.   (17.8/BC)                                                          4977
    FIXED INTERVAL ANALYSIS AND FRACTILE ANALYSIS.
    CONTRIB. STATIST. (MAHALANOBIS VOLUME), 449–470.

SETHURAMAN, J. 1965.   (17.4/17.6)                                                        4978
    ON A CHARACTERIZATION OF THE THREE LIMITING TYPES OF THE EXTREME.
    SANKHYA SER. A, 27, 357–364.   (MR34-2044)


SETHURAMAN, J.    SEE ALSO
    4742. RUBIN, HERMAN + SETHURAMAN, J. (1965)

    4743. RUBIN, HERMAN + SETHURAMAN, J. (1965)

SEVERO, NORMAN C.; MONTZINGO, LLOYD J., JR. + SCHILLO, PAUL J. 1965.   (2.1/14.2/AU)    4979
    CHARACTERISATION OF THE ASYMPTOTIC DISTRIBUTION OF A TRANSFORMED
    NORMAL VARIATE.
    SANKHYA SER. A, 27, 417–422.

SEVERO, NORMAN C.    SEE ALSO
    6071. ZELEN, MARVIN + SEVERO, NORMAN C. (1960)

SHAH, B.K. 1963.   (2.5/14.5)                                                             4980
    DISTRIBUTION OF DEFINITE AND OF INDEFINITE QUADRATIC FORMS FROM A
    NON-CENTRAL NORMAL DISTRIBUTION.
        (AMENDED BY NO. 4982)
    ANN. MATH. STATIST., 34, 186–190.   (MR26-1953)

SHAH, B.K. 1966.   (18.3/B)                                                               4981
    ON THE BIVARIATE MOMENTS OF ORDER STATISTICS FROM A LOGISTIC DISTRIBUTION.
    ANN. MATH. STATIST., 37, 1002–1010.

SHAH, B.K. 1968.  (2.5/14.5)                                                          4982
    CORRECTION TO: DISTRIBUTION OF DEFINITE AND OF INDEFINITE QUADRATIC FORMS FROM
    A NON-CENTRAL NORMAL DISTRIBUTION.
        (AMENDMENT OF NO. 4980)
    ANN. MATH. STATIST., 39, 289.   (MR26-1953)

SHAH, B.K. + KHATRI, C. G. 1961.  (2.5/14.5)                                          4983
    DISTRIBUTION OF A DEFINITE QUADRATIC FORM FOR NON-CENTRAL NORMAL VARIATES.
        (AMENDED BY NO. 4984)
    ANN. MATH. STATIST., 32, 883-887.  (MR23-A4196)

SHAH, B.K. + KHATRI, C. G. 1963.  (2.5/14.5)                                          4984
    CORRECTION TO: DISTRIBUTION OF A DEFINITE QUADRATIC FORM FOR
    NON-CENTRAL NORMAL VARIATES.
        (AMENDMENT OF NO. 4983)
    ANN. MATH. STATIST., 34, 673.

SHAH, D. C.   SEE
    4462. RAO, C. RADHAKRISHNA + SHAH, D. C. (1948)

SHAKUN, MELVIN F. 1965.  (5.1/14.5/C)                                                 4985
    MULTIVARIATE ACCEPTANCE SAMPLING PROCEDURES FOR GENERAL SPECIFICATION
    ELLIPSOIDS.
    J. AMER. STATIST. ASSOC., 60, 905-913.   (MR32-3240)

SHANAWANY, M. R. EL  SEE  EL SHANAWANY, M. R.

SHANBHAG, D. N. 1966.  (2.5)                                                          4986
    ON THE INDEPENDENCE OF QUADRATIC FORMS.
    J. ROY. STATIST. SOC. SER. B, 28, 582-583.

SHANNON, C. F. 1948.  (17.9)                                                          4987
    THE MATHEMATICAL THEORY OF COMMUNICATION.
    BELL SYSTEM TECH. J., 27, 379-423, 623-656.

SHARPE, G. E. + STYAN, GEORGE P. H. 1965.  (20.2)                                     4988
    CIRCUIT DUALITY AND THE GENERAL NETWORK INVERSE.
    IEEE TRANS. CIRCUIT THEORY,CT-12, 22-27.

SHARPE, G. E. + STYAN, GEORGE P. H. 1965.  (20.2)                                     4989
    A NOTE ON THE GENERAL NETWORK INVERSE.
    IEEE TRANS. CIRCUIT THEORY,CT-12, 632-633.

SHAW, LIAN CHENG  SEE  CHENG, SHAO-LIEN

SHELUPSKY, DAVID 1962.  (20.4)                                                        4990
    AN INTRODUCTION OF SPHERICAL COORDINATES.
    AMER. MATH. MONTHLY, 69, 644-646.

SHEN, EUGENE 1924.  (4.4/4.8)                                                         4991
    THE STANDARD ERROR OF CERTAIN ESTIMATED COEFFICIENTS OF CORRELATION.
    J. EDUC. PSYCHOL., 15, 462-465.

SHENTON, L. R. + JOHNSON, WHITNEY L. 1965.  (2.6)                                     4992
    MOMENTS OF A SERIAL CORRELATION COEFFICIENT.
    J. ROY. STATIST. SOC. SER. B, 27, 308-320.

SHEPARD, HERBERT A.   SEE
    319. BARRETT, F. DERMOT + SHEPARD, HERBERT A. (1953)

SHEPARD, ROGER N. 1962.  (15.5)                                                       4993
    THE ANALYSIS OF PROXIMITIES - MULTIDIMENSIONAL SCALING WITH AN
    UNKNOWN DISTANCE FUNCTION.  I.
    PSYCHOMETRIKA, 27, 125-140.  (MR25-3798)

SHEPARD, ROGER N. 1962.  (15.5)                                                       4994
    THE ANALYSIS OF PROXIMITIES:  MULTIDIMENSIONAL SCALING WITH AN
    UNKNOWN DISTANCE FUNCTION.  II.
    PSYCHOMETRIKA, 27, 219-246.  (MR30-3555)

SHEPARD, ROGER N. 1964.  (15.5)                                                       4995
    ATTENTION AND THE METRIC STRUCTURE OF THE STIMULUS SPACE.
    J. MATH. PSYCHOL., 1, 54-87.

SHEPARD, ROGER N. 1966.  (15.5)                                    **4996**
    METRIC STRUCTURES IN ORDINAL DATA.
    J. MATH. PSYCHOL., 3, 287-315.

SHEPARD, ROGER N. + CARROLL, J. DOUGLAS 1966.  (15.5/E)            **4997**
    PARAMETRIC REPRESENTATION OF NONLINEAR DATA STRUCTURES.
    MULTIVARIATE ANAL. PROC. INTERNAT. SYMP. DAYTON (KRISHNAIAH), 561-592.

SHEPPARD, WILLIAM F. 1889.  (20.4)                                 **4998**
    ON SOME EXPRESSIONS OF A FUNCTION OF A SINGLE VARIABLE IN TERMS OF
    BESSEL'S FUNCTIONS.
    QUART. J. PURE APPL. MATH., 23, 223-260.

SHEPPARD, WILLIAM F. 1898.  (2.1/9.1/14.3/B)                       **4999**
    ON THE GEOMETRICAL TREATMENT OF THE "NORMAL CURVE" OF STATISTICS,
    WITH SPECIAL REFERENCE TO CORRELATION AND TO THE THEORY OF ERROR.
    PROC. ROY. SOC. LONDON, 62, 170-173.

SHEPPARD, WILLIAM F. 1899.  (2.1/T)                                **5000**
    ON THE APPLICATION OF THE THEORY OF ERROR TO CASES OF NORMAL
    DISTRIBUTION AND NORMAL CORRELATION.
    PHILOS. TRANS. ROY. SOC. LONDON SER. A, 192, 101-167.

SHEPPARD, WILLIAM F. 1899.  (2.1/4.1/T)                            **5001**
    ON THE STATISTICAL REJECTION OF EXTREME VARIATIONS, SINGLE OR
    CORRELATED.  (NORMAL VARIATION AND NORMAL CORRELATION.)
    PROC. LONDON MATH. SOC., 31, 70-99.

SHEPPARD, WILLIAM F. 1900.  (2.7/BT)                               **5002**
    ON THE CALCULATION OF THE DOUBLE-INTEGRAL EXPRESSING NORMAL CORRELATION.
    TRANS. CAMBRIDGE PHILOS. SOC., 19, 23-68.

SHEPPARD, WILLIAM F. 1912.  (20.4)                                 **5003**
    SUMMATION OF THE COEFFICIENTS OF SOME TERMINATING HYPERGEOMETRICAL SERIES.
    PROC. LONDON MATH. SOC. SER. 2, 10, 469-478.

SHERIN, RICHARD J. 1966.  (15.2/20.2)                             **5004**
    A MATRIX FORMULATION OF KAISER'S VARIMAX CRITERION.
    PSYCHOMETRIKA, 31, 535-538.  (MR34-6941)

SHERMAN, JACK + MORRISON, WINIFRED J. 1950.  (20.1/20.3)          **5005**
    ADJUSTMENT OF AN INVERSE MATRIX CORRESPONDING TO A CHANGE IN ONE
    ELEMENT OF A GIVEN MATRIX.
    ANN. MATH. STATIST., 21, 124-127.  (MR11,P.693)

SHERMAN, SEYMOUR 1955.  (20.5)                                    **5006**
    A THEOREM ON CONVEX SETS WITH APPLICATIONS.
    ANN. MATH. STATIST., 26, 763-767.  (MR17,P.655)

SHIMADA, SHOZO 1954.  (14.1/U)                                    **5007**
    POWER OF THE R-CHART.
    REP. STATIST. APPL. RES. UN. JAPAN. SCI. ENGRS., 3, 70-74.  (MR16,P.727)

SHIMBEL, ALFONSO 1951.  (20.1/E)                                  **5008**
    APPLICATIONS OF MATRIX ALGEBRA TO COMMUNICATION NETS.
    BULL. MATH. BIOPHYS., 13, 165-178.  (MR13,P.371)

SHIMIZU, RYOICHI 1962.  (2.2)                                     **5009**
    CHARACTERIZATION OF THE NORMAL DISTRIBUTION. II.
    ANN. INST. STATIST. MATH. TOKYO, 14, 173-178.  (MR26-7073)

SHOMAN, G. N. 1935.  (4.6)                                        **5010**
    PROCEDURE FOR RAPID CALCULATION OF MULTIPLE CORRELATION COEFFICIENTS.
    J. AGRIC. RES., 50, 59-70.

SHRIVASTAVA, M.P. 1941.  (5.3)                                    **5011**
    ON THE $D^2$-STATISTIC.
    BULL. CALCUTTA MATH. SOC., 33, 71-86.  (MR4,P.23)

SHRIVASTAVA, M.P. 1941.  (18.1/B)                                 **5012**
    BI-VARIATE CORRELATION SURFACES.
    SCI. AND CULT., 6, 615-616.  (MR5,P.126)

SHUKLA, G. K. 1965.  (8.1/14.3)                                           5013
    MULTIVARIATE REGRESSION ESTIMATE.
    J. INDIAN STATIST. ASSOC., 3, 202-211.  (MR34-2136)


SHUKLA, G. K. 1966.  (18.6)                                               5014
    AN ALTERNATIVE MULTIVARIATE RATIO ESTIMATE FOR FINITE POPULATION.
    CALCUTTA STATIST. ASSOC. BULL., 15, 127-134.

SHULTZ, EDNA L.   SEE
    4021. NORTON, KENNETH A.; SHULTZ, EDNA L. + YARBROUGH, HELEN (1952)

SIBUYA, MASAAKI 1960.  (17.6/B)                                           5015
    BIVARIATE EXTREME STATISTICS, I.
    ANN. INST. STATIST. MATH. TOKYO, 11, 195-210.  (MR22-6043)

SIBUYA, MASAAKI 1962.  (18.6/R)                                           5016
    A METHOD FOR GENERATING UNIFORMLY DISTRIBUTED POINTS ON N-DIMENSIONAL SPHERES.
    ANN. INST. STATIST. MATH. TOKYO, 14, 81-85.


SIBUYA, MASAAKI   SEE ALSO
    2462. HIGUTI, ISAO + SIBUYA, MASAAKI (1965)


SICHEL, HERBERT S. 1952.  (15.4)                                          5017
    THE SELECTIVE EFFICIENCY OF A TEST BATTERY.
    PSYCHOMETRIKA, 17, 1-39.

SIDDIQUI, M.M. 1958.  (16.1/20.3)                                         5018
    ON THE INVERSION OF THE SAMPLE COVARIANCE MATRIX IN A STATIONARY
    AUTOREGRESSIVE PROCESS.
    ANN. MATH. STATIST., 29, 585-588.  (MR20-2070)

SIDDIQUI, M.M. 1958.  (8.1)                                               5019
    COVARIANCES OF LEAST-SQUARES ESTIMATES WHEN RESIDUALS ARE CORRELATED.
    ANN. MATH. STATIST., 29, 1251-1256.  (MR20-7374)

SIDDIQUI, M.M. 1960.  (8.3)                                               5020
    TESTS FOR REGRESSION COEFFICIENTS WHEN ERRORS ARE CORRELATED.
    ANN. MATH. STATIST., 31, 929-938.  (MR22-10140)

SIDDIQUI, M.M. 1960.  (17.6/B)                                            5021
    DISTRIBUTION OF QUANTILES IN SAMPLES FROM A BIVARIATE POPULATION.
    J. RES. NAT. BUR. STANDARDS SECT. B, 64, 145-150.  (MR25-4591)

SIDDIQUI, M.M. 1965.  (2.5/A)                                             5022
    APPROXIMATIONS TO THE DISTRIBUTION OF QUADRATIC FORMS.
    ANN. MATH. STATIST., 36, 677-682.  (MR31-1730)

SIEGEL, CARL LUDWIG 1935.  (20.1)                                         5023
    UBER DIE ANALYTISCHE THEORIE DER QUADRATISCHEN FORMEN.
    ANN. MATH. SER. 2, 36, 527-606.

SIEGEL, CARL LUDWIG 1937.  (20.1)                                         5024
    UBER DIE ANALYTISCHE THEORIE DER QUADRATISCHEN FORMEN III.
    ANN. MATH. SER. 2, 38, 212-291.

SIEGEL, LAURENCE + CURETON, EDWARD E. 1952.  (4.4/15.4)                   5025
    NOTE ON THE COMPUTATION OF BISERIAL CORRELATIONS IN ITEM ANALYSIS.
    PSYCHOMETRIKA, 17, 41-43.


SIEGERT, ARNOLD J. F. 1957.  (16.4)                                       5026
    A SYSTEMATIC APPROACH TO A CLASS OF PROBLEMS IN THE THEORY OF NOISE
    AND OTHER RANDOM PHENOMENA -- PART II, EXAMPLES.
    IRE TRANS. INFORMATION THEORY, IT-3, 38-43.

SIEGERT, ARNOLD J. F. 1958.  (16.4)                                       5027
    A SYSTEMATIC APPROACH TO A CLASS OF PROBLEMS IN THE THEORY OF NOISE
    AND OTHER RANDOM PHENOMENA--PART III, EXAMPLES.
    IRE TRANS. INFORMATION THEORY, IT-4, 4-14.  (MR23-A2246)

SIEGERT, ARNOLD J. F.   SEE ALSO
    1201. DARLING, D. A. + SIEGERT, ARNOLD J. F. (1957)

SIGIURA, NARIAKI 1965.  (17.8)                                                        5028
    MULTISAMPLE AND MULTIVARIATE NONPARAMETRIC TESTS BASED ON U STATISTICS
    AND THEIR ASYMPTOTIC EFFICIENCIES.
    OSAKA J. MATH., 2, 385–426.  (MR33–832)

SIGMON, HOWARD   SEE
    5238. STEVENS, CHARLES D. + SIGMON, HOWARD (1952)

SILBERSTEIN, LUDWIK 1946.  (17.9)                                                     5029
    ON TWO ACCESSORIES OF THREE-DIMENSIONAL COLORIMETRY.  I.  THE PROBABLE ERROR OF
    COLORIMETRIC TENSOR COMPONENTS AS DERIVED FROM A NUMBER OF COLOR MATCHINGS.
    II. THE DETERMINATION OF THE PRINCIPAL COLORIMETRIC AXES AT ANY POINT OF THE
    COLOR THREEFOLD.
    J. OPT. SOC. AMER., 36, 464–468.  (MR8,P.44)

SILER, WILLIAM   SEE
    6011. WOODBURY, MAX A. + SILER, WILLIAM (1966)

SILLITTO, G. P. 1947.  (4.8)                                                          5030
    THE DISTRIBUTION OF KENDALL'S $\tau$ COEFFICIENT OF RANK CORRELATION IN
    RANKINGS CONTAINING TIES.
    BIOMETRIKA, 34, 36–40.  (MR8,P.475)

SILVERMAN, S.   SEE
    314. BARNES, R. B. + SILVERMAN, S. (1934)

SILVERSTEIN, A. B. + MC LAIN, RICHARD E. 1964.  (15.4)                                5031
    INTERNAL STRUCTURE AND ITEM DIFFICULTY:  REPLY TO CAMPBELL.
    PSYCHOL. REP., 15, 838.

SILVEY, SAMUEL D. 1964.  (4.9)                                                        5032
    ON A MEASURE OF ASSOCIATION.
    ANN. MATH. STATIST., 35, 1157–1166.  (MR29–2839)

SILVEY, SAMUEL D.   SEE ALSO
    41. AITCHISON, JOHN + SILVEY, SAMUEL D. (1958)

    85. ALI, S. M. + SILVEY, SAMUEL D. (1965)

    86. ALI, S. M. + SILVEY, SAMUEL D. (1965)

    87. ALI, S. M. + SILVEY, SAMUEL D. (1965)

    88. ALI, S. M. + SILVEY, SAMUEL D. (1966)

SIMAIKA, J.B. 1941.  (4.7/5.2)                                                        5033
    ON AN OPTIMUM PROPERTY OF TWO IMPORTANT STATISTICAL TESTS.
    BIOMETRIKA, 32, 70–80.  (MR2,P.236)

SIMAIKA, J.B. 1946.  (4.9)                                                            5034
    NOTE ON M. FRECHET INDEX OF CORRELATION.
    PROC. MATH. PHYS. SOC. EGYPT, 3(2) 21–22.  (MR8,P.592; MR11,P.870)

SIMMONS, V. L.   SEE
    709. BRIER, GLENN W.; SCHOOT, R. G. + SIMMONS, V. L. (1940)

SIMON, HERBERT A. 1943.  (5.2)                                                        5035
    SYMMETRIC TESTS OF THE HYPOTHESIS THAT THE MEAN OF ONE NORMAL
    POPULATION EXCEEDS THAT OF ANOTHER.
    ANN. MATH. STATIST., 14, 149–154.  (MR5,P.128)

SIMON, HERBERT A. 1954.  (4.1)                                                        5036
    SPURIOUS CORRELATION – A CAUSAL INTERPRETATION.
    J. AMER. STATIST. ASSOC., 49, 467–479.

SIMONSEN, WILLIAM 1937.  (7.2/7.3/8.2)                                                5037
    ON THE DISTRIBUTIONS OF CERTAIN FUNCTIONS OF SAMPLES FROM A
    MULTIVARIATE INFINITE POPULATION.
    SKAND. AKTUARIETIDSKR., 20, 200–219.

SIMONSEN, WILLIAM 1944.  (4.7/7.3/8.2)                                                5038
    ON DISTRIBUTIONS OF FUNCTIONS OF SAMPLES FROM A NORMALLY DISTRIBUTED
    INFINITE POPULATION.
    SKAND. AKTUARIETIDSKR., 27, 235–261.  (MR7,P.212)

SIMONSEN, WILLIAM 1945.   (7.3/20.4)                                5039
    ON DISTRIBUTIONS OF FUNCTIONS OF SAMPLES FROM A NORMALLY DISTRIBUTED
    INFINITE POPULATION.
    SKAND. AKTUARIETIDSKR., 28, 20–43.   (MR7,P.212)

SIMPSON, PAUL B. 1951.   (17.5/B)                                   5040
    NOTE ON THE ESTIMATION OF A BIVARIATE DISTRIBUTION FUNCTION.
    ANN. MATH. STATIST., 22, 476–478.   (MR13,P.142)

SINGH, BAIKUNTH NATH 1958.   (17.8)                                 5041
    ON THE APPLICATIONS OF THE STATISTICS W'S AND T'S FOR TESTING TWO SAMPLES.
    J. INDIAN SOC. AGRIC. STATIST., 10, 107–130.   (MR22–1965)

SINGH, DALJIT 1958.   (17.9/20.4/T)                                 5042
    TABLES OF ORTHOGONAL POLYNOMIALS WHEN THE INDEPENDENT VARIABLE X IS
    IN THE GEOMETRIC PROGRESSION.
    J. INDIAN SOC. AGRIC. STATIST., 10, 131–140.   (MR22–300)

SINGH, NAUNIHAL 1960.   (3.2)                                       5043
    ESTIMATION OF PARAMETERS OF A MULTIVARIATE NORMAL POPULATION FROM
    TRUNCATED AND CENSORED SAMPLES.
    J. ROY. STATIST. SOC. SER. B, 22, 307–311.   (MR22–6039)

SINGLETON, HENRY E. 1950.   (20.3)                                 5044
    A DIGITAL ELECTRONIC CORRELATOR.
    PROC. INST. RADIO ENGRS., 38, 1422–1428.   (MR12,P.362)

SINHA, GURUDAS 1949.   (4.5)                                       5045
    A NOTE ON THE EXPRESSION FOR THE SAMPLE ESTIMATE OF THE COEFFICIENT
    OF PARTIAL CORRELATION.
    BULL. CALCUTTA MATH. SOC., 41, 159–161.   (MR11,P.259)

SIOTANI, MINORU 1951.   (19.3/A)                                   5046
    ASYMPTOTIC PROPERTIES OF MAXIMUM LIKELIHOOD ESTIMATES IN THE CASE OF SEVERAL
    UNKNOWN PARAMETERS.
    RES. MEM. INST. STATIST. MATH. TOKYO, 7, 467–477.

SIOTANI, MINORU 1956.   (5.3)                                      5047
    ON THE DISTRIBUTIONS OF THE HOTELLING'S $T^2$-STATISTICS.
        (TRANSLATION OF NO. 5048)
    ANN. INST. STATIST. MATH. TOKYO, 8, 1–14.   (MR18,P.243)

SIOTANI, MINORU 1956.   (5.3)                                      5048
    ON THE DISTRIBUTIONS OF THE HOTELLING'S $T^2$-STATISTICS.
        (IN JAPANESE, WITH SUMMARY IN ENGLISH, TRANSLATED AS NO. 5047)
    PROC. INST. STATIST. MATH. TOKYO, 4(1) 33–42.   (MR18,P.243)

SIOTANI, MINORU 1957.   (8.1/12.1)                                5049
    EFFECT OF THE ADDITIONAL VARIATES ON THE CANONICAL CORRELATION COEFFICIENTS.
        (IN JAPANESE, WITH SUMMARY IN ENGLISH)
    PROC. INST. STATIST. MATH. TOKYO, 5, 52–57.

SIOTANI, MINORU 1958.   (8.4/8.5/A)                               5050
    NOTE ON THE UTILIZATION OF THE GENERALIZED STUDENT RATIO IN THE
    ANALYSIS OF VARIANCE OR DISPERSION.
    ANN. INST. STATIST. MATH. TOKYO, 9, 157–171.

SIOTANI, MINORU 1959.   (8.4/T)                                   5051
    THE EXTREME VALUE OF THE GENERALIZED DISTANCES OF THE INDIVIDUAL
    POINTS IN THE MULTIVARIATE NORMAL SAMPLE.
        (AMENDED BY NO. 5054)
    ANN. INST. STATIST. MATH. TOKYO, 10, 183–208.   (MR21–5251)

SIOTANI, MINORU 1959.   (6.5/C)                                   5052
    ON THE RANGE IN MULTIVARIATE CASE.
        (IN JAPANESE, WITH SUMMARY IN ENGLISH)
    PROC. INST. STATIST. MATH. TOKYO, 6, 155–165.

SIOTANI, MINORU 1960.   (3.1/8.4/8.5/CT)                          5053
    NOTES ON MULTIVARIATE CONFIDENCE BOUNDS.
    ANN. INST. STATIST. MATH. TOKYO, 11, 167–182.   (MR22–5091)

SIOTANI, MINORU 1960.  (8.4/T)                                                              5054
    ERRATA TO: THE EXTREME VALUE OF THE GENERALIZED DISTANCES OF THE INDIVIDUAL
    POINTS IN THE MULTIVARIATE NORMAL SAMPLE.
        (AMENDMENT OF NO. 5051)
    ANN. INST. STATIST. MATH. TOKYO, 11, 220.  (MR21–5251)

SIOTANI, MINORU 1960.  (8.1/C)                                                              5055
    A NOTE ON THE INTERVAL ESTIMATION RELATED TO THE REGRESSION MATRIX.
    ANN. INST. STATIST. MATH. TOKYO, 12, 147–149.  (MR23–A723)

SIOTANI, MINORU 1961.  (5.3/14.2)                                                           5056
    THE EXTREME VALUE OF GENERALISED DISTANCES AND ITS APPLICATIONS.
    BULL. INST. INTERNAT. STATIST., 38(4) 591–599.

SIOTANI, MINORU 1961.  (1.1/1.2)                                                            5057
    THE RECENT PROGRESS IN THE THEORY OF MULTIVARIATE ANALYSIS.
        (IN JAPANESE, WITH SUMMARY IN ENGLISH)
    PROC. INST. STATIST. MATH. TOKYO, 8, 95–142.  (MR24–A2486)

SIOTANI, MINORU 1964.  (14.4)                                                               5058
    TOLERANCE REGIONS FOR A MULTIVARIATE NORMAL POPULATION.
    ANN. INST. STATIST. MATH. TOKYO, 16, 135–153.  (MR30–2613)

SIOTANI, MINORU 1964.  (18.1/19.3/T)                                                        5059
    INTERVAL ESTIMATION FOR LINEAR COMBINATIONS OF MEANS.
    J. AMER. STATIST. ASSOC., 59, 1141–1164.

SIOTANI, MINORU + HAYAKAWA, TAKESI 1964.  (4.5/4.6/7.3/A)                                    5060
    ASYMPTOTIC DISTRIBUTIONS OF FUNCTIONS OF WISHART MATRIX.
        (IN JAPANESE, WITH SUMMARY IN ENGLISH)
    PROC. INST. STATIST. MATH. TOKYO, 12, 191–198.  (MR33–1918)

SIOTANI, MINORU + KAWAKAMI, HISAKO 1963.  (8.8/T)                                            5061
    SIMULTANEOUS CONFIDENCE INTERVAL ESTIMATION ON REGRESSION COEFFICIENTS.
        (IN JAPANESE, WITH SUMMARY IN ENGLISH)
    PROC. INST. STATIST. MATH. TOKYO, 10, 79–98.  (MR27–3053)

SIRAZDINOV, S. H. + ABDURAHMANOV, T. 1964.  (18.4/18.5/A)                                    5062
    STATISTICAL ACCEPTANCE CONTROL BY SEVERAL INDICATIONS FOR N→∞.
        (IN RUSSIAN)
    THEORY PROB. MATH. STATIST. (IZDAT. NAUKA UZBEK. SSR), 13–25.  (MR33–6797)

SITARAMAN, B. 1940.  (4.1)                                                                  5063
    ON CORRELATION CONSTANTS IN MINGLED RECORDS.
    MATH. STUDENT, 8, 73–75.  (MR2,P.234)

SITGREAVES, ROSEDITH 1952.  (6.2/7.3)                                                       5064
    ON THE DISTRIBUTION OF TWO RANDOM MATRICES USED IN CLASSIFICATION PROCEDURES.
    ANN. MATH. STATIST., 23, 263–270.  (MR15,P.239)

SITGREAVES, ROSEDITH 1961.  (15.4/T)                                                        5065
    A STATISTICAL FORMULATION OF THE ATTENUATION PARADOX IN TEST THEORY.
    STUD. ITEM ANAL. PREDICT. (SOLOMON), 17–28.  (MR24–B1693A)

SITGREAVES, ROSEDITH 1961.  (15.4)                                                          5066
    OPTIMAL TEST DESIGN IN A SPECIAL TESTING SITUATION.
    STUD. ITEM ANAL. PREDICT. (SOLOMON), 29–45.  (MR24–B1693B)

SITGREAVES, ROSEDITH 1961.  (15.4)                                                          5067
    FURTHER CONTRIBUTIONS TO THE THEORY OF TEST DESIGN.
    STUD. ITEM ANAL. PREDICT. (SOLOMON), 46–63.  (MR24–B1693C)

SITGREAVES, ROSEDITH 1961.  (6.2)                                                           5068
    SOME RESULTS ON THE DISTRIBUTION OF THE W-CLASSIFICATION STATISTIC.
    STUD. ITEM ANAL. PREDICT. (SOLOMON), 241–251.  (MR23–A2998)

SITGREAVES, ROSEDITH    SEE ALSO
    688. BOWKER, ALBERT H. + SITGREAVES, ROSEDITH (1961)
    5380. TEICHROEW, DANIEL + SITGREAVES, ROSEDITH (1961)

    1545. ELFVING, GUSTAV: SITGREAVES, ROSEDITH + SOLOMON, HERBERT (1959)

    1546. ELFVING, GUSTAV: SITGREAVES, ROSEDITH + SOLOMON, HERBERT (1961)

SIXTL, F. + WENDER, K. 1964.   (15.1/15.2/15.5)                                    5069
     DER ZUSAMMENHANG ZWISCHEN MULTI-DIMENSIONALEN SKALIEREN UND FAKTORENANALYSE.
   BIOM. Z., 6, 251-261.

SKITOVIC, V.P. 1953.   (2.2)                                                       5070
     ON A PROPERTY OF A NORMAL DISTRIBUTION.
        (IN RUSSIAN).
   DOKL. AKAD. NAUK SSSR NOV. SER., 89, 217-219.   (MR14,P.1098)

SKITOVIC, V.P. 1954.   (2.2)                                                       5071
     LINEAR FORMS OF INDEPENDENT RANDOM VARIABLES AND THE NORMAL DISTRIBUTION LAW.
        (IN RUSSIAN)
   IZV. AKAD. NAUK SSSR SER. MAT., 18, 185-200.   (MR16,P.52)

SKLAR, A. 1959.   (17.3)                                                           5072
     FONCTIONS DE REPARTITION A N DIMENSIONS ET LEURS MARGES.
   PUBL. INST. STATIST. UNIV. PARIS, 8, 229-231.   (MR23-A2899)

SKLAR, A.    SEE ALSO
   4941. SCHWEITZER, BERTHOLD + SKLAR, A. (1960)

SLATER, LUCY J. 1962.   (1.1)                                                      5073
     REGRESSION ANALYSIS.
   COMPUT. J., 4, 287-291.   (MR26-7094)

SLATER, PATRICK 1947.   (15.1/15.2/E)                                              5074
     THE FACTOR ANALYSIS OF A MATRIX OF 2 X 2 TABLES.
   J. ROY. STATIST. SOC. SUPP., 9, 114-127.

SLATER, PATRICK 1951.   (15.1/20.2)                                                5075
     THE TRANSFORMATION OF A MATRIX OF NEGATIVE CORRELATIONS.
   BRITISH J. PSYCHOL. STATIST. SECT., 4, 9-20.

SLATER, PATRICK 1953.   (15.1)                                                     5076
     THE FACTOR ANALYSIS OF NEGATIVE CORRELATIONS.
   BRITISH J. STATIST. PSYCHOL., 6, 101-106.

SLATER, PATRICK 1956.   (15.4)                                                     5077
     WEIGHTING RESPONSES TO ITEMS IN ATTITUDE SCALES.
   BRITISH J. STATIST. PSYCHOL., 9, 41-48.

SLATER, PATRICK 1958.   (15.1)                                                     5078
     THE GENERAL RELATIONSHIP BETWEEN TEST FACTORS AND PERSON FACTORS -
     APPLICATION TO PREFERENCE MATRICES.
   NATURE, 181, 1225-1226.

SLATER, PATRICK    SEE ALSO
   4463. RAO, C. RADHAKRISHNA + SLATER, PATRICK (1949)

SLEPIAN, DAVID 1962.   (16.4)                                                      5079
     ON THE ONE-SIDED BARRIER PROBLEM FOR GAUSSIAN NOISE.
   BELL SYSTEM TECH. J., 41, 463-501.   (MR24-A3017)

SLEPIAN, DAVID    SEE ALSO
   2888. KAC, MARK + SLEPIAN, DAVID (1959)

   2129. GRENANDER, ULF; POLLAK, H.O. + SLEPIAN, DAVID (1959)

SLUCKII, E. F. 1916.   (4.8)                                                       5080
     CONCERNING AN ERROR IN THE APPLICATION OF FORMULAS FROM CORRELATION THEORY.
        (IN RUSSIAN)
   STATIST. VESTNIK, 3, 18-19.

SLUCKII, E. F. 1923.   (4.8)                                                       5081
     UBER EINIGE KORRELATIONSSCHEMEN UND UBER DEN SYSTEMATISCHEN FEHLER
     DES EMPIRISCHEN KORRELATIONSKOEFFIZIENTEN.
        (IN RUSSIAN)
   VESTNIK STATIST., 13, 31-50.

SLUCKII, E. F. 1929.   (4.2/A)                                                     5082
     SUR L'ERREUR QUADRATIQUE MOYENNE DU COEFFICIENT DE CORRELATION DANS LE
     CAS DES SUITES DES EPREUVES NON-INDEPENDANTES.
   C.R. ACAD. SCI. PARIS, 189, 612-614.

SLUCKII, E. F. 1930.  (16.5)                                                    5083
     ON THE QUADRATIC ERROR OF THE CORRELATION COEFFICIENT IN HOMOGENEOUS
     INTERRELATED SERIES.
          (IN RUSSIAN, WITH SUMMARY IN ENGLISH)
     TRUDY KONJUNKTURNYI INST. SER. 2, 2, 64-101.

SLUCKII, E. F. 1932.  (16.5)                                                    5084
     ON THE DISTRIBUTION OF ERRORS OF THE CORRELATION COEFFICIENT IN
     HOMOGENEOUS CONNECTED SERIES.
          (IN RUSSIAN, WITH SUMMARY IN ENGLISH)
     ZUR. GEOFIZ. INST. MOSK., 2, 66-98.

SLUTSKY, E. F.  SEE  SLUCKII, E. F.

SMILEY, MALCOLM F. 1964.  (2.5/8.8/20.2)                                        5085
     THE SPECTRAL THEOREM FOR FINITE MATRICES AND COCHRAN'S THEOREM.
     ANN. MATH. STATIST., 35, 443-444.  (MR28-1687)

SMITH, B. BABINGTON  SEE  BABINGTON SMITH, B.

SMITH, BRADFORD BIXBY 1925.  (16.5)                                             5086
     THE ERROR IN ELIMINATING SECULAR TREND AND SEASONAL VARIATION BEFORE
     CORRELATING TIME SERIES.
     J. AMER. STATIST. ASSOC., 20, 543-545.

SMITH, BRADFORD BIXBY 1926.  (16.5)                                             5087
     COMBINING THE ADVANTAGES OF FIRST-DIFFERENCE AND DEVIATION-FROM-TREND
     METHODS OF CORRELATING TIME SERIES.
     J. AMER. STATIST. ASSOC., 21, 55-59.

SMITH, BRADFORD BIXBY 1929.  (4.6)                                              5088
     ANOTHER ATTEMPT TO EXPLAIN MULTIPLE CORRELATION IN SIMPLE TERMS.
     J. AMER. STATIST. ASSOC., 24, 61-65.

SMITH, CEDRIC A.B. 1947.  (6.3/E)                                               5089
     SOME EXAMPLES OF DISCRIMINATION.
     ANN. EUGENICS, 13, 272-282.  (MR8,P.593)

SMITH, CEDRIC A.B. 1953.  (4.8/6.5)                                             5090
     THE LINEAR FUNCTION MAXIMIZING INTRACLASS CORRELATION.
     ANN. EUGENICS, 17, 286-292.

SMITH, CEDRIC A.B. 1957.  (4.8)                                                 5091
     ON THE ESTIMATION OF INTRACLASS CORRELATION.
     ANN. HUMAN GENET., 21, 363-373.  (MR19,P.1027)

SMITH, CEDRIC A.B.   SEE ALSO
     1662. FIELLER, E. C. + SMITH, CEDRIC A.B. (1951)

SMITH, CLARENCE DEWITT 1930.  (17.2/U)                                          5092
     ON GENERALISED TCHEBYCHEFF INEQUALITIES IN MATHEMATICAL STATISTICS.
     AMER. J. MATH., 52, 109-126.

SMITH, CLARENCE DEWITT 1958.  (4.1)                                             5093
     ON THE MATHEMATICS OF SIMPLE CORRELATION.
     MATH. MAG., 32, 57-69.  (MR20-5537)

SMITH, CLARENCE DEWITT 1959.  (8.1)                                             5094
     SOME FURTHER NOTES ON THE THEORY OF CORRELATION.
     MATH. MAG., 32, 269-270.  (MR21-2336)

SMITH, CYRIL STANLEY + GUTTMAN, LESTER 1953.  (18.6)                            5095
     MEASUREMENT OF INTERNAL BOUNDARIES IN THREE-DIMENSIONAL STRUCTURES
     BY RANDOM SECTIONING.
     J. METALS, 5, 81-87.

SMITH, E.H. + STONE, D. E. 1961.  (2.7/14.4/B)                                  5096
     A NOTE ON THE EXPECTED COVERAGE OF ONE CIRCLE BY ANOTHER, THE CASE OF
     OFFSET AIM.
     SIAM REV., 3, 51-53.  (MR23-A2897)

SMITH, GEORGE MILTON, JR. 1933.  E(11.1/15.2)                                   5097
     GROUP FACTORS IN MENTAL TESTS SIMILAR IN MATERIAL OR IN STRUCTURE.
     ARCH. OF PSYCHOL. (COLUMBIA), 24(156) 1-56.

SMITH, H. FAIRFIELD 1936.   (6.7/E)                                      5098
     A DISCRIMINANT FUNCTION FOR PLANT SELECTION.
          (AMENDED BY NO. 5099)
     ANN. EUGENICS, 7, 240-250.

SMITH, H. FAIRFIELD 1936.   (6.7/E)                                      5099
     ERRATA TO:  A DISCRIMINANT FUNCTION FOR PLANT SELECTION.
          (AMENDMENT OF NO. 5098)
     ANN. EUGENICS, 7, 240A.

SMITH, H. FAIRFIELD 1957.   (8.5)                                        5100
     INTERPRETATION OF ADJUSTED TREATMENT MEANS AND REGRESSIONS IN ANALYSIS
       OF COVARIANCE.
     BIOMETRICS, 13, 282-308.

SMITH, H. FAIRFIELD 1958.   (8.5)                                        5101
     A MULTIVARIATE ANALYSIS OF COVARIANCE.
     BIOMETRICS, 14, 107-127.   (MR19,P.1095)

SMITH, HARRY, JR.; GNANADESIKAN, R. + HUGHES, J.B. 1962.   (8.5)         5102
     MULTIVARIATE ANALYSIS OF VARIANCE (MANOVA).
     BIOMETRICS, 18, 22-41.   (MR27-2047)

SMITH, JOHN H. 1942.   (17.7)                                            5103
     WEIGHTED REGRESSIONS IN THE ANALYSIS OF ECONOMIC SERIES.
     STUD. MATH. ECON. ECONOMET. (SCHULTZ VOLUME), 151-164.

SMITH, KIRSTINE 1922.   (4.8)                                            5104
     THE STANDARD DEVIATIONS OF FRATERNAL AND PARENTAL CORRELATION COEFFICIENTS.
     BIOMETRIKA. 14, 1-22.

SMITH, R.C.T. 1953.   (20.4/T)                                           5105
     CONDUCTION OF HEAT IN THE SEMI-INFINTE SOLID, WITH A SHORT TABLE OF
       AN IMPORTANT INTEGRAL.
     AUSTRAL. J. PHYS., 6, 127-130.   (MR15,P.64)

SMITH, STEVENSON    SEE
     2578. HORST, PAUL + SMITH, STEVENSON (1950)

SMOLIAKOW, P.T. 1933.   (4.1/E)                                          5106
     DIE FECHNERSCHE KORRELATIONSFORMEL.
     METEOROL. Z., 50, 87-93.

SNEDECOR, GEORGE W.    SEE
     5749. WALLACE, H. A. + SNEDECOR, GEORGE W. (1925)

SNELL, J. L.    SEE
     2974. KEMENY, J. G. + SNELL, J. L. (1966)

SNEYERS, R. 1966.   (17.7/E)                                             5107
     ON THE NOTION OF CLIMATOLOGICAL INDEPENDENCE.  (IN FRENCH)
     REV. STATIST. APPL., 14(2) 31-36.

SNOW, BARBARA 1963.   (4.8)                                              5108
     THE DISTRIBUTION OF KENDALL'S TAU FOR SAMPLES OF FOUR FROM A NORMAL
       BIVARIATE POPULATION WITH CORRELATION $\rho$.
     BIOMETRIKA. 50, 538-539.

SNOW, BARBARA    SEE ALSO
     4167. PEARSON, EGON SHARPE + SNOW, BARBARA (1962)

SNOW, E.C. 1911.   E(4.6)                                                5109
     THE APPLICATION OF THE METHOD OF MULTIPLE CORRELATION TO THE
       ESTIMATION OF POST-CENSAL POPULATIONS.
          (WITH DISCUSSION)
     J. ROY. STATIST. SOC., 74, 575-629.

SNOW, E.C. 1911.   (17.9)                                               5110
     ON RESTRICTED LINES AND PLANES OF CLOSEST FIT TO SYSTEMS OF POINTS IN
       ANY NUMBER OF DIMENSIONS.
     PHILOS. MAG. SER. 6, 21, 367-386.

SNOW, F.C. 1912.   (4.4)
    THE APPLICATION OF THE CORRELATION COEFFICIENT TO MENDELIAN DISTRIBUTIONS.
    BIOMETRIKA, 8, 420–424.                                                                        5111

SOBEL, MILTON 1954.   (20.5)
    ON A GENERALIZATION OF AN INEQUALITY OF HARDY,LITTLEWOOD,AND POLYA.
    PROC. AMER. MATH. SOC., 5, 596–602.   (MR16,P.118)                                             5112

SOBEL, MILTON    SEE ALSO
    311. BARNDORFF-NIELSEN, O. + SOBEL, MILTON (1966)

    312. BARNDORFF-NIELSEN, O. + SOBEL, MILTON (1966)

    413. BECHHOFER, ROBERT E. + SOBEL, MILTON (1954)

    820. CACOULLOS, THEOPHILOS + SOBEL, MILTON (1966)

    1420. DUNNETT, CHARLES W. + SOBEL, MILTON (1954)

    1421. DUNNETT, CHARLES W. + SOBEL, MILTON (1955)

    2216. GUPTA, SHANTI S. + SOBEL, MILTON (1957)

    2217. GUPTA, SHANTI S. + SOBEL, MILTON (1962)

    4078. OLKIN, INGRAM + SOBEL, MILTON (1965)

    414. BECHHOFER, ROBERT E.; DUNNETT, CHARLES W. + SOBEL, MILTON (1954)

SOCIVKO, V. P. 1966.   (6.4)
    PATTERN RECOGNITION BY MEANS OF COMPUTING MACHINES.                                             5113
        (IN RUSSIAN)
    THEORY PROB. MATH. STATIST. (IZDAT.NAUKA UZBEK. SSR), 55–99.   (MR36–1232)

SOKAL, ROBERT R. 1958.   (15.3)
    THURSTONE'S ANALALYTICAL METHOD FOR SIMPLE STRUCTURE AND A MASS                                 5114
    MODIFICATION THEREOF.
    PSYCHOMETRIKA, 23, 237–257.

SOKAL, ROBERT R. 1961.   (6.5)
    DISTANCE AS A MEASURE OF TAXONOMIC SIMILARITY.                                                  5115
    SYSTEMATIC ZOOL., 10, 70–79.

SOKAL, ROBERT R.    SEE ALSO
    4600. ROHLF, F. J. + SOKAL, ROBERT R. (1962)

SOKOLOV, S.N.    SEE
    3056. KLEPIKOV, N. P. + SOKOLOV, S.N. (1957)

    3057. KLEPIKOV, N. P. + SOKOLOV, S.N. (1957)

SOLAND, RICHARD M.    SEE
    697. BRACKEN, JEROME + SOLAND, RICHARD M. (1966)

SOLOMON, HERBERT 1953.   (17.3/8)
    DISTRIBUTION OF THE MEASURE OF A RANDOM TWO-DIMENSIONAL SET.                                    5116
    ANN. MATH. STATIST., 24, 650–656.   (MR15,P.329)

SOLOMON, HERBERT 1956.   (6.4/15.4)
    PROBABILITY AND STATISTICS IN PSYCHOMETRIC RESEARCH:  ITEM ANALYSIS                             5117
    AND CLASSIFICATION TECHNIQUES.
    PROC. THIRD BERKELEY SYMP. MATH. STATIST. PROB., 5, 169–184.   (MR18,P.955)

SOLOMON, HERBERT 1960.   (6.4/ET)
    CLASSIFICATION PROCEDURES BASED ON DICHOTOMOUS RESPONSE VECTORS.                                5118
        (REPRINTED AS NO. 5121)
    CONTRIB. PROB. STATIST. (HOTELLING VOLUME), 414–423.   (MR22–11466)

SOLOMON, HERBERT 1960.   (15.1)
    A SURVEY OF MATHEMATICAL MODELS IN FACTOR ANALYSIS.                                             5119
    MATH. THINK. MEAS. BEHAV. (SOLOMON), 269–314.   (MR23–B2078)

SOLOMON, HERBERT 1961. (2.5/14.4/14.5/N) 5120
ON THE DISTRIBUTION OF QUADRATIC FORMS IN NORMAL VARIATES.
PROC. FOURTH BERKELEY SYMP. MATH. STATIST. PROB., 1, 645-653.

SOLOMON, HERBERT 1961. (6.4/ET) 5121
CLASSIFICATION PROCEDURES BASED ON DICHOTOMOUS RESPONSE VECTORS.
(REPRINT OF NO. 5118)
STUD. ITEM ANAL. PREDICT. (SOLOMON), 177-186. (MR22-11472)

SOLOMON, HERBERT + ROSNER, BENJAMIN 1954. (1.2/15.1) 5122
FACTOR ANALYSIS.
REV. EDUC. RES., 24, 421-438.

SOLOMON, HERBERT    SEE ALSO
1735. FORTIER, JEAN J. + SOLOMON, HERBERT (1966)

2100. GRAD, ARTHUR + SOLOMON, HERBERT (1955)

1545. ELFVING, GUSTAV; SITGREAVES, ROSEDITH + SOLOMON, HERBERT (1959)

1546. ELFVING, GUSTAV; SITGREAVES, ROSEDITH + SOLOMON, HERBERT (1961)

SOMERMEIJER, W.H. 1957. (4.1/8.1/E) 5123
SUBSTITUUT-VARIABLEN IN CORRELATIEBEREKINGEN.
(WITH SUMMARY IN ENGLISH)
STATISTICA NEERLANDICA, 11, 153-160. (MR19,P.780)

SOMERS, ROBERT H. 1962. (4.8) 5124
A NEW ASYMMETRIC MEASURE OF ASSOCIATION FOR ORDINAL VARIABLES.
AMER. SOCIOL. REV., 27, 799-811.

SOMERS, ROBERT H. 1962. (4.8) 5125
A SIMILARITY BETWEEN GOODMAN AND KRUSKAL'S TAU AND KENDALL'S TAU, WITH
A PARTIAL INTERPRETATION OF THE LATTER.
J. AMER. STATIST. ASSOC., 57, 804-812.

SOMERS, ROBERT H. 1965. (16.2) 5126
A CAPITAL INTENSIVE APPROACH TO THE SMALL SAMPLE PROPERTIES OF VARIOUS
SIMULTANEOUS EQUATION ESTIMATORS.
ECONOMETRICA, 33, 1-41. (MR31-5684)

SOMERVILLE, PAUL N. 1954. (19.1) 5127
SOME PROBLEMS OF OPTIMUM SAMPLING.
BIOMETRIKA, 41, 420-429. (MR16,P.604)

SOMMERVILLE, D. M. Y. 1927. (20.5) 5128
THE RELATIONS CONNECTING THE ANGLE-SUMS AND VOLUME OF A POLYTOPE IN
SPACE OF N DIMENSIONS.
PROC. ROY. SOC. LONDON SER. A, 115, 103-119.

SONDHI, MAN MOHAN 1961. (2.7) 5129
A NOTE ON THE QUADRIVARIATE NORMAL INTEGRAL.
BIOMETRIKA, 48, 201-203. (MR23-B2596)

SONQUIST, JOHN A.    SEE
3882. MORGAN, JAMES N. + SONQUIST, JOHN A. (1963)

SOOM, ERICH 1956. (4.1) 5130
DIE ANWENDUNG DER KORRELATIONSRECHNUNG IM ARBEITS- UND ZEITSTUDIENWESEN.
ALLGEMEIN. STATIST. ARCH., 40, 347-351.

SOPER, H.E. 1913. (4.2) 5131
ON THE PROBABLE ERROR OF THE CORRELATION COEFFICIENT TO A SECOND APPROXIMATION.
BIOMETRIKA, 9, 91-115.

SOPER, H.E. 1914. (4.4) 5132
ON THE PROBABLE ERROR OF THE BISERIAL EXPRESSION FOR THE CORRELATION
COEFFICIENT.
BIOMETRIKA, 10, 384-390.

SOPER, H.E. 1926. (18.1) 5133
THE MOMENTS OF THE HYPERGEOMETRIC SERIES.
J. ROY. STATIST. SOC., 89, 326-328.

SOPER, H.E. 1929. (4.7)                                                                                          5134
    THE GENERAL SAMPLING DISTRIBUTION OF THE MULTIPLE CORRELATION COEFFICIENT.
    J. ROY. STATIST. SOC., 92, 445–447.

SOPER, H.E.; YOUNG, ANDREW W.; CAVE, BEATRICE M.; 1917. (4.2)                                                     5135
    LEE, ALICE + PEARSON, KARL
    ON THE DISTRIBUTION OF THE CORRELATION COEFFICIENT IN SMALL SAMPLES. APPENDIX II
    TO THE PAPERS OF "STUDENT" AND R.A. FISHER.  A COOPERATIVE STUDY.
    BIOMETRIKA, 11, 328–413.

SPARRE ANDERSEN, ERIK  SEE  ANDERSEN, ERIK SPARRE

SPEARMAN, CHARLES 1904. (4.8)                                                                                     5136
    THE PROOF AND MEASUREMENT OF ASSOCIATION BETWEEN TWO THINGS.
    AMER. J. PSYCHOL., 15, 72–101.

SPEARMAN, CHARLES 1904. (15.1)                                                                                    5137
    "GENERAL INTELLIGENCE", OBJECTIVELY DETERMINED AND MEASURED.
    AMER. J. PSYCHOL., 15, 201–293.

SPEARMAN, CHARLES 1906. (4.1)                                                                                     5138
    A FOOTRULE FOR MEASURING CORRELATION.
    BRITISH J. PSYCHOL., 2, 89–108.

SPEARMAN, CHARLES 1907. (15.1)                                                                                    5139
    DEMONSTRATION OF FORMULAE FOR TRUE MEASUREMENT OF CORRELATION.
    AMER. J. PSYCHOL., 18, 161–169.

SPEARMAN, CHARLES 1910. (4.1)                                                                                     5140
    CORRELATION CALCULATED FROM FAULTY DATA.
    BRITISH J. PSYCHOL., 3, 271–295.

SPEARMAN, CHARLES 1911. (4.8)                                                                                     5141
    EINE NEUE KORRELATIONSFORMEL.
    KONG. EXPER. PSYCHOL. BER., 4, 189–191.

SPEARMAN, CHARLES 1914. (15.1)                                                                                    5142
    THE THEORY OF TWO FACTORS.
    PSYCHOL. REV., 21, 101–115.

SPEARMAN, CHARLES 1920. (15.1)                                                                                    5143
    MANIFOLD SUB-THEORIES OF "THE TWO FACTORS".
    PSYCHOL. REV., 27, 159–172.

SPEARMAN, CHARLES 1922. (15.1)                                                                                    5144
    RECENT CONTRIBUTIONS TO THE THEORY OF "TWO FACTORS".
    BRITISH J. PSYCHOL., 13, 26–30.

SPEARMAN, CHARLES 1922. (15.1)                                                                                    5145
    CORRELATIONS BETWEEN ARRAYS IN A TABLE OF CORRELATIONS.
    PROC. ROY. SOC. LONDON SER. A, 101, 94–100.

SPEARMAN, CHARLES 1923. (15.1)                                                                                    5146
    A FURTHER NOTE ON THE THEORY OF TWO FACTORS.
    BRITISH J. PSYCHOL., 13, 266–270.

SPEARMAN, CHARLES 1928. (15.1)                                                                                    5147
    PEARSON'S CONTRIBUTION TO THE THEORY OF TWO FACTORS.
    BRITISH J. PSYCHOL., 19, 95–101.

SPEARMAN, CHARLES 1930. (15.1)                                                                                    5148
    HETEROGENEITY AND THE THEORY OF FACTORS.
    AMER. J. PSYCHOL., 42, 645–646.

SPEARMAN, CHARLES 1930. (15.1/15.2/E)                                                                             5149
    LA THEORIE DES FACTEURS.
    ARCH. DE PSYCHOL. GENEVA, 22, 313–327.

SPEARMAN, CHARLES 1931. (15.1/15.2)                                                                               5150
    THE THEORY OF "TWO FACTORS" AND THAT OF SAMPLING.
    BRITISH J. EDUC. PSYCHOL., 1, 140–161.

SPEARMAN, CHARLES 1931.  (15.1)                                                      5151
    WHAT THE THEORY OF FACTORS IS NOT.
    J. EDUC. PSYCHOL., 22, 112-117.

SPEARMAN, CHARLES 1931.  (15.2)                                                      5152
    SAMPLING ERROR OF TETRAD DIFFERENCES.
    J. EDUC. PSYCHOL., 22, 388.

SPEARMAN, CHARLES 1933.  (15.1)                                                      5153
    THE UNIQUENESS AND EXACTNESS OF G.
    BRITISH J. PSYCHOL., 24, 106-108.

SPEARMAN, CHARLES 1933.  (15.1)                                                      5154
    THE FACTOR THEORY AND ITS TROUBLES.  II.  GARBLING THE EVIDENCE.
    J. EDUC. PSYCHOL., 24, 521-524.

SPEARMAN, CHARLES 1933.  (15.1)                                                      5155
    THE FACTOR THEORY AND ITS TROUBLES.  III.  MISREPRESENTATION OF THE THEORY.
    J. EDUC. PSYCHOL., 24, 591-601.

SPEARMAN, CHARLES 1934.  (15.2)                                                      5156
    ANALYSIS OF ABILITIES INTO FACTORS BY THE METHOD OF LEAST SQUARES.
    BRITISH J. EDUC. PSYCHOL., 4, 183-185.

SPEARMAN, CHARLES 1934.  (15.1)                                                      5157
    THE FACTOR THEORY AND ITS TROUBLES.  IV.  UNIQUENESS OF G.
    J. EDUC. PSYCHOL., 25, 142-153.

SPEARMAN, CHARLES 1934.  (15.1)                                                      5158
    THE FACTOR THEORY AND ITS TROUBLES. V. ADEQUACY OF PROOF.
    J. EDUC. PSYCHOL., 25, 310-319.

SPEARMAN, CHARLES 1934.  (15.1)                                                      5159
    THE FACTOR THEORY AND ITS TROUBLES.  CONCLUSION.  SCIENTIFIC VALUE.
    J. EDUC. PSYCHOL., 25, 383-391.

SPEARMAN, CHARLES 1934.  (15.1)                                                      5160
    PROFESSOR TRYON ON FACTORS.
    PSYCHOL. REV., 41, 306-307.

SPEARMAN, CHARLES 1937.  (4.1)                                                       5161
    ABILITIES AS SUMS OF FACTORS OR AS THEIR PRODUCTS.
    J. EDUC. PSYCHOL., 28, 629-631.

SPEARMAN, CHARLES 1939.  (15.3)                                                      5162
    THE FACTORIAL ANALYSIS OF ABILITY.  II.  DETERMINATION OF FACTORS.
    BRITISH J. PSYCHOL., 30, 78-83.

SPEARMAN, CHARLES 1946.  (15.1)                                                      5163
    THEORY OF GENERAL FACTOR.
    BRITISH J. PSYCHOL., 36, 117-131.

SPEARMAN, CHARLES + HOLZINGER, KARL J. 1924.  (15.2)                                 5164
    THE SAMPLING ERROR IN THE THEORY OF TWO FACTORS.
        (AMENDED BY NO. 5165)
    BRITISH J. PSYCHOL., 15, 17-19.

SPEARMAN, CHARLES + HOLZINGER, KARL J. 1925.  (15.2)                                 5165
    NOTE ON THE SAMPLING ERROR OF TETRAD DIFFERENCES.
        (AMENDMENT OF NO. 5164)
    BRITISH J. PSYCHOL., 16, 86-88.

SPEARMAN, CHARLES + HOLZINGER, KARL J. 1930.  (15.2)                                 5166
    THE AVERAGE VALUE FOR THE PROBABLE ERROR OF TETRAD DIFFERENCES.
    BRITISH J. PSYCHOL., 20, 368-370.

SPEARMAN, CHARLES    SEE ALSO
    3156. KRUEGER, F. + SPEARMAN, CHARLES (1906)

SPENCER, JOHN 1906.  (20.5)                                                          5167
    SOME PRACTICAL HINTS ON TWO-VARIABLE INTERPOLATION.
    J. INST. ACTUAR., 40, 293-304.

SPIEGELMAN, MORTIMER    SEE
   1393. DUBLIN, LOUIS I.; LOTKA, ALFRED J. + SPIEGELMAN, MORTIMER (1935)

SPITZER, F. 1958.    (16.4/B)                                                                    5168
   SOME THEOREMS CONCERNING TWO-DIMENSIONAL BROWNIAN MOTION.
   TRANS. AMER. MATH. SOC., 87, 187-197.    (MR21-3051)

SPJOTVOLL, EMIL 1966.    (8.8/20.2)                                                              5169
   A MIXED MODEL IN THE ANALYSIS OF VARIANCE.
   SKAND. AKTUARIETIDSKR., 49, 1-38.

SPOERL, CHARLES A. 1943.    (20.3)                                                               5170
   A FUNDAMENTAL PROPOSITION IN THE SOLUTION OF SIMULTANEOUS LINEAR EQUATIONS.
   TRANS. ACTUAR. SOC. AMER., 44, 276-288.    (MR5,P.161)

SPOERL, CHARLES A. 1944.    (20.3)                                                               5171
   ON SOLVING SIMULTANEOUS LINEAR EQUATIONS.    (WITH DISCUSSION)
   TRANS. ACTUAR. SOC. AMER., 45, 18-32, 67-69.    (MR6,P.50)

SPRENT, P. 1965.    (8.1)                                                                        5172
   FITTING A POLYNOMIAL TO CORRELATED EQUALLY SPACED OBSERVATIONS.
   BIOMETRIKA, 52, 275-276.    (MR34-5223)

SPRENT, P. 1966.    (17.7)                                                                       5173
   A GENERALIZED LEAST-SQUARES APPROACH TO LINEAR FUNCTIONAL RELATIONSHIPS.
   J. ROY. STATIST. SOC. SER. B, 28, 278-297.

SPRINGER, M. D. + THOMPSON, W. E. 1966.    (17.5)                                                5174
   THE DISTIRBUTION OF PRODUCTS OF INDEPENDENT RANDOM VARIABLES.
   SIAM J. APPL. MATH., 14, 511-526.

SPURR, WILLIAM A. 1951.    (4.1)                                                                 5175
   A SHORT-CUT MEASURE OF CORRELATION.
   J. AMER. STATIST. ASSOC., 46, 89-94.

SPURRELL, D. J. 1963.    E(11.1)                                                                 5176
   SOME METALLURGICAL APPLICATIONS OF PRINCIPAL COMPONENTS.
   APPL. STATIST., 12, 180-188.

SREEDHARAN PILLAI, K. C.    SEE   PILLAI, K.C.SREEDHARAN

SRIKANTAN, K. S. 1961.    (8.3/8.4/TU)                                                           5177
   TESTING FOR THE SINGLE OUTLIER IN A REGRESSION MODEL.
   SANKHYA SER. A, 23, 251-260.

SRINATH, MANDYAM D.    SEE
   1816. FU, YUMIN; SRINATH, MANDYAM D. + YEN, ANDREW T. (1965)

SRINIVASAN, R. 1965.    (18.1/B)                                                                 5178
   THE BIVARIATE GAMMA DISTRIBUTION AND THE RANDOM WALK PROBLEM.
   PROC. INDIAN ACAD. SCI. SECT. A, 62, 358-366.    (MR32-3150)

SRINIVASAN, S. K.    SEE
   4399. RAMAKRISHNAN, ALLADI + SRINIVASAN, S. K. (1956)

SRINIVASIENGAR, C.N. 1932.    (4.1)                                                              5179
   REMARKS ON SPURIOUS CORRELATION.
   J. INDIAN MATH. SOC., 19, 251-252.

SRIVASTAVA, A.B.L. 1960.    (8.2/18.3/B)                                                         5180
   THE DISTRIBUTION OF REGRESSION COEFFICIENTS IN SAMPLES FROM BIVARIATE
   NON-NORMAL POPULATIONS.  I. THEORETICAL INVESTIGATION.
   BIOMETRIKA, 47, 61-68.    (MR22-10072)

SRIVASTAVA, H. M. 1966.    (20.4)                                                                5181
   SOME EXPANSIONS IN PRODUCTS OF HYPERGEOMETRIC FUNCTIONS.
   PROC. CAMBRIDGE PHILOS. SOC., 62, 245-247.

SRIVASTAVA, H. M. 1966.    (20.4)                                                                5182
   THE INTEGRATION OF GENERALIZED HYPERGEOMETRIC FUNCTIONS.
   PROC. CAMBRIDGE PHILOS. SOC., 62, 761-764.

SRIVASTAVA, J. N. 1964.  (8.3)                                                    5183
    ON THE MONOTONICITY PROPERTY OF THE THREE MAIN TESTS FOR MULTIVARIATE
    ANALYSIS OF VARIANCE.
    J. ROY. STATIST. SOC. SER. B, 26, 77-81.  (MR30-5433)

SRIVASTAVA, J. N. 1965.  (8.1)                                                    5184
    A MULTIVARIATE EXTENSION OF THE GAUSS-MARKOV THEOREM.
    ANN. INST. STATIST. MATH. TOKYO, 17, 63-66.  (MR30-5413)

SRIVASTAVA, J. N. 1966.  (8.5)                                                    5185
    SOME GENERALIZATIONS OF MULTIVARIATE ANALYSIS OF VARIANCE.
    MULTIVARIATE ANAL. PROC. INTERNAT. SYMP. DAYTON (KRISHNAIAH), 129-145.  (MR35-7496)

SRIVASTAVA, J. N. 1966.  (10.2)                                                   5186
    ON TESTING HYPOTHESES REGARDING A CLASS OF COVARIANCE STRUCTURES.
    PSYCHOMETRIKA, 31. 147-164.  (MR33-6771)

SRIVASTAVA, J. N. 1966.  (8.5)                                                    5187
    INCOMPLETE MULTIRESPONSE DESIGNS.
    SANKHYA SER. A, 28, 377-388.  (MR35-5095)

SRIVASTAVA, J. N.    SEE ALSO
    4702. ROY, S.N. + SRIVASTAVA, J. N. (1964)

SRIVASTAVA, M. S. 1965.  (7.1)                                                    5188
    ON THE COMPLEX WISHART DISTRIBUTION.
    ANN. MATH. STATIST., 36, 313-315.  (MR30-2620)

SRIVASTAVA, M. S. 1965.  (10.2)                                                   5189
    SOME TESTS FOR THE INTRACLASS CORRELATION MODEL.
    ANN. MATH. STATIST., 36, 1802-1806.

SRIVASTAVA, M. S. 1966.  (19.1)                                                   5190
    ON A MULTIVARIATE SLIPPAGE PROBLEM I.
    ANN. INST. STATIST. MATH. TOKYO, 18, 299-305.  (MR36-2268)

SRIVASTAVA, M. S. 1966.  (19.1/A)                                                 5191
    SOME ASYMPTOTICALLY EFFICIENT SEQUENTIAL PROCEDURES FOR RANKING AND SLIPPAGE PROBLEMS.
    J. ROY. STATIST. SOC. SER. B, 28, 370-380.  (MR34-2132)

SRIVASTAVA, OM PRAKASH; HARKNESS, W. L. + BARTOO, J. B. 1964.  (14.1/A)           5192
    ASYMPTOTIC DISTRIBUTION OF DISTANCES BETWEEN ORDER STATISTICS FROM
    BIVARIATE POPULATIONS.
    ANN. MATH. STATIST., 35, 748-754.  (MR28-4617)

SRIVASTAVA, SURENDRA K. 1965.  (17.7)                                             5193
    AN ESTIMATE OF THE MEAN OF A FINITE POPULATION USING SEVERAL AUXILIARY
    VARIABLES.
    J. INDIAN STATIST. ASSOC., 3, 189-194.

SRIVASTAVA, SURENDRA K. 1966.  (17.7)                                             5194
    PRODUCT ESTIMATOR.
    J. INDIAN STATIST. ASSOC., 4, 29-37.

SRIVASTAVA, SURENDRA K. 1966.  (17.7)                                             5195
    ON RATIO AND LINEAR REGRESSION METHODS OF ESTIMATION WITH SEVERAL AUXILIARY VARIABLES.
    J. INDIAN STATIST. ASSOC., 4, 66-72.

SRIVASTAVA, TEJ NARAIN 1966.  (3.3)                                               5196
    USE OF RATIO ESTIMATE IN SINGLE SAMPLING PLAN FOR CORRELATED VARIABLE.
    J. INDIAN STATIST. ASSOC., 4, 51-57.

SRIVASTAVA, V. K.    SEE
    4648. ROY, A. R. + SRIVASTAVA, V. K. (1965)

STALNAKER, JOHN M.    SEE
    4542. RICHARDSON, MOSES W. + STALNAKER, JOHN M. (1933)

STANDISH, CHARLES 1956.  (17.3)                                                   5197
    N-DIMENSIONAL DISTRIBUTIONS CONTAINING A NORMAL COMPONENT.
    ANN. MATH. STATIST., 27, 1161-1165.  (MR18,P.957)

STANGE, K. 1948.   (2.1)                                                              5198
    UBER DIE VERTEILUNGSDICHTE DER MESS- ODER BEOBACHTUNGSFEHLER EINES
    DREIDIMENSIONALEN PUNKTRAUMES.
    Z. ANGEW. MATH. MECH., 28, 235-243.   (MR10,P.134)

STANTON, RALPH G. 1946.   (4.8/E)                                                     5199
    FILIAL AND FRATERNAL CORRELATIONS IN SUCCESSIVE GENERATIONS.
        (AMENDED BY NO. 5200)
    ANN. EUGENICS, 13, 18-24.

STANTON, RALPH G. 1947.   (4.8/E)                                                     5200
    ERRATA TO: FILIAL AND FRATERNAL CORRELATIONS IN SUCCESSIVE GENERATIONS.
        (AMENDMENT OF NO. 5199)
    ANN. EUGENICS, 14, V-A.

STARKEY, DAISY M. 1939.   (4.7/16.5)                                                  5201
    THE DISTRIBUTION OF THE MULTIPLE CORRELATION COEFFICIENT IN
    PERIODOGRAM ANALYSIS.
    ANN. MATH. STATIST., 10, 327-336.   (MR1,P.152)

STEAD, H. G. 1923.   (15.1/15.2/F)                                                    5202
    THE CORRECTION OF CORRELATION COEFFICIENTS.
    J. ROY. STATIST. SOC., 86, 412-419.

STECK, GEORGE P. 1958.   (2.7/T)                                                      5203
    A TABLE FOR COMPUTING TRIVARIATE NORMAL PROBABILITIES.
        (AMENDED BY NO. 5204)
    ANN. MATH. STATIST., 29, 780-800.   (MR20-323)

STECK, GEORGE P. 1959.   (2.7/T)                                                      5204
    CORRECTION TO: A TABLE FOR COMPUTING TRIVARIATE NORMAL PROBABILITIES.
        (AMENDMENT OF NO. 5203)
    ANN. MATH. STATIST., 30, 1297.   (MR20-323)

STECK, GEORGE P. 1962.   (2.7)                                                        5205
    ORTHANT PROBABILITIES FOR THE EQUICORRELATED MULTIVARIATE NORMAL DISTRIBUTION.
    BIOMETRIKA, 49, 433-445.   (MR28-1702)

STECK, GEORGE P. + OWEN, DONALD B. 1962.   (2.7)                                      5206
    A NOTE ON THE EQUICORRELATED MULTIVARIATE NORMAL DISTRIBUTION.
    BIOMETRIKA, 49, 269-271.   (MR26-4446)

STECK, GEORGE P.    SEE ALSO
    4119. OWEN, DONALD B. + STECK, GEORGE P. (1962)

    2218. GUPTA, SHANTI S.; PILLAI, K.C.SREEDHARAN + STECK, GEORGE P. (1964)

STEECE, H. M.    SEE
    2678. IMMER, F. R.; TYSDAL, H. M. + STEECE, H. M. (1939)

STEEL, ROBERT G. D. 1951.   (7.2/9.1)                                                 5207
    MINIMUM GENERALIZED VARIANCE FOR A SET OF LINEAR FUNCTIONS.
    ANN. MATH. STATIST., 22, 456-460.   (MR13,P.144)

STEFFENSEN, J. F. 1922.   (4.1/E)                                                     5208
    A CORRELATION-FORMULA.
    SKAND. AKTUARIETIDSKR., 5, 73-91.

STEFFENSEN, J. F. 1925.   (17.2)                                                      5209
    ON A GENERALIZATION OF CERTAIN INEQUALITIES BY TSCHEBYCHEF AND JENSEN.
    SKAND. AKTUARIETIDSKR., 8, 137-147.

STEFFENSEN, J. F. 1934.   (4.8)                                                       5210
    ON CERTAIN MEASURES OF DEPENDENCE BETWEEN STATISTICAL VARIABLES.
    BIOMETRIKA, 26, 251-255.

STEFFENSEN, J. F. 1941.   (4.9)                                                       5211
    ON THE COEFFICIENT OF CORRELATION FOR CONTINUOUS DISTRIBUTIONS.
        (AMENDED BY NO. 5213)
    SKAND. AKTUARIETIDSKR., 24, 1-12.   (MR3,P.5)

STEFFENSEN, J. F. 1941.   (4.8)                                                       5212
    ON THE ω TEST OF DEPENDENCE BETWEEN STATISTICAL VARIABLES.
    SKAND. AKTUARIETIDSKR., 24, 13-33.   (MR3,P.5)

STEFFENSEN, J. F. 1941.   (4.9)                                                    5213
     ERRATA TO:  ON THE COEFFICIENT OF CORRELATION FOR CONTINUOUS DISTRIBUTIONS.
          (AMENDMENT OF NO. 5211)
     SKAND. AKTUARIETIDSKR., 24, 232.   (MR3,P.5)


STEIN, CHARLES M. 1956.   (5.1/5.2)                                                5214
     THE ADMISSIBILITY OF HOTELLING'S $T^2$-TEST.
     ANN. MATH. STATIST., 27, 616-623.   (MR18,P.243)


STEIN, CHARLES M. 1956.   (3.1)                                                    5215
     INADMISSIBILITY OF THE USUAL ESTIMATOR FOR THE MEAN OF A
     MULTIVARIATE NORMAL DISTRIBUTION.
     PROC. THIRD BERKELEY SYMP. MATH. STATIST. PROB., 1, 197-206.   (MR18,P.948)


STEIN, CHARLES M. 1960.   (8.1)                                                    5216
     MULTIPLE REGRESSION.
     CONTRIB. PROB. STATIST. (HOTELLING VOLUME), 424-443.   (MR22-11467)


STEIN, CHARLES M. 1962.   (3.1/C)                                                  5217
     CONFIDENCE SETS FOR THE MEAN OF A MULTIVARIATE NORMAL DISTRIBUTION.
          (WITH DISCUSSION)
     J. ROY. STATIST. SOC. SER. B, 24, 265-296.   (MR26-5692)


STEIN, CHARLES M.   SEE ALSO
     2765. JAMES, W. + STEIN, CHARLES M. (1961)

     1999. GIRI, N. C.; KIEFER, J. + STEIN, CHARLES M. (1963)


STEIN, S. + STORER, J. E. 1956.   (2.1/R)                                          5218
     GENERATING A GAUSSIAN SAMPLE.
     IRE TRANS. INFORMATION THEORY, IT-2, 87-90.


STEINBERG, LEON   SEE
     3147. KRISHNAIAH, PARUCHURI R.; HAGIS, PETER,JR. + STEINBERG, LEON (1963)

     3148. KRISHNAIAH, PARUCHURI R.; HAGIS, PETER,JR. + STEINBERG, LEON (1965)


STEINER, L. 1931.   (4.9)                                                          5219
     ZUR DEUTUNG DES QUADRATS DES KORRELATIONSKOEFFIZIENTEN.
     METEOROL. Z., 48, 350-353.


STEPHENS, M. S.   SEE  STEPHENS, MICHAEL A.


STEPHENS, MICHAEL A. 1962.   (18.1/20.2/T)                                         5220
     EXACT AND APPROXIMATE TESTS FOR DIRECTIONS.  I.
     BIOMETRIKA, 49, 463-477.   (MR28-2591)


STEPHENS, MICHAEL A. 1962.   (18.1/20.2/T)                                         5221
     EXACT AND APPROXIMATE TESTS FOR DIRECTIONS.  II.
     BIOMETRIKA, 49, 547-552.   (MR28-2592)


STEPHENS, MICHAEL A. 1965.   (18.6)                                                5222
     APPENDIX TO: EQUATORIAL DISTRIBUTIONS ON A SPHERE.
     BIOMETRIKA, 52, 200-201.


STEPHENSON, WILLIAM 1934.   (4.8)                                                  5223
     A NOTE ON CORRELATIONS.
     BRITISH J. PSYCHOL., 24, 335-338.


STEPHENSON, WILLIAM 1935.   (4.5/15.1)                                             5224
     A NOTE ON FACTORS AND THE PARTIAL CORRELATION PROCEDURE.
          (AMENDED BY NO. 5287)
     BRITISH J. PSYCHOL., 25, 399-401.


STEPHENSON, WILLIAM 1935.   (15.1)                                                 5225
     ON THOMSON'S THEOREM FOR MEASURING G BY OVERLAPPING TESTS.
     BRITISH J. PSYCHOL., 25, 490-493.


STEPHENSON, WILLIAM 1935.   (15.1)                                                 5226
     A NOTE ON THE "PURIFICATION" TECHNIQUE IN TWO-FACTOR ANALYSIS.
     BRITISH J. PSYCHOL., 26, 196-198.

STEPHENSON, WILLIAM 1935.   (15.1/E)
  CORRELATING PERSONS INSTEAD OF TESTS.
  CHARACTER AND PERSONALITY, 4, 17-24.                                           5227

STEPHENSON, WILLIAM 1936.   (15.3)
  THE INVERTED FACTOR TECHNIQUE.
  BRITISH J. PSYCHOL., 26, 344-361.                                             5228

STEPHENSON, WILLIAM 1936.   (15.1)
  SOME RECENT CONTRIBUTIONS TO THE THEORY OF PSYCHOMETRY.
  CHARACTER AND PERSONALITY, 4, 294-304.                                        5229

STEPHENSON, WILLIAM 1936.   (15.1/E)
  INTRODUCTION TO INVERTED FACTOR ANALYSIS WITH SOME APPLICATIONS TO STUDIES.
  J. EDUC. PSYCHOL., 27, 353-367.                                              5230

STEPHENSON, WILLIAM 1936.   (15.1)
  THE FOUNDATIONS OF PSYCHOMETRY:  FOUR FACTOR SYSTEMS.
  PSYCHOMETRIKA. 1, 195-209.                                                    5231

STEPHENSON, WILLIAM 1939.   (15.1/15.2)
  TWO CONTRIBUTIONS TO THE THEORY OF MENTAL TESTING.  I.  A NEW
  PERFORMANCE TEST FOR MEASURING ABILITIES AS CORRELATION COEFFICIENTS.
  BRITISH J. PSYCHOL., 30, 19-35.                                              5232

STEPHENSON, WILLIAM 1939.   (15.1)
  THE FACTORIAL ANALYSIS OF ABILITY.  IV.  ABILITIES DEFINED AS
  NON-FRACTIONAL FACTORS.
  BRITISH J. PSYCHOL., 30, 94-104.                                             5233

STEPHENSON, WILLIAM 1940.   (15.1/15.2)
  TWO CONTRIBUTIONS TO THE THEORY OF MENTAL TESTING.  II.  A
  STATISTICAL REGARD OF PERFORMANCE.
  BRITISH J. PSYCHOL., 30, 230-247.                                            5234

STEPHENSON, WILLIAM 1952.   (15.1)
  SOME OBSERVATIONS ON Q TECHNIQUE.
  PSYCHOL. BULL., 49, 483-498.                                                 5235

STEPHENSON, WILLIAM + BROWN, WILLIAM 1933.   (15.1)
  PROFESSOR GODFREY THOMSON'S NOTE.
  BRITISH J. PSYCHOL., 24, 209-212.                                            5236

STEPHENSON, WILLIAM   SEE ALSO
  806. BURT, CYRIL + STEPHENSON, WILLIAM (1939)

STERN, CURT 1933.   (4.8)
  FAKTORENKOPPELUNG UND FAKTORENAUSTAUSCH.
  HANDB. VERERBUNGSWISS., 19, 1-331.                                           5237

STEVENS, CHARLES D. + SIGMON, HOWARD 1952.   (4.1)
  INSPECTION OF SIMPLE RELATIONS AMONG A NUMBER OF VARIABLES.
  SCIENCE NEW SER., 116, 368.                                                  5238

STEVENS, CHARLES D.   SEE ALSO
  5561. TURNER, MALCOLM E. + STEVENS, CHARLES D. (1959)

STEVENS, K. N. 1950.  E(16.3)
  AUTOCORRELATION ANALYSIS OF SPEECH SOUNDS.
  J. ACOUST. SOC. AMER., 22, 769-771.                                          5239

STEVENS, W. L. 1945.   (2.1/6.2/6.3/BE)
  ANALISE DISCRIMINANTE.
  QUESTOES MET. INST. ANTROP. COIMBRA, 7, 5-54.                                5240

STEVENS, W. L.   SEE ALSO
  2693. IRWIN, J. O.; COCHRAN, WILLIAM G.; FIELLER, E. C. +   (1936)
  STEVENS, W. L.

STEWART, NAOMI   SEE
  754. BRYAN, MIRIAM M.; BURKE, PAUL J. + STEWART, NAOMI (1952)

STEYN, H.S. 1951.    (7.1)                                                          5241
    THE WISHART DISTRIBUTION DERIVED BY SOLVING SIMULTANEOUS LINEAR
    DIFFERENTIAL EQUATIONS.
  BIOMETRIKA, 38, 470–472.   (MR13,P.662)

STEYN, H.S. 1951.    (18.1)                                                         5242
    ON DISCRETE MULTIVARIATE PROBABILITY FUNCTIONS.
  KON. NEDERL. WETENSCH. PROC. SER. A, 54, 23–30.   (MR12,P.722)

STEYN, H.S. 1955.    (18.3)                                                         5243
    ON DISCRETE MULTIVARIATE PROBABILITY FUNCTIONS OF HYPERGEOMETRIC TYPE.
  KON. NEDERL. WETENSCH. PROC. SER. A, 58, 588–595.   (MR17,P.634)

STEYN, H.S. 1956.    (18.3)                                                         5244
    ON THE UNIVARIABLE SERIES $F(t) \equiv F(a; b_1, b_2, ..., b_k; c; t, t^2, ..., t^k)$
    AND ITS APPLICATIONS IN PROBABILITY THEORY.
  KON. NEDERL. WETENSCH. PROC. SER. A, 59, 190–197.   (MR17,P.981)

STEYN, H.S. 1960.    (18.2)                                                         5245
    ON REGRESSION PROPERTIES OF MULTIVARIATE PROBABILITY FUNCTIONS OF
    PEARSON'S TYPES.
  KON. NEDERL. WETENSCH. PROC. SER. A, 63, 302–311.   (MR22-5095)

STEYN, H.S. 1963.    (18.1)                                                         5246
    ON APPROXIMATIONS FOR THE DISTRIBUTIONS OBTAINED FROM MULTIPLE EVENTS.
  KON. NEDERL. WETENSCH. PROC. SER. A, 66, 85–96.   (MR26-5661)

STEYN, H.S. + WILD, A.J.B. 1958.    (18.1)                                          5247
    ON EIGHTFOLD PROBABILITY FUNCTIONS.
  KON. NEDERL. WETENSCH. PROC. SER. A, 61, 129–138.   (MR20-3618)

STIELTJES, T.S. 1889.    (2.1/20.2/20.5)                                            5248
    EXTRAIT D'UNE LETTRE ADDRESSEE A M. HERMITE.
  BULL. SCI. MATH. SER. 2, 13, 170–172.

STIGUM, B. P.    SEE
  3002. KESTEN, HARRY + STIGUM, B. P. (1966)

  3003. KESTEN, HARRY + STIGUM, B. P. (1966)

STOLLER, DAVID S.    SEE
  4907. SCHEUER, ERNEST M. + STOLLER, DAVID S. (1962)

STONE, CHARLES 1965.    (17.1)                                                      5249
    A LOCAL LIMIT THEOREM FOR NONLATTICE MULTI-DIMENSIONAL DISTRIBUTION FUNCTIONS.
  ANN. MATH. STATIST., 36, 546–551.

STONE, D. E.    SEE
  5096. SMITH, E.H. + STONE, D. E. (1961)

STONE, J. R. N. 1947.    (16.1)                                                     5250
    PREDICTION FROM AUTOREGRESSIVE SCHEMES AND LINEAR STOCHASTIC DIFFERENCE SYSTEMS.
      (AMENDED BY NO. 5251)
  BULL. INST. INTERNAT. STATIST., 31(5) 29–38.   (MR12,P.512)

STONE, J. R. N. 1951.    (16.1)                                                     5251
    ERRATUM TO:  PREDICTION FROM AUTOREGRESSIVE SCHEMES AND LINEAR
    STOCHASTIC DIFFERENCE SYSTEMS.
      (AMENDMENT OF NO. 5250)
  ECONOMETRICA, 19, 227.   (MR12,P.512)

STONE, MERVYN 1964.    (14.2)                                                       5252
    COMMENTS ON A POSTERIOR DISTRIBUTION OF GEISSER AND CORNFIELD.
  J. ROY. STATIST. SOC. SER. B, 26, 274–276.   (MR30-2605)

STONE, RICHARD 1947.    (15.1/E)                                                    5253
    ON THE INTERDEPENDENCE OF BLOCKS OF TRANSACTIONS.
    (WITH DISCUSSION)
  J. ROY. STATIST. SOC. SUPP., 9, 1–45.

STORER, J. E.    SEE
  5218. STEIN, S. + STORER, J. E. (1956)

STOUFFER, SAMUEL A. 1934.  (4.5/E)                                                5254
    A COEFFICIENT OF "COMBINED PARTIAL CORRELATION" WITH AN EXAMPLE FROM
    SOCIOLOGICAL DATA.
    J. AMER. STATIST. ASSOC., 29, 70–71.

STOUFFER, SAMUEL A. 1936.  (4.5)                                                  5255
    EVALUATING THE EFFECT OF INADEQUATELY MEASURED VARIABLES IN PARTIAL
    CORRELATION ANALYSIS.
    J. AMER. STATIST. ASSOC., 31, 348–360.

STOUFFER, SAMUEL A. 1936.  (15.4)                                                 5256
    RELIABILITY COEFFICIENTS IN A CORRELATION MATRIX.
    PSYCHOMETRIKA, 1(2) 17–20.

STRASSEN, V. 1965.  (17.4)                                                        5257
    THE EXISTENCE OF PROBABILITY MEASURES WITH GIVEN MARGINALS.
    ANN. MATH. STATIST., 36, 423–439.

STRAUGHAN, J. H.   SEE
    1573. ESTES, WILLIAM K. + STRAUGHAN, J. H. (1954)

    1574. ESTES, WILLIAM K. + STRAUGHAN, J. H. (1963)

STRAUS, ERNST G.   SEE
    1732. FORSYTHE, GEORGE E. + STRAUS, ERNST G. (1955)

STRECKER, GRACE 1935.  (4.5)                                                      5258
    ON EVALUATING A COEFFICIENT OF PARTIAL CORRELATION.
    ANN. MATH. STATIST., 6, 143–145.

STRIEBEL, C. T. 1961.  (16.4/B)                                                   5259
    EFFICIENT ESTIMATION OF A REGRESSION PARAMETER FOR CERTAIN SECOND
    ORDER PROCESSES.
    ANN. MATH. STATIST., 32, 1299–1313.  (MR26-856)

STROTZ, ROBERT H. 1960.  (17.7)                                                   5260
    INTERDEPENDENCE AS A SPECIFICATION ERROR (PART II OF A TRIPTYCH ON CAUSAL CHAIN SYSTEMS).
    ECONOMETRICA, 28, 428–442.  (MR22-10792)

STROTZ, ROBERT H. + WOLD, HERMAN O. A. 1960.  (17.7)                              5261
    RECURSIVE VS. NONRECURSIVE SYSTEMS.  AN ATTEMPT AT SYNTHESIS  (PART I
    OF A TRIPTYCH ON CAUSAL CHAIN SYSTEMS).
    ECONOMETRICA, 28, 417–427.  (MR22-10791)

STRUIK, DIRK J. 1930.  (4.1/4.5)                                                  5262
    CORRELATION AND GROUP THEORY.
    BULL. AMER. MATH. SOC., 36, 869–878.

STUART, ALAN 1958.  (2.7)                                                         5263
    EQUALLY CORRELATED VARIATES AND THE MULTINORMAL INTEGRAL.
    J. ROY. STATIST. SOC. SER. B, 20, 373–378.  (MR20-4898)

STUART, ALAN 1962.  (14.2/B)                                                      5264
    GAMMA-DISTRIBUTED PRODUCTS OF INDEPENDENT VARIABLES.
    BIOMETRIKA, 49, 564–565.  (MR27-6317)

STUART, ALAN   SEE ALSO
    1431. DURBIN, JAMES + STUART, ALAN (1951)

"STUDENT" 1908.  (14.5/18.3/EU)                                                   5265
    THE PROBABLE ERROR OF A MEAN.
    BIOMETRIKA, 6, 1–25.

"STUDENT" 1908.  (4.2)                                                            5266
    PROBABLE ERROR OF A CORRELATION COEFFICIENT.
    BIOMETRIKA, 6, 302–310.

"STUDENT" 1913.  (4.8)                                                            5267
    THE CORRECTION TO BE MADE TO THE CORRELATION RATIO FOR GROUPING.
    BIOMETRIKA, 9, 316–320.

"STUDENT" 1914.   (4.1)
    THE ELIMINATION OF SPURIOUS CORRELATION DUE TO POSITION IN TIME OR SPACE.
    BIOMETRIKA. 10, 179-180.

5268

"STUDENT" 1921.   E(4.8)
    AN EXPERIMENTAL DETERMINATION OF THE PROBABLE ERROR OF DR SPEARMAN'S
    CORRELATION COEFFICIENTS.
    BIOMETRIKA, 13, 263-282.

5269

"STUDENT" 1931.   (5.1)
    ON THE 'Z' TEST.
    BIOMETRIKA. 23, 407-408.

5270

STUIT, D. B.   SEE
    3859. MONROE, WALTER S. + STUIT, D. B. (1935)

STUMPERS, F. LOUIS H. M. 1953.   (1.2)
    A BIBLIOGRAPHY OF INFORMATION THEORY. COMMUNICATION THEORY CYBERNETICS.
    TRANS. IRE PROF. GROUP INFORMATION THEORY,PGIT-2, 1-60.   (MR15,P.638)

5271

STURGES, ALEXANDER 1934.   (16.5)
    NOTE ON CORRELATION IN RANDOM SERIES.
    J. AMER. STATIST. ASSOC., 29, 421-422.

5272

STURGES, HERBERT A. 1927.   (4.9/E)
    THE THEORY OF CORRELATION APPLIED IN STUDIES OF CHANGING ATTITUDES.
    AMER. J. SOCIOL., 33, 269-275.

5273

STYAN, GEORGE P. H.   SEE
    4988. SHARPE, G. E. + STYAN, GEORGE P. H. (1965)

    4989. SHARPE, G. E. + STYAN, GEORGE P. H. (1965)

    5997. WOLF, ROBERT L.; MENDLOWITZ, MILTON; ROBOZ, JULIA;   (1966)
    STYAN, GEORGE P. H.; KORNFELD, PETER + WEIGL, ALFRED

SUBRAHMANIAN, K. 1966.   (18.2)
    A TEST FOR "INTRINSIC CORRELATION" IN THE THEORY OF ACCIDENT PRONENESS.
    J. ROY. STATIST. SOC. SER. B, 28, 180-189.   (MR34-3711)

5274

SUBRAHMANIAN, K. 1966.   (17.3/17.5/U)
    SOME CONTRIBUTIONS TO THE THEORY OF NON-NORMALITY--I.   (UNIVARIATE CASE).
    SANKHYA SER. A, 28, 389-406.

5275

SUBRAMANIAN, S 1935.   (4.5)
    ON A PROPERTY OF PARTIAL CORRELATION.
    J. ROY. STATIST. SOC., 98, 129.

5276

SUBRAMANIAN, S + VENKATACHARI, S. 1934.   (4.5)
    ON CERTAIN PROPERTIES OF THE CORRELATION COEFFICIENT.
    J. ANNAMALAI UNIV., 3, 202-204.

5277

SUGAR, GEORGE R. 1954.   (4.3)
    ESTIMATION OF CORRELATION COEFFICIENTS FROM SCATTER DIAGRAMS.
    J. APPL. PHYS., 25, 354-357.

5278

SUGAWARA, M.   SEE
    5343. TAKASHIMA, MICHIO + SUGAWARA, M. (1951)

SUGIYAMA, T. 1965.   (13.2/B)
    ON THE DISTRIBUTION OF THE LATENT VECTORS FOR PRINCIPAL COMPONENT ANALYSIS.
    ANN. MATH. STATIST., 36, 1875-1876.   (MR32-557)

5279

SUGIYAMA, T. 1966.   (11.2/13.2)
    ON THE DISTRIBUTION OF THE LARGEST LATENT ROOT AND THE CORRESPONDING LATENT
    VECTOR FOR PRINCIPAL COMPONENT ANALYSIS.
    ANN. MATH. STATIST., 37, 995-1001.   (MR34-897)

5280

SUKHATME, B. V.   SEE
    2087. GOSWAMI, J. N. + SUKHATME, B. V. (1965)

SUKHATME, SHASHIKALA 1962.   (8.8/17.8)                                    5281
    SOME NON-PARAMETRIC TESTS FOR LOCATION AND SCALE PARAMETERS IN A MIXED
    MODEL OF DISCRETE AND CONTINUOUS VARIABLES.
  J. INDIAN SOC. AGRIC. STATIST., 14, 121–137.


SULANKE, ROLF 1965.   (18.6)                                               5282
    SCHNITTPUNKTE ZUFALLIGER GERADEN.
  ARCH. MATH., 16, 320–324.



SULANKE, ROLF    SEE ALSO
   4517. RENYI, ALFRED + SULANKE, ROLF (1963)

   4518. RENYI, ALFRED + SULANKE, ROLF (1964)


SUMMERFIELD, A. + LUBIN, ARDIE 1951.   (8.1/20.3)                          5283
    A SQUARE ROOT METHOD OF SELECTING A MINIMUM SET OF VARIABLES IN
    MULTIPLE REGRESSION.
  PSYCHOMETRIKA, 16, 271–284, 425–437.


SUMNER, F. C. + DEHANEY, K. G. 1943.   (4.1)                               5284
    SIZE AND PLACEMENT OF INTERVALS AS INFLUENCING A PEARSON PRODUCT-MOMENT
    CORRELATION COEFFICIENT OBTAINED BY THE SCATTER-DIAGRAM PROCEDURE.
  J. PSYCHOL., 15, 27–30.


SUTCLIFFE, J. P. 1965.   (6.1)                                             5285
    A PROBABILITY MODEL FOR ERRORS OF CLASSIFICATION.   I. GENERAL CONSIDERATIONS.
  PSYCHOMETRIKA, 30, 73–96.   (MR33–838)


SUTCLIFFE, J. P. 1965.   (6.1/6.4/15.5)                                    5286
    A PROBABILITY MODEL FOR ERRORS OF CLASSIFICATION.   II. PARTICULAR CASE.
  PSYCHOMETRIKA, 30, 129–155.


SUTHERLAND, JOHN D. 1935.   (4.5/15.1)                                     5287
    A REPLY TO DR STEPHENSON'S NOTE ON FACTORS AND PARTIAL CORRELATION PROCEDURE.
    (AMENDMENT OF NO. 5224)
  BRITISH J. PSYCHOL., 25, 402–403.


SUTHERLAND, JOHN D. 1951.   (15.3)                                         5288
    FACTOR ROTATION BY THE METHOD OF EXTENDED VECTORS.
  BRITISH J. PSYCHOL. STATIST. SECT., 4, 21–30.


SVERDRUP, ERLING 1947.   (7.1)                                             5289
    DERIVATION OF THE WISHART DISTRIBUTION OF THE SECOND ORDER SAMPLE
    MOMENTS BY STRAIGHTFORWARD INTEGRATION OF A MULTIPLE INTEGRAL.
  SKAND. AKTUARIETIDSKR., 30, 151–166.   (MR9,P.453)


SVERDRUP, ERLING 1965.   (16.5/E)                                          5290
    ESTIMATES AND TEST PROCEDURES IN CONNECTION WITH STOCHASTIC MODELS FOR DEATHS,
    RECOVERIES AND TRANSFERS BETWEEN DIFFERENT STATES OF HEALTH.
  SKAND. AKTUARIETIDSKR., 48, 184–211.


SVERDRUP, ERLING 1966.   (19.1/19.2/19.4)                                  5291
    THE PRESENT STATE OF DECISION THEORY AND THE NEYMAN-PEARSON THEORY.
  REV. INST. INTERNAT. STATIST., 34, 309–333.


SWAIN, A. K. P. C. 1964.   (17.7)                                         5292
    THE USE OF SYSTEMATIC SAMPLING IN RATIO ESTIMATE.
  J. INDIAN STATIST. ASSOC., 2, 160–164.


SWANSON, Z.   SEE
   3314. LAWLEY, D. N. + SWANSON, Z. (1954)


SWERLING, PETER 1959.   (16.5/E)                                          5293
    FIRST ORDER ERROR PROPAGATION IN A STAGEWISE SMOOTHING PROCEDURE FOR
    SATELLITE OBSERVATIONS.
  J. ASTRONAUTICAL SCI., 6, 46–52.


SWINEFORD, FRANCES 1936.   (15.4)                                         5294
    VALIDITY OF TEST ITEMS.
  J. EDUC. PSYCHOL., 27, 68–78.

SWINEFORD, FRANCES 1936.  (4.8)                                                          **5295**
    BISERIAL R VERSUS PEARSON R AS MEASURES OF TEST-ITEM VALIDITY.
    J. EDUC. PSYCHOL., 27, 471-472.

SWINEFORD, FRANCES 1941.  (15.1)                                                         **5296**
    SOME COMPARISONS OF THE MULTI-FACTOR AND THE BI-FACTOR METHODS OF ANALYSIS.
    PSYCHOMETRIKA, 6, 375-382.

SWINEFORD, FRANCES 1946.  (17.9)                                                         **5297**
    GRAPHICAL AND TABULAR AIDS FOR DETERMINING SAMPLE SIZE WHEN PLANNING EXPERIMENTS
    WHICH INVOLVE COMPARISONS OF PERCENTAGES.
    PSYCHOMETRIKA, 11, 43-49.

SWINEFORD, FRANCES 1948.  (18.2)                                                         **5298**
    A TABLE FOR ESTIMATING THE SIGNIFICANCE OF THE DIFFERENCE BETWEEN
    CORRELATED PERCENTAGES.
    PSYCHOMETRIKA, 13, 23-25.  (MR9,P.363)

SWINEFORD, FRANCES 1949.  (4.8/T)                                                        **5299**
    FURTHER NOTES ON DIFFERENCES BETWEEN PERCENTAGES.
    PSYCHOMETRIKA, 14, 183-187.

SWINEFORD, FRANCES 1952.  (1.2/15.1/15.4)                                                **5300**
    SELECTED REFERENCES ON STATISTICS, THE THEORY OF TEST CONSTRUCTION,
    AND FACTOR ANALYSIS.
    SCHOOL REV., 60, 491-497.

SWINEFORD, FRANCES 1953.  (1.2/15.1/15.4)                                                **5301**
    SELECTED REFERENCES ON STATISTICS, THE THEORY OF TEST CONSTRUCTION,
    AND FACTOR ANALYSIS.
    SCHOOL REV., 61, 491-499.

SWINEFORD, FRANCES 1954.  (1.2/15.1/15.4)                                                **5302**
    SELECTED REFERENCES ON STATISTICS, THE THEORY OF TEST CONSTRUCTION,
    AND FACTOR ANALYSIS.
    SCHOOL REV., 62, 488-496.

SWINEFORD, FRANCES 1955.  (1.2/15.1/15.4)                                                **5303**
    SELECTED REFERENCES ON STATISTICS, THE THEORY OF TEST CONSTRUCTION,
    AND FACTOR ANALYSIS.
    SCHOOL REV., 63, 448-452.

SWINEFORD, FRANCES 1956.  (1.2/15.1/15.4)                                                **5304**
    SELECTED REFERENCES ON STATISTICS, THE THEORY OF TEST CONSTRUCTION,
    AND FACTOR ANALYSIS.
    SCHOOL REV., 64, 370-374.

SWINEFORD, FRANCES + FAN, CHUNG-TEH 1957.  (15.4)                                        **5305**
    A METHOD OF SCORE CONVERSION THROUGH ITEM STATISTICS.
    PSYCHOMETRIKA, 22, 185-188.  (MR19,P.333)

SWINEFORD, FRANCES + HOLZINGER, KARL J. 1933.  (1.2/15.1/15.4)                           **5306**
    SELECTED REFERENCES ON STATISTICS AND THE THEORY OF TEST CONSTRUCTION.
    SCHOOL REV., 41, 462-466.

SWINEFORD, FRANCES + HOLZINGER, KARL J. 1934.  (1.2/15.1/15.4)                           **5307**
    SELECTED REFERENCES ON STATISTICS AND THE THEORY OF TEST CONSTRUCTION.
    SCHOOL REV., 42, 459-465.

SWINEFORD, FRANCES + HOLZINGER, KARL J. 1935.  (1.2/15.1/15.4)                           **5308**
    SELECTED REFERENCES ON STATISTICS AND THE THEORY OF TEST CONSTRUCTION.
    SCHOOL REV., 43, 462-467.

SWINEFORD, FRANCES + HOLZINGER, KARL J. 1936.  (1.2/15.1/15.4)                           **5309**
    SELECTED REFERENCES ON STATISTICS, THE THEORY OF TEST CONSTRUCTION,
    AND FACTOR ANALYSIS.
    SCHOOL REV., 44, 462-469.

SWINEFORD, FRANCES + HOLZINGER, KARL J. 1937.  (1.2/15.1/15.4)                           **5310**
    SELECTED REFERENCES ON STATISTICS, THE THEORY OF TEST CONSTRUCTION,
    AND FACTOR ANALYSIS.
    SCHOOL REV., 45, 459-469.

SWINEFORD, FRANCES + HOLZINGER, KARL J. 1938.  (1.2/15.1/15.4)                                    5311
    SELECTED REFERENCES ON STATISTICS, THE THEORY OF TEST CONSTRUCTION,
    AND FACTOR ANALYSIS.
    SCHOOL REV., 46, 463-469.

SWINEFORD, FRANCES + HOLZINGER, KARL J. 1939.  (1.2/15.1/15.4)                                    5312
    SELECTED REFERENCES ON STATISTICS, THE THEORY OF TEST CONSTRUCTION,
    AND FACTOR ANALYSIS.
    SCHOOL REV., 47, 459-466.

SWINEFORD, FRANCES + HOLZINGER, KARL J. 1940.  (1.2/15.1/15.4)                                    5313
    SELECTED REFERENCES ON STATISTICS, THE THEORY OF TEST CONSTRUCTION,
    AND FACTOR ANALYSIS.
    SCHOOL REV., 48, 460-466.

SWINEFORD, FRANCES + HOLZINGER, KARL J. 1941.  (1.2/15.1/15.4)                                    5314
    SELECTED REFERENCES ON STATISTICS, THE THEORY OF TEST CONSTRUCTION,
    AND FACTOR ANALYSIS.
    SCHOOL REV., 49, 461-467.

SWINEFORD, FRANCES + HOLZINGER, KARL J. 1942.  (1.2/15.1/15.4)                                    5315
    SELECTED REFERENCES ON STATISTICS, THE THEORY OF TEST CONSTRUCTION,
    AND FACTOR ANALYSIS.
    SCHOOL REV., 50, 456-465.

SWINEFORD, FRANCES + HOLZINGER, KARL J. 1943.  (1.2/15.1/15.4)                                    5316
    SELECTED REFERENCES ON STATISTICS, THE THEORY OF TEST CONSTRUCTION,
    AND FACTOR ANALYSIS.
    SCHOOL REV., 51, 369-374.

SWINEFORD, FRANCES + HOLZINGER, KARL J. 1944.  (1.2/15.1/15.4)                                    5317
    SELECTED REFERENCES ON STATISTICS, THE THEORY OF TEST CONSTRUCTION,
    AND FACTOR ANALYSIS.
    SCHOOL REV., 52, 370-375.

SWINEFORD, FRANCES + HOLZINGER, KARL J. 1945.  (1.2/15.1/15.4)                                    5318
    SELECTED REFERENCES ON STATISTICS, THE THEORY OF TEST CONSTRUCTION,
    AND FACTOR ANALYSIS.
    SCHOOL REV., 53, 364-368.

SWINEFORD, FRANCES + HOLZINGER, KARL J. 1946.  (1.2/15.1/15.4)                                    5319
    SELECTED REFERENCES ON STATISTICS, THE THEORY OF TEST CONSTRUCTION,
    AND FACTOR ANALYSIS.
    SCHOOL REV., 54, 364-367.

SWINEFORD, FRANCES + HOLZINGER, KARL J. 1947.  (1.2/15.1/15.4)                                    5320
    SELECTED REFERENCES ON STATISTICS, THE THEORY OF TEST CONSTRUCTION,
    AND FACTOR ANALYSIS.
    SCHOOL REV., 55, 363-368.

SWINEFORD, FRANCES + HOLZINGER, KARL J. 1948.  (1.2/15.1/15.4)                                    5321
    SELECTED REFERENCES ON STATISTICS, THE THEORY OF TEST CONSTRUCTION,
    AND FACTOR ANALYSIS.
    SCHOOL REV., 56, 361-366.

SWINEFORD, FRANCES + HOLZINGER, KARL J. 1949.  (1.2/15.1/15.4)                                    5322
    SELECTED REFERENCES ON STATISTICS, THE THEORY OF TEST CONSTRUCTION,
    AND FACTOR ANALYSIS.
    SCHOOL REV., 57, 315-320.

SWINEFORD, FRANCES + HOLZINGER, KARL J. 1950.  (1.2/15.1/15.4)                                    5323
    SELECTED REFERENCES ON STATISTICS, THE THEORY OF TEST CONSTRUCTION,
    AND FACTOR ANALYSIS.
    SCHOOL REV., 58, 489-493.

SWINEFORD, FRANCES + HOLZINGER, KARL J. 1951.  (1.2/15.1/15.4)                                    5324
    SELECTED REFERENCES ON STATISTICS, THE THEORY OF TEST CONSTRUCTION,
    AND FACTOR ANALYSIS.
    SCHOOL REV., 59, 489-497.

SWINEFORD, FRANCES    SEE ALSO
    2534. HOLZINGER, KARL J. + SWINEFORD, FRANCES (1932)

    2535. HOLZINGER, KARL J. + SWINEFORD, FRANCES (1937)

SYLVESTER, J. J. 1867.  (20.2)                                              5325
    THOUGHTS ON INVERSE ORTHOGONAL MATRICES, SIMULTANEOUS SIGN-SUCCESSIONS, AND
    TESSELLATED PAVEMENTS IN TWO OR MORE COLOURS, WITH APPLICATION TO NEWTON'S RULE,
    ORNAMENTAL TILE-WORK, AND THE THEORY OF NUMBERS.
    PHILOS. MAG. SER. 4, 34, 461-475.

SYMONDS, PERCIVAL M. 1930.  (4.4/T)                                         5326
    COMPARISON OF STATISTICAL MEASURES OF OVERLAPPING WITH CHARTS FOR
    ESTIMATING THE VALUE OF BI-SERIAL R.
    J. EDUC. PSYCHOL., 21, 586-596.

TACK, P.I.   SEE
    392. BATEN, WILLIAM DOWELL; TACK, P.I. + BAEDER, HELEN A. (1958)

TAGA, YASUSHI 1959.  (20.3)                                                 5327
    LINEAR COMPUTATIONS IN STATISTICAL ANALYSIS.
        (IN JAPANESE, WITH SUMMARY IN ENGLISH)
    PROC. INST. STATIST. MATH. TOKYO, 7, 109-122.

TAGA, YASUSHI 1963.  (19.3)                                                 5328
    ON UNBIASED RATIO AND REGRESSION ESTIMATORS.
        (IN JAPANESE, WITH SUMMARY IN ENGLISH)
    PROC. INST. STATIST. MATH. TOKYO, 11, 1-6.

TAGA, YASUSHI 1966.  (16.5)                                                 5329
    OPTIMUM TIME SAMPLING IN STOCHASTIC PROCESSES.
        (IN JAPANESE)
    PROC. INST. STATIST. MATH. TOKYO, 14, 59-61.

TAGA, YASUSHI + KOMAZAWA, TSUTOMU 1960.  (20.3)                             5330
    SOME REMARKS ON THE SOLUTIONS OF ALGEBRAIC EQUATIONS BASED ON THE METHOD FOR
    UNITARY TRANSFORMATIONS OF THE COMPANION MATRICES.
        (IN JAPANESE, WITH SUMMARY IN ENGLISH)
    PROC. INST. STATIST. MATH. TOKYO, 8, 39-44.   (MR25-1643)

TAGIURI, RENATO   SEE
    3531. LUCE, R. DUNCAN; MACY, JOSIAH, JR. + TAGIURI, RENATO (1955)

TAKACS, LAJOS 1957.  (16.4/18.1)                                            5331
    ON SECONDARY STOCHASTIC PROCESSES GENERATED BY A MULTIDIMENSIONAL
    POISSON PROCESS.
    MAGYAR TUD. AKAD. MAT. KUTATO. INT. KOZL., 2, 71-80.   (MR20-7356)

TAKACS, LAJOS 1961.  (1.2)                                                  5332
    CHARLES JORDAN, 1871-1959.
    ANN. MATH. STATIST., 32, 1-11.   (MR22-6686)

TAKACS, LAJOS 1964.  (3.1/17.2/C)                                           5333
    AN APPLICATION OF A BALLOT THEOREM IN ORDER STATISTICS.
    ANN. MATH. STATIST., 35, 1356-1358.

TAKAHASHI, MASAYUKI   SEE
    4038. OGASAWARA, TOZIRO + TAKAHASHI, MASAYUKI (1951)

    4039. OGASAWARA, TOZIRO + TAKAHASHI, MASAYUKI (1953)

    4040. OGASAWARA, TOZIRO + TAKAHASHI, MASAYUKI (1953)

TAKAHASI, KOITI 1965.  (18.1)                                              5334
    NOTE ON THE MULTIVARIATE BURR'S DISTRIBUTION.
    ANN. INST. STATIST. MATH. TOKYO, 17, 257-260.   (MR31-5294)

TAKAHASI, KOITI + KOMAZAWA, TSUTOMU 1966.  (4.8/NT)                         5335
    SOME NUMERICAL RESULTS OF RANK CORRELATIONS AND THE ORDERED COMPONENT
    OF SUM OF TWO RANKS.
        (IN JAPANESE, WITH SUMMARY IN ENGLISH)
    PROC. INST. STATIST. MATH. TOKYO, 14, 119-126.

TAKAKURA, SETSUKO 1962.  (6.4/E)                                           5336
    SOME STATISTICAL METHODS OF CLASSIFICATION BY THE THEORY OF QUANTIFICATION.
        (IN JAPANESE, WITH SUMMARY IN ENGLISH)
    PROC. INST. STATIST. MATH. TOKYO, 9, 81-105.   (MR27-3063)

TAKANO, KINSAKU 1953.  (17.3)                                                  5337
     ON THE MANY-DIMENSIONAL DISTRIBUTION FUNCTIONS.
     ANN. INST. STATIST. MATH. TOKYO, 5, 41-58.  (MR15,P.329)

TAKANO, KINSAKU 1954.  (16.4)                                                  5338
     NOTE ON WIENER'S PREDICTION THEORY.
     ANN. INST. STATIST. MATH. TOKYO, 5, 67-72.  (MR16,P.151)

TAKANO, KINSAKU 1954.  (17.1/17.3)                                             5339
     ON SOME LIMIT THEOREMS OF PROBABILITY DISTRIBUTIONS.
          (AMENDED BY NO. 5340)
     ANN. INST. STATIST. MATH. TOKYO, 6, 37-113.  (MR16,P.149)

TAKANO, KINSAKU 1954.  (17.1/17.3)                                             5340
     ERRATA TO: ON SOME LIMIT THEOREMS OF PROBABILITY DISTRIBUTIONS.
          (AMENDMENT OF NO. 5339)
     ANN. INST. STATIST. MATH. TOKYO, 6, 125.  (MR16,P.149)

TAKANO, KINSAKU 1956.  (17.1)                                                  5341
     MULTIDIMENSIONAL CENTRAL LIMIT CRITERION IN THE CASE OF BOUNDED VARIANCES.
     ANN. INST. STATIST. MATH. TOKYO, 7, 81-93.  (MR18,P.156)

TAKANO, KINSAKU 1956.  (17.1)                                                  5342
     CENTRAL CONVERGENCE CRITERION IN THE MULTIDIMENSIONAL CASE.
     ANN. INST. STATIST. MATH. TOKYO, 7, 95-102.  (MR18,P.156)

TAKASHIMA, MICHIO + SUGAWARA, M. 1951.  (17.3)                                 5343
     ON THE SET OF SUCH RANDOM VARIABLES THAT CORRELATION COEFFICIENT BETWEEN EACH
     PAIR OF ELEMENTS IS A FUNCTION OF THE DISTANCE BETWEEN THEM.
          (IN JAPANESE)
     SUGAKU, 3, 109-110.

TAKAYAMA, T.   SEE
     2855. JUDGE, G. G. + TAKAYAMA, T. (1966)

TAKEDA, M.   SEE
     3078. KONDO, H. + TAKEDA, M. (1956)

TALACKO, J. 1962.  (1.2)                                                       5344
     PROFESOR RONALD A. FISHER:  SU VIDA, TRABAJOS Y PUBLICACIONES.
     TRABAJOS ESTADIST., 13, 155-172.

TALLIS, G.M. 1961.  (2.4)                                                      5345
     THE MOMENT GENERATING FUNCTION OF THE TRUNCATED MULTI-NORMAL DISTRIBUTION.
     J. ROY. STATIST. SOC. SER. B, 23, 223-229.  (MR23-A1394)

TALLIS, G.M. 1962.  (4.8)                                                      5346
     THE MAXIMUM LIKELIHOOD ESTIMATION OF CORRELATION FROM CONTINGENCY TABLES.
     BIOMETRICS, 18, 342-353.  (MR26-3143)

TALLIS, G.M. 1962.  (18.1/18.2)                                                5347
     THE USE OF A GENERALIZED MULTINOMIAL DISTRIBUTION IN THE ESTIMATION
     OF CORRELATION IN DISCRETE DATA.
     J. ROY. STATIST. SOC. SER. B, 24, 530-534.  (MR26-7084)

TALLIS, G.M. 1963.  (2.4)                                                      5348
     ELLIPTICAL AND RADIAL TRUNCATION IN NORMAL POPULATIONS.
     ANN. MATH. STATIST., 34, 940-944.  (MR27-2061)

TALLIS, G.M. 1965.  (2.4)                                                      5349
     PLANE TRUNCATION IN NORMAL POPULATIONS.
     J. ROY. STATIST. SOC. SER. B, 27, 301-307.  (MR33-6677)

TALLIS, G.M. + YOUNG, S. S. Y. 1962.  (3.2/18.2/BU)                            5350
     MAXIMUM LIKELIHOOD ESTIMATION OF PARAMETERS OF THE NORMAL, LOG-NORMAL,
     TRUNCATED NORMAL AND BIVARIATE NORMAL DISTRIBUTIONS FROM GROUPED DATA.
     AUSTRAL. J. STATIST., 4, 49-54.  (MR31-6307)

TAMURA, RYOJI 1966.  (17.8/A)                                                  5351
     MULTIVARIATE NONPARAMETRIC SEVERAL-SAMPLE TESTS.
     ANN. MATH. STATIST., 37, 611-618.  (MR33-833)

TANAKA, HIROSHI    SEE
   3701. MARUYAMA, GISHIRO + TANAKA, HIROSHI (1959)

TANAKA, K.    SEE
   3957. NAKAGAMI, MINORU; TANAKA, K. + KANEHISA, M. (1957)

TANAKA, SADAKO    SEE
   197. AOYAMA, HIROJIRO + TANAKA, SADAKO (1957)

TANG, P.C. 1938.  (8.4/TU)                                                    5352
   THE POWER FUNCTION OF THE ANALYSIS OF VARIANCE TESTS WITH TABLES AND
   ILLUSTRATIONS OF THEIR USE.
   STATIST. RES. MEM. LONDON, 2, 126-157.

TANIMOTO, TAFFEE T.    SEE
   4595. ROGERS, DAVID J. + TANIMOTO, TAFFEE T. (1960)

TAO, TSUNG-YING    SEE
   943. CHENG, SHAO-LIEN; TAO, TSUNG-YING + OTHERS (SIC) (1962)

   944. CHENG, SHAO-LIEN; TAO, TSUNG-YING + OTHERS (SIC) (1962)

TAPPAN, M. 1927.  (4.5/4.6/4.8)                                               5353
   ON PARTIAL MULTIPLE CORRELATION COEFFICIENTS IN A UNIVERSE OF
   MANIFOLD CHARACTERISTICS.
   BIOMETRIKA, 19, 39-44.

TARVER, M.G. + SCHENCK, A.M. 1957.  (4.6/8.1/E)                               5354
   THE SQC APPROACH TO MULTIPLE CORRELATION AND REGRESSION IN FOOD
   PACKAGING RESEARCH.
   FOOD TECH., 11, 558-562.

TATCHELL, J. B. 1965.  (20.2)                                                 5355
   LIMITATION THEOREMS FOR TRIANGULAR MATRIX TRANSFORMATIONS.
   J. LONDON MATH. SOC., 40, 127-136.

TATE, ROBERT F. 1954.  (4.4)                                                  5356
   CORRELATION BETWEEN A DISCRETE AND A CONTINUOUS VARIABLE.
   POINT-BISERIAL CORRELATION.
   ANN. MATH. STATIST., 25, 603-607.  (MR16,P.271)

TATE, ROBERT F. 1955.  (4.4/4.8)                                              5357
   THE THEORY OF CORRELATION BETWEEN TWO CONTINUOUS VARIABLES WHEN ONE
   IS DICHOTOMIZED.
   BIOMETRIKA, 42, 205-216.  (MR17,P.54)

TATE, ROBERT F. 1955.  (4.4/4.8)                                              5358
   APPLICATIONS OF CORRELATION MODELS FOR BISERIAL DATA.
   J. AMER. STATIST. ASSOC., 50, 1078-1095.

TATE, ROBERT F. 1966.  (18.1/18.2/18.3)                                       5359
   CONDITIONAL-NORMAL REGRESSION MODELS.
   J. AMER. STATIST. ASSOC., 61, 477-489.  (MR34-8567)

TATE, ROBERT F.    SEE ALSO
   2333. HANNAN, J. F. + TATE, ROBERT F. (1965)

   4079. OLKIN, INGRAM + TATE, ROBERT F. (1961)

   4080. OLKIN, INGRAM + TATE, ROBERT F. (1965)

TATSUOKA, M.M. + TIEDEMAN, DAVID V. 1954.  (1.2/6.2/8.5)                      5360
   DISCRIMINANT ANALYSIS.
   REV. EDUC. RES., 24, 402-420.

TAUSSKY, OLGA 1950.  (20.3)                                                   5361
   NOTES ON NUMERICAL ANALYSIS.  2. NOTE ON THE CONDITION OF MATRICES.
   MATH. TABLES AIDS COMPUT., 4, 111-112.  (MR12,P.361)

TAUSSKY, OLGA 1958.  (20.1)                                                   5362
   ON A MATRIX THEOREM OF A. T. CRAIG AND H. HOTELLING.
   KON. NEDERL. WETENSCH. PROC. SER. A, 61, 139-141.  (MR20-3165)

TAUSSKY, OLGA 1964.  (20.1)                                                        5363
    ON THE VARIATION OF THE CHARACTERISTIC ROOTS OF A FINITE MATRIX UNDER
    VARIOUS CHANGES OF ITS ELEMENTS.
    RECENT ADVANC. MATRIX THEORY (SCHNEIDER), 125-138.


TAUSSKY, OLGA + MARCUS, MARVIN 1962.  (20.1)                                       5364
    EIGENVALUES OF FINITE MATRICES.
    SURVEY NUMER. ANAL. (TODD), 279-297.  (MR24-A1918)


TAUSSKY, OLGA + TODD, JOHN 1952.  (20.1)                                           5365
    SYSTEMS OF EQUATIONS, MATRICES, AND DETERMINANTS.
    MATH. MAG., 26, 9-20.  (MR14,P.715)


TAUSSKY, OLGA + ZASSENHAUS, H. 1959.  (20.1)                                       5366
    ON THE SIMILARITY TRANSFORMATION BETWEEN A MATRIX AND ITS TRANSPOSE.
    PACIFIC J. MATH., 9, 893-896.


TAYLOR, ERWIN K.   SEE
    5843. WHERRY, ROBERT J. + TAYLOR, ERWIN K. (1946)


TAYLOR, H. C. + RUSSELL, J. T. 1939.  E(15.4)                                      5367
    THE RELATIONSHIP OF VALIDITY COEFFICIENTS TO THE PRACTICAL
    EFFECTIVENESS OF TESTS IN SELECTION:  DISCUSSION AND TABLES.
    J. APPL. PSYCHOL., 23, 565-578.


TAYLOR, JOHN G.   SEE
    4520. REYBURN, H. A. + TAYLOR, JOHN G. (1943)


TAYLOR, S. J. 1953.  (16.4)                                                        5368
    THE HAUSDORFF $\alpha$-DIMENSIONAL MEASURE OF BROWNIAN PATHS IN N-SPACE.
    PROC. CAMBRIDGE PHILOS. SOC., 49, 31-39.  (MR14,P.663)


TAYLOR, S. J.   SEE ALSO
    1438. DVORETZKY, ARYEH; ERDOS, PAUL; KAKUTANI, SHIZUO + (1957)
    TAYLOR, S. J.


TAYLOR, WILLIAM F. 1953.  (19.3/A)                                                 5369
    DISTANCE FUNCTIONS AND REGULAR BEST ASYMPTOTICALLY NORMAL ESTIMATES.
    ANN. MATH. STATIST., 24, 85-92.  (MR14,P.996)


TAYLOR, WILSON L. + FONG, CHING 1963.  (4.8/T)                                     5370
    SOME CONTRIBUTIONS TO AVERAGE RANK CORRELATION METHODS AND TO THE
    DISTRIBUTION OF THE AVERAGE RANK CORRELATION COEFFICIENT.
    J. AMER. STATIST. ASSOC., 58, 756-769.  (MR27-2035)


TCHEN, CHAN-MOU 1952.  (4.5/16.4/E)                                                5371
    RANDOM FLIGHT WITH MULTIPLE PARTIAL CORRELATIONS.
    J. CHEM. PHYS., 20, 214-217.


TCHOUPROV, AL. A.   SEE  CUPROV, ALEX. A.


TEATINI, UGO 1950.  (4.8)                                                          5372
    DEGLI INDICI DI CORRELAZIONE TRA CARATTERI CICLICI.
    STATISTICA, BOLOGNA, 10, 46-67.  (MR12,P.38)


TEDESCHI, BRUNO 1957.  (17.1/17.3)                                                 5373
    SULLE LIMITAZIONI PIU CONVENIENTI DELLA PROBABILITA CHE UNA VARIABILE CASUALE A
    PIU DIMENSIONI ASSUMA UN VALORE APPARTENENTE A UN CAMPO ASSEGNATO.
    SCR. MAT. FILIPPO SIBIRANI, 261-279.  (MR19,P.326)


TEGHEM, J. 1959.  (17.7)                                                           5374
    L ANALYSE DE LA CONFLUENCE.
    CAHIERS CENTRE ETUDES RECH. OPERAT., 1(3) 5-34.  (MR21-6660)


TEH, FAN CHUNG   SEE  FAN, CHUNG-TEH


TEICHER, HENRY 1954.  (18.1)                                                       5375
    ON THE MULTIVARIATE POISSON DISTRIBUTION.
    SKAND. AKTUARIETIDSKR., 37, 1-9.  (MR17,P.983)


TEICHER, HENRY 1956.  (17.7)                                                       5376
    IDENTIFICATION OF A CERTAIN STOCHASTIC STRUCTURE.
    ECONOMETRICA, 24, 172-177.  (MR17,P.1219)

TEICHER, HENRY 1965.   (17.1)                                                5377
    ON RANDOM SUMS OF RANDOM VECTORS.
    ANN. MATH. STATIST., 36, 1450-1458.


TEICHER, HENRY 1965.   (14.1/A)                                              5378
    ON THE MAXIMUM OF A STATIONARY GAUSSIAN SEQUENCE.
    SANKHYA SER. A, 27, 422-424.

TEICHER, HENRY   SEE ALSO
    951. CHERNOFF, HERMAN + TEICHER, HENRY (1965)

    1442. DWASS, MEYER + TEICHER, HENRY (1957)

TEICHROEW, DANIEL 1955.   (17.8/TU)                                          5379
    NUMERICAL ANALYSIS RESEARCH UNPUBLISHED STATISTICAL TABLES.
    J. AMER. STATIST. ASSOC., 50, 550-556.   (MR16,P.1037)

TEICHROEW, DANIEL + SITGREAVES, ROSEDITH 1961.   (6.2/7.1/RT)               5380
    COMPUTATION OF AN EMPIRICAL SAMPLING DISTRIBUTION FOR THE
    W-CLASSIFICATION STATISTIC.
    STUD. ITEM ANAL. PREDICT. (SOLOMON), 252-275.   (MR23-A2994)

TEISSIER, GEORGES 1938.   (15.1/ET)                                          5381
    UN ESSAI D'ANALYSE FACTORIELLE.   LES VARIANTS SEXUELS DE MAIA SQUINADO.
    BIOTYPOLOGIE, 6, 73-97.

TELSER, LESTER G. 1964.   (8.1/20.3)                                         5382
    ITERATIVE ESTIMATION OF A SET OF LINEAR REGRESSION EQUATIONS.
    J. AMER. STATIST. ASSOC., 59, 845-862.   (MR29-4148)

TEODORESCU, C.C. 1932.   (4.1/B)                                             5383
    LA CORRELATION STATISTIQUE.
    BULL. SCI. ECOLE POLYTECH. TIMISOARA, 4, 202-218.

TEODORESCU, C.C. 1935.   (4.1)                                               5384
    NOTE SUR LA CORRELATION.
    MATHEMATICA, CLUJ, 9, 294-299.

TERPSTRA, T.J. 1953.   (17.8/AU)                                             5385
    THE EXACT PROBABILITY DISTRIBUTION OF THE T STATISTIC FOR TESTING
    AGAINST TREND AND ITS NORMAL APPROXIMATION.
    KON. NEDERL. WETENSCH. PROC. SER. A, 56, 433-437.   (MR15,P.452)

TERRAGNO, PAUL J.   SEE
    2156. GUENTHER, WILLIAM C. + TERRAGNO, PAUL J. (1964)

TEUGELS, J. 1963.   (18.5)                                                   5386
    SELECTIVE SAMPLING IN K-DIMENSIONAL DISTRIBUTIONS.
    SIMON STEVIN WIS- EN NATUURK. TIJDSCHR., 37, 55-70.

THANH, LUU-MAU-   SEE   LUU-MAU-THANH

THEIL, H. 1950.   (8.1/17.7/17.8/C)                                          5387
    A RANK-INVARIANT METHOD OF LINEAR AND POLYNOMIAL REGRESSION ANALYSIS, I.
    KON. NEDERL. AKAD. WETENSCH. PROC. SECT. SCI., 53, 386-392.   (MR12,P.117)

THEIL, H. 1950.   (8.1/17.7/17.8/C)                                          5388
    A RANK-INVARIANT METHOD OF LINEAR AND POLYNOMIAL REGRESSION ANALYSIS, II.
    KON. NEDERL. AKAD. WETENSCH. PROC. SECT. SCI., 53, 521-525.   (MR12,P.117)

THEIL, H. 1950.   (8.1/17.8/C)                                              5389
    A RANK-INVARIANT METHOD OF LINEAR AND POLYNOMIAL REGRESSION ANALYSIS, III.
    KON. NEDERL. AKAD. WETENSCH. PROC. SECT. SCI., 53, 1397-1412.   (MR12,P.725)

THEIL, H. 1954.   (16.1)                                                     5390
    ESTIMATION OF PARAMETERS OF ECONOMETRIC MODELS.
    BULL. INST. INTERNAT. STATIST., 34(2) 122-129.   (MR16,P.1040)

THEIL, H. 1963.   (8.1)                                                      5391
    ON THE USE OF INCOMPLETE PRIOR INFORMATION IN REGRESSION.
    J. AMER. STATIST. ASSOC., 58, 401-414.

THEIL, H. 1963.  (17.7)                                                              **5392**
    ON THE SPECIFICATION OF MULTIVARIATE RELATIONS AMONG SURVEY DATA.
    MEAS. ECON. (GRUNFELD VOLUME), 293-313.


THEIL, H. 1965.  (16.5/E)                                                             **5393**
    THE ANALYSIS OF DISTURBANCES IN REGRESSION ANALYSIS.
    J. AMER. STATIST. ASSOC., 60, 1067-1079.


THEIL, H. + BOOT, J. C. G. 1962.  (16.2)                                              **5394**
    THE FINAL FORM OF ECONOMETRIC EQUATION SYSTEMS.
    REV. INST. INTERNAT. STATIST., 30, 136-152.  (MR28-3518)


THEIL, H. + KLOEK, T. 1959.  (12.1/16.1)                                              **5395**
    THE STATISTICS OF SYSTEMS OF SIMULTANEOUS ECONOMIC RELATIONSHIPS.
    STATISTICA NEERLANDICA, 13, 65-89.  (MR26-7436)


THEIL, H.   SEE ALSO
    2543. HOOPER, JOHN W. + THEIL, H. (1958)

    6076. ZELLNER, ARNOLD + THEIL, H. (1962)


THEILER, G.   SEE
    3820. MIHOC, GH. + THEILER, G. (1961)


THEISS, EDE (EDOUARD) 1937.  (4.8)                                                    **5396**
    STATISZTIKAI KORRELACIO ES KERESLETI TORVENY.
        (TRANSLATED AS NO. 5397)
    KOZGAZDASAGI SZEMLE, 80, 1-53.


THEISS, EDE (EDOUARD) 1938.  (4.8)                                                    **5397**
    LA CORRELATION STATISTIQUE ET LA LOI DE LA DEMANDE.
        (TRANSLATION OF NO. 5396)
    J. SOC. HONGR. STATIST., 16, 240-262.


THEODORESCU, ANA 1964.  (17.3)                                                        **5398**
    ON MULTIDIMENSIONAL DISTRIBUTION FUNCTIONS.  (IN ROUMAINIAN)
    GAZETA MAT. SER. A, 69, 250-253.  (MR35-7377)


THEODORESCU, RADU 1955.  (16.4)                                                       **5399**
    SUR LES RELATIONS CARACTERISTIQUES DES CHAINES DE MARKOFF CONTINUES DE
    MULTIPLICITE P.
        (IN ROUMAINIAN, WITH SUMMARY IN RUSSIAN AND FRENCH)
    ACAD. REPUB. POP. ROMINE BUL. STI. MAT. FIZ., 7(3) 763-774.  (MR17,P.635)


THEODORESCU, RADU 1955.  (16.4)                                                       **5400**
    PROCESSUS STOCHASTIQUES DE MULTIPLICITE P.
        (IN ROUMAINIAN, WITH SUMMARY IN RUSSIAN AND FRENCH)
    ACAD. REPUB. POP. ROMINE BUL. STI. MAT. FIZ., 7(3) 775-794.  (MR17,P.636)


**THEODORESCU, RADU 1956.  (16.4)**                                                   5401
    **AN ERGODIC THEOREM FOR P-MULTIPLE CONTINUOUS STOCHASTIC PROCESSES.**
    AN. UNIV. BUCURESTI SER. STI. NATUR., 5(10) 23-24.  (MR20-3598)


THIELE, T. N. 1931.  (4.1/8.1)                                                        **5402**
    THE THEORY OF OBSERVATIONS (REPRINT OF A BOOK, OUT OF PRINT AT THE TIME).
    ANN. MATH. STATIST., 2, 165-308.


THOMAS, JOHN B. + WOLF, J.K. 1962.  (16.4)                                            **5403**
    ON THE STATISTICAL DETECTION PROBLEM FOR MULTIPLE SIGNALS.
    IRE TRANS. INFORMATION THEORY, IT-8, 274-280.  (MR25-4943)


THOMAS, JOHN B. + ZADEH, LOTFI A. 1961.  (16.4/16.5)                                  **5404**
    NOTE ON AN INTEGRAL EQUATION OCCURRING IN THE PREDICTION, DETECTION,
    AND ANALYSIS OF MULTIPLE TIME SERIES.
    IRE TRANS. INFORMATION THEORY, IT-7, 118-120.  (MR27-2071)


THOMAS, JOHN B.   SEE ALSO
    6005. WONG, E. + THOMAS, JOHN B. (1962)


THOMAS, LEON L. 1952.  E(6.6)                                                         **5405**
    A CLUSTER ANALYSIS OF OFFICE OPERATIONS.
    J. APPL. PSYCHOL., 36, 238-242.

THOMPSON, J. RIDLEY 1919.  (15.1)                                    5406
    THE ROLE OF INTERFERENCE FACTORS IN PRODUCING CORRELATION.
    BRITISH J. PSYCHOL., 10, 81-100.

THOMPSON, J. RIDLEY 1928.  (15.1)                                    5407
    BOUNDARY CONDITIONS FOR CORRELATION COEFFICIENTS BETWEEN THREE AND
    FOUR VARIABLES.
    BRITISH J. PSYCHOL., 19, 77-94.

THOMPSON, J. RIDLEY 1929.  (15.1)                                    5408
    THE LIMITS OF CORRELATION BETWEEN THREE VARIABLES.
    BRITISH J. PSYCHOL., 19, 239-252.

THOMPSON, J. RIDLEY 1929.  (15.1/E)                                  5409
    THE GENERAL EXPRESSION FOR BOUNDARY CONDITIONS AND THE LIMITS OF CORRELATIONS.
    PROC. ROY. SOC. EDINBURGH, 49, 65-71.

THOMPSON, ROBERT C.   SEE
    3661. MARCUS, MARVIN + THOMPSON, ROBERT C. (1963)

THOMPSON, W. A., JR. 1962.  (3.1/7.1/C)                              5410
    ESTIMATION OF DISPERSION PARAMETERS.
    J. RES. NAT. BUR. STANDARDS SECT. B, 66, 161-164.   (MR26-5686)

THOMPSON, W. A., JR. 1963.  (8.6/8.9)                                5411
    PRECISION OF SIMULTANEOUS MEASUREMENT PROCEDURES.
    J. AMER. STATIST. ASSOC., 58, 474-479.   (MR27-2027)

THOMPSON, W. E.   SEE
    5174. SPRINGER, M. D. + THOMPSON, W. E. (1966)

THOMSON, GODFREY H. 1916.  (15.1)                                    5412
    A HIERARCHY WITHOUT A GENERAL FACTOR.
    BRITISH J. PSYCHOL., 8, 271-281.

THOMSON, GODFREY H. 1919.  (15.2)                                    5413
    ON THE DEGREE OF PERFECTION OF HIERARCHICAL ORDER AMONG CORRELATION
    COEFFICIENTS.
    BIOMETRIKA, 12, 355-366.

THOMSON, GODFREY H. 1919.  (15.1)                                    5414
    THE PROOF OR DISPROOF OF THE EXISTENCE OF GENERAL ABILITY.
    BRITISH J. PSYCHOL., 9, 321-336.

THOMSON, GODFREY H. 1919.  (15.1)                                    5415
    THE HIERARCHY OF ABILITIES.
    BRITISH J. PSYCHOL., 9, 337-344.

THOMSON, GODFREY H. 1919.  (15.1)                                    5416
    ON THE CAUSE OF HIERARCHICAL ORDER AMONG THE CORRELATION COEFFICIENTS
    OF A NUMBER OF VARIATES TAKEN IN PAIRS.
    PROC. ROY. SOC. LONDON SER. A, 95, 400-408.

THOMSON, GODFREY H. 1920.  (15.1)                                    5417
    THE GENERAL FACTOR IN PSYCHOLOGY.
    BRITISH J. PSYCHOL., 10, 319-326.

THOMSON, GODFREY H. 1923.  (15.2)                                    5418
    ON HIERARCHICAL ORDER AMONG CORRELATION COEFFICIENTS.
    BIOMETRIKA, 15, 150-160.

THOMSON, GODFREY H. 1927.  (15.1/15.2)                               5419
    THE TETRAD-DIFFERENCE CRITERION.
    BRITISH J. PSYCHOL., 17, 235-255.

THOMSON, GODFREY H. 1927.  E(4.8/15.2)                               5420
    A WORKED OUT EXAMPLE OF THE POSSIBLE LINKAGES OF FOUR CORRELATED
    VARIABLES ON THE SAMPLING THEORY.
    BRITISH J. PSYCHOL., 18, 68-76.

THOMSON, GODFREY H. 1932.  (4.5/8.1)                                 5421
    ON THE COMPUTATION OF REGRESSION EQUATIONS, PARTIAL CORRELATIONS, ETC.
    BRITISH J. PSYCHOL., 23, 64-68.

THOMSON, GODFREY H. 1934. (15.1)                                  5422
   THE MEANING OF 'I' IN THE ESTIMATE OF 'G'.
  BRITISH J. PSYCHOL., 25, 92-99.

THOMSON, GODFREY H. 1934. (15.3)                                  5423
   HOTELLING'S METHOD MODIFIED TO GIVE SPEARMAN'S 'G'.
  J. EDUC. PSYCHOL., 25, 366-374.

THOMSON, GODFREY H. 1934. (15.1/15.2)                            5424
   THE ORTHOGONAL MATRIX TRANSFORMING SPEARMAN'S TWO-FACTOR EQUATIONS
   INTO THOMSON'S SAMPLING EQUATIONS IN THE THEORY OF ABILITY.
  NATURE, 134, 700.

THOMSON, GODFREY H. 1935. (15.2)                                  5425
   ON COMPLETE FAMILIES OF CORRELATION COEFFICIENTS, AND THEIR TENDENCE TO ZERO
  TETRAD-DIFFERENCES: INCLUDING A STATEMENT OF THE SAMPLING THEORY OF ABILITIES.
  BRITISH J. PSYCHOL., 26, 63-92.

THOMSON, GODFREY H. 1935. (15.1/15.3)                            5426
   THE DEFINITION AND MEASUREMENT OF 'G' GENERAL INTELLIGENCE.
  J. EDUC. PSYCHOL., 26, 241-262.

THOMSON, GODFREY H. 1936. (15.3)                                  5427
   SOME POINTS OF MATHEMATICAL TECHNIQUE IN THE FACTORIAL ANALYSIS OF ABILITY.
  J. EDUC. PSYCHOL., 27, 37-54.

THOMSON, GODFREY H. 1936. (15.1)                                  5428
   BOUNDARY CONDITIONS IN THE COMMON-FACTOR SPACE, IN THE FACTORIAL
   ANALYSIS OF ABILITY.
  PSYCHOMETRIKA, 1, 155-163.

THOMSON, GODFREY H. 1938. (2.4/15.1)                             5429
   THE INFLUENCE OF UNIVARIATE SELECTION ON THE FACTORIAL ANALYSIS OF ABILITY.
  BRITISH J. PSYCHOL., 28, 451-459.

THOMSON, GODFREY H. 1938. (15.2)                                  5430
   THE ESTIMATION OF SPECIFIC AND BI-FACTORS.
  J. EDUC. PSYCHOL., 29, 355-362.

THOMSON, GODFREY H. 1939. (2.4/15.1)                             5431
   THE INFLUENCE OF MULTIVARIATE SELECTION OF THE FACTORIAL ANALYSIS OF
   ABILITY. SECTIONS I-V.
  BRITISH J. PSYCHOL., 29, 288-297.

THOMSON, GODFREY H. 1939. (15.1)                                  5432
   THE FACTORIAL ANALYSIS OF ABILITY. III. THE PRESENT POSITION AND
   THE PROBLEMS CONFRONTING US.
  BRITISH J. PSYCHOL., 30, 71-77, 105-108.

THOMSON, GODFREY H. 1940. (15.4)                                  5433
   WEIGHTING FOR BATTERY RELIABILITY AND PREDICTION.
  BRITISH J. PSYCHOL., 30, 357-366.

THOMSON, GODFREY H. 1944. (15.1)                                  5434
   THE APPLICABILITY OF KARL PEARSON'S SELECTION FORMULAE IN FOLLOW-UP EXPERIMENTS.
  BRITISH J. PSYCHOL., 34, 105.

THOMSON, GODFREY H. 1947. (12.1/15.4)                           5435
   THE MAXIMUM CORRELATION OF TWO WEIGHTED BATTERIES. (HOTELLING'S MOST PREDICTABLE CRITERION
  BRITISH J. PSYCHOL. STATIST. SECT., 1, 27-34.

THOMSON, GODFREY H. 1947. (1.2)                                   5436
   CHARLES SPEARMAN.
  OBIT. NOTICES FELLOWS ROY. SOC., 5, 373-385.

THOMSON, GODFREY H. 1949. (15.2/15.3)                           5437
   ON ESTIMATING OBLIQUE FACTORS.
  BRITISH J. PSYCHOL. STATIST. SECT., 2, 1-2.

THOMSON, GODFREY H.   SEE ALSO
  1851. GARNETT, J. C. MAXWELL + THOMSON, GODFREY H. (1919)

THOMSON, KENNETH F.    SEE
   1485. EDGERTON, HAROLD A. + THOMSON, KENNETH F. (1942)

THORNBER, H.    SEE
   6077. ZELLNER, ARNOLD + THORNBER, H. (1966)

THORNDIKE, EDWARD L. 1937.   (4.1)                                    5438
   ON CORRELATIONS BETWEEN MEASUREMENTS WHICH ARE NOT NORMALLY DISTRIBUTED.
   J. EDUC. PSYCHOL., 28, 367-370.

THORNTON, G. R. 1943.   (4.8)                                         5439
   THE SIGNIFICANCE OF RANK DIFFERENCE COEFFICIENTS OF CORRELATION.
   PSYCHOMETRIKA, 8, 211-222.

THOULESS, ROBERT H. 1936.   (15.4)                                    5440
   TEST UNRELIABILITY AND FUNCTION FLUCTUATION.
   BRITISH J. PSYCHOL., 26, 325-343.

THOULESS, ROBERT H. 1939.   (4.1/17.7)                                5441
   THE EFFECT OF ERRORS OF MEASUREMENT ON CORRELATION COEFFICIENTS.
   BRITISH J. PSYCHOL., 29, 383-403.

THRALL, ROBERT M.; COOMBS, CLYDE H. + CALDWELL, WILLIAM V. 1958.   (6.6)   5442
   LINEAR MODEL FOR EVALUATING COMPLEX SYSTEMS.
   NAVAL RES. LOGISTICS QUART., 5, 347-361.

THUMB, NORBERT 1935.   (15.1)                                         5443
   ZUR PROBLEMATIK DER FAKTORENTHEORIEN.
   C. R. CONF. INTERNAT. PSYCHOTECH., 8, 712-721.

THURSTONE, L.L. 1919.   (15.4)                                        5444
   A SCORING METHOD FOR MENTAL TESTS.
   PSYCHOL. BULL., 16, 235-240.

THURSTONE, L.L. 1925.   (15.4/15.5)                                   5445
   A METHOD OF SCALING PSYCHOLOGICAL AND EDUCATIONAL TESTS.
   J. EDUC. PSYCHOL., 16, 433-451.

THURSTONE, L.L. 1928.   (15.5)                                        5446
   SCALE CONSTRUCTION WITH WEIGHTED OBSERVATIONS.
   J. EDUC. PSYCHOL., 19, 441-453.

THURSTONE, L.L. 1931.   (15.1)                                        5447
   MULTIPLE FACTOR ANALYSIS.
   PSYCHOL. REV., 38, 406-427.   (MR12,P.38)

THURSTONE, L.L. 1934.   (15.1)                                        5448
   THE VECTORS OF MIND.
   PSYCHOL. REV., 41, 1-32.

THURSTONE, L.L. 1937.   (15.1)                                        5449
   CURRENT MISUSE OF THE FACTORIAL METHODS.
   PSYCHOMETRIKA, 2, 73-76.

THURSTONE, L.L. 193 .   (15.3)                                        5450
   A NEW ROTATIONAL METHOD IN FACTOR ANALYSIS.
   PSYCHOMETRIKA, 3, 199-218.

THURSTONE, L.L. 1940.   (15.1)                                        5451
   CURRENT ISSUES IN FACTOR ANALYSIS.
   PSYCHOL. BULL., 37, 189-236.

THURSTONE, L.L. 1944.   (15.3/20.2/20.3)                              5452
   GRAPHICAL METHOD OF FACTORING THE CORRELATION MATRIX.
   PROC. NAT. ACAD. SCI. U.S.A., 30, 129-134.   (MR6,P.6)

THURSTONE, L.L. 1944.   (15.1)                                        5453
   SECOND-ORDER FACTORS.
   PSYCHOMETRIKA, 9, 71-100.

THURSTONE, L.L. 1945.   (15.3/20.2)                                   5454
   A MULTIPLE GROUP METHOD OF FACTORING THE CORRELATION MATRIX.
   PSYCHOMETRIKA, 10, 73-78.   (MR7,P.20)

THURSTONE, L.L. 1945.  (15.1)                                              5455
    THE EFFECTS OF SELECTION IN FACTOR ANALYSIS.
    PSYCHOMETRIKA, 10, 165-198.


THURSTONE, L.L. 1946.  (15.3)                                              5456
    A SINGLE PLANE METHOD OF ROTATION.
    PSYCHOMETRIKA, 11, 71-79.  (MR8,P.54)


THURSTONE, L.L. 1949.  (15.1)                                              5457
    NOTE ABOUT THE MULTIPLE GROUP METHOD.
    PSYCHOMETRIKA, 14, 43-45.


THURSTONE, L.L. 1951.  (15.1)                                             5458
    L'ANALYSE FACTORIELLE:  METHODE SCIENTIFIQUE.
    ANNEE PSYCHOL., 50, 61-75.


THURSTONE, L.L. 1954.  (15.3)                                             5459
    AN ANALYTICAL METHOD FOR SIMPLE STRUCTURE.
    PSYCHOMETRIKA, 19, 173-182.


THURSTONE, L.L. 1964.  (15.1)                                             5460
    MULTIPLE-FACTOR ANALYSIS.
    MATH. PSYCHOL. (MILLER), 236-246.


TIAGO DE OLIVEIRA, J. 1960.  (4.8/18.1/18.2/B)                            5461
    LES DISTRIBUTIONS BIVARIEES DE MARGES DONNEES ET L'ESTIMATION DE SES PARAMETRES.
    REV. FAC. CI. UNIV. LISBOA SER. 2A CI. MAT., 8, 339-347.  (MR26-4383)


TIAGO DE OLIVEIRA, J. 1962.  (17.6/B)                                     5462
    LA REPRESENTATION DES DISTRIBUTIONS EXTREMALES BIVARIEES.
    BULL. INST. INTERNAT. STATIST., 39(2) 477-480.  (MR29-683)


TIAGO DE OLIVEIRA, J. 1962.  (17.6/B)                                     5463
    STRUCTURE THEORY  OF BIVARIATE EXTREMES; EXTENSIONS.
    ESTUDOS MAT. ESTATIST. ECONOMET., 7, 165-195.


TIAGO DE OLIVEIRA, J. 1964.  (17.6)                                       5464
    L'INDEPENDANCE DANS LES DISTRIBUTIONS EXTREMALES BIVARIEES.
    PUBL. INST. STATIST. UNIV. PARIS, 13, 137-141.  (MR30-2534)


TIAGO DE OLIVEIRA, J. 1965.  (17.6/B)                                     5465
    STATISTICAL DECISION FOR BIVARIATE EXTREMES.
    PORTUGAL. MATH., 24, 145-154.  (MR35-2437)


TIAO, GEORGE C. + GUTTMAN, IRWIN 1965.  (7.4/18.1/20.4)                   5466
    THE INVERTED DIRICHLET DISTRIBUTION WITH APPLICATIONS.
    J. AMER. STATIST. ASSOC., 60, 793-805.  (MR32-3177)


TIAO, GEORGE C. + ZELLNER, ARNOLD 1964.  (8.1)                           5467
    ON THE BAYESIAN ESTIMATION OF MULTIVARIATE REGRESSION.
    J. ROY. STATIST. SOC. SER. B, 26, 277-285.  (MR31-1743)


TIEDEMAN, DAVID V. 1951.  (6.2/6.3)                                       5468
    THE UTILITY OF THE DISCRIMINANT FUNCTION IN PSYCHOLOGICAL AND
    GUIDANCE INVESTIGATIONS.
    HARVARD EDUC. REV., 21(4) 71-80.


TIEDEMAN, DAVID V. 1966.  (6.6/E)                                         5469
    A GEOMETRIC MODEL FOR THE PROFILE PROBLEM.
    TEST. PROBLEMS PERSPECTIVE (ANASTASI), 331-354.


TIEDEMAN, DAVID V.   SEE ALSO
    5360. TATSUOKA, M.M. + TIEDEMAN, DAVID V. (1954)


TIHONOV, A. N. 1930.  (20.5)                                              5470
    UBER DIE TOPOLOGISCHE ERWEITERUNG VON RAUMEN.
    MATH. ANN., 102, 544-561.


TIKKIWAL, B.D. 1960.  (18.5)                                              5471
    ON THE THEORY OF CLASSICAL REGRESSION AND DOUBLE SAMPLING ESTIMATION.
    J. ROY. STATIST. SOC. SER. B, 22, 131-138.  (MR22-3070)

TIKKIWAL, B.D. + KABE, D. G. 1958.  (17.9/20.3)                    5472
    ON THE PROOF OF THE LEMMA USED IN DERIVING THE DISTRIBUTION OF THE
    SECOND ORDER SAMPLE MOMENTS FOR A SYMMETRIC P-VARIATE POPULATION.
    J. KARNATAK UNIV., 3(1) 113-115.  (MR22-3075)

TIKU, M. L. 1963.  (8.8/14.5)                    5473
    A LAGUERRE PRODUCT SERIES APPROXIMATION TO ONE-WAY CLASSIFICATION
    VARIANCE RATIO DISTRIBUTION.
    J. INDIAN SOC. AGRIC. STATIST., 15, 223-231.

TIKU, M. L. 1965.  (14.5/NT)                    5474
    SERIES EXPANSIONS FOR THE DOUBLY NON-CENTRAL F-DISTRIBUTION.
    AUSTRAL. J. STATIST., 7, 78-89.

TIKU, M. L. 1966.  (14.5/T)                    5475
    A NOTE ON APPROXIMATING TO THE NON-CENTRAL F DISTRIBUTION.
    BIOMETRIKA, 53, 606-610.

TILDESLEY, M. L. 1921.  E(6.5/17.9)                    5476
    A FIRST STUDY OF THE BURMESE SKULL.
    BIOMETRIKA, 13, 247-251.

TINBERGEN, J. 1947.  (16.2/16.5/E)                    5477
    THE USE OF CORRELATION ANALYSIS IN ECONOMIC RESEARCH.
    EKON. TIDSKR., 49, 173-192.

TINTNER, GERHARD 1939.  (2.5)                    5478
    THE DISTRIBUTION OF SYMMETRIC QUADRATIC FORMS IN NORMAL AND
    INDEPENDENT VARIABLES.
    IOWA STATE COLL. J. SCI., 13, 231-233.

TINTNER, GERHARD 1945.  (8.1/8.4/17.7)                    5479
    A NOTE ON RANK, MULTICOLLINEARITY AND MULTIPLE REGRESSION.
    ANN. MATH. STATIST., 16, 304-308.  (MR7,P.132)

TINTNER, GERHARD 1946.  (8.1/17.7)                    5480
    MULTIPLE REGRESSION FOR SYSTEMS OF EQUATIONS.
    ECONOMETRICA, 14, 5-36.  (MR7,P.318)

TINTNER, GERHARD 1946.  (1.1/E)                    5481
    SOME APPLICATIONS OF MULTIVARIATE ANALYSIS TO ECONOMIC DATA.
    J. AMER. STATIST. ASSOC., 41, 472-500.  (MR8,P.397)

TINTNER, GERHARD 1950.  (17.7)                    5482
    SOME FORMAL RELATIONS IN MULTIVARIATE ANALYSIS.
    J. ROY. STATIST. SOC. SER. B, 12, 95-101.  (MR12,P.513)

TINTNER, GERHARD 1950.  (17.7)                    5483
    A TEST FOR LINEAR RELATIONS BETWEEN WEIGHTED REGRESSION COEFFICIENTS.
    J. ROY. STATIST. SOC. SER. B, 12, 273-277.

TINTNER, GERHARD 1952.  (1.2/16.2)                    5484
    ABRAHAM WALD'S CONTRIBUTIONS TO ECONOMETRICS.
    ANN. MATH. STATIST., 23, 21-28.  (MR13,P.613)

TINTNER, GERHARD 1952.  (8.7/17.7)                    5485
    DIE ANWENDUNG DER VARIATE DIFFERENCE METHODE AUF DIE PROBLEME DER
    GEWOGENEN REGRESSION UND DER MULTIKOLLINEARITAT.
    MITT. MATH. STATIST., 4, 159-162.  (MR14,P.297)

TINTNER, GERHARD 1961.  (1.2)                    5486
    THE STATISTICAL WORK OF OSKAR ANDERSON.
    J. AMER. STATIST. ASSOC., 56, 273-280.  (MR22-9432)

TINTNER, GERHARD 1962.  (14.3/20.1/T)                    5487
    EIN TEST FUR DIE SINGULARITAT EINER MATRIX.
    IFO-STUDIEN, 8, 1-4.

TINTNER, GERHARD 1964.  (8.7)                    5488
    A NOTE ON THE RELATION BETWEEN MAHALANOBIS DISTANCE AND WEIGHTED REGRESSION.
    CONTRIB. STATIST. (MAHALANOBIS VOLUME), 481-484.  (MR35-5080)

TINTNER, GERHARD + NARAYANAN, R. 1966.  (16.4)                                        5489
    A MULTI-DIMENSIONAL STOCHASTIC PROCESS FOR THE EXPLANATION OF
    ECONOMIC DEVELOPMENT.
    METRIKA, 11, 85-90.


TODD, JOHN   SEE
    5365. TAUSSKY, OLGA + TODD, JOHN (1952)


TODD-TAUSSKY, OLGA  SEE  TAUSSKY, OLGA


TOLEDO TOLEDO, MANUEL 1964.  (1.1)                                                    5490
    DISTRIBUCION NORMAL MULTIVARIANTE.
    ESTADIST. ESPAN., 23, 32-44.  (MR30-2623)


TOLLEY, H. R. + EZEKIEL, MORDECAI J. B. 1923.   (4.6)                                 5491
    A METHOD OF HANDLING MULTIPLE CORRELATION PROBLEMS.
    J. AMER. STATIST. ASSOC., 18, 993-1003.


TOLLEY, H. R. + EZEKIEL, MORDECAI J. B. 1927.   (4.6/20.3)                            5492
    THE DOOLITTLE METHOD FOR SOLVING MULTIPLE CORRELATION EQUATIONS
    VERSUS THE KELLEY-SALISBURY "ITERATION" METHOD.
    J. AMER. STATIST. ASSOC., 22, 497-500.


TOLMAN, EDWARD C.   SEE
    1120. CRUTCHFIELD, RICHARD S. + TOLMAN, EDWARD C. (1940)


TOMAN, W. 1949.  (15.1)                                                               5493
    DIE FAKORENANALYSE IN DER PSYCHOLOGIE.
    STATIST. VIERTELJAHRESSCHR., 2, 57-68.


TOOPS, HERBERT A. 1927.  (4.1)                                                        5494
    STATISTICAL CHECKS ON THE ACCURACY OF INTERCORRELATION COMPUTATIONS.
    RES. BULL. EDUC. TEST. SERVICE, 6, 385-391.


TOOPS, HERBERT A.   SEE ALSO
    25. ADKINS, DOROTHY C. + TOOPS, HERBERT A. (1937)

    115. ANDERSON, L. DEWEY + TOOPS, HERBERT A. (1928)

    116. ANDERSON, L. DEWEY + TOOPS, HERBERT A. (1929)

    1486. EDGERTON, HAROLD A. + TOOPS, HERBERT A. (1928)


TORGERSON, ERIK N. 1965.  (19.4)                                                      5495
    MINIMAL SUFFICIENCY OF THE ORDER STATISTIC IN THE CASE OF TRANSLATION
    AND SCALE PARAMETERS.
    SKAND. AKTUARIETIDSKR., 48, 16-21.


TORGERSON, WARREN S. 1952.  (15.5)                                                    5496
    MULTIDIMENSIONAL SCALING. I. THEORY AND METHOD.
    PSYCHOMETRIKA, 17, 401-419.  (MR14,P.889)


TORGERSON, WARREN S. 1956.  (4.8)                                                     5497
    A NONPARAMETRIC TEST OF CORRELATION USING RANK ORDERS WITHIN SUBGROUPS.
    PSYCHOMETRIKA, 21, 145-152.


TORGERSON, WARREN S. 1965.  (15.5)                                                    5498
    MULTIDIMENSIONAL SCALING OF SIMILARITY.
    PSYCHOMETRIKA, 30, 379-393.


TORNQVIST, LEO 1946.  (17.1/17.3)                                                     5499
    ON THE DISTRIBUTION FUNCTION FOR A FUNCTION OF N STATISTIC VARIABLES AND THE
    CENTRAL LIMIT THEOREM IN THE MATHEMATICAL THEORY OF PROBABILITY.
    SKAND. AKTUARIETIDSKR., 29, 206-229.  (MR8,P.389)


TORNQVIST, LEO 1957.  (8.1/20.3)                                                      5500
    A METHOD FOR CALCULATING CHANGES IN REGRESSION COEFFICIENTS AND INVERSE
    MATRICES CORRESPONDING TO CHANGES IN THE SET OF AVAILABLE DATA.
    SKAND. AKTUARIETIDSKR., 40, 219-226.  (MR20-4906)


TORTRAT, ALBERT 1949.  (16.3/U)                                                       5501
    SUR LES FONCTIONS DE CORRELATION DES PROCESSUS DE MARKOFF.
    C.R. ACAD. SCI. PARIS, 228, 1559-1561.  (MR10,P.720)

TOWNSEND, M. W.   SEE
    1078. COX, D. R. + TOWNSEND, M. W. (1948)

TRANQUE, T. GARCIA   SEE   GARCIA TRANQUE, TOMAS

TRANQUILLI, GIOVANNI BATTISTA 1966.   (17.5)                5502
    SU UN GENERICO TEST CARATTERISTICO DI NORMALITA E DI OMOSCHEDASTICITA
    PER PIU VARIABILI CASUALI.
        (WITH SUMMARY IN FRENCH AND ENGLISH)
    GIORN. IST. ITAL. ATTUARI, 29, 64-96.

TRAPERO, JESUS BERNARDO PENA   SEE   PENA TRAPERO, JESUS BERNARDO

TRAVERS, R.M.W. 1938.   E(15.4)                5503
    THE ELIMINATION OF THE INFLUENCE OF REPETITION ON THE SCORE OF A
    PSYCHOLOGICAL TEST.
    ANN. EUGENICS, 8, 303-318.

TRAWINSKI, IRENE MONAHAN + BARGMANN, ROLF E. 1964.   (3.2)                5504
    MAXIMUM LIKELIHOOD ESTIMATION WITH INCOMPLETE MULTIVARIATE DATA.
    ANN. MATH. STATIST., 35, 647-657.   (MR31-6320)

TRENCH, WILLIAM F. + SCHEINOK, PERRY A. 1966.   (20.2/20.3)                5505
    ON THE INVERSION OF A HILBERT TYPE MATRIX.
    SIAM REV., 8, 57-61.

TRICOMI, FRANCESCO G. 1951.   (18.3)                5506
    APPLICAZIONE DELLA FUNZIONE GAMMA INCOMPLETA ALLO STUDIO DELLA SOMMA
    DI VETTORI CASUALI.
    BOLL. UN. MAT. ITAL., 6(3) 189-194.   (MR14,P.59)

TRISKA, KAREL 1934.   (4.2)                5507
    HOW TO COMPUTE THE STANDARD ERROR OF A CORRELATION WITH $\rho$.
    AKTUAR. VEDY, 4, 134-137.

TRISKA, KAREL 1937.   (4.4)                5508
    UCELNEJSI UPRAVA KELLEYOVA VZORCE PRO BISERIALNI KORELACI.
        (WITH SUMMARY IN FRENCH)
    STATIST. OBZOR, 18, 239.

TRISKA, KAREL 1947.   (15.4)                5509
    O MERENI SPOLEHLIVOSTI TESTU.
        (WITH SUMMARY IN ENGLISH AND FRENCH)
    STATIST. OBZOR, 27, 199-215.

TROCONIZ, ANTONIA FZ. DE   SEE   FZ. DE TROCONIZ, ANTONIA

TROSKIE, C. G. 1966.   (7.4/8.4/18.1)                5510
    NIE-SENTRALE MEERVERANDERLIKE BETAVERDELINGS.
    TYDSKR. NATUURWETENSK., 6, 58-71.

TRUAX, DONALD R. 1953.   (19.1)                5511
    AN OPTIMUM SLIPPAGE TEST FOR THE VARIANCES OF K NORMAL DISTRIBUTIONS.
    ANN. MATH. STATIST., 24, 669-674.   (MR15,P.727)

TRUAX, DONALD R.   SEE ALSO
    2932. KARLIN, SAMUEL + TRUAX, DONALD R. (1960)

TRUKSA, LADISLAV 1929.   (20.5)                5512
    APPLICATION OF BESSEL COEFFICIENTS IN APPROXIMATIVE EXPRESSING OF COLLECTIVES.
    AKTUAR. VEDY, 1, 2-13.

TRUKSA, LADISLAV 1931.   (20.4)                5513
    HYPERGEOMETRIC ORTHOGONAL SYSTEMS OF POLYNOMIALS.
    AKTUAR. VEDY, 2, 65-84, 113-144, 177-203.

TRYBULA, STANISLAW 1958.   (19.3)                5514
    SOME PROBLEMS OF SIMULTANEOUS MINIMAX ESTIMATION.
    ANN. MATH. STATIST., 29, 245-253.   (MR20-376)

TRYBULA, STANISLAW 1958.   (18.2)                5515
    ON THE MINIMAX ESTIMATION OF THE PARAMETERS IN A MULTINOMIAL DISTRIBUTION.
    POLSKA AKAD. NAUK ZASTOS. MAT., 3, 307-322.   (MR21-393)

TRYBULA, STANISLAW 1965.  (17.3)
   ON THE PARADOX OF  N  RANDOM VARIABLES.
  POLSKA AKAD. NAUK ZASTOS. MAT., 8, 143-156.  (MR32-8380)
         5516

TRYON, ROBERT C. 1929.  (4.1)
   THE INTERPRETATION OF THE CORRELATION COEFFICIENT.
  PSYCHOL. REV., 36, 419-445.
         5517

TRYON, ROBERT C. 1935.  (15.1)
   DISCUSSION:  INTERPRETATION OF PROFESSOR SPEARMAN'S COMMENTS.
  PSYCHOL. REV., 42, 122-125.
         5518

TRYON, ROBERT C. 1935.  (15.1)
   A THEORY OF PSYCHOLOGICAL COMPONENTS--AN ALTERNATIVE TO 'MATHEMATICAL FACTORS'.
  PSYCHOL. REV., 42, 425-454.
         5519

TRYON, ROBERT C. 1939.  (6.6)
   COMPARATIVE CLUSTER ANALYSIS.
  PSYCHOL. BULL., 36, 645-646.
         5520

TRYON, ROBERT C. 1957.  (6.6/15.1/15.4)
   COMMUNALITY OF A VARIABLE - FORMULATION BY CLUSTER ANALYSIS.
  PSYCHOMETRIKA, 22, 241-260.  (MR19,P.1027)
         5521

TRYON, ROBERT C. 1958.  (6.6/15.1/E)
   CUMULATIVE COMMUNALITY CLUSTER ANALYSIS.
  EDUC. PSYCHOL. MEAS., 18, 3-35.
         5522

TRYON, ROBERT C. 1958.  (6.6/15.1)
   GENERAL DIMENSIONS OF INDIVIDUAL DIFFERENCES:  CLUSTER ANALYSIS VS. MULTIPLE
   FACTOR ANALYSIS.  (SYMPOSIUM:  THE FUTURE OF FACTOR ANALYSIS.)
  EDUC. PSYCHOL. MEAS., 18, 477-495.
         5523

TRYON, ROBERT C. 1959.  (6.6/15.1)
   DOMAIN SAMPLING FORMULATION OF CLUSTER AND FACTOR ANALYSIS.
  PSYCHOMETRIKA, 24, 113-135.  (MR21-947)
         5524

TSCH, C. HSU LEE  SEE  HSU, LEETZE CHING

TSCHEPOURKOWSKY, E. 1905.  (4.1/E)
   CONTRIBUTIONS TO THE STUDY OF INTERRACIAL CORRELATION.
  BIOMETRIKA, 4, 286-312.
         5525

TSCHUPROW, ALEX A.  SEE  CUPROV, ALEX. A.

TSIAN, TSE-PEI  SEE  JIANG, ZE-PEI

TSUDA, TAKAO + MATSUMOTO, HIROSHI 1966.  (20.5)
   A NOTE ON LINEAR EXTRAPOLATION OF MULTIVARIABLE FUNCTIONS BY THE
   MONTE CARLO METHOD.
  J. ASSOC. COMPUT. MACH., 13, 143-150.
         5526

TSUJIOKA, BIEN  SEE
  891. CATTELL, RAYMOND B.; COULTER, MALCOLM A. + TSUJIOKA, BIEN (1966)

TSUKIBAYASHI, SHOMEI 1962.  (17.5)
   ESTIMATION OF BIVARIATE PARAMETERS BASED ON RANGE.
  REP. STATIST. APPL. RES. UN. JAPAN. SCI. ENGRS., 9, 10-23.  (MR25-667)
         5527

TSUNG-YING, TAO  SEE  TAO, TSUNG-YING

TUCKER, HOWARD G.  SEE
  4744. RUBIN, HERMAN + TUCKER, HOWARD G. (1959)

TUCKER, LEDYARD R. 1940.  (15.1/15.2)
   THE ROLE OF CORRELATED FACTORS IN FACTOR ANALYSIS.
  PSYCHOMETRIKA, 5, 141-152.
         5528

TUCKER, LEDYARD R. 1940.  (20.2)
   A MATRIX MULTIPLIER.
  PSYCHOMETRIKA, 5, 289-294.  (MR2,P.240)
         5529

TUCKER, LEDYARD R. 1944.  (15.3)
   A SEMI-ANALYTICAL METHOD OF FACTORIAL ROTATION TO SIMPLE STRUCTURE.
  PSYCHOMETRIKA, 9, 43-68.
         5530

TUCKER, LEDYARD R. 1944.   (20.2)                                 5531
     THE DETERMINATION OF SUCCESSIVE PRINCIPAL COMPONENTS WITHOUT
     COMPUTATION OF TABLES OF RESIDUAL CORRELATION COEFFICIENTS.
  PSYCHOMETRIKA, 9, 149-153.   (MR6,P.51)

TUCKER, LEDYARD R. 1946.   (15.4)                                 5532
     MAXIMUM VALIDITY OF A TEST WITH EQUIVALENT ITEMS.
  PSYCHOMETRIKA, 11, 1-13.   (MR7,P.463)

TUCKER, LEDYARD R. 1947.   (15.3)                                 5533
     SIMPLIFIED PUNCHED CARD METHODS IN FACTOR ANALYSIS.
  PROC. RES. FORUM, 9-19.

TUCKER, LEDYARD R. 1948.   (4.1)                                  5534
     A NOTE ON THE COMPUTATION OF A TABLE OF INTER-CORRELATIONS.
  PSYCHOMETRIKA, 13, 245-250.

TUCKER, LEDYARD R. 1949.   (15.4)                                 5535
     A NOTE ON THE ESTIMATION OF TEST RELIABILITY BY THE KUDER-RICHARDSON
     FORMULA (20).
  PSYCHOMETRIKA, 14, 117-119.

TUCKER, LEDYARD R. 1955.   (15.1)                                 5536
     THE OBJECTIVE DEFINITION OF SIMPLE STRUCTURE IN LINEAR FACTOR ANALYSIS.
  PSYCHOMETRIKA, 20, 209-225.

TUCKER, LEDYARD R. 1958.   (15.1)                                 5537
     DETERMINATION OF PARAMETERS OF A FUNCTIONAL RELATION BY FACTOR ANALYSIS.
  PSYCHOMETRIKA, 23, 19-23.

TUCKER, LEDYARD R. 1958.   (15.1)                                 5538
     AN INTER-BATTERY METHOD OF FACTOR ANALYSIS.
  PSYCHOMETRIKA, 23, 111-136.   (MR20-6175)

TUCKER, LEDYARD R. 1960.   (15.4/15.5)                            5539
     INTRA-INDIVIDUAL AND INTER-INDIVIDUAL MULTIDIMENSIONALITY.
  PSYCHOL. SCALING THEORY APPL. (GULLIKSEN + MESSICK), 155-167.

TUCKER, LEDYARD R. 1963.   (15.1/15.5/E)                          5540
     IMPLICATIONS OF FACTOR ANALYSIS OF THREE-WAY MATRICES FOR MEASUREMENT OF CHANGE.
  PROBLEMS MEAS. CHANGE (HARRIS), 122-137.

TUCKER, LEDYARD R. 1963.   (12.1/12.2/E)                          5541
     FORMAL MODELS FOR A CENTRAL PREDICTION SYSTEM.
  PSYCHOMET. MONOG., 10, 1-61.

TUCKER, LEDYARD R. 1964.   (15.1)                                 5542
     THE EXTENSION OF FACTOR ANALYSIS TO THREE-DIMENSIONAL MATRICES.
  CONTRIB. MATH. PSYCHOL. (FREDERIKSEN + GULLIKSEN), 109-127.

TUCKER, LEDYARD R. 1964.   (8.1/14.4/E)                           5543
     A SUGGESTED ALTERNATIVE FORMULATION IN THE DEVELOPMENTS BY HURSCH,
     HAMMOND, AND HURSCH, AND BY HAMMOND, HURSCH, AND TODD.
  PSYCHOL. REV., 71, 528-530.

TUCKER, LEDYARD R. 1966.   (15.1/15.2/15.5/E)                     5544
     LEARNING THEORY AND MULTIVARIATE EXPERIMENT:   ILLUSTRATION BY
     DETERMINATION OF GENERALIZED LEARNING CURVES.
  HANDB. MULTIVARIATE EXPER. PSYCHOL. (CATTELL), 476-501.

TUCKER, LEDYARD R. 1966.   (15.1/15.2)                            5545
     SOME MATHEMATICAL NOTES ON THREE-MODE FACTOR ANALYSIS.
  PSYCHOMETRIKA, 31, 279-311.

TUCKER, LEDYARD R. + MESSICK, SAMUEL J. 1963.   (15.5/E)          5546
     AN INDIVIDUAL DIFFERENCES MODEL FOR MULTIDIMENSIONAL SCALING.
  PSYCHOMETRIKA, 28, 333-367.

TUCKER, LEDYARD R.    SEE ALSO
  2192. GULLIKSEN, HAROLD + TUCKER, LEDYARD R. (1951)

   1344. DIEDERICH, GERTRUDE W.; MESSICK, SAMUEL J. + TUCKER, LEDYARD R. (1957)

   2429. HELM, C. E.; MESSICK, SAMUEL J. + TUCKER, LEDYARD R. (1961)

TUCKERMAN, L. B. 1941.  (20.3)                                                    5547
    ON THE MATHEMATICALLY SIGNIFICANT FIGURES IN THE SOLUTION OF
    SIMULTANEOUS LINEAR EQUATIONS.
    ANN. MATH. STATIST., 12, 307-316.  (MR3,P.154)


TUKEY, JOHN W. 1948.  (17.8/U)                                                    5548
    NONPARAMETRIC ESTIMATION.  III.  STATISTICALLY EQUIVALENT BLOCKS AND
    MULTIVARIATE TOLERANCE REGIONS--THE DISCONTINUOUS CASE.
    ANN. MATH. STATIST., 19, 30-39.  (MR9,P.453)



TUKEY, JOHN W. 1949.  (8.5/B)                                                     5549
    DYADIC ANOVA, AN ANALYSIS OF VARIANCE FOR VECTORS.
    HUMAN BIOL., 21, 65-110.


TUKEY, JOHN W. 1954.  (4.8/17.7)                                                  5550
    CAUSATION, REGRESSION, AND PATH ANALYSIS.
    STATIST. MATH. BIOL. (KEMPTHORNE ET AL.), 35-66.


TUKEY, JOHN W. 1962.  (1.1)                                                       5551
    THE FUTURE OF DATA ANALYSIS.
        (AMENDED BY NO. 5552)
    ANN. MATH. STATIST., 33, 1-67.  (MR24-A3761)


TUKEY, JOHN W. 1962.  (1.1)                                                       5552
    CORRECTION TO: THE FUTURE OF DATA ANALYSIS.
        (AMENDMENT OF NO. 5551)
    ANN. MATH. STATIST., 33, 812.  (MR24-A3761)


TUKEY, JOHN W. 1966.  (1.2/16.3/16.5/E)                                           5553
    USE OF NUMERICAL SPECTRUM ANALYSIS IN GEOPHYSICS.
        (WITH DISCUSSION)
    BULL. INST. INTERNAT. STATIST., 41, 267-307.


TUKEY, JOHN W. + WILK, MARTIN B. 1966.  (17.9)                                    5554
    DATA ANALYSIS AND STATISTICS, AN EXPOSITORY OVERVIEW.
    FALL JOINT COMP. CONF. PROC., 29, 695-710.


TUKEY, JOHN W. + WILKS, S.S. 1946.  (8.4/10.1/20.4)                               5555
    APPROXIMATION OF THE DISTRIBUTION OF THE PRODUCT OF BETA VARIABLES BY
    A SINGLE BETA VARIABLE.
    ANN. MATH. STATIST., 17, 318-324.  (MR8,P.162)

TUKEY, JOHN W.   SEE ALSO
    3530. LUCE, R. DUNCAN + TUKEY, JOHN W. (1964)

    4081. OLMSTEAD, P. S. + TUKEY, JOHN W. (1947)

    3214. KURTZ, T. F.; LINK,R.F.; TUKEY, JOHN W. +  (1966)
    WALLACE, D. L.


TUMANYAN, S.H. 1955.  (17.1/18.1/A)                                               5556
    ASYMPTOTIC INVESTIGATION OF THE MULTINOMIAL PROBABILITY DISTRIBUTION.
        (IN RUSSIAN, WITH SUMMARY IN ARMENIAN)
    DOKL. AKAD. NAUK ARMJAN. SSR, 20, 65-74.  (MR17,P.47)


TUMURA, YOSIRO 1965.  (13.1/13.2)                                                 5557
    THE DISTRIBUTIONS OF LATENT ROOTS AND VECTORS.
    T. R. U. MATH., 1, 1-16.


TUNIS, C. J.   SEE
    2142. GRIFFIN, JOHN S., JR.; KING, J. H., JR. + TUNIS, C. J. (1963)


TURIN, GEORGE L. 1960.  (2.5/14.5)                                                5558
    THE CHARACTERISTIC FUNCTION OF HERMITIAN QUADRATIC FORMS IN COMPLEX
    NORMAL VARIABLES.
    BIOMETRIKA, 47, 199-201.  (MR22-8546)


TURING, ALAN M. 1947.  (20.3)                                                     5559
    ROUNDING-OFF ERRORS IN MATRIX PROCESSES.
    QUART. J. MECH. APPL. MATH., 1, 287-308.  (MR10,P.405)

TURNBULL, HERBERT W. 1927. (20.1)                                              5560
    ON DIFFERENTIATING A MATRIX.
  PROC. EDINBURGH MATH. SOC. SER. 2, 1, 111-128.

TURNER, MALCOLM E. + STEVENS, CHARLES D. 1959. (17.7/CE)                       5561
    THE REGRESSION ANALYSIS OF CAUSAL PATHS.
  BIOMETRICS, 15, 236-258. (MR21-2329)

TURNER, MALCOLM E.; MONROE, ROBERT J. + HOMER, L. D. 1963. (16.5)             5562
    GENERALIZED KINETIC REGRESSION ANALYSIS: HYPERGEOMETRIC KINETICS.
  BIOMETRICS, 19, 406-428. (MR28-3509)

TURNER, MALCOLM E.; MONROE, ROBERT J. + LUCAS, HENRY L., JR. 1961. (17.7)     5563
    GENERALIZED ASYMPTOTIC REGRESSION AND NON-LINEAR PATH ANALYSIS.
  BIOMETRICS. 17, 120-143. (MR22-12651)

TURRI, TULLIO 1934. (4.8)                                                      5564
    CORRELAZIONI PROIETTIVAMENTE DISTINTE LE CUI OMOGRAFIE QUADRATE SONO
    PROIETTIVAMENTE IDENTICHE.
  REND. CIRCOLO MAT. PALERMO, 58, 175-189.

TUTUBALIN, V. N. 1965. (17.1)                                                  5565
    ON LIMIT THEOREMS FOR PRODUCTS OF RANDOM MATRICES.
        (IN RUSSIAN, TRANSLATED AS NO. 5566)
  TEOR. VEROJATNOST. I PRIMENEN., 10, 19-32.

TUTUBALIN, V. N. 1965. (17.1)                                                  5566
    ON LIMIT THEOREMS FOR THE PRODUCT OF RANDOM MATRICES.
        (TRANSLATION OF NO. 5565)
  THEORY PROB. APPL., 10, 15-27.

TWERY, RAYMOND J.; SCHMID, JOHN, JR. + WRIGLEY, CHARLES F. 1958. E(15.2)       5567
    SOME FACTORS IN JOB SATISFACTION: A COMPARISON OF THREE METHODS OF ANALYSIS.
  EDUC. PSYCHOL. MEAS., 18, 189-201.

TYCHNOFF, A. SEE TIHONOV, A. N.


TYLER, FRED T. 1952. (5.2/6.2/E)                                               5568
    SOME EXAMPLES OF MULTIVARIATE ANALYSIS IN EDUCATIONAL AND
    PSYCHOLOGICAL RESEARCH.
  PSYCHOMETRIKA, 17, 289-296.

TYSDAL, H. M. SEE
  2678. IMMER, F. R.; TYSDAL, H. M. + STEECE, H. M. (1939)

UCHIDA, GINZO + PEARSON, KARL 1904. (4.1/E)                                    5569
    ON THE CORRELATION BETWEEN AGE AND THE COLOUR OF HAIR AND EYES IN MAN.
  BIOMETRIKA, 3, 462-466.

UD-DIN, M. ZIA SEE ZIA UD-DIN, M.

UEMATU, TOSIO 1959. (8.5/12.1/20.3)                                            5570
    NOTE ON THE NUMERICAL COMPUTATION IN THE DISCRIMINATION PROBLEM.
  ANN. INST. STATIST. MATH. TOKYO, 10, 131-135. (MR21-4506)

UEMATU, TOSIO 1964. (6.1/18.2)                                                 5571
    ON A MULTIDIMENSIONAL LINEAR DISCRIMINANT FUNCTION.
  ANN. INST. STATIST. MATH. TOKYO, 16, 431-437. (MR31-842)

UENO, TADASHI 1956. (14.4)                                                     5572
    A METHOD IN THE STUDY OF MUTUALLY DEPENDENT RANDOM VARIABLES.
  SUGAKU, 8, 1-9.

UHLENBECK, G. E. SEE
  5755. WANG, MING CHEN + UHLENBECK, G. E. (1945)

UKITA, YOSHIMASA 1955. (20.2/E)                                                5573
    CHARACTERIZATION OF 2-TYPE DIAGONAL MATRICES, WITH AN APPLICATION TO ORDER
    STATISTICS.
        (IN JAPANESE, WITH SUMMARY IN ENGLISH)
  J. HOKKAIDO GAKUGEI UNIV. SECT. B, 6, 66-75.

UKITA. YUKICHI 1955. (4.5/8.1)                                            5574
    ON THE VECTOR LINEAR MEAN REGRESSION AND THE PARTIAL CORRELATION.
    J. HOKKAIDO GAKUGEI UNIV. **SECT. B,** 6, 26-34.


ULLMAN. JOSEPH L. 1944. (20.3)                                            5575
    THE PROBABILITY OF CONVERGENCE OF AN ITERATIVE PROCESS OF INVERTING A MATRIX.
    ANN. MATH. STATIST., 15, 205-213. (MR6,P.51)


ULMO. JEANINE 1962. (8.1)                                                 5576
    ETUDE GEOMETRIQUE DE LA REGRESSION LINEAIRE MULTIPLE.
    C.R. ACAD. SCI. PARIS, 254, 61-63. (MR24-A1165)


UNGRIA. A. DIAZ  SEE  DIAZ UNGRIA, A.


UVEN. M. J. VAN  SEE  VAN UVEN, M.J.


VACEK. MILOS 1932. (18.1)                                                 5577
    SUR LA LOI DE POLYA REGISSANT LES FAITS CORRELATIFS.
    AKTUAR. VEDY. 3, 18-28, 49-61.


VADUVA. I. 1963. (17.8/A)                                                 5578
    ESTIMATION OF A DENSITY OF DISTRIBUTION IN K-DIMENSIONS.
        (IN ROUMAINIAN, WITH SUMMARY IN RUSSIAN AND FRENCH)
    STUDII CERC. MAT., 14, 653-660. (MR32-1807)


VAIL. RICHARD W.   SEE
    1803. FREUND. RUDOLF J.; VAIL, RICHARD W. + CLUNIES ROSS, CHARLES W. (1961)

    1804. FREUND. RUDOLF J.; VAIL, RICHARD W. + CLUNIES ROSS, CHARLES W. (1961)


VAN BEAUMONT. R. C.   SEE
    5588. VAN NAERSSEN, R. F. + VAN BEAUMONT, R. C. (1966)


VAN BODA. STEPHAN 1936. (4.1)                                             5579
    VEREINFACHTE KORRELATIONSRECHNUNG.
    INDUST. PSYCHOTECH., 13, 243-246.


VAN BOVEN. ALICE 1947. (4.6/8.1/20.3)                                     5580
    A MODIFIED AITKEN PIVOTAL CONDENSATION METHOD FOR PARTIAL REGRESSION
    AND MULTIPLE CORRELATION.
    PSYCHOMETRIKA, 12, 127-133. (MR9.P.47)


VAN DE PANNE. C.   SEE
    3985. NEUDECKER, H. + VAN DE PANNE, C. (1966)


VAN DER STOK. J. P. 1916. (2.1/4.3/8.8/E)                                 5581
    DE METHODE DER CORRELATIE EN HARE TOEPASSINGEN (ANWENDUNGEN).
    ARCH. VERZEKERINGS-WETENSCH. AANVERW. VAKKEN, 15, 27-67.


VAN DER VAART. H. ROBERT 1953. (2.7)                                      5582
    THE CONTENT OF CERTAIN SPHERICAL POLYHEDRA FOR ANY NUMBER OF DIMENSIONS.
    EXPERIENTIA, 9, 88-89. (MR14,P.1007)


VAN DER VAART. H. ROBERT 1955. (2.7/20.4/20.5)                            5583
    THE CONTENT OF SOME CLASSES OF NON-EUCLIDEAN POLYHEDRA FOR ANY
    NUMBER OF DIMENSIONS, WITH SEVERAL APPLICATIONS.  I, II.
        (AMENDED BY NO. 5584)
    KON. NEDERL. WETENSCH. PROC. SER. A, 58, 199-221. (MR17,P.401)


VAN DER VAART. H. ROBERT 1955. (2.7/20.4/20.5)                            5584
    THE CONTENT OF SOME CLASSES OF NON-EUCLIDEAN POLYHEDRA FOR ANY NUMBER
    OF DIMENSIONS, WITH SEVERAL APPLICATIONS.  I. AND II.  ERRATA.
        (AMENDMENT OF NO. 5583)
    KON. NEDERL. WETENSCH. PROC. SER. A, 58, 564. (MR17,P.401)


VAN DER VAART. H. ROBERT 1961. (13.1/17.5)                                5585
    ON CERTAIN CHARACTERISTICS OF THE DISTRIBUTION OF THE LATENT ROOTS
    OF A SYMMETRIC RANDOM MATRIX UNDER GENERAL CONDITIONS.
    ANN. MATH. STATIST., 32, 864-873. (MR24-A608)

VAN DER VAART, H. ROBERT 1965.  (6.2)                                          **5586**
    A NOTE ON WILKS' INTERNAL SCATTER.
    ANN. MATH. STATIST., 36, 1308-1312.  (MR31-2790)


VAN GOOL, J.   SEE
    1261. DE FROE, A.; HUIZINGA, J. + VAN GOOL, J. (1947)


VAN KAMPEN, F. R.   SEE
    2889. KAC, MARK + VAN KAMPEN, E. R. (1939)


VAN NAERSSEN, R. F. 1966.  (15.4)                                              **5587**
    DE FOUT RIJ HET GEBRUIK VAN EEN TWEETAL BENADERINGSFORMULES.
        (WITH SUMMARY IN ENGLISH)
    STATISTICA NEERLANDICA, 20, 251-256.


VAN NAERSSEN, R. F. + VAN BEAUMONT, R. C. 1966.  (4.4/15.4)                    **5588**
    THE LONG COEFFICIENT AS AN APPROXIMATION OF THE BISERIAL $r$  IN THE
    ITEM ANALYSIS OF TESTS.
        (IN DUTCH)
    NEDERL. TIJDSCHR. PSYCHOL. GRENSGEB., 21, 308-316.


VAN NESS, J.   SEE
    1799. FREIBERGER, WALTER F.; ROSENBLATT, MURRAY + VAN NESS, J. (1962)


VAN RYZIN, J. R. 1966.  (19.1)                                                 **5589**
    THE COMPOUND DECISION PROBLEM WITH M×N FINITE LOSS MATRIX.
    ANN. MATH. STATIST., 37, 412-424.


VAN RYZIN, J. R. 1966.  (19.1)                                                 **5590**
    THE SEQUENTIAL COMPOUND DECISION PROBLEMS WITH  M×N  FINITE LOSS MATRIX.
    ANN. MATH. STATIST., 37, 954-975.


VAN RYZIN, J. R. 1966.  (6.4/17.8/A)                                           **5591**
    BAYES RISK CONSISTENCY OF CLASSIFICATION PROCEDURES USING DENSITY
    ESTIMATION.
    SANKHYA SER. A, 28, 261-270.  (MR35-1158)


VAN UVEN, M.J. 1914.  (2.1)                                                    **5592**
    THE THEORY OF BRAVAIS (ON ERRORS IN SPACE) FOR POLYDIMENSIONAL SPACE,
    WITH APPLICATIONS TO CORRELATION.
        (TRANSLATION OF NO. 5595)
    KON. AKAD. WETENSCH. AMSTERDAM PROC. SECT. SCI., 16, 1124-1135.


VAN UVEN, M.J. 1914.  (2.1)                                                    **5593**
    THE THEORY OF BRAVAIS (ON ERRORS IN SPACE) FOR POLYDIMENSIONAL SPACE,
    WITH APPLICATIONS TO CORRELATION.  (CONTINUATION).
        (TRANSLATION OF NO. 5595)
    KON. AKAD. WETENSCH. AMSTERDAM PROC. SECT. SCI., 17, 150-156.


VAN UVEN, M.J. 1914.  (8.8/20.5)                                               **5594**
    THE THEORY OF THE COMBINATION OF OBSERVATIONS AND THE DETERMINATION
    OF THE PRECISION, ILLUSTRATED BY MEANS OF VECTORS.
        (TRANSLATION OF NO. 5596)
    KON. AKAD. WETENSCH. AMSTERDAM PROC. SECT. SCI., 17, 490-500.


VAN UVEN, M.J. 1914.  (2.1)                                                    **5595**
    DE THEORIE VAN BRAVAIS (OVER DE FOUTEN IN DE RUIMTE) VOOR DE
    MEERDIMENSIONALE RUIMTE, MET TOEPASSINGEN OP DE CORRELATIE.
        (TRANSLATED AS NO. 5592, TRANSLATED AS NO. 5593)
    VERSLAG AFD. NATUURK. KON. AKAD. WETENSCH., 22, 1075-1086, 1265-1271.


VAN UVEN, M.J. 1914.  (8.8/20.5)                                               **5596**
    FOUTENVEREFFENING EN BEPALING VAN DE NAUWKEURIGHEID MET BEHULP VAN VECTOREN.
        (TRANSLATED AS NO. 5594)
    VERSLAG AFD. NATUURK. KON. AKAD. WETENSCH., 23, 300-310.


VAN UVEN, M.J. 1925.  (2.1/4.1/18.1/B)                                         **5597**
    ON TREATING SKEW CORRELATION.
        (TRANSLATION OF NO. 5599)
    KON. AKAD. WETENSCH. AMSTERDAM PROC. SECT. SCI., 28, 797-811.

VAN UVEN, M.J. 1925.   (2.1/4.1/18.1/B)                                                                    5598
    ON TREATING SKEW CORRELATION.   (CONTINUATION).
        (TRANSLATION OF NO. 5600)
    KON. AKAD. WETENSCH. AMSTERDAM PROC. SECT. SCI., 28, 919-935.

VAN UVEN, M.J. 1925.   (2.1/4.1/18.1/B)                                                                    5599
    OVER HET BEWERKEN VAN SCHEEVE CORRELATIE.
        (TRANSLATED AS NO. 5597, AMENDED BY NO. 5603)
    VERSLAG AFD. NATUURK. KON. AKAD. WETENSCH., 34, 787-802.

VAN UVEN, M.J. 1925.   (2.1/4.1/18.1/B)                                                                    5600
    OVER HET BEWERKEN VAN SCHEEVE CORRELATIE. -(VERVOLG.)
        (TRANSLATED AS NO. 5598, AMENDED BY NO. 5603)
    VERSLAG AFD. NATUURK. KON. AKAD. WETENSCH., 34, 965-982.

VAN UVEN, M.J. 1926.   (4.1/18.1/BE)                                                                       5601
    ON TREATING SKEW CORRELATION.   (THE END).
        (TRANSLATION OF NO. 5602)
    KON. AKAD. WETENSCH. AMSTERDAM PROC. SECT. SCI., 29, 580-590.

VAN UVEN, M.J. 1926.   (4.1/18.1/BE)                                                                       5602
    OVER HET BEWERKEN VAN SCHEEVE CORRELATIE.   (SLOT).
        (TRANSLATED AS NO. 5601)
    VERSLAG AFD. NATUURK. KON. AKAD. WETENSCH., 35, 129-141.

VAN UVEN, M.J. 1926.   (2.1/4.1/18.1/B)                                                                    5603
    ERRATA: OVER HET BEWERKEN VAN SCHEEVE CORRELATIE.
        (AMENDMENT OF NO. 5599, AMENDMENT OF NO. 5600)
    VERSLAG AFD. NATUURK. KON. AKAD. WETENSCH., 35, 137.

VAN UVEN, M.J. 1927.   (4.5/8.1)                                                                           5604
    METHOD OF CALCULATING THE MEAN ERRORS OF THE STANDARD DEVIATIONS, COEFFICIENTS
    OF CORRELATION AND COEFFICIENTS OF REGRESSION, WITH 'N' LINEARLY CORRELATED VARIABLES.
        (TRANSLATION OF NO. 5606)
    KON. AKAD. WETENSCH. AMSTERDAM PROC. SECT. SCI., 30, 823-838.

VAN UVEN, M.J. 1927.   (17.7)                                                                              5605
    LINEAR ADJUSTMENT OF A SET OF PAIRS OF NUMBERS $(X_k, Y_k)$.
    KON. AKAD. WETENSCH. AMSTERDAM PROC. SECT. SCI., 30, 1021-1038.

VAN UVEN, M.J. 1927.   (4.5/8.1)                                                                           5606
    METHODE TER BEREKENING VAN DE MIDDELBARE FOUTEN DER STANDAARDAFWIJKINGEN,
    CORRELATIECOEFFICIENTEN EN REGRESSIECOEFFICIENTEN, BIJ N LINEAIR-GECORRELATEERDE
    VERANDERLIJKEN.
        (TRANSLATED AS NO. 5604)
    VERSLAG AFD. NATUURK. KON. AKAD. WETENSCH., 36, 875-890.

VAN UVEN, M.J. 1929.   (4.2)                                                                               5607
    SCHEEVE CORRELATIE.
    HANDEL. NEDERL. NATUUR- EN GENEESK. CONG., 22, 100-102.

VAN UVEN, M.J. 1929.   (4.1/18.1/BE)                                                                       5608
    SCHEEVE CORRELATIE TUSSCHEN TWEE VERANDERLIJKEN.
        (WITH SUMMARY IN ENGLISH)
    KON. AKAD. WETENSCH. AMSTERDAM PROC. SECT. SCI., 32, 408-413.

VAN UVEN, M.J. 1929.   (2.1/3.1)                                                                           5609
    SKEW CORRELATION BETWEEN THREE AND MORE VARIABLES, I.   SKEW CORRELATION
    BETWEEN THREE VARIABLES.
    KON. AKAD. WETENSCH. AMSTERDAM PROC. SECT. SCI., 32, 793-807.

VAN UVEN, M.J. 1929.   (2.1/3.1)                                                                           5610
    SKEW CORRELATION BETWEEN THREE AND MORE VARIABLES, II.   SKEW CORRELATION
    BETWEEN $n$ VARIABLES.
    KON. AKAD. WETENSCH. AMSTERDAM PROC. SECT. SCI., 32, 995-1007.

VAN UVEN, M.J. 1929.   (2.1/3.1)                                                                           5611
    SKEW CORRELATION BETWEEN THREE AND MORE VARIABLES, III.   SKEW CORRELATION
    BETWEEN $n$ VARIABLES.
    KON. AKAD. WETENSCH. AMSTERDAM PROC. SECT. SCI., 32, 1085-1103.

VAN UVEN, M.J. 1930.   (8.8/20.5)                                                                          5612
    ADJUSTMENT OF N POINTS (IN $n$-DIMENSIONAL SPACE) TO THE BEST LINEAR
    $(n-1)$-DIMENSIONAL SPACE. I.
    KON. AKAD. WETENSCH. AMSTERDAM PROC. SECT. SCI., 33, 143-157.

VAN UVEN, M.J. 1930.  (8.8/20.5)                                          5613
    ADJUSTMENT OF N POINTS (IN n-DIMENSIONAL SPACE) TO THE BEST LINEAR
    (n-1)-DIMENSIONAL SPACE.  II.
KON. AKAD. WETENSCH. AMSTERDAM PROC. SECT. SCI., 33, 307-326.

VAN UVEN, M.J. 1939.  (3.3)                                               5614
    ADJUSTMENT OF A RATIO.
ANN. EUGENICS, 9, 181-202.

VAN UVEN, M.J. 1943.  (2.1/B)                                             5615
    CORRELATIE I., II.
LANDBOUWK. TIJDSCHR. WAGENINGEN, 55, 622-638.

VAN UVEN, M.J. 1947.  (18.1/B)                                            5616
    EXTENSION OF PEARSON'S PROBABILITY DISTRIBUTIONS TO TWO VARIABLES. I.
KON. NEDERL. AKAD. WETENSCH. PROC. SECT. SCI., 50, 1063-1070.  (MR9,P.363)

VAN UVEN, M.J. 1947.  (18.1/B)                                            5617
    EXTENSION OF PEARSON'S PROBABILITY DISTRIBUTIONS TO TWO VARIABLES.  II.
KON. NEDERL. AKAD. WETENSCH. PROC. SECT. SCI., 50, 1252-1264.  (MR9,P.363)

VAN UVEN, M.J. 1948.  (18.1/B)                                            5618
    EXTENSION OF PEARSON'S PROBABILITY DISTRIBUTIONS TO TWO VARIABLES. III.
KON. NEDERL. AKAD. WETENSCH. PROC. SECT. SCI., 51, 41-52.  (MR9,P.452)

VAN UVEN, M.J. 1948.  (18.1/B)                                            5619
    EXTENSION OF PEARSON'S PROBABILITY DISTRIBUTIONS TO TWO VARIABLES.  IV.
KON. NEDERL. AKAD. WETENSCH. PROC. SECT. SCI., 51, 191-196.  (MR9,P.452)

VAN WIJNGAARDEN, A.   SEE
    2861. KAAPSFMAKER, L. + VAN WIJNGAARDEN, A. (1953)

VARADARAJAN, V.S. 1958.  (17.1)                                           5620
    A USEFUL CONVERGENCE THEOREM.
SANKHYA, 20, 221-222.  (MR21-6015)

VARADARAJAN, V.S.   SEE ALSO
    4402. RANGA RAO, R. + VARADARAJAN, V.S. (1960)

    4464. RAO, C. RADHAKRISHNA + VARADARAJAN, V.S. (1963)

    4465. RAO, C. RADHAKRISHNA + VARADARAJAN, V.S. (1964)

VARADY, PAUL V.   SEE
    1417. DUNN, OLIVE JEAN + VARADY, PAUL V. (1966)

VARANGOT, V. 1951.  (4.6)                                                 5621
    GRAFISCHE BEPALING VAN DE TWEEVOUDIGE CORRELATIECOEFFICIENT.
STATISTICA. LEIDEN, 5, 145-147.

VARBERG, DALE F. 1966.  (2.5/17.1)                                        5622
    CONVERGENCE OF QUADRATIC FORMS IN INDEPENDENT RANDOM VARIABLES.
ANN. MATH. STATIST., 37, 567-576.  (MR32-8394)

VARGA, R. S. 1965.  (20.3)                                                5623
    ITERATIVE METHODS FOR SOLVING MATRIX EQUATIONS.
AMER. MATH. MONTHLY, 72 (2), 67-74.

VARMA. K.BHASKARA 1951.  (10.1)                                           5624
    ON THE EXACT DISTRIBUTION OF WILKS  $L_{mvc}$ AND $L_{vc}$ CRITERIA.
BULL. INST. INTERNAT. STATIST., 33(2) 181-214.  (MR16,P.841)

VARMA, R. S. 1952.  (2.1/20.4)                                            5625
    ON THE PROBABILITY FUNCTION IN A NORMAL MULTIVARIATE DISTRIBUTION.
QUART. J. MECH. APPL. MATH., 5, 361-362.  (MR14,P.293)

VASWANI, SUNDRI P. 1947.  (4.1/18.1)                                      5626
    A PITFALL IN CORRELATION THEORY.
NATURE, 160, 405-406.  (MR9,P.294)

VASWANI, SUNDRI P. 1950.  (4.4)                                                    5627
     ASSUMPTIONS UNDERLYING THE USE OF THE TETRACHORIC CORRELATION COEFFICIENT.
     SANKHYA, 10, 269-276.  (MR12,P.429)

VELLANDO, GONZALO ARNAIZ  SEE  ARNAIZ, GONZALO

VENKATACHARI, S.  SEE
     3153. KRISHNASWAMI, G. V. + VENKATACHARI, S. (1934)

     5277. SUBRAMANIAN, S + VENKATACHARI, S. (1934)

VENTTSEL', A. D. 1959.  (16.4/17.1/A)                                              5628
     ON BOUNDARY CONDITIONS FOR MULTIDIMENSIONAL DIFFUSION PROCESSES.
          (IN RUSSIAN, WITH SUMMARY IN ENGLISH, TRANSLATED AS NO. 5629)
     TEOR. VEROJATNOST. I PRIMENEN., 4, 172-185.  (MR21-5246)

VENTTSEL', A. D. 1959.  (16.4/17.1/A)                                              5629
     ON BOUNDARY CONDITIONS FOR MULTIDIMENSIONAL DIFFUSION PROCESSES.
          (TRANSLATION OF NO. 5628)
     THEORY PROB. APPL., 4, 164-177.  (MR21-5246)

VERDOOREN, L. R.  SEE
     3004. KEULS, M. + VERDOOREN, L. R. (1964)

VERGOTTINI, MARIO DE  SEE  DE VERGOTTINI, MARIO

VERHAGEN, A. M. W. 1961.  (8.1/17.8)                                               5630
     THE ESTIMATION OF REGRESSION AND ERROR-SCALE PARAMETERS WHEN THE JOINT
     DISTRIBUTION OF ERRORS IS OF ANY CONTINUOUS FORM AND KNOWN APART FROM
     A SCALE PARAMETER.
     BIOMETRIKA, 48, 125-132.  (MR25-1598)

VERMA, M. C. + GHOSH, MAHINDRA NATH 1963.  (8.3/CT)                                5631
     SIMULTANEOUS TESTS OF LINEAR HYPOTHESES AND CONFIDENCE INTERVAL ESTIMATION.
     J. INDIAN SOC. AGRIC. STATIST., 15, 194-212.

VERNON, P. E. 1948.  (15.4)                                                        5632
     INDICES OF ITEM CONSISTENCY AND VALIDITY.
     BRITISH J. PSYCHOL. STATIST. SECT., 1, 152-166.

VESSEREAU, A.  SEE
     922. CHARBONNIER, A.; CYFFERS, B.; SCHWARTZ, D. +  (1957)
     VESSEREAU, A.

VICKERY, C. W. 1939.  (20.5)                                                       5633
     SPACES OF UNCOUNTABLY MANY DIMENSIONS.
     BULL. AMER. MATH. SOC., 45, 456-462.

VIDAL, ANDRE 1955.  (1.1)                                                          5634
     L'ANALYSE DES GROUPES A VARIABLES MULTIPLES OU ANALYSE TYPOLOGIQUE.
     REV. STATIST. APPL., 3(4) 87-94.

VILKAUSKAS, L. 1961.  (17.1)                                                       5635
     ZONES OF NORMAL CONVERGENCE IN THE HIGHER-DIMENSIONAL CASE.
          (IN RUSSIAN, WITH SUMMARY IN LITHUANIAN AND ENGLISH)
     LITOVSK. MAT. SB., 1, 25-40.  (MR26-5604)

VILKAUSKAS, L. 1965.  (17.1)                                                       5636
     LARGE DEVIATIONS OF LINNIK TYPE IN THE MULTI-DIMENSIONAL CASE ON
     CERTAIN REGIONS.
          (IN RUSSIAN, WITH SUMMARY IN ENGLISH AND LITHUANIAN)
     LITOVSK. MAT. SB., 5, 25-43.  (MR35-4971)

VILLARS, D.S. + ANDERSON, T. W. 1943.  (8.3/10.2/17.7/B)                           5637
     SOME SIGNIFICANCE TESTS FOR NORMAL BIVARIATE DISTRIBUTIONS.
     ANN. MATH. STATIST., 14, 141-148.  (MR5,P.127)

VILLE, JEAN-A. 1943.  (17.8)                                                       5638
     SUR UN CRITERE D'INDEPENDANCE.
     C.R. ACAD. SCI. PARIS, 216, 552-553.  (MR5,P.206)

VILLE, JEAN-A. 1943.  (17.8)                                                    5639
    SUR L'APPLICATION, A UN CRITERE D'INDEPENDANCE, DU DENOMBREMENT DES
    INVERSIONS PRESENTEES PAR UNE PERMUTATION.
    C.R. ACAD. SCI. PARIS, 217, 41-42.  (MR6,P.8)

VILLE, JEAN-A. 1963.  (16.2)                                                    5640
    LE ROLE DES MATHEMATIQUES DANS LA FORMATION DE LA PENSEE ECONOMIQUE.
    CAHIERS INST. SCI. ECON. APPL. SER. E, 2, 15-30.

VILLE, M. J. 1955.  (20.1)                                                      5641
    PRINCIPES D'ANALYSE MATRICIELLE.
    PUBL. INST. STATIST. UNIV. PARIS, 4, 141-217.  (MR17,P.1044)

VILLEGAS, CESAREO 1961.  (17.7)                                                 5642
    MAXIMUM LIKELIHOOD ESTIMATION OF A LINEAR FUNCTIONAL RELATIONSHIP.
    ANN. MATH. STATIST., 32, 1048-1062.  (MR24-A1767)

VILLEGAS, CESAREO 1962.  (17.7)                                                 5643
    ON THE LEAST SQUARES ESTIMATION OF A LINEAR RELATION.
        (REPRINTED AS NO. 5644)
    BOL. FAC. INGEN. AGRIMENS. MONTEVIDEO, 8, 47-63.  (MR28-2607)

VILLEGAS, CESAREO 1963.  (17.7)                                                 5644
    ON THE LEAST SQUARES ESTIMATION OF A LINEAR RELATION.
        (REPRINT OF NO. 5643)
    UNIV. REPUB. FAC. INGEN. MONTEVIDEO PUBL. INST. MAT. ESTADIST., 3, 189-204.  (MR28-2606)

VILLEGAS, CESAREO 1964.  (17.7/C)                                               5645
    CONFIDENCE REGION FOR A LINEAR RELATION.
    ANN. MATH. STATIST., 35, 780-788.  (MR29-697)

VINCENT, DOUGLAS F. 1953.  (15.1)                                               5646
    THE ORIGIN AND DEVELOPMENT OF FACTOR ANALYSIS.
    APPL. STATIST., 2, 107-117.

VINCENT, NORMAN L. 1962.  (15.1/E)                                              5647
    A NOTE ON STOETZEL'S FACTOR ANALYSIS OF LIQUOR PREFERENCES.
    J. ADV. RES., 2(1) 24-27.

VINCENZ, S.A. + BRUCKSHAW, J. MC G. 1960.  (17.3)                               5648
    NOTE ON THE PROBABILITY DISTRIBUTION OF A SMALL NUMBER OF VECTORS.
    PROC. CAMBRIDGE PHILOS. SOC., 56, 21-26.  (MR22-999)

VINCZE, ISTVAN 1960.  (17.8/AB)                                                 5649
    ON THE DEVIATION OF TWO-VARIATE EMPIRICAL DISTRIBUTION FUNCTIONS.
    MAGYAR TUD. AKAD. MAT. OSZT. KOZL., 10, 361-372.  (MR25-692)

VINCZE, ISTVAN  SEE ALSO
    4843. SARKADI, KAROLY; SCHNELL, EDIT + VINCZE, ISTVAN (1962)

VINOGRADE, BERNARD 1950.  (20.1)                                                5650
    CANONICAL POSITIVE DEFINITE MATRICES UNDER INTERNAL LINEAR TRANSFORMATIONS.
    PROC. AMER. MATH. SOC., 1, 159-161.  (MR11,P.637)

VINOKUROV, V. G. 1959.  (16.4)                                                  5651
    ON PROBABILITY PROCESSES GIVEN IN CO-ORDINATE SPACE.
        (IN RUSSIAN, WITH SUMMARY IN UZBEKISH)
    IZV. AKAD. NAUK UZSSR. SER. FIZ.-MAT. NAUK, 1959(5) 42-48.

VISONHALER, JOHN F. + MEREDITH, WILLIAM 1966.  (15.4)                           5652
    A STOCHASTIC MODEL FOR REPEATED TESTING.
    MULTIVARIATE BEHAV. RES., 1, 461-524.

VITALI, ORNELLO 1964.  (11.2)                                                   5653
    IL METODO DELLE COMPONENTI PRINCIPALI E LE SUE POSSIBILITA DI APPLICAZIONE.
    STATISTICA, BOLOGNA, 24, 253-294.  (MR32-1837)

VOAK, HELMUT 1952.  (15.1)                                                      5654
    SPORTLEISTUNGEN IM LICHTE DER MULTIPLEN FAKTORENANALYSE.
    STATIST. VIERTELJAHRESSCHR., 5, 23-31.

VOLKONSKII, V. A. 1957. (17.1)
A MULTIDIMENSIONAL LIMIT THEOREM FOR MARKOV CHAINS WITH A COUNTABLE
SET OF STATES.
(IN RUSSIAN, WITH SUMMARY IN ENGLISH, TRANSLATED AS NO. 5656)
TEOR. VEROJATNOST. I PRIMENEN., 2, 230-255. (MR19,P.1089)

5655

VOLKONSKII, V. A. 1957. (17.1)
A MULTI-DIMENSIONAL LIMIT THEOREM FOR HOMOGENEOUS MARKOV CHAINS WITH A
COUNTABLE NUMBER OF STATES.
(TRANSLATION OF NO. 5655)
THEORY PROB. APPL., 2, 221-244. (MR19,P.1089)

5656

VOLODIN, I. N. 1964. (17.9)
TESTING THE HYPOTHESIS OF NORMALITY OF A DISTRIBUTION BY SMALL SAMPLES
(MULTIVARIATE CASE).
(IN RUSSIAN)
KAZAN. GOS. UNIV. UCEN. ZAP., 124(2) 21-25. (MR33-6761)

5657

VON HOLDT, R. E. 1956. (20.3)
AN ITERATIVE PROCEDURE FOR THE CALCULATION OF THE EIGENVALUES AND
EIGENVECTORS OF A REAL SYMMETRIC MATRIX.
J. ASSOC. COMPUT. MACH., 3, 223-238. (MR18,P.418)

5658

VON MISES, HILDA GEIRINGER SEE GEIRINGER, HILDA

VON MISES, RICHARD 1945. (6.1)
ON THE CLASSIFICATION OF OBSERVATION DATA INTO DISTINCT GROUPS.
ANN. MATH. STATIST., 16, 68-73. (MR6,P.235)

5659

VON NEUMANN, JOHN 1937. (20.5)
UBER EIN OKONOMISCHES GLEICHUNGSSYSTEM UND EINE VERALLGEMEINERUNG DES
BROUWER'SCHEN FIXPUNKTSATZES.
ERGEB. MATH. KOLLOQ., 8, 73-83.

5660

VON NEUMANN, JOHN 1941. (2.6)
DISTRIBUTION OF THE RATIO OF THE MEAN SQUARE SUCCESSIVE DIFFERENCE TO
THE VARIANCE.
ANN. MATH. STATIST., 12, 367-395. (MR4,P.21)

5661

VON NEUMANN, JOHN + GOLDSTINE, HERMAN H. 1947. (20.3)
NUMERICAL INVERTING OF MATRICES OF HIGH ORDER.
BULL. AMER. MATH. SOC., 53, 1021-1099. (MR9,P.471)

5662

VON NEUMANN, JOHN SEE ALSO
918. CHANDRASEKHAR, S. + VON NEUMANN, JOHN (1942)

919. CHANDRASEKHAR, S. + VON NEUMANN, JOHN (1943)

2049. GOLDSTINE, HERMAN H. + VON NEUMANN, JOHN (1951)

2377. HART, B. I. + VON NEUMANN, JOHN (1942)

VON NIESSL, G. 1894. (6.6)
UEBER DIE WAHRSCHEINLICHSTE BAHNFORM FUR DIE AUS DEM WELTENRAUM IN
UNSERE BEOBACHTUNGSSPHARE GELANGENDEN KORPER.
ASTRONOM. NACHR., 135, 137-150.

5663

VON SCHELLING, HERMANN 1931. (16.1/B)
DIE WIRTSCHAFTLICHEN ZEITREIHEN ALS PROBLEM DER KORRELATIONSRECHNUNG.
MIT BESONDERER BERUCKSICHTIGUNG DER "LAG" KORRELATION.
VEROFF. FRANKFURTER GES. KONJUNKTURF., 11, 1-64.

5664

VON SCHELLING, HERMANN 1932. (4.8)
KORRELATIONSMESSUNG AUF GRUND DER ANORDRUNG DER BEOBACHTUNGEN.
Z. ANGEW. MATH. MECH., 12, 377-380.

5665

VON SCHELLING, HERMANN 1949. (20.4)
A FORMULA FOR THE PARTIAL SUMS OF SOME HYPERGEOMETRIC SERIES.
ANN. MATH. STATIST., 20, 120-122. (MR10,P.454)

5666

VON SZELISKI, VICTOR S. 1929. E(16.5)
EXPERIMENTS IN THE CORRELATION OF TIME SERIES.
J. AMER. STATIST. ASSOC., 24(SP) 241-247.

5667

VOSS, A. 1896.  (20.1)                                                    5668
     SYMMETRISCHE UND ALTERNIRENDE LOSUNGEN DER GLEICHUNG SX = XS'.
SITZUNGSBER. MATH. PHYS. CL. KAIS. AKAD. WISS. MUNCHEN, 26, 273–281.

VOTAW, DAVID F., JR. 1933.  (15.4/T)                                      5669
     GRAPHICAL DETERMINATION OF PROBABLE ERROR IN VALIDATION OF TEST ITEMS.
J. EDUC. PSYCHOL., 24, 682–686.

VOTAW, DAVID F., JR. 1946.  (17.3)                                        5670
     THE PROBABILITY DISTRIBUTION OF THE MEASURE OF A RANDOM LINEAR SET.
ANN. MATH. STATIST., 17, 240–244.  (MR8,P.281)

VOTAW, DAVID F., JR. 1948.  (10.2)                                        5671
     TESTING COMPOUND SYMMETRY IN A NORMAL MULTIVARIATE DISTRIBUTION.
ANN. MATH. STATIST., 19, 447–473.  (MR10,P.387)

VOTAW, DAVID F., JR. 1952.  (6.7)                                         5672
     METHODS OF SOLVING SOME PERSONNEL-CLASSIFICATION PROBLEMS.
PSYCHOMETRIKA, 17, 255–266.

VOTAW, DAVID F., JR.; KIMBALL, ALLYN W. + RAFFERTY, J.A. 1950.  (10.2/E)  5673
     COMPOUND SYMMETRY TESTS IN THE MULTIVARIATE ANALYSIS OF MEDICAL EXPERIMENTS.
BIOMETRICS, 6, 259–281.  (MR12,P.271)

VOTAW, DAVID F., JR.; RAFFERTY, J.A. + DEEMER, WALTER L. 1950.  (3.2)     5674
     ESTIMATION OF PARAMETERS IN A TRUNCATED TRIVARIATE NORMAL DISTRIBUTION.
PSYCHOMETRIKA, 15, 339–347.  (MR13,P.367)

VRANCEANU, G. G. 1965.  (16.4)                                           5675
     THEORIE GEOMETRIQUE DES CHAINES PROBABILISTIQUES.
ACAD. ROY. BELG. BULL. CL. SCI. SER. 5, 51, 1158–1167.

VRANIC, VLADIMIR 1950.  (4.1)                                            5676
     O PRIMJENI DUALITETA NA PROMATRANJE KORELACIONIH ODNOSA U STATISTICI.
          (WITH SUMMARY IN ENGLISH)
GLASNIK MAT. FIZ. ASTRONOM. SER. 2, 5, 166–174.

VRANIC, VLADIMIR 1963.  (4.6/8.1)                                        5677
     ON THE USE OF DUALITY IN THE THEORY OF CORRELATION.
RAD JUGOSLAV. AKAD. ZNAN. UMJET. ODJEL MAT. FIZ. TEH. NAUKE, 325, 165–187.  (MR27-3054)

VRANIC, VLADIMIR 1964.  (4.6/8.1)                                        5678
     ON AN APPLICATION OF DUALITY IN REPRESENTING CORRELATION.
REV. INST. INTERNAT. STATIST., 32, 65–71.  (MR30-5434)

WAGNER, ASCHER 1910.  (8.1/U)                                            5679
     UBER DEN EINFLUSS DES MITTLEREN FEHLERS AUF DIE WAHRSCHEINLICHSTE
     BEZIEHUNG ZWISCHEN ZWEI VERANDERLICHEN.
METEOROL. 7., 27, 542–549.

WAGNER, ASCHER 1929.  (18.6)                                             5680
     ZUR THEORIE DER HAUFIGKEITSVERTEILUNG VON FEHLERN IN DER EBENE MIT
     BESONDERER BERUCKSICHTIGUNG DER WINDVEKTOREN.
Z. GEOPHYS., 5, 366–371.

WAGNER, HARVEY M. 1958.  (17.7)                                          5681
     A MONTE CARLO STUDY OF ESTIMATES OF SIMULTANEOUS LINEAR STRUCTURAL EQUATIONS.
ECONOMETRICA, 26, 117–133.  (MR19,P.897)

WAGNER, HARVEY M. 1962.  (8.8/20.3)                                      5682
     NON-LINEAR REGRESSION WITH MINIMAL ASSUMPTIONS.
J. AMER. STATIST. ASSOC., 57, 572–578.

WAHLUND, STEN 1928.  (4.9)                                               5683
     ZUSAMMENSETZUNG VON POPULATIONEN UND KORRELATIONSERSCHEINUNGEN VOM
     STANDPUNKT DER VERERBUNGSLEHRE AUS BETRACHTET.
HEREDITAS GENET. ARK. LUND, 11, 65–106.

WAHLUND, STEN 1935.  (4.4/4.8/E)                                         5684
     A NEW METHOD OF DETERMINING CORRELATION FROM TETRACHORIC GROUPINGS.
UPPSALA LANTBRUKSHOGSK. ANN., 2, 181–242.

WAITE, WARREN C. 1932.   (8.8)                                                                    5685
     SOME CHARACTERISTICS OF THE GRAPHIC METHOD OF CORRELATION.
          (AMENDED BY NO. 1597)
  J. AMER. STATIST. ASSOC., 27, 68-70.

WAITE, WARREN C. 1932.   (8.1)                                                                    5686
     II. REJOINDER.
  J. AMER. STATIST. ASSOC., 27, 185.

WALD, ABRAHAM 1933.   (20.2/20.3)                                                                 5687
     UBER DIE VOLUMSDETERMINANTE.
  ERGEB. MATH. KOLLOQ., 4, 25-28.

WALD, ABRAHAM 1933.   (20.5)                                                                      5688
     UBER DAS VOLUMEN DER EUKLIDISCHEN SIMPLEXE.
  ERGEB. MATH. KOLLOQ., 4, 32-33.

WALD, ABRAHAM 1933.   (20.1)                                                                      5689
     VEREINFACHTER BEWEIS DES STEINITZSCHEN SATZES UBER VEKTORENREIHEN IM $R_n$.
  ERGEB. MATH. KOLLOQ., 5, 10-13.

WALD, ABRAHAM 1933.   (17.1)                                                                      5690
     BEDINGT KONVERGENTE REIHEN VON VEKTOREN IM $R_\omega$.
  ERGEB. MATH. KOLLOQ., 5, 13-14.

WALD, ABRAHAM 1933.   (20.5)                                                                      5691
     KOMPLEXE UND INDEFINITE RAUME.
  ERGEB. MATH. KOLLOQ., 5, 32-42.

WALD, ABRAHAM 1935.   (20.5)                                                                      5692
     SUR LA COURBURE DES SURFACES.
  C.R. ACAD. SCI. PARIS, 201, 918-920.

WALD, ABRAHAM 1936.   (20.5)                                                                      5693
     BEGRUNDUNG EINER KOORDINATENLOSEN DIFFERENTIALGEOMETRIE DER FLACHEN.
  ERGEB. MATH. KOLLOQ., 7, 24-46.

WALD, ABRAHAM 1938.   (17.2)                                                                      5694
     GENERALIZATION OF THE INEQUALITY OF MARKOFF.
  ANN. MATH. STATIST., 9, 244-255.

WALD, ABRAHAM 1939.   (19.2/19.3)                                                                 5695
     CONTRIBUTIONS TO THE THEORY OF STATISTICAL ESTIMATION AND TESTING HYPOTHESES.
          (REPRINTED AS NO. 5712)
  ANN. MATH. STATIST., 10, 299-326.   (MR1,P.152)

WALD, ABRAHAM 1940.   (8.1/17.7)                                                                  5696
     THE FITTING OF STRAIGHT LINES IF BOTH VARIABLES ARE SUBJECT TO ERROR.
          (REPRINTED AS NO. 5713)
  ANN. MATH. STATIST., 11, 284-300.   (MR2,P.108)

WALD, ABRAHAM 1941.   (19.2/A)                                                                    5697
     ASYMPTOTICALLY MOST POWERFUL TESTS OF STATISTICAL HYPOTHESES.
          (REPRINTED AS NO. 5714)
  ANN. MATH. STATIST., 12, 1-19.   (MR3,P.8)

WALD, ABRAHAM 1942.   (8.3/8.8/14.5)                                                              5698
     ON THE POWER FUNCTION OF THE ANALYSIS OF VARIANCE TEST.
          (REPRINTED AS NO. 5715)
  ANN. MATH. STATIST., 13, 434-439.   (MR5,P.129)

WALD, ABRAHAM 1943.   (17.8)                                                                      5699
     AN EXTENSION OF WILKS' METHOD FOR SETTING TOLERANCE LIMITS.
          (REPRINTED AS NO. 5716)
  ANN. MATH. STATIST., 14, 45-55.   (MR4,P.222)

WALD, ABRAHAM 1943.   (19.2/A)                                                                    5700
     TESTS OF STATISTICAL HYPOTHESES CONCERNING SEVERAL PARAMETERS WHEN
     THE NUMBER OF OBSERVATIONS IS LARGE.
          (REPRINTED AS NO. 5717)
  TRANS. AMER. MATH. SOC., 54, 426-482.   (MR7,P.20)

WALD, ABRAHAM 1944.  (6.1/6.2)                                    5701
    ON A STATISTICAL PROBLEM ARISING IN THE CLASSIFICATION OF AN
    INDIVIDUAL INTO ONE OF TWO GROUPS.
        (REPRINTED AS NO. 5718)
    ANN. MATH. STATIST., 15, 145-162.   (MR6,P.9)

WALD, ABRAHAM 1945.  (17.3)                                       5702
    SOME GENERALIZATIONS OF THE THEORY OF CUMULATIVE SUMS OF RANDOM VARIABLES.
    ANN. MATH. STATIST., 16, 287-293.   (MR7,P.209)

WALD, ABRAHAM 1947.  (19.1)                                       5703
    AN ESSENTIALLY COMPLETE CLASS OF ADMISSIBLE DECISION FUNCTIONS.
    ANN. MATH. STATIST., 18, 549-555.   (MR9,P.364)

WALD, ABRAHAM 1947.  (8.6/CU)                                     5704
    A NOTE ON REGRESSION ANALYSIS.
        (REPRINTED AS NO. 5720)
    ANN. MATH. STATIST., 18, 586-589.   (MR9,P.364)

WALD, ABRAHAM 1947.  (17.1)                                       5705
    LIMIT DISTRIBUTION OF THE MAXIMUM AND MINIMUM OF SUCCESSIVE
    CUMULATIVE SUMS OF RANDOM VARIABLES.
        (REPRINTED AS NO. 5719)
    BULL. AMER. MATH. SOC., 53, 142-153.   (MR8,P.471)

WALD, ABRAHAM 1948.  (19.3/A)                                     5706
    ASYMPTOTIC PROPERTIES OF THE MAXIMUM LIKELIHOOD ESTIMATE OF AN
    UNKNOWN PARAMETER OF A DISCRETE STOCHASTIC PROCESS.
        (REPRINTED AS NO. 5721)
    ANN. MATH. STATIST., 19, 40-46.   (MR9,P.454)

WALD, ABRAHAM 1948.  (19.3)                                       5707
    ESTIMATION OF A PARAMETER WHEN THE NUMBER OF UNKNOWN PARAMETERS
    INCREASES INDEFINITELY WITH THE NUMBER OF OBSERVATIONS.
        (REPRINTED AS NO. 5722)
    ANN. MATH. STATIST., 19, 220-227.   (MR10,P.135)

WALD, ABRAHAM 1949.  (19.3/A)                                     5708
    NOTE ON THE CONSISTENCY OF THE MAXIMUM LIKELIHOOD ESTIMATE.
        (REPRINTED AS NO. 5723)
    ANN. MATH. STATIST., 20, 595-601.   (MR11,P.261)

WALD, ABRAHAM 1950.  (16.2)                                       5709
    NOTE ON THE IDENTIFICATION OF ECONOMIC RELATIONS.
        (REPRINTED AS NO. 5724)
    COWLES COMMISS. MONOG., 10, 238-244.

WALD, ABRAHAM 1950.  (16.1/16.2/C)                                5710
    REMARKS ON THE ESTIMATION OF UNKNOWN PARAMETERS IN INCOMPLETE SYSTEMS
    OF EQUATIONS.
        (REPRINTED AS NO. 5725)
    COWLES COMMISS. MONOG., 10, 305-310.

WALD, ABRAHAM 1952.  (19.1)                                       5711
    BASIC IDEAS OF A GENERAL THEORY OF STATISTICAL DECISION RULES.
        (REPRINTED AS NO. 5726)
    PROC. INTERNAT. CONG. MATH., 1950(1) 231-243.

WALD, ABRAHAM 1955.  (19.2/19.3)                                  5712
    CONTRIBUTIONS TO THE THEORY OF STATISTICAL ESTIMATION AND TESTING HYPOTHESES.
        (REPRINT OF NO. 5695)
    SELECT. PAPERS STATIST. PROB. WALD, 87-114.   (MR1,P.152)

WALD, ABRAHAM 1955.  (8.1/17.7)                                   5713
    THE FITTING OF STRAIGHT LINES IF BOTH VARIABLES ARE SUBJECT TO ERROR.
        (REPRINT OF NO. 5696)
    SELECT. PAPERS STATIST. PROB. WALD, 136-152.   (MR2,P.108)

WALD, ABRAHAM 1955.  (19.2/A)                                     5714
    ASYMPTOTICALLY MOST POWERFUL TESTS OF STATISTICAL HYPOTHESES.
        (REPRINT OF NO. 5697)
    SELECT. PAPERS STATIST. PROB. WALD, 154-172.   (MR3,P.8)

WALD, ABRAHAM 1955.  (8.3/8.8/14.5)                                      5715
    ON THE POWER FUNCTION OF THE ANALYSIS OF VARIANCE TEST.
        (REPRINT OF NO. 5698)
    SELECT. PAPERS STATIST. PROB. WALD, 241-246.  (MR5,P.129)

WALD, ABRAHAM 1955.  (17.8)                                             5716
    AN EXTENSION OF WILKS' METHOD FOR SETTING TOLERANCE LIMITS.
        (REPRINT OF NO. 5699)
    SELECT. PAPERS STATIST. PROB. WALD, 247-257.  (MR4,P.222)

WALD, ABRAHAM 1955.  (19.2/A)                                           5717
    TESTS OF STATISTICAL HYPOTHESES CONCERNING SEVERAL PARAMETERS WHEN
    THE NUMBER OF OBSERVATIONS IS LARGE.
        (REPRINT OF NO. 5700)
    SELECT. PAPERS STATIST. PROB. WALD, 323-379.  (MR7,P.20)

WALD, ABRAHAM 1955.  (6.1/6.2)                                          5718
    ON A STATISTICAL PROBLEM ARISING IN THE CLASSIFICATION OF AN
    INDIVIDUAL INTO ONE OF TWO GROUPS.
        (REPRINT OF NO. 5701)
    SELECT. PAPERS STATIST. PROB. WALD, 391-408.  (MR6,P.9)

WALD, ABRAHAM 1955.  (17.1)                                             5719
    LIMIT DISTRIBUTION OF THE MAXIMUM AND MINIMUM OF SUCCESSIVE
    CUMULATIVE SUMS OF RANDOM VARIABLES.
        (REPRINT OF NO. 5705)
    SELECT. PAPERS STATIST. PROB. WALD, 474-485.  (MR8,P.471)

WALD, ABRAHAM 1955.  (8.6/CU)                                           5720
    A NOTE ON REGRESSION ANALYSIS.
        (REPRINT OF NO. 5704)
    SELECT. PAPERS STATIST. PROB. WALD, 493-496.  (MR9,P.364)

WALD, ABRAHAM 1955.  (19.3/A)                                           5721
    ASYMPTOTIC PROPERTIES OF THE MAXIMUM LIKELIHOOD ESTIMATE OF AN UNKNOWN
    PARAMETER OF A DISCRETE STOCHASTIC PROCESS.
        (REPRINT OF NO. 5706)
    SELECT. PAPERS STATIST. PROB. WALD, 497-503.  (MR9,P.454)

WALD, ABRAHAM 1955.  (19.3)                                             5722
    ESTIMATION OF A PARAMETER WHEN THE NUMBER OF UNKNOWN PARAMETERS
    INCREASES INDEFINITELY WITH THE NUMBER OF OBSERVATIONS.
        (REPRINT OF NO. 5707)
    SELECT. PAPERS STATIST. PROB. WALD, 513-520.  (MR10,P.135)

WALD, ABRAHAM 1955.  (19.3/A)                                           5723
    NOTE ON THE CONSISTENCY OF THE MAXIMUM LIKELIHOOD ESTIMATE.
        (REPRINT OF NO. 5708)
    SELECT. PAPERS STATIST. PROB. WALD, 541-547.  (MR11,P.261)

WALD, ABRAHAM 1955.  (16.2)                                             5724
    NOTE ON THE IDENTIFICATION OF ECONOMIC RELATIONS.
        (REPRINT OF NO. 5709)
    SELECT. PAPERS STATIST. PROB. WALD, 569-575.

WALD, ABRAHAM 1955.  (16.1/16.2/C)                                      5725
    REMARKS ON THE ESTIMATION OF UNKNOWN PARAMETERS IN INCOMPLETE SYSTEMS
    OF EQUATIONS.
        (REPRINT OF NO. 5710)
    SELECT. PAPERS STATIST. PROB. WALD, 576-581.

WALD, ABRAHAM 1955.  (19.1)                                             5726
    BASIC IDEAS OF A GENERAL THEORY OF STATISTICAL DECISION RULES.
        (REPRINT OF NO. 5711)
    SELECT. PAPERS STATIST. PROB. WALD, 656-668.

WALD, ABRAHAM 1955.  (5.2/U)                                            5727
    TESTING THE DIFFERENCE BETWEEN THE MEANS OF TWO NORMAL POPULATIONS
    WITH UNKNOWN STANDARD DEVIATIONS.
    SELECT. PAPERS STATIST. PROB. WALD, 669-695.

ENTRY NO. 5728 APPEARS BETWEEN ENTRY NOS. 5991 & 5992.

**WALD**, ABRAHAM + BROOKNER, RALPH J. 1941.  (9.2/T)                                   5729
    ON THE DISTRIBUTION OF WILKS' STATISTIC FOR TESTING THE INDEPENDENCE
    OF SEVERAL GROUPS OF VARIATES.
        (REPRINTED AS NO. 5730)
  ANN. MATH. STATIST., 12, 137-152.  (MR3,P.9)

**WALD**, ABRAHAM + BROOKNER, RALPH J. 1955.  (9.2/T)                                   5730
    ON THE DISTRIBUTION OF WILKS' STATISTIC FOR TESTING THE INDEPENDENCE
    OF SEVERAL GROUPS OF VARIATES.
        (REPRINT OF NO. 5729)
  SELECT. PAPERS STATIST. PROB. WALD, 173-188.  (MR3,P.9)


**WALD**, ABRAHAM + WOLFOWITZ, J. 1943.  (17.8)                                         5731
    AN EXACT TEST FOR RANDOMNESS IN THE NON-PARAMETRIC CASE BASED ON
    SERIAL CORRELATION.
        (REPRINTED AS NO. 5733)
  ANN. MATH. STATIST., 14, 378-388.  (MR5,P.211)

**WALD**, ABRAHAM + WOLFOWITZ, J. 1944.  (17.8)                                         5732
    STATISTICAL TESTS BASED ON PERMUTATIONS OF THE OBSERVATIONS.
        (REPRINTED AS NO. 5734)
  ANN. MATH. STATIST., 15, 358-372.  (MR6,P.163)

**WALD**, ABRAHAM + WOLFOWITZ, J. 1955.  (17.8)                                         5733
    AN EXACT TEST FOR RANDOMNESS IN THE NON-PARAMETRIC CASE BASED ON
    SERIAL CORRELATION.
        (REPRINT OF NO. 5731)
  SELECT. PAPERS STATIST. PROB. WALD, 380-390.  (MR5,P.211)

**WALD**, ABRAHAM + WOLFOWITZ, J. 1955.  (17.8)                                         5734
    STATISTICAL TESTS BASED ON PERMUTATIONS OF THE OBSERVATIONS.
        (REPRINT OF NO. 5732)
  SELECT. PAPERS STATIST. PROB. WALD, 417-431.  (MR6,P.163)

**WALD**, ABRAHAM    SEE ALSO
    3639. MANN, HENRY B. + WALD, ABRAHAM (1942)

    3640. MANN, HENRY B. + WALD, ABRAHAM (1943)

    3641. MANN, HENRY B. + WALD, ABRAHAM (1943)

    3642. MANN, HENRY B. + WALD, ABRAHAM (1955)

    3643. MANN, HENRY B. + WALD, ABRAHAM (1955)

    3644. MANN, HENRY B. + WALD, ABRAHAM (1955)

    1436. DVORETZKY, ARYEH; WALD, ABRAHAM + WOLFOWITZ, J. (1951)

    1437. DVORETZKY, ARYEH; WALD, ABRAHAM + WOLFOWITZ, J. (1955)


**WALKER**, A.M. 1958.  (16.1)                                                          5735
    THE EXISTENCE OF BARTLETT-RAJALAKSHMAN GOODNESS OF FIT G-TESTS FOR MULTIVARIATE
    AUTOREGRESSIVE PROCESSES WITH FINITELY DEPENDENT RESIDUALS.
  PROC. CAMBRIDGE PHILOS. SOC., 54, 225-232.  (MR21-414)

**WALKER**, A.M.    SEE ALSO
    173. ANDERSON, T. W. + WALKER, A.M. (1964)

**WALKER**, GILBERT T. 1909.  (4.1/E)                                                   5736
    CORRELATION IN SEASONAL VARIATION OF CLIMATE.
  MEM. INDIAN METEOROL. DEPT., 20, 117-124.

**WALKER**, GILBERT T. 1910.  (4.1/E)                                                   5737
    CORRELATION IN SEASONAL VARIATIONS OF WEATHER, II.
  MEM. INDIAN METEOROL. DEPT., 21(2) 22-45.

**WALKER**, GILBERT T. 1914.  (4.1/E)                                                   5738
    CORRELATION IN SEASONAL VARIATIONS OF WEATHER, III.  ON THE CRITERION
    FOR THE REALITY OF RELATIONSHIPS OR PERIODICITIES.
  MEM. INDIAN METEOROL. DEPT., 21(9) 13-15.

WALKER, GILBERT T. 1915.  (4.1/E)                                                      5739
    CORRELATION IN SEASONAL VARIATIONS OF WEATHER, IV.  SUNSPOTS AND RAINFALL.
MEM. INDIAN METEOROL. DEPT., 21(10) 17-59.

WALKER, GILBERT T. 1915.  (4.1/E)                                                      5740
    CORRELATION IN SEASONAL VARIATIONS OF WEATHER, V.  SUNSPOTS AND TEMPERATURE.
MEM. INDIAN METEOROL. DEPT., 21(11) 61-90.

WALKER, GILBERT T. 1915.  (4.1/E)                                                      5741
    CORRELATION IN SEASONAL VARIATIONS OF WEATHER, VI.  SUNSPOTS AND PRESSURE.
MEM. INDIAN METEOROL. DEPT., 21(12) 91-118.

WALKER, GILBERT T. 1915.  (4.1/E)                                                      5742
    CORRELATION OF RAINFALL AND THE SUCCEEDING CROPS WITH SPECIAL
    REFERENCE TO THE PUNJAB.
MEM. INDIAN METEOROL. DEPT., 21(14) 130-146.

WALKER, GILBERT T. 1941.  (4.1/4.6)                                                    5743
    SELECTION OF FACTORS IN STATISTICAL INVESTIGATIONS.
QUART. J. ROY. METEOROL. SOC., 67, 261-262.

WALKER, GILBERT T. 1950.  (16.5)                                                       5744
    APPARENT CORRELATION BETWEEN INDEPENDENT SERIES OF AUTOCORRELATED OBSERVATIONS.
BIOMETRIKA, 37, 184-185.

WALKER, HELEN M. 1928.  (4.1)                                                          5745
    THE RELATION OF PLANA AND BRAVAIS TO THE THEORY OF CORRELATION.
ISIS, 10(34) 466-484.

WALKER, HELEN M. 1958.  (1.1)                                                          5746
    THE CONTRIBUTIONS OF KARL PEARSON.
J. AMER. STATIST. ASSOC., 53, 11-22.   (MR22-A9)

WALKER, J. F. 1930.  (4.1/T)                                                           5747
    SHORT METHOD FOR FINDING ZERO ORDER COEFFICIENTS OF CORRELATIONS.
J. EDUC. PSYCHOL., 21, 65-67.

WALL, F. J.   SEE
    3570. MC HUGH, RICHARD B. + WALL, F. J. (1962)

WALLACE, D. L.   SEE
    3214. KURTZ, T. E.; LINK, R.F.; TUKEY, JOHN W. + (1966)
    WALLACE, D. L.

WALLACE, E. W. 1965.  (20.2)                                                           5748
    ON MATRICES OF COFACTORS.
AMER. MATH. MONTHLY, 72, 144-148.

WALLACE, H. A. + SNEDECOR, GEORGE W. 1925.  (4.1/4.5/4.6)                              5749
    CORRELATION AND MACHINE CALCULATION.
IOWA STATE COLL. OFF. PUBL., 23(35) 1-47.

WALLIS, W. ALLEN 1939.  (4.8)                                                          5750
    THE CORRELATION RATIO FOR RANKED DATA.
J. AMER. STATIST. ASSOC., 34, 533-538.

WALSH, JAMES A. 1964.  (15.3)                                                          5751
    AN IBM 709 PROGRAM FOR FACTOR ANALYZING THREE MODE MATRICES.
EDUC. PSYCHOL. MEAS., 24, 669-673.

WALSH, JAMES A. 1965.  (15.1)                                                          5752
    METHODOLOGICAL NOTE ON FACTOR ANALYSIS AS AN EXPERIMENTAL TECHNIQUE.
PSYCHOL. REP., 16, 1099-1100.

WALSH, JOHN E. 1947.  (3.2/4.8/C)                                                      5753
    CONCERNING THE EFFECT OF INTRACLASS CORRELATION ON CERTAIN SIGNIFICANCE TESTS.
ANN. MATH. STATIST., 18, 88-96.   (MR8,P.476)

WALSH, JOHN E. 1963.  (6.4)                                                            5754
    SIMULTANEOUS CONFIDENCE INTERVALS FOR DIFFERENCES OF CLASSIFICATION
    PROBABILITIES.
BIOM. Z., 5, 231-234.

WALTERS, LILLIE C.    SEE
    2794. JOHLER, J. RALPH + WALTERS, LILLIE C. (1959)

WANG, MING CHEN + UHLENBECK, G. E. 1945.   (16.4)                5755
    ON THE THEORY OF THE BROWNIAN MOTION.   II.
    REV. MODERN PHYS., 17, 323-342.   (MR7,P.130)


WARBURTON, F.W. 1954.   (15.1)                                   5756
    THE FULL FACTOR ANALYSIS.
    BRITISH J. STATIST. PSYCHOL., 7, 101-106.

WARD, D. J.    SEE
    3278. LARMAN, D. G. + WARD, D. J. (1966)

WARNER, STANLEY L. 1963.   (14.4)                                5757
    MULTIVARIATE REGRESSION OF DUMMY VARIATES UNDER NORMALITY ASSUMPTIONS.
    J. AMER. STATIST. ASSOC., 58, 1054-1063.   (MR27-6340)

WARRINGTON, WILLARD G.    SEE
    1115. CRONBACH, LEE J. + WARRINGTON, WILLARD G. (1952)

WARTMANN, ROLF 1951.   (6.1)                                     5758
    DIE STATISTISCHE TRENNUNG SICH IN MEHREREN MERKMALEN UBERLAPPENDER
    INDIVIDUENGRUPPEN (DISKRIMINANZANALYSE).
    Z. ANGEW. MATH. MECH., 31, 256-257.

WATANABE, S. 1959.   (4.8)                                       5759
    CORRELATION INDICES.
    NUOVO CIMENTO, 13, 576-582.   (MR22-1951)

WATANABE, S. 1960.   (4.9)                                       5760
    INFORMATION THEORETICAL ANALYSIS OF MULTIVARIATE CORRELATION.
    IBM J. RES. DEVELOP., 4, 66-82.   (MR22-641)

WATANABE, TOSHIO 1952.   (14.4)                                  5761
    ON THE AMOUNT OF INFORMATION.
        (IN JAPANESE)
    RES. MEM. INST. STATIST. MATH. TOKYO, 8, 293-307.

WATKINS, G. P. 1933.   (4.8/E)                                   5762
    AN ORDINAL INDEX OF CORRELATION.
    J. AMER. STATIST. ASSOC., 28, 139-151.

WATSON, F. R. 1964.   (8.1/20.3)                                 5763
    A NEW METHOD FOR SOLVING SIMULTANEOUS LINEAR EQUATIONS ASSOCIATED
    WITH MULTIVARIATE ANALYSIS.
    PSYCHOMETRIKA, 29, 75-86.   (MR29-2909)

WATSON, G. N. 1933.   (14.5/20.4)                                5764
    NOTES ON GENERATING FUNCTIONS OF POLYNOMIALS:(1) LAGUERRE POLYNOMIALS.
    J. LONDON MATH. SOC., 8, 189-192.

WATSON, G. N. 1933.   (20.5)                                     5765
    NOTES ON GENERATING FUNCTIONS OF POLYNOMIALS:(2) HERMITE POLYNOMIALS.
    J. LONDON MATH. SOC., 8, 194-199.

WATSON, G.S. 1955.   (2.6/14.5)                                  5766
    THE DISTRIBUTION OF THE RATIO OF TWO QUADRATIC FORMS.
    AUSTRAL. J. PHYS., 8, 402-407.   (MR17,P.503)

WATSON, G.S. 1955.   (18.2)                                      5767
    ANALYSIS OF DISPERSION ON A SPHERE.
    MONTHLY NOTICES ROY. ASTRONOM. SOC. GEOPHYS. SUPP., 7, 153-159.   (MR18,P.769)


WATSON, G.S. 1956.   (18.1)                                      5768
    A NOTE ON THE CIRCULAR MULTIVARIATE DISTRIBUTION.
    BIOMETRIKA, 43, 467.   (MR18,P.522)

WATSON, G.S. 1960.   (18.6)                                      5769
    MORE SIGNIFICANCE TESTS ON THE SPHERE.
    BIOMETRIKA, 47, 87-91.   (MR22-1960)

WATSON, G.S. 1961. (18.6)                                                    5770
   GOODNESS-OF-FIT TESTS ON A CIRCLE.
   BIOMETRIKA. 48, 109-114.


WATSON, G.S. 1964. (3.1)                                                     5771
   A NOTE ON MAXIMUM LIKELIHOOD.
   SANKHYA SER. A, 26, 303-304.


WATSON, G.S. 1965. (18.6)                                                    5772
   EQUATORIAL DISTRIBUTIONS ON A SPHERE.
   BIOMETRIKA. 52, 193-201.


WATSON, G.S. 1966. (18.1/18.2/18.3)                                          5773
   THE STATISTICS OF ORIENTATION DATA.
   J. GEOL., 74, 786-797.


WATSON, G.S. + DURBIN, JAMES 1951. (2.6/16.5)                               5774
   EXACT TESTS OF SERIAL CORRELATION USING NON-CIRCULAR STATISTICS.
   ANN. MATH. STATIST., 22, 446-451. (MR13,P.144)


WATSON, G.S. + WILLIAMS, E.J. 1956. (18.2/18.3)                             5775
   ON THE CONSTRUCTION OF SIGNIFICANCE TESTS ON THE CIRCLE AND THE SPHERE.
   BIOMETRIKA. 43, 344-352. (MR18,P.521)


WATSON, G.S.   SEE ALSO
   547. BINET, F. E. + WATSON, G.S. (1956)

   5832. WHEELER, S. + WATSON, G.S. (1964)


WATSON, H. E. 1956. (6.1/15.5)                                               5776
   AGREEMENT ANALYSIS. A NOTE ON PROFESSOR MCQUITTY'S ARTICLE.
   BRITISH J. STATIST. PSYCHOL., 9, 17-20.


WATTERSON, GEOFFREY A. 1959. (3.2)                                          5777
   LINEAR ESTIMATION IN CENSORED SAMPLES FROM MULTIVARIATE NORMAL POPULATIONS.
   ANN. MATH. STATIST., 30, 814-824. (MR22-3071)


WAUGH, FREDERICK V. 1935. (8.8/20.3)                                        5778
   A SIMPLIFIED METHOD OF DETERMINING MULTIPLE REGRESSION CONSTANTS.
   J. AMER. STATIST. ASSOC., 30, 694-700.


WAUGH, FREDERICK V. 1936. (8.3)                                             5779
   THE ANALYSIS OF REGRESSION IN SUBSETS OF VARIABLES.
   J. AMER. STATIST. ASSOC., 31, 729-730.


WAUGH, FREDERICK V. 1942. (12.1/E)                                          5780
   REGRESSIONS BETWEEN SETS OF VARIATES.
   ECONOMETRICA. 10, 290-310.


WAUGH, FREDERICK V. 1943. (8.8)                                             5781
   CHOICE OF THE DEPENDENT VARIABLE IN REGRESSION ANALYSIS.
   J. AMER. STATIST. ASSOC., 38, 210-214.


WAUGH, FREDERICK V. 1945. (20.1)                                            5782
   A NOTE CONCERNING HOTELLING'S METHOD OF INVERTING A PARTITIONED MATRIX.
   ANN. MATH. STATIST., 16, 216-217. (MR7,P.84)


WAUGH, FREDERICK V. 1946. (4.5)                                             5783
   THE COMPUTATION OF PARTIAL CORRELATION COEFFICIENTS.
   J. AMER. STATIST. ASSOC., 41, 543-546. (MR8,P.282)


WAUGH, FREDERICK V. 1950. (16.2)                                            5784
   INVERSION OF THE LEONTIEF MATRIX BY POWER SERIES.
   ECONOMETRICA, 18, 142-153. .


WAUGH, FREDERICK V. 1961. (17.7)                                            5785
   THE PLACE OF LEAST SQUARES IN ECONOMETRICS.
   ECONOMETRICA, 29, 386-396.


WAUGH, FREDERICK V. 1962. (16.2)                                            5786
   FURTHER COMMENT.
   ECONOMETRICA, 30, 568-569.

WAUGH, FREDERICK V. 1964. (16.2/E)                                                      5787
    COBWEB MODELS.
  J. FARM ECON., 46, 732-750.


WAUGH, FREDERICK V. + DWYER, PAUL S. 1945. (20.3)                                       5788
    COMPACT COMPUTATION OF THE INVERSE OF A MATRIX.
  ANN. MATH. STATIST., 16, 259-271. (MR7,P.218)


WAUGH, FREDERICK V. + FOX, KARL A. 1957. (4.6/T)                                        5789
    GRAPHIC COMPUTATION OF $R_{1.23}$.
        (AMENDED BY NO. 5790)
  J. AMER. STATIST. ASSOC., 52, 479-481.


WAUGH, FREDERICK V. + FOX, KARL A. 1958. (4.6/T)                                        5790
    CORRIGENDA:  GRAPHIC COMPUTATION OF $R_{1.23}$.
        (AMENDMENT OF NO. 5789)
  J. AMER. STATIST. ASSOC., 53, 1031.


WAUGH, FREDERICK V.   SEE ALSO
  1469. DWYER, PAUL S. + WAUGH, FREDERICK V. (1953)

  1470. DWYER, PAUL S. + WAUGH, FREDERICK V. (1953)


WEBB, E. L. R. 1962. (17.3/U)                                                           5791
    NOTE ON THE PRODUCT OF RANDOM VARIABLES.
  CANAD. J. PHYS., 40, 1394-1396. (MR26-792)


WEBB, H. A. + AIREY, JOHN R. 1918. (20.4)                                               5792
    THE PRACTICAL IMPORTANCE OF THE CONFLUENT HYPERGEOMETRIC FUNCTION.
  PHILOS. MAG. SER. 6, 36, 129-141.


WEBSTER, HAROLD 1957. (15.4)                                                            5793
    ITEM SELECTION METHODS FOR INCREASING TEST  HOMOGENEITY.
  PSYCHOMETRIKA, 22, 395-403.



WEGGE, L. L. 1965. (17.7)                                                               5794
    IDENTIFIABILITY CRITERIA FOR A SYSTEM OF EQUATIONS AS A WHOLE.
  AUSTRAL. J. STATIST., 7, 67-76.


WEGNER, PETER 1963. (8.1/20.5)                                                          5795
    RELATIONS BETWEEN MULTIVARIATE STATISTICS AND MATHEMATICAL PROGRAMMING.
  APPL. STATIST., 12, 146-150.


WEIBULL, MARTIN 1953. (18.3)                                                            5796
    THE DISTRIBUTION OF T- AND F-STATISTICS AND OF CORRELATION AND REGRESSION
    COEFFICIENTS IN STRATIFIED SAMPLES FROM NORMAL POPULATIONS WITH DIFFERENT MEANS.
  SKAND. AKTUARIETIDSKR., 36(SP) 1-106. (MR15,P.725)


WEICHELT, JOHN A. 1946. (4.3)                                                           5797
    A FIRST-ORDER METHOD FOR ESTIMATING CORRELATION COEFFICIENTS.
  PSYCHOMETRIKA, 11, 215-221. (MR8,P.282)


WEIDA, FRANK M. 1926. (4.1)                                                             5798
    ON THE CORRELATION BETWEEN TWO FUNCTIONS.
  AMER. MATH. MONTHLY, 33, 440-444.


WEIDA, FRANK M. 1928. (4.1/4.8/4.9)                                                     5799
    ON VARIOUS CONCEPTIONS OF CORRELATION.
  ANN. MATH. SER. 2, 29, 276-312.


WEIDA, FRANK M. 1934. (4.8)                                                             5800
    ON MEASURES OF CONTINGENCY.
  ANN. MATH. STATIST., 5, 308-319.


WEIGL, ALFRED   SEE
  5997. WOLF, ROBERT L.; MENDLOWITZ, MILTON; ROBOZ, JULIA;  (1966)
  STYAN, GEORGE P. H.; KORNFELD, PETER + WEIGL, ALFRED


WEILER, H. 1959. (2.4/B)                                                                5801
    MEANS AND STANDARD DEVIATIONS OF A TRUNCATED NORMAL BIVARIATE DISTRIBUTION.
  AUSTRAL. J. STATIST., 1, 73-81. (MR22-271)

WEILER. H.    SEE ALSO
    5932. WILLIAMS, JEAN M. + WEILER. H. (1964)


WEINBERG, W. 1916.  (4.8)                                                    5802
    UBER KORRELATIONSMESSUNG.
    DEUTSCHE STATIST. ZENTRALBLATT. 8. 145–158.


WEINER. JOHN M. + DUNN. OLIVE JEAN 1966.  (6.2/E)                            5803
    ELIMINATION OF VARIATES IN LINEAR DISCRIMINATION PROBLEMS.
    BIOMETRICS. 22. 268–275.


WEINGARTEN. HARRY + DI DONATO, A. R. 1961.  (14.5/T)                         5804
    A TABLE OF GENERALIZED CIRCULAR ERROR.
    MATH. COMPUT.. 15. 169–173.  (MR23–B608)


WEINSTEIN. A. S. 1962.  (20.1)                                               5805
    A NECESSARY AND SUFFICIENT CONDITION IN THE MAXIMUM–MINIMUM THEORY OF
    EIGENVALUES.
    STUD. MATH. ANAL. RELATED TOPICS (POLYA VOLUME), 429–434.


WEISS. GEORGE H. + KIMURA. MOTOO 1965.  (16.4)                               5806
    A MATHEMATICAL ANALYSIS OF THE STEPPING STONE MODEL OF GENETIC CORRELATION.
    J. APPL. PROB.. 2, 129–149.


WEISS. LIONEL 1958.  (17.8)                                                  5807
    A TEST OF FIT FOR MULTIVARIATE DISTRIBUTIONS.
    ANN. MATH. STATIST.. 29. 595–599.  (MR20–381)


WEISS. LIONEL 1960.  (17.8)                                                  5808
    TWO–SAMPLE TESTS FOR MULTIVARIATE DISTRIBUTIONS.
    ANN. MATH. STATIST.. 31. 159–164.  (MR22–10071)


WEISS. LIONEL 1962.  (18.2)                                                  5809
    A SEQUENTIAL TEST OF THE EQUALITY OF PROBABILITIES IN A MULTINOMIAL
    DISTRIBUTION.
    J. AMER. STATIST. ASSOC.. 57, 769–774.  (MR25–4596)


WEISS. LIONEL 1962.  (17.5/B)                                               5810
    A SEQUENTIAL TEST OF FIT FOR MULTIVARIATE DISTRIBUTIONS.
    SANKHYA SER. A, 24, 377–384.  (MR30–682)


WEISS. LIONEL 1964.  (17.6/A)                                               5811
    ON THE ASYMPTOTIC JOINT NORMALITY OF QUANTILES FROM A MULTIVARIATE DISTRIBUTION.
    J. RES. NAT. BUR. STANDARDS SECT. B, 68, 65–66.  (MR29–4139)


WEISS. LIONEL 1966.  (19.3/A)                                               5812
    THE RELATIVE MAXIMA OF THE LIKELIHOOD FUNCTION. II.
    SKAND. AKTUARIETIDSKR.. 49, 119–121.


WEISS. MARIE–CLAUDE 1966.  (2.2)                                            5813
    DETERMINATION D'UNE VARIABLE GAUSSIENNE A PLUSIEURS DIMENSIONS AU MOYEN DE LA
    FONCTION CARACTERISTIQUE.
    J. SOC. STATIST. PARIS. 107, 135–136.


WEITZMAN, R. A.    SEE
    4642. ROSS. JOHN + WEITZMAN, R. A. (1964)


WELCH. B.L. 1935.  (8.3/B)                                                  5814
    SOME PROBLEMS IN THE ANALYSIS OF REGRESSION AMONG K SAMPLES OF TWO VARIABLES.
    BIOMETRIKA, 27, 145–160.


WELCH. B.L. 1937.  (8.5/U)                                                  5815
    ON THE Z TEST IN RANDOMIZED BLOCKS AND LATIN SQUARES.
    BIOMETRIKA, 29, 21–52.


WELCH. B.L. 1938.  (10.2)                                                   5816
    ON TESTS FOR HOMOGENEITY.
    BIOMETRIKA. 30, 149–158.


WELCH. B.L. 1939.  (6.1/6.2)                                                5817
    NOTE ON DISCRIMINANT FUNCTIONS.
    BIOMETRIKA. 31, 218–220.  (MR1,P.154)

WELCH, B.L. 1947. (3.2/5.2)                                                    5818
    THE GENERALIZATION OF "STUDENT'S" PROBLEM WHEN SEVERAL DIFFERENT
    POPULATION VARIANCES ARE INVOLVED.
    BIOMETRIKA, 34, 28-35. (MR8,P.394)

WELCH, B.L.    SEE ALSO
    2820. JOHNSON, N. L. + WELCH, B.L. (1940)

WELCH, PETER D. + WIMPRESS, RICHARD S. 1961. E(6.2/6.4/20.3)                   5819
    TWO MULTIVARIATE STATISTICAL COMPUTER PROGRAMS AND THEIR APPLICATION
    TO THE VOWEL RECOGNITION PROBLEM.
    J. ACOUST. SOC. AMER., 33, 426-434. (MR23-B3115)

WELD, L. D. H. 1932. (4.1)                                                     5820
    USE OF CORRELATION IN THE MEASUREMENT OF SALES POTENTIALS.
    J. AMER. STATIST. ASSOC., 27(SP) 202-205.

WELDON, W. F. R. 1897. (17.3/AE)                                               5821
    NOTE, JANUARY 13, 1897.
    PROC. ROY. SOC. LONDON, 60, 498.

WELKER, E.L. 1951. (8.1/A)                                                     5822
    CORRELATION AND REGRESSION ANALYSIS.
    PROC. INDUST. COMPUT. SEM. IBM NEW YORK, 36-43. (MR13,P.366)

WELKER, E.L. + WYND, F. L. 1943. (4.1/4.3/4.8/E)                               5823
    INFLUENCE OF UNKNOWN FACTORS ON THE VALIDITY OF MATHEMATICAL
    CORRELATIONS OF BIOLOGICAL DATA.
    PLANT PHYSIOL., 18, 498-507.

WELKER, E.L.    SEE ALSO
    2298. HALL, D. M.; WELKER, E.L. + CRAWFORD, ISABELLE (1945)

WELLMAN, R. H. + MC CALLAN, S. E. A. 1943. E(4.8/17.9)                         5824
    CORRELATION WITHIN AND BETWEEN LABORATORY SLIDE-GERMINATION, GREENHOUSE TOMATO
    FOLIAGE DISEASE, AND WHEAT SMUT METHODS OF TESTING FUNGICIDES.
    CONTRIB. BOYCE THOMPSON INST., 13, 143-169.

WELLS, W. T.; ANDERSON, R. L. + CELL, JOHN W. 1962. (18.1/U)                   5825
    THE DISTRIBUTION OF THE PRODUCT OF TWO CENTRAL OR NON-CENTRAL
    CHI-SQUARE VARIATES.
    ANN. MATH. STATIST., 33, 1016-1020. (MR25-5563)

WELSH, D.J.A.    SEE
    3883. MORGAN, R.W. + WELSH, D.J.A. (1965)

WELSH, GEORGE S. 1955. (4.4/T)                                                 5826
    A TABULAR METHOD OF OBTAINING TETRACHORIC R WITH MEDIAN-CUT VARIABLES.
    PSYCHOMETRIKA, 20, 83-85.

WENDEL, JAMES G. 1957. (2.1)                                                   5827
    INVARIANCE OF NORMAL DISTRIBUTIONS.
    MICHIGAN MATH. J., 4, 173-174. (MR20-1344)

WENDER, K.    SEE
    5069. SIXTL, F. + WENDER, K. (1964)

WENTZELL, A. D.  SEE  VENTTSEL', A. D.

WERENSKIOLD, W. 1930. (4.1)                                                    5828
    ON THE COMPUTATION OF THE CORRELATION COEFFICIENT.
    AVH. NORSKE VIDENS. AKAD. OSLO, 8, 1-16.

WESLER, OSCAR 1959. (6.4)                                                      5829
    A CLASSIFICATION PROBLEM INVOLVING MULTINOMIALS.
    ANN. MATH. STATIST., 30, 128-133. (MR21-4502)

WESLER, OSCAR    SEE ALSO
    4745. RUBIN, HERMAN + WESLER, OSCAR (1958)

WEST, VINCENT I. 1952. (8.1/20.3/E)                                            5830
    REPLACING VARIABLES IN CORRELATION PROBLEMS.
    J. AMER. STATIST. ASSOC., 47, 185-190.

WESTWICK, ROY    SEE
    3662. MARCUS, MARVIN; MOYLS, V.N. + WESTWICK, ROY (1957)

WEYL, HERMANN 1949.  (20.1/20.3)                                                    5831
    INEQUALITIES BETWEEN THE TWO KIND OF EIGENVALUES OF A LINEAR TRANSFORMATION.
    PROC. NAT. ACAD. SCI. U.S.A., 35, 408-411.

WHEELER, S. + WATSON, G.S. 1964.  (17.8/BT)                                         5832
    A DISTRIBUTION-FREE TWO SAMPLE TEST ON THE CIRCLE.
    BIOMETRIKA, 51, 256-257.

WHERRY, ROBERT J. 1931.  (4.6)                                                      5833
    A NEW FORMULA FOR PREDICTING THE SHRINKAGE OF THE COEFFICIENT OF
    MULTIPLE CORRELATION.
    ANN. MATH. STATIST., 2, 440-457.

WHERRY, ROBERT J. 1932.  (20.3)                                                     5834
    A MODIFICATION OF THE DOOLITTLE METHOD.  A LOGARITHMIC SOLUTION.
    J. EDUC. PSYCHOL., 23, 455-459.

WHERRY, ROBERT J. 1935.  (15.4)                                                     5835
    THE SHRINKAGE OF THE BROWN-SPEARMAN PROPHECY FORMULA.
    ANN. MATH. STATIST., 6, 183-189.

WHERRY, ROBERT J. 1940.  (4.6)                                                      5836
    AN APPROXIMATION METHOD FOR OBTAINING A MAXIMIZED MULTIPLE CRITERION.
    PSYCHOMETRIKA, 5, 109-116.

WHERRY, ROBERT J. 1946.  (15.4)                                                     5837
    TEST SELECTION AND SUPPRESSOR VARIABLES.
    PSYCHOMETRIKA, 11, 239-247.

WHERRY, ROBERT J. 1947.  (4.8/15.4)                                                 5838
    MULTIPLE BI-SERIAL AND MULTIPLE POINT BI-SERIAL CORRELATION.
    PSYCHOMETRIKA, 12, 189-195.  (MR9,P.47)

WHERRY, ROBERT J. 1949.  (15.3)                                                     5839
    A NEW ITERATIVE METHOD FOR CORRECTING ERRONEOUS COMMUNALITY ESTIMATES
    IN FACTOR ANALYSIS.
    PSYCHOMETRIKA, 14, 231-241.

WHERRY, ROBERT J. 1959.  (15.3)                                                     5840
    HIERARCHIAL FACTOR SOLUTIONS WITHOUT ROTATION.
    PSYCHOMETRIKA, 24, 45-51.

WHERRY, ROBERT J. + GAYLORD, RICHARD H. 1943.  (15.4)                               5841
    THE CONCEPT OF TEST AND ITEM RELIABILITY IN RELATION TO FACTOR PATTERN.
    PSYCHOMETRIKA, 8, 247-264.

WHERRY, ROBERT J. + GAYLORD, RICHARD H. 1944.  (15.4)                               5842
    FACTOR PATTERN OF TEST ITEMS AND TESTS AS A FUNCTION OF THE CORRELATION
    COEFFICIENT:  CONTENT, DIFFICULTY, AND CONSTANT ERROR FACTORS.
    PSYCHOMETRIKA, 9, 237-244.

WHERRY, ROBERT J. + TAYLOR, ERWIN K. 1946.  (4.8)                                   5843
    THE RELATION OF MULTISERIAL ETA TO OTHER MEASURES OF CORRELATION.
    PSYCHOMETRIKA, 11, 155-161.  (MR8,P.41)

WHERRY, ROBERT J. + WINER, BEN J. 1953.  (15.3)                                     5844
    A METHOD FOR FACTORING LARGE NUMBERS OF ITEMS.
    PSYCHOMETRIKA, 18, 161-179.

WHERRY, ROBERT J.; CAMPBELL, JOEL T. + PERLOFF, ROBERT 1951.  (15.3)                5845
    AN EMPIRICAL VERIFICATION OF THE WHERRY-GAYLORD ITERATIVE FACTOR
    ANALYSIS PROCEDURE.
    PSYCHOMETRIKA, 16, 67-74.

WHERRY, ROBERT J.; NAYLOR, JAMES C.; WHERRY, ROBERT J., JR. + 1965.  (17.9)         5846
    FALLIS, ROBERT F.
    GENERATING MULTIPLE SAMPLES OF MULTIVARIATE DATA WITH ARBITRARY
    POPULATION PARAMETERS.
    PSYCHOMETRIKA, 30, 303-313.

WHERRY, ROBERT J., JR.   SEE
    5846. WHERRY, ROBERT J.; NAYLOR, JAMES C.; WHERRY, ROBERT J., JR. + (1965)
    FALLIS, ROBERT F.

WHIPPLE, F. J. W. 1926.  (8.8/11.1)                                            5847
    ON THE BEST LINEAR RELATION CONNECTING THREE VARIABLES.
    PHILOS. MAG. SER. 7, 1, 378–384.

WHIPPLE, F. J. W.   SEE ALSO
    3552. MC CREA, WILLIAM H. + WHIPPLE, F. J. W. (1940)

WHITCOMB, MILTON A. 1956.  (15.2/E)                                            5848
    EVALUATION OF A METHOD FOR THE CONSTRUCTION OF FACTOR-PURE APTITUDE TESTS.
    SYMP. AIR FORCE HUMAN ENGRG. PERS. TRAIN. RES. (FINCH + CAMERON), 298–305.

WHITE, PAUL A. 1958.  (20.3)                                                   5849
    THE COMPUTATION OF EIGENVALUES AND EIGENVECTORS OF A MATRIX.
    J. SOC. INDUST. APPL. MATH., 6, 393–437.  (MR20-6783)

WHITELEY, M. A. + PEARSON, KARL 1899.  E(17.9)                                 5850
    DATA FOR THE PROBLEM OF EVOLUTION IN MAN.  I.  A FIRST STUDY OF THE
    VARIABILITY AND CORRELATION OF THE HAND.
    PROC. ROY. SOC. LONDON, 65, 126–151.

WHITFIELD, J. W. 1947.  (4.8)                                                  5851
    RANK CORRELATION BETWEEN TWO VARIABLES, ONE OF WHICH IS RANKED, THE
    OTHER DICHOTOMOUS.
    BIOMETRIKA, 34, 292–296.  (MR9,P.453)

WHITFIELD, J. W. 1949.  (4.8)                                                  5852
    INTRA-CLASS RANK CORRELATION.
    BIOMETRIKA, 36, 463–467.

WHITNEY, D. RANSOM 1951.  (17.8/B)                                             5853
    A BIVARIATE EXTENSION OF THE U STATISTIC.
    ANN. MATH. STATIST., 22, 274–282.  (MR12,P.840)

WHITTINGHILL, MAURICE   SEE
    4345. POTTHOFF, RICHARD F. + WHITTINGHILL, MAURICE (1966)

WHITTLE, PETER 1952.  (11.1/15.2)                                             5854
    ON PRINCIPAL COMPONENTS AND LEAST SQUARE METHODS OF FACTOR ANALYSIS.
    SKAND. AKTUARIETIDSKR., 35, 223–239.  (MR17,P.872)

WHITTLE, PETER 1953.  (16.5)                                                  5855
    THE ANALYSIS OF MULTIPLE STATIONARY TIME SERIES.
    J. ROY. STATIST. SOC. SER. B, 15, 125–139.  (MR15,P.143)

WHITTLE, PETER 1957.  (16.3/16.5)                                             5856
    CURVE AND PERIODOGRAM SMOOTHING (SYMPOSIUM ON SPECTRAL APPROACH TO TIME SERIES).
        (WITH DISCUSSION)
    J. ROY. STATIST. SOC. SER. B, 19, 38–63.  (MR19,P.1098)

WHITTLE, PETER 1958.  (17.2)                                                  5857
    A MULTIVARIATE GENERALIZATION OF TCHEBICHEV'S INEQUALITY.
    QUART. J. MATH. OXFORD SECOND SER., 9, 232–240.  (MR20-6754)

WHITTLE, PETER 1958.  (17.2)                                                  5858
    CONTINUOUS GENERALIZATIONS OF CHEBYSHEV'S INEQUALITY.
        (REPRINTED AS NO. 5859)
    TEOR. VEROJATNOST. I PRIMENEN., 3, 386–394.  (MR21-6619)

WHITTLE, PETER 1958.  (17.2)                                                  5859
    CONTINUOUS GENERALIZATIONS OF CHEBYSHEV'S INEQUALITY.
        (REPRINT OF NO. 5858)
    THEORY PROB. APPL., 3, 358–366.  (MR21-6619)

WHITTLE, PETER 1959.  (18.1/A)                                                5860
    QUADRATIC FORMS IN POISSON AND MULTINOMIAL VARIABLES.
    J. AUSTRAL. MATH. SOC., 1, 233–240.

WHITTLE, PETER 1960.  (2.5/14.5/17.3)                                         5861
    BOUNDS FOR THE MOMENTS OF LINEAR AND QUADRATIC FORMS IN INDEPENDENT VARIABLES.
        (REPRINTED AS NO. 5862)
    TEOR. VEROJATNOST. I PRIMENEN., 5, 331–335.  (MR24-A3673)

WHITTLE, PETER 1960.  (2.5/14.5/17.3)                                              5862
    BOUNDS FOR THE MOMENTS OF LINEAR AND QUADRATIC FORMS IN INDEPENDENT VARIABLES.
        (REPRINT OF NO. 5861)
  THEORY PROB. APPL., 5, 302-305.  (MR24-A3673)

WHITTLE, PETER 1962.  (17.9)                                                       5863
    TOPOGRAPHIC CORRELATION, POWER-LAW COVARIANCE, FUNCTIONS, AND DIFFUSION.
  BIOMETRIKA, 49, 305-314.

WHITTLE, PETER 1963.  (16.1)                                                       5864
    ON THE FITTING OF MULTIVARIATE AUTOREGRESSIONS, AND THE APPROXIMATE
    CANONICAL FACTORIZATION OF A SPECTRAL DENSITY MATRIX.
  BIOMETRIKA, 50, 129-134.  (MR28-4635)

WHITTLE, PETER 1964.  (16.4)                                                       5865
    STOCHASTIC PROCESSES IN SEVERAL DIMENSIONS.  (WITH DISCUSSION)
  BULL. INST. INTERNAT. STATIST., 40, 974-994, 1009-1010.  (MR30-3500)

WICKSELL, S.D. 1916.  (4.8)                                                        5866
    SOME THEOREMS IN THE THEORY OF PROBABILITY, WITH SPECIAL REFERENCE TO
    THEIR IMPORTANCE IN THE THEORY OF HOMOGRADE CORRELATION.
  SVENSKA AKTUARIEFOREN. TIDSKR., 3, 165-213.

WICKSELL, S.D. 1917.  (18.1/18.3)                                                  5867
    THE CORRELATION FUNCTION OF TYPE A AND THE REGRESSION OF ITS CHARACTERISTICS.
        (REPRINTED AS NO. 5868)
  KUNGL. SVENSKA VETENSKAPSAKAD. HANDL., 58(3) 1-48.

WICKSELL, S.D. 1917.  (18.1/18.3)                                                  5868
    THE CORRELATION FUNCTION OF TYPE A, AND THE REGRESSION OF ITS CHARACTERISTICS.
        (REPRINT OF NO. 5867)
  MEDD. LUNDS ASTRONOM. OBS. SER. 2, 17, 1-48.

WICKSELL, S.D. 1917.  (2.1/4.1/BE)                                                 5869
    ON LOGARITHMIC CORRELATION, WITH AN APPLICATION TO THE DISTRIBUTION OF
    AGES AT FIRST MARRIAGE.
        (REPRINT OF NO. 5872, TRANSLATED AS NO. 5873)
  MEDD. LUNDS ASTRONOM. OBS. SER. 2, 84, 1-46.

WICKSELL, S.D. 1917.  (18.1)                                                       5870
    THE APPLICATION OF SOLID HYPERGEOMETRICAL SERIES TO FREQUENCY
    DISTRIBUTIONS IN SPACE.
  PHILOS. MAG. SER. 6, 33, 389-394.

WICKSELL, S.D. 1917.  (2.1/B)                                                      5871
    THE CONSTRUCTION OF THE CURVES OF EQUAL FREQUENCY IN CASE OF TYPE A CORRELATION.
  SVENSKA AKTUARIEFOREN. TIDSKR., 4, 122-140.

WICKSELL, S.D. 1917.  (2.1/4.1/BE)                                                 5872
    ON LOGARITHMIC CORRELATION, WITH AN APPLICATION TO THE DISTRIBUTION OF
    AGES AT FIRST MARRIAGE.
        (TRANSLATED AS NO. 5873, REPRINTED AS NO. 5869)
  SVENSKA AKTUARIEFOREN. TIDSKR., 4, 141-161.

WICKSELL, S.D. 1918.  (2.1/4.1/BE)                                                 5873
    DAS HEIRATSALTER IN SCHWEDEN 1891-1910.  EINE KORRELATIONSSTATISTISCHE
    UNTERSUCHUNG.
        (TRANSLATION OF NO. 5869, TRANSLATION OF NO. 5872)
  LUNDS UNIV. ARSSKR. N. F. AVD. 2, 14(18) 1-46.

WICKSELL, S.D. 1918.  (18.1)                                                       5874
    ON THE CORRELATION OF ACTING PROBABILITIES.
  SKAND. AKTUARIETIDSKR., 1, 98-135.

WICKSELL, S.D. 1919.  (2.1/18.1)                                                   5875
    MULTIPLE CORRELATION AND NON-LINEAR REGRESSION.
        (REPRINTED AS NO. 5876)
  ARK. MAT. ASTRONOM. FYS., 14(10) 1-18.

WICKSELL, S.D. 1919.  (2.1/18.1)                                                   5876
    MULTIPLE CORRELATION AND NON-LINEAR REGRESSION.
        (REPRINT OF NO. 5875)
  MEDD. LUNDS ASTRONOM. OBS. SER. 1, 5(91) 1-18.

WICKSELL, S.D. 1919. (2.3) 5877
    A GENERAL FORMULA FOR THE MOMENTS OF THE NORMAL CORRELATION FUNCTION
    OF ANY NUMBER OF VARIATES.
    PHILOS. MAG. SER. 6, 37, 446-452.

WICKSELL, S.D. 1921. (4.1) 5878
    AN EXACT FORMULA FOR SPURIOUS CORRELATION.
    METRON, 1(4) 33-40.

WICKSELL, S.D. 1930. (8.1/U) 5879
    REMARKS ON REGRESSION.
    ANN. MATH. STATIST., 1, 3-13.

WICKSELL, S.D. 1933. (18.1/B) 5880
    ON CORRELATION FUNCTIONS OF TYPE III.
    BIOMETRIKA, 25, 121-133.

WICKSELL, S.D. 1934. (8.1) 5881
    ANALYTICAL THEORY OF REGRESSION.
    LUNDS UNIV. ARSSKR. N. F. AVD. 2, 30(1) 1-32.

WIENER, NORBERT + AKUTOWICZ, E. J. 1959. (20.1) 5882
    A FACTORIZATION OF POSITIVE HERMITIAN MATRICES.
    J. MATH. AND MECH., 8, 111-120. (MR21-2158)

WIENER, NORBERT + MASANI, PESI R. 1957. (16.4) 5883
    THE PREDICTION THEORY OF MULTIVARIATE STOCHASTIC PROCESSES. I. THE
    REGULARITY CONDITION.
    ACTA MATH., 98, 111-150. (MR20-4323)

WIENER, NORBERT + MASANI, PESI R. 1958. (16.4) 5884
    THE PREDICTION THEORY OF MULTIVARIATE STOCHASTIC PROCESSES. II. THE
    LINEAR PREDICTOR.
    ACTA MATH., 99, 93-137. (MR20-4325)

WIENER, NORBERT + MASANI, PESI R. 1958. (16.4) 5885
    SUR LA PREVISION LINEAIRE DES PROCESSUS STOCHASTIQUES VECTORIELS A DENSITE
    SPECTRALE BORNEE, ET LA DETERMINATION DE LA FONCTION GENERATRICE.
    C.R. ACAD. SCI. PARIS, 246, 1492-1495. (MR20-4324A)

WIENER, NORBERT + MASANI, PESI R. 1958. (16.4) 5886
    SUR LA PREVISION LINEAIRE DES PROCESSUS STOCHASTIQUES VECTORIELS A
    DENSITE SPECTRALE BORNEE.
    C.R. ACAD. SCI. PARIS, 246, 1655-1656. (MR20-4324B)

WIENER, NORBERT SEE ALSO
    3710. MASANI, PESI R. + WIENER, NORBERT (1959)

    3711. MASANI, PESI R. + WIENER, NORBERT (1959)

    3712. MASANI, PESI R. + WIENER, NORBERT (1959)

WIER, J.B. DE V. 1966. (5.2/NTU) 5887
    TABLE OF 0.1 PERCENTAGE POINTS OF BEHRENS'S d.
    BIOMETRIKA, 53, 267-268.

WIERWILLE, WALTER W. 1965. (4.8/E) 5888
    A THEORY AND METHOD FOR CORRELATION ANALYSIS OF NONSTATIONARY SIGNALS.
    IEEE TRANS. ELECTRONIC COMPUT.,EC-14, 909-919.

WIESEN, J.M. SEE
    4120. OWEN, DONALD B. + WIESEN, J.M. (1959)

WIESNER, JEROME B. SEE
    3349. LEE, Y. W. + WIESNER, JEROME B. (1950)

    3350. LEE, Y. W.: CHEATHAM, T. P., JR. + WIESNER, JEROME B. (1950)

WIGGINS, RALPH A. + ROBINSON, ENDERS A. 1965. (16.4) 5889
    RECURSIVE SOLUTION TO THE MULTICHANNEL FILTERING PROBLEM.
    J. GEOPHYS. RES., 70, 1885-1891. (MR32-589)

WIGNER, EUGENE P. 1955.   (20.1/20.2)                                                    5890
      CHARACTERISTIC VECTORS OF BORDERED MATRICES WITH INFINITE DIMENSIONS.
            (REPRINTED AS NO. 5894)
      ANN. MATH. SER. 2. 62, 548-564.   (MR17,P.1097)

WIGNER, EUGENE P. 1957.   (13.1/20.2)                                                    5891
      CHARACTERISTIC VECTORS OF BORDERED MATRICES WITH INFINITE DIMENSIONS.II.
            (REPRINTED AS NO. 5895)
      ANN. MATH. SER. 2. 65, 203-207.   (MR18,P.771)

WIGNER, EUGENE P. 1958.   (13.1/17.3)                                                    5892
      ON THE DISTRIBUTION OF. THE ROOTS OF CERTAIN SYMMETRIC MATRICES.
            (REPRINTED AS NO. 5897)
      ANN. MATH. SER. 2, 67, 325-327.   (MR20-2029)

WIGNER, EUGENE P. 1959.   (13.1/20.1)                                                    5893
      STATISTICAL PROPERTIES OF REAL SYMMETRIC MATRICES WITH MANY DIMENSIONS.
            (REPRINTED AS NO. 5896)
      PROC. CANAD. MATH. CONG., 4, 174-184.

WIGNER, EUGENE P. 1965.   (20.1/20.2)                                                    5894
      CHARACTERISTIC VECTORS OF BORDERED MATRICES WITH INFINITE DIMENSIONS.
            (REPRINT OF NO. 5890)
      STATIST. THEOR. SPECTRA FLUCTUATIONS (PORTER), 145-161.   (MR17,P.1097)

WIGNER, EUGENE P. 1965.   (13.1/20.1)                                                    5895
      CHARACTERISTIC VECTORS OF BORDERED MATRICES WITH INFINITE DIMENSIONS. II.
            (REPRINT OF NO. 5891)
      STATIST. THEOR. SPECTRA FLUCTUATIONS (PORTER), 176-180.   (MR18,P.771)

WIGNER, EUGENE P. 1965.   (13.1/20.1)                                                    5896
      STATISTICAL PROPERTIES OF REAL SYMMETRIC MATRICES WITH MANY DIMENSIONS.
            (REPRINT OF NO. 5893)
      STATIST. THEOR. SPECTRA FLUCTUATIONS (PORTER), 188-198.

WIGNER, EUGENE P. 1965.   (13.1/17.3)                                                    5897
      ON THE DISTRIBUTION OF THE ROOTS OF CERTAIN SYMMETRIC MATRICES.
            (REPRINT OF NO. 5892)
      STATIST. THEOR. SPECTRA FLUCTUATIONS (PORTER), 226-227.   (MR20-2029)

WIGNER, EUGENE P. 1965.   (13.1)                                                         5898
      DISTRIBUTION LAWS FOR THE ROOTS OF A RANDOM HERMITEAN MATRIX.
      STATIST. THEOR. SPECTRA FLUCTUATIONS (PORTER), 446-461.

WIJSMAN, ROBERT A. 1957.   (5.3/7.1/7.2)                                                 5899
      RANDOM ORTHOGONAL TRANSFORMATIONS AND THEIR USE IN SOME CLASSICAL
      DISTRIBUTION PROBLEMS IN MULTIVARIATE ANALYSIS.
      ANN. MATH. STATIST., 28, 415-423.   (MR18,P.955)

WIJSMAN, ROBERT A. 1959.   (4.2/4.7/7.1)                                                 5900
      APPLICATIONS OF A CERTAIN REPRESENTATION OF THE WISHART DISTRIBUTION.
      ANN. MATH. STATIST., 30, 597-601.   (MR21-3069)

WILCOXON, FRANK   SEE
      698. BRADLEY, RALPH A.; MARTIN, DONALD C. + WILCOXON, FRANK (1965)

WILD, A.J.B.   SEE
      5247. STEYN, H.S. + WILD, A.J.B. (1958)

WILDE, DOUGLASS J.   SEE
      425. BEIGHTLER, CHARLES S. + WILDE, DOUGLASS J. (1966)

WILDER, CARLTON E.   SEE
      568. BITTNER, REIGN H. + WILDER, CARLTON E. (1946)

WILDER, MARIAN A. 1931.   (4.1/20.5)                                                     5901
      CORRELATION COEFFICIENTS AND TRANSFORMATION OF AXES.
      AMER. MATH. MONTHLY, 38, 64-66.

WILK, MARTIN B. 1955.   (8.5/17.9)                                                       5902
      THE RANDOMIZATION ANALYSIS OF A GENERALIZED RANDOMIZED BLOCK DESIGN.
            (AMENDED BY NO. 5903)
      BIOMETRIKA. 42, 70-79.   (MR16,P.943)

WILK, MARTIN B. 1956. (8.5/17.9)                                              5903
    ERRATA TO: THE RANDOMIZATION ANALYSIS OF A GENERALIZED RANDOMIZED BLOCK DESIGN.
        (AMENDMENT OF NO. 5902)
    BIOMETRIKA, 43, 235. (MR16,P.943)

WILK, MARTIN B. + GNANADESIKAN, R. 1961. (17.9)                              5904
    GRAPHICAL ANALYSIS OF MULTI-RESPONSE EXPERIMENTAL DATA USING ORDERED DISTANCES.
    PROC. NAT. ACAD. SCI. U.S.A., 47, 1209-1212. (MR23-A3005)

WILK, MARTIN B. + GNANADESIKAN, R. 1964. (17.9)                             5905
    GRAPHICAL METHODS FOR INTERNAL COMPARISONS IN MULTIRESPONSE EXPERIMENTS.
    ANN. MATH. STATIST., 35, 613-631. (MR29-6585)

WILK, MARTIN B.   SEE ALSO
    5554. TUKEY, JOHN W. + WILK, MARTIN B. (1966)

WILKIE, D. 1965. (20.4/T)                                                   5906
    COMPLETE SET OF LEADING COEFFICIENTS, $\lambda(r,n)$, FOR ORTHOGONAL
    POLYNOMIALS UP TO N = 26.
    TECHNOMETRICS, 7, 644-648.

WILKINSON, J. H.   SEE
    2056. GOLUB, GENE H. + WILKINSON, J. H. (1966)

WILKS, S.S. 1932. (3.1/3.2/17.5/A)                                          5907
    MOMENTS AND DISTRIBUTIONS OF ESTIMATES OF POPULATION PARAMETERS FROM
    FRAGMENTARY SAMPLES.
    ANN. MATH. STATIST., 3, 163-195.

WILKS, S.S. 1932. (4.7)                                                     5908
    ON THE SAMPLING DISTRIBUTION OF THE MULTIPLE CORRELATION COEFFICIENT.
    ANN. MATH. STATIST., 3, 196-203.

WILKS, S.S. 1932. (7.2/7.3/8.4)                                             5909
    CERTAIN GENERALIZATIONS IN THE ANALYSIS OF VARIANCE.
    BIOMETRIKA, 24, 471-494.

WILKS, S.S. 1932. (4.2/8.2/14.2)                                            5910
    ON THE DISTRIBUTIONS OF STATISTICS IN SAMPLES FROM A NORMAL
    POPULATION OF TWO VARIABLES WITH MATCHED SAMPLING OF ONE VARIABLE.
    METRON, 9(3) 87-126.

WILKS, S.S. 1932. (7.3/15.2)                                                5911
    THE STANDARD ERROR OF A TETRAD IN SAMPLES FROM A NORMAL POPULATION OF
    INDEPENDENT VARIABLES.
    PROC. NAT. ACAD. SCI. U.S.A., 18, 562-565.

WILKS, S.S. 1934. (7.2)                                                     5912
    MOMENT-GENERATING OPERATORS FOR DETERMINANTS OF PRODUCT MOMENTS IN
    SAMPLES FROM A NORMAL SYSTEM.
    ANN. MATH. SER. 2, 35, 312-340.

WILKS, S.S. 1935. (9.1)                                                     5913
    ON THE INDEPENDENCE OF K SETS OF NORMALLY DISTRIBUTED STATISTICAL VARIABLES.
    ECONOMETRICA, 3, 309-326.

WILKS, S.S. 1935. (1.1)                                                     5914
    TEST CRITERIA FOR STATISTICAL HYPOTHESES INVOLVING SEVERAL VARIABLES.
    J. AMER. STATIST. ASSOC., 30, 549-560.

WILKS, S.S. 1936. (7.1/8.4)                                                 5915
    THE SAMPLING THEORY OF SYSTEMS OF VARIANCES, COVARIANCES AND
    INTRACLASS COVARIANCES.
    AMER. J. MATH., 58, 426-432.

WILKS, S.S. 1937. (8.3)                                                     5916
    THE ANALYSIS OF VARIANCE FOR TWO OR MORE VARIABLES.
    REP. THIRD ANNUAL CONF. ECON. STATIST., 82-85.

WILKS, S.S. 1938. (7.2/15.4)                                                5917
    WEIGHTING SYSTEMS FOR LINEAR FUNCTIONS OF CORRELATED VARIABLES WHEN
    THERE IS NO DEPENDENT VARIABLE.
    PSYCHOMETRIKA, 3, 23-40.

WILKS, S.S. 1946. (10.1)                                                                      5918
    SAMPLE CRITERIA FOR TESTING EQUALITY OF MEANS, EQUALITY OF VARIANCES,
    AND EQUALITY OF COVARIANCES IN A NORMAL MULTIVARIATE DISTRIBUTION.
    ANN. MATH. STATIST., 17, 257–281.   (MR8,P.162)

WILKS, S.S. 1959. (17.8)                                                                      5919
    NON-PARAMETRIC STATISTICAL INFERENCE.
    PROB. STATIST. (CRAMER VOLUME), 331–354.

WILKS, S.S. 1960. (1.1)                                                                       5920
    MULTIDIMENSIONAL STATISTICAL SCATTER.
    CONTRIB. PROB. STATIST. (HOTELLING VOLUME), 486–503.   (MR22-11470)

WILKS, S.S. 1963. (14.4)                                                                      5921
    MULTIVARIATE STATISTICAL OUTLIERS.
    SANKHYA SER. A, 25, 407–426.   (MR30-3547)

WILKS, S.S. + DALY, JOSEPH F. 1939. (17.5/19.3/C)                                             5922
    AN OPTIMUM PROPERTY OF CONFIDENCE REGIONS ASSOCIATED WITH THE
    LIKELIHOOD FUNCTION.
    ANN. MATH. STATIST., 10, 225–235.   (MR1,P.64)

WILKS, S.S.    SEE ALSO
    2193. GULLIKSEN, HAROLD + WILKS, S.S. (1950)

    4168. PEARSON, EGON SHARPE + WILKS, S.S. (1933)

    5555. TUKEY, JOHN W. + WILKS, S.S. (1946)

WILLERS, FRIEDRICH A. 1931. (4.5)                                                             5923
    KORRELATION ZWISCHEN DREI VERANDERLICHEN.
    SKAND. AKTUARIETIDSKR., 14, 158–166.

WILLIAMS, CLYDE M.   SEE
    4115. OVERALL, JOHN E. + WILLIAMS, CLYDE M. (1961)

WILLIAMS, E.J. 1952. (11.1/E)                                                                 5924
    APPLICATIONS OF COMPONENT ANALYSIS TO THE STUDY OF PROPERTIES OF TIMBER.
    AUSTRAL. J. APPL. SCI., 3, 101–118.

WILLIAMS, E.J. 1952. (8.4/8.5/13.1/E)                                                         5925
    SOME EXACT TESTS IN MULTIVARIATE ANALYSIS.
    BIOMETRIKA, 39, 17–31.   (MR14,P.299)

WILLIAMS, E.J. 1952. (4.8)                                                                    5926
    USE OF SCORES FOR THE ANALYSIS OF ASSOCIATION IN CONTINGENCY TABLES.
    BIOMETRIKA, 39, 274–289.   (MR17,P.641)

WILLIAMS, E.J. 1953. (8.3)                                                                    5927
    TESTS OF SIGNIFICANCE FOR CONCURRENT REGRESSION LINES.
    BIOMETRIKA, 40, 297–305.   (MR17,P.641)

WILLIAMS, E.J. 1955. (8.4/8.6/8.7)                                                            5928
    SIGNIFICANCE TESTS FOR DISCRIMINANT FUNCTIONS AND LINEAR FUNCTIONAL
    RELATIONSHIPS.
    BIOMETRIKA, 42, 360–381.   (MR17,P.381)

WILLIAMS, E.J. 1961. (12.2)                                                                   5929
    TESTS FOR DISCRIMINANT FUNCTIONS.
    J. AUSTRAL. MATH. SOC., 2, 243–252.   (MR25-1608)

WILLIAMS, E.J. + KLOOT, N. H. 1953. (16.5)                                                    5930
    INTERPOLATION IN A SERIES OF CORRELATED OBSERVATIONS.
    AUSTRAL. J. APPL. SCI., 4, 1–17.   (MR14,P.1104)

WILLIAMS, E.J.   SEE ALSO
    5775. WATSON, G.S. + WILLIAMS, E.J. (1956)

WILLIAMS, J. S. 1966. (17.3/B)                                                                5931
    AN EXAMPLE OF THE MISAPPLICATION OF CONDITIONAL DENSITIES.
    SANKHYA SER. A, 28, 297–300.

WILLIAMS, JEAN M. + WEILER, H. 1964.   (2.4/T)                    5932
     FURTHER CHARTS FOR THE MEANS OF TRUNCATED NORMAL BIVARIATE DISTRIBUTIONS.
     AUSTRAL. J. STATIST., 6, 117-129.   (MR32-4765)

WILLIAMS, W. T.   SEE
     3583. MACNAUGHTON-SMITH, P.; WILLIAMS, W. T.; DALE, M. G. +   (1964)
     MOCKETT, L. G.

WILLIS, R. H. 1959.   (4.1)                                       5933
     LOWER BOUND FORMULAS FOR THE INTERCORRELATION COEFFICIENT.
     J. AMER. STATIST. ASSOC., 54, 275-280.

WILSON, EDWIN B. 1928 ((15.1/15.2/E)                              5934
     ON HIERARCHICAL CORRELATION SYSTEMS.
     PROC. NAT. ACAD. SCI. U.S.A., 14, 283-291.

WILSON, EDWIN B. 1929.   (4.2)                                    5935
     PROBABLE ERROR OF CORRELATION RESULTS.
     J. AMER. STATIST. ASSOC., 24(SP) 90-93.

WILSON, EDWIN B. 1931.   (4.5/4.8/8.8)                            5936
     CORRELATION AND ASSOCIATION.
     J. AMER. STATIST. ASSOC., 26(SP) 250-257.

WILSON, EDWIN B. 1932.   (4.1)                                    5937
     A CORRELATION CURIOSITY.
     SCIENCE, 76, 515-516.

WILSON, EDWIN B. 1933.   (15.1)                                   5938
     TRANSFORMATIONS PRESERVING THE TETRAD EQUATIONS.
     PROC. NAT. ACAD. SCI. U.S.A., 19, 882-884.

WILSON, EDWIN B. 1951.   (4.8/E)                                  5939
     NOTE ON ASSOCIATION OF ATTRIBUTES.
     PROC. NAT. ACAD. SCI. U.S.A., 37, 696-704.

WILSON, EDWIN B. + WORCESTER, JANE 1939.   (15.1/15.2/E)          5940
     NOTE ON FACTOR ANALYSIS.
          (AMENDED BY NO. 2968)
     PSYCHOMETRIKA, 4, 133-148.

WILSON, EDWIN B. + WORCESTER, JANE 1942.   (4.8)                  5941
     THE ASSOCIATION OF THREE ATTRIBUTES.
     PROC. NAT. ACAD. SCI. U.S.A., 28, 384-390.   (MR4,P.106)

WILSON, L. A.   SEE
     6013. WOODROW, H. + WILSON, L. A. (1936)

WILSON, THURLOW R.   SEE
     2942. KATZ, LEO + WILSON, THURLOW R. (1956)

WIMP, JET   SEE
     1659. FIELDS, JERRY L. + WIMP, JET (1961)

WIMPRESS, RICHARD S.   SEE
     5819. WELCH, PETER D. + WIMPRESS, RICHARD S. (1961)

WINE, R.L. + FREUND, JOHN E. 1957.   (19.1)                       5942
     ON THE ENUMERATION OF DECISION PATTERNS INVOLVING N MEANS.
     ANN. MATH. STATIST., 28, 256-259.   (MR18,P.832)

WINER, BEN J. 1955.   (15.5)                                      5943
     A MEASURE OF INTERRELATIONSHIP FOR OVERLAPPING GROUPS.
     PSYCHOMETRIKA, 20, 63-68.

WINER, BEN J.   SEE ALSO
     5844. WHERRY, ROBERT J. + WINER, BEN J. (1953)

WINKLER, C. A.   SEE
     299. BARDWELL, J. + WINKLER, C. A. (1949)

     300. BARDWELL, J. + WINKLER, C. A. (1949)

     301. BARDWELL, J. + WINKLER, C. A. (1949)

WINTNER, AUREL 1935.   (17.1/17.3)                                                                5944
    GAUSSIAN DISTRIBUTIONS AND CONVERGENT INFINITE CONVOLUTIONS.
    AMER. J. MATH., 57, 821-838.

WINTNER, AUREL   SEE ALSO
    2391. HARTMAN, PHILIP + WINTNER, AUREL (1940)

WIRTH, W. 1925.   (4.8)                                                                            5945
    DIE PSYCHOTECHNISCHE BRAUCHBARKEIT DES SPEARMAN'SCHEN
    RANGKORRELATIONS-KOEFFIZIENTEN, ZUMAL FUR AUGENMASSPRUFUNGEN.
    INDUST. PSYCHOTECH., 2, 22-31.

WISE, M. E. 1963.   (17.8/18.1/A)                                                                  5946
    MULTINOMIAL PROBABILITIES AND THE $\chi^2$ AND $X^2$ DISTRIBUTIONS.
        (AMENDED BY NO. 5947)
    BIOMETRIKA, 50, 145-154.   (MR27-5307)

WISE, M. E. 1963.   (17.8/18.1/A)                                                                  5947
    CORRIGENDA.
        (AMENDMENT OF NO. 5946)
    BIOMETRIKA, 50, 546.   (MR27-5307)

WISHART, JOHN 1928.   (7.1)                                                                        5948
    THE GENERALIZED PRODUCT MOMENT DISTRIBUTION IN SAMPLES FROM A NORMAL
    MULTIVARIATE POPULATION.
        (AMENDED BY NO. 5949)
    BIOMETRIKA, 20(A) 32-52.

WISHART, JOHN 1928.   (7.1)                                                                        5949
    NOTE ON THE PAPER BY DR J. WISHART.
        (AMENDMENT OF NO. 5948)
    BIOMETRIKA, 20(A) 424.

WISHART, JOHN 1928.   (15.2)                                                                       5950
    SAMPLING ERRORS IN THE THEORY OF TWO FACTORS.
    BRITISH J. PSYCHOL., 19, 180-187.

WISHART, JOHN 1928.   (4.6/4.8/E)                                                                  5951
    LE TRAITEMENT CORRECT DES PROBLEMES DE CORRELATION MULTIPLE EN
    METEOROLOGIE ET EN AGRICULTURE.
    C. R. ASSOC. FRANC. AVANCE. SCI. CONF., 52, 188-191.

WISHART, JOHN 1928.   (4.7)                                                                        5952
    ON ERRORS IN THE MULTIPLE CORRELATION COEFFICIENT DUE TO RANDOM SAMPLING.
    MEM. ROY. METEOROL. SOC., 2, 29-37.

WISHART, JOHN 1928.   (4.7/T)                                                                      5953
    TABLE OF SIGNIFICANT VALUES OF THE MULTIPLE CORRELATION COEFFICIENT.
    QUART. J. ROY. METEOROL. SOC., 54, 258-259.

WISHART, JOHN 1929.   (7.1)                                                                        5954
    A PROBLEM IN COMBINATORIAL ANALYSIS GIVING THE DISTRIBUTION OF CERTAIN
    MOMENT STATISTICS.
    PROC. LONDON MATH. SOC. SER. 2, 29, 309-321.

WISHART, JOHN 1929.   (14.2)                                                                       5955
    THE CORRELATION BETWEEN PRODUCT MOMENTS OF ANY ORDER IN SAMPLES FROM
    A NORMAL POPULATION.
    PROC. ROY. SOC. EDINBURGH, 49, 78-90.

WISHART, JOHN 1931.   (4.7)                                                                        5956
    THE MEAN AND SECOND MOMENT COEFFICIENT OF THE MULTIPLE CORRELATION
    COEFFICIENT IN SAMPLES FROM A NORMAL POPULATION.
    BIOMETRIKA, 22, 353-361.

WISHART, JOHN 1932.   (4.8)                                                                        5957
    A NOTE ON THE DISTRIBUTION OF THE CORRELATION RATIO.
    BIOMETRIKA, 24, 441-456.

WISHART, JOHN 1947.   (14.5/A)                                                                     5958
    THE CUMULANTS OF THE Z AND OF THE LOGARITHMIC $\chi^2$ AND T DISTRIBUTIONS.
        (AMENDED BY NO. 5959)
    BIOMETRIKA, 34, 170-178.   (MR8,P.474,P.709)

WISHART, JOHN 1947.　(14.5)　　　　　**5959**
　　CORRIGENDA.
　　　　(AMENDMENT OF NO. 5958)
　BIOMETRIKA, 34, 374.　(MP8,P.474,P.709)

WISHART, JOHN 1948.　(7.1)　　　　　**5960**
　　PROOFS OF THE DISTRIBUTION LAW OF THE SECOND ORDER MOMENT STATISTICS.
　　　　(AMENDED BY NO. 5961)
　BIOMETRIKA, 35, 55–57.　(MR9,P.600)

WISHART, JOHN 1948.　(7.1)　　　　　**5961**
　　NOTE ON "PROOFS OF THE DISTRIBUTION LAW OF THE SECOND ORDER MOMENT STATISTICS".
　　　　(AMENDMENT OF NO. 5960)
　BIOMETRIKA, 35, 422.　(MR9,P.600)

WISHART, JOHN 1949.　(18.1)　　　　　**5962**
　　CUMULANTS OF MULTIVARIATE MULTINOMIAL DISTRIBUTIONS.
　BIOMETRIKA, 36, 47–58.　(MR11,P.528)

WISHART, JOHN 1952.　(17.3)　　　　　**5963**
　　MOMENT COEFFICIENTS OF THE K-STATISTICS IN SAMPLES FROM A FINITE POPULATION.
　BIOMETRIKA, 39, 1–13.　(MR14,P.296)

WISHART, JOHN 1955.　(1.1)　　　　　**5964**
　　MULTIVARIATE ANALYSIS.
　APPL. STATIST., 4, 103–116.　(MR16,P.1040)

WISHART, JOHN 1957.　(14.5)　　　　　**5965**
　　AN APPROXIMATE FORMULA FOR THE CUMULATIVE Z-DISTRIBUTION.
　ANN. MATH. STATIST., 28, 504–510.　(MR19,P.471)

WISHART, JOHN + BARTLETT, MAURICE S. 1932.　(7.1)　　　　　**5966**
　　THE DISTRIBUTION OF SECOND ORDER MOMENT STATISTICS IN A NORMAL SYSTEM.
　PROC. CAMBRIDGE PHILOS. SOC., 28, 455–459.

WISHART, JOHN + BARTLETT, MAURICE S. 1933.　(7.1)　　　　　**5967**
　　THE GENERALISED PRODUCT MOMENT DISTRIBUTION IN A NORMAL SYSTEM.
　PROC. CAMBRIDGE PHILOS. SOC., 29, 260–270.

WISNER, ROBERT J.　SEE
　2743. JACOBSON, BERNARD + WISNER, ROBERT J. (1965)

WISNIEWSKI, JAN K. 1934.　(4.8)　　　　　**5968**
　　PITFALLS IN THE COMPUTATION OF THE CORRELATION RATIO.
　J. AMER. STATIST. ASSOC., 29, 416–417.

WISNIEWSKI, JAN K. 1934.　(4.8)　　　　　**5969**
　　KILKA UWAG O MIARACH KORELACJI.
　　　　(WITH SUMMARY IN FRENCH)
　KWART. STATYST. GLOWNY URZAD STATYST., 11, 535–543.

WITTENBERG, HELEN 1964.　(17.1/U)　　　　　**5970**
　　LIMITING DISTRIBUTIONS OF RANDOM SUMS OF INDEPENDENT RANDOM VARIABLES.
　Z. WAHRSCHEIN. VERW. GEBIETE, 3, 7–18.

WOLD, HERMAN O. A. 1934.　(17.3)　　　　　**5971**
　　SHEPPARD'S CORRECTION FORMULAE IN SEVERAL VARIABLES.
　SKAND. AKTUARIETIDSKR., 17, 248–255.

WOLD, HERMAN O. A. 1936.　(1.1)　　　　　**5972**
　　ON QUANTITATIVE STATISTICAL ANALYSIS.
　SKAND. AKTUARIETIDSKR., 19, 281–284.

WOLD, HERMAN O. A. 1943.　(20.1/20.3)　　　　　**5973**
　　ON INFINITE, NON-NEGATIVE DEFINITE, HERMITIAN MATRICES, AND
　　CORRESPONDING LINEAR EQUATION SYSTEMS.
　ARK. MAT. ASTRONOM. FYS., 29A(19) 1–13.　(MR5,P.30)

WOLD, HERMAN O. A. 1945.　(4.9/8.1/17.7)　　　　　**5974**
　　A THEOREM ON REGRESSION COEFFICIENTS OBTAINED FROM SUCCESSIVELY
　　EXTENDED SETS OF VARIABLES.
　SKAND. AKTUARIETIDSKR., 28, 181–200.　(MR7,P.317)

WOLD, HERMAN O. A. 1946.  (4.1)                                              5975
    A COMMENT ON SPURIOUS CORRELATION.
    FORSAKRINGSMAT. STUD. TILL. F. LUNDBERG, 278-285.  (MR8,P.393)


WOLD, HERMAN O. A. 1947.  (17.7)                                            5976
    STATISTICAL ESTIMATION OF ECONOMIC RELATIONSHIPS.
    BULL. INST. INTERNAT. STATIST., 31(5) 1-22.  (MR13,P.481)



WOLD, HERMAN O. A. 1950.  (8.1/16.5)                                        5977
    ON LEAST SQUARE REGRESSION WITH AUTOCORRELATED VARIABLES AND RESIDUALS.
    BULL. INST. INTERNAT. STATIST., 32(2) 277-289.  (MR13,P.261)

WOLD, HERMAN O. A. 1951.  (16.1)                                            5978
    DYNAMIC SYSTEMS OF THE RECURSIVE TYPE--ECONOMIC AND STATISTICAL ASPECTS.
    SANKHYA, 11, 205-216.

WOLD, HERMAN O. A. 1952.  (4.9)                                             5979
    UBER GLEICHMASSIGE INTERKORRELATION.
    MITT. MATH. STATIST., 4, 163-166.  (MR14,P.297)

WOLD, HERMAN O. A. 1953.  (15.1)                                            5980
    SOME ARTIFICIAL EXPERIMENTS IN FACTOR ANALYSIS.
    NORDISK PSYKOL. MONOG. NR. 3 (UPPSALA SYMP. PSYCHOL. FACT. ANAL.), 43-64.  (MR16,P.731)

WOLD, HERMAN O. A. 1955.  (16.1/16.2)                                       5981
    POSSIBILITES ET LIMITATIONS DES SYSTEMES A CHAINE CAUSALE.
    CAHIERS SEM. ECONOMET., 3, 81-101.

WOLD, HERMAN O. A. 1956.  (1.1)                                             5982
    CAUSAL INFERENCE FROM OBSERVATIONAL DATA, A REVIEW OF ENDS AND MEANS.
    J. ROY. STATIST. SOC. SER. A, 119, 28-50.

WOLD, HERMAN O. A. 1958.  (11.2)                                            5983
    A CASE STUDY OF INTERDEPENDENT VERSUS CAUSAL CHAIN SYSTEMS.
    REV. INST. INTERNAT. STATIST., 26, 5-25.

WOLD, HERMAN O. A. 1959.  (16.1/16.2/16.5)                                  5984
    ENDS AND MEANS IN ECONOMETRIC MODEL BUILDING.  BASIC CONSIDERATIONS REVIEWED.
    PROB. STATIST. (CRAMER VOLUME), 355-434.  (MR21-7800)

WOLD, HERMAN O. A. 1960.  (17.7)                                            5985
    A GENERALIZATION OF CAUSAL CHAIN SYSTEMS (PART III OF A TRIPTYCH ON
    CAUSAL CHAIN SYSTEMS).
    ECONOMETRICA, 28, 443-463.  (MR22-10793)

WOLD, HERMAN O. A. 1961.  (1.2)                                             5986
    OSKAR ANDERSON, 1887-1960.
    ANN. MATH. STATIST., 32, 651-660.  (MR23-A2301)


WOLD, HERMAN O. A. 1961.  (16.2/E)                                          5987
    CONSTRUCTION PRINCIPLES OF SIMULTANEOUS EQUATION MODELS IN ECONOMETRICS.
    BULL. INST. INTERNAT. STATIST., 38(4) 111-138.

WOLD, HERMAN O. A. 1963.  (16.1/16.2/16.5)                                  5988
    FORECASTING BY THE CHAIN PRINCIPLE.
        (REPRINTED AS NO. 5990)
    PROC. SYMP. TIME SERIES ANAL. BROWN UNIV. (ROSENBLATT), 471-497.  (MR27-1302)

WOLD, HERMAN O. A. 1963.  (8.8/A)                                           5989
    ON THE CONSISTENCY OF LEAST SQUARES REGRESSION.
    SANKHYA SER. A, 25, 211-215.

WOLD, HERMAN O. A. 1964.  (16.1/16.2/16.5)                                  5990
    FORECASTING BY THE CHAIN PRINCIPLE.
        (REPRINT OF NO. 5988)
    ECONOMET. MODEL BUILDING (WOLD). 5-36.  (MR27-1302)

WOLD, HERMAN O. A. 1964.  (1.1/E)                                           5991
    A LETTER REPORT TO PROFESSOR P.C. MAHALANOBIS.
    ESSAYS ECONOMET. PLANNING (MAHALANOBIS VOLUME), 309-320.

WOLD, HERMAN O.A. 1965.  (16.2)                                                    5728
     TOWARD A VERDICT ON MACROECONOMIC SIMULTANEOUS EQUATIONS. (WITH DISCUSSION)
     SEMAINE D'ETUDE ROLE ANAL. ECONOMET. PONT. ACAD. SCI., 115-185.

WOLD, HERMAN O. A. 1966.  (16.1)                                                   5992
     A FIX-POINT THEOREM WITH ECONOMETRIC BACKGROUND. PART I. THE THEOREM.
     ARK. MAT., 6, 209-220.  (MR33-6772)

WOLD, HERMAN O. A. 1966.  (16.1)                                                   5993
     A FIX-POINT THEOREM WITH ECONOMETRIC BACKGROUND. PART II. ILLUSTRATIONS. FURTHER
     DEVELOPMENTS.
     ARK. MAT., 6, 221-240.  (MR33-6773)

WOLD, HERMAN O. A. 1966.  (11.2/15.2/E)                                            5994
     ESTIMATION OF PRINCIPAL COMPONENTS AND RELATED MODELS BY ITERATIVE
     LEAST SQUARES.
     MULTIVARIATE ANAL. PROC. INTERNAT. SYMP. DAYTON (KRISHNAIAH), 391-420.   (MR36-3457)

WOLD, HERMAN O. A. 1966.  (12.1/16.2/17.7)                                         5995
     NONLINEAR ESTIMATION BY ITERATIVE LEAST SQUARE PROCEDURES.
     RES. PAPERS STATIST. FESTSCHR. NEYMAN (DAVID), 411-444.  (MR35-1144)

WOLD, HERMAN O. A. + FAXER, P. 1957.  (17.7)                                       5996
     ON THE SPECIFICATION ERROR IN REGRESSION ANALYSIS.
     ANN. MATH. STATIST., 28, 265-267.  (MR19,P.74)

WOLD, HERMAN O. A.   SEE ALSO
     467. BENTZEL, R. + WOLD, HERMAN O. A. (1946)

    1104. CRAMER, HARALD + WOLD, HERMAN O. A. (1936)

    1843. GALVENIUS, I. + WOLD, HERMAN O. A. (1947)

    4583. ROBINSON, ENDERS A. + WOLD, HERMAN O. A. (1963)

    5261. STROTZ, ROBERT H. + WOLD, HERMAN O. A. (1960)

WOLF, J.K.   SEE
     5403. THOMAS, JOHN B. + WOLF, J.K. (1962)

WOLF, ROBERT L.; MENDLOWITZ, MILTON; ROBOZ, JULIA; 1966.  E(8.6)                   5997
     STYAN, GEORGE P. H.; KORNFELD, PETER + WEIGL, ALFRED
     TREATMENT OF HYPERTENSION WITH SPIRONOLACTONE:  DOUBLE-BLIND STUDY.
     J. AMER. MED. ASSOC., 198, 1143-1149.

WOLFLE, DAEL 1940.  (1.2/15.1)                                                     5998
     FACTOR ANALYSIS TO 1940.
     PSYCHOMET. MONOG., 3, 1-69.

WOLFLE, DAEL 1956.  (1.2)                                                          5999
     LOUIS LEON THURSTONE, 1887-1955.
     AMER. J. PSYCHOL., 69, 131-134.

WOLFOWITZ, J. 1952.  (17.7)                                                        6000
     CONSISTENT ESTIMATORS OF THE PARAMETERS OF A LINEAR STRUCTURAL RELATION.
     SKAND. AKTUARIETIDSKR., 35, 132-151.  (MR14,P.776)

WOLFOWITZ, J. 1954.  (17.1)                                                        6001
     GENERALIZATION OF THE THEOREM OF GLIVENKO-CANTELLI.
     ANN. MATH. STATIST., 25, 131-138.  (MR15,P.808)

WOLFOWITZ, J. 1954.  (16.1/U)                                                      6002
     ESTIMATION BY THE MINIMUM DISTANCE METHOD IN NON-PARAMETRIC
     STOCHASTIC DIFFERENCE EQUATIONS.
     ANN. MATH. STATIST., 25, 203-217.  (MR15,P.808

WOLFOWITZ, J. 1957.  (19.3)                                                        6003
     THE MINIMUM DISTANCE METHOD.
     ANN. MATH. STATIST., 28, 75-88.  (MR19,P.472,P.1432)

WOLFOWITZ, J. 1960.  (17.1)                                                        6004
     CONVERGENCE OF THE EMPIRIC DISTRIBUTION FUNCTION ON HALF- SPACES.
     CONTRIB. PROB. STATIST. (HOTELLING VOLUME), 504-507.

WOLFOWITZ, J.   SEE ALSO
   3040. KIEFER, J. + WOLFOWITZ, J. (1952)

   3041. KIEFER, J. + WOLFOWITZ, J. (1956)

   3042. KIEFER, J. + WOLFOWITZ, J. (1959)

   3375. LEVENE, HOWARD + WOLFOWITZ, J. (1944)

   5731. WALD, ABRAHAM + WOLFOWITZ, J. (1943)

   5732. WALD, ABRAHAM + WOLFOWITZ, J. (1944)

   5733. WALD, ABRAHAM + WOLFOWITZ, J. (1955)

   5734. WALD, ABRAHAM + WOLFOWITZ, J. (1955)

   1434. DVORETZKY, ARYEH; KIEFER, J. + WOLFOWITZ, J. (1953)

   1435. DVORETZKY, ARYEH; KIEFER, J. + WOLFOWITZ, J. (1953)

   1436. DVORETZKY, ARYEH; WALD, ABRAHAM + WOLFOWITZ, J. (1951)

   1437. DVORETZKY, ARYEH; WALD, ABRAHAM + WOLFOWITZ, J. (1955)

   2890. KAC, MARK; KIEFER, J. + WOLFOWITZ, J. (1955)

WONG, F. + THOMAS, JOHN B. 1962.  (17.3)                                    6005
   ON POLYNOMIAL EXPANSIONS OF SECOND-ORDER DISTRIBUTIONS.
   J. SOC. INDUST. APPL. MATH., 10, 507-516.

WONG, Y.K. 1937.  (4.6)                                                     6006
   ON THE ELIMINATION OF VARIABLES IN MULTIPLE CORRELATION.
   J. AMER. STATIST. ASSOC., 32, 357-360.

WOO, T.L. 1929.  (4.8/T)                                                    6007
   TABLES FOR ASCERTAINING THE SIGNIFICANCE OR NON-SIGNIFICANCE OF
   ASSOCIATION MEASURED BY THE CORRELATION RATIO.  (WITH INTRODUCTION BY
   KARL PEARSON)
   BIOMETRIKA, 21, 1-66.  (MR13,P.478)

WOO, T.L. + MORANT, G. M. 1932.  E(6.4/6.5)                                 6008
   A PRELIMINARY CLASSIFICATION OF ASIATIC RACES BASED ON CRANIAL MEASUREMENTS.
   BIOMETRIKA, 24, 108-134.

WOOD, FRANCES   SEE
   736. BROWN, J. W.; GREENWOOD, MAJOR + WOOD, FRANCES (1914)

WOODBURY, MAX A. 1940.  (4.8)                                               6009
   RANK CORRELATION WHEN THERE ARE EQUAL VARIATES.
   ANN. MATH. STATIST., 11, 358-362.  (MR2,P.110)

WOODBURY, MAX A. + LORD, FREDERIC M. 1956.  (15.4)                          6010
   THE MOST RELIABLE COMPOSITE WITH A SPECIFIED TRUE SCORE.
   BRITISH J. STATIST. PSYCHOL., 9, 21-28.

WOODBURY, MAX A. + SILER, WILLIAM 1966.  (15.2)                             6011
   FACTOR ANALYSIS WITH MISSING DATA.
   ANN. NEW YORK ACAD. SCI., 128(3) 746-754.

WOODING, R.A. 1956.  (14.5)                                                 6012
   THE MULTIVARIATE DISTRIBUTION OF COMPLEX NORMAL VARIATES.
   BIOMETRIKA, 43, 212-215.  (MR17,P.978)

WOODROW, H. + WILSON, L. A. 1936.  (15.3)                                   6013
   A SIMPLE PROCEDURE FOR APPROXIMATE FACTOR ANALYSIS.
   PSYCHOMETRIKA, 1, 245-258.

WOODWARD, R. S. 1884.  (8.8)                                                6014
   THE SPECIAL TREATMENT OF CERTAIN FORMS OF OBSERVATION-EQUATIONS.
      (REPRINTED AS NO. 6015)
   BULL. PHILOS. SOC. WASHINGTON, 6, 156-157.

WOODWARD, R. S. 1888.  (8.8)                                                    6015
    THE SPECIAL TREATMENT OF CERTAIN FORMS OF OBSERVATION-EQUATIONS.
        (REPRINT OF NO. 6014)
    SMITHSONIAN MISC. COLLECT., 33(1) 156-157.


WOOLLEY, ELLIOTT B. 1941.  (8.8/17.7)                                           6016
    THE METHOD OF MINIMIZED AREAS AS A BASIS FOR CORRELATION ANALYSIS.
    ECONOMETRICA, 9, 38-62.  (MR2,P.235)


WORCESTER, JANE   SEE
    5940. WILSON, EDWIN B. + WORCESTER, JANE (1939)

    5941. WILSON, EDWIN B. + WORCESTER, JANE (1942)


WORKING, HOLBROOK 1921.  (4.1/T)                                               6017
    A USE FOR TRIGONOMETRIC TABLES IN CORRELATION.
    QUART. PUBL. AMER. STATIST. ASSOC., 17, 765-769.


WREN, F. L. 1937.  (20.1/20.3)                                                 6018
    NEO-SYLVESTER CONTRACTIONS AND THE SOLUTION OF SYSTEMS OF LINEAR EQUATIONS.
    BULL. AMER. MATH. SOC., 43, 823-834.


WREN, F. L. 1938.  (4.5/4.6)                                                   6019
    THE CALCULATION OF PARTIAL AND MULTIPLE COEFFICIENTS OF REGRESSION
    AND CORRELATION.
    J. EDUC. PSYCHOL., 29, 695-700.


WRIGHT, F. M. J.; MANNING, W. H. + DU BOIS, PHILIP H. 1959.  (4.6/20.2)        6020
    DETERMINANTS IN MULTIVARIATE CORRELATION.
    J. EXPER. EDUC., 27, 195-202.


WRIGHT, SEWALL 1917.  (4.1/E)                                                  6021
    THE AVERAGE CORRELATION WITHIN SUB-GROUPS OF A POPULATION.
    J. WASHINGTON ACAD. SCI., 7, 532-535.


WRIGHT, SEWALL 1918.  (4.5/15.1/E)                                             6022
    ON THE NATURE OF SIZE FACTORS.
    GENETICS, 3, 367-374.


WRIGHT, SEWALL 1921.  (4.5/4.6/17.7/E)                                         6023
    CORRELATION AND CAUSATION.
    J. AGRIC. RES., 20, 557-585.


WRIGHT, SEWALL 1923.  (4.6/4.8/17.7/E)                                         6024
    THE THEORY OF PATH COEFFICIENTS - A REPLY TO NILES'S CRITICISM.
    GENETICS, 8, 239-255.


WRIGHT, SEWALL 1932.  (4.5/15.1/E)                                             6025
    GENERAL, GROUP AND SPECIAL SIZE FACTORS.
    GENETICS, 17, 603-619.


WRIGHT, SEWALL 1954.  (8.1/17.7/E)                                             6026
    THE INTERPRETATION OF MULTIVARIATE SYSTEMS.
    STATIST. MATH. BIOL. (KEMPTHORNE ET AL.), 11-33.


WRIGLEY, CHARLES F. 1957.  (15.1)                                             6027
    THE DISTINCTION BETWEEN COMMON AND SPECIFIC VARIANCE IN FACTOR THEORY.
    BRITISH J. STATIST. PSYCHOL., 10, 81-98.


WRIGLEY, CHARLES F. 1958.  (15.1)                                            6028
    OBJECTIVITY IN FACTOR ANALYSIS.
    EDUC. PSYCHOL. MEAS., 18, 463-476.


WRIGLEY, CHARLES F. 1959.  (15.2)                                            6029
    THE EFFECT UPON THE COMMUNALITIES OF CHANGING THE ESTIMATE OF THE
    NUMBER OF FACTORS.
    BRITISH J. STATIST. PSYCHOL., 12, 34-54.


WRIGLEY, CHARLES F. + NEUHAUS, JACK O. 1952.  (15.3)                         6030
    A RE-FACTORIZATION OF THE BURT-PEARSON MATRIX WITH THE ORDVAC
    ELECTRONIC COMPUTER.
    BRITISH J. PSYCHOL. STATIST. SECT., 5, 105-108.

WRIGLEY, CHARLES F. + NEUHAUS, JACK O. 1955. (15.3)                                    6031
    THE USE OF AN ELECTRONIC COMPUTER IN PRINCIPAL AXES FACTOR ANALYSIS.
    J. EDUC. PSYCHOL., 46, 31–41.

WRIGLEY, CHARLES F.: SAUNDERS, DAVID R. + NEUHAUS, JACK O. 1958. (15.3)                6032
    APPLICATION OF THE QUARTIMAX METHOD OF ROTATION TO THURSTONE'S
    PRIMARY MENTAL ABILITIES STUDY.
    PSYCHOMETRIKA, 23, 151–170.

WRIGLEY, CHARLES F.: CHERRY, CHARLES N.; LEE, MARILYN C. + 1957. (15.2/E)              6033
    MC QUITTY, LOUIS L.
    USE OF THE SQUARE-ROOT METHOD TO IDENTIFY FACTORS IN THE JOB
    PERFORMANCE OF AIRCRAFT MECHANICS.
    PSYCHOL. MONOG., 71(1) 1–28.

WRIGLEY, CHARLES F.    SEE ALSO
    3986. NEUHAUS, JACK O. + WRIGLEY, CHARLES F. (1954)

    5567. TWERY, RAYMOND J.; SCHMID, JOHN, JR. + WRIGLEY, CHARLES F. (1958)

WYKES, ELIZABETH CROSSLEY    SEE
    4275. PETERS, CHARLES C. + WYKES, ELIZABETH CROSSLEY (1931)

    4276. PETERS, CHARLES C. + WYKES, ELIZABETH CROSSLEY (1931)

WYLD, C.    SEE
    4491. RAY, W.D. + WYLD, C. (1965)

WYND, F. L.    SEE
    5823. WELKER, E.L. + WYND, F. L. (1943)

YAGLOM, A. M.    SEE    JAGLOM, A. M.

YAMAMOTO, SUMIHARU 1951. (14.2/U)                                                      6034
    ON THE DISTRIBUTION OF THE RATIO OF TWO NORMAL VARIATES.
    (IN JAPANESE)
    RES. MEM. INST. STATIST. MATH. TOKYO, 7, 366–376.

YAMAZAKI, H.    SEE
    4063. OGAWARA, MASAMI + YAMAZAKI, H. (1952)

YAMAZAKI, KAZUHIDE    SEE
    3744. MATUFUJI, HAJIME; YAMAZAKI, KAZUHIDE + KOMAZAWA, TSUTOMU (1966)

YAO, YING 1965. (5.2/T)                                                                6035
    AN APPROXIMATE DEGREES OF FREEDOM SOLUTION TO THE MULTIVARIATE
    BEHRENS-FISHER PROBLEM.
    BIOMETRIKA, 52, 139–147.

YARBROUGH, HELEN    SEE
    4021. NORTON, KENNETH A.; SHULTZ, EDNA L. + YARBROUGH, HELEN (1952)

YARDI, M. R. 1946. E(6.2)                                                              6036
    A STATISTICAL APPROACH TO THE PROBLEM OF CHRONOLOGY OF SHAKESPEARE'S PLAYS.
    SANKHYA, 7, 263–268.

YASTREMSKII, B. S.    SEE    JASTREMSKII, B. S.

YASUKAWA, KAZUTARO 1925. (17.3/17.9/18.3/T)                                            6037
    ON THE MEANS, STANDARD DEVIATIONS, CORRELATIONS AND FREQUENCY
    DISTRIBUTIONS OF FUNCTION OF VARIATES.
    BIOMETRIKA, 17, 211–237.

YATES, FRANK 1939. (8.3/B)                                                             6038
    TESTS OF SIGNIFICANCE OF THE DIFFERENCES BETWEEN REGRESSION
    COEFFICIENTS DERIVED FROM TWO SETS OF CORRELATED VARIATES.
    PROC. ROY. SOC. EDINBURGH, 59, 184–194.    (MR1,P.23)

YATES, FRANK 1952. (1.2)                                                               6039
    GEORGE UDNY YULE.
    OBIT. NOTICES FELLOWS ROY. SOC., 8, 309–323.

YATES, FRANK 1966.   (1.1)                                                          6040
     COMPUTERS, THE SECOND REVOLUTION IN STATISTICS (THE FIRST FISHER
     MEMORIAL LECTURE).
     BIOMETRICS, 22, 233-251.


YATES, FRANK + MATHER, K. 1963.   (1.2)                                             6041
     RONALD AYLMER FISHER.
     BIOGRAPH. MEM. FELLOWS ROY. SOC., 9, 91-129.


YEN, ANDREW T.   SEE
     1816. FU, YUMIN; SRINATH, MANDYAM D. + YEN, ANDREW T. (1965)


YLVISAKER, N. DONALD 1964.   (16.5)                                                 6042
     LOWER BOUNDS FOR MINIMUM COVARIANCE MATRICES IN TIME SERIES REGRESSION PROBLEMS.
     ANN. MATH. STATIST., 35, 362-368.   (MR28-4636)


YOSHIMURA, ISAO 1964.   (17.3)                                                      6043
     UNIFIED SYSTEM OF CUMULANT RECURRENCE RELATIONS.
     REP. STATIST. APPL. RES. UN. JAPAN. SCI. ENGRS., 11, 1-8.   (MR30-5391)


YOSHIMURA, ISAO 1964.   (17.3)                                                      6044
     A COMPLEMENTARY NOTE ON THE MULTIVARIATE MOMENT RECURRENCE RELATION.
     REP. STATIST. APPL. RES. UN. JAPAN. SCI. ENGRS., 11, 9-12.   (MR30-5392)


YOSIDA, KOSAKU 1949.   (16.4)                                                       6045
     BROWNIAN MOTION ON THE SURFACE OF THE 3-SPHERE.
     ANN. MATH. STATIST., 20, 292-296.   (MR10,P.721)


YOUNG, ANDREW W.   SEE
     4248. PEARSON, KARL + YOUNG, ANDREW W. (1918)

     5135. SOPER, H.E.; YOUNG, ANDREW W.; CAVE, BEATRICE M.;   (1917)
     LEE, ALICE + PEARSON, KARL


YOUNG, D. H.   SEE
     2821. JOHNSON, N. L. + YOUNG, D. H. (1960)


YOUNG, GALE 1937.   (15.1/20.1)                                                     6046
     MATRIX APPROXIMATIONS AND SUBSPACE FITTING.
     PSYCHOMETRIKA, 2, 21-25.


YOUNG, GALE 1937.   (15.1/20.1)                                                     6047
     MATRIX APPROXIMATION CRITERIA.
     PSYCHOMETRIKA, 2, 71-72.


YOUNG, GALE 1939.   (6.6/15.1)                                                      6048
     FACTOR ANALYSIS AND THE INDEX OF CLUSTERING.
     PSYCHOMETRIKA, 4, 201-208.


YOUNG, GALE 1941.   (15.2)                                                          6049
     MAXIMUM LIKELIHOOD ESTIMATION AND FACTOR ANALYSIS.
     PSYCHOMETRIKA, 6, 49-53.   (MR2,P.235)


YOUNG, GALE + HOUSEHOLDER, ALSTON S. 1938.   (15.1/20.1)                            6050
     DISCUSSION OF A SET OF POINTS IN TERMS OF THEIR MUTUAL DISTANCES.
          (AMENDED BY NO. 6051)
     PSYCHOMETRIKA, 3, 19-22.


YOUNG, GALE + HOUSEHOLDER, ALSTON S. 1938.   (15.1/20.1)                            6051
     ERRATA TO: DISCUSSION OF A SET OF POINTS IN TERMS OF THEIR MUTUAL DISTANCES.
          (AMENDMENT OF NO. 6050)
     PSYCHOMETRIKA, 3, 126.


YOUNG, GALE + HOUSEHOLDER, ALSTON S. 1940.   (15.1)                                 6052
     FACTORIAL INVARIANCE AND SIGNIFICANCE.
     PSYCHOMETRIKA, 5, 47-56.   (MR2,P.110)


YOUNG, GALE + HOUSEHOLDER, ALSTON S. 1941.   (15.5)                                 6053
     A NOTE ON MULTIDIMENSIONAL PSYCHOPHYSICAL ANALYSIS.
     PSYCHOMETRIKA, 6, 331-333.

YOUNG. GALE    SEE ALSO
   1480. ECKART. CARL + YOUNG. GALE (1936)

   1481. ECKART. CARL + YOUNG. GALE (1939)

   2615. HOUSEHOLDER. ALSTON S. + YOUNG. GALE (1938)

YOUNG. J. E. 1954.  (16.4)
   CORRELATION FUNCTIONS FOR NOISE FIELDS.
   J. ACOUST. SOC. AMER.. 26. 788–789.                                        6054

YOUNG. S. S. Y.   SEE
   5350. TALLIS. G.M. + YOUNG. S. S. Y. (1962)

YUAN. PAE-TSI 1933.  (18.1/B)                                                 6055
   ON THE LOGARITHMIC FREQUENCY DISTRIBUTION AND THE SEMI-LOGARITHMIC
   CORRELATION SURFACE.
   ANN. MATH. STATIST.. 4. 30–74.

YULE. G. UDNY 1897.  (4.9)                                                    6056
   ON THE THEORY OF CORRELATION.
   J. ROY. STATIST. SOC.. 60. 812–854.

YULE. G. UDNY 1897.  (8.1)                                                    6057
   ON THE SIGNIFICANCE OF BRAVAIS' FORMULAE FOR REGRESSION,  &  C., IN
   THE CASE OF SKEW CORRELATION.
   PROC. ROY. SOC. LONDON. 60. 477–489.

YULE. G. UDNY 1903.  (4.9)                                                    6058
   NOTES ON THE THEORY OF ASSOCIATION OF ATTRIBUTES IN STATISTICS.
   BIOMETRIKA. 2. 121–134.

YULE. G. UDNY 1906.  (4.9)                                                    6059
   ON A PROPERTY WHICH HOLDS GOOD FOR ALL GROUPINGS OF A NORMAL DISTRIBUTION OF
   FREQUENCY FOR TWO VARIABLES. WITH APPLICATIONS TO THE STUDY OF
   CONTINGENCY-TABLES FOR  THE INHERITANCE OF UNMEASURED QUALITIES.
   PROC. ROY. SOC. LONDON SER. A, 77. 324–336.

YULE. G. UDNY 1907.  (4.1/4.5/4.6)                                            6060
   ON THE THEORY OF CORRELATION FOR ANY NUMBER OF VARIABLES. TREATED BY
   A NEW SYSTEM OF NOTATION.
   PROC. ROY. SOC. LONDON SER. A, 79. 182–193.

YULE. G. .UDNY 1909.  (4.1/4.5/8.8/E)                                         6061
   LES APPLICATIONS DE LA METHODE DE CORRELATION AUX STATISTIQUES
   SOCIALES ET ECONOMIQUES.
       (TRANSLATED AS NO. 6062)
   BULL. INST. INTERNAT. STATIST.. 18(1) 265–277, 537–548.

YULE. G. UDNY 1909.  (4.1/4.5/8.1/E)                                          6062
   THE APPLICATION OF THE METHOD OF CORRELATION TO SOCIAL AND ECONOMIC STATISTICS.
       (TRANSLATION OF NO. 6061)
   J. ROY. STATIST. SOC.. 72. 721–730.

YULE. G. UDNY 1910.  (4.1)                                                    6063
   ON THE INTERPRETATION OF CORRELATIONS BETWEEN INDICES OR RATIOS.
   J. ROY. STATIST. SOC.. 73. 644–647.

YULE. G. UDNY 1912.  (4.8)                                                    6064
   ON THE METHODS OF MEASURING ASSOCIATION BETWEEN TWO ATTRIBUTES.
       (WITH DISCUSSION)
   J. ROY. STATIST. SOC.. 75. 579–652.

YULE. G. UDNY 1921.  (4.8/16.5)                                               6065
   ON THE TIME-CORRELATION PROBLEM. WITH ESPECIAL REFERENCE TO THE
   VARIATE-DIFFERENCE CORRELATION METHOD.  (WITH DISCUSSION)
   J. ROY. STATIST. SOC.. 84. 497–537.

YULE. G. UDNY 1926.  (4.1/16.5)                                              6066
   WHY DO WE SOMETIMES GET NONSENSE-CORRELATIONS BETWEEN TIME-SERIES?--A
   STUDY IN SAMPLING AND THE NATURE OF TIME-SERIES.  (WITH DISCUSSION)
   J. ROY. STATIST. SOC.. 89. 1–69.

6067

YULE, G. UDNY 1938. (2.1/B)
    ON SOME PROPERTIES OF NORMAL DISTRIBUTIONS, UNIVARIATE AND BIVARIATE,
    BASED ON SUMS OF SQUARES OF FREQUENCIES.
    BIOMETRIKA, 30, 1-10.


YULE, G. UDNY   SEE ALSO
    2118. GREENWOOD, MAJOR + YULE, G. UDNY (1920)

    2539. HOOKER, R. H. + YULE, G. UDNY (1906)

    4249. PEARSON, KARL; BEETON, MARY + YULE, G. UDNY (1900)


ZADEH, LOTFI A.    SEE
    5404. THOMAS, JOHN B. + ZADEH, LOTFI A. (1961)

3235

ZAHLEN, J. P. 1966. (19.2)
    ABOUT THE FOUNDATIONS OF THE THEORY OF PARAMETRIC HYPOTHESIS TESTING.
    (IN GERMAN)
    STATIST. HEFTE, 7. 148-174.

6068

ZAHN, SAMUEL 1966. (17.1/U)
    BOUNDS FOR THE CENTRAL LIMIT THEOREM ERROR.
    SIAM J. APPL. MATH., 14, 1225-1245.


ZAKHAROV, V. K.    SEE
    4862. SARMANOV, O. V. + ZAKHAROV, V. K. (1960)

    4863. SARMANOV, O. V. + ZAKHAROV, V. K. (1960)

    4864. SARMANOV, O. V. + ZAKHAROV, V. K. (1961)


ZALOKAR, JULIA    SEE
    2307. HALPERIN, MAX; GREENHOUSE, SAMUEL W.; CORNFIELD, JEROME +   (1955)
    ZALOKAR, JULIA


ZAREMBA, S. K.    SEE
    2037. GODWIN, H. J. + ZAREMBA, S. K. (1961)


ZASSENHAUS, H.    SEE
    5366. TAUSSKY, OLGA + ZASSENHAUS, H. (1959)

6069

ZASUHIN, V. 1941. (16.4)
    ON THE THEORY OF MULTIDIMENSIONAL STATIONARY RANDOM PROCESSES.
    C. R. ACAD. SCI. URSS NOUV. SER., 33, 435-437.   (MR5,P.102)

6070

ZELEN, MARVIN 1962. (2.5/8.8)
    LINEAR ESTIMATION AND RELATED TOPICS.
    SURVEY NUMER. ANAL. (TODD), 558-584.   (MR24-A3747)

6071

ZELEN, MARVIN + SEVERO, NORMAN C. 1960. (2.7/B)
    GRAPHS FOR BIVARIATE NORMAL PROBABILITIES.
    ANN. MATH. STATIST., 31, 619-624.   (MR23-B1653)


ZELEN, MARVIN    SEE ALSO
    2048. GOLDMAN, A. J. + ZELEN, MARVIN (1964)

6072

ZELLNER, ARNOLD 1963. (17.8)
    ESTIMATORS FOR SEEMINGLY UNRELATED REGRESSION EQUATIONS:   SOME EXACT
    FINITE SAMPLE RESULTS.
    J. AMER. STATIST. ASSOC., 58, 977-992.   (MR28-673)

6073

ZELLNER, ARNOLD + CHETTY, V. KARUPPAN 1965. (8.1/14.4)
    PREDICTION AND DECISION PROBLEMS IN REGRESSION MODELS FROM THE
    BAYESIAN POINT OF VIEW.
        (AMENDED BY NO. 6074)
    J. AMER. STATIST. ASSOC., 60, 608-616.

6074

ZELLNER, ARNOLD + CHETTY, V. KARUPPAN 1968. (8.1/14.4)
    CORRECTION TO: PREDICTION AND DECISION PROBLEMS IN REGRESSION MODELS FROM THE
    BAYESIAN POINT OF VIEW.
        (AMENDMENT OF NO. 6073)
    J. AMER. STATIST. ASSOC., 63, 1551.

ZELLNER, ARNOLD + HUANG, DAVID S. 1962.  (8.1/8.2)
    FURTHER PROPERTIES OF EFFICIENT ESTIMATORS FOR SEEMINGLY UNRELATED
    REGRESSION EQUATIONS.
    INTERNAT. ECON. REV., 3, 300-313.                                           6075

ZELLNER, ARNOLD + THEIL, H. 1962.  (17.7)
    THREE-STAGE LEAST SQUARES:  SIMULTANEOUS ESTIMATION OF SIMULTANEOUS EQUATIONS.    6076
    ECONOMETRICA, 30, 54-78.  (MR31-4122)


ZELLNER, ARNOLD + THORNBER, H. 1966.  (16.2/E)
    COMPUTATIONAL ACCURACY AND ESTIMATION OF SIMULTANEOUS EQUATION                6077
    ECONOMETRIC MODELS.
    ECONOMETRICA, 34, 727-729.

ZELLNER, ARNOLD   SEE ALSO
    2544. HOOPER, JOHN W. + ZELLNER, ARNOLD (1961)

    5467. TIAO, GEORGE C. + ZELLNER, ARNOLD (1964)

ZEMACH, RITA   SEE
    4963. SEIDEN, ESTHER + ZEMACH, RITA (1966)

ZEMANIAN, A. H. 1965.  (20.4)
    A CHARACTERIZATION OF THE INVERSE LAPLACE TRANSFORMS OF RATIONAL             6078
    POSITIVE-REAL MATRICES.
    J. SOC. INDUST. APPL. MATH., 13, 463-468.

ZIA-UD-DIN, M. 1940.  (20.5/T)
    TABLES OF SYMMETRIC FUNCTIONS FOR STATISTICAL PURPOSES.                      6079
    PROC. NAT. ACAD. SCI. INDIA SECT. A, 10, 53-60.  (MR8,P.191)

ZIA-UD-DIN, M. 1951.  (20.1)
    NOTE ON EQUATION OF THE SQUARED DIFFERENCES OF THE CONJUGATE MATRIX ROOTS.   6080
    PROC. PAKISTAN STATIST. CONF., 1, 62.

ZIA-UD-DIN, M. 1951.  (20.1)
    GROUPS AND CONJUGATE MATRIX SOLUTIONS OF EQUATIONS.                          6081
    PROC. PAKISTAN STATIST. CONF., 1, 63-65.  (MR13,P.621)

ZIA-UD-DIN, M. 1962.  (17.3)
    DEVELOPMENT OF SYMMETRIC FUNCTIONAL AND K-STATISTICS.                        6082
    BULL. INST. INTERNAT. STATIST., 39(2) 501-508.

ZIA-UD-DIN, M. 1966.  (2.3)
    EXPRESSION OF GENERALIZED K-STATISTICS IN TERMS OF POWER-SUM SYMMETRIC FUNCTIONS.    6083
    BULL. INST. INTERNAT. STATIST., 41, 81-83.

ZIEGLER, JAMES 1954.  (15.4)
    A NOTE ON ITEM ANALYSIS WITH AN ELECTRONIC COMPUTOR.                         6084
    PSYCHOMETRIKA, 19, 261-262.


ZIMERING, SHIMSON 1965.  (20.2)
    MATRICES LIMITES D'UNE MATRICE TRIANGULAIRE ET LEUR APPLICATION AUX THEOREMES    6085
    MERCERIENS.
    C.R. ACAD. SCI. PARIS, 260, 3248-3250.

ZIMMERMAN, WAYNE S. 1946.  (15.3)
    A SIMPLE GRAPHICAL METHOD FOR ORTHOGONAL ROTATION OF AXES.                   6086
    PSYCHOMETRIKA, 11, 51-55.  (MR7,P.338)

ZINGER, A.A. 1956.  (2.5/14.5/17.3)
    ON INDEPENDENCE OF POLYNOMIAL AND QUASI-POLYNOMIAL STATISTICS.               6087
        (IN RUSSIAN)
    DOKL. AKAD. NAUK SSSR, 110, 319-322.  (MR19,P.991)

ZINGER, A.A. 1965.  (2.2/2.5)
    ON THE DISTRIBUTION OF POLYNOMIAL STATISTICS IN SAMPLES FROM NORMAL          6088
    AND RELATED POPULATIONS.
        (IN RUSSIAN, TRANSLATED AS NO. 6089)
    TRUDY MAT. INST. STEKLOVA, 79, 150-159.  (MR34-6898)

ZINGER, A.A. 1966.  (2.2/2.5)                                                6089
     ON THE DISTRIBUTION OF POLYNOMIAL STATISTICS IN SAMPLES FROM NORMAL
     AND RELATED POPULATIONS.
          (TRANSLATION OF NO. 6088)
   PROC. STEKLOV INST. MATH., 79, 167-177.  (MR34-6898)

ZIRM, R. R.    SEE
     4122. PAGE, P. M.; BRODZINSKY. A. + ZIRM, R. R. (1953)

ZOLOTAREV, V. M. 1966.  (17.2/17.3)                                          6090
     A MULTI-DIMENSIONAL ANALOG OF THE BERRY-ESSEEN INEQUALITY FOR SETS WITH
     BOUNDED DIAMETER.
          (IN RUSSIAN, TRANSLATED AS NO. 6091)
   TEOR. VEROJATNOST. I PRIMENEN., 11, 507-514.  (MR34-844)

ZOLOTAREV, V. M. 1966.  (17.2/17.3)                                          6091
     A MULTI-DIMENSIONAL ANALOG OF THE BERRY-ESSEEN INEQUALITY FOR SETS WITH
     BOUNDED DIAMETER.
          (TRANSLATION OF NO. 6090)
   THEORY PROB. APPL., 11, 447-454.  (MR34-844)

ZUBIN, JOSEPH 1934.  (15.4)                                                  6092
     THE METHOD OF INTERNAL CONSISTENCY FOR SELECTING TEST ITEMS.
   J. EDUC. PSYCHOL., 25, 345-356.

ZUBIN, JOSEPH 1938.  (15.5/E)                                                6093
     A TECHNIQUE FOR MEASURING LIKE-MINDEDNESS.
   J. ABNORM. SOC. PSYCHOL., 33, 508-516.

ZUBIN, JOSEPH    SEE ALSO
     1822. FULCHER, JOHN S. + ZUBIN, JOSEPH (1942)

ZUBRZYCKA, L.    SEE
     3103. KOPOCINSKI. B. + ZUBRZYCKA. L. (1964)

ZUCKERMAN, HERBERT S.    SEE
     563. BIRNBAUM, Z. W.; RAYMOND, JOSEPH L. + ZUCKERMAN, HERBERT S. (1947)

## III.3   NUMBER OF PAPERS PER YEAR

| Year | No. | Year | No. | Year | No. |
|------|-----|------|-----|------|-----|
| 1811 | 1 | 1890 | 1 | 1930 | 44 |
| 1812 | 1 | 1892 | 3 | 1931 | 55 |
| 1813 | 1 | 1893 | 7 | 1932 | 67 |
|      |   | 1894 | 3 | 1933 | 72 |
| 1821 | 1 | 1895 | 2 | 1934 | 72 |
| 1823 | 1 | 1896 | 4 | 1935 | 95 |
| 1828 | 2 | 1897 | 8 | 1936 | 93 |
|      |   | 1898 | 4 | 1937 | 87 |
| 1832 | 1 | 1899 | 6 | 1938 | 76 |
|      |   |      |   | 1939 | 97 |
| 1846 | 1 | 1900 | 4 |      |   |
|      |   | 1901 | 4 | 1940 | 91 |
| 1852 | 1 | 1902 | 4 | 1941 | 108 |
| 1853 | 2 | 1903 | 3 | 1942 | 82 |
| 1858 | 3 | 1904 | 6 | 1943 | 63 |
|      |   | 1905 | 7 | 1944 | 76 |
| 1862 | 1 | 1906 | 10 | 1945 | 71 |
| 1866 | 1 | 1907 | 10 | 1946 | 86 |
| 1867 | 1 | 1908 | 10 | 1947 | 121 |
|      |   | 1909 | 17 | 1948 | 109 |
| 1871 | 1 |      |   | 1949 | 121 |
| 1872 | 2 | 1910 | 16 |      |   |
| 1874 | 1 | 1911 | 7 | 1950 | 135 |
| 1875 | 2 | 1912 | 9 | 1951 | 154 |
| 1876 | 1 | 1913 | 16 | 1952 | 118 |
| 1877 | 1 | 1914 | 22 | 1953 | 128 |
| 1879 | 2 | 1915 | 15 | 1954 | 163 |
|      |   | 1916 | 13 | 1955 | 186 |
| 1880 | 4 | 1917 | 16 | 1956 | 195 |
| 1881 | 2 | 1918 | 13 | 1957 | 173 |
| 1882 | 3 | 1919 | 17 | 1958 | 175 |
| 1883 | 2 |      |   | 1959 | 144 |
| 1884 | 2 | 1920 | 16 |      |   |
| 1885 | 1 | 1921 | 18 | 1960 | 191 |
| 1886 | 2 | 1922 | 18 | 1961 | 211 |
| 1887 | 3 | 1923 | 28 | 1962 | 236 |
| 1888 | 5 | 1924 | 22 | 1963 | 219 |
| 1889 | 2 | 1925 | 36 | 1964 | 255 |
|      |   | 1926 | 29 | 1965 | 372 |
|      |   | 1927 | 26 | 1966 | 446 |
|      |   | 1928 | 44 |      |   |
|      |   | 1929 | 51 | TOTAL | 5786 |

Numbers of papers are computed according to the numbers
of entries in §III.2 not including reprints, translations
and amendments.

## III.4 NUMBER OF PAPERS ACCORDING TO LANGUAGE

|  |  |
|---|---|
| Afrikaans | 1 |
| Chinese | 7 |
| Czech | 12 |
| Danish | 3 |
| Dutch | 20 |
| English | 4774 |
| French | 312 |
| German | 200 |
| Greek | 2 |
| Hebrew | 1 |
| Hungarian | 6 |
| Italian | 134 |
| Japanese | 61 |
| Latin | 4 |
| Polish | 8 |
| Portuguese | 5 |
| Roumanian | 8 |
| Russian | 166 |
| Serbo-Croatian | 5 |
| Spanish | 45 |
| Swedish | 4 |
| Ukrainian | 8 |
| TOTAL | 5786 |

Numbers of papers are computed according to the numbers of entries in §III.2 not including reprints, translations and amendments.

## III.5  NUMBERS OF AUTHORS AND NUMBERS OF PAPERS

| Number of authors who have written | Number of papers |
|:---:|:---:|
| 1889 | 1 |
| 435 | 2 |
| 189 | 3 |
| 102 | 4 |
| 67 | 5 |
| 38 | 6 |
| 50 | 7 |
| 33 | 8 |
| 14 | 9 |
| 16 | 10 |
| 6 | 11 |
| 8 | 12 |
| 7 | 13 |
| 7 | 14 |
| 11 | 15 |
| 2 | 16 |
| 5 | 17 |
| 4 | 18 |
| 1 | 19 |
| 4 | 20 |
| 4 | 21 |
| 2 | 22 |
| 2 | 23 |
| 2 | 25 |
| 3 | 26 |
| 1 | 27 |
| 1 | 29 |
| 1 | 30 |
| 2 | 31 |
| 2 | 33 |
| 1 | 35 |
| 1 | 37 |
| 2 | 38 |
| 1 | 39 |
| 1 | 49 |
| 1 | 61 |
| 1 | 86 |
| 2916 | |

Of the 2916 authors cited in §III.2, 1889 have each written one paper, 435 two, 189 three, ..., and one author (Karl Pearson) has written 86 papers. Of the 5786 papers listed in §III.2, 4816 are by a single author alone; 849 have two authors who are given alphabetically in 538 papers. Two papers have 5 authors, and two have 6; the authorship is not alphabetical in any of these four papers.

| Number of authors of a single paper | Number of papers | |
|:---:|:---:|:---:|
| | Total | Alphabetical authorship |
| 1 | 4816 | – |
| 2 | 849 | 538 |
| 3 | 104 | 51 |
| 4 | 13 | 4 |
| 5 | 2 | 0 |
| 6 | 2 | 0 |
| TOTAL | 5786 | |

Numbers of papers are computed according to the numbers of entries in §III.2 not including reprints, translations and amendments.

# III.6  LIFE SPANS AND BIBLIOGRAPHIES OF
## PROMINENT DECEASED AUTHORS

| | | | |
|---|---|---|---|
| AITKEN, A. C. | 1895–1967 | 22 | WHITTAKER & BARTLETT (1968) *below*. |
| EDGEWORTH, F. Y. | 1845–1926 | 15 | BOWLEY (1928) in §I.2.  No. 690 in §III.2. |
| FISHER, Ronald A. | 1890–1962 | 20 | Nos. 182, 358, 2415, 4450, 5344, 6041 in §III.2. |
| GINI, Corrado | 1884–1965 | 18 | Nos. 182, 879, 871 in §III.2. |
| GUMBEL, Emile J. | 1891–1966 | 15 | CAPPELLI EDITORE (1959) in §I.2. 12 pp. mimeographed list (Columbia University). |
| HOLZINGER, Karl J. | 1892–1954 | 38 | SWINEFORD (1967) *below*. |
| PEARSON, Karl | 1857–1936 | 86 | MORANT (1939) in §I.2. |
| ROMANOVSKII, Vsevolod I. | 1879–1954 | 15 | No. 4866 in §III.2. |
| ROY, S. N. | 1906–1964 | 49 | No. 186 in §III.2. |
| SPEARMAN, Charles | 1863–1945 | 31 | Nos. 2529, 5436 in §III.2. |
| THOMSON, Godfrey H. | 1881–1955 | 27 | No bibliography found. |
| THURSTONE, L. L. | 1887–1955 | 17 | Nos. 24, 2168, 5999 in §III.2. |
| VAN UVEN, M. J. | 1878–1960 | 20 | No. 3004 in §III.2. |
| WALD, Abraham | 1902–1950 | 33 | Nos. 184, 185, 5484 in §III.2. |
| WILKS, S. S. | 1906–1964 | 19 | No. 158 in §III.2.  ANDERSON (1967) *below*. |
| WISHART, John | 1898–1956 | 17 | No. 4166 in §III.2. |

The numbers in the third column are the numbers of papers computed according to the numbers of entries in §III.2 not including reprints, translations and amendments.

## REFERENCES

ANDERSON, T. W. *editor* (1967)  *S. S. Wilks: Collected Papers. Contributions to Mathematical Statistics.*  John Wiley & Sons, Inc., New York.  xxxii + 693 pp.

SWINEFORD, Frances (1967) Factor analysis by Karl J. Holzinger. *The School Review*, 75, 93–104.

WHITTAKER, J. M., & BARTLETT, M. S. (1968) Alexander Craig Aitken: 1895–1967. *Biographical Memoirs of the Fellows of the Royal Society*, 14, 1–14.

CHAPTER IV.    INDEX BY SUBJECT-MATTER CODE
               TO AUTHORS OF RESEARCH PAPERS


Listed under each subject-matter code *(SMC)*

are the authors of entries in Chapter III;

the number of entries by each author is

given.  Descriptions of the subject-matter

codes, as well as an index by topic, appear

in Chapter V.

*SMC = 1.1*

| | |
|---|---|
| ANDERSON, HARRY E., JR. | ... 1 |
| BARTLETT, MAURICE S. | ... 4 |
| BENNETT, B. M. | ... 1 |
| BLOOMERS, PAUL | ... 1 |
| BOCK, R. DARRELL | ... 2 |
| BOSTWICK, ARTHUR E. | ... 1 |
| BURT, CYRIL | ... 2 |
| CASTELLANO, VITTORIO | ... 1 |
| CHAMBERLIN, T. C. | ... 1 |
| CHAMBERS, E. G. | ... 1 |
| COCHRAN, WILLIAM G. | ... 3 |
| CRAIG, CECIL C. | ... 1 |
| CRAMER, ELLIOT M. | ... 1 |
| DALENIUS, TORE | ... 1 |
| DUNLAP, JACK W. | ... 1 |
| EDGEWORTH, F. Y. | ... 1 |
| ELFVING, GUSTAV | ... 1 |
| FERAUD, LUCIEN | ... 1 |
| FIELLER, E. C. | ... 2 |
| FINNEY, D. J. | ... 1 |
| FISHER, FRANKLIN M. | ... 1 |
| FUNKHOUSER, H. G. | ... 1 |
| GATTY, RONALD | ... 1 |
| GINI, CORRADO | ... 2 |
| HALDANE, J. B. S. | ... 1 |
| HALPERIN, MAX | ... 1 |
| HANNAN, E. J. | ... 1 |
| HARTLEY, H. O. | ... 1 |
| HIGUTI, ISAO | ... 1 |
| HOEL, PAUL G. | ... 1 |
| HORST, PAUL | ... 1 |
| HOTELLING, HAROLD | ... 3 |
| HUSSON, R. | ... 1 |
| IRWIN, J. O. | ... 5 |
| KABE, D. G. | ... 1 |
| KEMPTHORNE, OSCAR | ... 1 |
| KENNEY, J. F. | ... 1 |
| KSHIRSAGAR, A. M. | ... 1 |
| KUIPER, NICOLAAS H. | ... 1 |
| KULLBACK, SOLOMON | ... 1 |
| LA GARRIGUE, V. ROUQUET | ... 1 |
| LINDBLAD, TORD | ... 1 |
| MADOW, WILLIAM G. | ... 1 |
| MODER, JOSEPH J., JR. | ... 1 |
| MORGAN, JAMES N. | ... 1 |

| | |
|---|---|
| OZAWA, MASARU | ... 1 |
| POMPILJ, GIUSEPPE | ... 1 |
| RAO, C. RADHAKRISHNA | ... 3 |
| REGAN, MARY C. | ... 1 |
| REYMENT, R. A. | ... 1 |
| RIETZ, H. L. | ... 1 |
| ROY, S. N. | ... 1 |
| SIOTANI, MINORU | ... 1 |
| SLATER, LUCY J. | ... 1 |
| SONQUIST, JOHN A. | ... 1 |
| STEVENS, W. L. | ... 1 |
| TINTNER, GERHARD | ... 1 |
| TOLEDO TOLEDO, MANUEL | ... 1 |
| TUKEY, JOHN W. | ... 2 |
| VIDAL, ANDRE | ... 1 |
| WALKER, HELEN M. | ... 1 |
| WILKS, S. S. | ... 2 |
| WISHART, JOHN | ... 1 |
| WOLD, HERMAN O. A. | ... 3 |
| YATES, FRANK | ... 1 |

*SMC = 1.2*

| | |
|---|---|
| ADKINS, DOROTHY C. | ... 1 |
| AFIFI, A. A. | ... 1 |
| AGHEVLI, MAHMOUD | ... 1 |
| ALLEN, R. G. D. | ... 1 |
| ANDERSON, OSKAR | ... 1 |
| ANDERSON, T. W. | ... 3 |
| ANONYMOUS | ... 8 |
| ARROW, K. J. | ... 1 |
| BALAGANGADHARAN, K. | ... 1 |
| BARRETT, F. DERMOT | ... 1 |
| BARTLETT, MAURICE S. | ... 3 |
| BENINI, R. | ... 1 |
| BLACKWELL, DAVID H. | ... 1 |
| BOCK, R. DARRELL | ... 1 |
| BODIO, LUIGI | ... 3 |
| BOWKER, ALBERT H. | ... 1 |
| BOWLEY, A. L. | ... 1 |
| CASTELLANO, VITTORIO | ... 2 |
| CHESSIN, P. L. | ... 1 |
| COCHRAN, WILLIAM G. | ... 3 |
| CRAMER, ELLIOT M. | ... 1 |
| CRAMER, HARALD | ... 1 |
| DEMING, LOLA S. | ... 5 |
| DOLEZAL, E. | ... 1 |

SMC = 4.1 (cont.)

| | |
|---|---|
| GADDIS, L. WESLEY | ... 1 |
| GALTON, FRANCIS | ... 3 |
| GEARY, ROBERT C. | ... 2 |
| GEBELEIN, HANS | ... 1 |
| GEHLKE, C. E. | ... 1 |
| GEIRINGER, HILDA | ... 1 |
| GERSMAN, S. G. | ... 2 |
| GIBSON, WINIFRED | ... 1 |
| GILI, ADOLFO | ... 1 |
| GOLDIE, N. | ... 1 |
| GOTTSCHICK, J. | ... 1 |
| GRAVELIUS, H. | ... 1 |
| GREENWOOD, MAJOR | ... 2 |
| GRIFFIN, HAROLD D. | ... 1 |
| GULDBERG, ALF. | ... 2 |
| GUMBEL, EMILE J. | ... 3 |
| HALDANE, J. B. S. | ... 1 |
| HARRIS, J. ARTHUR | ... 8 |
| HEILMAN, J. D. | ... 1 |
| HEKKER, TH. | ... 1 |
| HINCIN, A. JA. | ... 1 |
| HOLWERDA, A. O. | ... 1 |
| HOLZINGER, KARL J. | ... 1 |
| HOPE, A. E. | ... 1 |
| HORST, PAUL | ... 1 |
| HOSTINSKY, BOHUSLAV | ... 1 |
| HOTELLING, HAROLD | ... 1 |
| HOWELLS, W. W. | ... 1 |
| HUIZINGA, J. | ... 1 |
| HULL, CLARK L. | ... 1 |
| HUNTINGTON, EDWARD V. | ... 1 |
| HURON, ROGER | ... 1 |
| IRWIN, J. O. | ... 1 |
| ISSERLIS, L. | ... 2 |
| JACKSON, DUNHAM | ... 2 |
| JAMES, A. N. | ... 1 |
| JASTREMSKII, B. S. | ... 1 |
| JOHNSON, HELMER G. | ... 1 |
| JONES, H. GERTRUDE | ... 1 |
| JONES, ROBERT A. | ... 1 |
| JORDAN, KAROLY | ... 1 |
| JULIN, A. | ... 2 |
| KAERSNA, ALFRED | ... 1 |
| KAISER, HENRY F. | ... 1 |
| KAPROCKI, STANISLAW | ... 1 |
| KAPTEYN, J. C. | ... 1 |

| | |
|---|---|
| KEEN, JOAN | ... 2 |
| KENDALL, MAURICE G. | ... 1 |
| KEYS, NOEL | ... 1 |
| KING, WILLFORD I. | ... 1 |
| KLAUBER, L. M. | ... 1 |
| KOGA, Y. | ... 1 |
| KOSSACK, CARL F. | ... 1 |
| KRAUSCH, F. | ... 1 |
| KRAVCHUK, M. | ... 1 |
| KRUEGER, F. | ... 1 |
| KUBOTA, ISAO | ... 1 |
| KUNREUTHER, HOWARD | ... 1 |
| LAMOTTE, EDITH M. M. DE G. | ... 1 |
| LASKA, V. | ... 1 |
| LASKI, H. J. | ... 1 |
| LATICHEVA, K. JA. | ... 1 |
| LEE, ALICE | ... 2 |
| LEVY, PAUL | ... 1 |
| LI, CHEN-NAN | ... 1 |
| LINCOLN, EDWARD A. | ... 1 |
| LINDEBERG, J. W. | ... 1 |
| LINEHAN, THOMAS P. | ... 1 |
| LOPES QUEIROZ, A. | ... 1 |
| LUDWIG, ROLF | ... 1 |
| MC INTYRE, FRANCIS | ... 1 |
| MAHALANOBIS, P. C. | ... 4 |
| MARAIS, HENRI | ... 1 |
| MARCANTONI, ALESSANDRO | ... 2 |
| MARCH, M. LUCIEN | ... 2 |
| MATSUMURA, SOJI | ... 1 |
| MEACHAM, ALAN D. | ... 1 |
| MICHAEL, WILLIAM B. | ... 1 |
| MITROPOLSKY, A. K. | ... 1 |
| MONROE, WALTER S. | ... 3 |
| MORAN, P. A. P. | ... 2 |
| MORANT, G. M. | ... 1 |
| MORTARA, GIORGIO | ... 1 |
| MUNZNER, HANS-FRIEDRICH | ... 1 |
| MUSHAM, M. V. | ... 1 |
| NEIFELD, M. R. | ... 1 |
| NEPRASH, JERRY A. | ... 1 |
| NYGAARD, P. H. | ... 1 |
| OKONENKO, A. | ... 1 |
| ONICESCU, OCTAV | ... 1 |
| O'TOOLE, A. L. | ... 1 |
| PAGE, D. J. | ... 1 |

SMC = 7.4 (cont.)

| | |
|---|---|
| GUTTMAN, IRWIN | ... 1 |
| HAGIS, PETER, JR. | ... 1 |
| HAYAKAWA, TAKESI | ... 1 |
| HOFFMAN, WILLIAM C. | ... 1 |
| HOGG, ROBERT V. | ... 1 |
| KHATRI, C. G. | ... 5 |
| KRISHNAIAH, PARUCHURI R. | ... 1 |
| KULLBACK, SOLOMON | ... 1 |
| MANTEL, NATHAN | ... 1 |
| MILLER, KENNETH S. | ... 4 |
| NAKAGAMI, MINORU | ... 1 |
| NARAYAN, RAM DEVA | ... 1 |
| OLKIN, INGRAM | ... 1 |
| PARK, JOHN H., JR. | ... 1 |
| STEINBERG, LEON | ... 1 |
| TIAO, GEORGE C. | ... 1 |
| TROSKIE, C. G. | ... 1 |

SMC = 8.1

| | |
|---|---|
| AITCHISON, JOHN | ... 1 |
| AITKEN, A. C. | ... 5 |
| ALBERT, ARTHUR | ... 1 |
| AMATO, VITTORIO | ... 1 |
| ANDERSON, OSKAR | ... 4 |
| ARNAIZ, GONZALO | ... 1 |
| BAETSLE, P.-L. | ... 1 |
| BAKST, AARON | ... 1 |
| BALESTRA, PIETRO | ... 1 |
| BANERJEE, DURGA PROSAD | ... 1 |
| BARGMANN, ROLF E. | ... 1 |
| BARTLETT, MAURICE S. | ... 1 |
| BARTON, D. E. | ... 1 |
| BEAN, LOUIS H. | ... 1 |
| BEAUCHAMP, JOHN J. | ... 1 |
| BECKMAN, F. S. | ... 1 |
| BENNETT, B. M. | ... 2 |
| BERGSTROM, A. R. | ... 1 |
| BERKSON, JOSEPH | ... 1 |
| BERNSTEIN, FELIX | ... 1 |
| BHATTACHARYA, P. K. | ... 2 |
| BJERHAMMAR, ARNE | ... 1 |
| BONFERRONI, CARLO EMILIO | ... 1 |
| BOSE, MRS. CHAMELI | ... 1 |
| BOX, GEORGE E. P. | ... 2 |
| BRIDGER, CLYDE A. | ... 1 |

| | |
|---|---|
| BROGDEN, HUBERT E. | ... 2 |
| BUCK, S. F. | ... 1 |
| BURKET, GEORGE R. | ... 1 |
| CAMP, BURTON H. | ... 1 |
| CARLSON, PHILLIP G. | ... 1 |
| CARPMAEL, C. | ... 1 |
| CARTER, A. H. | ... 1 |
| CASLEY, D. J. | ... 1 |
| CHARNLEY, F. | ... 1 |
| CHATTERJEE, SHOUTIR KISHORE | ... 1 |
| CHETTY, V. KARUPPAN | ... 2 |
| CHIPMAN, JOHN S. | ... 1 |
| CHOPRA, A. S. | ... 1 |
| CHOW, GREGORY C. | ... 1 |
| CLUNIES-ROSS, CHARLES W. | ... 1 |
| COHEN, JOZEF | ... 1 |
| CONGARD, ROGER | ... 1 |
| CORNELL, RICHARD G. | ... 1 |
| CORNISH, E. A. | ... 1 |
| CORSTEN, L. C. A. | ... 1 |
| COWDEN, DUDLEY J. | ... 2 |
| COX, D. R. | ... 1 |
| CURETON, EDWARD E. | ... 1 |
| DARMOIS, GEORGES | ... 1 |
| DE CAROLIS, LINDA V. | ... 1 |
| DEMING, W. EDWARDS | ... 1 |
| DENT, B. | ... 1 |
| DIEULEFAIT, CARLOS E. | ... 2 |
| DIVISIA, F. J. | ... 1 |
| DRAPER, NORMAN R. | ... 2 |
| DUNCAN, DAVID B. | ... 1 |
| DURBIN, JAMES | ... 1 |
| DWYER, PAUL S. | ... 4 |
| DYKSTRA, OTTO, JR. | ... 1 |
| EDDINGTON, ARTHUR S. | ... 1 |
| EDGEWORTH, F. Y. | ... 1 |
| EHRENBERG, A. S. C. | ... 1 |
| ELFVING, GUSTAV | ... 2 |
| ELSTON, R. C. | ... 1 |
| EZEKIEL, MORDECAI J. B. | ... 2 |
| FERON, ROBERT | ... 3 |
| FISHER, WALTER D. | ... 1 |
| FLEISS, JOSEPH L. | ... 1 |
| FOURGEAUD, CLAUDE | ... 1 |
| FRASER, A. R. | ... 1 |
| FREUND, RUDOLF J. | ... 1 |

SMC = 10.2 (cont.)

| | |
|---|---|
| LAWLEY, D. N. | ... 2 |
| LORD, FREDERIC M. | ... 1 |
| MATUSIRA, KAMEO | ... 1 |
| MAUCHLY, JOHN W. | ... 2 |
| MORGAN, W. A. | ... 1 |
| MAKHERJEE, BISHWA NATH | ... 1 |
| MURTHY, V. KRISHNA | ... 1 |
| NANDI, HARIKINKAR K. | ... 1 |
| NEYMAN, JERZY | ... 1 |
| PITMAN, E. J. G. | ... 1 |
| PLACKETT, R. L. | ... 2 |
| POTTHOFF, RICHARD F. | ... 1 |
| RAFFERTY, J. A. | ... 1 |
| RAO, V. V. NARAYAN | ... 1 |
| ROY, J. | ... 3 |
| ROY, S. N. | ... 1 |
| SCHAIE, K. WARNER | ... 1 |
| SCHEFFE, HENRY | ... 1 |
| SRIVASTAVA, J. N. | ... 1 |
| SRIVASTAVA, M. S. | ... 1 |
| STEINBERG, LEON | ... 1 |
| VILLARS, D. S. | ... 1 |
| VOTAW, DAVID F., JR. | ... 2 |
| WELCH, B. L. | ... 1 |

SMC = 11.1

| | |
|---|---|
| ABRAHAM, T. P. | ... 1 |
| BAILEY, D. W. | ... 1 |
| BRANNSTROM, B. | ... 1 |
| BURT, CYRIL | ... 1 |
| CASSIE, R. MORRISON | ... 1 |
| DARMOIS, GEORGES | ... 1 |
| DARROCH, J. N. | ... 1 |
| DEEMING, TERENCE J. | ... 1 |
| FERIBERGER, W. | ... 1 |
| FLOOD, MERRILL M. | ... 1 |
| GIRSHICK, M. A. | ....1 |
| GLAHN, HARRY R. | ... 1 |
| GOODMAN, LEO A. | ... 1 |

| | |
|---|---|
| GRENANDER, ULF | ... 1 |
| GUPTA, R. P. | ... 1 |
| GUTTMAN, LOUIS | ... 3 |
| HAITOVSKY, YOEL | ... 1 |
| HARMAN, HARRY H. | ... 1 |
| HOLZINGER, KARL J. | ... 1 |
| HOTELLING, HAROLD | ... 2 |
| JACKSON, J. EDWARD | ... 2 |
| JEFFERS, J. N. R. | ... 1 |
| JOLICOEUR, PIERRE | ... 3 |
| KELLOGG, CHESTER E. | ... 2 |
| KHOSLA, R. K. | ... 1 |
| KNOTT, VIRGINIA | ... 1 |
| KULLBACK, SOLOMON | ... 1 |
| MC CLOY, C. H. | ... 1 |
| MATSUDA, K. | ... 1 |
| METHENY, ELEANOR | ... 1 |
| MISRA, R. K. | ... 1 |
| MORRIS, R. H. | ... 1 |
| MOSIMANN, JAMES E. | ... 1 |
| ORLOCI, L. | ... 1 |
| PEARSON, KARL | ... 1 |
| PINEAU, H. | ... 1 |
| RAO, C. RADHAKRISHNA | ... 2 |
| REYMENT, R. A. | ... 2 |
| RICHARDSON, LEWIS F. | ... 1 |
| ROHLF, F. J. | ... 1 |
| ROUVIER, R. | ... 1 |
| SMITH, GEORGE MILTON, JR. | ... 1 |
| SPURRELL, D. J. | ... 1 |
| WHIPPLE, F. J. W. | ... 1 |
| WHITTLE, PETER | ... 1 |
| WILLIAMS, E. J. | ... 1 |

SMC = 11.2

| | |
|---|---|
| ABRAHAM, T. P. | ... 1 |
| ANDERSON, T. W. | ... 3 |
| BARTLETT, MAURICE S. | ... 2 |
| BLACKITH, R. E. | ... 1 |
| GUPTA, R. P. | ... 1 |

SMC = 15.4 (cont.)

| | |
|---|---|
| CLEMENS, WILLIAM V. | ... 1 |
| CONRAD, HERBERT S. | ... 1 |
| COTTON, JOHN W. | ... 1 |
| COX, GERTRUDE M. | ... 1 |
| CRONBACH, LEE J. | ... 6 |
| CURETON, EDWARD E. | ... 8 |
| DAS GUPTA, SOMESH | ... 1 |
| DAVIS, FREDERICK B. | ... 1 |
| DEEMER, WALTER L. | ... 1 |
| DENNEY, H. R. | ... 1 |
| DIEDERICH, GERTRUDE W. | ... 1 |
| DINGMAN, HARVEY F. | ... 1 |
| DRESSEL, PAUL L. | ... 1 |
| DU BOIS, PHILIP H. | ... 2 |
| DUNLAP, JACK W. | ... 6 |
| EAGLE, ALBERT R. | ... 1 |
| EDGERTON, HAROLD A. | ... 1 |
| ELFVING, GUSTAV | ... 6 |
| EMMETT, W. G. | ... 1 |
| EWART, EDWIN | ... 1 |
| FAN, CHUNG-TEH | ... 2 |
| FEDER, DANIEL D., LT. COMM. ( U.S.N.R.) | ... 1 |
| FERGUSON, GEORGE A. | ... 6 |
| FINCH, J. H. | ... 1 |
| FLANAGAN, JOHN C. | ... 2 |
| FORLANO, GEORGE | ... 1 |
| FRENCH, JOHN W. | ... 1 |
| FULCHER, JOHN S. | ... 1 |
| GAGE, N. L. | ... 1 |
| GAYLORD, RICHARD H. | ... 2 |
| GHISELLI, EDWIN E. | ... 2 |
| GLESER, GOLDINE C. | ... 1 |
| GREENE, JOEL E. | ... 1 |
| GROSSMAN, DAVID P., SGT. | ... 1 |
| GROSSNICKLE, LOUISE T. | ... 1 |
| GUILFORD, J. P. | ... 7 |
| GULLIKSEN, HAROLD | ... 2 |
| GUTTMAN, LOUIS | ... 3 |
| HANDY, UVAN | ... 1 |
| HIRSCHSTEIN, BERTHA | ... 1 |
| HOLZINGER, KARL J. | .. 20 |
| HORST, PAUL | ... 9 |
| HOUSE, J. MILTON | ... 1 |
| HOYT, CYRIL J. | ... 2 |
| HULL, CLARK L. | ... 1 |

| | |
|---|---|
| IKEDA, TOYOJI | ... 1 |
| JACKSON, ROBERT W. B. | ... 4 |
| JEEVES, T. A. | ... 1 |
| JOHNSON, A. P. | ... 1 |
| JOHNSON, H. M. | ... 2 |
| KAITZ, HYMAN B. | ... 2 |
| KARSLAKE, R. | ... 1 |
| KEATS, JOHN A. | ... 1 |
| KELLEY, TRUMAN L. | ... 4 |
| KHAN, F. | ... 1 |
| KRISTOF, WALTER | ... 1 |
| KROLL, ABRAHAM | ... 1 |
| KRUSKAL, J. B. | ... 1 |
| KUDER, G. F. | ... 2 |
| KUZNETS, GEORGE | ... 1 |
| LAWLEY, D. N. | ... 2 |
| LAWSHE, C. H., JR. | ... 1 |
| LAZARSFELD, PAUL F. | ... 3 |
| LENTZ, THEODORE F., JR. | ... 2 |
| LEV, JOSEPH | ... 1 |
| LOEVINGER, JANE | ... 1 |
| LOHNES, PAUL R. | ... 1 |
| LONG, JOHN A. | ... 1 |
| LONG, W. F. | ... 1 |
| LORD, FREDERIC M. | .. 21 |
| LUCE, R. DUNCAN | ... 1 |
| LYONS, THOBURN C. | ... 1 |
| MAC EWAN, CHARLOTTE | ... 2 |
| MC GEE, VICTOR E. | ... 1 |
| MC LAIN, RICHARD E. | ... 1 |
| MC QUITTY, J. V. | ... 1 |
| MAHMOUD, A. F. | ... 1 |
| MALONE, R. DANIEL | ... 1 |
| MARTINS, OCTAVIO A. LINS | ... 1 |
| MAY, MARK A. | ... 1 |
| MEREDITH, WILLIAM | ... 2 |
| MERRIL, WALTER W., JR. | ... 1 |
| MESSICK, SAMUEL J. | ... 1 |
| MILLER, WILBUR C. | ... 1 |
| MORRISON, ALEXANDER W. | ... 1 |
| MOSIER, CHARLES I. | ... 5 |
| PEEL, E. A. | ... 2 |
| PINTNER, RUDOLF | ... 1 |
| POTTHOFF, RICHARD F. | ... 1 |
| RAIFFA, HOWARD | ... 1 |
| RAND, GERTRUDE | ... 1 |
| RASCH, G. | ... 3 |

# CHAPTER V  THE SUBJECT-MATTER CODES

# V.1    INTRODUCTION

Each entry in Chapter III is annotated according to subject-matter. Each subject-matter code consists of two integers separated by a period. The first integer ranges from 1 to 20 identifying the following areas of multivariate statistical analysis:

1. Expository and Bibliography.
2. The Multivariate Normal Distribution.  General Properties.
3. Estimation of the Mean Vector and Covariance Matrix of a Multivariate Normal Distribution.
4. Correlation.
5. Tests of Hypotheses about One or Two Mean Vectors of Multivariate Normal Distributions and Hotelling's $T^2$.
6. Classification of Observations.
7. Distributions of Random Matrices and Related Quantities connected with the Multivariate Normal Distribution.
8. The Multivariate General Linear Model and Multivariate Regression.
9. Dependence of Sets of Variates.
10. Inference about Covariance Matrices and Inference Simultaneously about Mean Vectors and Covariance Matrices in Multivariate Normal Distributions.
11. Principal Components.
12. Canonical Correlations and Canonical Variates with Both Sets Random.
13. Distributions of Characteristic Roots and Vectors of Certain Matrices related to the Multivariate Normal Distribution.
14. Miscellaneous Topics related to the Multivariate Normal Distribution.
15. Factor Analysis.  Multivariate Statistical Methods in Psychology and Related Areas.
16. Multiple Time Series.
17. General Results specifically derived for Multivariate Statistical Analysis.
18. Specific Multivariate Statistical Analysis not based on the Multivariate Normal Distribution.
19. General Results which apply to but were not specifically derived for Multivariate Statistical Analysis.
20. Mathematical Methods applicable to Multivariate Statistical Analysis.

Sections 1 to 13 correspond approximately to the first thirteen chapters of the book by Anderson (1958).

Up to three subject-matter codes are given for each entry, followed at times by one or more letters which qualify the subject-matter as follows:

| | | | |
|---|---|---|---|
| *A* | Asymptotic | *N* | Numerical Analysis |
| *B* | Bivariate | *R* | Random Number Generation |
| *C* | Confidence Regions | *T* | Tables and Charts |
| *E* | Examples and Numerical Applications | *U* | Univariate. |

The letter *E* precedes the subject-matter code(s) when the main theme of the entry is an example or numerical application.

The subject-matter codes are listed in §V.2 with explanatory notes which were used in coding the entries. These notes, however, are not intended to be complete descriptions of the codes.

It is not easy to formulate a set of codes which describe perfectly the content of each of the entries listed. Moreover, the title of a paper is sometimes misleading and its content may not conform to the categorization adopted by us. We have, however, tried to assign the codes to describe the entries as accurately as possible. The original source material has been inspected whenever possible, though inevitably some entries had to be coded from the title, an abstract or a review. Reprints, translations and amendments appear as separate entries coded identically to the originals.

We have concentrated on papers which contain statistical inference for multivariate analysis or probabilistic aspects of multivariate distribution theory. In the following areas we have made a selection from the available papers:

Rank correlation and similar nonparametric measures of dependence *(4.8)*.

Univariate linear models *(8.8)*, univariate distributions *(14.5)* and other aspects of univariate statistical analysis (suffix *U*).

Applications of factor analysis and associated model building in psychology and related areas. (Section 15)

Multivariate stochastic processes. *(16.4)*

General results related to multivariate statistical analysis. (Section 17)

General theory of multiple decision procedures and statistical inference which applies to multivariate statistical analysis. (Section 19)

Mathematical methods applicable to multivariate statistical analysis. (Section 20)

Examples and numerical applications of existing theory. (Prefix or suffix *E*)

Our selection procedures were based in part on the relative importance and historical significance of the papers.

The table in §V.3 gives the numbers of papers (entries in Chapter III not including reprints, translations and amendments) and associated pages for each subject-matter code.  Of the 5786 such papers, 4177 have one subject-matter code, 1221 have two, and 388, three.  The table in §V.4 is a cross-tabulation of subject-matter code against year of publication; the subject-matter codes are pooled within Sections 2 up to 20.  The entries in this table are the average or total numbers of papers per year; the average number (to the nearest integer) is given for 1811-1919, 1920-9, 1930-9, 1940-9, 1950-9 and the total number for 1960, 1961,..., 1966.

Chapter IV is an index by subject-matter code to the authors of all the entries in Chapter III; §V.5 is an index by topic to the subject-matter codes.

## V.2   AN ANNOTATED LIST OF THE SUBJECT-MATTER CODES

1.  **EXPOSITORY AND BIBLIOGRAPHY.**

    1.1 *Expository and General.*
        Surveys of more than one topic in multivariate analysis.

    1.2 *Bibliography.*
        Bibliographies on statistical literature which include some
        multivariate entries.  Obituaries which contain bibliography.

2.  **THE MULTIVARIATE NORMAL DISTRIBUTION.   GENERAL PROPERTIES.**

    2.1 *Descriptive Properties.*
        The spherical normal distribution, distributions of linear
        combinations of normal variables, conditional distributions.
        Random normal deviate tables and the methods to generate them
        (see also *17.9*).

    2.2 *Characterization and Other Derivations.*
        Independence of the sample mean vector and sample covariance
        matrix as characterization of multivariate normality.  Uses of
        polynomial forms to characterize the multivariate normal
        distribution.  For other studies of polynomial forms see *2.5*.

    2.3 *Characteristic Function, Moments, Cumulants (Semi-invariants).*
        Moments and cumulants as population quantities.

    2.4 *Truncated and Censored Distributions.  Mixtures.*
        Effects of truncation on some variables.  Convex mixtures
        of multivariate normal densities.  Pearson-type selection.

    2.5 *Quadratic and Bilinear Forms.  Polynomial Forms.*
        Distributions of $\underset{\sim}{x}'A\underset{\sim}{x}$ and of $\underset{\sim}{x}'B\underset{\sim}{y}$ where $\underset{\sim}{x}$ and $\underset{\sim}{y}$ are
        multivariate normal.  Theorems of Cochran, Craig and Sakamoto
        and related developments.  For polynomial forms used to
        characterize the multivariate normal distribution see *2.2*.

    2.6 *Distribution of Functions of Quadratic Forms and Related Topics.*
        Radial error, circular probable error.  Ratios of quadratic
        forms, serial correlation.  Coverage and hit problems based on
        radial error (see also *14.4, 14.5*).

    2.7 *Methods for Numerical Evaluation of the Cumulative Distribution.*
        Computational procedures and computer programs.  Methods for
        evaluating integrals of the multivariate normal (cf. Gupta,
        Shanti S. (1963), no. 2212).

3.  ESTIMATION OF THE MEAN VECTOR AND COVARIANCE MATRIX OF A MULTIVARIATE
        NORMAL DISTRIBUTION.

    3.1  *Mean Vector and Covariance Matrix Estimation without Side
          Conditions.*
          Both separate and simultaneous estimation of the mean vector
    and of the covariance matrix.  Admissibility and minimax
    properties of the estimators (cf. Stein, Charles M. (1956),
    no. 5215).  For special structures of covariance matrices see
    *3.2;* for hypotheses about the mean vector when the covariance
    matrix is known, or unknown, see Section 5 (case of single
    degree of freedom) and Section 8 (more than one degree of
    freedom).

    3.2  *Mean Vector and Covariance Matrix Estimation with Side
          Condititions.*
          Estimation when special structure is assumed (e.g., mean
    vector with all components equal, covariance matrix with all
    diagonal elements equal and all off-diagonal elements equal).
    Missing observations and censored distributions.  For esti-
    mation of correlations see *4.3;* for tests concerning mean
    vectors see Section 5 and covariance matrices, Section 10.

    3.3  *Estimation of Other Quantities, such as Cumulants.*
          Estimation of functions of means and covariances, e.g., ratio
    of two means.  For sampling distributions see *14.2.*


4.  CORRELATION.

    A parent multivariate normal distribution is not necessarily
    assumed.  For sampling distributions and inference when a specific
    parametric nonnormal multivariate distribution is assumed see *18.2,
    18.3.*

    4.1  *Definition, Calculation and Description of Product-Moment
          Correlation.*
          Geometric interpretations.  Correlation between $x/y$ and $z/y$.
    Applications leading to new theory, new calculation algorithms,
    or further usefulness of product-moment correlation.

    4.2  *Distribution of Product-Moment Correlation.*
          Fisher's $z$-transformation.  For joint distributions of all
    the elements of a correlation matrix see *7.1.*

    4.3  *Inference about Product-Moment Correlation.*
          Tests of hypotheses and estimation.  For estimation along
    with other parameters under normality see *3.1.*

4.    CORRELATION.    (cont.)

   *4.4   Tetrachoric, Biserial, Point-Biserial Correlation Coefficients
          and Similar Measures of Correlation.*

   *4.5   Definition, Calculation and Description of Partial Correlation.
          Distribution of and Inference about Partial Correlation.*
          Correlation in a conditional multivariate distribution.  Tests
          of hypotheses and estimation in a conditional multivariate
          distribution.

   *4.6   Definition, Calculation and Description of Multiple Correlation.*
          Maximum correlation between one random variable and a linear
          combination of other random variables.  Selected results when
          the other variables are held fixed as in univariate multiple
          regression.  See also *4.7*.

   *4.7   Distribution of and Inference about Multiple Correlation.*
          Tests of hypotheses and estimation.  See also *4.6*.

   *4.8   Other Specific Measures of Association and Dependence.*
          Selected results on rank correlation, parabolic correlation,
          transvariation, correlation ratio, measures of association in
          contingency tables, filial correlation, intraclass correlation,
          genetic correlation, path analysis, "concordanza" (cf. Gini,
          Corrado (1916), no. 1977).  For serial correlation see *2.6* and
          for tests of independence see *9.1, 9.2*.

   *4.9   General Theory of Measures of Association and Dependence.*
          Different uses of measures of association.  For measures of
          association between vectors see *9.3*.

5.    TESTS OF HYPOTHESES ABOUT ONE OR TWO MEAN VECTORS OF MULTIVARIATE
      NORMAL DISTRIBUTIONS AND HOTELLING'S $T^2$.

   *5.1   Tests of Hypotheses about One Mean Vector.*
          Cases of known and unknown covariance matrix with or without
          side conditions.  Properties of test procedures.  Stein's two-
          sample procedure in the multivariate case.  Various kinds of
          hypotheses on the mean vector and its components.  For cases
          with more than a single linear combination see Section 8.

5.  TESTS OF HYPOTHESES ABOUT ONE OR TWO MEAN VECTORS OF MULTIVARIATE
       NORMAL DISTRIBUTIONS AND HOTELLING'S $T^2$.  (cont.)

   5.2  *Tests of Hypotheses of Equality of the Mean Vectors of Two*
        *Populations.*
           Behrens-Fisher problem in the multivariate case.  Distribu-
        tions of relevant statistics involved in testing equality of
        the mean vectors of two multivariate normal populations; for $T^2$
        see *5.3*.  For cases with more than a single linear combination
        see Section 8.  See also *6.2*.

   5.3  *Distribution and Properties of Hotelling's $T^2$.*
           Distribution of Studentized $D^2$.  See also *5.1, 5.2, 8.4, 14.2*.


6.  CLASSIFICATION OF OBSERVATIONS.

   6.1  *General Theory of Classification for Both Multivariate Normal*
        *and Other Multivariate Distributions.*

   6.2  *Linear Classification Functions for Two Multivariate Normal*
        *Distributions with Common Covariance Matrix.*
           Distribution, uses and properties of linear classification
        statistics.  Inference on the coefficients of the linear
        classification functions of Fisher, Wald and Anderson.
        "Discriminant analysis" treating classification problems; for
        hypothesis-testing problems see *5.1, 5.2*.

   6.3  *Other Classification Procedures for Multivariate Normal*
        *Distributions.*
           Two multivariate normal populations with different covariance
        matrices; more than two multivariate normal populations.
        Growth-invariant discriminant functions.

   6.4  *Classification Procedures for assigning an Observation to One*
        *of a Finite Number of Multivariate Distributions other than*
        *Normal.*

   6.5  *Measures of Distance between Populations.*
           Measures such as those of Pearson and of Mahalanobis.  Co-
        efficient of racial likeness.  Distributions of measures of
        distance; Mahalanobis distance.  A parent multivariate normal
        distribution is not necessarily assumed.  For distribution of
        Hotelling's $T^2$ and of Studentized $D^2$ see *5.3*.

   6.6  *Clustering and Group Constellations.*
           "Classification" as used by biologists.  Clustering in the
        sense of grouping populations or individuals based on
        internally-defined criteria.  Profile similarity; pattern
        recognition.

6.   CLASSIFICATION OF OBSERVATIONS.   (cont.)

6.7   *Ranking and Selection.   Allocation.*
         Ranking and selection of populations or of individuals.
     Personnel selection.


7.   DISTRIBUTIONS OF RANDOM MATRICES AND RELATED QUANTITIES CONNECTED
        WITH THE MULTIVARIATE NORMAL DISTRIBUTION.

7.1   *The Central and Non-Central Wishart Distribution.*
         Distributions of matrices of variances and covariances,
     and of matrices of correlation coefficients.   For distribu-
     tions of a single correlation coefficient see *4.2.*

7.2   *Scatter and Generalized Variance.*
         Descriptions, distributions.

7.3   *Matrices which are Functions of Wishart Matrices.*
         Distribution of $F'AF$, where $A$ and $B = (FF')^{-1}$ are both
     Wishart matrices, distribution of $T$, where $A = TT'$ ($T$ not
     necessarily triangular); of $A^{-1}$ ("inverted" Wishart), and of
     minors of $A$.   For distribution of a single principal minor see
     *7.2.*

7.4   *Other Topics related to the Wishart Distribution.*
         Ratios of sample standard deviations and of sample variances.
     Studies of $X'AX$ where $X' = (x_1, \ldots, x_N)$ is the $p \times N$
     matrix of observations $(p > 1)$.   Rayleigh distribution (joint
     distribution of sample variances).   Matrix analogue of the
     Dirichlet distribution (see also *18.1*).


8.   THE MULTIVARIATE GENERAL LINEAR MODEL AND MULTIVARIATE REGRESSION.

The linear model with more than one dependent variable: $E(X) = A\Gamma B$,
where $A$ and $B$ are known matrices and $\Gamma$ is unknown ($B$ is often the
identity matrix), and $X' = (x_1, \ldots, x_N)$ is the $p \times N$ matrix of N
independent observations with common unknown $p \times p$ covariance
matrix $\Sigma$ $(p > 1)$.   A parent multivariate normal distribution is
assumed in *8.3, 8.4.*

8.1   *General Linear Model:   Estimation of Parameters.   Definition
      and Properties of the Sample Regression Coefficients.*
         Unbiasedness, admissibility and minimax properties of
     estimators; Bayes estimates.   Selected applications.   For
     selected results with $p = 1$ or $N = 1$ see *8.8.*

8.  THE MULTIVARIATE GENERAL LINEAR MODEL AND MULTIVARIATE REGRESSION.
        (cont.)

8.2   *General Linear Model: Distribution of the Sample Regression Coefficients.*
        Large sample approximations *(8.2/A)*.

8.3   *General Linear Model: Test Procedures and their Properties.*
        Invariant tests such as those based on Wilks' $\Lambda$, Lawley-Hotelling trace criterion, Roy maximum-root statistic, and Pillai trace criterion. Admissibility and power monotonicity properties. Associated confidence regions *(8.3/C)*. Tests of hypotheses concerning the matrix $\Gamma$ of regression coefficients. For the single degree of freedom linear hypothesis see Section 5. For specializations of $\underset{\sim}{A}$ to experimental designs see *8.5*.

8.4   *General Linear Model: Distributions of the Test Criteria and their Large Sample Approximations.*
        Exact and approximate distributions; asymptotic distributions *(8.4/A)*. Evaluation of power functions. Maximum of several $T^2$ (see also *14.2*).

8.5   *Multivariate Analysis of Variance: Fixed Effects (Model I).*
        Specialized results from *8.1 - 8.4* for experimental designs. Equality of *k* mean vectors. Multivariate analysis of covariance.

8.6   *Multivariate Analysis of Variance: Random Effects (Model II). Mixed Models.*
        Multivariate components of variance.

8.7   *Testing Rank of the Unknown Parameter Matrix, Linear Restrictions on the Mean Vectors or Regression Coefficients. Canonical Correlations and Canonical Variates with One Set Fixed.*
        See also Section 12.

8.8   *General Linear Model: Univariate Case.*
        Selected results for the case where $\underset{\sim}{X}$ is a vector ($p = 1$ or $N = 1$); classical least squares, estimation and tests of hypotheses, univariate analysis of variance.

8.9   *General Linear Model: Other Topics.*
        Prediction by regression methods, residual analysis (including outliers, if appropriate). Errors in forecasting; for time series analysis see *16.1, 16.5*.

9. DEPENDENCE OF SETS OF VARIATES.

> 9.1 *Criteria for testing Independence in Multivariate Normal Distributions: Description and Properties.*
> Likelihood-ratio tests.  Step-down procedures.

> 9.2 *Criteria for testing Independence in Multivariate Normal Distributions:  Distributions of Criteria.*
> Large sample chi-square approximations (9.2/A).

> 9.3 *Measures of Dependence between Vectors.*
> Descriptions and properties.  General distributions including the multivariate normal.

10. INFERENCE ABOUT COVARIANCE MATRICES AND INFERENCE SIMULTANEOUSLY ABOUT MEAN VECTORS AND COVARIANCE MATRICES IN MULTIVARIATE NORMAL DISTRIBUTIONS.

> 10.1 *Hypothesis of Equality of Covariance Matrices.  Hypothesis of Equality of Mean Vectors and Covariance Matrices.*
> The hypotheses are:  $\underset{\sim}{\Sigma}_1 = \ldots = \underset{\sim}{\Sigma}_q$ ;  $\underset{\sim}{\mu}_1 = \ldots = \underset{\sim}{\mu}_q$ and $\underset{\sim}{\Sigma}_1 = \ldots = \underset{\sim}{\Sigma}_q$ .  Distributions of test criteria.  Confidence regions for the characteristic roots of $\Sigma_1\Sigma_2^{-1}$ (10.1/C).  Selected univariate cases (10.1/U).  For testing mean vectors only see Section 5 (q = 2) and Section 8 (q > 2).

> 10.2 *Hypotheses of Special Forms of Covariance Matrices in One Population.  Hypotheses of Mean Vectors and Covariance Matrices other than Equality.*
> Sphericity, compound symmetry and proportionality of covariance matrices (inference about $\underset{\sim}{\Sigma}$ in each case).  Tests of $\underset{\sim}{\mu} = \underset{\sim}{\mu}_0$, $\underset{\sim}{\Sigma} = \underset{\sim}{\Sigma}_0$; and $\underset{\sim}{\Sigma} = \underset{\sim}{\Sigma}_0$ ($\underset{\sim}{\mu}$ unspecified).  $L_{mvc}$, $L_{vc}$, and $L_c$ tests.  Distributions of test criteria.  Confidence regions for the characteristic roots of $\underset{\sim}{\Sigma}$ (10.2/C).  See also Section 3.

11. PRINCIPAL COMPONENTS.

> 11.1 *Definitions and Descriptions.*
> Algebra of the procedure for both population and sample.  For the factor analysis model $\underset{\sim}{X} = \underset{\sim}{\Lambda}\underset{\sim}{Z} + \underset{\sim}{U}$ see Section 15.

> 11.2 *Inference.*
> For distributions of sample quantities see Section 13.

12.  CANONICAL CORRELATIONS AND CANONICAL VARIATES WITH BOTH SETS RANDOM.

12.1  *Definitions and Properties.*
Algebra of the procedure for both population and sample.
Description of the pattern of dependence. For one set of
variates fixed see *8.7.*

12.2  *Estimation.*
For distributions of sample quantities see Section 13.

13.  DISTRIBUTIONS OF CHARACTERISTIC ROOTS AND VECTORS OF CERTAIN
MATRICES RELATED TO THE MULTIVARIATE NORMAL DISTRIBUTION.

13.1  *Distributions that do not depend on the Parameters.*
Case of one Wishart matrix; case of two Wishart matrices
with same parameter covariance matrix.

13.2  *Distributions that depend on the Parameters.*
Case of two Wishart matrices with different parameter
covariance matrices and/or with nonzero noncentrality parameters.

14.  MISCELLANEOUS TOPICS RELATED TO THE MULTIVARIATE NORMAL DISTRIBUTION.

14.1  *Distributions of Order Statistics and Extreme Values from the*
*Multivariate Normal.*
For extreme values of general multivariate distributions see
*17.6;* for extreme values of specific nonnormal distributions
see Section 18.

14.2  *Sampling Distributions not included elsewhere.*
Distribution of ratio of two means. Fiducial and posterior
sampling distributions. Maximum of several $T^2$ (see also *8.4*).

14.3  *Tests of Hypotheses and Confidence Regions not included else-*
*where.*
Tests and confidence regions based on fiducial and posterior
distributions. For the estimation problem *per se* see Section 3.

14.4  *All Other Aspects and Discussions related to the Multivariate*
*Normal.*
Multivariate tolerance regions. Linear equations with coefficients distributed normally. Probit analysis. Information
theory. Coverage and hit problems (not included in *2.6, 14.5*).
Probability inequalities (without normality assumption see
*17.2*). Characterizations (see also *2.2, 7.1*). See also *17.9.*

14.  MISCELLANEOUS TOPICS RELATED TO THE MULTIVARIATE NORMAL DISTRIBUTION.
        (cont.)

   14.5  *Some Results on Standard Univariate Distributions useful for*
         *Multivariate Analysis.*
            Chi-square, *F-*, *t-*, *z*-distributions. Mixtures, approximations.
         Use of Laguerre polynomials (see also *20.5*). Coverage and hit
         problems involving mixtures of chi-square distributions (see
         also *2.6*, *14.4*).


15.  FACTOR ANALYSIS.  MULTIVARIATE STATISTICAL METHODS IN PSYCHOLOGY
        AND RELATED AREAS.

   The factor analysis model $X = \Lambda Z + U$.  Theoretical results,
   especially those dealing with different models.  Selected results
   on related topics, few which deal principally with applications.
   For principal components see Section 11.

   15.1  *Factor Analysis:  Descriptions and Models.*
            Algebraic expositions.  Some criticisms and verbal
         descriptions.

   15.2  *Factor Analysis:  Statistical Inference, Statistical Procedures*
         *and Distributions of Test Criteria.*
            Tetrad differences.  A parent multivariate normal distribu-
         tion is assumed in most cases.

   15.3  *Factor Analysis:  Computational Methods and Computer Programs.*
            Computational algorithms and procedures.

   15.4  *Models of Psychological Tests:  Item Analysis, Attenuation*
         *and Split-Half Tests.*
            Parallel forms.  Reliability.  Selected results from the
         psychological literature.

   15.5  *Other Psychometric and Sociometric Procedures.*
            Theoretical and other selected results.  Latent structure
         analysis.  Multidimensional scaling, psychological scaling;
         dimensional analysis (cf. Luce, R. Duncan (1963), no. 3526).
         Method of quantification (cf. Hayashi, Chikio (1961), no. 2404).

16.  MULTIPLE TIME SERIES.

Time series with observations on a vector random variable.

16.1  *Autoregressive Processes.*
      Autoregressive schemes for vector processes and moving
averages.  Estimation, hypothesis testing and prediction for
autoregressive models.  Stochastic difference equations without structural aspects (for structural aspects see *17.7*).
The model:  $\underset{\sim}{X}_t = \underset{\sim\sim}{B}\underset{\sim}{X}_{t-1} + \underset{\sim}{U}_t$ (see also *16.2*).

16.2  *Simultaneous Stochastic Equations used in Econometrics.*
      Maximum-likelihood limited-information estimation.  Selected
studies in economics.  The model:  $\underset{\sim\sim}{B}\underset{\sim}{X}_t = \underset{\sim\sim}{\Gamma}\underset{\sim}{X}_{t-1} + \underset{\sim}{U}_t$ (see also
*16.1*).

16.3  *Spectral Analysis; Statistical Inference.*
      Estimation of multivariate spectral densities; distributions
of the estimators.  Spectral density estimation related to
filtering and periodogram analysis (see also *16.5*).  For
spectral analysis of a process see *16.4*.

16.4  *Multivariate Stochastic Processes.*
      Selected results relevant to statistical inference; restricted
to stochastic processes where the range space and the index set
are finite-dimensional Euclidean spaces.  Prediction with the
process known; representations.  Characterizations of processes
through multivariate distributions.  Gaussian processes.
Ergodic theorems.  Stochastic dynamic programming.  Spectral
analysis of a process.  Stochastic control theory.

16.5  *Other Multiple Time Series Analysis.*
      Statistical inference for multiple time series not covered
by *16.1* - *16.3*.  Multivariate trend analysis including linear
hypotheses and estimation with a stochastic process residual.
Filtering and periodogram analysis (see also *16.3*).  Prediction
in multiple time series not based on autoregressive models.


17.  GENERAL RESULTS SPECIFICALLY DERIVED FOR MULTIVARIATE STATISTICAL
        ANALYSIS.

Selected results not specifically derived for the multivariate
normal distribution but which are definitely multivariate in nature
or have some direct bearing on multivariate statistical analysis.
For specific nonnormal multivariate distributions see Section 18;
for more general results which are not specifically multivariate in
nature see Section 19.

17.  GENERAL RESULTS SPECIFICALLY DERIVED FOR MULTIVARIATE STATISTICAL
        ANALYSIS.  (cont.)

17.1  *General Multivariate Limit Theorems.*
       Results with finite dimensional vectors.

17.2  *Probability Inequalities in Multivariate Cases.*
       Čebyčev-type inequalities.  Bounds for multiple integrals.
       For multivariate normal distributions see *14.4.*

17.3  *General Results in Multivariate Distribution Theory.*
       Properties, characteristic functions, moments and cumulants
       (semi-invariants), conditional distributions, and character-
       ization of general multivariate distributions.  For the
       multivariate normal distribution see Sections 2 and 14;
       for specific nonnormal multivariate distributions see Section
       18.

17.4  *Joint Distributions with Given Marginals or Given Conditional
       Distributions.*
       No specific distribution is assumed; for the multivariate
       normal see Section 2 and for specific nonnormal multivariate
       distributions see Section 18.

17.5  *General Results for Inference and Sampling Distributions in
       Multivariate Analysis.*
       Tests for normality.  Ratio estimates.  Hypothesis testing,
       estimation, confidence regions *(17.5/C)* for general
       multivariate distributions.  For the multivariate normal see
       *14.2, 14.3;* for specific nonnormal multivariate distributions
       see *18.2, 18.3.*

17.6  *Order Statistics and Extreme Value Analysis for General
       Multivariate Distributions.*
       Order statistics for general multivariate distributions.
       For the multivariate normal see *14.1;* for specific nonnormal
       multivariate distributions see Section 18.

17.7  *Structural Relations.  Confluence Analysis.*
       Estimation, identification and description.  Estimation of
       $\gamma$ such that $\gamma'\mu_i = 0$, where $x_i$ has mean vector $\mu_i$ and covariance
       matrix $\Sigma$.  Regression and estimation of the measure of
       dependence when all variables are subject to error.  Confluence
       analysis through correlations.  Wright's path coefficients.

17.  GENERAL RESULTS SPECIFICALLY DERIVED FOR MULTIVARIATE STATISTICAL
      ANALYSIS.   (cont.)

   17.8  *Nonparametric Multivariate Statistical Analysis.  Inference
         and Procedures.  Distributions of Test Criteria.*
            Estimation of density functions and tolerance regions.  For
      Wilk-Gnanadesikan and similar extensions of Daniel's half-
      normal plot see *17.9;* for rank and similar measures of corre-
      lation see *4.8.*

   17.9  *All Other Aspects of Multivariate Statistical Analysis which
         are General and specifically derived for the Multivariate Case.*
            Equations with random terms, data analysis, stochastic
      approximation.  Heuristic and graphical multivariate analysis
      (cf. Wilk, Martin B., and Gnanadesikan, R. (1964), no. 5905).
      Tables of generated random samples (see also *2.1).*  Information
      theory.  See also *14.4.*

18.  SPECIFIC MULTIVARIATE STATISTICAL ANALYSIS NOT BASED ON THE
      MULTIVARIATE NORMAL DISTRIBUTION.

   18.1  *Descriptions and Properties.  Characterizations.*
            Multivariate *t*-distribution, multivariate Poisson.  Dirichlet
      distribution (for its matrix analogue see *7.4).*  Selected
      results for the multinomial distribution.

   18.2  *Statistical Inference.*

   18.3  *Sampling Distributions of Statistics.*
            Distributions of test criteria, moments, cumulants.  For
      correlation, regardless of parent distribution, see Section 4.

   18.4  *Other Aspects of Specific Nonnormal Parametric Multivariate
         Distributions and Statistical Analysis.*

   18.5  *Multivariate Theory for Finite Populations.*

   18.6  *Geometrical Probability.*
            Measures of random geometrical figures and allied problems.

19.  GENERAL RESULTS WHICH APPLY TO BUT WERE NOT SPECIFICALLY DERIVED
     FOR MULTIVARIATE STATISTICAL ANALYSIS.

     Selected results which have direct relevance to multivariate
     statistical analysis.

     *19.1  Multiple Decision Procedures.*
          General decision theory and multiple decision theory; some
          illustrations.  For classification problems see Section 6.

     *19.2  Testing of Statistical Hypotheses.*

     *19.3  Estimation, Confidence Regions and Tolerance Regions.*

     *19.4  Sufficient Statistics.  Invariance.  Completeness.*

20.  MATHEMATICAL METHODS APPLICABLE TO MULTIVARIATE STATISTICAL
     ANALYSIS.

     Selected results which have direct relevance to multivariate
     statistical analysis.

     *20.1  Theory of Linear Algebra applicable to Multivariate Statistical
          Analysis.  General Results.*
          Transformations of linear systems; Jacobians.  Diagonalization
     of matrices; singular value and characteristic root factor-
     izations.  Bounds for singular values and characteristic roots.
     Inequalities and equalities for trace and for rank; approx-
     imations by matrices of lower rank.  General theory of deter-
     minants and generalized inverses.  Positive definite, positive
     semi-definite and nonnegative definite matrices (for corre-
     lation and covariance matrices see 20.2).  Determinants,
     inverses and generalized inverses of partitioned matrices (for
     patterned matrices see 20.2).  Quadratic and bilinear forms
     (for related statistical distribution theory see 2.5, 2.6).
     For special cases see 20.2; for computational procedures see
     20.3.

     *20.2  Theory of Linear Algebra applicable to Multivariate Statistical
          Analysis.  Special Cases.*
          Bordered, patterned, tri-diagonal matrices; Vandermonde and
     other special determinants.  Generalized inverses of matrices
     with special structure.  Idempotent matrices (see also 2.5).
     Correlation and covariance matrices (see also 7.1, 7.4, 20.1).
     For general results see 20.1; for computational procedures
     see 20.3.

20. MATHEMATICAL METHODS APPLICABLE TO MULTIVARIATE STATISTICAL ANALYSIS. (cont.)

20.3 *Numerical Analysis. Computational Procedures and Computer Programs for Linear Algebra applicable to Multivariate Statistical Analysis.*

Algorithms for inversion of matrices and solution of systems of simultaneous equations. Computational techniques for evaluation of determinants, characteristic roots and vectors, singular values.

20.4 *Integrals and Special Functions.*

Bessel and hypergeometric-type functions.

20.5 *Other Mathematics relevant to Multivariate Statistical Analysis.*

Convex sets, multidimensional geometry. Multivariate interpolation specific to multivariate statistical analysis. General theory of integration and differential equations; Haar integrals. Methods for solving general mathematical equations. Methods for fitting a surface to a set of points in a finite dimensional geometry. Laguerre polynomials (see also *14.5*).

| | | | | | | |
|------|-----|------|---|------|-----|------|
| 1.1  | 74  | 1593 | | 10.1 | 48  | 598  |
| 1.2  | 121 | 2835 | | 10.2 | 55  | 636  |
| 2.1  | 126 | 1948 | | 11.1 | 46  | 861  |
| 2.2  | 35  | 288  | | 11.2 | 25  | 427  |
| 2.3  | 31  | 447  | | 12.1 | 41  | 667  |
| 2.4  | 51  | 518  | | 12.2 | 17  | 269  |
| 2.5  | 90  | 751  | | 13.1 | 46  | 557  |
| 2.6  | 25  | 270  | | 13.2 | 27  | 411  |
| 2.7  | 61  | 734  | | 14.1 | 35  | 453  |
| 3.1  | 55  | 634  | | 14.2 | 61  | 754  |
| 3.2  | 35  | 389  | | 14.3 | 22  | 308  |
| 3.3  | 9   | 106  | | 14.4 | 62  | 803  |
| 4.1  | 316 | 4651 | | 14.5 | 85  | 823  |
| 4.2  | 71  | 889  | | 15.1 | 383 | 5555 |
| 4.3  | 57  | 648  | | 15.2 | 162 | 2253 |
| 4.4  | 69  | 711  | | 15.3 | 76  | 876  |
| 4.5  | 93  | 1195 | | 15.4 | 254 | 3148 |
| 4.6  | 125 | 1759 | | 15.5 | 154 | 2537 |
| 4.7  | 40  | 424  | | 16.1 | 47  | 813  |
| 4.8  | 354 | 5657 | | 16.2 | 90  | 1991 |
| 4.9  | 109 | 1852 | | 16.3 | 49  | 809  |
| 5.1  | 36  | 411  | | 16.4 | 242 | 3657 |
| 5.2  | 71  | 781  | | 16.5 | 133 | 2116 |
| 5.3  | 33  | 362  | | 17.1 | 157 | 1901 |
| 6.1  | 58  | 847  | | 17.2 | 47  | 507  |
| 6.2  | 86  | 1082 | | 17.3 | 212 | 2903 |
| 6.3  | 52  | 847  | | 17.4 | 36  | 532  |
| 6.4  | 58  | 822  | | 17.5 | 44  | 715  |
| 6.5  | 77  | 1940 | | 17.6 | 34  | 345  |
| 6.6  | 65  | 1255 | | 17.7 | 143 | 2107 |
| 6.7  | 31  | 443  | | 17.8 | 119 | 1519 |
| 7.1  | 48  | 486  | | 17.9 | 108 | 1627 |
| 7.2  | 26  | 318  | | 18.1 | 225 | 2743 |
| 7.3  | 22  | 286  | | 18.2 | 51  | 570  |
| 7.4  | 27  | 245  | | 18.3 | 62  | 909  |
| 8.1  | 223 | 3466 | | 18.4 | 8   | 88   |
| 8.2  | 30  | 425  | | 18.5 | 25  | 259  |
| 8.3  | 118 | 1556 | | 18.6 | 35  | 433  |
| 8.4  | 100 | 1292 | | 19.1 | 72  | 1077 |
| 8.5  | 88  | 1565 | | 19.2 | 46  | 771  |
| 8.6  | 20  | 241  | | 19.3 | 64  | 854  |
| 8.7  | 32  | 413  | | 19.4 | 8   | 111  |
| 8.8  | 123 | 2045 | | 20.1 | 179 | 2467 |
| 8.9  | 23  | 298  | | 20.2 | 101 | 941  |
| 9.1  | 17  | 239  | | 20.3 | 206 | 2172 |
| 9.2  | 10  | 71   | | 20.4 | 114 | 1661 |
| 9.3  | 16  | 201  | | 20.5 | 90  | 1274 |

Numbers of papers (column 2) and associated pages (column 3) are computed
for each subject-matter code (column 1) according to the numbers of entries
in Chapter III not including reprints, translations and amendments.

| | Average number of papers per year | | | | | Total number of papers per year | | | | | | |
|------|-----------|--------|--------|--------|--------|------|------|------|------|------|------|------|
| | 1811–1919 | 1920–9 | 1930–9 | 1940–9 | 1950–9 | 1960 | 1961 | 1962 | 1963 | 1964 | 1965 | 1966 |
| 1.1 | 0 | 0 | 2 | 1 | 1 | 2 | 4 | 5 | 3 | 4 | 5 | 10 |
| 1.2 | 0 | 0 | 2 | 2 | 4 | 7 | 5 | 4 | 7 | 6 | 8 | 5 |
| 2 | 0 | 2 | 5 | 7 | 11 | 13 | 18 | 21 | 13 | 15 | 23 | 29 |
| 3 | 0 | 0 | 0 | 1 | 3 | 2 | 5 | 9 | 7 | 7 | 13 | 11 |
| 4 | 2 | 19 | 27 | 22 | 22 | 24 | 16 | 17 | 23 | 18 | 17 | 35 |
| 5 | 0 | 0 | 2 | 3 | 4 | 3 | 6 | 4 | 9 | 3 | 7 | 23 |
| 6 | 0 | 1 | 3 | 5 | 12 | 17 | 20 | 34 | 29 | 29 | 38 | 52 |
| 7 | 0 | 0 | 2 | 2 | 3 | 3 | 3 | 10 | 10 | 11 | 8 | 10 |
| 8 | 0 | 2 | 7 | 11 | 22 | 29 | 26 | 40 | 37 | 37 | 52 | 77 |
| 9 | 0 | 0 | 1 | 1 | 1 | 2 | 2 | 4 | 1 | 3 | 2 | 1 |
| 10 | 0 | 0 | 1 | 2 | 3 | 3 | 3 | 10 | 8 | 4 | 11 | 7 |
| 11 | 0 | 0 | 1 | 0 | 2 | 2 | 3 | 4 | 7 | 4 | 10 | 9 |
| 12 | 0 | 0 | 0 | 1 | 1 | 1 | 5 | 4 | 5 | 2 | 6 | 8 |
| 13 | 0 | 0 | 0 | 1 | 2 | 4 | 4 | 2 | 5 | 6 | 11 | 5 |
| 14 | 0 | 1 | 3 | 4 | 7 | 9 | 16 | 19 | 17 | 16 | 19 | 20 |
| 15 | 0 | 4 | 23 | 22 | 31 | 21 | 25 | 25 | 23 | 34 | 24 | 61 |
| 16 | 0 | 1 | 1 | 8 | 21 | 23 | 31 | 23 | 39 | 37 | 49 | 38 |
| 17 | 0 | 3 | 8 | 12 | 24 | 49 | 39 | 39 | 32 | 51 | 77 | 100 |
| 18 | 0 | 3 | 5 | 4 | 9 | 13 | 21 | 24 | 24 | 20 | 41 | 35 |
| 19 | 0 | 0 | 1 | 2 | 6 | 5 | 13 | 9 | 10 | 10 | 15 | 43 |
| 20 | 0 | 2 | 8 | 12 | 19 | 22 | 16 | 22 | 20 | 35 | 61 | 53 |

Numbers of papers are computed according to the numbers of entries in Chapter III not including reprints, translations and amendments. Subject-matter codes are pooled within Sections 2 up to 20.

# ADDENDA

*Modifications of entries in §II.2.*

ANWEND. MATRIZ. WIRTSCH. STATIST. PROBLEME (ADAM ET AL.) is Volume 9 of
 EINZELSCHR. DEUTSCHEN STATIST. GES.

BULL. AMER. MATH. SOC. Volume 72, Number 1, Part II "Norbert Wiener
 1894-1964" is paginated 1-145 in addition to rest of Volume 72 which
 is paginated 1-1100.

CONG. INTERNAT. PHILOS. SCI. is Volume 1146 of ACTUAL. SCI. INDUST.

KON. NEDERL. WETENSCH. PROC. SER. A should be KON. NEDERL. AKAD. WETENSCH.
 PROC. SER. A (also in §III.2).

*Additions to §II.2.*

COLLECT. MATH. *Collectanea Mathematica*. (Barcelona) [Consejo Superior
 de Investigaciones Cientificas. Universidad de Barcelona, Seminario
 Matematico]

ECONOMET. ANAL. NAT. ECON. PLANNING (HART, MILLS, + WHITAKER) *Econometric
 Analysis for National Economic Planning*. Edited by P. E. Hart, G.
 Mills, & J. K. Whitaker. [Volume XVI of the Colston Papers. Proceedings of the Sixteenth Symposium of the Colston Research Society
 held in the University of Bristol, April 6th-9th, 1964. Pub. Butterworths, London, 1964]

ESSAYS PROB. STATIST. (BARTLETT REPRINTS) *Essays on Probability and
 Statistics*. By M. S. Bartlett. [Methuen & Co., Ltd., London, &
 John Wiley & Sons, Inc., New York, 1962]

GESAMMELTE ABHANDL. ERNST ABBE *Gesammelte Abhandlungen von Ernst Abbe*.
 [Zweiter Band: "Wissenschaftliche Abhandlungen aus verschiedenen
 Gebieten. Patentschriften. Gedächtnisreden", 1906. Pub. Verlag
 von Gustav Fischer, Jena]

MEM. MET. STATIST. UNIV. ROMA *Memorie di Metodologia Statistica*. [Vol-
 ume Secondo, "Transvariazione", edited by Corrado Gini (a cura di
 Giuseppe Ottaviani), 1959. Università degli Studi di Roma, Facoltà
 di Scienze Statistiche, Demografiche ed Attuariali. Pub. Libreria
 Goliardica, Roma]

STUDENT'S COLLECTED PAPERS *"Student's" Collected Papers*. Edited by E.
 S. Pearson & John Wishart. [Cambridge University Press, Cambridge,
 1942. Published for the Biometrika Trustees. Reprinted 1947, 1958]

TECH. MODELES SCI. HUMAINES MODEL BUILDING HUMAN SCI. (WOLD) *La Technique
 des Modèles dans les Sciences Humaines. Model Building in the Human
 Sciences*. Scientific Organiser: Herman O. A. Wold. [Entretiens de
 Monaco en Sciences Humaines, Session 1964. Pub. Éditions "Sciences
 Humaines", Union Européenne D'éditions, 17, rue de Millo, Monaco.
 No date of publication given]

*Modifications of entries in* §III.2.

183   This may be the same paper as No. 4452.
1176  German title = WAHRSCHEINLICHKEIT UND AUSMASS DER TRANSVARIATION
          IM N-DIMENSIONALEN RAUM.
1374  Change SER. 2 to SER. 3 in journal abbreviation.
3285  Delete URSR in journal abbreviation.
3589  Change year to 1967 and save for *Abomsa II.*
3630  Amendment of No. 3629.
3708  Add Part number 1/II.
4170  Amended by No. 5821.
4452  This may be the same paper as No. 183.
4618  Add Part number 1/II. This paper is unsigned, but follows a paper
          by William L. Root.
5401  In Roumanian, with summary in Russian and French.
5821  Amendment of No. 4170.
(IN ROUMAINIAN) should be (IN ROUMANIAN).

Additional references to *Mathematical Reviews:*

| | | |
|---|---|---|
| 626:  MR30-5339 | 2276:  MR32-6522 | 3842:  MR29-1217 |
| 1034:  MR30-5394 | 2501:  MR31-845 | 3843:  MR34-2587 |
| 1039:  MR34-6935 | 2502:  MR33-3417 | 3876:  MR35-4958 |
| 1245:  MR29-6592 | 3025:  MR32-556 | 3897:  MR34-3604 |
| 1246:  MR30-5440 | 3029:  MR36-2264 | 4154:  MR34-3669 |
| 1247:  MR32-4762 | 3104:  MR34-6907 | 4281:  MR37-5907 |
| 1248:  MR32-4762 | 3105:  MR34-6907 | 4296:  MR34-5220 |
| 1313:  MR29-4084 | 3331:  MR34-6909 | 4744:  MR22-1056 |
| 1819:  MR30-3542 | 3681:  MR31-2747 | 4963:  MR33-5061 |
| 2023:  MR30-2643 | 3708:  MR32-4473 | 5377:  MR31-4074 |
| 2041:  MR33-59 | 3794:  MR35-1058 | 5630:  MR25-1598 |
| 2069:  MR35-1106 | | |

*Modifications of names of authors cited in* §III.2.

CRATHORNE, Arthur R.
HANNAN, Edward James
H'DOUBLER, Francis Todd
HERZEL, Amato
HUBER, Hans
JOHNS, Milton Vernon, Jr.
NANDI, Hari Kinkar
REED, William Gardner
REINSCHKE, Kurt

ROBINSON, Warren S.
ROHLF, F. James
SNOW, Barbara A. S.
STRIEBEL, Charlotte T.
THEIL, Henri
TINBERGEN, Jan
TOLLEY, Howard R.
ZUBRZYCKA, Ludmila

# *Additions to* §III.2.

ABBE, ERNST 1906.   (2.5/2.6)                                                          A1
     UEBER DIE GESETZMASSIGKEIT IN DER VERTHEILUNG DER FEHLER BEI BEOBACHTUNGSREIHEN.
     GESAMMELTE ABHANDL. ERNST ABBE, 2, 55-81.

BALAKRISHNAN, A. V. 1959.   (17.3)                                                     A2
     ON A CHARACTERIZATION OF COVARIANCES.
     ANN. MATH. STATIST., 30, 670-675.   (MR21-6676)

BARTLETT, MAURICE S. 1962.   (15.1/15.2)                                               A3
     FACTOR ANALYSIS IN PSYCHOLOGY AS A STATISTICIAN SEES IT.   (REPRINT OF NO. 350)
     ESSAYS PROB. STATIST. (BARTLETT REPRINTS), 37-48.

CASTELLANO, VITTORIO 1959.   (4.8)                                                     A4
     SULLO SCARTO QUADRATICO MEDIO DELLA PROBABILITA DI TRANSVARIAZIONE.
        (REPRINT OF NO. 867)
     MEM. MET. STATIST. UNIV. ROMA, 2, 56-117.

CROUSE, C. F. 1966.   (17.8/B)                                                         A5
     DISTRIBUTION FREE TESTS BASED ON THE SAMPLE DISTRIBUTION FUNCTION.
     BIOMETRIKA, 53, 99-108.   (MR33-5047)

DAGUM, CAMILO 1959.   (4.8)                                                            A6
     TRANSVARIAZIONE FRA PIU DI DUE DISTRIBUZIONI.   (TRANSLATED AS NO. 1173)
     MEM. MET. STATIST. UNIV. ROMA, 2, 608-647.

DAGUM, CAMILO 1960.   (4.8/E)                                                          A7
     TEORIA DE LA TRANSVARIACION. SUS APPLICACIONES A LA ECONOMIA.
     METRON, 20, 45-250.

DE LUCIA, LUIGI 1959.   (4.8)                                                          A8
     TRANSVARIAZIONE TRA CARATTERI CONNESSI.   (PARTE PRIMA, TRANSVARIAZIONE NELLE
     VARIABILI CASUALI.   PARTE SECONDA, TRANSVARIAZIONE NEI CAMPIONI.)
        (REPRINT OF NO. 1287)
     MEM. MET. STATIST. UNIV. ROMA, 2, 281-358.

DE NOVELLIS, MIRELLA 1959.   (4.8)                                                     A9
     PROCEDIMENTI E DISTRIBUZIONI DEI CALCOLI PER LA DETERMINAZIONE DELLE INTENSITA
     DI TRANSVARIAZIONE TRA DUE GRUPPI.
     MEM. MET. STATIST. UNIV. ROMA, 2, 183-209.

DE NOVELLIS, MIRELLA 1959.   (4.8)                                                     A10
     ULTERIORI CONTRIBUTI ALLA TEORIA DELLA TRANSVARIAZIONE.   GLI INDICI DI
     TRANSVARIAZIONE NEL CASO DI DISTRIBUZIONI CHE DIANO LUOGO A CURVE IPERBOLICHE.
     MEM. MET. STATIST. UNIV. ROMA, 2, 648-681.

DE NOVELLIS, MIRELLA 1959.   (4.8/E)                                                   A11
     APPLICAZIONI DELLA TEORIA DELLA TRANSVARIAZIONE ALLO STUDIO DI ALCUNI PROBLEMI
     DI ANTROPOMETRIA E BIOLOGIA.
     MEM. MET. STATIST. UNIV. ROMA, 2, 682-694.

EMANUELLI, FILIPPO 1959.   (4.8/B)                                                     A12
     SULLA DETERMINAZIONE DELLA PROBABILITA E DELL'INTENSITA DI TRANSVARIAZIONE TRA
     DUE VARIABILI CASUALI NEL CASO CHE QUESTE SEGUANO UNA LEGGE DI DISTRIBUZIONE
     DOVUTA ALL'EDGEWORTH.
     MEM. MET. STATIST. UNIV. ROMA, 2, 719-730.

FISHER, FRANKLIN M. 1965.   (16.2)                                                     A13
     DYNAMIC STRUCTURE AND ESTIMATION IN ECONOMY-WIDE ECONOMETRIC MODELS.
        (WITH DISCUSSION)
     SEMAINE D'ETUDE ROLE ANAL. ECONOMET. PONT. ACAD. SCI., 385-464.

FRASER, DONALD A. S. 1966.   (6.7/19.4)                                                A14
     SUFFICIENCY FOR SELECTION MODELS.
     SANKHYA SER. A, 28, 329-334.   (MR35-1119)

GINI, CORRADO 1959.   (4.8/E)                                                          A15
     IL CONCETTO DI "TRANSVARIAZIONE" E LE SUE PRIME APPLICAZIONI.
     MEM. MET. STATIST. UNIV. ROMA, 2, 1-55.

GINI, CORRADO 1959.   (4.8)                                                            A16
     PER LA DETERMINAZIONE DELLE PROBABILITA DI TRANSVARIAZIONE TRA PIU GRUPPI.
     MEM. MET. STATIST. UNIV. ROMA, 2, 210-215.

GINI, CORRADO 1959.   (4.8)                                                          A17
    DELLA MISURA SINTETICA DELLA TRANSVARIAZIONE RISPETTO AD N CARATTERI.
    MEM. MET. STATIST. UNIV. ROMA, 2, 434-476.

GINI, CORRADO 1959.   (4.8/6.7)                                                      A18
    DELLA SELEZIONE INDIRETTA.
    MEM. MET. STATIST. UNIV. ROMA, 2, 695-697.

GINI, CORRADO + LIVADA, GREGORIO 1959.   (4.8/T)                                     A19
    TRANSVARIAZIONE A PIU DIMENSIONI.   (REPRINT OF NO. 1988)
    MEM. MET. STATIST. UNIV. ROMA, 2, 216-253.

GINI, CORRADO + LIVADA, GREGORIO 1959.   (4.8)                                       A20
    NUOVI CONTRIBUTI ALLA TEORIA DELLA TRANSVARIAZIONE.
    MEM. MET. STATIST. UNIV. ROMA, 2, 254-280.

GINI, CORRADO + SONNINO, G. 1959.   (2.6/4.8)                                        A21
    CONTRIBUTO ALLA TEORIA DELLA TRANSVARIAZIONE FRA SERIAZIONI CORRELATE.
    MEM. MET. STATIST. UNIV. ROMA, 2, 389-433.

GINI, CORRADO; VITERBO, C.; BENEDETTI, CARLO + HERZEL, AMATO 1959.   (4.8)           A22
    PROBLEMI DI TRANSVARIAZIONE INVERSA.
    MEM. MET. STATIST. UNIV. ROMA, 2, 520-593.

HERZEL, AMATO 1958.   (4.8)                                                          A23
    INFLUENZA DEL RAGGRUPPAMENTO IN CLASSI SULLA PROBABILITA E SULL'INTENSITA
    DI TRANSVARIAZIONE.
        (REPRINTED AS NO. A24)
    METRON, 19 (1) 199-242.   (MR21-357)

HERZEL, AMATO 1959.   (4.8)                                                          A24
    INFLUENZA DEL RAGGRUPPAMENTO IN CLASSI SULLA PROBABILITA E SULL'INTENSITA
    DI TRANSVARIAZIONE.
        (REPRINT OF NO. A23)
    MEM. MET. STATIST. UNIV. ROMA, 2, 477-519.   (MR21-357)

HOLGATE, P. 1964.   (17.4/B)                                                         A25
    BIVARIATE GENERALIZATIONS OF NEYMAN'S TYPE A DISTRIBUTION.
    BIOMETRIKA, 53, 241-245.   (MR33-3389)

HULTQUIST, ROBERT A. 1966.   (2.5)                                                   A26
    DIAGRAMMING THEOREMS RELATIVE TO QUADRATIC FORMS.
    AMER. STATIST., 20 (5) 31.

IYER, P. V. KRISHNA 1945.   (5.3)                                                    A27
    A NOTE ON HOTELLING'S $T^2$.
    CURRENT SCI., 14, 173-175.

JACKSON, J. EDWARD 1956.   (5.2/8.3/8.4/BEC)                                         A28
    QUALITY CONTROL METHODS FOR TWO RELATED VARIABLES.
    INDUST. QUAL. CONTROL, 12 (7) 4-8.

KAISER, HENRY F. + DICKMAN, KERN 1962.   (15.2/R)                                    A29
    SAMPLE AND POPULATION SCORE MATRICES AND SAMPLE CORRELATION MATRICES FROM AN
    ARBITRARY POPULATION CORRELATION MATRIX.
    PSYCHOMETRIKA, 27, 179-182.   (MR25-4618)

KLEIN, LAWRENCE R. 1964.   (16.2)                                                    A30
    PROBLEMS IN THE ESTIMATION OF INTERDEPENDENT SYSTEMS.
    TECH. MODELES SCI. HUMAINES MODEL BUILDING HUMAN SCI. (WOLD), 51-87.

LEIPNIK, ROY B. 1958.   (2.6)                                                        A31
    NOTE ON THE CHARACTERISTIC FUNCTION OF A SERIAL-CORRELATION DISTRIBUTION.
    BIOMETRIKA, 45, 559-562.

LEIPNIK, ROY B. 1962.   (2.6/A)                                                      A32
    LIMIT DISTRIBUTIONS OF THE CIRCULAR SERIAL CORRELATION COEFFICIENT.
    SANKHYA SER. A, 24, 395-408.   (MR30-1582)

MAJINDAR, KULENDRA N. 1966.   (20.1)                                                 A33
    ON CERTAIN INEQUALITIES FOR THE CHARACTERISTIC ROOTS OF HERMITIAN MATRICES.
    AMER. MATH. MONTHLY, 73, 268-270.

MENGES, G. + DIEHL, H. 1964.  (17.7)                                         A34
    TIME STABILITY OF STRUCTURAL PARAMETERS.  (WITH DISCUSSION)
    ECONOMET. ANAL. NAT. ECON. PLANNING (HART, MILLS, + WHITAKER), 299-320.

MIANI CALABRESE, DONATO 1959.  (4.8/E)                                       A35
    LA TRANSVARIAZIONE RISPETTO AL SESSO DEI CARATTERI FISICI DELL'INFANZIA.
    MEM. MET. STATIST. UNIV. ROMA, 2, 130-182.

OSTROVSKII, I. V. 1966.  (17.3)                                              A36
    DECOMPOSITION OF MULTI-DIMENSIONAL PROBABILISTIC LAWS.
        (TRANSLATION OF NO. 4102)
    SOVIET MATH. DOKL., 7, 1052-1055.  (MR34-3619)

OTTAVIANI, GIUSEPPE 1959.  (18.1)                                            A37
    SULLA PROBABILITA CHE UNA PROVA SU DUE VARIABILI CASUALI X E Y VERIFICHI LA
    DISUGUAGLIANZA X < Y E SUL CORRISPONDENTE SCARTO QUADRATICO MEDIO.
        (REPRINT OF NO. 4110)
    MEM. MET. STATIST. UNIV. ROMA, 2, 118-129.  (MR1,P.340)

OTTAVIANI, GIUSEPPE 1959.  (4.8)                                             A38
    SULLA TRANSVARIABILITA.
    MEM. MET. STATIST. UNIV. ROMA, 2, 594-607.

OTTAVIANI, GIUSEPPE 1959.  (4.8)                                             A39
    ALCUNE CONSIDERAZIONI SUL CONCETTO DI TRANSVARIAZIONE.
        (REPRINTED AS NO. 4112)
    MEM. MET. STATIST. UNIV. ROMA, 2, 698-718.

PASTERNACK, BERNARD + LIUZZI, ANTHONY 1965.  (2.6/8.9)                       A40
    PATTERNS IN RESIDUALS:  A TEST FOR REGRESSION MODEL ADEQUACY IN RADIONUCLIDE
    ASSAY.
    TECHNOMETRICS, 7, 603-621.

SALES VALLES, FRANCISCO DE A. 1959.  (6.5/17.3)                             A41
    ON THE CONTINUITY OF THE COVARIANCE OF RANDOM FUNCTIONS OF SECOND ORDER.
        (IN SPANISH)
    COLLECT. MATH., 11, 69-75.  (MR22-3024)

SALVEMINI, TOMMASO 1959.  (4.8/6.2)                                         A42
    TRANSVARIAZIONE E ANALISI DISCRIMINATORIA.
    MEM. MET. STATIST. UNIV. ROMA, 2, 731-742.

SARGAN, JOHN D. 1964.  (16.2/20.3/E)                                        A43
    WAGES AND PRICES IN THE UNITED KINGDOM:  A STUDY IN ECONOMETRIC METHODOLOGY.
        (WITH DISCUSSION)
    ECONOMET. ANAL. NAT. ECON. PLANNING (HART, MILLS, + WHITAKER), 25-63.

SIDDIQUI, M. M. 1958.  (2.6)                                                A44
    DISTRIBUTION OF A SERIAL CORRELATION COEFFICIENT NEAR THE ENDS OF THE RANGE.
    ANN. MATH. STATIST., 29, 852-861.  (MR20-362)

STERIOTIS, PIETRO J. 1954.  (4.8)                                           A45
    PROBABILITY OF TRANSVARIATION IN THE REGULAR DISTRIBUTIONS OF GAUSS.
        (IN GREEK, WITH SUMMARY IN ENGLISH)
    BULL. SOC. MATH. GRECE, 28, 1-36.  (MR16,P.493)

STERIOTIS, PIETRO J. 1959.  (4.8)                                           A46
    CONTRIBUTO ALLA TRANSVARIAZIONE DELLE DISTRIBUZIONI GAUSSIANE.
    MEM. MET. STATIST. UNIV. ROMA, 2, 359-388.

"STUDENT" 1942.  (14.5/18.3/EU)                                             A47
    THE PROBABLE ERROR OF A MEAN.
        (REPRINT OF NO. 5265)
    STUDENT'S COLLECTED PAPERS, 11-34.

"STUDENT" 1942.  (4.2)                                                      A48
    PROBABLE ERROR OF A CORRELATION COEFFICIENT.
        (REPRINT OF NO. 5266)
    STUDENT'S COLLECTED PAPERS, 35-42.

"STUDENT" 1942.   (4.8)                                                                    A49
      THE CORRECTION TO BE MADE TO THE CORRELATION RATIO FOR GROUPING.
            (REPRINT OF NO. 5267)
      STUDENT'S COLLECTED PAPERS, 53-57.

"STUDENT" 1942.   (4.1)                                                                    A50
      THE ELIMINATION OF SPURIOUS CORRELATION DUE TO POSITION IN TIME OR SPACE.
            (REPRINT OF NO. 5268)
      STUDENT'S COLLECTED PAPERS, 58-60.

"STUDENT" 1942.   E(4.8)                                                                   A51
      AN EXPERIMENTAL DETERMINATION OF THE PROBABLE ERROR OF DR SPEARMAN'S
      CORRELATION COEFFICIENTS.
            (REPRINT OF NO. 5269)
      STUDENT'S COLLECTED PAPERS, 70-89.

"STUDENT" 1942.   (5.1)                                                                    A52
      ON THE 'Z' TEST.
            (REPRINT OF NO. 5270)
      STUDENT'S COLLECTED PAPERS, 179-180.

WATSON, G. S. 1956.   (2.6)                                                                A53
      ON THE JOINT DISTRIBUTION OF THE CIRCULAR SERIAL CORRELATION COEFFICIENTS.
      BIOMETRIKA, 43, 161-168.   (MR18,P.79)

WOLD, HERMAN O. A. 1964.   (16.1/16.2/17.7)                                                A54
      THE APPROACH OF MODEL BUILDING.  CROSSROADS OF PROBABILITY THEORY, STATISTICS,
      AND THEORY OF KNOWLEDGE.
      TECH. MODELES SCI. HUMAINES MODEL BUILDING HUMAN SCI. (WOLD), 1-38.